Alcamo's
Fundamentals
of Microbiology

Jones and Bartlett Titles in Biological Science

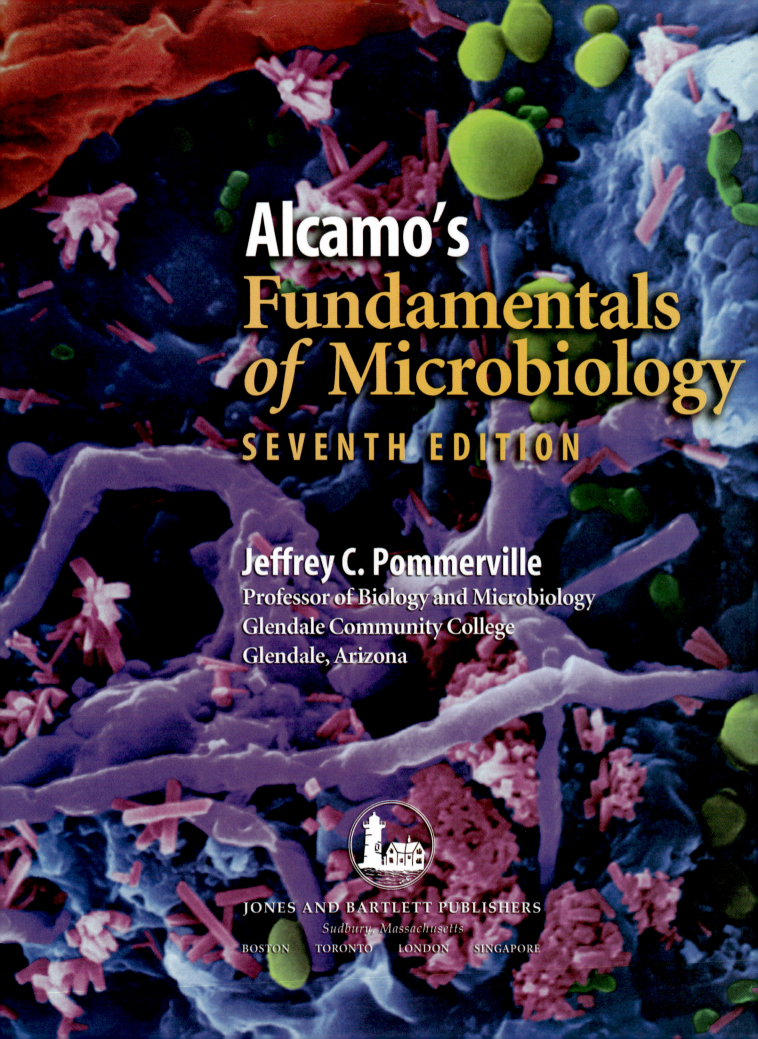

Alcamo's
Fundamentals
of Microbiology

SEVENTH EDITION

Jeffrey C. Pommerville

Professor of Biology and Microbiology
Glendale Community College
Glendale, Arizona

JONES AND BARTLETT PUBLISHERS
Sudbury, Massachusetts

BOSTON   TORONTO   LONDON   SINGAPORE

WORLD HEADQUARTERS

Jones and Bartlett Publishers
40 Tall Pine Drive
Sudbury, MA 01776
978-443-5000
info@jbpub.com
www.jbpub.com

Jones and Bartlett Publishers Canada
2406 Nikanna Road
Mississauga, ON L5C 2W6
CANADA

Jones and Bartlett Publishers International
Barb House, Barb Mews
London W6 7PA
UK

Library of Congress Cataloging-in-Publication Data
Pommerville, Jeffery.
    Alcamo's fundamentals of microbiology / Jeffery
Pommerville.—7th ed.
        p. cm.
    Previously published: Fundamentals of Microbiology / Edward I.
Alcamo. 6th ed.
    Sudbury, Mass.: Jones and Bartlett, 2001.
    Includes index.
    ISBN 0-7637-0067-3
    1. Microbiology. 2. Medical microbiology. I. Alcamo, I.
Edward. Fundamentals of microbiology. II. Title
    QR41.2.A43 2004
    579—dc22
2003066202

Printed in the United States of America
08 07 06 05 04   10 9 8 7 6 5 4 3 2 1

PRODUCTION CREDITS

Chief Executive Officer: Clayton Jones
Chief Operating Officer: Don W. Jones, Jr.
President: Robert W. Holland, Jr.
V.P., Design and Production: Anne Spencer
V.P., Sales and Marketing: William Kane
V.P., Manufacturing and Inventory Control: Therese Bräuer
Executive Editor, Science: Stephen L. Weaver
Managing Editor, Science: Dean W. DeChambeau
Associate Editor, Science: Rebecca Seastrong
Senior Production Editor: Louis C. Bruno, Jr.
Marketing Manager: Matthew Bennett
Associate Marketing Manager: Matthew Payne
Text Design and Composition: Seventeenth Street Studios
Cover Design: Anne Spencer
Copy Editor: Shellie Newell
Indexing: Deborah Patton
Photo Researcher: Kimberly Potvin
Art Work: Elizabeth Morales, Seventeenth Street Studios,
    Jim Bryant
Printing and Binding: Courier Kendallville
Cover Printing: Lehigh Press

ABOUT THE COVER

The image on the cover is a false color scanning electron microscope view of rod-shaped bacteria (pink), yeast cells (green), and filamentous fungi (purple) on a kitchen sponge. Dishrags and sponges from home kitchens can harbor large numbers of potentially disease-causing bacteria, including *Escherichia coli* and strains of *Salmonella*, *Pseudomonas*, and *Staphylococcus*. The moist environment and nutrients from wiped-up food scraps enable these microorganisms to survive for days, even weeks. If not cleaned, these same germs can be spread around the kitchen as you wipe other surfaces. It is best to frequently wash sponges, dishcloths, and dishtowels in the hot cycle of your washer. Using clean paper towels to wipe up messy spills also can help prevent spreading such germs throughout your house.

Photo Credit: © 2001 Dennis Kunkel Microscopy, Inc./Dennis Kunkel

Brief Contents

Contents

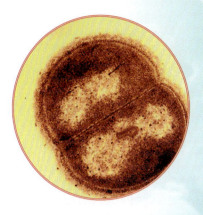

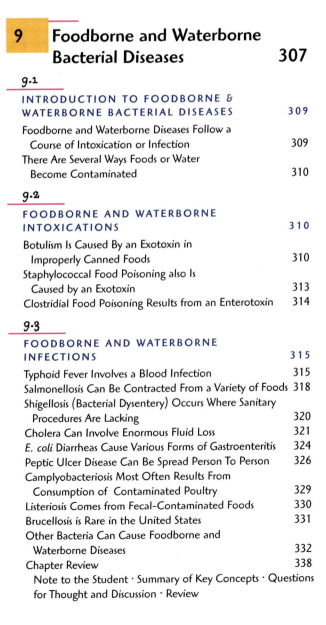

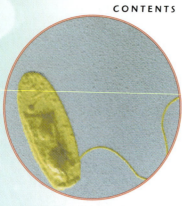

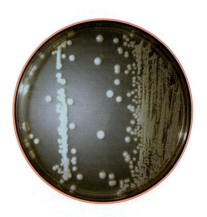

Preface

A TRADITION CONTINUES

In December 2002, Ed Alcamo passed away after a battle with cancer. Since the publication of the first edition of *Fundamentals of Microbiology* in 1983, Ed had devoted much of his time and effort to microbiology teaching and producing a textbook that students would find challenging yet enjoyable to read. He was both a colleague and friend, and I miss seeing Ed at conferences and meetings, having a beer or two while discussing microbiology education or more mundane topics, and, yes, listening to him play the harmonica. All of us in microbiology miss his passion, enthusiasm, and humor.

Ed's philosophy with *Fundamentals of Microbiology* was to share with students in a friendly and accessible way what he knew about the subject and what he considered important for them to know. He believed students must be informed and knowledgeable citizens in their community. As the new author of *Fundamentals,* I can say that his philosophy and tradition remain the backbone of the new seventh edition; in fact, notice that the new edition is titled *Alcamo's Fundamentals of Microbiology.* His influence and spirit remain—the tradition continues.

The times in which we live are progressing so rapidly that it is hard to keep up with the world around us. In microbiology, newly emerging infectious diseases, such as West Nile fever and severe acute respiratory syndrome (SARS), the spread of antibiotic resistance, and the threat of bioterrorism are just some of the issues we face today. *Alcamo's Fundamentals of Microbiology* continues to provide you, the student, with up-to-date content so you can read newspaper or magazine articles about microorganisms with understanding, or talk to your doctor with confidence about infectious diseases. In addition, if you are a nursing student or allied health major, the text will give you the information and understanding of microbiology that you need to perform well in your chosen health field.

The new seventh edition contains new pieces of artwork and many new photographs to improve learning and understanding. Jones and Bartlett Publishers remains committed to producing the best microbiology textbook available today, and I believe Ed would be proud of the new edition. He may not be with us physically, but his ideas and thoughts remain integral to the new edition of *Alcamo's Fundamentals of Microbiology.*

AUDIENCE

Alcamo's Fundamentals of Microbiology, Seventh Edition, is written for introductory microbiology courses having an emphasis on the biology of human disease. It is geared toward students in health and allied health science curricula such as nursing, dental hygiene, medical assistance, sanitary science, and medical laboratory technology. It also will be an asset to students studying food science, agriculture, environmental science, and health administration. In addition, the text should provide a firm foundation for advanced programs in biological sciences, as well as medicine, dentistry, and other health professions.

OBJECTIVES

Alcamo's Fundamentals of Microbiology presents a broad coverage of microorganisms. Looking at the thickness of this text, you might believe that learning and understanding the fundamentals of microbiology are going to be an enormous task. Honestly, there is a substantial body of material to learn and it will require significant effort on your part. As such, one of my objectives as author has been to make sure the fundamentals are presented in an organized and logical way. Also, realize that a substantial portion of the textbook contains figures, tables, diagrams, photographs, and other study aids that are designed to help you more easily envision the concepts being described and discussed. In this regard, make sure you read the section **To the Student** (p. xxiii) because I give you many tips on how best to use and read this textbook to master the fundamentals. One of my most important objectives, as was Ed's, is to ensure the material presented is accessible. I have tried to build on Ed's framework of storytelling (that was how people learned in past centuries) mentioning the men and women who succeeded in advancing our knowledge; and telling you why something is important, not just that it is important. I hope you find the microbial world you are about to enter interesting if not downright fascinating. Enjoy!

ORGANIZATION

Alcamo's Fundamentals of Microbiology, Seventh Edition, is divided into seven major areas of concentration. These areas use basic principles as frameworks to provide the unity and diversity of microbiology. Among the principles explored are the variations in structure and growth of microorganisms, the basis for infectious disease and resistance, and the beneficial effects microorganisms have on our lives.

Part 1 deals with the foundations of microbiology. It includes chapters on the origins of microbiology and the universal concepts that underpin the science. Part 2 then concentrates on the bacteria, the microorganisms most of us think of when we say "germs." The discussions carry over to Part 3, where the spectrum of bacterial diseases is surveyed. Part 4 looks at the significance of other microorganisms, including viruses, fungi, protozoa, and the multicellular parasites

In Part 5 of the text, the emphasis turns to infectious disease and the body's resistance through the immune system. Here, we study the reasons for disease and the means for surviving it. A logical segue is in Part 6, which presents various mechanisms for controlling microorganisms using such things as heat, radiation, disinfectants, and antibiotics. Part 7 closes the text with brief discussions of how public health measures interrupt epidemics. Some key insights also are given on the positive effects microorganisms exert through biotechnology.

WHAT'S NEW

Today, microbiology is in a third Golden Age. It is one of the most dynamic disciplines of science and information is changing almost exponentially. This has necessitated a comprehensive updating of several chapters.

New Infectious Disease Coverage. Since the last edition, several newly emerging diseases have made themselves known (either naturally or through bioterrorism) while others have spread in geographic location. New material, therefore, has been added to Chapter 12 on **mad cow disease. Severe acute respiratory syndrome (SARS)** as well as material on **smallpox** and **monkeypox** has been added to Chapter 13. Chapter 14 has been updated with new information on **AIDS** and new material on **West Nile fever.**

Protein Synthesis Has Moved. The material on protein synthesis that was in Chapter 5 of the previous edition is now in Chapter 6, Bacterial Genetics. Chapter 6 represents a more traditional chapter, covering DNA replication, transcription and translation, mutations, and gene regulation.

New Coverage of Bacterial Genomics. Rapid advances in genetic engineering and bacterial genomics has necessitated adding a new chapter entitled **Genetic Engineering and Bacterial Genomics** (Chapter 7) to cover the exciting discoveries being made.

New MicroInquiry Exercises. One of the major movements in science education today involves bringing a more inquiry approach to learning and engaging students in critical reasoning. To this end, embedded in each chapter is a **MicroInquiry** box (exercise), which require critical reasoning on your part and an understanding of material presented in that chapter. These "real life" exercises include such activities as analyzing experiments and experimental data, diagnosing the signs and symptoms of an infectious disease in a case study, understanding the steps and decisions required to test and bring an antibiotic to market, and determining how to decontaminate a building after a bioterrorism incident. Answers to each MicroInquiry are provided in Appendix E.

SPECIAL FEATURES

To help you achieve your learning goals and to reduce the anxiety that comes with approaching a new subject, the publisher and I have incorporated several features into this book. We hope that these features will enhance the accessibility of the key concepts and generate real enthusiasm for microbiology.

Career Essays, titled "Microbiology Pathways," open each part of the text and discuss various career options available.

Chapter Introductions provide a stimulating thought or historical perspective to set the tone for the chapter.

Marginal Definitions present succinct definitions of notable terms as they enter the discussion.

Boldface Terms highlight important terms and ideas in the text.

Mid-Chapter Summaries, titled "To This Point," allow you to pause and summarize the previous few pages and preview the next section.

MicroFocus Boxes explore interesting topics concerning microbiology or microorganisms.

Serotypes:
variants of organisms that differ according to the antibodies they elicit from the immune system.

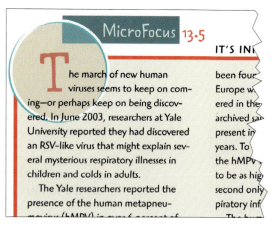

MicroFocus 13.5

IT'S IN[

The march of new human viruses seems to keep on coming—or perhaps keep on being discovered. In June 2003, researchers at Yale University reported they had discovered an RSV–like virus that might explain several mysterious respiratory illnesses in children and colds in adults.

The Yale researchers reported the presence of the human metapneu-

been fou[
Europe w[
ered in the[
archived sa[
present in[
years. To[
the hMPV[
to be as hig[
second only[
piratory inf[

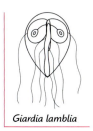

Giardia lamblia

WHY PATHOGENS?

Microorganisms perform many useful services for humans when they produce food products, manufacture organic materials in industrial plants, and recycle such elements as carbon and nitrogen. The emphasis of this book, however, is on the tiny, but important percentage of microorganisms that cause human disease, the so-called pathogens. Why do we emphasize pathogens? Here are several reasons:

■ Pathogens have regularly altered the course of human history.

■ Pathogens are familiar to audiences of microbiology.

■ Pathogens add drama to an invisible world of microorganisms.

■ Pathogens illustrate ecological relationships between humans and microorganisms.

■ Pathogens point up the diversity of microorganisms.

Moreover, the study of pathogens makes basic science relevant and shows how microbiology interfaces with other disciplines such as sociology, economics, history, politics, and geography. Finally, the study of pathogens helps us to understand contemporary newspaper articles, magazine headlines, and stories on the late news. And in the end, that makes us better citizens. Indeed, the famous essayist Thomas Mann has written, "All interest in disease is only another expression of interest in life."

Note to the Student

MicroInquiry Boxes allow you to investigate (usually interactively) some important aspect of the chapter being studied.

Textbook Cases help you understand pathogens and present contemporary disease outbreaks in pictorial format.

Summary Tables pull together the similarities and differences of topics discussed in the chapter.

Notes to the Student at the end of each chapter present an editorial comment within the framework of the chapter.

Summaries of Key Concepts pull together the major ideas discussed in the chapter.

Questions for Thought and Discussion encourage you to use the text to resolve thought-provoking problems with contemporary relevance.

Chapter Reviews contain questions of a somewhat unconventional type to assist review of the chapter contents.

The Glossary contains concise definitions for the major microbiological terms and concepts.

Pronouncing Microorganism Names (inside back cover) helps you correctly pronounce difficult microorganism names.

A List of the Textbook Cases

SUPPLEMENTS TO THE TEXT

We have developed a number of ancillaries that further amplify the concepts of this text:

For the Student

The web site we have developed exclusively for the seventh edition of this text, **http://microbiology. jbpub.com**, offers a variety of resources to enhance understanding of microbiology. The site contains eLearning, a free on-line study guide with chapter outlines, chapter essay questions, key term reviews, and short study quizzes.

The *Study Guide* that accompanies this textbook contains over 3000 practice exercises and study questions of various types to help you learn and retain the information in the text.

Laboratory Fundamentals of Microbiology, Seventh Edition, is a series of 30 multipart laboratory exercises that provide basic training in the handling of microorganisms and reinforce ideas and concepts described in the textbook.

An anthology called *Encounters with Microbiology* brings together "Vital Signs" articles from *Discover Magazine* in which health professionals use their knowledge of microbiology in their medical cases.

For the Instructor

Compatible with Windows and Macintosh platforms, the **Instructor's Tool Kit—CD-ROM** provides instructors with the following traditional ancillaries:

- The *Instructor's Manual,* provided as a text file, includes chapter summaries, complete chapter lecture outlines, and answers to all the Questions for

Thought and Discussion.

- The *Test Bank* is available as straight text files and as part of the Diploma™ Test Generator software included on the CD. The Test Generator software helps choose an appropriate variety of questions, create multiple versions of tests, even administer and grade tests on-line.

- The *PowerPoint™ Lecture Outline Slides* presentation package provides lecture notes, and images for each chapter of *Alcamo's Fundamentals of Microbiology*. Instructors with the Microsoft PowerPoint™ software can customize the outlines, art, and order of presentation.

The *Kaleidoscope Media Viewer* provides a library of all the art and tables in the text to which Jones and Bartlett Publishers holds the copyright or has digital reprint rights. The Kaleidoscope Media Viewer uses your browser—Internet Explorer or Netscape Navigator—to enable you to project images from the text in the classroom, insert images into PowerPoint presentations, or print your own acetates.

A set of *Acetate Overhead Transparencies* includes 100 color figures and 75 electron micrographs from the textbook.

Acknowledgments

Without an experienced and capable publication team, a textbook such as this would be impossible. At Jones and Bartlett Publishers, Stephen Weaver, Executive Editor, Science, signed me up. I have had the privilege of working with a great group of production professionals. Lou Bruno supervised the production process with the skill of an orchestra leader; Dean DeChambeau once more contributed his experience and project management skills to the project; Shellie Newell was the "eagle-eye" copy editor; Elizabeth Morales of Elizabeth Morales Illustration Services once again lent her expert experience in translating my primitive drawings into first-class artwork; Kimberly Potvin tracked down many of the great photos that embellish these pages; Deborah Patton read every page and created the index; Rebecca Seastrong coordinated the the building of the student web site and the processing of the ancillary materials; and again Seventeenth Street Studio set each page with care.

The book benefited from the expertise of several fellow microbiologists and biologists. I wish to thank Gerald Goldstein, Paul Gulig, John Roth, and Stephen Williams for their input during the writing of this edition. I also want to thank Christopher Woolverton for his excellent review of Chapter 7.

After more than 20 years of university and college instruction, I must thank all my former students who kept me on my toes in the classroom and to be always prepared. Your suggestions and evaluations have encouraged me to continually assess my instruction so it can be easily understood. I salute you, and I hope those of you who read this text also will let me know what works and what still needs improvement to make your learning efficient and still enjoyable.

I wish to thank the sabbatical committee at the Maricopa Community Colleges for providing me with a sabbatical so this project could be finished in record time.

Finally, I could not finish without acknowledging the most important person of all—my wife Yvonne. This project could not have been completed without her support, enthusiasm, and patience. After more than 30 years of marriage, she has been not only my loving wife but also my best friend and motivator, keeping me on track during wavering moments.

J. Pommerville
Scottsdale, AZ
Spring, 2004

To the Student—Read This

When I was an undergraduate student, I hardly ever read the "To the Student" section (if indeed one existed) in my textbooks because the section rarely contained any information of importance. This one does, so please read on.

In college, I was a mediocre student until my junior year. Why? Mainly because I did not know how to study properly, and, important here, I did not know how to read a textbook effectively. My textbooks were filled with underlined sentences (highlighters hadn't been invented yet!) without any plan on how I would use this "emphasized" information. In fact, most textbooks *assume* you know how to properly read a textbook. I didn't and you might not, either.

Reading a textbook is difficult if you are not properly prepared. So you can take advantage of what I learned and have learned from instructing thousands of students; I have worked hard to make this text user friendly with a reading style that is not threatening or complicated to understand. Still, there is a substantial amount of information to learn and understand, so having the appropriate reading and comprehension skills is critical. Therefore, I encourage you to spend 30 minutes to read this section as I am going to give you several tips and suggestions to acquire the skills to read *Alcamo's Fundamentals of Microbiology.* Let me show you how to be an active reader.

BE A PREPARED READER

Before you jump into reading a section of a chapter in this text, prepare yourself by finding the place and time and having the tools for study.

Place. Where are you right now as you read these lines? Are you in a quiet library or at home? If at home, are there any distractions, such as loud music, a blaring television, or screaming kids? Is the lighting adequate to read? Are you sitting at a desk or lounging on the living room sofa? Get where I am going? When you read for an educational purpose—that is, to learn and understand something—you need to maximize the environment for reading. Yes, it should be comfortable but not to the point that you will doze off.

Time. All of us have different times during the day when we perform some skill, be it exercising or reading, the best. The last thing you want to do is read when you are tired or simply not "in tune" for the job that needs to be done. You cannot learn and understand the information if you fall asleep or lack a positive attitude. I have kept the chapters in this text to about the same length so you can estimate the time necessary for each and plan your reading accordingly. If you have done your preliminary survey of the chapter or chapter section, you can determine about how much time you will need. If 40 minutes is needed to read—and comprehend (see below)—a section of a chapter, find the place and time that will give you 40 minutes of uninterrupted study. Brain research suggests that most people's brains cannot spend more than 45 minutes in concentrated, technical reading. Therefore, I have avoided lengthy presentations and instead have focused on smaller sections, each with its own heading. These should accommodate shorter reading periods.

Reading Tools. Lastly, as you read this, what study tools do you have at your side? By that I mean, do you have highlighter or pen for emphasizing or underlining important words or phrases? Notice that the text has wide margins, which allow you to make notes or to indicate something that needs further clarification. Do you have a pencil or pen handy to make these notes? Or, if you do not want to "deface" the text, make your notes in a notebook, which you have with you, right? Lastly, some students find having a ruler is useful to prevent your eyes from wandering on the page and to read each line without distraction.

BE AN EXPLORER BEFORE YOU READ

When you sit down to read a section of a chapter, do some preliminary exploring. Look at the section head and subheadings to get an idea of what is discussed. Preview any diagrams, photographs, tables, graphs, or other visuals used. They give you a better idea of what is going to occur. We have used a good deal of space in the text for these features, so use them to your advantage. They will help you learn the written information and comprehend its meaning. Do not try to understand all the visuals, but try to generate a mental "big picture" of what is to come. Familiarize yourself with any symbols or technical jargon that might be used in the visuals.

The end of each chapter contains a **Summary of Key Concepts** for that chapter. It is a good idea to read the summary before delving into the chapter. That way you will have a framework of the chapter before filling in the nitty-gritty information.

BE A DETECTIVE AS YOU READ

Reading a section of a textbook is not the same as reading a novel. With a textbook, you need to uncover the important information (the terms and concepts) from the forest of words on the page. So, the first thing to do is read the complete paragraph. When you have determined the main ideas, highlight or underline them. In my years of instruction, I have seen students' highlighting where the entire paragraph is yellow, including every *a, the,* and *and.* This is an example of highlighting before knowing what is important. Now, I have helped you out somewhat. Important terms and concepts are in **bold face** followed by the definition (or the definition might be in the margin). So only highlight or underline with a pen essential ideas and key phrases—not complete sentences, if possible. By the way, the important microbiological terms and major concepts also are in the **Glossary** at the back of the text.

What if a paragraph or section has no boldfaced words? How do you find what is important here? From an English course, you may know that often the most important information is mentioned first in the paragraph. If it is followed by one or more examples, then you can backtrack and know what was important in the paragraph. In addition, I have written section summaries (called **To this point...**) to let you reorient yourself before getting too far ahead in the material. These summaries also are clues to what was important in the section you just read. Several chapters contain stories of local disease outbreaks expressed as pictorial representations, a medium that may help you remember them better. We call them **Textbook Cases.**

At the end of each chapter, you will find a **Note to the Student.** A textbook writer has to be objective, and one soon gets tired of "just the facts." I, therefore, have added a personal thought or two, which I offer for your consideration.

BE A REPETITIOUS STUDENT

Brain research has shown that each individual can only hold so much information in short-term memory. If you try to hold more, then something else needs to be removed—sort of like a full computer disk. So that you do not lose any of this important information, you need to transfer it to long-term memory—to the hard drive if you will. In reading and studying, this means retaining the term or concept; so, write it out in your notebook *using your own words.* Memorizing a term does not mean you have learned the term or understood the concept. By actively writing it out in your own words, you are forced to think and actively interact with the information. This repetition reinforces your learning.

BE A PATIENT STUDENT

In technical academic textbooks, you cannot read at the speed that you read your e-mail or a magazine story. There are unfamiliar details to be learned and understood—and this requires being a patient, slower reader. Actually, if you are not a fast reader to begin with, as I am, it may be an advantage in your learning process. Identifying the important information from a textbook chapter requires you to *slow down* your reading speed. Speed-reading is of no value here.

KNOW THE WHAT, WHY, AND HOW

Have you ever read something only to say, "I have no idea what I read!" As I've already mentioned, reading a microbiology text is not the same as reading *Sports Illustrated* or *People* magazine. In these entertainment magazines, you read passively for leisure or perhaps amusement. In *Alcamo's Fundamentals of Microbiology,* you must read actively for learning and understaming—that is, for *comprehension.* This can quickly lead to boredom unless you engage your brain as you read—that is, be an active reader. Do this by knowing the *what, why,* and *how* of your reading.

- *What* is the general topic or idea being discussed? This often is easy to determine because the section heading might tell you. If not, then it will appear in the first sentence or beginning part of the paragraph.

- *Why* is this information important? If I have done my job, the text section will tell you why it is important or the examples provided will drive the importance home. These surrounding clues further explain why the main idea was important.

- *How* do I "mine" the information presented? This was discussed under being a detective.

A MARKED UP READING EXAMPLE

So let's put words into action. The following page contains a passage from the text. I have marked up the passage as if I were a student reading it for the first time. It uses many of the hints and suggestions I have provided. Remember, it is important to read the passage slowly, and concentrate on the main idea (concept) and the special terms that apply.

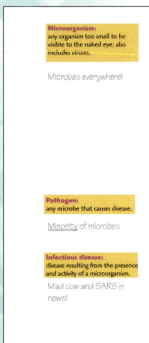

Microorganism:
any organism too small to be visible to the naked eye; also includes viruses.

Microbes everywhere!

Pathogen:
any microbe that causes disease.

Minority of microbes

Infectious disease:
disease resulting from the presence and activity of a microorganism.

Mad cow and SARS in news!

Microorganisms (or **microbes** for short) are present in vast numbers in nearly every environment and habitat on Earth, not just the Sargasso Sea. Microbes are found in Antarctica, on top of the tallest mountains, in the deepest parts of the oceans, and even miles down within the crust of the earth. In all, by weight they make up about two-thirds of Earth's living material. It is probably safe to say that if a microbe can match its metabolism to a habitat or environment on Earth, microorganisms will be found there.

Microbes inhabit the soil and flourish in oceans and freshwater lakes where they recycle nutrients and help control our climate. Microbes survive in, and might be purposely put in, many of the foods we eat. They also are found in the air we breathe and, at times, in the water we drink. Even closer to home, microorganisms are on our skin and grow in our mouth, ears, nose, throat, and digestive tract. Fortunately, the majority of these microbes are benign or may actually be beneficial in keeping the dangerous microbes out.

When most of us hear the word bacteria or virus though, we think infection or disease. Such **pathogens** periodically have carved out great swatches of humanity as epidemics passed over the land. Some diseases—such as plague, cholera, and smallpox—have become known as "slate-wipers," a reference to the barren towns they left in their wake (**MicroFocus 1.1**); others—such as measles, polio, and AIDS—have been more insidious in their progress but no less dangerous.

Infectious disease often left people thunderstruck with terror. Up to the 1850s, the fear of disease was compounded by ignorance because no one knew the cause of diseases, much less how to deal with them. Many strange brews and potions were offered as cures, but few were reliable, and some made an already bad situation even worse. Even today, with antibiotics and vaccines to cure and prevent many infectious diseases, they still bring concern and sometimes panic. Just think about the scares that AIDS, mad cow disease, and, more recently, severe acquired respiratory syndrome (SARS) have caused worldwide (**FIGURE 1.1**).

HAVE A DEBRIEFING STRATEGY

After reading the material, be ready to debrief. Verbally summarize what you have learned. This will start moving the short-term information into the long-term memory storage—that is, *retention*. Any notes you made concerning confusing material should be discussed as soon as possible with your instructor. For microbiology, allow time to draw out diagrams. Again, repetition makes for easier learning and better retention.

In many professions, such as sports or the theater, the name of the game is practice, practice, practice. The hints and suggestions I have given you form a skill that requires practice to perfect and use efficiently. Be patient, things will not happen overnight. But perseverance and willingness will pay off with practice. You might also check with your college or university academic (or learning) resource center. These folks may have more ways to help you to read a textbook better and to study well overall.

SEND ME A NOTE

In closing, I would like to invite you to write me and let me know what is good about this textbook so I can build on it and what is bad so I can change or eliminate it. Also, I would be pleased to hear about any news of microbiology in your community, and I'd be happy to help you locate any information not covered in the text. I can be reached at the Department of Biology, Glendale Community College, 6000 W. Olive Avenue, Glendale, Arizona 85302. Feel free to e-mail me at: jeffrey.pommerville@gcmail.maricopa.edu

I wish you great success in your microbiology course. Welcome! Let's now plunge into the wonderful and sometimes awesome world of microorganisms.

—Dr. P.

About the Author

Although today I am a microbiologist, researcher, and science educator, in high school in Santa Barbara, California I wanted to play professional baseball, study the stars, and own a "66" Corvette. None of these desires would come true—my batting average was miserable, I hated the astronomy correspondence course I took, and I never bought that Corvette.

I found an interest in biology at Santa Barbara City College. After squeaking through college calculus (I impressed my instructor so much, he told me, "You're certainly smarter than your sister-in-law"), I transferred to the University of California at Santa Barbara (UCSB) where I received a B.S. in Biology and stayed on to pursue a Ph.D. degree in the lab of Ian Ross studying cell communication and sexual pheromones in a water mold. After receiving my doctorate in Cell and Organismal Biology, my graduation was written up in the local newspaper as a native son who was a fungal sex biologist—an image that was not lost on my three older brothers!

While in graduate school at UCSB, I rescued a secretary in distress from being licked to death by a German Shepard. Within a year, we were married (the secretary and I). I then spent several years as a post-doctoral fellow at the University of Georgia. Worried that I was involved in too many research projects, a faculty member told me "Jeff, it's when you can't think of a project that you need to worry." When the post-doctoral position ended, I accepted a biology faculty position at Texas A&M University, where I spent eight years in teaching and research—and telling Aggie jokes. Toward the end of this time, after publishing over 30 peer-reviewed papers in national and international research journals, I realized that I had a real interest in teaching and education. Leaving the sex biologist nomen behind (but taking my wife of now 16 years), I headed farther west to Arizona to join the biology faculty at Glendale Community College where I teach introductory biology and microbiology.

I have been part of several educational research projects and have been honored with two of my colleagues with a Team Innovation of the Year Award by the League of Innovation in the Community Colleges, which recognizes outstanding contributions to teaching and learning in American community colleges. In 2000, I became project director and lead principal investigator for a National Science Foundation grant to improve student outcomes in science through changes in curriculum and pedagogy. Within the two-year period of the grant, I coordinated more than 60 science faculty members in designing and field testing 18 interdisciplinary science units. I was honored in 2003 to receive the Gustav Ohaus Award (College Division) for Innovations in Science Teaching from the National Science Teachers Association.

Currently, I am an editorial board member for *Microbiology Education*, the journal of the American Society for Microbiology (ASM) devoted to educational research in microbiology. In 2004, I was co-chair for the ASM Conference for Undergraduate Educators entitled "Facilitating Student Learning in Diverse Environments."

With the recent changes in world events, I have presented numerous seminars and workshops to colleges, universities, businesses, medical, and social organizations on understanding and responding to bioterrorism. Along the career road, life sometimes offers opportunities in unexpected ways. With the untimely passing of my friend and professional colleague Ed Alcamo, I have taken over the authorship of *Fundamentals of Microbiology.* It is an undertaking I relish as I (along with the wonderful folks at Jones and Bartlett) try to evolve a new breed of microbiology textbook that reflects the pedagogy change occurring in science classrooms today. And, hey, who knows—maybe that "66" Corvette could be in my garage yet.

DEDICATION

I dedicate this book to Ian K. Ross, Professor Emeritus and Research Professor

As a senior undergraduate, I enrolled in a mycology class taught by Ian at the University of California, Santa Barbara. Little did I know those many years ago that his course, his teaching, and eventual mentorship in research would define my career in biology and my fascination for the microbial world.

Alcamo's
Fundamentals of Microbiology

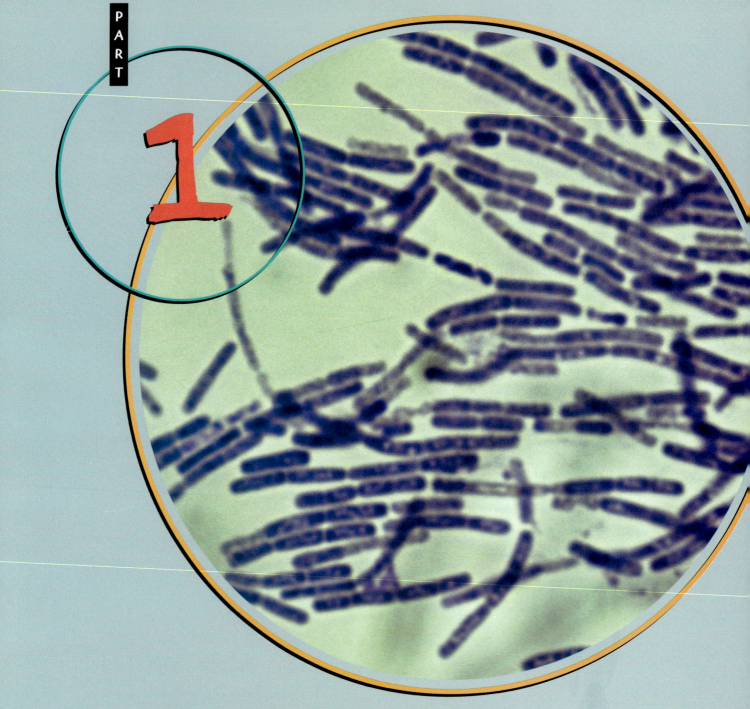

Foundations

of

Microbiology

I n 1676, a century before the Declaration of Independence, a Dutch merchant named Anton van Leeuwenhoek sent a noteworthy letter to the Royal Society of London. Writing in the vernacular of his home in the Netherlands, Leeuwenhoek described how he used a primitive microscope to observe vast populations of minute creatures. His reports opened a chapter of science that would evolve into the study of microscopic organisms and the discipline of microbiology. At that time, few people, including Leeuwenhoek, attached any practical significance to the microorganisms, but during the next three centuries, scientists learned how profoundly they influence the quality of our lives.

We shall begin our study of the microorganisms by exploring the grassroots developments that led to the establishment of microbiology as a science. These developments are surveyed in Chapter 1, where we focus on the individuals who stood at the forefront of discovery. Although the case is made for three Golden Ages for microbiology, Chapter 1 primarily describes events up to the end of the classical Golden Age in the early 1900s. The remaining chapters of this book describe in more detail the discoveries of microbiology since that time. Indeed, we are now in the midst of a third Golden Age and our understanding of microorganisms continues to grow even as you read this book.

Part 1 also contains a chapter on basic chemistry, inasmuch as microorganisms are chemical machines. Moreover, their activities are all related to chemistry. The third and final chapter in Part 1 will set down some basic concepts that apply to all microorganisms, much as the alphabet applies to word development. In succeeding chapters, we shall formulate words into sentences and sentences into ideas, as we survey the different groups of microorganisms and concentrate on their importance to public health and human welfare.

■ *Cells of* Bacillus anthracis, *the causative agent of anthrax.*

BEING A SCIENTIST

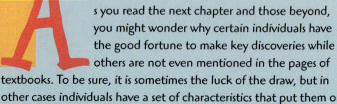

As you read the next chapter and those beyond, you might wonder why certain individuals have the good fortune to make key discoveries while others are not even mentioned in the pages of textbooks. To be sure, it is sometimes the luck of the draw, but in other cases individuals have a set of characteristics that put them on the trail to success.

Robert S. Root-Bernstein attempted to identify some of these characteristics in an article published in *The Sciences* (May/June 1988). Root-Bernstein points out that many prominent scientists like to goof around, play games, and surround themselves with a type of chaos aimed at revealing the unexpected. Their labs may appear to be a disorderly mess, but they know exactly where every tube or bottle belongs. Scientists also identify intimately with the things or creatures they study (it is said that Louis Pasteur actually dreamed about microorganisms), and this identification brings on an intuition, a sense about what an organism will do. Plus, there is the ability to recognize patterns that might eventually reveal hidden truths. (Pasteur had studied art as a teenager, and therefore he had an appreciation of patterns.)

In 1992, Professor Alcamo received a letter from a student at Red Deer College in Alberta, Canada, asking why he became a microbiologist. Essentially, he answered, "It was because I enjoyed my undergraduate microbiology course (thanks, Dr. Marazzella), and when I needed to select a graduate major, microbiology seemed like a good idea. I also think I had some of the characteristics described by Root-Bernstein: I loved to try out different projects; my corner of the world qualified as disaster area; still I was a nut on organization, insisting that all the square pegs fit into the square holes." If all this sounds familiar, then maybe you fit the mold of a scientist. Why not consider pursuing a career in microbiology? Some possibilities are listed in this book, but you should also visit with your instructor. Simply stop by the cafeteria, buy two cups of coffee, and you're on your way.

Microbiology: Then and Now

Science contributes to our culture in many ways, as a creative intellectual activity in its own right, as the light which has served to illuminate man's place in the universe, and as the source of understanding of man's own nature.

—John F. Kennedy (1917–1963)

W HEN WE HEAR news stories about the Bermuda or Devil's Triangle, we might think of the ships, boats, and aircraft that have disappeared in this region apparently without a trace. This imaginary area, located off the southeastern US coast, also makes up the southwestern part of the Sargasso Sea, often called an ocean desert ecosystem because of the lack of nutrient turnover.

In 1990, microbiologist Stephen Giovannoni of Oregon State University identified in the Sargasso Sea what is perhaps the most abundant organism on the planet. Called SAR11 (SAR for Sargasso), this bacterium, which has been renamed *Pelagibacter ubique,* has been identified across the oceans of the world. What makes it significant is its population size, which Giovannoni estimates to be 2.4×10^{28} cells. (Compare this to the 6 billion (6×10^9) humans and 2.4×10^{17} termites on Earth and you see how big a number this is!) Since there are an estimated 1.2×10^{29} bacteria living in the open oceans, SAR11 alone would account for 20 percent of all oceanic bacteria! Add in all the terrestrial bacteria (5×10^{30}) and SAR11 still accounts for 0.5 percent of all bacteria on Earth.

With this number of SAR11 cells, it is amazing that the bacteria were not observed until 2002. The reason is that, until recently, they could not be grown in the lab. SAR11 was originally discovered using techniques that look for the prevalence of specific DNA fragments in water samples. Such methods can detect specific microorganisms without visually seeing them. Happily, genomics and cell culture now have come together. Giovannoni's group has cultured the SAR11 bacteria and discovered that these crescent-shaped bacteria are among the smallest known.

SAR11's success story suggests the bacterium must have a significant impact on the planet. Although such roles remain to be identified, Giovannoni believes that they are responsible for up to 10 percent of all nutrient recycling on the planet, influencing the cycling of carbon, and even affecting global warming.

Also plumbing the depths of the Sargasso Sea is J. Craig Venter and his biotech team at the Institute for Biological Energy Alternatives. Fresh from his success in sequencing the human genome, Venter's team plans on reading the DNA of every organism—microbial at least—in the Sargasso Sea. Over a period of about one year, they will isolate the microbial DNAs, fragment them into small, random pieces, and use a large array of sequencing machines to piece the DNA molecules back together to their respective microbes. Reading all the DNA sequences is routine. Trying to match DNA sequences to the correct microbe however might be like taking a pile of tens of thousands of jigsaw puzzle pieces, sorting them by puzzle type, and reassembling them into their correct "picture." Why would Venter's group wish this on themselves? Venter believes that by sequencing the Sargasso microbes, his institute can discover novel metabolic pathways or species that can be used for alternative energy sources and to help curb global warming.

So, unlike the Bermuda Triangle, the Sargasso Sea is giving up some of its secrets—well, microbial at least.

Microorganism:
any organism too small to be visible to the naked eye; also includes viruses.

Microorganisms (or **microbes** for short) are present in vast numbers in nearly every environment and habitat on Earth, not just the Sargasso Sea. Microbes are found in Antarctica, on top of the tallest mountains, in the deepest parts of the oceans, and even miles down within the crust of the earth. In all, by weight they make up about two-thirds of Earth's living material. It is probably safe to say that if a microbe can match its metabolism to a habitat or environment on Earth, microorganisms will be found there.

Microbes inhabit the soil and flourish in oceans and freshwater lakes where they recycle nutrients and help control our climate. Microbes survive in, and might be purposely put in, many of the foods we eat. They also are found in the air we breathe and, at times, in the water we drink. Even closer to home, microorganisms are on our skin and grow in our mouth, ears, nose, throat, and digestive tract. Fortunately, the majority of these microbes are benign or may actually be beneficial in keeping the dangerous microbes out.

Pathogen:
any microbe that causes disease.

When most of us hear the word bacteria or virus though, we think infection or disease. Such **pathogens** periodically have carved out great swatches of humanity as epidemics passed over the land. Some diseases—such as plague, cholera, and small-pox—have become known as "slate-wipers," a reference to the barren towns they left in their wake (MicroFocus 1.1); others—such as measles, polio, and AIDS—have been more insidious in their progress but no less dangerous.

Infectious disease:
disease resulting from the presence and activity of a microorganism.

Infectious disease often left people thunderstruck with terror. Up to the 1850s, the fear of disease was compounded by ignorance because no one knew the cause of diseases, much less how to deal with them. Many strange brews and potions were offered as cures, but few were reliable, and some made an already bad situation even worse. Even today, with antibiotics and vaccines to cure and prevent many infectious diseases, they still bring concern and sometimes panic. Just think about the scares that AIDS, mad cow disease, and, more recently, severe acquired respiratory syndrome (SARS) have caused worldwide (FIGURE 1.1).

MicroFocus 1.1

THE TRAGEDY OF EYAM

Each year a group of English pilgrims gather in the English countryside outside the village of Eyam, to pay homage to the townsfolk who gave their lives three centuries before so that others might live. The pilgrims pause, bow their heads, and remember.

Bubonic plague erupted in Eyam in the spring of 1666. The rich were the first to flee the town, and soon the commonfolk also considered leaving. However, they knew that by doing so they would probably spread the disease to nearby communities. At this point, the village rector made a passionate plea that they stay. After some deep soul-searching, most resolved to remain. They marked off a circle of stones outside the village limits, and people from the adja-

cent towns nervously brought food and supplies to the barrier, leaving it there for the self-quarantined villagers. In the end, 259 of the town's 350 residents succumbed to the plague.

The memorial service has a poignant moment as the pilgrims somberly recite a rhyme traced to that period:

> *Ring-a-ring of rosies*
> *A pocketful of posies*
> *Achoo! Achoo!*
> *We all fall down.*

There is no laughter in the group; indeed, some are moved to tears. The ring of rosies refers to the rose-shaped splotches on the chest and armpits of plague victims. Posies were tiny flowers the people hoped would ward off the

evil spirits. "Achoo!" refers to the fits of sneezing that accompanied the disease. And the last line, the saddest of all, suggests the death that befell so many.

■ *A dance in a graveyard to ward off the plague.*

FIGURE 1.1

Recent Infectious Diseases

(a) The agent of mad cow disease, scientifically called bovine spongiform encephalopathy (BSE), can be transmitted to humans who consume infected processed meats. (b) People wearing masks in an attempt to avoid being infected with the SARS virus.

(a)

(b)

A major focus of this introductory chapter is to give you an introspective "first look" at microbiology—then and now. Although the study of microorganisms began in earnest with the work of Pasteur and Koch, they were not the first to report microorganisms. To begin our story, we reach back to the 1600s, where we encounter some equally inquisitive individuals.

1.1

The Beginnings of Microbiology

In 1665, Robert Hooke, an English naturalist (the term scientist was not coined until 1833), published a work called *Micrographia*. The book contained, among other things, a description of his microscope and its uses. Hooke's writings and his illustrations, such as those shown in FIGURE 1.2, awakened the learned of Europe to the world of the very small. He also described a slice of cork and suggested that it was composed of tiny compartments, which he called "cella." With this first account of "cells," he secured a place in the history of biology.

The discoveries by Hooke and others showed that the microscope was an important tool for unlocking the secrets of nature. It is not surprising, therefore, that naturalists would be interested in Anton van Leeuwenhoek's descriptions of microorganisms.

ANTON VAN LEEUWENHOEK DESCRIBES "ANIMALCULES"

Anton van Leeuwenhoek was a draper, haberdasher, and owner of a dry goods business in Delft, the Netherlands. He was head of the City Council, inspector of weights and measures, and court surveyor. In his spare time, he ground pieces of glass into fine lenses, placing them between two silver or brass plates riveted together,

---

FIGURE 1.2

Hooke's *Micrographia* and Microscope

(a) In 1665, Robert Hooke published his observations and drawings in a volume called *Micrographia*. Shown here is his drawing of thin shavings of cork as he saw them through his microscope. (b) A drawing of the microscope used by Hooke.

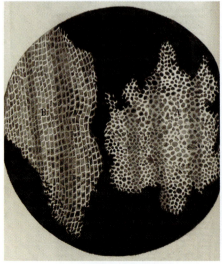
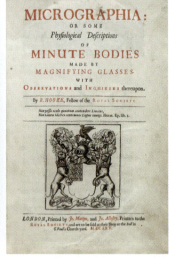

(a)   (b)

as shown in FIGURE 1.3. Gradually, he developed a skill that would remain unmatched for many generations.

Leeuwenhoek's lenses were no larger than the head of a pin, but by most modern accounts, they could magnify an object nearly 300 times. Initially he used the lenses to inspect the quality of cloth, but as his fascination with microscopic objects developed, his attention turned to observing living objects, including hair fibers, skin scales, blood cells—even samples of his own feces! As one anonymous writer penned:

He worked instead of tending store,
At grinding special lenses for
A microscope. Some of the things
He looked at were: mosquito wings,
The hairs of sheep, the legs of lice,
The skin of people, dogs, and mice.

The process of observation is an important skill for all scientists, including microbiologists—and Leeuwenhoek was an obsessed observer. He would look at the same sample with his microscopes repeatedly before he would accept his observations as accurate. He believed only sound observation and experimentation could be trusted—a requirement that remains a cornerstone of all sciences today.

Leeuwenhoek's work did not go unnoticed. Probably the only scientific group of that period was the Royal Society of London, but correspondence with this group was tenuous because England and the Netherlands were bitter rivals for commercial treasures of the East Indies. Nevertheless, through Royal Society contacts, Leeuwenhoek soon was sending along his illustrations and observations.

FIGURE 1.3

Anton van Leeuwenhoek

(a) Leeuwenhoek at work in his study. Using a primitive microscope, Leeuwenhoek was able to achieve magnifications of nearly 300 times and describe various biological specimens, including numerous types of microorganisms. (b) Details of Leeuwenhoek's lens system. The object is placed on the point of the specimen mount. The mount is adjusted by turning the focusing screw and the elevating screw. Light is reflected from the specimen through the lens, thereby magnifying the specimen.

(a)

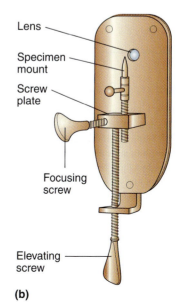

Lens

Specimen mount

Screw plate

Focusing screw

Elevating screw

(b)

In 1674, Leeuwenhoek examined a sample of the greenish, cloudy water from a marshy lake outside Delft. Placing the sample before his lens, he saw hundreds of what he thought were tiny animals, which he called **animalcules**. His curiosity aroused, Leeuwenhoek soon located animalcules in rainwater, and in material from his own teeth and feces. The creatures astonished him at first, then delighted and perplexed him as he pondered their origin and purpose.

In 1676, Leeuwenhoek sent his eighth letter to the Royal Society. This letter has special significance to microbiology because it contains the first description of what were probably bacteria:

> *May the 26th, I took about ⅓ of an ounce of whole pepper and having pounded it small, I put it into a Thea-cup with 2½ ounces of Rainwater upon it, stirring it about, the better to mingle the pepper with it, and then suffering the pepper to fall to the bottom. After it had stood an hour or two, I took some of the water, before spoken of, wherein the whole pepper lay, and wherein were so many several sorts of little animals; and mingled it with this water, wherein the pounded pepper had lain an hour or two, and observed that, when there was much of the water of the pounded pepper, with that other, the said animals soon died, but when little they remained alive.*
>
> *June 2, in the morning, after I had made divers observations since the 26th of May, I could not discover any living thing, but saw some creatures, which tho they had the figures of little animals, yet could I perceive no life in them how attentively I beheld them.…*

Leeuwenhoek's sketches were elegant in detail and clarity. His letters outlined structural details of protozoa and described threadlike fungi and microscopic algae. A particularly noteworthy letter, written in 1683, included drawings of what we believe were living bacteria from teeth scrapings (**FIGURE 1.4**). His observations, performed exclusively with single-lens microscopes, opened the door to a completely new world of living things.

Leeuwenhoek was elected to fellowship in the Royal Society in 1680, and, with Isaac Newton and Robert Boyle, he became one of the most famous men of his time. Peter the Great of Russia and Queen Anne of England came to peer through his microscopes. In all, Leeuwenhoek bombarded the Royal Society with over 200 letters

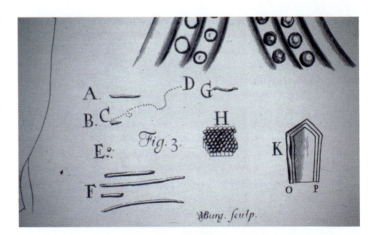

FIGURE 1.4

Leeuwenhoek's Drawings of Bacteria

Examples from a letter written by Leeuwenhoek to the Royal Society dated September 17, 1683. Both A and B represent rod forms, with C and D indicating the pathway of motion. A spherical form is shown in E and longer types are depicted in F. The type shown in G appears to be a spiral form, and a cluster of spheres is shown in H. Leeuwenhoek found many of these organisms between his teeth.

MicroFocus 1.2

THAT WAS THEN . . .

Warning: Do not read this box if you are squeamish, if you've just finished eating, or if you get upset easily. We are going to tell you what life was like in Anton van Leeuwenhoek's time when no one had any inkling that microorganisms are the agents of infectious disease.

At that period in history, it was customary for a person to be bathed three times during his or her life: at birth, on getting married, and after death. This applied to people of nobility as well as peasants. Skin boils and abscesses were common.

There was no indoor plumbing, and chamber pots were often emptied out the window. The contents landed on the street below and remained there. In other places, flies moved easily from outdoor cesspools to the surfaces of foods, and when the cesspool flooded, its runoff mixed with the community drinking water. Typhoid fever, cholera, and dysentery were rampant.

There were no refrigerators, so food decomposed almost overnight. This left people the choice of either eating rotten food or starving. Food poisoning and botulism broke out often.

Rats patrolled the streets and were everywhere, carrying their fleas to all corners of the community. No one was without lice, and during the hot months, mosquitoes were as thick as fog. Epidemics of malaria, plague, and typhus fever occurred regularly.

In 2004, the life expectancy of an average American is estimated to be roughly 78 years. In Europe, during Leeuwenhoek's time (1775), the life expectancy was 42. Small wonder.

about his findings. He died in 1723 at the age of 90. His longevity was a notable achievement in itself considering what life was like at that time (**MicroFocus 1.2**).

Leeuwenhoek was a very suspicious and secretive person. He invited no one to work with him, nor did he show anyone how to grind lenses or construct a microscope. Such secrecy stalled the advancement of the knowledge of microorganisms because, after his death, it would take over 100 years to grind lenses and construct a microscope equal to or better than those of Leeuwenhoek. Also, without these instruments, naturalists could not repeat his observations or verify his results, which are key components of scientific inquiry (see MicroInquiry 1). Another reason, and perhaps of more significance, is that natural philosophers of the 1700s saw microbes simply as curiosities of nature. Certainly, they could not cause disease in so lofty a creature as the human being.

To this point . . .

The discoveries made by Robert Hooke and Anton van Leeuwenhoek awakened the world to a previously invisible world of microorganisms. Leeuwenhoek used his hand lenses (microscopes) to observe and describe many of the protozoa, fungi, and bacteria that we commonly recognize today. After his death, studies of the animalcules diminished because other naturalists could not replicate Leeuwenhoek's microscopes. In addition, they could not conceive how these tiny creatures could cause sickness and death.

In the next section, we will see that during this lull in understanding life and the animalcules, arguments over the spontaneous generation of life occupied many natural philosophers. However, other naturalists would try to understand better the nature of infectious disease.

The Transition Period

In the 1700s, few naturalists continued to explore the microscopic world, although Louis Joblot published a review of protozoa in 1718. Most naturalists did not believe such microscopic organisms could cause infection. They hypothesized that infectious diseases, such as plague and malaria, spread by an altered chemical quality of the atmosphere or a poisoning of the air, an entity called **miasma**. (The word malaria comes from *mala aria,* meaning "bad air.") Miasma also might arise from decaying or diseased bodies known as miasms. The miasma idea figured prominently in medical thinking well into the 1800s.

As the years unfolded, some naturalists began to scrutinize the laws of nature and to question the origin of living things, as they exist today. Importantly, such investigations required experimentation as a way of knowing and explaining divergent observations.

DOES LIFE GENERATE ITSELF SPONTANEOUSLY?

Spontaneous generation: the doctrine that held that non-living matter could spontaneously give rise to living organisms.

In the fourth century B.C., Aristotle wrote that flies, worms, and other small animals could arise from decaying matter (i.e., without the need of parent organisms). His observations laid the basis for the doctrine of **spontaneous generation**. Indeed, in the early 1600s, the eminent Flemish physician Jan Baptista van Helmont lent credence to the belief when he observed that rats "originate" from wheat bran and old rags. Common people embraced the idea, for even they could see what appeared to be slime that produced toads and meat that generated wormlike maggots. Leeuwenhoek was one who opposed this idea. His observations on reproduction of grain weevils suggested that they did not arise from wheat grains, but rather they came from tiny eggs in the grain that he saw with his microscopes. Such divergent observations required careful clarification through experimentation.

Among the first to dispute the theory of spontaneous generation using experimentation was the Florentine naturalist, Francesco Redi. Noting Leeuwenhoek's descriptions, Redi reasoned that because flies had reproductive organs, they could land on pieces of exposed meat and lay their invisible eggs, which then hatch into visible maggots. To test this idea, in 1668 Redi performed one of history's first experiments in biology (**FIGURE 1.5**). He set up a series of tests in which he covered some jars of meat with fine lace, thereby preventing the entry of flies. So protected, the meat should not produce maggots. Redi's experiments supported his idea and temporarily put to rest the notion of spontaneous generation.

Reports of animalcules were becoming widespread during the 1700s, and in 1748, a British clergyman named John Needham suggested they arise spontaneously in tubes of mutton gravy. Needham even boiled several tubes of gravy and sealed the tubes with corks, as Redi had sealed his jars. Still, the tiny creatures appeared. The Royal Society of London was duly impressed and elected Needham to membership.

It was an Italian cleric and naturalist, Lazzaro Spallanzani, who first criticized Needham's work. In 1767, Spallanzani boiled meat and vegetable broths for longer periods and then sealed the necks by melting the glass. As control experiments, he left some flasks open to the air, stoppered some loosely with corks, and boiled some briefly, as Needham had done. After two days, he found the open flasks swarming

FIGURE 1.5

Francesco Redi: Disputing Spontaneous Generation

In the 1670s, Francesco Redi (a) attempted to disprove the belief that maggots (fly larvae) arise from decaying meat. (b) He placed a piece of meat in an open jar and showed that maggots originate as eggs laid by flies. People of that period believed that the maggots arose spontaneously from the decaying meat. He then placed a second meat sample in a jar covered with lace. The flies could not reach the meat, and maggots did not appear on the surface. This relatively simple experiment refuted spontaneous generation and was among the first ever performed in biology.

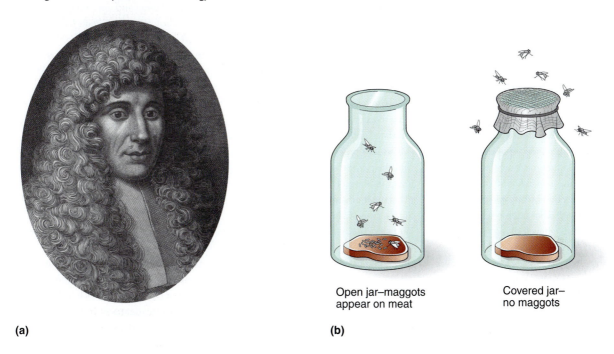

Open jar—maggots appear on meat

Covered jar— no maggots

(a)

(b)

with microorganisms, but the sealed flasks contained none. Spallanzani's experiment did not settle the issue, though. Needham countered that Spallanzani had destroyed the "vital force" of life with excessive heating. Other naturalists also suggested that the air necessary for life had been excluded by sealing the flasks.

The controversy over spontaneous generation continued into the mid-1800s when the idea of spontaneous generation finally was debunked through an elegant series of experiments carried out by Louis Pasteur. His well-thought-out experiments and the process of scientific inquiry are explained in MicroInquiry 1. Thus, Pasteur's work brought to an end the long and tenacious debate on spontaneous generation begun two centuries earlier.

EPIDEMIOLOGY LENDS SUPPORT TOWARD UNDERSTANDING DISEASE TRANSMISSION

While spontaneous generation was being debated and investigated, other naturalists studied how infectious disease could be transmitted. People had realized for centuries that diseases could be contagious and quarantines were used as early as the 1300s to combat the spread of diseases like plague. In 1546, the Italian poet and naturalist Girolamo Fracostoro wrote: "Contagion is an infection that passes from one thing to another." Fracostoro recognized three forms of transmission: through direct contact; from lifeless objects like clothing and eating utensils; and through the air.

MicroInquiry 1

EXPERIMENTATION AND SCIENTIFIC INQUIRY

Science certainly is a body of knowledge as you can see from the thickness of this textbook! However, science also is a process—a way of learning. Often we accept and integrate into our understanding new information because it appears consistent with what we believe is true. But, are we confident our beliefs are always in line with what is actually true? To test or challenge current beliefs, scientists must present logical arguments supported by well-designed and carefully executed **experiments**.

In this chapter, we are discovering that the idea that microorganisms cause infectious disease was hard to accept because infection by tiny microbes did not agree with what people believed. It was only through well thought out experiments by Pasteur, Koch, and others that physicians finally were convinced that microorganisms could cause disease.

THE COMPONENTS OF SCIENTIFIC INQUIRY

There are many ways of finding out the answer to a problem. In science, **scientific inquiry**—or what has been called the "scientific method"—is the way problems are investigated. Let's understand how scientific inquiry works by following the logic of the experiments Louis Pasteur published in 1862 to refute the idea of spontaneous generation (see **FIGURE A** "Pasteur and the Spontaneous Generation Controversy" in this box).

When studying a problem, the inquiry process usually begins with **observation**. For spontaneous generation, Pasteur's earlier observations suggested that organisms do not appear from nonliving matter (see text discussion of the early observations supporting spontaneous generation).

Next comes the **question**, which can be asked in many ways but usually as a "what," "why," or "how" question. For example, What accounts for the generation of flies on rotten meat? Or, How can flies develop spontaneously from

rotten meat? There is more than one way to ask a valid scientific question.

From the question, various hypotheses are proposed that might answer the question. A **hypothesis** is a provisional but *testable* explanation for an observed phenomenon. In almost any scientific question, several hypotheses can be proposed to account for the same observation. However, previous work or observations usually bias which hypothesis looks most promising, and scientists then put their "pet hypothesis" to the test first.

Pasteur's previous work suggested that the purported examples of life arising spontaneously in mutton gravy or other meat or vegetable broths were simply cases of airborne microorganisms landing on a suitable substance and then multiplying in such profusion that they could be seen.

PASTEUR'S EXPERIMENTS

Pasteur set up a series of experiments to prove the hypothesis that *Life only arises from other life* (see Figure A).

Experiment 1a and 1b: Pasteur sterilized a meat broth in glass flasks by heating. He then either left the neck open to the air (a) or sealed the glass neck (b). Organisms only appeared (turned the broth cloudy) in the open flask.

Experiment 2a and 2b: Pasteur sterilized a meat broth in necked glass flasks by heating. The glass neck was either heated (a) or left unheated in swan-neck flasks (b), so named because their S-shaped necks resembled a swan's neck. No organisms appeared in either case, even after several days.

ANALYSIS OF PASTEUR'S EXPERIMENTS

Let's analyze the experiments. Pasteur had a preconceived notion of the truth and designed experiments to test his hypothesis. In his experiments, only one **variable** (an adjustable condition) changed. In experiment 1, the neck was open or closed; in experiment 2, the

neck was heated or it wasn't heated. Pasteur kept all other factors the same; that is, the broth was the same in each experiment; it was heated the same length of time; and similar flasks were used. Thus, the experiments had rigorous **controls** (the comparative condition): in experiment 1, the control was the flask left open; in experiment 2, the control was the unheated swan-neck. Such controls are pivotal when explaining an experimental result. Pasteur's finding that no life appeared in the swan-necked flask (experiment 2b) is interesting, but tells us very little by itself. We only learn something by comparing this finding to the result in the control condition: that life did grow when the flask neck was straight (experiment 1a) or in an extension of the experiment when the neck was cut off.

Also note that the idea of spontaneous generation could not be dismissed by just one experiment (see "His critics" in Figure A). Pasteur's experiments required the accumulation of many experiments, all of which pointed to the same conclusion.

HYPOTHESIS AND THEORY

When does a hypothesis become a theory? The answer is that there is no set time or amount of evidence that specifies the change from hypothesis to theory. A **theory** is defined as a hypothesis that has been tested and shown to be correct every time by many separate investigators. So, at some point, sufficient evidence exists to say a hypothesis is now a theory. However, theories are not written in stone. They are open to further experimentation and so can be refuted.

As a side note, a theory often is used incorrectly in everyday speech and in the news media. In these cases, a theory is equated incorrectly with a hunch or belief—whether or not there is evidence to support it. In science, a theory is a general set of principles, supported by large amounts of experimental evidence.

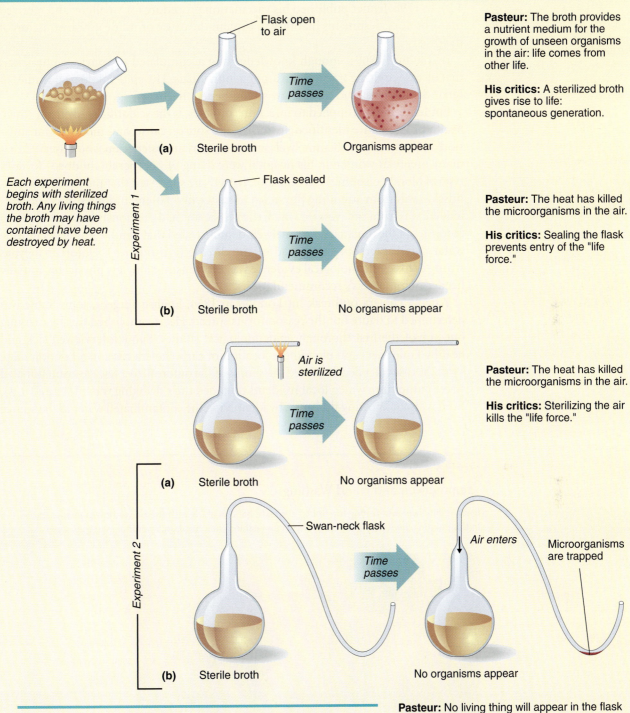

Pasteur: The broth provides a nutrient medium for the growth of unseen organisms in the air: life comes from other life.

His critics: A sterilized broth gives rise to life: spontaneous generation.

Each experiment begins with sterilized broth. Any living things the broth may have contained have been destroyed by heat.

Flask open to air

(a) Sterile broth → *Time passes* → Organisms appear

Flask sealed

(b) Sterile broth → *Time passes* → No organisms appear

Pasteur: The heat has killed the microorganisms in the air.

His critics: Sealing the flask prevents entry of the "life force."

Air is sterilized

(a) Sterile broth → *Time passes* → No organisms appear

Pasteur: The heat has killed the microorganisms in the air.

His critics: Sterilizing the air kills the "life force."

Swan-neck flask

Air enters   Microorganisms are trapped

(b) Sterile broth → *Time passes* → No organisms appear

Experiment 1

Experiment 2

Pasteur: No living thing will appear in the flask because microorganisms will not be able to reach the broth.

His critics: If the "life force" has free access to the flask, life will appear, given enough time.

Some days later the flask is still free of any living thing. Pasteur has refuted the doctrine of spontaneous generation.

FIGURE A

Pasteur and the Spontaneous Generation Controversy

(1a) When a flask of sterilized broth is left open to the air, organisms appear. (1b) When a flask of sterilized broth is boiled and sealed, no living things appear. (2a) When air entering a flask of sterilized broth is heated with a flame, no living things appear. (2b) Broth sterilized in a swan–neck flask is left open to the air. The curvature of the neck traps dust particles and microorganisms, preventing them from reaching the broth.

Although naturalists were slow to accept microorganisms as the agents of contagion, Edward Jenner's work to develop a method of inoculation against smallpox did show that contagion, however transmitted, could be interrupted (MicroFocus 1.3). In 1798, Jenner published his vaccination procedure for smallpox and received many honors, despite his inability to explain the cause or mode of transmission of the disease.

By the mid-1800s, enough knowledge had accumulated to convince physicians that disease could be transmitted among individuals. The epidemiological studies of Ignaz Semmelweis and John Snow strengthened this belief.

Epidemiology, as applied to infectious diseases, is the scientific study from which health problems are identified, including the source, cause, and mode of transmission of diseases. Ignaz Semmelweis was a Hungarian obstetrician who observed that many pregnant women in his hospital were dying of puerperal (childbed) fever (a type of blood poisoning) during labor. In1847, he suggested that the mode of transmission was from attending physician to maternity patient and the source, he contended, was from cadavers on which the physicians had previously been performing autopsies in the mortuary. Semmelweis demonstrated that hand washing in chlorine water could interrupt the spread of the disease (FIGURE 1.6). Unfortunately, as strong as his epidemiological evidence was for transmission, few physicians initially heeded Semmelweis's recommendations.

When a cholera outbreak hit London in 1854, English surgeon John Snow was determined to figure out the reason for its spread. He accomplished this by carrying out one of the first thorough epidemiological studies. Snow interviewed sick and healthy Londoners and plotted the location of each cholera case on a London map. With the cases mapped, he traced the source of the outbreak to a sewage-contaminated street pump from which local residents obtained their drinking water (FIGURE 1.7). Even without knowing the cause, by identifying the contaminated water as the source

FIGURE 1.6

Semmelweis and Hand Washing

Ignaz Semmelweis was among the first to correlate hand washing with the prevention of disease spread. In this painting, he is seen overseeing physicians washing their hands in a hospital maternity ward.

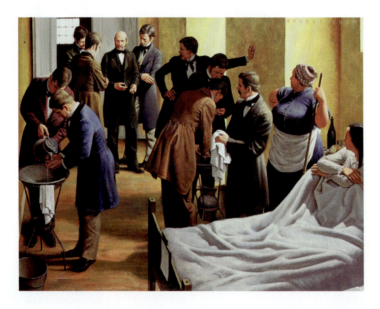

AHEAD OF HIS TIME

In the 1700s, smallpox was hardly ever absent from Europe. In England, epidemics were so severe that one-third of the children died before reaching the age of three. Many victims were blinded by smallpox, and most people were left pockmarked for life.

But for those who survived the pox, immunity lasted forever. People therefore sought a way of contracting a mild form of the disease. From the Far East came word of inoculation parties. A doctor would make a small wound in the arm and insert a few drops of pus from a smallpox skin lesion. A walnut was then tied over the area. Though mild smallpox often resulted, there was the danger of developing a deadly form.

Then came Edward Jenner. In the 1700s, while apprenticed to a country surgeon, Jenner learned that people who experienced cowpox were apparently immune to smallpox. Cowpox was a disease of the udders of cows. It was common in farmers and milkmaids, and it expressed itself as a mild, smallpox-like disease.

For years, Jenner pondered whether intentionally giving cowpox to people would protect them against smallpox.

He finally decided to put the matter to the test. In 1796, a dairy maid named Sarah Nelmes came to his office, the lesions of cowpox evident on her hand. Jenner took material from her lesions and scratched it into the skin of a boy named James Phipps. Several weeks later he inoculated young Phipps with material from a smallpox lesion. Within days, the boy developed a reaction at the site but failed to show any sign of smallpox.

Jenner continued his experiments for two years, switching to infected cows as a source of cowpox material. In 1798, he published an historic pamphlet on his work that generated considerable interest. Prominent physicians confirmed his findings, and within a few years, Jenner's method of "vaccination" spread through the world (*vacca* is Latin for "cow"). It is estimated that by 1801, some 100,000 people in England were vaccinated. In Russia, the first child vaccinated was renamed Vaccinor and was educated by the state. And President Thomas Jefferson wrote to Jenner: "You have erased from the calendar of human afflictions one of its great-

est. Yours is the comfortable reflection that mankind can never forget that you have lived."

A hundred years passed before scientists realized that the milder cowpox viruses were setting up a defensive mechanism in the body against the deadlier smallpox viruses. It is thus remarkable that Jenner accomplished what he did. His experimentation methods and interpretation of the vaccination results might well serve as models for accurate and careful laboratory work in modern microbiology.

■ *A cartoon from Jenner's time showing skepticism for his new method.*

FIGURE 1.7

Cholera and the Broad Street Pump

John Snow was able to correlate the spread of cholera with a contaminated water pump on Broad Street in London. This cartoon shows "death" pumping water to neighborhood children and adults.

of the disease, Snow instituted the first known example of a public health measure to interrupt disease transmission—he had the street pump handle removed!

In their work, both Semmelweis and Snow also drew attention to the fact that a poison or unseen object in the environment, not a miasma, was responsible for disease. Today, we have a good grasp of disease transmission mechanisms as we will discuss in Chapter 18. This understanding has allowed public health systems to produce better protective measures against the spread of disease. However, even with the advances that have been made in sanitation and public health, cholera remains a public health concern in many parts of the developing world. In addition, more than 150 years after Semmelweis's suggestions, a lack of hand washing by hospital staff and physicians, even in developed nations, ranks as a major mechanism for disease transmission from patient to patient. The simple mechanics of washing one's hands with soap and water still could reduce substantially disease transmission in hospitals (FIGURE 1.8).

TABLE 1.1 is a summary of the early observations and conclusions regarding spontaneous generation and disease transmission.

To this point . . .

We have described spontaneous generation and showed how the controversy surrounding this belief stirred naturalists to question the nature of living things. Redi, Needham, and Spallanzani achieved distinction during the controversy, though Pasteur's experiments were needed to finally discredit the idea. Importantly, the debate focused attention on the fundamental nature of life and encouraged naturalists to undertake scientific experimentation.

We next focused on disease transmission. The epidemiological work of Semmelweis and Snow showed how naturalists were coming to believe that something tangible in nature was causing disease. This contrasted with the miasma idea, which pointed to a vague, intangible disease cause.

In the next segment, we consider how Louis Pasteur and Robert Koch sought to establish the principle that microorganisms cause infectious disease. We will discuss their experiments and observe how the proofs for the germ theory of disease evolved. As the foundations of microbiology were strengthened, many other scientists expanded the work of Pasteur and Koch, and worldwide interest in microorganisms emerged.

FIGURE 1.8

The Importance of Clean Hands

Clean hands save lives as this Centers for Disease Control and Prevention (CDC) poster illustrates. Better hand hygiene can prevent outbreaks of diseases from within hospitals and other health care facilities.

TABLE 1.1

Some Early Observations in Microbiology

INVESTIGATOR	TIME FRAME	OBSERVATIONS/CONCLUSIONS
Aristotle	Fourth century B.C.	Living things do not need parents; spontaneous generation apparently occurs
Fracostoro	Mid-1500s	"Contagion" passes among individuals, objects, and air
Kircher	Mid-1600s	"Microscopic worms" are present in blood of plague victims
Redi	Mid-1600s	Fly larvae do not arise by spontaneous generation
Leeuwenhoek	Late 1600s	Microscopic organisms are present in numerous environments
Fabricius	Early 1700s	Fungi cause rust and smut diseases in plants
Joblot	Early 1700s	Various forms of protozoa exist
Needham	Mid-1700s	Microorganisms in broth arise by spontaneous generation
Spallanzani	Mid-1700s	Heat destroys microorganisms in broth
Jenner	Late 1700s	Recoverers from cowpox do not contract smallpox
Semmelweis	Mid-1800s	Chlorine disinfection prevents disease spread
Snow	Mid-1800s	Water is involved in disease transmission

1.3

The Classical Golden Age of Microbiology (1854–1914)

Beginning around 1854, microbiology blossomed through the work of Louis Pasteur and continued into the 20th century until the advent of World War I. During these 60 years, numerous branches of microbiology were established, and the foundations were laid for the maturing process that has led to modern microbiology. We shall refer to this period as the first, or classical, Golden Age of microbiology because we will see that in fact three such epochs exist.

LOUIS PASTEUR PROPOSES THAT GERMS CAUSE INFECTIOUS DISEASE

In a world ravaged by plague, tuberculosis, typhoid fever, and diphtheria, neither royalty nor commonfolk were immune to disease. There were virtually no cures for disease, and no one had yet confirmed the roles of microbes as the cause of infectious disease.

Such were the times in which Louis Pasteur studied at the French school, the École Normale Supérieure (**FIGURE 1.9**). In 1848, he achieved distinction in organic chemistry for his discovery of two different types of tartaric acid crystals. Using a microscope, Pasteur successfully separated the crystals and developed a skill that would aid his later studies of microorganisms. In 1854, at the age of 32, he was appointed Professor of Chemistry at the University of Lille in northern France.

Extending from President John Kennedy's quote at the top of this chapter, microbiology always has focused on its applications to health, nutrition, and the environment. Indeed, Pasteur was among the first scientists who believed that the discoveries of science should have practical applications. He therefore grasped the opportunity in 1857 to try to unravel the mystery of why local wines were turning sour. The prevailing theory held that wine fermentation resulted from the chemical breakdown of grape juice to alcohol. No living thing seemed to be involved. But, Pasteur's microscope consistently revealed large numbers of tiny yeast cells in wine that were overlooked by other scientists. Importantly, he observed that only soured wines also contained populations of barely visible sticks and rods, known then and now as bacteria.

In another classic series of experiments, Pasteur clarified the role of yeasts in fermentation and hypothesized that the contaminating bacteria were responsible for sour wine. To test his hypothesis, he set up a series of experiments. First, Pasteur removed all traces of yeasts from a flask of grape juice and set the juice aside. Nothing happened. Next, he added back the yeasts, and soon the fermentation was proceeding normally. He then found that if he could remove all bacteria from the grape juice, the wine would not turn sour.

Pasteur's work shook the scientific community. His experimental results demonstrated that yeast cells and bacteria are tiny, living factories in which important chemical changes take place. His work also drew attention to microorganisms as agents of change, because bacteria appeared to make the wine "sick."

For years, physicians had interpreted bacteria as an *effect* of disease; that is, they were thought to arise in the body during illness. Pasteur's work seemed to indicate that microbes could be a cause of disease because if they could sour wine, perhaps they also could make people sick. In 1857, Pasteur published a short paper on wine souring by bacteria. In the paper, he implied that **germs** were related to human illness. Five years later, he formulated the **germ theory of disease**, which holds that microorganisms are responsible for infectious disease.

Pasteur also recommended a practical solution to the sour wine problem. He suggested that grape juice be heated to destroy all the evidence of life, after which yeasts should be added to begin the fermentation. An alternative was to heat the

Germ:
a common word for a microorganism.

(a)

(b)

wine after fermentation and before aging, when the bacteria soured it. Acceptance of his heating technique, known as **pasteurization**, gradually ended the problem and made Pasteur famous. His elation was tempered with sadness, however. In 1859, his daughter Jeanne died of typhoid fever.

Pasteurization:
a mild heating of liquids to kill potential pathogens present.

PASTEUR'S WORK CONTRIBUTES TO DISEASE IDENTIFICATION AND CONTROL

Pasteur's interest in microorganisms rose as he learned more about them. He found bacteria in soil, water, air, and the blood of disease victims. Extending his germ theory of disease, he reasoned that if microorganisms were acquired from the environment, their spread could be controlled and the chain of disease transmission broken.

Pasteur showed that where disease was rampant, the air was full of microorganisms; but where disease was uncommon, the air was cleaner. He opened flasks of nutrient-rich broth to air from the crowded city, then from the countryside, and next from a high mountain. In each succeeding experiment, fewer flasks became contaminated with microorganisms. Such experiments made Pasteur confident that one's health was influenced by the germs in the air in which one lived. He believed that the environment, what he called the "terrain," might be as important to the control of infectious disease as the microorganisms themselves. Pasteur's belief in the germ theory again was strengthened.

By the early 1860s, Pasteur was a national celebrity. Once more, though, tragedy entered his life; in 1865, his two-year-old daughter Camille developed a tumor and died of blood poisoning. Pasteur realized that he was no closer to solving the riddle of disease.

That same year cholera engulfed Paris, killing 200 people a day. Pasteur attempted to capture the responsible bacterium by filtering the hospital air and trapping the bacteria in cotton. Unfortunately, Pasteur was unable to cultivate one bacterium apart from the others because he was using a **broth** that allows microorganisms to mix freely. (In later experiments, Koch would solve the problem by using solid culture media instead of broth media.) Although Pasteur demonstrated that bacterial inoculations made animals ill, he could not pinpoint an exact cause. Some of his critics claimed that a poison, or toxin, in the broth was responsible for the disease.

Broth:
a liquid medium containing nutrients for microbial growth.

In an effort to help French industry again, Pasteur turned his attention to pébrine, the disease of silkworms. Late in 1865, he identified a protozoan infesting the silkworms and the mulberry leaves fed to them. By separating the healthy silkworms from the diseased silkworms and their food, he managed to quell the spread of disease. The achievement strengthened the germ theory of disease. For Pasteur, however, it was another time of grief. Cecille, his third daughter, succumbed at the age of 12 to typhoid fever. Again, he returned to the study of human disease.

Although Pasteur failed to relate a specific microorganism to a specific human disease, his work stimulated others to investigate the nature of microorganisms and to ponder their association with disease. For example, Gerhard Hansen, a Norwegian physician, identified bacteria in the tissues of leprosy patients in 1871, and Otto Obermeier of Germany described bacteria in the blood of relapsing fever patients in 1873. Another German bacteriologist, Ferdinand Cohn, discovered that bacteria multiply by dividing into two cells, suggesting that infecting bacteria could grow and multiply in number.

In England, Joseph Lister (**FIGURE 1.10a**) was sufficiently impressed with Pasteur's writings on disease to suggest that microbes were responsible for postsurgical gangrene and other surgical complications. To block transmission, he employed carbolic acid, a disinfectant used at the time to clean sewers. Use of the acid to clean wounds and surgical instruments dramatically lowered the rate of surgical infections.

ROBERT KOCH FORMALIZES STANDARDS TO IDENTIFY GERMS WITH INFECTIOUS DISEASE

The definitive verification of the germ theory of disease was carried out by Robert Koch (**FIGURE 1.10b**), a country doctor from East Prussia (now part of Germany). Koch's primary interest was anthrax, a deadly bacterial disease in cattle and sheep. Anthrax was a threat to farmers because it ravaged their herds periodically and seemed to reappear time and again in the same district without warning.

Koch was determined to learn more about anthrax. In 1875, in a makeshift laboratory in his home, he injected mice with the blood of diseased sheep and cattle. He then performed meticulous autopsies and noted the same symptoms that had appeared in the sheep and cattle. Next, he isolated from the blood a few rod-shaped bacteria (called bacilli) in the aqueous humor of an ox's eye. Koch watched for hours as the bacilli multiplied, formed tangled threads, and finally reverted to highly resistant spores. At this point, he took several spores on a sliver of wood and injected them into healthy mice. The symptoms of anthrax appeared within hours. Koch autopsied the animals and found their blood swarming with bacilli. He reisolated the bacilli in fresh aqueous humor. The cycle was now complete. The bacilli definitely caused anthrax.

A year later Koch presented his work at the University of Breslau. Scientists there were astonished. Here was the verification of the germ theory of disease that had eluded Pasteur and for which many scientists were waiting. Koch's procedures came to be known as **Koch's postulates**. These techniques, illustrated in **FIGURE 1.11**, were quickly adopted by others as the formalized standards for relating a specific organism to a specific disease.

Koch's postulates:
a series of procedures by which a specific microorganism can be related to a specific infectious disease.

FIGURE 1.10

Joseph Lister and Robert Koch

(a) Joseph Lister was a British physician who was impressed with the work of Pasteur and who promulgated the germ theory of disease. In the 1860s, Lister introduced the principles of sterile surgery to his practice. (b) Koch was one of the first to relate a specific organism to a specific disease, and thus verify the germ theory of disease.

(a)

(b)

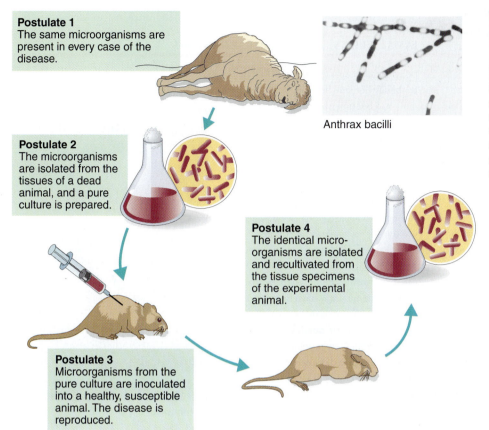

Postulate 1
The same microorganisms are present in every case of the disease.

Anthrax bacilli

Postulate 2
The microorganisms are isolated from the tissues of a dead animal, and a pure culture is prepared.

Postulate 4
The identical micro-organisms are isolated and recultivated from the tissue specimens of the experimental animal.

Postulate 3
Microorganisms from the pure culture are inoculated into a healthy, susceptible animal. The disease is reproduced.

FIGURE 1.11

A Demonstration of Koch's Postulates

Koch's postulates are used to relate a single microorganism to a single disease. The photo shows the rods of the anthrax bacillus as Koch observed them. Many rods are swollen with spores (white ovals).

KOCH DEVELOPS PURE CULTURE TECHNIQUES

After the sensation at Breslau subsided, Koch returned to his laboratory and developed numerous staining methods for bacteria. In 1880, he accepted an appointment to the Imperial Health Office, and while there, he happened upon the cultivation techniques that sparked the further development of microbiology.

Koch chanced to observe that a slice of potato contained small masses of bacteria, which he termed **colonies**. Since bacteria could grow and multiply on solid surfaces, Koch added gelatin to his broth to prepare a solid culture medium in a culture (Petri) dish. He then inoculated bacteria on the surface and set the medium aside to incubate. Next morning, visible colonies were present on the surface, which grew together forming a **pure culture**. Koch could now inoculate laboratory animals with a pure culture of bacteria and be certain that only one species of bacterium was involved. His work proved that bacteria, not toxins in the broth, were the cause of disease. MicroFocus 1.4 details a further advance in cultivation techniques.

In 1881, Koch outlined his methods and demonstrated his pure culture techniques to the International Medical Congress, meeting at Lister's laboratory in London. Pasteur and several of his coworkers were present. Several days later, Koch received a personal letter of congratulations from Pasteur.

Colony:
a viable mass of microorganisms, usually of a single type.

Pure culture:
an accumulation of one type of microorganism formed by the growth of colonies of that organism.

COMPETITION FUELS THE STUDY OF INFECTIOUS DISEASE

At another point in history, Pasteur and Koch might have become friends and colleagues in their search for the agents of disease. However, the years following the 1870 Franco-Prussian war were accompanied by fierce national pride. Both France

MicroFocus 1.4

JAMS, JELLIES, AND MICROORGANISMS

One of the major developments in microbiology was Robert Koch's use of a solid culture medium on which bacteria would grow. He accomplished this by solidifying beef broth with gelatin. When inoculated onto the surface of the nutritious medium, bacteria grew vigorously at room temperature and produced discrete, visible mounds of cells.

On occasion, however, Koch was dismayed to find that the gelatin turned to liquid. It appeared that certain types of bacteria were producing a chemical substance to digest the gelatin. Moreover, gelatin liquefied at the high incubator temperatures commonly used to cultivate certain bacteria.

Walther Hesse, an associate of Koch's, mentioned the problem to his wife and laboratory assistant, Fanny Eilshemius. She had a possible solution. For years, she had been using a seaweed-derived powder called agar (pronounced ah'gar) to solidify her jams and jellies. The formula had been passed to her by her mother, who learned it from Dutch friends living in Java. Agar was valuable because it mixed easily with most liquids and once gelled, it did not liquefy, even at high temperatures.

Hesse was sufficiently impressed to recommend agar to Koch. Soon Koch was using it routinely in his culture media, and in 1884 he first mentioned agar in his paper on the isolation of the tubercle bacillus. It is noteworthy that Fanny Eilshemius may have been among the first Americans to make a significant contribution to microbiology (she was originally from New Jersey).

Another point of interest: The common Petri dish was also invented about this time by Julius Petri, another of Koch's assistants.

■ *Fanny Hesse*

and Germany were undergoing unification, and heroes, including scientists, played an important role in the spirit of nationalism. A competition arose that would last into the next century (TABLE 1.2). However, Pasteur's laboratory was more interested in the mechanism of infection and immunity, while Koch's group focused on isolation, cultivation, and identification of specific pathogens.

Koch's verification of the germ theory was presented in 1876. Within two years, Pasteur had verified the proof and gone a step further. He reported that the bacilli causing anthrax were temperature sensitive. Chickens did not acquire anthrax at their normal body temperature of 42°C, but did so when the animals were cooled down to 37°C. Pasteur also recovered anthrax spores from the soil and suggested that dead animals should be burned or buried deeply in soil unfit for grazing.

One of Pasteur's more remarkable discoveries was made in 1880. For months, he had been working on ways to weaken the bacteria of chicken cholera using heat, different growth media, successive inoculations in animals, and virtually anything that might damage the bacteria. Finally, he developed a weak strain of bacteria. The trick, according to his notebooks, was to suspend the bacteria in a mildly acidic medium and allow the culture to remain undisturbed for a long period. When the bacteria were inoculated into chickens and later followed by a dose of lethal bacilli, the animals did not develop cholera. This principle is the basis for many **vaccines** for immunity. Pasteur applied the principle to anthrax in 1881 and found he could protect sheep against the disease (FIGURE 1.12).

Pasteur's experiments put France in the forefront of science until Koch's lab made several discoveries. These included pure culture techniques for growing bacteria and

Vaccine:
a preparation of modified microorganisms, treated toxins, or parts of microorganisms used for immunization purposes.

TABLE 1.2

A Comparison of Louis Pasteur and Robert Koch

CHARACTERISTIC	LOUIS PASTEUR	ROBERT KOCH
Country of origin	France	Germany (Prussia)
Preparatory education	Chemistry	Medicine
Initial investigations	Milk souring; beer, wine fermentations	Cause of anthrax
Accomplishments	Proposed germ theory of disease	Verified germ theory of disease
	Disproved theory of spontaneous generation	Developed cultivation methods for bacteria
	Developed immunization techniques	Isolated bacterium that causes tuberculosis
	Resolved pébrine problem of silkworms	Developed staining methods for bacteria
	Developed rabies vaccine	Investigated cholera, malaria, sleeping sickness
Associates	Roux, Yersin, Metchnikoff	Gaffky, Löeffler, von Behring, Kitasato
Nobel Prize	No	Yes

isolation of the bacilli that cause tuberculosis, typhoid fever, and diphtheria. Soon, news was forthcoming from France that Émile Roux and Alexandre Yersin of Pasteur's group had linked diphtheria to a toxin produced in the body. In later years, Koch's coworker, Emil von Behring, successfully treated diphtheria by injecting an **antitoxin**, a preparation of antibodies obtained from animals immunized against diphtheria.

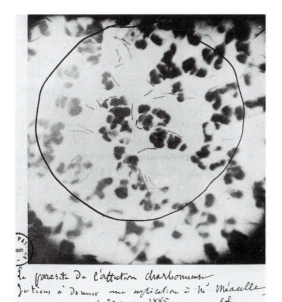

FIGURE 1.12

The Anthrax Bacillus as Photographed by Pasteur

A photomicrograph of the anthrax bacillus taken by Louis Pasteur. Pasteur circled the bacilli (the tiny rods) in tissue and annotated the photograph "the parasite of Charbonneuse." (Charbonneuse is the French equivalent of anthrax.) He requested that the photograph be brought to the attention of Monsieur Méderlle, dated it 20 March, 1885, and initialed it in the lower right.

There also was an international flavor in the French and German laboratories. Shibasaburo Kitasato of Japan studied with Koch and successfully cultivated the tetanus bacillus, an organism that grows only in the absence of oxygen. One of Pasteur's associates was Elie Metchnikoff, a native of Ukraine. In 1884, Metchnikoff published an account of **phagocytosis**, an immunological defensive process in which the body's white blood cells engulf and destroy microorganisms.

In 1885, Pasteur reached the zenith of his career when he successfully immunized young Joseph Meister against the dreaded disease rabies. Although he never saw the causative agent of rabies, Pasteur was able to cultivate it in the brain of animals and inject the boy with bits of the tissue (MicroFocus 1.5). The experiment was a triumph because it fulfilled his dream of applying the principles of science to practical problems. Many monetary rewards followed, including a generous gift from the Russian government after Pasteur immunized 20 peasants against rabies. The funds helped establish the Pasteur Institute in Paris, one of the world's foremost scientific establishments. Pasteur presided over the Institute until his death in 1895.

Koch also reached the height of his influence in the 1880s. In 1883, he interrupted his work on tuberculosis to lead groups studying cholera in Egypt and India. In both countries, Koch isolated a comma-shaped bacillus and confirmed the suspicion first raised by John Snow 30 years earlier that water is the key to transmission. In 1891, Koch became director of Berlin's Institute for Infectious Diseases. At various times, he studied malaria, plague, and sleeping sickness, but his work with tuberculosis ultimately gained him the 1905 Nobel Prize in Physiology or Medicine (MicroFocus 1.6). He died of a stroke in 1910 at the age of 66.

OTHER GLOBAL PIONEERS CONTRIBUTE TO MICROBIOLOGY

With the aging of Pasteur and Koch, a new generation of international scientists stepped in to expand the work on infectious disease. These microbiologists and their accomplishments are listed in TABLE 1.3.

Amid the burgeoning interest in medical microbiology, other scientists devoted their research to the environmental importance of microorganisms. The Russian sci-

MicroFocus 1.5

THE PRIVATE PASTEUR

The notebooks of Louis Pasteur had been an enduring mystery of science ever since the scientist himself requested his family not to show them to anyone. But in 1964, Pasteur's last surviving grandson donated the notebooks to the National Library in Paris, and after soul-searching for a decade, the directors made them available to a select group of scholars. Among the group was Gerald Geison of Princeton University. What Geison found stripped away part of the veneration conferred on Pasteur and showed another side to his work.

In 1881, Pasteur conducted a trial of his new anthrax vaccine by inoculating half a flock of animals with the vaccine, then exposing the entire flock to the disease. When the vaccinated half survived, Pasteur was showered with accolades. However, Pasteur's notebooks, according to Geison, reveal that he had prepared the vaccine not by his own method, but by a competitor's. (Coincidentally, the competitor suffered a nervous breakdown and died a month after the experiment ended.)

Pasteur also apparently sidestepped established protocols when he inoculated two boys with a rabies vaccine before it was tested on animals. Fortunately, the two boys survived, possibly because they were not actually infected or because the vaccine was, indeed, safe and effective. Nevertheless, the untested treatment should not have been used, says Geison. His book, *The Private Science of Louis Pasteur,* places the scientist in a more realistic light and shows that today's pressures to succeed in research are little different than they were more than a century ago.

MicroFocus 1.6

THE NOBEL PRIZE

The Nobel Prizes are among the world's most venerated awards. They were first conceived as a gesture of peace by a man whose discovery had unintentionally added to the destructive forces of warfare.

Alfred Bernhard Nobel was the third son of a Swedish munitions expert. As a young engineer, he developed an interest in nitroglycerine, the oily substance 25 times more explosive than gunpowder. In 1863, Nobel obtained a patent for a detonator of mercury fulminate, and within 4 years he used it with solid nitroglycerine mixed with a type of sandy clay. The mixture was called dynamite, from the Greek *dynamis* meaning "power."

Dynamite had a clear advantage over other explosives because it could be transported easily and handled without fear. It became an overnight success and was adapted to applications in mining,

tunnel construction, and bridge and road building. Before long, it was being used in armaments on the battlefield.

Nobel soon amassed a fortune through the control of several European companies that produced dynamite. However, toward the end of his life he became a pacifist and began speaking out against the use of dynamite in warfare. In 26 lines of his handwritten will, Nobel directed that the bulk of his estate should be used to award prizes that would promote peace, friendship, and service to humanity.

After his death in 1896, the governments of Sweden and Norway established Nobel Prizes in five categories: chemistry, physics, physiology or medicine, literature, and peace. A sixth category, economics, was added in 1969. Every year, Nobel laureates assemble in Oslo or Stockholm on December 10, the

anniversary of Nobel's death. Each laureate receives a medallion, a scroll, and all or part of a cash award currently valued at about $1 million per category.

The first Nobel Prize winners were announced in 1901. Among the recipients were Wilhelm K. Roentgen, the discoverer of X-rays; Jean Henri Durant, the founder of the Red Cross; and Emil von Behring, the developer of the diphtheria antitoxin.

entist Sergei Winogradsky, for example, discovered that certain bacteria could metabolize sulfur while other bacteria could use carbon dioxide to synthesize carbohydrates, much as plants do in photosynthesis (Chapter 5). Martinus Beijerinck, a Dutch investigator, isolated bacteria that trap nitrogen in the soil and make it available to plants for growth (Chapter 26). Together with Winogradsky, he developed many of the laboratory media essential for the study of environmental microbiology, while adding to the growing list of beneficial microorganisms.

Although the list of identified microbes was growing, the agents responsible for diseases such as measles, mumps, smallpox, and yellow fever continued to elude microbiologists. Then, Pasteur's group developed a way to filter solutions that effectively removed bacteria from the liquid. In 1882, a Russian scientist, Dimitri Ivanowsky, used the filter method to trap what he thought were bacteria responsible for tobacco mosaic disease, which produces mottled and stunted tobacco leaves. Surprisingly, Ivanowsky discovered that when he applied the liquid that passed through the filter to healthy tobacco plants, the filtered liquid caused infection—tobacco plants produced mottled, stunted leaves. Ivanowsky assumed bacteria somehow had slipped through the filter.

Unaware of Ivanowsky's work, Beijerinck repeated the experiments in 1895 and confirmed that tobacco mosaic disease was a "contagious, living liquid," dissimilar from bacteria and acting like a poison or virus (*virus* = "poison"). In 1898, scientists isolated the first virus from animals suffering from hoof-and-mouth disease, and in 1901 the virus that causes yellow fever was isolated. When viruses were finally visualized in the 1940s, they were unlike anything that scientists had imagined. We shall explore how this took place in Chapter 12.

TABLE 1.3

Some Notable Figures and Their Accomplishments During the Golden Age of Microbiology, 1854–1914

INVESTIGATOR	COUNTRY	ACCOMPLISHMENT
Gerhard Hansen	Norway	Observed bacteria in leprosy patients
Otto Obermeier	Germany	Observed bacteria in relapsing fever patients
Ferdinand Cohn	Germany	Described life cycle of certain bacteria
Joseph Lister	Great Britain	Developed the principles of aseptic surgery
Georg Gaffky	Germany	Cultivated the typhoid bacillus
Friedrich Löeffler	Germany	Isolated diphtheria bacillus
Emile Roux and Alexandre Yersin	France	Identified the diphtheria toxin
Emil von Behring	Germany	Developed the diphtheria antitoxin; won Nobel Prize
Shibasaburo Kitasato	Japan	Isolated the tetanus bacillus
Elie Metchnikoff	Ukraine	Described phagocytosis; shared Nobel Prize
Ernst Karl Abbé	Germany	Developed the oil-immersion lens and Abbé condenser
Charles Nicolle	France	Proved that lice transmit typhus fever; won Nobel Prize
Albert Calmette	France	Developed immunization process for tuberculosis
Jules Bordet	France	Isolated the pertussis bacillus
Richard Pfeiffer	Germany	Identified a cause of meningitis
Paul Ehrlich	Germany	Synthesized a "magic bullet" for syphilis; shared Nobel Prize
Ronald Ross	Great Britain	Showed mosquitoes can transmit malaria; won Nobel Prize
David Bruce	Great Britain	Proved that tsetse flies transmit sleeping sickness
Almroth Wright	Great Britain	Described opsonins to assist phagocytosis
Masaki Ogata	Japan	Discovered that rat fleas transmit plague
Kiyoshi Shiga	Japan	Isolated a cause of bacterial dysentery
Daniel E. Salmon	United States	Studied swine plague
Theobald Smith	United States	Proved that ticks transmit Texas fever
Howard Ricketts	United States	Showed that ticks transmit Rocky Mountain spotted fever
William Welch	United States	Isolated the gas gangrene bacillus
Walter Reed	United States	Studied mosquito transmission of yellow fever in Cuba
Sergius Winogradsky	Russia	Studied the biochemistry of soil bacteria
Martinus Beijerinck	Netherlands	Developed the discipline of environmental microbiology and provided some of the first clues for viruses as infectious agents
Dimitri Ivanowsky	Russia	Studied tobacco mosaic disease from which he isolated a filterable agent

The advent of World War I in 1914 signaled a dramatic pause in microbiology research and brought the classical Golden Age of microbiology to an end. The Pasteur Institute was closed as Paris came under siege and the German laboratories focused on developing antibacterial serums for war-related diseases. Nevertheless, the discoveries made about infectious disease would set the stage for their accurate diagnoses, prevention, and cure.

Today the search for microorganisms continues. In fact, it has been suggested that less than 2 percent of all microorganisms on Earth have been identified and fewer have been cultured or studied closely. There is still a lot to discover!

MICROBES REPRESENT A DIVERSE GROUP OF ORGANISMS

With the end of the classical Golden Age of microbiology, the diversity of known microbes had grown considerably. Let's briefly catalog these four groups (FIGURE 1.13).

■ FUNGI. The organisms in the fungal group primarily include mushrooms, puffballs, truffles, morels, molds, and yeasts. About 70,000 species of fungi have been described; however, there may be as many as 1.5 million species in nature. They grow best in warm, moist places and secrete digestive enzymes that break down nutrients. Fungi thus live in their own food supply. Sometimes that food supply is humans and disease may result. Many fungi provide useful products including some antibiotics, such as penicillin. Others are used in the food industry to impart distinctive flavors in foods such as Roquefort cheeses. Together with the bacteria, many fungi play a major role as **decomposers,** serving to bring about the rotting and decay of dead matter, and the recycling of this matter, in the environment.

■ PROTISTA. The protista consist of single-celled protozoa and algae. Of the estimated 200,000 species, some, such as the protozoan *Paramecium* or *Amoeba*, may be familiar to you. Some are free living on land or in water, or live in association with other plants and animals. Locomotion is achieved by flagella or cilia, or by a crawling

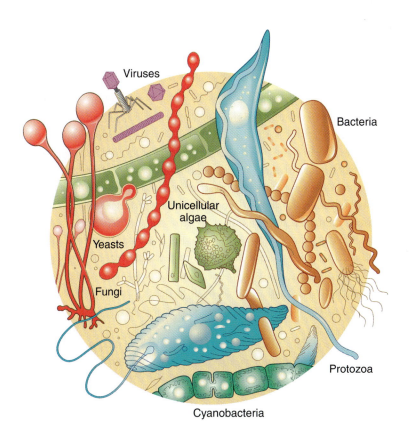

Viruses · Bacteria · Unicellular algae · Yeasts · Fungi · Protozoa · Cyanobacteria

FIGURE 1.13

A Microbial Menagerie

A menagerie of microorganisms as conceived by an artist. Many of these organisms were studied for the first time during the classical Golden Age of microbiology. The notable exception are the viruses, which awaited the electron microscope for visualization. All of these microorganisms are discussed in later chapters of the text.

movement. Different protista obtain nutrients in different ways. Protozoa either absorb nutrients from the surrounding environment or ingest small plants or animals. Algae make their own food through photosynthesis. Most protozoa are helpful in that they are important in lower levels of the food chain, providing food for living things such as snails, clams, and sponges. Some protozoa are capable of causing diseases in humans and other animals. Some human diseases caused by protozoa are malaria, sleeping sickness, and several types of diarrhea.

■ **BACTERIA**. There may be more than 10 million species of bacteria. They are single-celled and have been divided recently into two major domains, the *Eubacteria* and the *Archaea*. Both groups are metabolically more diverse than the other microbes. Most eubacteria and archaea absorb their food from the environment, but some, like the cyanobacteria, can carry out photosynthesis. Bacteria come in three different shapes: bacillus (rod-shaped), coccus (spherical), and spirillum (spiral). Although many eubacteria, along with the fungi, are decomposers, there are species responsible for many animal and human diseases, including the historical killers described in this chapter: tuberculosis, cholera, typhoid fever, and plague. Some bacteria also are responsible for food spoilage while others are useful in the food industry in giving foods like yoghurts and cheeses their distinctive texture or flavor.

■ **VIRUSES**. Currently, there are more than 3,600 known types of viruses. Viruses are structurally simpler than the other microbes because they are not made of cells. Viruses have a core of nucleic acid (DNA or RNA) surrounded by a protein coat. Among the features used to identify viruses are morphology (size, shape), genetic material (RNA, DNA), and biological properties (host infected, transmission). Outside of their **host** cells, viruses are inactive. However, inside living cells, viruses can reproduce. Of the known viruses, only a small percentage cause disease in humans. Smallpox, polio, measles, the flu, AIDS, and SARS are some examples.

Host:
the organism on or in which a microorganism lives and reproduces.

To this point . . .

The classical Golden Age witnessed a series of discoveries unparalleled in medicine. Pasteur, Koch, and their contemporaries forged the link between disease and microorganisms. Neither saw the connection in a blinding flash of light; instead, both worked long, arduous hours to prove a germ theory of disease they believed to be true. Pasteur formulated the germ theory of disease through his work on wine fermentation and silkworm disease. Pasteurization, a heating method to destroy bacteria, also was developed along with a vaccine for rabies. Koch confirmed the germ theory through the formulation of Koch's postulates and identified the cause of many infectious diseases.

As their notoriety grew, scientists flocked to them from all corners of the world. This explosion of research and discovery led to an understanding and control of many diseases, a dream anticipated for centuries. Although cures for established diseases would not come until the 1940s, Pasteur, Koch, their contemporaries, and their students became aware that infectious disease is caused by microorganisms and that the chains of transmission can be broken.

The microbial contributions to science though had just begun. The next part briefly describes a second Golden Age, the age of molecular biology that in great part was catalyzed and supported by experiments using microbes, notably viruses, bacteria, and a few fungi.

1.4

The Second Golden Age of Microbiology (1943–1970)

The 1940s brought the birth of molecular genetics to biology. Many biologists focused on understanding the genetics of organisms, including the nature of the genetic material and its regulation. During these years, microbiology and molecular biology were much the same discipline.

MOLECULAR GENETICS DEPENDS ON MICROORGANISMS

In 1943, the Italian-born microbiologist Salvador Luria and the German physicist Max Dulbrück carried out a series of experiments with bacteria and viruses that began the second Golden Age of microbiology. They used the bacterium *Escherichia coli* to address a basic question regarding evolutionary biology: do mutations occur spontaneously or are they induced by the environment? Luria and Dulbrück showed that bacteria could develop spontaneous mutations that generate resistance to viral infection. Besides the significance of their findings to microbial genetics, their use of *E. coli* as a microbial **model system** showed to other researchers that these relatively simple microorganisms could be used to study general principles of biology.

Model system: a research organism that is amenable to experimental manipulation.

Biologists were quick to jump on the "microbial bandwagon." The experiments carried out by Americans George Beadle and Edward Tatum, using the fungus *Neurospora,* showed that one gene codes for one enzyme. Oswald Avery, Colin MacLeod, and Maclyn McCarty, working with the bacterium *Streptococcus pneumoniae,* suggested that deoxyribonucleic acid (DNA) is the genetic material in cells. In 1953, American biochemist Alfred Hershey and geneticist Martha Chase, using bacterial viruses, provided irrefutable evidence that DNA is the substance of the genetic material. These experiments and discoveries, which will be discussed in more detail in Chapter 6, placed microbiology in the middle of the molecular genetics revolution.

TWO TYPES OF CELLULAR ORGANIZATION ARE REALIZED

The small size of bacteria hindered scientists' abilities to confirm that bacteria were "cellular" in organization. In the 1940s and 1950s, a new type of microscope, the electron microscope, was being developed that could magnify objects and cells thousands of times better than typical light microscopes. With the electron microscope, for the first time bacteria were seen as being cellular like all other microbes, plants, and animals. However, studies showed that they were organized in a fundamentally different way from other organisms.

It was known that animal and plant cells contained a cell nucleus that houses the genetic instructions in the form of chromosomes and was separated physically from other cell structures by a membrane envelope (**FIGURE 1.14a**). This type of cellular organization is called eukaryotic (*eu* = true; *karyon* = kernel, nucleus). Microscope observations of the protista and fungi had revealed that these organisms also had a eukaryotic organization. Thus, not only are all plants and animals **eukaryotes,** so are the microorganisms that make up the fungi and protists.

Eukaryote: an organism possessing DNA enclosed by a membrane envelope.

Studies with the electron microscope revealed that bacterial cells had few of the cellular structures typical of eukaryotic cells. They lacked a cell nucleus, indicating the bacterial chromosome was not surrounded by a membrane envelope (**FIGURE 1.14b**).

FIGURE 1.14

False Color Images of Eukaryotic and Prokaryotic Cells

(a) A scanning electron micrograph of a eukaryotic cell. All eukaryotes including the protozoa, algae, and fungi have their DNA (pink) enclosed in a cell nucleus with a membrane envelope. (Bar = 3 μm.)
(b) A transmission electron micrograph of a dividing prokaryotic cell. Bacteria lack a cell nucleus. The DNA (orange) is not surrounded by a membrane. (Bar = 0.5 μm.)

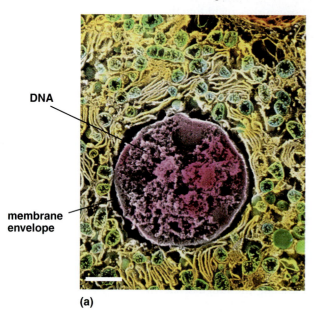

(a)

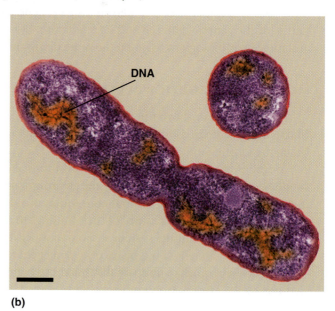

(b)

Prokaryote:
an organism possessing naked DNA not enclosed in a membrane envelope.

Therefore, bacteria have a prokaryotic (*pro* = primitive) type of cellular organization. *Eubacteria* and *Archaea*, then, are **prokaryotes**. In Chapters 3 and 4, we will spend time identifying other major features that define eukaryotic and prokaryotic cells. By the way, viruses do not have a cellular organization and, therefore, are neither prokaryotes nor eukaryotes.

ANTIBIOTICS ARE USED TO CURE INFECTIOUS DISEASE

As a prelude to the second Golden Age, Lister's use of carbolic acid represented one of the first examples of how chemicals could combat infectious disease. In 1910, another coworker of Koch's, Paul Ehrlich, synthesized the first "magic bullet"—a chemical that would search out and destroy pathogens. Called Salvarsan, Ehrlich showed that the compound was very effective at curing syphilis, a sexually transmitted disease that was becoming widespread at the time.

In 1929, Alexander Fleming, a Scottish scientist, went on vacation leaving several culture plates of bacteria on the lab bench. When he returned, he found a mold growing in one of the bacterial cultures. On further inspection, Fleming observed that the mold, a species of *Penicillium*, killed the bacteria that were near it (FIGURE 1.15). He named the antimicrobial substance penicillin and developed an **assay** for its production. In 1940, British biochemists Howard Florey and Ernst Chain purified penicillin and carried out clinical trials that showed the antimicrobial potential of penicillin.

Assay:
an evaluation or test for a substance.

FIGURE 1.15

Fleming and the Discovery of Penicillin

(a) Fleming in his laboratory. (b) Fleming's notes on the inhibition of bacterial growth by the fungus *Penicillium*.

(a)

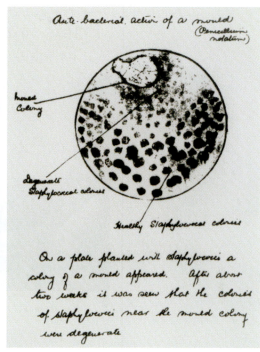

(b)

Additional antimicrobial substances were discovered prior to the second Golden Age. For example, the German chemist Gerhard Domagk discovered the first sulfa drug in 1932. Then, in the early years of the second Golden Age, Selman Waksman discovered the antibacterial chemicals actinomycin and streptomycin. He coined the term "**antibiotic**" to refer to those antimicrobial substances produced by molds and bacteria that inhibit growth or kill other microorganisms.

During the second Golden Age, the push to market effective antibiotics was stimulated by a need to treat infections in casualties of World War II (**FIGURE 1.16**). By the 1950s, penicillin and several additional antibiotics had established treatments in the medical practice. In fact, the growing arsenal of antibiotics convinced many that the age of infectious disease was waning. In 1969, then–US Surgeon General William Stewart confidently declared to Congress that it was time to "close the books on infectious diseases."

Partly due to the perceived benefits of antibiotics, interest in microbes was waning by the end of the 1960s. Research funding became harder to obtain and the knowledge gained from bacterial studies was being applied to eukaryotic organisms, especially animals. What was ignored was the mounting evidence that bacteria were becoming resistant to antibiotics.

FIGURE 1.16

Penicillin—One of the Magic Bullets

A World War II poster touting the benefits of penicillin and illustrating the great enthusiasm in the United States for treating infectious diseases.

To this point . . .

The second Golden Age of microbiology was characterized by the use of microbial organisms to answer questions of significance throughout biology. Organisms, such as the bacterium E. coli, *became very amenable to experimental manipulation and were used to delve into the secrets of molecular genetics.*

The use of the electron microscope allowed biologists to confirm that bacteria were true cells, but that their organization was fundamentally different from other organisms. Animals, plants, and the fungi and protista were eukaryotes because they had a true cell nucleus. Bacteria were identified as prokaryotes since they lacked a cell nucleus.

In an attempt to find the magic bullet, many groups of antibiotics were developed that were effective at curing infectious diseases. However, by the 1950s, evidence already was present that some bacterial pathogens were becoming resistant to the drugs.

In the last part of this chapter, we will briefly examine the reasons why microbiology is currently in yet another Golden Age, this time ushered in by the technological revolution.

The Third Golden Age of Microbiology—Now

Once again, the tides of science and research have changed and microbiology finds itself on the world stage in yet a third Golden Age. The new Golden Age arose in part from the biotechnology advances made in the latter part of the 20th century. During this period, genetic engineers were able to manipulate microorganisms so that the microbes could act as tiny factories to produce needed human proteins, such as insulin or human growth factor, or be used to produce new synthetic vaccines, such as the hepatitis B vaccine. In the latest Golden Age, the biotechnology advances have blossomed into the field of **genomics**, which we will discuss in more detail in Chapter 7.

Genomics:
the study of an organism's DNA.

Infectious disease is still with us. In fact, in the first years of the 21st century, infectious disease is the second leading cause of death (mortality) and the leading cause of disability (morbidity) worldwide. More than 11 million people die each year just from tuberculosis, malaria, lower respiratory infections, diarrheal diseases, and AIDS. The HIV/AIDS, Tuberculosis, and Malaria Act of 2003 signed by President George W. Bush is but one example of the continuing concern over infectious disease. The Act has committed $15 billion over five years in the fight against HIV/AIDS, tuberculosis, and malaria in Africa and the Caribbean. Health policy makers are counting on the microbiological and biological science communities to help achieve global health through the eradication and effective control of infectious disease.

MICROBIOLOGY AGAIN IS FACING MANY CHALLENGES

The third Golden Age of microbiology must try to answer several challenges. Among these are the spread of antimicrobial drug resistance, the emergence of new or resurgent infectious diseases, and the threat of bioterrorism.

■ **ANTIBIOTIC RESISTANCE.** One major challenge concerns the increasing inability to fight infectious disease. In large part, this is due to pathogens evolving resistance to antibiotics and other antimicrobial drugs. Ever since the general microbiology community and governmental health organizations realized pathogens could mutate into "supermicrobes" resistant to many drugs, a crusade has been undertaken to restrain the inappropriate use of these drugs by doctors and to educate patients not to demand them in uncalled-for situations.

There are further challenges facing microbiologists and interested drug companies. There is a need to find new and effective antibiotics before the current arsenal is completely useless. In developing such new drugs, another challenge is to identify antimicrobial drugs to which pathogens will not quickly develop resistance. On the near horizon is the information coming from microbial genomics. One benefit from sequencing the DNA of bacteria and other microbes is to discover a molecular "Achilles heel" or sensitive spot where the organisms would be vulnerable to antimicrobial drugs or to which effective vaccines could be generated. There is still much to be learned here.

■ **EMERGING AND REEMERGING INFECTIOUS DISEASES.** From time to time, strange new disease-causing microbes seem to pop up from nowhere and threaten the health of populations and even whole continents. Such occurrences are nothing new though. For example, Europe experienced epidemics of plague in the 14th and 17th centuries, and cholera pandemics have occurred several times throughout history. With today's growing and highly mobile world population, it is important to quickly identify a disease outbreak and control it before it can spread to epidemic or pandemic proportions.

Microbiologists and epidemiologists today especially are concerned with emerging and reemerging infectious diseases. **Emerging infectious diseases** are those that have recently surfaced in a population. Among the more newsworthy have been AIDS, hantavirus pulmonary syndrome, Lyme disease, mad cow disease, and most recently, SARS. **Reemerging infectious diseases** are ones that have existed in the past but are now showing a resurgence in incidence or geographic range. Often the cause for the reemergence is antibiotic resistance or an increase in susceptible individuals. Among the more prominent reemerging diseases are cholera, tuberculosis, dengue fever, and, for the first time in the Western Hemisphere, West Nile fever. Again, these diseases are of concern to public health officials because lack of identification or control measures could bring on a serious health crisis (**FIGURE 1.17**). In fact, the West Nile virus has now spread across all the lower 48 states, bringing with it an increasing death toll. Since additional emerging or reemerging infectious diseases are inevitable in the future, now is the time for microbiology to devise new and effective measures to detect and minimize such threats.

■ **BIOTERRORISM.** One threat to society is bioterrorism. **Bioterrorism** involves the intentional or threatened use of biological agents to cause fear in or actually inflict death or disease upon a large population. Most of the recognized biological agents are microorganisms or microbial toxins that are bringing diseases like anthrax, smallpox, and plague back into the human psyche (**FIGURE 1.18**). To minimize bioterrorism threats, the challenge is to eliminate the threat of using biological agents. The scientific community and microbiologists need to improve the ways that bioterror agents are detected, discover effective measures to protect the public, and

FIGURE 1.17

West Nile Virus—An Emerging Disease Threat

Ways to protect yourself from mosquitoes that spread the West Nile virus (WNV). WNV is just one of several microorganisms responsible for emerging or reemerging diseases.

FIGURE 1.18

Bioterrorism

Combating the threat of bioterrorism often requires special equipment and protection, because many organisms seen as possible bioweapons are spread through the air.

develop new and effective treatments for individuals or whole populations. If there is anything good to come out of such challenges, it is that we will be better prepared for potential natural emerging infectious disease outbreaks, which initially would be difficult to tell apart from a bioterrorist attack.

Clinical microbiology, being at the center of these challenges, has at its disposal the diagnostic tools of the technological revolution. This means that advances in genomics and **bioinformatics** can produce better and more rapid detection methods for microorganisms in both clinical and environmental samples. These technologies also will be critical for the development of new diagnostics, vaccines, and therapies. The field of **molecular epidemiology** will be critical for understanding disease susceptibility by microbial pathogens.

Bioinformatics:
the use of computers and statistical techniques to manage biological information.

Molecular epidemiology:
the study of the sources, causes, and mode of transmission of diseases using molecular diagnostic techniques.

MICROBIAL ECOLOGY AND EVOLUTION ARE HELPING DRIVE THE NEW GOLDEN AGE

Advances in microbial ecology and understanding microbial evolution also are responsible for boosting microbiology into another Golden Age.

■ **MICROBIAL ECOLOGY.** Since the time of Pasteur, microbiologists have wanted to know how microbes interact, survive, and thrive. For example, advances in microbial ecology indicate that microbes do not act as individual entities; rather, in nature they survive in complex communities called a **biofilm**. Bacteria in biofilms act very differently than individual cells and are difficult to treat when biofilms cause infectious disease. If you or someone you know has had a middle ear infection, the cause was a bacterial biofilm. A potentially deadly biofilm is formed by the bacterium *Pseudomonas aeruginosa* when the organism infects individuals suffering from cystic fibrosis (FIGURE 1.19). On the good side, microbial ecologists are understanding how biofilms can absorb pollutants, reduce toxic chemicals in the environment, and break down and transform many chemicals found in sewage and wastewaters.

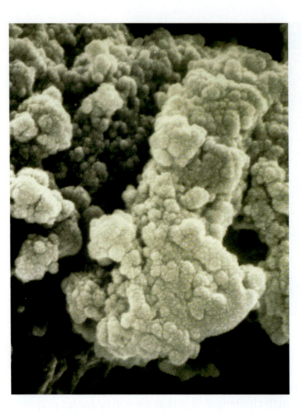

Traditional methods of microbial ecology required that organisms from an environment be cultivated in the laboratory so that they can be characterized and identified. However, up to 99 percent of microorganisms do not grow well in the lab (if at all) and therefore could not be studied. Today, many microbial ecologists, armed with genomic and other technological tools, can study and characterize these unculturable microbes. Such genomic inventories make it possible to even learn something about the microbial biochemical processes that might be going on in the environment. SAR11 and the plans of Craig Venter, mentioned in the chapter's opening piece, were but two examples.

The research of microbial ecologists is being applied to problems that have the potential to benefit humankind. **Bioremediation** is one example where the understanding of microbial ecology has produced a useful outcome (**FIGURE 1.20**). Other microbes hold potential to solve ecological impacts caused by toxic wastes, fertilizers, and pesticides released into the environment.

Bioremediation:
the use of microorganisms to remove or degrade dangerous environmental wastes.

■ **MICROBIAL EVOLUTION**. The technological advances in genomics are telling us much about the origins of life and the evolutionary relationships between microbes and the other organisms making up the tree of life (**FIGURE 1.21**). As of 2003, many more microbes have had their genetic information sequenced than other organisms. This is in part because of their smaller **genome**. All this amassed genetic information can be "mined" to dig out information of practical use. In fact, the explosion of information is what allowed an entire new domain of microbes, the *Archaea*, to be assembled. Studies in microbial evolution are not only tracing the evolution of microbes and identifying their place in the tree of life, but also providing us with an appreciation of our relationships with them.

Genome:
the complete set of genes in an organism.

(a)

(b)

(c)

FIGURE 1.20

Bioremediation

Microbes can be used to help clean up toxic spills.
(a) A shoreline coated with oil from an oil spill can
be sprayed with microorganisms that degrade oil.
(b) and (c) Such remediation processes often are
important agents helping clean such shorelines.

MICROORGANISMS ARE THE "INVISIBLE EMPERORS" OF THE WORLD

The material you have read in this chapter should be giving you an idea of the tremendous role microbes play in us and around us. As emphasized, they certainly affect us when it comes to infectious disease. Nevertheless, it is important to recognize that their beneficial contributions globally far outweigh their negative attributes, and the advances being made by microbiologists are helping humanity better see its own nature and its relations with other organisms, including microbes. As scientists and microbiologists continue to explore this microbial world, it is becoming more apparent that microbes are "invisible emperors" that rule the world! By the time you finish this course, I suspect you too will have this conviction.

FIGURE 1.21

Taxonomic Tree of Life

Four of the six groups of organisms are composed of microorganisms. The fungi and protista are eukaryotic microbes, while the eubacteria and archaea make up the prokaryotic groups. The eukaryotic groups probably evolved from an ancestral eukaryote that evolved from a combination of prokaryotic cells.

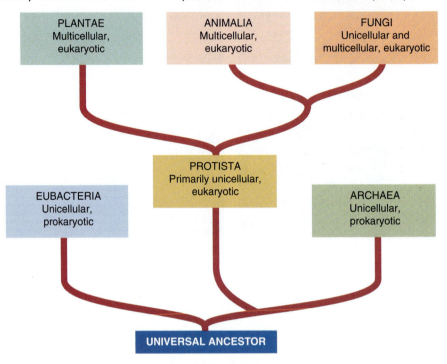

Note to the Student

   Through this opening chapter, I have tried to highlight some of the excitement of the times, the importance of scientific inquiry and experimentation, and the variety of studies taking place that have placed microbiology in the limelight on more than one occasion.

   Pasteur and Koch started the first Golden Age of microbiology, but it would be overly simplistic to believe these two scientists were the only ones of their era to think that microorganisms cause infectious disease. Quite the contrary, a substantial number of scientists sensed that different diseases were due to microbial agents, but few were bold enough to say it publicly without proof. Blaming disease on microorganisms was breaking with centuries-old traditions rooted in beliefs and dogmas. That is why the works of Pasteur and Koch are so special: Pasteur's studies brought microorganisms to the public eye, and Koch provided proof of their involvement in disease. It is amazing that microbiology has gone through two additional Golden Ages since then. It shows the usefulness of microorganisms as model organisms for study of important biological principles as well as their potential to help us live in a better world. The veritable explosion in microbiology knowledge currently going on worldwide suggests that amazing discoveries are yet to be made.

Summary of Key Concepts

1.1 THE BEGINNINGS OF MICROBIOLOGY

- **Anton van Leeuwenhoek Describes "Animalcules."** In the late 1600s, Anton van Leeuwenhoek reported the existence of microorganisms and sparked interest in a previously unknown world of microscopic life.

1.2 THE TRANSITION PERIOD

- **Does Life Generate Itself Spontaneously?** Although strong interest in the microbial world did not continue after Leeuwenhoek's death, the controversy concerning spontaneous generation drew scientists to explore the origin and nature of living things. The need for accurate scientific experimentation came into general practice.

- **Epidemiology Lends Support toward Understanding Disease Transmission.** In the 1860s, scientists such as Semmelweis and Snow believed that infectious disease could be caused by something transmitted from the environment and that the transmission could be interrupted.

1.3 THE CLASSICAL GOLDEN AGE OF MICROBIOLOGY (1854–1914)

- **Louis Pasteur Proposes that Germs Cause Infectious Disease.** The fermentation experiments of Pasteur, begun in the 1850s, were innovative and imaginative because they focused attention on the chemical changes that microorganisms induce. Gradually, Pasteur was led to the possibility that human disease could be due to chemical changes brought about by microorganisms in the body, and he proposed the germ theory of disease.

- **Pasteur's Work Contributes to Disease Identification and Control.** Although Pasteur demonstrated that bacteria could make animals ill, he was unable to cultivate microorganisms in pure culture and verify his germ theory. Influenced by Pasteur's work, Joseph Lister advocated handwashing as a way to break the transmission cycle.

- **Robert Koch Formalizes Standards to Identify Germs with Infectious Disease.** Robert Koch's work with anthrax allowed him to formalize the methods for relating a single microorganism to a single disease (Koch's postulates).

- **Koch Develops Pure Culture Techniques.** Koch invented several staining methods for bacteria. His methods for culturing bacteria on solid media confirmed that bacteria are the cause of infectious disease.

- **Competition Fuels the Study of Infectious Disease.** An intense rivalry developed between the labs of Pasteur and Koch as they hunted down the microorganisms of infectious disease. Pasteur's lab studied the mechanisms for infection and developed vaccines for chicken cholera and human rabies. Koch's lab focused on isolation, cultivation, and identification of pathogens responsible for tuberculosis, typhoid fever, and diphtheria.

- **Other Global Pioneers Contribute to Microbiology.** Scientists throughout the world continued to study infectious disease. In addition, Winogradsky and Beijerinck began examining the role of noninfectious microorganisms in the soil. Ivanowsky and Beijerinck provided the first evidence for viruses as infectious agents.

- **Microbes Represent a Diverse Group of Organisms.** With the arrival of the 20th century, microbes were recognized as making up a growing menagerie of organisms. It includes the fungi (yeasts and molds), protista (protozoa and algae), bacteria (eubacteria and archaea), and viruses.

1.4 THE SECOND GOLDEN AGE OF MICROBIOLOGY (1943–1970)

- **Molecular Genetics Depends on Microorganisms.** The advances made on understanding molecular genetics and general principles in biology were based in great measure on microbial model systems. It was with such systems that DNA was shown to be the genetic material in cells.

- **Two Types of Cellular Organization Are Realized.** With the advent of the electron microscope, microbiologists realized that bacteria lack an organized cell nucleus with a membrane envelope. All organisms were assigned to either the eukaryotic or prokaryotic group.

- **Antibiotics Are Used to Cure Infectious Disease.** Based on the early work of Lister, Ehrlich, and Fleming, antibiotics were developed as "magic bullets" to cure infectious disease. Few were aware that bacteria could quickly develop resistance to the drugs.

1.5 THE THIRD GOLDEN AGE OF MICROBIOLOGY —NOW

■ **Microbiology Again Is Facing Many Challenges.** In the 21st century, increasing antibiotic resistance, emerging and reemerging infectious diseases, and bioterrorism are challenges facing microbiology and society. Advances in genomics offer new ways to approach these challenges.

■ **Microbial Ecology and Evolution Are Helping Drive the New Golden Age.** Microbial ecology is providing new clues to the roles of microorganisms in the environment. Biofilms are recognized as the

dominant form of organization of microbial communities. The vast number of unculturable microbes can be studied and characterized with genomic tools. The understanding of microbial evolution has advanced with the use of genomic technologies and has provided new perspectives on the relationships between microorganisms.

■ **Microorganisms Are the "Invisible Emperors" of the World.** Microorganisms play more roles than simply causing infectious disease. The majority of microbes are seen as emperors of the world because of their essential and important beneficial roles that can provide humanity with an even better and more healthful existence.

Questions for Thought and Discussion

Answers to selected questions can be found in Appendix C.

1. Abu-Bakr Muhammed al-Razi, better known to history as Rhazes, was a famous Arabic doctor during the 900s. On moving to Baghdad to establish a hospital, he picked a site by hanging pieces of meat throughout the city and choosing the place where the last piece of meat turned rotten. How does this insight show an early appreciation of the relationship between disease and something present in the air?

2. In her biography of Louis Pasteur, Patrice Debré describes Pasteur's 1857 paper on lactose fermentation of milk as "the birth certificate of microbiology." Why do you suppose she thinks so highly of the paper? Further, she writes that as a result of the "Pasteurian revolution," medicine could no longer do without science, and hospitals could no longer be mere hospices. What does she have in mind?

3. Suppose you were a research microbiologist in the year 1900 and you had a consuming interest in the organism that causes measles. What direction would your research be taking, and what would be its fate?

4. Controversy in scientific thought often generates what one author has described as "dueling experiments." In this model, a scientific experiment is conducted to counter or refute another experiment. How many examples of dueling experiments can you find in this chapter?

5. Many people are fond of pinpointing events that alter the course of history. In your mind, which single event in this chapter had the greatest influence on the development of microbiology? What event would be in second place?

6. Suppose spontaneous generation were an accepted tenet of medicine and science. How would our view of infectious disease be affected? Now consider how our view is affected by belief in the germ theory of disease.

7. One of the foundations of proper experimental design is the use of controls. What is the role of a control in an experiment? For each of the experiments described in the classical Golden Age of microbiology, identify the control(s) and explain how the interpretation of the experimental results would change without such controls.

8. In 1878, after studying bacteriology in Germany, William Welch returned to the United States, eager to apply the principles he had learned to his medical practice. Welch approached administrators at New York's Bellevue Hospital Medical College for space and money to conduct research, and he was awarded a paltry $25 for three kitchen tables to serve as laboratory benches. Why do you believe Welch was rebuffed?

9. Suppose uncooked hamburger meat becomes smelly after some days in the refrigerator. You bring a sam-

ple to the laboratory, where microorganisms are revealed. Design an experiment, different from the one used by Francesco Redi, that would refute the view that microorganisms arise by spontaneous generation.

10. Why do you suppose there was no follow-up work on smallpox after Edward Jenner's development of a smallpox vaccine? Why might you enjoy learning the art of persuasion from Jenner?

11. At the beginning of the 20th century, many types of microorganisms, including fungi, protozoa, algae, and bacteria, had been seen with the microscope. Provide some ways that microbiologists in 1900 would have assigned microorganisms to the correct group.

12. In downtown Mexico City, at the crossroads of Insurgents Avenue and Paseo de la Reforma, there is an area called Plaza Louis Pasteur. In the plaza stands an elegant statue of Pasteur given to Mexico City by the French residents of Mexico in 1910. That year marked the centennial of the start of the Mexican War of Independence. What connection can you see among these facts and events?

13. As a microbiologist in the 1940s, you are interested in discovering new antibiotics that will kill bacteria. You have been given a liquid sample of a chemical substance you need to test to see if it kills bacteria. Drawing on the culture techniques of Robert Koch, design an experiment that would allow you to determine the killing properties of the sample substance.

14. One of the key discoveries during the second Golden Age of microbiology was the discovery that eukaryotic microorganisms—the fungi, protozoa, and algae—have a true cell nucleus that is surrounded by a membrane envelope. Prokaryotic microorganisms—the eubacteria and archaea—were observed to lack the envelope. Both have DNA as the genetic material though. Propose a reason why the eubacteria and archaea lack a membrane envelope. In other words, why must eukaryotic cells have a membrane envelope?

15. One reason for the rapid advance in knowledge concerning molecular genetics during the second Golden Age of microbiology was because many researchers switched to using microorganisms as model systems. Why would something like the bacterium *Escherichia coli* be more advantageous to use for research at the time than, say, rats or guinea pigs?

16. Textbooks, like this one, usually take about one year for publication. So, from the time that these pages were written for this edition of *Alcamo Fundamentals of Microbiology,* it is about one year before the textbook actually appears in print—and it may have been longer before you bought this copy. Can you identify any newly emerging or reemerging infectious diseases that have been identified since the textbook was published? [SARS (severe acute respiratory syndrome), which first appeared in the spring of 2003, was the last emerging infectious disease mentioned in this text.]

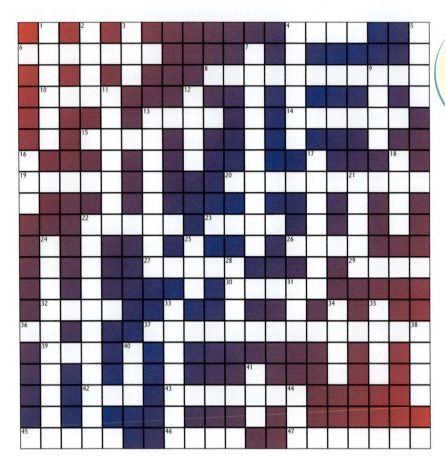

P art of this chapter has explored the development of microbiology up to World War I (Classical Golden Age). To test your knowledge of this period, complete the following cross-word puzzle. The answers are listed in Appendix D.

■ ACROSS

4. Insect that transmits plague
6. Microorganisms poorly studied during the classical Golden Age
8. Leeuwenhoek's term for micro-organisms
10. Pasteur's flasks had the neck of a

   _____

13. Related diphtheria to a toxin
14. Mass of bacteria isolated by Koch
15. Attempted to disprove spontaneous generation by covering meat
19. Isolated the gas gangrene bacillus
20. Used chlorine water for hand disinfection in hospital
22. Related a specific insect to Texas fever

23. Environment where anthrax spores are located
26. Souring attracted attention of Pasteur
27. Isolated the bacterium that causes dysentery
29. Believed by Pasteur to be the cause of spoilage in wine
30. Proved that tsetse flies transmit African sleeping sickness
32. First described by Robert Hooke
36. Tetanus bacilli grow where there is _____ oxygen
37. Dutch investigator who reported the existence of microorganisms
39. Needed by algae and cyanobacteria for photosynthesis
42. In the classical Golden _____ , microbiology experienced substantial growth
43. Searched for a magic bullet to treat syphilis
45. Produced by the bacterium in cases of diphtheria
46. Studied the role of mosquitoes in yellow fever epidemics
47. Causative agent studied by Kitasato

■ DOWN

1. Vague airborne principle believed to cause disease
2. Country in which Reed performed his studies
3. Koch's solid culture medium had a _____ -like consistency
4. Pasteur's country
5. Fabricius suggested this fungus caused plant disease
7. Disease agent isolated by Löffler
9. Leeuwenhoek's microscope had a single _____
11. His broth experiments supported spontaneous generation
12. Type of culture obtained by Koch
13. Related ticks to Rocky Mountain spotted fever
16. Leeuwenhoek's letters number over _____ hundred
17. Disease of silkworms studied by Pasteur
18. Insect that can transmit Texas fever
21. Environment in which Leeuwenhoek observed microorganisms

22. Broth experiments opposed spontaneous generation
24. Recognized three forms by which contagion passed
25. Believed by Pasteur to be a source of disease microorganisms
28. Developed the condenser for use in microscopy
31. Animal infected by anthrax bacillus
33. One of the founders of the science of microbiology
34. Studied cholera during an outbreak in London
35. Won the Nobel Prize for work on malaria and mosquitoes
38. Proved that a specific microorganism causes a specific disease
40. Source of aqueous humor used by Koch
41. Used by Abbé to increase microscope magnification
44. Article of clothing sold by Leeuwenhoek

http://microbiology.jbpub.com

The site features **eLearning,** an on-line review area that provides quizzes and other tools to help you study for your class. You can also follow useful links for in-depth information, or just find out the latest microbiology news.

The Chemical Building Blocks of Life

The significant chemicals in living tissue are rickety and unstable, which is exactly what is needed for life.

—Isaac Asimov (1920–1992)

The words "microbial life" often bring to mind images of microorganisms cavorting in a droplet of pond water, molds growing on a piece of stale bread, or bacteria multiplying in a festering wound. But, microbiologists generally go a step beyond. To them, such images would conjure up such questions as, *Why* do microorganisms act as they do? *How* are they organized? or *What* are they made of? Their questions reflect a curiosity about aspects of microbial life that are less visible, but perhaps more fundamental.

Untold generations ago, humans first pondered the composition of living things. They reasoned that because living organisms differ so greatly from nonliving things, there must be a corresponding difference in their construction. Then, during the early 1800s, chemists identified a group of basic substances found in the earth and atmosphere, including carbon, hydrogen, and oxygen. Surprisingly, the cells of living organisms proved to be made of the same materials; the only apparent difference was how the substances were organized. Chemists began referring to the materials associated with living things as "organic substances," while all others were termed "inorganic substances." Organic substances could be converted to inorganic substances easily enough, but the reverse appeared impossible. Living things seemed to be unique.

In 1821, an important discovery was made by the German chemist Friedrich Wöhler. Wöhler was investigating the properties of cyanides, a group of chemicals generally accepted as inorganic. While heating ammonium cyanate, he found, to his amazement, that crystals of urea were forming. Urea is a major component of urine and an organic compound. Wöhler's work showed that an inorganic substance could be converted to an organic substance. More fundamentally, it indicated that it was possible to synthesize the substances of living things. His work encouraged other scientists to tackle the synthesis of the organic substances, and soon scientists realized that knowledge of the chemicals of life was within their grasp.

During the late 1800s and through the 1900s, the distinction between living and nonliving things continued to evaporate as chemists produced numerous **organic compounds** and began to understand life processes in terms of chemical reactions. Today it is clear that all biology, including microbiology, has a chemical basis. Biology and chemistry are inseparable if scientists are to find answers to the questions of *why*, *how*, and *what*.

Organic compound: a compound associated with living things.

In this chapter, we will review the fundamental concepts of chemistry that will form a foundation for the chapters ahead. We shall identify the elements that make up all known substances and show how these elements combine to form the four major groups of organic compounds found in virtually all forms of life. Your understanding of chemistry will make subsequent chapters easier and prepare you for a rewarding learning experience as we study microorganisms.

2.1

The Elements of Life

As far as scientists know, all matter in the physical universe is composed of substances called **elements**. Ninety-two naturally occurring elements have been discovered and characterized. Each is designated by one or two letters standing for its English or Latin name. For example, H is the symbol for hydrogen, O for oxygen, Cl for chlorine, and Mg for magnesium. Some Latin abbreviations include Na from *natrium* (which translates to "sodium"), K from *kalium* (which is Latin for "potassium"), and Fe from *ferrum* (the Latin word for "iron").

Element: a basic substance that cannot be broken into simpler substances by ordinary means.

Of the 92 naturally occurring elements, only about 25 are essential to living organisms (TABLE 2.1). Note that just six of these elements—carbon, hydrogen, nitrogen, oxygen, phosphorus, and sulfur—make up 99 percent of a bacterium's weight. (The acronym CHNOPS is helpful in remembering these six important elements.)

ATOMS ARE COMPOSED OF SUBATOMIC PARTICLES

An **atom** is the smallest unit having the properties of that element and cannot be broken down further without losing the quality of that element. Simply stated, carbon consists of carbon atoms, oxygen of oxygen atoms, and so forth. If you split a carbon atom into simpler substances, it no longer has the properties of carbon.

Every atom consists of a positively charged core, the **atomic nucleus** (FIGURE 2.1a). The atomic nucleus contains most of the atom's mass and is composed of two kinds of tightly packed subatomic particles called **protons** and **neutrons**. Although these two particles have about the same mass, protons bear a positive electrical charge, while neutrons have no charge. MicroFocus 2.1 examines these subatomic particles a little closer. Surrounding the atomic nucleus is a negatively charged array of **electrons**. In any atom, the number of electrons is equal to the number of protons,

TABLE 2.1

The Major Elements of Living Organisms

ELEMENT	SYMBOL	PERCENTAGE BY WEIGHT IN BODY	ATOMIC NUMBER	ATOMIC WEIGHT
Oxygen	O	65	8	16
Carbon	C	18	6	12
Hydrogen	H	10	1	1
Nitrogen	N	3	7	14
Calcium	Ca	2	20	40
Phosphorus	P	1	15	31
Potassium	K	0.9	19	39
Sulfur	S	0.9	16	32
Chlorine	Cl	0.9	17	35
Sodium	Na	0.9	11	23
Magnesium	Mg	0.9	12	24
Iron	Fe	0.9	26	56
SOME TRACE ELEMENTS (NEEDED AT 0.1 PERCENT OR LESS)				
Manganese	Mn	0.1	25	55
Copper	Cu	0.1	29	64
Iodine	I	0.1	53	127
Cobalt	Co	0.1	27	59
Zinc	Zn	0.1	30	65
Boron	B	0.1	5	11
THE MAJOR ELEMENTS OF A BACTERIUM				
Carbon	C	12.14		
Hydrogen	H	9.94		
Nitrogen	N	3.04		
Oxygen	O	73.68		
Phosphorus	P	0.60		
Sulfur	S	0.32		

so that an atom has no net electrical charge. The number of protons in an atom defines each element. For example, carbon atoms always have six protons. If there are seven protons, it is no longer carbon but rather the element nitrogen.

The number of protons represents the **atomic number** of the atom. Thus, carbon with six protons has an atomic number of 6. The **mass number** is the number of protons and neutrons combined. Since carbon atoms have six protons and usually six neutrons in the atomic nucleus, the mass number of carbon is 12. FIGURE 2.1b provides a simple diagram of the structures, atomic numbers, and mass numbers of five atoms essential to life.

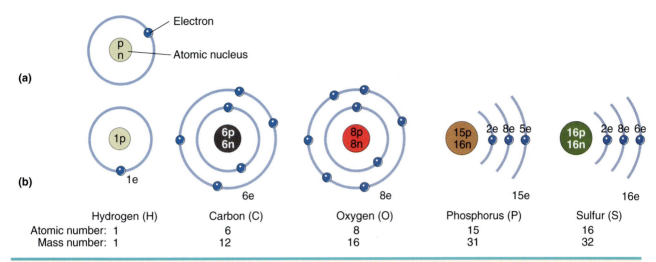

FIGURE 2.1

The Atomic Structure of Five Important Elements

Note that the number of protons is equal to the number of electrons (though not necessarily equal to the number of neutrons). Also note that the electrons are arranged in concentric shells about the nucleus. The atomic number is the proton number, and the mass number is the sum of the numbers of protons and neutrons.

The **atomic weight** of an atom is nearly the same as its mass number. Atomic weights are determined by adding together the weights of the protons, neutrons, and electrons in the atom. For example, the mass number for carbon is 12 where, in reality, the atomic weight is 12.01. For our purposes, the atomic weight of an atom is close enough to be the same as the mass number.

ISOTOPES VARY IN THE NUMBER OF NEUTRONS PRESENT

Although the number of protons is the same for all atoms in an element, the number of neutrons may vary. Consequently, the mass number of atoms of an element may vary. Most carbon atoms, for example, have a mass number of 12, but some carbon atoms have eight neutrons, rather then six, in the atomic nucleus and, hence, a

MicroFocus 2.1

THREE QUARKS IN A PROTON

You might not have thought about it before (and after this brief discussion, you might be just as glad), but have you ever wondered what is inside a proton? For a long time physicists thought nothing was inside, there were no smaller parts. But then physicists devised mathematical models suggesting that protons contain smaller particles. Physicist Murray Gell-Mann named these smaller components "quarks" after the chapter, "Three quarks for Muster Mark" in James Joyce's *Finnegans Wake*. The statement probably has a liquor connotation, as in "Three quarts for Mister Mark," which is how you might feel after finishing this piece.

Anyway, such particles now have been detected and, as Gell-Mann implied, there are in fact three quarks in a proton: two "up" quarks and one "down" quark. This should not be confused with a neutron, which has one "up" quark and two "down" quarks! But that's not all. Further investigations have revealed additional particles. Not only are there quarks, there are gluon particles that hold the quarks together as well as antiquarks. And all this activity is occurring at dizzying rates close to the speed of light.

I don't know about you, but I think I'll stick with microorganisms.

mass number of 14. Atoms of the same element that have different numbers of neutrons are called **isotopes**. Therefore, carbon-12 and carbon-14 (symbolized as ^{12}C and ^{14}C) are isotopes of carbon.

Some isotopes are unstable and give off energy in the form of radiation. Such **radioisotopes** are useful in research and medicine. C-14, for example, can be incorporated into an organic compound and used as a radioactive tracer to follow the fate of a substance, as MicroInquiry 2 demonstrates near the end of this chapter.

ATOMS CAN BECOME ELECTRICALLY CHARGED

Atoms are uncharged when they contain equal numbers of electrons and protons. Atoms do not gain or lose protons, but a gain or loss of electrons is possible. When this takes place, the atom is converted to an electrically charged atom, called an **ion**. For example, the addition of one electron adds an additional negative electrical charge to the atom and converts it to a negatively charged ion, called an **anion**. By contrast, the loss of one electron leaves the atom with one extra proton and yields a positively charged ion, called a **cation**. As we will see, ion formation is important in bonding between atoms and molecules.

ELECTRON PLACEMENT DETERMINES CHEMICAL REACTIVITY

In 1913, the Danish physicist Niels Bohr suggested one of the first models of the atom. In his model, the atom is represented as a miniature solar system, with the electrons revolving around the atomic nucleus in concentric orbits (FIGURE 2.2a). This simplified model has been greatly modified by contemporary physicists. Definite electron pathways have been eliminated, and the location of electrons now is regarded as a three-dimensional region of space around the nucleus, as shown in FIGURE 2.2b. This cloudlike area, where electrons are most likely to be found, is referred to as an **orbital**.

The distance of an electron from the nucleus is a function of the electron's energy. Electrons with higher energy are farther from the nucleus than those with lower energy. The energy level in an atom in which an electron is found most often is called the **electron shell**. In illustrations of the atom, the shell and the orbital are commonly represented by a ring, with the electron shown as a spot on the ring, as done in Figure 2.1. The model, though technically wrong, is useful for interpreting chemical phenomena.

Each shell can hold a maximum number of electrons. The shell closest to the nucleus can accommodate two electrons, while the second shell can hold eight.

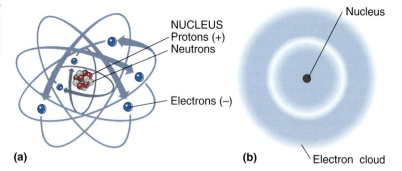

FIGURE 2.2

Two Models of the Atom's Structure

(a) An early model of the atom shows electrons "in orbit" about the nucleus. (b) A modern view of the atom has electrons occupying distinct cloud-like areas of three-dimensional space around the nucleus.

NUCLEUS
Protons (+)
Neutrons

Electrons (−)

Nucleus

Electron cloud

(a)   (b)

Other shells also have maximum numbers but usually have no more than 18 in the outer shell. Since the 25 essential elements are of lower atomic weight, only the first few shells are of significance to life. Inner shells are filled first and, if there are not enough electrons to fill all the shells, the outermost shell is left incompletely filled.

Atoms with unfilled outer shells are unstable but can become stable by interacting with another unstable atom. A carbon atom, with six electrons, has two electrons in its first shell and only four in the second. For this reason, carbon is extremely reactive and, as we will see, forms innumerable combinations with other elements. Therefore, only atoms with an unfilled outer shell will participate in a chemical reaction.

The shells of a few elements normally are filled completely and these elements are chemically stable. Such an element is said to be an **inert element**. Helium is stable because it has two electrons in its only electron shell. Neon is another inert element. Its outer (second) shell contains the maximum eight electrons.

COMPOUNDS AND MOLECULES FORM THROUGH CHEMICAL BONDING

Unstable atoms interact in specific ways to fill their outer shells. When atoms of two or more elements interact with one another to achieve stability, they form a substance called a **compound**. In a compound, the atoms of the elements combine in specific proportions with a particular pattern of linkages involving chemical bonding. Each compound, like each element, has a definite formula and set of properties that distinguish it from its components. For example, sodium (Na) is an explosive metal and chlorine (Cl) is a poisonous gas, but the compound they form is edible table salt (NaCl).

A **molecule** is the smallest part of a compound that retains the chemical properties of that compound. Molecules may be composed of only one kind of atom, as in oxygen gas (O_2), or they may consist of different kinds of atoms in substances such as water (H_2O), carbon dioxide (CO_2), and the simple sugar glucose ($C_6H_{12}O_6$). As shown by these examples, the kinds and amounts of atoms (the subscript) in a molecule are called the **molecular formula**.

To appreciate the relative size of a molecule, it is valuable to know its **molecular weight**. This is determined by adding together the atomic weights of all the atoms in the molecule and expressing the sum in units called **daltons**. The molecular weight of a water molecule therefore is 18 daltons, while the molecular weight of a glucose molecule is 180 daltons (**FIGURE 2.3**). Some molecular weights reach astonishing proportions. For instance, the antibodies produced by the body's immune system may have a molecular weight of 150,000 daltons and the toxic poison secreted by the bacterium that causes botulism has a molecular weight of over 900,000 daltons.

The term **mole** is used to express the quantity of a substance whose weight in grams is numerically equivalent to its molecular weight: A mole of water weighs 18 grams, a mole of glucose weighs 180 grams, and a mole of botulinum toxin weighs 900,000 grams. The mole concept is important in Chapter 5, where the energy content of a mole of certain substances is discussed.

Dalton:
a measure of mass equal to one hydrogen atom.

FIGURE 2.3

The Relationship between Atomic Weights and Molecular Weights

Atomic weights refer to the weights of the protons and the neutrons of the individual atoms, while the molecular weight is the sum of the atomic weight in a molecule, such as water, carbon dioxide, or glucose.

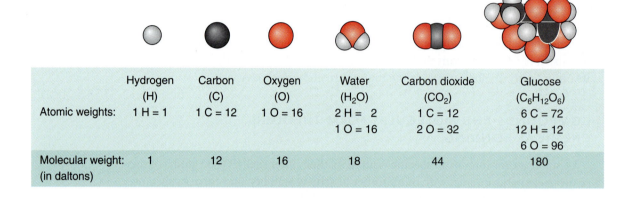

	Hydrogen (H)	Carbon (C)	Oxygen (O)	Water (H_2O)	Carbon dioxide (CO_2)	Glucose ($C_6H_{12}O_6$)
Atomic weights:	1 H = 1	1 C = 12	1 O = 16	2 H = 2 1 O = 16	1 C = 12 2 O = 32	6 C = 72 12 H = 12 6 O = 96
Molecular weight: (in daltons)	1	12	16	18	44	180

To this point . . .

We began studying the principles of chemistry by noting why chemistry is essential to the study of living things. We then outlined the elements of the physical universe and reduced the elements to the level of atoms, because atoms join to form the chemical substances of life. To understand how the joining takes place, we explored atomic structure, emphasizing electrons and their placement in the atom. As we shall see in the next section, atoms may combine in a number of ways to yield the familiar substances of all living organisms—be they bacteria, trees, or humans.

The section also reviewed important concepts of chemistry that will recur in some later chapters. For example, isotopes are commonly used in diagnostic procedures for disease, and ions are participants in the chemical processes of most microorganisms. We also shall encounter molecules and compounds in the chemistry of life processes, and, in many places, we will mention molecular formulas and molecular weights. It therefore is essential to grasp the meaning of these terms as a prelude to later use.

We now turn to the types of bonding that occur in chemical compounds. We will examine chemical reactions from a general standpoint and learn the basis for the chemical processes that occur in all life forms. A short explanation of acids and bases is included because they also are very significant in the microbial world.

Chemical Bonding

Isaac Asimov's opening quote that chemicals are "rickety and unstable" is validated when considering how atoms interact. Linkage requires that the two unstable atoms come close enough for their electron orbitals to overlap. At this point, an energy exchange takes place, and each of the participating atoms assumes an electron configuration more stable than its original configuration. The rearrangement can occur in one of two major ways: atoms can lose one or more electrons to another atom, or each atom can share electrons with one or more other atoms. When two or more atoms are linked together, the force that holds them is called a **chemical bond**. So, chemical bonds are the result of these rickety, unstable atoms filling their outer electron shells. **FIGURE 2.4** and **TABLE 2.2** compare the different types of chemical bonds we will discuss next.

FIGURE 2.4

Chemical Bonding

(a) The transfer of an electron from an atom of sodium to an atom of chlorine creates oppositely charged ions. The electrical attraction between these ions creates an ionic bond. The resulting molecule is sodium chloride. (b) A covalent bond involves the sharing of electrons between atoms, the example shown here being the simple organic compound methane. (c) Hydrogen bonds form when unequal sharing of electrons creates polarized "ends" with-in a molecule. Hydrogen atoms are typicaly involved, as is the case with the hydrogen bonding that occurs between water molecules.

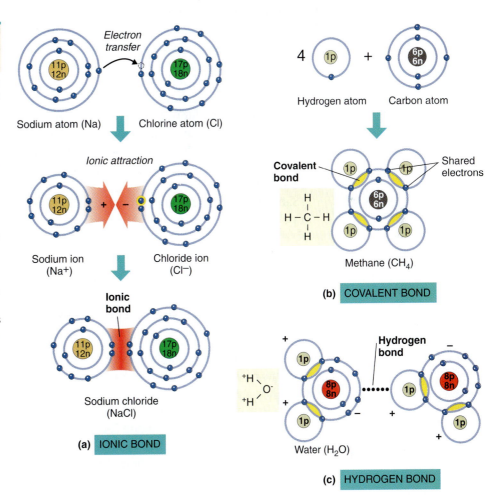

TABLE 2.2

Three Types of Chemical Bonds in Organic Compounds

TYPE	CHEMICAL BASIS	STRENGTH	EXAMPLE
Ionic	Attraction between oppositely charged ions	Weak	Sodium chloride
Covalent	Sharing of electron pairs between atoms	Strong	Glucose
Hydrogen	Attraction of a hydrogen nucleus (a proton) to negatively charged oxygen or nitrogen atoms in the same or neighboring molecules	Weak	Water molecules

IONIC BONDS FORM BETWEEN OPPOSITELY CHARGED IONS

Ionic bond:
the force that holds ions together in a compound.

In **ionic bonds** one atom gives up its outermost electrons to another. The reaction between sodium and chlorine is illustrative (Figure 2.4a). Sodium has an atomic number of 11, with electrons arranged in groups of 2, 8, and 1. Chlorine, with an atomic number of 17, has its electrons in groups of 2, 8, and 7. To achieve stability, sodium atoms only need to lose one electron and chlorine atoms only need to gain one electron. Thus, when atoms of the two elements are brought together, the sodium atoms donate one electron to the chlorine atoms. Both atoms now have their outermost shell filled.

The transfer of electrons leads to ion formation. By acquiring one electron, chlorine atoms become chloride ions (Cl^-). By contrast, sodium atoms now have an extra proton and therefore are sodium ions (Na^+). Since opposite electrical charges attract each other, the chloride ions and sodium ions come together to form stable sodium chloride (NaCl). Under cellular conditions, ionic bonds are relatively weak and break quite easily.

Salt:
a compound formed by ionic bonds between ions.

Salts are typically formed through ionic bonding. Besides sodium and chloride, important salts are formed from other ions, including calcium (Ca^{+2}), potassium (K^+), magnesium (Mg^{+2}), and iron (Fe^{+2} or Fe^{+3}). As we will see, ionic bonds also play important roles in protein structure and the reactions between antigens and antibodies in the immune response (Chapter 19).

COVALENT BONDS SHARE ELECTRONS

Covalent bond:
the force resulting from the sharing of electrons among the atoms in a molecule or between molecules.

Atoms also can achieve stability by sharing electrons through the formation of **covalent bonds**. Such bonds are very important in biology because the CHNOPS elements of life almost always enter into covalent bonds with themselves or one another.

Covalent bonding occurs frequently in carbon because this element has four electrons in its outer shell (Figure 2.4b). The carbon atom is not strong enough to acquire four additional electrons, but it is sufficiently strong to retain the four that it has. It therefore enters into a huge variety of covalent bonds with four other atoms, ions, or groups of atoms. The vast array of carbon compounds that can be formed is responsible for the chemistry of life.

Many of the microbes that reside in the ruminant stomach produce methane or natural gas (CH_4) as a by-product of cellulose digestion. This gas is a good example to illustrate covalent bonding between carbon and hydrogen. A carbon atom shares each of its four outer shell electrons with the electron of a hydrogen atom, forming four single covalent bonds (Figure 2.4b).

Scientists use one or more lines between chemical symbols to represent covalent bonds and the **structural formulas** for molecules. In Figure 2.4b, each line between carbon and hydrogen (C—H) represents a single covalent bond between a pair of shared electrons. Other molecules, such as carbon dioxide (CO_2), share two pairs of electrons and therefore two lines are used to indicate the double covalent bond: O=C=O. However, in all cases, the atoms now are stable because the outer electron shell of each atom is filled through this sharing.

The simplest derivatives of carbon are the **hydrocarbons**, so named because they consist solely of hydrogen and carbon. Methane is the most fundamental hydrocarbon. Other hydrocarbons consist of chains of carbon atoms and, in some cases, the chains may be closed to form a ring.

Methane (CH_4)

Propane (C_3H_8)

Benzene (C_6H_6)

When atoms bond together to form a molecule, they establish a definite geometric relationship determined largely by the electron configuration. Notice in the hydrocarbons drawn above that the covalent bonds are distributed equally around each carbon atom. Each of these examples, where there is an equal sharing of electron pairs, represents a **nonpolar molecule**—there are no electrical charges (poles) present.

HYDROGEN BONDS FORM BETWEEN POLAR GROUPS OR MOLECULES

Not all molecules are nonpolar. Indeed, one of the most important molecules to life is water, which is a **polar molecule**—it has electrically charged poles. In a water molecule, the two hydrogen atoms are attached to one side of the oxygen atom (Figure 2.4c). As a result, the protons gather at this side and give the molecule a slightly positive charge. By contrast, the other side of the molecule carries a slightly negative charge owing to the accumulation of electrons. The water molecule therefore has poles and is said to be polar (Figure 2.4c).

Hydrogen bonds hold polar molecules together. They generally involve attraction of a partially positive hydrogen atom on one polar molecule toward either a partially negative oxygen or nitrogen atom on another polar molecule. Hydrogen bonds may last only a brief instant, being much weaker than covalent bonds. Still, hydrogen bonds are extremely important in shaping the structures of proteins and nucleic acids, two of the components of living cells.

Hydrogen bond: the force resulting from the attractions between oppositely charged poles of adjacent molecules.

CHEMICAL REACTIONS CHANGE BONDING PARTNERS

A **chemical reaction** is a process in which atoms or molecules interact to form new bonds, thereby undergoing a change through electron rearrangement. Different combinations of atoms or molecules result from the reaction; that is, bonding partners change. However, the total number of interacting atoms remains constant.

For chemical reactions, an arrow is used to separate the original and final substances in a chemical reaction and indicate in which direction the reaction will proceed. The atoms or molecules drawn to the left of the arrow are the **reactants** and those to the right are the **products**. In some reactions, two reactants *combine* to form a product:

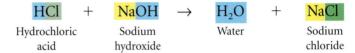

$$Na^+ \quad + \quad Cl^- \quad \rightarrow \quad NaCl$$
Sodium ion   Chloride ion   Sodium chloride

In other cases, reactions involve an *exchange* of reactant atoms with two new products formed:

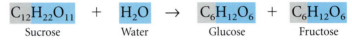

$$HCl \quad + \quad NaOH \quad \rightarrow \quad H_2O \quad + \quad NaCl$$
Hydrochloric   Sodium   Water   Sodium
acid   hydroxide   chloride

Hydrolysis:
a decomposition chemical reaction in which a molecule is split by the addition of a water molecule.

Other reactions involve a *decomposition* in which water functions as a reactant. Such reactions are called **hydrolysis** reactions:

$$C_{12}H_{22}O_{11} \quad + \quad H_2O \quad \rightarrow \quad C_6H_{12}O_6 \quad + \quad C_6H_{12}O_6$$
Sucrose   Water   Glucose   Fructose

Note that in all cases, new products have been formed but the number of atoms in the reactants is equal to those in the products. In forming new products, only the bonding partners have changed. A fourth type of reaction, a dehydration synthesis will be discussed shortly.

To this point . . .

We have progressed in our understanding of chemical bonding by knowing why chemical bonds are formed between unstable atoms. Chemical bonding is an attempt to fill the outermost electron shells of the unstable interacting atoms. This can be accomplished by forming ionic bonds or covalent bonds.

In ionic bonds, electrons are lost and gained by the interacting atoms. This generates oppositely charged ions that are electrically attracted to one another. The result is the formation of a salt. Covalent bonds involve the sharing of electron pairs between the interacting atoms. In both cases, the chemical bonds formed produce atoms with filled outer electron shells.

In covalent bond formation, some electron pairs are shared equally, producing nonpolar molecules that lack any electrical charge. In other cases, electron pairs are shared unequally. This is especially true in hydrogen-oxygen and hydrogen-nitrogen atom interactions. The resulting molecules, such as water, are polar, meaning they have partial positive and negative charges. The electrical attractions between opposite electrical poles can generate hydrogen bonds.

Finally, we looked at some specific examples of chemical reactions where reactants change bonding partners in the process of forming new products.

Next, we need to briefly examine water, the medium in which all chemical reactions occur. We also will explore acid-base relationships because of their influence on chemical reactions and significance in microbiology.

Water and pH

All living organisms are composed primarily of water. In fact, most eukaryotic organisms are about 90 percent water while bacteria are about 70 percent water. No organism, not even the bacteria, can develop and grow without water.

WATER IS THE UNIVERSAL SOLVENT OF LIFE

Liquid water is the medium in which cellular chemical reactions must occur. Being polar, water molecules are attracted to other polar molecules and act as the universal **solvent** in cells. Take for example what happens when you put a **solute** like salt in water. The solute dissolves into separate sodium ions and chloride ions because water molecules break the weak ionic bonds and surround each ion in a sphere of water molecules. The metabolism in cells also requires many molecules to be soluble in water, so they can dissociate and participate in other chemical reactions.

Water also participates as a reactant in many chemical reactions. The example of the hydrolysis reaction shown on the previous page involved water in splitting sucrose into separate molecules of glucose and fructose. Also, in exchange reactions, water often is a product of the reaction.

The polar nature of water molecules means that they tend to be attracted to one another through hydrogen bonding. These bonds are strong enough to keep water liquid and prevent it from easily becoming ice or steam. By forming a large number of hydrogen bonds between water molecules, it takes a large amount of heat to increase the temperature of water. Likewise, a large amount of heat must be lost before water decreases temperature. So, by being 70 to 90 percent water, cells are bathed in a solvent that maintains a more consistent temperature even when the environmental temperatures change.

ACIDS AND BASES MUST BE BALANCED IN CELLS

Many solutes in cells act as acids or bases, which can adversely affect cell function. For our purposes, an **acid** is a chemical substance that donates hydrogen ions (H^+); that is, protons to water or other solution. By contrast, a **base** (or **alkali**) is a substance that accepts hydrogen ions in solution. Since the acceptance of a hydrogen ion increases the concentration of hydroxyl ions (OH^-) in solution, a base also may be considered a substance that increases the amount of hydroxyl ions.

Acids are distinguished by their sour taste. Some common examples are acetic acid in vinegar, citric acid in citrus fruits, and lactic acid in sour milk products. Strong acids are those that donate large numbers of hydrogen ions to a solution. Hydrochloric acid (HCl), sulfuric acid (H_2SO_4), and nitric acid (HNO_3) are examples. Weak acids, typified by carbonic acid (H_2CO_3), donate a smaller number of hydrogen ions.

Bases have a bitter taste. Strong bases take up numerous hydrogen ions from a solution and leave it with a high concentration of hydroxyl ions. Potassium hydroxide (KOH), a material used to make soap, is among them. Weak bases take up smaller amounts of hydrogen ions and are typical of compounds such as ammonia that contain an amino (—NH_2) group.

Solvent:
a liquid doing the dissolving.

Solute:
a substance dissolved in a solvent.

Acids and bases frequently react with each other because of their opposing chemical characteristics. Such a reaction neutralizes both the acid and the base to form water and a salt. The exchange reaction involving hydrochloric acid (HCl) and sodium hydroxide (NaOH) described earlier was one example, which produced water (H_2O) and ordinary table salt (NaCl).

Substances differ in their degree of acidity or alkalinity. To indicate the degree to which a substance is acidic or basic, the Danish chemist Søren P. L. Sørensen introduced the symbol **pH** and the **pH scale** to refer to the strength of an acid or base. The scale extends from 0 to 14 and is based on actual calculations of the number of hydrogen ions present when a substance mixes with water. The strongest acid has a pH of 0; the strongest base has a pH of 14. A substance with a pH of 7, such as water, is said to be neutral. Most, but not all, bacteria live in near neutral pH environments, although there are some spectacular exceptions (MicroFocus 2.2). Many fungi also can tolerate a more acidic environment, often causing food spoilage in acidic foods such as sour cream, yogurt, and citrus fruit products. FIGURE 2.5 summarizes the pH values of several common substances.

BUFFERS PREVENT pH SHIFTS

All organisms must balance the acids and bases in cells because chemical reactions and organic compounds are very sensitive to pH shifts. Proteins are especially vulnerable, as we will soon see. However, many chemical reactions in cells normally produce hydrogen ions or absorb them, meaning that the pH could fluctuate. To prevent such changes, cells contain **buffers**, which are compounds that prevent pH swings from occurring and keep the internal pH constant. Most biological buffers consist of a weak acid and a weak base. If an excessive number of hydrogen ions are produced (potential pH drop), the base can "absorb" them through chemical reactions. Alternatively, if there is a decrease in the hydrogen ion concentration (potential pH increase), the weak acid can dissociate, replacing the lost hydrogen ions.

FIGURE 2.5

A Sample of pH Values for Some Common Substances

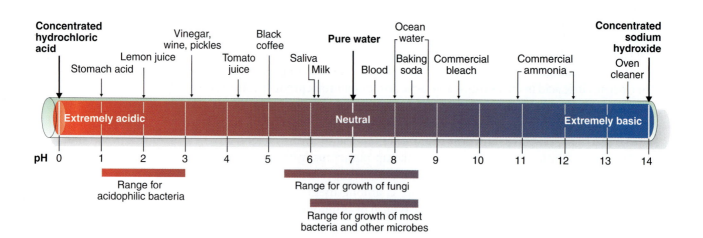

To this point . . .

We discovered water is an essential component of cells and has unique properties due to the polar nature of the molecules. Water serves as a solvent in all cells, allowing polar molecules (solutes) to dissociate (dissolve) so they can be used in cell metabolism. Water also moderates temperature fluctuations in cells and organisms.

The discussion finished with a brief review of acids and bases. We reviewed the pH scale used to measure the relative strength or weakness of an acid or base and learned that chemical reactions are sensitive to even the slightest change in pH. Cells contain buffers to prevent potential pH shifts.

In the final section, we shall focus on the four important classes of organic compounds: carbohydrates, lipids, proteins, and nucleic acids. We will examine the structures and functions of these compounds and touch on their importance in microbiology.

MicroFocus 2.2

LIFE IN STRANGE PLACES BENEFITS CHICKENS

Life as a whole for organisms on planet Earth is fairly comfortable. Most microorganisms make their living in environments that have moderate temperatures and neutral pHs. Then there are microorganisms that live in strange, seemingly inhospitable places where the pH is about that of your car battery. Such microbes are acidophilic (acid loving).

The domain *Archaea* that we mentioned in Chapter 1 is one of the two groups of prokaryotic microorganisms. Some of its members live in unusually exotic places. For example, take the genus *Sulfolobus*. This microbe normally grows in hot sulfur springs and volcanic vents where the pH is 2 to 3, and it easily tolerates a pH of 1. The genus *Picrophilus* is the most acidophilic prokaryote yet discovered. It grows best at a pH of 0.7 and will continue to grow at pHs below 0.

In these examples, it is the environmental pH that is so acid. The internal pH of the organisms remains near neutrality or perhaps a few pH units to the acid side. So, how can these organisms keep the excessive hydrogen ions out? It's a role for the cell membrane surrounding the organism. These bacteria have evolved a membrane that contains an unusual complement of lipids that are impermeable to hydrogen ions. If the ions cannot enter the cell, they cannot damage the cell's inner workings. However, place the organisms in a growth medium at pH 4 and their membranes become leaky and the cells rapidly disintegrate.

Many of these organisms also have enzymes that work best at acidic pHs. So, even if the cell interior has an acidic pH, chemical reactions are not hindered and the microbes live on. This ability to produce acid-tolerant enzymes has been captured for practical use. Agricultural companies now isolate and produce these acid-tolerant enzymes as a feed supplement for chickens. When eaten, the enzymes go to the chicken's stomach. Being acid-tolerant, they maximize the digestion of otherwise indigestible fibrous materials normally found in feed. Being able to digest the fibrous material represents better nutrition for chickens and promotes growth that is more rapid.

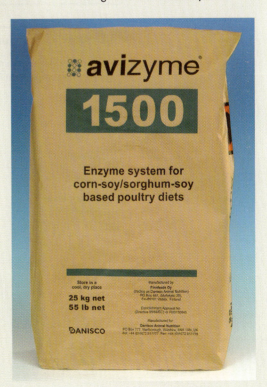

Major Organic Compounds of Living Organisms

As mentioned in the last section, a typical bacterial cell is about 70 percent water and 30 percent chemicals. If we evaporate away all the water in a bacterial cell—or any living cell—the predominant "dry weight" remaining consists of organic compounds (FIGURE 2.6). There are four distinct classes of organic compounds: the carbohydrates, lipids, proteins, and nucleic acids. Except for the lipids, each class represents a type of **polymer** built from a large number of separate building blocks called **monomers**.

Polymer:
a long chain of covalently linked similar subunits.

FUNCTIONAL GROUPS DEFINE MOLECULAR BEHAVIOR

Before we look at the major classes of organic compounds, we need to address one question. The monomers that build these polymers are essentially stable molecules because their outer shells have been filled through covalent bonding. So why should these molecules take part in chemical reactions?

Well, in fact, these molecules are not completely stable. Rather, projecting from the hydrocarbon scaffoldings are groups of atoms called **functional groups**. These groups represent points where further chemical reactions can occur if facilitated by a specific enzyme. There are only a small number of functional groups but their

FIGURE 2.6

Organic Compounds in Cells

Macromolecules are abundant in cells. The approximate composition of these in a bacterial cell is shown. The composition in other microbes, except viruses, is similar.

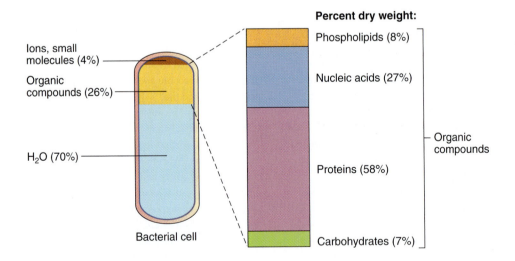

Ions, small molecules (4%)
Organic compounds (26%)
H₂O (70%)
Bacterial cell

Percent dry weight:
Phospholipids (8%)
Nucleic acids (27%)
Proteins (58%)
Carbohydrates (7%)
Organic compounds

differences and placement on monomers makes possible a large variety of chemical reactions.

The important functional groups in living organisms, including microbes, include the hydroxyl group (—OH), the amino group (—NH$_2$), the carboxyl group (—COOH), the carbonyl group (—CHO), the sulfhydryl group (—SH), and the phosphate group (—OPO$_3^{-2}$). Functional groups interact and build bigger molecules through a **dehydration synthesis** reaction. In a dehydration synthesis (also called condensation reaction), a water molecule (H and OH) is removed. This leaves the remaining atoms in each molecule unstable. However, by forming a covalent bond between these molecules, stability is restored.

Dehydration synthesis: a chemical reaction in which two molecules become covalently bonded through the removal of a water molecule.

$$A - OH + HO - B \longrightarrow [A - OH \quad HO - B] \longrightarrow [A \cdots O \cdots B]$$

Molecule A with a hydroxyl group   Molecule B with a hydroxyl group   H$_2$O   (unstable)

A — O — B
Molecule AB

We will see specific examples of these groups and how they interact as we now visit each of the four classes of organic compounds.

CARBOHYDRATES PROVIDE ENERGY AND BUILDING MATERIALS

Carbohydrates are organic compounds composed of carbon, hydrogen, and oxygen atoms that build sugars and starches. In simple sugars, like glucose (C$_6$H$_{12}$O$_6$), the ratio of hydrogen to oxygen is 2 to 1, the same as in water. For this reason, the carbohydrates often are considered "hydrated carbon," hence their name. However, the atoms are not present as water molecules bound to carbon, but rather carbon covalently bonded to hydrogen and hydroxyl groups (H—C—OH). This combination occurs frequently in carbohydrate structures.

Carbohydrates function as energy sources in cells. They also function as structural molecules in cell walls and nucleic acids. Often the carbohydrates are termed **saccharides** (*saccharon,* meaning "sugar") and, based on size (molecular weight), are divided into three groups: monosaccharides, disaccharides, and polysaccharides (FIGURE 2.7).

Monosaccharides (*mono* = "one") are the simplest carbohydrates; they represent the monomers for disaccharides and polysaccharides. Monosaccharides have three to seven carbon atoms. Glucose and fructose are both six-carbon sugars and are among the most widely encountered monosaccharides. Both have a molecular formula typical of carbohydrates: C$_6$H$_{12}$O$_6$. However, their structural formulas are different (Figure 2.7a). Such molecules are called **isomers**. Isomers react differently and usually have different properties because of the arrangement of the functional groups. Also, fructose tastes sweeter than glucose.

Isomers: molecules with identical molecular formulas but different structural formulas.

Glucose serves as the basic supply for energy in the world. Estimates vary, but many scientists suggest that half the world's carbon exists as glucose. Such sugars are synthesized from water and carbon dioxide through the process of photosynthesis. Algae and cyanobacteria have the chemical machinery for this process, which is described in Chapter 5.

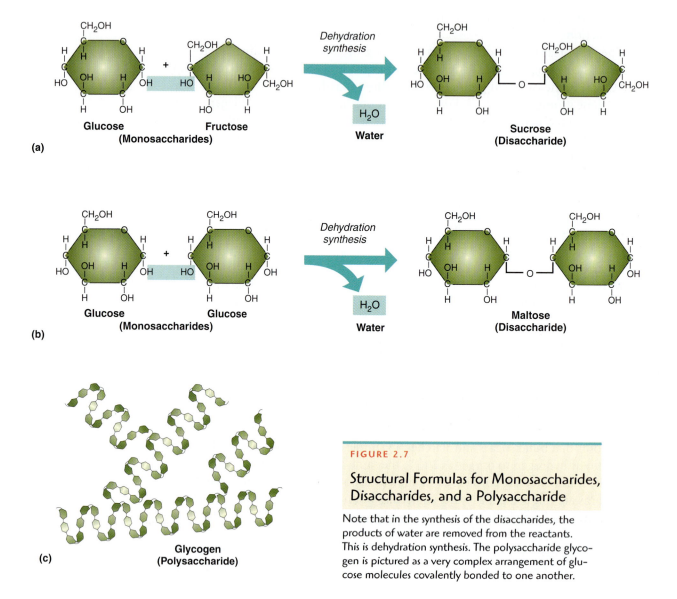

(a)

Glucose   Fructose
(Monosaccharides)

Dehydration synthesis

H_2O
Water

Sucrose
(Disaccharide)

(b)

Glucose   Glucose
(Monosaccharides)

Dehydration synthesis

H_2O
Water

Maltose
(Disaccharide)

(c)

Glycogen
(Polysaccharide)

FIGURE 2.7

Structural Formulas for Monosaccharides, Disaccharides, and a Polysaccharide

Note that in the synthesis of the disaccharides, the products of water are removed from the reactants. This is dehydration synthesis. The polysaccharide glycogen is pictured as a very complex arrangement of glucose molecules covalently bonded to one another.

Disaccharides (*di* = "two") are double sugars. They are composed of two monosaccharides held together by a covalent bond. Sucrose (table sugar) is an example (Figure 2.7a). This disaccharide is constructed from a glucose and fructose molecule by dehydration synthesis. Sucrose is a starting point in wine fermentations, and it is often involved in tooth decay (MicroFocus 2.3). Maltose is another disaccharide but is composed of two glucose monomers (Figure 2.7b). Maltose occurs in cereal grains, such as barley, and is fermented by yeasts for energy. An important by-product of the fermentation is the formation of alcohol in beer. Lactose is a third disaccharide. It is composed of the monosaccharides glucose and galactose. Lactose is known as milk sugar because it is the principal carbohydrate in milk. Under controlled industrial conditions, microorganisms digest the lactose for energy; in the process, they produce the acid in yogurt, sour cream, and other sour dairy products.

MicroFocus 2.3

SUGARS, ACID, AND DENTAL CAVITIES

At least once in our lives, each of us probably has feared the dentist's drill after a cavity has been detected. Dental cavities usually result from eating too much sugar or sweets. These sugars contribute to cavity formation only in an indirect way. The real culprits are bacteria.

What happens in a cavity is this: There are many species of bacteria that normally inhabit the mouth. Some of these bacteria along with saliva and food debris form a gummy layer called dental plaque, a biofilm mentioned in Chapter 1. If not removed, plaque will accumulate on the grooved chewing surfaces of back molars and at the gum line.

Plaque starts to accumulate within 20 minutes after eating. For example, as the bacteria multiply they digest sucrose (table sugar) in sweets for energy. The metabolism of sucrose has two consequences. Some plaque bacteria produce dextran, an adhesive polysaccharide that increases the thickness of plaque. They also produce lactic acid as a by-product of sugar metabolism. Being trapped under the plaque, the acid is not neutralized by the saliva. When the pH drops to 5.5 or lower, the hydrogen ions start to dissolve or demineralize the dental enamel.

Over time, a depression or cavity forms. When the soft dental tissues underneath the enamel are reached, toothache pain results from the exposure of the sensitive nerve endings in the soft tissues.

Good oral hygiene, which includes brushing, flossing, and regular professional dental cleaning, can keep plaque to a minimum. At home, watch what you eat. Consuming sugary foods with a meal or for dessert is less likely to cause cavities because the increased saliva produced while eating helps wash food debris off the tooth surface and neutral-ize any acids produced. More important is the nature and frequency of sugar intake. Unfortunately, if you frequently snack on sugary foods or eat sugary foods that are sticky, like caramel, toffee, dried fruit, or candies, the food debris will cling to teeth for a longer time, causing more plaque and providing continuous acid attack on your teeth. No wonder cavities are one of the most prevalent infectious diseases, second only to the common cold. More useful information on preventing tooth decay is provided in MicroFocus 11.5.

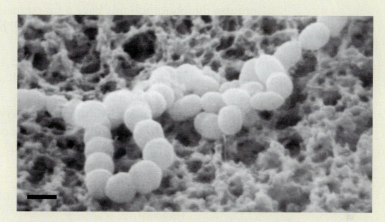

■ *A highly magnified view of streptococci from the surface of the teeth. Organisms such as these exist in the dental plaque and produce the acid that leads to cavities. (Bar = 2 μm.)*

In most bacteria, the cell wall is composed of carbohydrate and protein, forming a peptidoglycan (also called murein). A disaccharide is the building block of the peptidoglycan. In Chapter 4, we will examine the cell wall in more detail.

Polysaccharides (*poly* = "many") are complex carbohydrates. These polymers generally are large compounds formed by joining hundreds of thousands of glucose monomers. Covalent bonds resulting from dehydration synthesis link the units. Starch, a common polysaccharide, is used by some eukaryotic microbes as an energy source. Glycogen, the common storage polysaccharide in humans, functions as a stored energy source in bacteria (Figure 2.7c). Cellulose, another polysaccharide, is a component of the cell walls of plants and many algae.

LIPIDS STORE ENERGY AND ARE COMPONENTS OF MEMBRANES

The **lipids** are a broad group of nonpolar organic compounds that generally do not dissolve in water. Like carbohydrates, lipids are composed of carbon, hydrogen, and oxygen, but the proportion of oxygen is much lower.

The best-known lipids are the fats. They serve living organisms as important stored energy sources. As shown in FIGURE 2.8, **fats** consist of a three-carbon glycerol molecule and up to three long-chain fatty acids. Each fatty acid is a long nonpolar hydrocarbon chain containing between 16 and 18 carbon atoms. Bonding of fatty acids to the glycerol molecule occurs by dehydration synthesis between the hydroxyl and carboxyl functional groups.

There are two major types of fatty acids. **Saturated fatty acids** contain the maximum number of hydrogen atoms extending from the carbon backbone, while **unsaturated fatty acids** contain less than the maximum due to covalent double bonds between a few carbon atoms.

In bacteria, lipids are not stored as fats but rather, along with proteins, are a component of cell membranes. The membrane lipids are called **phospholipids** because in place of the third fatty acid chain there is a phosphate group. By having phosphate groups, the phospholipids have a functional group that is polar and can actively interact with other polar molecules. We will have more to say about phospholipids and membranes in the next chapter.

Other types of lipids are the waxes and sterols. Waxes are composed of long chains of fatty acids and form part of the cell wall in *Mycobacterium tuberculosis*, the bacterium that causes tuberculosis. Sterols, such as cholesterol, are very different from lipids and are included with lipids solely because they too are nonpolar molecules. Sterols are composed of several rings of carbon atoms with side chains, and stabilize membranes of fungi and the bacterium *Mycoplasma*.

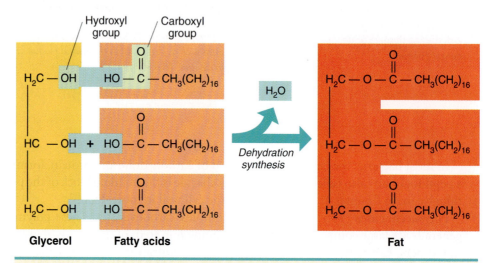

FIGURE 2.8

The Molecular Structures of Fat Components and the Synthesis of a Fat

Fats consist of fatty acids and glycerol. Each fatty acid contains numerous carbon and hydrogen atoms. Fat formation occurs by dehydration synthesis, as the components of water are removed from the reactants and covalent bonds are formed.

PROTEINS ARE THE WORKHORSES OF CELLS

Proteins are by far the most abundant organic compounds in microorganisms and all living organisms, making up about 58 percent of the cell's dry weight. They are composed of carbon, hydrogen, oxygen, nitrogen, and usually sulfur atoms. The high percentage of protein indicates their essential and diverse roles in organisms. Many proteins function as structural components of cells and cell walls, as transport agents in membranes, and as toxins (FIGURE 2.9). A large number of proteins serve as **enzymes**, a group of biological compounds that catalyze chemical reactions. In the human body, the cornea of the eye is composed primarily of protein, and, as such, it may supply nutrients to encourage the growth of disease-causing (pathogenic) bacteria, as FIGURE 2.10 indicates.

FIGURE 2.9

A Case of Salmonella Food Poisoning

This outbreak occurred at a restaurant during October 1991. *Salmonella enteritidis* was isolated from stools of all 13 patrons of the restaurant who sought medical attention. *Salmonella* food poisoning is caused by a protein toxin produced by the bacterium (Chapter 9).

1. During the morning of October 17, a restaurant employee prepared dressing for Caesar salad, cracking fresh eggs into a large bowl containing olive oil. The eggs were probably contaminated with *Salmonella* bacteria.

3. The warm water raised the temperature of the mixture slightly and encouraged bacteria to grow. Although the dressing was placed in the refrigerator, the bacteria probably had an opportunity to grow before the dressing cooled.

2. Anchovies, garlic, and warm water then were mixed into the egg and oil.

4. Later that day, the Caesar dressing was placed at the salad bar in a cooled compartment having a temperature of about 60°F (comfortable room temperature is a slightly higher 70–75°F). There the dressing remained until the restaurant closed, a period of 8 to 10 hours. During that time, many patrons helped themselves to the Caesar salad.

5. Within three days, fifteen restaurant patrons fell ill with diarrhea, fever, abdominal cramps, nausea, and chills. Thirteen sought medical care, and eight required intravenous rehydration.

TEXTBOOK CASES

Proteins are built from nitrogen-containing monomers called **amino acids**. At the center of each amino acid is a carbon atom (FIGURE 2.11a). Attached to this atom are two functional groups: an amino group (—NH$_2$) and a carboxyl group (—COOH). Also attached to the carbon is a side chain, called the **R group**. Each of the 20 amino acids differs only by the atoms composing the R group (FIGURE 2.11b). Therefore, these side chains, many being functional groups, are essential in determining the final shape, and therefore function, of the protein.

In protein formation, amino acids (sometimes called peptides) are joined together by covalent bonds when each amino group is linked to the carboxyl through dehydration synthesis (FIGURE 2.12). The chain that results is known as a **polypeptide**

FIGURE 2.10

A Case of Corneal Infection Caused by a Pathogenic Bacterium

This incident happened in Georgia during the winter of 1989.

1. On the morning of January 11, a woman scratched the cornea of her left eye while applying mascara. The scratch was painful, but she thought little else of the incident and continued her routine as she prepared to go to work.

2. A day later her eye was painfully swollen and red, the eye was very sensitive to light. She decided to seek medical help.

3. That day, the woman went to an ophthalmologist and received treatment with gentamicin ointment. Unfortunately, the symptoms worsened, and on January 14 she was admitted to the hospital with a severe corneal abscess.

4. Technologists at the hospital isolated the bacterium *Pseudomonas aeruginosa* from the abscessed cornea. They found the same organism in her mascara and concluded that the scratch had introduced bacteria to her eye tissues. The woman was treated with a series of antibiotics and recovered.

TEXTBOOK CASES

FIGURE 2.11

Examples of Amino Acids

(a) The generalized structure for an amino acid. In cells, amino acids usually exist in an ionized form. (b) Four amino acids that differ by the molecular nature of the R group. Cysteine contains a sulfhydral functional group and glutamic acid has a carboxyl functional group as the R group.

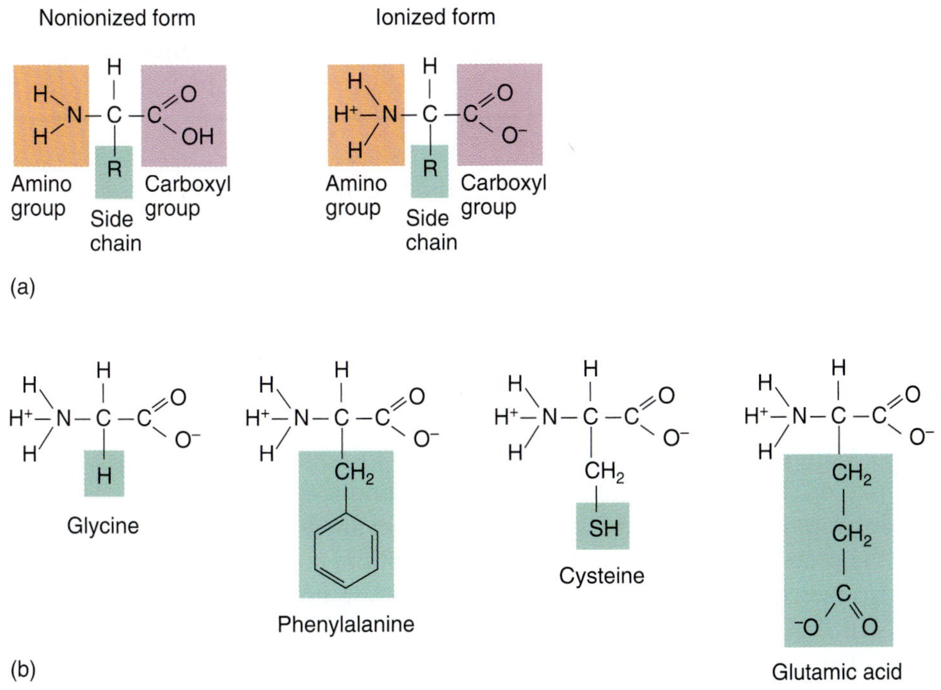

and the bond therefore is called a **peptide bond**. How the amino acids are slotted into position building protein synthesis is a complex process discussed in Chapter 6. It should be noted that the sequence of amino acids is of the utmost importance because a single amino acid improperly positioned may change the three-dimensional shape and function of the protein.

Since proteins have tremendously diverse roles, they come in many sizes and shapes. The final shape of a protein depends on several factors associated with the amino acids. The chain of amino acids in the protein represents the **primary structure** (FIGURE 2.13a). Each protein that has a different function will have a different primary structure. However, the sequence of amino acids alone is not sufficient to confer function.

Proteins have regions in the polypeptide that fold into a corkscrew shape or alpha helix. These regions represent part of the protein's **secondary structure** (FIGURE 2.13b). Hydrogen bonds between amino groups (-NH) and carbonyl groups (-C=O) on nearby amino acids maintain this structure. A secondary structure may form when the hydrogen bonds cause portions of the polypeptide chain to line up alongside one another, forming a pleated sheet. Other regions may not interact and remain in a random coil.

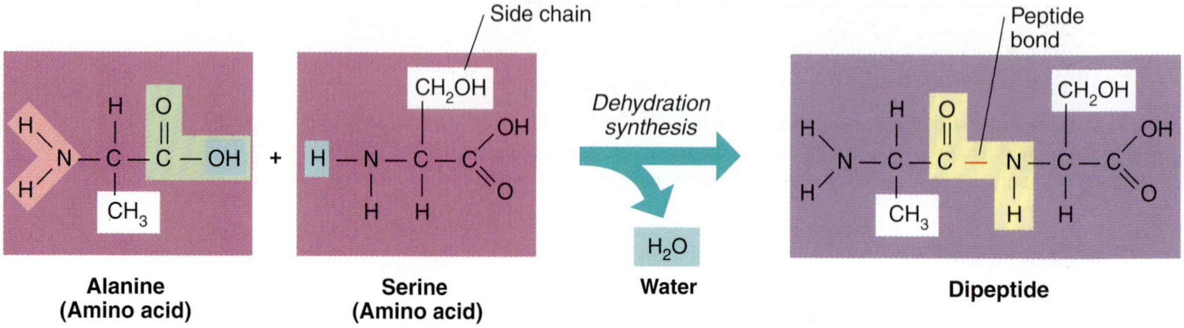

FIGURE 2.12

Formation of a Dipeptide by Dehydration Synthesis

The amino acids alanine and serine are shown, with the differences in white. The —OH from the carboxyl group of alanine combines with the —H from the amino group of serine to form water. The open bonds then link together, yielding a peptide bond. The process is repeated hundreds of times, adding additional amino acids to form a polypeptide.

Many proteins have a **tertiary structure** (FIGURE 2.13c). In this case, the protein is folded back on itself much like a spiral telephone cord folded on a table. Ionic and hydrogen bonds between R groups on adjacent amino acids help form and maintain the protein in its tertiary structure. In addition, covalent bonds, called **disulfide bridges**, between sulfur atoms in R groups are important in stabilizing tertiary structure.

The shape of some proteins requires tertiary structure for functional activity. However, many proteins contain two or more polypeptide chains that form the complete and functional protein. Such association of polypeptides represents the **quaternary structure** (FIGURE 2.13d). Each polypeptide chain is folded into its tertiary structure and the unique association between separate polypeptides produces the quaternary structure. The same types of chemical bonds are involved as in tertiary structure.

The ionic and hydrogen bonds helping hold a protein in its functional shape are relatively weak associations. Such weak interactions in a protein are influenced by environmental conditions. When subjected to heat, pH changes, or certain chemicals, many of the non-covalent bonds holding a protein together break. This loss of a functional tertiary structure is referred to as **denaturation**. Since most enzymes are tertiary or quaternary proteins, heat, pH, or chemicals may be used to denature them. For example, the white of a boiled egg is denatured egg protein (albumen) and cottage cheese is denatured milk protein. Should enzymes be denatured, the important chemical reactions they control will be interrupted and death of the organism may result. Now you should understand the importance of buffers in cells; they prevent protein denaturation. Viruses also can be destroyed by denaturing the proteins found in the viral protein coat.

FIGURE 2.13

Protein Folding

The function of a protein is dependent on how it folds. Folding depends on the primary structure (a), which dictates all higher levels of folding. Secondary structure (b) is dependent on the interactions between amino and carbonyl groups on amino acids, while tertiary structure (c) depends on bonding interactions between R groups. Some proteins consist of more than one polypeptide (d), and this quaternary structure also is determined by the bonding interactions between each polypeptide.

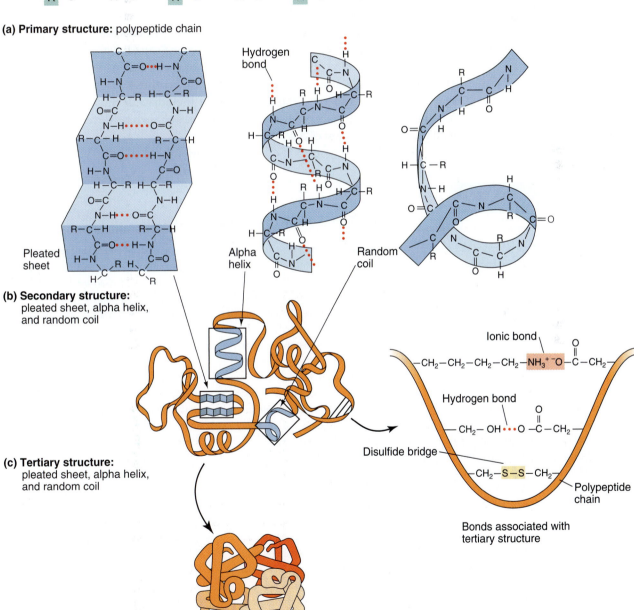

(a) Primary structure: polypeptide chain

(b) Secondary structure: pleated sheet, alpha helix, and random coil

(c) Tertiary structure: pleated sheet, alpha helix, and random coil

(d) Quaternary structure: two or more folded polypeptides

Bonds associated with tertiary structure

NUCLEIC ACIDS STORE, TRANSPORT, AND CONTROL HEREDITARY INFORMATION

Gene:
a segment of DNA specifying a protein or RNA.

The **nucleic acids**, among the largest organic compounds found in organisms, are composed of carbon, hydrogen, oxygen, nitrogen, and phosphorus atoms. Two types function in all living things (except viruses): **deoxyribonucleic acid** (**DNA**) and **ribonucleic acid** (**RNA**). DNA is the genetic material that stores and encodes hereditary information in the form of **genes**, while RNA functions in the construction of proteins and gene control.

Both DNA and RNA are composed of repeating monomers called **nucleotides** (**FIGURE 2.14a**). Each nucleotide has three components: a carbohydrate molecule, a phosphate group, and a nitrogenous base. The carbohydrate molecule in DNA is deoxyribose, while in RNA it is the closely related sugar ribose. The **nitrogenous bases** are nitrogen-containing compounds. In DNA, the bases are adenine (A), guanine (G), cytosine (C), and thymine (T). In RNA, adenine, guanine, and cytosine also are present, but uracil (U) is found instead of thymine. The nucleotides in one strand are covalently joined through dehydration synthesis reactions between the sugar of one nucleotide and the phosphate of the adjacent nucleotide to form a polynucleotide (**FIGURE 2.14b**). **MicroInquiry 2** uses these base differences to investigate the mode of action of a potential antibiotic.

MicroFocus 2.4

CREATING LIFE IN A TEST TUBE

In Chapter 1 we discussed the idea that life could arise from inanimate objects. This idea of spontaneous generation was debunked by Louis Pasteur in 1861. Today, a somewhat similar idea concerning life has arisen. The idea concerns the ability to create "new" life in a test tube. As we will see in Chapter 7, some organisms, including many microorganisms, have had their complete genetic makeup sequenced and, in some cases, much of the regulation of the genetic machinery understood. Therefore, the suggestion has been made that it should be possible to mix genes (inanimate chemicals?) to generate a new life form—essentially making life from new gene combinations.

Scientists in several labs around the world are trying to discover the minimal number of genes required to produce a living cell. The bacterium *Mycoplasma genitalium*, which infects the urinary tract, has one of the smallest genomes of any independently living cell. It has 450 protein-coding genes, about 300 of which appear to be essential. From a construction view, identifying the genes might be the easiest part. The genetic engineering techniques also are available, so what would we need in our test tube to create life?

The genes for all RNAs that convert genetic information into proteins would need to be included. Perhaps a few hundred genes for reproduction and growth would be essential and, as a result, our new organism must have the capability to produce a cell membrane.

All this gene-coding information would exist as DNA, so techniques would be required to make long DNA molecules (artificial chromosomes). As a first step, one plan is to add the artificial chromosome to an *M. genitalium* cell from which its DNA had been removed. The ultimate question then is, Will the components of the bacterial cell "turn on" the genes in the new chromosome?

Importantly, these ideas have more use than simply trying to build new life. Scientists see the research as an opportunity to expand evolution's repertoire by designing organisms that are better at doing certain jobs. Can we, for example, design microorganisms that are better at bioremediation, providing new energy resources, or helping eliminate global pollution on our planet?

Such ideas are not without concern for safety. Certainly, it is not the intention to produce "new bugs" that could cause human or animal disease. But, in an age of bioterrorism, the capabilities of creating life could be used for other nefarious purposes.

However, creating new life in a test tube has not been perfected yet—or has it? On July 11, 2002, scientists at the State University of New York, Stony Brook reported that they had created a virus from scratch. They reconstructed a "live" poliovirus by assembling the poliovirus genes. Although many do not consider viruses to be "living" microbes, this certainly qualifies as the closest thing yet to creating life in a test tube.

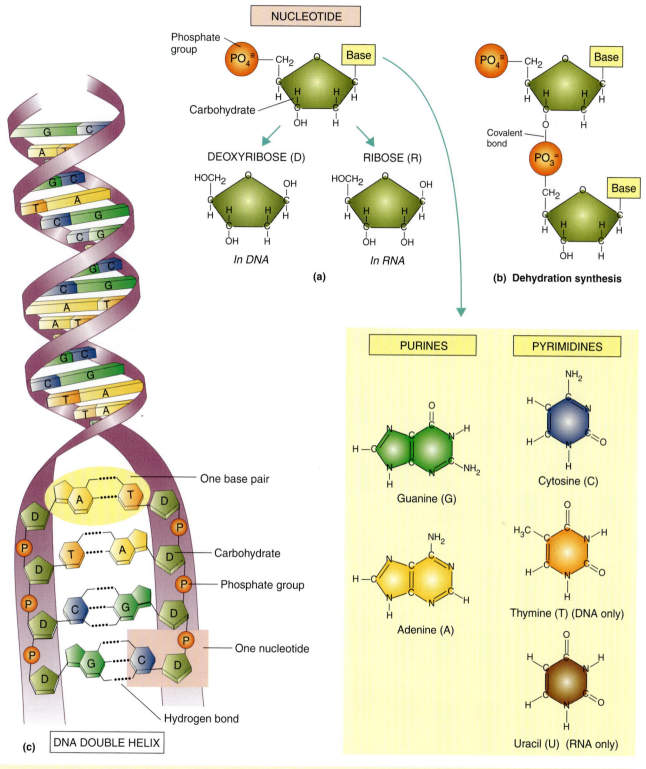

FIGURE 2.14

The Molecular Structures of Nucleotide Components and the Construction of DNA

(a) The carbohydrates in nucleotides are ribose and deoxyribose. These two sugars are identical except at one oxygen atom. The nitrogenous bases include adenine and guanine, which are large purine molecules, and thymine, cytosine, and uracil, which are smaller pyrimidine molecules. Note the similarities in the structures of these bases and the differences in the side groups. (b) Nucleotides are bonded together by dehydration synthesis reactions. (c) The two polynucleotides of DNA are bonded together by hydrogen bonds between adenine (A) and thymine (T) or guanine (G) and cytosine (C) to form a double helix.

MicroInquiry 2

ISOTOPES AS TRACERS IN DRUG DISCOVERY

Several biologically important elements have isotopes that are radioactive. The table below lists several of those commonly used in research.

Let's examine the use of these isotopes in drug development. Note: today, there are more sophisticated ways to accomplish the same thing.

Scenario: As a researcher for a drug company, you are given a chemical that seems to kill the bacterium *Escherichia coli*. However, your job is to discover how the potential antibacterial chemical works; that is, what organic compound (DNA, RNA, or protein) is affected that brings about the death of the bacteria?

2a. What bases are unique to RNA and DNA?

2b. How could isotopes be used to label radioactively these nucleic acids so they can be identified separately?

2c. Which of the three elements is specific to proteins?

Here is one way we could approach the problem. We can purchase the following items from science supply houses:

^{3}H–uridine (that is, radioactive uracil, the base found only in RNA); ^{14}C–thymine (the base found only in DNA); and ^{35}S–methionine (one of the 20 amino acids that contains sulfur in its structure). So, we have specific "labels" that would be incorporated into RNA, DNA, and protein.

The experiment will involve growing the bacterium *E. coli* in each of three broth cultures. To these actively growing cultures, we will add ^{3}H–uridine to one culture, ^{14}C–thymine to another, and ^{35}S–methionine to the third. If our potential antimicrobial chemical is not added to the cultures, the following occurs: as DNA is copied, the molecules should become radioactive with ^{14}C–thymine; as RNA is synthesized, the

ELEMENT	COMMON FORM	RADIOACTIVE FORM
Hydrogen	^{1}H	^{3}H (tritium)
Carbon	^{12}C	^{14}C
Sulfur	^{32}S	^{35}S

James Watson and Francis Crick showed in 1953 that a complete DNA molecule consists of two polynucleotide strands that oppose each other in a ladder-like arrangement (FIGURE 2.14c). Guanine and cytosine line up opposite one another, and thymine and adenine oppose each other in the two strands. The complementary base pairs in the double-stranded DNA molecule are held together by hydrogen bonds. The double strand then twists to form a spiral arrangement called the DNA **double helix**.

As with proteins, the nucleic acids cannot be altered without injuring the organism or killing it. Ultraviolet light damages DNA, and thus it can be used to control bacteria on an environmental surface. Chemicals, such as formaldehyde, alter the nucleic acids of viruses and can be used in the preparation of vaccines. Certain antibiotics interfere with DNA or RNA function and thereby kill bacteria. In Chapters 22 to 24, we shall encounter these and other examples where tampering with nucleic acids is the basis for controlling microorganisms. Chapter 6 is devoted to the role of nucleic acids in the genetics of bacteria.

The four major classes of organic compounds are summarized in TABLE 2.3. In Chapter 1, we discussed the doctrine of spontaneous generation, which alleged that life could arise from nonliving things. MicroFocus 2.4 looks at the attempts to "create life in a test tube" using the types of compounds described in this chapter.

OK to rewrite so MicroFocus 2.4 reference is on page 68?

RNAs should become radioactive with ³H-uridine; and as proteins are synthesized, they will become radioactive with ³⁵S-methionine. We can assume we have instruments to measure the radioactivity.

To these actively growing cultures containing the isotope, we add our chemical to be tested. With time, we remove samples from each of the cultures and determine the level of radioactivity in the DNA, RNA, and proteins. The results of the experiment are shown in the graph to the right.

2d. What do these isotope experiments indicate about the target for the chemical? Which of the three organic compounds is affected?

Answers can be found in Appendix E.

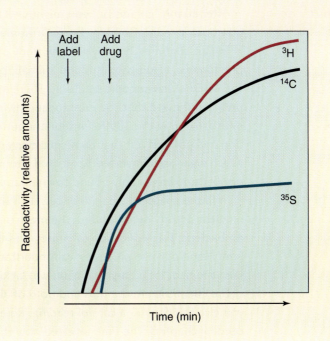

TABLE 2.3

Major Organic Compounds of Microorganisms

ORGANIC COMPOUND	MONOMER	SOME MAJOR FUNCTIONS	EXAMPLES
Carbohydrate:			
Monosaccharides	–	Energy source	Glucose, fructose
Disaccharides	Monosaccharides	Energy source; physical structure	Lactose, sucrose; bacterial walls
Polysaccharides	Monosaccharides	Energy storage; physical structure	Starch, glycogen; cellulose, cell walls
Lipid:			
Fats	Fatty acids and glycerol	Energy source	Fat; oil
Phospholipids	Fatty acids, glycerol, phosphate	Foundation for membranes	Cell membranes
Waxes	Fatty acids	Physical structures	*Mycobacterium* cell walls
Sterols	–	Membrane stability	Cholesterol
Protein	Amino acids	Enzymes; toxins; physical structures	Antibodies; viral surface; flagella; membranes
Nucleic acid	Nucleotides	Inheritance; instructions for protein synthesis	DNA, RNA

Note to the Student

After all this "chemistry," you might well ask whether this is going to be a chemistry book or a microbiology book. Trust me—it is a microbiology book.

In past centuries, biology was the study of nature. Naturalists described natural phenomena, classified plants and animals, performed dissections, and marveled at the world of microscopic creatures. Then, in 1828, Friedrich Wöhler synthesized urea and showed that the substances of living organisms could be produced by scientists. Later, in the 1850s, Pasteur announced his observations about fermentation and pointed out that those microorganisms were tiny, chemical factories in which sugar was converted to alcohol. From these and other advances, microbiology became irrevocably bound to chemistry.

In future chapters, we shall discuss such topics as the metabolism of microorganisms, the activity of antibodies, the mechanisms of disease, the action of antibiotics on bacteria, the replication of viruses, and the chemical control of microorganisms. None of these processes was understood 100 years ago, partly because knowledge of their chemical bases was lacking. Such is not the case today.

The chemicals are the nuts and bolts of all living organisms. I recommend that you give this chapter a careful reading. Refer back to these pages as you read succeeding chapters. The time invested now though will be well worth the effort.

Summary of Key Concepts

2.1 THE ELEMENTS OF LIFE

■ **Atoms Are Composed of Subatomic Particles.** Ninety-two naturally occurring elements comprise the substance of all known living and nonliving organisms. The simplest unit of these elements is the atom, a particle consisting of an atomic nucleus (with protons and neutrons) and a cloud of electrons.

■ **Isotopes Vary in the Number of Neutrons Present.** Isotopes of an element have different numbers of neutrons. Some unstable ones, called radioisotopes, are useful in research and medicine.

■ **Atoms Can Become Electrically Charged.** If an atom gains or loses electrons, it becomes an electrically charged ion. Anions have gained one or more electrons while cations have lost one or more electrons. Many ions are important in microbial metabolism.

■ **Electron Placement Determines Chemical Reactivity.** Electrons are arranged outside the nucleus in electron shells, with each shell holding a maximum number of electrons. Interactions occur between atoms to fill the outer shells with electrons.

■ **Compounds and Molecules Form through Chemical Bonding.** The interactions between two or more different atoms result in compounds. The atoms are held together by chemical bonds. A molecule shares electrons in chemical bonds between interacting atoms. The particular atoms and their numbers in a substance are determined by its molecular formula while the size is determined by its molecular weight.

2.2 CHEMICAL BONDING

■ **Ionic Bonds Form between Oppositely Charged Ions.** Ionic bonds result from the attraction of oppositely charged ions (cation and anion). Salts are a common result, which in cells are relatively weak bonds.

■ **Covalent Bonds Share Electrons.** Most atoms achieve stability by filling the outer electron shells through a sharing of electrons between atoms, forming covalent bonds. Hydrocarbons are the simplest compounds with covalent bonds. The equal sharing of electrons produces molecules that are nonpolar (no electrical charge).

- **Hydrogen Bonds Form between Polar Groups or Molecules.** Atomic interactions between hydrogen and oxygen or nitrogen produce unequal sharing of electrons, which generate polar molecules (have electrical charges). Separate polar molecules, like water, are electrically attracted to one another and form hydrogen bonds, again involving positively charged hydrogen atoms and negatively charged oxygen atoms.

- **Chemical Reactions Change Bonding Partners.** In a chemical reaction, the atoms in the reactant change bonding partners in forming one or more products. Examples include combination reactions, exchange reactions, hydrolysis reactions, and dehydration synthesis. The number of atoms is the same in the reactants and products.

2.3 WATER AND pH

- **Water Is the Universal Solvent of Life.** All chemical reactions in organisms occur in liquid water. Being polar, water has unique properties. These include its role as a solvent, as a chemical reactant, as a cohesive molecule that stays liquid, and as a solvent that maintains a fairly constant temperature.

- **Acids and Bases Must Be Balanced in Cells.** Among the important compounds in living organisms are acids and bases. Acids are compounds that donate hydrogen ions to a solution; bases are compounds that remove hydrogen ions from a solution and, in so doing, increase the number of hydroxyl ions. The pH scale indicates the number of hydrogen ions in a solution and indicates the relative acidity or alkalinity of a solution. Shifts in the acid/base balance can lead to a disruption in cell or organism function.

- **Buffers Prevent pH Shifts.** Buffers are a mixture of a weak acid and a weak base that maintains acid/base balance in cells. Excess hydrogen ions can be "absorbed" by the base and too few hydrogen ions can be provided by the acid.

2.4 MAJOR ORGANIC COMPOUNDS OF LIVING ORGANISMS

- **Functional Groups Define Molecular Behavior.** The building of large organic compounds depends on the functional groups found on the building blocks, called monomers. Functional groups on monomers interact through dehydration synthesis reactions to produce water in the course of forming a covalent bond between monomers.

- **Carbohydrates Provide Energy and Building Materials.** One important compound in living organisms is the carbohydrates. These compounds contain carbon, hydrogen, and oxygen. They are used primarily as energy sources for life processes, and they include monosaccharides such as glucose and fructose; disaccharides such as sucrose, lactose, and maltose; and polysaccharides such as glycogen and cellulose.

- **Lipids Store Energy and Are Components of Membranes.** Lipids have the same types of atoms as carbohydrates, but the proportion of oxygen is much lower. Like carbohydrates, lipids serve as energy sources, but their major role is as phospholipids in the construction of cell membranes.

- **Proteins Are the Workhorses of Cells.** Another group of compounds in living organisms is the proteins. These are chains of amino acids connected by peptide bonds, a type of covalent bond. Proteins are used as enzymes and as structural components of cells. Primary, secondary, and tertiary structures form the functional shape of proteins. Many proteins consist of more than one polypeptide, producing the protein's quaternary structure.

- **Nucleic Acids Store, Transport, and Control Hereditary Information.** The genetic instructions for living organisms are located in compounds called nucleic acids. A nucleic acid is composed of carbohydrate molecules, phosphate groups, and a series of nitrogenous bases. Two nucleic acids are important in biological systems: deoxyribonucleic acid (DNA), which stores and encodes the hereditary information and ribonucleic acid (RNA) that transmits the information to make proteins. RNA also controls genes and genetic activity.

Questions for Thought and Discussion

Answers to selected questions can be found in Appendix C.

1. Calcium atoms have two electrons in their outer shell; magnesium atoms also have two. Would you expect that these elements will interact through bonding when their atoms are brought together? Explain.

2. Certain detergent-based disinfectants are known to dissolve lipids. How would this activity affect bacterial cells?

3. Why do you think organic molecules tend to be so large?

4. Why is Wöhler's synthesis of urea in 1828 considered a landmark achievement in bridging the gap between biology and chemistry?

5. Bacteria do not grow on bars of soap even though the soap is wet and covered with bacteria after one has washed. Explain this observation.

6. Oxygen comprises about 65 percent of the weight of a living organism. This means that a 120-pound person contains 78 pounds of oxygen. How can this be so?

7. Suppose you had the choice of destroying one class of organic molecules in bacteria in order to prevent their spread. Which would you choose? Why?

8. Milk production typically has the bacterium *Lactobacillus* added to the milk before it is delivered to market. This bacterium produces lactic acid. (1) Why would this bacterium be added to the milk and (2) why was this bacterium chosen?

9. You want to grow a species of bacterium that is acid loving; that is, it grows best in very acid environments. Would you want to grow it in a culture that has a pH of 2.0, 6.8, or 11.5? Explain.

10. You are given two beakers of a broth growth medium. However, only one of the beakers of broth is buffered. How could you determine which beaker contains the buffered broth solution?

11. Carbon is so common in living organisms that the chemistry of carbon is often said to be the chemistry of life. What property of carbon is responsible for its ability to enter a wide variety of chemical combinations?

12. The toxin associated with the foodborne disease botulism is a protein. To avoid botulism, home canners are advised to heat preserved foods to boiling for at least 12 minutes before tasting or consuming them. How does the heat help?

13. Water is generally considered the universal solvent of life. In how many places in this chapter does water live up to its reputation?

14. Proteins are made up of chains of amino acids, yet the proteins themselves are not acidic. Why do you think this is so?

15. Four classes of organic compounds are built from monomers that contain functional groups. Identify the type of functional group(s) present in the monomers for each organic compound.

16. Lipids are the only organic compounds that are not considered a polymer, even though they are built from monomers. Explain this statement.

17. Explain how Isaac Asimov's quote, "The significant chemicals in living tissue are rickety and unstable, which is exactly what is needed for life," applies to microorganisms and the organic compounds described in this chapter.

Review

This chapter has focused on the elements of living things, how the elements combine to form molecules, and the major compounds of microorganisms and other life forms. To test your knowledge of the chapter contents, rearrange the scrambled letters to spell out the correct word for the available space. The answers are listed in Appendix D.

1. During _____ bonding, an electron or a series of electrons is transferred between atoms.

   O C I I N

2. The chemistry of living things is _____ chemistry.

   R N C I O G A

3. An acid is a chemical substance that donates _____ ions to a solution such as water.

   Y R N H G O D E

4. The carbohydrate _____ contains the basic supply of the world's energy to living things.

   U C E S L G O

5. The best-known lipids are the _____.

   A S T F

6. In all living organisms, proteins function both as structural materials and as _____.

   E E S Y M N Z

7. Both DNA and RNA are composed of repeating units called _____.

   U L S T E N I E C O D

8. The _____ structure of a protein consists of the amino acid chain.

   I A M R Y P R

9. In _____ synthesis, the products of water are removed during the formation of covalent bonds.

   Y D A I R E O D H T N

10. The pH scale relates the measure of _____ of a chemical substance.

    A D Y T I C I

11. The smallest part of a compound that retains the property of the compound is a _____.

    L U C O E L E M

12. _____ are atoms of the same element that have different numbers of neutrons.

    T I P E S S O O

13. A functional group designated —COOH is known as a _____ group.

    A B X C R O L Y

14. Examples of disaccharides include lactose, sucrose, and _____.

    T A M E O L S

15. The _____ bond is a weak bond that exists between poles of adjacent molecules.

    G R N H E O D Y

http://microbiology.jbpub.com

The site features **eLearning**, an on-line review area that provides quizzes and other tools to help you study for your class. You can also follow useful links for in-depth information, or just find out the latest micro-biology news.

Basic Concepts of Microbiology

It's as if [he] lifted a whole submerged continent out of the ocean.

—A prominent biologist speaking of Carl Woese, whose work led to the three-domain vision of living things

E D ALCAMO REMEMBERS the spring of 1954.

I was a lad of thirteen growing up in the Bronx and looking forward to a carefree summer. But I could feel the tension in my parents' voices as they anticipated the months ahead, for summer was the dreaded polio season.

And sure enough, by early July the tension had turned to outright fear. I was told to avoid the public pool and the lusciously cool air-conditioned movie house. I had to report any cough or stiff neck promptly. 'Stick with your old friends,' my father told me. 'You've already got their germs.' Most of the time I was indoors, and the only baseball I got to play was in my imagination, as I listened to the Yankees every afternoon on my portable radio.

Our family was one of the lucky few to have a television, and each night we watched row upon row of iron lungs, and we saw the faces of the kids whose bodies were captured forever in their iron prisons. (Iron lungs, I was told, help you breathe when paralysis affects the respiratory muscles.) We heard and read about the daily toll from polio, where the victims lived, and how many kids had died.

But there was hope. My mom and her friends were out collecting dimes to fight polio (they called it the Mothers' March Against Polio), and the National Foundation of Infantile Paralysis said it had 75 million dimes to help fund the tests of a new vaccine—Dr. Salk's vaccine. Two million children would be getting shots. Maybe next year would be different.

Boy, was next year ever different! On April 12, 1955, at a televised news confer-ence, Dr. Jonas Salk declared, 'The vaccine works!' The celebration was wild. Our school closed for the day. And the church bells rang, even though it was a Thursday. I could tell my mother was relieved—we had steak for dinner that night.

By summertime, things were back to normal. I was now fourteen and eager to show off my baseball skills to any girl who cared to watch. Down at the neighborhood pool I was learning how to dive (when no one was watching). And for a quarter, I got to cheer for the cavalry at the Saturday afternoon movie. Summer was back."

Once again, we encounter how microorganisms have influenced the way people think and act. Virtually no one has escaped infectious disease, and the search con-tinues daily for the causes of microbial disorders, as well as for their treatment and prevention (MicroFocus 3.1). In the quality-control laboratory, inspectors are on con-stant alert to interrupt disease transmission in food and dairy products, and in water-purification and sewage-treatment facilities. Highly sophisticated technologies also are used to prevent the spread of microorganisms through the fluids we drink.

However, microbiology also has many positive aspects. Microbial ecologists study how microorganisms help in the natural recycling of minerals such as carbon and nitrogen. Evolutionary microbiologists look to the microorganisms to learn more about the ancestors of contemporary forms, and biochemists use microorganisms as miniature laboratories to discover the chemical framework that underlies all living

MicroFocus 3.1

EPIDEMIC

Patrons of fast food restaurants usually are safe from infectious disease, surprisingly safe in view of the huge volume of food served and the speed of preparation.

But, occasionally things go wrong, and in the winter of 1993 things went very wrong. That January, close to 500 people in several northwestern states became seriously ill after eating at their local Jack-in-the-Box restaurants. Three children died during the epidemic, and many others required hospitalization, including some who needed kidney dialysis.

The problem came to light when sev-eral patrons in Washington called their physicians to report the onset of bloody diarrhea. Most were experiencing gut-wrenching stomach cramps, and many were sick with fever. Similar reports were soon received from patients in Nevada, Oregon, Idaho, and California.

Local health departments moved quickly and contacted the Centers for Disease Control and Prevention, the federal agency for dealing with problems of this magnitude. Health officials were soon on the scene, taking food samples at several restaurants and sending the samples to laboratories for bacterial testing. Within days, the laboratories reported isolation of a common bac-terium called *Escherichia coli*. Normally a benign inhabitant of the human and animal intestines, this particular strain of *E. coli* was the notorious toxin-producing strain O157:H7. The toxin was destroying cells of the intestinal lining, which led to the bleeding; and it was causing hemor-rhages in the kidneys. All signs pointed to the hamburger meat as the source. Most likely, it had been contaminated with feces from the intestines of cattle during slaughter.

Decisive action followed. The man-agement of Jack-in-the-Box restaurants ordered its employees to increase the cooking time for hamburgers by 12.5 percent. Federal guidelines were insti-tuted requiring ground beef to be cooked at 155 degrees Fahrenheit (instead of the usual 140 degrees). Physicians throughout the country were alerted to watch for additional cases, and infected children were not permit-ted to return to day-care centers until two successive tests proved they had no residual bacteria in their intestine. Consumers were warned to avoid rare hamburgers. The operative expression became, "If it's gray, it's OK."

organisms. Geneticists study the hereditary material in microorganisms and then apply that knowledge to more complex organisms. Even behavioral psychologists find value in studying how certain microorganisms respond to environmental stimuli.

Our study of microbiology begins with a review of the position microorganisms occupy in the world of living organisms. After surveying the major groups of microorganisms, we will explore the methods used to catalog microorganisms and discuss the origin of their names, their sizes and shapes, and the techniques used to view them. These basic considerations provide a foundation for studying the microorganisms in detail in later chapters.

3·1 A Brief Survey of Microorganisms

Despite their microscopic size, microorganisms (except for the viruses) exhibit several features that are common to all living organisms. For example, DNA is the hereditary material in microorganisms as it is in all living things. Many of the biochemical patterns of growth and energy usage also are similar, if not identical. However, a number of significant differences set microorganisms apart. Before we explore these differences and unique features, we shall examine the total spectrum of living organisms by distinguishing between the two broad groups of organisms, the prokaryotes and eukaryotes. Once the characteristics of these groups are defined, we shall see how microorganisms fit into the pattern.

PROKARYOTES AND EUKARYOTES HAVE SOME STRUCTURES THAT ARE UNIVERSAL

In Chapter 1, we saw that the observations made in the 1940s and 1950s led to one of the most important generalizations in biology—that all organisms could be categorized as either prokaryotes or eukaryotes. Thus, prokaryotes lack a cell nucleus, whereas eukaryotes possess a well-defined cell nucleus. Accordingly, eubacteria and archaea are prokaryotes, while fungi, protista (protozoa and unicellular algae), plants, and animals (including humans) are eukaryotes. Knowing this can have practical significance because we can define many characteristics of organisms simply by knowing whether their cells have a prokaryotic or eukaryotic organization.

A few characteristics are universal to all prokaryotic and eukaryotic organisms. Both prokaryotic and eukaryotic cells have their hereditary characteristics stored in molecules of DNA organized in chromosomes. In prokaryotic cells, the chromosome consists of a single, circular DNA molecule, while in eukaryotic cells, the DNA occurs in multiple, linear chromosomes arranged within the membrane envelope of the cell nucleus.

All living cells also have a **cell membrane** (also known as the **plasma membrane**) that forms the border between the exterior and the gel-like cytoplasm. This membrane is composed primarily of proteins and lipids, especially phospholipids. The phospholipids form two layers, referred to as a bilayer, which is quite fluid. Membrane proteins, which form a mosaic in the bilayer, appear to float within or on the bilayer, and carry out most of the transport functions of the membrane. Therefore, the membrane is constantly in motion and is referred to as a **fluid mosaic structure**.

The cytoplasm represents everything surrounded by the cell membrane and, in eukaryotic cells, exterior to the cell nucleus. If all the cell structures are removed from the cytoplasm, what remains is called the **cytosol**. The cytosol represents the water, salts, ions, and organic compounds referred to in the previous chapter.

The third structure common to all prokaryotes and eukaryotes is the **ribosomes**, the RNA-protein bodies that participate in protein synthesis (Chapter 6). Although functionally identical, prokaryotic ribosomes are smaller than their counterparts in eukaryotic cells. This distinction will be important in targeted antibiotic chemotherapy, described in Chapter 24. Moreover, whereas prokaryotic ribosomes exist free in the cytosol, eukaryotic ribosomes can be free or bound to membranes of the endoplasmic reticulum.

PROKARYOTES AND EUKARYOTES ARE STRUCTURALLY DISTINCT

Distinctions between prokaryotes and eukaryotes can be appreciated by looking deeper into their cells. Eukaryotic cells, including eukaryotic microbes, have a variety of structurally discrete compartments called **organelles** that are absent in prokaryotes (FIGURE 3.1).

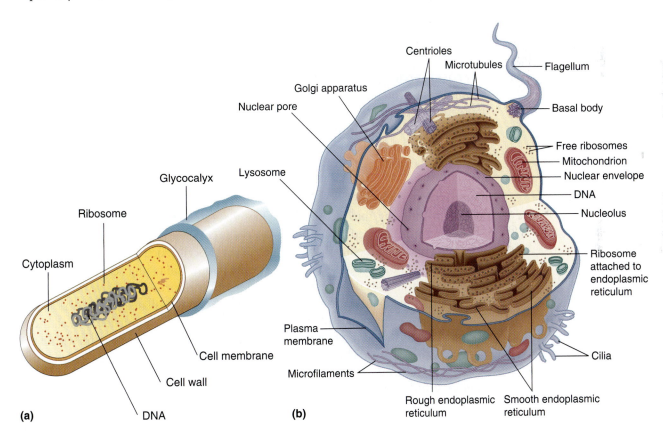

(a) Cytoplasm · Ribosome · Glycocalyx · Cell membrane · Cell wall · DNA

(b) Nuclear pore · Golgi apparatus · Centrioles · Microtubules · Flagellum · Basal body · Lysosome · Free ribosomes · Mitochondrion · Nuclear envelope · DNA · Nucleolus · Ribosome attached to endoplasmic reticulum · Plasma membrane · Microfilaments · Rough endoplasmic reticulum · Smooth endoplasmic reticulum · Cilia

FIGURE 3.1

A Comparison of Prokaryotic and Eukaryotic Cells

(a) A stylized bacterial cell as an example of a prokaryotic cell. Relatively few structures are present. (b) An animal cell as a typical eukaryotic cell. Note the variety and large size of the cellular features, many of which are discussed in the text. The presence of a nucleus is a primary feature distinguishing the eukaryotic cell from the prokaryotic cell.

■ **ENDOMEMBRANE SYSTEM.** Eukaryotic microbes have a series of membrane-enclosed organelles in the cytosol that compose the cell's **endomembrane system**. These include the **endoplasmic reticulum** (ER), a series of ribosome-attached membranes (rough ER) and tube-like membranes (smooth ER) extending throughout the cell. The ER is involved in protein and lipid synthesis. The **Golgi apparatus** is a group of independent stacks of flattened membranes and vesicles where the proteins and lipids from the ER are processed, sorted, and packaged for transport. **Lysosomes**, somewhat circular sacs containing digestive (hydrolytic) enzymes, are vital to the killing of many pathogens by white blood cells.

■ **ENERGY ORGANELLES.** Eukaryotic microbes have one or two organelles that function in energy transformations. All eukaryotes have one or more **mitochondria** (sing., mitochondrion). The function of these membrane-enclosed organelles is to convert chemical energy into cellular energy needed for cellular work (which is why mitochondria are called the "powerhouses of the cell"). Prokaryotes also must carry out these transformations; however, the reactions occur on the cell membrane. Eukaryotic microorganisms (primarily the green algae) contain membrane-bound **chloroplasts**, in which the process of photosynthesis converts sunlight into chemical energy. Certain prokaryotes, such as the cyanobacteria and green-sulfur bacteria also carry out these energy transformations. Again, the cell membrane is the chemical workbench for the process.

■ **CYTOSKELETON.** Another system unique to eukaryotic cells is the **cytoskeleton**, a complex of proteins that are not membrane bound. These proteins are organized into an interconnected system of fibers, threads, and interwoven molecules that give structure to the cell and assist in the transport of materials throughout the cell. The main components of the cytoskeleton are microtubules, microfilaments, and intermediate filaments, each assembled from different protein subunits.

■ **MOTILITY.** Many prokaryotic and eukaryotic cells have flagella or cilia. **Flagella** (sing., flagellum) are long, thin protein projections that extend from the cell to provide a type of cell movement called motility. In prokaryotic cells, the naked flagella rotate like the propeller of a motorboat (Chapter 4). In eukaryotic cells, such as certain protozoa and algae, the membrane-enveloped flagella whip about. **Cilia** (sing., cilium) are shorter and more numerous than flagella. In some motile protozoa, they wave in synchrony and propel the cell forward. *Paramecium* is a well-known ciliated protozoan.

Many prokaryotic and eukaryotic cells contain structures outside the cell membrane. The most prominent is the **cell wall**, which differs in composition within and between the groups. The fungal cell wall contains a polysaccharide called chitin (Chapter 15), while the major cell wall component of many unicellular algae is cellulose. With only a few exceptions, all prokaryotes have a rigid cell wall containing varying amounts of a different polysaccharide that is a major component of peptidoglycan (Chapter 4). All cell walls provide support for the cells, give them shape, and help them resist mechanical pressures. A summary of the differences between prokaryotes and eukaryotes is presented in TABLE 3.1.

It is important to remember that all living organisms are either prokaryotes or eukaryotes, and microbes are found in both groups. Note, however, that viruses are neither, since they are not living organisms. Viruses are fragments of nucleic acid packaged in a protein shell (Chapter 12). Nevertheless, we consider them "microorganisms" because of their extremely small size and infectious nature.

TABLE 3.1

A Comparison of Prokaryotes and Eukaryotes

CHARACTERISTIC	PROKARYOTES	EUKARYOTES
Nucleus	Absent	Present with nuclear membrane
DNA structure	Single circular chromosome	Multiple linear chromosomes in nucleus
Membranes	Cell membrane only	Cell and organelle membranes
Organelles	Absent	Present in a variety of forms
Ribosomes	Smaller than eukaryotic ribosomes Free in cytoplasm	Larger than prokaryotic ribosomes Free or bound to ER membranes
Cytoskeleton	Absent	Present
Cell walls	Generally present Complex chemical composition	Present in fungi, algae, plants Complex chemical composition
Flagella	Rotating movement	Whipping movement
Cilia	Absent	In some cells
Examples	Eubacteria, Archaea	Fungi, protista, plants, animals, including humans

In the following sections, we shall summarize other significant properties of the major groups of microorganisms.

BACTERIA ARE AMONG THE MOST ABUNDANT ORGANISMS ON EARTH

The term **bacteria** is a plural form of the Latin *bacterium,* meaning "staff" or "rod." The vast majority play a positive role in nature: They digest sewage into simple chemicals, they extract nitrogen from the air and make it available to plants for protein production, they break down the remains of all that die and recycle the carbon and other elements, and they produce foods for human consumption and products for industrial technology. It is safe to say that life as we know it would be impossible without the bacteria.

Of course, we know from Chapter 1 and personal experience that some bacteria are harmful. Certain species multiply within the human body, where they disrupt tissues or produce toxins that result in disease. Other bacteria infect plant crops and animal herds. Indeed, disease-causing bacteria (i.e., pathogenic bacteria) are a global threat to all forms of life.

Bacteria more than any other group of organisms have adapted to the diverse environments on Earth. They inhabit the air, soil, and water, and they exist in enormous numbers on the surfaces of virtually all plants and animals. They can be isolated from Arctic ice, thermal hot springs, the fringes of space, and the tissues of animals (FIGURE 3.2). Some can withstand the intense acid in volcanic ash, the crushing pressures of ocean trenches, and the powerful activity of digestive enzymes. Other bacteria survive in oxygen-free environments, boiling water, or extremely dry locations. Bacteria have so completely colonized every part of the Earth that the mass of bacterial cells is estimated to outweigh the mass of all plants and animals

FIGURE 3.2

Mixed Bacteria

An electron microscopic view of mixed bacteria from the gastrointestinal tract. Rod and spherical forms in various sizes, shapes, and arrangements are visible. (Bar = 1 μm.)

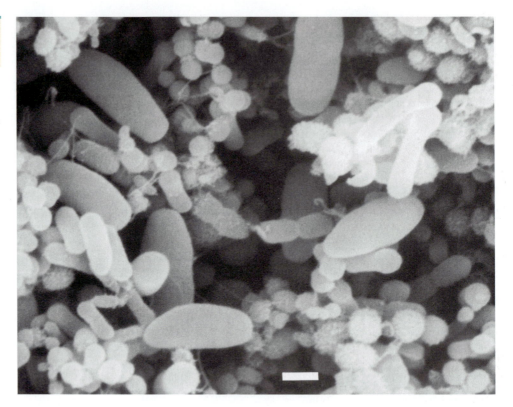

combined (MicroFocus 3.2). Chapters 4 to 7 are devoted to the structure, growth, biochemistry, and genetics of bacteria, and Chapters 8 through 11 discuss the diseases they cause.

Let's look briefly at a few unusual examples of bacteria.

The rickettsiae, chlamydiae, and mycoplasmas once were considered separate from the bacteria, but contemporary microbiologists now classify them as "small bacteria." Unfortunately, old habits linger and some microbiologists still think of them as distinct groups, which is why we separate them here.

Rickettsiae (sing., rickettsia) were first described by Howard Taylor Ricketts in 1909. These tiny bacteria can barely be seen with the most powerful light microscope. They are transmitted among humans primarily by arthropods such as ticks and lice, and are cultivated only in living tissues such as fertilized eggs. Different species cause a number of important diseases, including Rocky Mountain spotted fever and typhus fever. Chapter 10 contains a more thorough description of their properties.

Roughly half the size of the rickettsiae, the **chlamydiae** (sing., chlamydia) are so small that they cannot be seen with the light microscope. They can be cultivated only within living cells, and one species causes the gonorrhea-like disease known as chlamydia. Chlamydial diseases are described in Chapters 8 and 11.

Even smaller than chlamydiae, and possibly the smallest bacteria, are the **mycoplasmas**. They can be cultivated in artificial laboratory media, and, although they are prokaryotes, they do not have cell walls like other bacteria. One form of pneumonia is caused by a mycoplasma (Chapter 8) and one sexually transmitted disease is a mycoplasmal illness (Chapter 11). We have mentioned this group already when we discussed the possibility of "creating life in the test tube" in Chapter 2.

AN UNKNOWN WORLD

Practically everyone is familiar with the word *bacteria*, but it may surprise you that hardly anyone really knows the bacteria. Yes, bacteria have been thoroughly studied in medicine, ecology, and molecular genetics, but the vast majority of bacteria remain unknown to science.

Hold a pinch of rich soil in the palm of your hand. You are now face to face with an estimated billion bacteria representing some 10,000 different species. If you were to try cultivating them using the most sophisticated methods available, perhaps 500 species might grow. That leaves 9,500 species unknown until either the right combinations of nutrients and environmental conditions are established in the laboratory or their genomes are isolated as described in the opener to Chapter 1.

Traditionally, *Bergey's Manual* has been the most authoritative guide to bacterial species. The few thousand species listed in this book are a far cry from the thousands more believed to exist. Those who believe there are no more worlds to conquer should take note.

Cyanobacteria once were known as blue-green algae. Today, these microorganisms are considered bacteria because their structural and biochemical properties are similar to typical bacteria. However, cyanobacteria are unique among prokaryotes because they carry out photosynthesis similar to unicellular algae (Chapter 5). Cyanobacteria possess light-trapping pigments that function in photosynthesis. Many of the pigments are blue, but some are black, yellow, green, or red. The periodic redness of the Red Sea, for example, is due to "blooms" of cyanobacteria whose members contain large amounts of red pigment.

Cyanobacteria may occur as unicellular or filamentous forms (FIGURE 3.3a). Many species can incorporate ("fix") atmospheric nitrogen into organic compounds

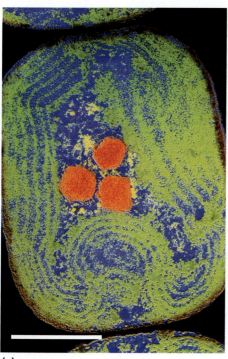

(a)

(b)

FIGURE 3.3

Cyanobacteria

(a) False color transmission electron micrograph of the cyanobacterium *Pseudanabaena*. Note the numerous membranes (green) on which photosynthesis occurs. (Bar = 1 µm.) (b) Thick mats of cyanobacteria in the Baltic Sea. These bacteria often form thick masses or blooms in nutrient-rich waters.

useful to plants, thereby filling an important ecological niche. When lakes or bays contain a rich supply of nutrients, the organisms may "bloom" (FIGURE 3.3b) and convert the water to a pea-soup green with a foul odor. Swimming pools and aquaria experience this problem if algicide is not used regularly.

PROTISTA (PROTOZOA AND UNICELLULAR ALGAE) ARE VERY DIVERSE EUKARYOTIC MICROBES

The protista introduced in Chapter 1 consist of the protozoa and single-celled algae. **Protozoa** (sing., protozoan) are single-celled eukaryotes, most of which lack cell walls, ingest food particles, and move about freely. Many protozoa join with fungi and bacteria to decompose dead organisms and recycle nutrients. Indeed, numerous species are important links in the food chain as they convert nutritionally inaccessible materials into substances easily digested by other organisms. In some plant-eating animals (e.g., cattle, goats, and other ruminants), protozoa live in the digestive tract and, along with bacteria, enable the animals to use grass and other high-cellulose foods they could not otherwise digest. In Chapter 2, we mentioned that in these animals methane gas was a by-product of the microbial metabolism of cellulose.

Protozoa exhibit a bewildering assortment of shapes, sizes, and structural components (FIGURE 3.4). A few species have the necessary pigments for photosynthesis. However, most species must obtain nutrients from preformed organic matter.

One way of grouping protozoa is according to how they move. Some have one or more flagella (Figure 3.4a, b), others possess cilia, and still others move by means of cytoplasmic extensions called pseudopodia ("false feet") (Figure 3.4c).

Although most protozoa are harmless to humans, a few species are pathogenic. Among the more notorious pathogens are those that cause malaria and sleeping sickness. Other species are transmitted in water and cause such diseases as giardiasis, amoebic dysentery, and cryptosporidiosis (FIGURE 3.5). We shall study protozoal diseases in depth in Chapter 16.

The **unicellular algae** also are a diverse group of eukaryotic microbes. The word algae (sing., alga) refers to any plant-like organisms that carry out photosynthesis and differ structurally from typical land plants such as mosses, ferns, and seed plants. Two types of unicellular algae, the diatoms and the dinoflagellates, have important ecological roles. Together, they compose the **phytoplankton**, which is a major source of food for many aquatic and marine animals (FIGURE 3.6).

Phytoplankton:
diatoms and dinoflagellates that float near the surface of lakes and oceans.

Diatoms are an important source of food in the world's oceans. Through photosynthesis, they trap the sun's energy and manufacture carbohydrates, which are passed on to other marine organisms in the food chain. The cell walls of diatoms are impregnated with silicon dioxide, a glass-like substance. When they die, their glassy remains accumulate on the seafloor as diatomaceous earth. The latter is gathered and used to produce filtering agents and mild abrasives found in toothpastes and other products.

Dinoflagellates are another group of unicellular algae in which the cells are encased in hard cellulose shells. Dinoflagellates also are important members of the world's food chains. Under certain environmental conditions, some species cause the periodic red tides occurring in the oceans. During these periods, paralytic shellfish poisoning (a respiratory paralysis) can occur in humans who eat shellfish contaminated with the dinoflagellates. Both diatoms and dinoflagellates are discussed in Chapter 26.

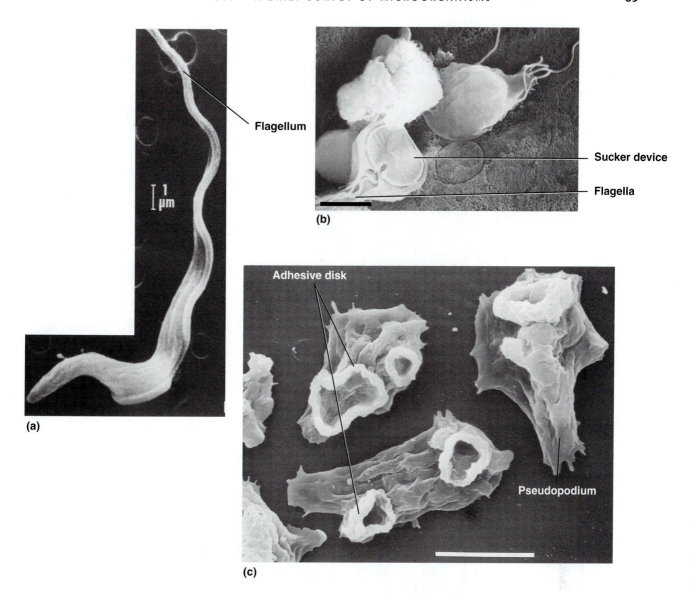

Flagellum

1 μm

(a)

Sucker device

Flagella

(b)

Adhesive disk

Pseudopodium

(c)

FIGURE 3.4

Three Species of Protozoa

Electron micrographs of three species of protozoa illustrating differences in shape, movement, and size. (a) A protozoan called a trypanosome. The elongated shape and hair-like appendage (flagellum) are apparent. This organism causes African sleeping sickness. (Bar = 1μm.) (b) Another flagellated protozoan, *Giardia lamblia*. The organism has a flat shape with multiple flagella extending toward the rear. The protozoan on the left shows the sucker device on its lower surface for holding fast to tissue. *Giardia* causes diarrhea in humans. (Bar = 2μm.) (c) The amoeba *Naegleria fowleri*, a cause of meningitis in humans. Note the irregular shape of this organism, which can move about using pseudopodia. The adhesive disks are used to attach to a surface. (Bar = 5 μm.)

FIGURE 3.5

An Outbreak of Intestinal Disease at a Summer Day Camp

This outbreak occurred in Florida during July and August 1996. Health officials recommended measures to restrict access to the hose believed to be the source of the outbreak.

1. During the summer, a group of 98 children and 6 counselors attended a day camp in Alachua County, Florida, on the grounds of a public elementary school. The attendees ranged in age from 4 to 12 years.

2. The days were hot, and occasionally the children took a break from play to take a drink from a nearby hose with a spray nozzle. There was also water available from coolers, refilled often from the same garden hose.

3. The hose also was used to clean the garbage cans. Occasionally, camp personnel would leave the hose lying on the ground unattended. Feces of unknown origin were sometimes seen near the hose.

4. That summer, 72 children and 5 of the 6 counselors reported symptoms of intestinal disease, including abdominal pain, nausea, vomiting, and three or more watery stools each day.

5. Health officials visited the camp and obtained samples from various food and water sources. The outdoor faucet, its hose, and the spray nozzle were found to harbor *Cryptosporidium parvum*, a protozoan that causes intestinal illness.

FUNGI HAVE A DISTINCT GROWTH STYLE

Because of the presence of a cell wall, **fungi** once were considered members of the plant kingdom (indeed, many botany books and courses still discuss the fungi). However, researchers have found the polysaccharide chitin in fungal cell walls but not in plant walls. Moreover, fungi do not carry out photosynthesis; instead, they absorb and use preformed organic matter from the environment as their nutritional source. These characteristics, among others, separate them from plants.

Fungi are eukaryotic microorganisms often spoken of in two broad groups: the yeasts and the molds. **Yeasts** are unicellular organisms larger than most bacteria (**FIG-URE 3.7a**). They play a vital role in industry for the fermentation of wine and beer

FIGURE 3.6

Marine Phytoplankton

(a) Scanning electron micrograph of marine phytoplankton. These microbes drift around in oceans, providing the basis for the entire marine food chain. They inhabit only the very top layers of the water where sunlight is able to penetrate. (Bar = 10 μm.) (b) Satellite image of the Gulf of Mexico, showing the blooms of phytoplankton in the water. The shallow, plankton-rich coastal waters support large fisheries. The nutrient-rich waters flowing out of the Mississippi delta (center right) and along the Louisiana and Texas coasts produce phytoplankton plumes that can be hundreds of kilometers in length.

(a)

(b)

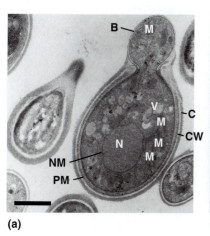

(a)

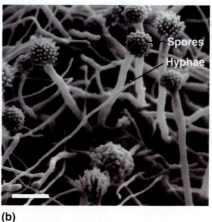

(b)

(c)

FIGURE 3.7

Three Views of Fungi

(a) A transmission electron micrograph of a yeast cell showing typical eukaryotic features. Note the cell wall (CW) and plasma membrane (PM) on the cell surface. In the cytoplasm, clear bodies called vacuoles (V) and mitochondria (M) may be seen. The nucleus (N) is surrounded by a nuclear membrane (NM). In the upper portion of the photograph, a reproductive structure, the bud (B), is growing out from the parent cell. (Bar = 3 μm.) (b) A scanning electron micrograph of hyphae of a mold. The network of hyphae form a mycelium on which spores are formed. (Bar = 30 μm.) (c) A similar mold growing on a grapefruit. The mycelium is covered with blue-green spores.

and the production of bread. **Molds** are long chains of cells often seen as fuzzy or fluffy masses on bread and other food products, especially acidic products. The molds often assume vivid colors from the pigments they produce in spores for reproductive purposes.

One of the unique structural features of the molds is how they grow. The body of most molds consists of fuzzy masses of multicellular filaments called **hyphae** (sing., hypha). Usually the hyphae form dense networks called a **mycelium** (FIG-URE 3.7b, c). Cells of each hypha are highly efficient exploiters of available nutrients—during periods of peak growth, a mold can grow more than a half-mile of new hyphae in a single day.

FIGURE 3.8

A Case of Rabies Occurring in Warren County, New Jersey, in October 1997

This was the first case of human rabies in New Jersey since 1971.

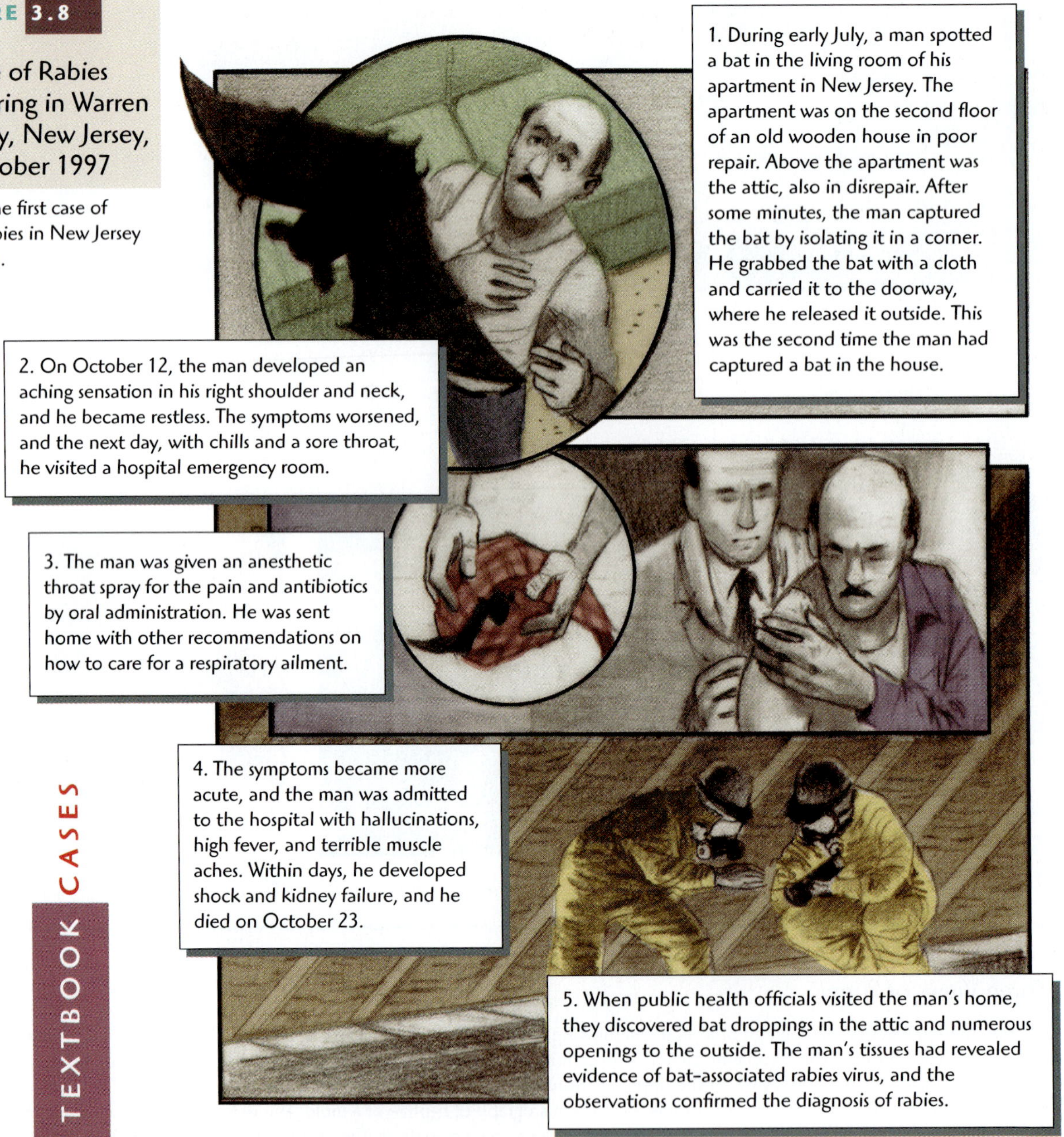

TEXTBOOK CASES

1. During early July, a man spotted a bat in the living room of his apartment in New Jersey. The apartment was on the second floor of an old wooden house in poor repair. Above the apartment was the attic, also in disrepair. After some minutes, the man captured the bat by isolating it in a corner. He grabbed the bat with a cloth and carried it to the doorway, where he released it outside. This was the second time the man had captured a bat in the house.

2. On October 12, the man developed an aching sensation in his right shoulder and neck, and he became restless. The symptoms worsened, and the next day, with chills and a sore throat, he visited a hospital emergency room.

3. The man was given an anesthetic throat spray for the pain and antibiotics by oral administration. He was sent home with other recommendations on how to care for a respiratory ailment.

4. The symptoms became more acute, and the man was admitted to the hospital with hallucinations, high fever, and terrible muscle aches. Within days, he developed shock and kidney failure, and he died on October 23.

5. When public health officials visited the man's home, they discovered bat droppings in the attic and numerous openings to the outside. The man's tissues had revealed evidence of bat-associated rabies virus, and the observations confirmed the diagnosis of rabies.

Fungi are among the major decomposers of organic matter in the world. Yet, a few fungal species do not wait for an organism to die before they start consuming it. These are the pathogenic fungi, which cause some of the most common and persistent plant diseases. Human illnesses caused by fungi represent challenges to modern medicine. More than 20 percent of the world's population suffers from fungal disease, ranging from irritating maladies of the skin (e.g., athlete's foot) to life threatening diseases such as valley fever and cryptococcosis. Chapter 15 describes the fungi and human fungal diseases in more detail.

VIRUSES ARE THE SMALLEST MICROORGANISMS

Viruses are neither prokaryotes nor eukaryotes, as we have noted previously. Many microbiologists even question whether viruses are living organisms because they are noncellular. As independent entities, they do not grow, nor do they display any metabolic activity. Viruses have no observable activity except replication, and they accomplish this function only within living cells. Indeed, outside a living cell, a virus is no more alive than a grain of sand.

Most viruses are smaller than the tiniest bacteria, so small that an electron microscope is necessary to see them. The simplest viruses consist of nothing more than nucleic acid, either DNA or RNA, and are packed inside a protein shell. When a virus penetrates an appropriate host cell, the genetic information is released. By commandeering the host cell's structures and enzymes, the virus replicates itself hundreds of times, often destroying the host cell in the process. As nearby cells are penetrated and destroyed, the tissues are destroyed. Such tissue destruction is common to many viral diseases, including influenza, AIDS, hepatitis, chickenpox, herpes infections, and rabies (FIGURE 3.8). We shall pay considerable attention to the structure and physiology of viruses in Chapter 12 and survey their diseases in Chapters 13 and 14.

TABLE 3.2 summarizes the classification and features of the microorganisms. It is worthwhile at this time to understand how some of these microorganisms could be used as agents for bioterrorism. MicroInquiry 3 is a primer on this topic.

TABLE 3.2

An Overview of the Microorganisms

MICROORGANISMS	CLASSIFICATION	DISTINGUISHING CHARACTERISTICS
Bacteria	Prokaryotic	Extremely abundant; microscopic; many positive roles in nature; some cause disease; rickettsiae and chlamydiae multiply only in host cells; mycoplasmas have no cell walls; cyanobacteria similar to true bacteria except carry out photosynthesis
Protozoa	Eukaryotic	Animal-like; classified by type of motion; no cell walls; usually not photosynthetic
Unicellular algae	Eukaryotic	Plant-like; photosynthetic; most marine forms; include diatoms and dinoflagellates
Fungi	Eukaryotic	Molds and yeasts; usually filamentous; not photosynthetic; unique cell walls
Viruses		Noncellular inert particles; fragment of nucleic acid (RNA or DNA) enclosed in protein; replicate only in living host cells; ultramicroscopic

MicroInquiry 3   BIOTERRORISM: THE WEAPONIZATION AND PURPOSEFUL DISSEMINATION OF HUMAN PATHOGENS

The anthrax attacks that occurred on the East Coast in October 2001 confirmed what many health and governmental experts had been saying for over 10 years—it is not *if* bioterrorism would occur but *when* and *where*. **Bioterrorism** represents the intentional or threatened use of primarily microorganisms or their toxins to cause fear in or actually inflict death or disease upon a large population for political, religious, or ideological reasons. This MicroInquiry serves as a primer to better understand bioterrorism and why microorganisms are often the weapons of choice.

IS BIOTERRORISM SOMETHING NEW?

Bioterrorism is not new, and several MicroFocus boxes in this text (MicroFocus 9.8, 10.4, and 10.6) mention historical examples. Such bioterrorism agents also have been used as biowarfare agents. In the United States, during the aftermath of the French and Indian Wars (1754–1763) British forces, under the guise of goodwill, gave smallpox-laden blankets to rebellious tribes sympathetic to the French. The disease decimated the Native Americans, who had never been exposed to the disease before and had no immunity. Between 1937 and 1945, the Japanese established Unit 731 to carry out experiments designed to test the lethality of several microbiological weapons as biowarfare agents on Chinese soldiers and civilians. In all, some 10,000 "subjects" died of bubonic plague, cholera, anthrax, and other diseases.

In 1973, the United States, the Soviet Union, and more than 100 other nations signed the Biological and Toxin Weapons Convention, which prohibited nations from developing, deploying, or stockpiling biological weapons. Unfortunately, the treaty provided no way to monitor compliance. As a result, in the 1980s the Soviet Union developed and stockpiled many microbiological agents, including the smallpox virus, and anthrax and plague bacteria.

After the 1991 Gulf War, the United Nations Special Commission (UNSCOM) analysts reported that Iraq had produced 8,000 liters of concentrated anthrax solution and more than 20,000 liters of botulinum toxin solution. In addition, anthrax and botulinum toxin had been loaded into SCUD missiles.

In the United States, several biocrimes have been committed. *Biocrimes* are the intentional introduction of biological agents into food or water, or by injection, to harm or kill groups of individuals. The most well known biocrime occurred in Oregon in 1984 when the Rajneeshee religious cult, in an effort to influence local elections, intentionally contaminated salad bars of several restaurants with the bacterium *Salmonella*. The unsuccessful plan sickened over 750 citizens and hospitalized 40. Whether biocrime or bioterrorism, the 2001 events concerning the anthrax spores mailed to news offices and to two US congressmen only increases our concern over the use of microorganisms or their toxins as bioterror agents.

WHAT MICROORGANISMS ARE CONSIDERED BIOTERROR AGENTS?

A considerable number of human pathogens and toxins have potential as many microbiological agents, including the smallpox virus, and anthrax and plague bacteria.

microbiological weapons. These "select agents" include bacteria, bacterial toxins, fungi, and viruses. The seriousness of the agent depends on the severity of the disease it causes (virulence) and the ease with which it can be disseminated. The pathogens of most concern, called the Category A Select Agents, are those that can be spread by aerosol contact, such as anthrax and smallpox, and toxins that can be added to food or water supplies, such as the botulinum toxin (TABLE A).

WHY USE MICROORGANISMS?

At least 17 nations are believed to have the capability of producing bioweapons from microorganisms. Such microbiological weapons offer clear advantages to these nations and terrorist organizations in general. Perhaps most important, biological weapons represent "The Poor Nation's Equalizer." Microbiological weapons are cheap to produce compared to chemical and nuclear weapons and provide those nations with a deterrent every bit as dangerous and deadly as the nuclear weapons possessed by other nations. With biological weapons, you get high impact and the most "bang for the buck."

In addition, microorganisms can be deadly in minute amounts to a defenseless (nonimmune) population. They are odorless, colorless, and tasteless, and unlike conventional and nuclear

TABLE A

Category A Select Agents and Perceived Risk of Use

TYPE OF MICROBE	DISEASE (MICROBE SPECIES OR VIRUS NAME)	PERCEIVED RISK
Bacteria	Anthrax (*Bacillus anthracis*)	High
	Plague (*Yersinia pestis*)	Moderate
	Tularemia (*Francisella tularensis*)	Moderate
Viruses	Smallpox (Variola)	Moderate
	Hemorrhagic fevers (Ebola, Marburg, Lassa, Machupo)	Low
Toxins	Botulinum toxin (*Clostridium botulinum*)	Moderate

weapons, microbiological weapons do not damage infrastructure, yet they can contaminate such areas for extended periods. Without rapid medical treatment, most of the select agents can produce high numbers of casualties that would overwhelm medical facilities. Lastly, the threatened use of microbiological agents creates panic/anxiety that represents the very nature of terrorism.

HOW WOULD MICROBIOLOGICAL WEAPONS BE USED?

All known microbiological agents (except smallpox) represent organisms naturally found in the environment. For example, the bacterium that causes anthrax is found in soils around the world (FIGURE A). Assuming one has the agent, the microorganisms can be grown (cultured) easily in large amounts. However, most of the select agents must be "weaponized"; that is, they must be modified into a form that is deliverable, stable, and has increased infectivity and/or lethality. Nearly all of

the microbiological agents in category A are infective as an inhaled aerosol. Weaponization, therefore, requires that the agents be small enough in size that inhalation would bring the organism deep into the respiratory system and prepared so that the particles do not stick together or form clumps that would not stay suspended in the air. Several of the anthrax letters of October 2001 involved such weaponized spores.

Dissemination of biological agents by conventional means would be a difficult task. Aerosol transmission, the most likely form for dissemination, exposes microbiological weapons to environmental conditions to which they are usually very sensitive. Excessive heat, ultraviolet light, and oxidation would limit the potency and persistence of the agent in the environment. Although anthrax spores are relatively resistant to typical environmental conditions, the bacterial cells causing tularemia become ineffective after just a few minutes in sunlight.

The possibility also exists that some nations have developed or are developing more lethal bioweapons though genetic engineering and biotechnology. The former Soviet Union may have done so. Commonly used techniques in biotechnology could create new, never before seen bioweapons, making the resulting "designer diseases" true doomsday weapons.

CONCLUSIONS

In May 2000, Ken Alibek, a scientist and defector from the Soviet bioweapons program, testified before the House Armed Services Committee that the best biodefense is to concentrate on developing appropriate medical defenses that will minimize the impact of bioterrorism agents. If these agents are ineffective, they will cease to be a threat; therefore, the threat of using human pathogens or

toxins for bioterrorism, like that for emerging diseases such as SARS and West Nile fever, is being addressed by careful monitoring of sudden and unusual disease outbreaks. Extensive research studies are being carried out to determine the effectiveness of various antibiotic treatments (FIGURE B) and how best to develop effective vaccines or administer antitoxins. To that end, vaccination perhaps offers the best defense. As of 2003, the United States has stockpiled sufficient smallpox vaccine to vaccinate the entire population if a smallpox outbreak occurred. Other vaccines for other agents are in the development stages.

This primer is not intended to scare or frighten; rather, it is intended to provide an understanding of why microbiological agents have been developed as weapons for bioterrorism. We cannot control the events that occur in the world, but by understanding bioterrorism, we can control how we should react to those events—should they occur in the future.

FIGURE A

Anthrax Bacteria

Light micrograph of Gram-stained *Bacillus anthracis*, the causative agent of anthrax. There is concern that terrorists could release large quantities of anthrax spores in a populated area, which potentially could cause many deaths.

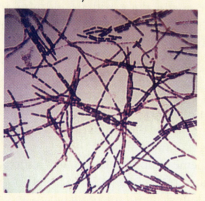

FIGURE B

Anthrax Antibiotics Research

Antibiotic drugs in paper discs are used to test the sensitivity of anthrax bacteria (*Bacillus anthracis*) cultured on an agar growth medium. The clear zone surrounding each disc indicates the bacterium is sensitive to the antibiotic.

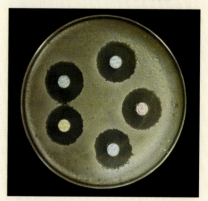

To this point . . .

We have begun our study of microbiology by placing microorganisms into the panorama of living things. Knowing whether a microorganism is a prokaryote or a eukaryote is important because it enables us to see microorganisms in relation to plants and animals, while giving us a glimpse of their properties. We then developed brief sketches of different microorganisms that form the foundation for our studies in the chapters ahead. Viruses were included as microorganisms because they are important agents of disease, even though they are neither prokaryotes nor eukaryotes.

We now focus on how microorganisms are classified and named. This study will serve as a prelude to studying individual characteristics of the groups of microorganisms in other chapters.

3.2 Cataloging Microorganisms

If you open any catalog, items are separated by types, styles, or functions. For example, shoes are separated from watches and, within the watches, men's and women's styles are separated from one another. Organisms, including the microorganisms, also are cataloged; that is, there are ways of grouping organisms based on specific characteristics. In this section, we shall explore the principles on which these methods are based.

CLASSIFICATION ATTEMPTS TO CATALOG ORGANISMS

The science of classification, or **taxonomy**, deals with the systematized arrangement of related microorganisms and other living organisms into logical categories. Taxonomy is essential to understanding the relationships among living organisms. It focuses on unifying concepts among organisms, while providing a basis for communication among biologists.

One of the first taxonomists was the Greek philosopher Aristotle. In the fourth century B.C., Aristotle categorized living forms in the world around him and described over 500 species of plants and animals according to their appearance and habits—without reference to books or instruments to help him.

Modern taxonomy was devised by the Swedish botanist Carl von Linné, better known to history as Carolus Linnaeus (FIGURE 3.9a). In his *Systema Naturae,* published in several editions between 1735 and 1759, Linnaeus inspired an unprecedented worldwide program of specimen hunting and sent his students around the globe in search of new life forms. As a result, he named thousands of plants and animals and classified them in the kingdoms Plantae and Animalia.

Linnaeus also popularized the binomial scheme of nomenclature, thus resurrecting the use of Latin as a language of the learned (MicroFocus 3.3). In his otherwise exacting system, Linnaeus reflected the scant knowledge of microorganisms and his general disinterest by grouping them separately under the heading of Vermes (as in vermin) in the category Chaos (confusion). In the years after Linnaeus, however,

FIGURE 3.9

Early Taxonomists

(a) Modern taxonomy began with the observations of Carolus Linnaeus who named thousands of plants and animals collected by him and his students. (b) Ernst Haeckel added a third kingdom, the Protista, to encompass all organisms that were not plants or animals; that is, fungi, protozoa, algae, and bacteria.

(a)

(b)

MicroFocus 3.3

"WHAT WAS THAT NAME?"

The binomial system of nomenclature appears to be so obvious that it hardly needed to be invented. However, before Carolus Linnaeus popularized the scheme in 1735, scientists could not agree on scientific names for the organisms they studied. Different names were invented by different writers to serve as both designation and description. Confusion was rampant.

Linnaeus's great simplifying decision was to use the genus name and a modifying adjective for the label and description of an organism. He had to work quickly lest other naturalists use the same name for different organisms. In a monumental task of linguistic invention, Linnaeus ransacked his Latin for enough terms to make up thousands of labels. Some names he took from an organism's manner of growth, others from the discoverer, others from classical heroes, and still others from vernacular names. Any parent who has had to name a child can appreciate what he was up against.

In a 1753 book on plants, Linnaeus supplied binomial names for over 5,900 plant species known at that time. In his tenth edition of *Systema Naturae* (1759), he extended the scheme to animals. Within decades, his binomial names were adopted by European scientists and were reaching across the world. And just in time, because explorers were returning to Europe from distant corners of the globe with newly discovered life forms (penguins, tobacco, potato, manatees, kangaroos, and others).

How important is the binomial system of nomenclature? It is clear that, to be understood, scientists must avoid ambiguity, and this is what the binomial system accomplishes. In the United States, for example, corn refers to a tall plant that produces yellow kernels on a cob. However, in England, the same plant is called maize. English "corn" is any number of different cereal grains, such as wheat, rye, and barley. Thus, to eliminate confusion, scientists use the term *Zea mays* when they refer to American corn or English maize. For scientists, the binomial system contains more than a grain of truth.

some microorganisms, such as the bacteria, algae, and fungi, were considered plants while others, such as the protozoa, were categorized as animals.

In 1866, the German naturalist Ernst H. Haeckel (FIGURE 3.9b) disturbed the tidiness of the plant and animal kingdoms by proposing a new system to separate the microorganisms. Haeckel was not satisfied with mushrooms being included with plants because mushrooms do not carry out photosynthesis. Realize that Haeckel was living at the time that Pasteur and Koch were advancing the idea of the germ theory of disease, so there was a plethora of newly described "in-between" organisms, including many forms of protozoa, microscopic algae, and bacteria. Haeckel therefore coined the term *protist* for a microorganism, and he placed all bacteria, protozoa, algae, and fungi in a new, third kingdom, the **Protista**. Protista came to include not only fungi, protists, and bacteria, but virtually all organisms that share plant and animal characteristics but are not plants or animals. Some microbiologists, however, were not eager to accept Haeckel's view, and they continued to classify bacteria and fungi with plants because all three have cell walls, while placing protozoa with the animals.

THE FIVE-KINGDOM SYSTEM RECOGNIZES PROKARYOTIC AND EUKARYOTIC DIFFERENCES

During the twentieth century, advances in cell biology and interest in evolutionary biology led scientists to question the two- or three-kingdom classification schemes. In 1969, Robert H. Whittaker of Cornell University proposed a system that quickly gained wide acceptance in the scientific community. Further expanded in succeeding years by Lynn Margulis of the University of Massachusetts, the system recognizes five kingdoms of living things: Monera (bacteria), Protista (also known as Protoctista by Margulis's supporters), Fungi, Plantae, and Animalia (FIGURE 3.10).

Microorganisms comprise three of these five kingdoms. Bacteria clearly should be classified together in the kingdom **Monera**, also called **Procaryotae**. These organisms are the only true prokaryotes, and they differ significantly from members of the four eukaryotic kingdoms in the cellular details described earlier.

The second kingdom, **Protista**, now is limited to the protozoa and the unicellular algae. Members of the kingdom generally have flagella at some time in their lives, and even though there are some larger forms (e.g., complex algae), the tissue level of organization typical of plants and animals is absent. Many members of the kingdom are "taxonomic misfits" because they do not appear to fit in other kingdoms. However, they share certain characteristics with plants and animals, and some species appear to be plant and animal ancestors.

In Whittaker's system, the kingdom **Fungi** includes nongreen, nonphotosynthetic eukaryotic organisms whose cell walls differ chemically from those in bacteria, algae, and plants. Also, there is a mingling of cytoplasms of adjacent cells in the fungal hyphae. This means that the cytosol and some organelles can move from one cell to another in the filament, so true multicellularity does not exist, as compared to plants and animals. Fungi absorb dissolved organic matter. Finally, remember how early taxonomists placed fungi with plants? New molecular evidence suggests that, although fungi are not ancestors of animals, they appear to have close relationships with animals.

The final two kingdoms, Plantae and Animalia, are the traditional multicellular plants and animals. Plants use photosynthesis to synthesize their organic molecules; animals ingest their food through some form of mouth, and then use digestive enzymes to break food particles into absorbable fragments.

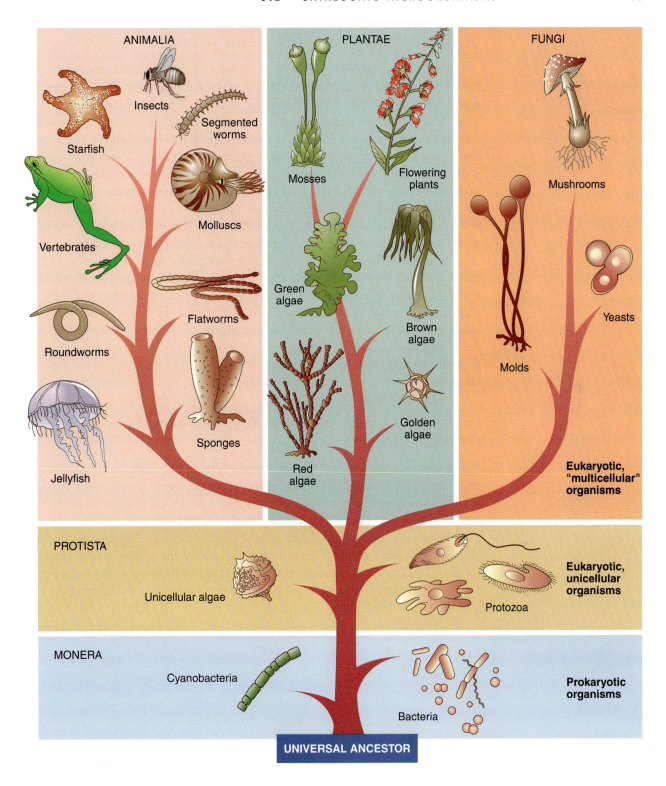

FIGURE 3.10

Whittaker's Five-Kingdom System

Devised by Robert H. Whittaker, this system implies an evolutionary lineage, beginning with the Monera and extending to the Protista. Certain of the Protista are believed to be ancestors of the Plantae, Fungi, and Animalia. Divergence at each level is based on the three modes of nutrition: photosynthesis, absorption, and ingestion. Unicellular and multicellular organization also are key features in the system, although the fungi are not truly multicellular.

CLASSIFICATION USES A HIERARCHICAL SYSTEM

Species:
the fundamental rank in the classification system.

The mechanics of the classification system have remained consistent since first outlined by Linnaeus. The fundamental and least inclusive rank is the **species** (pl., species). For bacteria, the members of a species have 70 percent biochemical similarity and they differ significantly from other species. (For most eukaryotes, a species is a group of individuals in a population that can breed with one another). Two or more similar species are grouped together as a **genus** (pl., genera). A collection of similar genera makes up a **family**, and families with similar characteristics make up an **order**. Orders are placed together as a **class**, and classes are assembled into a **phylum** (pl., phyla) or **division** (in bacteriology and botany). Two or more similar phyla/divisions are grouped as a **kingdom,** the most inclusive rank in classification. TABLE 3.3 outlines the classification schemes for three organisms in the five-kingdom system.

In bacteria, an organism may belong to a rank below the species level to indicate that a special characteristic exists within a subgroup of the species. Such ranks have no official standing in nomenclature, but they have practical usefulness in helping to identify an organism. For example, two biotypes of the cholera bacillus, *Vibrio cholerae,* are known to exist: *Vibrio cholerae* classic and *Vibrio cholerae* El Tor. Other designations of ranks include subspecies, serotype, strain, morphotype, and variety. Some bacteriologists recommend changing all these subspecies ranks to "varieties" and then using names such as biovar and serovar.

THE THREE-DOMAIN SYSTEM PLACES THE MONERA IN SEPARATE LINEAGES

The view that five kingdoms alone represent the natural lines of division among living things has been modified further by the development of the three domains, or superkingdoms system. First proposed in the 1970s by Carl Woese (FIGURE 3.11) and his coworkers at the University of Illinois, the three domain system is based on new techniques in molecular biology and biochemistry. It also encompasses new knowledge about a group of bacteria formerly called the archaebacteria (*archae* = "ancient"). Many of these bacterial forms are known for their ability to live under extremely harsh environments, similar to ancient ones on Earth (Chapter 4).

TABLE 3.3

Taxonomic Classification of Three Species of Organisms

	HUMAN BEING	BACTERIUM	YEAST
Kingdom	Animalia	Monera	Fungi
Phylum (Division)	Chordata	Gracilicutes	Ascomycota
Class	Mammalia	Scotobacteria	Saccharomycotina
Order	Primates	Spirochaetales	Saccharomycetales
Family	Hominidae	Leptospiraceae	Saccharomycetaceae
Genus	*Homo*	*Leptospira*	*Saccharomyces*
Species	*H. sapiens*	*L. interrogans*	*S. cerevisiae*

In Woese's **three-domain system**, one domain includes the archaebacteria and is called the domain **Archaea**. The second encompasses all the remaining true bacteria and is called the domain **Eubacteria** (similar to Monera but without archaebacteria). So, why two separate domains? Biochemical and molecular evidence indicate that the archaea and eubacteria differ significantly in the base sequence of the RNA in their ribosomes, the composition of their cell walls, the types of lipids in their membranes, and their sensitivity to certain antibiotics. These were such unexpected differences that it is not surprising the chapter opening quote was made about Woese's discoveries.

FIGURE 3.11

Carl Woese

Much of the early evidence for the three-domain system of classification is based on Carl Woese's investigations.

The third domain includes the four remaining kingdoms of Whittaker (i.e., Protista, Fungi, Plantae, and Animalia) and is called the domain **Eukarya** (FIGURE 3.12).

Scientists initially were reluctant to accept the three-domain system of classification, and many deemed it a threat to the tenet that all living things are either prokaryotes or

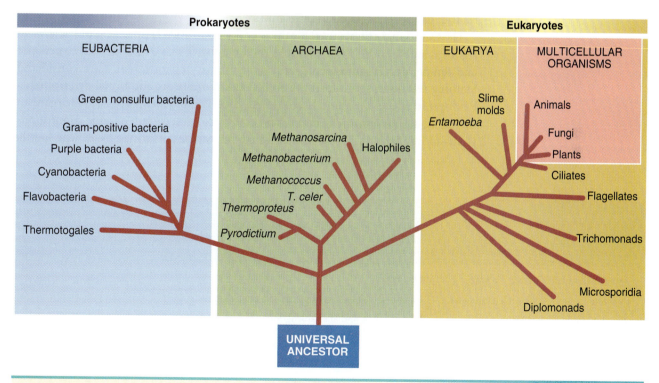

FIGURE 3.12

The Three-Domain System

Fundamental differences in genetic endowments are the basis for the three domains of all organisms on Earth. The relationships are determined from the sequences of nitrogenous bases in ribosomal RNA. The line length between any two groups is proportional to their genetic differences.

eukaryotes. Then in 1996, Craig Venter and his coworkers deciphered the DNA base sequence of the archaean *Methanococcus jannaschii* and showed that almost two-thirds of its genes are different from those of a common eubacterium. They also found that proteins replicating the DNA and involved in the RNA synthesis of archaea have no counterpart in the eubacteria. Additional recent genomic evidence has persuaded more scientists to accept the scheme. It appears that the three-domain system is on firm ground.

TABLE 3.4 summarizes the "evolution" of taxonomic thinking.

BACTERIAL TAXONOMY NOW INCLUDES MOLECULAR CRITERIA

We shall consider the taxonomy of most groups of microorganisms in their respective chapters. Bacterial taxonomy, however, merits special attention because bacteria occupy an important position in microbiology and have had a complex system of taxonomy. One of the first systems of classification for the bacteria was devised in 1923 by David Hendricks Bergey. His book *Bergey's Manual of Determinative Bacteriology* was updated and greatly expanded in the decades that followed and exists today as the five-volume *Bergey's Manual of Systematic Bacteriology*. **Bergey's Manual**, as it is commonly known, is considered the official listing of all recognized bacteria. It also is intended to be a guide to identification. Since the first edition in 1984, bacterial taxonomy has gone through tremendous changes. For example, the second edition (2001) has added more than 2,200 new species and 390 new genera. (Note: A single volume entitled *Bergey's Manual of Determinative Bacteriology* is available for use as a quick, handy guide to identifying unknown bacteria that have been isolated in pure culture.)

The traditional criteria used for the identification and classification of bacteria are rigorous and thorough. For example, the organism's shape and size, its oxygen, pH, and temperature requirements, and its laboratory characteristics are considered. Staining reactions, spore-forming ability, and motility are additional determinants. Pathogenic effects on animals are noted also. The biochemistry of the organism, including its photosynthetic nature and ability to digest certain organic compounds, yields additional data. Tests are mentioned that can be performed to see whether the

TABLE 3.4

A Summary of Microbial "Classifiers"

INVESTIGATOR	TIME FRAME	PROPOSAL
Aristotle	Fourth century B.C.	Described 500 species of plants and animals
Carolus Linnaeus	1750s	Devised plant and animal kingdoms Developed classification system for plants and animals Grouped microorganisms as "Chaos"
Ernst Haeckel	1860s	Separated microorganisms into third kingdom (Protista)
Ferdinand Cohn	1880s	Placed bacteria with plants because of presence of cell wall
David Bergey	1923	Developed a classification of bacteria in *Bergey's Manual*
Robert Whittaker	1969	Devised five-kingdom classification; microorganisms occupy kingdoms Monera, Protista, and Fungi
Carl Woese	1976	Devised three-domain classification system with Archaea, Eubacteria, and Eukarya

organism interacts with known antibodies. Even a laser beam can be used to identify an organism. In 1993, researchers discovered that each species of bacterium has an "optical fingerprint" and reflects or absorbs laser light in its own characteristic way.

In 1984, the editors of *Bergey's Manual* noted that there is no "official" classification of bacteria and that the closest approximation to an official classification is the one most widely accepted by the community of microbiologists. The editors stated that a comprehensive classification might one day be possible. Today, the fields of molecular genetics and genomics have advanced the analysis and sequencing of nucleic acids. Results of this work are reflected in the new second edition of *Bergey's Manual* and have made it clear that many of the original criteria have no bearing in classifying particular taxonomic groups. This has given rise to a new era of molecular taxonomy.

Molecular taxonomy is based on the universal presence of ribosomes in all living organisms. In particular, it is the RNAs that build the ribosome, called ribosomal RNA (rRNA), that are of most interest and were the primary basis of Woese's construction of the three-domain system. Many scientists today believe the rRNA molecule is the ultimate "molecular chronometer" that allows for precise bacterial classification in all taxonomic classes. The ultimate aim of molecular taxonomy is to reconstruct the main evolutionary events that have occurred (as suggested in Figure 3.12) and influenced life on Earth over the past 4 billion years. The pioneering work by Woese and others is an example of that goal.

NOMENCLATURE GIVES SCIENTIFIC NAMES TO ORGANISMS

In addition to giving the descriptions and properties of microorganisms, taxonomies like the one in *Bergey's Manual* provide the genus name by which scientists refer to specific microorganisms. All organisms have a double name, usually from Latin or Greek stems. The name consists of the **genus** to which the organism belongs and a **specific epithet** (a descriptive adjective) that further describes the genus name. Together these two words make up the species name (see Table 3.3). This system of nomenclature is called the **binomial system** (*binomial* = "two names"). First suggested by Linnaeus, the binomial system gives biologists throughout the world an international language for life forms and eliminates incalculable amounts of confusion.

When a species of an organism is written, only the first letter of the genus name is capitalized. The remainder of the genus name and specific epithet is written in lowercase letters. Both words should be printed in italics, but if this is not possible, they should be underlined. For example, a species of intestinal bacteria discovered in 1885 by Theodor Escherich is written as *Escherichia coli* or Escherichia coli (**Micro-Focus 3.4**).

Scientists often abbreviate binomial names by writing the first letter of the genus name, or some accepted substitution, together with the full specific epithet, after the first time the species has been spelled out. The abbreviation should also be italicized or underlined. Thus, *Escherichia coli* becomes *E. coli*, and *Bacillus subtilis* is written *B. subtilis*. Where similarities occur, international committees on nomenclature have intervened.

The rules for assigning names to protozoa are listed in the *International Code of Zoological Nomenclature*, for fungi in the *International Code for Botanical Nomenclature*, and for bacteria in the *Bacteriological Code*. Viruses have no universally accepted binomial names (Latinized binomials are not used). Virus hierarchy is determined by The International Committee on Taxonomy of Viruses.

MicroFocus 3.4

NAMING NAMES

As you read this book, you will come across many scientific names for microbes, where a species name is a combination of the genus and specific epithet. Not only are many of these names tongue twisting to pronounce (they are all listed with their pronunciation inside the back cover), but how in the world did the organisms get those names? Here are a few examples.

GENERA NAMED AFTER INDIVIDUALS:

Escherichia coli: named after Theodore Escherich who isolated the bacterium from infant feces in 1885. Being in feces, it commonly is found in the colon.

Neisseria gonorrhoeae: named after Albert Neisser who discovered the bacterium in 1879. As the specific epithet points out, the disease it causes is gonorrhea.

GENERA NAMED AFTER THE MICROBE'S SHAPE:

Vibrio cholerae: vibrio means "comma-shaped," which describes the shape of the bacteria that cause cholera.

Staphylococcus epidermidis: staphylo means "clusters" and coccus means "spheres." So, these bacteria form clusters of spheres that are found on the skin surface (epidermis).

GENERA NAMED AFTER AN ATTRIBUTE OF THE MICROBE:

Saccharomyces cerevisiae: in 1837, Theodor Schwann observed yeast cells and called them *Saccharomyces* (meaning "sugar fungus") because the yeast converted grape juice (sugar) into alcohol; *cerevisiae* (from *cervisia* = "beer") refers to the use of yeast to make beer since ancient times.

Myxococcus xanthus: myxo means "slime," so there are slime-producing spheres that grow as yellow (*xanthos* = yellow) colonies on agar.

Thiomargarita namibiensis: see MicroFocus 3.5.

To this point . . .

We have discussed the beginnings of modern taxonomy through the work of Linnaeus and Haeckel. The currently accepted system is the one devised by Whittaker, then modified by Woese. Microorganisms occupy three of the five kingdoms in Whittaker's system and all three domains in Woese's. In both systems, organisms can be ranked from species to kingdom.

We then looked at how bacteria are classified. The classification system is complex and is outlined in Bergey's Manual of Systematic Bacteriology. To provide a glimpse of the taxonomist's work, we listed several criteria used to classify and identify a bacterium. Of importance was molecular taxonomy, which has brought about a reordering of organisms, as exemplified by the three-domain classification system.

The discussion then turned to nomenclature, where the binomial system is the method employed for all living things. A binomial name consists of the genus to which an organism belongs and a specific epithet. Both parts of the binomial name are written in italics.

In the final section of this chapter, we will survey the methods used for observing microorganisms. After describing how cell size is measured, we will discuss the common light microscope and then review three specialized types of microscopy. Some remarks on electron microscopy will complete the chapter.

3·3

Microscopy

By now you should be aware that microorganisms usually are very small. Before we examine the instruments used to "see" these tiny creatures, we need to be familiar with the units of measurement.

MOST MICROORGANISMS ARE IN THE MICROMETER SIZE RANGE

An important property in the study of microorganisms is their size. In biology, the unit of length is the **micrometer** (**μm**), which is equivalent to a millionth (micro-) of a meter. To appreciate how small a micrometer is, consider this: Comparing a micrometer to an inch is like comparing a housefly to New York City's Empire State Building, 1,472 feet high.

Microorganisms range in size from the relatively large, almost visible protozoa (100 μm) down to the incredibly tiny viruses (0.01 μm), ten-thousand times smaller (**FIGURE 3.13**). Molds consist of intertwined filaments so long and twisted that they are visible (such as a mushroom), but the individual cells measure only about 40 μm by 10 μm. The yeasts are about 8 μm in diameter. Most bacteria are about 1 μm to 5 μm in length, but the largest ones reach approximately 20 μm in length, although notable exceptions have been discovered (**MicroFocus 3.5**). The smaller bacteria we encountered earlier in this chapter, such as the rickettsiae, may be only about 0.4 μm long. The chlamydiae and mycoplasmas are a scant 0.25 μm in length.

To avoid using decimals with micrometers, scientists often express the size of viruses in nanometers. A **nanometer** (**nm**) is equivalent to a billionth (nano-) of a meter; that is, 1/1,000 of a μm. Using nanometers, the size of a smallpox virus may be written as 250 nm, rather than 0.25 μm. The poliovirus, among the smaller viruses, measures 20 nm in diameter. Other objects measured in nanometers include the wavelength of radiant energy, such as visible or ultraviolet light, and the size of certain large molecules.

LIGHT MICROSCOPY USES VISIBLE LIGHT TO RESOLVE OBJECTS

The basic microscope system used in the microbiology laboratory is the **light microscope**, in which visible light passes directly through the lenses and specimen (**FIGURE 3.14a**). Such an optical configuration is called **bright-field microscopy**. More specifically, visible light is projected through a substage condenser lens, which focuses the light into a sharp cone (**FIGURE 3.14b**). The light then passes through the opening in the stage, into the slide, where the light is reflected, refracted, or passes through the specimen. Next, light passing through the specimen enters the objective lens to form a magnified intermediate image inverted from that of the specimen. This intermediate image becomes the object magnified by the ocular lens (eyepiece) and seen by the observer.

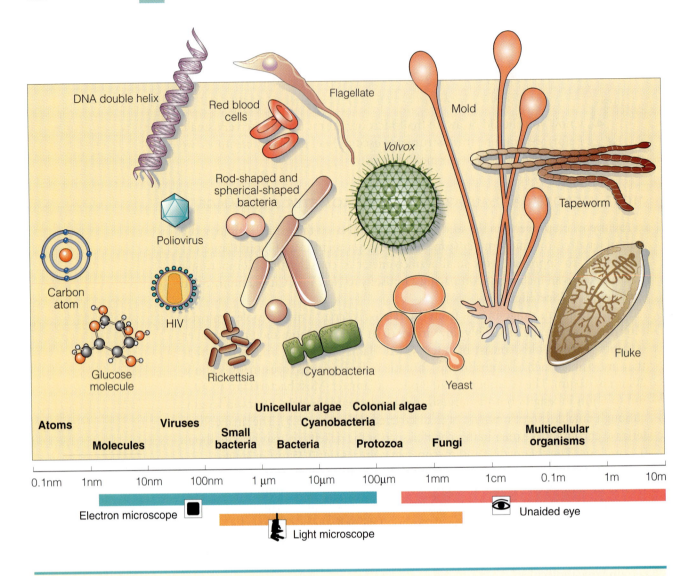

FIGURE 3.13

Size Comparisons among Various Atoms, Molecules, and Microorganisms (not drawn to scale)

MAGNIFICATION AND RESOLUTION DEFINE THE LIMITS OF WHAT IS VISIBLE

A light microscope usually has at least three objective lenses: the low-power, high-power, and oil-immersion lenses. In general, these lenses magnify an object 10, 40, and 100 times, respectively. (Magnification is represented by the multiplication sign, ×.) This intermediate image is remagnified by the ocular lens, as previously noted. With a standard 10× ocular lens, the **total magnification** achieved is 100×, 400×, and 1,000×, respectively.

For an object to be seen distinctly, the lens system must have good **resolving power**; that is, it must transmit light without variation and allow closely spaced objects to be clearly distinguished. For example, a car seen in the distance at night may appear to have a single headlight because at that distance the unaided eye lacks

Total magnification: the magnification of the ocular multiplied by the magnification of the objective lens.

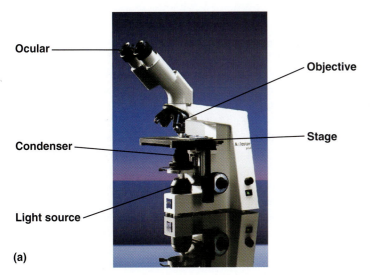

Ocular

Objective

Condenser

Stage

Light source

(a)

FIGURE 3.14

The Light Microscope

(a) This is the familiar light microscope used in many instructional and clinical laboratories. Note the important features of the microscope that contribute to the visualization of the object.
(b) Image Formation in Light Microscopy. Light passes through the objective lens, forming an intermediate image. This image serves as an object for the ocular lens, which remagnifies the image and forms the final image the eye perceives.

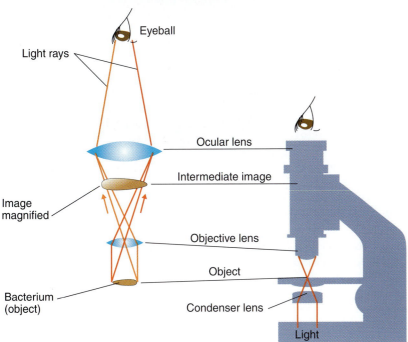

Eyeball

Light rays

Ocular lens

Intermediate image

Image magnified

Objective lens

Object

Bacterium (object)

Condenser lens

Light

(b)

resolving power. However, using binoculars, the two headlights can be seen clearly as the resolving power of the eye increases.

The resolving power (RP) of a lens system is important in microscopy because it denotes the size of the smallest object that can be seen clearly. The resolving power varies for each objective lens and is calculated using the following formula:

$$RP = \frac{\lambda}{2 \times NA}$$

In this formula, the Greek letter λ (lambda) represents the wavelength of light and is usually set at 500 nm, the halfway point between the limits of visible light. The symbol NA stands for the **numerical aperture** of the lens and refers to the size of the

MicroFocus 3.5

A BIOLOGICAL OXYMORON

An English professor would define an oxymoron as two mutually exclusive juxtaposed terms. To ordinary people, an oxymoron is simply two words that do not go together, as they seem to refer to separate things. Some typical oxymorons are jumbo shrimp, holy war, old news, negative attraction, and sweet sorrow.

In 1993, researchers at Indiana University reported their discovery of a "visible bacterium," and headlines proclaimed it a biological oxymoron. Why? Because a bacterium, by definition, is an invisible organism. And yet, here was a visible bacterium so large that a microscope was not needed to see it. The spectacular giant measures over 0.6 mm in length (that's 600 μm compared to 2 μm for *Escherichia coli*) and even dwarfs the protozoan *Paramecium.* Found in the gut of the surgeonfish near Australian reefs, the bacterium was the largest ever observed—until 1999.

While on an expedition off the coast of Namibia (western coast of southern Africa), Heide Schultz and teammates from the Max Planck Institute for Marine Microbiology in Bremen, Germany, found quite a surprise while looking at one of the sediment samples from the seafloor. Here were spherical cells about 0.1 to 0.3 mm in diameter but some as large as 0.75 mm—about the diameter of the period in this sentence. Their volume is about 3 million times greater than a typical bacterium like *E. coli.* The cells, shining white with enclosed sulfur granules, were held together in chains by a mucus sheath looking like a string of pearls. Thus, the bacterium was named *Thiomargarita namibiensis* (meaning "sulfur pearl of Namibia").

So, how does a prokaryotic cell survive in so large a size? The trick is to keep the cytoplasm as a thin layer plastered against the edge of the cell so materials do not need to travel (diffuse) far to get into or out of the cell. The rest of the cell is a giant "bubble," called a vacuole, in which nitrate and sulfur are stored. The bacterium uses the sulfur and nitrate for energy metabolism. Thus, the actual cytoplasmic layer is microscopic and as close to the surface as possible.

The bacterium grows quite happily in these Atlantic waters, which are very toxic to other life due to high levels of hydrogen sulfide. Without serious competition, the bacterium is not a rare organism because it can be found in concentrations of up to 47 grams per square meter. As we already have seen in the book, bacteria are amazing in many ways, including how they have evolved adaptations to most any environment—and in this case, their enormous size.

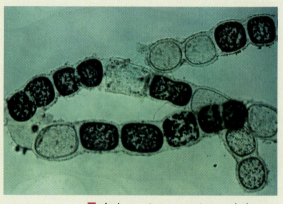

■ *A phase microscopy micrograph showing a chain of* Thiomargarita namibiensis *cells. It is the largest bacterium ever found and has shattered the idea that microorganisms have to be invisible.*

cone of light that enters the objective lens after passing through the specimen. This number generally is printed on the side of the objective lens. For a low-power objective with an NA of 0.25, the resolving power may be calculated as follows:

$$RP = \frac{500 \text{ nm}}{2 \times 0.25} = \frac{550}{0.50} = 1{,}000 \text{ nm or } 1.0 \text{ μm}$$

Since the resolving power for this lens system is 1.0 μm, any object smaller than 1.0 μm could not be seen as a clear, distinct object. An object larger than 1.0 μm would be resolved.

When switching from the low-power or high-power lens to the oil-immersion lens, one quickly finds that the image has become fuzzy. The object lacks resolution, and the resolving power of the lens system appears to be poor. This is because this one objective lens must be immersed in an optical oil (**FIGURE 3.15**). The system's

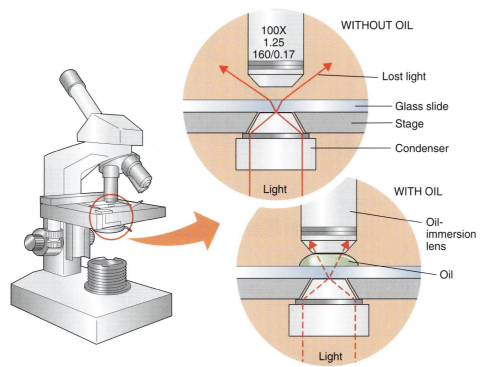

FIGURE 3.15

Aspects of Oil-Immersion Microscopy

When light rays enter the air (solid arrow), they miss the objective lens. However, they remain on a straight line (dashed arrow) in the oil. This pathway leads them directly into the lens. The resolution increases with increased light. On the barrel of the objective lens is engraved important information. For the oil immersion lens: 100× = magnification; 1.25 = numerical aperture; 160 = working distance (distance, in μm, between lens and cover glass); and 0.17 = optimal cover glass thickness (in mm).

resolving power is calculated with the lens suspended in oil rather than air, a factor that increases the numerical aperture to 1.25.

Both low-power and high-power objectives are wide enough to capture sufficient light for viewing. The oil-immersion objective, on the other hand, is so narrow that most light bends away and would miss the objective lens if oil were not used. The **index of refraction** (or refractive index) is a measure of the light-bending ability of a medium. Immersion oil has an index of refraction of 1.5, which is almost identical to the index of refraction of glass. Because the refractive index is the same for oil and glass, the light does not bend as it passes from the glass slide and the specimen into the oil. By comparison, air has an index of refraction of 1.0, which accounts for the abrupt bending as light enters it. The oil thus provides a homogeneous pathway for light from the slide to the objective, and the resolution of the object increases. With the oil-immersion lens having a numerical aperture of 1.25, the highest resolution possible with the light microscope is 0.2 μm (200 nm).

STAINING TECHNIQUES PROVIDE CONTRAST

Microbiologists commonly stain bacteria before viewing them because the cytoplasm of bacterial cells lacks color, making it hard to see the cells on a bright background. Several staining techniques have been developed to provide **contrast** for bright-field microscopy (**FIGURE 3.16**).

Contrast:
ability to see an object clearly from the background.

To perform the **simple stain technique**, a small amount of bacteria in a droplet of water or broth is smeared on a glass slide and the slide air-dried. Next, the slide is passed briefly through a flame in a process called **heat fixing**. This bonds the cells to the slide, kills the organisms that may still be alive, and increases stain absorption. Now the slide is flooded with a **basic** (cationic) **dye** such as crystal violet or methylene blue (Figure 3.16a). Since cationic dyes have a positive charge, the dye is

Simple stain technique:
a staining procedure that contrasts stained cells against a bright background.

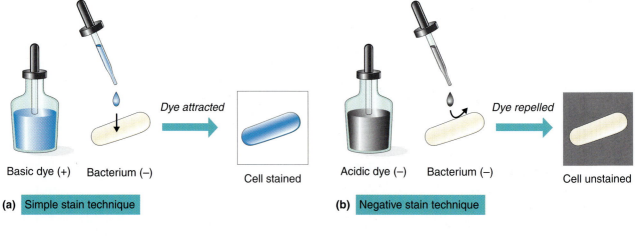

Dye attracted

Basic dye (+)   Bacterium (−)   Cell stained

(a) Simple stain technique

Dye repelled

Acidic dye (−)   Bacterium (−)   Cell unstained

(b) Negative stain technique

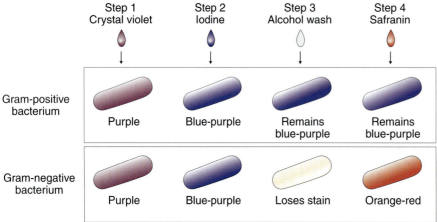

	Step 1 Crystal violet	Step 2 Iodine	Step 3 Alcohol wash	Step 4 Safranin
Gram-positive bacterium	Purple	Blue-purple	Remains blue-purple	Remains blue-purple
Gram-negative bacterium	Purple	Blue-purple	Loses stain	Orange-red

(c) Gram stain technique

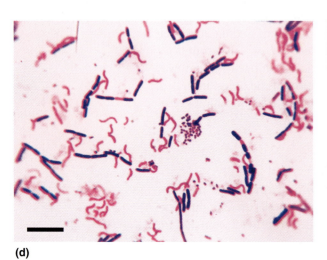

(d)

FIGURE 3.16

Important Staining Reactions in Microbiology

(a) In the simple stain technique, the positive-charged stain is attracted to the negative-charged bacteria, and staining takes place. (b) With the negative stain technique, negative-charged dye is repelled by the bacteria, and the cells remain unstained on a dark background. (c) The Gram stain technique is a differential procedure. All bacteria stain with the crystal violet and iodine, but only gram-negative bacteria lose the color when alcohol is applied. Subsequently, these bacteria stain with the safranin dye. Gram-positive bacteria remain blue purple. (d) This light micrograph demonstrates the staining results of a Gram stain. Not only can gram-positive and gram-negative cells be seen, but also cell size, cell shape, and arrangements can be determined. (Bar = 10 μm.)

attracted to the cytoplasm and cell wall, which have negative charges. By contrasting the purple or blue cells against the bright background, the staining procedure allows the observer to measure cell size and determine cell shape. It also can provide information about how cells are arranged with respect to one another (Chapter 4).

The **negative stain technique** works in the opposite manner (Figure 3.16b). Bacteria are mixed on a slide with an **acidic** (anionic) **dye** such as nigrosin (a black stain) or India ink (a black drawing ink). The mixture then is smeared across the face of the slide and allowed to air-dry. Because the anionic dye carries a negative charge, it is repelled from the cell wall and cytoplasm. The stain does not enter the cells and the microscopist observes clear or white cells on a black or gray background. Since this technique avoids chemical reactions and heat fixing, the cells appear less shriveled and less distorted than in a simple stain. They are closer to their natural condition.

Negative stain technique: a staining procedure that contrasts unstained cells against a stained background.

The **Gram stain technique** allows us to view stained cells while learning something about them. The technique is named for Christian Gram, the Danish physician who first perfected the technique in 1884. It is an example of a differential staining technique because it differentiates (separates) bacteria into two groups called gram-positive and gram-negative bacteria.

Gram stain technique: a staining procedure that differentiates bacteria into two separate groups, gram-positive and gram-negative.

The first two steps of the technique are straightforward. Air-dried, heat-fixed smears are stained with crystal violet (MicroFocus 3.6) and then with a special Gram's iodine solution (Figure 3.16c). All bacteria would appear blue-purple if the procedure was stopped and the sample viewed with the light microscope. Next, the smear is rinsed with a decolorizer, such as 95 percent alcohol or an alcohol-acetone mixture. Observed at this point, certain bacteria have lost their color and become transparent. These are the **gram-negative bacteria**. Other bacteria retain the blue-purple

---

MicroFocus 3.6

"A.O. MEANS WHAT?"

In modern bacteriology laboratories, the crystal violet solution used for Gram staining is prepared by mixing solid dye particles with ammonium oxalate. This procedure has not changed since 1929, when a graduate student named Thomas Hucker introduced it. How this "Hucker modification" came about is part of the folklore of microbiology.

Hucker was studying bacteriology at Yale University. Early in 1929, his advisor suggested that he contact several hospital and university laboratories to see how they were performing the Gram stain technique. Hucker was to report his findings in a paper presentation at an upcoming scientific meeting in Philadelphia. He dutifully sent out a series of letters and learned that the standard procedures were being used at all laboratories—all, that is, except Dartmouth's.

The reply from Dartmouth College piqued his interest. At the time, the usual procedure was to dissolve crystal violet in aniline oil. But Dartmouth bacteriologists apparently were using ammonium oxalate. Hucker tried ammonium oxalate and found that the stain improved with age and gave clearer results. He prepared his paper for the Philadelphia meeting and sent a draft to Dartmouth's biology department with a note of thanks. Soon thereafter he received a phone call from Dartmouth—they had never heard of ammonium oxalate for Gram staining. Hucker was perplexed.

In the days that followed, Hucker learned that a chemist had intercepted his survey letter and sent the reply. In writing out the method for crystal violet preparation, the chemist had read "A.O." on the bottle of stain and assumed that it meant the dye was dissolved in ammonium oxalate. Aniline oil simply did not occur to him. Moreover, he had not bothered to check with the biology department because it was inventory time and other things were on his mind. Thus, a case of badly interpreted bacteriological shorthand led to the Hucker modification. Hucker became famous; the chemist remained anonymous.

stain. These are the **gram-positive bacteria**. The last step is to use safranin, a red cationic dye, to contrast the gram-negative organisms' orange-red color. Thus, at the technique's conclusion, gram-positive bacteria are blue-purple while gram-negative organisms are orange-red (Figure 3.16d). By observing the color of the cells at the conclusion of the process, one may decide the group (gram-positive or gram-negative) to which the bacteria belong. Gram staining also allows the observer to determine size, shape, and arrangement of cells.

Knowing whether a bacterium is gram-positive or gram-negative is important for several reasons. Microbiologists and clinical technicians use the results from the Gram stain technique to help identify an unknown bacterium and classify it in *Bergey's Manual*. Gram-positive and gram-negative bacteria also differ in their susceptibility to chemical substances such as antibiotics (gram-positive bacteria are more susceptible to penicillin, gram-negatives to tetracycline). Gram-negative bacteria have more complex cell walls, as described in Chapter 4. In addition, gram-positive and gram-negative bacteria produce different types of toxic poisons called toxins.

One other differential staining procedure, the **acid-fast technique**, deserves mention. This technique is used to identify members of the genus *Mycobacterium*, one species of which causes tuberculosis. These bacteria are normally difficult to stain with the Gram stain because the cells have very waxy walls. However, the cell will stain red when treated with carbolfuchsin (red dye) and heat (or a lipid solubilizer). The cells then retain their color when washed with a dilute acid-alcohol solution. Other stained genera lose the red color easily during the acid-alcohol wash. The *Mycobacterium* species is therefore said to be acid resistant or "acid fast." (A blue counterstain is used to give color to non–acid-fast bacteria.) Because they stain red and break sharply when they reproduce, *Mycobacterium* species euphemistically are referred to as "red snappers."

> **Acid-fast technique:**
> a staining technique used to identify *Myobacterium* species in which stain is forced into the bacteria, then retained by the cells even on treatment with a dilute acid-alcohol solution.

LIGHT MICROSCOPY HAS OTHER OPTICAL CONFIGURATIONS

Without staining, bright-field microscopy provides little contrast (**FIGURE 3.17a**). However, the light microscope often contains other optical systems to improve contrast of prokaryotic and eukaryotic cells without staining. Three common methods are mentioned here.

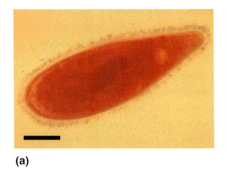 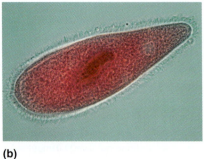 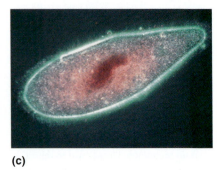

(a)   (b)   (c)

FIGURE 3.17

Variations in Light Microscopy

The same *Paramecium* specimen seen with three different optical configuratons. (Bar = 25 μm.) (a) Bright-field, (b) phase-contrast, and (c) dark-field. Note the different view that each microscopy technique affords.

Special condenser and objective lenses are used in **phase-contrast microscopy** (**FIGURE 3.17b**). The condenser lens on the light microscope splits a light beam and throws the light rays slightly out of phase. The separated beams of light then pass through and around the specimen, and small differences in the refractive index within the specimen show up as different degrees of brightness and contrast. With phase-contrast microscopy, microbiologists can see organisms alive and unstained. The structure of yeasts, molds, and protozoa is studied with this optical configuration.

In **dark-field microscopy**, the background remains dark, and the specimen is illuminated (**FIGURE 3.17c**). A special condenser lens mounted under the stage scatters the light and causes it to hit the specimen from the side. Only light bouncing off the specimen and into the objective lens makes the specimen visible, as the surrounding area appears dark because it lacks background light. Dark-field microscopy provides good resolution and often illuminates parts of a specimen not seen with bright-field optics.

Dark-field microscopy helps in the diagnosis of diseases caused by spiral bacteria because these organisms are near the limit of resolution of the light microscope and do not stain well. For example, syphilis, caused by the spiral bacterium *Treponema pallidum*, has a diameter of about 0.15 μm. This bacterium may be observed in scrapings taken from a lesion of a person who has the disease. Dark-field microscopy also is the preferred way to study motility of live prokaryotic and eukaryotic cells.

A more recent configuration is **fluorescence microscopy,** which has emerged as a major asset to diagnostic and research laboratories. The technique has been applied to the identification of many microorganisms and is a mainstay of modern microbial ecology and especially clinical microbiology (MicroFocus 3.7). With fluorescence

Phase-contrast microscopy: a microscopy system in which light rays are thrown out of phase by the condenser lens, and then used to illuminate details of living cells.

Dark-field microscopy: a microscopy system in which light rays hit an object from the side, producing images of reflected objects against a dark background.

Fluorescence microscopy: a microscopy system that uses one color (wavelength) of light to excite a dye coating an object to fluoresce a different color.

MicroFocus 3.7

THE CENTERS FOR DISEASE CONTROL AND PREVENTION

The Centers for Disease Control and Prevention (CDC) has its headquarters in Atlanta, Georgia, and is one of six major agencies of the US Public Health Service. Originally established as the Communicable Disease Center in 1946, the CDC was the first governmental health organization ever set up to coordinate a national control program against infectious diseases. At first, it was concerned with diseases spread from person to person, from animals to people, or from the environment to humans. Eventually, though, all communicable diseases came under its aegis. Atlanta was selected as the site for the CDC because it was a convenient central point for the study of malaria, which was then common in the South.

In April 1955, two weeks after release of the Salk vaccine for polio, the CDC received reports of six cases of polio in vaccinated children. Two days later, it established the Polio Surveillance Unit and began collecting data on polio occurrence and summarizing it for health professionals. More than 80 percent of vaccine-associated polio cases were related to a single manufacturer, and its vaccine was withdrawn at once. This incident established the role of the CDC in health emergencies, and soon it became a national resource for the development and dissemination of information on communicable disease. In 1960, the CDC moved to a new headquarters complex adjoining Emory University. The unassuming appearance of the facility belies its importance.

Reorganized with its current name in 1980 (the words "and Prevention" were added in 1992 but the CDC acronym was retained), the CDC is charged with protecting the public health of the US populace by providing leadership and direction in the prevention and control of infectious disease and other preventable conditions, such as cancer. It is concerned with urban rat control, quarantine measures, health education, and the upgrading and licensing of clinical laboratories. The CDC also provides international consultation on disease and participates with other nations in the control and eradication of communicable infections. It employs 3,500 physicians and scientists, the largest group in the world, and processes 170,000 samples of tissue annually. Its publication *The Morbidity and Mortality Weekly Report* is distributed each week to over 100,000 health professionals.

microscopy, objects fluoresce a certain color after light of another color (wavelength) shines on them.

Microorganisms are coated with a fluorescent dye, such as fluorescein, and then illuminated with ultraviolet (UV) light. The energy in UV light excites electrons in the dye, and they move to higher energy levels. However, the electrons quickly drop back to their original energy levels and give off the excess energy as visible light. The coated microorganisms thus appear to fluoresce; in the case of fluorescein, they glow a greenish yellow. Other dyes produce other colors (**FIGURE 3.18**).

An important application of fluorescence microscopy is the **fluorescent antibody technique** used to identify an unknown organism. In one variation of this procedure, fluorescein is chemically attached to antibodies, the protein molecules produced by the body's immune system. These "tagged" antibodies are mixed with a sample of the unknown organism. If the antibodies are specific for that organism, they will bind to it and coat the cells with the dye. When subjected to UV light, the organisms will fluoresce. If the organisms fail to fluoresce, the antibodies were not specific to that organism and a different tagged antibody is tried.

ELECTRON MICROSCOPY PROVIDES DETAILED IMAGES OF CELLS AND CELL PARTS

Electron microscopy:
a microscopy system in which a beam of electrons substitutes for the light energy used in other microscopes.

The **electron microscope** (**FIGURE 3.19a**) grew out of an engineering design made in 1933 by the German physicist Ernst Ruska (winner of the 1986 Nobel Prize in Physics). Ruska showed that electrons flow in a sealed tube if a vacuum is maintained to prevent electron scattering. Magnets, rather than glass lenses, pinpoint the flow onto an object, where the electrons are absorbed, deflected, or transmitted depending on the density of structures within the object (**FIGURE 3.19b**). When projected onto a screen underneath, the electrons form a final image that outlines the structures. As mentioned in Chapter 1, the early days of electron microscopy produced the images that showed bacteria indeed were cellular but their structure was different. This led to the development of the prokaryotic and eukaryotic groups of organisms.

FIGURE 3.18

Fluorescence Microscopy

Fluorescence microscopy of sporulating cells of *Bacillus subtilis*. DNA has been stained with a dye that fluoresces red and a sporulating protein with fluorescein (green). RNA synthesis activity of β-galactosidase is indicated by a dye that fluoresces blue. (Bar = 20 μm.)

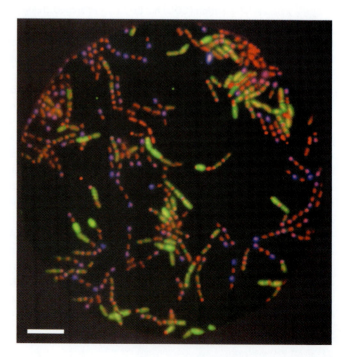

FIGURE 3.19

The Electron Microscope

(a) A transmission electron microscope (TEM). (b) A schematic of a TEM. A beam of electrons is emitted from the electron source and electromagnets are used to focus the beam on the specimen. The image is magnified by objective and projector lenses. The final image is projected on a screen, television monitor, or photographic film.

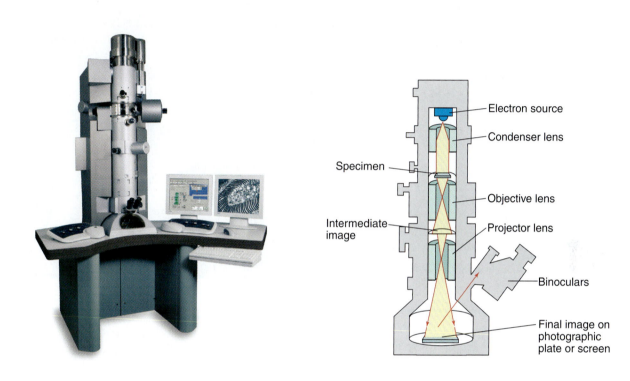

(a)   (b)

The key to electron microscopy is the extraordinarily short wavelength of the beam of electrons. Measured at 0.005 nm (compared to 500 nm for visible light), the short wavelength dramatically increases the resolving power of the system and makes possible the visualization of viruses and fine cellular structures, often called the *ultrastructure* of cells. The practical limit of resolution of biological samples with the electron microscope is about 2 nm, which is 100× better than the resolving power of the light microscope. The drawback of the electron microscope is that the preparation methods kill the cells or organisms.

Two types of electron microscopes are currently in use. The **transmission electron microscope** (**TEM**) is used to view and record detailed structures such as organelles within cells. Ultrathin sections of the object must be prepared because the electron beam can penetrate matter only a very short distance. After embedding the specimen in a suitable plastic mounting medium or freezing it, scientists cut the specimen into sections with a diamond knife. In this manner, a single bacterium can be sliced, like a loaf of bread, into hundreds of thin sections. Several of the sections are placed on a small grid and stained with heavy metals such as lead and osmium to provide contrast. The microscopist then inserts the grid into the vacuum chamber of the microscope

Transmission electron microscope: a microscope in which an electron beam passes through an ultrathin slice of an object.

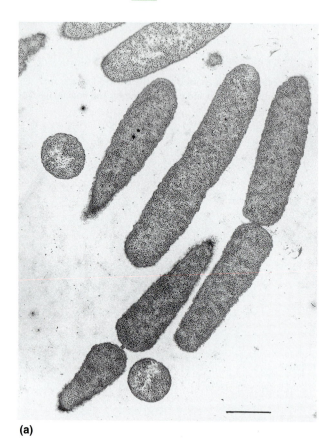

(a)

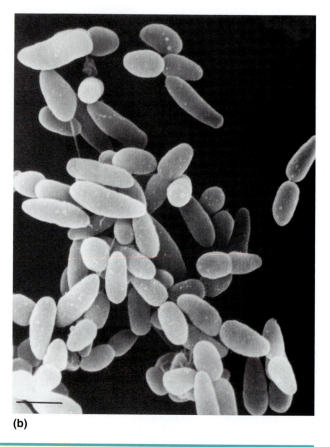

(b)

FIGURE 3.20

Transmission and Scanning Electron Microscopy Compared

The bacterium *Pseudomonas aeruginosa* as seen with two types of electron microscopy. (a) A view of sectioned cells seen with the transmission electron microscope. (Bar = 1.0 μm.) (b) A view of whole cells seen with the scanning electron microscope. (Bar = 2.0 μm.) The difference in perspective with the two microscopes is clear.

Scanning electron microscope:
a microscope in which an electron beam sweeps across the surface of an object.

and focuses a 100,000-volt electron beam on one portion of a section at a time. An image forms on the screen below or can be recorded on film. A photograph, called an *electron micrograph*, may be enlarged with enough resolution to achieve a total magnification of over 20 million ×.

The **scanning electron microscope** (**SEM**) was developed in the late 1960s to enable researchers to see the surfaces of objects in the natural state and without sectioning. The specimen is placed in the vacuum chamber and covered with a thin coat of gold. The electron beam then scans across the specimen and knocks loose showers of electrons that are captured by a detector. An image builds line by line, as in a television receiver. Electrons that strike a sloping surface yield fewer electrons, thereby producing a darker contrasting spot and a sense of three dimensions. The resolving power of the conventional SEM is about 7 nm and magnifications with the SEM are limited to about 50,000 ×. However, the instrument provides vivid and undistorted views of an organism's surface details. FIGURE 3.20 shows the same organism viewed with the TEM and SEM.

The electron microscope has added immeasurably to our understanding of the structure and function of microorganisms by letting us penetrate their innermost

TABLE 3.5

A Comparison of Various Types of Microscopy

TYPE OF MICROSCOPY	SPECIAL FEATURE	APPEARANCE OF OBJECT	MAGNIFICATION RANGE	OBJECTS OBSERVED
Bright-field	Visible light illuminates object	Stained microorganisms on clear background	100×–1,000×	Arrangement, shape, and size of killed microorganisms (except viruses)
Phase-contrast	Special condenser throws light rays "out of phase"	Unstained microorganisms with contrasted structures	100×–1,000×	Internal structures of live, unstained eukaryotic microorganisms
Dark-field	Special condenser scatters light	Unstained microorganisms on dark background	100×–1,000×	Live, unstained microorganisms (e.g., spirochetes)
Fluorescence	UV light illuminates fluorescent-coated objects	Fluorescing microorganisms on dark background	100×–1,000×	Outline of microorganisms coated with fluorescent-tagged antibodies
Transmission electron microscopy	Short-wavelength electron beam penetrates sections	Alternating light and dark areas contrasting internal cell structures	100×–200,000×	Ultrathin slices of microorganisms and internal components of eukaryotic and prokaryotic cells
Scanning electron microscopy	Short-wavelength electron beam knocks loose electron showers	Microbial surfaces	10×–50,000×	Surfaces and textures of microorganisms and cell components

secrets. In the chapters ahead, we will encounter many of the fine structures displayed by electron microscopy, and we will better appreciate microbial physiology as it is defined by microbial structures.

The various types of light and electron microscopy are compared in **TABLE 3.5**.

Note to the Student

As microbiology continues to capture headlines in the public media, it becomes increasingly common to see microbial names in print. And all too often the name is misspelled, miswritten, or misinterpreted. For example, a newspaper once reported that the bacterium *Staphylococcus aureus*, a well-known cause of skin infections, was a recurring problem in a local hospital. In the vernacular of our times, the disease is often called a "staph infection." However, the newspaper reported it as a "staff infection."

I believe that if one is to write or speak microbiology, one should try to do it correctly. I would recommend, therefore, that you pay particular attention to the microbial names as we encounter them and that you develop the habit of pronouncing them and writing them correctly. You should also insist that magazines and newspapers follow the accepted standards for writing the names of microorganisms. A letter to the editor does much to improve the publication's awareness of its error. You should not be reluctant to send one.

Summary of Key Concepts

3.1 A BRIEF SURVEY OF MICROORGANISMS

- **Prokaryotes and Eukaryotes Have Some Structures that Are Universal.** Bacteria are prokaryotes, while fungi, protozoa, and algae are eukaryotes. Some structural characteristics apply to all prokaryotes and eukaryotes. For example, universal structures include DNA organized as chromosomes, a cell membrane that surrounds the cytoplasm, and ribosomes to make proteins.

- **Prokaryotes and Eukaryotes Are Structurally Distinct.** Eukaryotes, including the fungi, protozoa, and algae, contain many organelles that are absent in the prokaryotes. These organelles include the endoplasmic reticulum, Golgi apparatus, and lysosomes. Prokaryotes also lack mitochondria and chloroplasts, although they may be able to carry out similar biochemical reactions on the cell membrane. The eukaryotes have a cytoskeleton. Although many prokaryotes and eukaryotes are motile, the structures are built differently between the two groups of organisms. Also, all microorganisms, except the protozoa, have cell walls whose composition differs for each group.

- **Bacteria Are among the Most Abundant Organisms on Earth.** Most bacteria play beneficial roles in the environment, although there are pathogenic species. Among the smallest of the bacteria are the rickettsiae, chlamydiae, and mycoplasmas. The cyanobacteria can carry out photosynthesis. Their overabundance can cause "blooms."

- **Protista (Protozoa and Unicellular Algae) Are Very Diverse Eukaryotic Microbes.** Many protozoa help decompose dead organisms and recycle nutrients. They come in a tremendous variety of shapes and sizes, some of which can carry out photosynthesis. A few species of protozoa are pathogenic. The unicellular algae carry out photosynthesis. The diatoms and dinoflagellates are part of the phytoplankton. During times of "red tides," dinoflagellates can cause paralytic shellfish poisoning in humans.

- **Fungi Have a Distinctive Growth Style.** The fungi include the yeasts and molds. A unique characteristic of the molds in their growth as networks of hyphae that form a mycelium. Many fungi are decomposers but some do cause plant or animal diseases.

- **Viruses Are the Smallest Microorganisms.** The viruses lack the structure seen in prokaryotic and eukaryotic cells, consisting of nucleic acid (DNA or RNA) and a protein shell. They can reproduce only when they infect living cells, often resulting in tissue destruction.

3.2 CATALOGING MICROORGANISMS

- **Classification Attempts to Catalog Organisms.** Many systems of classification have been devised to catalog organisms. The work of Aristotle, Linnaeus, and Haeckel represents early efforts.

- **The Five-Kingdom System Recognizes Prokaryotic and Eukaryotic Differences.** One of the currently accepted schemes is the five-kingdom system proposed by Whittaker. In this classification, bacteria are placed in the kingdom Monera, protozoa and algae in the kingdom Protista, and fungi in the kingdom Fungi.

- **Classification Uses a Hierarchical System.** Organisms are properly classified using a standardized hierarchical system from species (the least inclusive) to the kingdom (the most inclusive).

- **The Three-Domain System Places the Monera in Separate Lineages.** Based on several molecular and biochemical differences within the kingdom Monera, Woese proposed a three-domain system. The Kingdom Monera is separated into two domains, the Eubacteria and Archaea. The remaining kingdoms (Protista, Fungi, Plantae, and Animalia) are placed in the domain Eukarya.

- **Bacterial Taxonomy Now Includes Molecular Criteria.** *Bergey's Manual* is the standard reference to identify and classify bacteria. Criteria have included traditional characteristics, but modern molecular methods have led to a reconstruction of evolutionary events and organism relationships.

- **Nomenclature Gives Scientific Names to Organisms.** Part of an organism's binomial name is the genus name; the remaining part is an adjective, the specific epithet that describes the genus name. Thus, a species name consists of the genus and specific epithet.

3.3 MICROSCOPY

- **Most Microorganisms Are in the Micrometer Size Range.** Another criterion of a microorganism is its size, a characteristic that varies among members of different groups. The micrometer (μm) is used to measure the dimensions of bacteria, protozoa, and fungi. The nanometer (nm) is commonly used to express viral sizes.

- **Light Microscopy Uses Visible Light to Resolve Objects.** The instrument most widely used to observe microorganisms is the light microscope. Light passes through several lens systems that magnify and resolve the object being observed.

- **Magnification and Resolution Define the Limits of What Is Visible.** Although magnification is important, resolution is key. The light microscope can magnify up to $1,000\times$ and resolve objects as small as $0.2\ \mu m$.

- **Staining Techniques Provide Contrast.** For bacteria, staining generally precedes observation. The simple, negative, Gram, and other staining techniques can be used to impart contrast and determine structural or physiological properties.

- **Light Microscopy Has Other Optical Configurations.** Microscopes employing phase-contrast, dark field, and fluorescence optics have specialized uses in microbiology to contrast cells without staining.

- **Electron Microscopy Provides Detailed Images of Cells and Cell Parts.** To increase resolution and achieve extremely high magnification, the electron microscope employs a beam of electrons instead of a beam of light. To observe internal details (ultrastructure), the transmission electron microscope is most often used; to see whole objects, the scanning electron microscope is useful.

Questions for Thought and Discussion

Answers to selected questions can be found in Appendix C.

1. A student is asked on an examination to write a description of the protozoa. She blanks out. However, she remembers that protozoa are eukaryotes, and she recalls the properties of eukaryotes. How can she use this information to answer the question?

2. A local newspaper once contained an article about "the famous bacteria eecoli." How many things can you find wrong in this phrase? Rewrite the phrase correctly.

3. Microorganisms have been described as the most chemically diverse, the most adaptable, and the most ubiquitous organisms on Earth. Although your knowledge of microorganisms still may be limited at this point, try to add to this list of "mosts."

4. A student is performing the Gram stain technique in the laboratory. In reaching for the alcohol bottle in step 3, he inadvertently takes the water bottle and proceeds with the technique. What will be the colors of gram-positive and gram-negative bacteria at the conclusion of the technique?

5. Prokaryotes lack the cytoplasmic organelles commonly found in the eukaryotes. Provide a reason for this structural difference.

6. In 1997, a writer from *The New York Times* described microorganisms as the "New Yorkers of the living world: irrepressible, vilified, and able to reproduce in wildly inhospitable environments." From your general knowledge of microorganisms, what characteristics do they have in common with the residents of the city where you live?

7. While working in the lab, a student notices that there is no applicator in the bottle of immersion oil. But then she unscrews the cap and, as she lifts it from the oil, a glass applicator appears. Why did the applicator escape her ability to see it?

8. A new bacteriology laboratory is opening in your community. What is one of the first books that the laboratory director will want to purchase? Why is it important to have this book?

9. While scanning a menu in an Italian restaurant, you notice an entire section entitled "Fungi." Among the choices are spaghetti with fungi, stuffed fungi, and fungi parmigiana. What will you receive if you order any of these?

10. Would the best resolution with a light microscope be obtained using red light ($\lambda = 680$ nm), green light ($\lambda = 520$ nm), or blue light ($\lambda = 480$ nm)? Explain your answer.

11. Christian Gram, developer of the Gram stain technique, originally intended his technique for use in distinguishing bacteria from cellular nuclei in slides of patients' tissues. Develop a scenario in which he might have realized that it would be more useful for separating bacteria into groups.

12. In 1987, in a respected science journal, an author wrote, "Linnaeus gave each life form two Latin names, the first denoting its genus and the second its species." A few lines later, the author wrote, "Man

was given his own genus and species *Homo sapiens.*" What is conceptually and technically wrong with both statements?

13. A student of general biology observes a microbiology student using immersion oil and asks why the oil is used. "To increase the magnification of the microscope" is the reply. Would you agree? Why?

14. Assume that a small spherical bacterium has a diameter of 1 μm. A million of these bacteria therefore would fit in the space occupied by a meter (39.39 inches, or slightly more than 3 feet). Suppose you were to count each bacterium in this space at a rate of one bacterium per second. How long would it take you to count all the bacteria? Does this help you conceptualize how small a bacterium truly is?

15. A 1980s *Far Side* cartoon by Gary Larson was entitled "Single Cell Bar." Assuming you were the artist, how many different cell shapes and sizes would you include in your drawing?

16. Every state has an official animal, flower, or tree, but only Oregon has a bacterium named in its honor: *Methanohalophilus oregonese.* The species modifier *oregonese* is obvious, but can you decipher the meaning of the genus name?

http://microbiology.jbpub.com

The site features **eLearning,** an on-line review area that provides quizzes and other tools to help you study for your class. You can also follow useful links for in-depth information, or just find out the latest microbiology news.

Review

The types of microorganisms; their classification, nomenclature, and size; and the methods for observing microorganisms were the major themes of this chapter. To test your understanding of these themes, match the statement on the left to the term on the right by placing the letter of the term in the available space. Appendix D contains the correct answers.

_____ 1. System of nomenclature used for micro-organisms and other living things.

_____ 2. Unit of measurement used for viruses and equal to a billionth of a meter.

_____ 3. Major group of organisms whose cells have no nucleus or organelles in the cytoplasm.

_____ 4. Bacteria capable of photosynthesis.

_____ 5. Devised the five-kingdom system of classification in which microorganisms are placed.

_____ 6. Type of microscope that uses a special condenser to split the light beam.

_____ 7. Type of electron microscope for which cell sectioning is not required.

_____ 8. Eukaryotes classified into groups according to how they move.

_____ 9. Prokaryotic microorganisms that have no cell wall.

_____ 10. Neither prokaryotes nor eukaryotes.

_____ 11. Kingdom in which the bacteria are classified.

_____ 12. Staining technique that differentiates bacteria into two groups.

_____ 13. Author of an early system of classification for bacteria.

_____ 14. Category into which two or more species of bacteria are grouped.

_____ 15. Coined the name Protista for microorganisms.

_____ 16. Considered to be unicellular, eukaryotic algae.

_____ 17. Used to write the binomial name of microorganisms.

_____ 18. Unit of measurement for bacteria and equal to a millionth of a meter.

_____ 19. Type of microscopy using UV light to excite a dye-coated specimen.

_____ 20. Staining technique in which the background is colored and the cells are clear.

A. Viruses

B. Italics

C. Gram

D. Mycoplasmas

E. Simple

F. Genus

G. Binomial

H. Boldface

I. Diatoms

J. Micrometer

K. Prokaryotes

L. Cyanobacteria

M. Dark-field

N. Scanning

O. Haeckel

P. Bergey

Q. Negative

R. Fungi

S. Eukaryote

T. Nanometer

U. Phase-contrast

V. Monera

W. Protozoa

X. Whittaker

Y. Transmission

Z. Fluorescence

The Bacteria

We live at the center of a microbial universe. On all sides, microscopic organisms surround us and make their presence felt—for good or ill. The useful species outnumber the harmful ones by thousands and are so valuable we could not live without them. The remaining species are agents of disease and death.

Only since the mid–1800s have scientists linked microorganisms to events of human importance (not surprisingly, because microorganisms can be seen only with a microscope). Before then, disease processes now attributed to microorganisms seemed to happen almost spontaneously and without apparent cause. Even thinking about disease in terms of microorganisms could be dangerous because it was heresy to consider disease as anything other than supernatural.

In Part 2 of this text, we shall focus on one group of microorganisms, the bacteria. These microorganisms have traditionally occupied an important niche in microbiology because scientists probably know more about bacteria than any other organisms. Bacteria have been involved in the great plagues of history, and for centuries their effects have captured the imagination of scientists and writers. Bacteria are easily studied in the laboratory, and their chemical activities have been charted and well documented. Also, many helpful ones play key roles in industrial processes. We often mean bacteria when we talk about "germs," and we need only consider how often we use that word to appreciate the significance of bacteria in our lives.

Small as they are, bacteria are endowed with the ability to perform certain acts characteristic of all living things. They take in food, grow, excrete waste products, reproduce, and die. In addition, bacteria are sensitive to external agents and stimuli, and they respond in some fashion. In Chapter 4, we shall survey their structural frameworks and growth patterns, and in Chapter 5, we examine their biochemical activities. Chapter 6 is devoted to the genetics of bacteria, while Chapter 7 explores the modern fields of genetic engineering and bacterial genomics. The discussions in Part 2 have broad significance not only in medicine, research, and industry, but also in our daily lives.

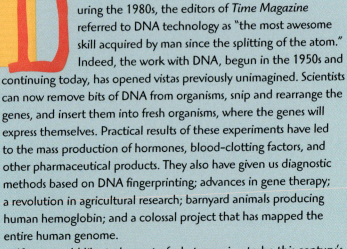

During the 1980s, the editors of *Time Magazine* referred to DNA technology as "the most awesome skill acquired by man since the splitting of the atom." Indeed, the work with DNA, begun in the 1950s and continuing today, has opened vistas previously unimagined. Scientists can now remove bits of DNA from organisms, snip and rearrange the genes, and insert them into fresh organisms, where the genes will express themselves. Practical results of these experiments have led to the mass production of hormones, blood-clotting factors, and other pharmaceutical products. They also have given us diagnostic methods based on DNA fingerprinting; advances in gene therapy; a revolution in agricultural research; barnyard animals producing human hemoglobin; and a colossal project that has mapped the entire human genome.

If you would like to be part of what promises to be this century's great technology, then microbiology is the place to be. You would be well advised to take a course in biochemistry, as well as one in genetics. Courses in physiology and neurobiology are also helpful. Employers will be looking for individuals with good laboratory skills, so be sure to take as many lab courses as you can. And don't be afraid to become a "lab-rat" (the scientific equivalent of basketball's "gym-rat").

You may enter the biotechnology field with an associate's, bachelor's, master's, or doctoral degree. This is because there are so many levels at which individuals are hired. An employer will be looking for work experience, which you can obtain by assisting a senior scientist, doing an internship, or working summers in a biotech firm (usually for slave wages). The campus research lab is another good place to obtain work experience. It also might be a good idea to sharpen your writing skills, since you will be preparing numerous reports. As Chapter 7 explains, the novel and imaginative research that established biotechnology was founded in microbiology, and it continues to call on microbiology for its continuing growth.

Bacterial Structure and Growth

Our planet has always been in the "Age of Bacteria," ever since the first fossils—bacteria of course—were entombed in rocks more than 3 billion years ago. On any possible, reasonable criterion, bacteria are—and always have been—the dominant forms of life on Earth.

—Paleontologist Stephen J. Gould (1941–2002)

IN A 1683 LETTER to the Royal Society of London, Anton van Leeuwenhoek described microscopic "streaks and threads" among his animalcules. The streaks and threads remained nameless until 1773, when the Danish scientist Otto Frederik Müller christened them "bacilli." Bacilli is the plural form of the Latin *bacillus,* meaning "little rod."

However not all bacilli were rods. Some were spiral and some were circular in shape, and the word *bacilli* would not do. Therefore, in the 1850s, the French investigator Casimir Davaine began calling the microscopic creatures "bacteria," even though this derivative of the Greek *bacterion* also means rod. Time has a way of sorting out confusion, and in the next few decades "bacteria" came to refer to all the microorganisms in that group, and the word "bacillus" was reserved for rod forms only.

The terminology problem was resolved just in time because, in the 1850s, bacteria were attracting considerable attention. At that time, Pasteur's work showed that bacteria are chemical factories capable of bringing about significant changes in nature, and in the 1870s, Koch's experiments verified their link to infectious disease. In the late 1800s, the rush to locate and isolate the bacterial causes of infectious disease was unlike anything previously experienced in medical science.

As it happened, none of these pioneers of microbiology could see what lay ahead. As the twentieth century unfolded, scientists found that bacteria have structures and growth patterns far beyond what had been imagined in the years before. With the development of the electron microscope in the 1940s and the revelations of the second Golden Age of microbiology, bacteria revealed themselves as more than simple sticks and rods. Scientists uncovered a wealth of microscopic and submicroscopic details in bacteria and showed how the very minute bacteria can be as complex as the very large, visible organisms. As we shall see in this chapter, a study of the structural features of bacteria (we will focus on the domain Eubacteria) provides a window to their activities and illustrates how bacteria relate to other living organisms.

4.1

The Shapes and Arrangements of Bacteria

Bacteria vary greatly in size, shape, and arrangement of cells. As described in Chapter 3, these characteristics are best studied by viewing stained cells with the light microscope. Such studies show that most bacteria, including the clinically significant ones, appear in one of three different shapes: the rod, the sphere, and the spiral (FIGURE 4.1).

BACILLI HAVE A CYLINDRICAL SHAPE

As suggested by Müller, the rod is known as a **bacillus** (pl., bacilli). In various species of rod-shaped bacteria, the cylindrical cell may be as long as 20 μm or as short as 0.5 μm. Certain bacilli are slender, such as those of *Salmonella typhi* that cause typhoid fever; others, such as the agent of anthrax (*Bacillus anthracis*), are rectangular with squared ends; still others, such as the diphtheria bacilli (*Corynebacterium diphtheriae*), are club shaped. Most rods occur singly, but some are arranged into long chains called **streptobacilli** (*strepto* = "chains") (Figure 4.1a). Realize that we are using the word *bacillus* in two different ways: to denote a rod-shaped bacterial cell, and as a genus name.

FIGURE 4.1

Variations in Bacterial Shape and Cell Arrangements

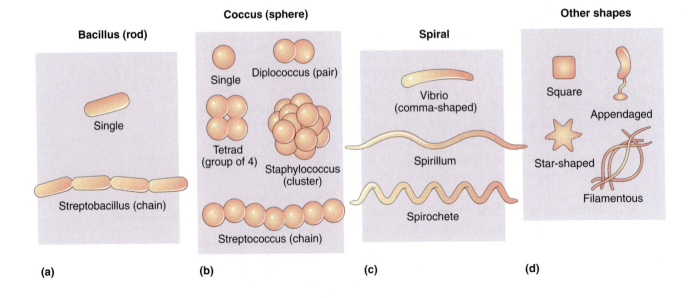

Bacillus (rod)
Single
Streptobacillus (chain)

Coccus (sphere)
Single   Diplococcus (pair)
Tetrad (group of 4)   Staphylococcus (cluster)
Streptococcus (chain)

Spiral
Vibrio (comma-shaped)
Spirillum
Spirochete

Other shapes
Square   Appendaged
Star-shaped   Filamentous

(a)   (b)   (c)   (d)

COCCI FORM A VARIETY OF ARRANGEMENTS

A spherically shaped bacterium is known as a **coccus** (pl., cocci), a term derived from the Greek *kokkos*, meaning, "berry." Cocci tend to be quite small, being only 0.5 μm to 1.0 μm in diameter. Although they are usually round, they also may be oval, elongated, or indented on one side.

Many bacterial species that are cocci stay together after division and take on cellular arrangements characteristic of the organism (Figure 4.1b). Those cocci that remain in pairs after reproducing are called **diplococci**. The organism that causes gonorrhea, *Neisseria gonorrhoeae*, and one type of bacterial meningitis (*N. meningitidis*) are examples. Cocci that remain in chains are called **streptococci** (FIGURE 4.2a). Certain species of streptococci are involved in strep throat (*Streptococcus pyogenes*) and tooth decay (*S. mutans*), while other species are harmless enough to be used for producing dairy products such as yogurt (*S. lactis*). Another arrangement of cocci is the **tetrad**, consisting of four cocci forming a square. A cube-like packet of eight cocci is called a **sarcina** (*sarcina* = "bundle"). *Micrococcus luteus*, a common inhabitant of the skin, is one example (FIGURE 4.2b). Other cocci may divide randomly and form an irregular grape-like cluster of cells called a **staphylococcus** (*staphyle* = "grape"). A well-known example, *Staphylococcus aureus,* is often a cause of food poisoning, toxic shock syndrome, and numerous skin infections (FIGURE 4.2c). The latter are known in the modern vernacular as "staph" infections. Notice again that the words *streptococcus* and *staphylococcus* can be used to describe cell shape and arrangement, or a genus of bacterium.

SPIRALS AND OTHER SHAPES ALSO EXIST

The third major shape of bacteria is the **spiral**, which can take one of three forms (Figure 4.1c). Certain spiral bacteria called **vibrios** are curved rods that resemble commas. The cholera-causing organism *Vibrio cholerae* is typical. Other spiral bacteria called **spirilla** (sing., spirillum) have a helical shape with a thick, rigid cell wall

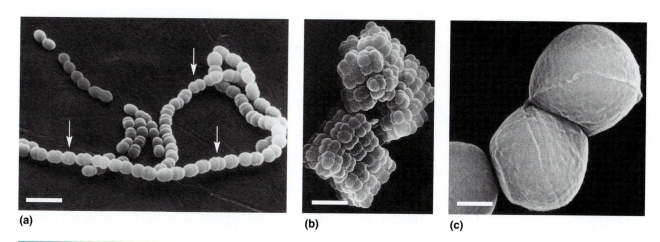

(a) (b) (c)

FIGURE 4.2

Scanning Electron Micrographs of Bacterial Cocci

(a) A streptococcus such as is found in the intestine. Note the characteristic string-of-beads appearance of the chain. The arrows indicate three of the many positions at which the cocci are undergoing division. (Bar = 10 μm.) (b) Clusters of sarcinae in cube-like packets of eight or more. (Bar = 10 μm.) (c) Two cocci from a cluster of *Staphylococcus aureus* cells. (Bar = 0.3 μm.)

FIGURE 4.3

Electron Micrograph of a Freshwater Spirillum

The spiral shape is seen clearly in this cell, and flagella are visible at the poles of the cell. (Bar = 10 μm.)

and flagella that assist movement (**FIGURE 4.3**). Those spiral-shaped bacteria known as **spirochetes** have a thin, flexible cell wall but no flagella in the traditional sense. Movement in these organisms occurs by contractions of axial filaments that run the length of the cell. The organism causing syphilis, *Treponema pallidum*, typifies a spirochete. The spiral-shaped bacteria can be from 1 μm to 100 μm in length.

In addition to the bacillus, coccus, and spiral shapes, other variations exist (Figure 4.1d). In the genus *Caulobacter*, there are appendaged bacteria; members of the genus *Nocardia* consist of branching filaments; and some archaea have square and star shapes.

Stained cells or live cells viewed with the light microscope do not show striking internal structure. When the electron microscope is used, however, resolution increases down to 2 nm allowing scientists to observe a level of bacterial detail not possible with the light microscope.

4.2 The Structure of Bacteria

Although bacteria observed with the light microscope appear to have little structure, they actually have a surprisingly complex structure. Some structures are necessary for growth and survival. Others aid in cell movement or enhance the ability of bacteria to cause disease.

We shall examine some of the common bacterial structures found in an idealized bacterium, as no single species contains all the structures (**FIGURE 4.4**). First, we will examine structures on or extending from the surface of the cell (bacterial flagella, pili, cell envelope). Then, we will explore the structures within the cytosol (ribosomes, granules, nucleoid, and plasmids).

BACTERIAL FLAGELLA PROVIDE MOTILITY

Numerous species of gram-positive and gram-negative rods and spirilla, and a few cocci, are capable of independent movement called motility. These movements are

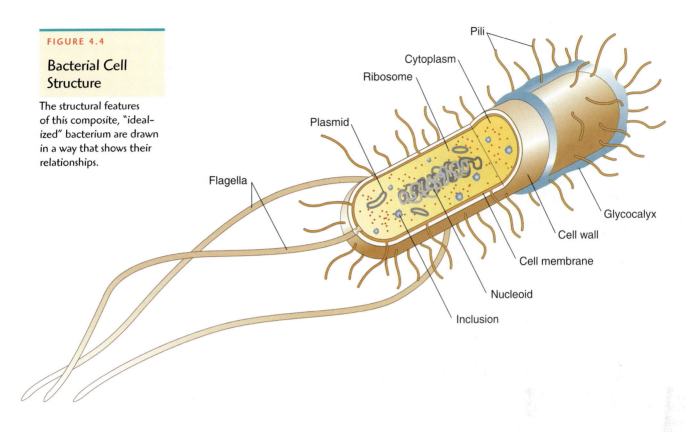

FIGURE 4.4

Bacterial Cell Structure

The structural features of this composite, "idealized" bacterium are drawn in a way that shows their relationships.

carried out using remarkable nanomachines called **flagella** (FIGURE 4.5). Each flagellum, composed of a filament, hook, and basal body, turns like a rotary engine to generate a propeller-like rotation.

The hollow filament is composed of long, rigid and helical strands of protein subunits called **flagellin**. By contrast, in the flagella in eukaryotic microorganisms, such as in some protozoa and algae, the strands are flexible, and the fibers, which are made of a different protein called tubulin, are elongated and able to slide past one another.

Electron microscopy reveals that the flagella are anchored in the cell membrane (Figure 4.5a). The filament attaches to a hook-like structure that is connected to a basal body. The **basal body** is an assembly of more than 20 different proteins that form a central rod and set of enclosing rings. Gram-positive bacteria have one ring embedded in the cell membrane and one in the cell wall, while gram-negative bacteria have a pair of rings embedded in the cell membrane and another pair in the cell wall.

The number and placement of flagella vary by species (Figure 4.5b). Bacteria can be **monotrichous**, possessing a single flagellum at one end; **lophotrichous**, having a group of two or more flagella at one end of the cell; **amphitrichous**, having a single or group of flagella at both ends; and **peritrichous**, having flagella at many locations over the cell surface. The arrangement of flagella is one that can be used to help classify bacterial species.

Bacterial flagella range in length from 10 μm to 20 μm and are many times longer than the diameter of the cell, as Figure 4.5c shows. However, the flagellum is only about 20 nm thick and cannot be seen with the light microscope unless stained. However, their existence can be inferred by watching live bacteria dart about with dark-field microscopy.

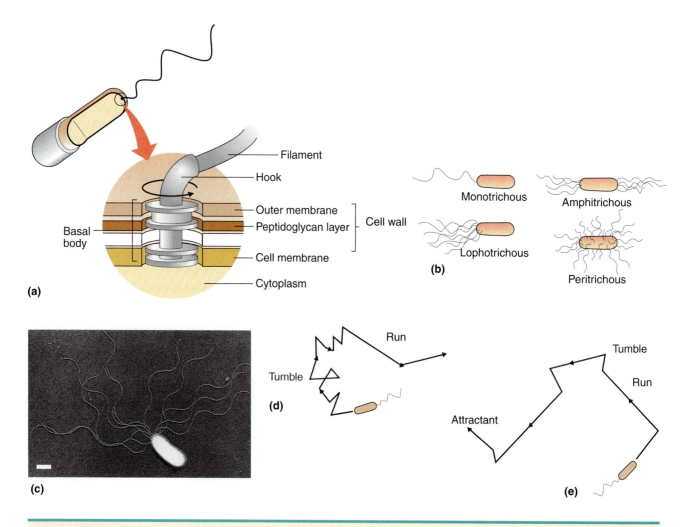

(a)

(b)
Monotrichous　　Amphitrichous
Lophotrichous　　Peritrichous

Filament
Hook
Outer membrane
Peptidoglycan layer — Cell wall
Cell membrane
Cytoplasm
Basal body

(c)

(d)
Run
Tumble

(e)
Tumble
Run
Attractant

FIGURE 4.5

Details of the Bacterial Flagellum and Chemotaxis

(a) In a gram-negative bacterium, shown here, the flagellum is attached to the cell wall and membrane by a complex mechanism of structures. (In a gram-positive bacterium, the arrangement is slightly different.) (b) Various configurations of flagella occur among bacteria. Monotrichous bacteria possess a single flagellum, amphitrichous bacteria have flagella at both poles of the cell, and lophotrichous organisms have them at one end. Peritrichous bacteria are surrounded by flagella. (c) A transmission electron micrograph of *Pseudomonas marginalis* showing polar flagella. (Bar = 1 μm.) Note that the flagella are many times the length of the bacillus and appear in a characteristic wavy format. This bacterium is lophotrichous. (d) Rotation of the flagellum counterclockwise causes the bacterium to "run," while rotation of the flagellum clockwise causes the bacterium to "tumble," as shown. (e) During chemotaxis to an attractant, such as sugar, flagellum behavior leads to longer runs and fewer tumbles, which will result in biased movement toward the attractant.

Most bacterial flagella rotate like a propeller, generating up to 1,500 rpm as they push the cell through its aqueous environment. When the flagellum rotates counterclockwise, the organism moves straight ahead in short bursts called "runs" (Figure 4.5d). These runs can last for a few seconds and the cells can move up to 10 body lengths per second (the fastest human can run about 5–6 body lengths per second). When the flagellum rotates clockwise, the bacterium "tumbles" randomly for a second. Then another run occurs. These movements require a considerable amount of energy.

What advantage is gained by cells having flagella? In nature, bacteria often are attracted to chemical nutrients, called attractants. The movement toward an attractant, called **chemotaxis**, is characterized by lengthened runs and shortened tumbles (Figure 4.5e). In some diseases like cholera, flagella enable the bacilli within the human body to move among the tissues and escape phagocytosis by white blood cells.

PILI ARE STRUCTURES USED FOR ATTACHMENT

Numerous short, thin hair-like fibers, called **pili** (sing., pilus), protrude from the surface of many gram-negative bacteria (**FIGURE 4.6**). These structures are shorter and straighter than bacterial flagella and are comprised primarily of protein subunits called **pilin**. The function of the pili is to attach the bacterial cell to specific surfaces, including animal tissues (**MicroFocus 4.1**). This requires that the pili have specialized proteins called **adhesins** located at their tips. For example, the adhesins on the pili of *Neisseria gonorrhoeae* cells specifically anchor the cells to surfaces of the urogenital tract. In this way, pili enhance a bacterium's ability to colonize an area and cause disease.

In 1997, researchers identified an adhesin of *E. coli* pili and used it to produce a vaccine. When injected into healthy mice, the adhesin induced the animal's immune system to produce antibodies, which when infected with *E. coli* united with the adhesin and prevented the cells from attaching. Without the chemical mooring line lashing the bacteria to the mouse cells, the bacterium could not infect the tissue. Antibodies might be used together with antibiotics to suppress attachment and growth in gram-negative bacteria such as *Neisseria gonorrhoeae* where resistance to antibiotics is an ongoing concern in gonorrhea therapy.

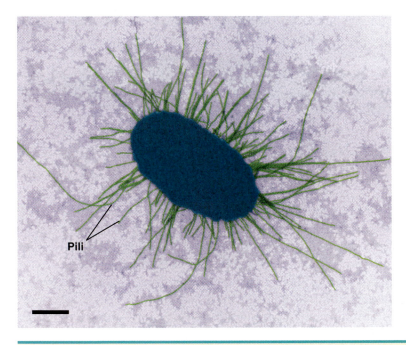

Pili

FIGURE 4.6

Example of Pili on Bacteria

False color transmission electron micrograph of a heavily piliated cell of *Escherichia coli*. The pili help the cell attach to surfaces. (Bar = 0.5 μm.)

MicroFocus 4.1

DIARRHEA DOOZIES

They gathered at the clinical research center at Stanford University to do their part for the advancement of science (and earn a few dollars as well). They were the "sensational sixty"—sixty young men and women who would spend three days and nights and earn $300 to help determine whether hair-like structures called pili have a significant place in disease.

A number of nurses and doctors were on hand to help them through their ordeal. The students would drink a fruit-flavored cocktail containing a special diarrhea-causing strain of *Escherichia coli*. Thirty cocktails had *E. coli* with normal pili, while thirty had *E. coli* with pili mutated beyond repair. The hypothesis was that bacteria with the thread-like

pili should latch onto intestinal tissue and cause diarrhea, while those with mutated pili should be swept away by the rush of intestinal movements and not cause intestinal distress. At least that's what the sensational sixty would either verify or prove false.

On that fateful day in 1997, the experiment began. Neither the students nor the health professionals knew who was drinking the diarrhea cocktail and who was getting the "free pass"; it was a so-called double-blind experiment. Then came the waiting. Some experienced no symptoms, but others felt the bacterial onslaught and clutched at their last remaining vestiges of dignity. For some it was three days of hell, with nausea, abdominal cramps, and numerous

bathroom trips; for others, luck was on their side, and investing in a lottery ticket seemed like a good idea.

When it was all over, the numbers appeared to bear out the hypothesis: The great majority of volunteers who drank the mutated bacteria experienced no diarrhea, while the great majority of those who drank the normal bacteria had attacks of diarrhea, in some cases real doozies. All appeared to profit from the experience: The scientists had some real-life evidence that pili contribute to infection; the students made their sacrifice to science and pocketed $300 each; and the local supermarket had a surge of profits from unexpected sales of toilet paper, Pepto-Bismol, and Immodium.

Besides these pili involved in attachment, certain bacteria produce **conjugation pili** that aid in the transfer of genetic material between bacterial cells (Chapter 7). These structures are longer and usually only one or a few are produced. It should be noted that some microbiologists use the word **fimbriae** (sing., fimbria) for the attachment pili and reserve the word pili for structures functioning in DNA transfer.

The last three external structures, the glycocalyx (if present), cell wall, and cell membrane, sometimes are referred to collectively as the **cell envelope**. Let's examine these structures next.

THE GLYCOCALYX IS A STICKY LAYER COATING MANY BACTERIA

Many gram-positive and gram-negative rods and cocci secrete an adhering layer of polysaccharides or polysaccharides and small proteins called the **glycocalyx**. The layer can be thick and tightly bound to the cell, in which case it is known as a **capsule**. When thinner, diffuse, and less tightly bound, it is referred to as a **slime layer**. The glycocalyx can be seen by light microscopy when observing a negative stain preparation or by transmission electron microscopy (FIGURE 4.7).

The glycocalyx (meaning "sweet coat") serves as a buffer between the cell and the external environment. Because of its high water content, the glycocalyx averts potential cell desiccation and traps nutrients. The glycocalyx also can contribute to the disease process. The glycocalyx of *V. cholerae*, for example, attaches the cells to the intestinal wall of the host. Bacteria that are encapsulated, such as those of *Streptococcus pneumoniae* (a principal cause of bacterial pneumonia) and *Bacillus anthracis*, cannot be easily engulfed by white blood cells during phagocytosis. When the capsule is experimentally removed, the organism is harmless because the

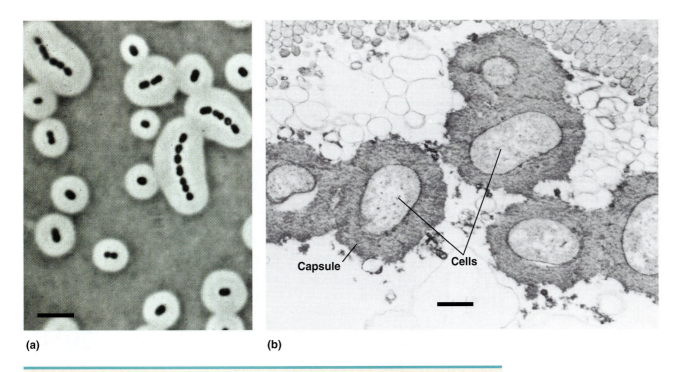

(a)   (b)

FIGURE 4.7

The Bacterial Glycocalyx

(a) Demonstration of the presence of a capsule in an *Acinetobacter* species by negative staining and observed by phase–contrast microscopy. (Bar = 10 µm.) (b) A transmission electron micrograph of *Escherichia coli* from the intestine of an animal. Each bacillus is surrounded by a thick capsule. (Bar = 1.0 µm.)

cells are easily engulfed and destroyed. Scientists believe that the repulsion between bacterium and white blood cell is due to strong negative charges in the capsule and white blood cell.

A slime layer usually contains a mass of tangled fibers of a polysaccharide called **dextran**. The fibers attach the bacterium to tissue surfaces. A case in point is *Streptococcus mutans,* an important cause of tooth decay that was discussed in MicroFocus 2.3.

A glycocalyx is a key element in the formation of biofilms. Introduced in Chapter 1, biofilms are embedded microcolonies of bacteria attached to such surfaces as industrial pipelines, sewage-treatment systems, medical instruments, or body tissues. In biofilms, a carbohydrate matrix binds the microcolonies together, and surrounding water channels deliver nutrients and remove waste. In the example for dental plaque, a slime layer about 10 µm thick lies at the base of the biofilm and attaches the bacteria to the surface. Living within such biofilms effectively shields bacteria from the body's immune defenses, as well as from antibiotics and other therapies. **FIGURE 4.8** details a medical consequence of a biofilm. Dental caries (cavities) and urinary tract infections also are consequences of biofilms (Chapter 11).

In food products, the slime-producing bacteria may cause an unsightly and distasteful experience. For instance, the glue-like slime produced by *Alcaligenes viscolactis* accumulates in milk, causing it to become thick and stringy. The result is "ropy milk." Bread may also become ropy if contaminated with slime-producing *Bacillus subtilis.*

TEXTBOOK CASES

FIGURE 4.8

An Outbreak of *Enterobacter cloacae* Infection Associated with Biofilms in the Waste Drainpipe of a Hemodialysis Machine

The incident points up the need for the correct maintenance of machines through which fluid flows.

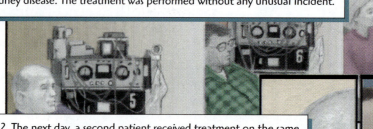

1. During September 1997, a patient at a hospital in Montreal, Canada received treatment on a hemodialysis machine to help relieve the effects of kidney disease. The treatment was performed without any unusual incident.

2. The next day, a second patient received treatment on the same hemodialysis machine. His treatment also went normally, and he returned to his usual activities after the session was completed.

3. In the following days, both patients experienced infection of the bloodstream. They had high fever, muscular aches and pains, sore throat, and impaired blood circulation. Because the symptoms were severe, the patients were hospitalized.

4. Microbiologists inspected the hemodialysis machine used by the patients and discovered biofilms containing *Enterobacter cloacae*. The identical gram-negative rod was isolated from samples taken from the patients' bloodstreams.

5. In the following days, public health officials visited with seven other adult patients who had used the hemodialysis machine. They discovered that all seven had been ill with the same type of bloodstream infection.

6. Further study indicated that hospital personnel were disinfecting the machines inadequately. Health officials began a hospital education program to ensure that further outbreaks of infection were curtailed.

THE CELL WALL PROVIDES SHAPE AND PROTECTION

The fact that all bacteria except the mycoplasmas have a cell wall suggests the critical role this structure must play. Indeed, the **cell wall** protects the cell and enhances its survival (MicroFocus 4.2). The cell wall also gives the cell its characteristic shape.

A major role for the cell wall is to prevent cells from bursting because of the internal pressure. Most microbes live in an environment where there are more dissolved materials inside the cell than outside. This **hypertonic** condition in the cell means that water diffuses inward and accounts for the increased internal pressure, which may be up to 20 times the external pressure. Without a cell wall, the cell would burst

Hypertonic: a condition where one of two solutions has a greater concentration of solutes.

MicroFocus 4.2

ASLEEP FOR 11,000 YEARS

It was 2 feet long, reddish brown, and tube shaped. It could have been a piece of rusted pipe, but it smelled so bad and had such a convoluted shape that the pipe theory was quickly discarded. Workers had found the object while building a new fourteenth hole at the Burning Tree Golf Course near Columbus, Ohio. While digging a mere 5 feet into the soil, they hit upon the skeletal remains of a mastodon, an elephant-like animal that lived 11,000 years ago. This tube-shaped object was near the animal's rib cage.

Paleontologists theorized that the object was probably part of the mastodon's intestinal tract. It so happened that Gerald Goldstein of Ohio Wesleyan University was visiting the site at the invitation of a friend involved in the excavation. He half jokingly suggested that something might still be alive in the intestinal contents, and he

was given a small bag of the smelly material. Back at his laboratory, he placed a sample of the intestinal contents on ordinary bacteriological medium and surprise, surprise...the next day the medium teemed with bacteria. The organism was *Enterobacter cloacae*, a well-known resident of the mammalian intestine.

In 1991, Goldstein announced his discovery to a skeptical scientific community. He determined that the *E. cloacae* was not a contaminant from the surrounding soil by searching for the organism in 12 samples from nearby sites. All 12 samples failed to yield *E. cloacae*. An independent "blind" analysis, in which scientists were not told the sources of the samples, confirmed the results.

Since then, several published research papers have demonstrated that the bacteria are definitely from the preserved intestinal contents, making the bacteria

the oldest known living bacteria recovered from nature (bacterial spores excluded). A 3-foot cap of clay had apparently sealed the site, and the chilly 7 °C temperature probably contributed to placing the bacteria in a state of suspended animation. As one writer suggested, "Move over Rip van Winkle. There's a new record for slumber time."

or undergo **lysis**. It is similar to blowing so much air into a balloon that the air pressure eventually bursts the balloon.

The eubacterial cell wall differs markedly from the walls of other microorganisms and plants in containing a network of **peptidoglycan** chains (FIGURE 4.9). This very large molecule, also called murein, is composed of alternating units of two amino-containing sugars, *N*-acetylglucosamine (NAG) and *N*-acetylmuramic acid (NAM). The carbohydrate backbone occurs in multiple layers connected by side chains of four amino acids and peptide cross-bridges (Figure 4.9a). Therefore, the many layers comprise one extremely large molecule. There are two predominant types of eubacterial cell walls, which provide the means to identify stained bacteria as gram-positive (blue-purple) and gram-negative (orange-red).

■ GRAM-POSITIVE BACTERIA. About 60 to 90 percent of the cell wall in gram-positive bacteria is peptidoglycan (Figure 4.9b). The abundance and thickness (25 nm) of this material may be one reason why they retain the crystal violet in Gram staining (Chapter 3). The peptidoglycan layer also contains an anionic polysaccharide derivative called **teichoic acid**. Teichoic acid helps link the peptidoglycan chains together.

The antibiotic penicillin interferes with the construction of the peptidoglycan of the cell wall in new cells, and they quickly lyse (FIGURE 4.10). That is why the antibiotic is used to treat strep throat infections caused by gram-positive *S. pyogenes* cells. However, penicillin has little effect on older cells whose walls already are formed. In these cases, lysozyme can destroy existing cells. **Lysozyme** is an enzyme

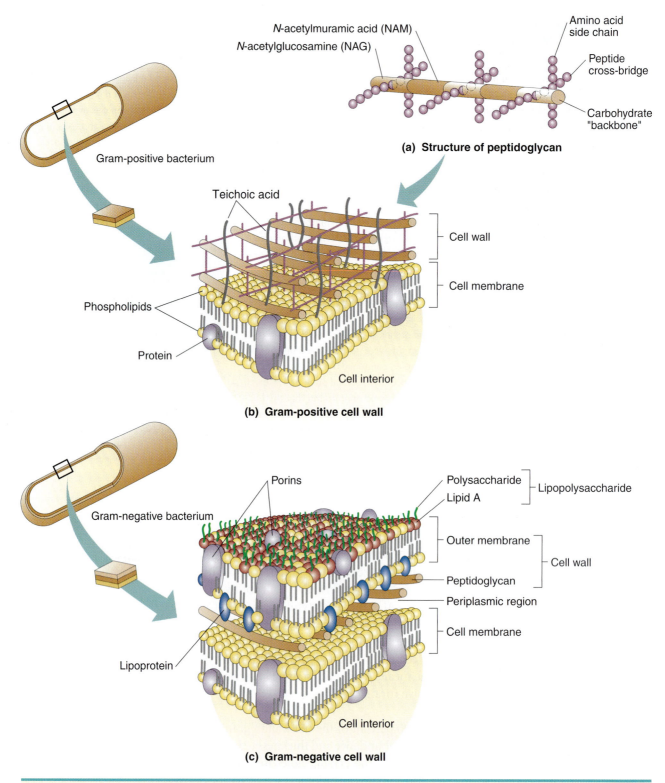

(a) **Structure of peptidoglycan**

(b) **Gram-positive cell wall**

(c) **Gram-negative cell wall**

FIGURE 4.9

A Comparison of the Cell Walls of Gram–Positive and Gram–Negative Bacteria

(a) The structure of peptidoglycan is shown as units of NAG and NAM joined laterally by amino acid cross–bridges and vertically by side chains of four amino acids. (b) The cell wall of a gram-positive bacterium is composed of peptidoglycan layers combined with teichoic acid molecules. (c) In the gram–negative cell wall, the peptidoglycan layer is much thinner, and there is no teichoic acid. Moreover, an outer membrane closely overlies the peptidoglycan layer so that the membrane and layer comprise the cell wall. Note the structure of the outer membrane in this figure. The outer half is unique in containing lipopolysaccharide and porin proteins.

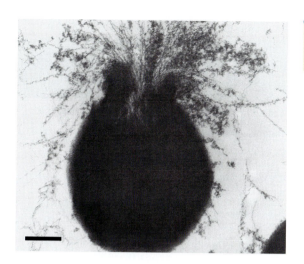

FIGURE 4.10

The Effect of Penicillin

A photomicrograph of a *Staphylococcus aureus* cell exploding on exposure to penicillin. (Bar = 0.25 μm.) The antibiotic has prevented construction of the peptidoglycan layer of the cell wall, and internal pressures have led to weakening and disruption of the cell membrane.

in human tears and saliva. It attacks the linkages between carbohydrates in the peptidoglycan layer, thus causing the cell wall to break down and the cell to lyse. In both cases, the effect is more dramatic in gram-positive bacteria because these organisms have more peptidoglycan.

■ **GRAM-NEGATIVE BACTERIA.** By contrast, the cell wall of gram-negative bacteria has no teichoic acid, and the thin (3 nm thick) peptidoglycan only accounts for about 10 percent of the cell wall (Figure 4.9c). This is one reason why it loses the crystal violet stain during Gram staining.

The outer portion of the wall consists of an **outer membrane** not found in gram-positive bacteria. This membrane bilayer consists of an inner layer of phospholipid and an outer layer of **lipopolysaccharide** (**LPS**) not found in any other organisms. In this unique molecule, the lipid portion (known as lipid A) is embedded in the outer half of the membrane bilayer. The lipid A portion is an **endotoxin**, causing fever and circulatory collapse when it is released from the bacterial cells into the bloodstream or digestive system of an infected host organism. *Salmonella* and *E. coli* are two endotoxin-producing bacteria.

Endotoxin:
a toxin retained in the outer membrane of certain gram-negative bacteria and released when they disintegrate.

The polysaccharide portion of the LPS is attached to the lipid A. Part of this polysaccharide, called the *O*-polysaccharide, is used to identify variants of a species (e.g., strain O157:H7 of *E. coli*).

The outer membrane also contains unique proteins called **porins**. These proteins form pores in the outer membrane through which small molecules pass into the **periplasmic region** that lies between the outer membrane and cell membrane. This area is filled with **periplasm**, a gel-like material containing digestive enzymes and transport proteins to speed entry of nutrients into the cell. Larger molecules cannot pass, accounting for the resistance of gram-negative cells to so many chemicals, including antimicrobial agents, dyes, disinfectants, and lysozyme.

TABLE 4.1 summarizes the major differences between the two types of bacterial cell walls.

THE CELL MEMBRANE IS A PERMEABILITY BARRIER

The **cell membrane** (also called the **plasma membrane** in eukaryotic cells) is the boundary layer between the cell cytoplasm and the cell's external environment (Figure 4.9b, c). It functions as a permeability barrier to keep cytoplasmic material inside

TABLE 4.1

A Comparison of Gram-Positive and Gram-Negative Cell Walls

CHARACTERISTIC	GRAM-POSITIVE	GRAM-NEGATIVE
Peptidoglycan	Yes, thick layer	Yes, thin layer
Teichoic acids	Yes	No
Outer membrane	No	Yes
Lipopolysaccharides	No	Yes
Porin proteins	No	Yes
Periplasmic region	No	Yes

the cell and many other substances out of the cell. Importantly, the cell membrane also functions in specifically transporting needed nutrients into the cell and eliminating waste materials. This membrane also anchors the DNA during replication.

The cell membrane of a bacterial cell is about 40 percent phospholipid and 60 percent protein. The phospholipid molecules are arranged in two parallel layers (a phospholipid bilayer) forming a nonpolar barrier that helps explain how it acts as a selectively permeable membrane. The proteins are embedded in the membrane, where some are enzymes functioning in cell wall synthesis. Others are enzymes used in energy metabolism, a factor that makes the cell membrane equivalent of the membranes of mitochondria in a eukaryotic cell. Some proteins extend from one surface of the membrane to the other and play important roles as transporters of charged nutrients, such as amino acids, simple sugars, nitrogenous bases, and ions.

In illustrations, the cell membrane appears very rigid. In reality, it is quite fluid, having the consistency of olive oil. This means the mosaic of phospholipids and proteins are not cemented in place but rather they can move laterally in the membrane surface. This dynamic model of membrane structure therefore is called the **fluid mosaic model**.

Several antimicrobial substances act on the cell membrane. The antibiotic polymyxin pokes holes in the membrane, while some detergents and alcohols dissolve the bilayer. Such action allows the cytoplasmic contents to leak out of the bacterial cells, resulting in death through cell lysis.

THE CYTOPLASM IS THE CENTER OF BIOCHEMICAL ACTIVITY

The cell membrane encloses the **cytoplasm**, which is the foundation substance of a cell and the center of its growth and metabolism. The cytoplasm consists of the cytosol, a gelatinous mass of proteins, amino acids, sugars, nucleotides, salts, vitamins, and ions—all dissolved in water—and a few bacterial structures, each with a specific function.

■ RIBOSOMES. One of the universal cell structures mentioned in Chapter 3 was the **ribosome**. There are hundreds of thousands of these particles in the bacterial cytoplasm, which gives it a granular appearance when viewed with the electron microscope. Ribosomes are built from RNA and protein and are composed of a small subunit and a large subunit. For proteins to be synthesized, the two subunits come together to form a functional ribosome (Chapter 6). Some antibiotics, such as streptomycin and tetracycline, prevent bacterial ribosomes from carrying out protein synthesis.

■ **INCLUSION BODIES.** Cytoplasmic structures, called **inclusion bodies**, are found in many species of bacteria. Many of these bodies store nutrients or the monomers for bacterial structures. For example, they can consist of aggregates or granules of polysaccharides (glycogen), elemental sulfur, or lipid. Other inclusion bodies can serve as important identification characters for disease-causing bacteria. One common example is the diphtheria bacilli that contain **metachromatic granules**, or volutin, which are phosphate depots. These granules stain deeply with dyes such as methylene blue. The **magnetosome**, another type of inclusion body, contains crystals of an iron-containing compound called magnetite (Fe_3O_4). Bacteria that contain magnetosomes can orient themselves to the environment by aligning themselves with the local magnetic field (**MicroFocus 4.3**).

■ **NUCLEOID.** The chromosome region in a bacterial cell is termed the **nucleoid** (**FIGURE 4.11**). The nucleoid does not contain a covering or membrane; rather it represents an area in the cytoplasm where the DNA aggregates and ribosomes are absent. The bacterial **chromosome** is a closed loop of DNA that contains the hereditary information or genes of the cell. Depending on the bacterial species, up to 3,500 genes may be present on the chromosome. These genes determine what proteins and enzymes the cell can make; that is, what metabolic reactions can be carried out. Unlike eukaryotic microorganisms and other eukaryotes, the nucleoid and chromosome do not undergo mitosis. Since there is only one copy of the bacterial genes, meiosis does not occur.

■ **PLASMIDS.** Smaller, nonessential molecules of DNA called **plasmids** exist apart from the nucleoid. About a tenth the size of a bacterial chromosome, these DNA molecules exist as closed loops containing only 5–100 genes. There can be one or more similar or different plasmids in a specific bacterial cell. Although they contain few genes and are not essential for bacterial growth, plasmids are significant in disease processes. Some plasmids possess genes for toxic substances and many carry genes for drug

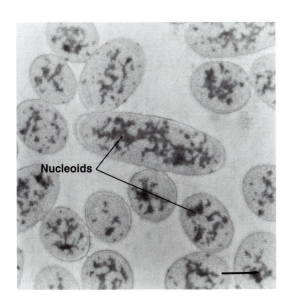

Nucleoids

FIGURE 4.11

The Nucleoids of Bacteria

In this transmission electron micrograph of *Escherichia coli*, nucleoids are seen in irregular coralline (coral shaped) forms. Nucleoids can be observed occupying a large area in a bacterium, which indicates how far the DNA of the bacterium is spread out. Both longitudinal and cross sections of *E. coli* are visible. (Bar = 0.5 μm.)

MicroFocus 4.3

A (NOT SO) FATAL ATTRACTION

To get from place to place, humans usually require the assistance of maps, compasses, and gas station attendants. In the microbial world, life is generally more simple, and traveling is no exception.

Consider the bacteria, for example. In the early 1980s, Richard P. Blakemore and his colleagues at the University of New Hampshire observed that certain mud-dwelling bacteria tend to gather at the north end of water droplets. On further study, they found that each bacterium had a chain of magnetic particles acting as a kind of bacterial compass directing the organism's movements. Bacteria possessing the particles swim toward the north in the Northern Hemisphere and toward the south in the Southern Hemisphere. Indeed, when genetic mutations are caused in the bacteria, they head in the wrong direction—and wind up in a hostile environment and die.

This last observation is particularly noteworthy because it appears to give rhyme and reason to the particles. The conventional wisdom is that the so-called magnetotactic bacteria use their traveling skills to locate a favorable environment. In 1992, Dennis A. Bazylinski of Iowa State University theorized how this hypothesis might work in nature. Certain magnetotactic bacteria are anaerobic; that is, they live in an oxygen-free environment. While swimming toward a pole (north or south), the bacteria also are oriented downward by Earth's magnetic field. The downward tilt pulls them away from the oxygen-rich water and toward the oxygen-poor mud below. On reaching their optimum environment, the bacteria reach a sort of biological nirvana and settle in for a life of anaerobic bliss.

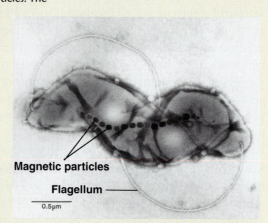

Magnetic particles

Flagellum

0.5μm

■ *The bacterial magnetosome is seen in this remarkable electron micrograph of a magnetotactic marine spirillum. Particles of magnetite in the magnetosome apparently influence the direction the bacteria move in the environment. The flagella of the spirillum are clearly visible.*

---

resistance. For this latter reason, they are often called **R plasmids** (R for resistance). Plasmids can be transferred between cells during recombination processes (Chapter 7) and multiply independently from cell reproduction. As we also will see in Chapter 7, plasmids are important in industrial technologies that use genetic engineering.

TABLE 4.2 summarizes the structural features of bacterial cells.

ENDOSPORES ARE DESIGNED FOR DORMANCY

A few gram-positive bacteria, especially members of the genera *Bacillus* and *Clostridium*, produce highly resistant structures called **endospores** or, simply, **spores** (**FIGURE 4.12**). Bacteria normally grow, mature, and reproduce as **vegetative cells**. However, when nutrients such as carbon or nitrogen are limiting, or other environmental pressures exist, species of *Bacillus* and *Clostridium* begin spore formation. The process begins when the bacterial chromosome replicates, a small amount of cytoplasm gathers with it, and then the cell membrane grows in to seal off the developing spore within the cytoplasm (**FIGURE 4.13**). Endospores may develop at the end of the cell, near the end, or at the center of the cell, depending on the species (the position is useful for identification purposes) (Figure 4.12a). In any case, thick layers of peptidoglycan form followed by a series of protein coats that protect the contents further (Figure 4.12b). The vegetative cell then disintegrates, and the spore is freed.

TABLE 4.2

A Summary of the Structural Features of Bacteria

STRUCTURE	CHEMICAL COMPOSITION	FUNCTION	COMMENT
Flagella	Protein	Motility	Present in many rods and spirilla; few cocci; vary in number and placement
Pili	Protein	Attachment to surfaces Genetic transfers	Found in many gram-negative bacteria
Glycocalyx	Polysaccharides and small proteins	Buffer to environment Contributes to disease Cell protection Attachment to surfaces	Capsule and slime layer Source of ropy milk and bread Found in plaque bacteria and biofilms
Cell wall	Gram-positives have much peptidoglycan, with teichoic acid Gram-negatives have little peptidoglycan and an outer membrane	Cell protection Shape determination Prevents cell lysis	Site of activity of penicillin and lysozyme Absent in mycoplasmas Gram-negatives release endotoxins
Cell membrane	Protein Phospholipid	Cell boundary Transport into/out of cell Site of enzymatic reactions	Conforms to fluid mosaic model Susceptible to detergents, alcohols, and some antibiotics
Cytosol	Water, amino acids, sugars, ions	Foundation substance of cell	Center of biochemistry and growth Semitransparent, gel-like
Ribosomes	RNA and protein	Protein synthesis	Inhibited by certain antibiotics
Inclusion bodies	Glycogen, sulfur	Nutrient storage	Used as nutrients during starvation periods
Metachromatic granules	Polyphosphate	Storage for ATP and nucleic acid synthesis	Found in diphtheria bacilli
Magnetosome	Magnetite	Cell orientation	Helps locate preferred habitat
Chromosome	DNA	Site of genetic code Site of inheritance	Exists as single, closed loop Located at nucleoid
Plasmids	DNA	Site of some genes	Contains R factors

When the environment is favorable for cell growth, the protective layers break down and each endospore germinates into a vegetative cell (Figure 4.12c). It should be noted that endospore formation is not a reproductive process. Rather, the spores represent a dormant stage in the life of the bacteria.

Endospores are probably the most resistant living things known. By containing little water, endospores are heat resistant and undergo very few chemical reactions. However, they do have some ribosomes and enzymes, and a large amount of **dipicolinic acid**, a unique organic substance that helps stabilize their proteins and DNA. This property makes them difficult to eliminate from contaminated medical materials and food products. For example, most vegetative bacteria die quickly in

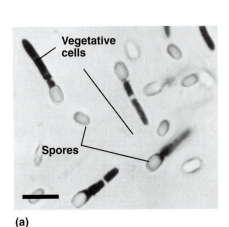

(a)

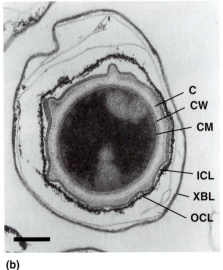

(b)

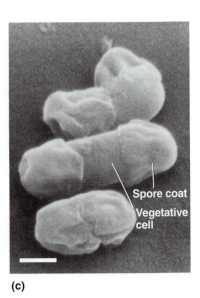

(c)

FIGURE 4.12

Three Different Views of Bacterial Spores

(a) A view of *Clostridium* with the light microscope, showing terminal spore formation. Note the characteristic drumstick appearance of the cells. (Bar = 2.0 µm.) (b) The fine structure of a *Bacillus thuringiensis* spore seen using the transmission electron microscope. The visible spore structures include the core membrane (CM), core wall (CW), cortex (C), inner coat layer (ICL), outer coat layer (OCL), and exosporium basal layer (XBL). These layers contribute to spore resistance. (Bar = 0.2 µm.) (c) A scanning electron microscope view of a germinating spore. Note that the spore coat divides equatorially along the long axis, and as it separates, the vegetative cell emerges. (Bar = 0.4 µm.)

water over 80°C. Endospores can remain viable in boiling water (100°C) for 2 hours or more. When placed in 70 percent ethyl alcohol, endospores have survived for 20 years. Humans can barely withstand 500 rems of radiation, but spores can survive one million rems. Desiccation has little effect on spores; living spores have been recovered from the intestines of Egyptian mummies. In 1983, archaeologists found spores alive in sediment lining Minnesota's Elk Lake. The sediment was 7,518 years old.

But all the records pale in comparison to the controversial discovery reported in 1996 by researcher Raul Cano of California Polytechnic State University. Cano found bacterial spores in the stomach of a fossilized bee caught in the resin flowing from a tree in the Dominican Republic (when the resin hardens it becomes amber). The fossilized bee was about 20 million years old. The equally ancient spores from its gut germinated when placed in laboratory nutrients and produced a culture strikingly similar to *Bacillus sphaericus*, which is found today in Dominican bees.

A few serious diseases in humans are caused by spore formers. The most newsworthy has been anthrax, the agent of the 2001 bioterror attack through the mail (MicroFocus 4.4). This potentially deadly disease, originally studied by Koch and Pasteur, is caused by *Bacillus anthracis* (Chapter 10).

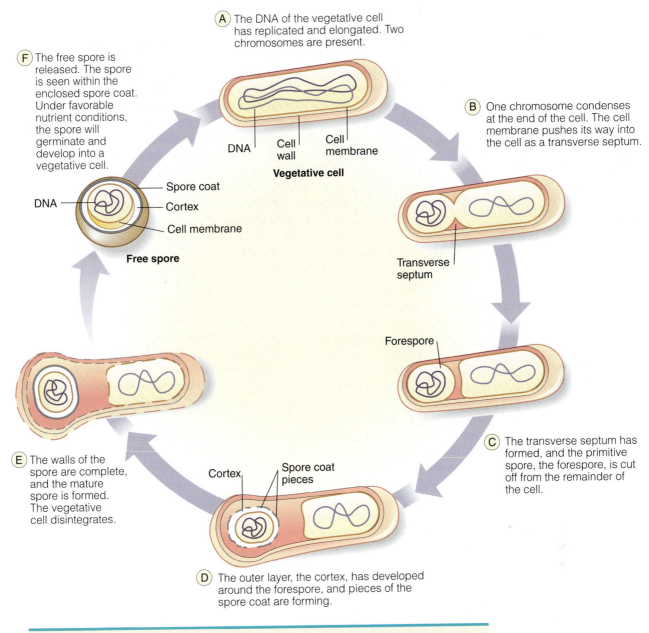

A The DNA of the vegetative cell has replicated and elongated. Two chromosomes are present.

DNA | Cell wall | Cell membrane

Vegetative cell

B One chromosome condenses at the end of the cell. The cell membrane pushes its way into the cell as a transverse septum.

Transverse septum

Forespore

C The transverse septum has formed, and the primitive spore, the forespore, is cut off from the remainder of the cell.

F The free spore is released. The spore is seen within the enclosed spore coat. Under favorable nutrient conditions, the spore will germinate and develop into a vegetative cell.

Spore coat
Cortex
Cell membrane

DNA

Free spore

E The walls of the spore are complete, and the mature spore is formed. The vegetative cell disintegrates.

Cortex | Spore coat pieces

D The outer layer, the cortex, has developed around the forespore, and pieces of the spore coat are forming.

FIGURE 4.13

The Formation of a Bacterial Spore

The cell metabolizes nutrients and multiplies for many generations as a vegetative cell. During periods of nutrient limitation (e.g., carbon, nitrogen), the cell enters the sporulaton cycle shown here.

Botulism, gas gangrene, and tetanus are diseases caused by different species of *Clostridium.* Clostridial endospores often are found in soil, as well as in human and animal intestines. However, the environment must be free of oxygen for the spores to germinate to vegetative cells. Dead tissue in a wound provides such an environment for the spore-forming bacteria that cause tetanus and gas gangrene (Chapter 10), and a vacuum-sealed can of food is suitable for the spore-forming bacteria of botulism (Chapter 9).

MicroFocus 4.4

ANTHRAX AND THE BIOTERRORISM EVENTS OF 2001

When the anthrax attacks occurred in Miami, Washington, D.C., and New York City in October 2001, it validated what many health and governmental experts had been saying for years—it is not *if* bioterrorism will occur, but *when* and *where*.

Bioterrorism represents the intentional or threatened use of biological agents to cause fear in or actually inflict death or disease upon a large population for political, religious, or ideological reasons. One of the most feared agents is the spore-forming bacterium *Bacillus anthracis* that causes anthrax. This deadly blood disease is the same one studied by Koch and Pasteur in the late 1800s. Anthrax endospores can be found in soil from around the world. There are a few naturally occurring cases, which are usually occupational. Between 1955 and 1999, 236 cases (mostly cutaneous) of occupational anthrax were reported in the United States.

Inhalational anthrax is the most likely form for bioterrorism and accounted for the five deaths and the necessity to make antibiotics available to 10,000 postal employees and government officials. The perpetrator(s) sent four anthrax-containing letters through the mail on the same day, addressed to NBC newscaster Tom Brokaw, the *New York Post*, and to United States Senators Tom Daschle and Patrick Leahy. This resulted in the closing of the Hart Senate Office Building and several postal sorting facilities for decontamination.

The resistance of the spores to chemicals is seen in the aftermath to the attacks. The cleanup in the Hart Building cost over $23 million and involved spraying noxious chlorine dioxide gas. The Bentwood postal sorting facility in Washington, D.C. and another postal facility cost more than $35 million. Cleanup crews had to wear full biohazard suits while spraying the chlorine

dioxide gas. These facilities could not reopen until the test standard indicates there are no *B. anthracis* colonies growing in cultures that originated from germinated endospores. The Brentwood facility did not resume full operation until early 2004, more than two years after the anthrax letter incident.

Letter to Tom Brokaw

Letter to Senator Daschle

To this point . . .

We have explored the structural features of bacteria and have noted the three major shapes that a bacterium can take: bacillus, coccus, and spiral. We then analyzed the fine details of a bacterium and discussed how bacterial structures are related to bacterial functions. We examined the external structures, including flagella and pili and discussed the three structures (glycocalyx, cell wall, and cell membrane) that make up the cell envelope.

The cytoplasm is the center of cell growth and the ground substance (cytosol) in which several structures, including ribosomes, inclusion bodies, DNA (nucleoid), and plasmids are located. The final structure discussed was the bacterial spore, a highly resistant body formed by species of Bacillus and Clostridium.

In the second half of this chapter, we turn to the growth patterns exhibited by bacteria. We will study how bacteria reproduce and then explore the dynamics of bacterial growth. We also will see how bacteria are cultivated in the laboratory and their numbers counted.

4.3

Bacterial Reproduction

Asexual reproduction in the eukaryotic organisms involves mitosis with the elaborate orchestration of microtubules and chromosomes for the precise events of nuclear division. The end of nuclear division overlaps and is followed by cytokinesis, the division of the cytoplasm into two daughter cells, each carrying the identical genetic instructions of the parent cell. Since there is but one chromosome in a bacterial cell, there is no need for such an elaborate system for DNA separation into daughter cells.

BACTERIA REPRODUCE BY BINARY FISSION

Most bacteria reproduce by an asexual process called **binary fission**. In this process, the cell elongates and the chromosome (DNA) replicates (**FIGURE 4.14**).

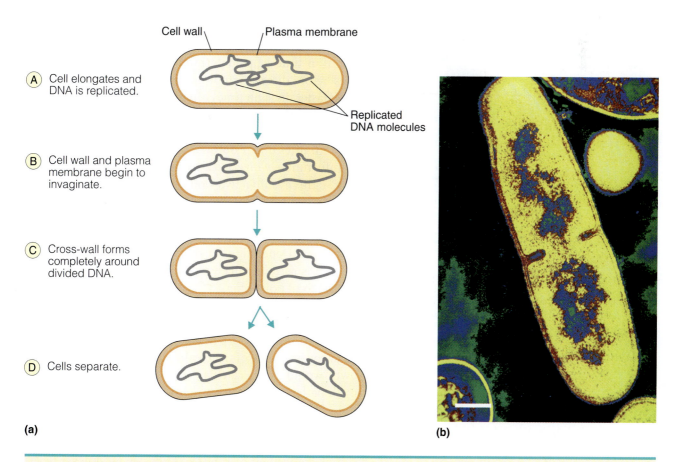

Cell wall   Plasma membrane

(A) Cell elongates and DNA is replicated.

Replicated DNA molecules

(B) Cell wall and plasma membrane begin to invaginate.

(C) Cross-wall forms completely around divided DNA.

(D) Cells separate.

(a)

(b)

FIGURE 4.14

Binary Fission of Bacteria

(a) As a result of DNA replication and binary fission, two cells are formed, each genetically identical to the parent cell. (b) A false color transmission electron micrograph of a cell of *Bacillus licheniformis* undergoing binary fission. The invagination of the cell membrane is evident. (Bar = 0.25 μm.)

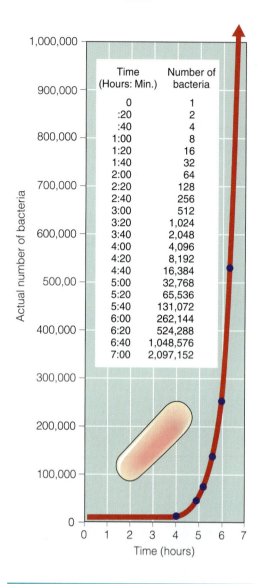

Time (Hours: Min.)	Number of bacteria
0	1
:20	2
:40	4
1:00	8
1:20	16
1:40	32
2:00	64
2:20	128
2:40	256
3:00	512
3:20	1,024
3:40	2,048
4:00	4,096
4:20	8,192
4:40	16,384
5:00	32,768
5:20	65,536
5:40	131,072
6:00	262,144
6:20	524,288
6:40	1,048,576
7:00	2,097,152

FIGURE 4.15

A Skyrocketing Bacterial Population

The number of bacteria progresses from 1 cell to 2 million cells in a mere 7 hours. The J-shaped growth curve gets steeper and steeper as the hours pass. Only a depletion of food, buildup of waste, or some other limitation will halt the progress of the curve.

Then, the two DNA molecules, attached to the cell membrane, are separated by new membrane growth. Finally, the cell membrane begins to pinch inward and the cell wall thickens between the invaginating cell membrane to separate the dividing cell into two genetically identical daughter cells. Reproduction by binary fission lends a certain immortality to bacteria because there is never a moment at which the first bacterium has died. Bacteria mature, undergo binary fission, and are young again.

It is interesting to see how fast such fission events can produce a large population of bacteria. Suppose you eat an undercooked hamburger contaminated with the pathogen *E. coli* O157:H7. It will infect human tissue and grow by binary fission. Here growth refers to an increase in the number of cells (population size). The interval of time between successive binary fissions of a cell or population of cells is known as the **generation time** (or **doubling time**). Under optimal growing conditions, some bacteria have a very short generation time; for others, it is quite long. For example, for *Staphylococcus aureus,* the generation time is about 30 minutes; for *Mycobacterium tuberculosis,* the agent of tuberculosis, it is approximately 15 hours; and for the syphilis spirochete, *Treponema pallidum,* it is a long 33 hours. The generation time helps determine the amount of time that passes before disease symptoms appear in an infected individual; faster division times often mean a shorter incubation period for a disease.

Escherichia coli has one of the shortest generation times, just 20 minutes under optimal conditions (**FIGURE 4.15**). If you unfortunately ingested a single rod of *E. coli* O157:H7 at 8:00 P.M. this evening, two would be present by 8:20, four by 8:40, and eight by 9:00. You would have sixty-four rods by 10:00 P.M. and five hundred twelve by 11:00 P.M. By 6:00 tomorrow morning, the culture would contain just over a billion rods. Depending on the response of one's immune system, it is certainly likely that by morning you will know you have food poisoning.

As another example, one enterprising mathematician has calculated that if binary fissions were to continue at their optimal generation time for 36 hours, there would be enough bacteria to cover the face of the Earth! Although such exponential growth would be spectacular, it does not reflect a possible growth pattern. Why? Such a reproductive potential of a bacterium could never be realized because of the limitation of nutrients and the ideal physical requirements in the external environment. The majority of the bacteria would starve to death or die in their own waste. Thus, we need never worry about being smothered with bacteria. Bacteria are subject to the same controls on growth as all other organisms on Earth, as we shall see next.

4.4

Bacterial Growth

In the previous example we saw how fast bacteria could grow under ideal circumstances. Let's look at the growth of bacterial populations in a little more detail.

A BACTERIAL GROWTH CURVE ILLUSTRATES THE DYNAMICS OF GROWTH

A typical **bacteria growth curve** for a population illustrates the events that occur over the course of time within a population of bacteria (FIGURE 4.16). If several bacteria infect the human respiratory tract or are transferred to a tube of fresh growth medium in the laboratory, four distinct phases of growth occur: the lag phase, the logarithmic phase, the stationary phase, and the decline phase.

The **lag phase** encompasses the first few hours of the curve. During this time, no cell divisions occur. Rather, bacteria are adapting to their new environment. In the respiratory tract, scavenging white blood cells may engulf and destroy some bacteria; in growth media, some organisms may die from the shock of transfer or the inability to adapt to the new environment. The actual length of the lag phase

Lag phase:
the phase of a bacterial growth curve prior to binary fission.

FIGURE 4.16

The Growth Curve for a Bacterial Population

(a) During the lag phase, the population numbers remain stable as bacteria prepare for division. (b) During the logarithmic (exponential growth) phase, the numbers double with each generation time. Environmental factors later lead to cell death, and (c) the stationary phase shows a stabilizing population. (d) The decline phase is the period during which cell death becomes substantial.

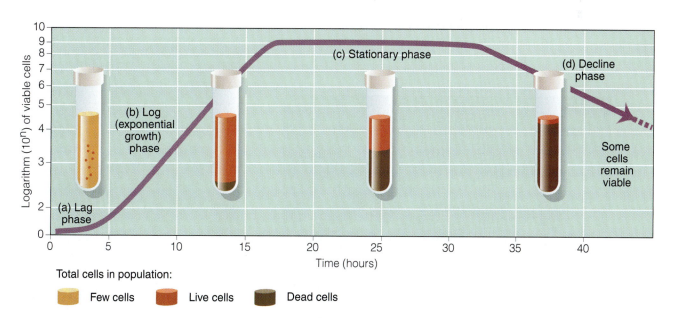

depends on the metabolic activity in the remaining bacteria. They must grow in size, store nutrients, and synthesize enzymes—all in preparation for binary fission.

The population then enters an active stage of growth called the **logarithmic phase** (or **log phase**). This is the exponential growth described above for *E. coli*. In the log phase, all cells are undergoing binary fission and the generation time is dependent on the species and environmental conditions present. As each generation time passes, the number of bacteria doubles and the graph rises in a straight line on a logarithmic scale.

In humans, disease symptoms usually develop during the log phase because the bacteria and their toxins are causing tissue damage. Coughing or fever may occur, and fluid may enter the lungs if the air sacs are damaged. If the bacteria produce toxins, tissue destruction may become apparent. But vulnerability to antibiotics is also highest at this active stage of growth. During the log phase in our broth tube, the medium becomes cloudy (turbid) due to increasing cell numbers. If plated on solid medium, bacterial growth may be so vigorous that visible colonies appear and each colony may consist of millions of organisms (**FIGURE 4.17**).

After some days (in an infection) or hours (in a culture tube), the vigor of the population changes and, as the reproductive and death rates equalize, the population enters a plateau, the **stationary phase**. In the respiratory tract, antibodies from the immune system are attacking the bacteria, and phagocytosis by white blood cells adds to their destruction. Perhaps the person was given an antibiotic to supplement the body's defensive measures. In the culture tube, available nutrients become scarce

Logarithmic phase:
the phase of a bacterial growth curve when reproduction and growth are at their highest rates.

Stationary phase:
the phase of a bacterial growth curve when the reproduction rate equals the death rate.

FIGURE 4.17

Two Views of Bacterial Colonies

(a) Colonies of *Bacillus macerans* isolated from sewage and growing in a medium of solidified soybean meal and casein peptone. These colonies are several millimeters in height; they have been euphemistically called "Rockies in a Petri dish." (b) A scanning electron micrograph of the surface of a colony of *Staphylococcus aureus* on a solid medium. (Bar = 5 µm.) Note the irregular nature of the surface of the colony, with numerous conical pits.

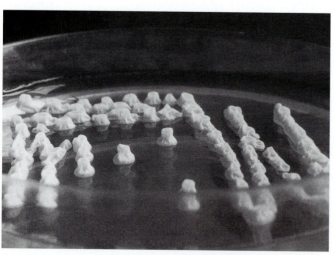

(a)

(b)

and waste products accumulate. Factors such as oxygen also may be in short supply. However, this is not a dormant period. If the bacterium is a species that secretes antibiotics, they are produced during the stationary phase. If the organism is a species of *Bacillus* or *Clostridium*, the vegetative cells will produce endospores.

If these conditions continue, the external environment will exert its limiting powers on the population and the **decline phase** (or **exponential death phase**) will ensue. Now the number of dying cells exceeds the number of new cells formed. A bacterial glycocalyx may forestall death by acting as a buffer to the environment, and flagella may enable organisms to move to a new location. For many species, though, the history of the population ends with the death of the last cell. When we discuss the progression of human diseases in Chapter 18, we will see a similar curve for the stages of the disease.

The bacterial growth curve we studied above was performed under ideal or optimal physical and nutrient conditions. Altering those conditions slows down or inhibits bacterial growth. Let's examine some of the major physical requirements for bacterial growth.

TEMPERATURE IS ONE OF THE MOST IMPORTANT FACTORS GOVERNING GROWTH

Different bacterial species grow at different temperatures. Each bacterial species has an optimal growth temperature and an approximate 30° range, from minimum to maximum, over which the microorganism will grow but at a slower rate (**FIGURE 4.18**). The reduced growth rate is due to slower reaction rates by the enzymes that

FIGURE 4.18

Growth Rates for Different Microorganisms in Response to Temperature

Temperature optima and ranges define the growth rates for different types of microorganisms. Notice that the growth rates decline quite rapidly to either side of the optimal growth temperature.

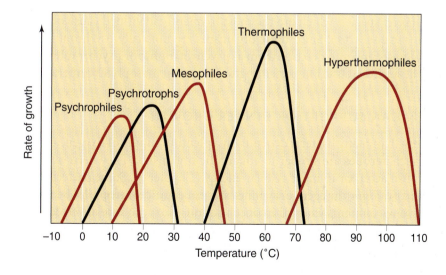

control metabolism. In general, bacteria can be assigned to one of three groups based on their optimal temperature for growth.

Bacteria that have their shortest generation times and their optimal growth rates in the range of 15°C but can still grow at 0°C to 20°C are called **psychrophiles**. True psychrophiles live in the ocean depths and in Arctic and Antarctic regions where the temperature may be 1–5°C. These bacteria contain unsaturated fatty acids in their phospholipids that allow the cell membrane to remain fluid at these extremely cold temperatures. In addition, they contain enzymes that are functional only at these low temperatures. Psychrophiles would not be pathogens in humans because they cannot grow at the warmer 37°C body temperature.

At the opposite extreme are the **thermophiles** that multiply best at temperatures around 60°C but still multiply from 40°C to 70°C. Thermophiles are present in compost heaps and hot springs, and are important contaminants in dairy products because they survive pasteurization temperatures. However, thermophiles pose little threat to human health because they do not grow well at the cooler temperature of the body. Opposite to the psychrophiles, thermophiles have highly saturated fatty acids in their cell membranes that stabilize these structures. They also contain heat-stable proteins and enzymes. However, surpassing their maximal growth temperature does denature critical proteins needed for growth.

There also are bacteria that grow optimally above 80°C. These **hyperthermophiles** have been isolated from seawater brought up from hot-water vents along rifts on the floor of the Pacific Ocean. Using high pressure to keep the water from boiling, some of these bacteria can grow at an astonishing 113°C.

Most of the best-characterized bacteria are **mesophiles** that thrive at the middle temperature range of 20°–45°C. This includes the pathogens that grow in the human body between 35°C and 42°C. Mesophiles that grow in the temperate and tropical latitudes of the world often can grow at temperatures substantially below their normal range. For example, refrigerated foods can harbor mesophiles that will grow very slowly and cause food spoilage. *Staphylococcus aureus* can contaminate improperly handled or prepared cold cuts, salads, or various leftovers. The slow growth of these bacteria at refrigeration temperature (5°C) can result in the deposit of toxins in the food products. When such foods are consumed without heating, the toxins may cause food poisoning. Other examples of mesophiles growing in the cold are *Campylobacter* species, which are the most frequently identified cause of infective diarrhea (**FIGURE 4.19**). Since these organisms are not truly psychrophilic, some microbiologists prefer to describe them as **psychrotrophic** or **psychrotolerant**; that is they will survive at 0°C but prefer to grow at typical mesophile temperatures.

OXYGEN CAN SUPPORT OR HINDER GROWTH

The growth of many bacteria depends on a plentiful supply of oxygen, and in this respect, such **aerobes** are similar to eukaryotic organisms. They need oxygen as a final electron acceptor to make cellular energy (Chapter 5). Some bacteria, such as *Treponema pallidum*, the agent of syphilis, are termed **microaerophiles** because they survive in environments where the concentration of oxygen is relatively low. In the body, certain microaerophiles cause disease of the oral cavity, urinary tract, and gastrointestinal tract. Conditions can be established in the laboratory to study these microbes (**FIGURE 4.20a**).

The **anaerobes**, by contrast, are bacteria that do not or cannot use oxygen. Many are **aerotolerant**, meaning they are insensitive to oxygen and will grow in environments with or without oxygen present. Others are **obligate anaerobes** that actually

FIGURE 4.19

An Outbreak of Campylobacteriosis Caused by *Campylobacter jejuni* Occurring in Oklahoma

This outbreak occurred between August 16 and 20, 1996.

TEXTBOOK CASES

1. On August 15, the cook began his day by cutting up raw chickens to be roasted for dinner.

2. He also cut up lettuce, tomatoes, cucumbers, and other salad ingredients on the same countertop. The countertop surface where he worked was unusually small.

3. For lunch that day, the cook prepared sandwiches on the same countertop. Most were garnished with lettuce.

4. Restaurant patrons enjoyed sandwiches for lunch and roasted chicken for dinner. Many patrons also had a portion of salad with their meal.

5. During the next three days, 14 people experienced stomach cramps, nausea, and vomiting. Public health officials learned that all the affected patrons had eaten salad with lunch or dinner. *Campylobacter*, a bacterial pathogen of the intestines, was located in their stools.

6. On inspection, microbiologists concluded that the chicken was probably contaminated with *Campylobacter*. Bacteria were deposited on the countertop, and they contaminated the lettuce, which was eaten raw. The chicken was not a source of illness because it was cooked well.

are killed if oxygen is present. This means they need other ways to make cell energy. Some anaerobic bacteria, such as *Thiomargarita namibiensis* discussed in MicroFocus 3.5, use sulfur in their metabolic activities instead of oxygen, and therefore they produce hydrogen sulfide (H_2S) rather than water (H_2O) as a waste product of their metabolism. Others we have already encountered, such as the ruminant archaea, produce methane as the by-product of the energy conversions. Both of these gases give putrid odors to marshes, swamps, and landfills. Petroleum is a product of anaerobic metabolism.

Some species of anaerobic bacteria cause disease in humans. For example, the *Clostridium* species that cause tetanus and gas gangrene multiply in the dead, anaerobic tissue of a wound and produce toxins that cause tissue damage. Another species of *Clostridium* multiplies in the oxygen-free environment of a vacuum-sealed can of food, where it produces the lethal toxin of botulism. In one bizarre incident, a restaurant owner died of botulism after tasting a piece of fish marinating under a layer of oil (MicroFocus 4.5).

A common way to test an organism's oxygen requirement is to use a **thioglycolate broth** that binds free oxygen so that only fresh oxygen entering at the top of the tube would be available. Among the most widely used methods to establish anaerobic conditions in the laboratory is the GasPak system, in which hydrogen reacts with oxygen in the presence of a catalyst to form water, thereby creating an oxygen-free atmosphere (FIGURE 4.20b).

Many bacteria are neither aerobic nor anaerobic, but **facultative**. Facultative bacteria grow in either the presence or the reduced concentration of oxygen. This group includes many staphylococci and streptococci, as well as members of the genus *Bacillus* and a variety of intestinal rods, among them *E. coli*. A facultative aerobe prefers anaerobic conditions (but grows aerobically), while a facultative anaerobe prefers oxygen-rich conditions (but grows anaerobically).

Finally, there are bacterial species, said to be **capnophilic**, that require an atmosphere low in oxygen but rich in carbon dioxide. The CO_2 content can be increased

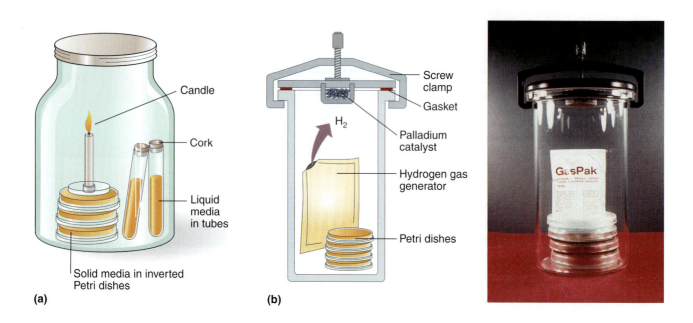

(a)   (b)

FIGURE 4.20

Bacterial Cultivation in Different Gas Environments

Two types of cultivation methods are shown for bacteria that grow poorly in an oxygen-rich environment. (a) A candle jar, in which microaerophilic bacteria grow in an atmosphere where the oxygen is reduced by the burning candle. (b) An anaerobic jar, in which hydrogen is released from a generator and then combines with oxygen through a palladium catalyst to form water and create an anaerobic environment.

MicroFocus 4.5

OF MARINATED FISH

It was September 1978, and the man was in serious condition. He had come to the hospital emergency room in Puerto Rico complaining of blurred vision, difficulty swallowing, erratic breathing, and numbness in his fingers and arms. Now the ER personnel were frantically trying to save his life. He owned a restaurant, he gasped. No, there were no other sick members in his house. No, he hadn't done anything unusual that he could remember. Slowly he was slipping away. By morning, he was gone.

Before long, the investigation was in progress. Public health officials visited

the man's restaurant and spoke with his wife. She was fond of marinating fish, and occasionally she would leave some in the restaurant to cure. Normally the fish cured in wide, shallow trays, but this time, she had no trays available so she used narrow-mouthed, screw-capped jars. Officials noted that a thick layer of oil had formed over the fish, and they surmised that the oil created anaerobic conditions underneath. (The narrow mouth of the jar also contributed to the lack of air movement.)

Now their suspicions were aroused. A sample of the fish was taken back to

the laboratory for bacteriological testing. At the same time, in another part of the lab, the man's tissues were being studied for evidence of bacterial toxins. The results came back almost simultaneously. The organism was *Clostridium botulinum*; the disease was botulism. The man had probably tasted the fish to see how it was coming along. He could not have foreseen what lay ahead.

---

in the laboratory by using a gas-generating apparatus or by burning a candle in a closed jar with the bacteria (Figure 4.20a). Members of the genera *Neisseria* and *Streptococcus* are capnophiles.

MOST BACTERIA PREFER TO GROW AT A NEUTRAL pH

The cytoplasm of most bacteria has a pH of about 7.0. This means the majority of species grow optimally at neutral pH. Human blood and tissues, with a pH of approximately 7.2 to 7.4, provide a suitable environment for the proliferation of disease-causing bacteria. Still bacteria have a pH range under which they will grow more slowly. The minimum to maximum range usually covers three pH units. However, some pH-hearty bacteria, such as *Vibrio cholerae*, can tolerate acidic conditions as low as pH 2.0 and alkaline conditions as high as pH 9.5.

Acid-tolerant bacteria called **acidophiles** are valuable in the food and dairy industries. For example, certain species of *Lactobacillus* and *Streptococcus* produce the acid that converts milk to buttermilk and cream to sour cream. These species pose no threat to good health even when consumed in large amounts. The "active cultures" in a cup of yogurt are actually acid-tolerant (acidophilic) bacteria. Extreme acidophiles are found among the archaea as we saw in the MicroFocus 2.2.

The majority of known bacterial species, however, do not grow well under acidic conditions. Thus, the acidic environment of the stomach helps deter disease in this organ, while providing a natural barrier to the organs beyond. In addition, you may have noted that certain acidic foods are hardly ever contaminated with bacteria. Examples are lemons, oranges, and other citrus fruits, as well as such vegetables as cabbage and rhubarb. Traditionally, tomatoes were too acidic to support bacterial growth. However, modern technologists have developed the "neutral tomato," along with a host of new problems for consumers, especially for those who grow and can their own tomatoes.

4.5

Culture Media and Growth Measurements

In this chapter, we have been discussing bacterial growth and the physical factors that control growth. To end this chapter, we need be familiar with some of the chemical media used to grow and separate specific bacteria, and with the measurements used to evaluate growth.

CULTURE MEDIA CONTAIN THE NUTRIENTS TO GROW BACTERIA

Since the time of Pasteur and Koch, microbiologists have tried to grow bacteria in laboratory cultures; that is, in ways that mimic the natural environment. Beef broth is one such **growth medium** used for the laboratory cultivation of bacteria. The modern form of this liquid medium, called **nutrient broth**, consists of water, beef extract, and peptone, a nitrogen preparation from plant or animal sources. When agar is added to solidify the medium, the product is called **nutrient agar**.

Agar is a polysaccharide derived from marine red algae. Introduced by the school of bacteriology of Robert Koch, agar is a unique colloid that remains liquid until cooled to below approximately 36°C. This allows for mixing of blood with culture media for determination of hemolytic reactions. The solidified medium can be used to cultivate bacteria, isolate pure cultures, or accomplish other tasks, such as a medium for measuring bacterial growth. Once solidified, agar will remain solid at room temperature. It will not melt until it reaches a temperature of 85°C, making the material excellent for growing thermophilic bacteria on nutrient agar. Agar adds no macronutrients to the nutrient medium. Sometimes it is valuable to use a semisolid medium, such as when testing bacterial motility. In this case, a lower concentration of agar is added to the medium to make it stiff but not as solid as nutrient agar.

Nutrient broth and nutrient agar media are examples of a chemically undefined medium, or **complex medium**. It is called complex because one cannot be certain of the exact components or their quantity. Beef broth is an example. One does not know precisely what carbon and energy sources are present nor what other factors necessary for growth are present. Complex media are commonly used in the teaching laboratory because you simply want to grow bacteria and are not concerned what specific nutrients are needed to accomplish this.

The other type of medium is the chemically defined, or **synthetic medium**. In this case, the chemical composition and amount of all components are known. This is the medium of choice for understanding the specific growth requirements of a bacterium.

TABLE 4.3 identifies some typical components of a complex and synthetic medium.

CULTURE MEDIA CAN BE DEVISED TO SELECT FOR AND DIFFERENTIATE BETWEEN BACTERIA

Most common bacteria grow well in common complex or synthetic media. Therefore, these growth media can be modified in one of three ways, depending on what one is trying to accomplish (TABLE 4.4).

A **selective medium** contains ingredients to inhibit the growth of certain bacteria in a mixture while allowing the growth of others. Another type of medium is the **differ-**

TABLE 4.3

Composition of Complex and Chemically Defined Growth Media

INGREDIENT	NUTRIENT SUPPLIED	AMOUNT
COMPLEX AGAR MEDIUM		
Peptone	Amino acids, peptides	5.0 g
Beef extract	Vitamins, minerals, other nutrients	3.0 g
Sodium chloride (NaCl)	Sodium and chloride ions	8.0 g
Agar		15.0 g
Water		1.0 liter
CHEMICALLY DEFINED BROTH MEDIUM		
Glucose	Simple sugar	5.0 g
Ammonium phosphate $[(NH_4)_2HPO_4]$	Nitrogen, phosphate	1.0 g
Sodium chloride (NaCl)	Sodium and chloride ions	5.0 g
Magnesium sulfate $(MgSO_4 \cdot 7H_2O)$	Magnesium ions, sulphur	0.2 g
Potassium phosphate (K_2HPO_4)	Potassium ions, phosphate	1.0 g
Water		1.0 liter

ential medium. This medium makes it easy to distinguish colonies of one organism from colonies of other organisms on the same culture plate. Although most common bacteria grow well in nutrient broth and nutrient agar, certain so-called fastidious bacteria may require an **enriched medium** containing special nutrients. MicroInquiry 4 looks closer at these methods used to identify or separate bacteria.

Other bacteria are simply impossible to cultivate in any laboratory culture medium yet devised. Rather, they require a living tissue medium to grow. Most rickettsiae and chlamydiae are examples of such bacteria. They must be grown in

TABLE 4.4

A Comparison of Bacterial Media

NAME	COMPONENTS	USES	EXAMPLES
Nutrient broth	Water, beef extract, peptone	General use	—
Nutrient agar	Water, beef extract, peptone, agar	General use	—
Selective medium	Growth stimulants Growth inhibitors	Selecting certain bacteria out of mixture	Mannitol salt agar for staphylococci
Differential medium	Dyes Growth stimulants Growth inhibitors	Distinguishing different bacteria in a mixture	MacConkey agar for gram-negative bacteria
Enriched medium	Growth stimulants	Cultivating fastidious bacteria	Blood agar for streptococci; chocolate agar for *Neisseria* species

MicroInquiry 4

IDENTIFICATION OF BACTERIA

It often is necessary to identify a bacterium or be able to tell the difference between similar-looking bacteria in a mixture. In microbial ecology, it might be necessary to isolate certain naturally growing bacteria from others in a mixture. In the clinical and public health setting, microbes might be pathogens associated with disease or poor sanitation. Identification can be accomplished by modifying the composition of a complex or synthetic growth medium. Let's go through several scenarios.

Selecting for a group of organisms. Suppose you are an undergraduate student in a marine microbiology course. On a field trip, you collect some seawater samples and, now back in the lab, you want to grow only photosynthetic microbes.

4a. How could you accomplish this using a complex or synthetic medium?

First, you know the photosynthetic organisms manufacture their own food, so their energy source will be sunlight, not the organic compounds typically found in nutrient media (see Table 4.3). So, you would need to use a synthetic medium but leave out the glucose. Also, knowing the salts typically in ocean waters, you would want to add them to the medium. Inoculate a sample of the collected material into a broth tube, place the tube in the light, and incubate for several days.

What you have used in this scenario is a **selective medium**; that is, one that will encourage the growth of photosynthetic microbes and suppress the growth of nonphotosynthetic microorganisms. Without the glucose source, nonphotosynthetic microbes probably will not grow.

Differentiating between similar species. You are working as a clinical technician and have a urine sample taken from a patient suffering a urinary tract infection. You remember from a microbiology course that both *E. coli* and *Enterobacter aerogenes* can cause such infections. So, after growing the bacterium on nutrient agar, you do a Gram stain and discover you do have

MicroFocus 4.6

SATISFYING KOCH'S POSTULATES

On July 21–23, 1976, some 5,000 legionnaires attended the Bicentennial Convention of the American Legion in Philadelphia, Pennsylvania. About 600 of the legionnaires stayed at the Bellevue Stratford Hotel. As the meeting was ending, several legionnaires who stayed at the hotel complained of flu-like symptoms. Four days after the convention ended, an Air Force veteran who had stayed at the Bellevue Stratford died. He would be the first of 34 legionnaires over several weeks to succumb to a lethal pneumonia that became known as Legionnaires' disease or legionellosis.

As with any new disease, epidemiological studies look for the source of the disease. The Centers for Disease Control and Prevention (CDC) had an easy time tracing the source back to the Bellevue Stratford Hotel. Epidemiological studies also try to identify the causative agent. Using Koch's postulates, CDC staff collected tissues from lung biopsies and sputum samples. However, no microbes could be detected on slides of stained material. By December 1976, they were no closer to identifying the infectious agent.

How can you verify Koch's postulates if you have no infectious agent? It was almost like being back in the times of Pasteur and Koch. Why was this bacterium so difficult to culture in bacteriological media? Was it a virus?

After trying 17 different culture media, the agent was finally cultured. It turns out that the bacterium, named *Legionella pneumophila*, has fastidious physiological needs. The initial medium used was called Mueller Hinton agar and contained a beef infusion, amino acids, and starch. When this medium was enriched with 1% hemoglobin and 1% isovitalex, small, barely visible colonies were seen after five days of incubation at 37°C. It was soon realized that the hemoglobin was supplying iron to the bacterium and the isovitalex was a source of the amino acid cysteine. Using these two chemicals in pure form, charcoal to absorb bacterial waste, a pH of 6.9, and an atmosphere of 2.5% CO_2, the growth of *L. pneumophila* was significantly enhanced. From these cultures, a gram-negative rod was confirmed.

short, pink rods. Unfortunately, both organisms are gram-negative and are short rods! A Gram stain was little help.

4b. If this were a pure culture of one of the two organisms, what can you do to identify the organism?

You can draw on the differences between the two bacteria. *E. coli* under fermentation conditions (without oxygen) produces mixed acids, but *E. aerogenes* does not produce the mixed acids. Therefore, you could simply take a complex medium, such as nutrient broth, and add a pH indicator, such as phenol red, to the medium. Next, you inoculate a sample of the bacteria into one of the red-colored broth tubes and seal the tubes. If acids are produced, the medium will turn yellow during the incubation period. If acids are not produced, the broth stays red. This method is an example of a **differential medium** because it allowed you to differentiate or distinguish between two very similar bacterial species.

You are working as a private food microbiologist for a small ham packing plant. Ham can become contaminated by *Staphylococcus aureus* during processing, so you need to constantly monitor the processing equipment and ham samples to make sure customers do not come down with staphylococcal food poisoning.

The typical medium to use is mannitol salt agar. This medium contains 7.5% NaCl, which is higher than typical nutrient agar (0.5%). It also contains mannitol as an energy source. Only *S. aureus* will grow under the high salt conditions and ferment mannitol. Other species of *Staphylococcus* do not. One of the products of mannitol fermentation is acid, so again phenol red is added as a pH indicator. This type of medium is both selective and differential. It is selective because only *Staphylococcus* species will grow in the high salt. It is differential because the ability of ferment mannitol and produce acid is specific to *S. aureus*.

Now you streak samples from the purported ham source on the agar and incubate for 48 to 72 hours at 37°C.

4c. If *S. aureus* is present, what should you find?

Answers can be found in Appendix E.

Being able to pure culture the bacterium meant that susceptible animals (guinea pigs) could be injected as required by Koch's postulates. Finally, *L. pneumophila* was recovered from infected guinea pigs, verifying the bacterium as the causative agent of Legionnaires' disease.

Today, we know that *Legionella* is found in many aquatic environments, both natural and artificial. At the Bellevue Stratford Hotel, epidemiological studies indicated guests were exposed to the bacterium as a fine aerosol emanating from the air-conditioning system. Through some type of leak, the bacteria gained access to the system from the water cooling towers.

Koch's postulates are still useful—it's just hard sometimes to satisfy the postulates without an isolated pathogen.

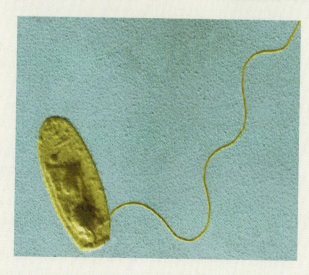

■ *False color transmission electron micrograph of an* L. pneumophila *cell. Note the single, polar flagellum.*

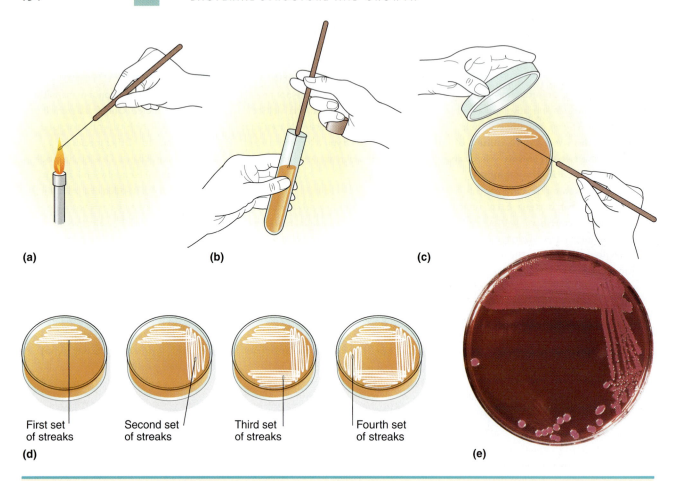

(a)   (b)   (c)

First set of streaks   Second set of streaks   Third set of streaks   Fourth set of streaks

(d)   (e)

FIGURE 4.21

The Streak-Plate Isolation Method

(a) A loop is sterilized, (b) a sample of bacteria is obtained, and (c) streaked along one edge of the plate of medium. (d) Successive streaks are performed, and the plate is incubated. (e) Well-isolated and defined colonies illustrate a successful isolation.

fertilized eggs, tissue cultures, animals, or under other conditions where living cells are found. The difficulty in cultivation often makes detection and study of such organisms a challenge (MicroFocus 4.6).

POPULATION MEASUREMENTS ARE MADE USING PURE CULTURES

Bacteria rarely occur in nature as a single species. Rather, they are mixed with other bacterial species, a so-called mixed culture. However, to work with bacteria, the research biologist and laboratory technologist must use a **pure culture**—that is, a population consisting of only one bacterial species. This is particularly important when trying to identify a pathogen, as Pasteur discovered when trying to discover the agent responsible for cholera.

If one has a mixed broth culture, how can the bacteria be isolated as pure colonies? Two standard methods are available. The first method, called the **streak plate isolation method**, uses a single plate of nutrient agar (FIGURE 4.21). An inoculum from a culture is removed with a sterile loop or needle, and a series of streaks is made on the surface of one area of the plate. The instrument is flamed, touched to the first area, and a second series is made in a second area. Similarly, streaks are made in the third and fourth areas, thereby spreading out the different bacteria so they can form discrete colonies on incubation. In a sense, the bacteria are being diluted to where there are individual bacterial cells on the agar that will then grow into pure colonies.

The second method is the **pour-plate isolation method**. Here, a sample of the mixed culture is diluted in several tubes of cooled, but still molten, agar medium. The agar then is poured into sterile Petri dishes and allowed to harden. During incubation, the bacteria will form discrete colonies where they have been diluted the most. In both methods, the researcher or technologist can then select samples of the colonies for further testing.

POPULATION GROWTH CAN BE MEASURED IN SEVERAL WAYS

To measure the amount (mass) of bacterial growth in a medium, there are numerous methods. For example, the cloudiness, or **turbidity**, of a broth culture may be determined using a spectrophotometer. This instrument detects the amount of light scattered by a suspension of cells. The amount of light scatter (optical density, OD) is a function of the cell number; that is, the more cells present, the more light is scattered and the higher the OD reading on the spectrophotometer. A standard curve can be generated that serves to measure cell numbers in other situations.

There are a number of ways to directly measure cell numbers. Scientists may wish to perform a **direct microscopic count** using a known sample of the culture on a specially designed slide (**FIGURE 4.22**). However, this procedure will count both live and dead cells. The **dry weight** of the bacteria gives an indication of the cell mass, and the **oxygen uptake** in metabolism can be measured as an indication of metabolic activity and therefore bacterial number.

Cell estimations also use indirect methods. Two common methods are the **most probable number test** and the **standard plate count procedure** (Chapter 26). In the

---

FIGURE 4.22

Direct Microscopic Counting Procedure Using the Petroff-Hausser Counting Chamber.

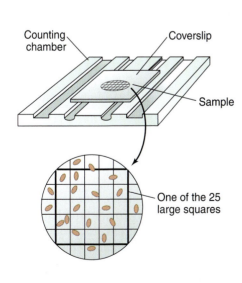

A The counting chamber is a specially marked slide containing a grid of 25 large squares of known area. The total volume of liquid held is 0.00002 ml (2×10^{-5} ml).

B The counting chamber is placed on the stage of a light microscope. The number of cells are counted in several of the large squares to determine the average number.

C Suppose the average was 14 cells per large square.
Calculation:
a. 14 cells $\times$ 25 squares = 350 cells
b. The total liquid volume = 2×10^{-5} ml
c. That calculates to 17,500,000 cells: there are 350 cells/2×10^{-5} ml = (1.75×10^7)/ml.
d. 1.75×10^7/ml $\times$ 10 ml = 1.75×10^8 cells.

former test, samples of bacteria are added to numerous lactose broth tubes, and the presence or absence of gas formed in fermentation gives a rough statistical estimation of the bacterial number. In the latter test, a bacterial culture is diluted, and samples of dilutions are placed in agar plates. Ideally, each cell will undergo multiple rounds of binary fission to produce a separate colony on the plate. Therefore, each cell is called a colony-forming unit (CFU). After incubation, the number of colonies will reflect the number of bacteria originally present. This test is desirable because it gives the **viable count** of bacteria (the living bacteria only), compared to a microscopic count or dry weight test that gives the **total bacterial count** (the living as well as dead).

Note to the Student

It is common to read in biology books about the "higher" and "lower" forms of life. Typically, humans are cast as higher forms, while insects, worms, microorganisms, and other creatures are considered "lower" forms. Discerning biologists would probably disagree with this concept. They would point out that each species of organism is a product of evolution and, as such, each is exquisitely adapted to its environment, and way of life. Just read again the opening quote by Stephen J. Gould.

So, consider the bacteria. Bacterial species thrive in environments ranging from ice to boiling-hot springs. Many species can live with or without oxygen. Large percentages make their own foods from chemicals in the soil. Bacteria have no built-in death age and they can double in number every hour or less. We humans, by contrast, must maintain a constant body temperature; we suffocate without oxygen; we eat complex foods; we reach a certain age, then die; and it takes a full 25 years to produce a generation.

It is difficult to believe that microorganisms are "lower" than any other form of life. Certainly, they are not "lower" than humans. Bacteria were here long before humans came on the scene, and they will undoubtedly be here long after we "higher" forms have vanished.

Summary of Key Concepts

4.1 THE SHAPES AND ARRANGEMENTS OF BACTERIA

- **Bacilli Have a Cylindrical Shape.** Many bacteria are rod-shaped (bacillus). When stained and viewed with the light microscope, rods generally appear singly or in chains (streptobacilli).

- **Cocci Form a Variety of Arrangements.** Spherical bacteria (cocci) occur in a number of arrangements, including the diplococcus, streptococcus, and staphylococcus.

- **Spirals and Other Shapes Also Exist.** The spiral bacteria can be curved rods (vibrios), spiral (spirochetes and spirilla). Spirals generally appear as single cells. Other bacteria have a stalked or filamentous shape.

4.2 THE STRUCTURE OF BACTERIA

- **Bacterial Flagella Provide Motility.** One or more flagella occur on many rods and provide for cell motility. In nature, flagella propel bacteria toward nutrient sources (chemotaxis). In causing human disease, flagella help the bacteria escape immune defenses.

- **Pili Are Structures Used for Attachment.** Pili are short hair-like appendages found on some gram-negative bacteria. Pili facilitate attachment to a surface.

■ **The Glycocalyx Is a Sticky Layer Coating Many Bacteria.** The glycocalyx is a sticky layer of polysaccharides that buffers a bacterium against the external environment. The glycocalyx can be thick and tightly bound to the cell (capsule) or thinner and loosely bound (slime layer).

■ **The Cell Wall Provides Shape and Protection.** The cell wall provides structure and protects against cell lysis. Gram-positive cells have a thick wall of peptidoglycan strengthened with teichoic acids. Gram-negative cells have a thinner layer of peptidoglycan and an outer membrane containing lipopolysaccharides and porin proteins.

■ **The Cell Membrane Is a Permeability Barrier.** The cell membrane represents a permeability barrier and the site of transfer for nutrients into and waste products out of the cell. The cell membrane reflects the fluid mosaic model for membrane structure in that the lipids are fluid and the proteins are a mosaic that can move laterally in the bilayer.

■ **The Cytoplasm Is the Center of Biochemical Activity.** The cytoplasm consists of the cytosol (solutes and water), which contains several structures. Ribosomes carry out protein synthesis while inclusion bodies store nutrients or structural building blocks. The DNA (bacterial chromosome) is located in an area called the nucleoid. Bacteria may contain one or more plasmids, circular pieces of nonessential DNA that replicate independently of the bacterial chromosome.

■ **Endospores Are Designed for Dormancy.** Highly resistant structures called endospores are produced by members of the genera *Bacillus* and *Clostridium*. These structures act as a dormant stage in the life of the bacteria when environmental conditions are not favorable for growth.

4.3 BACTERIAL REPRODUCTION

■ **Bacteria Reproduce by Binary Fission.** The reproduction of bacteria takes place by prokaryotic binary fission, a process wherein chromosomal duplication and cytoplasmic separation are major events. Binary fissions occur at intervals called the generation time, which for bacteria may be as short as 20 minutes.

4.4 BACTERIAL GROWTH

■ **A Bacterial Growth Curve Illustrates the Dynamics of Growth.** Although the potential for incalculable masses of bacteria is great, the dynamics of the bacterial growth curve show how a population grows exponentially, reaches a certain peak and levels off, and then may decline.

■ **Temperature Is One of the Most Important Factors Governing Growth.** Bacteria exhibit temperature ranges over which they grow. Psychrophiles grow at temperatures of 10°–20°C; mesophiles at 20°–40°C; and thermophiles at 40°–70°C. Proteins and enzymes are adapted to the psychrophilic and thermophilic temperature extremes so that denaturation does not occur.

■ **Oxygen Can Support or Hinder Growth.** Bacteria that require oxygen are aerobes or microaerophiles. Bacteria that grow without needing oxygen gas are aerotolerant while those that are killed by the presence of the gas are obligate anaerobes. Many bacteria are facultative, being able to grow with and without oxygen gas.

■ **Most Bacteria Prefer to Grow at a Neutral pH.** As with temperature, bacteria have a small pH range over which they will grow. Most prefer neutral conditions but many acid-tolerant bacteria (acidophiles) exist in acidic environments.

4.5 CULTURE MEDIA AND GROWTH MEASUREMENTS

■ **Culture Media Contain the Nutrients to Grow Bacteria.** Nutrient broth and nutrient agar contain the nutrients that most common bacteria require for growth. These media are examples of a complex medium because the precise form of the nutrients is not known.

■ **Culture Media Can Be Devised to Select for and Differentiate between Bacteria.** Complex or synthetic media can be modified to select for a desired bacterium, to differentiate between two similar bacterial species, or to enrich for bacteria that require special nutrients.

■ **Population Measurements Are Made Using Pure Cultures.** Pure cultures can be produced from a mixed culture by the streak-plate isolation method or the pour-plate isolation method. In both cases, discrete colonies can be identified that represent only one bacterial species.

■ **Population Growth Can Be Measured in Several Ways.** There is a variety of ways to measure growth. Direct measures include a direct microscopic count, dry weight, or oxygen uptake. Indirect methods include the most probable number test and the standard plate count procedure.

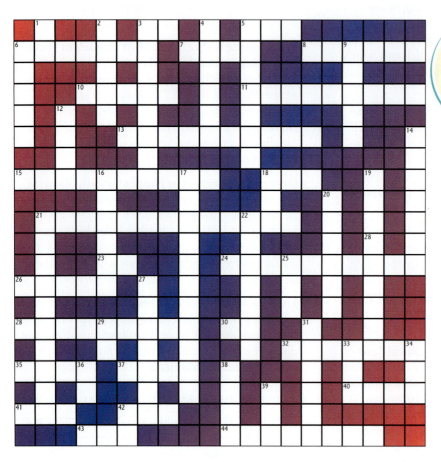

Review

One of the major topics of this chapter has been a survey of microscopic bacterial structures seen through the light and electron microscopes. To test your recall of these structures, fill in the following crossword puzzle. The answers to the puzzle are in Appendix D.

■ ACROSS

3. Glycocalyx-producing bacterial species that causes anthrax (initials)
5. The genetic material of a bacterium
6. A cluster of eight cocci
7. Staining allows one to determine the shape, ___ , and arrangement of bacterial cells
8. Contains the enzyme lysozyme for bacterial destruction
10. The ___ mosaic model describes the structure of the cell membrane
11. The cell membrane is the structure for nutrient ___ into the cell cytoplasm
13. It includes capsules and slime layers

Questions for Thought and Discussion

Answers to selected questions for thought and discussion can be found in Appendix C.

1. Suppose you were a bacterium and had the genetic potential to form a capsule, a flagellum, a pilus, or an endospore. Which would you choose and why?

2. The glycocalyx is a structure associated with many bacterial species. The term is relatively new to microbiology and is derived from the Greek words *glykos* for "sweet" and *kalyx*, referring to the cup of a flower. What do you think microbiologists had in mind when they conceived the word?

3. In reading a story about a bacterium that causes a human disease, the word bacillus is used. How would you know if the article is referring to a bacterial shape or a bacterial genus?

4. Consumers are advised to avoid stuffing a turkey the night before cooking, even though the turkey is refrigerated. A homemaker questions this advice and points out that the bacteria of human disease grow mainly at warm temperatures, not in the refrigerator. What explanation might you offer to counter this argument?

5. Extremophiles are of interest to industrial corporations, who see the bacteria as important sources of enzymes that function at temperatures of 100°C and pH levels of 10 (the enzymes have been dubbed "extremozymes"). What practical uses can you foresee for these enzymes?

6. In the fall of 1993, public health officials found that the water in a midwestern town was contaminated

15. Closed loop of DNA having the bacterial inheritance characteristics
18. Nucleic acid found in plasmids
21. The important chemical constituent of the bacterial cell wall
23. Unit of measurement (abbr) for viruses
24. Spiral bacteria with a flexible cell wall
26. A short hair-like projection for attachment and genetic transfer
28. Bodies of RNA and protein that function in protein synthesis
30. Bacterium (initials) that occurs as a sarcina
32. Closed loop of DNA apart from the chromosome in the cytoplasm
35. The slime of the capsule of *A. viscolactis* is _____ -like
37. Teichoic acid is present in the cell walls of ___-positive bacteria
40. Flagella permit a bacterium to _____
41. A monotrichous bacterium has _____ flagellum/a
42. One of the disaccharides in the bacterial cell wall (abbr)

43. The cell membrane has _____ important organic constituents
44. The name for the chromosomal region of a bacterium

■ **DOWN**
1. Bacterium (initials) that appears as a grape-like cluster
2. A pair of bacterial cocci is a _____-coccus
3. A bacterial rod
4. A curved rod that resembles a comma under the microscope
5. A polysaccharide existing as tangled fibers in the glycocalyx
6. A thin, loosely bound glycocalyx is called the _____ layer
9. About 40 percent of the cell membrane consists of _____
12. The side chains of peptidoglycan consist of _____ amino acids
14. The presence of a capsule contributes to the ability of a pathogen to cause _____
16. A capsule-containing cause of tooth decay is *Streptococcus* _____
17. A cytoplasmic body that helps a bacterium orient itself

18. Nucleic acid forming chromosomes
19. Many times the length of a bacterium but extremely thin; used for motility
20. A bacterial sphere
21. Antibiotic that interrupts construction of bacterial cell wall
22. Serves as a buffer between bacterium and external environment
25. Form displayed by typhoid, anthrax, and diphtheria bacilli
27. Alternative name for pilus
29. Bacterium (initials) well known for its capsule
31. Layer of bacteria and other materials on tooth surface
33. Microscope (abbr) used to visualize cell surfaces
34. Used to stain cells for bright-field microscopy
36. Unlike ___-karyotic cells, bacteria have no nuclei
38. Common site of infection by staphylococci
39. Type of cell (abbr) that cannot easily engulf encapsulated bacteria

with sewage bacteria. The officials suggested that homeowners boil their water for a couple of minutes before drinking it. (a) Would this treatment sterilize the water? Why? (b) Is it important that the water be sterile? Explain.

7. An organism is described as a peritrichous, anaerobic, mesophilic streptococcus. How might you translate this complex bacteriological language into a description of the organism?

8. Humans produce about 500 grams of feces per day. Researchers have estimated that, in broad terms, about one-third of human feces is composed of bacteria. If one *E. coli* cell weighs 1×10^{-12}g, how many bacteria are in a day's feces? That being the case, about how many grams of bacteria do we "produce" in a week? In a year? How can this be possible?

9. A thioglycolate tube has been inoculated with a bacterial species to determine its oxygen requirement.

After incubation for 24 hours, determine where the growth of the bacterium should be seen if (a) the bacterium is microaerophilic, (b) obligately anaerobic, (c) facultative, or (d) obligately aerobic.

10. Suppose this chapter on the structure and growth of bacteria had been written in 1940, before the electron microscope became available. Which parts of the chapter would probably be missing?

11. A bacterium has been isolated from a patient and identified as a gram-negative rod. Knowing that it is a human pathogen, what structures might it have? Explain your reasons for each choice.

12. Many people believe that bacteria do little more than cause human illness and infectious disease. How does the information in this chapter help you correct that misconception?

13. During the filming of the movie *Titanic*, researchers discovered at least 20 different species of bacteria

literally consuming the ship, especially a rather large piece of the midsection. What type of bacteria would you expect were at work on the ship?

14. Although thermophilic bacteria are presumably harmless because they do not grow at body temperatures, they may still present a hazard to good health. Can you think of a situation in which this might occur?

15. To prevent decay by bacteria and to display the mummified remains of ancient peoples, museum officials place the mummies in glass cases where oxygen has been replaced with nitrogen gas. Why do you think nitrogen is used? Will any bacteria-related decay occur in this environment?

http://microbiology.jbpub.com

The site features **eLearning,** an on-line review area that provides quizzes and other tools to help you study for your class. You can also follow useful links for in-depth information, or just find out the latest microbiology news.

Bacterial Metabolism

Life is like a fire; it begins in smoke and ends in ashes.

—Ancient Arab proverb connecting energy to life

CHARLIE SWAART had been a social drinker for years, but in 1945 he began a nightmare that would make medical history. One October day, while stationed in Tokyo after World War II, Swaart suddenly became drunk for no apparent reason. For years thereafter, the episodes continued—bouts of drunkenness and monumental hangovers without drinking so much as a beer. Doctors warned him not to drink for fear of damaging his liver. Swaart followed their advice to the letter; still, he experienced periods of drunkenness.

Twenty years passed before Swaart learned of a similar case in Japan. A Japanese businessman had endured years of social and professional disgrace before doctors found a yeast-like fungus fermenting carbohydrates to alcohol right there in his intestine. An antibiotic had worked to kill the yeast (known as *Candida albicans*). With this knowledge in hand, Swaart approached his doctor. Sure enough, lab tests showed massive colonies of *C. albicans* in Swaart's intestine (**FIGURE 5.1**). Having *Candida* in one's intestine is not uncommon; but finding a fermenting *Candida* was historic. The sugar in a cup of coffee or any carbohydrate in pasta, cake, or candy could bring on drunkenness.

Swaart's doctor prescribed an antibiotic, but the initial result was disappointing. After several tries, however, an effective antibiotic was found. Researchers believe that the atomic blasts of Hiroshima and Nagasaki may have caused a normal *Candida* to mutate to a fermenting form, which somehow found its way into Swaart's digestive system. Perhaps there

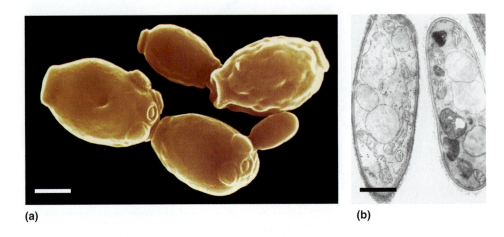

(a)   (b)

Candida albicans

(a) This false color scanning electron micrograph shows a group of *C. albicans* cells. Note their elongated shape. (Bar = 0.3 μm.) (b) Transmission electron micrograph of *C. albicans*. Mutant strains of this organism may ferment carbohydrates to ethyl alcohol. Note the internal structures typical of eukaryotic cells. (Bar = 0.2 μm.).

are thousands of other individuals who harbor a similar *C. albicans* and become fermentation vats. For Charlie Swaart, though, the nightmare was finally over.

The process of fermentation is but one aspect of the broad topic of microbial metabolism. **Metabolism** refers to the sum total of all biochemical processes taking place in living cells. These processes may be divided into two general categories: **Anabolism**, or the synthesis of chemical compounds; and **catabolism**, or the hydrolysis of chemical compounds. From an energy perspective, anabolic reactions form bonds, which require energy. Such energy-requiring reactions are called **endergonic reactions**. In contrast, catabolic reactions break bonds, releasing energy. These reactions are called **exergonic reactions. TABLE 5.1** compares anabolism and catabolism.

TABLE 5.1

A Comparison of Two Key Aspects of Cellular Metabolism

ANABOLISM	CATABOLISM
Buildup of small molecules	Breakdown of large molecules
Products are large molecules	Products are small molecules
Photosynthesis	Glycolysis, Krebs cycle
Mediated by enzymes	Mediated by enzymes
Energy generally is required (endergonic)	Energy generally is released (exergonic)

5.1

Enzymes and Energy in Metabolism

Living cells such as bacteria must have an adequate supply of enzymes for metabolic processes to take place. Therefore, we shall begin our study of metabolism with a detailed discussion of these proteins, which have been known only since the early 1900s (MicroFocus 5.1). In addition, this section will cover some general concepts relating to energy because enzyme catalyzed reactions of anabolism generally use energy, while those in catabolism often liberate energy. Since the discussions will deal with proteins, carbohydrates, and lipids, you may wish to review Chapter 2 as a refresher.

ENZYMES CATALYZE CHEMICAL REACTIONS

Enzymes are a group of organic molecules (usually proteins) that increase the rate of chemical reactions while themselves remaining unchanged. (MicroFocus 5.2 describes RNA molecules that act similar to enzymes.) They accomplish in fractions of a second what otherwise might take hours, days, or longer to happen spontaneously under normal biological conditions. For example, even though organic compounds like amino acids have functional groups, it is highly unlikely that they would randomly bump into one another in the precise way needed for a chemical reaction (dehydration synthesis) to occur and for a new peptide bond to be formed. Thus, the reaction rate would be very slow were it not for the activity of enzymes.

Enzymes are reusable. Once a chemical reaction has occurred, the enzyme is released to participate in another identical reaction, as illustrated in FIGURE 5.2.

MicroFocus 5.1

"HANS, DU WIRST DAS NICHT GLAUBEN!"

Louis Pasteur's discovery of the role of yeast cells in fermentation heralded the beginnings of microbiology because it showed that tiny organisms could bring about important chemical changes. However, it also opened debate on how yeasts accomplish fermentation. Soon, a lively controversy ensued among scientists. Some thought that sugars from grape juice entered yeast cells to be fermented; others believed that fermentation occurred outside the cells. The question would not be resolved until a fortunate accident happened in the late 1890s.

In 1897, two German chemists, Eduard and Hans Buchner, were prepar-ing yeast as a nutritional supplement for medicinal purposes. They ground yeast cells with sand and collected the cell-free "juice." To preserve the juice, they added a large quantity of sugar (as was commonly done at that time) and set the mixture aside. Several days later Eduard noticed an unusual alcoholic aroma coming from the mixture. Excitedly, he called to his brother, "Hans, you'll never believe this!" One taste confirmed their suspicion: The sugar had fermented to alcohol.

The discovery by the Buchner brothers was momentous because it demonstrated that a chemical substance inside yeast cells brings about fermentation, and that fermentation can occur without living cells. The chemical substance came to be known as an "enzyme," meaning "in yeast." In 1905, the English chemist Arthur Haden expanded the Buchner study by showing that "enzyme" is really a multitude of chemical compounds and should better be termed "enzymes." Thus, he added to the belief that fermentation is a chemical process. Soon, many chemists became biochemists, and biochemistry gradually emerged as a new scientific discipline.

Enzyme activity also is highly specific; an enzyme that functions in one chemical reaction usually will not participate in another type of reaction. That means there must be thousands of different enzymes to catalyze the different chemical reactions that occur in a microbial cell.

The substance acted upon by the enzyme is called the **substrate** and the product(s) formed appropriately are termed **products**. Since the reactions are usually reversible, enzymes may bring about synthesis as well as hydrolysis. This factor is important in metabolism because anabolism often occurs by a reversal of many steps in catabolism.

Enzymes originally were named for the chemical reactions they catalyze. For instance, pepsin (from the Greek *pepsis,* for "digestion") refers to the protein-digesting enzyme of the human gastrointestinal tract. Modern biochemists have adopted the *-ase* ending for many enzymes, which often are named for the substrate on which they act. Lactase, for example, is the enzyme that digests lactose, sucrase breaks down sucrose, and ribonuclease digests ribonucleic acid. Some groups of enzymes carry names corresponding to their activity. For example, the hydrolases operate in hydrolysis reactions in which hydroxyl and hydrogen ions from water are added to the substrate. Hydrolases include lipase, which breaks down lipids, and peptidase, which breaks the peptide bond between amino acids.

ENZYMES ACT THROUGH ENZYME-SUBSTRATE COMPLEXES

Enzymes function by aligning substrate molecules in such a way that a reaction is highly favorable. In the hydrolysis reaction shown in Figure 5.2, the jagged surface of

FIGURE 5.2

The Mechanism of Enzyme Action

Although this example shows an enzyme hydrolyzing a substrate (sucrose), enzymes also catalyze dehydration synthesis reactions, which, in this case, would combine glucose and fructose into sucrose.

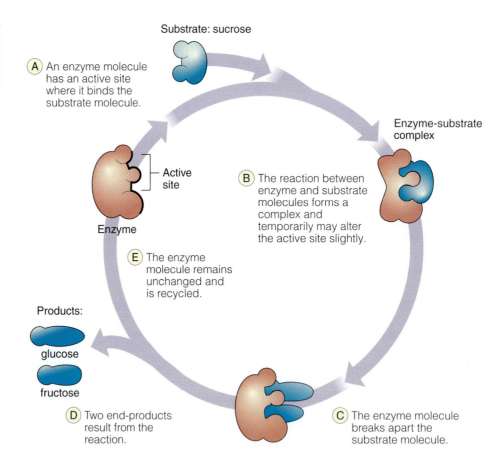

Substrate: sucrose

(A) An enzyme molecule has an active site where it binds the substrate molecule.

Enzyme-substrate complex

Active site

Enzyme

(B) The reaction between enzyme and substrate molecules forms a complex and temporarily may alter the active site slightly.

(E) The enzyme molecule remains unchanged and is recycled.

Products:

glucose

fructose

(D) Two end-products result from the reaction.

(C) The enzyme molecule breaks apart the substrate molecule.

MicroFocus 5.2   RIBOZYMES—TELLING US ABOUT OUR PAST AND HELPING WITH OUR FUTURE

Until the 1980s, one of the bedrock principles of biology held that nucleic acids (DNA and RNA) were the informational molecules responsible for directing the metabolic reactions in the cell. Proteins, specifically the enzymes, were the workhorses responsible for catalyzing the thousands of chemical reactions taking place in the cell. The dogma was "All enzymes are proteins."

In 1981, new research evidence suggested that RNA molecules could act as catalysts in certain circumstances. Today, scientists believe RNA acting by itself can trigger certain chemical reactions.

The seminal research on RNA was performed independently by Thomas R. Cech of the University of Colorado and Sidney Altman of Yale University. Altman had found an unusual enzyme in bacteria, an enzyme composed of RNA and protein. Initially, he thought the RNA was a contaminant, but when he separated the RNA from the protein, the bacterial enzyme could not function. After several years, Altman and his colleagues showed that RNA was the enzyme's key component because it could act alone. At about the same time, Cech discovered that RNA mole-

cules from *Tetrahymena*, a protozoan, could catalyze certain reactions under laboratory conditions. He showed that a molecule of RNA could cut internal segments out of itself and splice together the remaining segments.

Many biologists responded to the findings of Cech and Altman with disbelief. The implication of the research was that proteins and nucleic acids are not necessarily interdependent, as had been assumed. The research also opened the possibility that RNA could have evolved on Earth without protein. In fact, a number of scientists have proposed that life may have started in a primeval "RNA world." This world would have been swarming with self-catalyzing forms of RNA having the ability to reproduce and carry genetic information. In essence, there arose a whole new way of imagining how life might have begun on Earth. The Nobel Prize committee was equally impressed. In 1989, it awarded the Nobel Prize in Chemistry to Cech and Altman.

By 1990, these self-reproducing molecules of RNA had a name—ribozymes. They share many similarities with their protein counterparts, including the presence of binding pockets that, like

active sites on enzymes, recognize specific molecular shapes. Biochemists at Massachusetts General Hospital showed that one type of ribozyme could join together separate short nucleotide segments. The research was a step toward designing a completely self-copying RNA molecule.

Today, the understanding of catalytic ribozymes goes beyond the research laboratory. Several companies are using new molecular techniques to construct new catalytic ribozymes in what is termed "directed evolution." Development of these ribozymes may have uses in clinical diagnostics and as therapeutic agents. For example, in diagnostic applications ribozymes are being developed to identify potential new drugs. Other companies are using ribozymes as biosensors to detect viral contaminants in blood. These catalytic molecules also may be useful in fighting infectious diseases by inactivating RNA molecules in viruses or other pathogens.

So, ribozymes have much to offer in understanding our very distant past as well as providing for a more healthy future.

the enzyme molecule recognizes and holds the substrates in an **enzyme-substrate complex**. While in the complex, chemical bonds in the substrate are stretched or weakened by the enzyme, causing the bond to break. In a synthesis reaction, by contrast, the electron clouds of the substrates in the enzyme-substrate complex are forced to overlap in the spot where the chemical bond will form.

Thus, in a hydrolysis or synthesis reaction, recognition of the substrate(s) is key and not a random event. Each enzyme has a critical area on the surface, called the **active site**, where the substrate(s) is positioned and oriented to react. In Figure 5.2, the active site has a shape that closely matches the parts of the substrate that will interact in this hydrolysis reaction. The shape of an active site is determined by the levels of protein folding described in Chapter 2.

Looking at sucrose again, the bonds holding glucose and fructose together will not break spontaneously. The reason is that the bond between monosaccharides is stable and there is a substantial energy barrier preventing a reaction (FIGURE 5.3a). The job of sucrase is to bind the substrate and lower the energy barrier so that it is much

FIGURE 5.3

Enzymes and Activation Energy

Enzymes lower the activation energy barrier required for chemical reactions of metabolism. (a) The hydrolysis of sucrose is unlikely because of the high activation energy barrier. (b) When sucrase is present, the active site destablizes the covalent bond holding the two monosaccharides together. This destabilization effectively lowers the activation energy barrier, making the hydrolysis reaction highly likely.

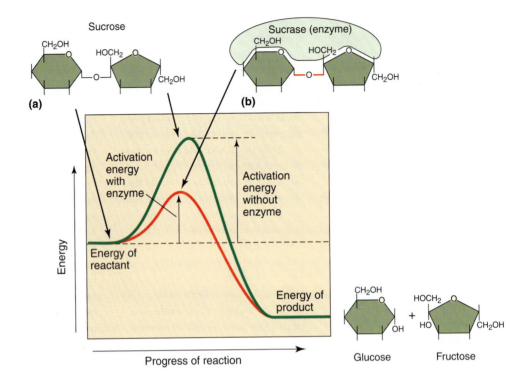

more likely that the reaction will occur. The bond holding glucose to fructose needs to be destabilized (i.e., stretched, weakened) by the enzyme (**FIGURE 5.3b**). The energy required to do this is called the **activation energy**. Enzymes, then, play a key role in metabolism because they lower the amount of activation energy required for a reaction to take place. They assist in the destabilization of chemical bonds and the creation of new ones by separating or joining atoms in a carefully orchestrated fashion.

Some enzymes are made up entirely of protein. An example is lysozyme, the enzyme in human tears and saliva that digests the cell walls of gram-positive bacteria. Other enzymes contain small, nonprotein substances that participate in the catalytic reaction. If the complementing substance is a metal ion, such as magnesium (Mg^{2+}), iron (Fe^{2+}), or zinc (Zn^{2+}), it is called a **cofactor**. When the nonprotein participant is a small organic molecule, it is referred to as a **coenzyme**. Examples of two important coenzymes are nicotinamide adenine dinucleotide (NAD^+) and flavin adenine dinucleotide (FAD). These coenzymes play a significant role as electron carriers in metabolism, and we shall encounter them in our ensuing study of bacterial metabolism.

ENZYMES OFTEN ACT IN METABOLIC PATHWAYS

There are many examples, such as the sucrose example, where an enzymatic reaction is a single substrate to product reaction. However, cells more often use metabolic pathways. A **metabolic pathway** is a sequence of chemical reactions, each reaction catalyzed by a different enzyme, in which the product of one reaction serves as a substrate for the next reaction (FIGURE 5.4). The pathway starts with the initial substrate and finishes with the final end product. The products of "in-between" stages are referred to as *intermediates*.

Metabolic pathways can be anabolic, where larger molecules are synthesized from smaller monomers. In contrast, other pathways are catabolic because they break larger molecules into smaller ones. Such pathways may be linear, branched, or cyclic. We will see many of these pathways in the microbial metabolism sections ahead.

FIGURE 5.4

Metabolic Pathways and Feedback Inhibition

In a metabolic pathway, a series of enzymes transforms an initiate substrate into a final end product. If excess final end product accumulates, it "feeds back" on the first enzyme in the pathway and inhibits its enzymatic action. As a result, the whole pathway becomes inoperative.

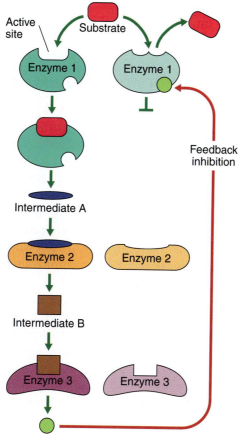

ENZYME ACTIVITY IS REGULATED AND CAN BE INHIBITED

The same chemical reaction does not occur in a cell all the time, even if the substrate(s) is present. Rather, cells regulate the enzymes so that they are present or active only at the appropriate time during metabolism.

One of the most common ways of regulating enzymes is for the final end product of a metabolic pathway to inhibit an enzyme in that pathway (Figure 5.4). If the first enzyme in the pathway is inhibited, then no product is available to "feed" the rest of the pathway, even though the enzymes are present and in an active form. Such **feedback inhibition** is typical of many metabolic pathways in cells. The final end product binds to a site on the enzyme, causing the shape of the active site to change so it can no longer bind substrate.

Active sites often contain sulfhydryl (—SH) groups, so any substance reacting with sulfhydryl groups will tie up the groups and inhibit enzyme activity. A heavy metal such as silver acts in this fashion, and, consequently, it is useful as a disinfectant in such forms as silver nitrate (Chapter 23). Another way of inhibiting an enzyme is by blocking its active site with a compound closely related to the normal substrate. **Sulfonamide drugs** operate in this way so that the enzyme cannot carry out its normal function (Chapter 24).

Since most enzymes are protein molecules, they are sensitive to any physical or chemical agents that injure proteins. Heat can be used to kill bacteria because heat alters the tertiary structure of enzymes. Chemicals, such as alcohol and phenol, precipitate enzyme proteins, as well as other proteins, and therefore act as disinfectants. Any antibiotic that interferes with protein synthesis automatically interferes with enzyme production.

ENERGY IN THE FORM OF ATP IS REQUIRED FOR METABOLISM

In many metabolic reactions, energy is needed, along with enzymes, for the reaction to occur. The cellular "energy currency" is a compound called **adenosine triphosphate**, or simply **ATP** (FIGURE 5.5a). In bacteria, the ATP is formed on the cell membrane, while in eukaryotic microbes the reactions occur primarily in the mitochondria.

An ATP molecule, which is quite similar to an RNA nucleotide, acts like a portable battery. It moves to any part of the cell where an energy-consuming reaction is taking place and provides energy. In a bacterium, ATP supplies energy for activities such as binary fission, flagellar motion, and spore formation. On a more chemical level, it fuels protein synthesis and carbohydrate breakdown. It is safe to say that a major share of bacterial functions depends on a continual supply of ATP. Should the supply be cut off, the cell dies very quickly.

ATP molecules are relatively unstable. In Figure 5.5a, notice that the three phosphate groups all have negative charges on oxygen atoms. Like charges repel, so the phosphate groups in ATP, being tightly packed together, are very unstable. Breaking the so-called "high-energy bond" holding the last phosphate group on the molecule produces a more stable adenosine diphosphate (ADP) molecule and a free phosphate group (FIGURE 5.5b). In breaking the bond, a single mole of ATP in a cell releases about 12,000 calories of energy. (A mole of ATP weighs 507 grams.) ATP hydrolysis is analogous to a spring compacted in a box. Open the box (hydrolyze the phosphate group) and you have a more stable spring (a more stable ADP molecule). The release of the spring (the freeing of a phosphate group) provides the means by which work can be done. Thus, the hydrolysis of the unstable phosphate groups in ATP molecules to a more stable condition is what drives

Feedback inhibition:
the end product of a metabolic pathway inhibits usually the first enzyme in the pathway.

ATP:
adenosine triphosphate, a high-energy molecule that serves as an immediate energy source for cells.

Calorie:
A unit of energy defined as the amount of heat required to raise one kilogram of water 1°C.

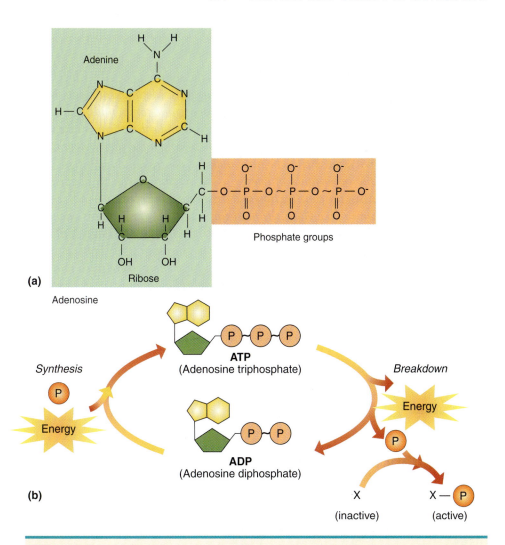

FIGURE 5.5

Adenosine Triphosphate and the ATP/ADP Cycle

Adenosine triphosphate (ATP) is a key immediate energy source for bacteria and other living things. (a) The ATP molecule is composed of adenine and ribose bonded to one another and to three phosphate groups. (b) When the ATP molecule breaks down (right), it releases a phosphate group and 12,000 calories of energy per mole; it becomes adenosine diphosphate (ADP); the freed phosphate can activate another chemical reaction through phosphorylation. For the synthesis of ATP (left), energy and a phosphate group must be supplied to an ADP molecule.

other energy-requiring reactions through the transfer of phosphate groups (Figure 5.5b). The addition of a phosphate group to another molecule is called **phosphorylation**.

Since ATP molecules are unstable, they cannot be stored. Therefore, microbial cells synthesize large organic compounds like glycogen or lipids for energy storage. As needed, the chemical energy in these molecules can be released in catabolic reactions and used to reform ATP from ADP and phosphate (Figure 5.5b). This ATP/ADP cycle occurs continuously in cells. It has been estimated that a typical bacterial cell must reform about 3 million ATP molecules per second from ADP and phosphate to supply its energy needs.

To this point . . .

We have introduced the concept of metabolism as the sum total of all the biochemical processes taking place in bacteria and other living cells. Metabolism is subdivided into two major categories: anabolism, the synthesis of organic compounds, and catabolism, the hydrolysis of these compounds. For anabolism and catabolism to take place, enzymes must be available. We therefore explored these vital molecules in depth, with emphasis on their activity, specificity, chemical makeup, control, and inhibition.

We also discussed the importance of energy as a governing factor in metabolism. Energy is required to assist the construction and destruction of chemical bonds. For the reactions of metabolism to take place, the most important energy source is adenosine triphosphate (ATP). This compound is not stored in the cell; as it is used up, it must be reformed, using the energy present in molecules such as glucose or lipid. In the next section, we shall follow the chemical events in which glucose energy is released and converted to ATP energy. As we proceed through this involved biochemistry, try to keep in mind that the ultimate goal of the process is to form ATP molecules.

5.2

The Catabolism of Glucose

One of the most thoroughly studied and best-understood aspects of metabolism is the catabolism of glucose. Since the early part of the twentieth century, the chemistry of glucose catabolism has been the subject of intense investigation by biochemists because glucose is a key source of energy for ATP production. Moreover, the process of glucose catabolism is very similar in all organisms, making this "metabolic interlock" one feature that unites all life. We therefore shall scrutinize this process closely.

GLUCOSE CONTAINS STORED ENERGY THAT CAN BE EXTRACTED

A mole of glucose (180 g) contains about 686,000 calories of energy. This fact can be demonstrated in the laboratory by setting fire to a mole of glucose and measuring the energy released. In a bacterium, however, not all the energy is set free from glucose, nor can the bacterium trap all that is released. The process accounts for the transfer of about 40 percent of the glucose energy to ATP energy; that is, chemical energy to cellular energy. To simplify matters, we shall follow the fate of one glucose molecule.

The catabolism of a glucose molecule does not take place in one chemical reaction, nor do ATP molecules form all at once. Rather, the energy will be extracted (converted) slowly to ATP. It is similar to the proverb quoted at the beginning of this chapter: "Life is like a fire; it begins in smoke and ends in ashes." The catabolism of glucose starts with a little energy being converted to ATP (the smoke), which builds to a point where large amounts of energy are converted to ATP (the fire), and the original glucose molecule has been depleted of its useful energy (the ashes).

The extraction of energy from glucose involves metabolic pathways. Should a reaction step in the pathway require energy, then a **coupled reaction** often takes place to supply the energy (**FIGURE 5.6**). Also, if a reaction happens to yield excess energy, then a coupled reaction may take place to trap and preserve the energy. Thus, these metabolic pathways involve a sequence of reactions, many of which are associated with coupled reactions. We will see specific examples of these reactions just ahead.

Coupled reaction:
an energy-requiring reaction linked to an energy-releasing reaction.

CELLULAR RESPIRATION IS A SERIES OF CATABOLIC PATHWAYS FOR THE PRODUCTION OF ATP

Virtually all cells make ATP by harvesting energy from exergonic metabolic pathways. Such a process is called **cellular respiration**. If cells consume oxygen in making ATP, the catabolic process is called **aerobic respiration**. In other instances, cells can make almost equally substantial amounts of ATP without using oxygen, in which case it is called **anaerobic respiration**. To begin our study of cellular respiration, we shall follow the process of aerobic respiration as it occurs in bacteria. There are several metabolic pathways for aerobic respiration, but the one we will discuss is represented by the following chemical formula:

$$C_6H_{12}O_6 \ + \ 6\,O_2 \ + \ 38\,ADP \ + \ 38\,P \rightarrow 6\,CO_2 \ + \ 6\,H_2O \ + \ 38\,ATP$$
Glucose   Oxygen   Carbon dioxide   Water

This straightforward equation summarizes a complex series of metabolic reactions conveniently divided into three processes: glycolysis, the Krebs cycle, and oxidative phosphorylation. Let's examine each of these in sequence.

GLYCOLYSIS IS THE FIRST STAGE OF ENERGY EXTRACTION

The chemical breakdown of glucose is called **glycolysis**, from *glyco-*, referring to glucose, and *lysis*, meaning "to break." The process occurs in the cytosol of bacteria and involves a metabolic pathway that converts glucose to a 3-carbon organic molecule called **pyruvate**. Between glucose and pyruvate, there are nine chemical reactions.

Pyruvate:
the final end product of glycolysis.

FIGURE 5.6

A Metabolic Pathway Coupled to the ATP/ADP Cycle

In this metabolic pathway, enzyme A catalyzes an energy-requiring reaction where the energy comes from ATP hydrolysis. Enzyme B converts the phosphorylated substrate to the end product. Being an energy releasing reaction, the free phosphate can be coupled to the reformation of ATP.

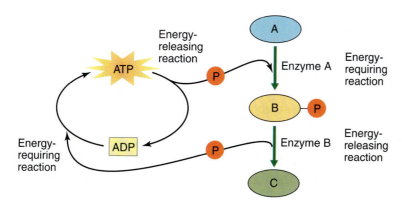

Each reaction is catalyzed by a specific enzyme. Even though glycolysis is the first step of aerobic respiration, it takes place in the absence of oxygen. FIGURE 5.7 illustrates the process. For easy referral, parenthetical numbers in the figure identify each reaction, and it would be helpful to refer to the figure as the discussion proceeds.

The first three steps of glycolysis actually require two molecules of ATP. Note in Figure 5.7 that one molecule of ATP is hydrolyzed in reaction (1) at the beginning of glycolysis and that a second ATP molecule is needed for reaction (3). In both cases, the phosphate group from ATP attaches to the product. Thus, reaction (1) gives us glucose-6-phosphate, and reaction (3) yields fructose-1,6-bisphosphate (*bis* means "two separate"; that is, two separate phosphate molecules).

An important split occurs in reaction (4). The fructose-1,6-bisphosphate molecule breaks apart to yield two molecules, each with three carbons. One is dihydroxyacetone phosphate (DHAP); the other is glyceraldehyde-3-phosphate (G3P). Note that an enzyme converts the DHAP molecule to another G3P molecule. This is important because we now have two molecules of G3P.

In the next series of reactions, each G3P molecule passes through an additional series of conversions and ultimately forms pyruvate. ATP formation occurs in reactions (6) and (9). In both steps, enough energy is released to synthesize an ATP molecule from ADP and phosphate in a coupled reaction. Thus, each time a G3P molecule is broken down to pyruvate through the sequence, two ATP molecules are formed. However, two G3P molecules are available. Therefore, as the second glyceraldehyde-3-phosphate molecule breaks down, two more ATP molecules result. The total is four molecules of ATP. Considering two ATP molecules were "invested" in reactions (1) and (3), the net gain from glycolysis is two molecules of ATP.

Before we proceed, take note of reaction (5). This enzymatic reaction causes two high-energy electrons and two protons (H^+) to be released and shuttled to the coenzyme NAD^+, forming two NADH molecules. This event will have great significance shortly as an additional source of energy.

THE KREBS CYCLE EXTRACTS MORE ENERGY FROM PYRUVATE

The **Krebs cycle** (or citric acid cycle) is a series of chemical reactions named for Hans A. Krebs, who won the 1953 Nobel Prize for the discovery of several participating substances in the cycle. This metabolic pathway is referred to as a cycle because the end product formed is identical to the initial substrate in the pathway. All the reactions are catalyzed by enzymes, and all take place along the cell membranes of bacteria. In eukaryotic microbes, including the protozoa, algae, and fungi, the Krebs cycle reactions occur in the mitochondria.

The Krebs cycle is somewhat like a constantly turning wheel. Each time the wheel comes back to the starting point, something must be added to spin it for another rotation. That something is the pyruvate molecule derived from glycolysis. FIGURE 5.8 shows the Krebs cycle. The reactions are identified by parenthetical capital letters to guide us through the cycle.

Before pyruvate molecules enter the Krebs cycle, they undergo a change, indicated in reaction (A). An enzyme removes a carbon atom from each of the two pyruvate molecules and releases the carbons as two carbon dioxide molecules ($2CO_2$). The remaining two carbon atoms of pyruvate are combined with a substance called coenzyme A (CoA) to form **acetyl CoA**. Equally important, this is another reaction associated with electron and proton transfer to NAD^+ to form NADH.

The two remaining carbons from pyruvate are now ready to enter the Krebs cycle. In reaction (B), each acetyl-CoA unites with a 4-carbon oxaloacetate to form

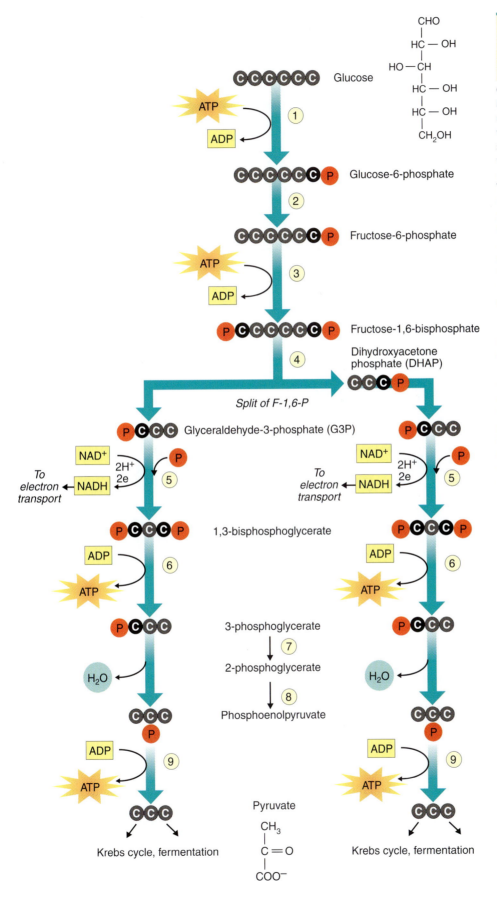

FIGURE 5.7

The Steps of Glycolysis

Carbon atoms are represented by circles. The dark circles represent carbon atoms bonded to phosphate groups. ATP is supplied to glucose in reaction (1) and to fructose-6-phosphate in reaction (3). The splitting of F-1,6-bisphosphate in reaction (4) yields G3P and DHAP. DHAP then converts to another G3P molecule. Both G3P molecules proceed through reactions (5) to (9) and yield two molecules of pyruvate. Two ATP molecules are generated in reaction (6) and two more in reaction (9). The total ATP produced is four molecules and, since two are used up in glycolysis, the net gain is two molecules of ATP. Also, in reaction (5), high-energy electrons and protons are captured by NAD^+ molecules. Note that the original six carbon atoms of glucose exist as two pyruvate molecules.

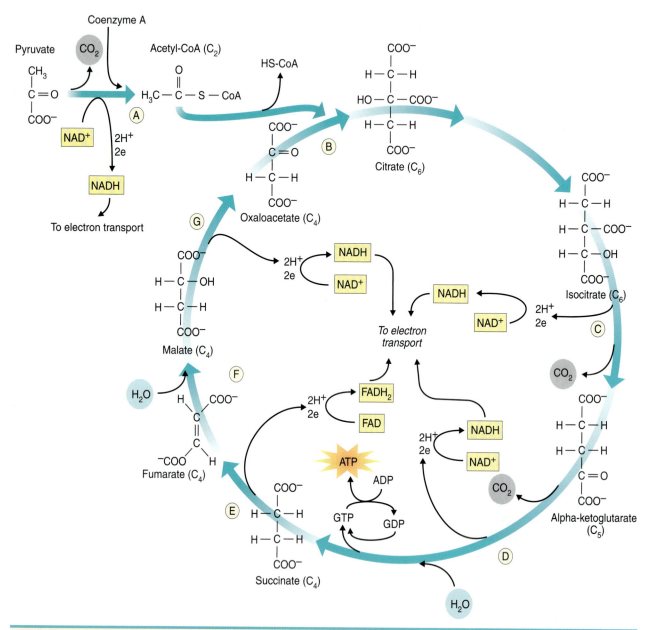

FIGURE 5.8

The Steps of the Krebs Cycle

Pyruvate from glycolysis combines with coenzyme A to form acetyl–CoA in reaction (A). This molecule then joins with oxaloacetate to form citrate (B). In reactions (C) through (F), citrate is converted to alpha-ketoglutarate and then to succinate, fumarate, and malate in succession. Malate regenerates oxaloacetate in the last reaction, (G). During the process, the three carbons of pyruvate are liberated as three molecules of carbon dioxide. ATP is formed in reaction (D) via GTP, and high–energy electrons and protons are liberated during reactions (A), (C), (D), (E), and (G). These are captured by NAD⁺ or FAD molecules, and the electrons are used for oxidative phosphorylation.

citrate, a 6-carbon molecule. (Citrate, or citric acid, may be familiar to you as a component of soft drinks.) Note in Figure 5.8 that isocitrate changes in reaction (C) to a substance having only five carbon atoms. The sixth atom has been released as CO_2. In the next step, (D), another carbon atom is released as CO_2. The 4-carbon molecule then undergoes a series of modifications (reactions E, F, and G) reforming oxaloacetate. The cycle is now complete, and oxaloacetate is ready to unite with another molecule of acetyl-CoA.

Several features of the Krebs cycle merit closer scrutiny. First, we shall follow the carbon. Pyruvate, with three carbon atoms, emerges from glycolysis, but after one turn of the Krebs cycle, its carbon atoms exist in three molecules of CO_2. There are two molecules of pyruvate from glycolysis, so when the second molecule enters the Krebs cycle, its carbon atoms also will form three CO_2 molecules. Remember that we began with a 6-carbon glucose molecule; six CO_2 molecules now have been produced. This fulfills part of the equation for aerobic respiration:

$$C_6H_{12}O_6 \; + \; 6\,O_2 \; + \; 38\,ADP \; + \; 38\,P \rightarrow 6\,CO_2 \; + \; 6\,H_2O \; + \; 38\,ATP$$

The second feature of the Krebs cycle that draws our attention is reaction (D). Here a molecule of guanosine triphosphate (GTP) forms in a coupled reaction. Its energy is immediately used to form an ATP molecule. Since we have two pyruvate molecules entering the cycle (per molecule of glucose), a second ATP molecule will form from GTP when the second pyruvate passes through the cycle. Combining the two ATPs with the net gain of two ATPs from glycolysis, the total gain rises to four ATPs. Impressive as this is, the major gain of ATP is still to come.

Last, and perhaps of most importance, are reactions C, D, E, and G. Reactions C, D, and G, like reaction 5 in glycolysis and in the conversion of pyruvate to acetyl CoA, are associated with NAD^+ and again produce NADH molecules. Two NADH molecules are produced in each step for a total of six. In addition, reaction E accomplishes much the same result except it is associated with another coenzyme, FAD. It too receives two electrons and two protons from the reaction. For the two pyruvates starting the process, two $FADH_2$ molecules are formed.

In summary, glycolysis and the Krebs cycle have extracted as much energy as possible from glucose and pyruvate (**FIGURE 5.9**). This has amounted to a small gain of ATP molecules formed from one glucose molecule. However, the 10 NADH and 2 $FADH_2$ molecules formed are significant. Let's see how.

OXIDATIVE PHOSPHORYLATION IS THE PROCESS BY WHICH MOST ATP MOLECULES FORM

Oxidative phosphorylation refers to a sequence of reactions in which two events happen: Pairs of electrons are passed from one chemical substance to another (electron transport), and the energy released during their passage is used to combine phosphate with ADP to form ATP (ATP synthesis). The adjective oxidative is derived from the term **oxidation**, which refers to the loss of electron pairs from molecules. Its counterpart is **reduction**, which refers to a gain of electron pairs by molecules. Phosphorylation, as we already have seen, implies adding a phosphate to another molecule. So, in oxidative phosphorylation, the loss and transport of electrons will enable ADP to be phosphorylated to ATP. Like the Krebs cycle, oxidative phosphorylation takes place at the cell membrane in bacteria and in the mitochondria of eukaryotic cells.

Oxidative phosphorylation is responsible for producing 34 molecules of ATP per glucose. The overall sequence involves the NAD^+ and FAD coenzymes that underwent reduction to NADH and $FADH_2$ during glycolysis and the Krebs cycle; remember, they gained two electrons in those metabolic pathways. In oxidative phosphorylation, the coenzymes will be reoxidized by passing off those two electrons to a series of electron carriers (**FIGURE 5.10**). These carriers, called **cytochromes**, are a set of protein pigments (*cyto* = "cell"; *chrome* = "color") containing iron ions that accept and release electron pairs. Together, the cytochromes form an **electron transport chain**. The last link in the chain is oxygen.

FIGURE 5.9

Summary of Glycolysis and the Krebs Cycle

Glycolysis and the Krebs cycle are metabolic pathways to extract chemical energy from glucose to make cellular energy (ATP).

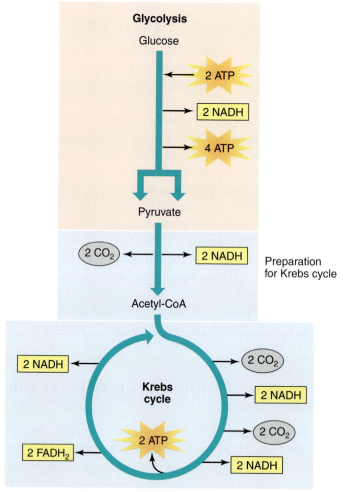

Summary: net 4 ATP (2 glycolysis, 2 Krebs cycle)
10 NADH (2 glycolysis, 8 Krebs cycle)
2 $FADH_2$ (Krebs cycle)

When oxidative phosphorylation is in operation (as in Figure 5.10), the two electrons with each NADH and $FADH_2$ are passed to the first cytochrome in the chain (A). The reoxidized coenzymes, NAD^+ and FAD, return to the cytosol (B) to be used again in glycolysis or the Krebs cycle. Like walking along stepping stones, each electron pair is passed from one cytochrome to the next down the chain until the electron pair is accepted by oxygen. Oxygen also acquires two protons (2 H^+) from the cytosol (C) and becomes water (H_2O). Oxygen's role is of great significance because if oxygen were not present, there would be no way for cytochromes to unload their electrons (but see MicroFocus 5.3), and the entire system would soon back up like a

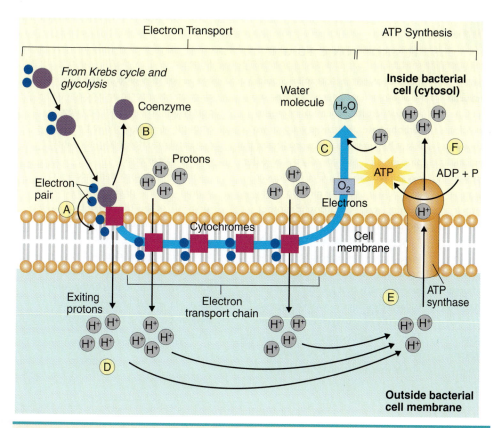

FIGURE 5.10

Oxidative Phosphorylation in Bacteria

(a) Originating in glycolysis or the Krebs cycle, coenzymes NADH and FADH$_2$ transport electron pairs to cytochromes in the cell membrane. (b) The NAD$^+$ or FAD coenzyme is regenerated for reuse. (c) Cytochromes transport the electron pairs, which combine with oxygen and protons to form water molecules. (d) As the electrons are transported, they release energy, which fuels the transport of protons (H$^+$) across the cell membrane at three points. (e) Protons then reenter the cytoplasm of the cell through a protein channel lined in the ATP synthase enzyme. (f) ADP molecules join with phosphates as protons move through the channel, producing the ATP molecules.

jammed conveyer belt and come to a halt. Oxygen's role also is reflected in the equation for aerobic respiration:

$$C_6H_{12}O_6 + 6\,O_2 + 38\,ADP + 38\,P \rightarrow 6\,CO_2 + 6\,H_2O + 38\,ATP$$

So, what is the importance of the electron transport chain since no ATP has been made? The actual mechanism for ATP synthesis comes from the pumping of protons through a process called **chemiosmosis** (*osmos* is Greek for "push"). First proposed by Nobel Prize winner Peter Mitchell, chemiosmosis uses the power of proton movement across a membrane to conserve energy for ATP synthesis.

What happens in chemiosmosis is this: As the electrons pass from cytochrome to cytochrome, the electrons gradually lose energy. The energy, however, is not lost in the sense that it is gone forever. Instead, the energy is used at three transition points to "pump" protons (H$^+$) across the membrane from the cytosol to the area

MicroFocus 5.3

"IT'S NOT TOXIC TO US!"

I t's hard to think of oxygen as a poisonous gas, but billions of years ago, oxygen was as toxic as cyanide. One whiff by an organism, and a cascade of highly destructive oxidation reactions was set into motion. Death followed quickly.

Difficult to believe? Not if you realize that ancient organisms relied on fermentation and anaerobic chemistry for their energy needs. They took organic materials from the environment and digested them to release the available energy. The atmosphere was full of methane, hydrogen, ammonia, carbon dioxide, and other gases. But no oxygen. And it was that way for hundreds of millions of years.

Then came the cyanobacteria and their ability to perform photosynthesis. Chlorophyll and chlorophyll-like pigments evolved, and organisms could now trap radiant energy from the sun and convert it to chemical energy in carbohydrates. But there was a down side: Oxygen was a waste product of the process—and it was deadly because the oxygen radicals (O_2^-, OH^-) produced could disrupt cellular metabolism by "tearing away" electrons from other molecules.

As millions of species died off in the toxic oceans and atmosphere, many organisms "escaped" to oxygen-free environments that are still in existence today. A few species survived by adapting to the new oxygen environment. They survived because they evolved the enzymes to safely tuck away oxygen atoms in a nontoxic form. That form was water. They used oxygen as a final electron acceptor in an electron transport system to tap foods for large amounts of energy. And so the Krebs cycle and oxidative phosphorylation came into existence.

Also coming into existence were millions of new species, some merely surviving, others thriving in the oxygen-rich environment. The face of planet Earth was changing as anaerobic and fermenting species declined and aerobic species proliferated. A couple of billion years would pass before one particularly well-known species of oxygen-breathing creature evolved: *Homo sapiens*.

---

outside of the membrane (D). Soon a large number of protons have built up outside the membrane, and because they cannot easily reenter the cell, they represent a large concentration of potential energy (much like a boulder at the top of a hill). The protons are positively charged, so there also is a buildup of charges outside the membrane.

Suddenly, a series of channels opens and the proton flow reverses (E). Each "channel" is contained within a large enzyme complex called **ATP synthase**. This enzyme complex has binding sites for ADP and phosphate. As the protons rush through the pore, they release their energy, and the energy is used to synthesize ATP molecules from ADP and phosphate ions (F), as MicroInquiry 5 explains. Three molecules of ATP can be synthesized for each pair of electrons originating from NADH. Two molecules of ATP are produced for each pair of electrons from $FADH_2$ because the coenzyme interacts further down the chain.

Chemiosmosis occurs only in structurally intact membranes. Indeed, if the membrane is damaged so proton movement cannot take place, the synthesis of ATP ceases even though electron transport through the cytochrome system continues. With the end of ATP production, the organism rapidly dies. This is one reason why damage to the bacterial membrane, such as with antibiotics or detergent disinfectants, is so harmful to a bacterium.

The ATP yield from aerobic respiration is summarized in **FIGURE 5.11**. The grand total of ATPs from the energy extracted from one glucose molecule is 38. It also completes the equation for aerobic respiration:

$$C_6H_{12}O_6 + 6\,O_2 + \textbf{38 ADP} + \textbf{38 P} \rightarrow 6\,CO_2 + 6\,H_2O + \textbf{38 ATP}$$

To this point . . .

We have discussed the catabolism of glucose by exploring the biochemical reactions of glycolysis, the Krebs cycle, and oxidative phosphorylation. In these processes, glucose first is converted to pyruvate, which then feeds into the Krebs cycle. The carbon of glucose ultimately is released as carbon dioxide. In addition, ATP molecules are formed from reactions that accompany the main metabolic pathway. The major percentage of ATP, however, results from oxidative phosphorylation (electron transport and chemiosmosis), as electrons from glycolysis and Krebs cycle reactions pass from coenzymes to cytochromes and eventually end up in oxygen. The energy released during the passages is used to form ATP. We saw that a total of 38 ATP molecules can be formed from the energy in a single glucose molecule through the reactions in metabolism.

Glucose is fundamental to the life of a bacterium because it provides the energy for chemical activities. However, glucose is not the only chemical substance used for energy. Numerous other carbohydrates, as well as proteins and fats, contribute to the energy metabolism of bacteria. In the next section, we shall see how these compounds are processed. In addition, we shall study mechanisms by which certain microorganisms obtain their energy even though oxygen is not available as an electron acceptor in the electron transport chain.

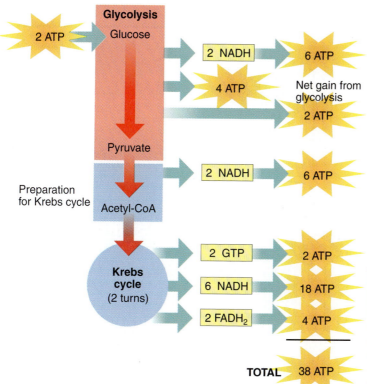

FIGURE 5.11

The ATP Yield from Aerobic Respiration

In a bacterial cell, 38 molecules of ATP normally result from the metabolism of a molecule of glucose. Each NADH molecule accounts for the formation of three molecules of ATP; each molecule of $FADH_2$ accounts for two ATP molecules.

MicroInquiry 5

HOW ATP IS SYNTHESIZED

Every day, an adult human weighing 160 pounds uses up about 80 pounds of ATP (about half of his or her weight). The ATP is changed to its two break-down products, ADP and phosphate, and huge amounts of energy are made available to do metabolic work.

However, the body's weight does not go down, nor does it change perceptibly because the cells are constantly regenerating ATP from the break-down products. Discovering how this is accomplished and how the recycling works were the seminal achievements of 1997's Nobel Prize winners in Chemistry.

THE CHEMIOSMOTIC BASIS FOR ATP SYNTHESIS

One of the three 1997 winners was Paul D. Boyer at the University of California at Los Angeles. Boyer's work expanded the pioneering work of Peter Mitchell, who developed the concept of chemiosmosis. Chemiosmosis proposes that electron transport between cytochromes provides the energy to "pump" protons (H^+) across the membrane; in the case of bacteria, from the cytosol to the environment. As explained in the text, this proton gradient provides the force or potential to drive the protons back into the cell through an enzyme called **ATP synthase**. This flow of rapidly streaming protons (H^+) brings together ADP and phosphate to form ATP.

HOW DOES PROTON FLOW CAUSE ATP SYNTHESIS?

The groundbreaking research as to how the ATP synthase works came from studies with *Escherichia coli* cells. Researchers knew that ATP synthase consisted of three functional regions (**FIGURE A**). Embedded in the cell membrane was an F_0 complex consisting of nine polypeptides of three different types. Extending from the cytoplasmic end of F_0 is a stalk made of three different proteins. The stalk connects the F_0 complex to the F_1 complex. The F_1 complex consists of six polypeptides, three called α polypeptides and three called β polypeptides. So, an ATP synthase consists of 18 polypeptides—a veritable nanomachine.

Boyer took the three complexes and hypothesized how they could manufacture ATP. His ideas plus newer findings have been merged into the current model (**FIGURES B** and **C**):

1. The flow of protons through F_0 causes F_0 and the stalk to spin as protons stream by (somewhat reminiscent of a turning waterwheel).

2. One of the polypeptides (γ) in the stalk extends into the F_1 complex and makes contact with each of the β-subunits as it rotates.

3. The rotation of the stalk causes three events to occur (**FIGURE C**).

   a. The open binding site loosely binds substrates [ADP + P] and the stalk rotates 120°.

   b. The binding site tightly binds the substrates [ADP + P]. The stalk rotates another 120°. In this tightly bound form, the substrates spontaneously interact to form ATP.

   c. The binding site releases the ATP. The stalk rotates another 120°, returning the binding site to its open shape.

Since there are three β-subunits, three ATP molecules are produced each time the stalk makes a complete rotation. Each β-subunit is like an active site in an enzyme. It recognizes and binds substrate [ADP and P], forms an enzyme-substrate complex, and releases the product (ATP).

So, like the rotation of the bacterial flagellum, here is another example of cells building and using a rotary engine, or nanomachine, to carry out very specific functions in the cell.

FIGURE A

Assembly of $F_0 F_1$ complex

FIGURE B

ATP synthesis

FIGURE C

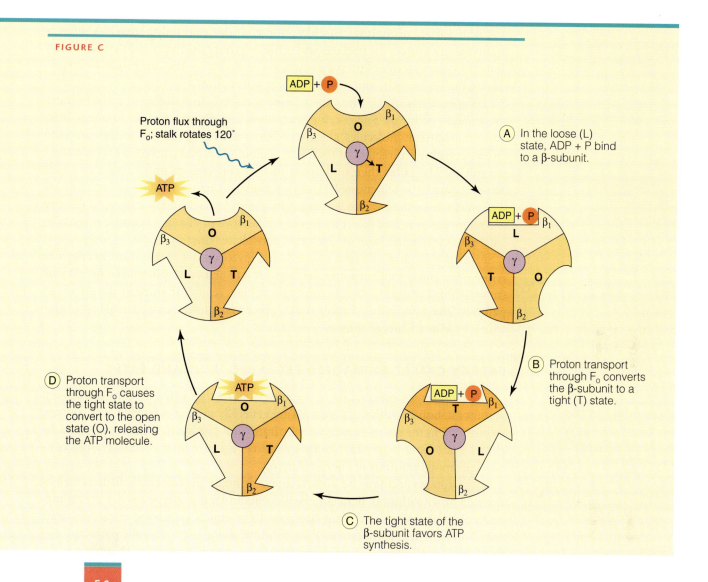

Proton flux through F_o; stalk rotates 120°

A In the loose (L) state, ADP + P bind to a β-subunit.

B Proton transport through F_o converts the β-subunit to a tight (T) state.

C The tight state of the β-subunit favors ATP synthesis.

D Proton transport through F_o causes the tight state to convert to the open state (O), releasing the ATP molecule.

5.3

Other Aspects of Catabolism

The catabolism of glucose is a complex process and, for the student of microbiology, it is a demanding aspect of metabolism to learn and understand. However, the process is central to the metabolism of bacteria, as well as a great many other living organisms, and it provides a glimpse of how living things obtain energy for life. Moreover, the process of glucose catabolism is a major thoroughfare to which many other biochemical pathways lead and from which numerous pathways extend. In this section, we shall examine how cells obtain energy from various other organic compounds by incorporating the compounds into the process of glucose catabolism. We also shall see how modifications of the basic scheme account for the use of glucose to make ATP without using oxygen as a final electron acceptor in electron transport. The use of one major pathway with multiple offshoots lends a modicum of economy to the metabolism of a cell.

OTHER CARBOHYDRATES REPRESENT POTENTIAL ENERGY SOURCES

A wide variety of monosaccharides, disaccharides, and polysaccharides serve as useful energy sources for bacteria and other microorganisms. All of these carbohydrates undergo a series of preparatory conversions before they are processed in glycolysis, the Krebs cycle, and oxidative phosphorylation.

In preparation for entry into the scheme of metabolism, different carbohydrates use different pathways. Sucrose, for example, is first digested by the enzyme sucrase into its constituent molecules, glucose and fructose. The glucose molecule enters the glycolysis pathway directly, but the fructose molecule is first converted to fructose-1-phosphate. The latter then undergoes further conversions and a molecular split before it enters the scheme as DHAP (reaction 4). Lactose, another disaccharide, is broken in two by the enzyme lactase to glucose and galactose. Glucose again directly enters the pathway, but galactose undergoes a series of changes before it is ready to enter glycolysis in the form of glucose-6-phosphate (reaction 2).

Stored polysaccharides, such as starch and glycogen, are metabolized as enzymes remove one glucose unit at a time and convert it to glucose-1-phosphate. An enzyme converts this compound to glucose-6-phosphate, ready for entry to the glycolysis pathway. The point is that carbohydrates other than glucose can be used just as glucose is used as an energy source.

ENERGY CAN BE EXTRACTED FROM PROTEINS AND FATS

The economy of metabolism is demonstrated further when we consider protein and fat catabolism. Although proteins are generally not considered energy sources, cells use them for energy when carbohydrates and fats are in short supply. Fats, by contrast, are extremely valuable energy sources because their chemical bonds contain enormous amounts of chemical energy.

Both proteins and fats are broken down through glucose catabolism as well as through other pathways. Basically, the proteins and fats undergo a series of enzyme-catalyzed conversions and form components normally occurring in carbohydrate metabolism. These components then continue along the metabolic pathways as if they originated from carbohydrates. Proteins are broken down to amino acids. Enzymes then convert many amino acids to pathway components by removing the amino group and substituting a carbonyl group. This process is called **deamination**. For example, alanine is converted to pyruvate and aspartic acid is converted to oxaloacetate. For certain amino acids, the process is more complex, but the result is the same: The amino acids become pathway intermediates of cellular respiration.

Fats consist of one or more fatty acids bonded to a glycerol molecule. To be useful for energy purposes, the fatty acids are separated from the glycerol by the enzyme lipase. Once this has taken place, the glycerol portion is converted to DHAP. For fatty acids, there is a complex series of conversions called **beta oxidation**, in which each long-chain fatty acid is broken by enzymes into 2-carbon units. Other enzymes then convert each unit to a molecule of acetyl CoA ready for the Krebs cycle. We previously noted that for each turn of the Krebs cycle, 16 molecules of ATP are derived. A quick calculation should illustrate the substantial energy output from a 16-carbon fatty acid (eight 2-carbon units).

ANAEROBIC RESPIRATION PRODUCES ATP USING OTHER FINAL ELECTRON ACCEPTORS

Certain anaerobic or facultative bacteria metabolize carbohydrates through **anaerobic respiration**, a process in which oxygen is not used as the final electron acceptor in electron transport. Still, the mechanism of ATP synthesis is dependent on oxidative phosphorylation. Instead of oxygen, anaerobic or facultative bacteria carrying out anaerobic respiration using a different inorganic molecule as a final electron acceptor. For example, *Escherichia coli* uses nitrate ions (NO_3^-) at the end of the cytochrome chain. Electrons combine with nitrate ions to form nitrite ions (NO_2^-). Microbiologists take advantage of this chemistry in a laboratory diagnostic test for identifying nitrite producers, such as *E. coli*.

Members of the genus *Desulfovibrio* use sulfate ions ($SO_4^=$) for anaerobic respiration. The sulfate combines with the electrons of the cytochrome chain and changes to hydrogen sulfide (H_2S). This gas gives a rotten egg smell to the environment (as in a tightly compacted landfill). A final example is exhibited by members of the genera *Methanobacterium* and *Methanococcus*. These archaeal cells use carbon dioxide as a final electron acceptor and, with hydrogen nuclei, they form large amounts of methane (CH_4). Some scientists believe that because of this chemistry, methane-producing bacteria existed on Earth when the only available gas was carbon dioxide. They postulate that methane actually may have entered the atmosphere for the first time through the activity of these bacteria.

FERMENTATION PRODUCES ATP USING AN ORGANIC FINAL ELECTRON ACCEPTOR

The chemical process of **fermentation** makes a few ATP molecules in the absence of cellular respiration. Fermentation is a unique process because the Krebs cycle and oxidative phosphorylation are shut down. Therefore, an organic molecule, usually an intermediary in a metabolic pathway, accepts the electrons rather than inorganic molecules like oxygen gas. For example, in the fermentation of glucose by certain bacteria and yeasts, an intermediary molecule accepts the electrons and protons from NADH formed in reaction (5) of glycolysis. This regenerates NAD^+ molecules for reuse as electron acceptors in glycolysis. Without oxidative phosphorylation running, NAD^+ exists in limited supply in the cytosol and must be continually regenerated so that glycolysis may proceed. Thus, fermentation is a way to recycle NAD^+ coenzymes so glycolysis can still make two ATP molecules for every glucose molecule consumed. Let's look at two examples.

The bacterium *Streptococcus lactis* carries out fermentation by using pyruvate to accept the electrons and protons from NADH (FIGURE 5.12a). An enzyme reaction converts the pyruvate to lactate (lactic acid). In the process, NAD^+ is reformed. In a dairy plant, fermentation by *S. lactis* is carefully controlled so the acid will curdle fresh milk to make buttermilk.

The fermentation chemistry in yeasts such as *Saccharomyces* is somewhat different because yeasts contain a different enzyme. In these cells, the pyruvate first is converted to acetaldehyde, a process in which carbon dioxide evolves (FIGURE 5.12b). Acetaldehyde then serves an acceptor for the electrons and protons of NADH. As the acetaldehyde is converted to ethyl alcohol, NAD^+ is reformed. The

FIGURE 5.12

The Relationship of Fermentation to Glycolysis

Fermentation is an anaerobic process that reoxidizes NADH to NAD+. The NAD+ is used to keep glycolysis running and producing two ATP molecules per glucose. (a) Lactic acid fermentation; (b) alcoholic fermentation.

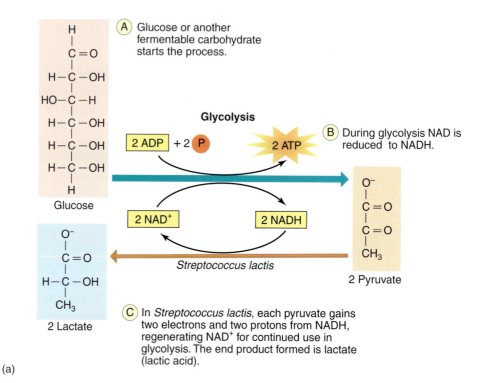

(a)

(b)

A Glucose or another fermentable carbohydrate starts the process.

B During glycolysis NAD is reduced to NADH.

C In *Streptococcus lactis*, each pyruvate gains two electrons and two protons from NADH, regenerating NAD+ for continued use in glycolysis. The end product formed is lactate (lactic acid).

D In yeast cells, pyruvate from glycolysis first is converted to acetaldehyde. This intermediary then gains two protons from NADH to regenerate the NAD+ for use in glycolysis. The end product formed is ethanol.

liquor industry uses the ethyl alcohol produced in fermentation to make alcoholic beverages such as beer and wine (Chapter 27). Fermentation of carbohydrates to alcohol by *C. albicans* also may take place in the human body, as explained in the opening of this chapter.

The energy benefits to *Streptococcus lactis* and yeasts are far less in fermentation than in cellular respiration. The only ATP produced is the net two molecules resulting from glycolysis. This is in sharp contrast to the 38 molecules evolving in cellular respiration. It is clear that cellular respiration is the better choice for energy conservation, but under the circumstances of an oxygen-free environment, there is little alternative if life for *Streptococcus lactis* or *Saccharomyces* is to continue.

In the food industry, fermentation results in a broad variety of useful products. Swiss cheese, for instance, develops its flavor partly from propionic acid resulting from fermentation and gets its holes from fermentation gases. Pickles and sauerkraut are sour because bacteria ferment the carbohydrates in cucumbers and cabbage, respectively. Sausage tastes like sausage because bacteria ferment the meat proteins. Thus, fermentation is useful not only to the microorganisms, but also to consumers who enjoy its products (Chapter 27).

To this point . . .

We have surveyed the process of cellular respiration through glucose metabolism, and we saw how numerous carbohydrates, proteins, and fats can be metabolized by bacterial cells through modifications of the scheme. In the latter cases, enzymes first convert the organic molecules to intermediary compounds of the process. The intermediates then proceed along the metabolic pathway and give up their energy to yield ATP. This chemistry illustrates a basic economy in cell metabolism because a centralized process is used, and various organic molecules fit into it.

The discussion then shifted gears to explore how energy may be obtained when oxygen is not used as a final electron acceptor. We described how other inorganic molecules are used as electron acceptors under anaerobic conditions so that the metabolism through the Krebs cycle and oxidative phosphorylation can continue. Some time was spent with fermentation, a unique process that uses an organic intermediary as an electron acceptor. Fermentation has great significance in the food and liquor industries.

We now shall turn our attention briefly to anabolism and discuss the methods by which carbohydrates are synthesized in bacterial cells. Our discussions will be relatively brief because we are more interested in an overview than in specific details. The processes of anabolism, together with those of catabolism, provide a window to the fundamental metabolism of bacterial life.

5.4

The Anabolism of Carbohydrates

Although the anabolism or synthesis of carbohydrates takes place through various mechanisms in bacteria, the unifying feature is the requirement for energy. In a few bacteria, energy comes from sunlight, but in most species, energy is derived from chemical reactions. On these bases, two general patterns of bacterial anabolism exist: photosynthesis and chemosynthesis. Chemosynthesis will be mentioned in the last section of this chapter.

PHOTOSYNTHESIS IS A PROCESS TO ACQUIRE CHEMICAL ENERGY

Photosynthesis is a process by which light energy is converted to chemical energy that is then stored as glucose or other organic compounds. In prokaryotes, the process takes place in the cell membrane, which contains chlorophyll or chlorophyll-like pigments. ATP is a key intermediary compound in the process, and glucose is a major end product. Among prokaryotes, photosynthesis occurs in the green sulfur bacteria, the purple sulfur bacteria, and the cyanobacteria (**FIGURE 5.13**). Among eukaryotes, photosynthesis occurs in the chloroplasts of such organisms as diatoms, dinoflagellates, and other algae. Our discussion will focus on the prokaryotes.

Cyanobacteria carry out photosynthesis in much the same manner as eukaryotic microorganisms and green plants. Light energy is absorbed by the green pigment **chlorophyll *a***, a magnesium-containing, lipid-soluble compound (**FIGURE 5.14a**). Chlorophylls and accessory pigments make up a light-receiving system called **photosystem II**. The light excites pigment molecules, and each molecule loses one electron (1). Dislodged electrons then are accepted by the first of a series of electron carriers (2). The electrons are passed along the series of cytochromes, and eventually the electrons are taken up by other chlorophyll pigments that form **photosystem I**.

As the electrons move down the cytochromes, energy is made available for proton pumping across the cell membrane of the cyanobacterium, followed by chemiosmosis. As described for oxidative phosphorylation, ATP is formed when protons pass back across the membrane and release their energy. Since light was involved in the formation of ATP, this process is called **photophosporylation**.

The electrons in photosystem I again are excited by light energy (3) and are boosted out of the pigment molecules to the first of another set of electron carriers, and finally to a coenzyme called **nicotinamide adenine dinucleotide phosphate** (**NADP⁺**). The coenzyme functions much like NAD⁺ in that NADP⁺ receives the electrons and also hydrogen ions from water molecules to form **NADPH** (4).

Hence, two major products, ATP and NADPH, result from this phase of photosynthesis. The phase is termed the **energy-fixing reactions** because light energy is trapped and converted to (or "fixed" as) chemical energy. It is important to note that electrons are replaced in the chlorophyll molecules by electrons from water molecules (which also supply hydrogen ions, as noted above). The residual portions of the water molecules recombine with one another and yield oxygen, the oxygen that fills the atmosphere and is used by aerobic organisms for cellular respiration. Organisms that produce oxygen in photosynthesis are said to be **oxygenic**.

In the second phase of photosynthesis, another cyclic metabolic pathway forms carbohydrates. The process is known as the **carbon-fixing reactions** because the carbon in carbon dioxide is trapped (**FIGURE 5.14b**). An enzyme bonds carbon dioxide to a 5-carbon organic substance called ribulose 1,5-bisphosphate (RuBP) (1). (The enzyme is called ribulose bisphosphate carboxylase.) The resulting 6-carbon molecule then splits to form two molecules of 3-phosphoglycerate (3GP). In the next step, the products of

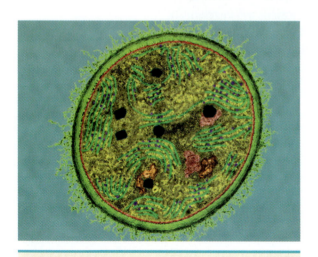

FIGURE 5.13

Cyanobacterial Membranes

False color transmission electron micrograph of a cyanobacterium displaying the membranes (green) along which photosynthetic pigments are located. These membranes are analogous to thylakoid membranes in the chloroplasts of algae and complex plant cells.

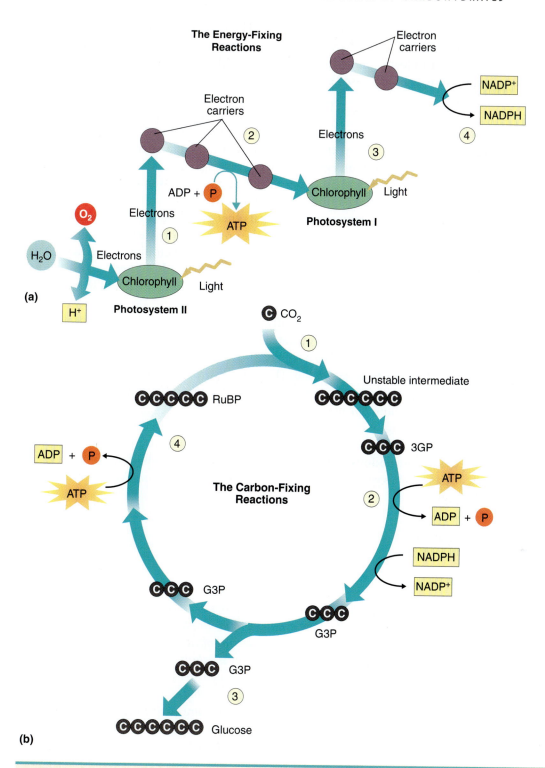

(a)

(b)

FIGURE 5.14

Photosynthesis in Microorganisms

(a) The energy-fixing reaction occurring along the membranes of a cyanobacterium. (1) Electrons in chlorophyll receive a boost in energy from light and (2) ATP is synthesized as the electrons pass among electron carriers. (3) In noncyclic photosynthesis, the electrons receive a second boost, and (4) the energy is used to form high-energy $NADPH_2$. The ATP and $NADPH_2$ are used in the carbon-fixing reactions. (b) The carbon-fixing reaction. (1) Carbon dioxide unites with ribulose bisphosphate (RuBP) to form an unstable 6-carbon molecule. (2) The latter splits to form two molecules of phosphoglyceric acid (PGA) then phosphoglyceraldehyde (PGAL). ATP and $NADPH_2$ from the light-fixing reaction are used in the conversion. (3) Condensations of two 3-carbon PGAL molecules yields glucose, and (4) the remainder is used to form RuBP to continue the process. ATP is used in the latter reaction.

the energy-fixing reactions, ATP and NADPH, drive the conversion of 3GP to glyceraldehyde-3-phosphate (G3P) (2). Two molecules of G3P then condense with each other to form a molecule of glucose (3). Thus, the overall formula for photosynthesis may be expressed as:

$$6\,CO_2 \;+\; 6\,H_2O \;+\; ATP \rightarrow C_6H_{12}O_6 \;+\; 6\,O_2 \;+\; ADP + P$$

Notice that this equation is the reverse of the equation for aerobic respiration. The fundamental difference is that aerobic respiration is an energy-yielding process, while photosynthesis is an energy-trapping process.

To finish off the cycle, G3P molecules also undergo a complex series of enzyme-catalyzed reactions that require ATP to reform RuBP (4).

In addition to the cyanobacteria, several other groups of prokaryotes trap energy by photosynthesis. Two such groups are the green sulfur bacteria and purple sulfur bacteria, so named because of the colors imparted by their pigments. These bacteria have chlorophyll-like pigments known as **bacteriochlorophylls**, to distinguish them from other chlorophylls. Bacteriochlorophylls *a* and *b* are found in purple sulfur bacteria. In the energy-fixing reactions, the organisms do not use water as a source of hydrogen ions. Consequently, no oxygen is liberated, and the bacteria are said to be **anoxygenic**. Instead of water, a series of inorganic or organic substances, such as fatty acids, are used as a source of hydrogen ions. Certain species of green sulfur bacteria use hydrogen sulfide (H_2S) as a hydrogen ion source. Thus, the green and purple sulfur bacteria commonly live under anaerobic conditions in environments such as sulfur springs and stagnant ponds.

Another variation of bacterial photosynthesis occurs in the archaea. Instead of the usual chlorophylls, the extreme **halophiles** of this group contain a pigment called **bacteriorhodopsin** (which is similar to the rhodopsin of the human eye). In the presence of oxygen, the extreme halophiles can synthesize ATP with the aid of this pigment.

Halophiles: bacteria that live in high-salt environments.

To this point . . .

We have examined the anabolism of carbohydrates by studying the process of photosynthesis in bacteria, specifically the cyanobacteria. In photosynthesis, light energy is converted into chemical energy in the form of carbohydrates. This was a two-phase process. In the energy-fixing reactions, light energy is captured by chlorophyll molecules. Excited electrons first are used to generate ATP through an electron transport chain and chemiosmosis. The electrons then are captured by another system of chlorophyll molecules. Light again excites the electrons, which now are used to reduce NADP$^+$ to NADPH. In the carbon-fixing reactions, the ATP and NADPH from the energy-fixing reactions are used to drive the synthesis of carbohydrates.

We briefly mentioned other types of photosynthesis among the green sulfur bacteria and the purple sulfur bacteria.

To conclude the chapter on metabolism, we will put together and review the patterns of metabolism that make up the metabolic diversity of the bacteria.

5·5

Patterns of Metabolism

Bacteria must meet certain nutritional requirements for growth. Besides water, which is an absolute necessity, bacteria need nutrients that can serve as energy sources and raw materials for the synthesis of cell components. These generally include proteins for structural compounds and enzymes, carbohydrates for energy, and a series of vitamins, minerals, and inorganic salts.

Two different patterns exist for satisfying an organism's metabolic needs. These patterns are called autotrophy and heterotrophy (TABLE 5.2). They are primarily based on the source of carbon used for making cell components.

AUTOTROPHS AND HETEROTROPHS GET THEIR ENERGY AND CARBON IN DIFFERENT WAYS

Organisms that synthesize their own foods from simple carbon sources are referred to as **autotrophs** (literally "self-feeding"). The organisms are said to be autotrophic. **Photoautotrophs**, such as the cyanobacteria, synthesize foods using light energy and carbon dioxide gas as described for photosynthesis.

Another group of autotrophs are the **chemoautotrophs**. During carbohydrate synthesis, these organisms also get their carbon from carbon dioxide gas. However, they use chemical reactions to obtain energy from inorganic compounds. Therefore, this process of carbohydrate anabolism is called **chemosynthesis** rather than photosynthesis. For example, species of *Nitrosomonas* convert ammonium ions (NH_4^+) into nitrite ions (NO_2^-) under aerobic conditions, thereby obtaining ATP. Bacteria of the genus *Nitrobacter* then convert the nitrite ions into nitrate ions (NO_3^-), also as an ATP-generating mechanism. In addition to providing energy to both species of bacteria, these reactions have great significance in the environment as a critical part of the **nitrogen cycle** (Chapter 26). By preserving nitrogen in the soil in the form of

Chemosynthesis: carbohydrate anabolism using chemical reactions as energy sources.

TABLE 5.2

A Nutritional Classification of Microorganisms

NUTRITIONAL TYPE	ENERGY SOURCE	CARBON SOURCE	EXAMPLES
Autotrophs			
Photoautotroph	Light	Carbon dioxide (CO_2)	Photosynthetic bacteria (green sulfur and purple sulfur bacteria), cyanobacteria, algae
Chemoautotroph	Inorganic compounds	Carbon dioxide (CO_2)	*Nitrosomonas, Nitrobacter*
Heterotrophs			
Photoheterotroph	Light	Organic compounds	Purple nonsulfur and green nonsulfur bacteria
Chemoheterotroph	Organic compounds	Organic compounds	Most bacteria; all fungi; and protozoa

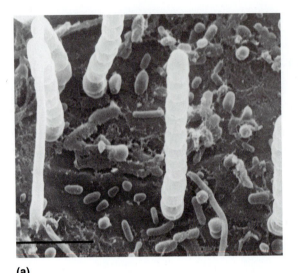

(a)   (b)

FIGURE 5.15

Chemoautotrophs

(a) A scanning electron micrograph of bacteria and other microorganisms attached to natural surfaces near cracks in the floor of the Pacific Ocean. These samples were retrieved from a depth of 2,550 meters (about 1.68 miles) near vents where the temperature was over 400°F. Hydrogen sulfide from the vent is a major source of energy for these marine bacteria. (Bar = 5 μm.) (b) The typical habitat of chemoautotrophic bacteria includes a diversity of other organisms, including red–tipped tubeworms.

nitrate, it can be used by green plants to form amino acids. Other bacteria living near ocean vents can obtain their energy from hydrogen sulfide (**FIGURE 5.15**). Marine bacteria also can use other electron acceptors.

The majority of microorganisms are **heterotrophs** (literally "other-feeders"). Such heterotrophic organisms obtain their energy and carbon in one of two ways. The **photoheterotrophs** use light as their energy source and preformed organic compounds such as fatty acids and alcohols as sources of carbon. Photoheterotrophs include certain green nonsulfur and purple nonsulfur bacteria.

The **chemoheterotrophs** use preformed organic compounds for both their carbon source and energy source. Glucose would be one example. Those chemoheterotrophic bacteria that feed exclusively on dead organic matter, such as rotting wood or decaying animal and plant matter, are commonly called **saprobes**. In contrast, chemoheterotrophs that feed on living organic matter, such as human tissues, are commonly known as **parasites**. The word **pathogen** (from the Greek *pathos* for "suffering") is used if the parasite causes disease in its host organism. We will certainly see many examples of this in upcoming chapters.

TABLE 5.3 summarizes the major pathways of bacterial metabolism.

TABLE 5.3

Some Characteristics of Major Pathways of Bacterial Metabolism

PATHWAY	LOCATIONS	REACTANTS	PRODUCTS
Glycolysis	Cytoplasm	Glucose	Pyruvate
Acetyl CoA formation	Bacterial membranes	Pyruvate	NADH, CO_2, acetyl CoA
Krebs cycle	Bacterial membranes	Acetyl CoA	ATP, NADH, $FADH_2$, CO_2
Electron transport and chemiosmosis	Bacterial membranes	NADH, $FADH_2$ O_2, NO_2^-, H_2S, CH_4	ATP, H_2O
Fermentation	Cytosol	Pyruvate	Ethanol, CO_2 (alcoholic) Lactic acid (lactic acid)
Photosynthesis (cyanobacteria) 1. Energy-fixing reactions 2. Carbon-fixing reactions	Bacterial membranes Cytosol	Light energy, H_2O CO_2, RuBP, ATP, NADPH	Oxygen gas, ATP, NADPH Glucose, H_2O
Chemosynthesis (chemoautotrophs)	Bacterial membranes Cytosol	H_2S, NH_4^+ NO_2^-	Elemental sulfur, NO_2^-, NO_3^-

Note to the Student

This chapter contains some of the most difficult concepts you will encounter in microbiology. The concepts, though, are among the most fundamental because biochemistry applies to all life, be it bacterial life, plant life, or human life. I hope you can step back and see the forest as well as the trees. Virtually all living things produce ATP by glucose catabolism, and all creatures depend ultimately on the carbohydrate produced in photosynthesis.

One of the corollary benefits of studying the metabolism of microorganisms is that you come away with a better understanding of living things in general. Nowhere is this more apparent than in this chapter. Bacteria and other microorganisms are mere tools used to discover the metabolic foundations that govern all living things. Beyond the window dressing of variation, there is an underlying kinship among all forms of life. To understand one form is to understand them all.

Summary of Key Concepts

5.1 ENZYMES AND ENERGY IN METABOLISM

■ **Enzymes Catalyze Chemical Reactions.** The two major themes of bacterial metabolism are catabolism (the breakdown of organic molecules) and anabolism (the synthesis of organic molecules). For either of these two processes to occur, bacteria and other living things use enzymes, a series of protein molecules that speed up a chemical change while themselves remaining unchanged.

■ **Enzymes Act through Enzyme-Substrate Complexes.** Enzymes bind to substrates by means of a three-dimensional surface area, called the active site. By binding the substrate(s) in an enzyme-substrate complex, the functional groups are destabilized and the activation energy is lowered.

■ **Enzymes Often Act in Metabolic Pathways.** Many metabolic processes in microbes occur by means of metabolic pathways, where a sequence of chemical reactions is catalyzed by different enzymes. Metabolic pathways comprise much of this chapter.

■ **Enzyme Activity Is Regulated and Can Be Inhibited.** One of the most common ways by which enzymes are regulated is through feedback inhibition, where the final end product of a metabolic pathway inhibits often the first enzyme in the pathway. Enzymes also can be inhibited by heavy metals and some antibiotics. Physical and chemical agents can denature enzymes and inhibit their action in cells.

■ **Energy in the Form of ATP Is Required for Metabolism.** Many metabolic reactions in cells require energy in the form of adenosine triphosphate (ATP). The breaking of the terminal phosphate produces enough energy to supply an endergonic reaction, and often involves the addition of the phosphate to another molecule (phosphorylation). Since ATP molecules are constantly being used in living cells, they must be resynthesized to maintain the reactions of metabolism.

5.2 THE CATABOLISM OF GLUCOSE

■ **Glucose Contains Stored Energy that Can Be Extracted.** One of the principal objectives of catabolism is to release the energy in organic molecules, such as glucose, for the synthesis of ATP. The breakdown of glucose with the release of energy occurs through several metabolic pathways. If excess energy is produced, the reaction is coupled to an endergonic

reaction. If energy is needed, it is coupled to an exergonic reaction.

■ **Cellular Respiration Is a Series of Catabolic Pathways for the Production of ATP.** The cellular reaction series in which energy is released to make ATP is called cellular respiration. It may require oxygen gas (aerobic respiration) or another inorganic molecule (anaerobic respiration).

■ **Glycolysis Is the First Stage of Energy Extraction.** The catabolism of glucose to pyruvate extracts some energy from which two ATPs and 2 NADH molecules result.

■ **The Krebs Cycle Extracts More Energy from Pyruvate.** The catabolism of pyruvate into carbon dioxide and water in the Krebs cycle extracts more energy as ATP, NADH, and $FADH_2$. Carbon dioxide gas is released.

■ **Oxidative Phosphorylation Is the Process by which Most ATP Molecules Form.** The process of oxidative phosphorylation involves the oxidation of NADH and $FADH_2$, the transport of freed electrons along a cytochrome chain, the pumping of protons across the cell membrane, and the synthesis of ATP from a reversed flow of protons. These last two steps are referred to as chemiosmosis.

5.3 OTHER ASPECTS OF CATABOLISM

■ **Other Carbohydrates Represent Potential Energy Sources.** Other carbohydrates, such as sucrose, lactose, and polysaccharides, represent energy sources that can be metabolized through cellular respiration.

■ **Energy Can Be Extracted from Proteins and Fats.** Besides carbohydrates, proteins and, especially, fats can be metabolized through the cellular respiratory pathways to produce ATP.

■ **Anaerobic Respiration Produces ATP Using Other Final Electron Acceptors.** The anaerobic respiration of glucose uses different final electron acceptors in oxidative phosphorylation. Glycolysis and the Krebs cycle still function and ATP synthesis occurs.

■ **Fermentation Produces ATP Using an Organic Final Electron Acceptor.** In fermentation, the catabolism of glucose can continue without a functional Krebs cycle or oxidative phosphorylation process. To maintain a steady supply of NAD^+ for glycolysis and ATP synthesis, pyruvate is redirected into other pathways that reoxidize NADH to NAD^+. End products include lactate or ethanol. Only the two ATP molecules of glycolysis are synthesized in fermentation from each molecule of glucose.

5.4 THE ANABOLISM OF CARBOHYDRATES

■ **Photosynthesis Is a Process to Acquire Chemical Energy.** The anabolism of carbohydrates can occur by photosynthesis. In this process, light energy is used to synthesize ATP, and the latter is then used to fix atmospheric carbon dioxide into carbohydrate molecules. Other bacteria carry out photosynthesis using other pigments and without water as a source of hydrogen ions.

5.5 PATTERNS OF METABOLISM

■ **Autotrophs and Heterotrophs Get Their Energy and Carbon in Different Ways.** Autotrophs that synthesize their own food from carbon dioxide and light energy (photosynthesis) are called photoautotrophs. Microbes that synthesize their own food from carbon dioxide and inorganic compounds (chemosynthesis) are chemoautotrophs. Heterotrophs that obtain their carbon from organic compounds and energy from light are referred to as photoheterotrophs, while those that obtain both carbon and energy from organic compounds are chemoheterotrophs.

Questions for Thought and Discussion

The answers to selected questions can be found in Appendix C.

1. Some years ago, a magazine cartoon pictured newspapers being carried into a "conversion plant" and beef cattle coming out. The cartoonist envisioned bacteria in the plant would convert the cellulose of the newspapers into the protein of beef cattle. Can you explain the chemistry behind this look into the future?

2. Citrase is the enzyme that converts citrate to α-ketoglutarate in the Krebs cycle. A chemical company has located a mutant microorganism that cannot produce this enzyme and proposes to use the microorganism to manufacture a particular product. What do you suppose the product is? How might this product be useful?

3. A student observes that during the process of respiration, a bacterium exhales before it inhales. What is he thinking?

4. A microbiology professor from a California college maintains that the most abundant enzyme in the world is ribulose 1,5-bisphosphate carboxylase. What do you think this enzyme accomplishes? On what basis does she make her claim?

5. One of the most important steps in the evolution of life on Earth was the appearance of certain organisms in which photosynthesis takes place. Why was this critical?

6. If ATP is such an important energy source in bacteria, why do you think it is not added routinely to the growth medium for these organisms?

7. Your lab partner maintains that organisms use proteins to synthesize enzymes. You counter that organisms use enzymes to synthesize proteins. Who is right? Why?

8. You have two flasks with broth media. One contains a species of cyanobacteria. The other flask contains *E. coli*. Both flasks are sealed and incubated under optimal growth conditions for two days. Assuming the cell volume and metabolic rate of the bacterial cells is identical in each flask, why would the carbon dioxide concentration be higher in the *E. coli* flask than in the cyanobacteria flask after the two-day incubation?

9. A microbiology professor from Virginia writes: "In the fall when the apples were ripe in our farm orchard, we had a cow that could not be stopped by any fence (shades of the old nursery rhyme). Topsy would stagger home to the barn every night doing a very good imitation of a drunken sailor. . . ." What two possibilities might you offer for this observation?

10. The formula for a growth medium for a particular bacterium stipulates that riboflavin must be added for the synthesis of a certain chemical compound of the catabolism process. Can you guess which compound? What would be the effect of omitting riboflavin?

11. A stagnant pond usually has a putrid odor because hydrogen sulfide has accumulated in the water. A microbiologist recommends that tons of green sulfur bacteria be added to remove the smell. What chemical process does the microbiologist have in mind? Do you think it will work?

12. You make up two batches of a synthetic agar medium. The only difference between them is the carbon course. Medium A has lactose as the energy and carbon source and medium B has protein as these sources. You streak *E. coli* over both plates and incubate them at 37°C for 48 hours. Which plate will show better growth of colonies? Explain your reasoning thoroughly.

13. A population of a *Bacillus* species is growing in a soil sample. Suppose glycolysis came to a halt in these bacterial cells. Would this mean that the Krebs cycle would also stop? Why?

14. An essential factor in the growth media for bacteria is phosphate. How many places in a cell can you cite where it is needed?

15. A student goes on a college field trip and misses the microbiology exam that covered the material on metabolism. Having made prior arrangements with the instructor for a make-up exam, he finds but one question on the exam: "Discuss the interrelationships between anabolism and catabolism." How might you have answered this question?

Review

When you have completed your study of bacterial metabolism, test your knowledge of its important facts and concepts by circling the choices that best complete each of the following statements. The answers are listed in Appendix D.

1. The sum total of all a bacterium's biochemical reactions is known as (catabolism, metabolism); it includes all the (synthesis, digestion) reactions called anabolism and all the break-down reactions known as (inactivation, catabolism).

2. Enzymes are a group of (carbohydrate, protein) molecules that generally (slow down, speed up) a chemical reaction by converting the (substrate, substart) to end products.

3. The aerobic respiration of glucose begins with the process of (oxidative phosphorylation, glycolysis) and requires that (amino acids, energy) be supplied by (ATP, GTP) molecules.

4. The process of (fermentation, Krebs cycle) takes place in the absence of (oxygen, magnesium) and begins with a molecule of (glucose, protein) and ends with molecules of (amino acid, alcohol).

5. In oxidative phosphorylation, pairs of (protons, electrons) are passed among a series of (chromosomes, cytochromes) with the result that (oxygen, energy) is released for (NAD, ATP) synthesis.

6. In the Krebs cycle, (glucose, pyruvate) undergoes a series of changes and releases its (carbon, nitrogen) as (nitrous oxide, carbon dioxide) and its electrons to (NAD^+, ATP).

7. For use as energy compounds, proteins are first digested to (uric, amino) acids, which then lose their (carboxyl, amino) groups in the process of (fermentation, deamination) and become intermediates of cellular respiration.

8. Ribulose 1,5-bisphosphate bonds with (carbon monoxide, carbon dioxide) molecules during (fermentation, photosynthesis), a process that ultimately results in molecules of (gluconic acid, glucose).

9. Chemoautotrophs use energy from (light, chemical reactions) to synthesize (carbohydrates, proteins) and are typified by species of (*Staphylococcus*, *Nitrosomonas*).

10. Fats are broken down to (fatty acids, coenzymes), which are converted through (beta oxidation, deamination) reactions to (glucose, two-carbon units) and eventually enter (cellular respiration, photosynthesis).

http://microbiology.jbpub.com

The site features **eLearning,** an on-line review area that provides quizzes and other tools to help you study for your class. You can also follow useful links for in-depth information, or just find out the latest microbiology news.

Bacterial Genetics

We wish to suggest a structure for the salt of deoxyribose nucleic acid (DNA). This structure has novel features which are of considerable biological interest.

—In the first 1953 paper by Watson and Crick describing the structure of DNA

I

N THE MODERN ERA, when time is measured in minutes and seconds, our minds find it difficult to imagine the colossal 4.5 billion years that the Earth has been in existence. It may help, however, to think of Earth's history as a single year. In our "historic" year, the Earth was a Mars-like, lifeless ball of rock until mid-July when bacteria, or something akin to bacteria, first appeared. These were the only creatures on Earth until mid-October, when multicellular organisms emerged. Not until the end of November did the first land plants come into being, and not until early December did animals move out of the sea onto the land. The dinosaurs were in existence from December 19 to December 25, and by December 27, the Earth bore a resemblance to modern Earth. Finally, on December 31, close to midnight, humans appeared.

We take this trek through geologic time to help us appreciate why bacteria have prospered genetically and in evolutionary terms. They have been successful primarily because they have been around the longest and have adapted well. Bacteria have been on Earth about 3.5 billion years (versus about 200,000 years for humans), as **FIGURE 6.1** shows. During this time, gene changes have been occurring regularly and nature has used the bacteria to test its newest genetic traits. The detrimental traits have been eliminated (together with the bacteria unlucky enough to have them), while the beneficial traits have thrived and have moved onto the next generation—and onto the present day.

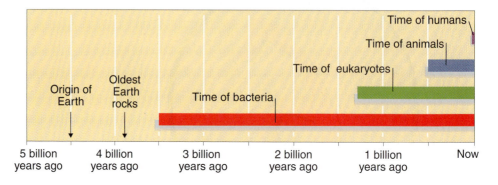

FIGURE 6.1

The Bacteria on Earth

This time line shows the relative amounts of time that various groups of organisms have existed on Earth. The bacteria have been in existence for a notably longer period than any other group, particularly humans. They have adapted well to Earth simply because they have had the longest opportunity to evolve through adaptation.

Modern bacteria, therefore, enjoy the fruits of genetic changes. Because of their diverse genes, bacteria can thrive in the diverse environments on Earth, whether it is the snows of the Arctic or the boiling hot volcanic vents of the ocean depths. No other organism can compare to bacteria in sheer numbers—a pinch of rich soil has more bacteria than all the people living in the United States today. Finally, consider a bacterium's multiplication rate—a new generation every half hour—and it is easy to see how a useful genetic change (such as drug resistance) can be propagated quickly in a stressful environment (one containing a drug).

Any one of these factors—time on Earth, sheer numbers, multiplication rate—would be sufficient to explain how bacteria have evolved to their current form. When taken together, the factors help us appreciate why bacteria have done very well in the evolutionary lottery—very well, indeed.

In this chapter, we shall examine **mutation**, one of the two processes that have brought ancient bacteria to the myriad forms we observe on Earth today. Mutation (**MicroFocus 6.1**) is an alteration in the bacterial chromosome by a permanent change in its DNA. The second process, **recombination**, is a chromosomal alteration by the acquisition of new DNA from another organism (Chapter 7). The two processes are the major subdivisions of bacterial genetics. To understand the material in these topics, we first must look at DNA replication and how the information in DNA is processed and regulated in making proteins.

6.1

The Bacterial DNA

In all organisms, except some viruses, the genetic information is **deoxyribonucleic acid** (**DNA**). However, before the 1950s, scientists were not sure if protein or nucleic acid was the genetic material. With 20 amino acids and only four nucleotides, it seemed that proteins could encode more information than nucleic acids. However, the work of Avery and his coworkers and the experiments

MicroFocus 6.1

MUTATING TO SURVIVE?

Giraffes got long necks by stretching into treetops for food when they were starving. Right? "No way," you say. "That's Lamarck's theory of acquired characteristics. It was discarded after Charles Darwin discovered natural selection." According to Darwin, a population of animals includes individuals with various neck lengths and when food at ground level is scarce, those with longer necks survive because they can feed a bit higher up in the trees.

Could there be exceptions that behave as Lamarck said? Some scientists believe mutations can be induced or even guided by selective stress. In 1988, John Cairns and colleagues at Harvard School of Public Health experimented with an *E. coli* strain that could not use lactose as an energy source. When these cells are placed on lactose medium, they should not grow. However, over several days of starvation, some cells did divide and colonies appeared. The conclusion was that in a few cells a mutation occurred that restored the ability to metabolize lactose. The shocking aspect of this result was that the mutants appeared directed—they only arose when lactose was available. No other mutation types appeared. The process of making directed "mutations on demand" has been called adaptive mutation.

Cairns's results were reinterpreted in 1995 by researchers who found evidence that stress seemed to induce general (rather than directed) mutations. They noticed that the lactose-utilizing mutants generated a variety of mutations during their time on the lactose medium. They, therefore, concluded that stress causes general and intense mutation events in a few cells in the population. Most of these cells die (of lethal mutations), but the few survivors are those that can metabolize lactose (and relieve the stress) before dying—so mutation only looks directed. These results contradicted the purest definition of the adaptive mutation theory (i.e., that stressed cells accumulate only useful mutations) but led to a new idea—when under stress, bacteria may undergo mutations throughout their whole genome to boost their chance of survival. Trying to improve your genome by random, wholesale mutation is like trying to fix your watch with a hammer. Could bacteria actually evolve a mechanism like this? Probably not.

Based on additional experiments, a new interpretation of Cairns's data proposes that mutants actually arise in slowly growing colonies that are too small to see. The strain Cairns used had a defective gene that produces too little enzyme for growth on lactose. In this strain, rare cells arise before plating on lactose that have two copies of this defective gene and, therefore, are just barely able to grow—very slowly—on lactose. Within these slow-growing clones, cells arise that have more copies of the defective gene, which allow better growth. When enough duplicated genes are present, the likelihood increases that a mutation will occur that repairs one copy of this gene. Now growth really accelerates. So, there does not have to be any increase in mutation rate because there are simply more gene copies susceptible to mutation. Given Cairns's assumption that mutations arise in the nongrowing cells, stress-induced mutation seemed like the only explanation. However, when cells grow under selection, Darwinian natural selection can operate (as it did for the giraffes) to slowly improve growth and generate more cells (and mutational targets).

This controversy is not settled. If it seems incredible that an apparently simple experiment done more than 15 years ago is still a source of argument, just remember—the question of whether selection causes mutations has remained unresolved since the time of Darwin. This fascinating system may provide insights about mutation under selection and perhaps even about how the giraffe got a long neck.

of Hershey and Chase described in Chapter 1 pointed to DNA as the genetic material. Then, in 1953 James Watson and Francis Crick worked out DNA's double helix structure based, in part, on the X-ray diffraction studies of Rosalind Franklin (MicroFocus 6.2).

In Chapter 2, we described the structure of the DNA molecule, so it might be helpful to review the chemical structure before proceeding too far in this chapter.

BACTERIAL DNA IS TIGHTLY PACKED WITHIN A SINGLE CHROMOSOME

Most of the genetic information in a bacterial (or prokaryotic) cell is contained within the **chromosome** as a single, usually circular molecule of DNA. The chromosome exists as thread-like fibers associated with some protein and is localized in

MicroFocus 6.2

THE TORTOISE AND THE HARE

We all remember the children's story of the tortoise and the hare. The moral of the story was that those who plod along slowly and methodically (the tortoise) will win the race over those who are speedy and impetuous (the hare). The names Watson and Crick are familiar to students of biology as the scientists who first proposed the correct structure of DNA. In doing so, they constructed a model of how hereditary material makes replicas of itself, and showed how genes could encode the synthesis of protein in a cell. It is a story of collaboration and competition—a science tortoise and the hare.

Rosalind Franklin (the tortoise) was 31 when she arrived at King's College in London in 1951 to work in J. T. Randell's lab. Having received a Ph.D. in physical chemistry from Cambridge University, she moved to Paris where she learned the art of X-ray crystallography. At King's College, Franklin was part of Maurice Wilkins's group and she was assigned the job of using X-ray crystallography to work out the structure of DNA fibers. Her training and constant pursuit of excellence allowed her to produce excellent, high-resolution X-ray photographs of DNA.

Meanwhile, at the Cavendish Laboratory in Cambridge, Watson (the hare) was in a rush for honor and greatness by figuring out the structure of DNA. His brash "bull in a china shop" attitude was in sharp contrast to Franklin's philosophy that you don't make conclusions until all of the experimental facts have been analyzed. Therefore, until she had all the facts, Franklin was reluctant to share her data with Wilkins—or anyone else for that matter.

Feeling left out, Wilkins was more than willing to help Watson and Crick. Since Watson thought Franklin was "incompetent in interpreting X-ray photographs" and he was better able to use the data, Wilkins shared with Watson an X-ray photograph and a report Franklin had filed. From these materials, it was clear that DNA was a helical molecule. It also seems clear that Franklin knew this as well but, perhaps being a physical chemist, she did not grasp its importance

because she was concerned with getting all the facts first and making sure they were absolutely correct. But, looking through the report that Wilkins shared, the proverbial "lightbulb" went on when Crick saw what Franklin had missed; that the two DNA strands were antiparallel. This knowledge, together with Watson's ability to work out the base pairing, led Watson and Crick to their "leap of imagination" and the structure of DNA.

In her book entitled *Rosalind Franklin: The Dark Lady of DNA* (HarperCollins, 2002), author Brenda Maddox suggests that it is uncertain if Franklin could have made that leap as it was not in her character to jump beyond the data in hand. In this case, the leap of intuition won out over the methodical, data collecting in research—the hare beat the tortoise this time. However, it cannot be denied that Rosalind Franklin's data provided an important key from which Watson and Crick made the historical proclamation.

In 1962, Watson, Crick, and Wilkins received the Nobel Prize in Physiology or Medicine for their work on the structure of DNA. Should Franklin have been included? Perhaps. However, the Nobel Prize committee does not make awards posthumously and Franklin had died four years earlier from ovarian cancer. The question remains: If she had lived, did she deserve to be included in the award?

---

the cytosol in a space called the **nucleoid**. In bacterial cells, remember that one of the unique features of the nucleoid area is the absence of a surrounding membrane envelope typical of eukaryotic cells. The eukaryotic chromosome is compared to the bacterial (prokaryotic) chromosome in **TABLE 6.1**.

The chromosome of *Escherichia coli* probably has been studied more thoroughly than any other single chromosome. The **genome** of *E. coli* has about 4,300 genes. Some viruses, by contrast, have as few as seven genes, while the human genome has some 35,000 genes. Distributed around the chromosome are individual sites to which genetic activity can be traced. Each site, called a **locus** (pl., loci), consists of one or several genes for that activity.

Genome:
the complete set of genetic information in a cell.

TABLE 6.1

Characteristics of Prokaryotic (Bacterial) and Eukaryotic Chromosomes

PROKARYOTIC (BACTERIAL) CHROMOSOME	EUKARYOTIC CHROMOSOME
Replicates just prior to binary fission	Replicates just prior to mitosis
Single molecule of DNA per genetic trait (haploid)	Two molecules of DNA per genetic trait (diploid); some organisms haploid
Chromosome is a closed loop	Chromosomes are linear
Little protein present	Histone protein present
No dominance or recessiveness in genes	Genes may be dominant or recessive, or have other inheritance patterns
Organized at the nucleoid	Organized in the nucleus with a membrane envelope
No introns present	Introns present
DNA also in plasmids	DNA also in mitochondria and chloroplasts (algae and plants)
Mutations occur in DNA	Mutations occur in DNA
Genetic recombinations occur	Genetic recombinations occur
4,000 genes in *E. coli* chromosome	35,000 genes in total of human chromosomes
About 1.5 mm in length	Tens or hundreds of millimeters in length
Replicates by semiconservative method Rolling circle method occurs	Replicates by semiconservative method No evidence of rolling circle method

In *E. coli*, the DNA occupies about half of the total volume of the cell, and when extended its full length, it is about 1.5 millimeters (mm) long. This is approximately 500 times the length of the bacterium. So, how can a 1.5 mm long chromosome fit into a 1.0 to 2.0 μm *E. coli* cell? The answer is supercoiling and tight packing, which accounts for the explosive release of DNA when the membrane of the cell is broken (FIGURE 6.2a). The DNA in the chromosome is **supercoiled**, meaning the DNA double helix is twisted on itself like a wound-up rubber band. The coils are folded further into loops of 50,000 to 100,000 bases and anchored to RNA and protein molecules in the nucleoid. An overall "flower" structure called the **looped domain structure** results (FIGURE 6.2b).

PLASMIDS CARRY NONESSENTIAL INFORMATION

Many bacteria contain relatively small, usually closed loops of DNA called **plasmids**. One or more plasmids may be found in most bacterial cells, although gram-negative bacteria are notable for the presence of plasmids. Plasmids exist as independent units in the cytosol (Figure 6.2a) and contain about 2 percent of the total genetic information of the cell. They can multiply independently of the chromosome and plasmids are easily transferred between cells.

Plasmids are not essential to the life of the bacterial cell but they can confer selective advantages for those organisms that have them. For example, some plasmids called **R factors** ("resistance" factors) carry genes for antibiotic resistance, while

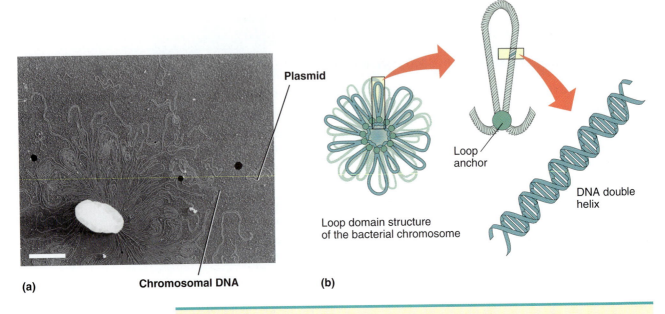

Plasmid

Loop anchor

DNA double helix

Loop domain structure
of the bacterial chromosome

(a)   Chromosomal DNA

(b)

FIGURE 6.2

Bacterial DNA Packing

(a) An electron micrograph of an *E. coli* cell immediately after disruption. The tangled mass is the organism's uncoiled DNA. A plasmid also is seen, showing the difference in size of the two DNA molecules. (Bar = 1μm.) (b) The loop domain structure of the chromosome, as seen head-on. The loops in DNA help account for the compacting of a large amount of DNA in a relatively small bacterial cell.

other plasmids allow bacteria to transfer their genetic material to receptive cells in a recombination process. Still other plasmids contain genes for the production of **bacteriocins**, a group of proteins that inhibit or kill other bacteria. There also are plasmids that contain genes that code for toxins that affect human cells and disease processes. We shall have much to say about these extrachromosomal units when we discuss recombination and genetic engineering in the next chapter.

To this point . . .

We have begun our study of bacterial genetics by examining the properties of the bacterial chromosome. This chromosome is a single molecule of DNA usually arranged as a closed loop that exists free in the cytoplasm without a surrounding membrane. To pack into a typical bacterial cell, the DNA must be supercoiled and arranged in loops.

Besides chromosomal DNA, most bacteria also contain smaller loops of DNA called plasmids. Although not essential for the normal survival of bacteria, they can carry genes that provide a selective advantage to the bacteria that contain them. Plasmids are key players in recombination and genetic engineering.

Knowing the organization of the bacterial chromosome leads us to investigate how the chromosome is replicated prior to binary fission.

6.2

DNA Replication

Watson and Crick's 1953 paper describing the structure of DNA also provided a glimpse into how DNA might be copied. They concluded, "It has not escaped our notice that the specific pairing we have postulated immediately suggests a possible copying mechanism for the genetic material."

DNA REPLICATION IS SEMICONSERVATIVE

The bacterial chromosome replicates just prior to the process of binary fission. At the beginning of the replication sequence, the DNA usually is anchored to a particular point on the cell membrane. The double helix then unwinds and enzymes synthesize a new polynucleotide strand of DNA for each of the two original old strands. This is accomplished by joining together nucleotides whose bases complement the bases in each original strand. Each new strand then combines with an old strand, and the replicated DNA twists to reform the double helix. This combination of a new and old strand was first observed in *E. coli* in 1958 by Matthew J. Meselson and Franklin W. Stahl. It is called the **semiconservative method of replication** because one strand of the replicate DNA is conserved in the new DNA molecule and one strand is newly synthesized.

DNA SYNTHESIS PROGRESSES FROM THE ORIGIN OF REPLICATION

The semiconservative method accounts for DNA replication, but it does not shed light on how a closed loop chromosome actually replicates. This problem perplexed microbiologists until 1962, when John Cairns and his coworkers clarified some of the details of the process.

Today we know that DNA starts to replicate at a fixed point called the **origin of replication**. A group of initiator proteins binds at the origin and, using the energy from ATP, starts to separate, or "unzip," the two polynucleotide strands. Since replication usually is bidirectional, two V-shaped **replication forks** form and move in opposite directions away from the origin (**FIGURE 6.3**).

Synthesis of DNA then occurs on each strand of each fork. Each strand represents a template for the synthesis of a new complementary strand. **DNA polymerase** enzymes move along each strand, catalyzing the insertion of new complementary nucleotides. Along one strand in each fork, new nucleotides are added in a *continuous* fashion, beginning at the origin. However, notice in the Figure 6.3 enlargement that along the other template strand in each fork, synthesis moves away from the fork and proceeds in a *discontinuous* fashion. Here, the new polynucleotide strand is synthesized in a series of segments that later are joined with the help of an enzyme called **DNA ligase**. These segments came to be known as **Okazaki fragments**, after Reiji Okazaki, who discovered them in 1968. Each of the new polynucleotide strands then combines with its old parent strand, as Meselson and Stahl postulated.

A second type of DNA replication is called the **rolling circle mechanism**. This process takes place in some viruses and in bacteria undergoing a type of recombination called conjugation. We will see this recombination mechanism in Chapter 7.

Replication fork:
point at which the double helix is unwound and new polynucleotide strands are synthesized.

DNA polymerase:
an enzyme that synthesizes a new DNA strand from a DNA template.

DNA ligase:
an enzyme that binds together short DNA fragments.

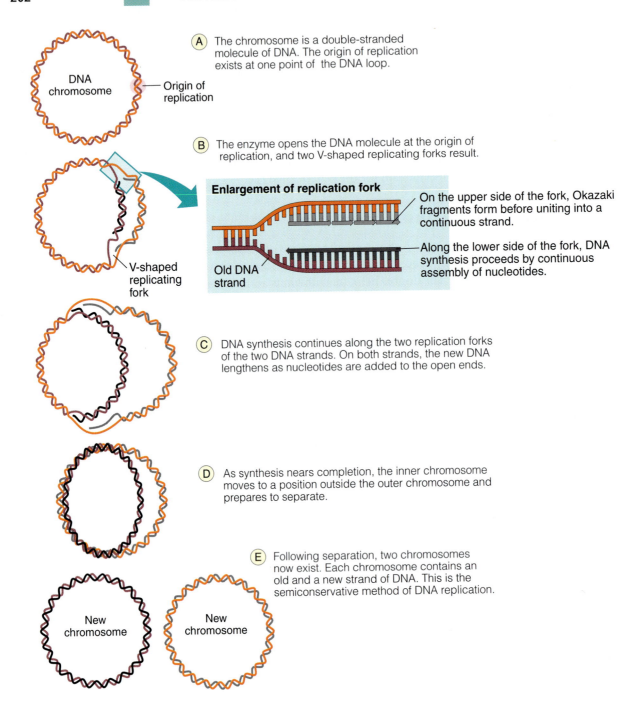

A The chromosome is a double-stranded molecule of DNA. The origin of replication exists at one point of the DNA loop.

DNA chromosome

Origin of replication

B The enzyme opens the DNA molecule at the origin of replication, and two V-shaped replicating forks result.

Enlargement of replication fork

On the upper side of the fork, Okazaki fragments form before uniting into a continuous strand.

Along the lower side of the fork, DNA synthesis proceeds by continuous assembly of nucleotides.

Old DNA strand

V-shaped replicating fork

C DNA synthesis continues along the two replication forks of the two DNA strands. On both strands, the new DNA lengthens as nucleotides are added to the open ends.

D As synthesis nears completion, the inner chromosome moves to a position outside the outer chromosome and prepares to separate.

E Following separation, two chromosomes now exist. Each chromosome contains an old and a new strand of DNA. This is the semiconservative method of DNA replication.

New chromosome

New chromosome

FIGURE 6.3

Replication of the Closed Loop Chromosome in *E. coli*

To this point . . .

We saw that the bacterial chromosome replicates by a semiconservative method prior to binary fission. Initiator proteins bind at the origin of replication and start to separate the two polynucleotide strands using the energy derived from ATP hydrolysis. DNA polymerase catalyzes the addition of new, complementary nucleotides to each template strand. Along each replication fork, one new strand is synthesized in a continuous fashion while the other is synthesized in a discontinuous fashion as a series of Okazaki fragments. DNA also can be replicated by the rolling circle method.

Next, we will investigate the process of protein synthesis, the anabolic process by which information in the genes is converted into amino acids that join to form those proteins the cell needs to carry out growth, metabolism, and reproduction.

6.3

Protein Synthesis

The discovery of the structure of DNA also provided a glimpse into understanding how a cell makes proteins. **Protein synthesis** is a process in which amino acids are precisely bound together in a sequence determined by the hereditary information, the genes, in the cell. The compounds resulting from protein synthesis are used as cellular enzymes, structural components, toxins, or other forms. The process requires not only DNA of the chromosome, but also ribonucleic acid (RNA). Both DNA and RNA were described in Chapter 2. A review of their structure is recommended so their functions can be fully comprehended.

THE CENTRAL DOGMA IDENTIFIES THE FLOW OF GENETIC INFORMATION

The central theme of protein synthesis holds that the genetic information first is expressed as RNA by a process called **transcription**. One type of RNA then functions as a messenger by carrying the code to areas of the cytosol where the ribosomes are located. There, amino acids are fitted together in a precise sequence to form the protein. This sequencing process, called **translation**, reflects the genetic information in the DNA. This **central dogma** of biology can be summarized as follows:

$$\text{DNA} \xrightarrow{\text{Transcription}} \text{RNA} \xrightarrow{\text{Translation}} \text{Protein}$$

As we will see in Chapter 12, a few viruses modify this rule.

TRANSCRIPTION COPIES GENETIC INFORMATION INTO RNA

In transcription, various types of RNA are produced according to the code of nitrogenous bases in the DNA molecule. (**TABLE 6.2** compares DNA and RNA.) The DNA thus serves as a template for new RNA molecules. The process begins with the

TABLE 6.2

A Comparison of DNA and RNA

DNA (DEOXYRIBONUCLEIC ACID)	RNA (RIBONUCLEIC ACID)
In prokaryotes, found in the nucleoid and plasmids; in eukaryotes, found in the nucleus, mitochondria, and chloroplasts	In prokaryotes and eukaryotes, found in the cytosol and in ribosomes; in eukaryotes, found in the nucleolus
Always associated with chromosome (genes); each chromosome has a fixed amount of DNA	Found mainly in combinations with proteins in ribosomes (ribosomal RNA) in the cytosol, as messenger RNA, and as transfer RNA
Contains a 5-carbon sugar called deoxyribose	Contains a 5-carbon sugar called ribose
Contains bases adenine, guanine, cytosine, thymine	Contains bases adenine, guanine, cytosine, uracil
Contains phosphorus (in phosphate groups) that connects deoxyribose sugars with one another	Contains phosphorus (in phosphate groups) that connects ribose sugars with one another
Functions as the molecule of inheritance	Functions in protein synthesis and gene regulation
Double stranded	Usually single stranded
Larger size	Smaller size

RNA polymerase:
an enzyme that synthesizes an RNA polynucleotide from the DNA template within a gene.

DNA double helix of a gene unwinding, followed by an uncoupling of the two DNA strands as the hydrogen bonds between opposing bases break (**FIGURE 6.4**).

An enzyme called **RNA polymerase** recognizes a sequence of bases called the **promoter** located on one of the two separated DNA strands. The polymerase binds to the promoter. It then moves along that strand of the DNA positioning complementary RNA nucleotides to the template strand; that is, guanine (G) and cytosine pair with one another and thymine (T) in the DNA template pairs with adenine (A) in the RNA. However, an adenine base on the DNA template pairs with a uracil (U) base in the RNA because RNA nucleotides contain no thymine bases. The RNA nucleotides then are linked together by the RNA polymerase. The sequence of bases in DNA thus forms a complementary image of itself in the RNA. The language, or genetic code, of DNA has been transcribed.

THREE TYPES OF RNA RESULT FROM TRANSCRIPTION

Transcription produces three types of RNA that are needed for protein synthesis.

Messenger RNA (mRNA). This RNA carries the message as to what protein will be synthesized. Each mRNA transcribed from a different gene carries a different message; that is, different information as to what protein will be synthesized. These messages are encoded in a series of three-base codes or **codons** found along the length of the mRNA. Each codon specifies an individual amino acid to be slotted into position during protein synthesis.

Ribosomal RNA (rRNA). Three rRNAs are transcribed from specific regions of the DNA. Together with protein, these RNAs serve as the framework of the **ribosomes**, which are the sites at which amino acids assemble into protein.

Transfer RNA (tRNA). The tRNAs are the smallest of the three RNA molecules discussed here (**FIGURE 6.5**). The conventional molecule is shaped roughly like a cloverleaf, with one point exhibiting a sequence of three nitrogenous bases (a triplet) that functions as an **anticodon**; that is, a sequence that complementary binds to an

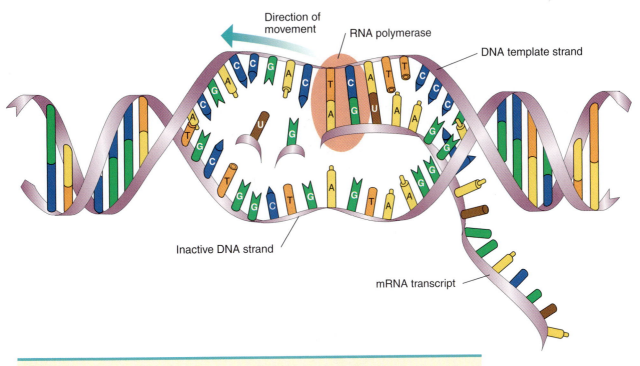

Direction of movement · **RNA polymerase** · **DNA template strand**

Inactive DNA strand

mRNA transcript

FIGURE 6.4

The Transcription Process

For transcription to occur, a gene must unwind and the base pairs separate. The enzyme RNA polymerase then moves along the template strand of the DNA and synthesizes a complementary molecule of RNA using the nitrogenous base code as a guide. The mRNA transcript will carry the genetic message of DNA to the ribosomes, where protein synthesis occurs. Note that the remaining strand of DNA is not transcribed.

mRNA codon. Each tRNA has a specific amino acid attached through an enzymatic reaction involving ATP. At least one type of tRNA exists for each of the 20 amino acids, and a high degree of specificity exists between the tRNA and its amino acid. For example, the amino acid alanine binds only to the tRNA specialized to transport alanine. Transfer RNA molecules deliver amino acids to the ribosome for assembly into proteins, as we shall see presently.

One of the startling discoveries of biochemistry is that the **genetic code** is redundant because in most cases there is more than one codon for each amino acid. Since there are four nitrogenous bases, mathematics tells us that 64 possible combinations can be made of the four bases, using three at a time. But, there are only 20 amino acids for which a code must be supplied. How do scientists account for the remaining 44 codes? It is now known that 61 of the 64 codons are used for an amino acid and that most amino acids have multiple codons (as shown in **TABLE 6.3**). For example, GCU, GCC, GCA, and GCG all code for alanine (ala). Also, one of the 64 codons, AUG, represents the *start codon* for making a protein. This codon codes for the amino acid methionine (met). The three codons that do not code for an amino acid (UGA, UAG, UAA) are called *stop codons* because they terminate the addition of amino acids to a growing polypeptide chain during translation. The genetic code of bases is nearly universal for all species, be they prokaryotic or eukaryotic. Thus, the GCU code for alanine in bacteria is also the code for alanine in human cells.

Genetic code:
the correspondence between the sequence of bases in a DNA or mRNA and the sequence of amino acids in a polypeptide.

(a)

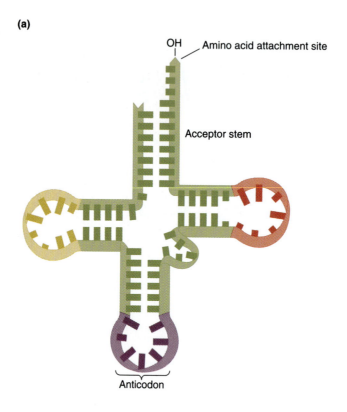

(b)

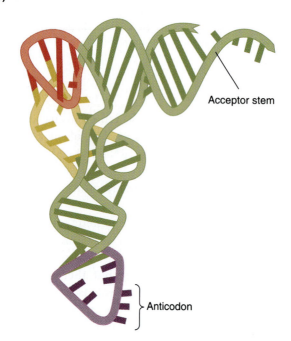

FIGURE 6.5

Structure of a tRNA

(a) The traditional "cloverleaf" configuration for tRNA. The anticodon will pair up with a complementary codon in the mRNA. The appropriate amino acid attaches to the end of the acceptor arm. (b) A schematic diagram of the more correct three-dimensional structure of a tRNA.

TABLE 6.3

The Genetic Code Decoder

The genetic code embedded in an mRNA is decoded by knowing which codon specifies which amino acid. On the far left column, find the first letter of the codon; then find the second letter from the top row; finally read up or down from the right-most column to find the third letter. The three-letter abbreviations for the amino acids are in parentheses.

SECOND LETTER

FIRST LETTER		U	C	A	G	THIRD LETTER
U		UUU UUC Phenylalanine (Phe) UUA UUG Leucine (Leu)	UCU UCC UCA UCG Serine (Ser)	UAU UAC Tyrosine (Tyr) UAA Stop condon UAG Stop condon	UGU UGC Cysteine (Cys) UGA Stop condon UGG Tryptophan (Trp)	U C A G
C		CUU CUC CUA CUG Leucine (Leu)	CCU CCC CCA CCG Proline (Pro)	CAU CAC Histidine (His) CAA CAG Glutamine (Gln)	CGU CGC CGA CGG Arginine (Arg)	U C A G
A		AUU AUC AUA Isoleucine (Ile) AUG Methionine; start codon	ACU ACC ACA ACG Threonine (Thr)	AAU AAC Asparagine (Asn) AAA AAG Lysine (Lys)	AGU AGC Serine (Ser) AGA AGG Arginine (Arg)	U C A G
G		GUU GUC GUA GUG Valine (Val)	GCU GCC GCA GCG Alanine (Ala)	GAU GAC Aspartic acid (Asp) GAA GAG Glutamic acid (Glu)	GGU GGC GGA GGG Glycine (Gly)	U C A G

There is one important difference between prokaryotes and eukaryotes in terms of the RNAs produced. In prokaryotes, all of the bases in a gene are transcribed and used to specify a particular protein. In eukaryotes, some are removed before the final RNA is produced. In 1977, Philip Sharp and his associates at the Massachusetts Institute of Technology found that certain portions of eukaryotic DNA are not part of the final RNA and must be removed from the RNA before the molecule can function. These intervening DNA segments removed after transcription are called **introns**, while the functioning, expressed segments are called **exons**. Bacteria do not have introns; the genes are entirely transcribed into the functional RNA.

TRANSLATION IS THE PROCESS OF MAKING THE PROTEIN

In the process of translation, the language of the genetic code (nucleotides) is translated into the language of proteins (amino acids). The process takes place at the ribosome, where the mRNA molecule meets tRNA molecules bound to their appropriate amino acids. As the process of translation is described, it would be helpful for you to follow along using FIGURE 6.6. In the figure, several of the early steps in the process have already occurred. A refresher on the structure of proteins and the peptide bonds that hold them together (Chapter 2) may be of value.

Translation takes place as the ribosome moves along the mRNA, with the codons in mRNA exposed to sites within the ribosome. The translation process occurs in three stages.

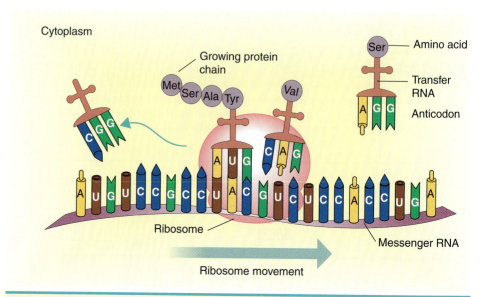

FIGURE 6.6

The Translation Process in Protein Synthesis

The messenger RNA moves to the ribosome, where it is met by transfer RNA molecules bonded to different amino acids. Within the ribosome, the tRNA molecules align themselves opposite the mRNA codon and bring the amino acids into position. A peptide bond forms between adjacent amino acids on the growing protein chain, after which the amino acid leaves the tRNA. The tRNA returns to the cytoplasm to bond with another molecule of the same amino acid. Note that the exceptionally short mRNA has a start and a stop codon.

■ **CHAIN INITIATION.** Translation began with the addition of the tRNA whose anticodon recognizes the start codon AUG. Therefore, the first amino acid was methionine (met).

■ **CHAIN ELONGATION.** The second tRNA is inserted into the ribosome. In this case, it was one that recognized the UCC codon in the mRNA. From the genetic code (Table 6.3), you know that the amino acid attached to the tRNA must have been serine (ser). Hydrogen bonds between the codon and anticodon bases temporarily hold the tRNA in position, while an enzyme attaches met to serine by a peptide bond. ATP and guanosine triphosphate (GTP) supply the energy for the reaction.

The first tRNA then breaks free, leaving its met molecule on the amino acid chain. The next codon GCC attracted a tRNA with the amino acid alanine (ala) attached. Again, an enzyme now transfers the dipeptide met-ser to alanine. The tRNA that carried serine now leaves the ribosome and the process continues.

In the step shown in Figure 6.6, the tRNA recognizing the GUC codon arrives. This tRNA carries the amino acid valine (val). Again, the enzyme transfers the growing polypeptide chain to val. The translation process then repeats itself with the next codon and on down the mRNA.

■ **CHAIN TERMINATION.** The process of adding tRNAs and transferring the elongating polypeptide to the entering amino acid/tRNA continues until the ribosome reaches a stop codon (either UAA, UAG, or UGA). There is no tRNA to recognize any of these codons. Rather, proteins called **releasing factors** bind where the tRNA would normally bind. This triggers the release of the polypeptide and a disassembly of the translation apparatus, which can be reassembled for translation of another mRNA.

During synthesis, the polypeptide already may start to twist into its secondary and tertiary structure. Groups of cytoplasmic proteins called **chaperones** ensure the folding process occurs correctly.

Cells typically make hundreds if not thousands of copies of each protein. Producing such large amounts of a protein can be done efficiently and quite quickly. Remember each bacterial cell contains thousands of identical ribosomes. Therefore, a single mRNA molecule can be translated simultaneously by several ribosomes (**FIGURE 6.7**). Once one ribosome has moved far enough along the mRNA, another ribosome can "jump on" and initiate translation. Such a cluster of ribosomes all translating the same message is called a **polyribosome**.

PROTEIN SYNTHESIS CAN BE CONTROLLED IN SEVERAL WAYS

In Chapter 5, we described how negative feedback can control enzyme activity. Another control mechanism simply is to not make the enzyme (or protein in general) when it is not needed. In the previous section, we saw that transcription is the first of the two steps leading to protein manufacture in cells. So, another way to control what proteins and enzymes are present in a cell is to regulate the mechanisms that induce ("turn on") or repress ("turn off") transcription of a gene or set of genes.

In 1961, two Pasteur Institute scientists, François Jacob and Jacques Monod, proposed such a mechanism for controlling protein synthesis. They suggested that segments of bacterial DNA are organized into functional units called **operons** (**FIGURE 6.8**). Their pioneering research along with more recent studies indicates that each operon consists of a cluster of **structural genes** that provides genetic codes for proteins having metabolically related functions. Adjacent to the structural genes is the **operator**, which is a sequence of bases that controls the expression (transcription)

FIGURE 6.7

Coupled Transcription and Translation in Bacteria

The electron micrograph shows transcription of a gene in *E. coli* and translation of the mRNA. The dark spots are ribosomes, which coat the mRNA. An interpretation of the electron micrograph is at the right. Each mRNA has ribosomes attached along its length. The large red dots are the RNA polymerase molecules; they are too small to be seen in the electron micrograph. The length of each mRNA is equal to the distance that each RNA polymerase has progressed from the transcription-initiation site. In this simplified drawing, the genetic information is being translated 23 times. For clarity, the polypeptides are not shown.

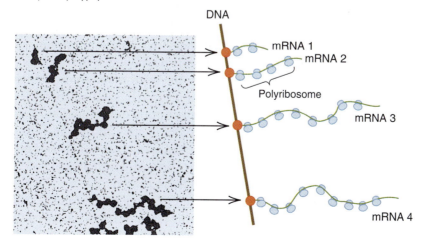

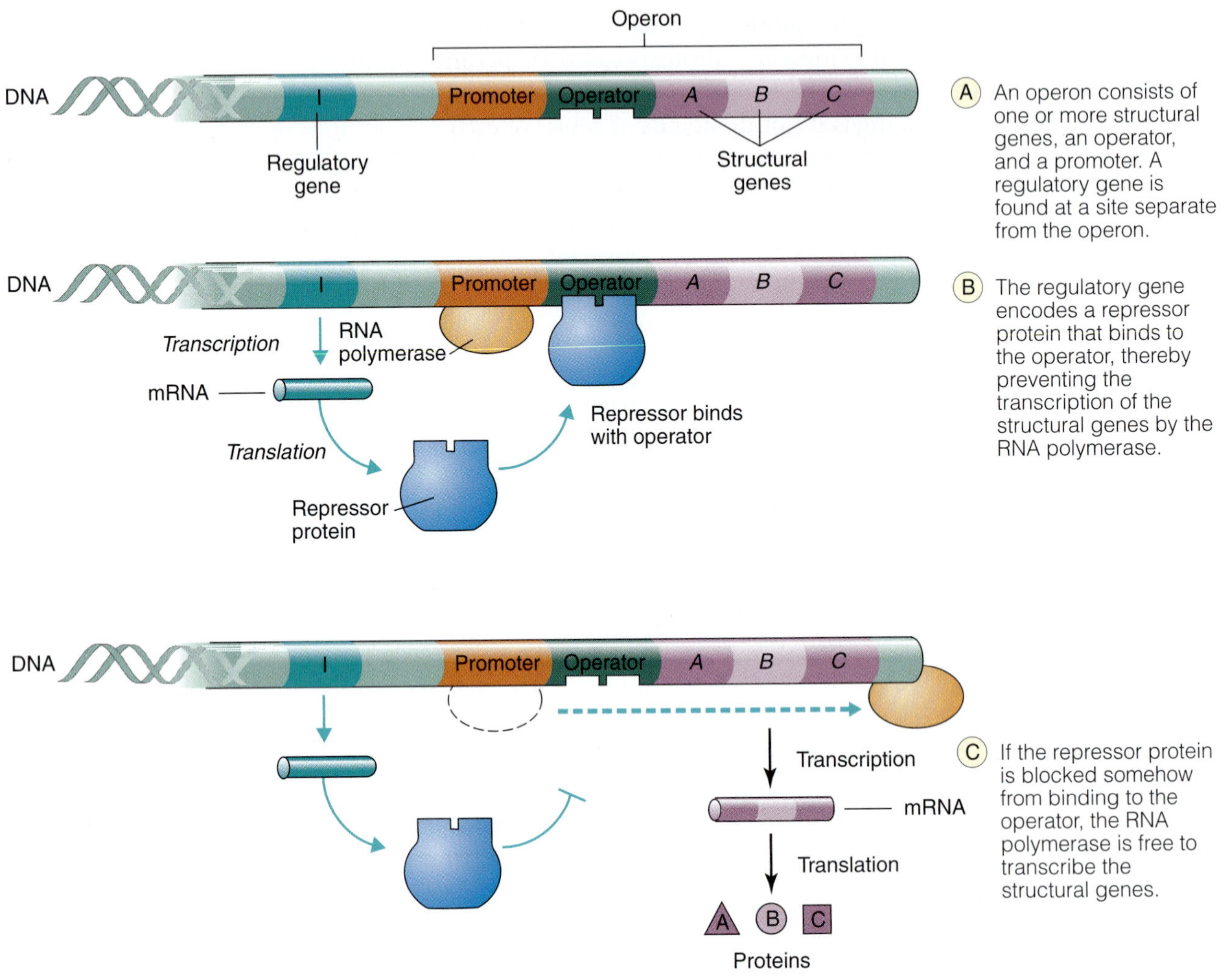

A — An operon consists of one or more structural genes, an operator, and a promoter. A regulatory gene is found at a site separate from the operon.

B — The regulatory gene encodes a repressor protein that binds to the operator, thereby preventing the transcription of the structural genes by the RNA polymerase.

C — If the repressor protein is blocked somehow from binding to the operator, the RNA polymerase is free to transcribe the structural genes.

FIGURE 6.8

The Operon Theory and Negative Control

The operon theory holds that control elements regulate the transcription of groups of structural genes. Since the role of the regulatory gene is to produce a repressor protein that inhibits transcription, such regulation is called negative control.

of the structural genes and a **promoter** next to the operator where the RNA polymerase binds that will transcribe the structural genes. Also important, but not part of the operon is a distant **regulatory gene** that codes for a **repressor protein**. Jacob and Monod proposed that when the repressor protein is bound to the operator, the RNA polymerase is blocked from moving down the operon and transcribing the structural genes. This is called **negative control** of protein synthesis because the repressor protein inhibits or "turns off" gene transcription in that operon.

MicroInquiry 6 presents two contrasting examples of how an operon works to induce or repress gene expression. Following the series of observations and explanations given in the MicroInquiry, you should have firm understanding of how bacterial cells can control protein synthesis through transcription.

Besides negative control, protein synthesis can be regulated through positive control of transcription as well as through varying levels of translational control.

MicroFocus 6.3

ANTISENSE MAKES SENSE

With the recent outbreak of fatal encephalitis caused by the West Nile virus and atypical pneumonia caused by the severe acute respiratory syndrome (SARS) coronavirus, scientists have been trying urgently to find ways to treat and cure these diseases. Being viral diseases, they are not amenable to antibiotics, so other strategies are needed.

One of the potential approaches being considered is the use of antisense molecules as therapeutic agents. Antisense molecules are fragments of nucleic acids that unite with and neutralize mRNA molecules carrying a specific genetic message for protein synthesis. Since about 1978, antisense molecules have been identified as a strategy to fight many types of diseases including infectious disease.

Antisense makes sense on paper because antisense molecules should have the ability to shut off the synthesis of unwanted or disease-causing proteins. To treat AIDS, for example, scientists have created a string of nucleotides that is complementary to specific mRNAs produced by the human immunodeficiency virus (HIV). In an infected individual, the antisense molecules should recognize and bind to these viral mRNAs. Existing now as a double-stranded RNA molecule, the mRNA could not be translated by the cell's ribosomes. Without these essential viral proteins, no new HIV particles could be formed (Chapter 14).

Although such strategies look promising on paper, they have yet to produce the successes that were hoped for in clinical trials. However, the research has made good progress recently in finding better ways of getting the antisense molecules into cells and keeping the drug's level constant. Of the biotech companies that are doing research with antisense molecules, some are now focusing attention on emerging infectious diseases, including the SARS coronavirus and West Nile virus. Both viruses have RNA genomes so they are susceptible to inhibition by antisense molecules. In the case of West Nile, one biotech company has identified a potential antisense drug for development. In animal trials, the antisense drug seems to work as it should. The same company is working on a SARS drug candidate as well. In both cases, if the antisense drugs can block the translation of critical viral mRNAs, then new virus particles cannot be made and the disease should be snuffed out.

The logic of antisense seems to make sense on paper. It is just a matter of getting the drugs to perform correctly in the body.

MicroFocus 6.3 describes how the understanding of protein synthesis has been used to block harmful proteins from being made.

Overall, the control of protein synthesis is an intricate series of recognitions, regulations, and biochemical changes at the very core of protein anabolism. Indeed, an understanding of control mechanisms is essential to an understanding of gene activity, and before microbial genetics or DNA technology could emerge as disciplines of science, the complexity of control systems had to be worked out. We shall explore the research that led to this breakthrough in the next chapter.

To this point . . .

In this section, we discussed the anabolism of proteins; that is, how protein synthesis occurs in cells. We saw that information passes from DNA to RNA through transcription, and then from RNA to protein through translation.

During the process of transcription, the information contained within a gene is transcribed into RNA. Three types of RNAs result: mRNA, which carries the information to specify a particular protein; rRNA, which forms part of the structure of the ribosome; and tRNA, which carries the amino acids to the ribosome. Transcription involves an RNA polymerase that recognizes a promoter sequence located on one of the two strands within a gene. We also learned that one of the differences between prokaryotic and eukaryotic RNAs is that the prokaryotic RNAs as transcribed from the

MicroInquiry 6

THE OPERON THEORY AND THE CONTROL OF PROTEIN SYNTHESIS

The best way to visualize and understand the operon model for control of protein synthesis is by working through a couple of examples.

The Lactose (*Lac*) Operon. Here is a piece of experimental data. The disaccharide lactose represents a potential energy source for *E. coli* cells if it can be broken into its monomers of glucose and galactose. One of the enzymes involved in the metabolism of lactose is ß-galactosidase. If *E. coli* cells are grown in the absence of lactose, ß-galactosidase activity cannot be detected as shown in the following graph.

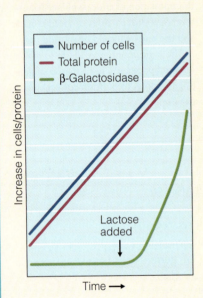

However, when lactose is added to the nutrient broth, very quickly enzyme activity is detected. How can this change from inhibition to expression be explained in the operon model?

Based on the operon theory, we would propose that when lactose is absent from the growth medium, the repressor protein for the *lac* operon binds to the operator and blocks passage of the RNA polymerase that is attached to the adjacent promoter (FIGURE A). Being unable to move past the operator, the polymerase cannot transcribe the structural genes, one of which codes for ß-galactosidase.

When lactose is added to the growth medium, lactose will be transported into the bacterial cell, where the disaccharide binds to the repressor protein and inactivates it. With the repressor protein inactive, it no longer can recognize and bind to the operator. The RNA polymerase now is not blocked and can translocate down the operon and transcribe the structural genes. Lactose is called an **inducer** because its presence has induced, or "turned on," structural gene transcription in the *lac* operon. It explains why ß-galactosidase activity enzyme forms when lactose was present.

Now let's see if you can figure out this scenario.

Tryptophan (*trp*) Operon. *E. coli* cells have a cluster of structural genes that code for five enzymes in the metabolic pathway for the synthesis of the amino acid tryptophan. Therefore, if *E. coli* cells are grown in a broth culture lacking the amino acid tryptophan, they continue to grow normally as shown below.

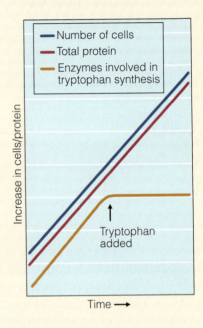

gene are functional molecules. In eukaryotic cells, the initial RNA must be processed and the introns removed before the remaining exons form the functional RNA.

We followed the process of translation, which involves the ribosome taking a sequence of codons in the mRNA and converting (translating) them into a chain of amino acids linked together by peptide bonds. Chain initiation involved the amino acid methionine. The genetic code of the mRNA specifies the sequencing of tRNAs that supply the amino acids to a growing polypeptide chain. This continues until a stop codon is reached, at which point the polypeptide chain is released.

However, as the graph shows, when tryptophan is added to the growth medium, new enzyme synthesis is repressed or "turned off" and cells use the tryptophan supplied in the growth medium.

How can enzyme repression be explained by the operon model? The solution is provided in Appendix E.

FIGURE A

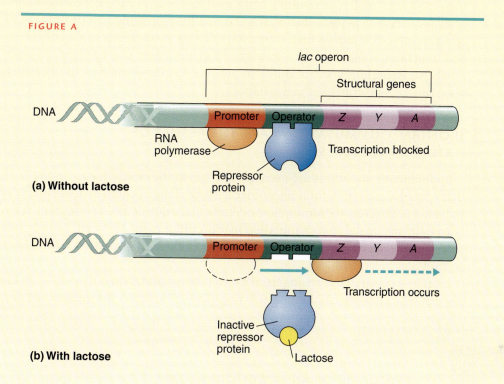

(a) Without lactose

(b) With lactose

We discovered that molecular chaperones were involved in guiding the correct folding of the polypeptide as translation proceeded. We also learned that many ribosomes simultaneously could translate the same mRNA, providing the thousands of copies of the protein required by the cell. We concluded the section with a MicroInquiry that explained how bacterial cells can regulate the synthesis of proteins through transcriptional induction or repression.

Next, we will examine the nature of mutations and discover how they develop and affect protein synthesis. We also will see how cells try to repair mutations or damage to DNA.

Bacterial Mutation

The information in a bacterial chromosome may be altered through a permanent change in the DNA called a **mutation**. In most cases, a mutation involves a disruption of the nitrogenous base sequence in the DNA molecule or the loss of significant parts of a gene. Often this leads to the production of a miscoded messenger RNA molecule, resulting in the insertion of one or more incorrect amino acids into the polypeptide during translation. Since proteins govern virtually all activities of a cell, it follows that a mutation will alter some aspects of the cell's metabolic functions. Some alterations have little effect, but others may be significant, such as when a bacterium loses its ability to produce a toxin or changes its chemistry. MicroFocus 6.4 describes such an instance.

SPONTANEOUS MUTATIONS ARE THE RESULT OF NATURAL PROCESSES

Spontaneous mutations are heritable changes to the base sequence in the DNA that occur because of natural phenomena. These changes could be from everyday radiation penetrating the atmosphere or errors made during DNA replication. It has been estimated that one such mutation may occur for every 10^6 to 10^{10} divisions of a bacterium. This implies that in a colony of a billion (10^9) bacteria, at least one mutant may be present.

Usually, a bacterial cell arising from a spontaneous mutation is masked by all the normal cells. However, should a selective agent be introduced, the mutant may sur-

MicroFocus 6.4

EVOLUTION OF AN INFECTIOUS DISEASE

Could the Black Death of the fourteenth century and the 25 million Europeans that succumbed to plague have been the result of a few genetic changes to a bacterium? Could the entire course of Western civilization have turned based on these changes?

Possibly so, maintain researchers from the federal Rocky Mountain Laboratory in Montana. In 1996, a research group led by Joseph Hinnebusch reported that three genes in the plague bacillus *Yersinia pestis* are not present in a fairly harmless form of the organism (*Y. pseudotuberculosis*) that causes mild food poisoning. Thus, it is possible that the entire story of plague's pathogenicity revolves around a small number of gene changes.

Bubonic, septicemic, and pneumonic plague are caused by *Y. pestis*, a rod-shaped bacterium transmitted by the rat flea (Chapter 10). In an infected flea, the bacteria eventually amass in its foregut and obstruct its gastrointestinal tract. Soon the flea is starving, and it starts biting victims (humans and rodents) uncontrollably and feeding on their blood. During the bite, the flea regurgitates some 24,000 plague bacilli into the bloodstream of the unfortunate victim.

At least three genes appear important in the evolution of plague. It appears that nonpathogenic plague bacilli have these genes, which encourage the bacilli to remain harmlessly in the midgut of the flea. Pathogenic plague bacilli, by contrast, do not have the genes. Free of control, the bacteria migrate from the midgut to the foregut and form a plug of packed bacilli that are passed on to the victim.

In 2002, Hinnebusch and colleagues published evidence that another gene, carried on a plasmid, codes for an enzyme that is required for the initial survival of *Y. pestis* bacilli in the flea midgut. By acquiring this gene from another organism, *Y. pestis* made a crucial jump in its host range. The bacterium now could survive in fleas and became adapted to relying on its blood-feeding host for transmission. So, these few genetic changes may have been a key force leading to the evolution and emergence of plague. This is just another example of the flexibility that many microbes have to repackage themselves constantly into new and, sometimes, more dangerous agents of infectious disease.

vive, multiply, and emerge as a dominant form. For many decades, for example, doctors have used penicillin to treat gonorrhea. Since 1976, however, a penicillin-resistant strain of *Neisseria gonorrhoeae* has been emerging in human populations. Many investigators believe that the resistant strain arose by spontaneous mutation at some unknown time, perhaps centuries ago, and as penicillin gradually eliminated susceptible strains, the resistant strain filled the niche.

INDUCED MUTATIONS ARE DELIBERATE MUTATIONS RESULTING FROM A MUTAGEN

Most of our understanding of mutations has come from planned experiments in which laboratory scientists subject bacteria to chemical or physical agents that cause mutations. These **mutagens** can cause a broad variety of **induced mutations**, depending on the type of mutagen used.

Mutations arising from treatment with **ultraviolet (UV) light** are among the best understood (**FIGURE 6.9**). UV light is a form of radiation not perceived by the human eye. When this energy is absorbed by DNA, it induces adjacent thymine (or cytosine) bases in the DNA to covalently link together forming dimers (**FIGURE 6.10**). If these dimers occur in a protein-coding gene that is to be transcribed, the inability to insert the correct bases in mRNA molecules occurs where the dimers are located. The radiation also can be used for disinfection purposes because it quickly kills bacteria (Chapter 22).

Nitrous acid is a chemical mutagen that converts DNA's adenine molecules to hypoxanthine molecules. Adenine will normally base pair with thymine during DNA replication, but hypoxanthine base pairs with cytosine. Later, when protein synthesis takes place, the new DNA molecule having cytosine substituted for thymine will code for guanine in the mRNA instead of the normal adenine (**FIGURE 6.11a**).

Mutations also can be induced by a series of **base analogs** that bear a chemical resemblance to nitrogenous bases. One such base analog, 5-bromouracil, is taken up

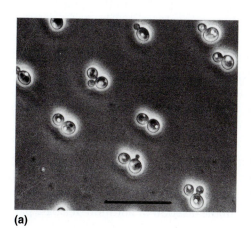
(a)

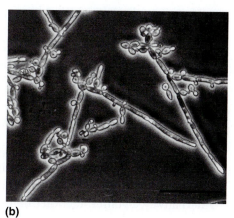
(b)

FIGURE 6.9

Two Morphological Forms of *Candida albicans*

Phase-contrast micrographs of *Candida albicans*, the agent of yeast infections in humans. (a) The oval yeast form commonly seen in vaginal infections. (Bar = 50 μm.) (b) The filamentous mold-like form often observed in invaded tissue. (Bar = 70 μm.) In 1990, investigators demonstrated that the yeast form could be converted to the mold-like form by treating the organism with ultraviolet light and nitrous acid, thereby mutating its genetic material.

FIGURE 6.10

The Formation of Thymine Dimers

When bacteria are irradiated with ultraviolet light, the radiations affect the DNA of the cell. (a) A normal DNA molecule is converted to (b) an abnormal DNA molecule as the UV light binds adjacent thymine molecules within the DNA to form (c) a thymine dimer. With its thymine molecules bound in dimers, the DNA molecule cannot function properly and cannot replicate.

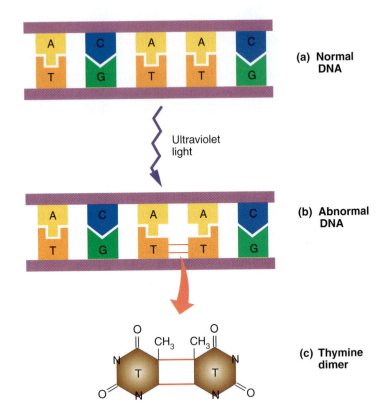

(a) Normal DNA

Ultraviolet light

(b) Abnormal DNA

(c) Thymine dimer

by cells and incorporated into DNA where thymine should be positioned (**FIGURE 6.11b**). The new DNA functions poorly with the analog in place. In the treatment of diseases caused by DNA viruses, a base analog is valuable because nucleic acid directs viral replication, and a virus with a functionless DNA molecule cannot replicate. The drug acyclovir is a base analog that works against herpesviruses (Chapter 13).

Mutations also can arise by chemicals that do not themselves become part of the DNA molecule. For example, benzopyrene, which is present in industrial soot and smoke, and aflatoxin, a fungal toxin found in certain animal products and foods, cause the loss or addition of nucleotides during replication. Such mutations have been associated with human cancers.

POINT MUTATIONS AFFECT ONE BASE PAIR IN A GENE

Regardless of the cause of the mutation, point mutations are one of the most common results. As the name suggests, **point mutations** affect just one point (base pair) in a gene. Such mutations may be a change to or substitution of a different base pair. Alternatively, a point mutation can result in the deletion or addition of a base pair.

■ BASE-PAIR SUBSTITUTIONS. If a point mutation causes a **base-pair substitution**, then the transcription of that gene will have one incorrect base in the mRNA sequence of codons. Perhaps one way to see the effects of such changes is using an English sentence made up of three-letter words (representing codons) where one letter has been changed. As three-letter words, the letter substitution still reads correctly, but the sentence makes less sense.

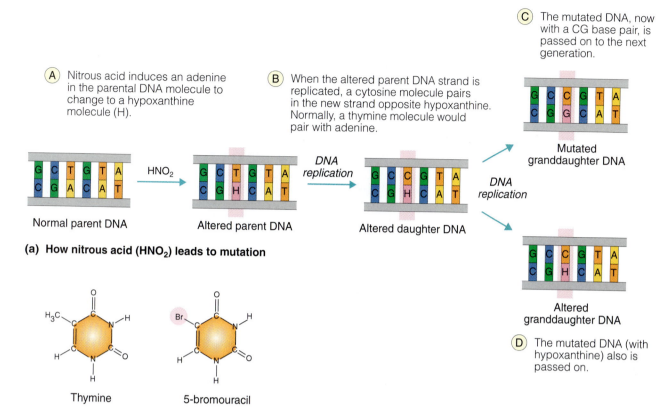

A — Nitrous acid induces an adenine in the parental DNA molecule to change to a hypoxanthine molecule (H).

B — When the altered parent DNA strand is replicated, a cytosine molecule pairs in the new strand opposite hypoxanthine. Normally, a thymine molecule would pair with adenine.

C — The mutated DNA, now with a CG base pair, is passed on to the next generation.

D — The mutated DNA (with hypoxanthine) also is passed on.

Normal parent DNA   Altered parent DNA   Altered daughter DNA

Mutated granddaughter DNA

Altered granddaughter DNA

(a) How nitrous acid (HNO₂) leads to mutation

Thymine   5-bromouracil

(b) A nitrogenous base and its mutation-causing analog

FIGURE 6.11

Mutations in Bacteria

(a) Nitrous acid causes an adenine to hypoxanthine change. After replication of the hypoxanthine containing strain, the granddaughter DNA has a mutated C–G base pair. (b) Base analogs induce mutations by substituting for nitrogenous bases in the synthesis of DNA. Note the similarity in chemical structure between thymine and the base analog 5-bromouracil.

Normal sequence: THE FAT CAT ATE THE RAT
Substitution: THE FAT CAN ATE THE RAT

As shown in **FIGURE 6.12a**, depending on the placement of the substituted base, when the mRNA is translated this may cause no change (**silent mutation**), lead to the insertion of the wrong amino acid (**missense mutation**) or generate a stop codon (**nonsense mutation**), prematurely terminating the polypeptide.

■ **BASE-PAIR DELETION OR INSERTION.** Point mutations also can cause the loss of a base or the addition of a base to the gene, resulting in an inappropriate number of bases. Again, using our English sentence we can see how a deletion or insertion of one letter affects the reading of the three-letter word sentence.

Normal sequence: THE FAT CAT ATE THE RAT
Deletion: THE F TC ATA TET HER AT
Insertion: THE FAT ACA TAT ETH ERA T

FIGURE 6.12

Categories and Results of Point Mutations

Mutations are permanent changes in DNA, but they are represented here as they are reflected in mRNA and its protein product. (a) Base-pair substitutions can produce silent, missense, or nonsense mutations. (b) Deletions or insertions shift the reading frame of the ribosome.

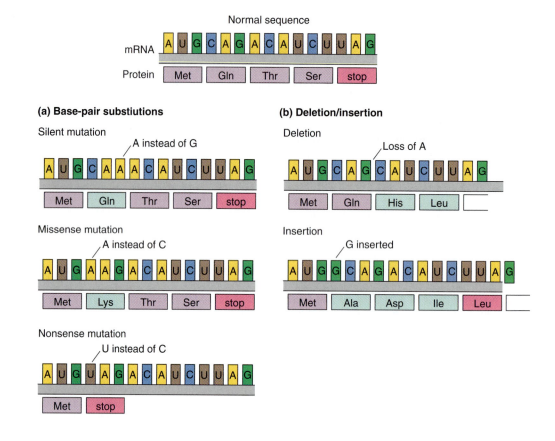

As you can see, the "sentence mutations" make little sense when reading the sentence as three-letter words. The same is true in a cell. Ribosomes always read three letters (one codon) at one time, generating potentially extensive mistakes in the amino acid sequence (FIGURE 6.12b). Thus, like our English sentence, the deletion or addition of a base will cause a **reading frameshift** because the ribosome always reads the genetic code in groups of three bases. Therefore, loss or addition of a base shifts the reading of the code by one base. The result is serious sequence errors in the amino acids that will probably produce an abnormal protein that will have lost its function in the cell or its role in metabolism.

REPAIR MECHANISMS ATTEMPT TO CORRECT MISTAKES OR DAMAGE IN THE DNA

During the life of a bacterial cell (indeed, of every prokaryotic and eukaryotic cell), the DNA undergoes a form of molecular punishment. Cellular DNA endures thousands of damage events every day. The damage comes from DNA replication errors and other base changes caused by exposure to everyday mutagens. If this damage is not repaired, there is a risk that normal metabolism will be disrupted. Although it is not perfect, cells try to maintain correct gene sequences in their DNA by using a variety of DNA repair mechanisms. These mechanisms rely on enzymes that literally patrol the DNA, locating and repairing alterations and distortions. (It's somewhat like driving with a mechanic in the back seat.) **MicroFocus 6.5** explains what may happen if repair is not effected.

The fact that DNA is a double-stranded molecule is not a fluke. This is because if one strand is damaged, the other can act as a template to correct the damage. One type of repair mechanism is called **mismatch repair**. In this case, the DNA polymerase that adds new bases to the DNA template strand during replication also "proofreads" DNA synthesis and removes nucleotides that it mismatched (**FIGURE 6.13a**). Scientists estimate that about 1 in 10,000 bases is incorrectly placed and subject to replacement. *E. coli* has about 4.6 million bases in its chromosome, so mathematics says over 460 replication errors will occur and have to be repaired every replication. But, considering the enzyme is catalyzing the addition of 50,000 bases every minute, that still is less than a 1 percent error rate.

Almost 100 different forms of nuclease enzymes are known to exist in *E. coli* cells. When DNA is damaged by a physical mutagen, such as UV light, several of these nucleases execute **excision repair**. First, nucleases cut out (excise) the damaged DNA (**FIGURE 6.13b**). Then, a different DNA polymerase from the one used in replication replaces the missing nucleotides with the correct ones. Finally, DNA ligase seals the new strand into the rest of the polynucleotide. In 1998, Texas A&M University researchers reported that plasmids are apparently the site of genes that

MicroFocus 6.5

"MAKE THOSE REPAIRS—PLEASE!"

As any homeowner knows, failure to repair the plumbing or fix the electrical wiring or patch the roof can make a day miserable. One would imagine that the same holds true for a bacterium. Unfortunately, the misery is often inflicted on its host.

Consider *Escherichia coli* O157:H7. This strain is associated with contaminated food and each year it infects thousands of people, killing some within days. Before 1982, *E. coli* O157:H7 apparently was not recognized. Then, this pathogenic strain emerged, and ever since, it has been a source of intestinal nightmares: the Jack-in-the-Box outbreak of 1993 (Chapter 3), the consumption of contaminated radish sprouts by thousands of Japanese people in 1996, and several recent incidents of severe diarrhea traced to unpasteurized apple cider.

Scientists believe that the new strain of *E. coli* can invade tissues and produce toxins unlike the traditionally harmless *E. coli* (Chapter 9). To do so, it must have genes that the traditional *E. coli* does not have. It is reasonably safe to assume that mutations in the chromosomal DNA brought about these new genes. But, scientists have wondered, why weren't the mutations corrected or discarded?

Researchers from the US Food and Drug Administration (FDA) think they have an answer. It all goes back to the repair mechanism, they say. Led by Thomas Cebula, FDA microbiologists examined numerous "new" strains of pathogens and compared their proteins to nonpathogenic strains of the same organism. In 1996, the researchers reported their findings: The new

pathogens lack the repair enzymes available in harmless strains. They simply cannot repair the faulty DNA sometimes produced during DNA replication. And, if the DNA encodes a toxin or protein that encourages tissue invasion, then the harmless bacterium potentially becomes a pathogen.

For many years, health officials believed that microbial pathogens had been brought under control, but their thinking has changed. Indeed, emerging pathogens such as *E. coli* O157:H7 are humbling reminders that bacteria can reinvent themselves and undergo a swift evolution as they adapt to new hosts, new conditions, and new pharmaceutical countermeasures. Something to think about as you figure out how to repair the plumbing.

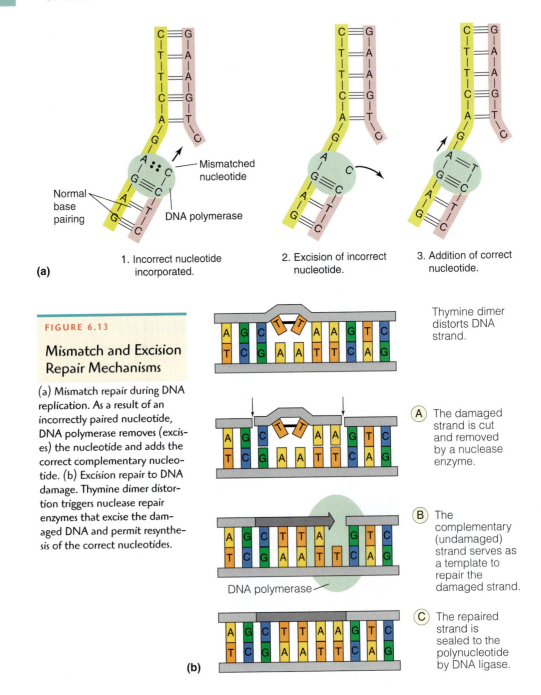

(a)

1. Incorrect nucleotide incorporated.

2. Excision of incorrect nucleotide.

3. Addition of correct nucleotide.

Mismatched nucleotide

Normal base pairing

DNA polymerase

FIGURE 6.13

Mismatch and Excision Repair Mechanisms

(a) Mismatch repair during DNA replication. As a result of an incorrectly paired nucleotide, DNA polymerase removes (excises) the nucleotide and adds the correct complementary nucleotide. (b) Excision repair to DNA damage. Thymine dimer distortion triggers nuclease repair enzymes that excise the damaged DNA and permit resynthesis of the correct nucleotides.

Thymine dimer distorts DNA strand.

(A) The damaged strand is cut and removed by a nuclease enzyme.

(B) The complementary (undamaged) strand serves as a template to repair the damaged strand.

DNA polymerase

(C) The repaired strand is sealed to the polynucleotide by DNA ligase.

(b)

encode excision repair enzymes. They discovered that when the number of plasmids increases, the damage due to ultraviolet light is repaired significantly faster.

TRANSPOSABLE GENETIC ELEMENTS CAN CAUSE MUTATIONS

Mutations of a different nature may be caused by fragments of DNA called **transposable genetic elements**. Two types are known in bacteria: insertion sequences and transposons. **Insertion sequences (IS)** are small segments of DNA with about 1,000 base pairs. They are found at one or more sites on the bacterial chromosome and plasmids. IS have no genetic information other than for the ability to insert into a chromosome. IS form copies of themselves, and the copies move into other areas of the chromosome. These events are rare, but when they occur they can interrupt the coding sequence in a gene, thereby inducing the wrong protein or, more likely, no protein to form. IS may be a prime force behind spontaneous mutation.

A second type of transposable genetic element is the **transposon**. These are the so-called "jumping genes" for which Barbara McClintock won the 1983 Nobel Prize in Physiology or Medicine (MicroFocus 6.6). Transposons are larger than IS and carry information, such as antibiotic resistance, that can be conferred to the bacterial cell. Like IS, they can interrupt the genetic code of a gene.

The movement of transposons appears to be nonreciprocal, meaning that an element moves ("jumps") away from its location and nothing takes its place. (This contrasts with insertion sequences, where copies move.) Transposons can move from plasmid to plasmid, from plasmid to chromosome, or from chromosome to plasmid. The presence of inverted repetitive base sequences at the ends of the element appears to be important in establishing the ability to move (FIGURE 6.14).

MicroFocus 6.6

JUMPING GENES

In the early 1950s, scientists assumed that genes were fixed elements, always found in the same position on the same chromosome. But in 1951, Barbara McClintock unveiled her research with corn plants at a symposium at Cold Spring Harbor Laboratory on Long Island, New York. McClintock described genes that apparently move from one chromosome to another. The audience listened in respectful silence. There were no questions after her talk, and only three people requested copies of her paper.

Like Gregor Mendel 100 years before, McClintock kept close watch over color changes in her plants. Whereas Mendel cultivated peas, McClintock grew Indian corn, or maize. In the 1940s, she noticed curious patterns of pigmentation on the kernels. Other scientists might have missed the patterns as random variations of nature, but McClintock's record keeping and careful analysis revealed a method to nature's madness. The pigment genes causing the splotches of color appeared to be switched on or off in particular generations. Still more remarkable, the "switches" seemed to occur at different places along the same

chromosome. Some switches even showed up in different chromosomes. Such "controlling elements," as McClintock called them, were available whenever needed to turn the genes on or off.

In the modern lexicon of molecular genetics, McClintock's elements are recognized as a two-gene system. One is an activator gene, the other a dissociation gene. The activator gene, for reasons unknown, can direct a dissociation gene to "jump" along the arm of the ninth chromosome in maize plants where color is regulated. When the jumping gene reinserts itself, it turns off the neighboring pigmentation genes, thereby altering the color of the kernel.

The jumping gene is identical to the transposon found in bacteria, and can cause important mutations and gene rearrangements. Many scientists suspect that jumping genes play significant roles in the development of a fertilized egg into a mature organism, while serving as a driving force in evolution.

For Barbara McClintock, recognition came 30 years after that symposium at Cold Spring Harbor. In 1981 (at the age of 79), she received eight awards, among them a $60,000-a-year lifetime grant

from the MacArthur Foundation and the $15,000 Lasker prize. In 1983, she was awarded the Nobel Prize in Physiology or Medicine. When informed of the Nobel award, she replied to an interviewer's question, "It seemed unfair to reward a person for having so much pleasure over the years, asking the maize plants to solve specific problems and then watching their response."

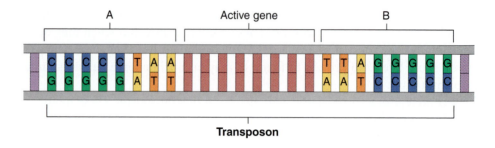

Transposon

FIGURE 6.14

Transposon Structure

A transposon contains one or more active genes bordered by inverted repetitive sequences. Note that the base sequence in A is the reverse and complement of the base sequence in B. Also note the presence of inverted repetitive base sequences (C–G and G–C) at the ends of the transposon. These sequences are important in conferring the ability of a transposon to move.

Of particular significance is the finding that many transposons contain genes for antibiotic resistance. If a plasmid containing a transposon moves from one bacterium to another, as plasmids are known to do, the transposon will move along with it, thus spreading the genes for antibiotic resistance among bacteria. Moreover, the movement of transposons among plasmids helps explain how a single plasmid acquires numerous genes for resistance to different antibiotics.

To this point . . .

With an understanding of protein synthesis, we examined how mutations arise. We discovered that mutations could occur spontaneously as the result of DNA replication errors or damage by normal atmospheric radiation. Mutations also could be induced by physical mutagens, such as UV light, which causes thymine dimer formation. Chemical mutagens included nitrous acid, which changes the chemical nature and pairing of a nitrogenous base and base analogs that resemble a nitrogenous base and cause DNA replication errors.

One of the most common types of changes to the DNA is a point mutation. Such a mutation could be a base-pair substitution for which the reading frame of the mRNA remains the same or a base-pair deletion/insertion, which shifts the reading frame (reading frameshift). The result of a mutation depends on the exact nature of the mutation in the DNA (silent, missense, or nonsense).

We saw that cells try to repair incorrect base pairing in or damage to the DNA. The repair mechanisms included mismatch repair where mismatched base pairs are correctly matched with complementary base pairs, and excision repair where a damaged region of DNA is removed and replaced with the correct base sequence.

Lastly, we learned that mutations also could arise from the insertion of transposable genetic elements (insertion sequences or transposons) into a gene sequence.

In the final section of this chapter, we will see how mutants can be identified through negative and positive selection techniques and how potential cancer-causing agents can be detected using bacterial mutations.

Identifying Mutants

A bacterium (or any organism) carrying a mutation is called a **mutant**, while the normal strain isolated from nature is the **wild type**. Some mutants are easy to identify because the phenotype (the physical appearance) of the bacteria or the colony have changed from that of the wild type. For example, some bacterial colonies appear red because they produce a red pigment. Treat the bacteria with a mutagen and, after plating on nutrient agar, mutants form colorless colonies. However, not all mutants can be identified solely by their phenotype.

PLATING TECHNIQUES SELECT FOR SPECIFIC MUTANTS OR CHARACTERISTICS

Suppose you want to find a nutritional mutant that is unable to grow without the amino acid histidine. This mutant (written his⁻) has lost the ability that the wild type strain (his⁺) has to make its own histidine. Such a nutritional mutant is called an **auxotroph**, while the wild type is a **prototroph**. Phenotypically, there is no difference between the two strains. However, you can identify visually the auxotroph using a **negative selection** plating technique (FIGURE 6.15a).

As another example, suppose you want to see if there are any bacteria in a hospital ward that are resistant to the antibiotic tetracycline. Again, phenotypically there is no difference between those strains that are sensitive to tetracycline and those that are resistant to the antibiotic. However, a **positive selection** plating technique permits visual identification of such tetracycline resistant mutants (FIGURE 6.15b).

Both techniques use of a *replica plating device* that consists of a sterile velveteen cloth mounted on a solid support (Figure 6.15). When an agar plate (*master plate*) with bacterial colonies is gently pressed against the surface of the velveteen, some cells from each colony stick to the velveteen. If another agar plate then is pressed against this velveteen cloth, some cells will be transferred (replicated) in the same pattern as on the master plate. The chemical composition of this transfer plate is key to visual identification of the colonies being hunted.

Negative selection: identification through replica plating of mutants that fail to grow on a minimal medium.

Positive selection: identification through replica plating of mutants that grow in the presence of a specific substance.

THE AMES TEST CAN IDENTIFY POTENTIAL MUTAGENS

Some years ago, scientists observed that about 90 percent of human **carcinogens**, agents causing cancer in humans, also induce mutations in bacteria. Working on this premise, Bruce Ames of the University of California developed a procedure to help identify human carcinogens by determining whether the agent can mutate bacteria. The procedure, called the **Ames test**, is a widely used, relatively inexpensive, rapid, and accurate screening test.

For the Ames test, an auxotrophic, histidine-requiring strain (his⁻) of *Salmonella typhimurium* is used. If inoculated onto a plate of nutrient medium lacking histidine, no colonies will appear because in this auxotrophic strain the gene inducing histidine synthesis is mutated and hence not active.

In preparation for the Ames test, the potential carcinogen is mixed with a liver enzyme preparation. The reason for doing this is because often chemicals only

FIGURE 6.15

Negative and Positive Selection of Bacterial Mutants

(a) Negative selection plating techniques can be used to detect nutritional mutants that fail to grow when replica plated on minimal medium. Comparison to replica plating on complete medium visually identifies the auxotrophic mutants. (b) Positive selection plating techniques are used to identify antibiotic resistant mutants that grow when replica plated onto complete medium plus antibiotic.

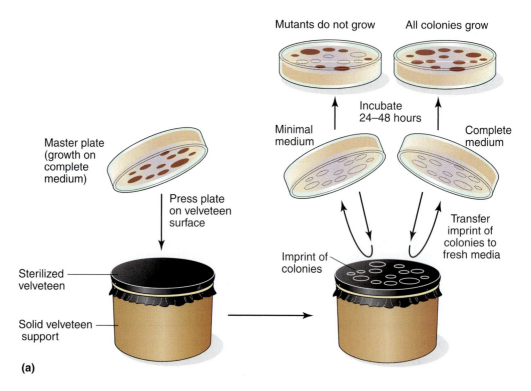

(a)

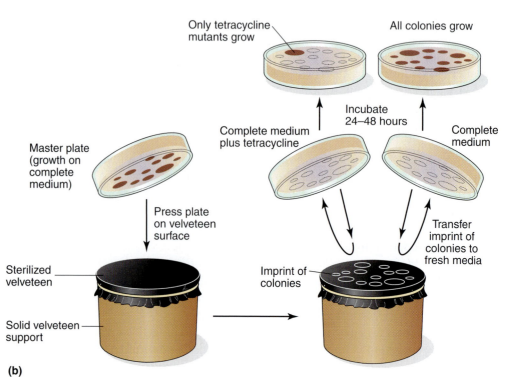

(b)

become cancer causing and mutagenic after they have been modified by liver enzymes in the human body.

To perform the Ames test, the his⁻ strain is inoculated onto an agar plate lacking histidine (**FIGURE 6.16**). A well is cut in the middle of the agar, and the potential carcinogen is added to the well where it diffuses into the agar. The plate is incubated for 24 to 48 hours. If bacterial colonies appear, one may conclude that the agent mutated the bacterial his⁻ gene back to the wild type (his⁺), so the cells could again encode the enzyme needed for histidine synthesis. Because the agent is a mutagen, it is therefore a possible carcinogen in humans. If bacterial colonies fail to appear, one assumes that no mutation took place. However, it is possible that the mutation did occur, but was repaired by a DNA repair enzyme. This possibility has been overcome by using bacterial strains known to be inefficient at repairing errors.

FIGURE 6.16

Using the Ames Test

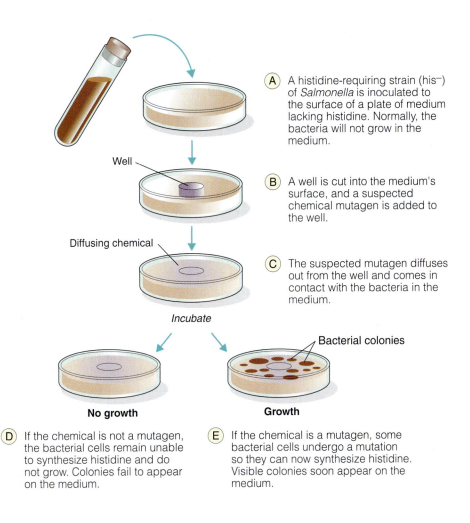

A) A histidine-requiring strain (his⁻) of *Salmonella* is inoculated to the surface of a plate of medium lacking histidine. Normally, the bacteria will not grow in the medium.

Well

B) A well is cut into the medium's surface, and a suspected chemical mutagen is added to the well.

Diffusing chemical

C) The suspected mutagen diffuses out from the well and comes in contact with the bacteria in the medium.

Incubate

Bacterial colonies

No growth

Growth

D) If the chemical is not a mutagen, the bacterial cells remain unable to synthesize histidine and do not grow. Colonies fail to appear on the medium.

E) If the chemical is a mutagen, some bacterial cells undergo a mutation so they can now synthesize histidine. Visible colonies soon appear on the medium.

Note to the Student

It has been almost 60 years since Avery, Hershey, Chase, and others proved that DNA is the genetic information in all cellular organisms. It has been just over 50 years since Watson and Crick worked out the structure of DNA. But, as we have seen in this textbook, science is a distinctly human way of learning. Science has a strong dependence on testable hypotheses, experimental data, logical reasoning, competition, collaboration—and even a bit of skepticism and arrogance. The collaborative work of Watson and Crick is a good example of such qualities.

Watson was a geneticist who knew nothing about the technique of X-ray crystallography that was used to elucidate the structure of DNA. Crick was a physicist who was well versed in X-ray crystallography, but who was more interested in theory than experiment. In the 16 months it took them to work out the structure of DNA, neither did an actual experiment. Rather, they built models for DNA structure that they thought best fit the experimental data of others. As we saw in MicroFocus 6.2, much of their evidence was taken from the X-ray crystallography work of Franklin and the research that Erwin Chargaff had carried out showing the base pairing (A—T, G—C) of DNA nucleotides. In fact, Chargaff stated, "I never met two men who knew so little—but aspired to so much." A bit of arrogance perhaps? However, as Watson said, "I solved it, I guess, because nobody else was paying full-time attention to the problem."

Summary of Key Concepts

6.1 BACTERIAL DNA

- **Bacterial DNA Is Tightly Packed within a Single Chromosome.** The DNA in a bacterium's chromosome is supercoiled and the supercoils are folded into a series of loops. Each loop consists of 50,000–100,000 bases anchored to RNA and protein in the nucleoid area.

- **Plasmids Carry Nonessential Information.** Bacteria often have one or more small closed loops of DNA called plasmids. These structures carry information that can confer selective advantages (e.g., antibiotic resistance, protein toxins) to the bacteria that contain them.

6.2 DNA REPLICATION

- **DNA Replication Is Semiconservative.** DNA copies (replicates) itself by a semiconservative mechanism where each strand of the original DNA molecule acts as a template to synthesize a new strand.

- **DNA Synthesis Progresses from the Origin of Replication.** DNA replication in bacteria starts at the origin of replication. Initiator proteins separate the strands, forming two replication forks that move in opposite directions around the chromosome. DNA polymerase enzymes move along the strands inserting the correct DNA nucleotide to complementary bind with the template strand. At each replication fork one of the two strands is synthesized in a continuous fashion while the other strand is formed in a discontinuous fashion, temporarily forming Okazaki fragments until the fragments are joined by a DNA ligase.

6.3 PROTEIN SYNTHESIS

- **The Central Dogma Identifies the Flow of Genetic Information.** The anabolism of proteins takes place by a complex mechanism in which the genetic information in DNA is first transcribed to a genetic message in RNA and then translated to a sequence of amino acids in the protein.

■ **Transcription Copies Genetic Information into RNA.** Transcription occurs from one of the two strands of a gene by the RNA polymerase binding to a promoter sequence on that strand. A careful ordering of codons on the mRNA, including start and stop codons, is essential to successful completion of the process.

■ **Three Types of RNA Result from Transcription.** Various forms of RNA, including mRNA, tRNA, and ribosomal RNA, are transcribed from the DNA and are important in the protein synthesis process.

■ **Translation Is the Process of Making the Protein.** Translation occurs on the ribosomes, which bring together mRNA and tRNAs. The ribosome "reads" the mRNA codons and inserts the correct tRNA to match codon and anticodon. Translation continues until the ribosome encounters a stop codon when the protein translation machinery disassembles and the protein is released. Chaperone proteins help the elongating polypeptide to fold properly during translation. A polyribosome is a string of ribosomes all translating the same mRNA.

■ **Protein Synthesis Can Be Controlled in Several Ways.** Different control factors influence protein synthesis and provide balance to the overall scheme of metabolism. The best understood is the operon model where binding of a repressor protein to the operon represses transcription. Other factors, such as inducers or corepressors, influence the ability of the repressor protein to bind to the operon.

6.4 BACTERIAL MUTATION

■ **Spontaneous Mutations Are the Result of Natural Processes.** Mutation is a permanent change in the cellular DNA. This can occur spontaneously in nature resulting from a replication error or the effects of natural radiation.

■ **Induced Mutations Are Deliberate Mutations Resulting from a Mutagen.** In the laboratory, ultraviolet light, chemicals, and base analogs represent mutagens that can induce mutations. Deletions or insertions involving the chromosome as well as transposons and insertion sequences also may be mutagenic.

■ **Point Mutations Affect One Base Pair in a Gene.** Base pairs in the DNA can change in one of two ways. There can be a base-pair substitution that does not change the reading frame of an mRNA by ribosomes. The result can be a silent, missense, or nonsense mutation. A point mutation also can occur from the loss or gain of a base pair. Such mutations change the reading frame and often lead to loss of function by the protein being synthesized.

■ **Repair Mechanisms Attempt to Correct Mistakes or Damage in the DNA.** Replication errors or other damage done to the DNA often can be repaired. Mismatch repair replaces an incorrectly matched base pair with the correct pair. Excision repair removes a section of damaged DNA and replaces it with the correctly paired bases.

■ **Transposable Genetic Elements Can Cause Mutations.** Two types of transposable genetic elements exist in many bacteria. Insertion sequences only carry information to copy the sequences and insert them into another location in the DNA. Transposons are genetic elements that "jump" from one location in the DNA and move to another location often in another DNA molecule.

6.5 IDENTIFYING MUTANTS

■ **Plating Techniques Select For Specific Mutants or Characteristics.** Auxotrophic mutants can be identified by negative selection plating techniques. Such mutants would grow as colonies on a complete medium but would not grow on a minimal medium lacking the growth factor. Positive selection can be used to identify bacteria that have certain attributes, such as antibiotic resistance. Only resistant bacteria would grow as colonies on plates containing the antibiotic.

■ **The Ames Test Can Identify Potential Mutagens.** Many mutagens also are carcinogens in humans. The Ames test is a method of using auxotrophic bacteria to identify such carcinogens. The test is based on the ability of a potential mutagen to revert an auxotrophic mutant to its prototrophic form.

Questions for Thought and Discussion

Answers to selected questions can be found in Appendix C.

1. Explain why the ribosome can be portrayed as a "cellular translator."

2. In hospitals, it is common practice to clear the air bubble from a syringe by expelling a small amount of the syringe contents into the air. One microbiologist estimates that this practice results in the release of up to 30 liters of antibiotic into a typical hospital's environment annually. How might this lead to the appearance of antibiotic-resistant mutants in hospitals?

3. One way to detect certain types of mutants is through a positive selection procedure. Explain how this form of identifying mutants is similar to a selective medium described in Chapter 5.

4. In an Ames test, often several apparently his⁻ colonies of *Salmonella* are seen on a control agar plate that contains no histidine. Explain why these colonies would appear on this plate.

5. The author of a general biology textbook writes in reference to the development of antibiotic resistance: "The speed at which bacteria reproduce ensures that sooner or later a mutant bacterium will appear that is able to resist the poison." How might this mutant bacterium appear? Do you agree with the statement? Does this bode ill for the future use of antibiotics?

6. Many viruses have double-stranded DNA as their genetic information while many others have single-stranded RNA as the genetic material. Which group of viruses do you believe is more likely to efficiently repair its genetic material? Explain.

7. Some scientists suggest that mutation is the single most important event in evolution. Do you agree? Why or why not?

8. In 1976, an outbreak of pulmonary infections among participants at an American Legion convention in Philadelphia led to the identification of a new disease, Legionnaires' disease. The bacterium responsible for the disease had never before been known to be pathogenic. From your knowledge of bacterial genetics, postulate how it might have acquired the ability to cause disease.

9. At this writing, the smallest known bacterium whose genome has been deciphered is the submicroscopic organism *Mycoplasma genitalium* (a possible cause of a sexually transmitted disease). This bacterium can survive on about 300 genes. (A human cell, by comparison, has about 35,000 genes.) What do you suppose are some of the proteins encoded by the genes in this minimal genome?

10. In 1994, the CDC reported that the percentage of antibiotic-resistant isolates of *Haemophilus influenzae*, which can cause meningitis in children, had risen from 4.5 percent to 28 percent over the previous five years. What factors might have accounted for this change?

11. In modern medicine, physicians are urged to prescribe an antibiotic that is specifically geared to the organism causing the present disease, rather than a broad-spectrum antibiotic that kills many different bacteria including the present organism. Why?

12. You are working for a bioremediation company that wants to develop a bacterial strain that will degrade toxic benzene found in many hazardous waste sites. How would you go about visually identifying a bacterium that has this characteristic?

13. A chemical is tested with the Ames test to see if the chemical is mutagenic and therefore possibly a cancer-causing chemical in humans. On the test plate containing the chemical, no his⁺ colonies are seen near the central well. However, many colonies are growing some distance from the well. If these colonies truly represent his⁺ colonies, why are there no colonies closer to the central well?

14. Why is supercoiling important to the structure of a bacterial chromosome?

15. What might happen if a mutation occurs in the promoter sequence of a gene such that a segment of the bases are deleted?

Review

Answer the following questions that pertain to (I) transcription and translation, and (II) mutations. Use the genetic code (Table 6.3) on page 207 as needed.

I. The following base sequence is a complete polynucleotide made in a bacterial cell.

—AUGGCGAUAGUUAAACCCGGAGGGUGA—

With this sequence, answer the following questions.

a. Provide the sequence of nucleotide bases found in the inactive strand of the DNA.

b. How many codons will be translated in the mRNA made from the template DNA strand?

c. How many amino acids are coded by the mRNA made and what are the specific amino acids?

d. How many nucleotide bases in the template DNA provide the information for this protein?

II. Use the sequence of bases in the box to answer the following questions about mutations in bacteria.

—TACACGATGGTTTTGAAGTTACGTATT—

a. Is the sequence in the box a single strand of DNA or RNA? Why?

b. Using the sequence in the box, show the translation result if a mutation results in a **C** replacing the **T** at base 12 from the left end of the sequence. Is this an example of a silent, missense, or nonsense mutation?

c. Using the sequence in the box, show the translation result if a mutation results in an **A** inserted between the **T** (base 12) and the **T** (base 13) from the left end of the sequence. Is this an example of a silent, missense, or nonsense mutation?

http://microbiology.jbpub.com

The site features **eLearning,** an on-line review area that provides quizzes and other tools to help you study for your class. You can also follow useful links for in-depth information, or just find out the latest microbiology news.

Genetic Engineering and Bacterial Genomics

Genetic engineering is the most powerful and awesome skill acquired by man since the splitting of the atom.

—The editors of *Time* magazine describing the potential for genetic engineering

W E OFTEN READ in the newspaper these days about this organism's genes being sequenced and that organism's DNA being mapped. That might seem well and fine to many of us because we perceive this as the advance of science. But, what is the underlying significance of such sequencing?

In late spring of 2003, a group of British scientists announced that they had mapped the genome of *Streptomyces coelicolor*, a common soil bacterium. The project began in 1997 and took six years in part because the bacterium is the largest ever sequenced. It has 8.7 million base pairs and some 7,825 genes. One of the scientists on the project said it is a fabulous resource for scientists. Why?

Well, here is where bacterial genomics shows its power. *S. coelicolor* and its relatives are responsible for producing over 65 percent of the naturally known antibiotics. This includes tetracycline and erythromycin. By analyzing the genome of *S. coelicolor* and other *Streptomyces* species, additional metabolic pathways may be discovered for the production of other unidentified and perhaps novel antibiotics. In fact, the researchers have identified 18 gene clusters that they suspect are involved with the production of antibiotics. If correct, knowing the genome and its organization might allow scientists to transform the bacterium into an antibiotics factory and add to the dwindling armada of usable antibiotics to which bacteria are not yet resistant. Using genetic engineering techniques, they could rearrange gene clusters and perhaps produce even more useful and potent antibiotics that do not even exist in nature. One example illustrating the need for newer antibiotics is multidrug-resistant *Staphylococcus aureus*, which causes several serious infections and several deaths each year. Unique antibiotics might be able to attack and destroy these bacteria that have become drug resistant.

But that is not all. *S. coelicolor* is a close relative of the tuberculosis, leprosy, and diphtheria bacilli. By comparing genomes, scientists hope to learn why the *Streptomyces* bacteria are not pathogenic, while the other three are pathogens. What is different about their genomes might be important in understanding the pathogenicity of their relatives and perhaps even designing new antibiotics through genetic engineering that will attack these pathogens.

Genetic engineering and bacterial genomics, two of the major topics for this chapter, are more than simply research procedures of interest to scientists. Their applications have far-reaching consequences for all of us. However, before we can explore these topics, we need to understand the process of genetic recombination, for it is this natural mechanism for DNA transfer between microorganisms that provided significant insight for genetic engineering and bacterial genomics.

7.1

Bacterial Recombination

Traditionally, when one thinks about the inheritance of genetic information, one envisions genes being passed from parent to offspring. This is what is referred to as **vertical gene transfer** (FIGURE 7.1a). However, imagine being able to transfer genes between members of your own family or between your close relatives. This ability to undergo **lateral (horizontal) gene transfer** (FIGURE 7.1b) might sound strange, but microbes are accomplished at doing both types of information transfer. MicroFocus 7.1 provides one scenario for a very extensive rate of genetic transformation through lateral gene transfer in the world's oceans.

GENETIC INFORMATION IN BACTERIA CAN BE TRANSFERRED VERTICALLY AND LATERALLY

In Chapter 6, we discussed **bacterial mutation**, which was one of the ways by which the genetic material in a cell can be permanently altered. Since the permanent change occurred in the parent cell, all future generations derived from the parent also will have the mutation because of vertical gene transfer.

This portion of the chapter concerns the second way by which genetic alterations can arise. This is through **genetic recombination**, which is the process of transferring plasmid DNA or chromosomal DNA fragments laterally from a donor cell to a recipient cell. If, for example, the recipient cell receives a plasmid from a donor cell, the plasmid exists independently in the recipient's cytosol and begins to multiply and encode proteins immediately. If the recipient obtains a chromosomal DNA fragment from the donor, the new DNA pairs with a complementary region of recipient DNA and replaces it. In this case, there is no change in quantity of the recipient's chromosomal DNA, but there may be a substantial change in its quality.

Although genetic recombination also occurs among some eukaryotic microbes, we will limit our discussion to lateral transfer in the prokaryotes. Three distinctive mechanisms mediate the lateral transfer of DNA between cells: *transformation, conjugation*, and *transduction*. We shall examine each in turn.

FIGURE 7.1

Gene Transfer Mechanisms

Genes can be transferred between bacteria in two ways. (a) In vertical transfer, a bacterium undergoes binary fission and the daughter cells of the next generation contain the identical genes found in the parent. (b) In lateral transfer, genes are transferred by various mechanisms to other bacteria of the same generation.

(a) Vertical gene transfer

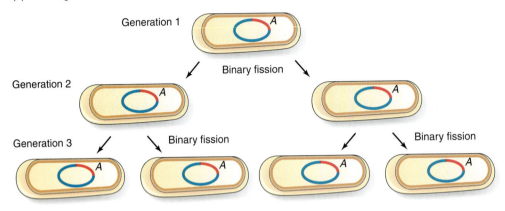

(b) Lateral gene transfer

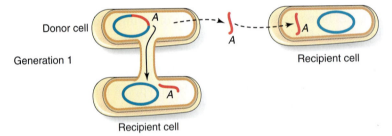

⬛ TRANSFORMATION IS THE UPTAKE AND EXPRESSION OF DNA IN A RECIPIENT CELL

Transformation is the uptake of a DNA fragment from the surrounding environment and the expression of that genetic information in the recipient cell; that is, the recipient has acquired a characteristic that it previously lacked. The process is best understood by examining a few experiments.

In 1928, an English bacteriologist named Frederick Griffith published the results of an interesting set of experiments with *Streptococcus pneumoniae*. This bacterium, referred to as a pneumococcus (pl., pneumococci), is a major cause of bacterial pneumonia (Chapter 8). At Griffith's time, the experimental results were considered unusual, but in retrospect, microbiologists note that they gave some of the first clues for genetic recombination.

Pneumococci occur in different strains. There are encapsulated strains, designated S, because the organisms grow in smooth colonies. These strains cause pneumonia.

MicroFocus 7.1

GENE SWAPPING IN THE WORLD'S OCEANS

Many of us are familiar with the accounts of microorganisms in and around us, but we are less familiar with the massive numbers of microbes in the world's oceans. For example, microbial ecologists estimate there are an estimated 10^{29} bacteria in the world's oceans. Also, there are some 10^{30} viruses called bacteriophages ("bacteria eaters") in the oceans that infect bacteria. These bacteriophages will be discussed later in the chapter. For now, the bacteriophages represent huge populations in the oceans that can infect bacteria.

Bacteriophages can carry pieces of the bacterial chromosome from one cell they have infected to another recipient cell. In the recipient cell, the new bacterial DNA can be swapped for an existing part of the bacterial chromosome. It is a fairly rare event, occurring only once in every 100 million (10^8) virus infections. That doesn't seem very significant until you now consider the number of bacteriophages and susceptible bacteria we mentioned that exist in the oceans. Working with these numbers and the potential number of virus infections, scientists suggest that if only one in every 100 million infections brings a fragment of bacterial DNA to a recipient cell, there are about 10 million billion (that's

10,000,000,000,000,000 or 10^{16}) such gene transfers *per second* in the world's oceans. That is about 10^{21} infections per day!

I think it is fair to say that there's an awful lot of gene swapping going on!

There also is an unencapsulated strain, designated R, because the colonies appear rough. Organisms in this strain are harmless. Griffith showed that mice injected with living S strain pneumococci die, while those injected with living R strain pneumococci live (**FIGURE 7.2**). This was what he expected. Also, he showed that mice injected with dead S strain organisms live. Again, this result was not unusual.

What happened next puzzled Griffith. He mixed heat-killed, dead S strain bacteria with live R strain bacteria and let the mixture incubate; then he injected the mixture into mice. The mice died. Griffith wondered how a mixture of live harmless bacteria (R strain) and debris from the dead pathogenic bacteria (S strain) could kill the mice. His answer came when he autopsied the animals: microscopic examination of their blood showed they were full of live S strain pneumococci. Knowing that spontaneous generation does not occur, Griffith reasoned that somehow the live R strain bacteria had been transformed to live S strain bacteria. Though he could not explain how this happened, he published his results but never tried to understand the nature of the transforming agent.

However, in 1944 Oswald T. Avery and his associates Colin M. MacLeod and Maclyn N. McCarty, of the Rockefeller Institute, purified and identified the transforming substance. These investigators found that the substance was not protein, as had been anticipated, but a then-obscure organic compound called **deoxyribonucleic acid** (**DNA**).

Most scientists were blind to Avery's discovery and were reluctant to accept DNA as a hereditary substance. Geneticists of that period were not trained as chemists, and Avery's experiments were difficult to repeat. Also, many scientists believed that

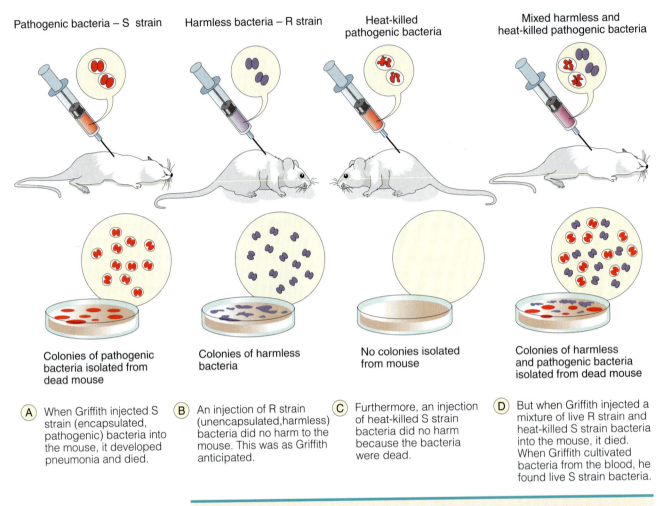

Pathogenic bacteria – S strain   Harmless bacteria – R strain   Heat-killed pathogenic bacteria   Mixed harmless and heat-killed pathogenic bacteria

Colonies of pathogenic bacteria isolated from dead mouse

Colonies of harmless bacteria

No colonies isolated from mouse

Colonies of harmless and pathogenic bacteria isolated from dead mouse

(A) When Griffith injected S strain (encapsulated, pathogenic) bacteria into the mouse, it developed pneumonia and died.

(B) An injection of R strain (unencapsulated, harmless) bacteria did no harm to the mouse. This was as Griffith anticipated.

(C) Furthermore, an injection of heat-killed S strain bacteria did no harm because the bacteria were dead.

(D) But when Griffith injected a mixture of live R strain and heat-killed S strain bacteria into the mouse, it died. When Griffith cultivated bacteria from the blood, he found live S strain bacteria.

FIGURE 7.2

The Transformation Experiments of Griffith

Griffith's experiments were the first to demonstrate bacterial transformation and the lateral transfer of genetic information.

experimental results from bacteria may not apply to eukaryotes. Moreover, preoccupation with World War II had restricted the dissemination and flow of scientific knowledge. Not until the early 1950s was DNA widely accepted as the molecule of heredity and transformation as a concept of bacterial recombination.

Modern scientists regard transformation as an important genetic recombination method even though it takes place in less than 1 percent of a bacterial population. Under natural conditions, both gram-positive and gram-negative bacteria are highly transformable when their DNA is very similar to the DNA being received, which generally implies cells of the same species. In natural environments, when bacterial cells die and lyse, the bacterial chromosome typically breaks apart into fragments of DNA composed of about 10 to 20 genes (**FIGURE 7.3a**). These fragments can be taken up by similar species of live bacteria and the recipient cells often are transformed.

The ability of a cell to be transformed depends on its **competence**, defined as the ability of a recipient bacterium to take up DNA from the environment. Competence

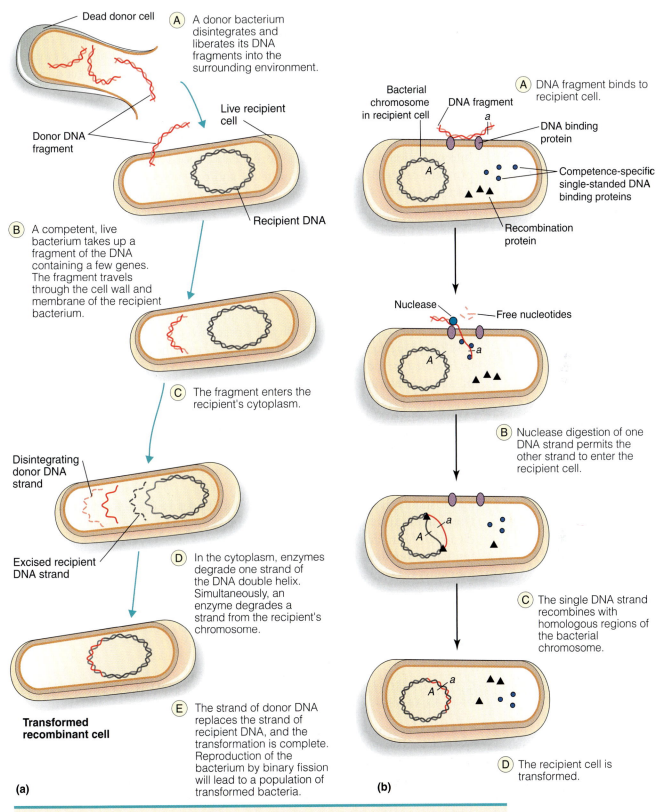

(A) A donor bacterium disintegrates and liberates its DNA fragments into the surrounding environment.

Dead donor cell

Donor DNA fragment

Live recipient cell

Recipient DNA

(B) A competent, live bacterium takes up a fragment of the DNA containing a few genes. The fragment travels through the cell wall and membrane of the recipient bacterium.

(C) The fragment enters the recipient's cytoplasm.

Disintegrating donor DNA strand

Excised recipient DNA strand

(D) In the cytoplasm, enzymes degrade one strand of the DNA double helix. Simultaneously, an enzyme degrades a strand from the recipient's chromosome.

Transformed recombinant cell

(E) The strand of donor DNA replaces the strand of recipient DNA, and the transformation is complete. Reproduction of the bacterium by binary fission will lead to a population of transformed bacteria.

(a)

Bacterial chromosome in recipient cell

DNA fragment

(A) DNA fragment binds to recipient cell.

DNA binding protein

Competence-specific single-stranded DNA binding proteins

Recombination protein

Nuclease

Free nucleotides

(B) Nuclease digestion of one DNA strand permits the other strand to enter the recipient cell.

(C) The single DNA strand recombines with homologous regions of the bacterial chromosome.

(D) The recipient cell is transformed.

(b)

FIGURE 7.3

Bacterial Transformation

(a) Transformation is the process in which a live bacterium acquires DNA fragments from the environment.
(b) The mechanism for DNA transfer by transformation involves binding of the DNA fragment, passage of a single-stranded molecule into the recipient, and incorporation into the bacterial chromosome.

is an intriguing property that varies among bacteria. For example, growing cells of *S. pneumoniae* secrete a competence factor that induces the competence state, while *Hemophilus influenzae* cells become competent when the culture is switched from a rich to a minimal growth medium. In both bacteria, several genes encode proteins for binding and uptake of DNA fragments. In fact, competent streptococci and *Bacillus* cells have approximately 50 surface receptors where DNA can bind before uptake.

A competent cell usually incorporates only one or at most a few DNA fragments. Internalization of DNA is an ATP-dependent process and requires DNA-binding proteins, cell wall degradation proteins, and cell membrane-transport proteins. Bacteria, such as *S. pneumoniae* and *Bacillus*, degrade one strand of the DNA fragment as it is being taken into the cell (FIGURE 7.3b). Such single-stranded DNA, if stably associated with a similar region of the bacterial chromosome, will replace a similar chromosome sequence. In *H. influenzae*, a double-stranded DNA fragment is internalized, but then one strand is digested by a nuclease before incorporation into the bacterial chromosome. MicroFocus 7.2 describes experiments that precisely located the insertion point for the new DNA.

Factors affecting the cell surface are important to competence, particularly changes in membrane permeability or surface receptors. *Escherichia coli* does not develop competence during normal growth. However, the uptake of DNA fragments by *E. coli* can be induced in the laboratory by chilling bacteria to 4°C in the presence of calcium chloride, and then quickly heating the cells to 42°C. This treatment apparently alters the membrane and encourages the passage of DNA strands.

One potential effect of transformation is to increase an organism's pathogenicity. In Griffith's pneumococci experiments, for example, the live R strain cells acquired the genes for capsule formation from the dead S strains, which allowed the organism to avoid body defenses and thus cause disease. Microbiologists also have demonstrated that when mildly pathogenic strains of bacteria take up DNA from other mildly pathogenic strains, there is a cumulative effect, and the degree of pathogenicity increases. Observations such as these may help explain why highly pathogenic strains of bacteria appear from time to time. Transformed bacteria also may display enhanced drug resistance from the acquisition of R plasmids. However, it does not appear that transformation is a significant contributor to the dispersal of antibiotic-resistance genes.

The significance of transformation as a means of genetic recombination under natural conditions continues to be studied. However, it appears certain that transformation occurs regularly where bacteria exist in crowded conditions, such as in rich soil or the human intestinal tract.

CONJUGATION INVOLVES CELL-TO-CELL CONTACT FOR DNA TRANSFER

In the recombination process called **conjugation**, two live bacteria come together and the donor cell directly transfers DNA to the recipient cell. This process was first observed in 1946 by Joshua Lederberg and Edward Tatum in a series of experiments with *E. coli*. Lederberg and Tatum mixed two different strains of bacteria and found that genetic traits could be transferred among them if contact occurred.

The process of conjugation requires a special conjugation apparatus called the **conjugation pilus** (FIGURE 7.4). It also has been called a *sex pilus*, but realize that conjugation is not a sexual process. For cell-to-cell contact, the donor cell, designated **F⁺**, produces the conjugation pilus that makes contact with the recipient cell, known as an **F⁻ cell**. The donor cell is called F⁺ because it contains an **F plasmid**, a double-stranded

MicroFocus 7.2

PROGRAMMING YOUR VCR

To the great majority of people, VCR stands for video cassette recorder. However, to Julian Davies and his coworkers at the University of British Columbia, VCR has a much more scientific meaning: *Vibrio cholerae* repeating sequences.

Davies is a molecular microbiologist. In 1998, he reported his research on repeating sequences in DNA in the cholera bacillus *Vibrio cholerae*. Repeating sequences are a type of genetic stutter; that is, stretches of DNA whose nitrogenous bases follow an identifiable pattern of repetition (e.g., note the five-base repeat in GTGGAGTGGAG TGGA...). Scientists have long known that repeating sequences border the genes for antibiotic resistance that insert into transformed bacteria. Davies and

his group performed experiments demonstrating that the sequences flank locations where other acquired genes insert as well. They extracted a *V. cholerae* gene and its adjacent repeating sequences and inserted the combination (a VCR cassette) into a plasmid. Then, they inserted the plasmid into *E. coli* cells and noted that the gene inserted into the recipient's DNA at a predetermined site. Soon the *V. cholerae* gene was expressing itself by encoding an identifiable protein.

The research evidence points up the possibility that acquired genes wind up at the same location in a recipient bacterium, regardless of whether they are antibiotic-resistance genes, toxin genes, adhesion genes, or any other genes. If so, the findings would improve the working

model for how bacteria acquire genes to enhance their virulence (virulent *E. coli* strains are an example), how bacteria acquire resistance to numerous antibiotics (multidrug-resistant *Staphylococcus aureus*, for instance), or how transformations in general take place.

Molecular geneticists have a name for the insertion site: the integron. They also have a name for the enzyme that inserts genes into the integron: integrase. Now they have an idea of where to find the integron and the site of integrase activity for all genes acquired by the cell: Just look for repeating sequences flanking the insertion site.

For Davies, the work with VCR was fruitful and significant. Finally, he could go home and catch a good movie. Where? On his VCR, of course.

FIGURE 7.4

Bacterial Conjugation in *E. Coli*

The direct transfer of DNA between live bacterial cells requires a conjugation pilus. In this false color transmission electron micrograph, the F⁺ cell on the left has produced a conjugation pilus that has contacted the F⁻ cell on the right. Notice the numerous fimbriae on the surface of the F⁺ cell. (Bar = 1 μm.)

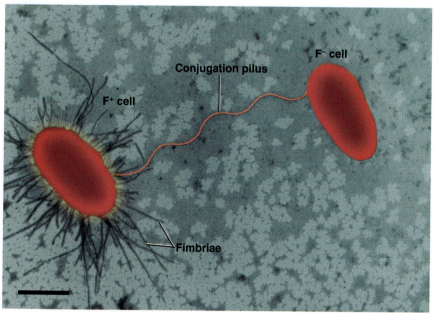

loop of DNA existing apart from the bacterial chromosome. The plasmid contains about 100 genes, most of which are associated with plasmid DNA replication and production of the conjugation pilus. Since the plasmid DNA is involved in conjugation and transfer to the F⁻ cell, the plasmid also is known as a **conjugative plasmid**. The F⁻ cell lacks an F plasmid.

Once contact is made, the pilus shortens to bring the two bacteria close together. The F plasmid then begins replicating by the rolling circle mechanism identified in Chapter 6. In **rolling circle replication**, while one strand of plasmid DNA remains in a closed loop, an enzyme nicks the other strand at a point called the **origin of transfer** (**oriT**) (FIGURE 7.5a). This broken strand then "rolls off" the loop and passes through the channel to the recipient cell. As the lateral transfer occurs, DNA synthesis produces a new complementary strand to replace the transferred strand. Once DNA transfer is complete, the two cells separate.

In the recipient cell, the new single-stranded DNA serves as a template for synthesis of a complementary polynucleotide strand, which then circularizes to reform an F plasmid. This completes the conversion of the recipient from F⁻ to F⁺ and this cell now can act as a donor cell (F⁺) with another F⁻ recipient. Transfer of the F factor does not involve the bacterial chromosome; therefore, the recipient does not acquire new genes other than those on the F plasmid.

The high efficiency of DNA transfer by conjugation shows that conjugative plasmids can spread rapidly, converting a whole population into plasmid-containing cells. Indeed, conjugation appears to be the major mechanism for antibiotic resistance transfer. In laboratory experiments, for example, bacterial strains carrying plasmid antibiotic-resistance genes were introduced into mice. Lateral gene transfer through conjugation of these R plasmids to native bacterial populations rapidly occurred. In nature, conjugation readily occurs between bacteria in soil and in water. MicroFocus 7.3 describes using conjugation as a mechanism for bioremediation.

CONJUGATION ALSO CAN TRANSFER CHROMOSOMAL DNA

Bacteria also can undergo a type of conjugation that accounts for the lateral passage of chromosomal material from donor to recipient cell. Bacteria that exhibit the ability to donate chromosomal genes are called **high frequency of recombination** (**Hfr**) strains.

In Hfr strains, the F plasmid attaches to the bacterial chromosome (FIGURE 7.5b). This attachment is a rare event that requires an insertion sequence to recognize the F plasmid. Once incorporated into the bacterial chromosome, the F plasmid no longer controls its own replication. The Hfr cell triggers conjugation just like an F⁺ cell. When a recipient cell is present, a conjugation pilus forms and attaches to the F⁻ cell. The two cells are brought together. One strand of the bacterial chromosome is nicked at oriT and a portion of the single-stranded chromosomal/plasmid DNA then passes into the recipient cell. Rolling circle replication in the Hfr cell replaces the DNA strand transferred so the donor cell remains an Hfr cell.

The oriT site in the bacterial chromosome is in the middle of the F plasmid genes. Therefore, in the recipient cell, the first genes to enter are only a part of the F plasmid and these genes are not the ones that would make the cell an F⁺ donor. Rather, the last genes that would be transferred to the recipient cell control the donor state. However, these rarely enter the recipient because conjugation usually is interrupted by movements that break the attachment between cells before complete transfer is accomplished. An estimated 100 minutes is required for the transfer of a complete *E. coli*

FIGURE 7.5

Conjugation in Bacteria

(a) Conjugation between an F⁺ cell and an F⁻ cell. When the F plasmid is transferred from a donor (F⁺) cell to a recipient (F⁻) cell, the F⁻ cell becomes an F⁺ cell as a result of the presence of an F plasmid.
(b) Conjugation between an Hfr and an F⁻ cell allows for the transfer of some chromosomal DNA from donor to recepient cell.

(a)

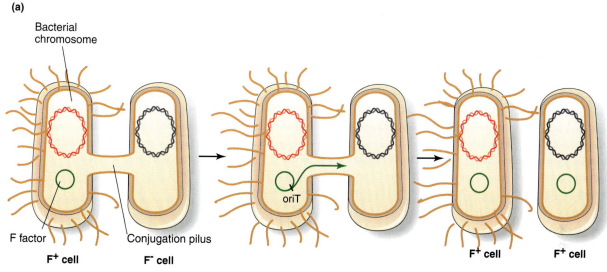

F⁺ cell	F⁻ cell	F⁺ cell F⁺ cell

(A) Conjugation pilus connects the F⁺ to the F⁻ cell.

(B) Rolling circle replication and transfer of F factor from the origin of transfer (oriT).

(C) Both cells now are F⁺.

(b)

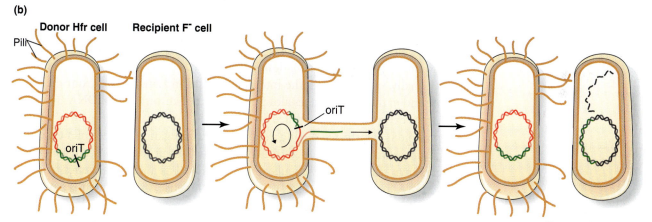

Recombined F⁻ cell

(A) The Hfr donor cell has the F plasmid (green) integrated into the donor's chromosome.

(B) A conjugation pilus forms between the donor and recipient cells, and replication of one strand of the donor's DNA begins at the oriT site and passes into the recipient cell.

(C) Usually only a portion of the donor DNA enters the recipient cell before the conjugation pilus breaks. The new DNA strand replaces a complementary portion of the recipient's DNA. The recombination is now complete, and the new genes can be expressed by the recipient.

MicroFocus 7·3

CONJUGATION-MEDIATED BIOREMEDIATION

In Chapter 1 we defined bioremediation as the use of microorganisms to remove or degrade dangerous environmental wastes. With the apparent rapid spread of conjugative plasmids between bacteria, scientists wondered if conjugation is a mechanism to introduce pollutant-degrading genes into bacterial populations that then could degrade the pollutant in a contaminated site.

A strain of the bacterium *Alcaligenes* has been isolated from a landfill in the Niagara River watershed, which contains a conjugative plasmid that carries the genes to degrade chlorobenzoate. Chlorobenzoate is a common pollutant in chemical waste dumps. To see if this

organism would grow and spread the pollutant-degrading genes to other *Alcaligenes* cells, bacteria containing the conjugative plasmid were introduced into an environment simulating a chlorobenzoate-contaminated river. Although the *Alcaligenes* cells did not survive well, the plasmid was transferred, presumably by conjugation, to a wide variety of indigenous bacterial species.

The conjugation process can spread genes over quite a large area. One of the most notable examples was in Estonia where an oil-shale mine caught on fire. Putting out the fire resulted in an accumulation of some 100,000 cubic meters of phenol-polluted waters below

ground. To try to clean up the accumulated phenol, a strain of *Pseudomonas putida* was used. The conjugative plasmid in this bacterium contained an operon for the degradation of phenol and polychlorinated compounds. Estonian scientists pumped a concentrated solution of the bacteria into the mine. The field trial was a success, with the concentration of phenols falling from some 25 mkg/l to 9 mkg/l. Interestingly, after six years, the original *P. putida* strain could not be detected. However, the operon encoding phenol degradation was associated with several other bacterial species up to 22 kilometers from the original site.

---

chromosome with plasmid genes—something that rarely occurs in nature. Thus, the F⁻ cell usually remains a recipient, although it now has some new chromosomal genes, depending on how many passed through before the two cells separated.

Should the entire chromosome be transferred to the recipient, the F plasmid usually detaches from the chromosome, and enzymes synthesize a strand of complementary DNA. The F plasmid now forms a loop to assume an existence as a plasmid, and the recipient becomes a donor (F⁺) cell.

Occasionally, in an Hfr cell, the integrated F plasmid breaks free from the chromosome and in the process takes along a fragment of chromosomal DNA. The plasmid with its extra DNA is now called an **F' plasmid** (pronounced "F-prime"). When the F' plasmid is transferred during a subsequent conjugation, the recipient acquires those chromosomal genes excised from the donor. This process results in a recipient having its own genes for a particular process, as well as additional genes from the plasmid DNA for that same process. In the genetic sense, the recipient is a partially diploid organism because there are two genes for a given function.

Conjugation has been demonstrated to occur between cells of various genera of bacteria. For example, conjugation occurs between such gram-negative bacteria as *Escherichia* and *Shigella*, *Salmonella* and *Serratia*, and *Escherichia* and *Salmonella*. **Intergenic transfer** has great significance in the transfer of antibiotic-resistance genes carried on plasmids. (MicroFocus 7.4 describes one case with serious medical overtones.) Moreover, when the genes are attached to transposons, the transposons may "jump" from ordinary plasmids to F plasmids, after which transfer by conjugation may occur.

Although conjugative pili are found only on some gram-negative bacteria, gram-positive bacteria also appear capable of conjugation. Microbiologists have experimented extensively with *Streptococcus mutans*, a common cause of dental caries. In this organism, conjugation appears to involve only plasmids, particularly those car-

Intergenic transfer:
the transfer of DNA between different genera.

TRANSFERABLE DRUG RESISTANCE

In 1968, an extremely serious form of bacterial dysentery broke out in Guatemala. Dysentery is a disease of the human intestine characterized by tissue erosion and considerable fluid loss. The responsible bacterium, *Shigella dysenteriae*, resisted treatment with chloramphenicol, tetracycline, streptomycin, and sulfanilamide antibiotics, any one of which is normally used in therapy. In the three years that the epidemic raged, 100,000 people were infected and 12,000 died.

This particular outbreak of drug-resistant dysentery points up the consequences when antibiotic treatment is stifled by resistant bacteria. How *Shigella* may have acquired the resistance was first shown in a remarkable set of experiments by a Japanese team of investigators.

The story began in 1955 when a Japanese woman suffered a case of

dysentery caused by bacteria resistant to the same quartet of drugs observed years later in Guatemala. Doctors at Tokyo University, led by Tomoichiro Akiba, investigated the case and found that drug-resistant *Shigella* strains were fairly widespread in Japan. They also noted a surprising coincidence: Patients with drug-resistant *Shigella* also had in their intestine a strain of *Escherichia coli* with resistance to the same four drugs. Since simultaneous mutations were highly unlikely, researchers postulated that the resistance had been transferred laterally between organisms.

Akiba and his colleagues devised a series of experiments to test this hypothesis. They mixed liquid suspensions of drug-resistant *E. coli* with laboratory-reared drug-sensitive *S. dysenteriae*. Then, they carefully isolated and tested the *Shigella*. The results were startling: *Shigella* was

now resistant to the same drugs as *E. coli*. A transfer indeed had taken place.

In the following years, numerous studies verified transferable drug resistance, and R factor plasmids were identified as the medium of transfer. Epidemics such as that in Guatemala soon broke out elsewhere, as drug-resistant bacteria began appearing in different human populations. By the 1990s, microbiologists had identified resistance in such diverse organisms as streptococci, gonococci, leprosy and tuberculosis bacilli, and malaria parasites. In the 1980s, an article in *Discover* magazine described a Detroit epidemic caused by a drug-resistant strain of *Staphylococcus aureus*. The strain was quickly dubbed "super staph," and the headline was an eye-grabbing "Bugs That Won't Die." Transferable drug resistance had made the popular media.

rying genes for antibiotic resistance. Moreover, the conjugation does not involve pili. Rather, the recipient cell apparently secretes substances encouraging the donor cell to produce **clumping factors** composed of protein. The factors bring together (clump) the donor and recipient cells, and pores form between the cells to permit plasmid transfer. Such observations have been made in *Clostridium* species. Chromosomal transfer has not been demonstrated.

TRANSDUCTION INVOLVES VIRUSES IN THE LATERAL TRANSFER OF DNA

Bacterial recombination by transduction was reported in 1952 by Joshua Lederberg and Norton Zinder. While working with *Salmonella* cells, Lederberg and Zinder observed recombination, but ruled out conjugation and transformation because the cells were separated by a thin membrane and DNA was absent in the extracellular fluid. Eventually, they discovered a virus in the fluid and uncovered the details of what was taking place.

Transduction requires a virus to carry the DNA fragment from donor to recipient cell. The virus that participates in transduction is called a **bacteriophage**, or simply **phage** (FIGURE 7.6). Though invisible at the time, the activity of a bacteriophage (literally "bacteria eater") was described in 1915 by Frederick Twort and two years later by Felix d'Herelle. Bacteriophages originally were assumed to be a type of poison

Bacteriophage:
a virus that infects and replicates within a bacterium.

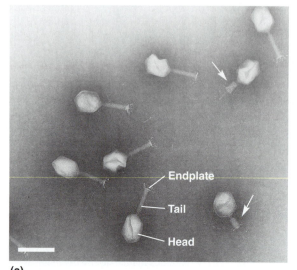

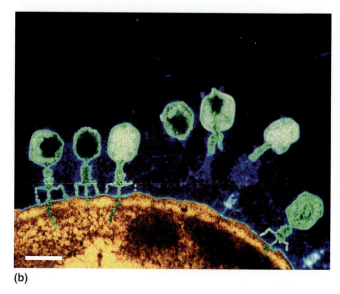

(a)   (b)

FIGURE 7.6

Bacteriophages

(a) A transmission electron microscope view of T2 bacteriophages. Note that each bacteriophage is composed of a head and a tail with a complex endplate. In two cases (arrows), the tail is contracted and a core may be seen protruding. This structure penetrates the bacterial cell wall and membrane to facilitate the transfer of viral DNA to the bacterial cell. (Bar = 130 nm). (b) False color transmission electron micrograph of bacteriophages adsorbed to receptor sites on a bacterial cell. Note the relative sizes of the viruses and the bacterium. (Bar = 80 nm.)

Lytic cycle:
a process in which phages replicate and lyse the host cell.

Lysogeny:
a process in which phage DNA integrates into host DNA without replicating or lysing the cell.

because they dissolve bacteria (Chapter 12). Today scientists recognize them as viruses composed of a core of DNA or RNA surrounded by a coat of protein. Phages that participate in transduction are called **transducing phages**. All the latter contain DNA.

In the replication cycle of a bacteriophage, the phage interacts with bacteria in either of two ways. In one way, the phage invades the bacterium, then replicates itself and destroys the bacterium as new phages are released. This cycle is called the **lytic cycle**. Phages that cause lysis are known as **virulent phages**.

The second way that phages interact with bacteria also involves invasion of the bacterium but without cell lysis. In this case, the phage DNA encodes a repressor protein that inhibits its own replication. The phage DNA often forms a closed circle that aligns next to the bacterial chromosome before integrating into the bacterial chromosome (as the F plasmid does in Hfr strains). This process is called **lysogeny** and the phages that participate in lysogeny are known as **temperate phages**.

Because two types of phage interact with bacteria, there are two ways phages can be involved in the transfer of bacterial genes: *generalized transduction* and *specialized transduction*. Generalized transduction is the more common form of lateral gene transfer between bacteria.

Generalized transduction is carried out by virulent phages that have a lytic cycle of infection (**FIGURE 7.7**). Two examples are phage P1, which infects *E. coli*, and P22, which infects *Salmonella typhimurium*. During virulent phage replication, the bacterial DNA is digested into small fragments. When the new phage DNA is produced, it quickly is enzymatically cut into smaller pieces that will become the DNA incorporated into new phage particles. During this process of packaging phage DNA into new phage particles, a fragment of bacterial DNA may accidentally be caught up in the packaging process and end up in a phage head rather than phage DNA. These

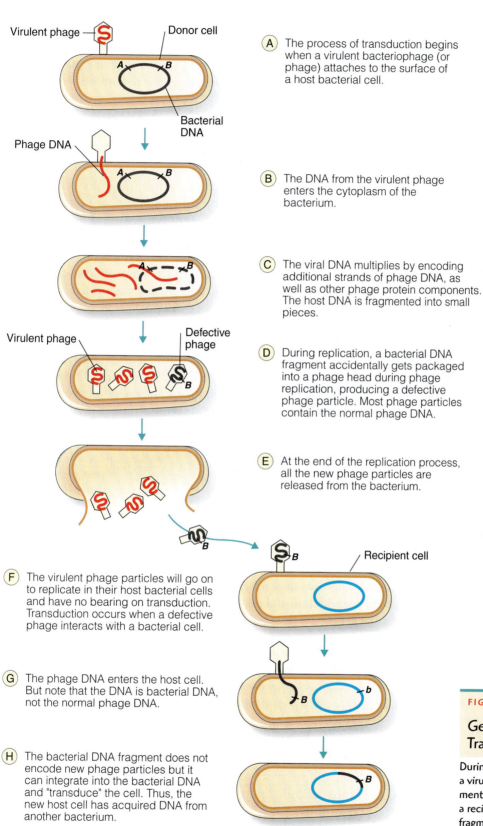

(A) The process of transduction begins when a virulent bacteriophage (or phage) attaches to the surface of a host bacterial cell.

(B) The DNA from the virulent phage enters the cytoplasm of the bacterium.

(C) The viral DNA multiplies by encoding additional strands of phage DNA, as well as other phage protein components. The host DNA is fragmented into small pieces.

(D) During replication, a bacterial DNA fragment accidentally gets packaged into a phage head during phage replication, producing a defective phage particle. Most phage particles contain the normal phage DNA.

(E) At the end of the replication process, all the new phage particles are released from the bacterium.

(F) The virulent phage particles will go on to replicate in their host bacterial cells and have no bearing on transduction. Transduction occurs when a defective phage interacts with a bacterial cell.

(G) The phage DNA enters the host cell. But note that the DNA is bacterial DNA, not the normal phage DNA.

(H) The bacterial DNA fragment does not encode new phage particles but it can integrate into the bacterial DNA and "transduce" the cell. Thus, the new host cell has acquired DNA from another bacterium.

Recombinant cell

FIGURE 7.7

Generalized Transduction

During generalized transduction, a virus carries random DNA fragments from a donor bacterium to a recipient bacterium. The DNA fragment may recombine with the recipient cell.

phages are fully formed though and they can infect another bacterial cell. However, they are defective because they carry no phage genes and cannot replicate themselves after infection. This event is considered rare (1 in every 100,000 new phages may have bacterial DNA) and represents the type of genetic recombination described in MicroFocus 7.1.

Now the transduction takes place. The transducing phage is released along with all the other normal phages. If the transducing phage attaches to a new (recipient) bacterium, it will inject its bacterial DNA into the recipient cell. Once released in the recipient, new genes can pair with a section of the recipient's DNA and replace the section in a fashion similar to conjugation. The recipient has now been transduced (changed) using genes from the donor bacterium and the phage intermediary.

Specialized transduction occurs as a result of lysogeny and unlike generalized transduction, results in the transfer of specific genes (FIGURE 7.8). One of the most studied temperate viruses is phage lambda, which infects *E. coli*. The site of phage incorporation (insertion site) into the bacterial chromosome varies for different phages and is often a region where an insertion sequence is located.

What happens is that the temperate phage DNA actually is integrated into the bacterial chromosome without replacing any bacterial genes. In the integrated state, the phage DNA is called a **prophage**, and the bacterium carrying the prophage is said to be **lysogenic**. The prophage can stay in this state for an extended period.

Prophage:
the phage DNA that has inserted into the host bacterium's chromosome.

At some time in the future, a mutagen such as UV light or a DNA inhibitor activates an enzyme complex that causes the prophage to excise itself from the bacterial chromosome and enter the lytic cycle. Most of the time the excision occurs precisely and the intact phage DNA is released. Sometimes, however, an imprecise excision occurs, and the excised prophage takes along a few flanking bacterial genes while leaving behind a few phage genes. At the conclusion of phage replication, multiple copies of the phage, each with some bacterial genes, are produced. Again, these phages would be defective since they are missing a few phage genes. Such a transducing phage can infect another bacterium and transfer its genes to a recipient, but the genes cannot encode a replicative cycle. Instead, the genes integrate into the bacterial chromosome, carrying the donor's bacterial genes with them. As before, the recipient bacterium has acquired genes from the original bacterium and the recipient is now considered transduced, as shown by Figure 7.8.

Generalized transduction and specialized transduction are summarized and compared in TABLE 7.1.

Specialized transduction is an extremely rare event in comparison to the generalized form, because genes do not easily break free from the bacterial chromosome and the types of chromosome sequences that are transferred are very restricted. The well-established correlation in clinical microbiology between prophage and pathogenesis demonstrates the potential results of recombination. For example, the diphtheria bacillus, *Corynebacterium diphtheriae*, harbors a prophage that provides the genetic code for a toxin that causes diphtheria. Other toxins encoded by prophages include staphylococcal enterotoxins in food poisoning, clostridial toxins in some forms of botulism, and streptococcal toxins in scarlet fever. Also, *Salmonella* cells carry prophages that encode lipopolysaccharides in the outer membrane. These lipopolysaccharides provide the basis for separating *Salmonella* into serological types (serotypes) rather than species.

It might seem strange that phage would transfer genes for human illnesses because these viruses only infect bacteria. However, phage-encoded toxins often cause the lysis of human host cells, which has advantages for the lysogenic bacteria. For example, many bacteria need iron for growth. However, "free iron" is very rare

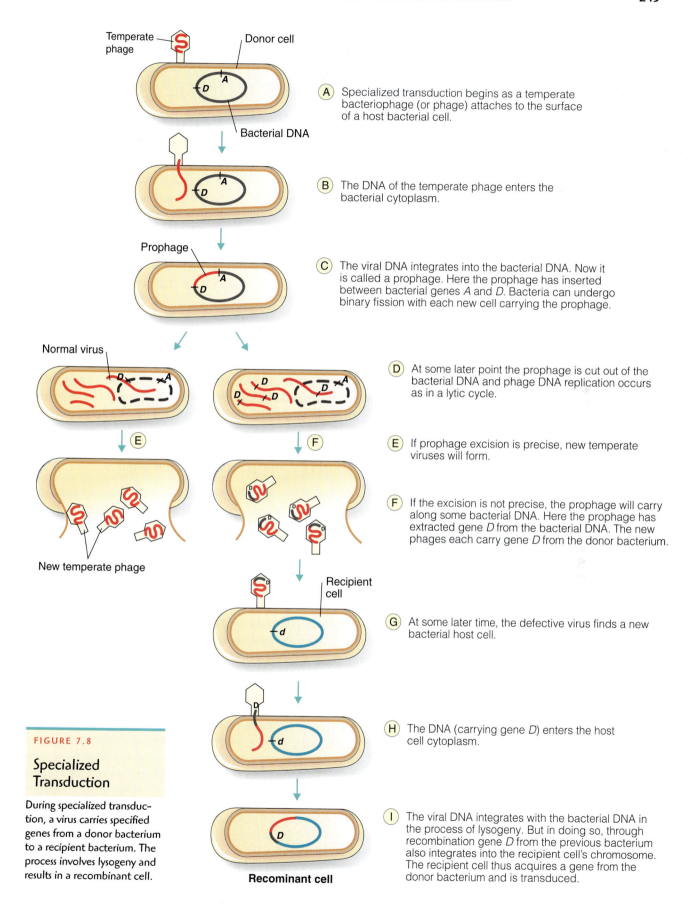

Temperate phage — **Donor cell**

Bacterial DNA

(A) Specialized transduction begins as a temperate bacteriophage (or phage) attaches to the surface of a host bacterial cell.

(B) The DNA of the temperate phage enters the bacterial cytoplasm.

Prophage

(C) The viral DNA integrates into the bacterial DNA. Now it is called a prophage. Here the prophage has inserted between bacterial genes A and D. Bacteria can undergo binary fission with each new cell carrying the prophage.

Normal virus

(D) At some later point the prophage is cut out of the bacterial DNA and phage DNA replication occurs as in a lytic cycle.

(E) If prophage excision is precise, new temperate viruses will form.

(F) If the excision is not precise, the prophage will carry along some bacterial DNA. Here the prophage has extracted gene D from the bacterial DNA. The new phages each carry gene D from the donor bacterium.

New temperate phage

Recipient cell

(G) At some later time, the defective virus finds a new bacterial host cell.

FIGURE 7.8

Specialized Transduction

During specialized transduction, a virus carries specified genes from a donor bacterium to a recipient bacterium. The process involves lysogeny and results in a recombinant cell.

(H) The DNA (carrying gene D) enters the host cell cytoplasm.

Recominant cell

(I) The viral DNA integrates with the bacterial DNA in the process of lysogeny. But in doing so, through recombination gene D from the previous bacterium also integrates into the recipient cell's chromosome. The recipient cell thus acquires a gene from the donor bacterium and is transduced.

TABLE 7.1

Generalized Transduction Compared to Specialized Transduction

GENERALIZED TRANSDUCTION	SPECIALIZED TRANSDUCTION
1. Viruses (phages) penetrate bacterial cell and enter lytic cycle.	1. Viruses (phages) penetrate bacterial cell and enter lysogenic cycle.
2. Viral DNA begins to replicate immediately in the bacterial cytoplasm.	2. Viral DNA incorporates into the bacterial chromosome as a prophage. Replication begins at a later time.
3. During replication, some bacterial DNA fragments may become mixed with phage DNA.	3. During excision from the bacterial chromosome, viral DNA accidentally carries along some bacterial DNA.
4. A bacterial DNA fragment may be randomly packaged into a new phage.	4. Bacterial genes and phage DNA are packaged into new a phage.
5. A phage having bacterial DNA is defective.	5. A phage having both viral DNA and bacterial DNA is defective.
6. Defective phage enters recipient bacterium.	6. Defective phage enters recipient bacterium.
7. Donor bacterial genes are incorporated into chromosome of recipient bacterium.	7. Donor bacterial genes are incorporated into chromosome of recipient bacterium together with phage DNA.

in the human body, so bacteria need some mechanism to acquire it. Therefore, in the case for *C. diphtheriae*, lysis of human cells is the answer because lysis frees iron from the host cell. So, there is an evolutionary advantage for these bacteria to carry the phages and it is probably safe to say that bacteria and phages have coevolved to the advantage of both.

TABLE 7.2 compares the three forms of genetic recombination in bacteria.

To this point . . .

We have described how the genetic information in a bacterium may be altered by the recombination processes. Transformation involved competent cells obtaining DNA fragments from the local environment and incorporating them into the bacterial chromosome. In the second method, conjugation, we observed how two cells come together and enable DNA to move from the donor cell to the recipient cell. The importance of the F plasmid was highlighted, and the Hfr strain was described. Plasmid transfer is common in conjugation, but chromosomal transfer is a rare event.

In the third type of recombination, transduction, a virus functions as an intermediary between cells. We began by describing the lytic cycle of viral replication and contrasted this with lysogeny. We then explained how in generalized transduction the virus might randomly incorporate fragments of bacterial DNA into a newly forming phage. The defective phage carries those genes to the next cell where they can be incorporated into the bacterial chromosome. This process contrasts with specialized transduction, where a few bacterial genes are excised from the chromosome with the phage DNA and replicated along with the virus.

In the next section, we shall focus on experiments that alter bacterial DNA in the process of genetic engineering. This is where the knowledge from genetic recombination is applied to the insertion of foreign genes into bacteria. We shall see how the process emerged and how the modern applications of genetic engineering yield products to enhance the quality of life.

TABLE 7.2

A Comparison of Transformation, Conjugation, and Transduction

CHARACTERISTIC	TRANSFORMATION	CONJUGATION	TRANSDUCTION
Method of DNA transfer	Soluble DNA moves across wall and membrane of recipient	Through cell-to-cell contact	By an intermediary virus
Amount of DNA transferred	Few genes	Variable	Few genes
Plasmid transferred	Yes	Yes	Not likely
Entire chromosome transferred	No	Sometimes	No
Virus required	No	No	Yes
Live bacteria required	Yes	Yes	Yes
Cell debris required	Yes	No	No
Used to acquire antibiotic resistance	Yes	Yes	Not likely

7.2

Genetic Engineering

Prior to the 1970s, bacteria that had special or unique metabolic properties were detected through mutant analysis or simply isolating bacteria that had certain metabolic talents (MicroFocus 7.5). Experiments in bacterial recombination entered a new dimension in the late 1970s, when it became possible to insert genes into bacterial DNA and thereby establish a cell line that would produce proteins from the inserted genes. The use of bacterial and microbial genetics, including the isolation, manipulation, and control of gene expression had far-reaching ramifications, leading to an entirely new field called **genetic engineering**.

Many of the products derived from genetic engineering have advanced the field of medicine and industrial production. **Biotechnology** is the name given to the commercial and industrial applications derived from genetic engineering.

GENETIC ENGINEERING WAS BORN IN THE 1970S

The science of genetic engineering surfaced in the early 1970s when the techniques became available to manipulate DNA. Among the first scientists to attempt genetic manipulation was Paul Berg of Stanford University. In 1971, Berg and his coworkers opened the DNA molecule from simian virus-40 (SV40) and spliced it into a bacterial chromosome. In doing so, they constructed the first **recombinant DNA molecule** (FIGURE 7.9). The process was extremely tedious though because the cut bacterial and viral DNAs had blunt ends, making sealing of the two DNAs difficult. Berg therefore had to use exhaustive enzyme chemistry to form staggered ends that would combine easily through complementary base pairing. While Berg was performing his experiments, an important development came from Herbert Boyer and his group at the University of California. Boyer isolated a **restriction endonuclease** enzyme that nicks a chromosome and leaves it with mortise-like staggered

Recombinant DNA:
a DNA molecule containing DNA from two different sources.

Restriction endonuclease:
an enzyme that recognizes and cuts specific short stretches of nucleotides.

MicroFocus 7.5

CLOSTRIDIUM ACETOBUTYLICUM AND THE JEWISH STATE

In 1999, scientists completed sequencing the genome of *Clostridium acetobutylicum*, a nonpathogenic bacterium. They hope that since some other species of *Clostridium* are major pathogens (one produces the food toxin that causes botulism, and another is responsible for tetanus) that DNA sequence comparisons will yield insights into what enables some species to become pathogens while others remain harmless. However, the bacterium's ability to convert starch into the organic solvents acetone and butanol is what has a prominent place in history.

In 1900, an outstanding chemist named Chaim Weizmann, a Russian-born Jew, completed his doctorate at the University of Geneva in Switzerland. He also was an active Zionist and advocated the creation of a Jewish homeland in Palestine. In 1904, Weizmann moved to Manchester, England, where he became a research fellow and senior lecturer at Manchester University. During this time, he was elected to the General Zionist Council.

Weizmann began working in the laboratory of Professor William Perkin, where he was trying to find an organism that would make butanol that was needed for rubber manufacture. Weizmann investigated the possibility of using fermentation to produce industrially useful substances. In particular, he noted that the bacterium *C. acetobutylicum*, which he isolated in 1914, converted starch to a mixture of ethanol, acetone, and butanol. Other than for rubber manufacture, the process seemed of no commercial value—until World War I broke out in 1914.

At that time, the favored propellant for rifle bullets and artillery projectiles was a material called cordite. To produce it, a mixture of cellulose nitrate and nitroglycerine was combined into a paste using acetone and petroleum jelly. Before 1914, acetone was obtained through the destructive distillation of wood. However, the supply was inadequate for wartime needs, and by 1915, there was a serious shell shortage, mainly due to the lack of acetone for making cordite.

After his inquiries to serve the government were not returned, a friend of Weizmann's went to Lloyd George, who headed the Ministry of Munitions. Lloyd George was told about Weizmann's work and that he might be able to synthesize acetone in a new way. The conversation resulted in a London meeting between Weizmann, Lloyd George, and Winston Churchill. After explaining the capabilities of *C. acetobutylicum*, Weizmann was given laboratory facilities and a distillery for the full-scale production of acetone from corn. Additional distilleries soon were added in Canada and India. The shell shortage ended.

After the war ended, British Prime Minister Lloyd George wished to honor Weizmann for his contributions to the war effort. Weizmann declined any honors but asked for support of a Jewish homeland in Palestine. Discussions with Foreign Minister Earl Balfour led to the Balfour Declaration of 1917, which committed Britain to help establish the Jewish homeland.

Weizmann went on to make significant contributions to science—he suggested other organisms be examined for their ability to produce industrial products and he laid the foundations for what would become the Weizmann Institute, one of Israel's leading scientific research centers. His political career also moved upward—he became the first President of Israel in 1948.

---

ends. The bits of single-stranded DNA extending out from the chromosome easily attached to a new fragment of DNA in recombinant experiments. Scientists quickly dubbed the single-stranded extensions "sticky ends."

Today, there is a vast array of restriction enzymes, each recognizing a different nucleotide sequence (**TABLE 7.3**). Each enzyme cuts both strands of the DNA because the recognition sequences are a **palindrome**. Each sequence has the same complementary set of nucleotide bases. Thus, restriction enzymes are "molecular scissors" that can be used to open a bacterial chromosome or plasmid at specific locations.

During that same period, Stanley Cohen, also at Stanford University, was accumulating data on the plasmids of *E. coli*. Cohen found that he could isolate plasmids from the bacterium and insert them into fresh bacteria by suspending the organisms in calcium chloride and heating them suddenly to achieve plasmid uptake via the

Palindrome:
a series of letters that reads the same left to right or right to left.

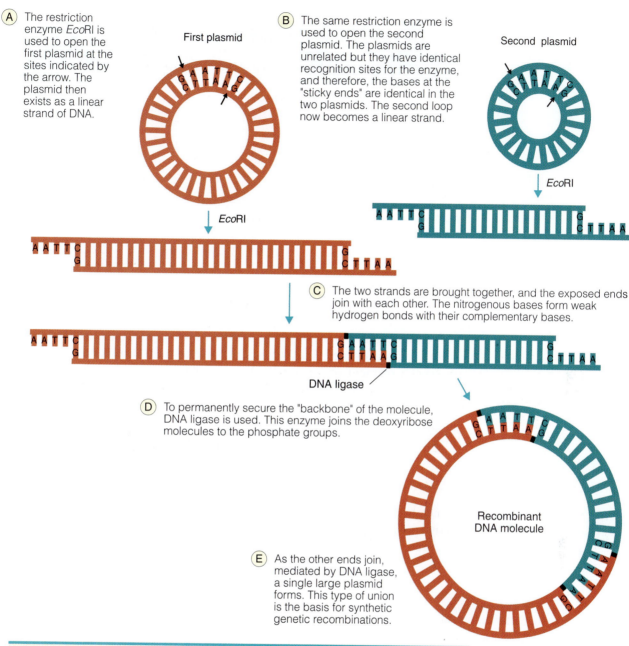

A The restriction enzyme *Eco*RI is used to open the first plasmid at the sites indicated by the arrow. The plasmid then exists as a linear strand of DNA.

First plasmid

B The same restriction enzyme is used to open the second plasmid. The plasmids are unrelated but they have identical recognition sites for the enzyme, and therefore, the bases at the "sticky ends" are identical in the two plasmids. The second loop now becomes a linear strand.

Second plasmid

*Eco*RI

*Eco*RI

C The two strands are brought together, and the exposed ends join with each other. The nitrogenous bases form weak hydrogen bonds with their complementary bases.

DNA ligase

D To permanently secure the "backbone" of the molecule, DNA ligase is used. This enzyme joins the deoxyribose molecules to the phosphate groups.

Recombinant DNA molecule

E As the other ends join, mediated by DNA ligase, a single large plasmid forms. This type of union is the basis for synthetic genetic recombinations.

FIGURE 7.9

Construction of a Recombinant DNA Molecule

In this construction, two unrelated plasmids (loops of DNA) are united to form a single plasmid.

transformation process. Once inside *E. coli* cells, the plasmids multiply independently and produce copies of themselves.

The final link to the process was provided by the **DNA ligases**. These enzymes function during the replication of DNA and the repair of broken DNA molecules. Essentially, they operate in a manner opposite that of endonucleases; they seal together DNA fragments.

TABLE 7.3

Examples of Restriction Endonuclease Recognition Sequences

ORGANISM	RESTRICTION ENZYME	RECOGNITION SEQUENCE*
Escherichia coli	*Eco*RI	G ↓ AATTC CTTAA ↓ G
Streptomyces albus	*Sal*I	G ↓ TCGAC CAGCT ↓ G
Haemophilus influenzae	*Hind*III	A ↓ AGCTT TTCGA ↓ A
Bacillus amyloliquefaciens	*Bam*HI	G ↓ GATC^{m}C CC^{m}TAG ↓ G
Providencia stuartii	*Pst*I	CTGCA ↓ G G ↓ ACGTC

*Arrows indicate where the restriction enzyme cuts the two strands of the recognition sequence; C^m = methylcytosine.

Progress came rapidly. Working together, Boyer and Cohen isolated plasmids from *E. coli* and opened them with restriction enzymes (MicroFocus 7.6). Next, they inserted a segment of foreign DNA and sealed the segment using DNA ligase. Then, they inserted the plasmids into fresh *E. coli* cells. By 1973, they had successfully spliced genes from *S. aureus* into *E. coli*. These genetic engineering experiments intrigued the scientific community because for the first time they could manipulate genes from a wide variety of species and splice them together.

HUMAN GENES CAN BE CLONED IN BACTERIA

The best way to understand how genetic engineering operates is to follow an actual procedure. MicroInquiry 7 describes one method to clone the human gene for the protein insulin in bacteria.

MicroFocus 7.6

OF CORNED BEEF AND PLASMIDS

In 1972, they met at a scientific conference in Hawaii—Stanley Cohen and Herbert Boyer. Cohen was there to lecture about his work with plasmids, the submicroscopic loops of DNA in many bacterial cells. Boyer was an expert on a restriction enzyme that could cut DNA—any DNA—at a precise point. As he sat in the audience and listened to Cohen, Boyer's mind stirred. Could his enzyme cut Cohen's plasmid and allow a foreign piece of DNA to attach?

Scientific conferences are the last place to talk about science, so Boyer invited Cohen to lunch at a local delicatessen in Waikiki. The corned beef was good that day, and the sandwiches hit the spot. The deli mustard was biting hot, and the beer was ice cold. The time was ripe to talk history—and talk history they did. They would collaborate on a set of genetic engineering experiments, the ones that, in retrospect, revolutionized the science of molecular genetics. As the afternoon wore on, the ideas flowed and the friendship took root. Only one thing about that historic lunch has remained a mystery: Who picked up the tab?

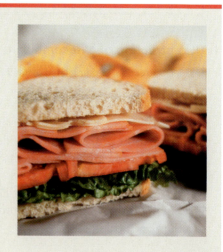

TABLE 7.4

Examples of Therapeutic Products Derived from Genetic Engineering

PRODUCT	TREATMENT/USE
Factors VIII, IX	Replace clotting factors missing in hemophiliacs
Tissue plasminogen activator (TPA)	Prevents blood clotting after heart attacks or strokes
Epidermal growth factor (EGF)	Promotes wound healing
Insulin	Used to stimulate glucose uptake from the blood in diabetics
Somatotropin	A growth hormone used to replace the missing hormone in people of short stature
Colony stimulating factor (CSF)	Used to stimulate white blood cell production in cancer and AIDS patients
α-Interferon	Used with other antiviral agents to fight viral infections
Vaccine proteins	Used to treat and prevent infectious diseases

The first diabetic patient in the world was injected with genetically engineered human insulin in 1980, and in 1986, the Eli Lilly Company began marketing the product as Humulin. Such commercial successes of biotechnology were a sign of things to come.

Besides insulin, a number of other proteins of important pharmaceutical value have been produced by genetically engineered microorganisms. Several noteworthy ones are listed in TABLE 7.4. Many of these proteins, such as the **interferons**, are produced in relatively low amounts in the body, making purification extremely costly. Therefore, the only economical solution to obtain significant amounts of the product is through genetic engineering.

Interferon:
a naturally produced human protein that interferes with viral replication.

GENETIC ENGINEERING HAS MANY COMMERCIAL AND PRACTICAL APPLICATIONS

By the year 2000, thousands of companies worldwide were working on the commercial and practical applications of genetic engineering and DNA technology, spurred in part by new technologies (MicroFocus 7.7). Some are research companies with special units for these studies, while others were established solely to pursue and develop new products by gene-splicing techniques.

■ **PHARMACEUTICAL APPLICATIONS.** The pharmaceutical products of DNA are numerous and diverse (Table 7.4). Many of the genetically engineered products are either the bacteria themselves or proteins expressed by recombinant DNA in bacterial clones (FIGURE 7.10).

■ **AGRICULTURAL APPLICATIONS.** Genetic engineering has extended into many realms of science. In agriculture, for example, genes for herbicide resistance have been transplanted from bacteria into tobacco plants, demonstrating that these **transgenic** plants better tolerate the herbicides used for weed control. For tomato growers, a notable advance was made when researchers at Washington University spliced genes

Transgenic:
referring to an organism that contains a gene from another organism that is incorporated in a stable condition.

MOLECULAR CLONING OF A HUMAN GENE IN BACTERIA

Genetic engineering has been used to produce pharmaceuticals of human benefit. One example concerns the need for insulin injections in people suffering from diabetes (an inability to produce the protein insulin to regulate blood glucose level). Prior to the 1980s, the only source for insulin was through a complicated and expensive extraction procedure from cattle or pig pancreases. But, what if you could isolate the human insulin gene and, through transformation, place it in bacterial cells? These cells would act as factories churning out large amounts of the pure protein that diabetics could inject.

To do molecular cloning of a gene, besides the bacterial cells, we need three ingredients: a cloning vector, the human gene of interest, and restriction enzymes. Plasmids are the **cloning vector**, a genetic element used to introduce the gene of interest into the bacterial cells. Human DNA containing the insulin gene must be obtained. We will not go into detail as to how the insulin gene can be "found" from among 35,000 human genes. Suffice it to say, there are standard procedures to isolate known genes. Restriction enzymes will cut open the plasmid and cut the gene fragment that contains the insulin gene, generating complementary sticky ends.

The following description and **FIGURE A** represent one procedure to genetically engineer the human insulin gene into cells of *Escherichia coli*. (There are other procedures, some more direct. However, this method reinforces many of the concepts we have learned in the last few chapters).

1. Plasmids often carry genes, such as antibiotic resistance. We are going to use the plasmid shown to the right because it contains genes for resistance to tetracycline (tet^R) and ampicillin (amp^R) antibiotics. This will be important for identification of clones that have been transformed. In addition, this plasmid has a single restriction sequence for the restriction enzyme *Sal*I (see Table 7.3). Importantly, notice that this cut site is within the tet^R gene. Also, the plasmid will replicate independently in *E. coli* cells and can be placed in the cells by transformation.

2. The plasmids and human DNA are cut with *Sal*I to produce complementary sticky ends on both the opened plasmid and the insulin gene. Plasmids and the insulin gene then are mixed together in solution with DNA ligase, which will covalently link the sticky ends. Some plasmids will be **recombinant plasmids**; that is, plasmids containing the insulin gene. Other plasmids will close back up without incorporating the gene.

3. The plasmids are placed in *E. coli* cells by transformation. The plasmids replicate independently in the bacterial cells, but as the bacterial cells multiply so do the plasmids. By allowing the plasmids to replicate, we have cloned the plasmids, including any that contain the insulin gene.

4. Since we do not know which plasmids contain the insulin gene, we need to screen the clones to identify the recombinant plasmids. This is why we selected a plasmid with tet^R and amp^R resistance genes.

Since the *Sal*I cut site is within tet^R, any recombinant plasmids will have a defective tetracycline resistance gene and the bacteria carrying those plasmids will not grow on a medium containing tetracycline. Plasmids without the insulin gene have an intact tet^R gene and can grow on the medium. In addition, there will be bacterial cells that did not take up a plasmid and will be sensitive to tetracycline and ampicillin. Thus, using the positive selection technique described in Chapter 6 and a selective medium as described in Chapter 4, we can identify which bacteria contain the recombinant plasmids.

Therefore, we plate all our bacteria onto a master plate—a nutrient medium without tetracycline. All bacterial cells should grow forming colonies. Using the replica plating technique mentioned in Chapter 6, we now replica plate all the colonies onto (a) a growth medium containing tetracycline and (b) onto a growth medium containing ampicillin.

7a. Explain which clones should grow on which plates. The answer can be found in Appendix E.

5. These colonies can now be isolated and grown in larger batches of liquid medium. These batch cultures then are inoculated into large "production vats," called **bioreactors**, in which the cells grow to massive numbers while secreting large quantities of insulin into the liquid.

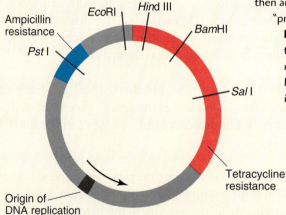

Ampicillin resistance

*Eco*RI

Hind III

*Bam*HI

Pst I

Sal I

Tetracycline resistance

Origin of DNA replication

FIGURE A

The Sequence of Steps to Engineer the Insulin Gene into *E. Coli* Cells

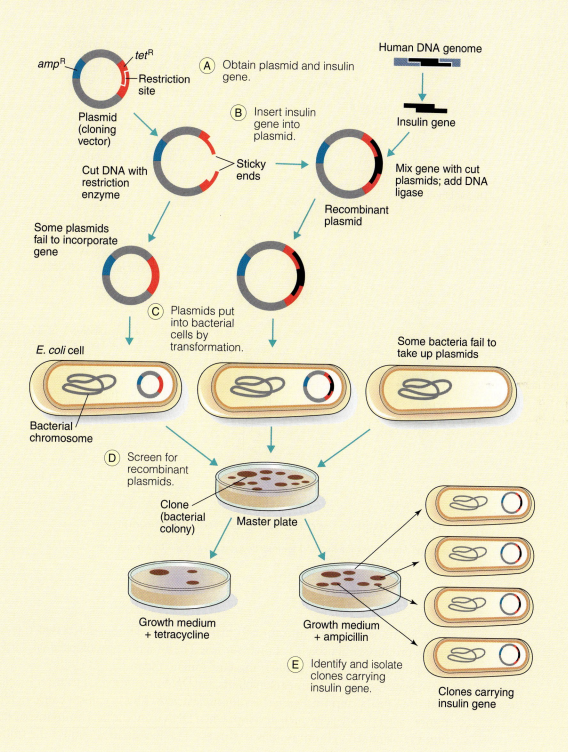

amp^R
tet^R
Restriction site
Plasmid (cloning vector)

Human DNA genome

Insulin gene

(A) Obtain plasmid and insulin gene.

(B) Insert insulin gene into plasmid.

Cut DNA with restriction enzyme

Sticky ends

Mix gene with cut plasmids; add DNA ligase

Recombinant plasmid

Some plasmids fail to incorporate gene

(C) Plasmids put into bacterial cells by transformation.

E. coli cell

Some bacteria fail to take up plasmids

Bacterial chromosome

(D) Screen for recombinant plasmids.

Clone (bacterial colony)

Master plate

Growth medium + tetracycline

Growth medium + ampicillin

(E) Identify and isolate clones carrying insulin gene.

Clones carrying insulin gene

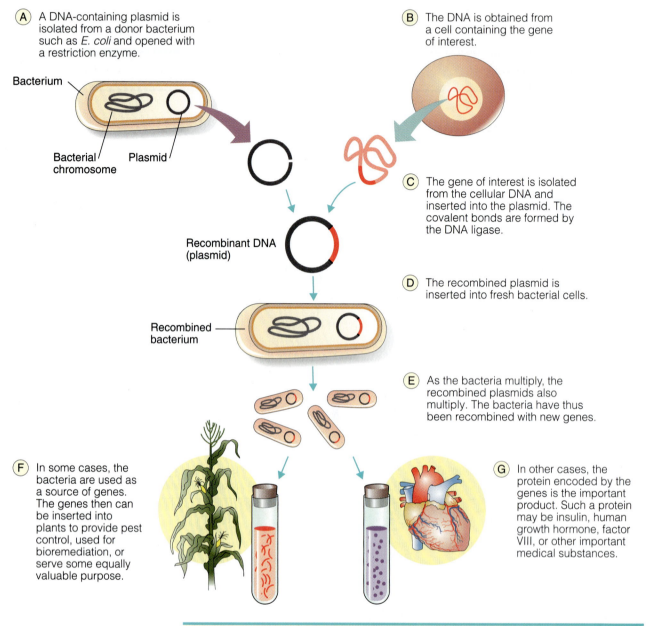

A A DNA-containing plasmid is isolated from a donor bacterium such as *E. coli* and opened with a restriction enzyme.

Bacterium

Bacterial chromosome   Plasmid

Recombinant DNA (plasmid)

Recombined bacterium

F In some cases, the bacteria are used as a source of genes. The genes then can be inserted into plants to provide pest control, used for bioremediation, or serve some equally valuable purpose.

B The DNA is obtained from a cell containing the gene of interest.

C The gene of interest is isolated from the cellular DNA and inserted into the plasmid. The covalent bonds are formed by the DNA ligase.

D The recombined plasmid is inserted into fresh bacterial cells.

E As the bacteria multiply, the recombined plasmids also multiply. The bacteria have thus been recombined with new genes.

G In other cases, the protein encoded by the genes is the important product. Such a protein may be insulin, human growth hormone, factor VIII, or other important medical substances.

FIGURE 7.10

Developing New Products Using Genetic Engineering

Genetic engineering is a method for inserting foreign genes into a bacterium and obtaining chemically useful products.

from a pathogenic virus into tomato plant cells and demonstrated that the cells would produce viral proteins at their surface. The viral proteins blocked viral infection, providing resistance to the transgenic tomato plants.

For gene transfer experiments in plants, the DNA often used for transfer is a plasmid from the bacterium *Agrobacterium tumefaciens*. This organism causes a plant tumor called crown gall, which develops when DNA from the bacterium inserts itself into the plant cell's chromosomes (**FIGURE 7.11**). Researchers remove the tumor-

MicroFocus 7·7

CUSTOMIZED PROTEINS

Before chemist Michael Smith began tinkering with genes, mutation studies were hit-and-miss. Scientists would expose cells to a mutagen such as ultraviolet light, and then forage among a crowd of mutated proteins, hoping to find a clue to where the mutation occurred. However, Smith had a better idea: He would control the mutation at a particular site and see what the organism produced.

Smith began by splicing the single-stranded DNA of a gene into the single-stranded DNA of a virus. (The viral DNA would later act as a carrier for the gene DNA and transport it into a cell.) Then, he synthesized a short strand of DNA complementary to the gene DNA, except at just one amino acid coding site (i.e., the mutation site). Next, he combined this mutated strand with the normal gene. The new, short strand bound tightly to the gene, except at the one site where the mutation existed. (The effect is somewhat like having a zipper with a small opening in the middle.) Smith then used an enzyme to complete the second strand, that is, the DNA complementary to the viral DNA. This formed a complete double-stranded DNA molecule.

Now, the DNA molecule was inserted into fresh bacteria to see what would happen. Smith was pleased to note that the normal version of the gene was encoding a normal protein, while the mutated version of the gene encoded a mutated protein. A customized gene was encoding a customized protein. He then set to work comparing the normal and new proteins.

Smith's work opened many new doors to microbial geneticists. For example, researchers now could determine whether a single amino acid mutation could induce a change in the function of the protein. In effect, does one amino acid influence a protein's activity? And if so, how? Moreover, they now could tailor enzyme proteins simply by adjusting the genetic code. Scientists were excited. They called the process site-directed mutagenesis. The Nobel Prize Committee had a better expression: Class A science. They awarded Smith a share of the 1993 Nobel Prize in Chemistry.

---

inducing gene from the plasmid and then splice the desired gene into the plasmid and allow the bacterium to infect the plant.

The dairy industry was the first to feel the dramatic effect of the new DNA technology. In the 1980s, researchers at Cornell University injected dairy cows with bovine growth hormone (BGH) produced by bacteria engineered with the BGH gene. They reported a 41 percent increase in milk from the experimental cows. Also being researched is a pig with more meat and less fat, a product of genetically engineered porcine growth hormone. Scientists at Auburn University have endowed young carp with extra copies of activated growth hormone genes, hoping to enable the fish to grow more efficiently in aquacultural surroundings.

■ **ANTIBIOTIC PRODUCTION.** The presence or threat of infectious disease represents a high demand for antibiotics. Therefore, many antibiotics today are produced commercially using microorganisms. Although antibiotics are produced in nature, the bacterium or fungus often does not produce these compounds in high yield. Furthermore, as no new class of antibiotics has been identified since the early 1990s, existing compounds need to be redesigned to reach their targets more efficiently. This means the microbes must be genetically engineered to produce larger quantities of antibiotics and/or to produce modified antibiotics to which infectious microbes have yet to show resistance.

■ **DETECTION AND DIAGNOSIS.** The genes of an organism contain the essential information that contributes to its behaviors and characteristics. Bacterial pathogens, for example, contain specific sequences of nucleotides that can confer the ability of the pathogen to infect and cause disease in the host. Because these nucleotide sequences

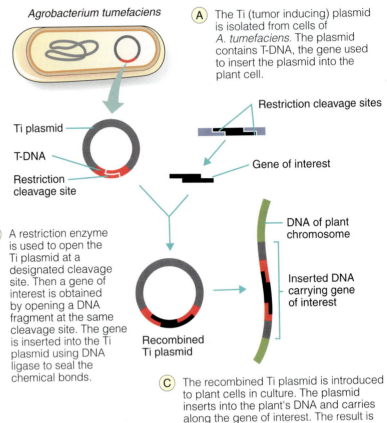

Agrobacterium tumefaciens

(A) The Ti (tumor inducing) plasmid is isolated from cells of *A. tumefaciens*. The plasmid contains T-DNA, the gene used to insert the plasmid into the plant cell.

Restriction cleavage sites

Ti plasmid

T-DNA

Gene of interest

Restriction cleavage site

(B) A restriction enzyme is used to open the Ti plasmid at a designated cleavage site. Then a gene of interest is obtained by opening a DNA fragment at the same cleavage site. The gene is inserted into the Ti plasmid using DNA ligase to seal the chemical bonds.

DNA of plant chromosome

Inserted DNA carrying gene of interest

Recombined Ti plasmid

(C) The recombined Ti plasmid is introduced to plant cells in culture. The plasmid inserts into the plant's DNA and carries along the gene of interest. The result is a transgenic plant.

FIGURE 7.11

The Ti Plasmid as a Vector in Plant Genetic Engineering

A. tumefaciens induces tumors in plants and causes a disease called crown gall. A lump of tumor tissue forms at the infection site, as the photograph shows. The catalyst for infection is a tumor-causing gene in the Ti plasmid. This plasmid, without the tumor-causing gene, is used to carry a gene of interest into plant cells.

are distinctive and often unique, if detectable, they can be used as a definitive diagnostic determinant.

In the medical laboratory, diagnosticians are optimistic about the use of **DNA probes**, single-stranded DNA molecules that recognize and bind to a distinctive and unique nucleotide sequence of a pathogen. The DNA probe binds (hybridizes) to its complementary nucleotide sequence from the pathogen, much like strips of Velcro stick together. To make a probe, scientists first identify the DNA segment (or gene) in the pathogen that will be the target of a probe. Using this segment, they construct the single-stranded DNA probe (**FIGURE 7.12**).

More than 100 DNA probes have been developed for the detection of pathogenic viruses, bacteria, fungi, and protozoa. As one example, a DNA probe exists for the early detection of malarial infections caused by the protozoan *Plasmodium falciparum*. DNA is isolated from the tissues of the suspected patient and fragmented into single-stranded DNA segments (Figure 7.12). These segments are attached to a solid support. The labeled DNA probe is added. If the DNA sample from the patient contains the target *P. falciparum* DNA segment, the probe binds to the complemen-

FIGURE 7.12

DNA Probes

Construction of a DNA probe and its use in disease detection and diagnosis.

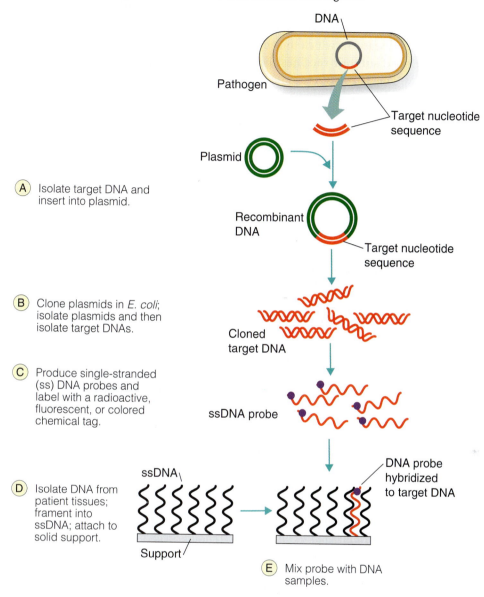

(A) Isolate target DNA and insert into plasmid.

(B) Clone plasmids in *E. coli*; isolate plasmids and then isolate target DNAs.

(C) Produce single-stranded (ss) DNA probes and label with a radioactive, fluorescent, or colored chemical tag.

(D) Isolate DNA from patient tissues; frament into ssDNA; attach to solid support.

(E) Mix probe with DNA samples.

tary nucleotide sequence. In doing so, it concentrates the label (radioactivity, fluorescence, or a colored chemical) at that site and indicates that a match has been made. Clearly, the use of DNA probes represent a reliable and rapid method for detecting and diagnosing many human infectious diseases.

■ ENVIRONMENTAL BIOLOGY. We already have mentioned in this chapter and previous ones the usefulness of bacteria as a source of genes. Bacteria represent a huge, mostly untapped gene pool representing metabolically diverse processes. Examples such as bioremediation have been discussed where genetically engineered or genetically recombined bacteria are provided with specific genes whose products will break down toxic pollutants, clean up waste materials, or degrade oil spills. We have barely scratched the surface to take advantage of the metabolic diversity that bacteria offer.

■ **VACCINE PRODUCTION.** Microbial vaccines traditionally have been either killed or attenuated (inactivated) preparations. Sometimes such microbial preparations are not completely killed or inactivated and a slight chance exists that the patient could develop the disease for which they have been vaccinated. Today, genetic engineering has developed **subunit vaccines** that are risk free of developing the disease when vaccinated. The hepatitis B vaccine, for example, is perfectly safe because the vaccine is made from only the protein coat subunits of the virus. The viral genes for the coat are cloned in yeast, and the vial protein harvested as the vaccine. Active hepatitis B viruses cannot arise from such vaccine material.

Sucg recombinant vaccines also are attractive because the precise genetic makeup of the vaccine is known, often making the administration of higher doses possible without the development of serious side effects. Such vaccines can be produced faster than traditional vaccines, although they are not always cheaper.

Also being developed are **DNA vaccines** where the genetic material (a gene) is the vaccine. Such gene vaccines often produce immunity in the individual in which the gene has been taken up. Again, they should be safe and cheaper to produce. Chapter 20 will describe more fully the nature of vaccines.

Lastly, it should be mentioned that genetically engineered products are not always a "no brainer" in terms of their development. This is no clearer than in the attempts to develop a vaccine for AIDS. Since 1987, scientists and genetic engineers have tried to identify viral subunits that can be used to develop an AIDS vaccine. However, it is not so much that genetic engineering can't be done as it is that the virus just seems to find ways to circumvent a vaccine and the immunity that develops in the patient. Still, hopes are high that eventually a genetically engineered vaccine will be developed that is safe and effective. We will have much more to say about AIDS and vaccines in Chapter 14.

Subunit vaccine: a vaccine that contains only a specific protein fragment from the pathogen.

To this point . . .

In this section of the chapter, we learned how genetic engineering got its start in the early 1970s. The discovery and use of restriction endonucleases were essential for the formation of recombinant DNA molecules. Today, genetic engineers have identified and isolated a large number of such enzymes that cut DNA molecules, most by forming sticky ends.

To appreciate the manipulations needed to insert genes into foreign DNA, MicroInquiry 7 illustrated the process of engineering a gene into a bacterial cell and a procedure for identifying recombinant DNA. Similar procedures have been designed to produce many human proteins for therapeutic and medical use. We finished this section by briefly outlining the major types of commercial products and practical applications that have been the result of genetic engineering.

In the last section, we will examine the new science of bacterial genomics and highlight some of the potential benefits and consequences coming from the field.

Microbial Genomics

In April 2003, exactly 50 years to the month after Watson and Crick announced the structure of DNA, a publicly financed, $3 billion international consortium of biologists, ethicists, industrial scientists, computer experts, and engineers completed perhaps the most ambitious project in the history of molecular genetics—indeed, in the history of biology. The Human Genome Project, as it was called, had succeeded in mapping the **human genome**—that is, the 3 billion nitrogenous bases in a human cell were identified and strung together in the correct order (sequenced). The completion of the project represents a scientific milestone with unimaginable health benefits.

MANY MICROBIAL GENOMES HAVE BEEN SEQUENCED

If the human genome were represented by a rope two inches in diameter, it would be 32,000 miles long. The genome of a bacterium like *E. coli* at this scale would be only 1,600 miles long or about 1/20 that of human DNA. So, being substantially smaller, microbial genomes are easier and much faster to sequence. By July 2003, over one hundred prokaryotic and eukaryotic genomes had been sequenced and published in the National Center for Biotechnology Information (NCBI) database (TABLE 7.5). Many more viral sequences have been deciphered.

In May 1995, the first complete genome of a free-living organism was sequenced: the 1.8 million base pairs (1.8 Mb) in the genome of the bacterium *Haemophilus influenzae* (MicroFocus 7.8). In a few short months, the genome for a second organism, *Mycoplasma genitalium* was reported. This reproductive tract pathogen has the smallest eubacterial genome, consisting of only 580,000 base pairs and 250 genes. In 1996, the genome for the yeast *Saccharomyces cerevisiae* was sequenced. In a field already littered with milestones, the sequencing of *S. cerevisiae* marked the first insight into the eukaryotic genome. Sixteen chromosomes were analyzed, 12 million bases were sequenced, and 6,000 genes were identified. The sequencing revealed many genes wholly new to biology.

Since then, many more microbial genomes have been sequenced, including a variety of pathogens. So, what do these sequences tell us and what use do they have?

MICROBIAL GENOMICS HAS PRODUCED SOME INTERESTING FINDINGS

Many microbial genomes had been sequenced before the human genome was deciphered. One of the interesting developments was the comparison of the human and bacterial DNA sequences. This comparison shows that about 200 of our 35,000 genes are essentially identical to those found in bacteria. However, these human genes did not come directly from bacteria, but rather were genes picked up by organisms that were early ancestors of humans. So important are these genes that they have been passed along from organism to organism throughout evolution.

Another discovery is that only about 5 percent of our DNA appears to code for proteins and regulatory RNAs. Scientists suggest that some of the non-gene DNA

TABLE 7.5

Microbial Genomes Sequenced and Representative Examples*

	GENOMES SEQUENCED	GENOME SIZE (Mb†)	COMMENTS
Viruses	>1,100		
HIV		0.009	Causative agent of AIDS
Poliovirus		0.007	Causative agent of polio
Rabies virus		0.012	Causative agent of rabies
SARS coronavirus		0.029	Causes an atypical pneumonia (SARS)
Variola virus		0.185	Causative agent of smallpox
West Nile virus		0.011	Causes a fatal encephalitis (inflammation of the brain)
Domain Archaea	17		
Methanococcus jannaschii		1.66	First archaean sequenced
Nanoarchaeum equitans		0.50	Smallest known cellular genome
Domain Eubacteria	128		
Bacillus anthracis		5.23	Causative agent of anthrax
Clostridium tetani		2.80	Causative agent of tetanus
E. coli O157:H7		5.59	Causes bloody diarrhea; hemolytic uremic syndrome
Escherichia coli		4.64	Common intestinal bacterium in humans
Mycobacterium tuberculosis		4.42	Causative agent of tuberculosis
Mycoplasma genitalium		0.58	Has the smallest known eubacterial genome
Staphylococcus aureus		2.82	Causes skin lesions, food poisoning, toxic shock syndrome
Treponema pallidum		1.14	Causative agent of syphilis
Vibrio cholerae		4.00	Causative agent of cholera
Yersinia pestis		4.65	Causative agent of plague
Domain Eukarya	21		
Kingdom Fungi	3		
Saccharomyces cerevisiae		12.07	Baker's yeast; used in fermentation
Kingdom Protista	7		
Plasmodium falciparum		22.90	Causative agent of malaria

* Data published in the Genomes OnLine Database. Available at: http://wit.integratedgenomics.com/GOLD/. Accessed November 17, 2003.

† Mb = 1,000,000 bases

may be "genetic debris" from viruses and bacteria that infected cells far back during cell evolution hundreds of millions of years ago. So, microbial genomes will have much to tell us about our past as comparisons continue.

DNA sequencing has produced some additional interesting results. In most of the microbial genomes that have been sequenced, nearly 50 percent of the identified genes that encode proteins have not yet been connected with a cellular function. Almost 30 percent of these proteins are unique to each species. What these proteins do remains a challenge to molecular biologists.

MicroFocus 7.8

PUTTING HUMPTY BACK TOGETHER AGAIN

Humpty Dumpty sat on a wall.
Humpty Dumpty had a
great fall.
All the king's horses,
And all the king's men,
Couldn't put Humpty together again.

We all remember this nursery rhyme, but it can be an analogy for the efforts required in sequencing the genome of an organism. To sequence a whole genome, you have to take thousands of small DNA fragments and, after sequencing them, try to "put the fragments together again." The good news in genomics is that you can "put Humpty together again."

Small fragments must be used when sequencing a whole genome because current methods will not work with the tremendously long stretches of DNA, even those shorter ones found in bacterial genomes. Therefore, one strategy is to break the genome into small fragments. These fragments then are sequenced using sequencing machines (see photo) and the fragments reassembled into the full genome. This technique is called the "whole-genome shotgun method." It can be extremely fast, but there are so many little pieces that it can be very difficult to put the whole genome together again.

The "shotgun" strategy first was used in 1995 by J. Craig Venter, Hamilton

Smith, and their colleagues to sequence the genome of a self-replicating, free-living organism—the bacterium *Haemophilus influenzae*. This gram-negative rod causes ear and respiratory infections and bacterial meningitis in children. With 1.8 million base pairs, the size of its genome is average for bacteria, but is about ten times larger than any virus that had been sequenced.

To sequence *H. influenzae*, Venter and Smith cut copies of the DNA into pieces of 1,600 to 2,000 base pairs. The segments then were partly sequenced at both ends, using automated sequencing machines. These base-pair sequences—with their many overlaps—became the sequence information that was entered into the computer. Using innovative computer software, the 24,304 DNA fragments generated from the *H. influenzae* genome were compared, clustered, and matched for assembling the genome.

Assembling the *H. influenzae* genome was a considerable achievement—and to some observers a surprise. The genome contains 1,830,137 base pairs, in which 1,749 genes are embedded. Once assembled, the genes could be located, compared to known genes, and a detailed map developed. Sequencing *H. influenzae* took about a year, but demonstrated that using "the king's horses" (supercomputers and shotgun sequencing) and "the king's men" (the large group of collaborators) could "put Humpty together again"—and with speed and accuracy. Note: Since 1995, great strides have been made in sequencing technology. If *H. influenzae* were to be sequenced today, it would take about five days, rather than an entire year.

Another interesting finding coming from DNA sequencing is that some strains of a bacterium may contain "genetic islands," where there can be up to 25 genes in a row that are absent from other strains of the same species. Many of these islands can be identified as having come from an altogether different species of bacterium. In fact, genomic analysis has discovered that some pathogens contain so-called "pathogenicity islands," representing a string of genes responsible for the infectious and disease-causing nature of the pathogen.

All this suggests that some form of lateral transfer, such as conjugation or transduction described earlier in this chapter, has occurred to account for the presence of these strings of genes. It is believed that the nonpathogenic bacterium *Thermotoga maritima* has acquired about 25 percent of its genome from lateral gene transfer (**FIGURE 7.13**). In addition, sequence analysis of its genes indicate they are a

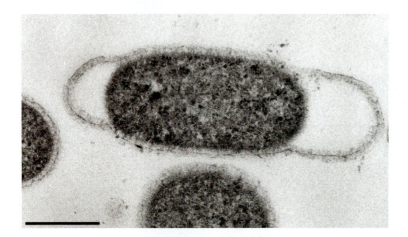

FIGURE 7.13

Thermotoga maritima

A transmission electron micrograph of *Thermotoga maritima*. This bacterium has genetic characteristics common to both the Eubacteria and Archaea domains. Note the balloon-like sheath that extends beyond the ends of the cell. (Bar = 0.5 μm.)

mixture of bacterial and archaeal genes. This suggests *T. maritima* evolved before the split of Eubacteria and Archaea domains.

Certainly, microbial DNA sequences have provided a treasure trove of information to better understand the metabolism and evolution of microorganisms. However, the information also has important potential for our future as resources for genetic engineering, energy production, environmental cleanup, and human health.

MICROBIAL GENOMICS WILL ADVANCE OUR UNDERSTANDING OF THE MICROBIAL WORLD

Microorganisms have existed on Earth for more than 3.8 billion years, although we have known about them for little more than 300 years. We know that over this long period of evolution, they have become established in almost every environment on Earth and, although they are the smallest organisms on the planet, they influence—if not control—some of the largest events. Yet, with few exceptions, we do not know a great deal about any of these microbes, and we have been able to culture and study in the laboratory less than 1 percent of all microorganism species.

However, our limited knowledge is changing. With the advent of **microbial genomics**, we have begun the third Golden Age of microbiology, a time when remarkable scientific discoveries will be made toward understanding the workings and interactions of the microbial world. Some potential consequences from the understanding of microbial genomes are outlined below.

Microbial genomics: the analysis of a complete set of genes (the genome) in a microorganism.

■ **BETTER FOOD PRODUCTION.** Many microorganisms are used in food production (Chapters 25 and 27). However, by understanding a bacterium's genome, better commercial applications can be developed. One example are those bacteria that carry out lactic acid fermentation. These lactic acid bacteria account for a $30 billion a year industry and are involved in everything from making cheddar cheese to wine. By understanding their genomes, food technologists can develop better-tasting and safer foods.

■ **A CLEANER ENVIRONMENT.** The applications of genetic engineering to environmental cleanup were described earlier. However, microbial genomics also has a strong contribution to make as the following example describes. One environmental concern is radioactive waste sites that remain from atomic weapons programs and the civilian nuclear power industry. These sites contain not only dangerous radioactive isotopes, but also high concentrations of mercury, toluene, and other toxic compounds. Although there are microbes, such as *E. coli* and *Pseudomonas* that can degrade toxic materials like mercury and toluene into less harmful products, the radioactivity issue has made it almost impossible to use bioremediation because microbes also are sensitive to the high levels of radiation and will develop lethal mutations.

In 1956, a bacterium was isolated from a can of ground beef that was thought to have been sterilized by radiation. The bacterium was called *Deinococcus radiodurans* (**FIGURE 7.14**). Scientists discovered that this bacterium could withstand 1,500 times the radiation exposure that would kill other bacteria or humans! It ends up that the organism contains a highly efficient DNA repair process that very quickly corrects the damage caused by radiation exposure. Its genome was sequenced in 1999.

The proposal is to engineer the genes for mercury and toluene degradation into the DNA of *D. radiodurans*. The organism then could be sprayed over a toxic site and the bacteria could clean up the toxic chemicals. Certainly, extensive laboratory experiments and other experimental work needs to be done before this idea actually is tried in the environment. The important point though is that it is possible using today's genomics information and genetic engineering capabilities. Since many toxic waste sites contain unique contaminants, knowing a bacterium's genome, it should be possible to develop "designer bacteria" with the appropriate ensemble of engineered genes capable of degrading whatever compounds are present.

■ **IMPROVED BIOSENSING.** The advent of the anthrax bioterrorism events of 2001 and the continued threat of bioterrorism has led many researchers to look for ways to more efficiently and more rapidly detect the presence of such bioweapons. Many of the diseases caused by these potential biological agents cause no symptoms for at least several days, and when symptoms appear, initially they are flu-like. The genomes of several microbes that represent potential bioweapons have been sequenced and that knowledge gained will be useful in designing rapid detection systems.

Beyond bioterrorism, biosensors could be developed to detect warfare agents. For example, bacteria could be used to detect the location of land mines. Genetically modified bacteria, engineered with the ability to detect trinitrotoluene (TNT), the explosive that leaks from land mines, could be sprayed over a suspect area. When monitored from the air, TNT leaks (land mines) could be detected by a fluorescent glow engineered into the bacteria and genetically programmed to be produced

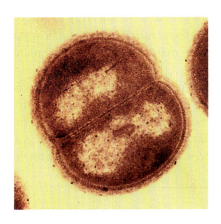

FIGURE 7.14

Deinococcus radiodurans

This false color transmission electron micrograph of *Deinococcus radiodurans* shows a cell undergoing binary fission. The bacterium is very resisitant to genetic damage caused by radiation.

when in contact with TNT. Similar scenarios could be devised for other pollutants and pathogens typically found in soil, food, or water.

■ **BETTER HUMAN HEALTH**. Microbial genomics and genetic engineering are developing many new products that are enhancing the quality of human life. Viruses are being used as gene delivery systems in attempts to cure genetic diseases (Chapter 21) and modified bacterial toxins, such as the botulinum toxin (Botox), are useful in muscle control. Heart attacks and strokes kill about 12 million people a year worldwide, about 1 million occurring in the United States. Genetic engineers have identified and developed "clot-busters," bacterial enzymes and human proteins that prevent blood clots and, thus, substantially decrease death and disability in patients suffering heart attacks. Streptokinase (Streptase) is one clot-buster produced by *Streptococcus pyogenes*. Another clot-buster is tissue plasminogen activator (t-PA; Activase), which is a human protein that participates in breaking down old blood clots. The human t-PA gene has been isolated and genetically engineered into bacteria, which become factories for producing the human enzyme. Both products effectively dissolve blood clots in patients who have suffered a heart attack, and prevent the formation of blood clots following open-heart surgery.

Sequencing the genomes of known pathogens can help in the fight to produce better vaccines and/or antibiotics against these agents. For example, the encapsulated bacterium *Streptococcus agalactiae* causes life-threatening infections. To understand why, researchers have sequenced the genome of *S. agalactiae* in the hope that by comparing its DNA sequence with that of other avirulent streptococci, they can determine why *S. agalactiae* is so virulent. Genes present in *S. agalactiae* but absent in other strains might represent virulence factors against which a vaccine could be developed.

MICROBIAL GENOMICS IS ONLY PART OF A GENETIC PROCESS

Sequencing the DNA bases of a microorganism (or any other organism) has and still does provide important information concerning the number of bases and genes comprising the organism. However, such sequences provide little understanding of how bacterial genes work together to run the metabolism of an organism. One needs to understand how the bacterium uses its genome to form the functioning organism. Sequencing is only the first part of that understanding.

Having sequenced a microbial genome, the next step is to discover the functions for the genes. Sequences need to be analyzed (annotated) to identify the location of the genes and the function of their RNA or protein products. As mentioned earlier, up to 50 percent of the genes discovered when sequencing an organism are not associated with a protein of known function. The field of **functional genomics** requires the expertise of computer scientists and other biological investigators to analyze a microbial sequence and to try to determine what those functions are and how those genes interact with others and the environment to maintain and allow the microbe to grow and reproduce.

One of the most important areas beyond DNA sequencing is the field of **comparative genomics** that compares the sequence from one microbe with that from another similar or dissimilar microorganism. Such comparisons can identify similar genes in different organisms, such as the 200 genes common to the human and microbial genomes. Comparing the sequences of similar genes indicates how genomes have evolved over time and provides clues to the relationships between microbes on the evolutionary tree of life (see MicroFocus 6.4). The discovery of novel genes opens the door to possible new metabolic pathways, which might have important industrial or applied uses. As described above, comparative genomics has great potential in identifying disease-causing genes and genes used to protect one microbe from another. Also, comparisons provide information on similar methods pathogens may use to enter cells and cause disease, which will be important information for the development of new vaccines and antibiotics. Elucidating the gene expression systems in microbial communities in nature also will provide a better understanding of how microbes survive and how they may be controlled or killed. Microbial genomics should prove exciting and challenging in the years ahead.

Note to the Student

Louis Pasteur once wrote, "There are science and the applications of science, separate yet bound together as the fruit to the tree which bears it." The truth of this statement is particularly apparent in genetic engineering. Genetic engineering and microbial genomics represent an apex in microbiology research. It is the equivalent of the engineer's dream of landing on the moon and the physicist's vision of peaceful uses of nuclear power.

The capabilities of genetic engineering and the potential of microbial genomics have turned microbiology from an analytical science into a synthetic science. By extracting genes from one species and inserting them into another, microbiologists have acquired the ability to make large quantities of proteins previously available only in minute amounts. Triumph has followed triumph. Rarely in scientific history have the discoveries of pure research had such immediate applications and implications upon the society of their time.

Before we get too big a head, it is worth remembering an important attribute of the microbial world, which is that microorganisms are found in most every environment on Earth. Therefore, environments exposed to high levels of radiation or containing antibiotics or other toxic substances have forced the resident microorganisms to undergo mutation and/or acquire new genes to survive in these stressful locations. They are the survivors of evolution's natural selection. Therefore, when we describe genetic engineering methods to transfer a gene into another organism or use comparative genomics to identify a microbe with special metabolic talents, we are simply exploiting or manipulating microbial processes that have been going on for millions of years. Still, it is a historic time for microbiology—a third Golden Age; and we should all share in the excitement. Louis Pasteur would have been proud to see this day.

Summary of Key Concepts

7.1 BACTERIAL RECOMBINATION

■ **Genetic Information in Bacteria Can Be Transferred Vertically and Laterally.** Recombination implies a lateral transfer of DNA fragments between bacteria and, thus, an acquisition of genes by the recipient cell. All three forms of recombination are characterized by the introduction of new genes to a recipient bacterium by lateral transfer.

■ **Transformation Is the Uptake and Expression of DNA in a Recipient Cell.** In transformation, a "competent" bacterium takes up DNA fragments from the local environment; this DNA has been left behind by a disrupted bacterium. The new DNA fragment displaces a segment of equivalent DNA in the recipient cell, and new genetic characteristics are assumed.

■ **Conjugation Involves Cell-to-Cell Contact for DNA Transfer.** In conjugation, a live donor cell contributes a portion of its DNA to a recipient cell. Plasmids commonly are transferred.

■ **Conjugation Also Can Transfer Chromosomal DNA.** In another form of conjugation, Hfr strains contribute a portion of the donor's chromosomal genes to the recipient cell.

■ **Transduction Involves Viruses in the Lateral Transfer of DNA.** In the third form of recombination, transduction, a virus enters a bacterium and later replicates within it. In generalized transduction a bacterial DNA fragment is mistakenly incorporated into an assembling phage, which transports that DNA to a new bacterium. In specialized transduction, the virus first incorporates itself into, then detaches from, the bacterial chromosome, taking a segment of bacterial DNA with it.

7.2 GENETIC ENGINEERING

■ **Genetic Engineering Was Born in the 1970s.** Genetic engineering is an outgrowth of studies in bacterial genetics. It was at this time that restriction endonucleases were discovered and isolated.

■ **Human Genes Can Be Cloned in Bacteria.** Plasmids are isolated from a bacterium, spliced with foreign genes, then inserted into fresh bacteria where the foreign genes are expressed as protein. Bacteria are used as the biochemical factories for the synthesis of such proteins as insulin, interferon, and human growth hormone.

■ **Genetic Engineering Has Many Commercial and Practical Applications.** Genetic engineering is only one branch of modern biotechnology. This technology uses DNA-based techniques to perform diagnoses, detect and treat genetic diseases, and spark innovative approaches to agriculture. The practical benefits of research in bacterial genetics have helped revolutionize myriad fields of human endeavor.

7.3 MICROBIAL GENOMICS

■ **Many Microbial Genomes Have Been Sequenced.** Since 1995, increasingly more microbial genomes have been sequenced; that is, the linear sequence of bases has been elucidated.

■ **Microbial Genomics Has Produced Some Interesting Findings.** A comparison of bacterial genomes with the human genome has shown that there are some 200 genes in common between these organisms. Comparisons between microbial genomes indicate that almost 50 percent of the identified genes have yet to be associated with a protein or function in the cell. Another finding is the presence of "genetic islands," which represent a contiguous series of genes that have been acquired through some form of genetic recombination.

■ **Microbial Genomics Will Advance Our Understanding of the Microbial World.** With the understanding of the relationships between sequenced microbial DNAs, comes the potential for better food production, a cleaner environment, improved monitoring of the environment for infectious diseases, and better human health.

■ **Microbial Genomics Is Only Part of a Genetic Process.** Sequencing is only the first step in understanding the behaviors and capabilities of microorganisms. Functional genomics attempts to determine the functions of the sequenced genes and how those genes interact with one another and with the environment. Comparative genomics compares the similarities and differences between microbial genome sequences. Such information provides a window on the evolutionary past and how microbes are related to one another. The discovery of novel genes provides hope that such genes can be applied to important industrial processes as well as vaccine and drug development.

Questions for Thought and Discussion

Answers to selected questions can be found in Appendix C.

1. Try to put yourself in Griffith's position in 1928. Genetics is poorly understood, DNA is virtually unknown, and bacterial biochemistry has not been clearly defined. How would you explain transformation?

2. Some geneticists maintain that the movement of transposons in a bacterial cell is a form of recombination—specifically, "illegitimate recombination." What arguments can be made for and against calling the movement a recombination, and why do you suppose it is labeled "illegitimate"?

3. Which of the recombination processes (transformation, conjugation, or transduction) would be most likely to occur in the natural environment? What factors would encourage or discourage your choice from taking place?

4. Although William Hayes and his colleagues established and outlined the method of bacterial conjugation in which DNA passes from a donor cell to a recipient cell, what other methods can you propose to explain the results of the bacterial recombination we now call conjugation?

5. In 1976, an outbreak of pulmonary infections among participants at an American Legion convention in Philadelphia led to the identification of a new disease, Legionnaires' disease. The bacterium responsible for the disease had never before been known to be pathogenic. From your knowledge of bacterial genetics, can you postulate how it might have acquired the ability to cause disease?

6. The development of genetic engineering has been hailed as the beginning of a second Industrial Revolution. Do you believe this label is justified? How many products of genetic engineering or applications of the process can you think of?

7. Since the 1950s, the world has been plagued by a broad series of influenza viruses that differ genetically from one another. For example, we have heard of swine flu, Hong Kong flu, Bangkok flu, and Victoria flu. How might the process of transduction help explain this variability?

8. It is not uncommon for students of microbiology to confuse the terms *reproduction* and *recombination.* How do the terms differ?

9. Using a diagram, illustrate how the gene for capsule formation was passed from donor to recipient cells in *Streptococcus pneumoniae.*

10. You are going to do a genetic engineering experiment, but the labels have fallen off the bottles containing the restriction endonucleases. One loose label says *Eco*RI and the other says *Pvu*I. How could you use the plasmid shown in MicroInquiry 7 to determine which bottle contains the *Pvu*I restriction enzyme?

11. Describe how comparative genomics can help in understanding the relationships between different groups of bacteria.

12. Discuss how the information coming from the sequencing of the human genome is helping us understand our relationships with the microorganisms on planet Earth.

13. As a research member of a genomics company, you are asked to take the lead on sequencing the genome of *Legionella pneumophila.* (a) Why is your company interested in sequencing this bacterium, and (b) what possible applications are possible from knowing the bacterium's DNA sequence?

14. Besides the example for *Deinococcus radiodurans,* how can knowing the sequence of a common soil bacterium help produce a cleaner environment?

15. While studying for the microbiology exam that covers the material in this chapter, your roommate asks you why genomics, and especially bacterial genomics, was emphasized. How would you answer this question?

Review

Use the following syllables to compose the term that answers each of the clues below. The number of letters in each term is indicated by the blank lines, and the number of syllables is shown by the number in parentheses. Each syllable is used only once, and the answers are listed in Appendix D.

ASE BAC CLE COC COM CON CUS DO DROME EN FER FITH GA GASE GE GE GRIF I IN IN JU LENT LI LI LY MIDS MO NOME NU NY O ON PAL PE PHAGE PHAGE PI PLAS PNEU PRO SO TENCE TER TER TION U VIR

1. Closed loops of DNA   (2) __ __ __ __ __ __ __ __

2. Restriction recognition sequence   (3) __ __ __ __ __ __ __ __ __ __

3. Transforming property   (3) __ __ __ __ __ __ __ __ __ __

4. Transduction virus   (5) __ __ __ __ __ __ __ __ __ __ __ __ __

5. Recombinant DNA enzyme   (5) __ __ __ __ __ __ __ __ __ __ __ __ __

6. Transformed bacterium   (4) __ __ __ __ __ __ __ __ __ __ __ __

7. Conjugation structures   (2) __ __ __ __ __

8. Viral nonreplication   (4) __ __ __ __ __ __ __ __ __

9. DNA linking enzyme   (2) __ __ __ __ __

10. Gene engineered drug   (4) __ __ __ __ __ __ __ __ __ __

11. Discovered transformation   (2) __ __ __ __ __ __ __ __

12. Type of recombination   (4) __ __ __ __ __ __ __ __ __ __ __

13. Phage that causes lysis   (3) __ __ __ __ __ __ __ __

14. Complete set of genes   (2) __ __ __ __ __ __

15. Virus DNA in a bacterial chromosome   (2) __ __ __ __ __ __ __

http://microbiology.jbpub.com

The site features **eLearning,** an on-line review area that provides quizzes and other tools to help you study for your class. You can also follow useful links for in-depth information, or just find out the latest micro-biology news.

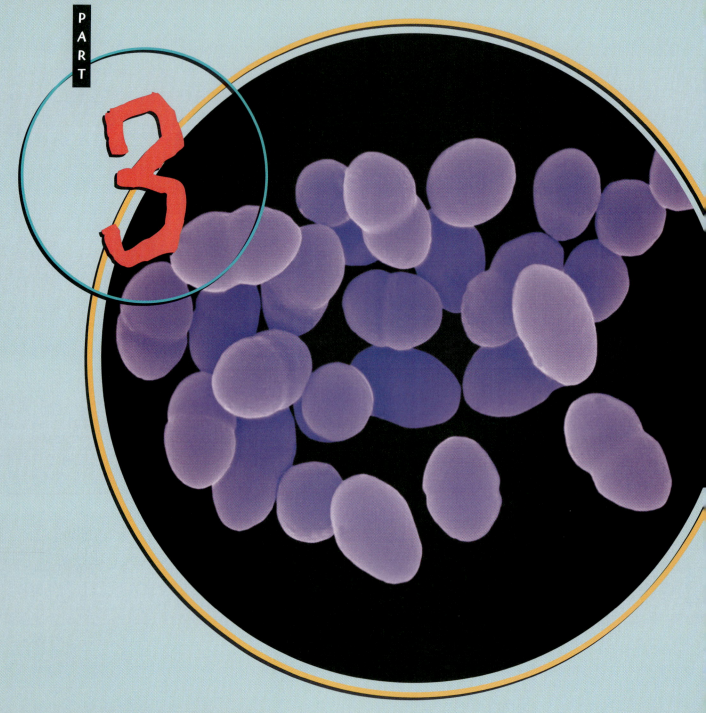

Bacterial Diseases of Humans

Throughout history, bacterial diseases have posed a formidable challenge to humans and have often swept through populations virtually unchecked. In the eighteenth century, for example, the first European visitors to the South Pacific found the islanders robust, happy, and well adapted to their environment. But the explorers introduced syphilis, tuberculosis, and pertussis (whooping cough), and soon these diseases spread like wildfire. Hawaii was struck with unusually terrible force. The population of the islands was about 300,000 when Captain Cook landed there in 1778; by 1860, it had been reduced to fewer than 37,000 people.

With equally devastating results, the Great Plague came to Europe from the Orient, and cholera spread eastward from India. Together with tuberculosis, diphtheria, and dysentery, these bacterial diseases ravaged European populations for centuries and insidiously wove themselves into the pattern of life. Infant mortality was particularly shocking: England's Queen Anne, who reigned in the early 1700s, lost 16 of her 17 babies to disease; and until the mid-1800s, only half the children born in the United States reached their fifth year.

Today, humans are better able to cope with bacterial diseases. Though credit often is given to wonder drugs, the major health gains have resulted from understanding disease and the body's resistance mechanisms, coupled with modern sanitary methods that prevent microorganisms from reaching their targets. Immunization also has played a key role in preventing disease. Indeed, very few people in our society die of the bacterial diseases that once accounted for the majority of all deaths.

In Part 3 of this text, we shall study the bacterial diseases of humans over the course of four chapters. The diseases have been grouped according to their major mode of transmission. Airborne diseases are discussed in Chapter 8; foodborne and waterborne diseases in Chapter 9; soilborne and arthropodborne diseases in Chapter 10; and sexually transmitted, contact, and miscellaneous diseases in Chapter 11. Many of the diseases we shall study are of historical interest and are currently under control. However, just as the garden always faces new onslaughts of weeds and pests, so, too, the human body is continually confronted with newly emerging or reemerging diseases. In this regard, the modern era is no different than Europe or the South Pacific islands of past centuries. On the fundamental level, disease has not changed. Only the pattern of disease has changed.

■ *False color scanning electron micrograph of Streptococcus pneumoniae.*

CLINICAL MICROBIOLOGY

One of the most famous books of the twentieth century was *Microbe Hunters* by Paul de Kruif. In his book, first published in the 1920s, de Kruif describes the joys and frustrations of Pasteur, Koch, Ehrlich, von Behring, and many of the original microbe hunters. The exploits of these scientists make for fascinating reading and help us understand how the concepts of microbiology were formulated. I would urge you to leaf through the book at your leisure.

Microbe hunters did not come to an end with Pasteur, Koch, and their contemporaries, nor did the stories of microbe hunters end with the publication of de Kruif's book. The men and women working in hospital, public, and private laboratories are today's detectives of microbiology. These individuals search for the pathogens of disease. Many travel to far corners of the world studying organisms, and many more remain close to home, identifying the pathogens in samples sent by physicians. Microbiologists even work in dental clinical labs, since many bacteria are involved in tooth decay and periodontal disease.

Clinical microbiology also offers an outlet for the talents of those who prefer to tinker with machinery. New instruments and laboratory procedures are constantly being researched in an effort to shorten the time between detection and identification of microorganisms. Many tests reflect human ingenuity. For example, there is a test that detects bacteria by its interference with the passage of light and its ability to scatter light at peculiar angles. Such modern devices as laser beams are used in this kind of instrumentation.

The microbe hunters have not changed materially in the past 100 years. The objectives of the search may be different, but the fundamental principles of the detective work remain the same. The clinical microbiologist is today's version of the great masters of a bygone era.

Airborne Bacterial Diseases

The captain of all these men of death that came against him to take him away was the Consumption, for it was that that brought him down to the grave.

—John Bunyan describing tuberculosis in
The Life and Death of Mr. Badman (1680)

THE PRODUCE LOOKED GOOD THAT DAY. The lettuce was green and crisp, and the carrots just seemed to ring out with good health. The broccoli was a deep, forest green and the cauliflower a lily white. It was probably due to the ultrasonic humidifier spraying its cool mist over the vegetable section. The fresh, clean mist made everything look terrific.

So it was in Bogaloosa, Louisiana, that November morning when officials from the local Department of Health walked into the supermarket. They had a disturbing story: 28 people were sick with Legionnaires' disease, and 2 had already died of pneumonia. Most of the patients, it seemed, had been in the grocery store within the previous ten days.

The investigators were friendly but firm. How long, they asked, did the air conditioner run each day? Were the vegetables washed before putting them out, and if so, for how long and where? Had any of the employees been sick recently? Who was the supermarket's supplier? And that humidifier—how long had it been there?

Three days later, the investigators were back with an answer. It was the humidifier! The lab had found *Legionella pneumophila* (FIGURE 8.1) in the water of its reservoir. Could the owner please remove the humidifier from the store and clean it out with disinfectant before reusing it? Within a day, the humidifier was gone. The broccoli didn't look quite as good anymore, and the lettuce leaves seemed to droop a bit. But, at least the air was safer to breathe, and that was important.

FIGURE 8.1

Legionella pneumophila

A false color transmission electron micrograph of *Legionella pneumophila*, the agent of Legionnaires' disease. The rod-shaped cells can have one or more long flagella. Legionnaires' disease is the only acute bacterial pneumonia known to occur in an outbreak form. (Bar = 0.5 µm.)

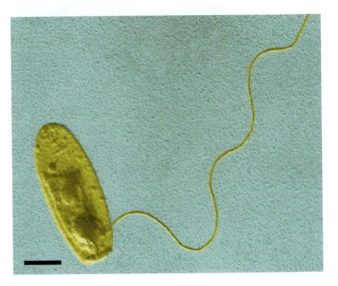

Though known only since 1976, Legionnaires' disease is recognized today throughout the world, and outbreaks, such as this 1989 incident in Louisiana, are reported regularly to the Centers for Disease Control and Prevention (CDC). In the United States alone, the CDC estimates that between 8,000 and 18,000 cases occur annually, though most go unreported (1,168 cases were reported in 2001). Legionnaires' disease is one of the contemporary airborne bacterial diseases that we shall discuss in this chapter. Such diseases are generally spread by airborne droplets where people crowd together.

We shall divide the airborne bacterial diseases into two general categories. The first category will include diseases of the upper respiratory tract, such as strep throat, scarlet fever, diphtheria, pertussis, and several forms of meningitis. The second category will include diseases of the lower respiratory tract: tuberculosis, pneumococcal pneumonia, primary atypical pneumonia, Legionnaires' disease, and others. As we proceed, note how the initial focus of infection is followed by spread to other organs. Also note that antibiotics are available for treating these diseases and that immunizations are used for protecting the community at large.

8.1 Diseases of the Upper Respiratory Tract

The diseases of the upper respiratory tract can be severe, as several diseases in this section illustrate. One reason is that the respiratory tract is a portal of entry to the blood, and from there, the disease can affect the more sensitive internal organs.

STREPTOCOCCAL DISEASES CAN BE MILD TO SEVERE

Streptococci are a large and diverse group of encapsulated, nonmotile, gram-positive cocci. They cause streptococcal sore throat (strep throat), scarlet fever, and several other diseases in humans. The bacteria divide in one plane and cling together to form chains of various lengths (FIGURE 8.2a). This observation is apparent in streptococci grown in liquid media, but it may not be obvious in streptococci from solid media.

Microbiologists classify the streptococci by several systems, two of which are widely accepted. The first system, developed by J. H. Brown in 1919, divides streptococci into hemolytic ("blood-digesting") groups, depending on how they affect red blood cells in blood agar. **Alpha-hemolytic streptococci** turn blood agar an olive-green color as they partially destroy red blood cells in the medium; colonies of **beta-hemolytic streptococci** are surrounded by clear, colorless zones due to the complete destruction of red blood cells; and **nonhemolytic streptococci** have no effect on red blood cells and thus cause no change in blood agar.

The second classification system, suggested by the American bacteriologist Rebecca Lancefield in 1933, is based on variants of a specific carbohydrate, the C substance, in cell walls of streptococci. Groups A through O streptococci have been distinguished. Most streptococcal diseases in humans are caused by species of group A streptococci; *Streptococcus pyogenes* is the most common species (FIGURE 8.2b). This beta-hemolytic organism is generally implied when physicians refer to "group A beta-hemolytic streptococci."

Streptococcus pyogenes causes **streptococcal pharyngitis**, popularly known as **strep throat**. The streptococci reach the upper respiratory tract within airborne

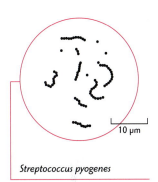

10 μm

Streptococcus pyogenes

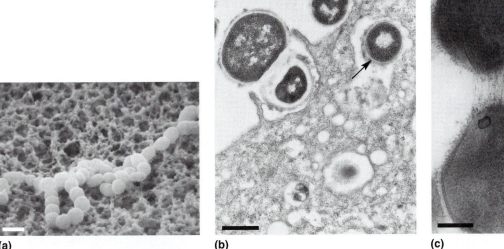

(a) (b) (c)

FIGURE 8.2

Streptococci and Their Host Cell

(a) A scanning electron micrograph of *Streptococcus mutans*, a species that causes many cases of tooth decay. (Bar = 1 μm.) (b) A transmission electron micrograph of *Streptococcus pyogenes* attaching to cells cultured from the human pharynx. The bacteria associate with microvilli of the cells. Surface proteins, especially the streptococcal M protein, function in this association. A streptococcal cell (arrow) has been internalized and is seen within the cytoplasm of the pharyngeal cell. (Bar = 0.5 μm.) (c) A close-up view of *S. pyogenes* cells showing strands of M protein protruding through the capsule. The M protein enhances virulence by encouraging attachment to the host cell. (Bar = 0.25 μm.)

droplets of mucus expelled by infected persons during coughing and sneezing. Patients experience a sore throat, high fever, coughing, swollen lymph nodes and tonsils, and a fiery red "beefy" appearance to pharyngeal tissues owing to tissue erosion. More than a million Americans, primarily children, suffer strep throat annually, many cases complicated by an infection of the middle ear, **otitis media**.

The pathogenicity of *S. pyogenes* is enhanced by a substance called **M protein** (**FIGURE 8.2c**). This protein, anchored in the cell wall and cell membrane, encourages adherence to the pharyngeal tissue and retards phagocytosis. Over 80 specific types (serotypes) of M protein have been identified, and complete immunity to streptococcal disease requires that a person produce antibodies against all 80 types.

Scarlet fever is strep throat accompanied by a skin rash. The rash develops as a pink-red blush, apparent on the neck, chest, and soft-skin areas of the arms. It results from blood leaking through the walls of capillaries damaged by a toxin. This toxin, called an **erythrogenic** ("red-forming") **toxin**, is produced only by strains of *S. pyogenes* that carry a prophage that encodes the toxin. Normally an individual experiences only one case of scarlet fever in a lifetime because the immune system produces special antibodies, called **antitoxins**, which circulate in the bloodstream and neutralize the toxins during succeeding episodes.

Treatment of streptococcal diseases is generally successful with antibiotics such as penicillin and erythromycin (**MicroFocus 8.1**). However, widespread resistance to antibiotics has been noted since the early 1970s, and research indicates that many strains of *S. pyogenes* carry prophages having antibiotic-resistance genes.

An important complication of streptococcal disease is **rheumatic fever**, which is most common in young school-age children. This condition, caused by specific M serotypes of group A streptococci, is characterized by fever and inflammation of the small blood vessels. Joint pain is common, but the most significant long-range effect is permanent scarring and distortion of the heart valves, a condition called **rheumatic heart disease**. The damage arises from a reaction of the body's antibodies to streptococcal antigens crossreacting with antigens on heart tissue. Another

Antitoxins:
highly specific antibodies that react with toxins.

MicroFocus 8.1

DANGER ON THE FARM

It's called tilapia (ti-lap'i-ah), and it's one of fish-farming's more popular products. Raised in huge tanks by modern methods of aquaculture, tilapia is appearing more and more in restaurants and supermarkets—but so are the bacteria that infect tilapia.

The lesson was driven home in 1996 when several people in Ontario, Canada, became ill after handling fresh tilapia. One patient punctured his hand with a fish bone while cutting the fish into filets; another punctured her finger with the dorsal fin of the fish while scaling it; yet another suffered a skin wound from a knife used to cut the fish. In all cases,

the patients developed infection with *Streptococcus iniae*, a known fish pathogen. The patients exhibited the symptoms of various diseases, including meningitis, transient arthritis, and endocarditis. All received antibiotic therapy and recovered successfully.

The outbreak underscores the possibility that fish pathogens can adapt to human tissues—in effect, jumping from one species to another. Health officials are concerned by this because it means a new human pathogen may emerge. Furthermore, aquaculture is a fast-growing industry and a promising source of foods for the future. However, it also represents

possible human exposure to new bacteria, and, as the cases in Canada illustrate, that can pose a formidable danger.

manifestation of this reaction may develop in the kidneys, where the condition is called **glomerulonephritis**. Rheumatic fever cases have been declining in the United States due to antibiotic treatment and, therefore, there has been a decline in the streptococcal strains that cause disease. The last year that the CDC required reporting of rheumatic fever cases was 1994. In that year, 112 cases were reported versus 10,000 reported cases in 1961. However, in developing nations, rheumatic fever remains a serious problem.

S. pyogenes is commonly present in the human nose and throat, and transmission to many other tissues is possible. For example, streptococci may cause disease in open wounds or skin abrasions, resulting in **erysipelas.** If invasive group A streptococci containing other specific prophages invade and infect the fascia over the muscles and beneath the skin, the phage-encoded toxins cause a condition called **necrotizing fasciitis** and an unsightly degeneration and dissolving of the skin tissues, as shown in FIGURE 8.3. (Tabloid newspapers and TV programs had a field day with this disease in 1994 as they heralded an outbreak of the "flesh-eating bacteria"; see MicroFocus 8.2.) If the tissue of the uterus becomes infected during the birth process, the mother may suffer **puerperal sepsis**, also known as **childbed fever.** Antibiotic use has been responsible for the elimination of the disease in the developed world. Perhaps you remember from Chapter 1 that this was the disease that prompted Ignaz Semmelweis to institute hand washing in the hospital maternity wards. Streptococci may infect the lower respiratory tract and cause streptococcal pneumonia as a complication to viral disease. In past generations, many died of **septicemia** or "blood poisoning," a disease now known to be due to several organisms, including streptococci. One of Louis Pasteur's daughters was a victim of this disease.

Sepsis:
The presence of pathogens or toxins in the blood.

DIPHTHERIA IS A DEADLY, TOXIN-CAUSED DISEASE

Diphtheria was first recognized as a clinical entity in 1826 by French pathologist Pierre F. Bretonneau. Bretonneau named the disease *la diphtherite*, from the Greek *diphthera* for "membrane," a reference to the exudate that appears in the throats of patients. In 1883, Edwin Klebs observed a bacterium in material from a patient's throat, and the next year, Friederich Löeffler successfully cultivated the organism. The so-called Klebs-Löeffler bacillus has the scientific name *Corynebacterium diphtheriae* because it resembles a club (*coryne* is Greek for "club").

Corynebacterium diphtheriae is a gram-positive, nonmotile bacillus containing numerous **metachromatic granules** that show up with a methylene blue stain as cytoplasmic dots. The bacteria remain close to one another after multiplying and

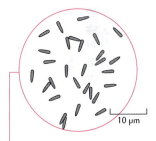

10 µm

Corynebacterium diphtheriae

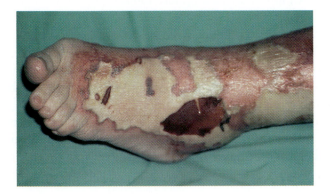

FIGURE 8.3

Necrotizing Fasciitis

This patient shows the symptoms of necrotizing fasciitis. This disease is caused by a species of *Streptococcus pyogenes*. The infection affects the fascia surrounding the muscle tissues and may result in rapid tissue degeneration as powerful prophage-encoded toxins interfere with cell metabolism. Prompt removal of dead tissue is necessary to prevent systemic toxicity. The disease has been known since ancient times but first attracted news-media interest in 1994.

MicroFocus 8.2

TOO GOOD TO BE TRUE

They needed a story for their tabloid newspaper. It had to be attention grabbing, current, just slightly scientific, and health related. Then they heard about the man in southwest England. It was too good to be true.

"A flesh-eating bug ate my face!" the tabloid screamed. Soon the stories rolled in: Limbs were rotting before people's eyes; facial features were melting away; doctors were making emergency amputations to save lives. It wasn't Armageddon—it was the flesh-eating bacteria!

The medical community scoffed. The "bug," they said, was a species of *Streptococcus* called *Streptococcus pyogenes*, a group A streptococcus, and the "new"

disease already had a name: necrotizing fasciitis. (Necrotizing is from the Greek *nekros* for "dead"; and fasciitis refers to the fascia that surrounds muscle tissue and attracts the streptococci.) There was nothing new about the disease—hospital physicians had seen it in contaminated wounds as early as the Civil War; and it was common in nursing home patients. They knew it as "hospital gangrene," "putrid ulcer," and by other descriptive expressions. Some even called it the Second Disease, since the skin often turned a deep red (see the discussion of Fifth Disease in Chapter 13).

But why was it flaring up now? Was it a publicity stunt, as some physicians sus-

pected? Apparently not, said the researchers. They presented evidence that the streptococci had become more virulent. They also linked contemporary cases to the new nonsteroidal anti-inflammatory drugs, such as ibuprofen. They pointed out that a significant number of patients were taking the drugs at the time of disease.

Whatever the source, the flesh-eating bacteria were here to stay. The tabloid had scored its coup, and the editors were ecstatic. Newspapers were selling faster than ever, and even the late news had picked up the story. Now, what would next week's story be?

Exotoxin:
a poisonous chemical substance produced by live bacteria and immediately excreted.

form a palisade layer, a picket-fence arrangement. Diphtheria is acquired by inhaling respiratory droplets from an infected person into the upper respiratory tract near the tonsils. Initial symptoms include a sore throat and moderate fever. The bacteria produce a potent **exotoxin** that inhibits protein synthesis in epithelial cells, the cube-like cells that line the skin and body cavities such as the respiratory tract. As dead tissue accumulates with mucus, white blood cells, and fibrous material, a leathery **pseudomembrane** forms ("pseudo" because it does not fit the definition of a true membrane) on the tonsils or pharynx. Respiratory blockage and death may follow, especially in children. In adults, the toxin often spreads to the bloodstream, where it causes neck swelling ("bullneck"), heart damage, and destruction of the fatty sheaths surrounding nerves. Treatment requires both antibiotics (penicillin or erythromycin) for the bacteria and antitoxins to neutralize the toxins.

At present, the number of cases of diphtheria in the United States is less than a dozen annually (two probable cases in 2001), but the disease remains a health problem in many regions of the world, especially parts of Asia. From 1990 to 1995, for example, a major outbreak occurred in 13 of the 14 Newly Independent and Baltic States of the former Soviet Union; it affected almost 150,000 people, with about 4,000 deaths.

Immunization against diphtheria may be rendered by an injection of diphtheria **toxoid** contained in the diphtheria-tetanus-acellular pertussis (DTaP) vaccine. A toxoid consists of toxin molecules treated with formaldehyde or heat to destroy their toxic qualities. The toxoid induces the immune system to produce antitoxins that circulate in the bloodstream throughout the person's life.

In 1961, V. J. Freeman discovered that a prophage integrated into the bacterial chromosome contains the genetic code for diphtheria **exotoxin** production. Strains of *C. diphtheriae* lacking the phage are harmless, but they may become **toxigenic** if the gene-carrying bacteriophage infects them. The same phage is carried by cells of

Toxigenic:
producing a toxin.

C. ulcerans, a related species found primarily in cattle. In 1997, a woman was diagnosed with diphtheria due to this organism.

PERTUSSIS (WHOOPING COUGH) IS HIGHLY CONTAGIOUS

Pertussis, also known as **whooping cough**, is one of the more dangerous and highly contagious diseases of childhood years (**MicroFocus 8.3**). Although the incidence of this disease has declined substantially since introduction of an effective vaccine in 1949, the number of cases in the United States still remains significant. For example, the CDC recorded about 7,580 cases in 2001, and the numbers were growing. Almost 25 percent occurred in children under six months of age. Recent studies also show that about 20 percent of adults with a persistent cough have undiagnosed pertussis.

Pertussis is caused by *Bordetella pertussis*, a small gram-negative, nonmotile rod first isolated by Jules Bordet and Octave Gengou in 1906. The bacillus is spread by respiratory droplets that adhere to and aggregate on the cilia of epithelial cells in the upper respiratory tract. Toxin production paralyzes the ciliated cells and impairs mucus movement, potentially causing pneumonia.

Typical cases of pertussis occur in three stages. The initial stage is marked by general malaise, low-grade fever, and increasingly severe cough. During the next stage, disintegrating cells and mucus accumulate in the airways and cause labored breathing. Patients experience multiple **paroxysms** of rapid-fire staccato coughs all in one exhalation, followed by a forced inhalation over a partially closed glottis. The rapid inhalation results in the characteristic "whoop" (hence, the name whooping

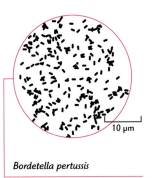

Bordetella pertussis

Paroxysm:
a severe spasm or convulsion.

MicroFocus 8.3

SOME PATHOGENS "AIN'T WHAT THEY USED TO BE"

Bordetella pertussis, the causative agent of whooping cough, is a highly contagious, strictly human pathogen. However, that is not how it used to be. In fact, some bacteria, including *B. pertussis*, have not become more mild over time but rather more pathogenic. How can that happen?

In March 2003, the genome of *B. pertussis* was sequenced by The Sanger Institute in Cambridge, England. The genome is 4,086,189 base pairs and the estimated number of genes is 3,816. Its sequence has been compared to that of *B. bronchiseptica* and *B. parapertussis*, both of which can survive in the environment outside a host. Comparative genomics indicates that *B. pertussis* has the smallest genome of the three related species. Closer comparative analysis suggests that *B. pertussis* has lost a large number of genes over a relatively short evolutionary period. So, how can losing genes cause an organism to become more virulent?

Researchers believe that the lost genes in *B. pertussis* once were needed to place constraints on virulence. In its relatives, these constraints remain and they ensure that the bacteria do not kill their host before the bacteria have a chance to reproduce and spread to other individuals; otherwise, the bacteria also will die. *B. pertussis* can afford to lose these controls because of its close contact with humans. Being spread in dense human populations, the bacterium has no problem spreading quickly from person to person before the host conceivably would die, so it can do without the controls.

The analysis of the *B. pertussis* genome indicates that its evolution has involved gene mutations through deletion and loss. Certainly, this is one reason why its genome is smaller than the genomes of its two close relatives. Also, a few base-pair changes have increased expression of the pertussis toxin gene. The toxin is responsible for paralyzing the host's white blood cells, so they cannot kill the bacterial cells.

Like all microbes, *B. pertussis* is good at taking advantage of opportunities presented by new environments. In this case, by losing or inactivating genes, the organism has become "leaner and meaner." It certainly "ain't what it used to be."

cough). Ten to fifteen paroxysms may occur daily, and exhaustion usually follows each. During the third stage, sporadic coughing continues for several weeks, even after the bacteria have vanished. (Doctors call it the "100-day cough.")

Eradication of the bacteria is generally successful when erythromycin is administered before the respiratory passageways become blocked. However, antibiotic treatment will not shorten the illness. Diagnosis is performed by obtaining swabs from the posterior pharyngeal wall and identifying *B. pertussis* on selective media. A fluorescent antibody test (Chapter 20) also is used. Although the bacterium produces several toxins, no one toxin is considered the prime factor in the disease. Convalescence depends on the speed at which the ciliated epithelium regenerates.

As with diphtheria, the declining incidence of pertussis stems partly from use of a pertussis vaccine. The older vaccine (diphtheria-pertussis-tetanus, or DPT) contained **merthiolate**-killed *B. pertussis* cells and was considered risky because about 1 in 300,000 vaccinees suffered high fevers and seizures. As of 1993, public health officials recommended the newer acellular pertussis vaccine prepared from *B. pertussis* chemical extracts. Combined with diphtheria and tetanus toxoids, the triple vaccine is given the acronym DTaP. Commercially, it is known as Tripedia.

MENINGOCOCCAL MENINGITIS CAN BE LIFE THREATENING

The term **meningitis** refers to several diseases of the meninges, the three membranous coverings of the brain and spinal cord. Meningitis may be caused by viruses, fungi, protozoa, or bacteria, and different forms of the disease have different mortality rates. In all forms, the meninges become inflamed, causing pressure on the spinal cord and brain. Patients experience intense headaches, neck aches, and lower backaches.

A particularly dangerous form of meningitis is **meningococcal meningitis**. This disease is caused by *Neisseria meningitidis*, a small gram-negative, encapsulated diplococcus commonly called the **meningococcus**. Meningococci enter the body by droplets, often from a **reservoir**. In most cases, the disease consists of an influenza-like upper respiratory infection. However, sepsis can occur where bacterial toxins may overwhelm the body in as little as two hours and cause death (**FIGURE 8.4**). This condition is called **meningococcemia**. In survivors, the meninges become inflamed, and patients experience a characteristic stiff, arched neck and pounding headache. A rash also appears on the skin, beginning as bright-red patches, which progress to blue-black spots. Fifty percent of untreated cases may be fatal.

Early diagnosis and treatment of meningococcal meningitis can prevent irreversible nerve damage or death. A principal criterion for diagnosis is the observation and/or cultivation of gram-negative diplococci in samples of spinal fluid obtained by a spinal tap. Treatment with penicillin, cefotaxime, or ceftriaxone drugs is usually recommended. A vaccine containing capsular polysaccharides from several groups of meningococci is available for immunization, but it is used only under special circumstances.

Cases of meningococcal meningitis are sometimes complicated by the formation of lesions in the adrenal glands and accompanying hormone imbalances. This condition, called the Waterhouse-Friderichsen syndrome, may be a manifestation of a hypersensitivity reaction taking place in the body (Chapter 21).

Neisseria meningitidis is a fragile organism that does not survive easily in the environment and must be maintained in nature by person-to-person transfer. Meningococcal meningitis is therefore prevalent where people are in close proximity for long

Merthiolate:
a heavy metal compound often used as an antiseptic and preservative.

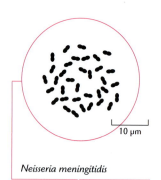

10 μm

Neisseria meningitidis

Reservoir:
living or nonliving source of a pathogen.

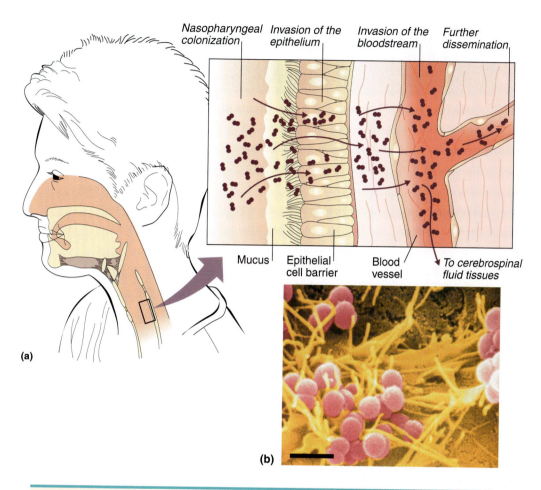

Nasopharyngeal colonization   Invasion of the epithelium   Invasion of the bloodstream   Further dissemination

Mucus   Epithelial cell barrier   Blood vessel   To cerebrospinal fluid tissues

(a)

(b)

FIGURE 8.4

Meningococcal Meningitis

Meningococcal meningitis is caused by the gram-negative diplococcus *Neisseria meningitidis*. (a) The bacteria colonize the nasopharynx, then invade the epithelium causing respiratory distress. They pass into the bloodstream, where they induce meningococcemia. Finally, they disseminate to tissues near the spinal cord, causing inflammation and meningitis. (b) A scanning electron micrograph of *N. meningitidis* at the surface of epithelial cells of the respiratory tract. (Bar = 2 μm.)

periods of time. Grade-school classrooms, military camps, and prisons are examples. Though most people suffer nothing worse than a respiratory disease, the CDC reported about 2,300 cases of *Neisseria*-linked meningitis in 2001.

TABLE 8.1 compares meningococcal meningitis and *Haemophilus* meningitis, which is discussed below.

HAEMOPHILUS MENINGITIS HAD BEEN PREVALENT IN PRESCHOOL CHILDREN

In 1892, Richard Pfeiffer isolated a small, gram-negative, nonmotile, encapsulated rod he thought was the cause of influenza. Because of this relationship and due to its attraction to blood ("hemo-philus"), he named the organism *Haemophilus influenzae*.

TABLE 8.1

A Comparison of Meningococcal Meningitis and *Haemophilus* Meningitis

CHARACTERISTIC	MENINGOCOCCAL MENINGITIS	HAEMOPHILUS MENINGITIS
Agent	*Neisseria meningitidis*	*Haemophilus influenzae* b
Description	Gram-negative diplococcus	Gram-negative rod
Transmission	Respiratory droplets	Respiratory droplets
Early symptoms	Respiratory disease	Respiratory disease
Blood symptoms	Meningococcemia, serious fever, malaise	Few symptoms
Nervous symptoms	Stiff, arched neck; headache	Stiff, arched neck; headache
Skin rash	Red to blue-black spots	Rare
Age group affected	General population	Young children
Mortality rate	High	Low
Vaccine available	Not to general public	Yes
Complications	Waterhouse-Friderichsen syndrome	Few

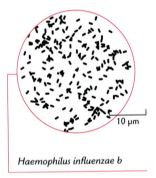

10 μm

Haemophilus influenzae b

However, during the great influenza epidemic of 1918 to 1919, Pfeiffer's bacillus was identified as a secondary cause of disease, and influenza was attributed to a virus rather than a bacterium (Chapter 13).

Haemophilus influenzae is now regarded as a cause of respiratory tract infections, occurring primarily as a complication to a previous disease. Moreover, *H. influenzae* type b (Hib) has attracted widespread attention as a cause of meningitis in children between the ages of six months and two years. The organism moves from the respiratory tract to the blood and then to the meninges, where it causes **Haemophilus meningitis**. Symptoms of disease include stiff neck, severe headache, and other evidence of neurological involvement such as listlessness, drowsiness, and irritability. Although *H. influenzae* is sensitive to a wide variety of antibiotics, ceftriaxone or cefotaxime is the first choice for treatment. The disease has a mortality rate of about 5 percent.

In 1986, about 18,000 cases of *Haemophilus* meningitis were occurring in the United States annually, and Hib was the most common cause of bacterial meningitis. By that time, however, a vaccine had been licensed by the FDA and the epidemic had peaked (FIGURE 8.5). The conjugated vaccine consists of polysaccharides from the organism's capsule joined to a bacterial protein. As of 1993, the vaccine was combined with the DTaP vaccine for distribution to children as Tetramune, and in 2001, the number of annual cases in the United States was down to 1,597. This disease and other airborne bacterial diseases of the upper respiratory tract are summarized in TABLE 8.2.

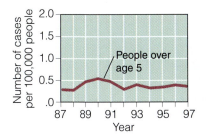

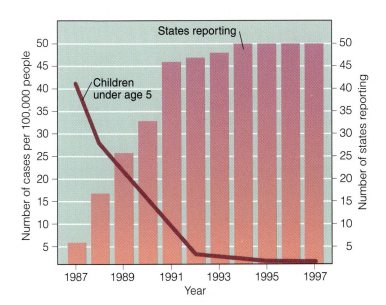

FIGURE 8.5

The Decline of *Haemophilus* Meningitis

This graph illustrates the declining number of cases of *Haemophilus* meningitis in the United States between 1987 and 1997 in children under age five. The decline is set against the increased number of states participating in the reporting mechanism. The graph indicates that the decline is due in part to the surveillance mechanism set up in the states. Note that the number of cases in people over age five (small box) has remained constant in this period. A very small number of cases occur in this group.

To this point . . .

We have studied a number of airborne bacterial diseases in which the initial focus of infection is in the upper respiratory tract. Symptoms generally include coughing, inflammation of the throat tissues, and the accumulation of mucus in the respiratory passageways. In diseases such as pertussis, this is the extent of disease, but in other diseases, bacteria invade the blood and cause serious damage elsewhere. This spread is evident in scarlet fever, diphtheria, and two forms of bacterial meningitis.

The next group of diseases involves the lower respiratory tract. Infection takes place in the lung tissue, and as the breathing capacity is reduced, a life-threatening situation is established. We shall see that lung damage occurs in tuberculosis, pneumococcal pneumonia, Legionnaires' disease, and other diseases surveyed. We also shall examine a number of respiratory diseases that do not spread easily from person to person, but develop from organisms already in the patient's respiratory tract. Our survey will conclude with diseases of the lower respiratory tract caused by rickettsiae, a group of small bacteria.

TABLE 8.2

A Summary of Airborne Bacterial Diseases of the Upper Respiratory Tract

DISEASE	CAUSATIVE AGENT	DESCRIPTION OF AGENT	ORGANS AFFECTED	CHARACTERISTIC SIGNS
Strep throat Scarlet fever	*Streptococcus pyogenes*	Gram-positive encapsulated streptococcus	Upper resp. tract Blood Skin	Sore throat Skin rash Septicemia
Diphtheria	*Corynebacterium diphtheriae*	Gram-positive rod	Upper resp. tract Heart, nerve fibers	Pseudomembrane Sore throat, moderate fever
Pertussis (whooping cough)	*Bordetella pertussis*	Gram-negative rod	Upper resp. tract	Mucous plugs Paroxysms of cough with "whoop"
Meningococcal meningitis	*Neisseria meningitidis*	Gram-negative encapsulated diplococcus	Upper resp. tract Blood Meninges	Toxemia Paralysis Skin spots
Haemophilus meningitis	*Haemophilus influenzae b*	Gram-negative encapsulated rod	Upper resp. tract Meninges	Respiratory symptoms Paralysis

8.2

Diseases of the Lower Respiratory Tract

In the lower respiratory tract, a number of bacterial diseases affect the lung tissues. As the latter are destroyed, fluid builds up in the lung cavity, and the space for obtaining oxygen and eliminating carbon dioxide is reduced. This is the basis for a possibly fatal pneumonia.

TUBERCULOSIS IS PRIMARILY A DISEASE OF THE LUNGS

At the turn of the century, **tuberculosis (TB)** was the world's leading cause of death from all causes, accounting for one fatality in every seven cases. Today's statistics, though improved, are still very high. In developing countries, public health officials report more deaths from TB than from any other bacterial disease. In the United States, the CDC reported 16,377 cases for 2000, half of which occurred in foreign-born persons. There were about 751 fatalities, most among individuals from minority groups. The World Health Organization (WHO) has referred to the global situation as a scandal and estimates 2 million died of the disease in 2001. The WHO also estimates that between 2002 and 2020 one billion people globally will become infected and 36 million will die of tuberculosis, unless control measures are strengthened. MicroFocus 8.4 recounts its existence in the Americas.

Tuberculosis is caused by *Mycobacterium tuberculosis*, the "tubercle" bacillus first isolated by Robert Koch in 1882. It is a small, nonmotile rod that enters the respiratory tract in aerosolized droplets (multiple exposures are generally necessary). People who live in urban ghettoes often contract TB because malnutrition and a

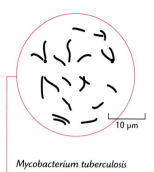

Mycobacterium tuberculosis

10 µm

TOXIN INVOLVED	TREATMENT ADMINISTERED	IMMUNIZATION AVAILABLE	COMMENT
Erythrogenic toxin	Penicillin Erythromycin	None	Rheumatic fever or glomerulonephritis possible Beta-hemolytic strains
Yes	Penicillin Antitoxin	Toxoid in DTaP	Bacteriophage involved Metachromatic granules
Yes	Erythromycin	Acellular vaccine available as DTaP	Vaccine may cause side effects Mucus movement impaired
Yes	Penicillin Cefotaxime Ceftriaxone	Capsule polysaccharide vaccine	High mortality if untreated Adrenal gland involvement
Not established	Cefotaxime Ceftriaxone	Conjugated vaccine to type b only	Dramatic decline in young children

generally poor quality of life contribute to the establishment of disease, and overcrowding increases the concentration of bacilli in the air.

About 10 percent of people who contract TB become ill within three months. They experience chronic cough, chest pain, and high fever, and they expel sputum, the thick matter accumulating in the lower respiratory tract. (Often the sputum is rust colored, signaling that blood has entered the lung cavity.) The remaining 90 percent exhibit mild symptoms, including malaise, fever, and weight loss. In these cases the body responds to the disease by forming a wall of white blood cells, calcium salts, and fibrous materials around the bacilli (FIGURE 8.6). As these materials accumulate in the lung, a hard nodule called a **tubercle** arises (hence the name tuberculosis). This

MicroFocus 8.4

"NOT GUILTY!"

Poor Christopher Columbus! Historians have accused you of bringing smallpox to the New World—and measles, whooping cough, tuberculosis, and almost every other conceivable disease. One can almost imagine that the stately Santa María was a hospital ship!

Well, rest easy, Chris, for scientists have cleared you of bringing at least one disease—tuberculosis. Your defense is based on 1995 research by Arthur

Aufderheide (pronounced off'der-hide) from the University of Minnesota. Some years before, Aufderheide was studying the remains of a mummified woman from Peru when he noticed in her lung tissues several lumps reminiscent of tuberculosis. He enlisted the help of a molecular biologist to extract DNA from the lumps and amplify it so there was enough to identify. The DNA turned out to be identical to that of

Mycobacterium tuberculosis, the tubercle bacillus.

Why was that important? Well, Chris, the mummy was a thousand years old—that's right, one thousand years. Apparently both the woman and the tuberculosis were already here hundreds of years before you arrived. It's even possible you might have taken some back with you to Europe. . . . Oops! Sorry!

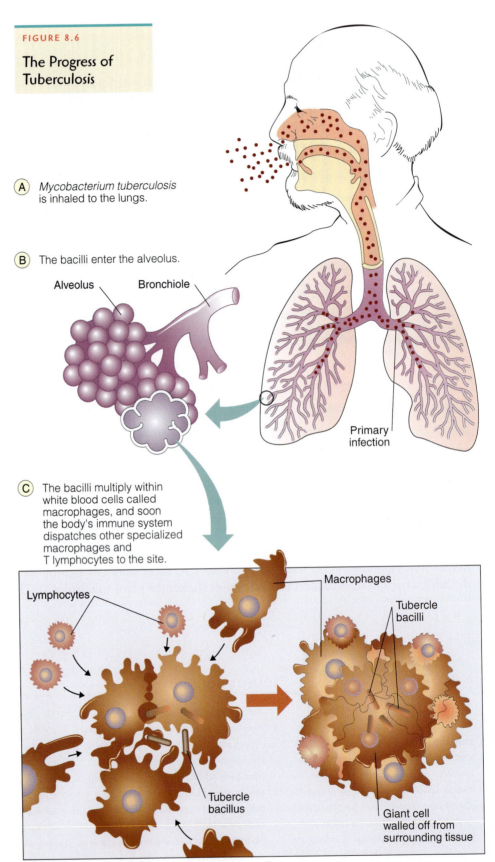

FIGURE 8.6

The Progress of Tuberculosis

(A) *Mycobacterium tuberculosis* is inhaled to the lungs.

(B) The bacilli enter the alveolus.

Alveolus

Bronchiole

Primary infection

(C) The bacilli multiply within white blood cells called macrophages, and soon the body's immune system dispatches other specialized macrophages and T lymphocytes to the site.

Lymphocytes

Macrophages

Tubercle bacilli

Tubercle bacillus

Giant cell walled off from surrounding tissue

(D) Multinucleated giant cells develop as the cells join together.

(E) A wall of cells, calcium salts, and fibrous materials eventually form around the giant cell. This is the tubercle.

tubercle may be visible in a chest X ray. Unfortunately, the bacilli are not killed, and the tubercle may expand as the lung tissue progressively deteriorates.

In many instances, the tubercle breaks apart and bacteria spread to other organs such as the liver, kidney, meninges, and bone. (Tuberculosis of the spine is called **Pott's disease**.) The disease is now called **miliary tuberculosis** (from the Latin *milium*, meaning "seed"). Tubercle bacilli produce no discernible toxins, but growth is so unrelenting that the tissues are literally consumed, a factor that gave tuberculosis its alternate name, **consumption**.

In its cell wall, *M. tuberculosis* contains a layer of waxy material that greatly enhances resistance to environmental pressures. In the laboratory, a sputum sample for staining must be accompanied by heat to penetrate this barrier, or a lipid-dissolving material must be used. Once stained, however, the organisms resist decolorization, even when subjected to a 5 percent acid-alcohol solution. Thus, the bacilli are said to be acid resistant, or **acid fast**. The acid-fast test (Chapter 3) is an important screening tool when used with the patient's sputum.

Early detection of tuberculosis is aided by the **tuberculin test**, a procedure that begins with the application of a purified protein derivative (PPD) of *M. tuberculosis* to the skin. One method of application, devised by Charles Mantoux in 1910, utilizes a superficial injection of PPD to the outer skin layers (the **Mantoux test**). If the patient has been exposed to tubercle bacilli, the skin becomes thick, and a raised, red welt develops within 48 to 72 hours. A positive test does not necessarily reflect the presence of tuberculosis, but may indicate a recent immunization, previous tuberculin test, or past exposure to the disease. It suggests a need for further tests.

Tuberculosis is an extremely stubborn disease. Physicians once recommended fresh air, but now they treat it with such drugs as isoniazid (INH), pyrazinamide, and rifampin (**MicroFocus 8.5**). The recent appearance of **multidrug-resistant *Mycobacterium tuberculosis*** (**MDR-TB**) has necessitated the use of ethambutol and streptomycin to help delay the emergence of resistant strains. In addition, drug therapy is intensive and must be extended over a period of six to nine months or more, partly because the organism multiplies at a very slow rate (its generation time is about 18 hours). Early relief, boredom, and forgetfulness often cause the patient to stop taking the medication, and the disease flares anew. In 1998, the FDA approved the drug rifapentine, which is taken only once a week. **MicroFocus 8.6** describes the WHO treatment strategy.

Tuberculosis is a particularly insidious problem to those who have **AIDS**. In these patients, the T lymphocytes, immune system cells that normally mount a response to *M. tuberculosis,* are being destroyed, and the patient cannot respond to the bacterial infection. HIV-infected patients face a mortality rate from tuberculosis of 70 to 90 percent, usually within one to four months of developing symptoms. Unlike most other tuberculosis patients, those with HIV usually develop tuberculosis in the lymph nodes, bones, liver, and numerous other organs. Ironically, AIDS patients often test negative for the tuberculin skin test because without T lymphocytes, they cannot produce the telltale red welt that signals infection. Tuberculosis often is the first disease to occur in the AIDS patient, even before any of the other opportunistic illnesses appear, and it is generally more intractable than in non-AIDS patients. The WHO estimates that worldwide, about 4.4 million people are coinfected with HIV and *M. tuberculosis*. AIDS is discussed in more detail in Chapter 14.

Immunization to tuberculosis may be rendered by injections of an **attenuated** (weakened) strain of *M. bovis*. This species causes tuberculosis in cows as well as humans. The attenuated strain is called **bacille Calmette Guérin**, or **BCG**, after Albert Calmette and Camille Guérin, the two French investigators who developed

Attenuated:
weakened by chemical or cultural processes.

MicroFocus 8.5

A BREATH OF FRESH AIR

Edward Livingston Trudeau, an American physician, understood the danger in nursing his brother, who was suffering from tuberculosis. It was 1865. Scientists believed that tuberculosis was passed among individuals, but the cause of the disease remained unknown. In Europe, Louis Pasteur was busy with his swan-neck flask experiments, and Robert Koch was still in medical school.

Several years later, Trudeau's worst fears became reality when he developed tuberculosis. His left lung was almost completely involved, and the disease had begun to appear in his right lung. Trudeau's doctor advised him to go south and exercise; hopefully, he would survive a few months. Instead, Trudeau resolved to spend his last days in the Adirondack Mountains of New York. In 1873, he left for Saranac Lake and a

precious last opportunity to hunt and fish in the mountains he loved.

However, death was not yet ready to claim Trudeau. While in the mountains, he found his tuberculosis slowly but steadily going into remission, and within months, his recovery was complete. Trudeau became a strong advocate for the open-air treatment of tuberculosis, and in 1884 he built two small cottages and established the sanatarium at Saranac Lake to care for tuberculosis

patients. Death did eventually come, but not until 1915—a full 42 years after he was told to expect it.

Today the Trudeau Institute at Saranac Lake remains a research facility for the study of human disease. It continues to be watched over by a Trudeau—the latest director, Francis Trudeau, is the grandson of Edward. (Francis Trudeau's son, incidentally, is Gary Trudeau, author of the comic strip *Doonesbury*.)

it in the 1920s (Chapter 20). Though the vaccine is used in parts of the world where the disease causes significant mortality and morbidity, many scientists oppose its use in the United States because they point to the success of early detection and treatment, and to the vaccine's occasional side effects. New vaccines consisting of subunits, molecules of DNA, and attenuated strains of mycobacteria are currently being developed, as Chapter 20 explores. In 1989, the CDC announced a strategic plan to eliminate tuberculosis from the United States by the year 2010. The plan emphasized more effective use of existing prevention and control methods, and the development of new technologies for diagnosis and treatment.

Several other species of *Mycobacterium* deserve a brief mention. The first, *M. cheloni*, is an acid-fast rod frequently found in soil and water. During the 1980s, microbiologists first recognized this fast-growing bacillus as a cause of lung diseases, wound infections, arthritis, and skin abscesses. *M. haemophilum* surfaced as a pathogen in 1991 when 13 cases occurred in immunocompromised individuals in New York City hospitals. Cutaneous ulcerating lesions and respiratory symptoms were observed in the patients. *M. abscessus*, another fast-growing species, caused an outbreak of abscesses at injection sites when it contaminated a hormone extract administered to 47 patients in Colorado in 1996. Another notable species is *M. avium-intracellulare*. This acid-fast rod may be the cause of lung disease, especially in people who have AIDS. In these patients, the bacteria are disseminated rapidly throughout the body. Infection comes from ingesting food or water contaminated with the bacterium.

MicroFocus 8.6

TRAGIC ENDINGS BUT HOPEFUL FUTURES

Tuberculosis (TB) is a contagious disease caused by *Mycobacterium tuberculosis*. When an infected person coughs, sneezes, or spits, they project TB bacilli into the air that can be inhaled by another person. In 2003, someone in the world was being newly infected with the bacillus every second. Overall, 33 percent of the world's population currently is infected. And TB is killing more people every year; in fact, in 2003 someone was dying of TB every 15 seconds. Among the reasons for the rise in TB cases is that infected individuals are not completing the full course of antibiotics once they start to feel better. Not only does this behavior fail to cure the disease, it also helps generate multidrug-resistant TB (MDR-TB).

To address these issues, the World Health Organization (WHO) has developed the most widely accepted TB treatment program available today. DOTS (Direct Observation Treatment System) is a program to detect and cure TB. Once a patient with infectious TB is identified by a sputum smear, their treatment follows the DOTS strategy. A physician, community worker, or trained volunteer observes and records the patient swallowing four basic medications over a six- to eight-month period. A sputum smear is repeated after two months to check progress and again at

the end of treatment. DOTS appears to be very effective when it is used.

Take for example the following case. In the early 1990s, a powerful strain of drug-resistant TB emerged in New York City. Affecting hundreds of people in hospitals and prisons, the outbreak killed 80 percent of the infected patients. The city responded with DOTS, which seemed to work since the outbreak subsided.

However, in 1997 it was back, this time in South Carolina. A New York patient (not on DOTS) moved to South Carolina. His TB had lingered and he infected three family members in his new community. Soon, another six members of the community were sick with TB. However, these six individuals had not had contact with the family; in fact, they did not even know the family. Investigators from the Centers for Disease Control and Prevention (CDC) were called in to investigate. They learned that one family member had been in the hospital, where he was examined with a bronchoscope (a lighted tube extended into the air passageways). Unfortunately, the bronchoscope was not disinfected properly after the examination and was used to examine these other individuals.

Many stories have happy endings, but this is not one of them. Of the six patients infected with the drug-resistant strain of *M. tuberculosis*, two died from

TB and three died from other causes while battling TB. Only one recovered.

Thus, one sees the importance and effectiveness of DOTS. The WHO says that DOTS produces a 95 percent cure rate and thus prevents new infections. DOTS also prevents the development of MDR-TB by making sure TB patients take the full course of treatment. Since DOTS was introduced in 1995, 10 million infectious patients have been treated successfully. In China, there has been a 96 percent cure rate and in Peru a 91 percent cure rate for new cases of TB. Overall, WHO has set a goal of detecting 70 percent of new infectious TB cases and curing 85 percent of the detected cases. In 2000, 10 of the participating 155 countries had met this goal. Meeting the goal in all countries remains a formidable task, but one that offers a hopeful future.

Another organism we shall mention is *M. scrofulaceum*. This rod causes tuberculosis of the neck tissues, a disease commonly known as **scrofula** (from the Latin *scrophula* for "glandular swelling"). Scrofula is characterized by the growth of mycobacteria in the lymph nodes of the neck and substantial swelling of the tissues. Its prevalence in the Elizabethan era may have influenced people to wear the high-neck collars characteristic of that period. FIGURE 8.7 shows an example.

PNEUMOCOCCAL PNEUMONIA PRIMARILY AFFECTS THE ELDERLY

The term **pneumonia** refers to microbial disease of the bronchial tubes and lungs. A wide spectrum of organisms, including many different viruses and bacteria, may cause pneumonia. Over 80 percent of bacterial cases are due to *Streptococcus pneumoniae*, a

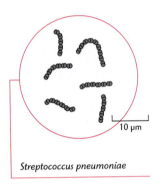

Streptococcus pneumoniae

gram-positive, encapsulated chain of diplococci traditionally known as the pneumococcus (**FIGURE 8.8**). The disease is commonly called **pneumococcal pneumonia**.

Pneumococcal pneumonia exists in all age groups, but the mortality rate is highest among the elderly and those with underlying medical conditions. There are more than 500,000 cases each year in the United States, resulting in approximately 40,000 deaths. *S. pneumoniae* is usually acquired by aerosolized droplets or contact, and the pneumococci exist in the respiratory tract of the majority of Americans. However, the natural resistance of the body is high, and disease usually does not develop until the defenses are compromised. Malnutrition, smoking, viral infections, and treatment with immune-suppressing drugs are typical conditions that may precede pneumonia.

Patients with pneumococcal pneumonia experience high fever, sharp chest pains, difficulty breathing, and rust-colored sputum. The color results from blood seeping into the alveolar sacs of the lung as bacteria multiply and cause the tissues to deteriorate. The involvement of an entire lobe of the lung is called **lobar pneumonia**. If both left and right lungs are involved, the condition is called **double pneumonia**. Scattered patches of infection in the respiratory passageways are referred to as **bronchopneumonia**.

The traditional drug of choice for pneumococcal pneumonia has been penicillin, with tetracycline and chloramphenicol as alternatives for people allergic to penicillin. To their chagrin, public health microbiologists have noted an increasing incidence of resistant strains of *S. pneumoniae*, and as early as 1981, the first bacterial strain resistant to all three antibiotics was reported. In such cases, erythromycin is useful.

Microbiologists have identified over 90 strains of *S. pneumoniae*, based on the presence of different capsule components. Unfortunately, recovery from one strain confers immunity to that strain alone. In 1983, the FDA licensed a capsuler polysaccharide vaccine for immunization to 23 strains of the organism. Since these strains are responsible for 87 percent of cases, there is hope that pneumococcal pneumonia may be controlled in high-risk patients.

PRIMARY ATYPICAL PNEUMONIA IS SPREAD BY CLOSE CONTACT

During the 1940s, a sudden increase in the number of cases of pneumonia prompted a search for a new infectious agent. At Harvard University, Monroe Eaton isolated a tiny agent from the respiratory tracts of patients and cultivated it

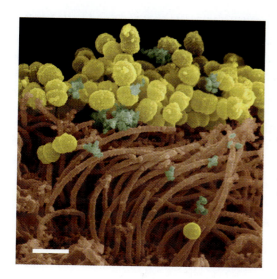

FIGURE 8.8

Streptococcus pneumoniae

False color scanning electron micrograph of *Streptococcus pneumoniae*, the cause of pneumococcal pneumonia. The gram-positive diplococci occur in short chains. Although found living harmlessly in the body, *S. pneumoniae* can cause dangerous pneumonia infections in the elderly or those with serious disease complications. (Bar = 5 μm.)

on media supplemented with blood (as shown in **FIGURE 8.9a**). The organism was subsequently named *Mycoplasma pneumoniae*, the Eaton agent. Its disease came to be known as **primary atypical pneumonia (PAP)**—"primary" because it occurs in previously healthy individuals (pneumococcal pneumonia is usually a secondary disease); "atypical" because the organism differs from the typical pneumococcus and because symptoms are unlike those in pneumococcal disease.

Mycoplasma pneumoniae is recognized as one of the smallest bacteria causing human disease. Mycoplasmas measure about 0.2 μm in size and are **pleomorphic**; that is, they assume a variety of shapes (**FIGURE 8.9b**). Because they have no cell wall, they have no Gram reaction or sensitivity to penicillin. *M. pneumoniae* is very fragile and does not survive for long outside the human or animal host. Therefore, it is maintained in nature by passage in droplets from host to host.

The symptoms of PAP resemble those of viral pneumonia (Chapter 13). The patient experiences fever, fatigue, and a characteristic dry, hacking cough. Research indicates that the organisms attach to and destroy the ciliated cells lining the respiratory tract.

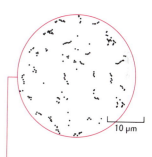

Mycoplasma pneumoniae

Pleomorphic:
occurring in a variety of shapes.

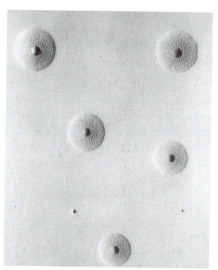

(a)

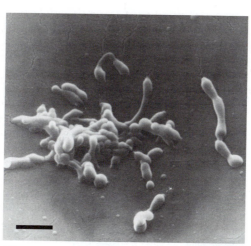

(b)

FIGURE 8.9

Mycoplasma pneumoniae

Two views of *Mycoplasma pneumoniae*, the agent of primary atypical pneumonia. (a) Colonies of *M. pneumoniae* on solid culture medium, showing the typical "fried egg" appearance. (b) A scanning electron micrograph of *M. pneumoniae*, demonstrating the pleomorphism exhibited by mycoplasmas. Note that the cells appear in multiple shapes, many in filamentous forms. (Bar = 2.5 μm.)

Blood invasion does not occur, and the disease is rarely fatal. Often it is called **walking pneumonia** (even though the term has no clinical significance). Epidemics are common where crowded conditions exist, such as in college dormitories, military bases, and urban ghettoes. Erythromycin and tetracycline are commonly used as treatments.

Research in the 1940s established that antibodies produced against *M. pneumoniae* agglutinate type O human red blood cells at 4°C but not at 37°C. This observation was used to develop the **cold agglutinin screening test** (**CAST**): A patient's serum is combined with red blood cells at cold temperatures, and the red cells are observed for clumping. Diagnosis is assisted also by isolation of the organism on blood agar and observation of a distinctive "fried-egg" colony appearance (Figure 8.9a).

Agglutinate:
an alternate expression for clump.

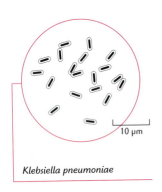

Klebsiella pneumoniae

KLEBSIELLA PNEUMONIA IS A HOSPITAL-ACQUIRED INFECTION

In 1882, Carl Friedländer isolated *Klebsiella pneumoniae*, an important cause of bacterial pneumonia. In the years thereafter, Friedländer's bacillus was related to about 5 percent of cases.

Klebsiella pneumoniae is a gram-negative, non-motile rod with a prominent capsule, as shown in **FIGURE 8.10**. The bacillus is acquired by droplets, and often it occurs naturally in the respiratory tracts of humans. ***Klebsiella* pneumonia** may be a primary disease or a secondary disease in alcoholics or people with impaired pulmonary function. As a primary pneumonia, it is characterized by sudden onset and a gelatinous reddish-brown sputum. The organisms grow over the lung surface and rapidly destroy the tissue, often causing death. In its secondary form, *Klebsiella* pneumonia occurs in already ill individuals and is a nosocomial, or hospital-acquired, disease spread by such routes as clothing, intravenous solutions, foods, and the hands of health-care workers.

SERRATIA PNEUMONIA ALSO IS A HOSPITAL-ACQUIRED INFECTION

For many decades, microbiologists considered *Serratia marcescens* a nonpathogenic bacillus and often used it as a test organism in their experiments (**MicroFocus 8.7** describes an example.) Today, they view this gram-negative, motile rod as a cause of

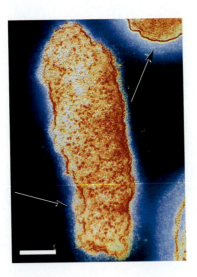

FIGURE 8.10

Klebsiella pneumoniae

False color transmission electron micrograph of *Klebsiella pneumoniae*, one of the agents of bacterial pneumonia. These gram-negative rods have a polysaccharide capsule (arrows) and are nonmotile. (Bar = 0.3 μm.)

respiratory disease in compromised patients. A patient may be predisposed to infection by such conditions as chronic illness, impaired immunity due to immunosuppressive therapy, radiation, or surgical treatments such as urinary catheterization, lumbar puncture, or blood transfusion.

Serratia **pneumonia** is accompanied by patches of bronchopneumonia and, in some cases, substantial tissue destruction in the lungs. The major clinical problem in treating the disease is resistance to antibiotic therapy, apparently due to R plasmids (Chapter 7), although gentamicin is quite reliable. The literature of the 1990s has also cited *Serratia marcescens* as a cause of eye infections, conjunctivitis, bone disease, arthritis, and meningitis. In addition, it is an agent of urinary tract diseases.

LEGIONNAIRES' DISEASE (LEGIONELLOSIS) HAS A MORTALITY RATE OF 15 TO 20 PERCENT

From July 21 to July 24, 1976, the Bellevue-Stratford Hotel in Philadelphia was the site of the 58th annual convention of Pennsylvania's chapter of the American Legion. Toward the end of the convention, 140 conventioneers and 72 other people in or near the hotel became ill with fever, coughing, and pneumonia. Eventually, 34 individuals died of the disease and its complications.

Initially some scientists suspected that the outbreak was the first wave of a swine flu epidemic predicted for that year, but as the weeks wore on, it became apparent that an unknown microorganism was responsible. By early December, the mystery had

MicroFocus 8.7

"KEEP IT SHORT, PLEASE!"

Defining, developing, and proving the germ theory of disease was one of the great triumphs of scientists in the late 1800s. Applying the theory to practical problems was another matter, however, because people were reluctant to change their ways. It would take some rather persuasive evidence to move them.

In the summer of 1904, influenza struck with terrible force among members of Britain's House of Commons. Soon the members began wondering aloud whether they should ventilate their crowded chamber. They decided to hire British bacteriologist Mervyn Henry Gordon to determine whether "germs" were being transferred through the air and whether ventilation would help the situation.

Gordon devised an ingenious (and, by today's standards, hazardous) experiment. He selected as his test organism *Serratia marcescens*, a bacterium that forms bright-red colonies in Petri dishes of nutrient agar. Gordon prepared a liquid suspension of the bacteria and gargled with it. (*S. marcescens* is now considered a pathogen in some individuals.) Gordon then stood in the chamber and delivered a two-hour oration consisting of selections from Shakespeare's *Julius Caesar* and *Henry V.* His audience was hundreds of open Petri dishes. The theory was simple: If bacteria were transferred during Gordon's long-winded speeches, then they would land on the agar plates and form red colonies.

And land they did. After several days, red colonies appeared on plates placed right in front of Gordon, as well as in distant reaches of the chamber. The members were impressed. They proposed a more constant flow of fresh air to the

chamber, as well as shorter speeches. No one was about to object to either solution, especially the latter.

deepened to the point that a writer from the respected journal *Science* described the "investigation that failed." Finally, in January 1977, CDC investigators announced the isolation of a bacterium from the lung tissue of one of the patients. The organism appeared responsible not only for the Legionnaires' disease (as the disease had come to be known), but also for a number of other unresolved pneumonia-like diseases. The writer from *Science* swallowed hard and acknowledged, "the investigation may be successful after all."

In contemporary microbiology, **Legionnaires' disease** (also called **legionellosis**) is known to be caused by the organism isolated in 1977, a gram-negative, motile rod named *Legionella pneumophila* (**FIGURE 8.11**). The bacillus exists where water collects, and apparently it becomes airborne in wind gusts and breezes. Cooling towers, industrial air-conditioning units, lakes, stagnant pools, and puddles of water have been identified as sources of bacteria. Humans breathe the contaminated aerosolized droplets into the respiratory tract, and disease develops within a week (recall the chapter opening).

The symptoms of Legionnaires' disease include fever, a dry cough with little sputum, and some diarrhea and vomiting. In addition, chest X rays show a characteristic pattern of lung involvement, and pneumonia is the most dangerous effect of the disease. Erythromycin is effective for treatment, and person-to-person transmission is uncommon.

After *L. pneumophila* was isolated in early 1977, microbiologists found that the organism was responsible for many epidemics of pneumonia in previous years. One such outbreak, called **Pontiac fever**, is an influenza-like illness that took place in Michigan in 1968. In the years after 1977, reports of Legionnaires' disease occurred

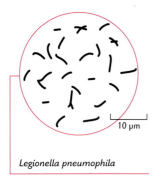

Legionella pneumophila

10 μm

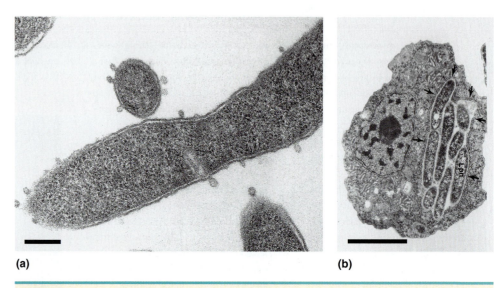

(a)   (b)

FIGURE 8.11

Legionella pneumophila

Two views of *Legionella pneumophila*, the agent of Legionnaires' disease. (a) A transmission electron micrograph of *L. pneumophila*. Both inner (cytoplasmic) and outer membranes can be seen, as well as several evaginations (blebs) at the outer membrane. (Bar = 0.1 μm.) (b) A transmission electron micrograph of the protozoan *Hartmanella vermiformis* infected with several cells of *L. pneumophila*. The arrows mark the boundary membrane of the vacuole enclosing the bacilli. Legionellae are now known to sequester themselves in protozoa like these and thereby escape environmental extremes. (Bar = 1 μm.)

throughout the world, and invariably, water was involved. For example, 66 cases in Sweden were linked to water collecting on the rooftop of a shopping center, and 23 cases in Italy were traced to well water used for bathing. In 1990 in South Dakota, 26 cases and 10 deaths were linked to the water in showers at a particular hospital, and in 1994, a notable outbreak occurred on a cruise ship (FIGURE 8.12).

Since the discovery of *Legionella* in water, microbiologists have been perplexed as to how such fastidious bacilli could survive in an aquatic environment that is often hostile. An answer was suggested by studies showing that the bacilli could live and grow within the protective confines of waterborne protozoa. In the South

FIGURE 8.12

Legionnaires' Disease on a Cruise Ship

This outbreak demonstrates how warm water can provide an incubator for infectious microorganisms.

TEXTBOOK CASES

1. On June 25, 1994, the cruise ship Horizon set sail from New York City with hundreds of passengers bound for Bermuda. Among the attractions that awaited them on deck were a set of Jacuzzi whirlpool baths. Passengers found them to be an exhilarating way to relax, the warm spray covering their faces.

2. The ship docked at Hamilton, Bermuda, where passengers enjoyed a three-day visit. On the return trip, the whirlpools were busy once again. The cruise ended on July 2, 1994.

3. Beginning on July 15, the New Jersey State Department of Health began receiving reports of passengers who had coughing, fever, and pneumonia. Some cases were quite severe, and some patients required hospitalization.

4. Cultures of respiratory secretions taken from the patients yielded *Legionella pneumophila*, the agent of Legionnaires' disease. Antibody tests confirmed the diagnosis.

5. Health investigators were drawn to the whirlpool baths, since most ill patients had used them. In the sand filters, they located the same strain of *L. pneumophila* as in the patients. Breathing the spray had apparently transmitted the bacteria.

Dakota episode cited on the previous page, amoebas were found in abundance in the hospital's water supply. The water was treated by heating and adding chlorine to help quell the spread of legionellae.

Q FEVER RESULTS FROM BREATHING THE AIRBORNE PARTICLES

Q fever is one of several diseases caused by a group of bacteria known as rickettsiae (sing., rickettsia). Once regarded as an intermediate type of organism between bacteria and viruses, the rickettsiae now are recognized as "small bacteria" below the resolving power of the normal light microscope and measuring about 0.45 µm. They have no flagella, pili, or capsules, and they reproduce by binary fission. With a few exceptions, rickettsiae are cultivated in the laboratory only within living tissue cultures such as fertilized eggs or animals. Arthropods are usually involved in their transmission.

The term **Q fever** was first used by E. H. Derrick in 1937 to describe an illness that broke out among workers at a meat-packing plant in Australia. The "Q" may have been derived either from "query," meaning unknown, or from Queensland, the province in which the disease occurred. The rickettsia that causes Q fever is named *Coxiella burnetii* (**FIGURE 8.13**), after H. R. Cox, who isolated it in Montana in 1935, and Frank Macfarlane Burnet, who studied its properties in the late 1930s.

Q fever is prevalent worldwide among livestock, especially in dairy cows, goats, and sheep, and being reservoirs for the bacterium, outbreaks may occur wherever these animals are raised, housed, or transported. Transmission among livestock

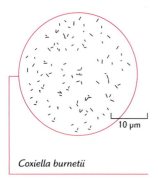

Coxiella burnetii

10 µm

MicroInquiry 8

INFECTIOUS DISEASE IDENTIFICATION

Below are several descriptions of infectious diseases based on material presented in this chapter. Read the case history and then answer the questions posed. Answers can be found in Appendix E.

Case 1: The patient, a 33-year-old male, arrives at a local health clinic complaining that he has felt "out of sorts," has a fever, and has lost over 10 percent of his body weight in the last month. He also has a cough that produced a rust-colored sputum. The patient is referred to a local hospital where a chest X ray and sputum sample are taken. Upon further questioning, the patient admits to having tested HIV-positive about one year ago. A tuberculin test also is ordered. Further questioning of the patient reveals that he had been living with two roommates

for two years after having lived for eight years with another roommate who had tested positive for tuberculosis about 6 months before the onset of the patient's symptoms. The sputum samples are negative for the two roommates, but both have a positive tuberculin test result. Both test negative for HIV.

8.1a. Why was a chest X ray ordered?

8.1b. Why was a sputum sample taken?

8.1c. What should a positive tuberculin skin test look like?

8.1d. What does such a test result indicate?

8.1e. Based on the symptoms and laboratory results, what infectious disease does the patient suffer? What is the agent?

8.1f. How did the patient contract the disease?

8.1g. Why is the infectious agent more virulent in HIV-infected patients?

Following diagnosis, the patient was placed on isoniazid (INH) for 12 months.

8.1h. Most treatment procedures call for a 6 to 8 month program. Why was the patient placed in INH for an extended period?

Case 2: The parents bring their 2-year-old daughter to the hospital emergency room. She appears to have an upper respiratory infection that her parents think started about one week previously. They say that their daughter had lost her appetite and appeared especially sleepy about four days ago. She complained of a sore throat. Examination indicates that she has a moderate fever but no chest congestion. Throat

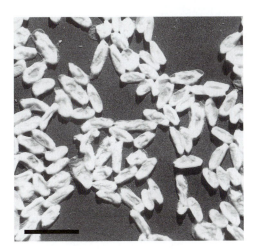

FIGURE 8.13

Coxiella burnetii

An electron micrograph of *Coxiella burnetii*, the agent of Q fever. Note the oval-shaped rods of the organism. (Bar = 1 μm.)

and to humans occurs primarily by inhaling infected dust particles or handling infected animals. In addition, humans may acquire the disease by consuming unpasteurized milk infected with *C. burnetii* or milk that has been improperly pasteurized. Patients experience severe headache, high fever, a dry cough, and occasionally, lesions on the lung surface. The mortality rate is low, and treatment with doxycline is effective for chronic infections. A vaccine is available for workers in high-risk environments.

and blood cultures are taken and their daughter is put on penicillin.

On returning to the hospital three days later, her throat culture shows gram-positive rods. The blood culture is negative. Her parents remark that this morning their daughter started breathing harder. Examination of her pharynx indicates the presence of a leathery membrane. On questioning the parents, it is discovered that the child has had no immunizations. The child is admitted to the hospital and treatment immediately begun.

8.2a. What infectious agent does the child have?

8.2b. The medical staff is concerned about the seriousness of the disease. How can the presence of bacteria in the throat and not in the blood cause such concern?

8.2c. How could this disease have been prevented?

8.2d. What is the prescribed treatment protocol?

Case 3: A 63-year-old retired steel worker who is a heavy smoker and alcoholic comes to the emergency room complaining of having a fever and shortness of breath for the last two days. This morning he had developed a cough with rust-colored sputum. A chest X ray is taken and shows involvement in the left lower lobe of the lungs. A sputum sample and blood sample are taken for Gram staining, and the patient is checked into the hospital where he is placed on penicillin. Two days later the patient is feeling much improved. His physician tells him that both sputum and blood cultures indicate the presence of gram-positive diplococci. The

patient is released from the hospital and recovers completely after finishing antibiotic therapy.

8.3a. What organism is responsible for the patient's infection?

8.3b. Why is this patient at high risk for becoming infected with the bacterium?

8.3c. How could the patient likely have prevented contracting the disease?

8.3d. What bacterial factors are responsible for virulence?

8.3e. If plated on blood agar, what type of hemolytic reaction should be seen?

PSITTACOSIS AND CHLAMYDIAL PNEUMONIA ARE AIRBORNE DISEASES

Psittacosis and chlamydial pneumonia are caused by a different species of chlamydia (pl., chlamydiae). Chlamydiae are among the smallest bacteria (0.25 µm) related to disease in humans. They are cultivated only in living human cells and have a complex life cycle that includes a number of different forms. Three species of chlamydiae are recognized: *Chlamydia trachomatis*, the cause of trachoma and two sexually transmitted diseases (Chapter 11); *Chlamydia psittaci*, the cause of psittacosis; and *Chlamydia pneumoniae*, the cause of chlamydial pneumonia.

Psittacosis affects parrots, parakeets, canaries, and other members of the psittacine family of birds (*psittakos* is the Greek word for "parrot"). The disease also occurs in pigeons, chickens, turkeys, and seagulls, and some microbiologists prefer to call it **ornithosis** to reflect the more widespread occurrence (*orni-* is from the Greek for "bird"). Humans acquire *C. psittaci* by inhaling airborne dust or dried droppings of infected birds. Sometimes the disease is transmitted by a bite from a bird or via the respiratory droplets from another human. The symptoms of psitta-

MicroFocus 8.8

THE BURDEN OF SUSPECTS

Here's one to startle the senses: the "germ theory of cardiovascular diseases."

And why not, suggest researchers? Coronary artery disease (CAD) remains a leading cause of death even with current medications and laser, angioplasty, or other innovative devices that are available. Perhaps antimicrobial drugs would help.

But, antibiotics for what? Why not start, they say, with *Chlamydia pneumoniae*? In several studies, *C. pneumoniae* has been associated with heart attack patients and in numerous males with CAD. Researchers suggest the bacteria could injure the blood vessels, triggering an inflammatory response where immune system cells attack the vessel walls and induce the formation of large, fibrous lesions, or plaques. When pieces of plaque break free, they start blood clots that clog the arteries and cause heart attacks (the condition is known as atherosclerosis). Also, inoculation of mice and rabbits with *C. pneumoniae* accelerates atherosclerosis, even more so if the animals are fed cholesterol-enriched diets.

A second suspect is cytomegalovirus (CMV), an animal virus of the herpes family. In fact, more than 70 percent of the CAD population is CMV-positive. Scientists have known that people infected with CMV respond very poorly to arterial cleaning, the technique of angioplasty, and the arteries quickly close up. CMV also accelerates atherosclerosis in mice and rats.

A weaker association with CAD is seen with *Streptococcus sanguis*, an agent of periodontal disease. Microbiologists believe that poor oral hygiene and bleeding gums give the bacterium access to the blood, where it produces blood-clotting proteins. It's no coincidence, they maintain, that people with unhealthy teeth and gums tend to have more heart trouble. Unfortunately, such groups also have other lifestyle factors that confound the association.

The final suspect is *Helicobacter pylori*, the cause of most peptic ulcers. Italian scientists have linked a virulent strain of *H. pylori* with increased incidence of heart disease. However, other trials do not find the bacterium as sig-

nificant in predicting CAD as the other agents.

To account for these observations and research findings, many medical researchers and microbiologists are proposing the concept of a total pathogen burden. First proposed in 2000, the concept suggests that while a single infectious agent may only minimally increase the risk of atherosclerosis, the burden of several agents could greatly increase the risk. In fact, several recent studies suggest that exposure to several microbial suspects does correlate with increased risk for CAD and, in established CAD cases, incident death. The studies propose that *C. pneumoniae* and CMV probably play a direct role in atherosclerosis while other agents, like *S. sanguis* and *H. pylori*, contribute indirectly to inflammation. The concept does not prove causality, but suggests avenues for further study, including the possible use of antibiotics, antivirals, or vaccines to prevent or cure CAD. So, before you run out to buy antibiotics, stick with the more likely anti-inflammatory factors of diet, exercise, and not smoking.

cosis resemble those of primary atypical pneumonia or influenza. Fever is accompanied by headaches, dry cough, and scattered patches of lung infection. Tetracycline is commonly used in therapy.

In 2001, the CDC reported 25 cases based on reports from less than 40 states. A notable series of cases occurred in 1992 after individuals came in contact with infected parakeets and cockatiels (FIGURE 8.14). The incidence of psittacosis in the United States remains low, partly because federal law requires a 30-day quarantine for imported psittacine birds. In addition, birds are given water treated with chlortetracycline hydrochloride (CTC) and CTC-impregnated feed.

FIGURE 8.14

Two Outbreaks of Psittacosis

These outbreaks of psittacosis were traced to birds distributed by a single supplier. This incident occurred in 1992. Coincidence linked the disease to the distributor's birds, even though there was no evidence of widespread disease among his stock.

TEXTBOOK CASES

1. On February 13, a distributor in Mississippi shipped a supply of parakeets and cockatiels to retail pet stores in Massachusetts and Tennessee. The birds had been supplied to him by domestic breeders.

2. Four days later, on February 17, a man purchased one of the parakeets from the Massachusetts store. The man noted that the bird became very "tired looking" some days after he brought it home.

3. On February 19, a family from Tennessee purchased a cockatiel sent by the distributor. The bird was taken home and kept in a bird cage where all family members could enjoy it. Some days later, family members observed that the bird was "irritable."

4. On March 1, the man from Massachusetts was hospitalized with fever, sore throat, and pneumonia. Two other members of his family were also sick.

5. Also on March 1, six members of the Tennessee family were experiencing fever, cough, and sore throat.

6. In both families a diagnosis of psittacosis was made. All recovered.

Chlamydial pneumonia is caused by *Chlamydia pneumoniae* (FIGURE 8.15), an organism formerly called **TWAR** because two of the original isolates of the organism were specified TW-183 and AR-39. The chlamydia is transmitted by respiratory droplets and causes a mild walking pneumonia, principally in young adults and college students. The disease is clinically similar to psittacosis and primary atypical pneumonia and is characterized by fever, headache, nonproductive cough, and infection of the lower lobe of the lung. Treatment with tetracycline or erythromycin hastens recovery from the infection. The organism was first observed in Seattle, Washington, in 1983 and now is believed to infect many thousands annually in the United States. Its relationship to cardiovascular disease is explored in MicroFocus 8.8.

The airborne bacterial diseases of the lower respiratory tract are summarized in TABLE 8.3. MicroInquiry 8 presents several case studies concerning some of the bacterial diseases discussed in this chapter.

TABLE 8.3

A Summary of Airborne Bacterial Diseases of the Lower Respiratory Tract

DISEASE	CAUSATIVE AGENT	DESCRIPTION OF AGENT	ORGANS AFFECTED	CHARACTERISTIC SIGNS
Tuberculosis	*Mycobacterium tuberculosis*	Acid-fast rod	Lungs, bones, other organs	Tubercle
Pneumococcal pneumonia	*Streptococcus pneumoniae*	Gram-positive encapsulated diplococcus in chains	Lungs	Rust-colored sputum
Primary atypical pneumonia	*Mycoplasma pneumoniae*	No cell wall Pleomorphic	Lungs	Dry cough
Klebsiella pneumonia	*Klebsiella pneumoniae*	Gram-negative encapsulated rod	Lungs	Pneumonia
Serratia pneumonia	*Serratia marcescens*	Gram-negative rod	Lungs	Pneumonia
Legionnaires' disease (legionellosis)	*Legionella pneumophila*	Gram-negative rod	Lungs	Pneumonia
Q fever	*Coxiella burnetii*	Rickettsia	Lungs	Influenza-like symptoms
Psittacosis	*Chlamydia psittaci*	Chlamydia	Lungs	Influenza-like symptoms
Chlamydial pneumonia	*Chlamydia pneumoniae*	Chlamydia	Lungs	Influenza-like symptoms

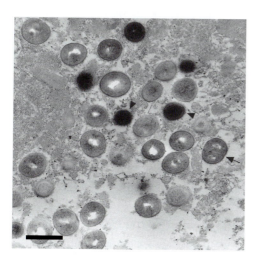

FIGURE 8.15

Chlamydia pneumoniae

An electron micrograph of *Chlamydia pneumoniae* in a glial cell from a patient suffering from Alzheimer's disease. Elementary bodies (a stage in the reproductive cycle of chlamydiae) are indicated by arrowheads, while reticulate bodies (another stage) are the less dense organisms. The arrow points to a chlamydia in binary fission. (Bar = 1 μm.)

TOXIN INVOLVED	TREATMENT ADMINISTERED	IMMUNIZATION AVAILABLE	COMMENT
None	Isoniazid Rifampin Pyrazinamide	Bacille Calmette Guérin (BCG)	Extended treatment necessary Diagnosis by tuberculin test Related to AIDS
Not established	Penicillin Tetracycline Erythromycin	Polysaccharide vaccine to 23 strains	Over 90 strains identified Natural resistance high Deterioration of alveoli
Not established	Erythromycin Tetracycline	None	Called walking pneumonia Diagnosis by CAST
Not established	Various antibiotics	None	Common in people with impaired pulmonary function
Not established	Gentamicin	None	Common nosocomial disease Resistance to antibiotic therapy
Not established	Erythromycin	None	Associated with airborne water droplets
Not established	Doxycline	Vaccine for high-risk workers	Reservoirs include dairy cows, goats, sheep Associated with raw milk
Not established	Tetracycline	None	Occurs in parrots and parrot-like birds
Not established	Tetracycline Erythromycin	None	Young adults affected

Note to the Student

In this text, we are surveying a broad group of diseases caused by an equally broad group of microorganisms. You will note, however, that other microorganisms generally do not induce the life-threatening situations posed by bacterial diseases. To be sure, there are nonbacterial diseases such as malaria, AIDS, and rabies that are deadly, but bacterial diseases are the ones that have ravaged humans for centuries. Tuberculosis, diphtheria, meningitis, and pneumonia are examples in this chapter. Typhoid fever, syphilis, cholera, and plague are examples encountered in other chapters.

Why do we not fear these bacterial diseases anymore in the United States, and why can we blithely discuss them as if they were occurring on another planet? How often, for example, do newspaper headlines trumpet diphtheria or pertussis epidemics?

The fact of the matter is that serious bacterial diseases do break out occasionally, especially in other parts of the world, and some of us may be affected by them. But, by and large we are insulated from widespread epidemics. Public health agencies understand bacterial diseases much better than in the past and can deal with them effectively by such means as sanitation measures, good hospital care, and a multitude of antibiotics. Perhaps this is a fitting place to pause and reflect on how much medical research has contributed to the quality of life. We might begin by trying to imagine what it might be like to fear infectious disease as our ancestors once did.

Summary of Key Concepts

This chapter surveyed a number of bacterial diseases of the respiratory tract, beginning with diseases of the upper tract and concluding with diseases of the lower tract and lungs. In many cases, the diseases are not confined to the respiratory organs, and a spread to distant organs can take place.

8.1 DISEASES OF THE UPPER RESPIRATORY TRACT

- **Streptococcal Diseases Can Be Mild to Severe.**
Streptococcal diseases of the upper respiratory tract are caused by *Streptococcus pyogenes*, a gram-positive, encapsulated coccus. Symptoms of streptococcal pharyngitis (strep throat) include sore throat, fever, cough, and swollen lymph nodes and tonsils. Toxins contribute to tissue erosion. The disease is transmitted by airborne droplets from infected individuals. Scarlet fever symptoms include fever, cough, swollen lymph nodes and tonsils, and a skin rash. Toxins contribute to the rash through damage to capillaries.

Treatment of streptococcal diseases is successful with penicillin and erythromycin. Complications include rheumatic fever that can develop into rheumatic heart disease. *S. pyogenes* infections in other parts of the body can cause necrotizing fasciitis.

- **Diphtheria Is a Deadly, Toxin-Caused Disease.**
Diphtheria is caused by those strains of *Corynebacterium diphtheriae* that harbor a prophage that codes for the diphtheria exotoxin. As dead tissue accumulates, a pseudomembrane forms in the throat or nasopharynx. Diphtheria is transmitted by respiratory droplets of infected persons. Treatment requires antibiotics (penicillin or erythromycin) and antitoxins. Prevention is accomplished by vaccination with the diphtheria toxoid that is part of the DTaP vaccine.

- **Pertussis (Whooping Cough) Is Highly Contagious.** *Bordetella pertussis*, a gram-negative rod, produces a toxin that paralyzes the epithelial cells of the upper respiratory tract, causing mucus congestion and a persistent cough. Symptoms of pertussis include fever, increasingly severe cough, and toxin-mediated paralysis of the ciliated epithelium. A narrowing of respiratory passages leads to the characteristic "whoop." Transmission of the bacillus is through respiratory droplets. Elimination of the bacteria can be successful with erythromycin treatment (does not shorten the illness) and prevention involves vaccination with the DTaP vaccine.

■ **Meningococcal Meningitis Can Be Life Threatening.** Meningococcal meningitis is an inflammation of the membranes surrounding the brain and spinal cord. The infection is caused by *Neisseria meningitidis*, a gram-negative diplococcus. The disease is accompanied by headache, neck ache, and lower backache. Serious problems arise after the bacteria pass through the blood and involve the meninges. Bacterial toxins can overwhelm the body causing death in a few hours. The bacteria are spread by airborne droplets. Treatment requires early diagnosis and penicillin, cefotaxime, or ceftriaxone antibiotics.

■ *Haemophilus* **Meningitis Had Been Prevalent in Preschool Children.** As a complication of a previous disease, *Haemophilus influenzae* can cause respiratory tract infections and meningitis in preschool children. Symptoms include a stiff neck, headache, and neurological involvement. Ceftriaxone or cefotaxime are the choices for antibiotic treatment. The Hib vaccine is available for prevention.

8.2 DISEASES OF THE LOWER RESPIRATORY TRACT

■ **Tuberculosis Is Primarily a Disease of the Lungs.** Tuberculosis (TB) is caused by *Mycobacterium tuberculosis*, an acid-fast bacillus. Inhalation into the lungs of bacilli from an infected person leads to the disease, which is characterized by a chronic cough, chest pain, and high fever. TB can be restricted to tubercle formation, or the bacteria can spread to other parts of the body. Treatment with isoniazid, pyrazinamide, and rifampin requires six to nine months. A BCG vaccine is available where the disease shows significant mortality and morbidity.

■ **Pneumococcal Pneumonia Primarily Affects the Elderly.** *Streptococcus pneumoniae*, a gram-positive diplococcus, is responsible for pneumococcal pneumonia. Patients experience high fever, sharp chest pains, breathing difficulties, and rust-colored sputum. It is acquired by aerosolized droplets from or contact with an infected person. The elderly and people with compromised immune systems are most susceptible to contracting the disease. Penicillin or erythromycin is the antibiotic of choice. A pneumococcal vaccine is available.

■ **Primary Atypical Pneumonia Is Spread by Close Contact.** *Mycoplasma pneumoniae* is responsible for primary atypical pneumonia (walking pneumonia). This pleomorphic bacterium generates symptoms similar to viral pneumonias. Patients have a fever and a dry, hacking cough. It is spread by droplets between individuals in close contact. Erythromycin or tetracycline is a useful antibiotic.

■ *Klebsiella* **Pneumonia Is a Hospital-Acquired Infection.** Another form of pneumonia is caused by *Klebsiella pneumoniae*, a gram-negative rod with a very prominent capsule. The primary disease involves sudden onset and reddish-brown sputum. It is community acquired through aerosolized droplets.

■ *Serratia* **Pneumonia Also Is a Hospital-Acquired Infection.** The gram-negative rods of *Serratia marcescens* cause disease in compromised patients, so usually it is a hospital-acquired infection. Substantial lung tissue can be destroyed. Gentamicin therapy is effective.

■ **Legionnaires' Disease (Legionellosis) Has a Mortality Rate of 15 to 20 Percent.** Legionnaires' disease is caused by the gram-negative rod, *Legionella pneumophila*. Symptoms include fever, dry cough with little sputum, and diarrhea and vomiting. The bacterium collects in water and can be transmitted through the air to contaminate other water sources or water-containing equipment. Aerosolized droplets can be transmitted to humans from such contaminated equipment. Person-to-person transmission is rare. Erythromycin treatment is effective.

■ **Q Fever Results from Breathing the Airborne Particles.** *Coxiella burnetii* is a member of the rickettsiae and the causative agent for Q fever. Reservoirs include dairy cows, goats, and sheep. Symptoms include severe headache, high fever, and dry cough. Transmission is through inhalation of contaminated dust particles or consuming unpasteurized or improperly pasteurized milk containing *C. burnetii*. Doxycline is the antibiotic of choice.

■ **Psittacosis and Chlamydial Pneumonia Are Airborne Diseases.** Psittacosis is a lower respiratory disease caused by *Chlamydia psittaci*. Symptoms include fever, headache, and dry cough. Humans acquire the disease by inhaling contaminated dust or dried droppings from infected birds (parrots, canaries, parakeets). Antibiotic therapy is most promising with tetracycline. Chlamydial pneumonia is caused by *Chlamydia pneumoniae*. The disease is characterized by fever, headache, and nonproductive cough. It is transmitted by respiratory droplets from infected individuals. Infections are treated with tetracycline or erythromycin.

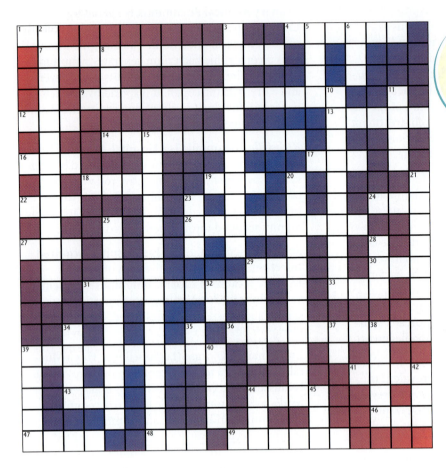

The main subject of this chapter has been respiratory diseases caused by bacteria. To test your recall of these diseases, fill in the following crossword puzzle. The answers to the puzzle are in Appendix D.

■ ACROSS

1. Scarlet fever is caused by a species of streptococc_____
4. Respiratory secretions sometimes containing blood
7. Caused by a species of *Mycobacterium*
9. Genus of acid-fast rods
12. Used for immunization against two respiratory diseases
13. Bacteria can _____ the body by respiratory droplets
14. Not visible with the light microscope
16. Shape of *Legionella* bacterium
17. Diphtheria immunization contains a tox _____

Questions for Thought and Discussion

Answers to selected questions can be found in Appendix C.

1. One of the remarkable public health stories of the ten-year period between 1986 and 1996 was the virtual elimination of *Haemophilus* meningitis as a concern to doctors and parents. Indeed, at the beginning of the period there were 18,000 cases annually in the United States, but in 1996, only 254 cases were reported. What factors probably contributed to the decline of the disease?

2. In 1997, in a widely publicized story, baseball pitcher Jason Isringhausen was diagnosed with tuberculosis. Doctors immediately put him on a regimen of four drugs. Which drugs do you suspect were prescribed? Why were four drugs needed? And why did the case generate so much press? (Postscript: Isringhausen was back in uniform for the 1999 season.)

3. A bacteriophage is responsible for the ability of the diphtheria bacillus to produce the toxin that leads to disease. Do you believe that having the virus is advantageous to the bacterium? Why or why not?

4. In New York City, in lower Manhattan, stands a notable building between Water Street and Franklin D. Roosevelt Drive. The building has strange rounded edges that make it appear like a weird circular planter. Constructed in 1901, the building was a hospital for immigrants, especially those with tuberculosis. Can you guess what the shape of the building had to do with the disease?

5. At present there is no licensed vaccine for the prevention of any streptococcal diseases, even though these are among the most commonly experienced bacterial diseases in the United States. Can you postulate why a vaccine, especially one composed of killed streptococci, might pose a threat to health?

18. Former name for *Chlamydia pneumoniae*
19. Incidence of diphtheria in the United States
22. Immunization to diphtheria is rendered using a _____-oid
24. The diphtheria toxin is a _____-tein
26. A species causes a form of meningitis
27. Used for immunization against tuberculosis
29. Global health group (abbr) concerned with infectious disease
30. In the CAST test, red blood cells agglutin_____
31. Species cause psittacosis and pneumonia
33. Streptococci can cause a blood disease called _____-ticemia
36. Transmitted by airborne droplets of water
39. Gram-negative rod that causes whooping cough
41. Certain strains of mycobacteria resist _____-biotics
43. Abbreviation for an acid-fast rod typically found in AIDS patients
44. Accompanies meningococcal meningitis

46. The counterstain in the Gram stain technique is _____ranine
47. Abrasions of the _____ permit streptococci to enter
48. Spreading tuberculosis is said to be _____-iary
49. *Haemophilus* meningitis is not common in _____

■ DOWN
2. Alpha, beta, and gamma hemolytic forms
3. Disease of parrots, parakeets, and canaries, as well as humans
5. *S. pneumoniae* is often called the _____-mococcus
6. Psittacosis may be carried by exo_____ birds
8. Cases of legionellosis are treated with _____-thromycin
10. May be due to *Neisseria* or *Haemophilus*
11. *Mycoplasma* species have no _____ wall
15. Causes pseudomembranes to form
20. Small gram-negative rod involved in meningitis in children

21. Rickettsia that causes Q fever
23. Drug (abbr) used to treat tuberculosis
25. Due to a species of *Corynebacterium*
28. A spinal _____ may be necessary to locate meningitis organisms
29. A symptom of meningitis may be the _____-house-Friderichsen syndrome
32. Q fever occurs in _____-mestic animals
34. The agent of pneumococcal pneumonia is _____-positive
35. Pneumonia may affect the termin_____ air sacs
37. *Klebsiella* may cause a _____-socomial infection
38. Organs affected by *Mycobacterium* species
39. Animals in which *C. psittaci* infects
40. An _____-fast rod is involved in cases of tuberculosis
42. A bout of _____-luenza may lead to streptococcal disease of the throat
44. Color of the pigment formed by *Serratia marcescens*
45. How a person feels when the fever is present in strep throat

6. It is a paradox of success that technology gives us appliances for better living but often provides breeding grounds for microorganisms. For example, *Legionella* probably lived undisturbed for millennia in ponds and lakes; but with the rise of air conditioning, powerful vents swept air over reservoirs of water and picked up droplets of *Legionella* to disperse to unsuspecting people who breathed the air. How many other examples of "technology" can you relate to the spreading problem of Legionnaires' disease?

7. Bacteria are generally designated gram-positive or gram-negative, but you may have noted that small bacteria such as rickettsiae and chlamydiae do not have this designation. Why do you think this is so? Also, why do you suppose the designation is lacking for members of the genus *Mycobacterium*?

8. It was February 1987. The patient was admitted to the hospital with high fever and a respiratory infection. Pneumococci and streptococci were eliminated as causes. Penicillin was ineffective. The most

unusual sign was a continually dropping count of red blood cells. Can you guess the final diagnosis?

9. "Be sure to dress warmly when you go out or you'll catch pneumonia." This precaution, or something like it, is familiar to almost every child. In what respect is it wrong? Why is it right?

10. In a report appearing in 1997, epidemiologists noted in children a rising incidence of otitis media (middle ear infections). They attributed the increase, in part, to the larger numbers of children in day care. How are these factors related?

11. Each year, a group of "mushers" gathers in Nome, Alaska, for a 770-mile dogsled tour to Nenana, an inland city. The two-week expedition is called the Serum 25. It commemorates the 1925 run, during which life-saving serum was delivered by dogsled to quell a diphtheria epidemic. At that time there were no antibiotics, but "serum" was then widely used. What was in the serum that was so valuable?

12. Now that *Haemophilus influenzae* is no longer considered to be the agent of influenza, would you agree that a name change is in order? If so, what name might you suggest?

13. One of the major world health stories of 1995 was the outbreak of diphtheria in the New Independent and Baltic States of the former Soviet Union. What factors might have contributed to this international public health emergency, and what do you think was the plan to help quell the spread of the diphtheria?

14. In this chapter we have encountered organisms that are commonly named for their discoverers. Examples are Friedländer's bacillus, Pfeiffer's bacillus, and the Bordet-Gengou bacillus. However, certain organisms, such as *Legionella* and *Chlamydia* are not often referred to by the names of their discoverers. Can you postulate why?

15. How many professions can you name in which workers might be exposed to the organism of Q fever? Why is it likely that many more cases have occurred than have been diagnosed?

16. In 1997, the incidence of pneumonia was recorded in various states of the United States. Massachusetts had a very high rate, but so did Georgia. North Dakota had the third lowest rate, Florida the lowest. The highest rate was in California. Suppose you were an epidemiologist. What explanation(s) might you offer for these observations?

17. In 1996, meningococcal meningitis swept through Nigeria (on Africa's west coast). Over 150,000 cases were reported, and deaths were estimated at 15,000. The epidemic arrived with the arid season when the air in Nigeria becomes exceptionally dry and dusty. It left with the rainy season. How do you think the climate related to the epidemic?

18. The CDC reports that an estimated 40,000 people in the United States die annually from pneumonia due to *Streptococcus pneumoniae*. Despite this high figure, only 30 percent of older adults who could benefit from the pneumonia vaccine are vaccinated (compared to over 50 percent who receive an influenza vaccine yearly). Suppose you were the epidemiologist in charge of bringing the pneumonia vaccine to a greater percentage of Americans. How would you proceed?

19. Manitou Springs is a small town at the foot of Pike's Peak in Colorado. The town is about 7,000 feet above sea level. Early in the twentieth century, people came to the town to drink from its mineral springs because the waters apparently helped cure tuberculosis. Many years later, scientists concluded there was no particular benefit to drinking the water. And yet, many patients had returned to good health after their visit to the town. How was this possible?

20. In 1998, Primary Children's Hospital in Salt Lake City reported a dramatic increase in the number of rheumatic fever cases. Doctors were alerted to start monitoring sore throats more carefully. Why do you suppose this prevention method was recommended?

http://microbiology.jbpub.com

The site features **eLearning,** an on-line review area that provides quizzes and other tools to help you study for your class. You can also follow useful links for in-depth information, or just find out the latest microbiology news.

9

Foodborne and Waterborne Bacterial Diseases

Scramble or gamble.

—A CDC official succinctly advising consumers
how to avoid *Salmonella* infection from eggs

TWENTY-FIVE MILLION POUNDS? You want us to recall twenty-five *million* pounds of hamburger meat? Is this a joke?"

It was no joke, assured the inspectors from the United States Department of Agriculture (USDA). They were standing in the office of Hudson Foods, Inc. in Columbus, Nebraska. And they were demanding a recall of the company's total production between June 5 and August 12, 1998. That amounted to 25 million pounds of hamburger meat.

No one is quite sure exactly who the first patient was, but one of the first was a young supermarket worker who reported to the emergency room at the St. Mary Corwin Hospital in Pueblo, Colorado, on July 15. The young man had bloody diarrhea and painful abdominal cramps. The doctor took a stool sample, gave him some medication, and sent him home. Then she sent the stool sample to the laboratory for routine testing. A day later, the results were ready: The patient was infected with *E. coli* O157:H7 (FIGURE 9.1).

City and county health departments routinely employ a communicable disease specialist who tracks outbreaks and reports them to the Centers for Disease Control and Prevention (CDC). That day, the person on duty in Pueblo was Sandra Gallegos. Gallegos recognized the *E. coli* strain; it was associated with extensive blood infection, severe kidney damage, and, in some cases, death. She called the patient and learned that he had eaten grilled hamburger patties for dinner on July 9. It didn't take long to make the connection— hamburger meat contaminated with cattle feces is a

307

FIGURE 9.1

Escherichia coli

A scanning electron micrograph of the surface of a colony of *Escherichia coli*. A strain of this organism causes serious intestinal disease and was involved in a widespread outbreak in 1998. The strain was *E. coli* O157:H7. (Bar = 10 μm.)

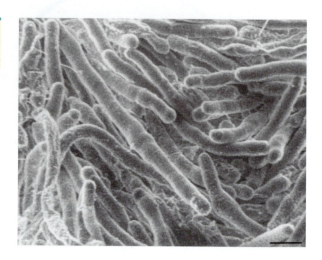

prime method for *E. coli* transmission. On her way home from work, she stopped by the patient's house and picked up the leftover frozen hamburgers. The next day, they were off to the USDA testing lab in Athens, Georgia.

By this time, other public health officials also were awakening to the fact that something unusual was going on. In Denver, for example, the city epidemiologist Pam Shillam noted an upsurge in the incidence of *E. coli* cases, and she was investigating two cases, both associated with a July 4 barbecue. Both people had eaten hamburgers, and both had bloody diarrhea yielding *E. coli*.

Almost simultaneously, the USDA report confirming *E. coli* in the Pueblo hamburgers arrived on Shillam's desk. Shillam checked the *E. coli* "fingerprint" in the report with that of the two local cases and found a match. By August 11, there were 16 more cases in Colorado, and all of them were connected to hamburger meat bearing the Hudson Foods label. By August 12, USDA inspectors were pulling into the parking lot at Hudson. They had some bad news.

The recall that followed was the largest ever for meat or poultry in the United States. As the shell-shocked company complied with federal mandates, supermarkets and fast-food restaurants around the country adapted as best they could (some Burger King outlets were left without burgers). Meanwhile, inspectors tried to pinpoint the origin of the epidemic, even though it would probably be impossible because infected cattle generally show no symptoms. Hamburger contamination is particularly difficult to trace because grinding spreads the bacteria throughout the meat, so they grow in multiple places. Moreover, freezing does not kill *E. coli*, and many people still like their hamburgers medium or rare. At this writing, the origin of the epidemic remains unknown.

The outbreak in Colorado illustrates the extreme measures that public health departments take to safeguard the quality of our lives. Although health inspectors are constantly vigilant to prevent disease transmission, sometimes their strongest precautions are not good enough, and outbreaks occur. (We shall see numerous examples in this chapter.) But, even when health officials cannot prevent an outbreak, it is comforting to know they react swiftly and strongly. In Colorado, many got sick—but no one died.

Introduction to Foodborne & Waterborne Bacterial Diseases

The principal foodborne and waterborne diseases in the United States have changed. At the turn of the 20th century, tuberculosis, typhoid fever, and cholera were common foodborne or waterborne diseases. However, with improvements in food safety and hygiene, including proper canning, pasteurization of milk and other beverages, and disinfection of water supplies, these diseases have been effectively eliminated at least in developed nations. Today, though, other foodborne and waterborne infections, such as staphylococcal food poisoning, salmonellosis, and shigellosis have taken their place.

FOODBORNE AND WATERBORNE DISEASES FOLLOW A COURSE OF INTOXICATION OR INFECTION

We shall categorize the foodborne and waterborne diseases as either intoxications or infections. **Intoxications** are diseases in which bacterial toxins, or poisons, are ingested in food or water. Examples are the toxins that cause botulism, staphylococcal food poisoning, and clostridial food poisoning. By contrast, **infections** refer to diseases in which live bacteria in food and water are ingested and subsequently grow in the body. Salmonellosis, shigellosis, and cholera are examples. Toxins may be produced, but they are the result of infection.

If an individual ingests and swallows a contaminated food or beverage, there is a delay, called the **incubation period**, before the symptoms appear. This period can range from hours to days, depending on the bacterial species and on how many of the organisms were swallowed. During the incubation period, the toxins or microbes pass through the stomach into the intestine. Toxins may directly affect gastrointestinal function or be absorbed into the bloodstream. Bacterial cells will attach to the lining of the intestinal walls and start dividing. Some will remain in the intestine, others will produce a toxin, and still others will penetrate the intestinal wall and invade deeper body tissues.

The symptoms produced depend on the specific toxin or microbe, and the number of toxins or cells ingested. Although the intoxications and infections have different symptoms, often nausea, abdominal cramps, vomiting, and diarrhea occur. Since these symptoms are so common, it often is difficult to identify the microbe causing a disease unless the disease is part of a recognized outbreak or laboratory tests are done to identify the microbe.

The Centers for Disease Control and Prevention (CDC) has estimated that there are over 76 million cases of foodborne disease each year in the United States. Most of these cases are mild and symptoms last only a few days. Some cases are more serious, leading to 325,000 hospitalizations and 5,000 deaths each year. The most severe cases occur in the elderly, the very young, those who have an illness already that reduces their immune system function, and in healthy individuals exposed to a very high dose of a toxin or microorganism.

THERE ARE SEVERAL WAYS FOODS OR WATER BECOME CONTAMINATED

There are many ways foods or water can become contaminated. Bacterial toxins vary in their sensitivity to heat. The toxin that causes staphylococcal food poisoning is not destroyed even if it is boiled. Fortunately, the potent toxin that causes botulism is completely inactivated by boiling.

Many foodborne microbes are present in healthy animals (usually in their intestines) raised for food. The carcasses of cattle and poultry can become contaminated during slaughter if they are exposed to small amounts of intestinal contents. Fresh fruits and vegetables can be contaminated if they are washed or irrigated with water that is contaminated with animal manure or human sewage.

Other foodborne microbes can be introduced from infected humans who handle the food, or by cross-contamination from some other raw agricultural product. For example, *Shigella* bacteria can contaminate foods from the unwashed hands of food handlers who are infected. In the home kitchen, microbes can be transferred from one food to another food by using the same knife, cutting board, or other utensil to prepare both without washing the surface or utensil between uses. A food that is fully cooked can become recontaminated if it touches raw foods or drippings from raw foods that contain pathogens.

Finally, some bacterial pathogens cause foodborne or waterborne bacterial diseases only when they are in large numbers. With warm, moist conditions and plenty of nutrients, lightly contaminated food left out overnight can be highly contaminated by the next day. *E. coli*, for instance, dividing every 30 minutes can produce 17 million progeny in 12 hours. If the food were refrigerated promptly, the bacteria would not multiply at all. However, two foodborne bacteria, *Listeria monocytogenes* and *Yersinia enterocolitica*, that we will discuss can grow at refrigerator temperatures.

9.2

Foodborne and Waterborne Intoxications

Intoxications are health problems in which a bacterial toxin is involved. Generally, there is a brief time between the entry of the toxin to the body and the appearance of symptoms (the **incubation period**), and the situation also resolves (for better or worse) in a relatively brief period. We shall observe this pattern in the diseases below.

BOTULISM IS CAUSED BY AN EXOTOXIN IN IMPROPERLY CANNED FOODS

Of all the foodborne intoxications in humans, none is more dangerous than **botulism**. The causative agent, *Clostridium botulinum*, produces a nerve toxin so powerful that one pint of the pure material could eliminate the entire population of the world; one ounce would kill all the people in the United States.

Clostridium botulinum is a gram-positive anaerobic bacillus that forms endospores. The spores exist in the intestines of humans as well as fish, birds, and barnyard animals. They reach the soil in manure, organic fertilizers, and sewage, and often, they cling to harvested products. When spores enter the anaerobic environment of cans or

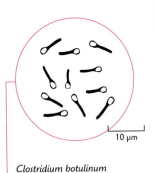

10 μm

Clostridium botulinum

jars, they germinate to vegetative bacilli, and the bacilli produce the toxin. The toxin, a protein of high molecular weight (900,000 daltons), is called an **exotoxin** because the bacilli release it to the food environment as it is synthesized. The bacteria themselves are of little consequence, but the toxin is lethal once absorbed into the bloodstream.

The symptoms of botulism develop within hours or a day after ingesting the toxin-contaminated food. Patients suffer blurred vision, slurred speech, difficulty swallowing and chewing, and labored breathing. The limbs lose their tone and become flabby, a condition called **flaccid paralysis**. These symptoms result from a complex process in which the toxin penetrates the ends of nerve cells and inhibits the release of the neurotransmitter acetylcholine into the junctions between nerves and muscles. Without acetylcholine, nerve impulses cannot pass into the muscles, and the muscles do not contract. Failure of the diaphragm and rib muscles to function leads to respiratory paralysis and death within a day or two.

Because botulism is a type of foodborne intoxication (or poisoning), antibiotics are of no value as a treatment. Instead, large doses of specific antibodies called **antitoxins** must be administered to neutralize the toxins. Life-support systems such as ventilators also are used. Frequently, the nature of the problem is not recognized because the patient has done nothing extraordinary in the previous hours. Although the annual number of cases in the United States is low (average of 110 cases per year), the number of deaths is about 8 percent. A vaccine consisting of botulism toxoid is available for laboratory workers.

Antitoxin: antibody molecule that binds specifically with toxin molecules.

Botulism can be avoided by heating foods before eating them, because the toxin is destroyed on exposure to temperatures of 90°C for 10 minutes. However, experience shows that most outbreaks are related to home-canned foods and food eaten cold. The largest episode recorded in the United States, for example, occurred in Michigan in 1977 when 58 people became ill after eating home-canned peppers at a restaurant. Another notable outbreak is illustrated in FIGURE 9.2. Foods linked to botulism include mushrooms, olives, salami, and sausage. (Indeed, the word *botulism* is derived from the Latin *botulus*, for "sausage.")

Scientists have identified seven types of *C. botulinum*, depending on the variant of toxin produced. Types A, B, and E cause most human disease. Knowing which type of *C. botulinum* caused the disease is important because antitoxin therapy must be type specific. In animals, botulism is manifested as **fodder disease**, acquired when cattle ingest toxin from silage and feed; in fowl, it is called **limberneck**. Massive fish kills also occur periodically from the consumption of botulism toxin in water.

Although foodborne botulism is very dangerous because many people can be poisoned by eating a contaminated food, other forms of botulism exist. **Wound botulism** is caused by toxins produced in the anaerobic tissue of a wound infected with *C. botulinum*. **Infant botulism** is the most common form of botulism in the United States, accounting for over 70 percent of the total annual cases reported. Unlike foodborne botulism where the botulinum toxin is ingested, infant botulism results from the ingestion of food containing *C. botulinum* endospores. Thus, parents are advised that they should not give their infant honey, which is the most common food triggering infant botulism. Spores in the honey germinate and grow in the intestinal tract where the botulinum cells release toxin. This form of botulism typically affects infants less than a year old because they have not established the normal balance of bowel microbes. Although hospitalization may be necessary, intoxication normally does not require antitoxin treatment. In addition, some cases of sudden infant death syndrome (SIDS) have been associated with infant botulism.

FIGURE 9.2

An Outbreak of Botulism

This incident occurred during April 1994 in El Paso, Texas. It was the largest outbreak of botulism in 11 years.

TEXTBOOK CASES

1. On April 7, a chef at a Greek restaurant wrapped a large number of potatoes in aluminum foil and baked them in the oven for about an hour, at roughly 400°F. Leaving them in their aluminum wrappings, he set them aside for use the next day. The baked potatoes were left at room temperature. A total of 18 hours would pass before they would be used.

2. The next day, April 8, the chef removed the potatoes from their wrappings and mashed them into a Greek dip called *skordalia*, which he then stored in a refrigerator.

3. That afternoon and evening, many restaurant patrons enjoyed the dip as an appetizer to their meals.

4. On April 10, a father and son, who had eaten at the restaurant, reported to the local hospital suffering from labored vision, difficult breathing, numbness, and general weakness. An alert physician diagnosed botulism.

5. Investigators found evidence of botulism toxin in the leftover dip. They concluded that *Clostridium botulinum* spores were on the potato skin and had germinated during the 18 hours of storage. Twenty-two cases were eventually found among 235 patrons. All recovered.

In recent years, the botulism toxin has been put to practical use. Scientists have found that, in minute doses, the toxin can relieve a number of movement disorders (the so-called dystonias) caused by involuntary sustained muscle contractions. For example, **Botox**® (botulism toxin type A), as the toxin is known, can be used to treat **strabismus**, or misalignment of the eyes, commonly known as cross-eye; it also is used against **blepharospasm**, or involuntarily clenched eyelids. The toxin may be valuable in relieving stuttering, uncontrolled blinking, and musician's cramp (the bane of the violinist). Botox® also has actual and potential cosmetic purposes. The toxin has been approved for use in the temporary relief of facial wrinkles and frown lines. Clinical studies are underway to use the toxin for temporary relief of migraine headaches, upper limb spasticity, and hyperhidrosis (excessive body sweating).

STAPHYLOCOCCAL FOOD POISONING ALSO IS CAUSED BY AN EXOTOXIN

Years ago it was common for people to complain of **ptomaine poisoning** shortly after eating contaminated food. (Indeed, the recovery of ptomaines from the intestinal tract appeared to justify their reputation.) Modern microbiologists, however, have exonerated the ptomaines and placed the blame for most food poisonings on the gram-positive, nonmotile bacterium *Staphylococcus aureus* (FIGURE 9.3a). Today, **staphylococcal food poisoning** ranks as the second most reported of all types of foodborne disease (*Salmonella*-related illnesses are first). Because most staphylococcal outbreaks probably go unreported, staphylococcal food poisoning could be the most common type.

Like botulism, staphylococcal food poisoning is caused by an exotoxin excreted in foods. Since the symptoms are restricted to the intestinal tract, the toxin is called an **enterotoxin** (*entero* refers to the intestines). Patients experience abdominal cramps, nausea, vomiting, prostration, and diarrhea as the toxin encourages the release of water. (The word *diarrhea* is derived from the Greek stems *dia*, meaning "through," and *rhein*, meaning "to flow"; hence, water "flows through"

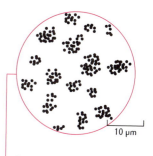

Staphylococcus aureus

Ptomaine:
a foul-smelling nitrogen compound often recovered from the intestine.

Enterotoxin:
a toxin whose effects are experienced in the intestine.

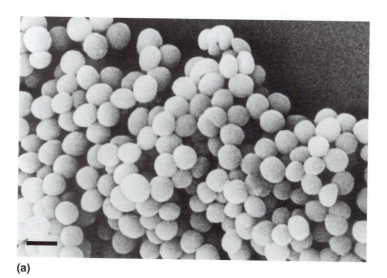

(a)

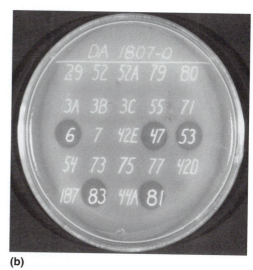

(b)

FIGURE 9.3

Staphylococcus aureus

(a) A scanning electron micrograph of *Staphylococcus aureus*, the most common cause of food poisoning in the United States. The typical grape-like cluster formation of the cocci can be seen. (Bar = 1 μm.) (b) Typing of *S. aureus* with bacteriophages. The plate of nutrient medium was seeded with the unknown strain of staphylococci, and numbered bacteriophages were then placed into different areas. The clear areas indicate which of the phages interacted specifically with the bacteria. In this case, the strain of *S. aureus* is one that interacts with phages 6, 47, 53, 81, and 83. Tests like this are important in relating a specific strain of *S. aureus* to an outbreak of food poisoning.

the intestines). The symptoms last for several hours, and recovery is usually rapid and complete.

The incubation period for staphylococcal food poisoning is a brief one to six hours. Often the individual can think back and pinpoint the source. Examples are spoiled meats and fish, as well as contaminated dairy products, cream-filled pastries, and salads such as potato salad and coleslaw. Foods containing *S. aureus* lack an unusual taste, odor, or appearance, and the only clues to possible contamination are factors such as moisture content, low acidity, and improper heating previous to arrival at the table. The staphylococcal enterotoxin is among the most heat resistant of all exotoxins. Heating at 100°C for 30 minutes will not destroy the toxin.

A key reservoir of *S. aureus* in humans is the nose. Thus, an errant sneeze may be the source of staphylococci in foods. Studies indicate, however, that the most common mode of transmission is from boils or abscesses on the skin that shed staphylococci. The staphylococci grow over the broad temperature range of 8°C to 45°C, and since refrigerator temperatures are generally set at about 5°C, refrigeration is not an absolute safeguard against contamination. Ham is particularly susceptible because staphylococci tolerate salt.

Staphylococcus aureus normally does not grow in human intestines because of competition by other organisms. Therefore, public health investigators usually are unable to locate the organisms in stool samples. Moreover, the contaminated food often has been consumed completely. Thus, case reports often are based on symptoms, patterns of outbreak, and type of food eaten. When investigators locate staphylococci, they can identify the organisms by growth on mannitol salt agar, Gram staining, and testing with bacteriophages to learn the strain involved, as **FIGURE 9.3b** illustrates.

CLOSTRIDIAL FOOD POISONING RESULTS FROM AN ENTEROTOXIN

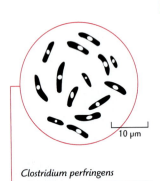

Clostridium perfringens

Since its recognition in the 1960s, **clostridial food poisoning** has risen to prominence as a common type of food poisoning. The causative organism, *Clostridium perfringens*, is also an agent of gas gangrene (Chapter 10). This gram-positive anaerobic spore former contaminates protein-rich foods such as meat, poultry, and beans. If the spores survive the cooking process, they germinate to vegetative cells and produce an enterotoxin. Consumption of the toxin leads to illness. Large numbers of vegetative cells also may be consumed into anaerobic pockets in the large intestine, where the enterotoxin is produced.

The incubation period for clostridial food poisoning is a relatively long 8 to 24 hours, a factor that distinguishes it from staphylococcal food poisoning. Moderate to severe abdominal cramping and watery diarrhea are common symptoms, as the enterotoxin encourages the outward movement of water from epithelial cells lining the intestinal tract. Recovery is rapid, often within 24 hours, and therapy is generally unnecessary.

To this point . . .

We have surveyed three foodborne diseases in which toxins are responsible for the characteristic symptoms. In botulism, the toxin affects the transmission of nerve impulses and causes a life-threatening situation, while in staphylococcal food poisoning and clostridial food poisoning, the toxins affect the intestinal lining and induce a loss of water by diarrhea. This water loss is less severe on the body than the nerve interference in botulism, but the incidence of food poisoning is much higher than for botulism.

We shall now move on to a series of foodborne and waterborne diseases in which bacteria enter the body and grow profusely. These diseases are more correctly called food infections than food intoxications. We shall begin with a serious health problem, typhoid fever, then survey other Salmonella-related diseases, and then focus on shigellosis and cholera, in which patient dehydration is a prime concern. The discussion will conclude with a series of intestinal disturbances that generally are not life threatening but have become widely recognized as detection methods continue to improve.

9.3
Foodborne and Waterborne Infections

Foodborne and waterborne infections have a longer incubation period than intoxications because bacteria must establish themselves in the body before symptoms develop. We shall see this pattern in the diseases that follow.

TYPHOID FEVER INVOLVES A BLOOD INFECTION

Typhoid fever is among the classical diseases (the "slate-wipers") that have ravaged human populations for generations. The disease captured the attention of microbiologists over a century ago and was studied by Karl Eberth and Georg Gaffky, both coworkers of Robert Koch. Eberth observed the causative organism, *Salmonella typhi*, in spleen tissue in 1880, and Gaffky isolated it in pure culture four years later.

Salmonella typhi, a gram-negative motile rod, displays high resistance to environmental conditions outside the body. This factor enhances its ability to remain alive for long periods of time in water, sewage, and certain foods. *S. typhi* is transmitted by the five Fs: flies, food, fingers, feces, and fomites. Humans are the only host for *S. typhi*.

The bacterium is acid resistant, and with the buffering effect of food and beverages, it survives passage through the stomach. In the small intestine it invades the tissues, causing deep ulcers and bloody stools. Blood invasion follows, and after several days, the patient experiences mounting fever, lethargy, and delirium. (The word *typhoid* is derived from the Greek *typhos*, for "smoke" or "cloud," a reference to the delirium.) The abdomen becomes covered with **rose spots**, an indication that blood is hemorrhaging in the skin.

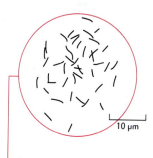

10 µm

Salmonella typhi

MicroFocus 9.1

TYPHOID MARY

By 1906, typhoid fever was claiming about 25,000 lives annually in the United States. During the summer of that year a puzzling outbreak occurred in the town of Oyster Bay on Long Island, New York. One girl died and five others contracted typhoid fever, but local officials ruled out contaminated food or water as sources. Eager to find the cause, they hired George Soper, a well-known sanitary engineer from the New York City Health Department.

Soper's suspicions centered on Mary Mallon, the seemingly healthy family cook. But she had disappeared three weeks after the disease surfaced. Soper was familiar with Robert Koch's theory that infections like typhoid fever could be spread by people who harbor the organisms. Quietly he began to search for the woman who would become known as Typhoid Mary.

Soper's investigations led him back over the ten years' time during which Mary Mallon cooked for several households. Twenty-eight cases of typhoid fever occurred in those households, and each time, the cook left soon after the outbreak. One epidemic in 1903 in Ithaca, New York, claimed 1,300 lives.

Ironically, Soper had gained his reputation during this episode.

Soper tracked Mary Mallon through a series of leads from domestic agencies and finally came face-to-face with her in March 1907. She had assumed a false name and was now working for a family in which typhoid had broken out. Soper explained his theory that she was a carrier, and pleaded that she be tested for typhoid bacilli. When she refused to cooperate, the police forcibly brought her to a city hospital on an island in the East River off the Bronx shore. Tests showed that her stools teemed with typhoid organisms, but fearing that her life was in danger, she adamantly refused the gall bladder operation that would eliminate them. As news of her imprisonment spread, Mary became a celebrity. Soon public sentiment led to a health department policy deploring the isolation of carriers. She was released in 1910.

But Mary's saga had not ended. In 1915, she turned up again at New York City's Sloane Hospital working as a cook under a new name. Eight people had recently died of typhoid fever, most of them doctors and nurses. Mary was taken back to the island, this time in

handcuffs. Still she refused the operation and vowed never to change her profession. Doctors placed her in isolation in a hospital room while trying to decide what to do. The weeks wore on.

Eventually Mary became less incorrigible and assumed a permanent residence in a cottage on the island. She gradually accepted her lot and began to help out with routine hospital work. However, she was forced to eat in solitude and was allowed few visitors. Mary Mallon died in 1938 at the age of 70 from the effects of a stroke. She was buried without fanfare in a local cemetery.

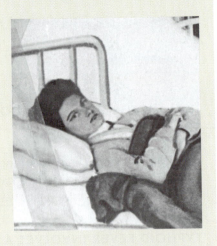

About 400 cases of typhoid fever are reported annually to the CDC, as FIGURE 9.4 indicates. However, 70 percent of the infections are acquired during international travel. Treatment is generally successful with the antibiotic chloramphenicol, except when antibiotic resistance due to R plasmids is noted (Chapter 7). About 5 percent of recoverers become **carriers** and continue to harbor and shed the organisms for a year or more. Because individuals such as food handlers can be carriers of disease to others, the public health department usually monitors the activities of carriers. The experiences of one of history's most famous carriers, Typhoid Mary, are recounted in MicroFocus 9.1.

Traditional vaccines for typhoid fever have consisted of dead *S. typhi* cells, but adverse reactions have introduced an element of risk. A newer oral Ty21a vaccine is composed of a weakened (attenuated) strain of *S. typhi*. Another injectable vaccine (ViCPS) consists of capsular polysaccharides from *S. typhi*. Its supporters point to a stronger response than that rendered by the Ty21a vaccine. Both vaccines lose effectiveness after a few years, so booster shots are needed for international travelers. MicroFocus 9.2 describes an interesting relationship between *S. typhi* and cystic fibrosis.

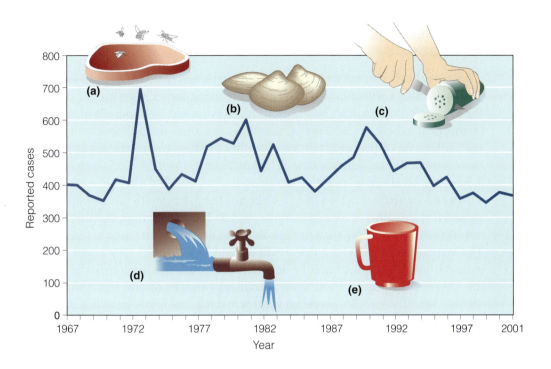

FIGURE 9.4

The Incidence of Typhoid Fever

Reported cases of typhoid fever in the United States by year, 1967 to 2001. For 2001, 368 cases were reported. Portrayed on the graph are the five Fs important in the transmission of typhoid fever: (a) flies, (b) food, (c) fingers, (d) feces, and (e) fomites.

MicroFocus 9.2

THE DOWNSIDE, THE UPSIDE

Cystic fibrosis is a gruesome disease. Abnormal amounts of thick, sticky mucus build up in the respiratory tracts of children and clog the air passageways. Parents have to slap children on the back repeatedly to help clear the clogging. Life-threatening respiratory infections often accompany the disease.

Scientists now know that cystic fibrosis is a genetic disease that develops when two mutant genes are inherited from the parents. With the mutation in place, a regulatory protein is not produced, and the sticky mucus accumulates. About 30,000 individuals in the United States suffer the misery of cystic fibrosis each year.

But there is a strange twist to this story. Harvard researchers have discovered that the regulatory protein is an attachment site for the typhoid bacillus *Salmonella typhi*. Apparently the regulatory protein protrudes from cells and binds to *S. typhi* at the start of the disease process. Now the bacilli are free to invade and destroy the tissues. However, if there is no protein, then there is no binding, and no disease.

Occasionally, Mother Nature does things that seem to defy explanation. Evidence suggests that the cystic fibrosis patient is immune to typhoid fever. Before the age of antibiotics, that would have been a worthwhile trade-off, since typhoid fever was a major killer. Indeed, in developing countries, typhoid fever is still a serious health problem. Is it possible that cystic fibrosis is another of those "diseases of civilization"? What do you think?

SALMONELLOSIS CAN BE CONTRACTED FROM A VARIETY OF FOODS

Salmonellosis currently ranks as the most reported of all foodborne diseases in the United States, with almost 40,000 cases occurring annually (TABLE 9.1 compares the disease to staphyloccal food poisoning.) It is caused by hundreds of **serotypes** (serological types) of *Salmonella*. Serotypes are used for *Salmonella* instead of species because of the uncertain relationships existing among the organisms. The three most common serotypes are *S. typhimurium* (FIGURE 9.5), *S. enteritidis*, and *S. newport*. All are gram-negative motile rods.

After an incubation period of 6 to 48 hours, the patient with salmonellosis experiences fever, nausea, vomiting, diarrhea, and abdominal cramps. Intestinal ulceration is usually less severe than in typhoid fever, and blood invasion is uncommon. The symptoms may last a week or more, and dehydration may occur in some patients; fluid replacement may be necessary. Diagnosis usually consists of isolating the *Salmonella* serotype from stool specimens or rectal swabs, using differential media. Often the infection is called **gastroenteritis,** or simply **enteritis**. A notable outbreak occurred in 1991, as MicroFocus 9.3 describes.

With increased awareness and modern methods of detection, salmonellosis has been linked to a broad variety of foods. Pasteurized milk was linked to 5,770 cases in the midwestern United States in 1985 (Chapter 25), and frozen pasta products were the source of dozens of cases in the northeastern states in 1986. Poultry products are particularly notorious because *Salmonella* serotypes commonly infect chickens and turkeys when the normal gut bacteria are absent (MicroFocus 9.4 explores a possible

TABLE 9.1

A Comparison of Staphylococcal Food Poisoning and Salmonellosis

CHARACTERISTIC	STAPHYLOCOCCAL FOOD POISONING	SALMONELLOSIS
Agent	*Staphylococcus aureus*	*Salmonella* serotypes
Description	Gram-positive cluster of cocci	Gram-negative rod
Toxin involved	Yes (enterotoxin)	No
Incubation period	1–6 hours	6–48 hours
Incidence in US	Possible millions annually	40,000 reported annually
Symptoms	Cramps, nausea, diarrhea; no ulceration; no fever	Cramps, nausea, diarrhea; some ulceration; fever
Duration of symptoms	Few hours	Few days
Foods involved	All	All, especially poultry/eggs
Animals involved	No	Yes (especially reptiles)
Treatment	None	Antibiotics possible
Source of bacteria	Skin infection, nose	Fecal contamination
Diagnosis	Isolate bacteria from food	Isolate bacteria from feces

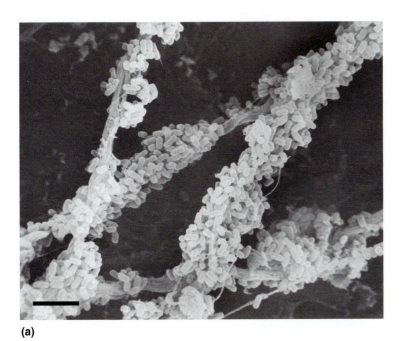

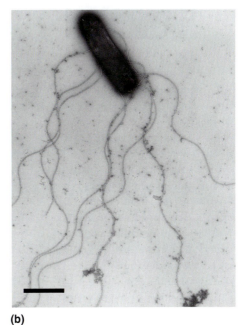

(a)   (b)

FIGURE 9.5

Salmonella Species

Two views of *Salmonella* involved in salmonellosis. (a) *S. typhimurium* observed on the collagen fibers of muscle tissue from an infected chicken. (Bar = 5 μm.) (b) A transmission electron micrograph of a *Salmonella* serotype showing peritrichous flagella. Note the long length of the flagella relative to the cell. (Bar = 1 μm.)

MicroFocus 9.3

YOU MAKE THE CALL

Poultry products and eggs are well known for their connections to *Salmonella* diseases, but the signs that summer did not point to the poultry and eggs; they pointed to the fruit salad.

There were 75 people at the party in New Jersey that June night in 1991. By the next week, 17 were ill with nausea, vomiting, diarrhea, abdominal cramps, and fever. Soon, investigators were visiting each one, asking questions and requesting stool samples. All 17 had eaten the fruit salad—the watermelon, cantaloupe, honeydew melon, strawberries, and grapes. The investigators scratched their heads because fruit salad and *Salmonella* just don't go together. Or do they? And if they do, which was the responsible fruit? They waited for additional cases to surface.

They did not have to wait long. In July, 20 cases of salmonellosis were reported in Minnesota, and several more occurred in Illinois, Pennsylvania, North Dakota, Missouri, and Michigan. Even Canada contributed 72 cases. All told, there were 400-plus cases of salmonellosis in 23 states and Canada during that June–July period. And the culprit seemed to be the cantaloupe.

At least the laboratory researchers suspected it was the cantaloupe, even though they could not isolate *Salmonella* from the fruit. But they recovered *Salmonella* from enough patients who had eaten cantaloupe; and they were able to pinpoint *Salmonella poona* as the epidemic's cause. In addition, they traced all the cantaloupes to a farm in south Texas where the soil was heavily

contaminated by *Salmonella* and where *S. poona* had probably infected the rinds. But people do not eat cantaloupe rinds; they eat the fruit. So how did the *Salmonella* get from the rind to the fruit? Any guesses?

MicroFocus 9.4

"SORRY, NO VACANCY"

In the old days, chicks hatched from their eggs, scrambled to their feet, and nestled under their mother's wing until it was safe to come out. By staying close to their mother hen, the chicks received protection, caring, and bacteria. Bacteria? Yes, bacteria—hundreds of strains of harmless organisms that filled the chicks' guts and prevented *Salmonella* serotypes from causing infection. With all the enterococci, fusobacteria, lactobacilli, and other strains, there simply was no room for *Salmonella*. The chicks remained healthy.

But mother hen is gone. The high-tech chicken farms of the modern era use machines to remove the eggs from the hen as they are produced. The chicks hatch and develop without ever seeing who made them. To be sure, that is sad. But microbiologically, the sadness is compounded by the absence of the harmless bacteria from the chick's gut (and the tendency of the chick to develop salmonellosis).

As of 1998, there was an answer. That year, the Food and Drug Administration approved a spray (called Pre-empt) that showers chicks with a mix of 29 species of harmless bacteria. The chicks pick the bacteria off their feathers and ingest them. As the harmless bacteria set up housekeeping in the gut, they compete with and exclude the dangerous ones. (Scientists call the process "competitive exclusion.") Numerous tests show that *Salmonella* serotypes are significantly reduced or completely eliminated. Once again, science has replaced mother hen.

method for replacement). The organisms may be consumed directly from the poultry or in pot pies, processed chicken roll, turkey roll, or chicken salads. With over 4 billion chickens and turkeys consumed annually in the United States (an average of 75 pounds per American), the possibilities for *Salmonella* passage are plentiful.

Eggs are another source of salmonellosis when used in foods such as custard pies, cream cakes, egg nog, ice cream, and mayonnaise (Chapter 25). In 1992, the CDC reported 38 cases associated with Caesar salad dressing made with raw eggs. In the past, salmonellosis was associated with cracked or contaminated egg shells, but researchers now believe that *Salmonella* serotypes infect the ovary of the hen and pass into the egg before the shell forms. If this is so, then even the highest-grade eggs may be contaminated. Consumers should store eggs in the main compartment of the refrigerator, refrigerate leftover egg dishes quickly in small containers (to accelerate cooling), and avoid "runny" or undercooked eggs. The current catchphrase at the CDC is "scramble or gamble." Indeed, in 1992, New Jersey outlawed "eggs over easy." (It later rescinded the ban.)

Live animals also are known to transmit salmonellosis, and many states now prohibit the sale of Easter chicks and ducklings for this reason. Moreover, some years ago, the FDA prohibited the distribution of pet turtles less than 4 inches long because a significant number of salmonellosis outbreaks were attributed to handling these animals. The sale of iguanas is restricted also, and a savannah monitor lizard was apparently the cause of an infant's disease in 1992.

SHIGELLOSIS (BACTERIAL DYSENTERY) OCCURS WHERE SANITARY PROCEDURES ARE LACKING

Members of the genus *Shigella* were first described by the Japanese investigator Kiyoshi Shiga in 1898, and two years later by the European microbiologist Simon

Flexner. *Shigella* species are small gram-negative nonmotile rods found mainly in humans and other primates. As few as 10 to 200 organisms can cause digestive disturbances ranging from mild diarrhea to a severe and sometimes fatal dysentery. (Dysentery is a syndrome manifested by waves of intense abdominal cramps and frequent passage of small-volume, bloody mucoid stools.) For many years, *Shigella* diseases were known as **bacterial dysentery**. However, watery diarrhea without blood or mucus is a more common symptom than dysentery, and the name **shigellosis** is preferred. About 14,000 cases are reported annually in the United States, which represents about 3 percent of the total cases.

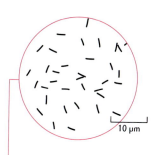

Shigella sonnei

Shigellosis in the United States is caused primarily by two species of *Shigella*: *S. sonnei* and *S. flexneri*. A third species, *S. dysenteriae*, causes deadly epidemics in the developing world. Humans ingest the organisms in contaminated water as well as in many foods, especially eggs, vegetables, shellfish, and dairy products. *Shigella* usually penetrates the epithelial cells lining the intestine and, after 2 to 3 days, it produces sufficient enterotoxins to encourage water release. Infection of the small intestine produces watery diarrhea, but infection of the large intestine results in bloody mucoid stools characteristic of dysentery. Indeed, the dysentery can be quite indisposing. It is said that at the Battle of Crécy in 1346, the English army was racked with dysentery. When the French attacked, they literally caught the English with their pants down.

Shigella species can be isolated from stools, and most cases of shigellosis subside within a week. Usually there are few complications, but patients who lose excessive fluids must be given salt tablets, oral solutions, or intravenous injections of salt solutions for rehydration. Antibiotics are sometimes effective, but many strains of *Shigella* are resistant because of R plasmids. Recoverers generally become carriers for a month or more and continue to shed the bacilli in their feces. Vaccines are not available, but 586 passengers aboard the cruise ship *Viking Serenade* wished they were. They were the unfortunate victims of a shigellosis outbreak in 1994 traced to bacteria in the water supply. Trapped hopelessly aboard ship, they could only wait for landfall and dream of a more pleasant "serenade" the next time.

CHOLERA CAN INVOLVE ENORMOUS FLUID LOSS

No diarrhea can compare with the extensive diarrhea associated with **cholera**. In the most severe cases, a patient may lose up to one liter of fluid every hour for several hours. The fluid is colorless and watery, with characteristic **rice-water stools** reflecting the conversion of the intestinal contents to a thin liquid like barley soup. The patient's eyes become gray and sink into their orbits. The skin is wrinkled, dry, and cold, and muscular cramps occur in the arms and legs. Despite continuous thirst, sufferers cannot hold fluids. The blood thickens, urine production ceases, and the sluggish blood flow to the brain leads to shock and coma. In untreated cases, the mortality rate may reach 70 percent. However, it is easily treated and prevented. Although prevalent in the United States in the 1800s, good sanitation and water treatment have eliminated the bacterium as a health threat.

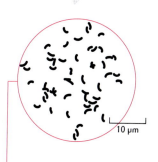

Vibrio cholerae

Cholera is caused by *Vibrio cholerae*, a curved gram-negative motile rod (**FIGURE 9.6a**) first isolated by Robert Koch in 1883. The bacilli enter the intestinal tract in contaminated water or food, such as raw oysters. *V. cholerae* is extremely susceptible to stomach acid. However, if high numbers are ingested, enough remain to colonize the intestines. As the bacilli move along the intestinal epithelium, they secrete an enterotoxin that stimulates the unrelenting loss of fluid (**MicroFocus 9.5**). Antibiotics such as tetracycline may be used to kill the bacteria, but the key treatment is restoration of the

MicroFocus 9.5

MEMORIES

It was supposed to be a simple evening together at a quiet, local restaurant—six women, dinner, some interesting scuttlebutt, and a pleasant memory or two. The August night was hot that summer in Maryland in 1991. All would go as planned, except for the memory. It would not be pleasant.

Two days later, one of the women developed vomiting and watery diarrhea so severe she had to be hospitalized. Then two others had acute diarrhea requiring medical attention. When blood samples from the three were analyzed for antibodies, the results raised the eyebrows of health officials—it was cholera.

The first thing that came to mind were the crabs served that night. But Marylanders are proud of their crabs, and health investigators were relieved to learn that crabs had been served to others at the restaurant with no apparent effect. What about the Thai-style rice pudding? It came with a topping made from imported coconut milk, and several unopened packages of the milk were still in the freezer. Did the sick women have the rice pudding with the topping? Yes. How about the other restaurant patrons that night? They had ordered rice pudding but without the topping. Aha! Could the health officials please have the unopened milk packages for testing?

During the next several days, emergency rooms were checked for additional cases of cholera, but none surfaced. Sewage collection points were tested for cholera bacilli, but samples turned up negative. Secondary contacts of the women were reached to see if they had any unusual symptoms; none reported any. By now the laboratory results on the coconut milk were ready. The milk was positive for *Vibrio cholerae* and a number of other bacteria. A voluntary product recall was issued by the distributor. Case closed.

Pandemic:
a worldwide epidemic.

body's water balance. Often, this entails intravenous injections of salt solutions. More commonly, patients can be treated with **oral rehydration solution (ORS)**, a solution of electrolytes and glucose that will restore the normal balances in the body.

Cholera has been observed for centuries in human populations, and seven **pandemics** have been documented since 1817. The current pandemic, caused by *V. cholerae* El Tor, began in 1961 in Indonesia and now involves about 35 countries (FIG-URE 9.6b). A major outbreak occurred in Peru and Ecuador early in 1991 and from there spread throughout the region, accounting for 731,000 cases by the end of 1992 and 6,300 deaths in 21 countries in the Western Hemisphere. By 1998, the epidemic was spreading throughout Africa at what the World Health Organization called a "catastrophic rate." MicroFocus 9.6 describes how the responsible strain may have emerged. A new epidemic emerged in India in 1992 and has spread across Asia. This strain, *V. cholerae* Bengal, may trigger the next cholera pandemic.

At present, travelers to cholera regions of the world are immunized with preparations of dead *V. cholerae*, thereby obtaining protection for about six months. This approach may change, however, because public health officials anticipate a genetically engineered cholera vaccine that will give longer-lasting protection without the risk of using dead pathogenic cells. In 1992, researchers at the University of Maryland identified the genes for enterotoxin production and successfully removed them from experimental *V. cholerae* cells. So treated, the vibrios became nonpathogenic but still alive (i.e., attenuated), and they could be used in a vaccine to stimulate an antibody response. Field trials for the new vaccine are ongoing at this writing.

For generations, scientists believed the cholera bacillus exists only within a human host. That principle was refuted by University of Maryland researchers led by Rita Colwell (current Director of the National Science Foundation). Colwell and her colleagues found *V. cholerae* in waters from the Chesapeake Bay, even though no cholera outbreaks were remotely close to the site. They postulated that

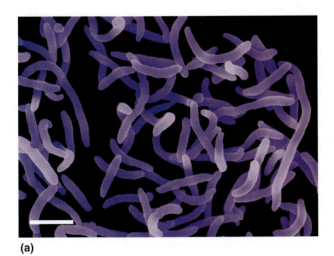

(a)   (b)

FIGURE 9.6

Vibrio cholerae and Disease Spread

False color scanning electron micrograph of the comma-shaped *Vibrio cholerae* (Bar = 1.5 μm.) (b) The spread of cholera is associated with poor sanitation and water supplies contaminated by feces. The Ganges River in India is considered sacred, and people wash in the river and drink from it. Cholera bacilli frequently inhabit the river and pass easily among unsuspecting bathers.

MicroFocus 9.6

HOW THE SHEEP GOT THE WOLF'S CLOTHES

The seventh pandemic of cholera began in Sulawesi, Indonesia, in 1961. Soon it spread to India, the Soviet Union, and the Middle East. By 1991, it had reached Latin America and affected all countries of South America, except Argentina. By 1995, over 5,000 people had died of cholera, and many hundreds of thousands had been terribly sick.

The pandemic was apparently due to a toxin-producing strain of *Vibrio cholerae* known as O1. The toxin binds to intestinal cells, setting off a cascade of reactions, and water pours out of the cells—up to 5 gallons of water per day. Where did the toxin come from? Microbiologists from Harvard think they have the answer: a virus.

Matthew Waldor and John Mekalanos had been studying cholera for many years. They were impressed by the toxicity of the O1 strain and wondered why other strains were far less lethal.

Their interest centered on the toxin gene in *V. cholerae*, and they speculated that the gene might have been delivered by a bacteriophage through the process of transduction (Chapter 7). To test their theory, they removed the entire toxin gene by sophisticated genetic engineering techniques, then they replaced it with an antibiotic-resistance gene. Now they cultured the new antibiotic-resistant cholera cells with normal cholera cells susceptible to antibiotics. Bingo! The susceptible cells became antibiotic resistant. Something (a phage?) seemed to be leaving the antibiotic-resistant cells and ferrying the resistance gene to the susceptible cells.

But, maybe the bacteria were conjugating and exchanging their genes directly. To test this theory, Waldor and Mekalanos passed the genetically engineered (antibiotic-resistant) cells through a filter that would trap everything except phages. They took the clear, cell-free liq-uid and added it to a fresh batch of normal cells. Double bingo! The cells became resistant to antibiotics. The phage theory strengthened.

Still another test: They treated the clear, cell-free fluid with enzymes that destroy free-floating nucleic acids but have no effect on viruses. (This would eliminate any molecular DNA or RNA that might transform cells.) Then they combined the fluid with normal cells. Once again, the cells became antibiotic resistant. And the coup de grace: Electron micrographs revealed long, stringy phages in the cell-free fluid.

To be sure, the cholera bacterium is not the first to have its toxicity associated with a phage (the diphtheria and botulism organisms are others), but the finding helps explain how an organism can suddenly become lethal. The sheep had acquired the wolf's clothing.

the organisms enter a spore-like dormant state that cannot be cultivated. (Highly sensitive antibody tests detected the organism when more traditional tests failed.) Apparently, in the cold, nutrient-poor water, the organism's metabolic rate diminishes, and it stops reproducing. This state, says Colwell, allows the cholera bacillus to survive in habitats and environments ranging from seawater to the human intestinal tract. In the sea, moreover, the bacillus appears to be a regular gut inhabitant of a tiny crustacean known as a copepod. The findings are novel, and they may signal a new outlook for cholera in the decades ahead.

E. COLI DIARRHEAS CAUSE VARIOUS FORMS OF GASTROENTERITIS

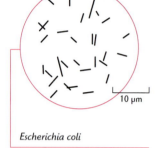

Escherichia coli

One of the major causes of **infantile diarrhea** is the gram-negative rod *Escherichia coli* (**FIGURE 9.7**). This organism may induce diarrhea by either of two mechanisms: Certain *enterotoxic* strains produce an enterotoxin similar to that in cholera, while other *enteroinvasive* strains penetrate the intestinal epithelium, as in shigellosis. The toxin causes fluid loss in the small intestine, while penetration occurs primarily in the large intestine. Both mechanisms lead to dehydration and salt imbalance substantial enough to be life threatening in infants. Antibiotic therapy and fluid replacement are usually effective. Often the infection is nosocomial (hospital acquired).

Traveler's diarrhea is a term usually applied to a disease in which the victim experiences diarrhea within two weeks of traveling to a tropical location; the diarrhea lasts 1 to 10 days. Sometimes this disease is called Montezuma's revenge, a reference to the leader of the Aztec nation decimated by smallpox and other diseases brought from Europe in the 1500s. A number of organisms including several types of bacteria, viruses, and protozoa may cause traveler's diarrhea, but recent studies point to *E. coli* as the principal agent. The bacilli adhere by pili to the intestinal lining and produce enterotoxins, which induce water loss. The volume lost is usually low, but occasionally it may be considerable, and dysentery may occur. The possibility of traveler's diarrhea may be reduced by careful hygiene and attention to the food and water consumed during visits to other countries.

As described in the introduction to this chapter, it is possible to contract a rather serious *E. coli* diarrhea without traveling. In 1993, for example, *E. coli* O157:H7, an

FIGURE 9.7

Escherichia coli

False color scanning electron micrograph of *Escherichia coli* bacteria inhabiting the surface of the intestines. *E. coli* are part of the normal intestinal microbiota of both humans and animals. When other strains of *E. coli* enter the intestines through contaminated food or water, infections such as gastroenteritis can occur. (Bar = 2 μm.)

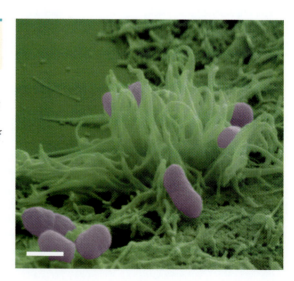

FIGURE 9.8

Outbreak of
E. coli O157:H7
Infection

This outbreak occurred in
New Haven County, Con-
necticut, during October
1996.

1. On October 6, 1996, a family from a large city in Connecticut decided to take a drive in the country. Along the way they stopped at a general store for a bite of lunch. The father and two children had apple cider; the mother had a soda instead.

2. Three days later, the father and children began to experience serious abdominal pains and vomitting. Moreover, there was blood in their stools. But the mother had no symptoms.

3. One of the children became worse. She had to be admitted to the hospital. The doctor said that the blood was infected and that the kidneys might be suffering. He advised dialysis to assist the kidney function.

4. Health officials were notified, and they began a telephone survey to find out if anyone else was similarly infected. Over two dozen cases were found. All had recently consumed apple cider sold by the same company.

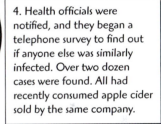

5. Inspectors visited the cider plant. The cider was unpasteurized. It was made from apples picked directly from the tree as well as apples picked from the ground, where contamination with cow manure might have occurred. They recommended pasteurization.

TEXTBOOK CASES

enterotoxic strain, was the cause of over 500 cases of serious illness in patrons of the Jack-in-the-Box fast-food chain (see MicroFocus 3.1). Also that year, mayonnaise contaminated with *E. coli* was implicated in numerous cases of bloody diarrhea contracted at The Sizzler restaurants in Oregon; and some years ago, the McDonald's chain faced an outbreak of *E. coli* infection traced to its hamburgers. Despite their acidity, orange juice, apple juice, and apple cider also have been connected to outbreaks, as **FIGURE 9.8** illustrates. When confined to the large intestine with grossly bloody diarrhea, the disease can lead to the complication known as **hemorrhagic colitis**. When the disease involves the kidney and leads to kidney failure, it is called hemolytic uremic syndrome (HUS); seizures, coma, colonic perforation, liver disorder, and heart muscle infections have been associated with HUS.

Epidemiologists at the CDC estimate that approximately 73,000 cases of *E. coli* O157:H7 infection occur annually in the United States, with about 61 deaths. In 2001, 28 states reported 202 cases of HUS. Most states require that *E. coli* O157:H7 isolates be reported to the state health department, and physicians are alerted to watch for cases of bloody diarrhea. A toxin similar to that produced by *Shigella* species has been identified in *E. coli* isolates (it is called the "shiga-like toxin"). Fresh-picked carrots in a garden salad were implicated in two incidents in 1993, and dry-cured salami was implicated in two 1994 outbreaks.

The prevailing wisdom is that *E. coli* O157:H7 exists in the intestines of cattle but causes no disease in these animals. (Researchers are considering an adjustment in cattle feeds to prevent the organism's proliferation.) Slaughtering brings *E. coli* to beef products, and excretion to the soil accounts for transfer to plants and fruits. The organism is particularly pathogenic because 100 bacilli are enough to establish infection; it produces toxins at an unusually high rate; and, since it colonizes the intestines, it can deliver toxins to this area efficiently. The organism ferments the alcoholic carbohydrate sorbitol very slowly, and this factor is useful in a diagnostic test: In MacConkey agar, the lactose is replaced by sorbitol, and *E. coli* O157:H7 produces white colonies, while other *E. coli* strains produce red or pink colonies.

In recent years, *E. coli* O111:NM and *E. coli* O104:H21 also have been identified as causes of intestinal illness. All told, over 100 serological types of *E. coli* have been implicated in hemolytic uremic syndrome and hemorrhagic colitis. Treatment regimens are not established, and in uncomplicated cases, the symptoms resolve within 5 to 10 days. Research continues on a vaccine for all *E. coli* diarrheas.

PEPTIC ULCER DISEASE CAN BE SPREAD PERSON TO PERSON

Approximately 25 million Americans suffer from peptic ulcers during their lifetime. One of the more remarkable discoveries of the modern era is that many cases of **peptic (stomach) ulcers** are caused by a bacterium. In past decades, scientists believed that all ulcers resulted from "excess acid" due to factors such as nervous stress, smoking, alcohol consumption, diet, and physiological dysfunction. However, the work of Barry Marshall and his coworkers has made it clear that the bacterium *Helicobacter pylori* is involved (MicroFocus 9.7). This gram-negative, motile, microaerophilic curved rod is responsible for 90 percent of duodenal ulcers and 80 percent of peptic ulcers. It is uncertain how *H. pylori* is transmitted. Most likely, it is spread person to person through contaminated food or water. Doctors have revolutionized the treatment of ulcers by prescribing antibiotics such as amoxicillin, tetracycline, or clarithromycin (Biaxin), along with omeprazole (Prilosec) for acid suppression. They have achieved cure rates of up to 94 percent; relapses are uncommon.

How *H. pylori* manages to survive in the intense acidity of the stomach is interesting. Apparently, the bacterium twists its way through the mucus coating of the stomach lining and attaches to the stomach wall. There it secretes the enzyme urease. Urease digests urea in the area and produces ammonia as an end product, as FIGURE 9.9 shows. The ammonia neutralizes acid in the stomach, and the organism begins its destruction of the tissue, supplemented by digestive enzymes normally found in the stomach tissue and secreted cytotoxins. In the stomach lining, a sore 0.6 to 12 cm in diameter appears (although some ulcers may be up to 30 cm in diameter). The pain is severe and is not relieved by food or an antacid (as is a duodenal ulcer).

MicroFocus 9·7

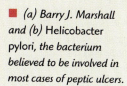

ULCERS ARE TRANSMISSIBLE?

The old claim "You're giving me an ulcer!" may be truer than anyone believed, if the latest laboratory evidence holds up. Why? The evidence indicates that ulcers are a transmissible disease.

The story began in 1982 when two Australian gastroenterologists, Barry J. Marshall and J. Robin Warren, identified bacteria living in the stomach lining of over 100 patients who had ulcers. The initial discovery was serendipitous because Marshall and Warren could not cultivate the bacterium under normal conditions. Only when they were swamped with work did they leave their culture plates in the incubator too long. And only then did the colonies of bacteria appear. Marshall (FIGURE a) and Warren identified the organism as the gram-negative curved rod *Campylobacter pyloridis*. Since then, the organism's name has been changed to *Helicobacter pylori* (FIGURE b).

Marshall's initial speculation that *H. pylori* causes ulcers aroused widespread skepticism and intensified research. It took ten years, but research has appeared to confirm Marshall's contention. In 1993, a study published in the *New England Journal of Medicine* indicated that 48 of 52 peptic ulcer patients could be cured of their ulcers in six weeks if treated with two antibiotics over a 12-day period. Another 52 patients received a placebo and 39 seemed to be cured, but a year later, the ulcers had returned in all 39 patients. By comparison, only 4 of the patients receiving antibiotics experienced a recurrence.

Marshall now conducts research at the University of Virginia Medical School and supervises an ongoing program to further elucidate the cause of ulcers. The bacterial cause of ulcers conflicts with the conventional wisdom, which says that stress, diet, or other factors trigger excess acid secretion and ulcer for-

mation. (Indeed, two of the top-selling drugs in the United States had been the antacids cimetidine [Tagamet] and ranitidine [Zantac], both used to control acid secretion.) Nevertheless, the evidence continues to mount that *H. pylori* is a major factor. In another study at Baylor University, researchers obtained results similar to Marshall's. The data have compelled some researchers to begin thinking of ulcers as a transmissible disease, thereby encouraging doctors to prescribe tetracycline together with Pepto-Bismol. Two biotechnology firms are even attempting to develop a vaccine against *H. pylori*.

And what of Marshall? It is said that once he drank a culture of *H. pylori* to prove his thesis that it causes peptic ulcers. Nowadays, he has adopted a more conservative approach to research, leaving the thesis-proving to his graduate students.

 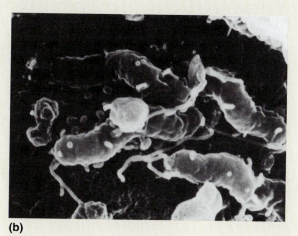

■ *(a) Barry J. Marshall and (b) Helicobacter pylori, the bacterium believed to be involved in most cases of peptic ulcers.* **(a)** **(b)**

In the past, physician biopsies of a patient's stomach tissue were used to detect *H. pylori*, but in 1996, a new and relatively simple, noninvasive **breath test** was approved by the FDA. The patient drinks a urea solution fortified with harmless carbon-13 isotopes. Because *H. pylori* breaks down urea rapidly, the carbon-13 is quickly expelled as CO_2 in the patient's breath, and it can be detected easily if the organism is present. If no isotope is detected, the organism is probably not present. The test can be used

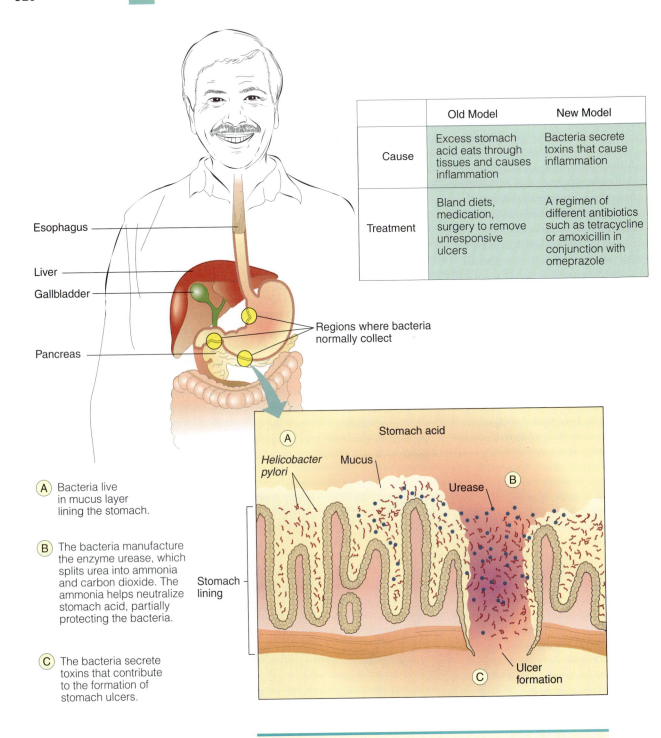

	Old Model	New Model
Cause	Excess stomach acid eats through tissues and causes inflammation	Bacteria secrete toxins that cause inflammation
Treatment	Bland diets, medication, surgery to remove unresponsive ulcers	A regimen of different antibiotics such as tetracycline or amoxicillin in conjunction with omeprazole

Esophagus

Liver

Gallbladder

Pancreas

Regions where bacteria normally collect

A Bacteria live in mucus layer lining the stomach.

B The bacteria manufacture the enzyme urease, which splits urea into ammonia and carbon dioxide. The ammonia helps neutralize stomach acid, partially protecting the bacteria.

C The bacteria secrete toxins that contribute to the formation of stomach ulcers.

Stomach acid

A

Helicobacter pylori   Mucus

Urease   B

Stomach lining

Ulcer formation   C

FIGURE 9.9

The Progression of Peptic Ulcers

Scientists now believe that the majority of peptic ulcers are caused by the bacterium *Helicobacter pylori*. This figure illustrates how they cause an ulcer and highlights the old and new approaches to an ulcer.

to document eradication of the organism after therapy, but recent antibiotic use also would yield a negative result.

The treatment of peptic ulcers with antibiotics has brought permanent relief to thousands of sufferers worldwide. Research is ongoing, because the modes of transmission must be studied, the treatments await refining, and the organism's relationship to stomach cancer is not yet clear. Still, these are historic times for patients and physicians alike.

CAMPYLOBACTERIOSIS MOST OFTEN RESULTS FROM CONSUMPTION OF CONTAMINATED POULTRY

Since the early 1970s, **campylobacteriosis** has emerged from an obscure disease in animals to being the most common bacterial cause of diarrhea in the United States, affecting over 2 million persons each year. The pathogen is *Campylobacter jejuni*, a curved (*campylo-* means "curved"), gram-negative motile rod (FIGURE 9.10). Reservoirs for the organism include the intestinal tracts of many animals, including dairy cattle, chickens, and turkeys. In fact, chicken raised commercially are colonized with *C. jejuni* by the fourth week of life. Contaminated water is also a source of infection.

The clinical symptoms of campylobacteriosis range from mild diarrhea to severe gastrointestinal distress, with fever, abdominal pains, and bloody stools. *Campylobacter jejuni* colonizes the small or large intestine, causing inflammation and occasional mild ulceration. However, the signs and symptoms of campylobacteriosis are not unique. Most patients recover in less than a week without treatment, but some have high fevers and bloody stools for prolonged periods. Erythromycin therapy hastens recovery.

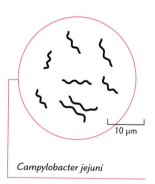

Campylobacter jejuni

FIGURE 9.10

Campylobacter jejuni

Two views of *Campylobacter jejuni*, the cause of campylobacteriosis. (a) A scanning electron micrograph of *C. jejuni* taken from a colony of cells. Note the curved shape of most organisms and the coccus shape of several. (Bar = 1 µm.) (b) A transmission electron micrograph of negatively stained cells, showing the flagellar arrangement of both types of cells. (Bar = 0.5 µm.)

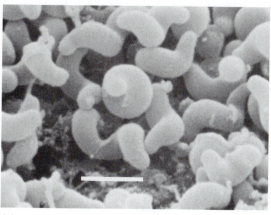

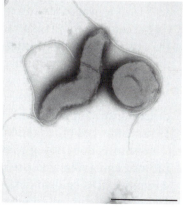

(a)   (b)

Some people may develop a rare nervous system disease several weeks after the diarrheal illness. This disease, called **Guillain-Barré syndrome**, results from the immune system attacking the body's own nerves. The resulting nerve damage can cause paralysis that lasts several weeks and usually requires intensive care. It is estimated that approximately 1 in every 1,000 reported cases of campylobacteriosis leads to Guillain-Barré syndrome and that up to 40 percent of Guillain-Barré syndrome cases in the United States might be caused by campylobacteriosis.

Raw *Campylobacter*-contaminated poultry often is the source of infection. A common route of infection is to cut chicken on a cutting board and then use the unwashed cutting board to prepare other raw foods or vegetables. Other foods also can be contaminated. In 1998, 79 persons at a summer camp were infected with *C. jejuni* through ingestion of contaminated tuna salad. Larger outbreaks due to *C. jejuni* usually result from drinking unpasteurized milk. For example, raw milk was identified as a source of *Campylobacter* in 18 outbreaks between 1981 and 1990, most of which involved school field trips to local dairy farms. Other species, including *C. coli* and *C. upsaliensis*, also have been identified as human pathogens, and as methods of detection improve, the number of reported cases of campylobacteriosis probably will continue to rise.

LISTERIOSIS COMES FROM FECAL-CONTAMINATED FOODS

Listeriosis is caused by *Listeria monocytogenes*, a small gram-positive rod that is motile at room temperature. At one time, this organism was dubbed the "Cinderella" of pathogenic bacteria because its ability to cause disease was not widely recognized. The bacillus is commonly found in the soil and in the intestines of many animals, including birds, fish, barnyard animals, dairy cattle, and household pets. It is transmitted to humans by food contaminated with fecal matter, as well as by the consumption of animal foods. Delicatessen cold cuts, as well as soft cheeses (e.g., Brie, Camembert, feta, and blue-veined cheeses), have been associated with a significant number of cases. Indeed, an outbreak of listeriosis associated with duck liver (used for *fois gras*) was among the first epidemics reported by the CDC in the twenty-first century.

Healthy adults experience few symptoms of *L. monocytogenes* ingestion. The disease primarily affects newborns, pregnant women, the elderly, and others with a weakened immune system.

Listeriosis occurs in many forms. One form, called **listeric meningitis**, is characterized by headaches, stiff neck, delirium, and coma. Another form is a blood disease accompanied by high numbers of white blood cells called **monocytes** (hence the organism's name) (**FIGURE 9.11**). A third form is characterized by infection of the uterus, with vague flu-like symptoms. If contracted during pregnancy, the disease may result in miscarriage of the fetus or mental damage in the newborn. Other individuals suffer respiratory distress, diarrhea, back pain, skin itching, and other nonspecific symptoms that make diagnosing the disease difficult. Penicillin, tetracycline, and other antibiotics are effective treatments.

Listeriosis does not appear to be transmissible among humans. Although the prompt and prolonged use of antibiotics is an effective treatment, relapses are common. CDC epidemiologists estimate that 2,500 cases of listeriosis occur in the United States annually and that 20 percent result in death. The causative organism has been isolated from vegetables, milk, poultry, sausages, and dairy products. The bacterium is considered psychrotrophic, meaning it is able to grow at refrigerator temperatures (Chapter 4).

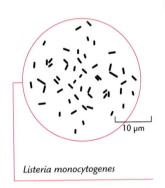

10 µm

Listeria monocytogenes

Monocyte:
a type of large white blood cell that functions in phagocytosis.

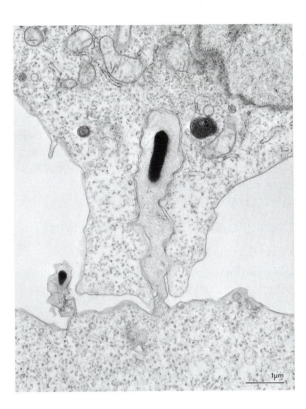

FIGURE 9.11

The Transmission of *Listeria monocytogenes*

A transmission electron micrograph of *Listeria monocytogenes* spreading from one monocyte to another. A *Listeria* cell within one pseudopod (bottom) has been engulfed by a pseudopod from the second cell (top). On pinching off, the bacterium will be inside the cytoplasm of the second cell. (Bar = 1 μm.)

A notable outbreak of listeriosis occurred in late 1998 and early 1999. Close to 100 cases of illness were reported in 22 states, all linked to hot dogs and deli meats distributed by Sara Lee, Ball Park, Hygrade, and other companies. Although health officials never found the bacterial source for certain, they postulated that dust from construction work on air-conditioning units at the production plant may have been involved. Fourteen adults died during the outbreak, and six pregnant women suffered miscarriages.

BRUCELLOSIS IS RARE IN THE UNITED STATES

Brucellosis (also known as Malta fever or Bang's disease) is an occupational hazard of farmers, veterinarians, dairy and meat plant workers, and others who work with large animals. Bacterial transmission can occur by splashing contaminated milk into the eye, by the accidental passage of contaminated fluids through skin abrasions, by contact with infected animals, and by the consumption of contaminated milk and other dairy products. In one example of food transmission, the CDC reported 29 cases in Mexican emigrants living in Houston, Texas. All had eaten goat cheese made from unpasteurized goat's milk. Human-to-human transmission is virtually unknown.

Among the organisms responsible for brucellosis are *Brucella abortus* from cattle, *B. suis* from swine, and *B. melitensis* from goats and sheep. Another species, *B. canis*, is associated with dogs and may be acquired by contact with pets. All are small, gram-negative, nonmotile rods. In animals, brucellosis manifests itself in several organs, especially the reproductive organs because the bacterium multiplies in the tissues of the uterus. Reproductive sterility is a common complication, and pregnant animals are known to abort their young. In veterinary literature, the disease is often referred to as **contagious abortion**.

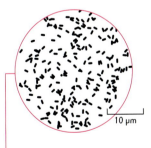

Brucella abortus

The major focus of brucellosis in humans is the blood-rich organs, such as the spleen and lymph glands. Patients experience flu-like weakness, as well as backache, joint pain, and a high fever (with drenching sweats) in the daytime and low fever (with chills) in the evening. This fever pattern gives the disease its alternate name, **undulant fever** (*undulat-* is Latin for "vary"). The disease seldom causes death in humans, and a combination of doxycyline and rifampin speeds recovery. The pasteurization of milk and livestock immunization have reduced the incidence of brucellosis in humans from 6,300 cases in 1947 to about 100 to 200 cases annually. However, a major outbreak of the disease has been taking place since 1997 among bison and elk herds in Yellowstone National Park (FIGURE 9.12).

OTHER BACTERIA CAN CAUSE FOODBORNE AND WATERBORNE DISEASES

A number of other bacterial organisms transmitted by food and water merit brief consideration in this chapter. Some have been recognized as pathogens only in recent years; others have been known for decades.

Among the recent concerns is *Vibrio parahaemolyticus* (FIGURE 9.13). This gram-negative rod is a major cause of foodborne infections in Japan and other areas of the world where seafood is the main staple of the diet. After about a 24-hour incubation period, patients experience acute abdominal pain, vomiting, diarrhea, and watery stools. Some years ago, an outbreak aboard an American cruise ship was linked to seafood salad, and another incident in Louisiana was traced to unrefrigerated cooked shrimp.

Another species of *Vibrio* also has caused recent concern. In 1996, the CDC reported an outbreak of intestinal illness due to *Vibrio vulnificus*. This organism occurs naturally in brackish and seawaters, where oysters and clams live. People who consume these molluscs raw are at risk, especially those with a compromised immune system and those who suffer from liver disease or low stomach acid. (Indeed, taking an antacid after a meal of any contaminated food may neutralize helpful stomach acid and facilitate the passage of the bacteria to the bloodstream.) The gastrointestinal infection involves fever, nausea, and severe abdominal cramps. Often it is accompanied by septicemia, necrotic skin lesions, and cellulitis (a diffuse inflammation of connective tissues of the skin). The mortality rate for the disease is over 50 percent. A notable outbreak related to eating raw seafood occurred in 1999 on New York's Long Island.

FIGURE 9.12

Brucellosis in Bison

Brucellosis poses a threat to bison in Yellowstone National Park, as well as to cattle they encounter outside the confines of the park. A vaccine has not yet been developed for wildlife like these, and resolution of the epidemic remains elusive.

FIGURE 9.13

FIGURE 9.13

Vibrio parahaemolyticus

A scanning electron micrograph of *Vibrio parahaemolyticus*, a gram-negative rod associated with seafood. This marine bacterium is widely distributed in natural aquatic environments around the world and is a well-known foodborne pathogen of the gastrointestinal tract. (Bar = 1 μm.)

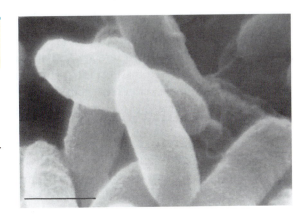

MicroFocus 9.8 FOODBORNE AND WATERBORNE BACTERIAL DISEASES AS BIOLOGICAL WEAPONS

There are several bacteria and bacterial toxins the Centers for Disease Control and Prevention (CDC) and the United States government have identified as potential biological weapons that could be used through foodborne or waterborne routes. Category A agents are of highest risk to national security because they could (1) be easily disseminated or transmitted person to person, (2) cause high mortality with a major public health impact, and (3) cause public panic and social disruption. The toxins of botulism fall into this category.

Category B agents are moderately easy to disseminate and cause moderate morbidity and low mortality. Several foodborne and waterborne pathogens are in this category, including *Clostridium perfringens*, *Staphylococcus aureus*, *Salmonella* species, *Shigella dysenteriae*, *Escherichia coli* O157:H7, *Vibrio cholerae*, and *Brucella* species.

Some of these agents actually have been used to commit biocrimes, the intentional introduction of a biological agent into food or water to sicken small groups of people. The first known biocrime in the United States occurred in The Dalles, Oregon, in 1984 when the Rajneeshee religious cult, in an effort to influence and win seats in the local election, intentionally contaminated salad bars and a city water tank with the bacterium *Salmonella typhimurium*. The result was unsuccessful, although the crime did sicken over 750 citizens and hospitalized 40. Another example occurred in 1996 at a Texas Medical Center. A disgruntled employee deliberately contaminated doughnuts and muffins in a hospital workroom with *Shigella dysenteriae*. All 12 employees who ate the food became ill and 4 were hospitalized.

Bioterrorism has similar aims but on a much larger scale, since the use of such biological agents would cause fear in or actually inflict death or disease upon a large population. Botulinum toxin is one such biological agent from the category A list that, disseminated in food or water, could be used as a bioterror weapon. The muscle-paralyzing botulinum toxins are among the most powerful toxins produced by living organisms.

If large amounts of the toxin could be produced and disseminated in food or water, the net result would be symptoms similar to typical foodborne botulism. A bioterror act should be relatively easy to identify as an act of bioterrorism rather than a natural outbreak because fewer than 200 cases of botulism are reported each year in the United States. Provided medical and health authorities act quickly, antitoxin treatment given before the onset of symptoms would be very effective against a botulism incident. Also, an investigational vaccine, available for high-risk exposures, is being tested.

The United States has more than 57,000 food processors and 1.2 million food retailers amounting to a $200 billion annual business. In May 2003, the Food and Drug Administration (FDA) proposed new regulations requiring food companies to keep better records for tracking foods involved in any future emergencies or terrorism-related contamination. The agency also plans to require advance information of food import shipments to intercept any contaminated products. In July 2003, the FDA began evaluating ways to prevent or reduce the risk of deliberate contamination of the nation's food supply. This might include chemical treatments, temperature controls, and technology intervention. The review should be completed by June 2004.

Still, for most of the agents identified at the top of this box, a large outbreak should signal bioterrorism because these agents normally do not produce large-scale outbreaks in the United States. Nevertheless, let's keep our fingers crossed that these potential bioterror agents remain a distant potential and not a reality.

Bacillus cereus is a gram-positive, spore-forming, motile bacillus that causes food poisoning in two distinct forms, both due to enterotoxins. The first form, *diarrheal*, is accompanied by diarrhea and abdominal pain, while the second form, *emetic*, is characterized by substantial vomiting, frequently experienced after consuming cooked rice. Neither form involves fever, and most patients recover within two days without treatment. In the 1980s, investigators traced a notable outbreak to a college cafeteria where a macaroni and cheese dish harbored a million bacilli per gram. Heat-resistant spores had apparently survived the cooking process. Fried rice prepared at a local restaurant was the cause of a 1993 outbreak in Virginia.

Another emerging cause of foodborne illness is *Yersinia enterocolitica* (FIGURE 9.14). This gram-negative, nonmotile rod is widely distributed in animals and in river and lake waters. The largest episode of disease thus far recorded in the United States occurred in 1982 and involved 172 patients in Arkansas, Tennessee, and Mississippi. Patients experienced fever, diarrhea, and abdominal pain, and most required hospitalization. Investigators believed that contaminated milk was the source. A 1990 outbreak in Georgia involved raw chitterlings (pork intestines). In this case, *Y. enterocolitica* was apparently transferred to a group of children by hand-to-hand contact with the individual preparing the food.

Still another cause of intestinal illness is *Plesiomonas shigelloides*, a gram-negative, facultatively anaerobic, nonmotile rod. *P. shigelloides* normally is limited to the tropical and subtropical climates of Asia and Africa, and it is commonly found in the gut of tropical fish. It can cause intestinal illness in people who eat raw seafood or who travel in the tropics and consume contaminated water or eat contaminated food. A 1990 case, however, appeared to fit neither description. The case occurred in a year-old child

MicroInquiry 9

FOODBORNE AND WATERBORNE DISEASE IDENTIFICATION

Below are several descriptions of foodborne and waterborne diseases based on material presented in this chapter. Read the case history and then answer the questions posed. Answers can be found in Appendix E.

Case 1: An 8-year-old male having diarrhea and abdominal distress is brought to a medical clinic by his mother. He had just returned from a weekend camping trip with friends where the abdominal pain began two days before his return home. Examination of the patient indicates he has a fever and fecal examination presents a bloody stool. A Gram stain reveals curved, gram-negative rods.

9.1a. Based on the examination and laboratory results, what organism most likely is responsible for the patient's symptoms?

9.1b. Propose several ways that the patient could have been exposed to this organism.

9.1c. Describe how foods or water can become contaminated with this organism.

9.1d. What complications can result from the diarrheal illness? What is the reason for these complications?

9.1e. In severe cases of the disease, what antibiotic could be given to shorten the duration of the symptoms?

Case 2: A 23-year-old woman comes to the emergency room complaining of lethargy and appears disoriented. She admits to abdominal pain and has a fever. She indicates that she recently returned from an Amazon River cruise where, on the last day, she ate

with the local people and drank local water. She had not been vaccinated for any infectious diseases before her trip. She exhibits a bloody stool and a blood smear indicates the presence of gram-negative rods.

9.2a. Given her travel history, identify three bacterial agents with which she has increased likelihood of being infected.

9.2b. Of the bacterial diseases presented in this chapter, which ones can be ruled out?

9.2c. Based on her symptoms and laboratory results, what bacterial disease is mostly likely responsible for her infection?

9.2d. Give two possible ways she could have contracted the disease.

9.2e. How might she have prevented getting the disease?

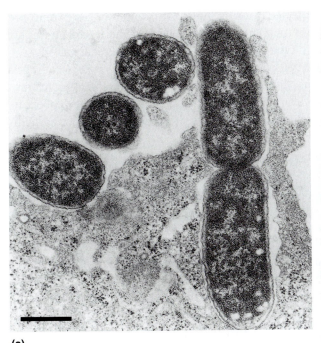

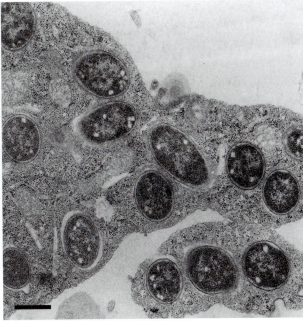

(a)   (b)

FIGURE 9.14

Yersinia enterocolitica

Transmission electron micrographs of invasive *Yersinia enterocolitica*. (a) A number of bacteria attached to the plasma membrane of infected cells. One bacterium appears to be undergoing cell division and entering the cell at the same time. (Bar = 1 μm.) (b) Several *Y. enterocolitica* cells observed within the cytoplasm of an infected cell. The bacteria are in membrane-enclosed vacuoles. Researchers postulate that a single gene is responsible for the invasive capacity displayed by this organism. (Bar = 2 μm.)

Case 3: A 27-year-old man arrives at the hospital emergency room complaining of headache, stiff neck, and irritability. He indicates that he had a kidney transplant about a year ago necessitated by an automobile accident. A blood culture is taken. A blood smear shows the presence of gram-positive rods and an assay for cell motility is positive. Asked about his recent eating habits, the patient says that he had purchased some feta cheese recently from an ethnic food market.

9.3a. Given the patient's symptoms and recent eating habits, what bacterial agent is most likely the cause of the disease?

9.3b. What is the significance of the headache, stiff neck, and irritability symptoms?

9.3c. What antibiotics should be given to the patient?

9.3d. What should immunocompromised individuals do to avoid this scenario and why doesn't it apply to healthy adults?

9.3e. Why is it significant that this agent is psychrotrophic?

Case 4: A previously healthy 12-year-old male comes to the emergency with his father. The son has exhibited bloody diarrhea and abdominal pain for the past 24 hours. The diarrhea is now worse, triggering the visit to the emergency room. His physical examination is unremarkable except for dehydration. Questioning the father, neither has been around anyone else with diarrhea and they have not traveled out of the country. His father does admit to cooking his son a hamburger using meat that had been sitting on the kitchen counter for "some time."

9.4a. Based on the symptoms and family activities, what organism and strain is the cause of the disease?

9.4b. What two clues lead you to this conclusion?

9.4c. Explain what other bacterial agents, although similar, would be ruled out.

9.4d. What is unique about this disease-causing strain?

9.4e. How can this strain of the bacterial species be culturally identified from other strains of the species?

who had apparently been infected by water in the bathroom tub. Her parents previously had poured water from their aquarium into the tub. The aquarium was home to a collection of piranhas, a type of tropical fish the family kept as pets.

Clinical reports in the 1980s documented that *Aeromonas* species, especially *Aeromonas hydrophila*, cause human gastrointestinal disease. The organisms are gram-negative, motile rods commonly found in soil and water. They appear to be transmitted by food. Both cholera-like and dysentery-like diarrheas, ranging from

TABLE 9.2

A Summary of Foodborne and Waterborne Bacterial Diseases

DISEASE	CAUSATIVE AGENT	DESCRIPTION OF AGENT	ORGANS AFFECTED	CHARACTERISTIC SIGNS
Botulism	*Clostridium botulinum*	Gram-positive spore-forming rod	Neuromuscular junction	Paralysis
Staphylococcal food poisoning	*Staphylococcus aureus*	Gram-positive staphylococcus	Intestine	Diarrhea Nausea Vomiting
Clostridial food poisoning	*Clostridium perfringens*	Gram-positive spore-forming rod	Intestine	Diarrhea Cramping
Typhoid fever	*Salmonella typhi*	Gram-negative rod	Intestine Blood Gall bladder	Ulcers Fever Rose spots
Salmonellosis	*Salmonella* serotypes	Gram-negative rods	Intestine	Fever Diarrhea Vomiting
Shigellosis	*Shigella* serotypes	Gram-negative rods	Intestine	Diarrhea Dysentery
Cholera	*Vibrio cholerae*	Gram-negative curved rod	Intestine	Rice-water stools Extreme diarrhea Shock
E. coli diarrheas	*Escherichia coli*	Gram-negative rod	Intestine	Diarrhea
Campylo-bacteriosis	*Campylobacter jejuni*	Gram-negative curved rod	Intestine	Diarrhea Fever
Listeriosis	*Listeria monocytogenes*	Gram-positive small rod	Intestine Meninges Monocytes	Diarrhea Meningitis
Brucellosis	*Brucella* species	Gram-negative rods	Spleen Lymph glands	Undulating fever Joint pain
Other foodborne and waterborne diseases	*Vibrio parahaemolyticus*	Gram-negative rod	Intestine	Diarrhea
	Vibrio vulnificus	Gram-negative rod	Intestine	Diarrhea
	Bacillus cereus	Gram-positive spore-forming rod	Intestine	Diarrhea
	Yersinia enterocolitica	Gram-negative rod	Intestine	Diarrhea
	Pleisomonas shigelloides	Gram-negative rod	Intestine	Diarrhea
	Aeromonas hydrophila	Gram-negative rod	Intestine	Diarrhea

mild to severe, have been reported in patients. An enterotoxin may be responsible for the symptoms.

The foodborne and waterborne bacterial diseases discussed in this chapter are summarized in TABLE 9.2. Several of the diseases mentioned in this chapter could be agents for bioterrorism, as MicroFocus 9.8 describes. Also, MicroInquiry 9 provides you with an opportunity to identify some of the diseases and bacterial pathogens discussed in this chapter.

TOXIN INVOLVED	TREATMENT ADMINISTERED	IMMUNIZATION AVAILABLE	COMMENT
Yes	Antitoxin	None	Most powerful toxin known Infant and wound botulism possible
Yes	None	None	Affects millions of Americans annually Due to an enterotoxin Brief incubation period
Yes	None	None	Common in protein-rich foods Spores survive cooking
Not established	Chloramphenicol	Vaccine of dead bacteria	Spread from carriers About 400 cases annually in US
Not established	Not recommended	None	Associated with poultry products Most reported foodborne disease in US About 40,000 cases annually
Yes	Rehydration Antibiotics	None	May be accompanied by dysentery R plasmids limit antibiotic use
Yes	Rehydration	Vaccine of dead bacteria	Danger from dehydration Rare in US Raw seafood involved
Yes	Antibiotics Rehydration	None	Symptoms due to enterotoxin Enteroinvasive strains observed Possible hemolytic uremic syndrome
Not established	Erythromycin	None	Associated with raw milk Symptoms mild to serious
Not established	Antibiotics	None	Dangerous in pregnancy Associated with animals Soft cheeses implicated
Not established	Doxycline Rifampin	None	Induces abortion in barnyard animals Hazardous to animal workers
Not established	Various antibiotics	None	Associated with seafood
Not established	Various antibiotics	None	Associated with shellfish
Yes	None	None	Spores survive in cooked foods
Not established	Various antibiotics	None	Widely found in animals
Not established	Various antibiotics	None	Occurs in tropical climates
Yes	Various antibiotics	None	Common in soil and water

Note to the Student

For over a half-century, *Escherichia coli* has been the tireless workhouse of biological research. In the 1940s, it was used as a host organism to determine the life cycle of viruses. Many of the important metabolic pathways, including the renowned Krebs cycle, were worked out first in this organism. In the 1950s, biochemists used *E. coli* to discover the three forms of microbial recombination. In the 1960s, it was the major research organism for deciphering the genetic code and learning how genes work. In the 1970s, *E. coli* became the guardian of public health as a valuable indicator of water pollution. It also emerged as an industrial giant for producing enzymes, growth factors, and vitamins. Since the 1980s, biochemists have used it as a living factory to produce an array of genetically engineered pharmaceuticals. And in the 1990s, *E. coli* continued to illustrate how bacteria can be put to work in the interest of science and for the betterment of humanity.

Then came *E. coli* O157:H7. To be sure, this strain has caused much human misery and pain because of its propensity to invade the intestinal tissues, pass to the blood, and cause serious injury to the kidneys. It has made us more careful of what we eat and has caused us to think twice about having a rare hamburger. Along the way it has become the "germ of the week" on *Dateline, 20–20*, and other television news programs, where it has been portrayed as the chief villain among a world of villains. Unfortunately, it also has made us forget all the good things that *E. coli* has done for us. And what a shame that is! Perhaps one bad apple can, indeed, spoil the whole barrel.

Summary of Key Concepts

A recurring theme of this chapter is that foodborne and waterborne bacterial diseases primarily affect the intestinal tract and are of two types: intoxications and infections.

9.1 INTRODUCTION TO FOODBORNE AND WATERBORNE BACTERIAL DISEASES

- **Foodborne and Waterborne Diseases Follow a Course of Intoxication or Infection.** Intoxications involve diseases caused by bacterial toxins while infections refer to diseases arising from the growth of bacteria in the body. An incubation period occurs before the symptoms appear. Symptoms depend on the specific toxin or microbe and the numbers ingested.

- **There Are Several Ways Foods or Water Become Contaminated.** Some bacterial toxins are resistant to heat. Cattle and poultry carcasses can become contaminated during slaughter. Fresh fruits and vegetables can become contaminated when washed or irrigated with water that is contaminated with animal manure or human sewage. Infected humans can contaminate food they handle, or there can be cross-contamination from other raw agricultural products. Some bacterial pathogens cause foodborne or waterborne bacterial diseases only when they are in large numbers.

9.2 FOODBORNE AND WATERBORNE INTOXICATIONS

- **Botulism Is Caused by an Exotoxin in Improperly Canned Foods.** Botulism is a severe form of food poisoning caused by a nerve toxin. Symptoms include blurred vision, slurred speech, and difficulty swallowing and breathing. Botulism is treated by administration of an antitoxin. Patients often need to be placed on a ventilator.

- **Staphylococcal Food Poisoning Also Is Caused by an Exotoxin.** *Staphylococcus aureus* is a pathogen causing food poisoning through ingestion of pre-formed enterotoxins. Abdominal cramps, nausea, vomiting, and diarrhea are common symptoms.

- **Clostridial Food Poisoning Results from an Enterotoxin.** *Clostridium perfringens* causes food poisoning through the ingestion of the enterotoxin. Symptoms include watery diarrhea and abdominal pain.

9.3 FOODBORNE AND WATERBORNE INFECTIONS

- **Typhoid Fever Involves a Blood Infection.** *Salmonella typhi* is an acid-resistance bacillus that invades the blood. The infection is characterized by fever, abdominal rose spots, lethargy, and delirium. Typhoid vaccines are available and chloramphenicol is the antibiotic of choice.

- **Salmonellosis Can Be Contracted from a Variety of Foods.** The infection is caused by the ingestion of a *Salmonella* serotype. Symptoms include fever, nausea, vomiting, diarrhea, and abdominal cramps.

- **Shigellosis (Bacterial Dysentery) Occurs Where Sanitary Procedures Are Lacking.** Ingestion of *Shigella* and the production of enterotoxins cause the disease. Symptoms include fever, abdominal cramps, and sometimes bloody diarrhea.

- **Cholera Can Involve Enormous Fluid Loss.** *Vibrio cholerae* infects the small intestines, producing an enterotoxin that causes profuse watery diarrhea and vomiting. Oral rehydration therapy restores electrolyte and glucose balance in the body.

- **E. coli Diarrheas Cause Various Forms of Gastroenteritis.** Several strains of *E. coli* can cause various forms of gastroenteritis. Watery diarrhea, caused by enterotoxin production, is typical of traveler's diarrhea. *E. coli* O157:H7 is a more serious form of diarrhea that can lead to complications called hemorrhagic colitis. Often this involves the kidneys, causing hemolytic uremic syndrome (HUS).

- **Peptic Ulcer Disease Can Be Spread Person to Person.** The presence of *Helicobacter pylori* is responsible for most cases of peptic ulcers. The exact transmission mechanism is not understood, although contaminated food or water is a likely candidate. High cure rates have been attained with antibiotics and acid suppressors.

- **Camplyobacteriosis Most Often Results from Consumption of Contaminated Poultry.** *Campylobacter jejuni* is the causative agent for this foodborne and waterborne infection. Symptoms include fever, abdominal pains, and bloody stools. If treatment is needed, erythromycin is the antibiotic used.

- **Listeriosis Comes from Fecal-Contaminated Foods.** Fecal-contaminated foods often contain *Listeria monocytogenes*. In individuals with a weakened immune system, listeric meningitis can occur. Infection in pregnant women can lead to miscarriage or mental damage to the newborn.

- **Brucellosis is Rare in the United States.** Several species of *Brucella* can cause brucellosis. Reproductive sterility in animals is a common complication of the infection.

- **Other Bacteria Can Cause Foodborne and Waterborne Diseases.** Several other bacteria can cause foodborne and waterborne diseases. These include *Vibrio parahaemolyticus*, *Vibrio vulnificus*, *Bacillus cereus*, *Yersinia enterocolitica*, *Plesiomonas shigelloides*, and *Aeromonas hydrophila*.

Questions for Thought and Discussion

Answers to selected questions can be found in Appendix C.

1. The story is told of an early nineteenth-century doctor in New York City who was an expert at diagnosing typhoid fever even before the symptoms of disease appeared. The doctor's forte was the tongue. He would go up and down the rows of hospital beds, feeling the tongues of patients and announcing that the patient was in the early stages of typhoid. Sure enough, a few days later the symptoms would surface. What do you think was the secret to his success?

2. In 1997, researchers in Boston reported that *Helicobacter pylori* accumulates in the gut of houseflies after the flies feed on food containing the bacteria. What are the implications of this research?

3. You have volunteered to do the supermarket shopping for the upcoming class barbecue. What are some precautions you can take to ensure that the event is remembered for all the right reasons?

4. In 1997, Swedish researchers provided evidence that *Salmonella* serotypes are infecting penguins on Bird Island, a remote island in the distant South Atlantic.

The organisms, not normally found in the birds, are presumably of human origin. Assuming humans had not visited the island in the previous decades, how might the bacteria have reached the birds?

5. "Finding the source of an outbreak is like trying to go back to a forest fire to find a spark." So said an exasperated University of Saskatchewan researcher explaining the difficulty of tracing the origin of a 1997 botulism outbreak in water fowl in which millions of birds died. The risk of botulism appears to rise when wetlands have a neutral pH and are relatively salt free. Also, it is known that water birds often acquire botulism by eating insect larvae that have concentrated the disease-related neurotoxin. With these facts in mind, develop a possible scenario in which such things as temperature, pH, salinity, pesticide use, and other wetland factors may contribute to a botulism epidemic.

6. Some years ago, the CDC noticed a puzzling trend: Reported cases of salmonellosis seemed to soar in the summer months, then drop radically in September. Can you venture a guess why this is so?

7. A laboratory instructor proposes to demonstrate the growth of soilborne organisms in canned food in the following way: A can is to be punctured with an ice pick and a small pinch of rich soil introduced to the can; the hole will then be sealed with candle wax and the can incubated. A fellow laboratory instructor hears of the experiment, reacts with concern, and advises the first instructor not to perform it. Why?

8. Between February 18 and 22, 1987, botulism was diagnosed in 11 patrons of a restaurant in Vancouver, B.C. The disease was subsequently traced to mushrooms bottled and preserved in the restaurant. What special cultivation practice enhances the possibility that mushrooms will be infected with the spores of the organism of botulism?

9. In preparation for a summer barbecue, a man cuts up chickens on a wooden carving board. After running the board under water for a few seconds, he uses it to cut up tomatoes, lettuce, peppers, and other salad ingredients. What sort of trouble is he asking for?

10. When a woman died of botulism in 1980, a public official was quoted in an interview as saying: "This death might have been prevented if any of the doctors involved had recognized and acted early on the symptoms of the disease they were dealing with." Do you believe the doctors were at fault? What might be your reply to the official?

11. During 1992, the North Carolina Department of Health received reports of illness in 18 workers at a local pork processing plant. All the affected employees worked on the "kill floor" of the plant. All had gram-negative rods in their blood. Their symptoms included fever, chills, fatigue, sweats, and weight loss. Which disease was pinpointed in the workers?

12. Three days ago, two hamburgers were purchased at the local market. One was frozen, the other remained chilled in the refrigerator section. Both are now placed on the grill. All other things being equal (size of hamburger, cooking time, source of meat, use of condiments, and so on), which hamburger might be safer to eat if you wish to avoid food poisoning?

13. Most physicians agree that the illness called "stomach flu" is not influenza at all. They maintain that the cramps, diarrhea, and vomiting can be due to a variety of bacteria and viruses. Which organisms in this chapter might be good candidates?

14. "You don't have to travel anymore to get traveler's diarrhea." So announced a state epidemiologist to thousands of attendees at a recent convention on emerging diseases. He was referring to the unusually high incidence of foodborne and waterborne diseases in the United States. What conditions in modern society would lead him to make this announcement?

15. In 1986, a New Rochelle, New York, frozen-food manufacturer recalled thousands of packages of jumbo stuffed shells and cheese lasagna after a local outbreak of salmonellosis. Which parts of the pasta products would attract the attention of inspectors as possible sources of *Salmonella*? Why?

16. Studies have shown that bismuth subsalicylate (Pepto-Bismol) can be used to prevent traveler's diarrhea, but the treatment is not easy to follow: 2 ounces or 2 tablets four times a day for 3 weeks before travel begins. Short of turning pink, what other measures can you use to prevent traveler's diarrhea while visiting other countries?

17. An estimated 25 million Americans will have an ulcer at some time of their lives, yet a recent study indicates that two-thirds still don't know that *Helicobacter pylori* causes 90 percent of ulcers. Suppose you were a public health official. What would you do to get the word out?

18. The CDC estimate that each year, 20,000 individuals suffer diarrhea from *E. coli* O157:H7, and that over 200 people die as a result of infection. Despite this prevalence, relatively few cases are diagnosed. What factors may contribute to misdiagnoses or nondiagnoses?

The preceding pages have summarized some of the major bacterial diseases transmitted by food and water. To test your knowledge of the chapter's contents, rearrange the scrambled letters and insert the correct word in each of the missing spaces. The answers are listed in Appendix D.

1. To treat patients who have botulism, large doses of _____ must be administered.

I I T A X N N T O

2. One of the most excessive diarrheas observed in an intestinal disease is associated with _____.

R L H E A C O

3. The fever pattern in brucellosis gives the disease its alternate name of _____ fever.

A U U T L N N D

4. Disease associated with *Shigella* species is usually accompanied by a syndrome of cramps and bloody stools called _____.

N S D E E Y Y T R

5. In cases of typhoid fever, the abdomen is often covered with a series of _____ spots.

E S R O

6. _____ food poisoning has a relatively short incubation period and is caused by a toxin deposited in food during aerobic bacterial growth.

O H L A Y S P C C T L O C A

7. Because *Salmonella* serotypes commonly infect _____, any products derived from these animals are potentially infected.

H C N K I E S C

8. A small percentage of those who recover from typhoid fever remain _____ and can spread the organism in their feces.

A C R E I R S R

9. *Escherichia coli* is a common gram-_____ rod that can be a cause of infantile and traveler's diarrhea.

G N V I A E T E

10. A spore-forming aerobic rod that can cause food-borne illness is *Bacillus* _____.

R C E S U E

11. A curved rod belonging to the genus _____ is known to cause mild to severe diarrhea in humans.

Y E A P O C C L A T R M B

12. In an animal such as a cow, infection with *Brucella* may result in the _____ of the fetus.

O T O A R I N B

13. A major symptom in patients experiencing botulism is _____ of the limbs and respiratory muscles.

A L S P A Y S R I

14. Many instances of staphylococcal food poisoning originate with staphylococci found in the _____.

O S E N

15. *Salmonella typhi* is able to reach the human intestinal tract because it is highly resistant to _____.

C I A D

16. One of the dangers of shigellosis is excessive _____ loss in the patient.

L I D U F

17. Outbreaks of cholera in the United States can best be described as _____.

A E R R

18. Consumers of seafood are particularly vulnerable to intestinal infection due to a species of _____.

O I B R V I

19. The most reported foodborne infection in the United States is that caused by types of _____.

A M E A S O L L L N

20. To avoid staphylococcal food poisoning, consumers should be sure to _____ all left-over foods before eating them.

E H T A

21. The organism *Clostridium perfringens* multiplies in foods only under _____ conditions.

E O I N R C A A B

22. Diagnosis of many of the intestinal diseases is assisted by the isolation of bacteria from _____ specimens.

O O T S L

23. Diseases such as shigellosis, typhoid fever, and cholera may be contracted by consuming contaminated _____.

O E F O D S A

24. In recent years, raw _____ has been implicated in many outbreaks of campylo-bacteriosis.

L K M I

25. Those who work with large _____ may be exposed to the bacterium that causes brucellosis.

M N L A A I S

http://microbiology.jbpub.com

The site features **eLearning,** an on-line review area that provides quizzes and other tools to help you study for your class. You can also follow useful links for in-depth information, or just find out the latest micro-biology news.

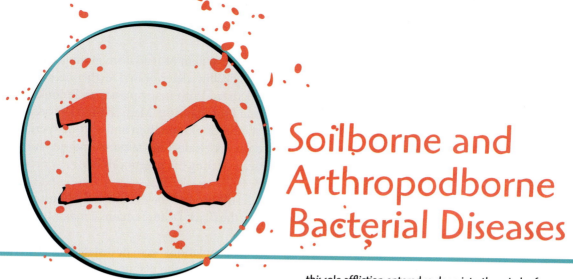

10 Soilborne and Arthropodborne Bacterial Diseases

. . . this sole affliction entered so deep into the minds of men and women that . . . fathers and mothers were found to abandon their own children, as if they had been strangers.

—Italian novelist Giovanni Boccaccio writing in *Decameron* about the Black Death of 1348

BUBONIC PLAGUE, COMMONLY known as "the Black Death," was probably the greatest catastrophe ever to strike Europe. It swept back and forth across the continent for almost a decade, each year increasing in ferocity. By 1348, two-thirds of the European population were stricken and half of the sick had died. Houses were empty, towns were abandoned, and a dreadful solitude hung over the land. The sick died too quickly for the living to bury them, and at one point, the Rhône River was consecrated as a graveyard for plague victims. Contemporary historians wrote that posterity would not believe such things could happen, because those who saw them were themselves appalled. The horror was almost impossible to imagine; to many people, it was the end of the world.

Before the century concluded, the Black Death visited Europe at least five more times in periodic reigns of terror. During one epidemic in Paris, an estimated 800 people died each day; in Siena, the population dropped from 42,000 to 15,000; and in Florence, almost 75 percent of the citizenry perished. Flight was the chief recourse for people who could afford it, but ironically, the escaping travelers spread the disease. Those who remained in the cities were locked in their homes until they succumbed or recovered.

The immediate effect of the plague was a general paralysis in Europe. Trade ceased and wars stopped. Bewildered peasants who survived encountered unexpected prosperity because landowners had to pay higher wages to obtain help. Land values declined and class relationships were upset, as the system of feudalism gradually crumbled. The authority of the clergy, already in decline, deteriorated further because the Church was helpless in the face of the disaster. With many priests and monks dead, a new order of reformers arose to found Protestantism. Medical practices became increasingly sophisticated, with new standards of sanitation and a 40-day period of detention (a "quarantine") imposed on vessels docking at ports. Indeed, the mechanical clock came into widespread use, reflecting the urgency of life.

The graveyard of plague left fertile ground for the renewal of Europe during the Renaissance. To many historians, the Black Death remains a major turning point in Western civilization.

Nor did it end there. European populations were devastated also by typhus and relapsing fever. Both of these diseases, like the plague, are transmitted by **arthropods**, and both can be interrupted by arthropod control. Neither is a major problem in our society, but other arthropodborne diseases—such as Lyme disease, tularemia, and Rocky Mountain spotted fever—occasionally are reported in the news media. We shall study each of these diseases in this chapter.

To begin, we shall examine a number of soilborne diseases where organisms enter the body through a cut, wound, or abrasion, or by inhalation. Among these are anthrax, a recent and feared disease in bioterrorism, and tetanus, a concern to anyone who has stepped on a nail or piece of glass. We also shall study other diseases receiving wider recognition as detection methods improve. The soilborne diseases, as well as the arthropodborne diseases, are primarily problems of the blood.

Arthropods:
insects and other animals with jointed appendages, segmented bodies, and outer skeletons of chitin.

10.1 Soilborne Bacterial Diseases

Soilborne bacterial diseases are those whose agents are transferred from the soil to the unsuspecting individual. To remain alive in the soil, the bacteria must resist environmental extremes, and often the bacteria form spores, as the first three diseases illustrate.

ANTHRAX PRODUCES THREE EXOTOXINS

Anthrax is primarily a disease of large hoofed mammals such as cattle, sheep, and goats. It is caused by *Bacillus anthracis*, a gram-positive spore-forming aerobic rod (**FIGURE 10.1**). Animals ingest the spores from the soil during grazing, and soon they are overwhelmed with vegetative bacteria as their organs fill with bloody black fluid (*anthrac* is the Greek word for "coal"; the disease name is thus a reference to the blackening of the blood). About 80 percent of untreated animals die, and since the bacteria remain in the dead body as spores, it is often necessary to cremate the carcass or bury it deeply in lime to prevent soil contamination. Burning the field also may be required.

Humans acquire anthrax in a number of ways. Workers who tan hides, shear sheep, or process wool may inhale the spores and contract inhalational (pulmonary) anthrax, often called **woolsorter's disease**. Consumption of contaminated meat may lead to gastrointestinal anthrax. Skin contact with spores may lead to cutaneous anthrax. Animal products related to past outbreaks of anthrax have included violin bows, shaving bristles, goatskin drums, and leather jackets.

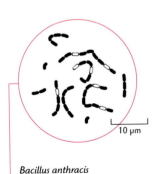

10 μm

Bacillus anthracis

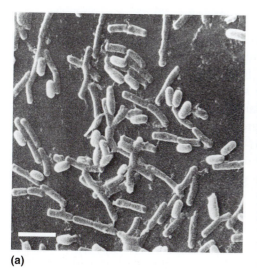

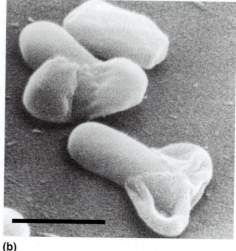

(a) (b)

Bacillus anthracis

Bacillus anthracis is the cause of anthrax. (a) Free spores and vegetative cells of *B. anthracis* visualized with the scanning electron microscope. Note the oval shape of the spores and the typical rod shape of the vegetative cells. (Bar = 5 μm.) (b) Anthrax spores in the process of germinating. The spore coat of the spore in the center of the photograph has divided and is beginning to separate. A vegetative cell may be seen emerging from the spore at the bottom. (Bar = 2 μm.)

Anthrax spores germinate rapidly on contact with human tissues. The thick protein capsule of the cells impedes phagocytosis, and the organisms produce three exotoxins that work together to cause disease. Capsule and toxins are coded by genes carried on two plasmids.

Inhalation anthrax initially resembles a common cold (fever, chills, cough, chest pain, headache, and malaise). After several days, the symptoms may progress to severe breathing problems and shock. Inhalation anthrax is usually fatal without early treatment. **Cutaneous anthrax** accounts for about 95 percent of all anthrax infections. Skin infection begins as a raised itchy bump that resembles an insect bite, but within one to two days it develops into a vesicle and then a painless ulcer, usually 1 to 3 cm in diameter, with a characteristic black necrotic (dying) area in the center (**FIGURE 10.2**). Lymph glands in the adjacent area may swell. About 20 percent of untreated cases of cutaneous anthrax will result in death. Deaths are rare with appropriate antimicrobial therapy. **Intestinal anthrax** is characterized by an acute inflammation of the intestinal tract. Initial signs include nausea, loss of appetite, vomiting, and fever. This is followed by abdominal pain, vomiting of blood, and severe diarrhea. Intestinal anthrax results in death in 25 to 60 percent of untreated cases.

In 2000 to 2002, the total number of anthrax cases in the United States was two per year. A vaccine prepared with an attenuated strain of *B. anthracis* has been successful

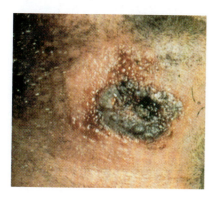

An Anthrax Lesion

This cutaneous lesion is a result of infection with anthrax bacilli. Lesions like this one develop when anthrax spores contact the skin, germinate to vegetative cells, and multiply.

in reducing outbreaks in domestic herds. A cell-free filtrate vaccine is available to veterinarians and others who work with livestock. Six inoculations over an 18-month period are required for immunity, which is 93 percent effective.

Anthrax also is considered a threat in bioterrorism and in biological warfare (MicroFocus 10.1). This is because the spores can be aerosolized in microscopic droplets for widespread distribution to large areas and spread to substantial numbers of people. Moreover, the bacillus can be grown fairly easily and in large quantities, and the spores are extraordinarily resistant to destruction. An aerosol of spores could be released unobtrusively and could drift through a large city without being detected. At present, there is no vaccine for civilian use, but since 1998, all military personnel receive the cell-free filtrate vaccine.

The seriousness of using biological agents as a means for bioterrorism was underscored in October 2001 when *B. anthracis* spores were distributed intentionally through the United States mail. In all, 22 cases of anthrax (11 inhalation and 11 cutaneous) were identified, making the case-fatality rate among patients with inhalation anthrax 45 percent (5/11). The six other individuals with inhalation anthrax and all the individuals with cutaneous anthrax recovered. Had it not been for antibiotic therapy, many more might have been stricken.

TETANUS PRODUCES HYPERACTIVE MUSCLE CONTRACTIONS

Tetanus is one of the most dangerous human diseases. *Clostridium tetani*, the bacillus that causes it, occurs everywhere in the environment, especially the soil.

MicroFocus 10.1

THE LEGACY OF GRUINARD ISLAND

In 1941, the specter of airborne biological warfare hung over Europe. Fearing that the Germans might launch an attack against civilian populations, British authorities performed a series of experiments to test their own biological weapons. The spot chosen was Gruinard Island, a mile-long patch of land off the coast of Scotland. Investigators placed 60 sheep on the island and exploded a bomb containing anthrax spores overhead. Within days, all the sheep were dead.

Warfare with biological weapons never came to reality in World War II, but the contamination of Gruinard Island remained. A series of tests in 1971 showed that anthrax spores were still alive at and below the upper crust of the soil, and that they could be spread by earthworms. Officials posted signs

warning people not to set foot on the island, but did little else.

Then a strange protest occurred in 1981. Activists demanded that the British government decontaminate the island. They backed their demands with packages of soil taken from the island. Notes led government officials to two 10-pound packages of spore-laden soil, and the writers threatened that 280 pounds were hidden elsewhere.

Partly because of the protests, the British government instituted a decontamination of the island in 1986. Technicians used a powerful brushwood killer, combined with burning and treatment with formalin in seawater. Finally, they managed to rid the soil of anthrax spores. By April 1987, sheep were once again grazing on the island. However, people were somewhat reluctant to return.

Gruinard Island remains a monument of sorts to the effects of biological warfare.

Spores enter a wound in very small numbers and revert to vegetative bacilli that produce the second most powerful toxin known to science (after the botulism toxin). The toxin provokes sustained and uncontrolled contractions of the muscles, and spasms occur throughout the body. Patients often experience violent deaths.

C. tetani was first isolated in 1889 by the Japanese bacteriologist Shibasaburo Kitasato. It is a gram-positive, anaerobic, spore-forming bacillus found in the intestines of many animals and humans. The bacillus possesses few invasive tendencies and does not cause intestinal disease. However, the spores are excreted in the feces to the soil, and later they enter the dead, oxygen-free tissue of a wound. The wound may result from a fracture, gunshot, animal bite, or puncture by a piece of glass, a thorn, or a needle. Rusty nails pose a threat because spores cling to the rough edges of the nail and the nail may cause extensive tissue damage as it penetrates. (The rust itself is of no consequence.)

Once inside the tissue, the spores germinate to vegetative cells that produce several toxins. The most important of these toxins appears to be **tetanospasmin**, an exotoxin of high molecular weight. At the synapse, this toxin blocks the relaxation pathway that follows muscle contraction. The toxin prevents the release of glycine (an amino acid) and other neurotransmitters needed to inhibit muscle contraction. Without any inhibiting influence, volleys of spontaneous impulses arise in the nerves, causing the muscles to contract continuously and without control (**FIGURE 10.3**).

Symptoms of tetanus develop rapidly, often within hours. A patient experiences generalized muscle stiffness, especially in the facial and swallowing muscles. Spasms of the jaw muscles cause the teeth to clench and bring on a condition called **trismus,** or **lockjaw.** Severe cases are characterized by a "fixed smile," arching of the back (**opisthotonus**), spasmodic inhalation, and seizures in the diaphragm and rib muscles, leading to reduced ventilation and death. Patients are treated with sedatives and muscle relaxants, and are placed in quiet, dark rooms. Physicians prescribe penicillin to destroy the organisms and tetanus antitoxin to neutralize the toxin.

Immunization to tetanus may be rendered by injections of tetanus toxoid in the diphtheria-tetanus-acellular pertussis (DTaP) vaccine. The toxoid, developed in 1933 by Gaston Ramon, is prepared by treating the toxin with formaldehyde to eliminate its toxic quality. Children usually receive the first of several injections at the age of two months. Booster injections of tetanus toxoid in the **Td vaccine** (a "tetanus shot") are recommended every 10 years to keep the level of immunity high.

The United States has had a steady decline in the incidence of tetanus, with 37 cases confirmed in 2001. Most cases occur in the young, who are more likely to have contact with soil, and in the elderly, in whom antibody levels have dropped or vaccination never took place. In other parts of the world, tetanus remains a major health

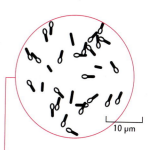

Clostridium tetani

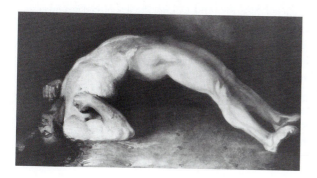

FIGURE 10.3

Tetanus

A photograph of a painting by Scottish surgeon Charles Bell showing a soldier dying of tetanus. The soldier was wounded at the battle of Corunna in 1809. Note that the muscles are fully contracting in virtually all parts of the body from head to toes. The soldier's face shows the clenched jaw and fixed ("sardonic") smile characteristic of lockjaw. Tetanus was once common in soldiers wounded in battle.

problem and is often related to traditional customs. In some countries, for example, unsanitary ear piercing is common, tattooing is widespread, and the umbilical stump of newborns is dressed with soil. In other cases, a simple splinter of wood may be the source of entry (as **MicroFocus 10.2** illustrates).

GAS GANGRENE CAUSES MASSIVE TISSUE DAMAGE

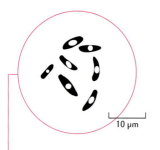

Clostridium perfringens

To understand gas gangrene as an infectious disease, it is important to understand the physiological term *gangrene*. Gangrene is a condition that develops when the blood flow ceases to a part of the body, usually as a result of blockage by dead tissue. The body part, generally an extremity, becomes dry and shrunken, and the skin color changes to purplish or black. The gangrene may spread as enzymes from broken cells destroy other cells, and the tissue may have to be excised (debrided) or the body part amputated. This form of gangrene is called dry gangrene (**FIGURE 10.4**).

Gas gangrene, or moist gangrene, occurs when soilborne bacteria invade the dead, anaerobic tissue. The organism responsible is *Clostridium perfringens*. These anaerobic, spore-forming, gram-positive rods multiply rapidly, while fermenting the muscle carbohydrates and putrefying the muscle proteins. Large amounts of gas result from this metabolism, causing a crackling sound, that tear the tissue apart. The gas also presses against blood vessels, thereby blocking the flow and forcing cells away from their blood supply. In addition, the organisms secrete lecithinase, an enzyme that dissolves cell membranes and releases toxic cellular enzymes. Two other bacterial enzymes, hyaluronidase and hemolysin, facilitate the passage of bacteria among the cells and destroy red blood cells, respectively. Neurotoxins also are released.

The symptoms of gas gangrene include intense pain and swelling at the wound site, as well as a foul odor. Initially the site turns dull red, then green, and finally blue black. Anemia is common, and bacterial toxins may damage the heart and

MicroFocus 10.2

SHOELACES

The untied shoelaces told the story—that and the difficult swallowing and distorted facial features. She had left the hospital the day before, but the symptoms were not gone—not yet, at least.

It began on Sunday, July 5, 1992. The 4th of July weekend was hot in Rutland, Vermont, and most of her friends were recovering from the previous day's celebrations. But she decided to catch up on her gardening. The ground was warm, and she took off her shoes to walk about the garden—nothing like good fertile soil, she must have been thinking.

Then it happened. "Ouch!" A splinter entered the base of her right big toe. No

matter. Take out the splinter and get on with the gardening. Gone and forgotten.

But it was not forgotten. Three days later, the pain on the left side of her face necessitated a visit to her family doctor. Probably a facial infection, she was told. Take the amoxicillin, and it should resolve.

It did not resolve—it worsened. Now it was July 12, and her jaw was so tight that she had not eaten for three days. The muscle spasms in her face were intense, and her friends nervously suggested that it looked like lockjaw. When she arrived at the hospital's emergency room, an alert doctor recognized the classic risus sardonicus (the grinning

expression caused by spasms of the facial muscles) and the trismus (the lockjaw caused by spasms of the chewing muscles). No question—she had tetanus.

Treatment was swift and aggressive—3,250 units of tetanus antitoxin, intravenous penicillin, and a tetanus booster, plus removing the traces of wood still in the wound. She was placed in a quiet room and given muscle relaxants. Fifteen days would pass before she was discharged. She was well on her way to recovery . . . except she still could not tie those darn shoelaces.

nervous system. Treatment consists of antibiotic therapy as well as debridement, amputation, or exposure in a hyperbaric oxygen chamber. The disease spreads rapidly, and death frequently results. Some microbiologists prefer the name **clostridial myonecrosis** ("muscle cell death") because gas is not always present in the early stages of disease.

LEPTOSPIROSIS CAUSES FEVER AND HEADACHE

Leptospirosis is a typical **zoonosis**—that is, a disease of animals that can spread to humans. The disease affects household pets such as dogs and cats, as well as rats, mice, and barnyard animals. Humans acquire it by contact with these animals or from soil, food, or water contaminated with their urine. Increasing rat populations in inner cities increase the risk for this disease through contact with standing water in parks.

The agent of leptospirosis is *Leptospira interrogans*, a small, delicate spirochete usually with a hook at one end that resembles a question mark, hence the name *interrogans* (**FIGURE 10.5**). The undulating movements of these organisms result from contractions of submicroscopic fibers called **axial filaments**. In infected animals, the spirochetes colonize the kidney tubules and are excreted in the urine to the soil. They enter the human body through the mucus membranes of the eyes, nose, and mouth, or through the skin, especially through abrasions and the soft parts of the feet. Patients experience flu-like symptoms, such as fever, aches, and muscle weakness. Five to ten percent of patients progress to the systemic form of leptospirosis, called **Weil's disease**. Here the spirochetes infect various organs, including the kidneys, liver, and meninges. Considerable jaundice may be present as bile seeps from the liver, and the patient may vomit blood from gastric hemorrhages. Despite the numerous tissues involved, the mortality rate from leptospirosis is low, and penicillin is generally used with success.

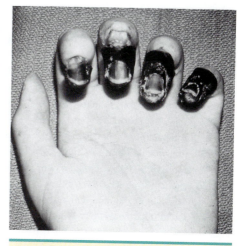

FIGURE 10.4

Dry Gangrene of the Tissues

This photograph shows blackening of the skin and the dry, shrunken nature of the tissue that characterizes dry gangrene. The photo was taken shortly before three digits were amputated. This form of gangrene is usually not due to bacterial infection.

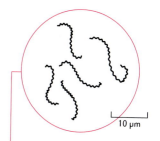

Leptospira interrogans

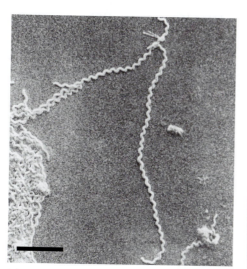

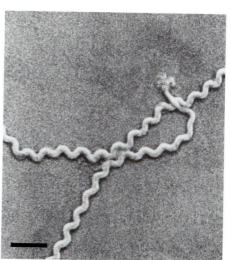

(a)   (b)

FIGURE 10.5

Leptospira interrogans

Details of *Leptospira interrogans*, the agent of leptospirosis. (a) A scanning electron micrograph. Note the tightly coiling spirals and the absence of flagella in this spirochete. (Bar = 2 μm.) (b) A closer view of the terminal hook of one spirochete. (Bar = 1 μm.)

MicroFocus 10.3

BAD DAY AT LAKE SPRINGFIELD

They swam a brisk mile and a half in Lake Springfield. They rode their bikes a long 45 miles over the Illinois hills. And to top it off, they ran a tough and painful 10 miles. It was the 1998 Springfield Ironhorse Triathlon. Participants expected to be tired—they didn't expect to get sick.

But they got sick in record numbers. Of about 1,000 athletes from 44 states and 7 countries who participated in the June 21 triathlon, 98 became ill and reported having chills, headache, diarrhea, muscle weakness, eye pain, or red eyes. Seventy-three sought medical attention, and 28 were hospitalized. Two others had their gallbladders needlessly removed because their doctors failed to recognize the disease. One of the worst-hit triathletes had such severe convulsions he had to be put on ice to bring down his 42°C fever. His doctors said he was close to death. If it had happened to one or two people, doctors might have passed it off as influenza, but dozens of athletes were sick. And athletes don't get sick like this—especially triathletes!

While the Illinois public health officials were publicizing the outbreak, another 248 recreational users of Lake Springfield came forward to report a similar illness. No, they hadn't all eaten at the same restaurant, been bitten by mosquitoes, or gone on a trip abroad together. Yes, they were all at Lake Springfield that fateful June day. Some were enjoying the beaches at the lake and swimming in the water; some were fishing; some were boating. And yes, they remembered watching the triathlon.

Eventually it all made sense: the symptoms, the bacterial isolations from the water, contact with the water, the blood tests. They all pointed to one disease—leptospirosis. The Springfield and state health departments on July 24 advised people not to swim, water ski, or use personal watercraft on Lake Springfield, which brought the active social environment on "the lake" to a grinding halt.

This was the largest outbreak of leptospirosis ever reported in the United States. What caused the outbreak of leptospirosis, you ask? Five years later, in 2003, no one knows what caused such an unusual disease to visit the lake in such a large scale. In the end, no one died that summer, but Lake Springfield would never be the same.

TABLE 10.1

A Summary of Soilborne Bacterial Diseases

DISEASE	CAUSATIVE AGENT	DESCRIPTION OF AGENT	ORGANS AFFECTED	CHARACTERISTIC SIGNS
Anthrax	*Bacillus anthracis*	Gram-positive spore-forming aerobic rod	Blood Lungs Skin	Hemorrhaged blood Boil-like lesions
Tetanus	*Clostridium tetani*	Gram-positive spore-forming anaerobic rod	Nerves at synapse	Spasms Tetanus
Gas gangrene	*Clostridium perfringens*	Gram-positive spore-forming anaerobic rod	Muscles Nerves Blood cells	Gangrene Swollen tissue
Leptospirosis	*Leptospira interrogans*	Spirochete	Kidney Liver Spleen	Jaundice Vomiting
Melioidosis	*Burkholderia pseudomallei*	Gram-negative rod	Heart Lung	Abscesses

Leptospirosis occurs throughout the United States, with 100 to 200 cases identified annually (50 percent of cases occur in Hawaii). Recognizing its symptoms, together with modern methods of diagnosis, has led to increased reports of its occurrence. In 1997, for example, a number of cases occurred in white-water rafters in Costa Rica, and in 1997, the disease broke out in triathletes in Wisconsin (MicroFocus 10.3). The disease is considered an occupational hazard of veterinarians, as well as tunnel diggers, sugarcane cutters, dockworkers, miners, and others who work in wet areas where rodents are present. Dogs may be immunized with a leptospirosis vaccine, usually combined with rabies, parvovirus, and distemper vaccines.

MELIOIDOSIS OCCURS IN TWO FORMS

Melioidosis is a disease that, until recently, has been rare in the United States. It is caused by a gram-negative rod named *Burkholderia pseudomallei*. This organism is common in soil and water, especially in Southeast Asia. Humans contract melioidosis by absorbing bacteria through soil-contaminated wounds, by inhalation, or by consuming contaminated food or water. The disease also affects horses, rodents, dogs, and cats. Person-to-person transmission can occur.

Melioidosis has been called a "medical time bomb" because of its propensity to lie dormant in the body for years. In *chronic melioidosis*, abscesses may occur in the heart, lungs, liver, or spleen, and symptoms may vary widely, depending on which organs are affected. A rare, *acute melioidosis* causes pneumonia and a blood infection. It usually is fatal. Tropical experts have known of melioidosis since the early 1900s, but Americans did not appear to be involved until the Vietnam War, when several cases surfaced.

This and other soilborne diseases are summarized in TABLE 10.1.

TOXIN INVOLVED	TREATMENT ADMINISTERED	IMMUNIZATION AVAILABLE	COMMENT
Yes	Ciprofloxacin	For animals	High mortality rate Affects many organs Rare disease in humans
Yes	Penicillin Antitoxin	Toxoid in DTaP	Second most powerful toxin Inhibits cholinesterase activity
Yes	Penicillin	None	Gas blocks flow of blood Called clostridial myonecrosis
Not established	Penicillin	For animals	Common in animals Typical zoonosis
Not established	Doxycycline	None	Symptoms develop slowly Occurs in Southeast Asia

To this point . . .

We have surveyed several bacterial diseases commonly transmitted from the soil. Diseases such as tetanus and gas gangrene involve the contamination of wounds with bacteria in the form of spores. As the bacilli multiply in the dead anaerobic tissue of the wound, the symptoms of disease surface. Tetanus bacilli produce powerful toxins that lead to sustained muscle contractions, while gas gangrene bacilli ferment and putrefy the organic compounds in muscle to produce large amounts of gas that lead to gangrene.

Anthrax, leptospirosis, and melioidosis also may be contracted from the soil, as well as by other means, such as contact with animals or consuming contaminated food or water. Anthrax occurs in a number of forms, including pulmonary, gastrointestinal, and skin forms, all of which may be fatal. Leptospirosis, often spread by domestic animals and household pets, affects many abdominal organs. Melioidosis has contemporary significance because many cases in Americans were contracted from the soil in Southeast Asia during the Vietnam War. You may note that in many of the soilborne diseases there is substantial involvement of the blood, since bacteria spread through this tissue from the initial site of invasion.

We shall now turn to a series of bacterial diseases transmitted by arthropods. In these diseases, infected arthropods introduce bacteria into the circulation, and the symptoms of disease commonly occur in the bloodstream. High fever is usually associated with the disease, and a general feeling of illness is common. Our survey will include bubonic plague, one of the most historically important diseases, as well as two diseases that were observed for the first time in the United States.

10.2

Arthropodborne Bacterial Diseases

Arthropods transmit diseases to humans usually by taking a blood meal and themselves becoming infected. Then they pass the organisms to another individual during the next blood meal. Arthropodborne diseases occur primarily in the bloodstream, and they often are characterized by a high fever and a body rash.

BUBONIC PLAGUE CAN BE A HIGHLY FATAL DISEASE

Few diseases have had the rich and terrifying history of **bubonic plague**, nor can any match the array of social, economic, and religious changes wrought by this disease.

The first documented pandemic of plague probably began in Africa during the reign of the Roman emperor Justinian in A.D. 542. It lasted 60 years, killed millions, and contributed to the downfall of Rome. The second pandemic was known as the **Black Death** because of the purplish-black splotches on victims and the terror it evoked in the 1300s (MicroFocus 10.4). The Black Death decimated the world and, by some accounts, killed an estimated 40 million people in Europe, almost one-third of the population of that continent, as the chapter introduction relates. A deadly epidemic

MicroFocus 10.4

CATAPULTING TERROR

One of the most horrendous emerging infectious diseases was starting to spread to Europe, North Africa, and the Near East in the mid-fourteenth century. It was the Black Death—bubonic plague—which historians believe moved out of the lands north of the Caspian and Black Seas.

Caffa (today Feodosija, Ukraine) was a port city situated on the north shore of the Black Sea. Through an agreement with the local Tartars (Mongols) who controlled the area, Caffa was placed under control of Genoa, Italy, and Christian merchants were allowed to trade goods with the Far East. In 1343, a group of Italian merchants from Genoa found themselves trapped behind the walls of Caffa after a brawl between the Italians and Tartars. The dreaded Tartars laid siege to the city and over the next five years, Genoa lost and regained control of the city several times.

During the siege of 1346, the Tartars were unable to drive the Italians and other Christians from the city, when plague broke out. Large numbers of Tartars started dying. Losing interest in the siege, the Tartars had the bodies of their dead plague victims placed in catapults and lobbed over the walls of Caffa into the city. The Tartars hoped the stench would kill everyone in Caffa. Soon plague was sweeping through Caffa. The townspeople were terrified: Either the plague would kill them inside the walls, or the Tartars would kill them outside the walls. But the Tartars were equally terrified of the plague, and they were withdrawing.

Sensing an opportunity to escape, the merchants ran for their ships and sailed off to Genoa, Venice, and other home-ports in the Mediterranean. Unfortunately, their voyage home would be a voyage of death. Many died of the plague onboard, and the survivors spread the disease wherever they stopped to replenish their supply of food and water.

Could such a tale be true? Could the dead diseased bodies catapulted into Caffa transmit plague? Almost certainly they could. City defenders would have carried away the dead, mangled bodies, which would spread the disease by contact. Poor sanitation and health of its citizens in Caffa would make transmission even easier and more widespread, especially if pneumonic plague broke out.

The attempted siege of Caffa in 1346 represents the most spectacular early episode of biological warfare, with the Black Death as its consequence. It demonstrates the very essence of terrorism as defined today—the intentional or threatened use of biological agents to cause fear in or actually inflict death or disease upon a large population for political, religious, or ideological reasons. The siege of Caffa also shows us the horrifying consequences that can come from the use of infectious disease as a weapon.

---

also occurred in London in 1665, where 70,000 people succumbed to the disease. Daniel Defoe's *Journal of the Plague Year* recounts how people reacted (**FIGURE 10.6**).

In the late 1800s, Asian warfare facilitated the spread of a Burmese focus of plague, and migrations brought infected individuals to China and Hong Kong. During this epidemic in 1894, the causative organism was isolated by Alexandre Yersin and, independently, by Shibasaburo Kitasato. Plague first appeared in the United States in San Francisco in 1900, carried by rats on ships from the Orient. The disease spread to ground squirrels, prairie dogs, and other wild rodents, and it is now endemic in the southwestern states, where it is commonly called **sylvatic plague**. **MicroFocus 10.5** recounts a case associated with wildlife.

Bubonic plague is caused by *Yersinia pestis*. This gram-negative, nonmotile rod stains heavily at the poles of the cell, giving it a safety-pin appearance and a characteristic called **bipolar staining**. As first demonstrated by Masaki Ogata in 1897, the bacillus is transmitted by the rat flea *Xenopsylla cheopis*. A living organism such as a flea that transmits disease agents is called a **vector**, from the Latin *vehere*, meaning "to carry." Normally, the fleas infest only rats, but as the rats die, the fleas jump to humans and feed in the skin, thus transferring the bacilli into the bloodstream.

Bubonic plague is a blood disease. The bacteria multiply in the bloodstream and localize in the lymph nodes, especially those of the armpits, neck, and groin.

Sylvatic:
occurring in nature; wild.

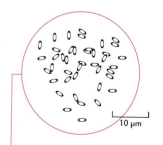

Yersinia pestis

FIGURE 10.6

Bubonic Plague

An engraving drawn during the London plague of 1665. Bodies heaped on "dead carts" are being pulled into huge pits by men dragging the corpses off the carts with a hook attached to a ten-foot pole. This practice is probably the origin of the saying, "I wouldn't touch it with a ten-foot pole." The pipes the men are smoking give off noxious fumes thought to ward off the disease.

Hemorrhaging in the lymph nodes causes substantial swellings called **buboes** (hence the name, bubonic plague). Dark, purplish splotches from hemorrhages also can be seen through the skin (the "rosies" in "Ring-a-ring of rosies"). Without treatment, mortality reaches 60 percent. From the buboes, the bacilli may spread to the bloodstream, where they cause **septicemic plague**. Nearly 100 percent of untreated cases of septicemic plague are fatal.

Human-to-human transmission of plague during epidemics has little to do with fleas. Instead, it is dependent upon respiratory droplets because *Y. pestis* localizes in the lungs of many of the first victims. In this form, the disease is called **pneumonic plague** and is highly contagious. Lung symptoms are similar to those in pneumonia,

---

MicroFocus 10.5

"WHAT A SHAME!"

Just an incredible day," he must have been thinking, as he trudged along. It was great to be out in the desert. Great to get away from the books for a while. Final exams were only a few weeks away, then summer, then college, then who knows? Maybe a career in physical education? Maybe a coaching job? Three miles completed, five miles to go. He had to keep moving to make it back home for dinner.

He didn't mind the solitude of the Arizona desert. It was cool up here in Flagstaff, just right for hiking. And there were plenty of animals to see and interesting plants to watch for. He would stop now and then for some water or to watch the horizon or feed a colony of prairie dogs as they bustled about the terrain. Then he continued on to home.

He enjoyed wrestling as well, and he was the team captain. Two days after the hike, he sustained a groin injury while wrestling. When the ache remained, he went to the doctor. The doctor noted the groin swelling and wondered whether it had anything to do with the fever the young man was experiencing. Was it just a coincidence—the ache, the swelling, the fever? This was no ordinary groin injury. "We'll watch it for a couple of days," the doctor said as he gave the young man a pain reliever.

Tragedy struck the next morning. All his mother remembered was a loud thud from the bathroom. She didn't remember her terrified shriek or dialing the emergency number. The emergency medical technicians were there in a flash, but it was too late. He was dead.

The investigation that followed took public health officials down many dead-ends. It wasn't the wrestling injury, they concluded. Still, the groin swelling made them suspicious. "That hiking trail," they asked his mother, "where is it?"

They set out in search of an elusive answer. About three and a half miles out, they came upon a colony of prairie dogs. The animals didn't look well. In fact, some were dead nearby. Carefully, they trapped a sick animal and carried it back to the lab. Two days later, the lab report was ready—the animal was sick with plague. Then the young man's tissues were tested—again, plague. The investigators shook their heads. "What a shame!"

with extensive coughing and sneezing. Hemorrhaging and fluid accumulation are common. Many suffer cardiovascular collapse, and death is common within 48 hours of the onset of symptoms. Mortality rates for pneumonic plague approach 100 percent. One 1992 death was linked to a cat and rodents (**FIGURE 10.7**), and a notable outbreak appears to have occurred in India in 1994. Overall, about 14 percent of all plague cases in the United States are fatal.

FIGURE **10.7**

An Episode of Plague

This incident demonstrates that plague occurs in modern times in the United States.

1. On August 19, 1992, a man reached under the crawlspace of a house in Chaffee County, Colorado, to help a cat trapped behind the fence.

2. The cat was apparently quite sick. It had abscesses under its jawbone and lesions around its mouth. The cat was lethargic and feverish.

3. The man returned home to Tucson, Arizona, the next day. Three days later he experienced high fever, abdominal cramps, and a severe cough.

4. Two days later he was hospitalized with septic shock, difficult breathing, and right lobar pneumonia. Despite aggressive antibiotic therapy, he died the next day.

5. When the laboratory tests on the man's tissues were complete, the laboratory director reported the presence of *Yersinia pestis*, the agent of plague. Antibody tests confirmed the diagnosis.

6. At the house, health investigators saw many dead rodents, and laboratory tests revealed *Yersinia pestis* in their tissues. The cat had died on August 19, the day it was rescued.

TEXTBOOK CASES

Plague may be treated with tetracycline or streptomycin when detected early, reducing mortality to about 18 percent. Diagnosis consists of the laboratory isolation of *Y. pestis*, together with tests for plague antibodies and typing with bacteriophages. The disease occurs sporadically in Native American populations and in travelers through the Southwest. Small-game hunters, taxidermists, veterinarians, zoologists, and others who handle small rodents must be aware of the possibility of contracting plague. About two dozen cases are reported to the CDC annually. A vaccine consisting of dead *Y. pestis* cells is available for high-risk groups.

TULAREMIA HAS MORE THAN ONE DISEASE PRESENTATION

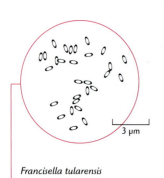

Francisella tularensis

Tularemia is one of several microbial diseases first recorded in the United States (others include St. Louis encephalitis, Rocky Mountain spotted fever, Lyme disease, and Legionnaires' disease). In 1911, a US Public Health Service investigator named George W. McCoy reported the plague-like disease in ground squirrels from Tulare County, California, and within a year he isolated the responsible bacillus. By 1920, researchers identified tularemia in humans, and Edward Francis assumed a detailed study of the disease. Over the next 25 years, Francis amassed data on over 10,000 cases, including his own. In the 1960s, when a name was coined for the causative organism, *Francisella tularensis* was selected to honor him and the county of first observation.

Francisella tularensis is a small gram-negative rod that displays bipolar staining. It is extremely virulent in that ten cells can cause disease. It occurs in a broad variety of wild animals, especially rodents, and it is particularly prevalent in rabbits (in which case it is known as rabbit fever). Cats and dogs may acquire the bacillus during romps in the woods, and humans are infected by arthropods from the fur of animals. Ticks are important vectors in this regard, as evidenced by an outbreak of 20 tickborne cases in South Dakota in the 1980s. Other methods of transmission include contact with an infected animal (such as skinning an animal), consumption of rabbit meat, splashing bacilli into the eye, and inhaling bacilli, even from laboratory specimens.

Various forms of tularemia exist, depending on where the bacilli enter the body. An arthropod bite, for example, may lead to a skin ulcer (**FIGURE 10.8**) and swollen salivary glands. Splashing into the eye may cause an eye lesion, and inhalation may lead to pulmonary symptoms. In all forms, the disease is difficult to recognize because the symptoms are mild and nonspecific. Various patients with tularemia have been mistakenly treated for strep throat, rickettsial disease, lymphoid cancer, and cat-scratch disease.

Tularemia usually resolves on treatment with tetracycline or streptomycin, and few people die of the disease. Epidemics are unknown, and evidence suggests that tularemia may not be communicable among humans despite the many modes of entry to the body. Physicians reported 129 cases to the CDC in 2001. MicroFocus 10.6 describes the potential of plague and tularemia as additional bioterrorism agents.

LYME DISEASE CAN BE DIVIDED INTO THREE STAGES

One of the major emerging infectious diseases is Lyme disease, currently the most commonly reported tickborne (indeed, arthropodborne) illness in the United States. Lyme disease has been reported, mostly from the northeastern and north-central states, and in 2001 it accounted for 17,029 US cases of infectious disease (**FIGURE 10.9**).

FIGURE 10.8

The Lesions of Tularemia

The lesions of tularemia occur where the bacilli enter the body. In (a), the patient acquired the disease by handling infected rabbit meat. In (b), infection was preceded by the bite of an infected arthropod. The crater-like form of the lesion can be seen in both cases.

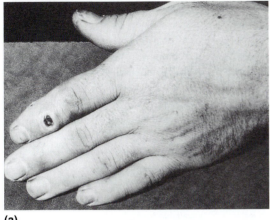

(a)

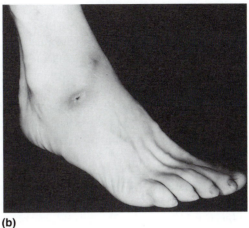

(b)

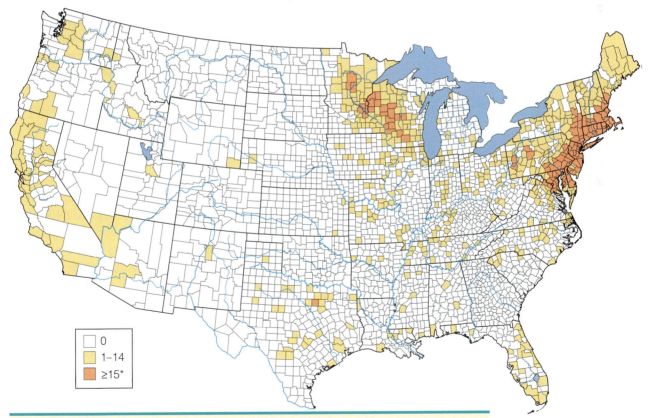

☐	0
▨	1–14
▨	≥15*

FIGURE 10.9

Lyme Disease, Reported Cases by County—United States, 2001

A total of 17,029 cases of Lyme disease were reported in 2001. Forty-three of the 48 lower states and the District of Columbia are involved in the epidemic. Ten northeastern states accounted for 93.9 percent of the reported cases.

*The total number of cases from these counties represented 90 percent of all cases reported in 2001.

MicroFocus 10.6

PLAGUE AND TULAREMIA AS WEAPONS OF BIOTERRORISM

Yersinia pestis, the causative agent of plague, is transmitted by fleas from infected rodents to humans. About 19,000 cases were reported worldwide between 1980 and 1994. In the United States, there are about a dozen cases each year, resulting from its endemic nature in small rodents in the Southwest and Pacific states. Lethal cases in the United States are very rare.

Of the three clinical forms of plague—bubonic, septicemic, and pneumonic—pneumonic plague is of most concern as a bioterror agent. However, it is important to know that pneumonic plague can develop from bubonic plague if *Y. pestis* spreads from the lymph nodes to the respiratory system. As a weapon of bioterrorism, the plague bacilli could be disseminated in an aerosol form and, after infection, transmission from person to person would occur simply through sneezing and coughing. Early administration of antibiotics (large doses of streptomycin) is critical because pneumonic plague is almost 100 percent fatal if

antibiotic therapy is delayed more than one day after the onset of symptoms. Penicillin is ineffective and no vaccine is currently available.

Like anthrax, the effects of plague are the result of the toxins produced by the living bacteria. Plague toxins are a set of six proteins called Yersinia outer proteins (Yops) that allow the bacterial cells to evade the host's immune defenses by crippling communication networks necessary for phagocytosis and other immune functions. The multiple effects of the toxins produce the same outcome as the anthrax toxins: to overcome the immune defense mechanisms of the host.

The causative agent of tularemia is the non–spore-forming bacterium *Francisella tularensis*. It is one of the most infectious pathogens known because as few as 10 to 25 cells can initiate disease. Humans acquire the disease accidentally by contact with diseased rodents and rabbits, through bites from infected ticks, mosquitoes, or deerflies, or by drinking water contaminated with the bacterium. *F.*

tularensis has an antiphagocytic capsule and, like most all gram-negative bacilli, produces an endotoxin. The bacterium is found worldwide. It causes about 200 disease cases per year in the United States, primarily in the south-central states (Arkansas, Missouri, Oklahoma) where it is endemic. Person-to-person transmission has not been reported.

An aerosol form of the bacterium most likely would be used by terrorists because inhalation of the aerosol can lead to pulmonary or septicemic tularemia, which has a kill rate of greater than 30 percent. Antibiotics are very effective with early treatment and an investigational vaccine is available for high-risk groups.

Both diseases are very rare in the United States, so as with all the potential bioterror agents, a major outbreak or epidemic should be realized very quickly as a bioterrorist attack. Because of the infectiousness of these two agents, ways to better detect an attack are being investigated.

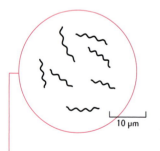

Borrelia burgdorferi

Lyme disease is named for Old Lyme, Connecticut, the suburban community where a cluster of cases was observed in 1975. That year, researchers led by Allan C. Steere of nearby Yale University traced the disease to ticks of the genus *Ixodes*. However, they were unable to locate a causative agent. Several years would pass until 1982, when a spirochete was observed in diseased tissue. Still another two years passed before the spirochete was isolated and cultivated (FIGURE 10.10). Researchers named it *Borrelia burgdorferi* for Willy Burgdorfer, the microbiologist who studied the spirochete in the gut of an infected tick.

Lyme disease has a variable incubation period that can be as long as six to eight weeks. One of the first signs is a slowly expanding red rash at the site of the tick bite. The rash is called **erythema** (red) **chronicum** (persistent) **migrans** (expanding), or **ECM**. Beginning as a small flat or raised lesion, the rash increases in diameter in a circular pattern over a period of weeks, sometimes reaching a diameter of 10 to 15 inches, as FIGURE 10.11 shows. It has an intense red border and a red center, and it resembles a bull's-eye (the **bull's-eye rash**). It can vary in shape and is usually hot to the touch, but it need not be present in all cases of disease. Indeed, about one-third of patients do not develop ECM. The tick bite can be distinguished from a mosquito bite because the latter itches, while a tick bite does not. Fever, aches and pains, and flu-like symptoms usually accompany the rash.

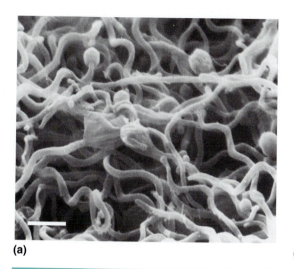

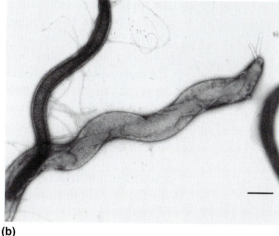

(a) (b)

FIGURE 10.10

Borrelia burgdorferi

Two views of *Borrelia burgdorferi*, the agent of Lyme disease. (a) A scanning electron micrograph showing an abundance of spirochetes. (Bar = 1 μm.) (b) A transmission electron micrograph in which the plasma membranes of two spirochetes have been disrupted, thereby releasing the axial filaments. These filaments run the length of the spirochete under the membrane and contribute to motility. (Bar = 1 μm.)

Lyme disease typically has three stages. The **early localized stage** begins about ten days after the tick bite. An ECM rash may develop at the site of the bite and persist for several weeks if untreated. The individual may experience a flu-like illness and lymph node swelling. Effective treatment can be rendered with penicillin or doxycycline.

An **early disseminated stage** begins weeks to months later with the spread of *B. burgdorferi* to the skin, heart, nervous system, and joints. On the skin, multiple smaller ECMs develop while invasion of the nervous system can lead to meningitis, facial palsy, and peripheral nerve disorders. Cardiac abnormalities are the most common, such as brief, irregular heartbeats. Joint and muscle pain also occur.

If left untreated, a **late stage** occurs weeks to years later. About 10 percent of patients develop chronic arthritis with swelling in the large joints, such as the knee. Although Lyme disease is not known to have a high mortality rate, the overall damage to the body can be substantial.

The tick that transmits most cases of Lyme disease in the Northeast and Midwest is the deer tick *Ixodes scapularis* (formerly *I. dammini*); in the West, the major vector is the western black-legged tick *I. pacificus*. In its human form, the tick is about the size of a pinhead or the period at the end of this sentence. It is smaller than the American dog tick, and it lives and mates in the fur of the white-tailed deer. Eventually it falls into the tall grass, where it waits for an unsuspecting dog, rodent, or human to pass by. The tick then attaches to its new host and penetrates into the skin. During the next 24 to 48 hours, it takes a blood meal and swells to the size of a small pea. While sucking the blood, it also defecates into

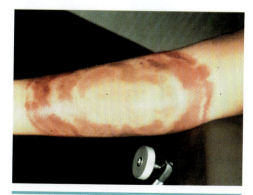

FIGURE 10.11

Erythema Chronicum Migrans

Erythema chronicum migrans (ECM) is the rash that accompanies two-thirds of cases of Lyme disease. Note that the rash consists of a large patch with an intense red border. It is usually hot to the touch, and it expands with time.

the wound, and, if the tick is infected, spirochetes will be transmitted (as the life cycle in FIGURE 10.12 indicates). If the tick is observed on the skin, it should be removed with a forceps or tweezers, and the area should be thoroughly cleansed with soap and water and an antiseptic applied.

A diagnosis of Lyme disease is usually based on symptoms, and the physician will often use a pen to mark off the border of the skin rash to see if it expands with time. The patient's recent activities are noted (hiking in the woods, living in a tick-infested area, camping), and a blood sample also may be taken for a test for spirochetal antibodies. The test, however, may not be accurate because a number of weeks are required for the body to produce enough antibodies to show up in the test. A newer test designed to detect spirochetal DNA has been developed, but it has not been standardized for routine diagnosis.

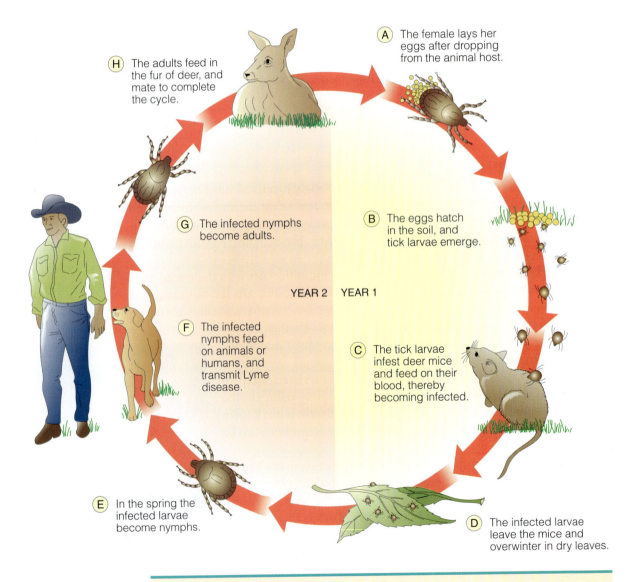

A The female lays her eggs after dropping from the animal host.

B The eggs hatch in the soil, and tick larvae emerge.

C The tick larvae infest deer mice and feed on their blood, thereby becoming infected.

D The infected larvae leave the mice and overwinter in dry leaves.

E In the spring the infected larvae become nymphs.

F The infected nymphs feed on animals or humans, and transmit Lyme disease.

G The infected nymphs become adults.

H The adults feed in the fur of deer, and mate to complete the cycle.

YEAR 2   YEAR 1

FIGURE 10.12

Life Cycle of the Tick

The life cycle of *Ixodes scapularis* and *I. pacificus*, the ticks that transmit Lyme disease.

Congenital transfers of *B. burgdorferi* have been documented, and pregnant women are encouraged to seek early diagnosis and treatment for the disease if they suspect they have been infected. A vaccine for dogs (LymeVax®)has been licensed and is in routine use. A vaccine for humans (LYMErix™) was approved by the US Food and Drug Administration (FDA) in January 1999. The FDA approved the vaccine for people 15 to 70 years old, since experimental trials did not involve children. It was intended for those who live or work in grassy or wooded areas where infected ticks are probably present. However, federal health authorities investigated whether some people who received the LYMErix™ vaccine later developed severe cases of arthritis and even the actual disease. The FDA also received reports of such problems, mainly from doctors and researchers in the Northeast. As of February 25, 2002, the manufacturer announced that the LYMErix™ was no longer available commercially.

Although Lyme disease is considered a "new" disease, there is evidence that it has existed for some decades, even though it had escaped recognition (MicroFocus 10.7).

RELAPSING FEVER IS CARRIED BY TICKS AND LICE

Relapsing fever is caused by a long spirochete that is responsible for two forms of the disease. **Endemic relapsing fever** often is caused by *Borrelia hermsii*. It is transmitted by the bite of the tick *Ornithodoros* from rodent hosts. Ticks normally inhabit the rodent burrows and nests, where the natural infection cycle proceeds without apparent disease in the rodents. Humans are incidental hosts, often bitten briefly and without notice at night by the ticks. Cabins in wilderness areas of the northwest and southwest United States are favorable nesting sites for infected rodents and their ticks, especially when food is made available by occupants of the cabin (as FIGURE 10.13 illustrates).

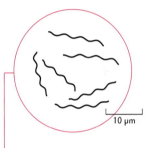

10 µm

Borrelia recurrentis

MicroFocus 10.7

NOT SO NEW AFTER ALL

Is Lyme disease a totally new disease, or has it been present but unrecognized in human populations for some time?

One of the hotspots for Lyme disease is Long Island, a 120-mile-long island off the coast of New York City. At the easternmost tip of Long Island is the town of Montauk, an old fishing village and a resort community. It seems that for decades local doctors have spoken of "Montauk knee," a disease accompanied by swollen and inflamed knee joints and a skin rash. Was Montauk knee what we now call Lyme disease?

Pathologist David Pershing of the Mayo Clinic attempted to find out. He and several colleagues sought out 140 specimens of ticks collected between 1924 and 1951, many from natural history collections. Then, they dissected

the ticks and removed the intestines. They amplified the DNA in the intestinal matter by a technique called the polymerase chain reaction. Next, they set out to identify any DNA that might belong to *Borrelia burgdorferi*, the cause of Lyme disease. The theory was simple: If the DNA of *B. burgdorferi* was present, then the spirochete must have been present also; if *B. burgdorferi* was present, Lyme disease would predate 1975. And present it was. Pershing's group identified the DNA of *B. burgdorferi* in 13 ticks. All 13 had been collected near Montauk.

FIGURE 10.13

An Outbreak of Tickborne Relapsing Fever

Although relapsing fever is not a well-known disease, outbreaks can occur, as this episode demonstrates.

TEXTBOOK CASES

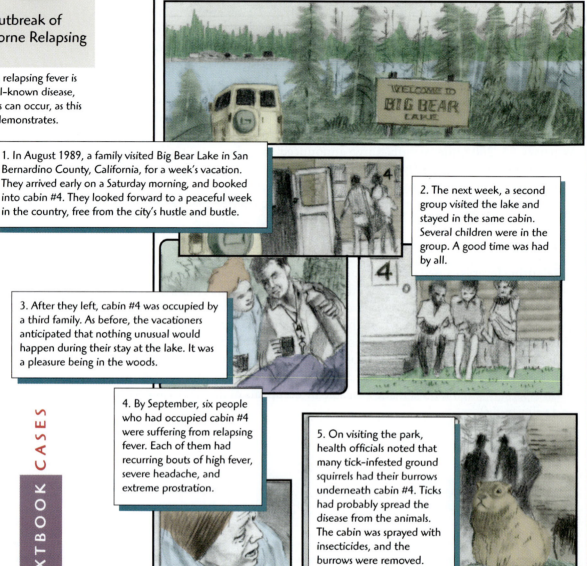

1. In August 1989, a family visited Big Bear Lake in San Bernardino County, California, for a week's vacation. They arrived early on a Saturday morning, and booked into cabin #4. They looked forward to a peaceful week in the country, free from the city's hustle and bustle.

2. The next week, a second group visited the lake and stayed in the same cabin. Several children were in the group. A good time was had by all.

3. After they left, cabin #4 was occupied by a third family. As before, the vacationers anticipated that nothing unusual would happen during their stay at the lake. It was a pleasure being in the woods.

4. By September, six people who had occupied cabin #4 were suffering from relapsing fever. Each of them had recurring bouts of high fever, severe headache, and extreme prostration.

5. On visiting the park, health officials noted that many tick-infested ground squirrels had their burrows underneath cabin #4. Ticks had probably spread the disease from the animals. The cabin was sprayed with insecticides, and the burrows were removed.

Epidemic relapsing fever is caused by *Borrelia recurrentis*, which is carried by the body louse *Pediculus humanus*. The disease is spread between humans by the lice, and infection occurs when the louse is crushed into the bite wound. No cases of epidemic relapsing fever have been recorded in the United States since 1906. It mainly is found in Africa, China, and parts of South America in overcrowded, poverty-stricken regions where the public health systems have failed. It also is a disease of war, such as in World War II when some 10 million people were infected. Another such epidemic descended upon Napoleon's lice-infested soldiers during the Russian campaign. Together with serious losses from another louseborne disease, typhus (discussed next), this disease so decimated the French army that the balance tipped in favor of the Russians. Peter Ilyich Tchaikovsky wrote the *1812 Overture* to celebrate the great Russian victory, one that might not have been possible without the intervention of microorganisms.

Cases of relapsing fever are characterized by substantial fever, shaking chills, headache, prostration, and drenching sweats. The symptoms last for a couple of days, then disappear for about eight days, then reappear up to ten times during the following weeks. Microbiologists believe that different serotypes of *B. recurrentis* emerge from the organs after each attack. Treatment with doxycycline or erythromycin hastens recovery.

To this point . . .

We have studied four arthropodborne diseases of varying significance. Bubonic plague has had substantial impact on the course of history because it is among the most prolific killers of humans. Plague is still present in today's world. Tularemia, by contrast, is a plague-like disease that has been recognized only in this century. Its symptoms are much milder than plague, and it often remains undiagnosed. Lyme disease has a history that is even more recent than tularemia's, having first been recorded in 1975. A distinctive lesion at the site of spirochete entry, and developing arthritis, are characteristic signs of the disease. In relapsing fever, disease is accompanied by recurring periods of fever.

In all these cases, the key element in control is the elimination of the arthropod vector. If the chain of transmission can be broken, the disease will not spread to the body. This is where the pest control operator plays a significant role in our system of public health. It is also where sanitation and personal hygiene contribute to good health.

In the final section of this chapter, we shall continue to examine arthropodborne diseases, emphasizing those diseases caused by rickettsiae, a group of small bacteria. Once again we shall encounter ticks, lice, and other arthropods as vectors, and we shall focus on their place in the disease process. Antibiotics are helpful in treatment of the diseases, but vector control is of prime importance.

10.3

Rickettsial Arthropodborne Diseases

In 1909, Howard Taylor Ricketts, a University of Chicago pathologist, described a new organism in the blood of patients with Rocky Mountain spotted fever and showed that ticks transmit the disease. A year later, he located a similar organism in the blood of animals infected with Mexican typhus, and discovered that fleas were the important vectors in this disease. Unfortunately, in the course of his work, Ricketts fell victim to the disease and died. When later research indicated that Ricketts had described a unique group of microorganisms, the name *rickettsiae* was coined to honor him. Originally, the rickettsiae were set apart from the bacteria, but microbiologists now consider them to be small, gram-negative, nonmotile bacteria that are obligate, intracellular parasites. All rickettsial disease can be treated effectively with doxycycline or chloramphenicol.

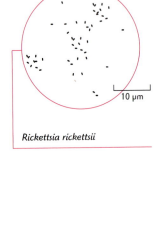

Rickettsia rickettsii

ROCKY MOUNTAIN SPOTTED FEVER BEGINS WITH A CHARACTERISTIC RASH

The agent of Rocky Mountain spotted fever (RMSF) is *Rickettsia rickettsii*. This organism is transmitted by ticks, especially those of the genera *Amblyomma* and *Dermacentor* (shown with other vectors in **FIGURE 10.14**). The hallmarks of RMSF are a high fever lasting for many days, severe headaches, and a skin rash reflecting damage to the small blood vessels. The rash begins as pink spots called **macules**, and progresses to pink-red pimple-like spots known as **papules**. Where the spots fuse, they form a **maculopapular rash**, which becomes dark red and then fades without evidence of scarring. The rash generally begins on the palms of the hands and soles of the feet and progressively spreads to the body trunk. Mortality rates of untreated cases are variable, with some outbreaks in Montana recording a rate as high as 75 percent. Antibiotic treatment reduces this rate significantly.

Accurate and rapid diagnosis is essential in treating RMSF. Evidence of a tick bite, progress of the rash, and the recent activities of the patient (camping, backpacking, and other outdoor activity) are taken into account. Traditionally, a procedure called the **Weil-Felix test** has been used as well. It is performed by mixing a sample of the patient's serum with the bacterium *Proteus* OX19. The bacteria clump together if the serum contains rickettsial antibodies. The test works because rickettsiae coincidentally possess antigens also located on *Proteus* cells. However, the test is nonspecific and relatively insensitive, so antibody-detection tests are advised.

Rocky Mountain spotted fever was first observed in early settlers to the American Northwest. About a thousand cases were reported annually in the United States in the early 1980s, but public education about the disease, along with improved methods of diagnosis and treatment, caused a drop to about 695 cases by 2001 (**FIGURE 10.15**). Contrary to its name, RMSF is not commonly reported in western states any longer, but it remains a problem in many southeastern and Atlantic Coast states. Children are its primary victims because of their contact with ticks.

Epidemic:
a sudden and rapid spread of an infectious disease through much of a population.

EPIDEMIC TYPHUS PRODUCES A POTENTIALLY FATAL ILLNESS

Epidemic typhus (also called **typhus fever**) is one of the most notorious of all bacterial diseases. It is considered a prolific killer of humans, and on several occasions it

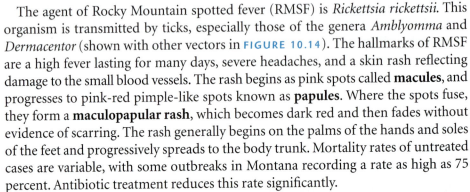

(a)   (b)   (c)

FIGURE 10.14

Three Arthropods that Transmit Bacterial Disease

(a) Rocky Mountain woodtick (*Dermacentor*). (b) The body louse (*Pediculus*). (c) The flea (*Xenopsylla*).

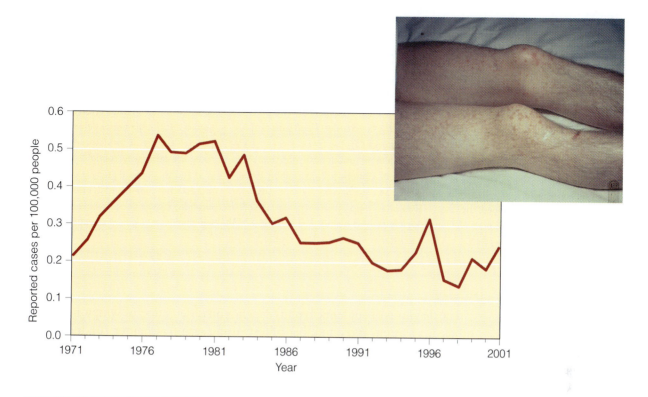

FIGURE 10.15

The Incidence and Symptoms of Rocky Mountain Spotted Fever

The incidence of Rocky Mountain spotted fever in the United States over the 30-year period preceding 2001. During the 1970s, this was the most prevalent tickborne disease, but insect control methods and antibiotic therapy have reduced the rate of incidence. Lyme disease is currently the most reported tickborne illness. Superimposed on the graph is a patient displaying the maculopapular rash associated with the disease.

has altered the course of history, as when it helped decimate the Aztec population in the 1500s. Historians report that Napoleon marched into Russia in 1812 with over 200,000 French soldiers but his forces were hit hard by typhus. Hans Zinsser's classic book *Rats, Lice, and History* describes events such as these and provides an engaging look at the effects of several infectious diseases on civilization (MicroFocus 10.8).

Epidemic typhus is caused by *Rickettsia prowazekii*. The bacteria are transmitted to humans by head and body lice of the genus *Pediculus* (Figure 10.14b). This was first noted by Charles Nicolle, winner of the 1928 Nobel Prize in Physiology or Medicine. Lice are natural parasites of humans. They flourish where sanitation measures are lacking and hygiene is poor. Often these conditions are associated with war, famine, poverty, and a generally poor quality of life.

The characteristic fever and rash of rickettsial disease are particularly evident in epidemic typhus. There is a **maculopapular rash**, but unlike the rash of RMSF, it appears first on the body trunk and progresses to the extremities. Intense fever, sometimes reaching 40°C, remains for many days as the patient hallucinates and becomes delirious. (Typhus, like typhoid, takes its name from the Greek word *typhos*, meaning "smoke" or "cloud.") Some patients suffer permanent damage to the blood vessels and heart, and over 75 percent of sufferers die in epidemics. Again, antibiotic therapy reduces this percentage substantially, and the disease is rare in the United States.

MicroFocus 10.8

RATS, LICE, AND HISTORY

The title of this box repeats the title of a book written by Hans Zinsser. Zinsser was born in New York City in 1878. He achieved fame for isolating the bacterium of epidemic typhus, but he is equally well known for his prose. Zinsser wrote some of the most uniquely personal, wise, and witty books of the twentieth century. Indeed, the unabridged title of his most famous book is (with tongue in cheek) *Rats,*

Lice, and History: Being a Study in Biography, Which, After Twelve Preliminary Chapters Indispensable for the Preparation of the Lay Reader, Deals with the Life History of Typhus Fever.

On a more serious note, here is a sample of Zinsser's writing from his 1935 book: "Soldiers have rarely won wars. They often mop up after the barrage of epidemics. And typhus, with its brothers and sisters—plague, cholera, typhoid,

dysentery—has decided more campaigns than Caesar, Hannibal, Napoleon, and all the other generals of history. The epidemics get the blame for defeat, the generals the credit for victory. It ought to be the other way 'round . . . ."

If you think you'd enjoy learning more about disease and its effect on civilization, Zinsser's book would be a worthwhile investment of your time.

The diagnosis of epidemic typhus depends on observation of symptoms, evidence of lice infestation, and a number of tests using serum (Chapter 20), including the Weil-Felix test. Control of the disease requires destruction of lice populations through good hygiene. The use of insecticide powders on the clothes and body surface are also helpful.

ENDEMIC TYPHUS IS TRANSMITTED BY INFECTED FLEAS

Endemic:
an infectious disease that exists continually in a population.

Endemic typhus is a second form of typhus. This disease occurs sporadically in human populations because the flea that transmits it, *Xenopsylla cheopis*, is not a natural parasite of humans (Figure 10.14c). However, the disease is prevalent in rodent populations where fleas abound (such as rats and squirrels), and thus it is called **murine typhus**, from the Latin *murine*, referring to "mouse." Cats and their fleas are involved also, and lice may harbor the bacilli.

The agent of endemic typhus is *Rickettsia typhi*. Both rodent and flea may harbor it for weeks without displaying any effects, but when an infected flea feeds in the human skin, it deposits the organism into the wound. Endemic typhus is usually characterized by a mild fever, persistent headache, and a maculopapular rash. Often the recovery is spontaneous, without the need of drug therapy. However, lice may transport the rickettsiae to other individuals and initiate an epidemic.

In the Southwest, endemic typhus is known as **Mexican typhus**, the disease from which Ricketts died. Since first described in 1570, it has also been called **tabardillo**, from *tabardo*, meaning "colored cloak," a reference to the mantle-like rash that covers the body.

OTHER RICKETTSIAE ALSO CAUSE DISEASE

Several other rickettsial diseases have microbiological significance because of their sporadic occurrence. These diseases are transmitted to humans by arthropod vectors, and most are characterized by fever and rash. The mortality rates are generally low.

Scrub typhus, found in Asia and the Southwest Pacific, is so named because it occurs in scrubland, where the soil is sandy or marshy and vegetation is poor. *Trombicula*, the mite that transmits the disease, lives in areas such as these. Scrub typhus also is called **tsutsugamushi fever**, from the Japanese words *tsutsuga* for "disease" and *mushi* for "mite." The causative agent, *Rickettsia tsutsugamushi*, enters the skin during mite infestations and soon causes fever and a rash, together with other typhus-like symptoms. Outbreaks may be significant, as evidenced by the 7,000 US servicemen affected in the Pacific during World War II.

Rickettsialpox was first recognized in 1946 in an apartment complex in New York City. Investigators traced the disease to mites in the fur of local mice, and named the disease rickettsialpox because the skin rash was similar to that of chickenpox. Rickettsialpox is now considered a benign disease. It is caused by *Rickettsia akari* (*acari* is Greek for "mite"). Fever and rash are typical symptoms, and fatalities are rare.

Except for the great influenza pandemic of 1918 to 1919 (Chapter 13), **trench fever** was the most widespread disease encountered during World War I. An estimated 1 million soldiers are thought to have been infected. The disease is caused by *Bartonella (Rochalimaea) quintana*, one of the few rickettsiae-like organisms cultivated outside of living cells. Transmission takes place by head and body lice. Symptoms include a maculopapular rash and a fever occurring at irregular intervals. Trench fever is prevalent where lice abound, and trench warfare provided a near-ideal setting. It also has been implicated in heart muscle inflammation in homeless individuals.

In various regions of the world, *Rickettsia conorii* causes a series of **tickborne fevers**. These diseases are known as boutonneuse fever, Marseilles fever, Nigerian typhus, and South African tick bite fever. All appear to be mild versions of RMSF, and all are accompanied by a fever and rash. In addition, a **black spot** (*la tache noire*) develops where the tick has fed in the skin. Wild rodents harbor the disease in nature.

In textbooks from the early 1900s, an illness called **Brill-Zinsser disease** was described as a mild form of typhus. The high fever was a distinguishing characteristic, and it was recognized that if a patient was infested with lice, an epidemic of typhus might ensue. Today microbiologists believe that Brill-Zinsser disease is a relapse of an earlier case of typhus in which *Rickettsia prowazekii* lay dormant in the patient for many years. However, symptoms are usually milder.

Ehrlichiosis, the final disease we shall consider, was first described in humans in 1986. Formerly believed to be confined to dogs, ehrlichiosis has been recognized in two forms in the United States: **human monocytic ehrlichiosis (HME)**, which is caused by the rickettsia *Ehrlichia chaffeensis* (because the first case was observed at Fort Chaffee, Arkansas); and **human granulocytic ehrlichiosis (HGE)**. The current thinking is that ehrlichiosis is identical with HME, and that HGE is a separate disease. Indeed, in 1996 the causative agent of HGE was first identified (**FIGURE 10.16**).

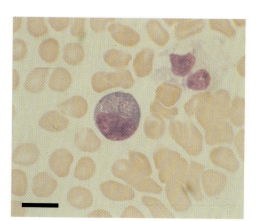

FIGURE 10.16

An *Ehrlichia* Species

The species of *Ehrlichia* that causes human granulocytic ehrlichiosis (HGE). This tickborne organism is seen multiplying within human white blood cells called granulocytes. The granulocytes are neutrophils. (Bar = 10 μm.)

It was found to be almost identical to *Ehrlichia phagocytophila* and to *Ehrlichia equi*, which causes ehrlichiosis in horses. As of this writing, the organisms are in the process of reclassification.

Patients with either ehrlichiosis (HME) or HGE suffer from headache, malaise, and fever, with some liver disease and, infrequently, a maculopapular rash. Ehrlichiosis (HME) is transmitted by the Lone Star tick (prevalent in the South), while HGE is transmitted by the dog tick and the deer tick, the same one that transmits Lyme disease (prevalent in the Northeast). Indeed, both HGE and HME are quite similar to Lyme disease, except that the symptoms come on faster in HGE and

TABLE 10.2

A Summary of Arthropodborne Bacterial Diseases

DISEASE	CAUSATIVE AGENT	DESCRIPTION OF AGENT	ORGANS AFFECTED	CHARACTERISTIC SIGNS
Bubonic plague	*Yersinia pestis*	Gram-negative bipolar staining rod	Lymph nodes Blood Lungs	Buboes Pneumonia Septicemia
Tularemia	*Francisella tularensis*	Gram-negative bipolar staining rod	Eyes Skin Blood	Eye lesion Skin ulcer Pneumonia
Lyme disease	*Borrelia burgdorferi*	Spirochete	Skin Joints Heart	Erythema chronicum migrans Arthritis
Relapsing fever	*Borrelia recurrentis*	Spirochete	Blood Liver	Fever Jaundice
Rocky Mountain spotted fever	*Rickettsia rickettsii*	Rickettsia	Blood Skin	Fever Rash
Epidemic typhus	*Rickettsia prowazekii*	Rickettsia	Blood Skin	Fever Rash
Endemic typhus	*Rickettsia typhi*	Rickettsia	Blood Skin	Fever Rash
Scrub typhus	*Rickettsia tsutsugamushi*	Rickettsia	Blood Skin	Fever Rash
Rickettsialpox	*Rickettsia akari*	Rickettsia	Blood Skin	Fever Rash
Trench fever	*Bartonella quintana*	Rickettsia	Blood Skin	Fever Rash
Tickborne fevers	*Rickettsia conorii*	Rickettsia	Blood Skin	Fever Rash
Ehrlichiosis (HME) (HGE)	*Ehrlichia chaffeensis* *E. equi* (?)	Rickettsia	Blood	Fever

HME, they clear more quickly, and the rash is infrequent. HME affects the body's monocytes (hence "monocytic"), while HGE affects the neutrophils (a type of granulocyte, hence "granulocytic"). Thus, a lowering of the white blood cell count (leukopenia) occurs in both diseases. A notable outbreak of HGE affected 68 patients in New York City in 1994.

The arthropodborne diseases covered in this chapter are summarized in TABLE 10.2. MicroInquiry 10 presents four cases for study involving soilborne and arthropodborne diseases.

VECTOR	TOXIN INVOLVED	TREATMENT ADMINISTERED	IMMUNIZATION AVAILABLE	COMMENT
Rat flea	Yes	Tetracycline Streptomycin	Killed bacteria for high risk	Great historic significance Septicemic and pneumonic forms Rodents infected
Flea Tick	Yes	Tetracycline Streptomycin	None	Resembles many other diseases Many modes of transmission No epidemics
Tick	Not established	Penicillin Doxycycline	Genetically engineered lipoprotein	Recognized since 1975 Most prevalent tickborne disease Three possible stages
Louse Tick	Not established	Doxycycline	None	Relapses common Rose spots develop on skin
Tick	Not established	Doxycycline	Killed bacteria for high risk	Most common rickettsial disease Rash first on extremities
Louse	Not established	Doxycycline Chloramphenicol	None	High mortality rate Rash first on body trunk
Flea	Not established	Doxycycline Chloramphenicol	None	Mild typhus-like disease Prevalent in rodents
Mite	Not established	Doxycycline Chloramphenicol	None	Occurs in scrubland Prevalent in Asia
Mite	Not established	Doxycycline Chloramphenicol	None	Rare disease Resembles chickenpox
Louse	Not established	Doxycycline Chloramphenicol	None	Common in World War I Cultivation in artificial medium
Tick	Not established	Doxycycline Chloramphenicol	None	Different forms
Tick	Not established	Doxycycline Chloramphenicol	None	Recognized since 1986

MicroInquiry 10

SOILBORNE AND ARTHROPODBORNE DISEASE IDENTIFICATION

Below are several descriptions of soilborne and arthropodborne bacterial diseases based on material presented in this chapter. Read the case history and then answer the questions posed. Answers can be found in Appendix E.

Case 1: An 18-year-old girl who had been in good health goes to her physician in Seattle, Washington, complaining of a headache and flu-like symptoms. Further questioning indicates she had developed a red rash on her thigh that enlarged but then disappeared after two weeks. During this time, she also had a fever. The physician discovers that the patient had been hiking in the hills east of Seattle ten days prior to developing the rash. The patient is placed on doxycycline and recovers.

10.1a. What important clues toward identifying the disease are indicated from the patient's case?

10.1b. What disease does she have and what organism is responsible for the disease?

10.1c. How was the organism responsible for the disease transmitted to the patient?

10.1d. What complications could occur if the patient had not visited her physician?

10.1e. How could this disease have been prevented?

Case 2: A 33-year-old indigent man comes to the emergency room of the county hospital. The emergency room nurse immediately notices that the man cannot open his mouth because of facial muscle spasms. The physician on duty detects right-sided face pain and trismus. She is able to ascertain from the patient that he has not been able to eat for two days because of jaw pain. Examination of his body shows necrotic, blackened areas on the bottom of his left foot. Based on the signs and physical exam, treatment is started.

10.2a. Based on these signs and the physical exam results, what organism has infected this man and what disease does he have?

10.2b. What treatment should be provided?

10.2c. Should treatment have waited until the diagnosis was confirmed by laboratory results? Explain.

10.2d. What is the significance of the necrotic areas of the patient's left foot?

10.2e. How might this indigent man have become infected?

Note to the Student

Several diseases discussed in this chapter have occurred in broadscale epidemics in past centuries and have accounted for widespread death. Bubonic plague, relapsing fever, and epidemic typhus are three examples. In today's world, microbiologists can quell the spread of these diseases by controlling arthropod populations.

But there is one disease in this chapter that may be spread by airborne spores. In the human body, its effects are so devastating that whole populations may be wiped out. The disease is anthrax. Its causative agent, *Bacillus anthracis,* is a potent weapon in bioterrorism.

Consider how easily bioterrorism could be carried out with *B. anthracis* spores. Spores released into the atmosphere may settle onto food and into water for human consumption. They also may be inhaled into the lungs or enter the blood through an abrasion in the skin. The terrifying symptoms of anthrax quickly follow any of these exposures, as the 2001 anthrax-containing letters demonstrated.

Can terrorism of this type come to pass? Apparently so, because for many years, several countries have explored its possibilities. To preclude biological warfare and bioterrorism, over 100 nations signed an international treaty in 1972 banning the

Case 3: A 10-year-old boy is brought to the emergency room in Charleston, South Carolina, by his mother and father. He had been in good health when the family went for a July Fourth holiday camping trip nine days ago in the Appalachian Mountains. The day after returning from the trip, a tick was discovered in his scalp and was removed. In the emergency room, the boy complains of a headache. He has a fever and the physician observes a pink rash on the palms of his hands and soles of his feet. His white blood cell count is slightly elevated. The physician starts the patient on erythromycin therapy and the boy eventually recovers completely.

10.3a. Identify two organisms that could produce the finding described in this case study.

10.3b. What is the agent responsible for the patient's disease? What were the clues for the diagnosis?

10.3c. What infectious bacterial diseases are spread by ticks?

10.3d. What test(s) might be used to confirm the diagnosis?

10.3e. How would organisms like *Ehrlichia chaffeensis* be eliminated as the agent responsible for the patient's disease?

Case 4: A 34-year-old woman arrives in the emergency room complaining of extreme pain in the right shin with limited mobility in the leg. Her breathing is normal. Examination of the lower leg indicates trauma and the skin is discolored a greenish blue. An additional finding was a crackling sound in her lower leg. The patient tells the physician that one week ago she had a severe mountain biking accident and had several deep cuts to her leg. She is given antibiotics and taken to the operating room where necrotic muscle is discovered. A biopsy

Gram stain from the tissue shows gram-positive rods.

10.4a. Based on the emergency room findings, what disease does the patient have and what bacterium is responsible for the infection?

10.4b. What physical conditions of growth should be used when incubating the blood culture?

10.4c. In the operating room, what should the surgeons do once they know the identity of the disease?

10.4d. To ensure a systemic infection does not develop, what physical treatment can be used to slow down or stop the growth of the infecting bacteria? Why would this treatment work?

development of biological weapons. Under one of its terms, all stockpiles of anthrax spores were to be destroyed, and all such weapons research was to cease. However, in 1979, observers reported an epidemic of anthrax in the city of Sverdlovsk in the Siberian region of Russia. The epidemic came only a few days after an explosion at a nearby military installation. Soviet officials denied that the two events were related but failed to provide a satisfactory explanation to critics. Then, in 1993, after the breakup of the Soviet Union, examination of autopsy notes and pathological specimens from victims confirmed that lesions in the lungs were characteristic of anthrax.

Many scientists believe that the threat of bioterrorism looms large, and the concern for the enemy's use of bacteriological weapons surfaced during the Persian Gulf War of 1992. The Iraqi authorities acknowledged in 1995 that they had produced 100 botulinum toxin and 50 anthrax bombs; armed 13 Scud missile warheads with botulinum toxin and 10 with anthrax; and filled 122-mm rockets with anthrax or botulinum toxin. The possibilities of genetically engineered biological weapons have added a new dimension to the issue and compounded the threat.

Summary of Key Concepts

10.1 SOILBORNE BACTERIAL DISEASES

- **Anthrax Produces Three Exotoxins.** Anthrax is an acute infectious disease caused by the gram-positive, aerobic endospore former *Bacillus anthracis*. Human contact can be by inhalation, consumption, or skin contact with spores. Inhalation produces symptoms of respiratory distress and causes a blood infection. Consumption and skin contacts lead to boil-like lesions. Ciprofloxacin has been the antibiotic of choice.

- **Tetanus Produces Hyperactive Muscle Contractions.** *Clostridium tetani is* the causative agent of tetanus. Symptoms of generalized muscle stiffness and trismus lead to convulsive contractions with an unnatural fixed smile. Antitoxin and antibiotics can be used to neutralize the toxin and kill the bacterial cells. A vaccine is available for prevention.

- **Gas Gangrene Causes Massive Tissue Damage.** Gas gangrene is caused by the anaerobic endospore-forming *Clostridium perfringens*. Symptoms include intense pain, swelling, and a foul odor at the wound site. Penicillin and removal of dead tissue help recovery.

- **Leptospirosis Causes Fever and Headache.** Leptospirosis is a disease spread from animals to humans (zoonosis). The organism, *Leptospira interrogans,* is a spirochete . Infected individuals have flulike symptoms. Up to 10 percent of patients experience a systemic form called Weil's disease. Penicillin is the antibiotic of choice.

- **Melioidosis Occurs in Two Forms.** *Burkholderia pseudomallei* can cause two forms of melioidosis. A chronic form develops abscesses in several body organs, while a rare form causes pneumonia and a blood infection.

10.2 ARTHROPODBORNE BACTERIAL DISEASES

- **Bubonic Plague Can Be a Highly Fatal Disease.** *Yersinia pestis*, a gram-negative rod, is the causative agent of plague. This highly fatal infectious disease is transmitted to humans by the bites of infected fleas. Bubonic plague is characterized by the formation of buboes, and without treatment, mortality can be as high as 60 percent. Spreading of the bacilli to the blood leads to septicemic plague. Localization in the lungs is characteristic of pneumonic plague, which can be spread person to person. Without treatment, septicemic and pneumonic plague are nearly 100 percent fatal.

- **Tularemia Has More than One Disease Presentation.** Tularemia is caused by *Francisella tularensis*, a gram-negative rod that is highly infectious at low doses. Various forms of the disease occur depending on where the bacilli enter the body. Skin ulcers, eye lesions, and pulmonary symptoms can result. Treatment is effective using tetracycline or streptomycin.

- **Lyme Disease Can Be Divided into Three Stages.** Tickborne Lyme disease results from an infection by the spirochete *Borrelia burgdorferi*. It involves three stages: the erythema chronicum migrans (ECM) rash; neurologic and cardiac disorders of the central nervous system; and migrating arthritis. Penicillin or doxycycline is effective.

- **Relapsing Fever Is Carried by Ticks and Lice.** Relapsing fever results in recurring attacks of high fever caused by ticks carrying *Borrelia hermsii* (endemic relapsing fever) or body lice carrying *Borrelia recurrentis* (epidemic relapsing fever). Symptoms include substantial fever, shaking chills, headache, and drenching sweats. Treatment with doxycycline or erythromycin speeds recovery.

10.3 RICKETTSIAL ARTHROPODBORNE DISEASES

- **Rocky Mountain Spotted Fever Begins with a Characteristic Rash.** RMSF is caused by *Rickettsia rickettsii*, a gram-negative, nonmotile, intracellular parasite. Carried by ticks, the symptoms include a high fever for several days, severe headache, and a maculopapular skin rash. Effective treatment requires doxycycline or chloramphenicol.

- **Epidemic Typhus Produces a Potentially Fatal Illness.** Epidemic typhus is a potentially fatal disease that occurs in unsanitary conditions and in overcrowded living conditions. It is caused by *Rickettsia prowazekii* that is carried by body lice. A maculopapular rash that progresses to the extremities, intense fever, hallucinations, and delirium are characteristic symptoms of the disease. In epidemics, mortality can be as high as 75 percent. Effective treatment requires doxycycline or chloramphenicol.

- **Endemic Typhus Is Transmitted by Infected Fleas.** Endemic typhus is transmitted by fleas. The causative agent, *Rickettsia typhi*, causes a mild fever, headache, and maculopapular rash. Recovery often is spontaneous, although doxycycline or chloramphenicol can be helpful.

- **Other Rickettsiae Also Cause Disease.** A number of other diseases caused by rickettsiae occur. This includes scrub typhus (*Rickettsia tsutsugamushi*), rickettsialpox (*Rickettsia akari*), trench fever (*Bartonella quintana*), tickborne fevers (*Rickettsia conorii*), Brill-Zinsser disease (recurring *Rickettsia prowazekii*), and ehrlichiosis (*Ehrlichia chaffeensis*).

Questions for Thought and Discussion

Answers to selected questions can be found in Appendix C.

1. In February 1980, a patient was admitted to a Texas hospital complaining of fever, headache, and chills. He also had greatly enlarged lymph nodes in the left armpit. A sample of blood was taken and Gram stained, whereupon gram-positive diplococci were observed. The patient was treated with cefoxitin, a drug for gram-positive organisms, but soon thereafter he died. On autopsy, *Yersinia pestis* was found in his blood and tissues. Why was this organism mistakenly thought to be diplococci, and what error was made in the laboratory? Why were the symptoms of plague missed?

2. There is a town outside of London known as Gravesend. The town apparently acquired its name during the 1660s in connection with a great medical upheaval. What was the name of that upheaval, and how do you suppose the name came about?

3. Although the tetanus toxin is second in potency to the toxin of botulism, many physicians consider tetanus to be a more serious threat than botulism. Would you agree? Why?

4. Some estimates place epidemic typhus among the all-time killers of humans; one listing even has it in third place behind malaria and plague. In 1997, during a civil war, an outbreak of epidemic typhus occurred in the African country of Burundi. The outbreak was estimated to be the worst since World War II. What conditions may have led to this epidemic?

5. On February 13, 1976, a newspaper article requested purchasers of a certain brand of Pakistani wool yarn to check with their local health departments because the wool was thought to be contaminated with anthrax spores. The article read as follows: "Spores of anthrax, which is a livestock disease, have been found on the yarn. The disease affects people as a skin ailment and is usually not fatal." Does the article minimize a potential medical emergency? If yes, write a letter to the editor in reply. If not, explain why not.

6. In 1993, a 57-year-old man died of tetanus in a Kansas hospital. His experience had begun on August 14 with a puncture wound to the foot and ended on September 16. Family members reported that the man had never received a vaccination with tetanus toxoid. Hospital costs for his therapy totaled $145,329. The administration of a dose of tetanus vaccine, by comparison, costs $3.30. Setting aside the value of a human life for the moment, what does this disparity of costs tell you?

7. At various times, local governments are inclined to curtail deer hunting. How might this lead to an increase in the incidence of Lyme disease?

8. In 1982, endemic typhus was observed in five members of a Texas household. On investigation, epidemiologists learned that family members had heard rodents in the attic, and two weeks previously they had used rat poison on the premises. Investigators concluded that both the rodents and the rat poison were related to the outbreak. Why?

9. Leptospirosis has been contracted by individuals working in such diverse locales as subway tunnels, gold mines, rice paddies, and sewage-treatment plants. What precautions might be taken by such workers to protect themselves against the disease?

10. A young woman was hospitalized with excruciating headache, fever, chills, nausea, muscle pains in her back and legs, and a sore throat. Laboratory tests ruled out meningitis, pneumonia, mononucleosis, toxic shock syndrome, and other diseases. On the third day of her hospital stay, a faint pink rash appeared on her arms and ankles. By the next day, the rash had become darker red and began moving from her hands and feet to her arms and legs. Can you guess what the diagnosis eventually was?

11. In Chapter 9 of the Bible, in the Book of Exodus, the sixth plague of Egypt is described in this way: "Then the Lord said to Moses and Aaron, 'Take a double handful of soot from a furnace, and in the presence of Pharaoh, let Moses scatter it toward the sky. It will

then turn into a fine dust over the whole land of Egypt and cause festering boils on man and cattle throughout the land.'" Which disease in this chapter is probably being described?

12. During the Civil War, some 92,000 soldiers died of battle wounds but 190,000 died of battle-related diseases. Which diseases in this chapter probably contributed to the enormous mortality rate? What conditions encouraged each disease cited?

13. In autumn, it is customary for homeowners in certain communities to pile leaves at the curbside for pickup. How might this practice increase the incidence of tularemia, Lyme disease, and Rocky Mountain spotted fever in the community?

14. At various times in past centuries, it was believed that cats were the medium through which witches spoke. Fearing cats, people would try to eliminate these animals from their neighborhood. Bubonic plague would occasionally break out shortly thereafter. Why?

15. Centuries ago, the habit of shaving one's head and wearing a wig probably originated in part as an attempt to reduce lice infestations in the hair. Why would this practice also reduce the possibilities of certain diseases? Which diseases?

16. An article in *Health* magazine (1996) opens with the following statement: "As pierce-o-mania sweeps the nation, it's bringing more ugly infections along with it." What are some possible infections that ear, nose, tongue, and other body piercings can lead to?

17. Even before October 2001, people from various government and civilian agencies were concerned about a terrorist attack using anthrax spores. In various scenarios, try to paint a picture of how such an attack might happen. Then, using your knowledge of microbiology, present your vision of how agencies might deal with such an attack.

http://microbiology.jbpub.com

The site features **eLearning,** an on-line review area that provides quizzes and other tools to help you study for your class. You can also follow useful links for in-depth information, or just find out the latest microbiology news.

Review

The bacterial diseases transmitted by soil and arthropods are the main focus of this chapter. To test your understanding of the chapter contents, match the statement on the left with the disease on the right by placing the correct letter in the available space. A letter may be used once, more than once, or not at all. Appendix D contains the answers.

_____ 1. Accompanied by erythema chronicum migrans.

_____ 2. Transmitted by lice; caused by *R. prowazekii*.

_____ 3. Affects neutrophils in the body; transmitted by ticks.

_____ 4. May be transmitted by contact with dogs.

_____ 5. Caused by a spore-forming rod that produces hemolysis and lecithinase.

_____ 6. Tickborne disease; caused by *R. rickettsii*.

_____ 7. Blood hemorrhaging in large animals such as cattle, sheep, and goats.

_____ 8. Also known as tabardillo and Mexican typhus.

_____ 9. Treated with antitoxins; caused by an anaerobic spore former.

_____ 10. Caused by *Borrelia burgdorferi*; transmitted by a tick.

_____ 11. Bubonic, septicemic, and pneumonic stages.

_____ 12. Maculopapular rash beginning on extremities and progressing to body trunk.

_____ 13. Caused by a spirochete that infects kidney tissues in pets and humans.

_____ 14. Occurs in small game animals, especially rabbits.

_____ 15. Transmitted by mites; also known as scrub typhus.

_____ 16. Caused by a gram-negative rod with bipolar staining; transmitted by rat flea.

_____ 17. Long-range complications include arthritis in large joints.

_____ 18. Up to ten attacks of substantial fever, joint pains, and skin spots; *Borrelia* involved.

_____ 19. Immunization rendered by the DTaP vaccine.

_____ 20. Pulmonary, intestinal, and skin forms possible; due to a *Bacillus* species.

_____ 21. A typical zoonosis; caused by a spiral bacterium.

_____ 22. Caused by a *Francisella* species; has multiple modes of transmission.

_____ 23. Epidemics where sanitation is poor; louseborne rickettsial disease.

_____ 24. Sustained and uncontrolled contractions of the body's muscles.

_____ 25. Most commonly reported tickborne disease in the United States.

A. Plague

B. Epidemic typhus

C. Anthrax

D. Melioidosis

E. Relapsing fever

F. Tularemia

G. Lyme disease

H. Rickettsialpox

I. Tsutsugamushi

J. Endemic typhus

K. Ehrlichiosis

L. Leptospirosis

M. Tetanus

N. Rocky Mountain spotted fever

O. Gas gangrene

11

Sexually Transmitted, Contact, and Miscellaneous Bacterial Diseases

I just froze. Then I closed the door, and went in my room and cried.

—A soft-spoken, 38-year-old woman recalling her reaction when she was visited by a health official and told she had syphilis

THE VICTORIAN ERA OF THE 1800s was notorious for its prudish attitude toward sex. Bulls were called "he-cows," and the legs of a piano were modestly covered with pantaloons. Women did not disrobe when visiting a physician; instead, they pointed to a chart to show where they hurt. An 1863 etiquette book stipulated that the works of male and female authors should be separated on bookshelves unless the authors were married. Even the biology books of that era mirrored the taboo on sex by carefully avoiding mention of the human reproductive system. It was as if the human species did not provide for the next generation.

Our era, by contrast, has been marked by broad sexual freedom, but it also has seen an alarming rise in diseases of the reproductive organs. Formerly, these diseases were called venereal diseases (VDs) from Venus, the Roman goddess of love. However, health departments now call them **sexually transmitted diseases** (**STDs**), a name that indicates the mode of transmission and avoids the "love" connotation. The STDs discussed in this chapter are of bacterial origin and include familiar names such as syphilis and gonorrhea, as well as emerging problems such as chlamydia. To underscore the seriousness of the STD problem, the Centers for Disease Control and Prevention (CDC) estimate that for each day in 1999, over 40,000 Americans contracted a sexually transmitted disease.

The increase in sexually transmitted diseases is but one example of how changing social patterns can affect the incidence of disease. Several other examples are evident in the diseases discussed in this chapter. For instance, the incidence of leprosy in the United States has risen because in the last decades, immigrant groups have brought the disease with them. Toxic shock syndrome was first recognized widely in 1980 when a new brand of high-absorbency tampon appeared on the commercial market. Finally, the mortality rate from nocardiosis rose when this disease was found to complicate cases of acquired immune deficiency syndrome (AIDS).

11.1

Sexually Transmitted Diseases

The sexually transmitted diseases of bacterial origin belong to a broad category of diseases that are transmitted by contact. The contact in this case is with the reproductive organs. This type of person-to-person transmission is necessary for bacterial survival because the bacteria usually cannot remain alive outside the body tissues. The list of sexually transmitted diseases is diverse, as the following examples will demonstrate. Other STDs are discussed in Chapters 14, 15, and 16.

SYPHILIS CAN PROGRESS THROUGH THREE STAGES IF UNTREATED

Over the centuries, Europeans have had to contend with four pox diseases: chickenpox, cowpox, smallpox, and the Great Pox, a disease now known as syphilis. The first European epidemic was recorded in the late 1400s, shortly after the conquest of Naples by the French army (MicroFocus 11.1). For decades the disease had various names, but by the 1700s, it had come to be called syphilis.

Syphilis is caused by *Treponema pallidum*, literally the "pale spirochete." This spiral bacterium moves by means of axial filaments. Humans are the only host for *T. pallidum*, so the organism must spread by human-to-human contact, usually during sexual intercourse. It penetrates the skin surface through the mucous membranes or via a wound, abrasion, or hair follicle. Then, it causes a disease that progresses in three stages, described below. The variety of clinical symptoms that accompany the stages, and their similarity to other diseases, have led some physicians to call syphilis the "great imitator."

The incubation period for syphilis varies greatly, but it averages about three weeks. **Primary syphilis** is the first stage to appear. This stage is characterized by the **chancre**, a painless circular, purplish ulcer with a raised margin and hard edges described as being like cartilage (FIGURE 11.1a). The chancre develops at the site of entry of the spirochetes, often the genital organs. However, any area of the skin may be affected, including the pharynx, rectum, or lips. (Centuries ago, people discovered that kissing could transmit syphilis, and even though they did not know what was causing the dreaded disease, they began to substitute hand-shaking and gentle kisses on the cheek for kissing on the lips.) The chancre teems with spirochetes. It persists for two to three weeks, and then it heals spontaneously. However, the infection has not been eliminated.

Several weeks later, the untreated patient experiences **secondary syphilis**. Symptoms include fever and a constitutional flu-like illness, as well as swollen lymph nodes reminiscent of infectious mononucleosis. The skin rash that appears may be mistaken for measles, rubella, or chickenpox (FIGURE 11.1b). It usually involves the palms, soles, face, and scalp. Loss of the eyebrows often occurs, and a patchy loss of hair results in "moth-eaten" areas commonly seen on the head. Involvement of the

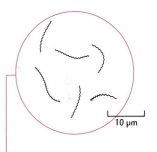

10 μm

Treponema pallidum

MicroFocus 11.1

THE ORIGIN OF A DISEASE

Among the more intriguing questions in medical history are how and why syphilis suddenly emerged in Europe in the late 1400s. Writers of that period tell of an awesome new disease that swept over Europe and on to India, China, and Japan. But where did the disease come from?

One oft-told story is that syphilis existed in the New World, and that members of Columbus's crew acquired it during stopovers in the Caribbean islands. Columbus returned to Palos in northern Italy in 1493, and some of his crew reportedly joined the army of Charles VII of France. In 1494, Charles's army attacked Naples in Italy, but mounting losses from the strange new sickness forced him to withdraw. His army of 30,000 French, German, Swiss, English, Polish, Spanish, and Hungarian troops returned to their native lands, and apparently brought the disease home with them.

A second theory holds that the disease first came to Spain and Portugal with slaves imported from Africa in the

mid-1400s. An African disease called yaws is very similar to syphilis in causative organism, transmission, and stages of development. Certain historians believe that some unknown factor caused yaws to flare up in the form of syphilis in the late 1400s. They speculate that the army of Charles VII provided a highly susceptible population of diverse men who spread the disease wherever they traveled. Indeed, recent studies of Native American burial grounds show no traces of syphilis before the arrival of Columbus. By contrast, remains of those dying after Columbus's arrival show signs that the disease was present in the community.

With its devastating effects, syphilis inspired a variety of epithets. The Italians called it the French disease (*morbus Gallicus*), while the French called it the Italian disease (*la maladie Italienne*); to the Japanese, it was the Chinese disease. The English impartially termed it the Great Pox. The name "syphilis" derives from the works of Girolamo Fracostoro, a sixteenth-century poet-scientist of

Verona. In 1530, Fracostoro wrote a long poem about a shepherd named Syphilus who momentarily left his pastoral responsibilities to commit a sexual indiscretion. The angered gods punished him with the horrible sores of the disease. In a later work on disease transmission, Fracostoro suggested that the illness be called syphilis after the mythical shepherd.

Syphilis was as international in effect as in name, and proved to be no respector of rank. Henry VIII of England, Napoleon of France, and Peter the Great of Russia all contracted the disease. Poets such as Keats, musicians such as Beethoven, and artists such as Gauguin also succumbed, as did millions of common people. For many generations, epidemics of syphilis swept back and forth across the world. As Lord Byron wrote in one of his poems:

The smallpox has gone out of late;
Perhaps it may be follow'd by the
Great.

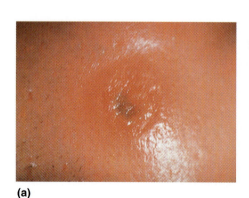

(a)

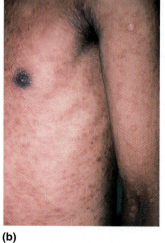

(b)

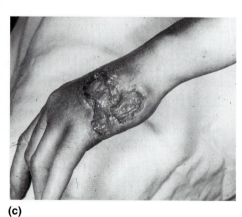

(c)

FIGURE 11.1

The Stages of Syphilis

Views of the skin lesion in the three stages of syphilis. (a) The chancre of primary syphilis as it occurs on the body surface. The chancre is circular with raised margins and is usually painless. It occurs where penetration of syphilis spirochetes has taken place. (b) The skin rash is characteristic of secondary syphilis. (c) The gumma that forms in tertiary syphilis. Note the granular, diffuse nature of this lesion compared with the primary chancre shown in (a).

liver may lead to jaundice and suspicion of hepatitis. In untreated patients, the symptoms last several weeks, and death may result. Most patients recover, but they bear pitted scars from the lesions and remain "pockmarked."

These individuals now enter a **latent stage**. Many patients will have relapses of secondary syphilis during which time they remain infectious. Within four years, the relapses cease and the disease is no longer infectious (except in pregnant females). Patients either remain asymptomatic or slowly progress to late-stage symptoms.

About one-third of untreated patients eventually develop **late** or **tertiary syphilis** over 6 to 40 years. This stage occurs in many forms, but most commonly it involves the skin, cardiovascular system, and nervous system. The hallmark of tertiary syphilis is the **gumma**, a soft, painless, gummy granular lesion that is noninfectious (**FIGURE 11.1c**). In the cardiovascular system, gummas weaken the major blood vessels, causing them to bulge and burst. In the spinal cord and meninges, gummas lead to degeneration of the tissues and paralysis. In the brain, they alter the patient's personality and judgment and cause insanity so intense that for many generations, people with tertiary syphilis were confined to mental institutions. Damage can be so serious as to cause death. It is conceivable that our ancestors failed to equate the chancre of primary syphilis with the horrible symptoms of tertiary syphilis because the stages were so distantly separated in time.

Syphilis is a serious problem in pregnant women because the spirochetes penetrate the placental barrier after the fourth month of pregnancy, causing **congenital syphilis** in the fetus. Syphilitic skin lesions and open sores may be apparent in the newborn, or symptoms may develop weeks after birth. Affected children often suffer poor bone formation, meningitis, or **Hutchinson's triad**, a combination of deafness, impaired vision, and notched, peg-shaped teeth. There were 441 cases reported to the CDC in 2001.

The cornerstone of syphilis control is the identification and treatment of the sexual contacts of patients. Penicillin is the drug of choice. *T. pallidum* multiplies very slowly in the tissues, partly because of its 33-hour generation time. This factor encourages successful therapy.

Since *T. pallidum* (**FIGURE 11.2**) was first observed in 1905 by Fritz R. Schaudinn and P. Erich Hoffman, exhaustive attempts have been made to cultivate it on laboratory media, but none have been successful. Diagnosis in the primary stage therefore depends on the observation of spirochetes from the chancre using darkfield microscopy. As the disease progresses, a number of tests to detect syphilis antibodies becomes useful, including the rapid plasma reagin test, the Venereal Disease Research Laboratory (VDRL) test, and others (Chapter 20).

Syphilis is currently among the most reported microbial diseases in the United States (**FIGURE 11.3**). Statistics indicate that over 32,000 people are afflicted with the disease annually, of whom about 6,000 are in the primary or secondary stage. Taken alone, these figures suggest the magnitude of the syphilis epidemic, but some public health microbiologists believe that for every case reported, as many as nine cases go unreported.

GONORRHEA IS AN INFECTION OF THE UROGENITAL TRACT

Gonorrhea is the second most frequently reported microbial disease in the United States, after chlamydial infections (Figure 11.3). During the 1960s, the incidence of gonorrhea rose dramatically; since 1975, several hundred thousand cases have been reported annually, the highest percentage being in persons under 24 years of age. Epidemiologists suggest that 3 to 4 million cases go undetected or unreported each

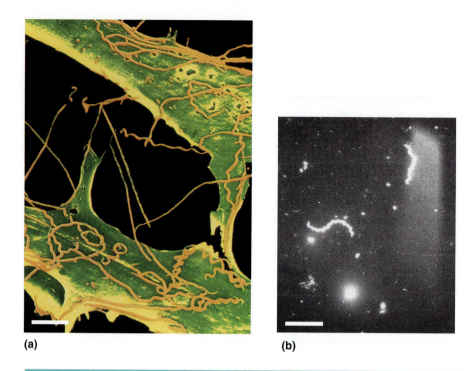

(a)   (b)

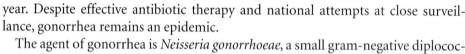

FIGURE 11.2

Treponema pallidum

(a) False color scanning electron micrograph of *Treponema pallidum*. (Bar = 5 μm.) (b) A dark-field microscope view of the bacteria seen in a sample taken from the chancre of a patient. (Bar = 10 μm.)

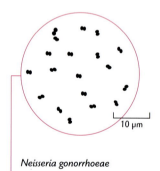

Neisseria gonorrhoeae

year. Despite effective antibiotic therapy and national attempts at close surveillance, gonorrhea remains an epidemic.

The agent of gonorrhea is *Neisseria gonorrhoeae*, a small gram-negative diplococcus named for Albert L. S. Neisser, who isolated it in 1879. The organism, commonly known as the **gonococcus**, has a characteristic double-bean shape exhibited by neisseriae (**FIGURE 11.4a**). *N. gonorrhoeae* is a very fragile organism susceptible to most antiseptics and disinfectants. It survives only a brief period outside the body and is rarely contracted from a dry surface such as a toilet seat. The great majority of cases of gonorrhea therefore are transmitted in person-to-person contact during sexual intercourse. (Gonorrhea is sometimes called "the clap," from the French *clap-poir* for "brothel.")

The incubation period for gonorrhea ranges from two to six days. In females, the gonococci invade the epithelial surfaces of the cervix and the urethra. The cervix may be reddened, and a discharge may be expressed by pressure against the pubic area. Patients often report abdominal pain and a burning sensation on urination, and the normal menstrual cycle may be interrupted.

In some females, gonorrhea also spreads to the fallopian tubes, which extend from the uterus to the ovaries. As these thin passageways become riddled with pouches and adhesions, the passage of egg cells becomes difficult. Complete blockage, or **salpingitis**, may take place. A condition of the pelvic organs such as this is called **pelvic inflammatory disease (PID)**. Sterility may result from scar tissue remaining after the disease has been treated, or a woman may experience an **ectopic pregnancy**.

FIGURE 11.3

Reported Cases of Microbial Diseases in the United States, 2001

For the year, chlamydia was considerably more prevalent than the next most common reportable disease, gonorrhea. AIDS was the third most common.

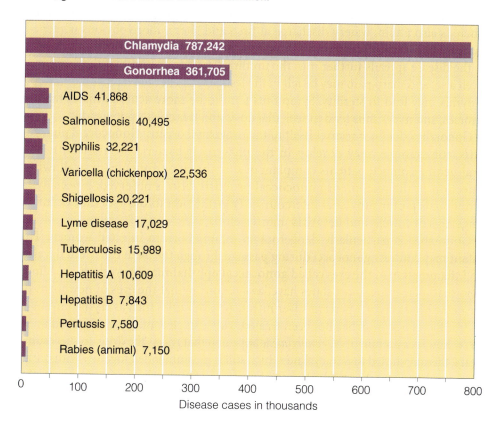

Disease	Cases
Chlamydia	787,242
Gonorrhea	361,705
AIDS	41,868
Salmonellosis	40,495
Syphilis	32,221
Varicella (chickenpox)	22,536
Shigellosis	20,221
Lyme disease	17,029
Tuberculosis	15,989
Hepatitis A	10,609
Hepatitis B	7,843
Pertussis	7,580
Rabies (animal)	7,150

Disease cases in thousands

(a)   (b)   (c)

FIGURE 11.4

Neisseria gonorrhoeae

Three views of *Neisseria gonorrhoeae*, the agent of gonorrhea. (a) An enlarged view of the cocci. Note the typical pairing of the cells and the dust-like granules on the cell surfaces. The significance of these granules is not clear. (Bar = 2 μm.) (b) A scanning electron micrograph of the edge of a colony of *Neisseria gonorrhoeae*. (Bar = 5 μm.) (c) A stained smear of discharge from the male urethra showing the diplococci of gonorrhea. Some of the diplococci can be seen in the cytoplasm of white blood cells. (Bar = 10 μm.)

It should be noted that symptoms are not universally observed in females, and that an estimated 50 percent of affected women exhibit no symptoms. Such asymptomatic women may spread the disease unknowingly.

In males, gonococci infect the mucus membranes of the urethra, the tube from the bladder that passes through the penis to the exterior. Onset usually is accompanied by a tingling sensation in the penis, followed in a few days by pain when urinating. There is also a thin, watery discharge at first, and later a more obvious whitened, thick fluid that resembles semen. Frequent urination and an urge to urinate develop as the disease spreads further into the urethra. The lymph nodes of the groin may swell also, and sharp pain may be felt in the testicles. Unchecked infection of the epididymis may lead to sterility. Symptoms tend to be more acute in males than in females, and males thus tend to seek diagnosis and treatment more readily.

Gonorrhea does not restrict itself to the urogenital organs. **Gonococcal pharyngitis**, for example, may develop in the pharynx if bacteria are transmitted by oral-genital contact; patients complain of sore throat or difficulty in swallowing. Infection of the rectum, or **gonococcal proctitis**, also is observed, especially in individuals performing anal intercourse. Transmission to the eyes may occur by fingertips or towels, and **keratitis** may develop.

Gonorrhea is particularly dangerous to infants born to infected women. The infant may contract gonococci during passage through the birth canal and develop an inflammation of the eyes called **gonococcal ophthalmia**. To preclude the blindness that may ensue, most states have laws requiring that the eyes of newborns be treated with drops of 1 percent silver nitrate or antibiotics such as erythromycin.

Traditionally, gonorrhea has been treated with a single large dose of penicillin. In 1976, however, a strain of *N. gonorrhoeae* appeared that resists the drug. In succeeding years, therapy shifted to tetracycline, which still is recommended together with ceftriaxone and cefixime. An attack of gonorrhea does not immunize one to future attacks, apparently because the immune system does not respond strongly enough to the first attack. No vaccine is available, but an antipili vaccine is in the experimental stage.

Gonorrhea can be detected by observing gram-negative diplococci in the discharge from the urogenital tract, as well as in colonies from swab samples cultivated on Thayer-Martin medium (**FIGURE 11.4b, c**). For an immunological test, physicians take a swab sample and dip the swab into an antibody solution. An immunoassay reaction takes place (Chapter 20), and within a few hours, a color reaction indicates the presence or absence of gonococci. The test allows doctors to detect gonorrhea early so that treatment can start immediately. A test to detect the DNA of gonococci is available also.

CHLAMYDIAL URETHRITIS CAN BE ASYMPTOMATIC

Chlamydia is the most frequently reported infectious disease in the United States. Also known as **chlamydial urethritis**, it is one of several diseases collectively known as **nongonococcal urethritis**, or **NGU**. Nongonococcal urethritis is a general term for a condition in which people without gonorrhea have a demonstrable infection of the urethra usually characterized by inflammation, and often accompanied by a discharge. Evidence is convincing that over 50 percent of cases of NGU are actually chlamydia (chlamydial urethritis). Another 25 percent of cases are believed to be ureaplasmal urethritis, and the remaining 25 percent are of unknown cause.

Chlamydia is a gonorrhea-like disease transmitted by sexual contact. The causative agent is *Chlamydia trachomatis*, a species of chlamydiae. *C. trachomatis* is an excep-

Keratitis:
a disease of the cornea of the eye.

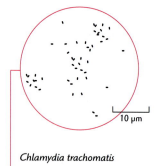

10 μm

Chlamydia trachomatis

tionally small organism, measuring about 0.35 μm in diameter. It is an obligate intracellular parasite; that is, it grows only in living tissue, such as fertilized chicken eggs and tissue cultures, and it has a complex reproductive cycle, as illustrated in FIG-URE 11.5. Humans appear to be the only host for the organism.

In 2001, over 780,000 new cases of chlamydia were reported to the CDC. This was the highest rate since reports began in the mid-1980s and the highest since mandatory

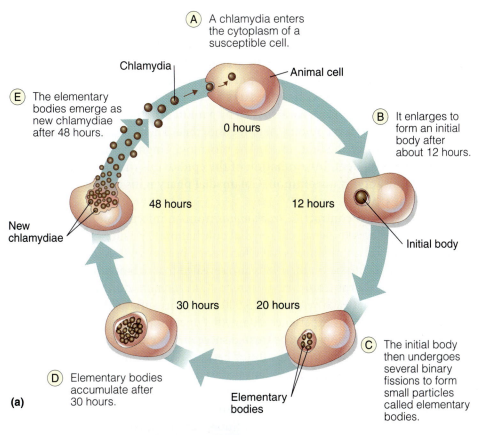

(A) A chlamydia enters the cytoplasm of a susceptible cell.

Chlamydia

Animal cell

(E) The elementary bodies emerge as new chlamydiae after 48 hours.

0 hours

(B) It enlarges to form an initial body after about 12 hours.

48 hours

12 hours

New chlamydiae

Initial body

30 hours

20 hours

(D) Elementary bodies accumulate after 30 hours.

Elementary bodies

(C) The initial body then undergoes several binary fissions to form small particles called elementary bodies.

(a)

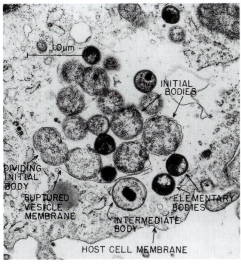

1.0 μm

INITIAL BODIES

DIVIDING INITIAL BODY

RUPTURED VESICLE MEMBRANE

ELEMENTARY BODIES

INTERMEDIATE BODY

HOST CELL MEMBRANE

(b)

FIGURE 11.5

The Chlamydiae

(a) The reproductive cycle of the chlamydiae. (b) A transmission electron micrograph of a microcolony of *Chlamydia* in the cytoplasm of a tissue cell after 48 hours of incubation. The various developmental forms are labeled. Note where the vesicular membrane has ruptured and chlamydiae are being released into the cytoplasm. (Bar = 1.0 μm.)

reporting in 1995. Seventy-five percent of these cases were in individuals under 25 years of age. Part of the reason for the increase could be due to the increased number of screening programs available and the development of better diagnostic tests.

The disease has an incubation period of about one to three weeks. However, about 75 percent of women and 50 percent of men have no symptoms and therefore do not seek health care. If symptoms do occur, they are remarkably similar to those of gonorrhea, although somewhat milder. Females often note a slight vaginal discharge, as well as inflammation of the cervix. Burning pain also is experienced on urination, reflecting disease in the urethra. In complicated cases, the disease may spread to the fallopian tubes, causing adhesions that block the passageways (**salpingitis**). Certain researchers report that **pelvic inflammatory disease (PID)** is a more likely consequence of chlamydia than of gonorrhea. (PID from gonorrhea and chlamydia is believed to affect about 50,000 women in the United States annually.) Often, however, there are few symptoms of disease before the salpingitis manifests itself, thus adding to the danger (**FIGURE 11.6**).

In males, chlamydia is characterized by painful urination and a discharge that is more watery and less copious than in gonorrhea. The discharge often is observed after urinating for the first time in the morning. Tingling sensations in the penis are generally evident. Inflammation of the epididymis may result in sterility, but this complication is uncommon. Chlamydial pharyngitis and proctitis are possible also.

Newborns may contract *C. trachomatis* from an infected mother and develop a disease of the eyes known as **chlamydial ophthalmia**. The silver nitrate used to prevent gonococcal ophthalmia is not effective as a preventative, so erythromycin therapy is required. Studies in the 1980s also revealed that **chlamydial pneumonia** may develop in newborns from an exposure to *C. trachomatis* during birth. Health officials estimate that each year, over 75,000 newborns suffer chlamydial ophthalmia and 30,000 newborns experience chlamydial pneumonia.

Chlamydial infections may be treated successfully with doxycycline. (If a woman is pregnant, erythromycin is substituted because doxycycline affects bone formation in newborns.) Since 1983, two relatively fast and simple laboratory tests have been available to detect *C. trachomatis*. In the first test, a physician takes a swab sample from the penis or the cervix (as in a Pap smear) and places the swab in a vial of fluid for transport to a local laboratory. A fluorescent antibody test using **monoclonal antibodies** (Chapter 20) is performed, and within 30 minutes the results are available. The second test is an immunoassay test, also performed with a swab sample. It is completed in the doctor's office and is similar to the test for gonorrhea. A test to detect the DNA of *C. trachomatis* is available as well. It uses a urine sample and is said to detect as few as five cells in a sample.

The big news of 1998 was the deciphering of the genome of *C. trachomatis*. (One researcher exclaimed, "It's like we were working in a room with the lights turned off, and now someone has turned on the lights.") The successful mapping of the 939 genes was achieved by Richard Stephens and his colleagues at the University of California at Berkeley. One interesting observation was a set of genes for synthesizing peptidoglycan. The organism was not thought to use this organic substance, but the presence of the genes indicates that it is employed somewhere in the cell, though no one is sure where. Also, the organism has genes for synthesizing ATP, allowing it to supplement the ATP obtained from its host for its energy needs. The organism has about 20 genes it apparently "borrowed" from a host cell sometime during its evolution millennia ago.

Monoclonal antibodies: antibodies experimentally produced against a single type of cell or substance.

FIGURE 11.6

A Case of Chlamydia

This case occurred in a 32-year-old professional woman. Tragic complications of the disease resulted because she and her doctor neglected to consider that she could be suffering from a sexually transmitted disease.

1. An educated, professional woman met a gentleman at a friend's house one evening. She was director of a New York law firm. The man was equally successful in his professional career.

2. The couple hit it off immediately. There were many evenings of quiet candlelit dinners, and soon, they became sexually intimate. Neither one used condoms or other means of protection.

3. Three days after having intercourse, the woman began experiencing fever, vomiting, and severe abdominal pains. She immediately made an appointment to see her doctor.

4. Assuming the illness was an intestinal upset, the physician prescribed appropriate medication. The woman was actually suffering from chlamydia, but the fact that she was sexually active did not come up during the examination.

5. Feeling better, the woman continued her normal routine. But six months later, with no apparent warning, she collapsed on a New York City sidewalk.

6. The woman was rushed to a local hospital, where doctors diagnosed a chlamydial infection of the peritoneum. They performed emergency surgery: her uterus and Fallopian tubes were badly scarred. She recovered, however, the scarring left her unable to have children.

UREAPLASMAL URETHRITIS PRODUCES MILD SYMPTOMS

Ureaplasmal urethritis is another type of NGU. It is caused by *Ureaplasma urealyticum*, a mycoplasma, so named because of its ability to digest urea in culture media. The organism often is referred to as a **T-mycoplasma** because "tiny" colonies of the organisms develop on laboratory media. At about 0.15 μm in size, *U. urealyticum* is one of the smallest known bacteria that cause human disease. Transmission is generally by sexual contact.

The symptoms of ureaplasmal urethritis are similar to those of gonorrhea and chlamydial urethritis. A distinction can be made between the diseases because in ureaplasmal urethritis, the discharge is variable in quantity, and the urethral pain is usually aggravated during urination. Symptoms are often very mild.

Penicillin cannot be used to treat ureaplasmal urethritis because *U. urealyticum* has no cell wall. Tetracycline is currently the drug of choice. Diagnosis often depends on eliminating gonorrhea or other types of NGU as possibilities, and for this reason, cases are not often recognized.

Infertility is one consequence of ureaplasmal urethritis because low sperm counts and poor movement of sperm cells have been observed in males. Salpingitis in females also has been described. Moreover, *Ureaplasma* is capable of colonizing the placenta during pregnancy, and reports have linked it to spontaneous abortions and premature births. As previously noted, 25 percent of NGU cases may be ureaplasmal urethritis.

CHANCROID CAUSES PAINFUL GENITAL ULCERS

Chancroid is a sexually transmitted disease believed to be more prevalent worldwide than gonorrhea or syphilis. The disease is endemic in many developing nations, and it is common in tropical climates and where public health standards are low; 38 cases were reported in the United States in 2001. However, it is difficult to culture and therefore the disease could be substantially underdiagnosed.

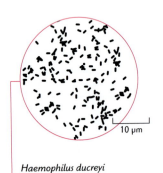

Haemophilus ducreyi

10 μm

The causative agent of chancroid is *Haemophilus ducreyi*, a small gram-negative rod named for Augusto Ducrey, who observed it in skin lesions in 1889. After a three- to five-day incubation period, a tender papule surrounded by a narrow zone of redness (erythema) forms at the entry site. The papule quickly becomes pus filled and then breaks down, leaving a shallow, saucer-shaped ulcer that bleeds easily and is painful. The ulcer has ragged edges and soft borders, a characteristic that distinguishes it from the primary lesion of syphilis. For this reason, the disease is often called **soft chancre**.

The lesions in chancroid most often occur on the penis in males and the labia or clitoris in females. Substantial swelling in the lymph nodes of the groin may be observed. However, the disease generally goes no further. The clinical picture in chancroid makes the disease recognizable, but definitive diagnosis depends on isolating *H. ducreyi* from the lesions.

The transmission of chancroid depends on contact with the lesion. Azithromycin, erythromycin, or ceftriazone drugs are useful for therapy, but the disease often disappears without treatment. However, treatment is essential because the open lesion increases the risk of HIV transmission.

OTHER SEXUALLY TRANSMITTED DISEASES ALSO EXIST

A number of other sexually transmitted diseases merit brief attention in this chapter. None of these diseases normally presents a life-threatening situation to humans.

Lymphogranuloma venereum (LGV) is caused by a serotype of *Chlamydia trachomatis* that is slightly different from those that cause chlamydial urethritis. LGV is more common in males than females, and is accompanied by fever, malaise, and swelling and tenderness in the lymph nodes of the groin. Females may experience infection of the rectum (proctitis), if the chlamydiae pass from the genital opening to the nearby intestinal opening. Approximately 500 cases are detected annually in the United States. Lymphogranuloma venereum is prevalent in Southeast Asia and Central and South America. Sexually active individuals returning from these areas may show symptoms of the disease, but treatment with doxycycline leads to rapid resolution.

Granuloma inguinale is a rare sexually transmitted disease in Europe and North America, but it remains an endemic problem in tropical and subtropical areas of the world, such as Caribbean countries and Africa. It is caused by *Calymmatobacterium granulomatis*, a small, gram-negative, encapsulated bacillus. The disease begins with a primary lesion starting as a nodule and progressing to a granular ulcer that bleeds easily. In most cases, this ulcer forms in the external genital organs but it may spread to other regions by contaminated fingers. The lymph nodes in the groin may swell, but fever and other body symptoms are usually absent, a factor that distinguishes the disease from LGV. Tissue samples reveal masses of bacteria called **Donovan bodies** within white blood cells in the lesion. The disease responds to various antibiotics, especially doxycycline.

Vaginitis is a general term for various mild infections of the vagina and sometimes the vulva. Up to 16 percent of pregnant women are infected. One cause of vaginitis is *Gardnerella vaginalis*, formerly called *Haemophilus vaginalis*. This small gram-negative rod usually lives uneventfully in the vagina, but flare-ups of disease may take place and transmission by sexual contact may occur during these times. A foul-smelling discharge is the most prominent symptom, and clindamycin or metronidazole therapy generally provides relief.

The final organism that we shall consider is *Mycoplasma hominis* (**FIGURE 11.7**). This mycoplasma causes **mycoplasmal urethritis**, a disease similar to ureaplasmal urethritis. In addition, the organism can colonize the placenta and cause spontaneous abortion or premature birth. *M. hominis* is distinguished from *Ureaplasma* by its inability to digest urea, its larger colony growth on agar, and its preference for anaerobic conditions during laboratory cultivation. Tetracycline is prescribed for active cases of mycoplasmal urethritis. The disease can also be caused by *Mycoplasma genitalium*, as reported first in 1993 by a joint American-British

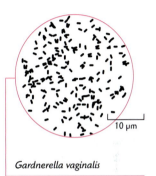

Gardnerella vaginalis

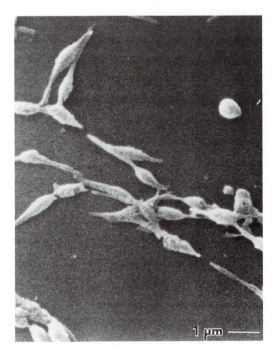

FIGURE 11.7

A Species of *Mycoplasma*

A species of *Mycoplasma* seen with the scanning electron microscope. A closely related species called *M. hominis* causes mycoplasmal urethritis, an STD that is similar to chlamydia. The pleomorphic (multishape) appearance of the organism is due to the absence of a cell wall.

research team. *M. genitalium* has another distinction in microbiology—it was the second organism of any kind (the first was *Haemophilus influenzae*) to have its entire genome deciphered, an achievement completed in 1995.

TABLE 11.1 summarizes the sexually transmitted bacterial diseases.

To this point . . .

We have studied a series of sexually transmitted diseases and have noted their prevalence in modern society. Syphilis is the only disease in the group that regularly spreads to other organs of the body. The remaining diseases are largely restricted to the urogenital organs. In addition, syphilis is the only STD that is life threatening. The other diseases generally are mild and may resolve even without treatment. However, with gonorrhea and chlamydia, there may be long-range complications, such as sterility or spontaneous abortion.

TABLE 11.1

A Summary of Sexually Transmitted Bacterial Diseases

DISEASE	CAUSATIVE AGENT	DESCRIPTION OF AGENT	ORGANS AFFECTED	CHARACTERISTIC SIGNS
Syphilis	*Treponema pallidum*	Spirochete	Skin Cardiovascular organs Nervous system	Chancre Skin lesions Gumma
Gonorrhea	*Neisseria gonorrhoeae*	Gram-negative diplococcus	Urethra, cervix Fallopian tubes Epididymis Eyes, pharynx	Pain on urination Discharge Salpingitis
Chlamydia (Chlamydial urethritis)	*Chlamydia trachomatis*	Chlamydia	Urethra, cervix Fallopian tubes Epididymis Eyes, pharynx	Pain on urination Watery discharge Salpingitis
Ureaplasma urethritis	*Ureaplasma urealyticum*	Mycoplasma	Urethra Fallopian tubes Epididymis	Pain on urination Variable discharge Salpingitis
Chancroid (Soft chancre)	*Haemophilus ducreyi*	Gram-negative rod	External genital organs Inguinal lymph nodes	Soft chancre Erythema Swollen inguinal lymph nodes
Lymphogranuloma venereum	*Chlamydia trachomatis*	Chlamydia	Lymph nodes of groin Rectum	Swollen lymph nodes Proctitis
Granuloma inguinale	*Calymmatobacterium granulomatis*	Gram-negative rod	External genital organs	Bleeding ulcer Swollen inguinal lymph nodes
Vaginitis	*Gardnerella vaginalis*	Gram-negative rod	Vagina	Foul-smelling discharge
Mycoplasmal urethritis	*Mycoplasma hominis* *M. genitalium*	Mycoplasma	Urethra Fallopian tubes Epididymis	Pain on urination Variable discharge Salpingitis

We also pointed out some sexually transmitted diseases that are less publicized, but that may occur widely in the United States. Chancroid and ureaplasmal urethritis are examples. It is conceivable that the incidence of lymphogranuloma venereum, granuloma inguinale, and Mycoplasma and Gardnerella infections may be equally high, but detection methods are currently lacking. The true extent of STDs may never be known.

We now shall turn our attention to a group of bacterial diseases usually acquired by contact, though generally not sexual contact. Some of these diseases have been long known and, left untreated, they may persist for years. Others were unknown a generation ago and are of short duration. Few are fatal, but all have contributed to our knowledge of the broad spectrum of bacterial diseases.

TOXIN INVOLVED	TREATMENT ADMINISTERED	IMMUNIZATION AVAILABLE	COMMENT
Not established	Penicillin	None	Primary, secondary, and tertiary stages The "Great Imitator" Congenital transmission
Not established	Tetracycline Ceftriaxone	None	One of the most reported US microbial diseases Complicated by pelvic inflammatory disease Eye infection in newborns
Not established	Doxycycline Erythromycin	None	Leads to infertility Antibody test for diagnosis
Not established	Tetracycline	None	Leads to infertility Linked to spontaneous abortion
Not established	Azithromycin Erythromycin Ceftriazone	None	Few complications Many immigrant cases
Not established	Doxycycline	None	Prevalent in Central and South America
Not established	Tetracycline	None	Donovan bodies seen in lesions
Not established	Tetracycline	None	Organism commonly found in the vagina
Not established	Tetracycline	None	Similar to ureaplasmal urethritis

Contact Bacterial Diseases

Numerous bacterial diseases are transmitted by contact other than sexual contact. Usually, some form of skin contact is required, as these diseases will illustrate.

LEPROSY (HANSEN'S DISEASE) IS A CHRONIC, SYSTEMIC INFECTION

For many centuries, **leprosy** was considered a curse of the damned. It did not kill, but neither did it seem to end. Instead, it lingered for years, causing the tissues to degenerate and deforming the body. In biblical times, the afflicted were required to call out "Unclean! Unclean!" and usually they were ostracized from the community. Among the more heroic stories of medicine is the work of Father Damien de Veuster, the Belgian priest who in 1870 established a hospital for leprosy patients on the island of Molokai in Hawaii. An equally heroic story was written more recently (MicroFocus 11.2).

The agent of leprosy is *Mycobacterium leprae*, an acid-fast rod related to the tubercle bacillus. This organism was first observed in 1874 by the Norwegian physician Gerhard Hansen. It is referred to as Hansen's bacillus, and leprosy is commonly

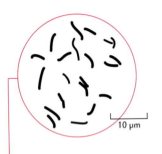

Mycobacterium leprae

10 μm

MicroFocus 11.2

THE "STAR" OF CARVILLE

On December 1, 1894, seven leprosy patients arrived at an old plantation on a crook in the Mississippi River. Soon thereafter, four nuns of the Order of St. Vincent de Paul joined them. Together this small band formed the nucleus of what was to become the National Hansen's Disease Center at Carville, Louisiana.

Change came slowly. In 1921, the United States Public Health Service acquired the institution, but it remained essentially a prison, patrolled by guards and surrounded by a cyclone fence with barbed wire. Then, in 1931, a leprosy patient named Stanley Stein arrived. Stanley Stein was not his real name—he had forsaken that for fear of bringing shame to his family. Soon, Stein instituted a weekly paper to bring a sense of community to the patients. Originally named *The Sixty-Six Star* (Carville was United States Marine Hospital Number

66), the name was eventually shortened to *The Star*.

As the circulation of *The Star* increased, Stein and others launched a campaign for change. In 1936, the patients acquired a telephone so they could hear the voices of their families. Three years later, the swamps were drained to reduce the incidence of malaria. Soon there came a better infirmary, a new recreation hall, and removal of the barbed wire. In 1946, the State of Louisiana allowed the patients to vote in local and national elections.

Through all these years, Stein's leprosy worsened. Originally he had tuberculoid leprosy, the form in which the nerves are damaged. Afterward, however, he developed lepromatous leprosy, which causes lesions to form on the face, ears, and eyes. Soon he was totally blind. Without feeling in his fingers, he could not even learn Braille.

But Stein was not finished. He and his newspaper tirelessly fought for a new post office and weekend passes for patients. In 1961, President Kennedy paid tribute to *The Star* on its thirtieth anniversary and singled out its indomitable editor for praise. Stanley Stein died in 1968. By that time, *The Star* had a circulation of 80,500 in all 50 states and 118 foreign countries.

■ *Stanley Stein stands next to the printing press as copies of* The Star *are printed.*

called **Hansen's disease**. Ironically, *M. leprae* has not yet been cultivated in artificial laboratory medium, and thus, Koch's postulates (Chapter 1) have not been fulfilled. In 1960, researchers at the CDC succeeded in cultivating the bacillus in the footpads of mice, and in 1969, scientists found that it would grow in the tissues of armadillos. Growth in apes was reported in 1982.

Leprosy is spread by multiple skin contacts, as well as by droplets from the upper respiratory tract. The disease has an unusually long incubation period of three to six years, a factor that makes diagnosis very difficult. Because the organisms are heat sensitive, the symptoms occur in the skin and peripheral nerves in the cooler parts of the body, such as the hands, feet, face, and earlobes. Severe cases also involve the eyes and the respiratory tract. Susceptibility is highest in childhood and decreases with age.

Patients with leprosy experience disfiguring of the skin and bones, twisting of the limbs, and curling of the fingers to form the characteristic **claw hand**. Loss of facial features accompanies thickening of the outer ear and collapse of the nose. Many tumor-like growths called **lepromas** form on the skin and in the respiratory tract (**lepromatous leprosy**). This is the most serious form of the disease because the immune system fails to react. However, the largest number of deformities develop from the loss of pain sensation due to nerve damage (**tuberculoid leprosy**). Inattentive patients can pick up a pot of boiling water without flinching, and they accidentally let cigarettes burn down and sear their fingers.

For many years, the principal drug for the treatment of leprosy was a sulfur compound known commercially as **dapsone** ("Zap it with dap"). In many cases, such as the one shown in FIGURE 11.8, the results were dramatic. However, studies indicate that *M. leprae* is becoming increasingly resistant to this drug, and alternative antibiotics, such as rifampin and clofazimine, are being used. In 1992, the WHO began a campaign to treat the disease with oflaxacin, and in 1998, thalidomide was approved as a therapy (MicroFocus 11.3). Some attempts have been made to immunize populations with BCG, the vaccine used for tuberculosis. Early diagnosis relies on a procedure called the **lepromin skin test**, performed in basically the same way as the tuberculin test (Chapter 8).

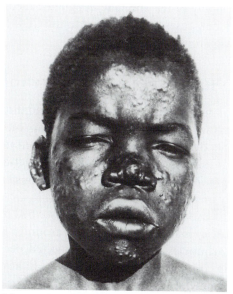

(a)

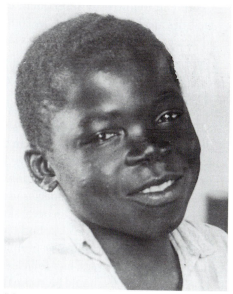

(b)

FIGURE 11.8

Treating Leprosy

The young boy with leprosy is pictured (a) before treatment with dapsone and (b) some months later, after treatment. Note that the lesions of the ear and face and the swellings of the lips and nose have largely disappeared.

MicroFocus 11.3

THE ROLLERCOASTER RIDE OF THALIDOMIDE

In 1998, the infamous drug thalidomide continued its long climb back to respectability, when the Food and Drug Administration (FDA) approved its use for treating a form of leprosy.

In the 1950s, thalidomide was administered to pregnant women as a treatment for morning sickness. By 1961, investigators realized—to their horror—that the drug was causing facial disfiguring in newborns and crippling birth defects. It was taken off the market. Over 12,000 children around the world were born with limbs that were either missing or malformed. Forty percent of those victims died within the first year of life although some 5,000 survive today.

Things began to change in 1996, when Celgene, a biotechnology company, filed

an application with the FDA to use thalidomide to treat a form of leprosy called erythema nodosum leprosum (ENL). This skin condition is accompanied by inflamed skin nodules. Extensive tests suggested it was effective in minimizing fever but had limited abilities to control nerve inflammation (neuritis). Still, the approval process was completed in 1998 to market thalidomide for moderate to severe ENL.

Approval was not rendered without restrictions. At the outset, the American Academy of Pediatrics lobbied strongly against the drug. Celgene was encouraged to provide a multistep education campaign and a program for safety. For example, doctors wishing to use the drug had to register with Celgene; a doctor had to confirm that a woman was

not pregnant before a prescription could be filled; women using the drug had to be tested for pregnancy weekly; and male partners had to be educated about contraceptive measures, including condom use.

Today, other drugs appear to be better at controlling the symptoms of ENL. Prednisolone is more effective at controlling ENL neuritis, and clofazimine prevents recurrent ENL reactions even when the patient is taken off the drug. Thalidomide patients experience recurrent reactions when not medicated.

Considering the health risks and a surveillance system that is not fool proof versus the benefits when other drugs are superior, it appears that thalidomide has had its ups and downs as a treatment medication.

The World Health Organization estimates there were 755,000 new cases in 2001, and there are over 10 million people with leprosy in the world today. Cases in the United States have risen during the last generation, in large measure because of infected immigrants. Approximately 7,000 patients currently are being treated in American hospitals. About 300 come each year to the National Hansen's Disease Programs in Baton Rouge, Louisiana, for two to six weeks of initial diagnosis and treatment, followed by subsequent care at any of ten specially designated medical facilities around the country. In 2001, 81 new cases were reported to the CDC and 110 cases to the National Hansen's Disease Programs.

STAPHYLOCOCCAL SKIN DISEASES HAVE SEVERAL MANIFESTATIONS

Staphylococci are normal inhabitants of the human skin, mouth, nose, and throat. Although they generally live in these areas without causing harm, they can initiate disease when they penetrate the skin barrier or the mucous membranes. Penetration is assisted by open wounds, damaged hair follicles, ear piercing, dental extractions, or irritation of the skin by scratching. *Staphylococcus aureus*, the grape-like cluster of gram-positive cocci, is the species usually involved in disease.

The hallmark of staphylococcal skin disease is the **abscess**, a circumscribed pus-filled lesion (**FIGURE 11.9**). A **boil** (**furuncle**) is a deeper skin abscess around a hair follicle, which often begins as a pimple. **Carbuncles** are a group of connected, deeper furuncles around several hair follicles. Skin contact with other people spreads the disease. Food handlers should be aware that staphylococci from boils and carbuncles can be transmitted to food, where they can cause food poisoning (Chapter 9).

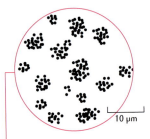

10 μm

Staphylococcus aureus

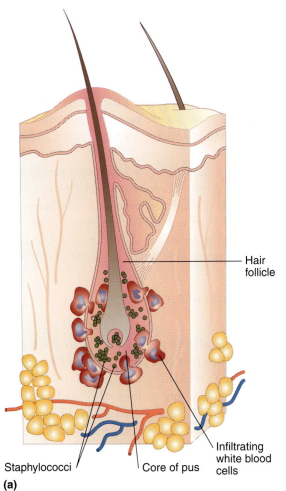

(a)

Staphylococci

Core of pus

Hair follicle

Infiltrating white blood cells

FIGURE 11.9

Staphylococci and Skin Abscesses

(a) Staphylococci at the base of a hair follicle during the development of a skin lesion. White blood cells that engulf bacteria have begun to collect at the site as pus accumulates, and the skin has started to swell. (b) A severe carbuncle on the head of a young boy. Abscesses often begin as trivial skin pimples and boils, but they can become serious as staphylococci penetrate to the deeper tissues.

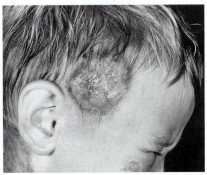

(b)

Another skin disease caused by another strain of *S. aureus* is the **scalded skin syndrome**, or **Ritter's disease**, which is occasionally seen in infants. The skin becomes red, wrinkled, and tender to the touch, with a sandpaper appearance. It may then peel off. Exotoxins produced by the staphylococci living at a point distant from the skin appear to be responsible for this condition. Mortality rates may be high in untreated cases due to bacterial invasion of the lungs or blood.

A more widespread and highly contagious staphylococcal skin disease is **impetigo contagiosum.** Here the infection is more superficial and involves patches of epidermis just below the outer skin layer. Impetigo first appears as thin-walled blisters that ooze a yellowish fluid and form yellowish-brown crusts. Usually the blisters occur on the exposed parts of the body, but they also may occur around the nose and upper lip after a child has had a cold with a runny nose, since the constant irritation provides a mechanism for penetration by the staphylococci. *Streptococcus pyogenes* (Chapter 8) also may cause impetigo.

Staphylococcal skin diseases and bloodborne infections are commonly treated with penicillin, but resistant strains of *S. aureus* are well known, and physicians may need to test a series of alternatives before an effective antibiotic is located. This problem was brought to public awareness during the 1990s, when reports of antibiotic resistance in staphylococci became widespread. Resistances to numerous antibiotics were reported, and **multidrug-resistant *Staphylococcus aureus* (MRSA)** soon appeared in many hospitals. It has been estimated that up to 100,000 people are hospitalized each year with MRSA infections. The last-resort antibiotic against the bacteria was vancomycin, a very expensive and somewhat toxic drug. However, by 1997, **vancomycin-resistant *S. aureus* (VRSA)** was detected in clinical settings in Japan. In 2002, the first two cases of VRSA were identified in patients in the United States. New drugs and new treatment approaches are being made available to the medical community to help stem the tide of MRSA and VRSA. Chapter 24 discusses the issue of antibiotic resistance in more detail.

A staphylococcal skin disease can be treated also by vigorously scrubbing the lesions with **Betadine** applied with 2-by-2-inch gauze pads under the direction of a physician. The disease should be treated with caution because staphylococci commonly invade the blood and penetrate to other organs. For example, staphylococcal blood poisoning (septicemia) may develop, as well as staphylococcal pneumonia, endocarditis, meningitis, or nephritis. A trivial skin boil is often the source.

TOXIC SHOCK SYNDROME IS A DISEASE CAUSED BY EXOTOXIN RELEASE

In 1978, James Todd of Children's Hospital in Denver, Colorado, coined the name **toxic shock syndrome (TSS)** for a blood disorder characterized by sudden fever and circulatory collapse. The name remained in relative obscurity until the fall of 1980, when a major outbreak occurred in menstruating women who used a particular brand of highly absorbent tampons. News of TSS dominated the media for about six months and led to a recall of the tampons. With that episode, toxic shock syndrome assumed a position of significance in modern medicine. Incidence rates are shown in **FIGURE 11.10**.

Toxic shock syndrome is caused by a toxin-producing strain of *S. aureus*. The earliest symptoms of disease include a rapidly rising fever, accompanied by vomiting and watery diarrhea. Patients then experience a sore throat, severe muscle aches, and a sunburn-like rash with peeling of the skin, especially on the palms of the hands and

FIGURE 11.10

Reported US Cases of Toxic Shock Syndrome, 1979 to 2001

The total number of TSS cases per year. The disease reached a peak in the fall of 1980; then, the incidence rates dropped with the recall of the highly absorbent tampons. Nonmenstrual episodes currently account for a significant number of cases. The photograph shows the peeling skin often associated with TSS.

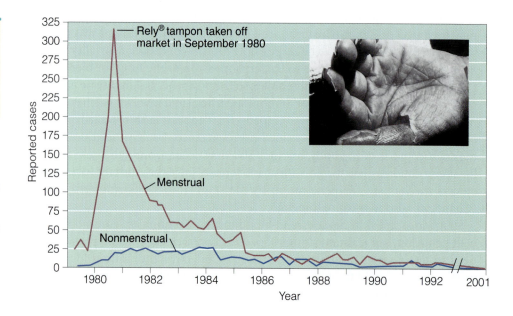

soles of the feet. A sudden drop in blood pressure also occurs, possibly leading to shock and heart failure. Antibiotics may be used to control the growth of bacteria, but measures such as blood transfusions must be taken to control the shock.

Although the staphylococci involved in TSS exist in various places in the body, the ones inhabiting the vagina have received the most attention. During the 1980 outbreak, scientists speculated that lacerations or abrasions of the tissue by tampon inserters gave the staphylococci access to the tissues. Others suggested that staphylococci grow in the warm, stagnant fluid during the long period that the tampon is in place. It appears certain that multiple factors play a role in TSS, because males, prepubertal girls, and postmenopausal women also have been stricken. Indeed, in 1994 one woman contracted TSS from a contaminated needle (MicroFocus 11.4). Over 100 cases of TSS are reported to the CDC annually.

TRACHOMA CAN LEAD TO BLINDNESS

Trachoma is a disease of the eyes. It occurs in hot, dry regions of the world and it is prevalent in Mediterranean countries, parts of Africa and Asia, and in the southwestern United States in Native American populations. Hundreds of millions throughout the world are believed to be afflicted by it (FIGURE 11.11).

The cause of trachoma is a different serotype of *Chlamydia trachomatis*, from those responsible for the chlamydia urethritis and lymphogranuloma venereum. Fingers, towels, optical instruments, and face-to-face contact are possible modes of transmission. The chlamydiae multiply in the **conjunctiva**, the thin membrane that covers the cornea and forms the inner eyelid. A series of tiny, pale **nodules** forms on this membrane, giving it a rough appearance. (The word *trachoma* is derived from the Greek *trachi-*, meaning "rough.") In serious cases, the upper eyelid turns in, causing abrasion of the cornea by the eyelashes. Blindness develops over 10 to 15 years from corneal abrasions and lesions.

MicroFocus 11.4

"IT SEEMED LIKE A GOOD IDEA!"

The idea of a tattoo seemed okay. All her friends had them, and a tattoo would add a sense of uniqueness to her personality. After all, she was already 22. It took some pushing from her friends, but she finally made it into the tattoo parlor that day in Fort Worth, Texas.

Two weeks later the pains started—first in her stomach, then all over. Her fever was high, and now a rash was breaking out; it looked like her skin was burned and was peeling away. There was one visit to the doctor, then immediately to the emergency room of the local hospital. The gynecologist guessed

it was an inflammation of the pelvic organs (pelvic inflammatory disease, they called it), so he gave her an antibiotic and sent her home.

But it got worse—the fever, the rash, the peeling, the pains. Back she went to the emergency room. This time they would admit her to the hospital, give her intravenous blood transfusions and antibiotics, keep her for 11 days, and discover a severe blood infection due to *Staphylococcus aureus*. And there was an unusual diagnosis: toxic shock syndrome. Don't women get that from tampons? Most do, she was told, but a few get it from staph entering a skin wound—a

wound that can be made by a contaminated tattooing needle.

Trachoma endemic countries

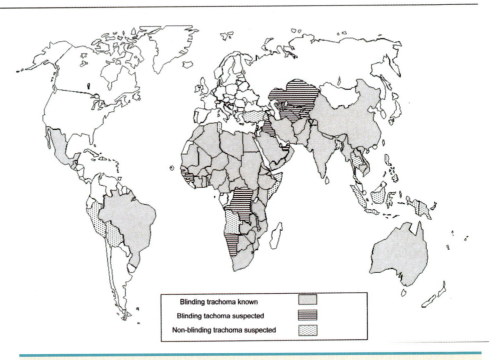

Blinding trachoma known

Blinding tachoma suspected

Non-blinding trachoma suspected

FIGURE 11.11

Worldwide Trachoma

Blinding or suspected blinding trachoma affects individuals worldwide, but especially in Africa, Southeast Asia, Mexico, and parts of South America.

Tetracycline and erythromycin help reduce the symptoms of trachoma, but in many patients, the relief is only temporary because chlamydiae reinfect the tissues. The World Health Organization maintains squads of trachoma nurses who travel from village to village, identifying trachoma cases in their early stages and treating them before they become serious. Trachoma is the world's leading cause of preventable blindness.

BACTERIAL CONJUNCTIVITIS (PINKEYE) IS VERY COMMON

Several microorganisms cause conjunctivitis, among them the bacterium *Haemophilus aegyptius*, the Koch-Weeks bacillus. This organism also is known in the literature as *Haemophilus influenzae* biotype III because of its close relationship to *H. influenzae*. The organism is a small gram-negative rod that grows in chocolate agar, a rich medium that contains disrupted red blood cells (*Haemophilus* means "blood-loving").

Conjunctivitis is a disease of the conjunctiva. When infected, the membrane becomes inflamed, a factor that imparts a brilliant pink color to the white of the eye (hence the name **pinkeye**). A copious discharge runs down the cheek in the waking hours and crusts the eyelids shut during sleep. The eyes are swollen and itch intensely, and vision in bright light is impaired (photophobia).

Conjunctivitis may be transmitted in a number of ways, including face-to-face contact and via airborne droplets. Contaminated optometric instruments, micro-

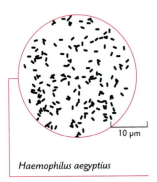

10 μm

Haemophilus aegyptius

scopes, and towels also may transmit the bacilli. The disease normally runs its course in about two weeks, and therapy usually is not required. Neomycin may be administered to hasten recovery. Conjunctivitis is extremely contagious, especially where people congregate.

In recent years, a variant of *H. aegyptius* also has been isolated from the blood of patients suffering from a disease called **Brazilian purpuric fever**. This life-threatening disease is accompanied by nausea, vomiting, fever, and hemorrhagic skin lesions. Many patients display conjunctivitis before the onset of more serious symptoms.

YAWS IS SPREAD PERSON TO PERSON

Yaws commonly occurs in tropical countries of Africa, South America, and Southeast Asia. It is caused by *Treponema pertenue*, a spirochete identical in appearance and similar in chemistry to the syphilis spirochete. Yaws is usually acquired by skin contact. A red, raised lesion called a **mother yaw** develops at the site of entry. Blood associated with the lesion gives it the appearance of a raspberry, and the disease is sometimes called **frambesia**, from *framboise*, the French word for "raspberry."

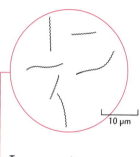

Treponema pertenue

In time the lesion disappears, but months later the patient develops numerous other yaws. Left untreated, these also disappear, only to reappear as soft granular lesions. The similarities have led investigators to postulate that syphilis may have originated as yaws, or vice versa. **Bejel** and **pinta** are two other diseases very similar to yaws.

TABLE 11.2 provides a summary of this and other contact bacterial diseases.

To this point . . .

We have discussed several bacterial diseases spread by contact by focusing on six examples. In leprosy, multiple contacts may be necessary for the transmission of the bacteria. Staphylococcal skin diseases also are spread by contact, although many episodes of infection begin by simply irritating the skin and allowing staphylococci to penetrate. Tissue irritation may be the source of the staphylococci involved in toxic shock syndrome also. Indeed, researchers who studied the tampon problem of 1980 concluded that the bacteria were already present in the body. Contact spreads the staphylococci to others.

In trachoma, conjunctivitis, and yaws, we discussed three other diseases spread by contact. Trachoma and yaws are prevalent in warm countries where less clothing is worn than in temperate climates. This yields a greater area of skin surface for contact with an infected individual. Bacterial conjunctivitis spreads where people congregate. You might note that the symptoms of the six diseases studied are largely restricted to the skin, where penetration has occurred. Except for the staphylococcal diseases, deeper organs seldom become involved, and life-threatening situations are uncommon. Also, we see the broad range of bacterial organisms that can cause disease including two rods, a staphylococcus, a chlamydia, and a spirochete.

In the final section of this chapter, we shall consider a number of miscellaneous bacterial diseases grouped into discrete categories. Certain of these diseases (the endogenous diseases) are caused by bacteria already in the body; some are fungal-like; other diseases (the animal bite diseases) are associated with bite wounds; still other diseases (oral diseases) occur in the mouth. A final group (nosocomial diseases) are contracted during a stay in the hospital. We also shall make brief mention of urinary tract and burn infections as we complete our survey.

TABLE 11.2

A Summary of Contact Bacterial Diseases

DISEASE	CAUSATIVE AGENT	DESCRIPTION OF AGENT	ORGANS AFFECTED	CHARACTERISTIC SIGNS
Leprosy (Hansen's disease)	*Mycobacterium leprae*	Acid-fast rod	Skin, bones Peripheral nerves	Tumor-like growths Skin disfigurement "Claw hand"
Staphylococcal skin diseases	*Staphylococcus aureus*	Gram-positive cluster of cocci	Skin	Abscess, boil Scalded-skin syndrome Impetigo contagiosum
Toxic shock syndrome	*Staphylococcus aureus*	Gram-positive cluster of cocci	Blood	Fever Watery diarrhea Sore throat Sunburn-like rash
Trachoma	*Chlamydia trachomatis*	Chlamydia	Eyes	Nodules on conjunctiva Scarring in eye
Bacterial conjunctivitis	*Haemophilus aegyptius*	Gram-negative rod	Eyes	Pinkeye Photophobia
Yaws	*Treponema pertenue*	Spirochete	Skin	Red, raised lesion

11.3

Miscellaneous Bacterial Diseases

The miscellaneous bacterial diseases include a diverse mix of diseases related to animal bites, poor oral hygiene, and hospital stays. Also in this group are diseases related to organisms already in the body, the so-called endogenous diseases. We shall discuss them along with fungal-like diseases first.

THERE ARE SEVERAL ENDOGENOUS BACTERIAL DISEASES

Endogenous bacterial diseases are caused by organisms that normally inhabit the body. Natural host resistance generally prevents proliferation of the causative organisms, but when the resistance is suppressed, disease may follow.

Species of *Bacteroides* are gram-negative anaerobic rods that inhabit the large intestine and the feces of most individuals (as **FIGURE 11.12** displays). These organisms enter the bloodstream when a person sustains an intestinal injury, and cause blood clots that clog the vessels, resulting in oxygen depletion and possible gangrene in the tissues. *Bacteroides fragilis* is the most common pathogenic species of the group, even though it lacks an outer cell membrane (no endotoxin production).

One of the side effects of excessive antibiotic use is the elimination of many species of intestinal bacteria that normally keep other species in check. Under the circumstances, an endogenous disease may develop from infection by *Clostridium difficile*, a gram-positive anaerobic rod. As other organisms disappear, the clostridia multiply and produce a series of exotoxins that induce a condition called

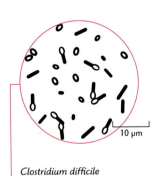

Clostridium difficile

10 µm

TOXIN INVOLVED	TREATMENT ADMINISTERED	IMMUNIZATION AVAILABLE	COMMENT
None	Dapsone Rifampin Clofazimine	BCG	Lepromin skin test for diagnosis Long incubation period
Probable	Penicillin Vancomycin	None	Antibiotic resistance in staphylococci Food handlers involved
Yes	Pencillin Blood transfusions	None	Occurs in all groups, especially menstruating women Over 100 cases annually
Not established	Tetracycline Erythromycin	None	Leading cause of preventable blindness
Not established	Neomycin	None	Extremely contagious Copious discharge
Not established	Penicillin	None	Found in tropical countries Related to bejel and pinta

pseudomembranous colitis. Yellowish-green membranous lesions cover the intestinal lining, and patients experience diarrhea with watery stools. Infants appear to be particularly susceptible to this condition, especially if a normal population of bacteria has not yet been established. Research in the 1980s also linked *C. difficile* and its toxins to sudden infant death syndrome (SIDS).

TWO FUNGAL-LIKE BACTERIAL DISEASES EXIST

An example of an endogenous disease that is fungal-like in growth is **actinomycosis.** This disease is caused by *Actinomyces israelii*, named for James A. Israel, who described the bacillus in 1878. It is a gram-positive, anaerobic, fungus-like rod often found in the gastrointestinal and respiratory tracts. Infection follows trauma to body tissues. For example, during a dental extraction, *A. israelii* can enter the gum tissues, multiply, and grow toward the facial surface, causing a red swelling that is lumpy and hard as wood. The condition, known as **lumpy jaw,** may develop into a skin problem with draining sinuses. Another form of actinomycosis involves **draining sinuses** of the chest wall, while a third form is characterized by **abdominal sinuses,** often as a complication of ulcers. Individuals who are HIV-positive are at increased risk.

Nocardiosis, another example of a disease that grows much like a fungus, is due to an acid-fast, filamentous rod called *Nocardia asteroides*. When inhaled, the disease strikes the

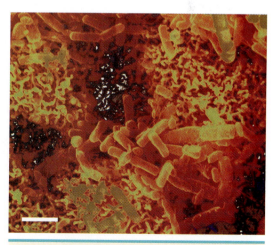

FIGURE 11.12

Endogenous Microorganisms

A scanning electron micrograph of the lining of the human large intestine. Endogenous disease may develop when bacteria like these invade the tissue during periods of suppressed host resistance. (Bar = 2 μm.)

lungs, where multiple abscesses form. Fever, coughing, and bloody sputum develop, and the symptoms may be mistaken for tuberculosis. Reports of death from nocardiosis have been linked to AIDS (Chapter 14). *N. asteroides* also may cause abscesses and swelling of the foot, a condition called **Madura foot**. This condition follows penetration from the soil via a wound.

ANIMAL BITE DISEASES OCCASIONALLY OCCUR

Public health officials estimate that each year in the United States, about 3.5 million people are bitten by animals. Most of these wounds heal without complications, but in certain cases, bacterial disease may develop.

An important cause of bite infections is *Pasteurella multocida*, a gram-negative rod. This organism is a common inhabitant of the pharynx of cats and dogs, where it causes a local disease called **pasteurellosis**. The bacterium causes most wound infections resulting from a dog or cat bite. In humans, the symptoms of pasteurellosis develop rapidly, with local redness, warmth, swelling, and tenderness at the site of the bite wound. Abscesses frequently form, especially if the wound has been sutured. Some patients may experience arthritis. The disease responds slowly to antibiotic therapy.

Although cats transmit few diseases to humans, a notable problem is **cat-scratch disease** (also known as **cat-scratch fever**). The disease affects an estimated 20,000 Americans each year, primarily children, and it is transmitted by a scratch, bite, or lick from a cat (or, in some cases, a dog). Symptoms include a papular or pustular lesion at the site of entry, followed by headache, malaise, and low-grade fever. Swollen lymph glands, generally on the side of the body near the bite, accompany the disease, as shown in **FIGURE 11.13a**. In rare cases, the brain and central nervous system may be involved. Most episodes of disease end after several days or weeks, and antibiotics such as rifampin hasten recovery.

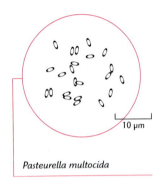

10 μm

Pasteurella multocida

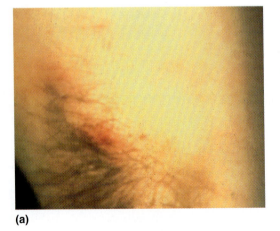

(a)

(b)

FIGURE 11.13

Cat-Scratch Disease

(a) A patient displaying in the armpit the considerable lymph node swelling that accompanies cat-scratch disease. (b) A photomicrograph of *Bartonella henselae*, considered by many researchers to be the causative agent of cat-scratch disease. (Bar = 10 μm.)

The causative agent of cat-scratch disease has not yet been isolated with certainty, but a leading candidate is *Bartonella henselae*, a rickettsia. Antibodies against this organism are usually present in people with cat-scratch disease, and the rickettsia is usually isolated from cats whose scratch has led to disease. *B. henselae* also appears to cause **bacillary angiomatosis**, a disease often found in individuals with lowered immune system function. The disease is characterized by tumors whose cells organize themselves to form blood vessels. Occurring in the body's cardiovascular system, bacillary angiomatosis resembles Kaposi's sarcoma (Chapter 14) and may be a late manifestation of cat-scratch disease. Another possible cause of cat-scratch disease is *Afipia felis*, a gram-negative rod.

Two different species of bacteria can cause **rat-bite fever**. One is *Streptobacillus moniliformis*, a gram-negative rod that occurs in long chains. It is found in the pharynx of rats and other rodents. Patients experience a lesion at the site of the bite, then a typical triad of fever, arthritis-like pain in the large joints, and skin rash. The second organism is *Spirillum minor*, a rigid spiral bacterium with polar flagella. A lesion occurs at the wound site, and a maculopapular rash spreads out from this point. In Japan and other parts of Asia, *Spirillum*-related rat-bite fever is known as **sodoku**. Antibiotic therapy is generally recommended for both forms of rat-bite fever.

TABLE 11.3 summarizes the endogenous, fungal-like, and animal bite bacterial diseases.

Streptobacillus moniliformis and Spirillum minor

ORAL DISEASES ARISE FROM BACTERIA NORMALLY FOUND IN THE MOUTH

The oral cavity is a type of ecosystem, with complex interrelationships among the members of the resident population of microorganisms and the oral environment. The cavity has various ecological niches, each with a different physical property and nutrient supply dictating the number and type of microorganisms that can survive. At least 20 different species of bacteria have been isolated from the normal oral environment, among them a variety of streptococci, diphtheria-like bacilli, lactobacilli, spirochetes, and filamentous bacteria. Scientists estimate that there are between 50 billion and 100 billion bacteria in the adult mouth at any one time. (To put the number in perspective, consider that about 78 billion people have lived on Earth since the beginning of time.)

The material that accumulates on the tooth surface is known by many terms, the most common of which is **dental plaque**. Plaque is essentially a biofilm, a deposit of dense gelatinous material consisting of protein, polysaccharide, and an enormous mass of bacteria. By some accounts there are more than a billion bacteria per gram of net weight of plaque. Most bacterial species have been cultivated in the laboratory, and two-thirds are either anaerobic or facultative species.

Dental caries, or tooth decay, takes its name from the Latin *cariosus*, meaning "rotten." In order for dental caries to develop, three elements must be present: a caries-susceptible tooth with a buildup of plaque; dietary carbohydrate, usually in the form of sucrose (sugar); and acidogenic (acid-producing) plaque bacteria (**FIGURE 11.14**). The bacteria produce acid that breaks down the calcium phosphate salts in hydroxyapatite, the major compound in the enamel and underlying dentin.

One of the primary bacterial causes of caries is acidogenic *Streptococcus mutans*. This gram-positive coccus has a high affinity for the smooth surfaces, pits, and fissures of a tooth. Its enzymes react with the glucose and fructose in sucrose and convert them to a long-chain carbohydrate called dextran. These materials give *S. mutans* its special adherence qualities. The bacilli then ferment dietary carbohydrates to

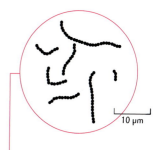

Streptococcus mutans

TABLE 11.3

A Summary of Miscellaneous Bacterial Diseases

DISEASE	CAUSATIVE AGENT	DESCRIPTION OF AGENT	ORGANS AFFECTED	CHARACTERISTIC SIGNS
Endogenous Diseases				
Bacteroides infection	*Bacteroides fragilis*	Gram-negative anaerobic rod	Intestine Blood	Blood clots
Pseudomembranous colitis	*Clostridium difficile*	Gram-positive spore-forming rod	Intestine	Intestinal lesions Diarrhea
Fungal-like Diseases				
Actinomycosis	*Actinomyces israelii*	Gram-positive anaerobic fungus-like rod	Gum tissues Chest wall Abdominal organs	Draining sinus
Nocardiosis	*Nocardia asteroides*	Acid-fast fungus-like rod	Lungs	Abscesses Rusty sputum Madura foot
Animal Bite Diseases				
Pasteurellosis	*Pasteurella multocida*	Gram-negative rod	Skin	Abscess at site of bite Arthritis
Cat-scratch disease	*Bartonella henselae* (?) *Afipia felis* (?)	Rickettsia Gram-negative rod	Skin	Lesion at site of bite Swollen lymph glands
Rat-bite fever	*Streptobacillus moniliformis*	Gram-negative streptobacillus	Skin	Lesion at site of bite Rash, fever
	Spirillum minor	Spirochete	Skin	Lesion at site of bite Rash, fever

FIGURE 11.14

Dental Caries

(a) Overlapping circles depicting the interrelationships of the three factors that lead to caries activity.
(b) *Streptococcus mutans*, a major cause of dental caries, as visualized by the scanning electron microscope. (Bar = 2 μm.)

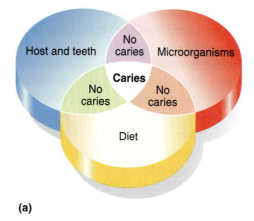

(a)

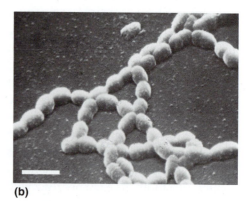

(b)

TOXIN INVOLVED	TREATMENT ADMINISTERED	IMMUNIZATION AVAILABLE	COMMENT
Not established	Clindamycin Chloramphenicol	None	May lead to gangrene Normal intestinal inhabitant
Possible	Withdrawal of causative antibiotic	None	Accompanies excessive antibiotic use
Not established	Penicillin Tetracycline	None	Lumpy jaw possible Associated with IUD use Normal lung inhabitant
Not established	Co-trimoxazole Sulfonamides	None	Associated with AIDS Endogenous disease Soil transmission via wound
Not established	Penicillin Doxycycline	None	Common in cats and dogs
Not established	Symptomatic management Various antibiotics	None	Treatment not always necessary
Not established	Penicillin Tetracycline	None	Typical fever, arthritis, and rash
Not established	Tetracycline Streptomycin	None	Common in Japan

lactic acid, with smaller amounts of acetic acid, formic acid, and butyric acid. The acids dissolve hydroxyapatite, after which protein-digesting enzymes break down any remaining organic materials. Other streptococcal species involved in caries include *S. sanguis, S. mitis,* and *S. salivarius.*

The prevention of dental caries relates to three principal areas: protecting the tooth, modifying the diet, and combatting cariogenic bacteria (Figure 11.14). Tooth protection may be accomplished by the ingestion and topical application of fluorides. These compounds displace hydroxyl ions in hydroxyapatite, thus reducing the solubility of the enamel. Teeth also can be protected by applying polymers to cover pits and fissures in the teeth, thereby preventing bacterial adhesion. Diet modification requires minimizing sucrose in foods. Some success has been observed by substituting **xylitol**, an alcoholic derivative of xylose, for the sugar in candy. Efforts to eliminate cariogenic bacteria focus on preventing the synthesis of dextrans by streptococci, as well as stimulating antibody production against the bacteria. Some novel approaches are explored in MicroFocus 11.5.

Caries is not the only form of dental disease. The teeth are surrounded by tissues that provide the support essential to tooth function. These tissues, called the periodontal tissues, may be the site of a **periodontal disease** called **acute necrotizing ulcerative gingivitis (ANUG)**. Microbiologists believe that ANUG is the result of an acute infection of the tissues by several bacteria, among which are *Porphyromonas*

MicroFocus 11.5

WATCH OUT *S. MUTANS*—YOUR DAYS MAY BE NUMBERED!

In the war against tooth decay, people have armed themselves with floss, toothpaste, toothbrushes, and many innovative variations of these. Still, the bacteria seem to win, especially *Streptococcus mutans*. Modifying the diet helps and adding fluoride to the water tips the balance still further. And newer innovations are just around the corner.

British researcher Charles G. Kelly and his research team have produced an antibacterial coating for the teeth. The preparation binds to the teeth surfaces where *S. mutans* attaches, and prevents bacterial attachment. In their studies, they smeared the synthetic preparation on the teeth of volunteers whose mouths were cleansed of bacteria. Control volunteers were treated with a placebo. Those receiving the synthetic preparation remained free of *S. mutans* for over three months while the bacteria appeared in the mouths of control volunteers after only three weeks.

Could a vaccine be the death knell for *S. mutans*? In London, Julian Ma and coworkers at Guy's Hospital are experimenting with a vaccine that contains antibodies that latch onto *S. mutans*, preventing the bacteria from binding to the tooth surface. Since the vaccine does not trigger an immune reaction, such treatments would need to be repeated every year or two.

In Boston, Martin Taubman, Daniel Smith, and colleagues at the Forsyth Institute have developed another vaccine against *S. mutans*. This vaccine blocks the enzyme responsible for synthesizing the long-chain bacterial carbohydrates that stick to the tooth surface. Taubman and Smith believe that "immunizing" children 18 months to 3 years of age would give lifelong protection. So, these three innovations come from different research directions but all would accomplish the same result: no attachment means no bacteria, which means no acid, which means no tooth decay.

Jeffrey Hillman and his colleagues at the University of Florida have come up with perhaps the most ingenious way to rid the mouth of *S. mutans*—chemically attack them. Hillman's goal is to replace the "bad" *S. mutans* that secrete lactic acid with "good" *S. mutans* that do not produce the acid. By collecting mouth samples from hundreds of patients, his group isolated a strain of *S. mutans* that produced a toxin that kills the other strains. The group then took this toxin-producing strain and genetically engineered it so that it would not secrete lactic acid. Using their "replacement therapy," this strain was squirted onto the teeth of rats. By producing the toxin, the genetically engineered bacteria killed and replaced the resident *S. mutans* population and the rats exhibited much reduced levels of cavities. Similarly, three human volunteers had their mouths rinsed for five minutes with the engineered strain. To this day, they either have no *S. mutans* in their mouth or only the engineered strain. Again, replacement therapy on this very limited group of volunteers worked. Clinical trials are beginning in 2003 and Hillman hopes that a commercial product will be available to dentists in 2006 or 2007. If so, a five-minute mouth rinse would do the trick.

For the present, the dental wars go on. It is comforting to know, however, that scientists have imaginative solutions that extend beyond a new flavor of toothpaste. Dental caries is the most widespread infectious disease in today's world. Putting it to an end would be a considerable feather in the scientific cap. So, watch out *S. mutans*, your days may be numbered!

---

gingivalis, a gram-negative rod; *Leptotrichia buccalis*, a long, thin gram-negative rod; *Treponema vincentii*, a spirochete; and species of *Eikenella*, another gram-negative rod. Indeed, in 1990 researchers at Mount Sinai Hospital in New York reported that *E. corrodens* can cause severe cellulitis and arthritis of the knee after entering the blood through trauma of the gingival tissues.

ANUG is characterized by punched-out ulcers that appear first along the gingival margin and interdental papillae, and then spread to the soft palate and tonsil areas. There often is severe pain and spontaneous bleeding. Infections in the latter area are sometimes called **Vincent's angina** after Jean Hyacinthe Vincent, who described the spirochete in 1892. A foul odor and bad taste come from gases produced by anaerobic bacteria. As the periodontal tissues decay, the teeth may become loosened and eventually dislodged completely. Diagnosis depends on the observation of gram-negative rods and spirochetes from the ulcers. The disease is sometimes called **fusospirochetal disease** because the rods have a long, thin fusiform shape and are mixed with spirochetes (**FIGURE 11.15**). Antibiotic washes may be used in therapy, and some physicians suggest painting the area with a traditional remedy of gentian violet.

ANUG has long been associated with conditions such as malnutrition, viral infection, excessive smoking, poor oral hygiene, and mental stress. The disease was common among soldiers in World War I and was known at that time as **trench mouth**. It is not considered a transmissible disease, but one that develops from bacteria already in the mouth. It is most common in teens and adults under 25.

Studies in the 1990s highlighted the hazards associated with dental equipment as a transfer mode for bacteria. Interest centered around **biofilms**, the populations of waterborne and airborne microorganisms that adhere to macromolecules on a solid surface. Among the pathogens found in biofilms are *Pseudomonas* species, *Legionella pneumophila*, and *Mycobacterium* species. At risk are dental patients with diminished resistance to infection, including elderly people, transplant recipients, and AIDS and cancer patients.

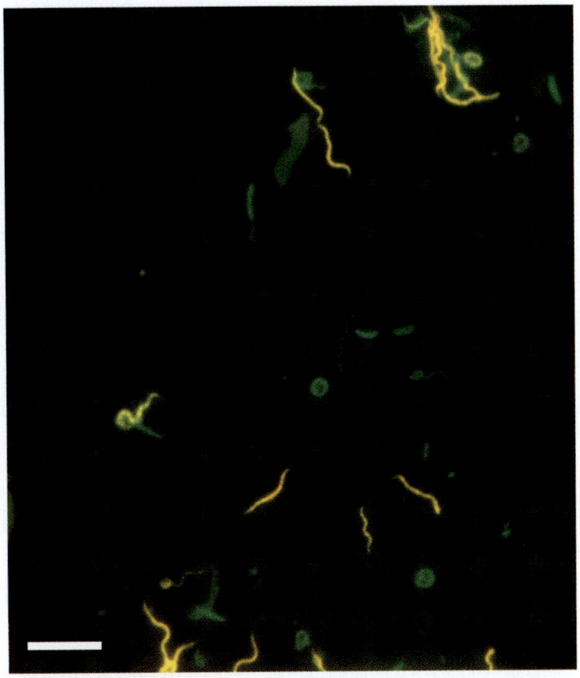

Of particular concerns are dental unit waterlines in hand pieces, air-water syringes, and ultrasonic scalers. Biofilms build up as microorganisms (most originating from the public water supply) colonize the narrow, smooth-walled waterlines. Counts as high as 1 million microorganisms per milliliter of water have been reported.

Although there is no established risk from biofilms in waterlines, the American Dental Association (ADA) has issued several recommendations to ensure safety. They recommend that independent water reservoirs be used, rather than the public supply; that chemical treatment regimens be put in place; that daily draining and air purging regimens be established; and that point-of-use filters be employed. The ADA recommends that the unfiltered output (water) of the dental unit contain no more than 200 colony forming units of aerobic mesophilic heterotrophic bacteria per milliliter. Through a proactive program, biofilms can be controlled and disease transmission can be interrupted.

FIGURE 11.15

Bacteria of the Oral Plaque

A photomicrograph of treponemes found in the subgingival plaque of a patient with periodontal disease. These spirochetes have been stained by fluorescence microscopy techniques. The organisms cannot be cultivated in laboratory media and must be studied directly as they come from the patient. (Bar = 10 μm.)

URINARY TRACT INFECTIONS ARE MORE COMMON IN WOMEN

Over 8 million episodes of **urinary tract infections (UTIs)** occur annually in the United States. Urinary tract infections are the second most frequent cause of visits to the doctor's office (after respiratory tract infections). Sufferers report abdominal discomfort, burning pain on urination, and frequent urges to urinate. *E. coli* often is the major cause of UTIs (**FIGURE 11.16a**). Other organisms of infection include *Chlamydia*, *Mycoplasma*, and *Proteus mirabilis* (**FIGURE 11.16b**).

Most infections develop in the urethra and cause a disease called **urethritis**. The bacteria then can move to the bladder, causing a bladder infection (**cystitis**). Left untreated, the infection may move up the ureters to infect the kidneys (**pyelonephritis**).

Women are apparently infected more often than men, most likely because of anatomical differences: The male urethra is about nine inches long, and infecting organisms must travel this lengthy distance from the environment before reaching

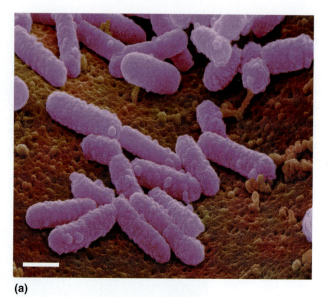

(a)

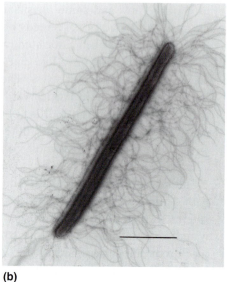

(b)

FIGURE 11.16

Pathogens of the Urinary Tract

False color scanning electron micrograph of *Escherichia coli* bacteria. These gram–negative bacilli are normal inhabitants of the human intestine and are usually harmless. However, under certain conditions they infect another part of the body and cause infection. For example, 80 percent of all urinary tract infections are caused by *E. coli*. (Bar = 2 μm.) (b) A transmission electron micrograph of *Proteus mirabilis*. This large cell contains hundreds of flagella that permit motion among the tissue cells of the urinary tract. (Bar = 5 μm.)

the bladder; the female urethra, by comparison, is only about one inch long. Furthermore, the female urethra opens in front of the vagina, and various microorganisms reside normally in the vaginal tract. In addition, the proximity of the urethra to the anus permits intestinal organisms to pass from the gastrointestinal tract to the urinary tract. Sexual intercourse increases the possibility of urinary tract infection because bacteria can enter the urethra during organ contact. Many women suffer from recurrent infections. Usually each of these involves a different strain of the bacterium from the previous infection.

Most urinary tract infections are successfully treated with sulfonamides or other antibiotics such as ofloxacin. Such practices as avoiding tight-fitting clothes and urinating soon after sexual intercourse also can reduce the possibility of infection. Studies have shown that cranberry juice and vitamin C may inhibit the bacteria by increasing the acidity of the urinary tract. Various cultivation methods are used in diagnosis, including one in which different bacteria give different color reactions, as FIGURE 11.17 shows.

Researchers have discovered that women who suffer from repeated UTIs often have lower antibody levels than those women who have normal levels of antibodies. Therefore, researchers are investigating the possibility of a vaccine against *E. coli* as a way of preventing urinary tract infections. Research groups have tested a genetically engineered, injectable vaccine in mice. The vaccine consists of proteins that trigger the immune system to produce antibodies against the bacterium's adhesion molecules of the pili. When coated with antibodies, the bacterium cannot adhere to the urinary tract tissues. A second group was experimenting with a different vaccine

FIGURE 11.17

Detecting Urinary Tract Pathogens

Various urinary tract pathogens are cultivated on a laboratory medium called CHROMagar. The different bacteria display different color reactions, which aid identification: 1) *Proteus mirabilis*, 2) *Enterobacter faecalis*, 3) *Klebsiella pneumoniae*, 4) *Pseudomonas aeruginosa*, 5) *Escherichia coli*, 6) *Staphylococcus aureus*.

composed of killed bacteria delivered to the vaginal tract by suppository. Local delivery of the vaccine would ensure a local immune response and avoid inflammation caused by injection at distant sites. Both vaccine studies look promising.

Mounting evidence has shown that **biofilms** can be a key factor in urinary tract infections. Bacteria sequestered in these slimy conglomerates are shielded from attack by the body's immune system, and they are difficult to kill with antibiotics. Biofilms in urinary catheters often provide starting points for bladder infections, as bacteria creep up the catheters. In males, biofilms have been implicated also in infections of the prostate gland, which are accompanied by chronic pain and sexual dysfunction. Researchers have demonstrated that biofilms are highly organized clumps of bacteria bound together by a carbohydrate matrix and surrounded by water channels for nutrient delivery and waste disposal. The carbohydrate shields the bacteria from defensive mechanisms, including antibodies and antibiotics. Finding ways to penetrate this barrier is important to stemming the tide of urinary tract infections.

MicroInquiry 11 presents four case studies covering some of the diseases described in this chapter.

NOSOCOMIAL DISEASES ARE ACQUIRED IN A HEALTH CARE SETTING

Nosocomial diseases are those acquired during hospitalization or in chronic care facilities. The CDC has estimated that up to 10 percent of all hospital patients may develop a nosocomial disease during their stay, with surgical patients particularly susceptible. Certain types of operations such as amputations and intestinal surgery are accompanied by an infection rate approaching 30 percent. Over 1 million patients may be involved annually, and an estimated $6 billion in hospital costs is spent to treat the nosocomial diseases.

Today though, the term "nosocomial infection" goes beyond the hospital environment. Patients in short-term care settings or chronic care facilities, such as nursing homes, also run the risk of being infected with the same pathogens as hospital patients. Therefore, identifying high-risk patients, instituting new or better

MicroInquiry 11   SEXUALLY TRANSMITTED, CONTACT, AND MISCELLANEOUS BACTERIAL DISEASE IDENTIFICATION

Below are several descriptions of sexually transmitted, contact, and miscellaneous bacterial diseases based on material presented in this chapter. Read the case history and then answer the questions posed. Answers can be found in Appendix E.

Case 1: The patient is a 17-year-old woman who comes to the clinic indicating that several days ago she started feeling nauseous but had not experienced any vomiting. She tells the physician that the day before coming to the clinic she had a fever and chills; she also has been urinating more frequently and the urine has a foul smell. She is diagnosed as having a urinary tract infection.

11.1a. What types of bacteria could be responsible for her illness?

11.1b. Why are these types of diseases more prevalent in women than they are in men?

11.1c. What types of urinary infections can occur?

11.1d. How could she attempt to avoid another UTI?

11.1e. What role do biofilms play in UTIs?

Case 2: A 19-year-old unwed mother arrives at the emergency room of the county hospital complaining of having cramps and abdominal pain for several days. She says she had never had a urinary tract infection and could not have gonorrhea, as she was treated and cured of that two years ago. She has not experienced nausea or vomiting. When questioned, she tells the emergency room nurse that she has a single male sexual partner and condoms always are used. Based on further examination, the patient is diagnosed with pelvic inflammatory disease (PID). An endocervical swab is used for preparing a tissue cul-

ture. Staining results indicate the presence of cell inclusions.

11.2a. What bacteria can be associated with PID? What disease does she most likely have?

11.2b. Why was a tissue culture inoculum ordered? Describe the reproductive cycle of this organism.

11.2c. What other tests could be ordered for the patient's infection?

11.2d. Why was the emergency room concerned about her sexual activity?

11.2e. What misconception does the patient have about her past gonorrhea infection?

Case 3: A 71-year-old man visits the local hospital emergency room at 6:00 PM after noticing a red infection streak running up his left forearm. He tells the physician that he was playing with his cat this morning when it bit him on the

---

Standard precautions: a method of infection control designed to minimize the risk of microorganism transmission between individuals in the hospital and health care facilities.

prevention strategies, and following **standard precautions** are critical toward disrupting the spread of infectious diseases in any health care setting.

All these settings represent high-density communities composed of unusually susceptible individuals. A variety of pathogenic bacteria abound, and new ones are continually being introduced as new patients arrive (**FIGURE 11.18**). In addition, the extensive use of antimicrobial agents contributes to the development of resistant strains of microorganisms (such as MRSA), and staff members become carriers of these strains (**FIGURE 11.19**). Many of the patients have already experienced some interference with their normal immune defenses, such as a breach of the skin barrier in surgery, radiation therapy for cancer, immunosuppressive medication, or indwelling apparatuses such as catheters and intravenous tubes. When all these factors meet, nosocomial diseases break out.

Up to 80 percent of nosocomial diseases are caused by microbes brought with the patient at the time of admission. These organisms are generally **opportunistic**—that is, they do not cause disease in normal humans, but they are dangerous in compromised individuals. Among the most common opportunistic bacteria are gram-negative rods such as *Escherichia coli*, *Serratia marcescens*, *Enterobacter aerogenes*, *Enterobacter cloacae*, *Klebsiella pneumoniae*, and *Proteus* species. Often these bacteria are the cause of urinary tract infections. *Staphylococcus aureus* is another important cause of nosocomial diseases, as are various types of streptococci.

left wrist. Thinking nothing of it and being an amateur photographer, he went about printing some photographs in his darkroom. At 3:00 PM, he finished and, in the daylight, noticed that his wrist was swollen and painful. The physician also notes that the patient experiences tenderness at the site. She then notices a small puncture wound on the wrist and a small abscess. A Gram stain smear from the abscess indicates the presence of gram-negative rods.

11.3a. What bacterium is responsible for the patient's illness?

11.3b. What clues lead you to identify this specific organism?

11.3c. Where is this organism normally found in cats?

11.3d. What could the patient have done to make it less likely that an infection occurred?

Case 4: A 17-year-old man comes to a free neighborhood clinic. He says that he noticed some white pus-like discharge and a tingling sensation in his penis. Since yesterday, he has had pain on urinating. He tells the physician that he has been sexually active with several female partners over the past eight months, but no one has had any sexually transmitted disease. Examination determines that there is no swelling of the lymph nodes in the groin or pain in the testicles. A Gram stain indicates the presence of gram-negative diplococci. The patient is given antibiotics, instructed to tell his female partners they should be medically examined, and then he is released.

11.4a. Based on the clinic findings, what disease does the patient have and what bacterium is responsible for the infection?

11.4b. Why is it important for his sexual partners to be medically examined, even if they experience no symptoms? What complications could arise if they are infected?

11.4c. For which other organisms is this patient at increased risk? Why?

11.4d. What significance can be drawn from the fact that the patient does not have any swelling of the lymph nodes in the groin or pain in the testicles?

11.4e. What antibiotics most likely would be given to the patient?

The other 20 percent of nosocomial infections result from contact with microbes within the hospital or health care facility. Contact often comes from the hands or instruments of health care workers, or contaminated equipment. Bacteria can be spread easily from one patient to another via the hands of a hospital worker. Person-to-person spread of infections also occurs by airborne droplets, and fecal-oral and bloodborne routes. To deal with nosocomial diseases, hospitals designate a specialist, usually a **nurse epidemiologist**, whose primary responsibility is to locate problem areas and report them to an infection-control committee. The local committee consists of nurses, doctors, dieticians, engineers, and housekeeping and laboratory workers who monitor equipment and procedures to interrupt the disease cycle.

One of the most basic ways to disrupt the disease cycle is by hand washing between patient contacts (MicroFocus 11.6). In ICUs, emergency rooms, urgent care clinics, and other areas such a simple procedure often is overlooked during the rush of crisis care. Also, awareness of protecting endotracheal tubes and other devices from contamination is crucial in decreasing disease transmission to or between patients. Approximately one-third of nosocomial infections are preventable. By identifying at-risk patients, hand washing, and protecting equipment and instruments through sterilization and disinfection procedures, control or elimination of nosocomial diseases is possible.

FIGURE 11.18

A Hospital Epidemic of *Burkholderia* Infection

This epidemic occurred over an extended period—from August 31, 1996, to June 1, 1998—when the causative organism was finally identified at its source.

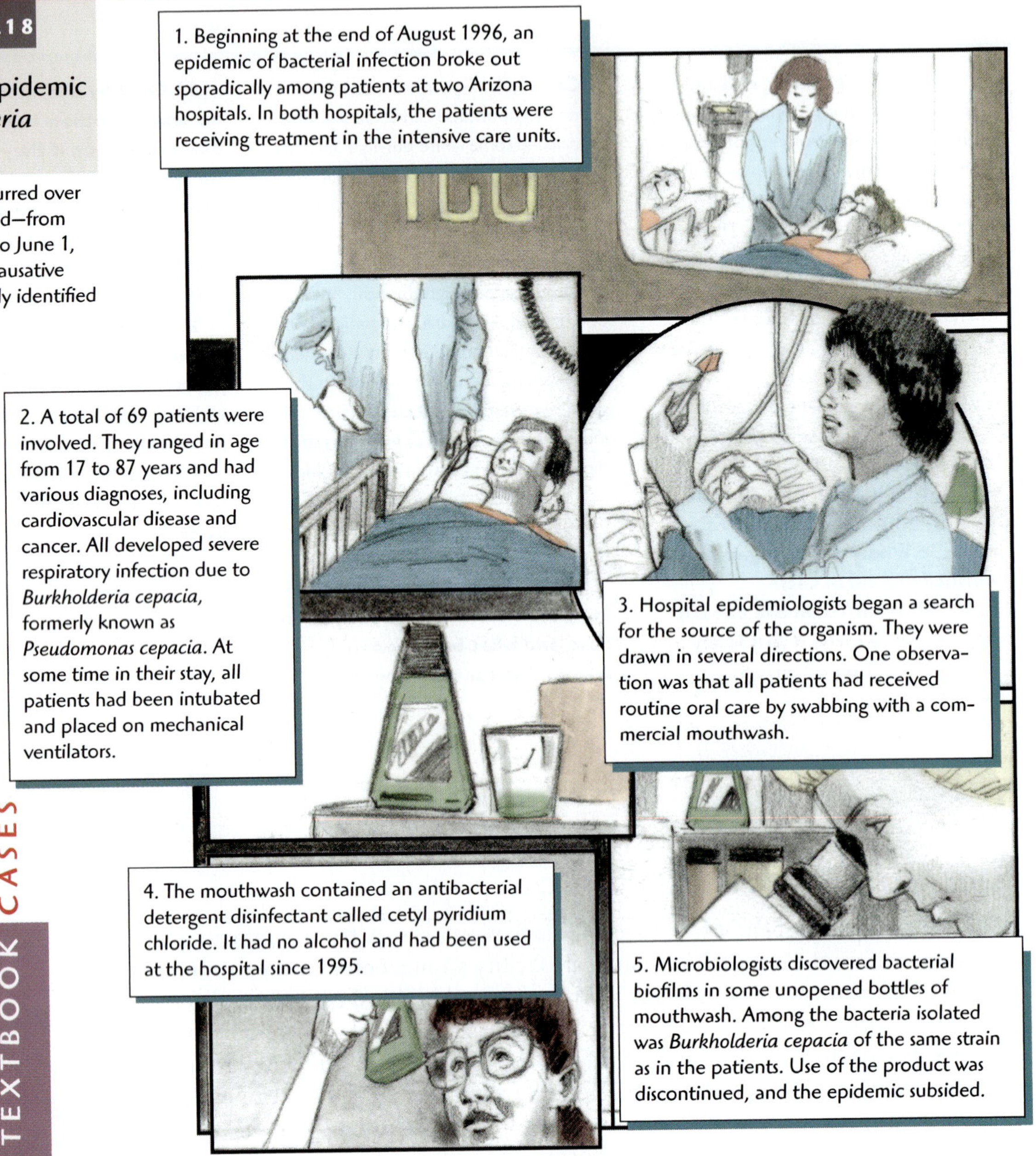

TEXTBOOK CASES

1. Beginning at the end of August 1996, an epidemic of bacterial infection broke out sporadically among patients at two Arizona hospitals. In both hospitals, the patients were receiving treatment in the intensive care units.

2. A total of 69 patients were involved. They ranged in age from 17 to 87 years and had various diagnoses, including cardiovascular disease and cancer. All developed severe respiratory infection due to *Burkholderia cepacia*, formerly known as *Pseudomonas cepacia*. At some time in their stay, all patients had been intubated and placed on mechanical ventilators.

3. Hospital epidemiologists began a search for the source of the organism. They were drawn in several directions. One observation was that all patients had received routine oral care by swabbing with a commercial mouthwash.

4. The mouthwash contained an antibacterial detergent disinfectant called cetyl pyridium chloride. It had no alcohol and had been used at the hospital since 1995.

5. Microbiologists discovered bacterial biofilms in some unopened bottles of mouthwash. Among the bacteria isolated was *Burkholderia cepacia* of the same strain as in the patients. Use of the product was discontinued, and the epidemic subsided.

FIGURE 11.19

Nonsocomial Infections Caused by MRSA

MRSA resistance in the ICU has continued to increase, surpassing the 50 percent mark in 1999. (Source: National Nosocomial Infections Surveillance data)

MicroFocus 11.6

YOU GOTTA WASH THOSE GERMS RIGHT OFF OF YOUR HANDS

All health experts agree that the most effective way to prevent the spread of infectious disease—be it in the hospital or at home—is hand washing with soap and water. Therefore, the following tips are important when you are out in public.

By frequently washing your hands, you wash away germs picked up as you touch people possibly infected with colds, flu, or any of the diseases discussed in the last several chapters of this textbook. Objects such as doorknobs, handkerchiefs, and dinnerware can be contaminated with infectious microbes—to say nothing about animals and animal waste. If you bring your germ-infested hands to your mouth, nose, or eyes, you can become infected. In fact, one of the best ways to catch a cold is to touch your nose or rub your eyes after your hands have been contaminated with a cold virus. The important thing to remember is that the spread of cold viruses as well as more serious diseases, such as hepatitis A, meningitis, and infectious diarrhea, can be prevented if you make hand washing a habit.

You should wash your hands often, which is probably more often than you do now. You can't see germs with the naked eye or smell them, so you do not really know where they are hiding. It is especially important to wash your hands before, during, and after you prepare food, and after you eat. It is estimated that one out of three people do not wash their hands after using the restroom, so make sure you do, whether at home or in public. Also, wash your hands more frequently when they are dirty or when someone in your home is sick.

Proper hand washing should be a normal routine.

■ Wet your hands and apply liquid or clean bar soap. Running water washes off many of the germs but you need soap to break up and loosen materials containing these microbes. Place the bar soap on a rack and allow it to drain.

■ Now rub your hands vigorously together and scrub all surfaces. Continue for 10 to 15 seconds. It is the combination of soap and scrubbing that dislodges and removes most microorganisms.

■ Rinse your hands well and dry them using a clean paper towel or warm-air hand drier. Note: cloth hand towels can harbor dangerous microbes since these towels often are shared by different people, especially in public restrooms.

Some people wonder how you can keep your hands clean when you have to shut off the sink faucet after you have finished washing your hands. Perhaps you are at an airport or other public restroom that has the high-tech sensor faucets that automatically turn on and off in response to the hand movements. If not, use a clean paper towel to turn the faucet on and off. The alternative is to act like a health professional—use your elbows to manipulate the faucet.

However you wash your hands, you are keeping yourself protected from infection and preventing the spread of potentially infectious microorganisms to others.

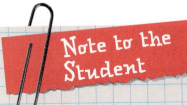

Note to the Student

We have surveyed numerous diseases in this chapter, but one group stands out and merits additional comment. I am referring to the sexually transmitted diseases.

Prior to World War II, doctors had to contend with five classical diseases transmitted by sexual contact. These were syphilis, gonorrhea, chancroid, lymphogranuloma venereum, and, rarely, granuloma inguinale. With the widespread use of antibiotics after the war, the annual incidence of syphilis and gonorrhea declined, and the other three diseases virtually disappeared.

Then came the 1960s and the sexual revolution. It was a time of affluence and defiance of traditional values. Birth control pills, vasectomies, and contraceptive devices offered sexual liberation to go along with social and economic freedoms. Not surprisingly, the incidence rates of STDs soared, and today, the United States is in the grip of an STD epidemic of unprecedented proportions.

The statistics are awesome: Public health officials estimate that one out of every four Americans between the ages of 15 and 55 will acquire an STD at some point in their life; 15 million Americans visit clinics and doctors' offices annually for treatment; and over $2 billion is spent each year in health care costs related to STDs. Moreover, the list of sexually transmitted diseases continues to expand, and at least 25 diseases are now involved, including the ones noted in this chapter, as well as genital warts, hepatitis B, shigellosis, genital herpes, candidiasis, AIDS, and numerous diseases discussed in upcoming chapters.

Is the epidemic likely to end? Some sociologists contend that the fear of getting an STD may be a motivating factor in limiting promiscuous sex. However, a more realistic view is that there has been a shift in attitude toward such things as premarital sex and sex in books and films. Condom use can do much to interrupt the transmission of an STD, but while values continue to be sorted out, it appears that the incidence of STDs will remain high, and that an end to the epidemic is still beyond expectation.

Summary of Key Concepts

11.1 SEXUALLY TRANSMITTED DISEASES

■ **Syphilis Can Progress through Three Stages if Untreated.** Syphilis is a sexually or congenitally transmitted disease caused by *Treponema pallidum*. Primary syphilis is characterized by a chancre at the site of infection. If left untreated, secondary syphilis occurs and presents symptoms of fever and flu-like illness along with a skin rash. Following a latent stage, late or tertiary syphilis occurs characterized by the gumma. Infected pregnant women can transmit syphilis to the fetus (congenital syphilis).

■ **Gonorrhea Is an Infection of the Urogenital Tract.** Gonorrhea is a disease of the urogenital tract caused by *Neisseria gonorrhoeae*. About 50 percent of women have symptoms of vaginal and urethral discharge of pus and burning during urination. In some women, salpingitis is a complication and pelvic inflammatory disease may develop. Males may experience urethral inflammation, a burning on urination, and discharge of pus. Infected women can pass *N. gonorrhoeae* to the newborn during birth. Untreated newborns can suffer gonococcal ophthalmia unless the eyes are treated with silver nitrate or antibiotics.

■ **Chlamydial Urethritis Can Be Asymptomatic.** *Chlamydia trachomatis* causes a form of urethritis that is the most prevalent of all infectious diseases. Symptoms are very similar to gonorrhea although they may be somewhat milder. However, about 75 percent of infected women and 50 percent of infected men are asymptomatic. Newborns can be infected during birth from an infected mother, leading to chlamydial ophthalmia or chlamydial pneumonia.

■ **Ureaplasmal Urethritis Produces Mild Symptoms.** *Ureaplasma urealyticum* causes yet another type of urethritis. Symptoms are mild and can be distinguished from chlamydial urethritis or gonorrhea by the presence of urethral pain during urination.

■ **Chancroid Causes Painful Genital Ulcers.** *Hemophilus ducreyi* causes genital ulcers called a soft chancre and swollen lymph nodes in the groin. The disease is self-limiting.

■ **Other Sexually Transmitted Diseases Also Exist.** Additional bacteria can cause STDs. Other serotypes of *Chlamydia trachomatis* can cause lymphogranuloma venereum that is prevalent in Southeast Asia, and Central and South America. *Calymmatobacterium granulomatis* causes a disease called granuloma inguinale characterized by an ulcer on the external genital organs. *Gardnerella vaginalis* is one bacterium that can cause vaginitis. Mycoplasmal urethritis can be caused by *Mycoplasma hominis*.

11.2 CONTACT BACTERIAL DISEASES

■ **Leprosy (Hansen's Disease) Is a Chronic, Systemic Infection.** Leprosy is a slowly progressive infection caused by *Mycobacterium leprae*. Patients suffer disfiguring of the skin and bones. Tumor-like growths called lepromas form on the skin and in the respiratory tract.

■ **Staphylococcal Skin Diseases Have Several Manifestations.** *Staphylococcus aureus* can cause boils (furuncles) or deeper infections of hair follicles called carbuncles. Other strains cause scalded skin syndrome. Yet other strains cause impetigo contagiosum. With *S. aureus*, concern exists of MRSA and VRSA development.

■ **Toxic Shock Syndrome Is a Disease Caused by Exotoxin Release.** A toxin-producing strain of *S. aureus* causes toxic shock syndrome, characterized by fever, vomiting, and diarrhea followed by sore throat, muscle aches, and skin peeling on the palms and the soles of the feet.

■ **Trachoma Can Lead to Blindness.** Other serotypes of *Chlamydia trachomatis* cause trachoma, an infection of the conjunctiva of the eye. Constant abrasions over a decade or longer by an in-turned upper eyelid can lead to blindness.

■ **Bacterial Conjunctivitis (Pinkeye) Is Very Common.** *Haemophilus aegyptius* is one organism that can cause conjunctivitis. Inflammation of the conjunctiva causes the characteristic "pinkeye."

■ **Yaws Is Spread Person to Person.** In tropical countries, a disease called yaws occurs that is caused by *Treponema pertenue*. The lesion formed eventually disappears but can reappear as a soft granular lesion.

11.3 MISCELLANEOUS BACTERIAL DISEASES

■ **There Are Several Endogenous Bacterial Diseases.** *Bacterioides fragilis* can enter the bloodstream and cause blood clots that can lead to gangrene. *Clostridium difficile* produces exotoxins responsible for pseudomembranous colitis.

■ **Two Fungal-like Bacterial Diseases Exist.** Two bacteria that have a filamentous growth form can cause human disease. *Actinomyces israelii* can cause lumpy jaw in the gum tissue of the mouth. Other forms of actinomycosis occur in the chest wall and abdomen. Nocardiosis is an abscess of the lungs caused by *Nocardia asteroids.*

■ **Animal Bite Diseases Occasionally Occur.** A dog or cat bite can transmit *Pasturella multocida.* The person bitten experiences redness, warmth, swelling, and tenderness at the bite site. Cat-scratch fever is caused by *Bartonella henselae*, a member of the rickettsiae. Symptoms are a lesion at the bite or scratch site, followed by headache, malaise, and fever. *Streptobacillus moniliformis* causes rat-bite fever characterized by a lesion at the bite site, followed by fever, arthritis-like pain in the joints, and a skin rash.

■ **Oral Diseases Arise from Bacteria Normally Found in the Mouth.** Dental caries is most often caused by *Streptococcus mutans.* Poor hygiene can lead to the development of a periodontal disease called acute necrotizing ulcerative gingivitis that is caused by a variety of gram-negative bacteria and spirochetes.

■ **Urinary Tract Infections Are More Common in Women.** Most urinary tract infections develop in the urethra and cause a disease called urethritis. The bacteria then can move to the bladder, causing a bladder infection (cystitis). Left untreated, the infection may move up the ureters to infect the kidneys (pyelonephritis). *Escherichia coli* is one of the most common causes of UTIs.

■ **Nosocomial Diseases Are Acquired in a Health Care Setting.** Nosocomial diseases are acquired in hospitals or other health care settings. About 80 percent of nosocomial infections are brought with the patient at the time of admission while 20 percent result from contact with microbes within the hospital or health care facility. Hand washing between patient contacts, recognizing at-risk patients, and protecting equipment and instruments from contamination are crucial in decreasing disease transmission to or between patients.

Questions for Thought and Discussion

Answers to selected questions can be found in Appendix C.

1. In 1995, the Rockefeller Foundation offered a $1 million prize to anyone who could successfully develop a simple and rapid test to detect chlamydia and/or gonorrhea. The test had to use urine as a test sample and be performed and interpreted by someone with a high school education. To date, no one has claimed the prize. Can you guess why?

2. One of the major problems of the current worldwide epidemic of AIDS is the possibility of transferring the human immunodeficiency virus (HIV) among those who have a sexually transmitted disease. Which diseases in this chapter would make a person particularly susceptible to penetration of HIV into the bloodstream? What explanation can you give for each example?

3. Studies indicate that most cases of *Staphylococcus*-related impetigo occur during the summer months. Why do you think this is the case?

4. Suppose a high incidence of leprosy existed in a particular part of the world. Why is it conceivable that there might be a correspondingly low level of tuberculosis?

5. It has been suggested that women should avoid vaginal douching because the practice can encourage the development of pelvic inflammatory disease (PID) if there is an underlying STD. How can you explain the connection to an inquisitive friend?

6. One day in the late 1980s, a Senegalese patient reported to a New York hospital with an upper lip swollen to about three times its normal size. Probing with a safety pin at facial points where major nerve endings terminate showed that the area to the left of the nose and above the lip was without feeling. When a biopsy of the tissue was examined, it revealed round reservoirs of immune system cells called granulomas within the nerves. On bacteriological analysis, acid-fast rods were observed in the tissue. What disease do all these data suggest?

7. On January 9, 1984, an undergraduate psychology student was bitten by a laboratory rat on the left index finger. Within 12 hours, her finger was swollen and throbbing. Soon thereafter she was hospitalized with swollen lymph nodes, a skin rash, fever, and exquisite sensitivity of the finger. Gram-negative branching rods were found in the tissue. What disease was she suffering from?

8. Researchers at the University of Maryland have suggested a "dip and brush" method of controlling oral diseases. The idea is to dip a toothbrush into an antiseptic several times while brushing to reduce the level of plaque bacteria. Do you think this method will reduce dental problems?

9. In some African villages, blindness from trachoma is so common that ropes are strung to help people locate the village well, and bamboo poles are laid to guide farmers planting in the fields. What measures can be taken to relieve such widespread epidemics as this?

10. Certain microscopes have the added feature of a small hollow tube that fits over the eyepiece or eyepieces. Viewers are encouraged to rest their eyes against the tube and thereby block out light from the room. Why is this feature hazardous to health?

11. A quote from the Book of Leviticus reads: "And the leper . . . shall be defiled; he is unclean; he shall dwell alone; without the camp shall his habitation be." Short of editing the Bible, how can this attitude toward leprosy be changed?

12. After a young man suffers an abrasion on the right arm, his affectionate cat licks the wound. Several days later, a pustular lesion appears at the site and a low-grade fever develops. He also experiences "swollen glands" on the right side of his neck. What disease has he acquired?

13. A woman suffers two miscarriages, each after the fourth month of pregnancy. She then gives birth to a child, but impaired hearing and vision become apparent as it develops. Also, the baby's teeth are shaped like pegs and have notches. What medical problem existed in the mother?

14. In the early 1990s, the CDC was reporting approximately 100,000 instances of ectopic pregnancy in the United States annually, a fivefold increase over the rate in 1970. What could account for this significant increase?

15. One of the diseases discussed in this chapter is probably the most widespread disease in all the world. Students go to special schools and earn a special degree to learn how to deal with it. Then they spend years developing their skill in treating it. What is the disease?

16. A 1998 report indicated that levels of syphilis in the United States were, until then, the lowest in history (about 70,000 new cases per year). The report suggested that the prospects for elimination may be emerging. Why do you think the incidence of syphilis has reached these low levels, and what can be done to hasten its elimination?

17. A recent column by Ann Landers carried the following letter: "I am a 34-year-old married woman who is trying to get pregnant, but it doesn't look promising. . . . When I was in college, I became sexually active. I slept with more men than I care to admit. . . . Somewhere in my wild days, I picked up an infection that left me infertile. . . . The doctor told me I have quite a lot of scar tissue inside my fallopian tubes." The young woman, who signed herself "Suffering in St. Louis," went on to implore readers to be careful in their sexual activities. What advice do you think the woman gave to readers?

18. At a specified hospital in New York City, hundreds of patients pay a regular visit to the "neurology ward." Some sign in with numbers; others invent fictitious names. All receive treatment for leprosy. Why do you think this disease still carries such a stigma?

19. While on their honeymoons, women are sometimes confronted with "honeymoon cystitis," a type of urinary tract infection. They have pain on urination, a frequent urge to urinate, and burning and abdominal discomfort. Why is this condition often related to one's honeymoon?

20. Cosmetic counters often have a number of eye makeup testers for consumers to sample. A recent survey of these testers revealed that over 50 percent were contaminated with bacteria involved in eye infections. What can consumers do to avoid exposure to the bacteria, and what should producers do to limit the danger to consumers?

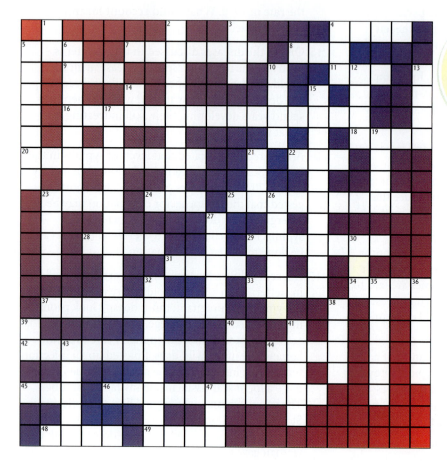

Review

A major topic of this chapter has been sexually transmitted diseases. To test your recall of these diseases, fill in the following crossword puzzle. The answers to the puzzle are in Appendix D.

■ ACROSS

4. The _____ of transmission for STDs is contact.

5. Health organization that charts epidemics of STDs.

7. Occurs during primary syphilis.

8. Number of exposures necessary to establish STD infection.

9. Living tissue for chlamydia cultivation.

11. Ureaplasmal urethritis is a type of _____.

16. Accompanied by swelling of lymph nodes in the groin.

18. Gender in which gonorrhea symptoms are more apparent.

20. Caused by *Haemophilus ducreyi.*

22. STD (abbr) caused by *Chlamydia trachomatis*.
23. Colloquial expression for gonorrhea.
24. Number of exposures necessary to establish STD infection.
25. Syphilis stage with skin rash, loss of hair, and flu-like symptoms.
28. Occurs during urination in gonorrhea patient.
29. Urine tube infected by *Neisseria gonorrhoeae*.
31. Syphilis stages are separated by substantial amounts of _____.
33. Container for sample transported to lab.
34. The _____ of choice for syphilis is penicillin.
37. Gonorrhea-like disease transmitted by sexual contact.
42. Caused by a gram-negative diplococcus discovered by Neisser.
44. Digested by T-mycoplasma in laboratory culture.
45. Possible complication of gonorrhea (abbr).
46. Blockage of fallopian tube as a result of gonorrhea.
48. Continent where lymphogranuloma is prevalent.
49. Can display symptoms of syphilis.

■ **DOWN**

1. Older name (initials) for a sexually transmitted disease.
2. Tubes invaded by gonococci during time of infection.
3. Gonorrhea is rarely contracted by exposure to a _____ surface.
4. Often display a discharge when gonorrhea is present.
5. Sexual _____ is generally required for transmission of syphilis spirochetes.
6. Absent in *Mycoplasma* species.
10. Odor of discharge from vaginitis patient.
12. Stain technique to identify gonococci.
13. Seventy-five percent of chlamydial cases occur in individuals under twenty-_____ years old.
14. Bacterium (initials) that causes syphilis.
15. Syphilis that passes from mother to child.
17. One species can cause a form of urethritis.
19. Chlamydiae do not grow on nutrient _____ media.
21. Organ that can be infected by gonococci and chlamydiae.

23. The organism that causes chlamydia is a very tiny _____.
26. Opening to uterus infected by gonococci.
27. Soft, granular lesion in tertiary syphilis.
30. Shape of *Haemophilus ducreyi*.
32. Caused by *Treponema pallidum*.
35. Organ infected by gonococci leading to proctitis.
36. Syphilis sometimes called the _____ imitator.
38. Object used to obtain sample from gonorrhea patient.
39. Bacterium (initials) that causes gonorrhea.
40. Gonococci that are resistant to penicillin (abbr).
41. Body region affected during cases of gonorrhea.
43. Swollen lymph _____ accompany LGU and are a major symptom.
47. STDs can also be transmitted by _____-sexual methods.

http://microbiology.jbpub.com

The site features **eLearning,** an on-line review area that provides quizzes and other tools to help you study for your class. You can also follow useful links for in-depth information, or just find out the latest microbiology news.

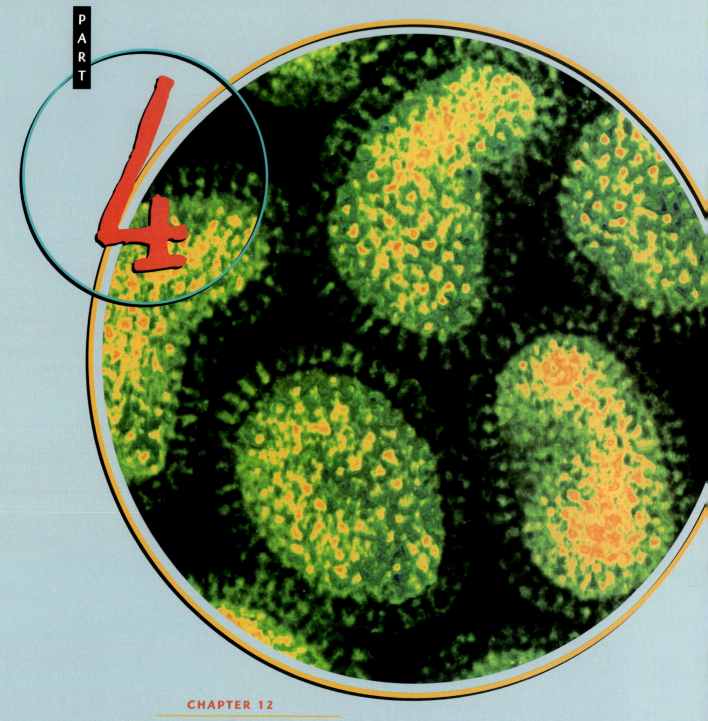

Other Microorganisms

Bacteria are but one of several groups of microorganisms interwoven with the lives of humans. Other prominent groups are the viruses, fungi, protozoa, and multicellular parasites. Knowledge of these microorganisms developed slowly during the early 1900s, partly because they were generally more difficult to isolate and cultivate than bacteria. (Indeed, viruses remained unseen and unphotographed until the late 1930s.) Another reason is that methods for research into bacteria were more advanced than for other microorganisms, and investigators often chose to build on established knowledge rather than pursue uncharted courses of study. Moreover, the urgency to learn about other microorganisms was not great because they did not appear to cause such great epidemics.

Much of that changed in the second half of the 1900s. Many bacterial diseases came under control with the advent of vaccines and antibiotics, and the increased funding for biological research allowed attention to shift to other infectious agents. The viruses finally were identified and cultivated, and microbiologists laid the foundations for their study. Fungi gained prominence as tools in biological research, and scientists soon recognized their significance in ecology and industrial product manufacturing. As remote parts of the world opened to trade and travel, public health microbiologists realized the global impact of protozoal disease. Moreover, as concern for the health of the world's peoples increased, observers expressed revulsion at the thought that hundreds of millions of human beings were infected by multicellular parasites.

In Part 4, we shall study four groups of microorganisms over the course of six chapters. Chapter 12 is devoted to a study of the viruses, and Chapters 13 and 14 outline the multiple diseases caused by these infectious particles. In Chapter 15, the discussion moves to fungi, while in Chapter 16, the area of interest is the protozoa, and in Chapter 17, we discuss the multicellular parasites. Throughout these chapters, the emphasis is on human disease. You will note some familiar terms, such as malaria, hepatitis, and chickenpox; as well as some less familiar terms, such as toxoplasmosis, giardiasis, and dengue fever. The spectrum of diseases continues to unfold as scientists develop new methods for the detection, isolation, and cultivation of microorganisms.

■ *False color transmission electron micrograph of influenza (flu) viruses.*

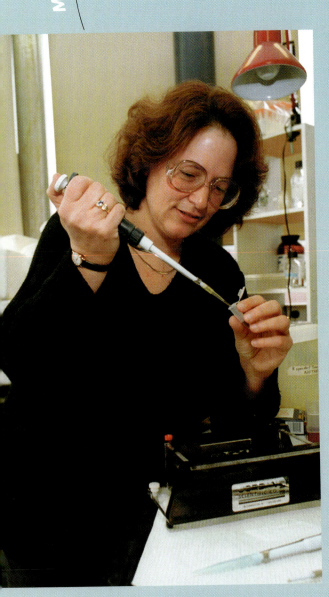

When Ed Alcamo was in college, he was part of a group of twelve biology majors. Each of them had a particular area of "expertise." John was going to be a surgeon, Jim was interested in marine biology, Walt was a budding dentist, and Ed was the local virologist. He was fascinated with viruses, the ultramicroscopic bits of matter, and at one time wrote a term paper summarizing arguments for the living or nonliving nature of viruses. (At the time, neither side was persuasive, and even his professor gracefully declined to place himself in either camp.)

Ed never quite made it to being a virologist, but if your fascination with these infectious particles is as keen as his, you might like to consider a career in virology. Virologists investigate dread diseases such as AIDS, polio, and rabies. They also study the beautiful variegations caused by viruses found in certain types of tulips. Some virologists concern themselves with many types of cancer, and others study the chemical interactions of viruses with various tissue culture systems and animal models. Virologists are working to replace agricultural pesticides with viruses that will destroy mosquitoes and other pests. Some virologists are inserting viral genes into plants and are hoping the plants will produce viral proteins to lend resistance to disease. One particularly innovative group is trying to insert genes from hepatitis B viruses into bananas. They hope that one day we can vaccinate ourselves against hepatitis B by having a banana for lunch.

If you wish to consider the study of viruses, there is one very important proviso you should know—virologists are chemists. I realize that to a biologist, "chemistry" ranks with "root canal," but the fact of the matter is that to do virology, you must be able to do chemistry. You must have general, organic, and, if possible, physical chemistry in your undergraduate program. Then, prepare for lots more chemistry in your graduate program.

The Viruses and Virus-like Agents

It's just a piece of bad news wrapped up in protein.

—Nobel laureate Peter Medawar (1915–1987) describing a virus

A SCENARIO SET IN THE NEAR FUTURE—
"Well, I did eat a meal of raw oysters at home yesterday that I purchased at a local seafood market," reported the patient in the emergency room. "I just felt bad, kind of like I was getting the flu, even though it is the middle of the summer. Then I started getting a fever. When these blood blisters showed up on my legs, I thought I'd better come to the hospital emergency room."

To the attending physician, these symptoms, eating raw oysters, a recorded fever of 102°F (39°C), and blood blisters (bullous lesions) were alarming enough. But what really caught her eye was on the admission chart. It indicated the patient had a history of heavy alcohol use and was suffering from alcoholic liver disease. Acting on her assumptions and the patient's medical history, she immediately transferred him to the ICU. From previous cases she had handled, lab tests were not needed to start treatment, but were ordered anyway. She knew the patient was suffering from an infection caused by eating raw oysters that were contaminated with the bacterium *Vibrio vulnificus*.

V. vulnificus is a gram-negative bacterium that causes a highly fatal blood and wound infection. Among healthy people, ingestion of seafoods contaminated with *V. vulnificus* is usually asymptomatic. However, immunocompromised individuals, such as our patient suffering chronic liver disease, are particularly at risk following ingestion of contaminated foods such as the raw oysters. The bloodstream infection in these individuals often leads to a severe and life-threatening illness characterized by fever and chills followed by decreased blood pressure (septic shock). The bullous lesions are formed by the bacteria growing in skin tissues and damaging blood vessels. Such infected individuals usually die because they present themselves at the hospital too late for effective antibiotic treatment. Thus, as in this case, something faster than antibiotics would be needed to save the patient.

"Hi, Carlos, this is Rebecca. I have a patient who is suffering from a very threatening *V. vulnificus* infection resulting from eating raw oysters. I'm afraid that antibiotics alone won't be good enough at this stage, and I suspect that phage therapy is his only hope." Although new to the medical profession, phage therapy uses naturally occurring bacterial viruses, called bacteriophages (or phages for short), to eradicate infections.

"Wow, this does sound very serious, Rebecca. We do have a 'phage cocktail,' which contains a mixture of phages that are specific to all known strains of *V. vulnificus*. It should work and I will have the cocktail sent over in short order."

The patient was given this "cocktail" intravenously. Amazingly, by late the next day, the patient was feeling much better and within a day no trace of the bacterium could be found in his blood or stool samples. Phage therapy had saved the patient's life.

Although this scenario is set in the near future, most US physicians today don't yet embrace phage therapy, which has been studied and used by physicians in the former Soviet Union for eight decades. Their pioneering work suggests that phage therapy has great potential because phages infect, reproduce within, and kill only the specific, targeted species of bacteria. They are harmless to humans and the beneficial bacteria normally found in the human gut. Moreover, once the targeted bacteria have been destroyed, the remaining phages are eliminated from the body by the immune system. Perhaps the day is not too far off when phage therapy will be a potential adjunct to antibiotic therapy.

In this chapter, we shall study the properties of bacteriophages and animal viruses, focusing on their unique mechanism for replication. We will see how they are classified, how they are eliminated outside the body, and how the body deals with them during a period of disease. The chapter also discusses even more bizarre virus-like agents that can cause disease in animals and plants.

You will note a simplicity in viruses that has led many microbiologists to question whether they are living organisms or fragments of genetic material leading an independent existence. Most of the information in this chapter has only been known since the 1950s, and the current era might be called the Golden Age of virology. Our survey will begin with a review of some of the events that led to this period.

12.1

Foundations of Virology

No single person discovered viruses. Instead, an appreciation of the viruses evolved in the late 1800s with the general understanding of the germ theory of disease. Different diseases had recognizable patterns, and although a bacterium, protozoan, fungus, or other agent could be isolated for most diseases, some diseases had no identifiable agent. Many of these diseases would turn out to be viral diseases.

MANY SCIENTISTS CONTRIBUTED TO THE EARLY UNDERSTANDING OF VIRUSES

In Chapter 1, we mentioned the work of the Russian pathologist Dimitri Ivanowsky (**FIGURE 12.1**). Studying **tobacco mosaic disease**, Ivanowsky filtered the crushed leaves of a diseased plant and found that the clear sap dripping from the filter (rather than the crushed leaves on the filter) contained the infectious agent. Unable to see any microorganism, Ivanowsky reported that a *filterable virus* was the agent of disease. This implied that the unseen agent, whatever its nature, would pass through a bacterial filter. Six years later, Martinus Beijerinck repeated Ivanowsky's work and tested the activity of many dilutions of the filtered viral fluid, and that it was inactivated by boiling. Beijerinck concluded that the disease agent was a "contagious living fluid" (*contagium vivum fluidum*) rather than a discrete object.

Other filterable agents also were discovered. In 1898, Paul Frosch and Friederich Löeffler reported that foot-and-mouth disease was caused by a filterable virus. This finding implied that "virus" could be transmitted among animals as well as plants. Three years later, Walter Reed and his group in Cuba wrote that **yellow fever** also is due to a filterable virus, and with this report, they established human involvement. For a while, the criterion of filterability remained significant, but scientists soon found that some viruses, such as those of rabies and cowpox, did not easily pass through filters. Hence, the agents came to be described simply as "viruses."

Tobacco mosaic disease:
a viral disease that causes tobacco leaves to shrivel and assume a mosaic appearance.

Yellow fever:
a mosquitoborne viral disease of the liver and blood.

(a)

(b)

(c)

FIGURE 12.1

The Investigator, the Disease, and the Virus

In 1892, the Russian pathologist Dimitri Ivanowsky (a) used an ultramicroscopic filter to separate the clear juice from crushed leaves of tobacco plants suffering from tobacco mosaic disease. He placed the juice on healthy leaves and reproduced the disease as the leaves (b) became shriveled (arrow) with a mosaic appearance. Ivanowsky had no idea what was causing the disease, since his microscope revealed no particles of any sort. He wrote that tobacco mosaic disease is caused by a "filterable virus," meaning some unknown poison that passes through a filter. Fifty years would pass before microbiologists finally saw (c) the particles that cause the disease. The particles are tobacco mosaic viruses. (Bar = 30 nm.)

A unique virus was discovered in 1915 by the English bacteriologist Frederick Twort, and independently in 1917 by the French scientist Felix d'Herrelle. The virus of Twort and d'Herrelle came to be called the **bacteriophage** ("bacteria-eater") for its ability to destroy bacteria (**MicroFocus 12. 1**). When a drop of virus was placed in a broth culture of bacteria, the bacteria disintegrated within minutes. D'Herrelle believed that bacteriophage (or, simply, phage) could be used to kill bacteria in the body, but investigators later found that the viruses are easily eliminated from the body and are highly specific for the bacterial strain they attack (**MicroFocus 12.2**). Bacteriophages have since become important tools in transduction research (Chapter 7) and in bacterial identification (Chapters 8 and 9).

By the 1930s, it was generally assumed that viruses were invisible microorganisms below the resolving power of available microscopes. As the years passed, a long list of viral diseases had developed, and work with plant viruses had been productive because of the ease of viral cultivation. For example, virologists learned that tobacco mosaic viruses were composed exclusively of nucleic acid and protein.

Cultivation in live animals was a difficult procedure until Alice M. Woodruff and Ernest W. Goodpasture published a paper in 1931 describing the use of fertilized chicken eggs as a nutrient for cultivating some viruses (**FIGURE 12.2**). The shell of the egg was a natural Petri dish for the nutrient medium, and viruses multiplied within the chick embryo tissues.

The assumption had been that viruses, though incredibly small, were living things, but Wendell M. Stanley of the Rockefeller Institute, in 1935, made the startling announcement that tobacco mosaic viruses could be crystallized. Many scientists suggested that viruses were as lifeless as any crystalline chemical molecule. Virologists pointed out, however, that the viruses replicate, cause fevers, and elicit antibody responses, properties not associated with chemical molecules. Stanley's work opened a debate on the living or nonliving nature of viruses that remains unresolved, and it cut across preconceived ideas that only living things could cause disease.

MicroFocus 12.1

WHY NOT TRY?

When bacteriophages were identified in 1915, some scientists came to believe they might be useful for curing the dreaded bacterial diseases then raging throughout the world. After all, bacteriophages could destroy bacteria in test tubes, so why not try them in the human body? Unfortunately, the bacteriophages turned out to be highly specific viruses, attacking only certain strains of bacteria. Moreover, no one knew much about them or had the means to work with them effectively.

Fast-forward to the modern era of electron microscopes, sophisticated laboratory analyses, and genetic engineer-ing. Technology had advanced substantially, and in the 1980s, the Polish microbiologist Stefan Slopek decided to try again. He identified a bacterium in the blood of an ill patient, then searched out and isolated a bacteriophage specific for that bacterium. He injected a solution of the phages into the patient and waited to see what would happen. Over a period of days, the infection gradually resolved. Slopek tried again—this time with 137 patients. Each patient benefited from the treatment, and several seemed completely cured.

Could bacteriophages be the answer to the growing problem of antibiotic resistance in bacteria? Perhaps so, but there is the potential problem that phages could convert benign or avirulent bacteria to toxin-producing strains (as for diphtheria or botulism), or that phages could alter the expression of host genes and promote the transfer of virulence genes. Still, some novel approaches to treating infectious disease must be found, and bacteriophages may be a candidate.

Here's a postscript: Martin Arrowsmith, the idealistic young physician of Sinclair Lewis's *Arrowsmith*, traveled to the West Indies to treat bubonic plague—with phage therapy.

The invention of the **electron microscope** revealed the nature of the viruses. By 1941, virologists were beginning to visualize the tobacco mosaic virus and other viruses, as FIGURE 12.3 shows.

The second key development of the 1940s was sparked by the national epidemic of poliomyelitis (or, simply, polio). Attempts at vaccine production were stymied by the inability to cultivate polioviruses outside the body, but John Enders, Thomas Weller, and Frederick Robbins of Children's Hospital in Boston solved that problem. Meticulously, they developed a test tube medium of nutrients, salts, and pH buffers in which living cells would remain alive. In the living cells, polioviruses replicated to huge numbers, and by the late 1950s, Jonas Salk and Albert Sabin had adapted the technique to produce massive quantities of virus for use in polio vaccines (Chapter 14). Enders, Weller, and Robbins did not invent tissue cultivation of viruses, but their work showed it had a practical value for vaccine production. In 1954 they shared the Nobel Prize in Physiology or Medicine.

Since viruses were first observed in the 1930s and 1940s, more than 1,500 viruses have been identified and studied. Amazingly, this is only a small proportion of the estimated 400,000 different viruses that are thought to exist. Electron microscopy and biochemistry paved the way for understanding the structure of viruses, which we will explore on the following pages.

FIGURE 12.2

Viral Cultivation

Inoculation of fertilized eggs by a technician in the virology laboratory. Techniques such as these are standard practice in virology research.

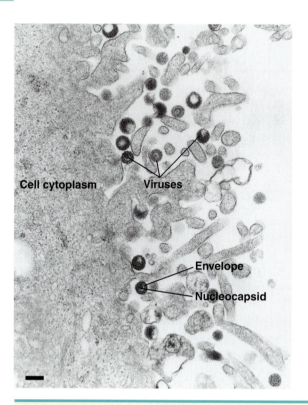

FIGURE 12.3

Cells and Viruses

A transmission electron micrograph of tissue cells and associated viruses. In this view at the margin of the cell, the viruses are the dark, round objects in the space outside the cell. The envelope of the virus is seen as a ring at the viral surface, and the nucleocapsid of the virus is the darker center. These viruses are herpesviruses. (Bar = 150 nm.)

To this point . . .

We have outlined some of the major events in the development of virology, beginning with the early concept of viruses in the late 1800s and continuing through to the 1940s. Ivanowsky was one of the first to note that something smaller than any known bacterium could cause disease. His work was developed further by Beijerinck, and expanded by the discoveries of Löeffler, Frosch, and Reed. In time, viruses were found to affect plants, animals, humans, and bacteria.

A surprising announcement of the 1930s was that viruses could be crystallized. This gave an insight into their simplicity and suggested that they might be nothing more than large chemical molecules. When the electron microscope was used to study viruses in the 1940s, scientists observed that viruses had structural details unlike anything seen in chemical compounds. An understanding of viral functions soon developed, as we shall see presently. Also, great strides were made toward the production of viral vaccines, the advances sparked by the tissue cultivation of polioviruses by Enders, Weller, and Robbins. Vaccines opened the way to the prevention of certain viral diseases.

In the next section we shall discuss the structure of viruses. You will note a level of structural simplicity matched by few other disease agents. These attributes lend uniqueness to viruses.

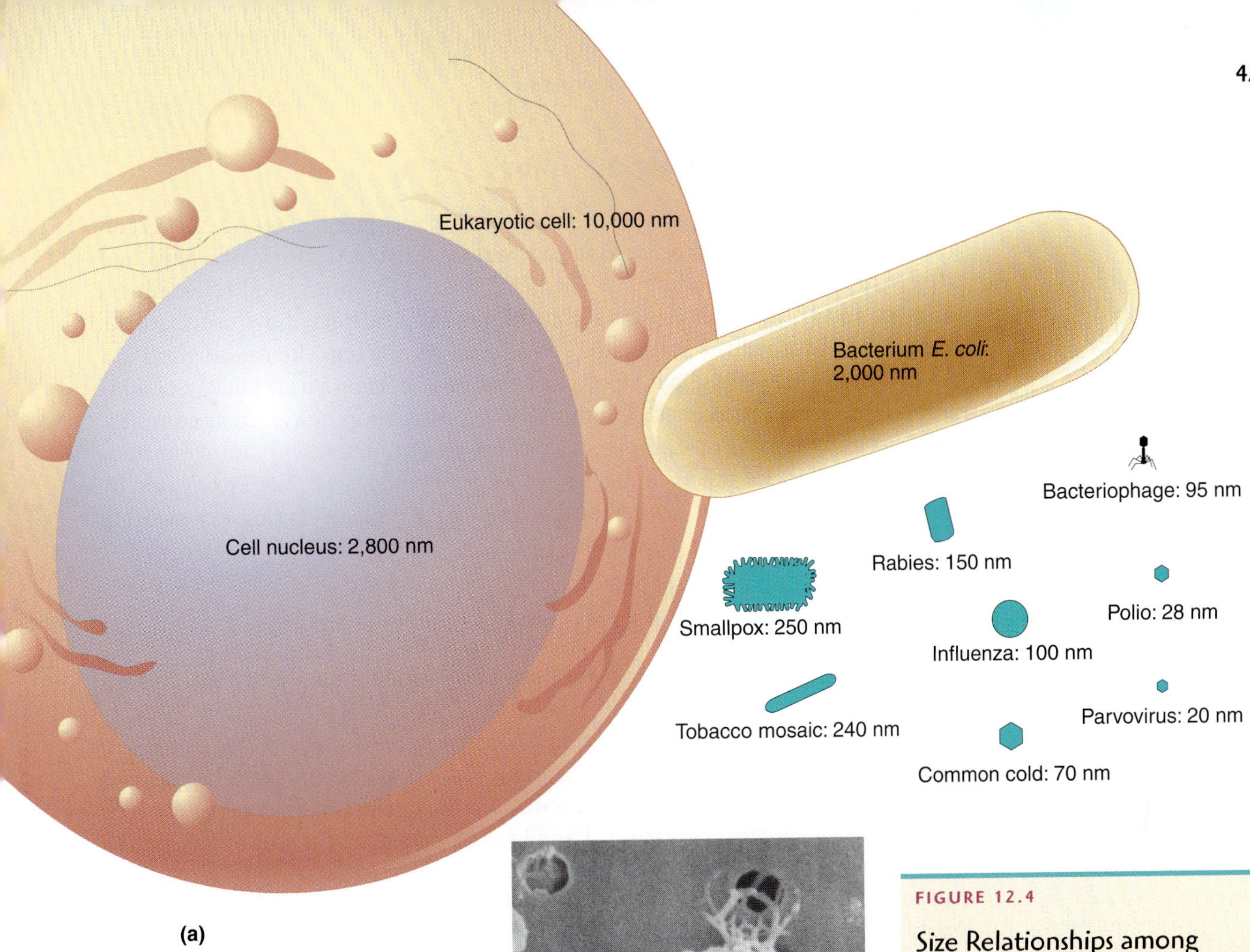

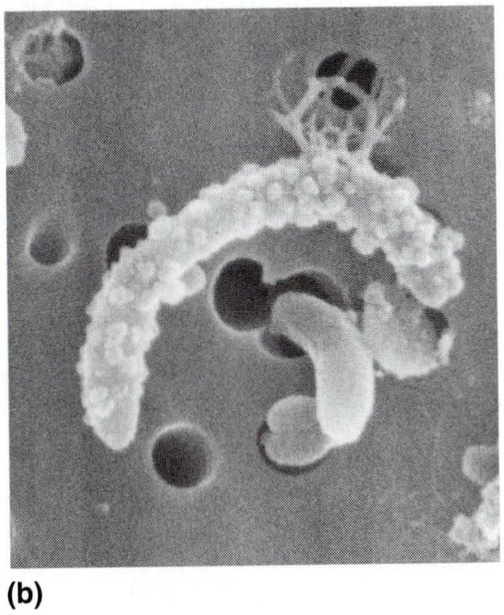

Eukaryotic cell: 10,000 nm

Bacterium *E. coli*:
2,000 nm

Bacteriophage: 95 nm

Cell nucleus: 2,800 nm

Rabies: 150 nm

Polio: 28 nm

Smallpox: 250 nm

Influenza: 100 nm

Parvovirus: 20 nm

Tobacco mosaic: 240 nm

Common cold: 70 nm

(a)

(b)

FIGURE 12.4

Size Relationships among Microorganisms

(a) The sizes of various viruses relative to a eukaryotic cell, a cell nucleus, and the bacterium *E. coli*. The smallpox viruses approximate the chlamydiae and mycoplasmas in size. (b) A scanning electron micrograph of bacterial rods isolated from a water sample on a sieve filter. The surface projections on the rods are bacteriophages. Note their size relative to the host bacterium.

12.2

The Structure of Viruses

Viruses are among the smallest agents able to cause disease in living things. They range in size from the large 250 nanometers (nm) of poxviruses to the 20 nm of parvoviruses (FIGURE 12.4). At the upper end of the spectrum, viruses approximate the size of the smallest bacterial cells, such as the chlamydiae and mycoplasmas; at the lower end, they have about the same diameter as a ribosome.

VIRUSES OCCUR IN VARIOUS SHAPES

Not only do viruses vary in size, they vary in shape as well (**FIGURE 12.5**). Certain viruses, such as rabies and tobacco mosaic viruses, exist in the form of a **helix** and are said to have helical symmetry. The helix is a tightly wound coil resembling a corkscrew or spring. Other viruses, such as herpes simplex and polioviruses, have the shape of an **icosahedron** and hence, icosahedral symmetry. The icosahedron is a polyhedron with 20 triangular faces and 12 corners. Certain viruses have a combination of helical and icosahedral symmetry, a construction described as **complex**. Some bacteriophages, for example, have complex symmetry, with an icosahedral head and a collar and tail assembly in the shape of a helical sheath. Poxviruses, by contrast, are brick shaped, with submicroscopic filaments occurring in a swirling pattern at the periphery of the virus.

VIRUSES LACK CELLULAR STRUCTURE

Unlike most other microbes, viruses have no organelles, no cytoplasm, and no cell nucleus.

All viruses consist of two basic components: a core of nucleic acid called the **genome**, and a surrounding coat of protein known as the **capsid**. The genome contains either DNA or RNA, but not both; and the nucleic acid occurs in double-stranded or single-stranded form. Usually the nucleic acid is unbroken, but in some instances (as in influenza viruses) it exists in separate segments. The genome is usually folded and condensed in icosahedral viruses, and coiled in helical fashion in helical viruses. **FIGURE 12.6** shows the components of a virus.

FIGURE 12.5

Various Viral Shapes

Viruses exhibit numerous variations in symmetry. (a) The nucleocapsid has helical symmetry in the tobacco mosaic, measles, and rabies viruses. The helix resembles a tightly coiled spiral. (b) Certain viruses, such as herpesviruses, polioviruses, and parvoviruses, exhibit icosahedral symmetry in their nucleocapsids. The icosahedron is a polyhedron having 20 triangular faces and 12 points. (c) In other viruses, neither helical nor icosahedral symmetry exists exclusively. The bacteriophage, for example, has an extended icosahedral "head" and a helical tail with extended fibers. The smallpox virus has a series of rod-like filaments embedded within the membranous envelope at its surface.

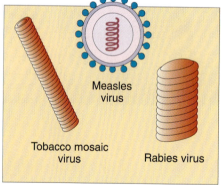

Measles virus

Tobacco mosaic virus   Rabies virus

(a) Helical viruses

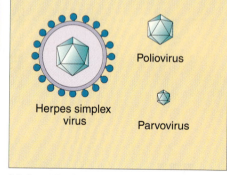

Poliovirus

Herpes simplex virus

Parvovirus

(b) Icosahedral viruses

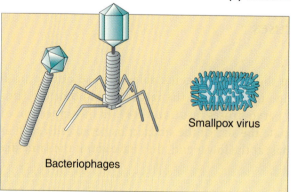

Smallpox virus

Bacteriophages

(c) Complex viruses

The capsid protects the genome. It also gives shape to the virus and is responsible for the helical, icosahedral, or complex symmetry. Generally, the capsid is subdivided into individual protein subunits called **capsomeres** (the organization of capsomeres yields the viral symmetry). The number of capsomeres is characteristic for a particular virus. For example, 162 capsomeres make up the capsid in herpesviruses, and 252 capsomeres compose the capsid in adenoviruses, one of the causes of the common cold.

The capsid provides a protective covering for the genome because the construction of its amino acids resists temperature, pH, and other environmental fluctuations. In some viruses, the capsid contains enzymes to assist cell penetration during replication. Also, the capsid is the structure that stimulates an immune response during periods of disease. The capsid plus the genome is called the **nucleocapsid** (though a better term is probably *genocapsid*, to maintain the structure-to-structure consistency). FIGURE 12.7 shows an icosahedral nucleocapsid.

Many viruses are surrounded by a flexible membrane known as an **envelope**. The envelope is composed of lipids and protein and is similar to the host cell membrane, except that it includes viral-specified components. It is acquired from the host cell during replication and is unique to each type of virus. In some viruses, such as influenza and measles viruses, the envelope contains functional projections known as **spikes**. The spikes often contain enzymes to assist viral attachment to host cells. Indeed, enveloped viruses lose their infectivity when the envelope is destroyed. Also, when the envelope is present, the symmetry of the capsid may not be apparent since the envelope is generally a loose-fitting structure. Some authors refer to viruses as spherical or cubical because the envelope gives the virus this appearance.

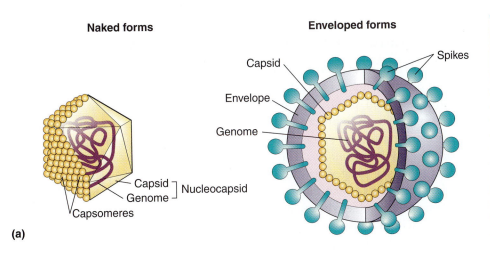

Naked forms   **Enveloped forms**

Capsid — Spikes

Envelope

Genome

Capsid — Nucleocapsid
Genome

Capsomeres

(a)

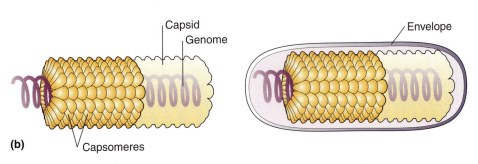

Capsid
Genome

Envelope

Capsomeres

(b)

FIGURE 12.6

The Components of Viruses

(a) An icosahedral virus in both naked and enveloped forms. Capsomere units are shown on one face of the capsid. The genome consists of either DNA or RNA and is folded and condensed. (b) A helical virus in both naked and enveloped forms. The genome winds in a helical fashion. The capsomeres are protein subunits that form the capsid cover.

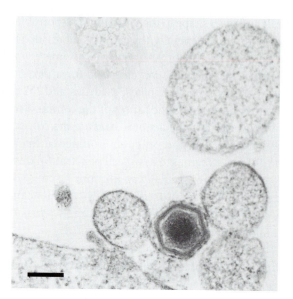

FIGURE 12.7

An Icosahedral Virus

A transmission electron micrograph of an iridovirus displaying icosahedral symmetry. The points of the icosahedron can be seen clearly since this virus has no envelope. The virus causes African swine fever. (Bar = 35 nm.)

A completely assembled and infectious virus outside its host cell is known as a **virion**. (We shall use the terms *virus* and *virion* interchangeably.) Compared to a prokaryote such as a bacterium, a virion is extraordinarily simple. As we have seen, it consists essentially of a segment of nucleic acid, a protein coat, and in some cases, an envelope. (Three variations are pictured in **FIGURE 12.8**.) Virions lack the chemical machinery for generating energy and synthesizing large molecules. Therefore, they must rely upon the structures and chemical components of their host cells for infection and replication.

VIRUS STRUCTURE DETERMINES HOST RANGE AND SPECIFICITY

Viruses can infect almost any cellular organism. There are viruses that specifically infect bacteria, other viruses that infect protozoa, and yet other viruses that specifically infect fungi, plants, or animals. A virus's **host range** refers to what organisms (hosts) that virus can infect and it is based on a virus's capsid structure. Most viruses have a very narrow host range. The phages described for phage therapy only infect the bacterium *Vibrio vulnificus*. They do not infect *Escherichia coli*, although there are other groups of phages that do. The smallpox virus only infects humans, while the poliovirus infects only humans and primates. A few viruses may have a more broad host range. The rabies viruses infect humans and many other warm-blooded animals.

Even within a host range, many viruses only infect certain cell types or tissues within a multicellular plant or animal. This limitation is called **tissue tropism** (tissue attraction). A few animal virus examples will demonstrate this specificity. The host range for the human immunodeficiency virus (HIV) is a human. In humans, HIV primarily infects a specific group of white blood cells called T helper cells because the envelope has protein spikes that recognize receptor molecules on these cells. The virus does not infect cells in other tissues or organs such as the heart or liver. Rabies virus is best at infecting cells of the nervous system and brain because its envelope contains proteins that recognize receptors on these tissues. Therefore, a

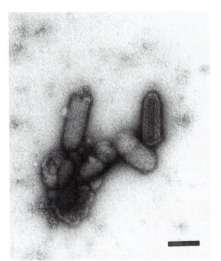
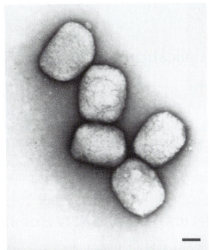
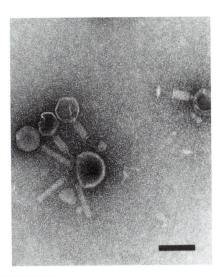

FIGURE 12.8

Viral Symmetry

Transmission electron micrographs of viruses displaying various forms of nucleocapsid symmetry. (a) The vesicular stomatitis virus, which causes skin sores in bovine animals. This virus has a helical nucleocapsid and appears in the shape of a bullet. It is related to the rabies virus. (Bar = 100 nm.) (b) The vaccinia virus, which causes cowpox. This viral nucleocapsid is rectangular with a series of rod-like fibers at its surface. The symmetry is described as complex. (Bar = 100 nm.) (c) A bacteriophage with an icosahedral head and an extended tail. This symmetry is also designated complex. (Bar = 100 nm.)

virus's host range and tissue tropism are linked to infectivity. If a potential host cell lacks the receptor or the virus lacks the complementary protein, the virus cannot bind and infection will not occur.

To this point . . .

We have begun the study of viruses by discussing their size and shape. We noted that different viruses have different forms of capsid symmetry. Structurally, all viruses consist of DNA or RNA, and a protein capsid. Some viruses are surrounded by a flexible membranous envelope. The complete virus is a virion.

Viral structure also determines its host range and tissue tropism. Most, but not all, viruses have a relatively narrow host range and, within the host, usually a specific tissue tropism. Thus, viruses have a specific strategy to ensure their survival. All viral genomes are packaged in a capsid that mediates transmission of the viral genome between cells or hosts. Lacking cellular structure, viral genomes must contain the information to initiate and complete viral replication using the cellular machinery of the susceptible host cell.

Next, we will examine viral replication, a process not encountered elsewhere in the biological world.

12.3

The Replication of Viruses

The process of viral replication is one of the most remarkable events in nature. A virion invades a living cell a thousand or more times its size, uses the metabolism of the cell, and produces copies of itself, often destroying the cell. The virion cannot replicate independently, but within the cell, the replication takes place with high efficiency.

THE REPLICATION OF BACTERIOPHAGES IS A FIVE-STEP PROCESS

Replication has been studied in a wide range of virions and their host cells. The best known process of replication is that carried on by **bacteriophages** of the T-even group (T for "type"). Bacteriophages T2, T4, and T6 are in this group. They are large, complex DNA virions with the characteristic head and tail of bacteriophages but without an envelope (**FIGURE 12.9**). These are virulent viruses, meaning they lyse the host cell, as described in Chapter 7, for the process of transduction.

It is important to note that the nucleic acid in a virion contains only a few of the many genes needed for viral synthesis and replication. It contains, for example, genes for synthesizing viral structural components, such as capsid proteins, and for a few enzymes used in the synthesis; but it lacks the genes for many other key enzymes, such as those used during nucleic acid synthesis. Therefore, its dependence on the host cell is substantial. We shall use phage replication in *E. coli* as a model for the viruses. An overview of the process is presented in **FIGURE 12.10**.

FIGURE 12.9

Bacteriophage Structure

(a) The structure of a bacteriophage consists of the head, inside of which is the nucleic acid, and the tail. The tail sheath is hollow to allow transfer of nucleic acid during infection. The tail fibers attach the phage to the host cell surface. (b) An electron micrograph of T–even bacteriophages (Bar = 40 μm.)

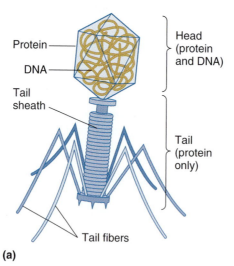

Protein
DNA
Tail sheath
Tail fibers

Head (protein and DNA)
Tail (protein only)

(a)

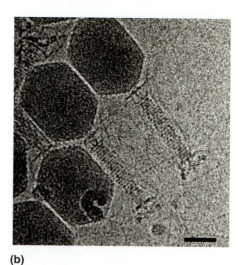

(b)

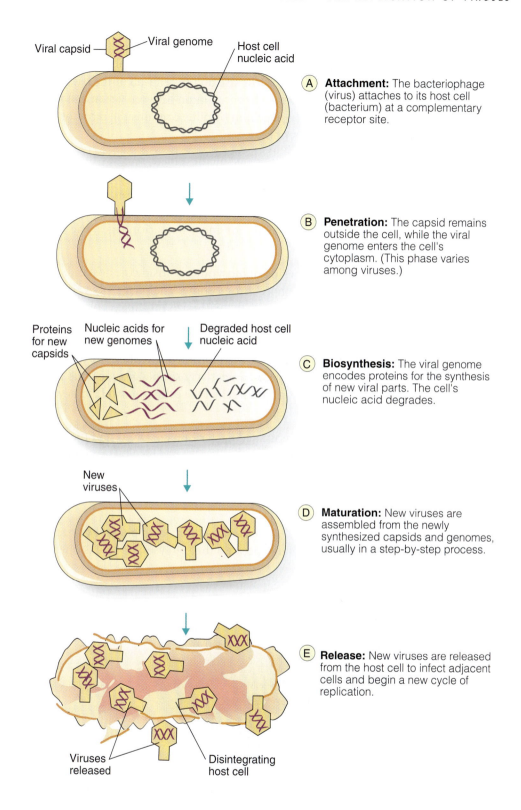

Viral capsid — **Viral genome**   **Host cell nucleic acid**

(A) **Attachment:** The bacteriophage (virus) attaches to its host cell (bacterium) at a complementary receptor site.

(B) **Penetration:** The capsid remains outside the cell, while the viral genome enters the cell's cytoplasm. (This phase varies among viruses.)

Proteins for new capsids   **Nucleic acids for new genomes**   **Degraded host cell nucleic acid**

(C) **Biosynthesis:** The viral genome encodes proteins for the synthesis of new viral parts. The cell's nucleic acid degrades.

New viruses

(D) **Maturation:** New viruses are assembled from the newly synthesized capsids and genomes, usually in a step-by-step process.

(E) **Release:** New viruses are released from the host cell to infect adjacent cells and begin a new cycle of replication.

Viruses released   **Disintegrating host cell**

FIGURE 12.10

Bacteriophage Replication

The pattern of replication in bacteriophages (bacterial viruses) has been studied for many generations. It serves as a model for the replication of other viruses.

1. Attachment The first phase in the replication of a bacteriophage is contact with its host cell. There is no long-distance chemical attraction between the two, so the collision is a chance event. For attachment to occur, sites on the phage's tail fibers must match with a complementary **receptor site** on the cell wall of the bacterium. The actual attachment consists of a weak chemical union between virion and receptor site. (In some cases, the bacterial flagellum or pilus contains the receptor site.) FIGURE 12.11 depicts the attachment phase.

2. Penetration In the next phase, the tail of the phage releases the enzyme lysozyme to dissolve a portion of the bacterial cell wall. Then the tail sheath contracts, and the tail core drives through the cell wall. As the tip of the core reaches the cell membrane below, the DNA passes through the tail core and on through the cell membrane into the bacterial cytoplasm. For most bacteriophages, the capsid remains outside (MicroFocus 12.3).

3. Biosynthesis Next comes the period of biosynthesis. Inside the host cell, phage genes code for the disruption of the host chromosome. The phage DNA then uses the bacterial nucleotides and cell enzymes to synthesize multiple copies of itself. Messenger RNA molecules transcribed from phage DNA appear in the cytoplasm, and the biosynthesis of phage enzymes and capsid proteins begins. Bacterial ribosomes, amino acids, and enzymes are all enlisted for the biosynthesis. Because viral capsids are repeating units of capsomeres, a relatively simple genetic code can be used over and over. For a number of minutes, called the **eclipse period**, no new viral capsids are present.

4. Maturation In this phase, the replicated bacteriophage DNAs and the capsids are assembled into complete virions. The enzymes encoded by viral genes guide the assembly in step-by-step fashion. In one area, phage heads and tails are assembled from protein subunits; in another, the heads are packaged with DNA; and in a third, the tails are attached to the heads.

5. Release The final phase of viral replication is the release phase. For bacteriophages, this also is called the **lysis stage** because the cell lyses, or breaks open. For

Viral Attachment

Scanning electron microscope view of the attachment of bacteriophages to the host cell. In this remarkable close-up view, the tail fibers are attached to the cell surface (arrow). (Bar = 65 nm.)

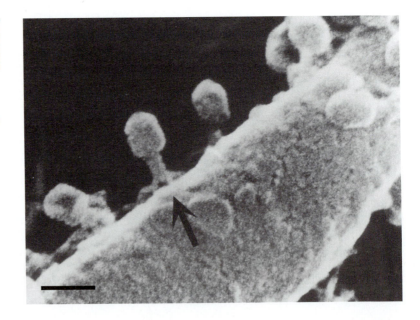

MicroFocus 12.3

EMPTY BOXES

What do you do after you've opened the box and removed the gift? You probably admire the gift, thank the giver, and think about how you'll use it. Perhaps you try it on or otherwise begin weaving it into your life. The poor box gets shunted aside until someone thinks to put it in the garbage pail (or the recycle bin).

It's not much different with viruses: The DNA or RNA genome enters the cell's cytoplasm and encodes new viruses, while the capsid gets broken down, or, in the case of bacteriophages, is left outside the cell. But things may be different in the future. Researchers are investigating viral capsids as miniature reaction chambers for designing ultramicroscopic wires or synthesizing crystals for microelectronic components. Because of a capsid's uniform size and shape, scientists see it as the ultimate small test tube.

Investigators at Montana State University have put the theory to work. Mark Young and his colleagues cultivated masses of viruses, separated the capsids from the genomes, then reassembled the capsid. They combined the protein shells with tungsten salts and found that tungsten molecules penetrate when the acidity level is varied to control the pore size. Soon they were dreaming of constructing computer chips inside a viral capsid—and perhaps a new meaning for the phrase "computer virus."

some phages, the important enzyme in this process is **lysozyme**, encoded by the bacteriophage genes late in the sequence of events. The enzyme degrades the bacterial cell wall, and the newly released bacteriophages are set free to infect other bacteria. The progressive disintegration of bacteria by lysis inspired Twort and d'Herrelle to name the viruses bacteriophages, or "bacteria-eaters." Now perhaps you can envision the scenario for phage therapy portrayed at the beginning of this chapter.

The time that passes from phage attachment to the release of new viruses is commonly referred to as the **burst time**. For bacteriophages, the burst time averages from 20 to 40 minutes. At the conclusion of the process, 50 to 200 new phages emerge from the host cell. This number is commonly called the **burst size**.

> **Lysozyme:**
> an enzyme that degrades bacterial cell walls, especially those of gram-positive species.

ANIMAL VIRUS REPLICATION HAS SIMILARITIES TO PHAGE REPLICATION

The method of replication displayed by T-even phages is similar to that in animal viruses, but with some notable exceptions. One example is the attachment phase. Like bacteriophages, animal viruses have attachment sites, but the receptor sites exist on the host plasma membrane rather than the cell wall. Furthermore, animal viruses have no tails, so the attachment sites often are the spikes distributed over the surface of the capsid or envelope. The sites themselves vary. For example, adenoviruses have small fibers at the corners of the icosahedron, while influenza viruses have spikes on the envelope surface. TABLE 12.1 summarizes some differences we shall discuss.

An understanding of the attachment phase can have practical consequences because the host's receptor sites are inherited characteristics. The sites vary from person to person, which may account for the susceptibility of different individuals to a particular virus. In addition, a drug aimed at an attachment site could conceivably bring an infection to an end. Many pharmaceutical scientists are investigating this approach to antiviral therapy. Indeed, a drug that prevents the attachment of influenza viruses to their host cells is now available (Chapter 13).

Penetration is also different. Phages inject their DNA into the host cell cytoplasm, but animal viruses usually are taken *in toto* into the cytoplasm. In some cases, the viral

TABLE 12.1

The Replication of Bacteriophages and Animal Viruses Compared

	BACTERIOPHAGE	ANIMAL VIRUS
ATTACHMENT	Precise attachment of special tail fibers to cell wall	Attachment of spikes, capsid, or envelope to cell surface receptors
PENETRATION	Phage tail enzymes dissolve cell wall; nucleic acid passes through	Whole virus enters cell, or viral envelope fuses with cell membrane; nucleic acid is released
BIOSYNTHESIS AND MATURATION	Occurs in cytoplasm Host cell activity ceases Viral DNA or RNA replicates and begins to function Viral components synthesized	Occurs in cytoplasm and or nucleus Host cell activity ceases Viral DNA or RNA replicates and begins to function Viral components synthesized
RELEASE FROM HOST CELL	Cell lyses when viral enzymes weaken it	Some cells lyse; enveloped viruses bud through host cell membrane
CELL DESTRUCTION	Immediate; some delayed	Immediate; some delayed

envelope fuses with the plasma membrane and releases the nucleocapsid into the cytoplasm. In other cases, the virion attaches to a small outfolding of the plasma membrane, and the cell then enfolds the virion within a vesicle and brings it into the cytoplasm like a piece of food during phagocytosis. FIGURE 12.12 illustrates both entry mechanisms. Uncoating takes place after the nucleocapsid has entered the cytoplasm. In this process, the protein coat is separated from the nucleic acid.

Now the process diverges once again because some animal viruses contain DNA, while some contain RNA. The DNA of a DNA virus supplies the genetic codes for enzymes that synthesize viral parts from available building blocks. A number of DNA viruses, such as poxviruses, replicate entirely in the host cell cytoplasm. Other DNA viruses employ a division of labor: DNA genomes are synthesized in the host cell nucleus, and capsid proteins are produced in the cytoplasm (FIGURE 12.13). The proteins then migrate to the nucleus and join with the nucleic acid molecules for assembly. Adenoviruses and herpesviruses follow this pattern.

RNA viruses follow a slightly different pattern. The RNA can act as a messenger RNA molecule (Chapter 6) and immediately begin supplying the codes for protein synthesis. Such a virus is said to have "sense"; it is called a **positive-stranded RNA virus**, or **sense virus**. In other RNA viruses, however, the RNA is used as a template to synthesize a complementary strand of RNA. The latter then is used as a messenger RNA molecule for protein synthesis. The original RNA strand is said to have "antisense," and the virus is therefore an **antisense virus** or a **negative-stranded RNA virus**. Usually the enzyme RNA polymerase is present in the virus to synthesize the complementary strand. Measles viruses are negative-stranded viruses, whereas polioviruses are positive-stranded viruses.

One RNA virus called the **retrovirus** has a particularly interesting method of replication. Retroviruses carry their own enzyme, called **reverse transcriptase**. The enzyme uses the viral RNA as a template to synthesize single-stranded DNA (the terms *reverse transcriptase* and *retrovirus* are derived from this reversal of the usual biochemistry). Once formed, the DNA serves as a template to form a complemen-

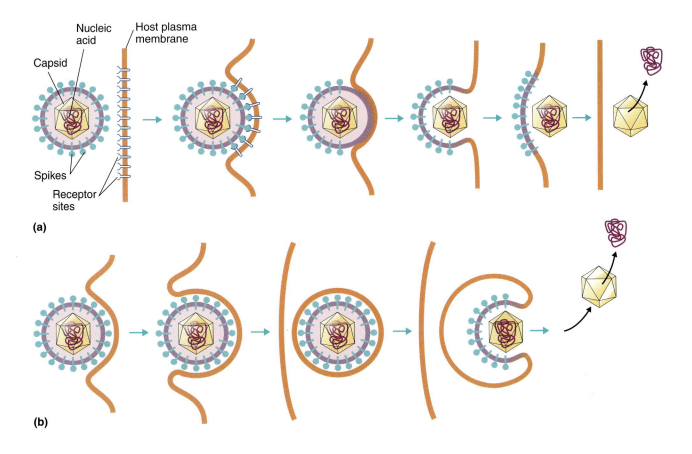

FIGURE 12.12

The Entry of Animal Viruses into Their Host Cells

Animal viruses enter their host cells by two major methods. (a) In the first method, an enveloped virus contacts the plasma membrane, and the spikes interact with receptor sites on the membrane surface. This action is highly specific for the viral spikes and receptor sites. The envelope fuses with the plasma membrane, and the nucleocapsid passes into the host cell's cytoplasm. Then the nucleic acid is released (uncoated). (b) In the second method, a specific interaction between spikes and receptor sites takes place, and a vesicle forms around the virus. The vesicle pinches off into the cytoplasm, then the envelope fuses with the vesicle membrane. This liberates the nucleocapsid into the cytoplasm, where the nucleic acid is uncoated.

tary DNA strand. The viral RNA is then destroyed, and the two DNA strands twist around each other to form a double helix. The DNA now migrates to the cell nucleus and integrates into one of the host cell's chromosomes, where it is known as a **provirus**, as shown in **FIGURE 12.14**. From this position, the DNA encodes new retroviruses. The process we have described here applies to certain leukemia viruses and to the human immunodeficiency virus (HIV), which causes AIDS (Chapter 14).

The final steps of viral replication may include the acquisition of an envelope. In this step, envelope proteins are synthesized and incorporated into a nuclear or cytoplasmic membrane, or the plasma membrane. Then, the virus pushes through the membrane, forcing a portion of the membrane ahead of it and around it, resulting in an envelope. This process, called **budding**, need not necessarily kill the cell during the virus's exit. The same cannot be said for unenveloped viruses,

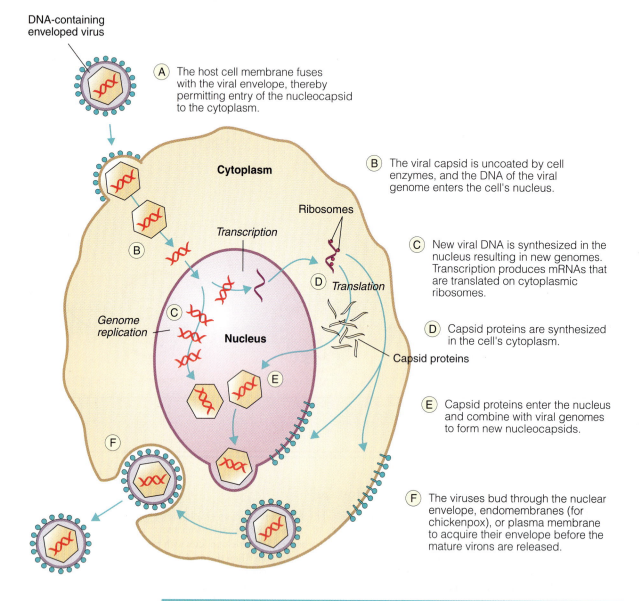

DNA-containing
enveloped virus

A The host cell membrane fuses
with the viral envelope, thereby
permitting entry of the nucleocapsid
to the cytoplasm.

Cytoplasm

Ribosomes

Transcription

B The viral capsid is uncoated by cell
enzymes, and the DNA of the viral
genome enters the cell's nucleus.

B

D *Translation*

C New viral DNA is synthesized in the
nucleus resulting in new genomes.
Transcription produces mRNAs that
are translated on cytoplasmic
ribosomes.

*Genome
replication*

C

Nucleus

D Capsid proteins are synthesized
in the cell's cytoplasm.

Capsid proteins

E

E Capsid proteins enter the nucleus
and combine with viral genomes
to form new nucleocapsids.

F

F The viruses bud through the nuclear
envelope, endomembranes (for
chickenpox), or plasma membrane
to acquire their envelope before the
mature virons are released.

FIGURE 12.13

Replication of a DNA Animal Virus

The virus illustrated here is a herpesvirus (such as one that might cause chickenpox), and the host cell is from human skin.

however. They leave the cell when the cell membrane ruptures, a process that generally leads to cell death.

SOME ANIMAL VIRUSES CAN EXIST AS PROVIRUSES

In the replication cycles, infection need not result in new viral particles or cell lysis. Rather, the virus may integrate its DNA or its RNA (via DNA) into a chromosome of the cell (as described in the previous paragraphs). When bacteriophages are involved, the phage DNA in the lysogenic state in a bacterial cell is called a **prophage**; when an animal virus (such as a retrovirus) is involved, the viral DNA is known as a **provirus**.

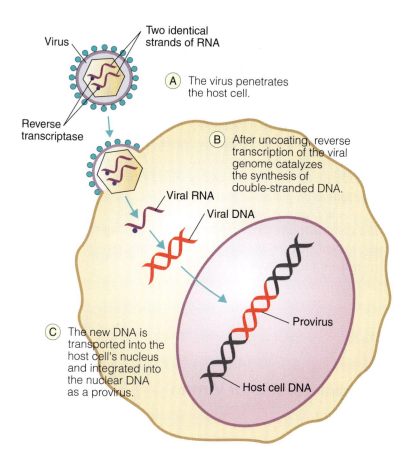

FIGURE 12.14

The Formation of a Provirus by HIV

In both cases, it appears that the viral genome is encoding a repressor protein that prevents activation of the genes necessary for replication. The virus, at least temporarily, is in a latent state.

Latency may have several implications. Proviruses, for example, are immune to body defenses since the body's antibodies cannot reach them (antibodies do not penetrate into cells). Moreover, the virus is propagated each time the cell's chromosome is reproduced, such as during mitosis in animal cells. The provirus, like the prophage, can confer new properties on the infected cell. A case in point is *Clostridium botulinum* (Chapter 9) and *Coryrebacterium diphteriae* (Chapter 8) bacteria whose lethal toxins are encoded by a prophage. In the case of HIV, the body's T lymphocytes harbor the virus as a provirus. Such an individual has **HIV infection**.

Another bacterial phenomenon is **specialized transduction**. In this process, a fragment of DNA from one cell is transferred to a second cell in combination with bacteriophage DNA (Chapter 7). A final implication involves cancer. As we shall see later in this chapter, some cancers may develop when certain animal viruses enter a cell and assume a provirus relationship with that cell. The proteins encoded by the virus often bring about the profound changes associated with this dreaded condition.

Latency:
a period during which viral replication is repressed in a host cell.

To this point . . .

We have studied the replication of viruses using the bacteriophage and its bacterial host as a model. The process begins with a union between phage and bacterium, followed by penetration of the viral DNA to the bacterial cytoplasm. Synthesis, maturation, and release of new virions are the final stages. Using the model as a basis, we noted how animal viruses differ in their replication mode, especially with respect to attachment and penetration. Differences are observed also in DNA and RNA viruses. The discussion concluded with the concept of lysogeny, where viruses remain in their host cells for long time periods as prophages or proviruses.

Our survey of viral characteristics continues as we see how viruses are classified and how viral diseases are detected. Inhibition of viruses using drugs inside the body will be surveyed, and we shall mention the types of viral vaccines. We also will see how viruses can be inactivated outside the body.

12.4

Other Characteristics of Viruses

Like all other microorganisms, viruses have characteristics that set them apart and help virologists understand their activities and deal with them effectively. In this section, we shall examine some of these characteristics.

NOMENCLATURE AND CLASSIFICATION DO NOT USE CONVENTIONAL TAXONOMIC GROUPS

A widely accepted classification system for viruses has not yet been devised, in part because sufficient data are not yet available to determine how different viruses relate to one another. Viruses therefore lack formal names (a fact that does not seem to bother students). Instead, they have acquired their names from several sources. Examples are the measles virus (after the disease), the adenovirus (after the adenoids, where it is commonly located), the Coxsackie virus (after Coxsackie, New York, where it was originally isolated), and the Epstein-Barr virus (after researchers who studied it). The situation is reminiscent of the late 1800s, when different bacteria were called the "tubercle bacillus," or the "diphtheria bacillus," or the "cholera bacillus."

This is not to say, however, that no classification system exists for viruses. Indeed, one classification scheme is loosely based on the body tissues affected by the virus. Though inexact, this scheme places viruses into four convenient groups, depending on whether they replicate in the respiratory, skin, visceral, or nervous tissue (TABLE 12.2). We shall use this scheme in Chapters 13 and 14.

A second classification system is in the process of evolving. At periodic meetings of the International Committee on Taxonomy of Viruses, a revised and updated viral classification scheme is presented. At this writing, no orders, divisions, or kingdom have been established for viruses, but virologists have prepared a working document in which a viral species is defined as a group of viruses sharing the same genetic information and ecological niche. Names have not yet been established for species,

TABLE 12.2

Classification of Human Viral Diseases by Tissue Affected

GROUP	TISSUES AFFECTED	IMPORTANT DISEASES
Pneumotropic	Respiratory system	Influenza, respiratory syncytial disease, rhinovirus infection, SARS
Dermotropic	Skin and subcutaneous tissues	Chickenpox, herpes simplex, measles, mumps, smallpox, molluscum contagiosum, rubella
Viscerotropic	Blood and visceral organs	Yellow fever, dengue fever, infectious mononucleosis, cytomegalovirus disease, viral fevers, viral gastroenteritis, hepatitis A, hepatitis B, AIDS
Neurotropic	Central nervous system	Rabies, polio, West Nile fever (encephalitis)

but the viruses have been categorized into genera; each genus name ends with the suffix *-virus* (e.g., *Herpesvirus*). The genera then have been organized into 73 families, each ending with *-viridae* (e.g., Herpesviridae). A selection of viral families affecting humans, together with some of their characteristics, is presented in **TABLE 12.3**. Note that due to the rapid changes taking place in viral taxonomy, some new names may be in use by the time you read this.

CULTIVATION AND DETECTION OF VIRUSES ARE CRITICAL TO VIRUS IDENTIFICATION

The methods used to detect viruses are more involved and considerably more time consuming than for bacteria and other microorganisms. Plant and animal tissues are difficult and expensive to maintain, and human pathogenic viruses often replicate only in human host cells, which causes additional complications. By contrast, bacteriophages are cultivated easily in bacterial cultures, and bacteriophages have been used as models for studying viral characteristics.

Before the advent of cell culture, one common method of cultivating viruses was to inoculate them into fertilized (embryonated) chicken eggs. A hole is drilled in the shell of the egg, and a suspension of viral material is introduced. Because different viruses replicate in different membranes or parts of the chick embryo, virologists must anticipate which virus is present (**FIGURE 12.15**). Today, only the influenza viruses are cultivated by this method, which produces high yields of viruses for vaccine production.

The most common method of cultivating and detecting viruses is to inoculate suspensions of material in **cell cultures**. To prepare the culture, cells are separated from a tissue with enzymes and suspended in a solution of nutrients, growth factors, pH buffers, and salts. The cells adhere to the wall of the container and reproduce to form a single layer, or **monolayer**. When viruses replicate in these cells, often a noticeable deterioration occurs. This is called a **cytopathic effect (CPE)**, which can be seen with a light or phase-contrast microscope (**TABLE 12.4**). The method can be used in the clinical laboratory where viruses are being identified from samples taken from patients.

TABLE 12.3

Major Families of Animal Viruses and Their Characteristics

FAMILY	STRAND TYPE*	CAPSID SYMMETRY	ENVELOPE OR NAKED VIRION	DIAMETER (NM)	DISEASE EXAMPLES
DNA Viruses					
Poxviridae	Double	Complex	Envelope	170–300	Smallpox, monkeypox
Herpesviridae	Double	Icosahedral	Envelope	150–200	Cold sores, genital herpes, chickenpox, shingles, infectious mononucleosis
Adenoviridae	Double	Icosahedral	Naked	70–90	Common cold, viral meningitis
Papovaviridae	Double	Icosahedral	Naked	45–55	Warts, genital warts, cervical cancer
Hepadnaviridae	Double (w/RNA intermediate)	Icosahedral	Envelope	42	Hepatitis B
Parvoviridae	Single	Icosahedral	Naked	18–26	Fifth disease
RNA Viruses					
Reoviridae	Double	Icosahedral	Naked	60–80	Gastroenteritis
Picornaviridae	Single (+)	Icosahedral	Naked	28–30	Polio, some colds, hepatitis A
Caliciviridae	Single (+)	Icosahedral	Naked	35–40	Gastroenteritis
Togaviridae	Single (+)	Icosahedral	Envelope	60–70	Rubella, encephalitis
Flaviviridae	Single (+)	Icosahedral	Envelope	40–50	Yellow fever, dengue fever, hepatitis C, West Nile fever encephalitis
Coronaviridae	Single (+)	Helical	Envelope	80–160	SARS
Filoviridae	Single (−)	Helical	Envelope	80–10,000	Ebola and Marburg hemorrhagic fevers
Bunyaviridae	Single (−)	Helical	Envelope	90–120	Hantavirus pulmonary syndrome
Orthomyxoviridae	Single (−)	Helical	Envelope	90–120	Influenza
Paramyxoviridae	Single (−)	Helical	Envelope	150–300	Mumps, measles
Rhabdoviridae	Single (−)	Helical	Envelope	70–380	Rabies
Arenaviridae	Single (−)	Helical	Envelope	50–300	Lassa fever
Retroviridae	Single (+) (w/DNA intermediate)	Icosahedral	Envelope	80–130	Human adult T-cell leukemia, AIDS

(+) = positive-stranded; (−) = negative-stranded

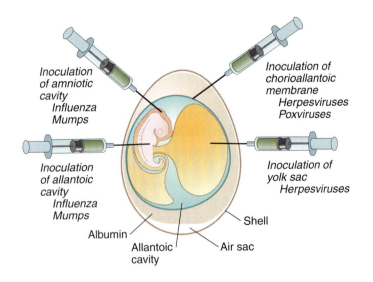

FIGURE 12.15

Viral Cultivation in Embryonated Chicken Egg

Inoculations to specific sites of the fertilized chicken egg is indicated for specific viruses.

TABLE 12.4

Examples of Virus Cytopathic Effects

VIRUS	CYTOPATHIC EFFECT
Changes in cell structure	
Picornaviridae	Shrinking of cell nucleus
Papovaviridae	Cytoplasmic vacuoles
Paramyxoviridae, Coronaviridae	Cell fusion
Herpesviridae	Chromosome breakage
Herpesviridae, Adenoviridae, Picornaviridae, Rhabdoviridae	Rounding and detachment of cells from culture
Cell inclusions	
Adenoviridae	Virions in nucleus
Rhabdoviridae	Virions in cytoplasm (Negri bodies)
Poxviridae	"Viral factories" in cytoplasm
Herpesviridae	Nuclear granules

An indirect method for detecting viruses is to search for viral antibodies in a patient's serum. This can be done by combining serum (the blood's fluid portion) with known viruses. In some **serological tests**, the viruses are attached to carrier particles, and the reaction results in a visible clumping. Certain viruses—such as those of influenza, measles, and mumps—have the ability to agglutinate (clump) red blood cells. This phenomenon, called **hemagglutination (HA)**, can be used for detection purposes. In addition, it is possible to detect antibodies against certain viruses because antibodies react with viruses and tie up the reaction sites that otherwise would bind to red blood cells, thereby inhibiting hemagglutination. Thus, a laboratory test for antibodies can be performed by combining the patient's serum with known viruses and red blood cells. Hemagglutination indicates that antibodies are absent from the serum, but the **hemagglutination-inhibition (HAI) test** points to the presence of serum antibodies (FIGURE 12.16). Such a finding implies that the patient has been exposed to the viruses. Chapter 20 explores gene probes and other contemporary approaches to viral detection.

Serological test:
a test that detects the presence of antibodies or antigens.

Viruses leave signs of their presence in the infected cells. For example, the brain cells of a rabid animal contain cytoplasmic granules called **Negri bodies**, and cells from herpes simplex patients have nuclear granules known as **Lipschütz bodies**. Moreover, a series of cellular or tissue modifications may signal the presence of viruses. Infectious mononucleosis, for example, is characterized by large numbers of lymphocytes with foamy, highly vacuolated cytoplasm.

In some cases, viral infections leave their mark on the infected individuals. Measles is accompanied by **Koplik spots**, a series of bright red patches with white pimple-like centers on the lateral mouth surfaces. Swollen salivary glands and teardrop-like skin lesions are associated with mumps and chickenpox, respectively. Blood clots are associated with certain viruses, as MicroFocus 12.4 indicates.

The most obvious method for detecting viruses is by direct observation with the electron microscope. In this procedure, tissue samples may be examined directly or

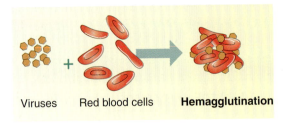

(a)

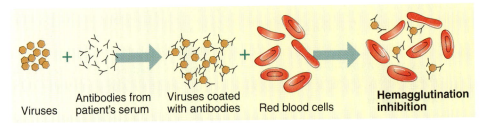

(b)

FIGURE 12.16

The Hemagglutination-Inhibition Test

(a) Viruses of certain diseases, such as measles and mumps, are able to agglutinate (clump) red blood cells. (b) In the hemagglutination-inhibition (HAI) test, known viruses are combined with serum from a patient. If the serum contains antibodies for that virus, the antibodies coat the virus, and when the viruses are then combined with red blood cells, no agglutination will take place. Since the nature of the virus is known, the type of antibody present can be determined.

after viral cultivation. Virologists often are able to identify unknown viruses by comparison to known viruses.

Bacteriophages may be detected by the formation of plaques. A **plaque** is a clear zone on a cloudy "lawn" of bacteria where bacteriophages have destroyed the cells. Technologists first cultivate the bacteria on an agar surface, then add the phages by spraying or other methods (**FIGURE 12.17**). If the phages are specific for that particular bacterium, they infect and replicate in the cells, thereby destroying them and forming plaques. This method also can be reversed, so that a known phage is used to detect an unknown bacterium. Epidemiological surveys of several bacterial diseases such as staphylococcal food poisoning are aided by this procedure, often called **phage typing**. An example was shown in Chapter 9 for *Staphylococcus aureus*.

Phage typing:
a method of identifying an unknown bacterium by its reaction with a known bacteriophage.

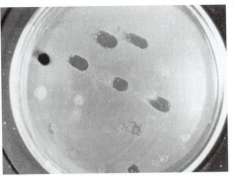

FIGURE 12.17

The Process of Plaque Formation

Susceptible bacteria are inoculated into plates of nutrient medium. Bacterial viruses are then sprayed onto the surface, and the plate is incubated. As viruses replicate in the bacteria, they destroy the cells and leave clear "moth-eaten" areas containing no bacteria. These areas are the plaques shown on the plate in the photograph.

Plaque assays also have been used in cell culture using a variety of animal viruses. MicroInquiry 12 uses this plaque assay to monitor intracellular and extracellular virus production.

The difficulty in detecting viruses has a bearing on relating a particular virus to a particular disease. In the classical sense, Koch's postulates cannot be applied to a viral disease because the viruses cannot be cultivated in pure culture. To circumvent this problem, Thomas M. Rivers in 1937 expanded Koch's postulates to include viruses as follows: Filtrates of the infectious material shown not to contain bacteria or other cultivatable organisms must produce the disease or its counterpart; or, the filtrates must produce specific antibodies in appropriate animals. This concept has come to be known as **Rivers's postulates**. MicroFocus 12.5 illustrates how these postulates have been used recently.

SYNTHETIC DRUGS CAN INHIBIT VIRUS REPLICATION

During periods of disease, the human body attempts to rid itself of viruses mainly by phagocytosis, neutralization with antibodies, and interactions with T lymphocytes. **Antibodies** are immune system protein molecules that react with viruses. Usually one antibody molecule unites specifically with a single virion, thereby preventing the virion from binding to its host cell. Antibodies also clump viruses into large

---

MicroFocus 12.4

ESCAPE

European rabbits had been a scourge in Australia ever since English settlers released a dozen of them in 1840. True to their reputation, the new arrivals bred prolifically, and by 1995, over 300 million rabbits dotted the Australian landscape.

But Australian scientists had a plan: They quarantined several thousand rabbits on Wardang Island off Australia's southern coast and infected them with calciviruses. (Calciviruses are a group of RNA viruses with cup-like projections on their surfaces. They quickly kill rabbits by causing blood clots in vital organs.) To reduce the spread of either viruses or rabbits, the scientists inspected the fences daily and tested the blood of wild rabbits outside the pens for evidence of calciviruses. If the experiment was successful on the island, they would consider releasing viruses on the Australian mainland.

Unfortunately, the viruses escaped. By September 1995, scientists were finding dead rabbits outside the pens and then on the mainland. The dead rabbits had evidence of calcivirus disease. There was no question: The virus was out of quarantine. (Research conducted in 1997 indicated that various insects carry the virus; in retrospect, that may have been the escape route.)

Now the scientists had some difficult days ahead. They would have to wait to learn whether the virus interrupts the breeding success of survivors; also, they would have to go without knowing how long the virus lasts in the environment. The escape was final and irrevocable. The genie was out of the bottle. They would have to go ahead with the general release.

Virologists and veterinarians acted swiftly. With government approval, they released infected rabbits at hundreds of sites on the mainland and hoped for the best. Soon, dead rabbits were everywhere. Within two years, the rabbit population was reduced by 95 percent in some areas; and many plant and animal species, preyed on by the rabbits, were rebounding after remaining unseen for generations. At this writing, the native wildlife remain unaffected, except that eagles and other predators of rabbits have declined in numbers. At first, scientists were discouraged at having to act in haste, but in this instance, the escape apparently had a beneficial twist. At least, so far.

MicroInquiry 12

THE ONE-STEP GROWTH CYCLE

In Chapter 4, we learned about the bacterial growth curve, which consisted of four phases (FIGURE A). The curve began with a short lag phase, when the cells are adapting to the growth medium and preparing for binary fission. This is followed by the log phase, a period of binary fission and the exponential increase in cell number. As nutrients become limiting, the cell population enters stationary phase, where the number of dividing cells equals the number of dying cells. Finally, there is a decline or death phase where the majority of cells are dying (not shown on graph).

In the research laboratory, we can follow the replication of animal viruses in a similar way by generating a one-step growth curve. Realize that we are not really looking at growth, but the replication and increase in number of virus particles. There are several periods associated with a virus growth cycle that you should remember because you will need to identify the periods in the growth curve. The **eclipse period** is the time when no virions can be detected inside cells and the **latent phase** is the time during which no extracellular virions can be detected. Also, remember the **burst size**, which is the number of virions released per infected cell.

To generate our growth curve, we will start by inoculating our viruses onto a sus-ceptible cell culture. There are 100,000 (10^5) cells in each of 10 cultures. We will add ten times the number of viruses (10^6) in a small volume of liquid to each culture to make sure all the cells will be infected rapidly. After 60 minutes incubation, we wash each culture to remove any viruses that did not attach to the cells and add fresh cell growth medium. Then, at 0 hour and every four hours after infection, we remove one culture, pour off growth medium, and lyse the cells. The virus titer in the growth medium and the lysed cells are determined. Since we cannot see viruses, we will measure any viruses in the growth medium or in the lysed cells using a plaque assay. We will

---

Phagocytosis:
a defensive measure of the body in which white blood cells engulf and destroy microorganisms.

masses efficiently removed by **phagocytosis**. In addition, viruses activate the complement system, a group of substances that encourage phagocytosis (Chapter 19). Finally, an important source of defense is the body's **T lymphocytes** (also called T cells). When activated, T lymphocytes migrate to virus-infected cells and interact directly with the cells, destroying the cells and the viruses within them (Chapter 19).

Normal body defenses against viral disease cannot be supplemented by common antibiotics because viruses lack the structures and metabolic machinery with which antibiotics interfere. For example, penicillin is useless for inhibiting viruses because viruses have no cell wall. However, several drugs have been proven effective against viruses. One drug, called **amantadine** (Symmetrel), prevents the attachment of influenza viruses to the host cell surface. Another drug, **vidarabine** (Vira-A), is used for treating herpes zoster (shingles) and encephalitis (brain disease) due to the herpes simplex virus. A third drug, **acyclovir** (Zovirax), has been available since 1985 as a topical ointment for genital herpes and, more recently, for chickenpox.

Acyclovir is typical of antimicrobial agents called **base analogs** (Chapter 6). These substances resemble nitrogenous bases and are erroneously incorporated into viral DNA. Other base analogs called **idoxuridine (IDU)** and **trifluridine** are taken up by herpesviruses in place of thymine, and the resulting genome cannot replicate itself. The base analog **azidothymidine (AZT)** has been used since 1987 to treat HIV infection and AIDS, and two other analogs called **dideoxyinosine (ddI)** and **dideoxycytidine (ddC)** are approved, also for treating AIDS. Another drug called **foscarnet** is used in patients having retinal disease (retinitis) caused by the cytomegalovirus (CMV). **Ganciclovir** also is used against CMV infection.

Another class of antiviral drug is called **reverse transcriptase inhibitors**. These drugs bind directly to reverse transcriptase and inhibit its activity, thereby preventing

assume that one plaque-forming unit (PFU) is equal to one virion.

The curve below plots the results from our growth experiment (**FIGURE B**).

Answer the following questions based on the one-step growth curve. Answers can be found in Appendix E.

12.1a. How long is the eclipse period for this viral infection? Explain what is happening during this period.

12.1b. How long is the latent period for this infection? Explain what is happening during this period.

12.1c. What is the burst size?

12.1d. Explain why the growth curve shows some intracellular PFUs between 4 and 11 hours.

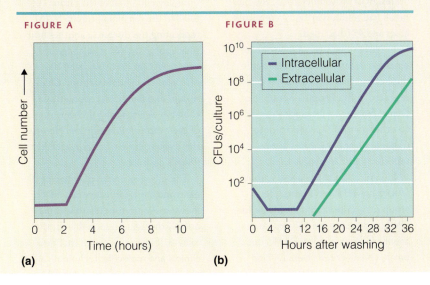

FIGURE A

FIGURE B

(a)

(b)

the synthesis of DNA in retroviruses. Used primarily against HIV, these medications include **nevirapine** and **delavirdine**. Still another class is **protease inhibitors**. These drugs react with protease, the enzyme that trims viral proteins down to working size for the construction of the capsid. Examples are **saquinavir** (Invirase), **indinavir** (Crixivan), and **ritonavir** (Norvir). The first fusion inhibitor **enfuvirtide** (Fuzeon) that blocks fusion of the virus with a host cell was approved in 2003. The protease and fusion inhibitors are used in patients with HIV infection and AIDS.

A final class of antiviral drug is the **neuraminidase inhibitors**. Neuraminidase is an enzyme in the spike of the influenza virus. It helps the virus infect cells and, after replication, helps the viruses spread to other cells (Chapter 13). The neuraminidase inhibitor drugs **zanamivir** (Relenza) and **oseltamivir** (Tamiflu) block the action of neuraminidase, preventing release of new virions—and thereby limiting disease spread in the body. The development of all these drugs is based on biochemical knowledge of virus function coupled with human ingenuity. Indeed, medical science one day may provide an antiviral drug to deal with obesity (MicroFocus 12.6).

INTERFERON PUTS CELLS IN AN ANTIVIRAL STATE

Interferon (**IFN**) represents one of the most optimistic approaches to inhibiting viruses. First identified in 1957 by Alick Isaacs and Jean Lindenmann, interferon is not a single substance but a group of over 20 substances designated alpha, beta, and gamma interferons. Each group has several members, and all appear to be proteins. IFN-alpha and IFN-beta are produced by viral infection of almost any type, be it DNA or RNA. The IFNs trigger a nonspecific reaction that protects against the stimulating virus, as well as many other viruses. In addition, some IFNs have anticancer

Interferon:
a group of cellular proteins that provide protection against viruses.

MicroFocus 12.5

DETECTING EMERGING VIRUSES

Recently, new diseases have emerged in the United States and globally. It has been critical to quickly identify the causative agents of these diseases so appropriate prevention measures and curative treatments can begin before large numbers of the population become ill or die.

In 1999, the New York City Health Department began hearing of citizens with a fatal arboviral (mosquitoborne) encephalitis. At first, health authorities thought it might be St. Louis encephalitis (SLE), which does occur in the United States during extremely dry summers as was occurring in New York City. Serological tests also pointed to SLE. However, birds and horses also were dying. Since other arboviruses cross-react with SLE in serological tests, alternate testing procedures were done. These tests, which analyzed the gene sequence of the virus, identified the agent as a West Nile–like virus, not SLE. As such, West Nile Virus (WNV) represented a new virus to the Western

Hemisphere. Since 1999, WNV has spread across the continental United States. As of November 2003, the outbreak had resulted in 12,775 reported cases and 491 had died.

In November 2002, a new disease emerged in Asia. Called severe acute respiratory syndrome (SARS), it first was reported in the People's Republic of China. From there, it quickly spread to other parts of Asia and Toronto, Canada. Over 8,000 cases were reported and 774 people died.

Identification of the agent causing SARS followed Rivers's postulates, a group of six criteria that should be fulfilled to identify a specific virus as the agent of a disease. For the SARS outbreak, several criteria were fulfilled early, including (1) isolating of the virus from people with the disease, (2) cultivating of the virus in cell culture, and (3) proving the virus was filterable. Detection of (4) a specific immune response to the virus also had been accomplished. These four criteria pointed to a previ-

ously unknown coronavirus (SARS–CoV) in SARS patients. However, it was not until July 2003 that investigators fulfilled the remaining criteria by (5) replicating the disease in another host (macaques) and (6) reisolating the virus from that host. Two bacteria, *Mycoplasma pneumoniae* and *Chlamydia,* also had been proposed as a causative agent during the early stages of the outbreak investigation. However, these agents have not been satisfied by Rivers's postulates.

These two examples of emerging diseases and the deaths they can inflict show how important it is to identify quickly the agent responsible for the disease. It is assuring to see that new clinical and molecular techniques available to investigators are making these identifications possible in very short periods. Incidentally, both viruses and their associated diseases will be described in more detail in Chapter 13.

properties. However, human IFN is the only one that will work in humans. Mouse, chicken, dog, or other animal IFNs are ineffective.

Unlike antibodies, IFN-alpha and IFN-beta do not interact directly with viruses, but with the cells they protect. For this reason, they have a broader inhibitory effect. These IFNs are produced when a virion releases its genome into the cell. High concentrations of double-strand RNA in the infected cells induces the cells to synthesize and secrete IFNs (**FIGURE 12.18**). These bind to specific receptor sites on the surfaces of adjacent cells and trigger the production of several proteins within those cells. Such cells are now in an **antiviral state**. The proteins inhibit viral replication by methods not completely understood, although it appears that cells in the antiviral state stop protein synthesis when infected. IFN-gamma appears to mobilize natural killer cells, which attack tumor cells (Chapter 21).

The inability to obtain sufficient concentrations of interferon stifled research for many years. (One approach was to inoculate huge batches of white blood cells with harmless viruses.) A breakthrough came in 1980 when Swiss and Japanese scientists deciphered the genetic code for interferon, and spliced the genes for interferon synthesis into *E. coli* plasmids (Chapter 7). Experiments showed that IFN from bacterial factories would reduce hepatitis symptoms, diminish the spread of herpes zoster, and shrink certain cancers. In 1984, a Swiss biotechnology firm

MicroFocus 12.6

OVERWEIGHT? TAKE AN ANTIBIOTIC

It's far out, to be sure, but as of 1997, University of Wisconsin researchers have linked a virus to obesity. And, they maintain, it may be possible one day to eliminate the virus and slim down quickly by taking an antiviral antibiotic.

The road to this startling conclusion began in 1990 with a veterinarian's chance remark that chickens infected with adenoviruses gain considerable weight. The conversation took place in Bombay, India, where Nikhil V. Dhurandhar, a nutritionist, was on his way to Wisconsin to do research. The veterinarian's observation was intriguing, and Dhurandhar resolved to pursue it.

At the medical school in Wisconsin, the fledgling researcher obtained a flock of chickens and a culture of adenoviruses (respiratory viruses often involved in the common cold). He soon confirmed that in chickens, the viruses are pathogenic and bring on rapid mortality—but not before the chickens become unusually obese, just as the vet had said.

But did it work that way with people? Dhurandhar solicited volunteers and got responses from 45 lean individuals and 154 obese people, each weighing over 250 pounds. The investigator was working with adenovirus strain 36 (AD-36), and antibodies against this virus would prove that it was currently in the body or that it had once been there. Dhurandhar took blood samples and analyzed them: 5 percent of the lean individuals had AD-36 antibodies, but 30 percent of the obese people had the telltale antibodies.

Dhurandhar, now a scientist at Wayne State University, also looked at a pair of identical twins who had about the same weight before they went to college. One twin gained considerable weight while the other did not. Blood tests indicated the twin that gained weight had the AD-36 antibodies.

Despite Dr. Dhurandhar's determination to prove his theory, most experts are not convinced. Most agree that lifestyle and eating habits are what counts. Overeating is a complex disorder though and Dhurandhar believes the virus may be important in some cases of obesity.

In the early 1980s, few people were willing to believe that gastric ulcers are related to a bacterium (Chapter 9), but the relationship is now an accepted tenet of medicine. Linking obesity to a virus may seem like a stretch. Still . . .

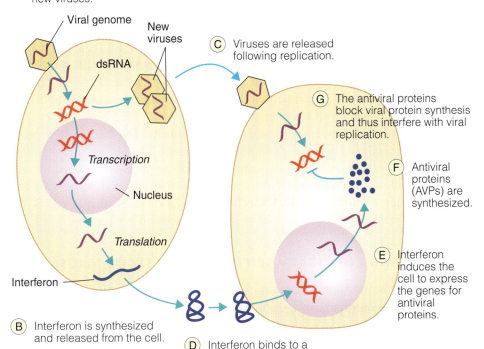

(A) Viral RNA stimulates the host to synthesize interferon while also directing replication of new viruses.

(C) Viruses are released following replication.

(G) The antiviral proteins block viral protein synthesis and thus interfere with viral replication.

(F) Antiviral proteins (AVPs) are synthesized.

(E) Interferon induces the cell to express the genes for antiviral proteins.

(B) Interferon is synthesized and released from the cell.

(D) Interferon binds to a neighboring cell.

Viral genome
New viruses
dsRNA
Transcription
Nucleus
Translation
Interferon

FIGURE 12.18

The Production and Activity of Interferon

Left: A host cell produces interferon following viral infection. Viruses replicate in the same cell. Right: The interferon reacts with receptors at the surface of a neighboring cell and induces the cell to produce antiviral proteins and the cell enters an activated state. The proteins interfere with viral replication in the cell, possibly by binding to messenger RNA molecules.

began marketing IFN-alpha using the trade name Intron. In 1986, the US Food and Drug Administration approved the sale of IFN-alpha for use against a form of leukemia; in 1988, it approved its use against genital warts; and in 1992, against chronic hepatitis B.

VIRAL VACCINES OFFER THE BEST PROTECTION AGAINST INFECTION

Despite the optimism inspired by the new drugs, the major approach to viral disease continues to be prevention through public education and the use of vaccines. Vaccines stimulate antibody production by the immune system and thereby induce specific defense. They also can stimulate the production of cytotoxic T cells (Chapter 19).

Several types of viral vaccines are currently in use. One type contains **inactivated viruses**, which are viruses treated with a physical agent (such as mild heat) or a chemical agent (such as formaldehyde). This treatment denatures the genome of the virus, thereby preventing replication, but it does not severely affect the capsid, which remains intact and stimulates antibody production. Opponents of inactivated viral vaccines point out that treatment may not reach all viruses, and that untreated viruses may remain active and cause disease. They also note that the vaccine must be given by injection to ensure an immune response, and that the immune response is low relative to that elicited by the second type of vaccine.

The second type of vaccine contains **attenuated viruses**, which are viruses that continue replicating in body cells but at an extremely low rate. Attenuated viruses stimulate the immune system for a longer period of time than inactivated viruses, and thus the immune response is higher. They are obtained by transferring viruses from culture to culture for a period of months or years, until a variant emerges with a greatly reduced replication rate. Opponents of attenuated viral vaccines suggest that the viruses may revert to their original form and cause disease, and that the viruses may infect cells other than the usual host cells or activate proviruses already in host cells. Moreover, attenuated viruses have been known to cause disease in rare cases.

Inactivated and attenuated viruses are sometimes called "dead" and "live" viruses, respectively, and the vaccines are therefore known by these terms (e.g., live measles vaccine). The words "dead" and "alive" are misleading, however, because they signify the life status of viruses, which remains uncertain. The **Salk vaccine** for polio typifies a vaccine made with inactivated viruses, while the **Sabin vaccine** for polio represents a vaccine made with attenuated viruses.

Another type of vaccine is composed of **viral subunits**. These are protein molecules, usually produced by genetic engineering methods. The vaccine for hepatitis B is typical. This vaccine (Chapter 14) contains capsid proteins synthesized by yeast cells genetically altered with the DNA from hepatitis B viruses. Researchers are currently pursuing a number of **DNA vaccines** for viral diseases such as AIDS. The spectrum of viral and other microbial vaccines is discussed in Chapter 20.

PHYSICAL AND CHEMICAL AGENTS CAN INACTIVATE VIRUSES

Viruses may be inactivated by many of the physical and chemical agents routinely used for other microorganisms. Among the physical agents are heat and ultraviolet light. Heat alters the structure of viral proteins and nucleic acids, causing them to unfold and denature. Sterilization temperatures reached in the **autoclave** (Chapter 22) will destroy all viruses, and boiling water for a few minutes will eliminate most viruses (with the notable exception of the hepatitis A virus, which requires a longer time). **Ultraviolet (UV) light** inactivates viruses by stimulating adjacent thymine or cytosine bases on DNA molecules to bind together and form pairs called **dimers** (Chapter 6). The dimers twist the molecule out of shape, and the distorted viral genome cannot replicate. Ultraviolet light is used in the preparation of some vaccines. **X rays** are another type of useful radiation. X rays cause breaks in the sugar-phosphate backbone of the nucleic acid.

Several **chemical agents** can be used outside the body to inactivate viruses, but not inside the body because of their toxic effects on the tissues. Examples are compounds such as chlorine and iodine derivatives; heavy metal compounds, such as mercury and silver derivatives; and phenol derivatives (Chapter 23). These compounds react strongly with protein, thereby altering the viral capsid. **Formaldehyde** is another useful chemical agent because it reacts with free amino groups on adenine, guanine, and cytosine molecules to modify the viral genome and prevent replication. The Salk polio vaccine is prepared with formaldehyde-inactivated viruses. Other valuable chemical agents are lipid solvents, such as ether, chloroform, and detergents, all of which dissolve the lipid in the envelope of viruses. Enzymes directed against viral proteins and nucleic acids also are useful. FIGURE 12.19 summarizes the activity of chemical and physical agents against viruses.

Autoclave:
a device that generates high-temperature steam under pressure for the destruction of microorganisms.

To this point . . .

This section of our study of viruses looked at the schemes used to name and classify viruses. We used the scheme of grouping viruses by their shared properties. Some disease examples also were given that will be described more fully in the next two chapters.

Cultivation and detection techniques to identify viruses were described. This included embryonated chicken eggs and cell culture. Cell culture can be used to detect various cytopathic effects caused by virus replication. Hemagglutination and hemagglutination-inhibition tests are two ways to detect viruses using red blood cells. Formation of plaques, clear areas on growth media, can be used to identify phage and animal virus infections.

A variety of synthetic antiviral drugs has been developed that interfere with viral replication. These drugs inhibit nucleic acid replication, interfere with new capsid biosynthesis, or prevent viral spread from host cells. Interferon represents a group of antiviral substances that are naturally produced by cells when infected. Although some success has come from the antiviral drugs and interferon, the best success in preventing infectious viral diseases has come from vaccines. Both inactivated viruses and attenuated viruses make up such vaccines. A variety of physical and chemical agents can inactivate viruses.

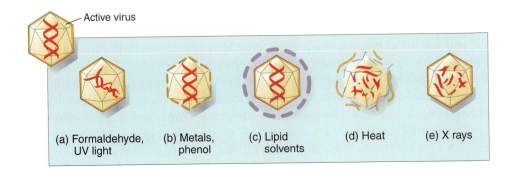

(a) Formaldehyde, UV light   (b) Metals, phenol   (c) Lipid solvents   (d) Heat   (e) X rays

FIGURE 12.19

Methods of Inactivating Viruses

(a) Formaldehyde combines with free amino groups on nucleic acid bases while ultraviolet light binds together thymine molecules in the genome and distorts the nucleic acid. (b) Metals and phenol react with the protein of the viral capsid. (c) Lipid solvents dissolve the envelope in enveloped viruses. (d) Heat denatures proteins of the capsid. (e) X rays break the DNA or RNA sugar–phosphate backbones.

12.5

Viruses and Cancer

Cancer is indiscriminate. It affects humans and animals, young and old, male and female, rich and poor. In the United States, over 450,000 people die of cancer annually, making the disease the second most common cause of death after cardiovascular disease. Worldwide, over 2 million people die of cancer each year.

CANCER IS AN UNCONTROLLED GROWTH OF CELLS

Cancer results from the uncontrolled reproduction of cells through the process of mitosis: The frequency of mitosis is greater for cancer cells than for normal cells. The cells escape controlling factors and as they continue to multiply, a cluster of cells soon forms. Eventually, the cluster yields an abnormal, functionless mass of cells. This mass is called a **tumor**.

Normally, the body will respond to a tumor by surrounding it with a capsule of connective tissue. Such a tumor is designated **benign**. If, however, the cells multiply too rapidly and break out of the capsule and spread to other parts of the body, the tumor is described as **malignant**. The individual now has **cancer**, a reference to the radiating spread of cells, which resembles a crab (the word is derived from the Greek *karkinos*, meaning "crab"). **Oncology**, the study of cancer, is derived from *onkos*, the Greek word for "tumor."

Cancer cells differ from normal cells in three major ways: They grow and undergo mitosis more frequently than normal cells, they stick together less firmly than normal cells, and they undergo dedifferentiation. **Dedifferentiation** means that normal cells revert to an early stage in their development. For example, when ciliated cells of the bronchi become cancer cells, they lose their cilia and dedifferentiate into formless cells that divide as rapidly as early embryonic cells. Moreover, abnormal cells often fail to exhibit **contact inhibition**; that is, they do not adhere tightly to one another, as normal cells do. Thus, they overgrow one another to form a tumor. Sometimes, they **metastasize**, or spread, to other body parts where new tumors begin. Also, they invade and grow in a broad variety of body tissues, since they do not stick to tissue cells. Through all this, the cells are evading or overcoming the natural body defenses centered in the immune system. There appears to be no boundary limiting the growth of a tumor.

How can such a mass of cells bring illness and misery to the body? By their sheer force of numbers, cancer cells invade and erode local tissues, thereby interrupting normal functions and choking organs to death. For example, a tumor in the kidney prevents kidney cells from performing their excretory function, a brain tumor cripples this organ by compressing the nerves and interfering with nerve impulse transmission, and a tumor in the bone marrow may block blood cell production.

In addition, tumor cells rob the body of vital nutrients to satisfy their own growth needs. Thus, the cancer patient will commonly experience weight loss even while maintaining a normal diet. Anemia and a general feeling of tiredness may develop when the tumor cells use up essential minerals for red blood cell production. Moreover, some tumor cells are known to produce hormones identical to those normally produced by the body's endocrine glands, thereby overloading the body with chemical regulators. Some tumors block air passageways; others interfere with the immune system so that microbial diseases take hold. Tumor cells often weaken the body until it fails.

VIRUSES ARE RESPONSIBLE FOR 20 PERCENT OF HUMAN CANCERS

Scientists are uncertain as to what triggers a normal cell to multiply without control (**FIGURE 12.20**). They know that certain chemicals are **carcinogens**, or cancer-causing

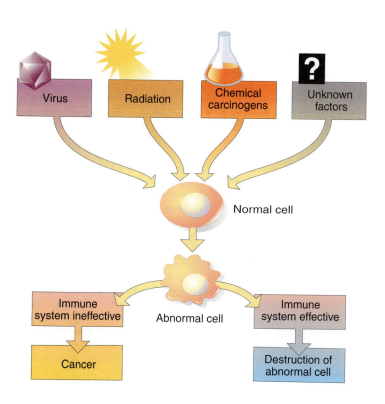

FIGURE 12.20

The Onset of Cancer

Viruses and other factors induce a normal cell to become abnormal. When the immune system is effective, it destroys the abnormal cell, and no cancer develops. However, when the abnormal cell evades the immune system, it develops first into a tumor and then may become a spreading cancer.

substances. The World Health Organization (WHO) estimates that carcinogens may be associated with 60 to 90 percent of all human cancers. Among the known carcinogens are the hydrocarbons found in cigarette smoke, as well as asbestos, nickel, certain pesticides and dyes, and environmental pollutants in high amounts. Physical agents such as UV light and X rays also are believed to be carcinogens.

There is considerable evidence that viruses are also carcinogens. Experiments with animals indicate that some viruses can induce tumor formation. Federal law, however, prohibits these experiments from being repeated with human volunteers, because of the serious consequences. Nevertheless, a number of viruses have been isolated from human cancers, and when these viruses are transferred to animals and tissue cultures, an observable transformation of normal cells to tumor cells takes place. Examples of such **oncogenic viruses** (cancer-causing viruses) are the herpesviruses associated with tumors of the human cervix and the Epstein-Barr virus, which is linked to Burkitt's lymphoma, a tumor of the jaw. A small number of viruses have been discovered that cause tumors in humans (TABLE 12.5). Of special note is cervical cancer, the second most common cancer in women under age 35. The human papilloma virus (HPV) subtypes (family *Papovaviridae*) have a strong correlation with this cancer. In one study, 71 percent of the women developing cervical cancer had HPV present in their Pap smears. Women with HPV 16 were

TABLE 12.5

Human Oncoviruses and Their Affects on Cell Growth

ONCOGENIC VIRUS	HUMAN TUMOR	EFFECT	RESULT
DNA Tumor Viruses			
Papovaviridae Human papilloma virus (HPV)	Cervical and penile cancer	Encodes genes that inactivate cell growth regulatory proteins	Loss of regulatory control leads to cell division and increased mutation rate
Herpesviridae Epstein–Barr virus (EBV)	Burkitt's lymphoma; nasopharyngeal carcinoma	Stimulates cell growth and activates a host cell gene that prevents cell death	Immortalization of B lymphocytes
Herpesvirus 8 (HV8)	Kaposi's sarcoma		Opportunistic infection in AIDS
Hepadnaviridae Hepatitis B virus (HBV)	Hepatocellular carcinoma	Stimulates overproduction of a transcriptional regulator	Liver cancer
RNA Tumor Viruses			
Flaviviridae Hepatitis C virus (HCV)	Hepatocellular carcinoma	Being investigated	Liver cancer
Retroviridae Human T-cell leukemia virus (HTLV-1)	Adult T-cell leukemia	Encodes a protein that activates growth-stimulating gene expression	Oncogenesis
		Integration of provirus near a cellular growth-stimulating gene	Cell division and growth stimulated
(HTLV-2)	Cervical cancer?		Cofactor of human cervical cancer?

more than 100 times as likely to develop cervical cancer while women with HPV 18, 31, or 33 had more than a 50-fold increased risk of developing cervical cancer than virus-free women.

One of the clearest virus-cancer links emerged in 1980 when a research team led by Robert T. Gallo of the National Cancer Institute isolated a virus that transforms normal T lymphocytes into the malignant T lymphocytes found in a rare cancer called **T-cell leukemia**. A year later, the same virus was found responsible for a relatively high rate of both T-cell leukemia and a form of lymphoma in Japan. Gallo identified the virus as an RNA-containing retrovirus and named it **HTLV-1 (human T-cell leukemia virus)**. His expertise with retroviruses proved valuable in 1984 when his research group set out to isolate the AIDS virus, also an RNA-containing retrovirus. Instead of transforming T lymphocytes, however, the AIDS virus destroyed them (Chapter 14).

In the 1990s, increasing evidence demonstrated that HTLV can cause leukemia, as well as neurological pain disorder and another condition marked by destruction of the sheaths that surround the nerve fibers. The virus appears to be bloodborne, and the primary mode of transport is through the use of illegal intravenous drugs. The growing incidence among patients also poses a threat to health-care workers who are exposed to patients' blood. A second virus, named **HTLV-2**, has been located also. Although the virus has not been definitely linked to any cancer, it is apparently common in patients suffering from a condition called **hairy-cell leukemia**, a type of leukemia in which the white blood cells develop long hair-like extensions to their cytoplasm at the surface.

ONCOGENIC VIRUSES TRANSFORM INFECTED CELLS

The mechanism by which viruses transform normal cells into tumor cells remained obscure until the **oncogene theory** was developed in the 1970s. First postulated by Robert Huebner and George Todaro in 1969, this theory suggests that transforming genes, the so-called **oncogenes**, normally reside in the chromosomal DNA of a cell. In the late 1970s, researchers J. Michael Bishop and Harold Varmus, of the University of California at San Francisco, located oncogenes in a wide variety of creatures from fruit flies to humans. Bishop and Varmus also made the astonishing discovery that practically the same genes exist in certain viruses, and they hypothesized that the genes could have been captured by the viruses from previous infections. It appeared that the oncogenes were not viral in origin but part of the genetic endowment of every living cell.

Oncogene:
a gene that can transform a normal cell to a tumor cell.

The discovery of oncogenes demonstrated that some forms of cancer have a genetic basis. As research continued, oncology researchers extracted DNA from tumors and used it to turn healthy cells into cancerous ones in the test tube. Moreover, they surmised that the transforming substance was in a small segment of the tumor cell DNA—probably a single gene. Finally, in 1981, three separate research groups isolated an oncogene residing in a human bladder cancer. At this writing, over 60 different oncogenes have been identified.

In recent years, the theory of oncogene activity has been revised slightly. Researchers now propose that normal genes, called **proto-oncogenes**, are the forerunners of oncogenes (FIGURE 12.21). Proto-oncogenes may have important functions as regulators of growth and mitosis. Indeed, research reported in 1985 linked proto-oncogenes to the production of cyclic adenosine monophosphate (cAMP), an organic substance central to many physiological processes. That proto-oncogenes exist in diverse forms of life argues for their important role in cell

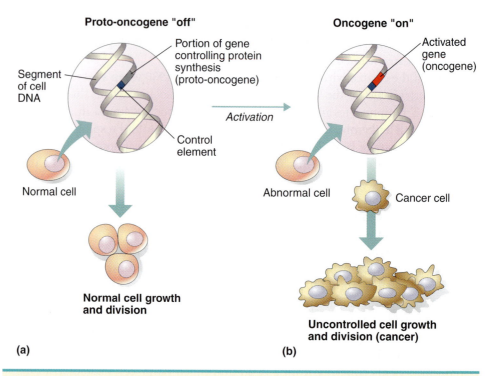

FIGURE 12.21

The Oncogene Theory

The oncogene theory helps explain the process of cancer development. (a) The normal cell grows and divides without complications. Within its DNA it contains proto-oncogenes that are "turned off." When the genes are activated by viruses or other factors, they revert to oncogenes, which "turn on." An abnormal cancer cell (b) results. The oncogenes encode proteins that regulate the transformation from a normal cell to a cancer cell and the development of a tumor.

metabolism, perhaps as growth regulators. (As one researcher notes, "They would not have survived through evolution just to make tumors.") It seems certain that proto-oncogenes can be converted to oncogenes by carcinogens, such as viruses, radiation, or chemicals, or by chromosomal breakage and rearrangement, after which tumor formation begins. Indeed, the bladder cancer oncogene differs from its counterpart proto-oncogene by only one nucleotide in 6,000.

How do viruses trigger the transformation? Virologists have observed that when some viruses enter a cell, they may become a provirus. Whether it is a retrovirus or a DNA virus, the double-stranded DNA inserts into the cell chromosome. Sometimes the integration of the provirus will be adjacent to a proto-oncogene. When virus replication is triggered, the viral DNA (provirus) not only replicates its own viral DNA, but also that of neighboring genes (**FIGURE 12.22**). Thus, when new viruses are produced, the viral DNA contains the proto-oncogene as part of its genome. Such proto-oncogenes "captured" in the viral genome are called **v-oncogenes**. The same scenario would hold for the retroviruses except the provirus produces RNA. Still, the viral RNA packaged into the virions will contain the viral information and the proto-oncogene information.

When these oncogenic viruses infect another cell and incorporate as a provirus, the v-oncogene is under control of the virus—not the host cell. So, similar to Figure 12.21, expression of the v-oncogene can influence cellular growth and mitosis

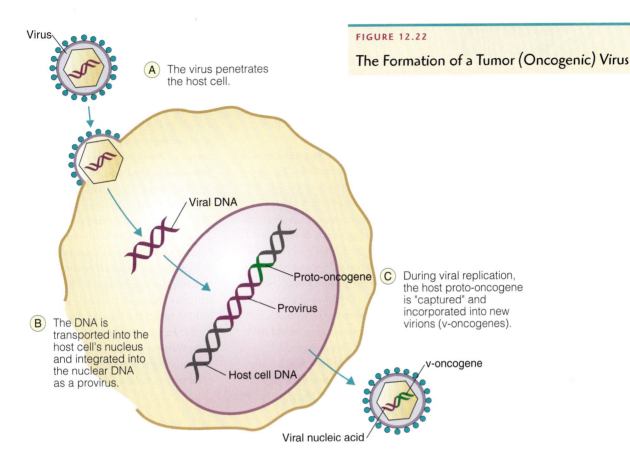

FIGURE 12.22

The Formation of a Tumor (Oncogenic) Virus

Virus

(A) The virus penetrates the host cell.

Viral DNA

Proto-oncogene

(C) During viral replication, the host proto-oncogene is "captured" and incorporated into new virions (v-oncogenes).

Provirus

(B) The DNA is transported into the host cell's nucleus and integrated into the nuclear DNA as a provirus.

Host cell DNA

v-oncogene

Viral nucleic acid

in several ways. Oncogenes, for example, may provide the genetic codes for growth factors that stimulate uncontrolled cell development and reproduction. Or the oncogenes may become incapable of encoding substances that turn off cell growth, a function the proto-oncogenes once had.

Another mechanism of transformation has been studied with the virus of **Burkitt's lymphoma**. In this cancer of the lymphoid connective tissues of the jaw, the viral genome inserts itself into a chromosome of B lymphocytes, the white blood cells important in immunity. The insertion triggers the proto-oncogene *c-myc* involved in cell growth to move from chromosome 8 to chromosome 14, far from the influence of its control genes. A segment of DNA from chromosome 14 replaces the proto-oncogene from chromosome 8. The proto-oncogene, now an oncogene, appears to produce elevated amounts of its protein product.

The oncogenic viruses described are involved in tumor formation and not true cancer development. The reason they are called oncogenic viruses is because the v-oncogene or activated oncogene is telling the infected cells, "Divide, divide, divide!" In a short period, a tumor forms. The excessive divisions provide an opportunity for mutations to occur that can lead the infected cells down the path of metastasis and cancer formation.

This ability of viruses to get inside cells, delivering their viral information, has been manipulated by medical research teams using gene engineering to deliver genes to cure genetic illnesses. MicroFocus 12.7 looks at the pros and cons of using viruses for this purpose.

MicroFocus 12.7

THE POWER OF THE VIRUS

Ever since the power of genetic engineering made it possible to transfer genes between organisms, scientists and physicians have wondered how they could use viruses to deliver good genes into individuals suffering from some form of genetic disease. For example, children suffering from severe combined immunodeficiency disease (SCID), lack a functioning immune system. These individuals have one gene that is not making the correct enzyme needed for immune function (Chapter 21). Commonly called the "bubble boy disease," many of these children must be kept in a sterile environment ("bubble") to prevent them from being infected by any of a number of disease-causing microorganisms.

Could a human virus be engineered that contains a copy of the "good gene" (therapeutic gene) for the missing enzyme? In other words, could one do gene therapy by infecting the body cells with an engineered virus that would supply those cells with the needed gene for the targeted illness? In 1990, this was accomplished by removing the lymphocytes from two SCID patients. The isolated cells were infected with a disease-disabled retrovirus containing a copy of the therapeutic gene, and, after growing more cells, the "infected" lymphocytes were replaced in the patient's body. Hopefully, the viral DNA would incorporate itself as a provirus in the patient's cells and the therapeutic gene would start pumping out its enzyme product. Although the technique worked, the patients also were being injected with the missing enzyme, so the gene therapy results were equivocal.

Since then, viral gene therapy has shown promise but it has had its misfires as well. In 1999, an 18-year-old patient, Jesse Gelsinger, died as a result of gene therapy. Although the gene therapy technique worked well, his body developed an immune response to the virus that resulted in the patient's death. In 2002, French scientists used gene therapy on four boys with SCID. Again, the therapy worked, but the inserted genes disrupted other cellular functions. In this case misplaced gene insertion led to leukemia (cancer of the white blood cells) in two of the boys. So, the idea of gene therapy certainly works. But, delivering the virus without complications and targeting the genes to the correct place remain to be worked out.

To this point . . .

Cancer was defined as an uncontrolled growth of cells that can be triggered by a variety of factors. Several viruses, called oncogenic viruses, were identified that are known to contribute to the development of human tumors, some of which can lead to cancer. These viruses can carry viral oncogenes or cause chromosomal translocations that activate genes that should be silent. About 20 percent of human cancers have a virus association.

In the next section, we briefly will discuss the emergence of "new" viruses and provide some ideas as to how viruses might have evolved.

12.6

Emerging Viruses and Virus Evolution

It may seem that of late viruses are appearing out of nowhere. Viruses that were not even heard of a few decades ago now seem almost commonplace: HIV, Ebola, Hantavirus, West Nile virus, and SARS virus, to name just a few.

These and other viruses represent **emerging viruses**; that is, viruses appearing for the first time in a population or rapidly expanding their host range with a corresponding increase in detectable disease (TABLE 12.6). In reality, many of these viruses are not new viruses. Some, like Hantavirus and West Nile Virus are just new to certain

TABLE 12.6

Examples of Emerging Viruses

VIRUS	FAMILY	EMERGENCE FACTOR
Influenza	*Orthomyxoviridae*	Mixed pig and duck agriculture, mobile population
Dengue fever	*Flaviviridae*	Increased population density, environments that favor breeding mosquitoes
Sin Nombre (Hantavirus)	*Bunyaviridae*	Large deer mice population and contact with humans
Ebola/Marburg	*Filoviridae*	Human contact with unknown host
HIV	*Retroviridae*	Increased host range, blood and needle contamination, sexual transmission, social factors
West Nile	*Flaviviridae*	Vector transported unknowingly to New York City
Machupo/Junin	*Arenaviridae*	Rodent contact via agricultural practices
Nipah/Hendra	*Paramyxoviridae*	Unknown
SARS-associated	*Coronaviridae*	Animal vector?

populations or geographical areas. Others, like HIV, have crossed host ranges and resulted in the development of new diseases. So, how do new viruses arise?

GENETIC RECOMBINATION AND MUTATION CREATE NEW VIRUSES

Almost every year a newly emerging influenza virus descends upon the human population. How can this happen? One way of creating new viruses is by the acquisition of new genetic information through **genetic recombination**. This is especially likely in viruses, like influenza, that have segmented genomes. Genetic recombination allows two different influenza viruses to reassort segments. The "bird flu" that broke out in Hong Kong in 1997 was the result of the reassortment of genome segments from a strain of avian flu virus with those of a human flu virus. The result was that the avian flu virus now had the capability of infecting humans. Not all the emerging viruses have segmented genomes. In those genomes that are a single linear or circular segment, genetic recombination still can occur.

HIV becomes resistant to new antiviral drugs quite rapidly. This can be explained by the second force that drives evolution, **mutation** (Chapter 6). Unlike cellular genes in the cell nucleus, when a single nucleotide is altered (**point mutation**) in an RNA virus genome in the host cytoplasm, there is no way to "proofread" or correct the mistake during replication. Although mutations are random, many point mutations will be damaging and result in defective virus particles; occasionally, one of these mutations will be advantageous. In the case of HIV, a beneficial mutation could confer resistance to an antiviral drug. Add to this the rapid replication rate and burst size for viral infections, and one can see that it does not take long for a beneficial mutation to establish itself within the viral population.

THE ENVIRONMENT AND SOCIAL CONDITIONS CAN PRODUCE NEW VIRUSES

Even if a new virus has emerged, it must encounter an appropriate host to replicate and spread. It is believed that smallpox and measles both evolved from cattle viruses, while flu probably originated in ducks and pigs. HIV almost certainly has evolved from a monkey (simian) immunodeficiency virus (SIV). Consequently, at some time such viruses had to make a species jump. What could facilitate such a jump?

Smallpox and measles are ancient diseases, so their cross to humans may extend back to the time when animals first were being domesticated and humans were becoming more sedentary. Similar scenarios may be true for other viral diseases as well. Our proximity to animals and pathogens makes the jump more probable. Today, population pressure is pushing the human population into new areas, to colonize and cut down forests where potentially virulent viruses may be lurking. This scenario is definitely true for the Machupo and Junin virus infections in humans. Increased agricultural practices in previously forested areas have brought infected rodents into contact with humans. New (emerging) viral diseases are a predictable outcome.

An increase in the size of the animal host population that carries a viral disease also can "explode" as an emerging viral disease. Spring 1993 in the American Southwest was a wet season, which provided ample food for deer mice. The explosion of the deer mice population brought them into closer contact with humans. Leaving behind mouse feces and dried urine containing the hantavirus made infection in humans likely. The deaths of 14 people with a mysterious respiratory illness in the Four Corners area that spring eventually were attributed to this newly recognized virus.

So, emerging viruses are not really new. They are simply evolving from existing viruses and, through human changes to the environment, are given the opportunity to spread or to increase their host range.

THERE ARE THREE HYPOTHESES FOR THE ORIGIN OF VIRUSES

You now should understand why viruses are obligate parasites—viruses absolutely depend on living host cells to replicate. In fact, they are *molecular parasites* because they use their host's molecular machinery to produce new virus parts and assemble those parts into new virions. Based on this definition, cells must have originated before viruses. But did they?

Although no one can know for certain how viruses originated, scientists and virologists have put forward three hypotheses for their origin.

- **Regressive evolution hypothesis.** Viruses are degenerate life-forms; that is, they are derived from intracellular parasites that have lost many functions that other organisms possess and have retained only those genes essential for their parasitic way of life.

- **Cellular origins hypothesis.** Viruses are derived from subcellular components and functional assemblies of macromolecules that have escaped their origins inside cells by being able to replicate autonomously in host cells.

- **Independent entities hypothesis.** Viruses coevolved with cellular organisms from the self-replicating molecules believed to have existed in the primitive prebiotic earth.

While each of these theories has its supporters, the topic generates strong disagreements among experts. In the end, it is not so much a matter of how viruses arose, as it is that we can become infected with them and we need to understand their behaviors better. However, practical implications derived from studying virus origins could increase our understanding of the properties and behavior of new viruses and help to develop new antiviral drugs to combat infection.

To this point . . .

In this section, we discovered that new viruses arise through genetic recombination and/or mutation. Emergence in humans often means that a virus needed to jump species, as HIV has done. Also, human penetration into newly settled areas often means exposure to viruses that have been present in that environment. We finished by examining three possible hypotheses explaining the origin of viruses.

In the final section of this chapter, we will study the virus-like agents, viroids and prions.

12.7 Virus-like Agents

When viruses were discovered, scientists believed they were the ultimate infectious particles. It was difficult to conceive that anything smaller than ultramicroscopic viruses could cause disease in plants, animals, and humans. In recent years, however, that perception has been revised, as scientists have researched a new class of disease agents—the subviral particles referred to as **virus-like agents**. We shall examine two types of virus-like agents in this section.

VIROIDS ARE INFECTIOUS RNA PARTICLES

Viroids are tiny fragments of nucleic acid known to cause several diseases in plants. Their discovery resulted from studies conducted in the 1960s at the US Department of Agriculture's Beltsville, Maryland, research center near Washington, D.C. A team of scientists led by Theodore O. Diener were investigating a suspected viral disease, **potato spindle tuber (PST)**, which results in long, pointed potatoes shaped like spindles. Nothing would destroy the disease agent except an RNA-degrading enzyme, and in 1971, the group postulated that a fragment of single-stranded RNA was involved. Diener called the agent a **viroid**, meaning "virus-like." The next year, a team led by Joseph Semancik at the University of California found a similar agent in a disease of citrus trees.

Currently, at least a dozen plant diseases have been related to viroids. The largest of these particles is about one-twentieth of the size of the smallest virus (FIGURE 12.23). The RNA chain of the PST viroid has a known molecular sequence (359 nucleotides), but it contains so few genetic codes that the replication cycle is not understood. Diener speculates that the viroids originated as introns, the sections of RNA spliced out of messenger RNA molecules before the messengers are able to function (Chapter 6). The similarity in size between introns and viroids, and the ring shape for both, have fueled the speculation. Semancik theorizes that viroids may be regulatory genes interacting with the host genome, because viroid diseases are characterized by interference with plant growth.

PRIONS ARE INFECTIOUS PROTEINS

In 1986, cattle in Great Britain began dying from a mysterious illness. The cattle experienced weight loss, became aggressive, lacked coordination, and were unsteady

FIGURE 12.23

Genome Relationships

The size relationships of a smallpox virus, poliovirus, and viroid. If the viroid were magnified 60,000 times, it would measure only one-eighth of an inch. By contrast, the genome of the smallpox virus would extend almost 14 feet and the genome of the poliovirus 4.5 inches. The genome of the potato spindle tuber viroid has 359 nucleotides, while that of the smallpox virus has almost 200,000.

on their hooves. These detrimental effects became known as **mad cow disease** and were responsible for the eventual death of these animals. A connection between mad cow disease and a similar human disease surfaced in Great Britain in the early 1990s when several young people died of a human brain disorder that resembled mad cow disease. Symptoms included dementia, weakened muscles, and loss of balance. Health officials proposed that the human disease was caused by eating beef that had been processed from cattle with mad cow disease. It appeared that the disease agent was transmitted from cattle to humans.

Since then, the disease has produced an economic disaster to the beef export industry in Britain and caused a political crisis throughout the European Union and beyond. In January 2001, over 1,000 head of cattle in Texas were feared to be the first American cattle to suffer mad cow disease. Luckily, tests conducted by the US Food and Drug Administration (FDA) indicated that there was no made cow disease present and no cause for alarm—until 2003. In May, the first mad cow was discovered in Alberta, Canada, and in December the first such cow was reported in the State of Washington. What is mad cow disease and how does it cause a lethal infection?

Mad cow disease, or **bovine spongiform encephalopathy** (**BSE**), is a fatal brain disorder (encephalopathy) that occurs in cattle (bovine). In autopsies, infected animals have sponge-like holes in the brain (spongiform) where brain cells were destroyed in large numbers (FIGURE 12.24a).

Besides BSE, similar neurologic degenerative diseases have been discovered and studied in other animals and humans. These include *scrapie* in sheep and goats, *wasting disease* in elk and deer, and *Creutzfeldt-Jakob disease* in humans. All are examples of a group of rare diseases called **transmissible spongiform encephalopathies** (**TSEs**) because, like BSE, they can be transmitted to other animals of the same species and possibly to other animal species, including humans. Many scientists

---

FIGURE 12.24

Prion Research and Disease

(a) A photomicrograph showing the vacuolar degeneration of gray matter characteristic of human and animal prion diseases. (b) This drawing shows the tertiary structure of the normal prion protein (PrP^C). The helical regions of secondary structure are green. (c) A misfolded prion (PrP^{SC}). This infectious form of the prion protein results from the helical regions in the normal protein unfolding and an extensive pleated sheet secondary structure (blue ribbons) forming within the protein. The misfolding allows the proteins to clump together and react in ways not completely understood that contribute to disease.

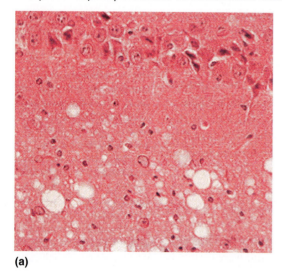

(a)

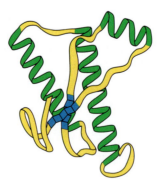

(b)

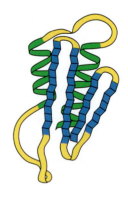

(c)

originally believed the agent was a new type of virus. However, to date no virus or nucleic acid characteristic of a virus has been reported. In the early 1980s, Stanley Prusiner and colleagues made a very interesting discovery while studying scrapie. Scrapie is the TSE that neurologically causes infected sheep to scrape their bodies raw on fences or other surfaces. Prusiner and colleagues isolated an unusual protein from scrapie-infected tissue that they thought represented the infectious agent. Prusiner called the proteinaceous infectious particle a **prion** (pronounced *pree'*on). The sequencing of the protein led to the identification of the coding gene, called *PrP*. The *PrP* gene is primarily expressed in the brain.

Based on more research and study, Prusiner and colleagues proposed the **protein-only hypothesis**. This hypothesis predicts that prions are composed solely of protein and contain no nucleic acid. The protein-only hypothesis further proposes there are two types of prion proteins. Normal cellular prions (PrP^C) are found in all uninfected brain cells (**FIGURE 12.24b**). Abnormal prions (PrP^{SC}) have a different shape (**FIGURE 12.24c**). They are found in infected cells and are the suspected infectious agents of diseases like BSE. With the wrong shape, PrP^{SC} proteins cannot organize properly in the cell membrane. Eventually cell function falters and cell death occurs. Based on his work with prions, Prusiner won the Nobel Prize in Physiology or Medicine in 1997. Still, some scientists believe a virus is somehow involved because the prion hypothesis seems to challenge the basic belief of DNA and inheritance.

If the protein-only hypothesis is correct, how can an infectious agent without a shred of nucleic acid spread and cause disease? Many researchers believe that prion diseases—that is, TSEs—are spread by the infectious PrP^{SC} binding to normal PrP^C, causing the latter to change shape. In a domino-like scenario, the newly converted PrP^{SC} proteins, in turn, would cause more PrP^C proteins to become abnormal. According to the hypothesis, all this occurs without any need for nucleic acid to replicate.

The 1990s outbreak of mad cow disease supports this idea. Scientists originally thought BSE made its way into cattle through feed that contained bone meal made from the carcasses of scrapie-infected sheep. However, in October 2000, the British BSE Inquiry concluded that BSE is a new disease, possibly arising from a mutation in one cow's genes. The inquiry concluded that the remains of this one cow were used in cattle feed, thus starting the infection of Britain's cattle herds.

Many people then ate beef or beef by-products that were unknowingly contaminated with nervous tissue, or from instruments that had contacted diseased nervous tissue, which contained the abnormally shaped prions. The PrP^{SC} proteins eventually crossed into the nervous system. Once the infectious agent enters the brain, it can lie dormant for several years (even as long as 10 to 15 years).

When the agent is active, the disease usually runs its course in less than one year. There remains little understanding of how abnormal PrP^{SC} proteins produce the clinical symptoms of the disease or the deadly pathogenesis that results. As mentioned, abnormal prions probably contact normal prion proteins and change them into abnormal ones, destroying the function of the normal prions. Abnormal prions then contact and change the shape of more normal prions in nerve cells. The infected nerve cells try to get rid of the abnormal prions by clumping them together into plaques for digestion by lysosomes. Because the nerve cells cannot digest the abnormal prions properly, they accumulate in the lysosomes. When the cell dies, the abnormal prions are released to infect other nerve cells. Large, sponge-like holes are left where groups of nerve cells have died. Death of the animal occurs from the numerous nerve cell deaths that lead to loss of brain function.

The human form of BSE appears similar to classical Creutzfeldt-Jakob disease (CJD). However, several characteristics separate the disease from classic CJD.

■ Prions are similar to those from the brains of BSE-infected cows and not to those prions found in victims of classic CJD.

■ Epidemiology studies have shown that victims had lived in areas where BSE outbreaks occurred. No victims were found in areas without BSE outbreaks.

■ Younger people are more susceptible to infection and the incubation period is shorter than for classical CJD.

Therefore, the new form of CJD was called variant CJD (vCJD). As of July 2003, the number of vCJD deaths worldwide was 129, although there have been more than 183,000 cases reported in Great Britain through the end of 2002. To protect the American beef industry, the federal government has banned the import of live cattle and animal parts from BSE-afflicted countries. However, with the December 2003 discovery, the US Department of Agriculture has implemented new protection measures. These include condeming those animals having signs of neurological illness and holding any cows suspected of having BSE until test results are known. In addition, "downer cattle"—those unable to walk on their own—cannot be used for human food or animal feed.

Keeping brain and spinal cord parts out of the food chain may not be enough though. In 2002, researchers from the University of California, San Francisco, reported that prions could accumulate in muscle—the animal part most of us eat. Since this discovery was made in mice, it remains to be shown that a similar situation would exist in animals typically eaten by humans; that is, cattle, sheep, deer, or elk (MicroFocus 12.8).

MicroFocus 12.8

BAMBI—STAY SAFE

In 2002, the Wisconsin Department of Natural Resources permitted the killing of some 18,000 deer in the state. The dead deer had their heads cut off and shipped in sealed plastic bags for testing. The carcasses were incinerated. What seemingly horrific act was going on here?

The kill was an attempt to keep an infectious disease called chronic wasting disease (CWD) from spreading through the state's deer population. Officials hoped the killing would thin out the deer population and slow the spread of the disease. The culling also would allow health examiners to determine how many of the killed animals had CWD.

CWD is a transmissible spongiform encephalopathy (TSE) of deer and elk. First recognized as a clinical "wasting" syndrome in 1967, the disease has been identified in wild and captive mule deer, white-tailed deer, North American elk, and in captive black-tailed deer in 12 states and two Canadian provinces.

Typical of TSEs, such as bovine spongiform encephalopathy (BSE) in cattle,

scrapie in sheep and goats, and variant Creutzfeldt-Jacob disease (vCJD) in humans, CWD produces a progressive, degenerative disease that affects infected animals. In deer and elk, clinical signs include chronic weight loss, behavioral changes, excessive salivation, and increased drinking and urination. CWD always is fatal in these animals. There is no known treatment or vaccine for CWD.

Being a TSE, the infectious agent that causes CWD is a prion. And that is cause for worry. Evidence suggests CWD is transmitted horizontally between animals by an unknown route. It appears that feed is not the transmission mechanism, although supplemental feeding of wild elk and deer could concentrate animals and make disease spread between animals easier. Although there is no known relationship between CWD and other TSEs of animals or people, in one experiment, CWD prions were mixed with normal prions from deer and humans. Less than 7 percent of the normal human prions (PrP^C) were induced to misfold into the abnormal form

(PrP^{SC}). However, the percentage of misfolding is about the same as that induced when BSE prions are mixed with human prions.

The CDC has issued this statement: "It is generally prudent to avoid consuming food derived from any animal with evidence of a TSE. To date, there is no evidence that CWD has been transmitted or can be transmitted to humans under natural conditions. However, there is not yet strong evidence that such transmissions could not occur."

The states with known cases of CWD in free-range animals are Colorado, Illinois, Nebraska, New Mexico, South Dakota, Utah, Wisconsin, and Wyoming. States with known CWD occurrences in farmed or captive animals are Arizona, Idaho, Iowa, Indiana, Kansas, Kentucky, Michigan, Minnesota, Missouri, Montana, Nevada, North Dakota, Oklahoma, and Texas. In April 2003, the Department of Agriculture made available $4 million to assist state wildlife agencies in addressing chronic wasting disease.

Note to the Student

Every science has its borderland where known and visible things merge with the unknown and invisible. Startling discoveries often come from this hazy, uncharted realm of speculation, and certain objects manage to loom large. In the borderland of microbiology, one curious and puzzling object is the virus.

Are viruses alive? At present, the tendency of many biologists is to sidestep the question. However, I shall make a suggestion. Although we have referred to viruses as microorganisms for the sake of convenience, I have avoided references to "live" or "dead" viruses. Instead, I have used the words "active" for replicating viruses and "inactive" for viruses unable to replicate. I suggest, therefore, that we consider viruses to be inert chemical molecules with at least one property of living things—the ability to replicate. Thus, viruses are neither totally inert nor totally alive, but somewhere in the threshold between. Perhaps viruses are transitional forms between inert molecules and living organisms. Indeed, a prominent virologist has suggested calling them "organules" or "molechisms," depending on one's preference.

Summary of Key Concepts

12.1 FOUNDATIONS OF VIROLOGY

■ **Many Scientists Contributed to the Early Understanding of Viruses.** The fundamental principles of this chapter are the structure and replication of viruses. Studying viruses was extremely difficult until the 1940s, when the electron microscope enabled scientists to see viruses and innovative methods allowed researchers to cultivate them.

12.2 THE STRUCTURE OF VIRUSES

■ **Viruses Occur in Various Shapes.** Viruses can have icosahedral, helical, or complex symmetry.

■ **Viruses Lack Cellular Structure.** Viruses are extraordinarily small particles composed of nucleic acid surrounded by a protein coat and, in some cases, a membrane-like envelope. The nucleic acid core known as the genome can be either DNA or RNA in either a single-stranded or double-stranded form. The protein coat, known as the capsid, is usually subdivided into smaller units called capsomeres. The envelope, when present, comes from the host

cell during replication and contains viral-specific proteins.

■ **Virus Structure Determines Host Range and Specificity.** A virus's host range refers to what organisms (hosts) that virus can infect and it is based on a virus's capsid structure. Many viruses only infect certain cell types or tissues within the host. This limitation represents a tissue tropism (tissue attraction).

12.3 THE REPLICATION OF VIRUSES

■ **The Replication of Bacteriophages Is a Five-Step Process.** Viral replication occurs only in living cells. The process involves attachment, penetration, biosynthesis, maturation, and release phases.

■ **Animal Virus Replication Has Similarities to Phage Replication.** Animal viruses go through the same five stages, but some of the processes vary according to the virus. For example, penetration can occur by a number of different methods, and the biochemistry of synthesis varies among DNA and RNA viruses.

■ **Some Animal Viruses Can Exist as Proviruses.** Many of the DNA viruses and the retroviruses can incorporate their viral DNA within a host chromosome. This viral DNA represents a provirus and is analogous to prophage formation in the bacteria.

12.4 OTHER CHARACTERISTICS OF VIRUSES

■ **Nomenclature and Classification Do Not Use Conventional Taxonomic Groups.** Viruses have no binomial names, but instead are named according to their disease (e.g., measles virus), their discoverer (e.g., Epstein-Barr virus), or another method. Classification schemes are based on physiological and biochemical characteristics, or on the tissue where replication takes place.

■ **Cultivation and Detection of Viruses Are Critical to Virus Identification.** Various detection methods for viruses are based on such things as characteristic changes in cell cultures, antibody responses, or pathological signs in cell culture or diseased tissue.

■ **Synthetic Drugs Can Inhibit Virus Replication.** A variety of synthetic antiviral drugs has been developed that interfere with viral replication. These drugs inhibit nucleic acid replication, interfere with new capsid biosynthesis, or prevent viral spread to other body cells.

■ **Interferon Puts Cells in an Antiviral State.** Interferon represents a group of antiviral substances that are naturally produced by cells when infected. Secretion of interferon stimulates uninfected neighboring cells to enter an antiviral state by producing antiviral proteins.

■ **Viral Vaccines Offer the Best Protection against Infection.** The primary public health response to viral disease is with vaccines. Vaccines consisting of inactivated viruses are available for polio and rabies, and vaccines containing weakened (attenuated) viruses are used to treat such viral diseases as measles, mumps, and rubella.

■ **Physical and Chemical Agents Can Inactivate Viruses.** Many of the physical and chemical methods routinely used on microorganisms can be employed to inactivate and destroy viruses.

12.5 VIRUSES AND CANCER

■ **Cancer Is an Uncontrolled Growth of Cells.** Cancer is a complex condition in which cells multiply without control.

■ **Viruses Are Responsible for 20 Percent of Human Cancers.** Among the many cancer-inducing agents are viruses. There are at least six carcinogenic viruses known to cause human tumors. Many of these tumors may develop into cancer.

■ **Oncogenic Viruses Transform Infected Cells.** Although the mechanisms of cancer induction are still unclear, viruses may bring about cancers by converting proto-oncogenes into cancer-causing oncogenes. If such genes are carried in a virus, they are called viral oncogenes.

12.6 EMERGING VIRUSES AND VIRUS EVOLUTION

■ **Genetic Recombination and Mutation Create New Viruses.** Similar to the way that bacteria can change and evolve, viruses use genetic recombination and mutations as mechanisms to evolve.

■ **The Environment and Social Conditions Can Produce New Viruses.** Human population expansion into new areas and increased agricultural practices in previously forested areas have exposed humans to existing viruses. Such viruses have to make the jump to humans and be able to replicate in the human host. If this occurs, new (emerging) viral diseases are a predictable outcome.

■ **There Are Three Hypotheses for the Origin of Viruses.** Three hypotheses have been proposed for the origin of viruses. The regressive evolution hypothesis suggests that viruses are derived from intracellular parasites. The cellular origins hypothesis proposes that viruses are derived from subcellular components that have escaped cells. The independent entities hypothesis puts forward the idea that viruses were self-replicating molecules that existed in the primitive prebiotic earth and coevolved with cellular organisms.

12.7 VIRUS-LIKE AGENTS

■ **Viroids Are Infectious RNA Particles.** Viroids are infectious particles made of RNA that infect a few plant species and cause disease. They lack protein and a capsid, but can replicate themselves inside the host.

■ **Prions Are Infectious Proteins.** Prions are infectious particles made of protein that are thought to cause a number of animal diseases, including mad cow disease in cattle and variant Creutzfeldt-Jakob disease (vCJD) in humans. Prions cause disease by folding improperly and in the misfolded shape, cause other prions to misfold. These and other similar diseases in animals are examples of a group of rare diseases called transmissible spongiform encephalopathies (TSEs).

Questions for Thought and Discussion

Answers to selected questions can be found in Appendix C.

1. If you were to stop 1,000 people on the street and ask if they recognize the term *virus*, all would probably respond in the affirmative. If you were then to ask the people to *describe* a virus, you might hear answers like "It's very small" or "It's a germ," or a host of other colorful but not very descriptive terms. As a student of microbiology, how would you describe a virus?

2. A textbook author referring to viruses once wrote: "Certain organisms seem to live only to reproduce, and much of their activity and behavior is directed toward the goal of successful reproduction." Would you agree with this concept? Can you think of any creatures other than viruses that fit the description?

3. Oncogenes have been described in the recent literature as "Jekyll and Hyde genes." What factors may have led to this label, and what does it imply? In your view, is the name justified?

4. Suppose the viroid turned out to be an infectious particle able to cause disease in humans. What difficulties might occur when dealing with this protein-free nucleic acid fragment both inside and outside the body?

5. Stanley's 1935 announcement that viruses could be crystallized stirred considerable debate about the living nature of viruses. Imagine that one day, a virus was discovered to contain both DNA and RNA. Might this stir an equal amount of controversy on the nature of viruses? Why?

6. Why was the cultivation of viruses in test tubes by Enders's group as significant as the cultivation of bacteria in test tubes by Koch? Why was each achievement critically important to the times? What type of follow-up experiment came after each foundation was established?

7. Researchers studying the bacteria that live in the oceans have long been troubled by the question of why bacteria have not saturated the oceanic environments. What might be a reason?

8. In broad terms, the public health approach to dealing with bacterial diseases is treatment. Can you guess the nature of the general public health approach to viral diseases? What evidence do you have to support your answer?

9. Bacteria can cause disease by using their toxins to interfere with important body processes; by overcoming body defenses, such as phagocytosis; by using their enzymes to digest tissue cells; or other similar mechanisms. Viruses, by contrast, have no toxins, cannot overcome body defenses, and produce no digestive enzymes. How, then, do viruses cause disease?

10. When Ebola fever broke out in Africa in 1994, the death toll was high, but the epidemic was short lived. By comparison, when influenza breaks out at the start of winter, the toll is low, but the epidemic lasts for six or more months. From the standpoint of the virus, what dynamics do you see in these two types of epidemics?

11. Many virologists believe that the agent of mad cow, scrapie, and other diseases is a prion. The term *prion* is derived from the words *pro*teinaceous *in*fectious particle. You will note that if these word parts are put together, the word is *proin*, not *prion*. How do you suppose the word got to be *prion*?

12. The eminent researcher Peter Medawar once described a virus using the derogatory words on the opening page of this chapter. How did the virus gain this questionable reputation? Explain whether it will ever change, and if so, how.

13. Many textbooks attempt to simplify biological concepts, and in doing so, they often oversimplify a particular thought. For example, it is not unusual to read that "viruses were discovered by Dimitri Ivanowsky." How would you react to this statement?

14. How have revelations from studies on viruses, viroids, and prions complicated some of the traditional views about the principles of biology?

15. When discussing the multiplication of viruses, virologists prefer to call the process replication, rather than reproduction. Why do you think this is so? Would you agree with virologists that *replication* is the better term?

http://microbiology.jbpub.com

The site features **eLearning,** an on-line review area that provides quizzes and other tools to help you study for your class. You can also follow useful links for in-depth information, or just find out the latest microbiology news.

Review

Use the following syllables to compose the term that answers each clue from virology. The number of letters in the term is indicated by the dashes, and the number of syllables in the term is shown by the number in parentheses. Each syllable is used only once. The answers are listed in Appendix D.

A A A AC AL AN AT AT BAC BO CAP CAP CEP CLO CO CO CY DE DERS DIES DINE DRON ED EN EN FER FORM GE GEN GENE HE HE HYDE I I I IN IN LET LEY LIX LY MAN MERES MOR NEG NOME O O OID ON ON ON ONS OPE PHAGE PRI RE RI SA SID SO SO STAN TA TED TEN TER TER TI TIV TOR TRA TU U UL VEL VI VIR VIR VIR Y

1. Viral protein coat   (2) ___ ___ ___ ___ ___ ___

2. Viral shape   (2) ___ ___ ___ ___ ___

3. Bacterial virus   (5) ___ ___ ___ ___ ___ ___ ___ ___ ___ ___

4. Neutralize viruses   (4) ___ ___ ___ ___ ___ ___ ___ ___

5. Rabies granules   (2) ___ ___ ___ ___ ___

6. Herpes drug   (4) ___ ___ ___ ___ ___ ___ ___ ___ ___

7. Natural antiviral   (4) ___ ___ ___ ___ ___ ___ ___ ___

8. Weakened virus   (5) ___ ___ ___ ___ ___ ___ ___ ___ ___ ___

9. Disease RNA fragment   (2) ___ ___ ___ ___ ___

10. Functionless cell mass   (2) ___ ___ ___ ___ ___

11. Virus-inactivating light   (5) ___ ___ ___ ___ ___ ___ ___ ___ ___

12. Cancer gene   (3) ___ ___ ___ ___ ___ ___

13. Site where virus attaches (3) ___ ___ ___ ___ ___ ___

14. Shape of poliovirus   (5) ___ ___ ___ ___ ___ ___ ___ ___ ___

15. Viral core   (2) ___ ___ ___ ___ ___

16. Cultivated poliovirus   (2) ___ ___ ___ ___ ___

17. Completely assembled virus (3) ___ ___ ___ ___ ___ ___

18. Virus incorporated to cell (4) ___ ___ ___ ___ ___ ___ ___ ___

19. Drug for influenza   (4) ___ ___ ___ ___ ___ ___ ___ ___

20. Surrounds the nucleocapsid (3) ___ ___ ___ ___ ___ ___

21. Crystallized viruses   (2) ___ ___ ___ ___ ___

22. Modifies viral genome   (4) ___ ___ ___ ___ ___ ___ ___ ___ ___ ___

23. Protein particles   (2) ___ ___ ___ ___ ___ ___

24. Capsid subunits   (3) ___ ___ ___ ___ ___ ___ ___

25. Vaccine virus   (5) ___ ___ ___ ___ ___ ___ ___ ___ ___

Pneumotropic and Dermotropic Viral Diseases

Medicine has never before produced any single improvement of such utility.

—Thomas Jefferson congratulating Edward Jenner on his development of a smallpox vaccine

HOW DOES ONE GO ABOUT eradicating a disease from the face of the Earth? How, indeed, when the disease at one time was killing one-third of English newborns; when the disease helped topple the highly civilized Aztec nation to a few hundred Spanish invaders; and when the ferocious disease was the major epidemic of the fledgling American colonies? How does one mount a global campaign against smallpox?

Apparently, the World Health Organization (WHO) was determined to try, and in 1966, it received $2.5 million to begin its campaign. The goal was to stamp out smallpox in ten years using global vaccination programs. Under the direction of Donald A. Henderson, vaccine-producing laboratories were established in countries where epidemics were raging. Wyeth Laboratories developed and donated a two-tined needle for administering the vaccine in 15 rapid jabs to the arm. By 1971, vaccination programs had been begun in 44 countries.

Next came surveillance containment. The WHO set up teams to improve the reporting and discovery of outbreaks, and every known contact of victims was vaccinated to break the chain of transmission. Vaccinators used persuasion and, in some cases, coercion to learn the whereabouts of people suffering from the disease. A favorite tool was the threat to withhold food ration cards. Rewards were offered for information, and people gradually came forward. By 1970, the number of smallpox countries had dropped to 17, and by 1973, only 6 countries were left.

469

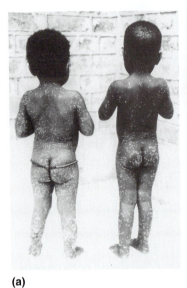

(a)

(b)

FIGURE 13.1

Smallpox

(a) Two young boys infected with smallpox. This photograph was taken during the Bangladesh epidemic in 1973. (b) Ali Maow Maalin, the Somali cook who had the last case of smallpox.

These countries included India, Pakistan, and Bangladesh, and 700 million people lived there (FIGURE 13.1). Now the WHO employed more rigorous techniques, including posting guards to prevent patients from leaving their homes, and vaccinating everyone within five miles of an infected village. By 1976, the job in Asia was done, and only Ethiopia remained. Despite a civil war, famine, several kidnappings, and torrential summer rains, the WHO pressed on. When victory seemed in sight, however, smallpox broke out in neighboring Somalia. Now, all efforts were directed at this African country, and a campaign involving 200,000 public health officials, 700 advisors, 55 countries, and untold millions of dollars eventually focused on a hospital cook named Ali Maow Maalin. Maalin developed smallpox on October 26, 1977. He was placed under guard, and vaccination was administered to 161 people previously in contact with him. None developed smallpox. In fact, no one has developed smallpox since that epic date.

In the minds of many scientists, the eradication of smallpox was the major medical event of the twentieth century. No claim of eradication had ever before been made for a disease, and nowhere have the potential benefits of applied public health been manifested better. Indeed, the eradication of smallpox probably has been public health's finest hour.

So, one might say, that's one less disease to worry about (and one less to learn for the test). Unfortunately not, because one of the most dangerous threats of bioterrorism comes from using smallpox as a biological weapon. Also, students need to know about the classical diseases of history in order to understand the contemporary ones. This is one reason why we study smallpox and numerous other diseases we probably will not encounter in our lifetimes. We also study viral diseases because, with the control of many bacterial diseases, viral diseases have become an important focus of attention in the medical community: For example, influenza continues to be an ongoing problem in the twenty-first century, and chickenpox is still among the most commonly reported diseases of childhood years. Another viral disease, genital herpes, has become so rampant that, by some estimates, 10 to 20 million Americans are currently infected. Indeed, the virus causing this disease may even be involved in clogged arteries (MicroFocus 13.1).

Scientists are learning to control many viral diseases even as we study them. For example, mumps and rubella were part of the fabric of life only a generation ago, but the annual case reports have dropped from hundreds of thousands to mere hundreds (and some officials are bold enough even to whisper the word "eradication").

In this chapter we shall focus on the **pneumotropic** viral diseases, which affect the respiratory system, and the **dermotropic** viral diseases, whose symptoms are found in the skin. As in Chapter 14, each disease is presented as an independent essay, so you can establish an order that best suits your needs. The pneumotropic and dermotropic divisions are an artificial classification simply for grouping convenience. Therefore, you may note that the symptoms go beyond the respiratory system or skin, respectively.

MicroFocus 13.1

A MISSING LINK?

During the late 1970s, Catherine Fabricant of Cornell University made the interesting observation that chickens infected with fowl herpesvirus develop a condition that looks suspiciously like atherosclerosis (fat accumulation on the inner walls of the coronary arteries, often resulting in heart disease). Her observation was intriguing, but there was more: The chickens were on a cholesterol-free diet. And more: Chickens whose diet included cholesterol suffered even greater fat accumulation. And still more: Chickens vaccinated against infection by fowl herpesvirus suffered no fat buildup. The link between herpesvirus and atherosclerosis appeared unmistakable.

Fabricant's work opened a set of inquiries that continues today. In 1993, for example, David Hajjar, also of Cornell, reported evidence that herpes-infected cells encourage blood clot formation while trapping fat deposits and accelerating fat buildup. These factors would speed up atherosclerosis and lead to heart attack. Hajjar's theory is that infected artery cells produce a glycoprotein that accumulates at the cell

surface. The glycoprotein unites with a clotting protein and sets the clotting mechanism into motion. Allied to clot formation is the arrival of white blood cells called monocytes. The monocytes collect fat and begin the process that will eventually lead to fat buildup and atherosclerosis.

Hajjar's theory is compelling not only because he has the evidence in molecular biology to support it, but because it also could explain why atherosclerosis occurs even though there are low cholesterol levels in certain patients.

The herpesviruses that might be involved are herpesvirus type II or cytomegalovirus (CMV). However, the association has remained equivocal since as many as 90 percent of the American population may carry CMV. The waters have been muddied more by other microbes being implicated. A Canadian heart study looked for the presence of antibodies to the four most cited microorganisms (CMV, *Chlamydia pneumoniae*, *Helicobacter pylori*, and hepatitis A virus) in more than 3,000 heart patients. The investigators concluded that exposure to CMV increased

the risk of a heart attack or stroke by 24 percent. Although the three other microorganisms showed no link with heart disease, people with antibodies in their blood to all four of the tested pathogens were 41 percent more likely to have a heart attack or stroke or die from cardiovascular disease.

Although the link between herpesviruses and heart disease is still speculative, some futurists are already thinking about immunizing against heart disease by using a herpes vaccine. It may seem unusual, but that is what futurists are for.

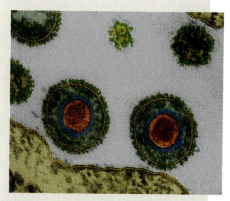

■ *Electron micrograph of herpesvirus*

13.1

Pneumotropic Viral Diseases

The pneumotropic viral diseases occur within the human respiratory tract. Certain ones, such as influenza, primarily affect the upper regions of the tract, while others, such as respiratory syncytial disease, involve the lungs and lead to viral pneumonia. The ubiquitous common colds and head colds that affect millions of Americans annually are in this group. In addition, recent emerging viruses, including the SARS coronavirus, produce respiratory-related diseases.

100 nm

influenza virus

INFLUENZA IS A HIGHLY COMMUNICABLE DISEASE

Influenza is a highly contagious **acute disease** of the upper respiratory tract transmitted by airborne droplets. The disease is believed to take its name from the Italian

Acute disease:
one that develops rapidly, exhibits substantial symptoms, and fades rather quickly.

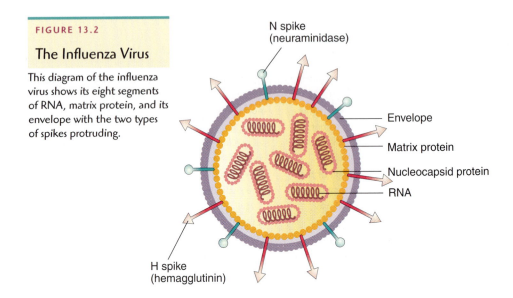

FIGURE 13.2

The Influenza Virus

This diagram of the influenza virus shows its eight segments of RNA, matrix protein, and its envelope with the two types of spikes protruding.

word for "influence," a reference either to the influence of heavenly bodies, or to the *influenza de freddo*, "influence of the cold." Since the first recorded epidemic in 1510, scientists have described 31 pandemics. The most notable pandemic of the twentieth century occurred in 1918 (MicroFocus 13.2); others took place in 1957 (the Asian flu) and in 1968 (the Hong Kong flu).

The influenza virion belongs to the *Orthomyxoviridae* family of viruses. It is composed of eight single-stranded (– strand) segments of RNA, each wound helically and associated with protein to form a nucleocapsid. Additional structural protein, the **matrix protein**, surrounds the core of segments, and an envelope lies outside the matrix protein. Projecting from the envelope is a series of **spikes** (FIGURE 13.2). One type of spike contains the enzyme **hemagglutinin (H)**, a substance that facilitates the attachment and penetration of influenza viruses into host cells. The second type contains another enzyme, **neuraminidase (N)**, a compound that assists the entry of the virion into the host cell for replication and exits from the host cell when replication is complete. Both enzymes are antigens.

Three types of influenza virus are recognized: type A, which strikes every year and causes most pandemics; type B, which also strikes every year and is less widespread than type A; and type C, which is rare. Each type is known for its **antigenic variation**, a process in which chemical changes occur periodically in hemagglutinin and neuraminidase, thereby yielding new strains of virus. The process also is called **antigenic drift** if it involves minor mutations, or **antigenic shift** if it involves major changes. These changes have practical consequences because as a result of antigenic drift, the antibodies produced during a previous attack of influenza fail to recognize the new strain, and a person may suffer another reasonably mild attack of disease. In addition, antigenic shift may give rise to new strains to which everyone is defenseless, and pandemics ensue. MicroInquiry 13 examines the differences in antigenic drift and antigenic shift. Antigenic variation precludes the development of a universally effective vaccine, although several have been developed over the years for particular strains. Moreover, the nomenclature for influenza viruses is based on the variant of antigen present (FIGURE 13.3). It is not uncommon to see references to strains such as A(H3N2), which predominated in the United States in the 2002–2003 influenza season.

MicroFocus 13.2

THE EPIDEMIC OF THE CENTURY

One theory suggests that it began when soldiers at Fort Riley, Kansas, were exposed to dust from burning manure piles during a March storm. Another theory implies that the source was an outbreak among hogs at an Iowa swine-breeders' show in September. Regardless of the origin, the epidemic that followed would be remembered as the most incredible of the twentieth century. The year was 1918; the disease was influenza.

At the time, influenza was called Spanish flu because it resembled an epidemic originating in Spain in 1893. American soldiers were the first to experience it, and as they traveled off to World War I in Europe, they carried the disease with them. Germany, France, and Great Britain soon were involved. Most of those affected were in their twenties and thirties—young, healthy, and vigorous.

The disease struck rich and poor alike, beginning with a cold. By the second day, victims were cyanotic (blue lips and face), and by the fourth or fifth day, they were dead from an overwhelming pneumonia as the lungs filled with fluid.

On October 6, 1918, the *Washington Post* recorded 658 fatalities from pneumonia, the most deaths from disease ever recorded in the United States in a single day. Indeed, October 1918 was the deadliest month in US history.

People adapted as best they could. Some wore camphor balls or bulbs of garlic; others consumed hot peppers or bootleg whiskey as medicinal agents. Public health officials fumigated street cars and trains daily with phenol and prohibited standing on a public conveyance. The police arrested people for not wearing masks or for spitting in the streets. Schools were closed, but homework assignments were printed in local newspapers, and students were expected to mail them in. On the sidewalks, girls skipped rope to this rhyme:

> I had a little birdie
> His name was enza
> I opened the window
> and in-flu-enza.

Medicinal remedies ranged from aspirin and laxatives to morphine and caffeine. Some people burned sulfur candles; others tried daily enemas. Many houses were trimmed with crepe to signify death—white for babies, purple for younger people, and black for the elderly. Survivors stacked coffins in cemeteries and behind funeral homes, and the army often had to be called in to bury the dead.

By late 1919, influenza had claimed 20 million lives throughout the world, a figure comparable to deaths from bubonic plague centuries before. Over 500,000 succumbed in the United States in a ten-month period. Some 24,000 American servicemen died of the disease (compared to a total of 34,000 World War I casualties). The notion of the Spanish flu remained until 1936 when Patrick Laidlaw, a British microbiologist, related the human disease to swine influenza. Memories of the great epidemic of 1918 to 1919 were vividly recalled when swine influenza was identified in 500 US Army recruits at Fort Dix, New Jersey, in the summer of 1976. The vaccination program that followed was the most ambitious ever mounted in the United States.

The onset of influenza is abrupt, with sudden chills, fatigue, headache, and pain most pronounced in the chest, back, and legs. Over a 24-hour period, body temperature can rise to 40°C, and a severe cough develops. Patients experience nasal congestion, dry throat, and tight chest, the latter a probable reflection of viral invasion of tissues of the trachea and bronchi. Despite these severe symptoms, influenza is normally short-lived and has a favorable prognosis. The disease is self-limiting and usually resolves in one week to ten days, although certain outbreaks raise public health concerns (MicroFocus 13.3). Secondary complications may occur if bacteria such as staphylococci or *Haemophilus influenzae* invade the damaged respiratory tissue.

A diagnosis of influenza is based on several factors, including the pattern of spread in the community, observation of disease symptoms, laboratory isolation of viruses, and hemagglutination of human type O red blood cells. Uncomplicated cases of type A influenza may be treated with bed rest, adequate fluid intake, and aspirin (or acetaminophen in children) for fever and muscle pain. In more serious cases, **amantadine** (Symmetrel), a drug believed to interfere with uncoating in the replication cycle, may be required. An alternate drug is **rimantadine** (Flumadine).

MicroInquiry 13

MIX AND MATCH INFLUENZA SEGMENTS

Influenza type A viruses are divided into subtypes based on two proteins, hemagglutinin (H) and neuraminidase (N), that extend from the virus envelope. The common nomenclature for influenza A viruses is to refer to the combinations of these two virus proteins. For example, the "Spanish flu" of 1918 may have been caused by influenza A(H1N1), while the "Hong Kong flu" of 1968 was due to influenza A(H3N2). Since 1977, these two influenza A virus subtypes, (H1N1) and (H3N2), have circulated widely in the human population. In fact, these two subtypes have been identified as the major subtypes during the 2002–03 season for which the 2002–03 flu vaccine was well matched.

After almost 70 years of research, how influenza causes yearly "flu" epidemics and less frequent pandemics is fairly well understood. The important factors are twofold: (1) influenza A virus not only infects humans, but also other animals, including birds and pigs; and (2) one or more of the eight RNA segments that make up the influenza A genome can be recombined with similar segments from another influenza A virus. For example, influenza A viruses that infect birds are not efficient at infecting humans and the influenza A viruses that infect humans do not efficiently infect birds. However, both often are efficient at infecting pigs. Consequently, pigs can be a "mixing pot" of avian and human influenza A viruses, producing new combinations of influenza A subtypes.

These subtypes can arise in two different ways. Small changes in the virus genome, called **antigenic drift**, happen continually over time to produce new virus strains that may not be recognized by the body's immune system. If a person is vaccinated with this year's flu vaccine, it might not provide protection against next year's strain. As newer virus strains appear, the antibodies against the older strains no longer recognize the "newer" virus, and another influenza A infection can occur. This is why people get the flu more than one time and people who want to be protected from flu need to get a flu shot every year.

Antigenic shift produces an abrupt, major change in the influenza A virus subtypes resulting in new hemagglutinin and/or neuraminidase proteins in influenza A viruses that infect humans. If an antigenic shift occurs, most people will have little or no protection against the new subtype. Influenza A viruses undergo both kinds of changes (drift and shift) while influenza B viruses change only by the more gradual process of antigenic drift.

With this background, see if you can predict the type of shift/drift and how the change occurred for the two cases presented. Answers can be found in Appendix E.

Case 1: On February 6, 2002, the WHO reported the identification of a new influenza virus strain, influenza A (H1N2), isolated from humans in England, Israel, and Egypt. In addition to the viruses reported by WHO, the CDC identified influenza A (H1N2) virus from patient specimens collected during the 2001–02 and 2002–03 seasons. Influenza A (H1N2) viruses had been identified as early as December 1988 and March 1989 when 19 influenza A (H1N2) viruses were identified in six cities in China. The virus strain apparently did not spread further.

13.1a. Is this an example of antigenic drift or antigenic shift? Explain.

13.1b. Provide an example of how the reassortment could have occurred.

13.1c. Explain how dangerous this subtype would be to people who received the 2002-03 flu vaccine.

Case 2: At the end of February 2003, the Netherlands reported outbreaks of avian influenza A (H7N7) in poultry on several farms. As of April 25, 2003, the National Influenza Center in the Netherlands had reported 83 confirmed cases of human H7N7 influenza virus infections in poultry workers and their families.

13.2a. Is this an example of antigenic drift or antigenic shift? Explain.

13.2b. Provide an example of how the reassortment could have occurred.

13.2c. Explain how dangerous this subtype would be to people comprising the 83 cases and to the general population who received the 2002-03 flu vaccine.

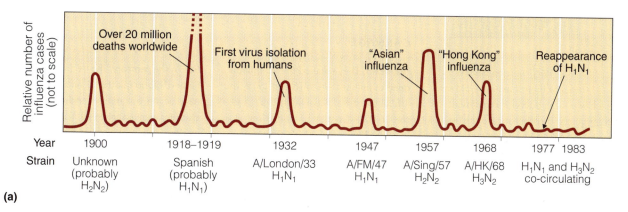

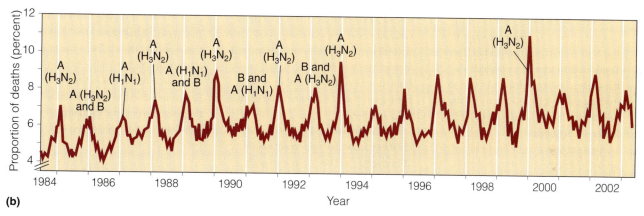

FIGURE 13.3

The Incidence of and Percentage of Deaths from Influenza in the United States

(a) Relative number of influenza cases 1900–1983. (b) Percent of all deaths since 1984 in 121 cities. Strains are labeled by type (A or B) and subtypes (hemagglutinin [H] and neuraminidase [N]) of type A. Source: Centers for Disease Control and Prevention. Available at www.cdc.gov. Accessed July 2003.

The best prevention for the flu is an annual flu vaccination. Each year's batch of vaccine (see MicroFocus 20.2) is based on the previous year's predominant influenza A and B viruses and is about 75 percent effective. Since the viruses are grown in chicken embryos (Chapter 12), people who are allergic to chickens or eggs should not be given the vaccine.

Recently, two antiviral drugs, **zanamivir** (Relenza) and **oseltamivir** (Tamiflu), have become available by prescription. These drugs target the neuraminidase spikes projecting from the influenza virus envelope and block the release of influenza A and B virus particles from infected cells. Being unable to leave cells limits the spread and transmission of the virus. If given to otherwise healthy adults and children early in disease onset, these drugs can reduce the duration of illness by one day and make complications less likely to occur. However, these drugs should not be taken in place of vaccination, which remains the best prevention strategy.

Two serious complications of influenza have surfaced during recent decades. One complication is **Guillain-Barré syndrome (GBS)**. This condition is characterized by nerve damage, polio-like paralysis, and coma. Some nonparalytic cases

Syndrome:
a collection of symptoms.

MicroFocus 13.3

CARNAGE

In May 1997, worried parents brought their three-year-old son to a Hong Kong doctor for treatment for influenza. The doctor gave the boy aspirin and an antibiotic and sent them home. But soon he was at Queen Elizabeth Hospital with a massive lung infection. Then his lungs collapsed, his liver shut down, and his kidneys failed. The boy died soon thereafter.

Baffled by his illness, doctors sent samples of the boy's tissues to experts in the Netherlands and the United States. Three months later, the diagnosis was complete: The boy's death was due to influenza virus A (H5N1), a virus fiercely pathogenic in chickens, but until then, unknown in humans.

An air of tension quickly developed among public health officials around the world. They recalled the 20 million dead of influenza in 1918 and 1919, the 100,000 dead of Asian flu in 1957, and the 36,000 who died of Hong Kong flu in 1968. They watched for other cases of the avian virus in humans (as a nervous Hong Kong community filled hospital emergency rooms with people having symptoms of routine colds). And they monitored the epidemic as it spread among the population of chickens: One moment a chicken stood contentedly in its cage; the next moment it leaned over and fell dead, blood oozing from its orifices. By November, 4 more human cases were confirmed; by Christmas, another 13.

On December 28, 1997, Hong Kong's chief executive decided to act swiftly: "No chicken will be allowed to walk free in the territory," he declared. Knives flashed, blood splattered, and the carnage began, as a million-and-a-half chickens were slaughtered. Thousands of workers were mobilized to stuff dead chickens into garbage bags and haul them to landfills. Carcasses rotted and rats picked apart bags left by the roadside. A thousand tons of chickens disappeared from the Hong Kong landscape.

In January 1998, the Chinese New Year came and went without the traditional poultry dishes. There were no live chickens or chicken dishes to offer to the gods or to honor Chinese ancestors. To be sure, it was a strange holiday. But there were no more deaths from influenza. It had been a close call.

---

are characterized by numbness in the arms and legs, and general weakness and shakiness. The second complication is **Reye's syndrome**, named for Ramon D. K. Reye, who reported it in 1963. Reye's syndrome (pronounced "rays" by some and "ryes" by others) usually makes its appearance when a child is recovering from influenza or chickenpox and has received aspirin (thus the use of acetaminophen for children). The fever rises, and repeated, protracted vomiting continues for a period of hours. The child may become lethargic, sleepy, and glassy eyed (or "starry eyed"), as well as disoriented, incoherent, and combative. Reye's syndrome, like GBS, is believed to be due to activity of the immune system because viruses cannot be found in sufficient numbers to explain the symptoms.

ADENOVIRUS INFECTIONS OFTEN ARE RESPONSIBLE FOR COLDS

Adenoviruses are a group of over 45 types of icosahedral virions having double-stranded DNA and belonging to the family *Adenoviridae* (**FIGURE 13.4**). They multiply in the nuclei of host cells and induce the formation of **inclusions**, a series of bodies composed of numerous virions arranged in a crystalline pattern. The viruses take their name from the adenoid tissue from which they were first isolated in 1953.

Adenoviruses are among the most frequent causes of upper respiratory diseases collectively called the **common cold**. Adenoviral colds are distinctive because the fever is substantial, the throat is very sore, and the cough is usually severe. In addition, the lymph nodes of the neck swell and a whitish-gray material appears over the throat surface.

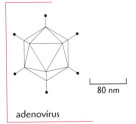

80 nm

adenovirus

One strain of adenovirus causes **keratoconjunctivitis**, an inflammation of the cornea (*kerato-*) and conjunctiva of the eye in adults. Patients experience reduced vision for several weeks, but recovery is usually spontaneous and complete. Transmission may be by respiratory droplets, contact, or ophthalmic instruments. Contaminated water also may transmit the viruses, as illustrated in an early 1970s outbreak where the chlorinator in a swimming pool malfunctioned, resulting in 44 cases of eye infection ("swimming pool conjunctivitis").

In recent years, adenoviruses have developed a more positive image as vectors (carriers) for genes during **gene therapy experiments**. In the 1990s, for example, researchers rendered adenoviruses incapable of replication, then reengineered them with a collection of genes to help control cystic fibrosis (CF). The genetically altered viruses ferried the helpful genes into the respiratory cells of CF patients, where they encoded proteins to help clear away the mucus and sticky material accumulating in the airways. Currently, a recombinant adeno-associated virus (parvovirus) has proven to be safe for gene delivery; however, its integration efficiency into the host DNA is low. The experiments continue as of this writing.

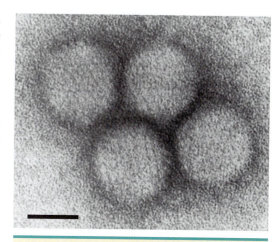

FIGURE 13.4

A Transmission Electron Micrograph of Adenoviruses

These viruses have no envelope, and the icosahedral symmetry of the nucleocapsid can be observed. The rough surface of the viruses is due to the presence of capsomeres. Adenoviruses are agents of common colds. (Bar = 100 nm.)

RESPIRATORY SYNCYTIAL DISEASE AFFECTS THE LOWER RESPIRATORY TRACT IN YOUNG CHILDREN

Since 1985, **respiratory syncytial (RS) disease** has been the most common lower respiratory tract disease affecting infants and children under one year of age. Infection takes place in the bronchioles and air sacs of the lungs, and the disease is often described as **viral pneumonia**. Maternal antibodies passed from mother to child probably provide protection during the first few months of life, but the risk of infection increases as these antibodies disappear. Indeed, researchers have successfully used preparations of antibodies (called immune globulin) to lessen the severity of established cases of RS disease. Aerosolized ribavirin also has been used with success.

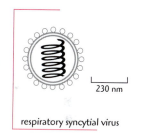

230 nm

respiratory syncytial virus

The respiratory syncytial virus (RSV) is an enveloped single-stranded (− strand) RNA virion of the *Paramyxoviridae* family. When the virus infects tissue cells, the latter tend to fuse (**FIGURE 13.5**) and form giant multinucleate cells called **syncytia** (sing., syncytium).

RS disease also can occur in adults, usually as an upper respiratory disease with influenza-like symptoms. Outbreaks occur yearly throughout the United States, but most cases are misdiagnosed or unreported. Some virologists believe that up to 95 percent of all children have been exposed to the disease by the age of five, and CDC epidemiologists estimate that 90,000 hospitalizations and 4,500 deaths occur in infants and children each year in the United States as a result of RS disease. A 1999 study linked the disease to a major proportion of middle-ear infections in children.

PARAINFLUENZA ACCOUNTS FOR 40 PERCENT OF ACUTE RESPIRATORY INFECTIONS IN CHILDREN

Parainfluenza is caused by another single-stranded (− strand) RNA virion of the *Paramyxoviridae* family of viruses (the RS and measles viruses are included in

FIGURE 13.5

Effects of the Respiratory Syncytial Virus (RSV)

(a) Uninfected cells. (b) Cells infected with RSV show the induction of cell fusion to produce giant, syncytial cells with several nuclei.

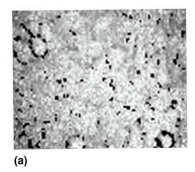 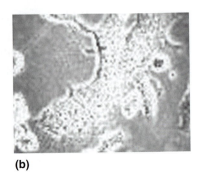

(a) (b)

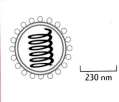

230 nm

parainfluenza virus

this family). Although as widespread as influenza, parainfluenza is a much milder disease and is transmitted by direct contact or aerosolized droplets. It is characterized by minor upper respiratory illness, often referred to as a cold. Bronchitis and croup may accompany the disease, which is most often seen in children under the age of six. The disease predominates in the late fall and early spring, and is seasonal, as **FIGURE 13.6** indicates. No specific therapy exists.

RHINOVIRUS INFECTIONS PRODUCE INFLAMMATION IN THE UPPER RESPIRATORY TRACT

Rhinoviruses are a broad group of over 100 different single-stranded (+ strand), naked RNA viruses with icosahedral symmetry. They belong to the family *Picornaviridae* (*pico-* means "small"; hence small-RNA-viruses). Rhinoviruses take their name from the Greek *rhinos*, meaning "nose," referring to their infection site. The viruses are among the major causes of **common colds**, also called head colds. They are transmitted through airborne droplets or by contact with an infected person or contaminated objects.

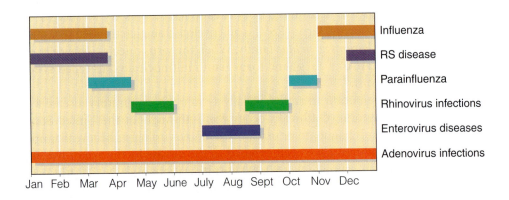

Jan Feb Mar Apr May June July Aug Sept Oct Nov Dec

FIGURE 13.6

The Seasonal Variation of Viral Respiratory Diseases

This chart shows the seasons associated with various viral diseases of the respiratory tract. Note that the influenza and RS disease seasons overlap and are the primary diseases of the winter season. Enteroviruses (Chapter 14) cause diseases of the gastrointestinal tract as well as respiratory disorders and are usually acquired from the environment.

A **head cold** involves a regular sequence of symptoms beginning with headache, chills, and a dry, scratchy throat. A "runny nose" and obstructed air passageways are the dominant symptoms, but the cough is variable and fever often is absent or slight. Some children suffer from **croup**. Antihistamines can sometimes be used to relieve the symptoms, which often are due to histamines released from damaged host cells. Another possible cause of symptoms is explored in MicroFocus 13.4.

Rhinoviruses thrive in the human nose, where the temperature is a few degrees cooler (33°C) than in the rest of the body. This may be one reason that the fumes from hot chicken soup appear to hasten recovery. Research on the use of **vitamin C** (ascorbic acid) as a preventive remains controversial. One study, for example, indicates that this vitamin induces the body to produce interferon, while another suggests that it encourages the formation of collagen to strengthen the "intercellular cement." High doses of vitamin C do not appear to prevent colds although high doses may lessen the length of cold symptoms.

Scientists have identified the receptor sites in nasal tissues where rhinoviruses attach. Furthermore, they have synthesized an antibody that binds to these sites and blocks viral attachment. Supposedly the antibody could be used as an anticold drug. Another group of researchers have used copies of the receptor sites as a drug to bind to the virus before it can find its host cell. And still another group conducted tests with a three-ply Kleenex tissue composed of two regular tissues sandwiched around a middle tissue impregnated with acidic compounds (the press dubbed them "killer Kleenexes").

The prospects for developing a cold vaccine are not promising, partly because many different viruses are involved. In addition to the rhinoviruses, the common cold viruses include adenoviruses and respiratory syncytial viruses, as well as coronaviruses, Coxsackie viruses, echoviruses, and reoviruses. A new nasal spray containing interferon has stimulated interest, but the side effects of this compound need to be understood before commercial products appear on the pharmacy shelf.

25 nm

rhinovirus

Croup:
hoarse coughing.

MicroFocus 13.4

BLAMING THE MESSENGER

Why do we feel so miserable when we develop a cold or the flu? Scientists think they may have an answer. The key, they suggest, is a protein called interleukin-6. Interleukin-6 is produced by white blood cells of the immune system (its T lymphocytes). Acting like a hormone, the protein travels to other white blood cells (the phagocytes), where it encourages an immune response to disease. When too much interleukin-6 is produced, however, tissue inflammation develops, and fever, achiness, and extreme exhaustion follow.

That's where the flu and common cold viruses come in. Working with student volunteers, scientists from various institutions have found that respiratory viruses stimulate the production of excessive amounts of interleukin-6 during infection. Their studies, reported in 1998, confirm that the amount of interleukin-6 in nasal washings is directly proportional to the amount of virus present and the severity of respiratory symptoms. Essentially, they point out, the body's response helps fight the infection but makes us miserable as well. The trick, researchers say, is for the body to produce enough interleukin-6 to combat the virus, but not so much that we have to suffer the symptoms of illness. Apparently the body has not yet learned the trick.

SEVERE ACUTE RESPIRATORY SYNDROME SPREADS THROUGH CLOSE PERSON-TO-PERSON CONTACT

Severe acute respiratory syndrome (**SARS**) is an emerging infectious disease of the respiratory system that was first reported in China in spring 2003 and quickly spread through Southeast Asia and to Canada. It is an example of how fast an emerging disease can spread (**FIGURE 13.7**).

Scientists at the CDC and other laboratories have identified in SARS patients a previously unrecognized coronavirus, which they named the SARS coronavirus

FIGURE 13.7

The Outbreak of SARS

This brief chronology for an atypical pneumonia outbreak shows how fast an emerging disease can spread—and be identified.

CHINA
Guangdong province   Hong Kong

1. On February 11, 2003, The Chinese Ministry of Health notifies the WHO of a mystery respiratory illness that has been occurring since November 2002 in Guangdong province in southern China. Officials of China refuse to allow WHO officials to investigate.

2. February 21, a 64-year-old doctor from Guangdong comes to Hong Kong to attend a wedding. He stays at a Hong Kong hotel, infecting 16 people who will spread the disease to Hanoi, Vietnam, Singapore, and Toronto, Canada.

3. February 23, a Canadian woman tourist checks out of the same Hong Kong hotel and returns to Toronto, where her family greets her. She dies 10 days later as five family members are hospitalized.

4. February 28, a WHO doctor in Hanoi, Carlo Urbani, treats one of the people infected in Hong Kong and realizes this is a new disease which he calls severe acute respiratory syndrome (SARS). He dies of the disease 29 days later.

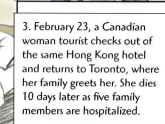

5. March 15, the WHO declares SARS a worldwide health threat.

6. April 16, the identification is made by 13 laboratories around the world that a new, previously unknown coronavirus causes SARS.

7. June 30, the WHO announces that there have been no new cases of SARS for two weeks. In all, more than 8,000 people from 32 countries were infected and 774 died.

TEXTBOOK CASES

(SARS-CoV). Being a member of the *Coronaviridae*, it is a single-stranded (+ strand) RNA virus with helical symmetry and a viral envelope (FIGURE 13.8). The mortality rate from SARS appears to be about 10 percent. As of November 2003, the WHO had received over 8,000 reported cases from 32 countries. WHO has reported 774 deaths, the majority being in China, Hong Kong, Taiwan, Canada, and Singapore.

SARS-CoV can be spread through close person-to-person contact by touching one's eyes, nose, or mouth after contact with the skin of someone with SARS. Spread also comes from contact with objects contaminated through coughing or sneezing with infectious droplets by a SARS-infected individual. Whether SARS can spread through the air or in other ways remains to be discovered.

Many people remain asymptomatic after contacting SARS-CoV. However, in affected individuals, moderate respiratory illness can occur and symptoms include fever (greater than 38°C), headache, an overall feeling of discomfort, and body aches. After two to seven days, SARS patients may develop a dry cough and have trouble breathing. In those patients progressing to a severe respiratory illness, pneumonia develops with insufficient oxygen reaching the blood. In 10 to 20 percent of cases, patients require mechanical ventilation.

Most of the cases of SARS in the United States in 2003 occurred among travelers returning from other parts of the world where SARS was present. Transmission of SARS to health care workers appears to have occurred after close contact with sick people before recommended infection control precautions were put into use.

Since this is a newly emerging disease, treatment options remain unclear. Antiviral therapy, including oseltamivir or ribavirin, has been tried as well as steroids given orally or intravenously in combination with ribavirin and other antimicrobials. The usefulness of these regimens remains unknown until controlled clinical

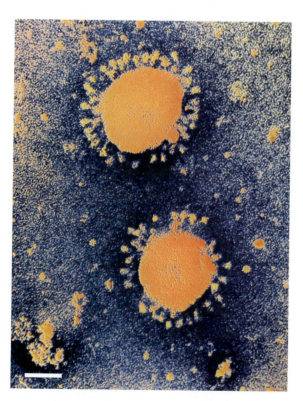

FIGURE 13.8

Coronaviruses

False color transmission electron micrograph of two human coronaviruses. In humans, coronavirus typically causes colds. The spikes can be seen clearly extending from the viral envelope. Viruses similar to these are responsible for severe acute respiratory syndrome (SARS). (Bar = 60 nm.)

MicroFocus 13.5

IT'S INFLUENZA, IT'S SARS, NO IT'S—METAPNEUMOVIRUS

The march of new human viruses seems to keep on coming—or perhaps keep on being discovered. In June 2003, researchers at Yale University reported they had discovered an RSV-like virus that might explain several mysterious respiratory illnesses in children and colds in adults.

The Yale researchers reported the presence of the human metapneumovirus (hMPV) in over 6 percent of retested lab samples originally taken from 296 children who reported mysterious flu-like and pneumonia-like symptoms in late 2001 and early 2002. In addition, researchers in Rochester, New York, identified hMPV in about 4 percent of similar retested samples from 1999 through 2001. The virus also has

been found in Canada, Australia, and Europe where it was originally discovered in the Netherlands in 2001. There, archived samples suggest hMPV has been present in the population for over 50 years. To date, the precise prevalence of the hMPV is not known but is suspected to be as high as 10 percent, making it second only to RSV in causing lower respiratory infections in children.

The human metapneumovirus is a member of the *Paramyxoviridae*, the same family that the respiratory syncytial and parainfluenza viruses belong. However, it seems to be milder than RSV and no one in the Yale and Rochester studies had died. Although many respiratory infections in children can be blamed on the influenza, respiratory syncytial,

parainfluenza, or cold viruses, over 30 percent of all respiratory infections never are identified as to cause. Perhaps hMPV is one of the culprits.

You might ask why it has taken so long to discover such a prevalent virus. The answer is that hMPV does not grow well in cell culture and so was not detected until better molecular methods were available.

More research is needed to determine the prevalence of hMPV and if a vaccine can be developed. Nevertheless, add another virus to the growing list of emerging viruses—in this case, one that has been around for decades but not detected. It presents yet another challenge to medical and microbiological researchers.

trials can be carried out. Early laboratory investigations suggest that ribavirin does not inhibit virus growth or prevent cell-to-cell spread. At the time of this writing, it is unknown whether SARS will establish itself as a seasonal disease. Perhaps by the time you read this, the answer will be known.

MicroFocus 13.5 presents yet another new pneumotropic virus that might explain a number of mysterious diseases.

The pneumotropic viral diseases are summarized in TABLE 13.1.

TABLE 13.1

A Summary of Pneumotropic Viral Diseases

DISEASE	CLASSIFICATION OF VIRUS	TRANSMISSION	ORGANS AFFECTED	VACCINE	SPECIAL FEATURES	COMPLICATIONS
Influenza	*Orthomyxoviridae*	Droplets	Respiratory tract	Available	Antigenic variation Strains A, B, C	Reye's syndrome Guillain-Barré syndrome
Adenovirus infections	*Adenoviridae*	Droplets Contact	Lungs, meninges Eyes	Not available	Common cold syndrome	Pneumonia Aseptic meningitis
Respiratory syncytial disease	*Paramyxoviridae*	Droplets	Respiratory tract	Not available	Syncytia of respiratory cells	Pneumonia
Parainfluenza	*Paramyxoviridae*	Droplets Contact	Upper respiratory tract	Not available	Common cold syndrome	None
Rhinovirus infections	*Picornaviridae*	Droplets Contact	Upper respiratory tract	Not available	Head-cold syndrome	None
SARS	*Coronaviridae*	Droplets Contact	Respiratory tract	None available	Newly emerged disease	Pneumonia

To this point . . .

We have surveyed a number of pneumotropic viral diseases that affect the respiratory tract. Our initial emphasis was on influenza, one of the most common diseases in our society. The influenza virus has a unique composition among viruses, with eight segments of RNA in its nucleocapsid. Antigenic variation in the virus accounts for the myriad strains that appear from year to year and that make resistance and vaccine development very difficult. Secondary infection, as well as Reye's and Guillain-Barré syndromes, can complicate cases of influenza.

We then concentrated on respiratory viruses that cause the familiar common cold. The adenoviruses were discussed as cold agents and their role in keratoconjunctivitis briefly was mentioned. We then turned to the respiratory syncytial (RS) virus, one of the most common causes of lower respiratory diseases in children. We described the rhinoviruses, a group of RNA viruses that cause the widely encountered head cold, and parainfluenza viruses, also the cause of colds. Throughout the discussions, the broad variety of viral strains and types that cause respiratory diseases were noted. This variety will probably preclude the development of a vaccine for many years. We ended by discussing the SARS virus, noting that it is the latest in newly emerging viruses.

We shall now turn our attention to the dermotropic viral diseases. These are viral diseases of the skin. In this group we shall discuss herpes simplex and chickenpox, two of the most widespread diseases in humans; measles and rubella, two viral diseases whose incidences are declining; and smallpox, a viral disease that is apparently extinct. In this diversity, we see the spectrum of viral diseases in the modern era.

13.2

Dermotropic Viral Diseases

The dermotropic viral diseases are a diverse collection of human maladies. Certain ones, such as herpes simplex, remain epidemic in contemporary times; while others, such as measles, mumps, and chickenpox, are being brought under control through effective vaccination programs. Indeed, there are relatively few antiviral drugs, and prevention programs remain a major course of action in dealing with these diseases.

HERPES SIMPLEX ARE WIDESPREAD, OFTEN RECURRENT INFECTIONS

Herpes simplex is an array of viral diseases caused by a large double-stranded DNA virion having icosahedral symmetry and an envelope with spikes. A member of the family *Herpesviridae*, it is one of the most common viruses in the environment. Indeed, some virologists contend that over 90 percent of Americans have been exposed to it by age 18. The virus passes among cells by intercellular bridges and remains in the nerve cells until something triggers it to multiply. Granules called **Lipschütz bodies** are seen in the cell nucleus.

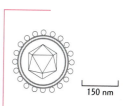

150 nm

herpes simplex virus

Lipschütz bodies: granules in the nucleus of cells infected with herpes simplex or other herpesviruses.

Physicians recognize many manifestations of herpes simplex infection, including **cold sores** (fever blisters), the unsightly lesions that form around the lips or nose (FIGURE 13.9); **herpes encephalitis**, a rare but potentially fatal brain disease; **neonatal herpes**, a life-threatening disease transmitted by mothers to newborns during childbirth; **gingivostomatitis**, a series of cold sores of the throat usually occurring in children; **herpes keratitis**, a disease of the eye and an important cause of blindness in young adults; and **genital herpes**, a troublesome sexually transmitted disease. The word *herpes* is Greek for "creeping," a reference to the spreading nature of herpes infections through the body after contact has been made (MicroFocus 13.6).

The sores and blisters of herpes simplex have been known for centuries. In ancient Rome, an epidemic was so bad that the Emperor Tiberius banned kissing; Shakespeare, in *Romeo and Juliet*, writes of "blisters o'er ladies lips"; and in the 1700s, genital herpes was so common that French prostitutes considered it a vocational disease. In current times, genital herpes is estimated to affect between 10 and 20 million Americans yearly, of whom about 500,000 are new cases. (The figures are inexact because genital herpes is not a reportable disease.) Signs generally appear within a few days of sexual contact, often as itching or throbbing in the genital area. This is followed by reddening and swelling of a small area where painful, thin blisters erupt. The blisters crust over and the sores disappear, usually within about three weeks. Usually a latent or dormant period then occurs during which time the virus remains in nerve cells near the blisters. In the majority of cases, however, the symptoms reappear, often in response to stressful triggers, such as sunburn, fever, menstruation, or emotional disturbance. FIGURE 13.10 shows the effect of ultraviolet light on the outbreak of infection. People with active herpes lesions are highly infectious and can pass the viruses to others during sexual contact. The cycle of latency and recurrent infections can occur three to eight times a year.

In the 1960s, scientists learned that the herpes simplex virus has two different forms: type I and type II. For reasons that remain unclear, **type I virus** often inhabits

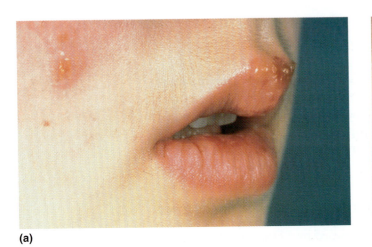

(a)

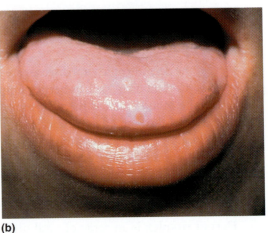

(b)

FIGURE 13.9

Two Manifestations of Herpes Simplex

(a) The cold sores (fever blisters) of herpes simplex erupting as tender, itchy papules and progressing to vesicles that burst, drain, and form scabs. Contact with the sores accounts for spread of the virus.
(b) Gingivostomatitis involving the oral mucosa, tongue, and cheeks with sores and blisters.

MicroFocus 13.6

GLADIATORS

The situation appeared normal: a camp for high school wrestlers in Minnesota from July 2 to July 28, 1989. The camp attracted 175 wrestlers from all over the United States. As they gathered that first day, they looked forward to daily sessions of grunting and wrestling their way to excellence. Three groups would participate: lightweights, middleweights, and heavyweights.

But this would be no ordinary experience. During the final week of camp, the first case of herpes simplex was observed. Then there was another, and another. Soon, there were too many infected participants for the camp to continue and, with two days remaining

on the schedule, the camp was suspended and the wrestlers sent home.

Subsequent contacts by the CDC revealed that 60 of the 175 participants had contracted herpes simplex during that two-week period. All experienced symptoms during the camp session or within one week of leaving. Lesions developed on the head or neck (73 percent of cases), the extremities (42 percent of cases), and the trunk (28 percent of cases). Five individuals experienced infection of the eye, and wrestlers in the heavyweight division were most frequently involved.

Herpes simplex in wrestlers and rugby players is called herpes gladiatorum ("herpes-of-the-gladiators," prob-

ably a whimsical term at first, but now technically acceptable). First described in the mid-1960s, herpes gladiatorum has broken out several times since then. Transmission occurs primarily by skin contact, and transmission on the fingertips can account for infection at several body sites. The disease illustrates another possible manifestation of herpes-related illness, and its swift passage by contact through a group of susceptible individuals demonstrates the ease of transfer.

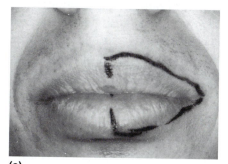

(a)

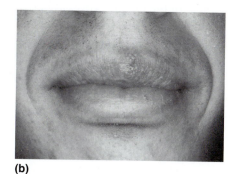

(b)

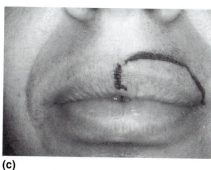

(c)

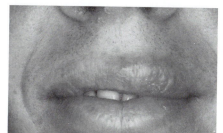

(d)

FIGURE 13.10

Ultraviolet Light and Herpes

An experiment showing the effects of ultraviolet (UV) light on the formation of herpes simplex sores of the lips. This patient usually experienced sores on the left upper lip. (a) The patient was exposed to UV light from a retail cosmetic sunlamp on the left upper and lower lips in the area designated by the line. The remainder of the face was protected by a sunscreen. (b) Sores formed on the left upper lip. (c) The patient was later exposed a second time to the sunlamp, but only on the left upper lip. (d) Sores formed and were larger than the previous ones. The results indicate that herpes sores can be experimentally stimulated by UV light, such as in sunlamps and sunlight.

areas above the waist (e.g., oral mucous membranes) and is the cause of herpes keratitis and most cold sores, while **type II virus** appears prevalent below the waist (e.g., genital area). This principle does not always hold true, however. Type II herpes simplex virus is especially worrisome because it is associated with **cervical cancer**, a disease that strikes over 15,000 American women annually. Though evidence is inconclusive, studies show that women who have suffered from genital herpes are several times more likely to develop cervical cancer than those who have not had herpes. Regular checkups, including Pap smears at frequent intervals, are recommended for women in this high-risk group.

Herpes encephalitis is a brain disease often accompanied by blindness, convulsions, or a range of neurological disorders, including mental impairment, and it may lead to death. The herpesviruses may be acquired by contact with an infected individual and may possibly reach the brain by the **olfactory nerves** after breathing into the nose. Herpes encephalitis also can occur in a newborn, where it is called **neonatal herpes**. In this case, the viruses infect the infant during passage through the birth canal. Indeed, if a woman has active genital herpes, her obstetrician may recommend birth by cesarean section. The passage of viruses across the placenta may lead also to neonatal herpes. In recent years, the acronym **TORCH** has been coined to focus attention on diseases with congenital significance: T for toxoplasmosis, R for rubella, C for cytomegalovirus, and H for herpes. O is for other diseases, such as syphilis.

Certain drugs have been approved by the Food and Drug Administration for the treatment of the various forms of herpes simplex. For example, **idoxuridine (IDU)** and **trifluridine** are used against herpes keratitis, and **vidarabine** is used to treat eye infections and herpes encephalitis, especially in newborns. Currently, the only drug approved for use against genital herpes is **acyclovir** (Zovirax). Acyclovir is a guanine derivative and a base analog that interferes with viral replication (**FIGURE 13.11**). Eye infections also respond to this drug.

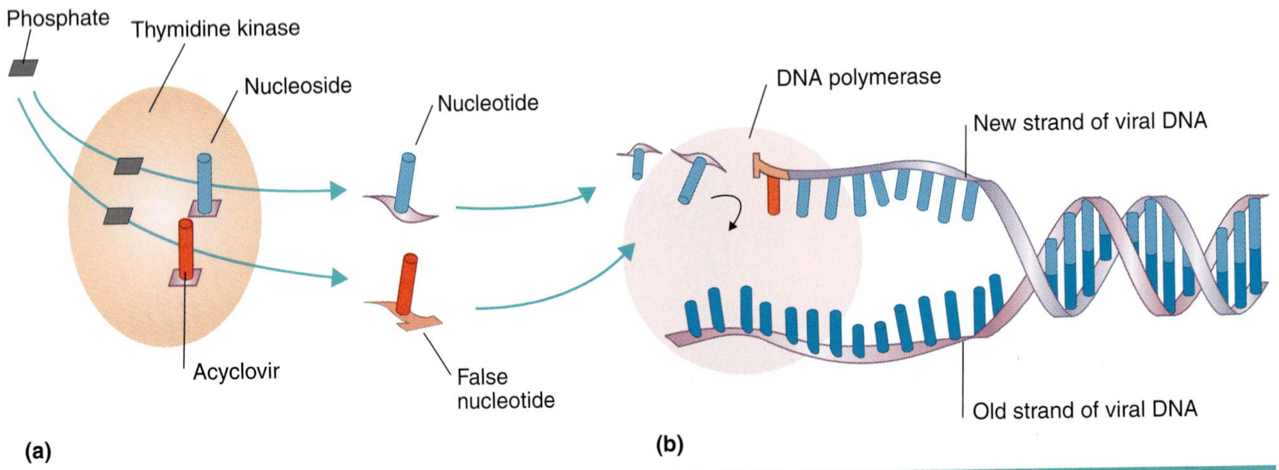

Olfactory nerves:
nerves used in the sense of smell.

(a)   (b)

FIGURE 13.11

The Mode of Action of Acyclovir

Acyclovir interferes with the replication of herpesviruses. (a) The enzyme thymidine kinase (circle, left) functions by combining phosphate groups with sugar-base combinations (nucleosides) to form nucleotides. Acyclovir resembles nucleosides, and the enzyme mistakenly adds phosphate groups to the acyclovir to form a false nucleotide. (b) During viral replication, another enzyme, DNA polymerase (circle, right) attaches the false nucleotide onto a developing DNA molecule. However, the false nucleotide lacks an attachment point for the next nucleotide. The elongation of DNA thus comes to a halt, and viral replication stops.

OTHER HERPESVIRUS INFECTIONS HAVE BEEN DETECTED RECENTLY

Although a direct cause-and-effect relationship has not been established, scientists have found that a type of herpesvirus called **human herpesvirus 6 (HHV-6)** bears a relationship to **multiple sclerosis (MS)**. Multiple sclerosis is a disease in which cells of the body's immune system attack myelin, the sleeve of tissue that surrounds nerve cells; the attack leads to the formation of numerous lesions called scleroses (hence the name). Muscle weakness, visual disturbances, and an array of other neurological impairments follow.

In 1998, researchers reported that the great majority of MS patients tested have antibodies against HHV-6 in their blood. The relationship was strengthened in 2003 when investigators identified HHV-6 genomes in patients with MS and in brain biopsies from patients who had acute MS. These signatures of the virus, together with the finding of herpes viral DNA in patients and HHV-6 in myelin lesions, have helped strengthen the relationship between the virus and the disease. Moreover, HHV-6 is known to be a cause of childhood **roseola**, a condition marked by fever and a red body rash. Researchers believe that HHV-6 may remain dormant in the body from the childhood years, then resurface to be part of the chain of events leading to multiple sclerosis. The relapsing, on-and-off progress of multiple sclerosis is reminiscent of the recurrent attacks that characterize herpesvirus infections. As many as 350,000 Americans suffer from multiple sclerosis.

Another herpesvirus called **human herpesvirus 8 (HHV-8)** now is regarded as the most probable cause of **Kaposi's sarcoma (KS)**, a highly **angiogenic** tumor most commonly seen in immunocompromised individuals, such as those with AIDS. Indeed, KS, which is marked by purple skin tumors, has become one of the most common tumors in parts of Africa where AIDS is endemic. The DNA of HHV-8 is present in most biopsies of tissue from KS patients, and antibodies against the virus are invariably detected in those with the disease or at risk of developing it. In 1997, two HHV-8 proteins were found to be instrumental in promoting blood vessel formation associated with the tumor.

Angiogenic: having many blood vessels.

Before leaving herpesviruses, we shall briefly mention a herpes-like illness called **B virus infection**. B virus infection is a relatively benign and common disease of Old World monkeys. Caused by a herpesvirus related to that of herpes simplex, human B virus infection is accompanied by serious neurological symptoms such as pain and numbness, with dizziness, local paralyses, and possible respiratory arrest. The disease is relatively rare in humans, but laboratory researchers who work with monkeys are at risk.

CHICKENPOX (VARICELLA) IS A COMMON, HIGHLY CONTAGIOUS DISEASE

In the centuries when pox diseases regularly swept across Europe, people had to contend with the Great Pox (syphilis), the smallpox, the cowpox, and the chickenpox. As of 2001, **chickenpox** was the sixth most reported disease in the United States, with over 22,000 cases reported annually. The causative agent is the **varicella-zoster virus**, another herpesvirus of the family *Herpesviridae*.

Chickenpox is a highly communicable disease. It is transmitted by respiratory droplets and skin contact, and it has an incubation period of two weeks. The disease begins in the respiratory tract, with fever, headache, and **malaise**. Viruses then pass into the bloodstream and localize in the peripheral nerves and skin. As they multiply in the cutaneous tissues, they trigger the formation of up to 500 small,

Malaise: a general feeling of illness.

teardrop-shaped, fluid-filled vesicles (**FIGURE 13.12a**). **Varicella**, the alternate name for chickenpox, is the Latin word for "little vessel."

The vesicles in chickenpox develop over three or four days in a succession of "crops." They itch intensely and eventually break open to yield highly infectious virus-laden fluid. Although many refer to the vesicles as pox, the latter term is more correctly reserved for the pitted scars of smallpox. In chickenpox, the vesicles form crusts that fall off without leaving a scar. **Acyclovir** has been shown to lessen the symptoms of chickenpox and hasten recovery. Development of this drug is described in **MicroFocus 13.7**. Varicella-zoster immune globulin also has been used.

The mortality rate from chickenpox is low, and the disease is considered benign. However, **Reye's syndrome** may occur during the recovery period, and public health officials have issued warnings against using aspirin to reduce fever. Other complications of chickenpox include pneumonia, encephalitis (brain inflammation), and bacterial infection of the skin. In pregnant women, the virus has been known to cross the placenta and cause damage in the fetus.

In 1995, the FDA licensed a vaccine for chickenpox. Known as Varivax, the vaccine consists of attenuated viruses administered subcutaneously and is recommended for all individuals over one year of age. One dose is given to children between ages 1 and 12, and two doses are given to adolescents and adults (pregnant women should not be immunized). The varicella vaccine is 85 percent effective in disease prevention.

Herpes zoster, or **shingles**, is an adult disease caused by the same virus that causes chickenpox. This is the reason why the virus is referred to as the varicella-zoster virus. The viruses multiply in ganglia (knots of nerve tissue) along the spinal cord, and travel down the nerves to the skin of the body trunk (**FIGURE 13.12b**). Here they cause blisters with blotchy patches of red that appear to encircle the trunk (*herpes* is Greek for "creeping," and *zoster* is Greek for "girdle"). Many sufferers also experience a series of headaches, as well as facial paralysis and sharp "ice-pick" pains described as among the most debilitating known. The condition can occur repeatedly and is linked to emotional and physical stress (such as radiation therapy), as well as to a suppressed immune system or aging.

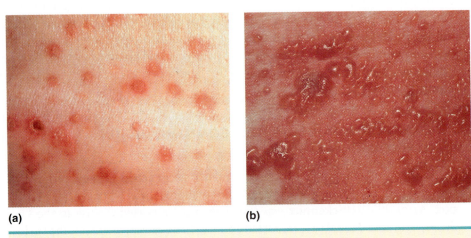

(a)   (b)

FIGURE 13.12

The Lesions of Chickenpox and Shingles

(a) A typical case of chickenpox. The lesions may be seen in various stages, with some in the early stage of development and others in the crust stage. (b) Dermal distribution of shingles lesions on the skin of the body trunk. The lesions contain less fluid than in chickenpox and occur in patches as red, raised blotches.

MicroFocus 13·7

THE PREFERRED WAY

Gertrude Belle Elion was getting dressed at 6:30 on the morning of October 17, 1988. Then the telephone rang. A moment later, Elion was speechless. She had won the Nobel Prize in Physiology or Medicine.

Only a few times in the century-long history of the Nobel Prize has the award been granted to researchers who developed drugs or worked for drug companies. This was one of those years. Gertrude B. Elion shared the award with George H. Hitchings, her former coworker at Burroughs Wellcome Research Laboratories in North Carolina, and with Sir James Black of King's College Medical School in London. The Nobel Committee named the three scientists for "their discovery of important principles of drug treatment" and for developing an intelligent method for designing new compounds based on an understanding of basic biochemical processes.

For Gertrude Elion, the award culminated a research career that almost did not happen. Even though she had a Bachelor of Science degree in biochemistry, Elion had difficulty obtaining a laboratory position because of her gender. She therefore accepted a job as a chemistry teacher. After World War II,

she went to Wellcome Laboratories, then to New York as an assistant to Hitchings. Although she never attained an advanced degree, Elion's technique and expertise were so respected that she soon came to be accepted as a colleague at the laboratory.

In 1944 Elion and Hitchings set out to learn how normal cell growth differs from that of abnormal cells, such as cancer cells. They hoped to find a way to destroy abnormal cells. The pair focused on differences in how various species metabo-

lize nucleic acid components, confining their studies mainly to nitrogenous bases of nucleic acids. In the 1950s, they developed antileukemia drugs called thioguanine and 6-mercaptopurine. Then, the biochemical clues led them to a series of other drugs including azathioprine (Imuran), which stalls the rejection mechanism in transplants; allopurinol, which is used to treat gout; and pyrimethamine and trimethoprim, for malaria and other diseases. In 1977, they developed acyclovir, now used widely against herpes simplex and more recently against chickenpox. Other colleagues, applying the basic ideas of Elion and Hitchings, synthesized AZT for AIDS patients.

Elion and Hitchings were part of the so-called "fundamentalist" world of chemotherapy. By concentrating on the fundamental physiology and biochemistry of cells, they came to understand essential cellular metabolic pathways and how to interfere with them. Other researchers, dubbed "screeners," preferred to bypass the cellular biochemistry and devote their efforts to screening a number of compounds, trusting their intuition and luck. It was the more rational approach that the Nobel Committee cited in its award.

There is substantial evidence that herpes zoster is caused by the same virus that caused chickenpox decades before in the individual. Most cases occur in people over age 50, and a person with an active case of herpes zoster can induce chickenpox (but not herpes zoster) in another susceptible person. AIDS patients may be susceptible to the disease because of their compromised immune systems (FIGURE 13.13). For herpes zoster, acyclovir therapy lessens the symptoms, but the immune globulin used to treat varicella has limited value.

MEASLES (RUBEOLA) IS ANOTHER HIGHLY COMMUNICABLE DISEASE

Measles is a highly contagious disease caused by a single-stranded (− strand) enveloped RNA virion first isolated in tissue culture by John Enders and Thomas Peebles in 1954. The virion has hemagglutinin spikes and is closely related to the mumps and RS viruses in the *Paramyxoviridae* family. Transmission usually occurs by respiratory droplets during the early stages of disease.

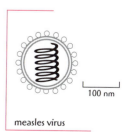

100 nm

measles virus

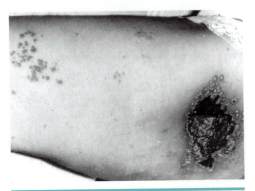

Herpes Lesions on a Patient's Leg

This patient suffered from disseminated herpes zoster associated with an immune deficiency due to HIV. The large lesion is a necrotic skin ulcer; the smaller lesions are herpetic pustules. The patient was treated intravenously with acyclovir, and the lesions healed. However, she later succumbed to the effects of HIV infection.

Measles symptoms commonly include a hacking cough, sneezing, nasal discharge, eye redness, sensitivity to light, and a high fever. Red patches with white grain-like centers appear along the gumline in the mouth two to four days after the onset of symptoms. These diagnostic patches are the **Koplik spots** first described in 1896 by Henry Koplik, a New York pediatrician.

The characteristic red rash of measles appears about two days after the first evidence of Koplik spots. Beginning as pink-red pimple-like spots (maculopapules), the rash breaks out at the hairline, then covers the face and spreads to the trunk and extremities (FIGURE 13.14). **Rubeola**, the alternative name for measles, is derived from the Latin *rube* for "red." Rashes resemble those in scarlet fever, but the severe sore throat of scarlet fever generally does not develop. Within a week, the rash turns brown and fades.

Measles usually is characterized by complete recovery. In some cases, however, bacterial disease may develop in the damaged respiratory tissue. Another possible problem is subacute sclerosing panencephalitis (SSPE), a rare brain disease characterized by a decrease in cognitive skills and loss of nervous function. Measles is a major cause of death in children in developing countries. There is also some evidence that the measles virus may be linked to multiple sclerosis and diabetes.

In 1978, the US Public Health Service launched a campaign to eliminate measles in the United States. The cornerstone of the campaign

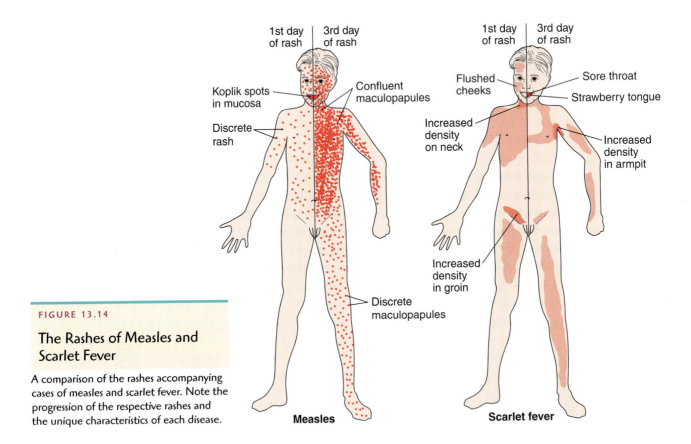

The Rashes of Measles and Scarlet Fever

A comparison of the rashes accompanying cases of measles and scarlet fever. Note the progression of the respective rashes and the unique characteristics of each disease.

was immunization of all school-age children with attenuated measles viruses in the **measles-mumps-rubella (MMR)** vaccine. Measles viruses have no known hosts other than humans, a factor that is helpful in containing the disease. By 1983, the total number of reported cases was 1,463, a reduction of 99.7 percent from the prevaccine era (**FIGURE 13.15**). However, the number skyrocketed to over 27,000 cases in 1990, reflecting outbreaks of measles in unimmunized children and college-age students inoculated with ineffective vaccines. Immunization of children entering grade school is now mandatory in all 50 states. By 1991, the number of cases had dropped to about 9,500, and by 1999, there were only 60 cases; the epidemic was over. In 2001, there were 116 reported cases, almost 50 percent of which occurred in immigrants.

RUBELLA (GERMAN MEASLES) IS AN ACUTE BUT MILDLY INFECTIOUS DISEASE

For generations, **rubella** was thought to be a mild form of measles. The distinction was not made until 1829 when Rudolph Wagner, a German physician, noted the differences in symptoms and suggested the two diseases were different. Thereafter, the

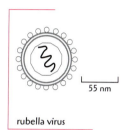

rubella virus

FIGURE 13.15

Reported US Cases of Measles (Rubeola), 1962–2001

Note the sharp drop-off in cases after licensing of the vaccine in the mid-1960s. Unfortunately, the immunity from this vaccine was not long lasting, and a new epidemic of measles broke out in the late 1980s. The inset shows the rise from a low of 1,400 cases in 1983 to over 27,000 cases in 1990. The recent reduction in cases is due to successful efforts to revaccinate susceptible individuals.

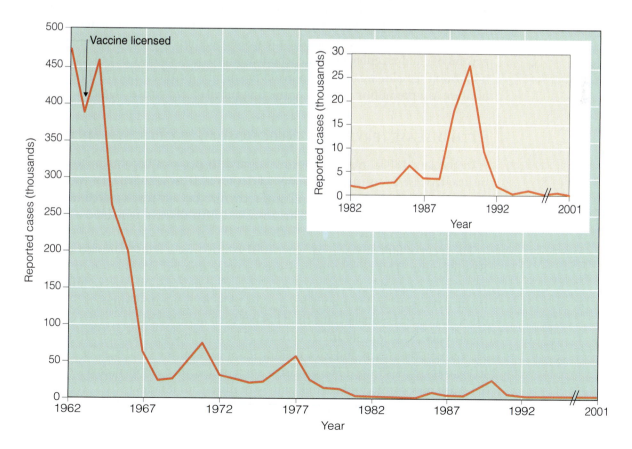

new disease was known as German measles (from Wagner's homeland) to distinguish it from measles. ("German" also may have been derived from the Latin *germanus,* meaning "akin"—in this case, akin to measles.) The name *rubella* ("small red") was suggested in the 1860s because the disease is accompanied by a slightly red rash.

Rubella is caused by a single-stranded (+ strand) RNA virus of the *Togaviridae* family. The virion is icosahedral with an envelope and spikes containing hemagglutinin. Viral transmission generally occurs by contact or respiratory droplets, and the disease is usually mild. It is accompanied by occasional fever with a variable, pale-pink **maculopapular** rash beginning on the face and spreading to the body trunk and extremities. The rash develops rapidly, often within a day, and fades after another two days. Recovery is usually prompt, but relapses appear to be more common than with other diseases, possibly because the viruses remain active within body cells.

Rubella is dangerous to the developing fetus in a pregnant woman. However, there were only three reported cases of this condition, called **congenital rubella syndrome**, in the United States in 2001 because of the successful vaccination program. Destruction of the fetal capillaries takes place, and blood insufficiency follows. The organs most often affected are the eyes, ears, and cardiovascular organs, and children may be born with **cataracts**, **glaucoma**, deafness, or heart defects. If rubella is contracted during the first month of pregnancy, the probability of damage to the fetus is about 50 percent. The probability declines sharply thereafter. In the 1965 to 1966 epidemic of rubella in the United States, over 50,000 instances of stillbirth and fetal deformity were recorded. The R in the TORCH group of diseases stands for rubella.

Since its introduction in 1969, the rubella vaccine has had a dramatic effect on the rate of incidence of the disease. That year, physicians reported 58,000 cases of rubella, but by 2001, the number was down to 23. The vaccine consists of attenuated viruses cultivated in human tissue cultures. It is combined with the measles and mumps vaccines (MMR) for subcutaneous inoculation of children. More than 95 percent of children entering grade school now provide evidence of rubella vaccination. Adult females are advised to avoid pregnancy for three months after immunization as a precaution against contracting rubella from viruses in the vaccine.

FIFTH DISEASE (ERYTHEMA INFECTIOSUM) PRODUCES A MILD RASH

In the late 1800s, numbers were assigned to diseases accompanied by skin rashes. Disease I was measles, II was scarlet fever, III was rubella, IV was Duke's disease (also known as **roseola** and now recognized as any rose-colored rash), and V was **erythema infectiosum**. This so-called **fifth disease** remained a mystery until the modern era.

The agent of fifth disease is parvovirus B19; fifth disease is therefore also known as **B19 infection**. This parvovirus is a small single-stranded DNA virion of the *Parvoviridae* family, having icosahedral symmetry. Community outbreaks of fifth disease occur worldwide, and transmission appears to be by respiratory droplets. Rubella is often suspected, especially if the child has not been immunized.

Fifth disease primarily affects children. The outstanding characteristic is a fiery red rash on the cheeks and ears, making it appear as if the child has been slapped, as **FIGURE 13.16** shows. (The disease is sometimes called **slapped-cheek disease**.) The rash may spread to the trunk and extremities, but it fades within several days, leaving a "lacy" rash on the skin. Recurrences during ensuing days or weeks are

Maculopapular:
referring to pink-red pimple-like spots that spread.

Cataract:
clouding of the lens of the eye.

Glaucoma:
visual defects due to the high pressures exerted by eye fluids.

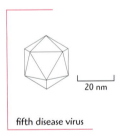

20 nm

fifth disease virus

Erythema:
reddening.

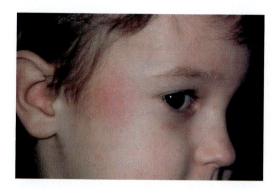

FIGURE 13.16

Fifth Disease

The fiery red rash of a child with fifth disease (erythema infectiosum). The confluent red rash makes it appear as if the child has been slapped.

related to bathing, sunlight, exercise, or stress. This characteristic, and the "slapped-cheek" appearance, are important to diagnosis.

Parvovirus B19 only infects humans (animal parvoviruses do not). However, fifth disease is not limited to children. Adults suffer from painful joints similar to the symptoms of rheumatoid arthritis, especially in the fingers, wrists, knees, and ankles. Infection of the bone marrow also may lead to anemia, and pregnant women may suffer miscarriage (but birth defects generally do not occur). Although antibody preparations (immune globulin) are available for treatment, the symptoms usually resolve spontaneously.

MUMPS IS SPREAD BY RESPIRATORY DROPLETS

Mumps takes its name from the English "to mump," meaning to be sullen or to sulk. The characteristic sign of the disease is enlarged jaw tissues arising from swollen salivary glands, especially the **parotid glands**. **Epidemic parotitis** is an alternate name for the disease.

The mumps virus is a single-stranded (− strand) RNA helical virion of the *Paramyxoviridae* family. Spikes with hemagglutinin are present in its envelope. The virus was among the first human viruses cultivated in fertilized chicken eggs, an achievement of Claud D. Johnson and Ernest Goodpasture in 1934.

Mumps is generally transmitted by droplets, contact, and contaminated objects; it is considered less contagious than measles or chickenpox. The virus is found in human blood, urine, and cerebrospinal fluid, even though its effects are observed primarily in the parotid glands. Obstruction of the ducts leading from the glands retards the flow of saliva, which causes the characteristic swelling. The skin overlying the glands is usually taut and shiny, and patients experience pain when the glands are touched.

In male patients, the mumps virus may pose a threat to the reproductive organs. As long ago as 1790, the Scottish physician Robert Hamilton observed swelling and damage to the testes and named the condition **orchitis**, from the Greek *orchi-*, referring to the testicles. The sperm count may be reduced, but sterility is not common. An estimated 25 percent of mumps cases in postadolescent males develop into orchitis.

The mumps vaccine, developed in 1967, consists of attenuated viruses and is usually combined with the measles and rubella vaccines (MMR). Although the campaign against mumps never attained the fame of the campaigns against measles or rubella, the reduction of mumps cases has been equally notable. Almost 200,000 cases were recorded in 1967, but only 266 cases occurred in 2001, the lowest number ever recorded. Humans are the only hosts for the virus.

Parotid gland:
the large salivary gland below the ear where the upper and lower jawbones come together.

Orchitis:
an infection of the male genital organs, a complication of mumps.

To this point . . .

We have surveyed a number of viral diseases whose symptoms occur largely on the skin surface. Herpes simplex infections are manifested as blister-like lesions in cold sores, genital herpes, and herpes keratitis. With chickenpox, the lesions are more like fluid-filled teardrops. Measles is not accompanied by lesions, but rather a blush-like rash that begins on the head and then spreads to the extremities. The rubella rash is similar, but it develops and fades rapidly and often does not occur at all. A fiery red rash on the cheeks and ears is a sign of fifth disease, and swollen parotid glands typify mumps.

None of the dermotropic diseases that we have studied is known to be life threatening, but the long-ranging effects may be consequential. For example, we have seen how SSPE is associated with measles, how orchitis may complicate mumps, how herpes zoster is an adult form of chickenpox, and how genital herpes recurs in a patient for many years. Also, a severe congenital problem is associated with rubella. Thus, many of the diseases formerly considered benign are currently viewed in a new light.

In the concluding section of this chapter, we shall study other dermotropic diseases, including smallpox, a viral disease that has been known to be potentially fatal for centuries. The remarkable feature of smallpox is that it has not been observed in humans for over 25 years. This claim cannot be made for any other disease. We also shall give brief mention to the skin warts caused by different viruses, and to Kawasaki disease, a malady not yet related to a virus with certainty.

13.3

Other Dermotropic Viral Diseases

Though dermotropic viral diseases tend to be benign, certain ones, such as smallpox, have exacted heavy tolls of human misery. Smallpox and other skin diseases are discussed in the final section of this chapter.

SMALLPOX (VARIOLA) IS A CONTAGIOUS AND SOMETIMES FATAL DISEASE

Smallpox has ravaged people around the world since prebiblical times. It moved swiftly across Europe and Asia, often doubling back on its path, and it was apparently brought to the New World in the 1500s by Cortez's troops. There it killed 3.5 million Native Americans and contributed to the collapse of the Inca and Aztec civilizations. Few people escaped the pitted scars that accompanied the disease, and children were not considered part of the family until they had survived smallpox.

Smallpox is caused by a brick-shaped double-stranded DNA virus of the *Poxviridae* family (**FIGURE 13.17**). It is one of the largest virions, approximately the size of chlamydiae. The nucleocapsid is surrounded by a series of fiber-like rods with an envelope. Transmission is by contact.

The earliest signs of smallpox are high fever and general body weakness. Pink-red spots, called **macules**, soon follow, first on the face and then on the body trunk. (In chickenpox, the spots appear randomly in crops.) The spots become pink pimples,

300 nm

smallpox virus

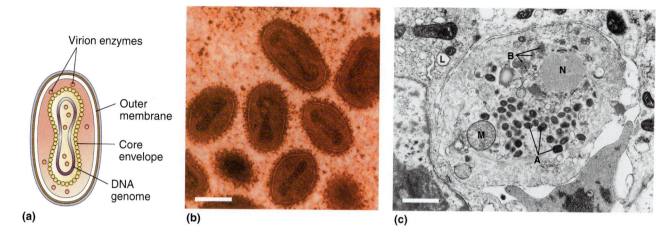

Virion enzymes

Outer membrane

Core envelope

DNA genome

(a)　　　(b)　　　(c)

FIGURE 13.17

The Smallpox Virus

(a) A drawing of the smallpox virus, showing its complex features. Fiber-like rods (not shown) are embedded in the outer membrane and envelope of the virus. (b) A false color transmission electron micrograph of the smallpox virus cultivated in cell culture. Note the brick-like shape of the virus and the characteristic rods at its surface. (Bar = 200 nm.) (c) A transmission electron micrograph of a cell infected with smallpox viruses. Rectangular mature virions can be observed (A) as well as immature virions (B). Also visible are the cell nucleus (N), mitochondrion (M), and lysosome (L). (Bar = 1440 nm.)

or **papules**, then fluid-filled **vesicles** so large and obvious that the disease is also called **variola**, from the Latin *varus* meaning "vessel" (**FIGURE 13.18**). The vesicles become deep **pustules**, which break open and emit pus. If the person survives, the pustules leave pitted scars, or **pocks**. These are generally smaller than the lesions of syphilis (the Great Pox) or varicella (chickenpox). **TABLE 13.2** summarizes the stages of smallpox.

Centuries ago, people discovered that they could survive smallpox if they were fortunate enough to experience the disease during a mild year. The custom thus arose of "buying the pox": One would approach a person who had a mild form and offer money to rub skin together. An Oriental custom of injecting oneself with pox fluid eventually spread to Europe, and later to North America. In 1721, a hospital was established in Boston for anyone interested in **variolation**, as the process was called.

In 1798, the English physician Edward Jenner noted that milkmaids contracted a mild form of smallpox named **cowpox**, or **vaccinia** (*vacca* is Latin for "cow"). Anyone who experienced cowpox apparently did not contract smallpox. Jenner therefore used material from a cowpox lesion for variolation and established the process of **vaccination**. His method was so successful that Napoleon ordered his entire army vaccinated in 1806. The effort to vaccinate the American population was led by an impressed President Thomas Jefferson, as the chapter-opening quote illustrates.

Cowpox:
a pox disease in animals that also may occur in humans.

Vaccination has been hailed as one of the greatest medical and social advances because it was the first attempt to control disease on a national scale. It was also the first effort to protect the community rather than the individual. A century passed before it was understood that the antibodies produced against the mild cowpox virus were equally effective in neutralizing the smallpox virus.

In 1966, the WHO received funding to attempt the global eradication of smallpox. Surveillance containment methods were used to isolate every known pox victim, and

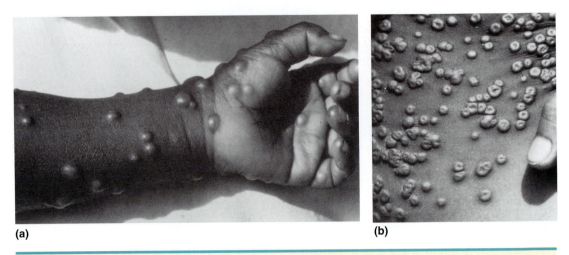

(a)   (b)

FIGURE 13.18

The Lesions of Smallpox

(a) The lesions are raised, fluid-filled vesicles similar to those in chickenpox. For this reason, cases of chickenpox have been misdiagnosed as smallpox. Later, the lesions will become pustules (b), and then form pitted scars, the pocks.

TABLE 13.2

Stages of Smallpox

STAGE	EXPLANATION	DURATION	CONTAGIOUS?
Incubation period	Following exposure to the virus, people do not have any symptoms and usually feel fine during the incubation period.	7–17 days (Average: 12–14 days)	No
Initial symptoms	First symptoms include high fever (38°–40°C), malaise, head and body aches, and sometimes vomiting. Affected individuals are too sick to carry on their normal activities.	2–4 days	No
Early rash	**Days 1 & 2:** A rash emerges as small red spots on the tongue and in the mouth, and develop into sores that break open and spread large amounts of the virus into the mouth and throat. About the time the sores in the mouth break down, a rash appears on the skin, starting on the face, and spreads to the arms and legs and then to the hands and feet. Usually the rash spreads to all parts of the body within 24 hours. As the rash appears, the fever usually falls and the person may start to feel better. **Day 3:** The rash becomes raised bumps. **Day 4:** The bumps fill with thick, opaque fluid and often have a depression in the center that looks like a belly button. (This is a major distinguishing characteristic of smallpox.) Fever often occurs again until scabs form over the bumps.	4 days	Very contagious
Pustular rash	The bumps become pustules—sharply raised, usually round and firm to the touch, as if there is a small round object under the skin. People often say the bumps feel like BB pellets embedded in the skin.	5 days	Yes
Pustules and scabs	By the end of the second week after the rash appears, the pustules begin to form a crust and then scab.	5 days	Yes
Resolving scabs	Within three weeks after the rash appears, the scabs begin to fall off, leaving marks on the skin that become pitted scars.	6 days	Yes
Scabs resolved	Scabs have fallen off.		No

Source: Centers for Disease Control. Available at: http://www.cdc.gov. Accessed July 2003.

all contacts were vaccinated (as noted in the chapter introduction). The eradication was aided by the fact that smallpox viruses apparently do not exist anywhere in nature except in humans. On October 26, 1977, health care workers reported isolation of the last case, and the WHO instituted a two-year waiting period to see if any new cases would appear. Finally, in 1979, the WHO announced worldwide smallpox eradication, the first such claim made for any disease.

Two of the key characteristics of smallpox are the person-to-person transmission and a kill rate of approximately 30 percent in exposed, susceptible individuals. Since vaccinations against smallpox stopped in the United States in 1972, a majority of Americans lacks immunity to the disease. This makes smallpox one of the most dangerous weapons of bioterrorism, even though a 2003 study suggests that those who were vaccinated prior to 1972 still have some level of immunity. In addition, the United States now has adequate stockpiles of smallpox vaccine to vaccinate every American if necessary.

In a purposeful attack, the virus most likely would be released in an aerosol form, where lack of surveillance could lead to a major outbreak. In another scenario, could a suicide "disease bomber" (terrorist) deliberately be infected with smallpox and fly to an American airport, infecting unknowing numbers of travelers who then spread it nationwide? It is unlikely that this scenario could occur because, as indicated in Table 13.2, smallpox is not communicable until the pox or pustules appear on the skin. By that time, the infected person would be so ill that it is highly unlikely the terrorist could get out of bed, yet fly on an airplane or roam the streets of a city spreading the disease.

There are two known stocks of live smallpox virus, one at the CDC in Atlanta and the other at a similar facility in Russia. However, the former Soviet Union produced massive amounts of smallpox virus during the "Cold War" years, so there may be stocks that "walked away" to rogue nations or terrorist organizations after the fall of the Soviet Union in 1991. The destruction of the remaining smallpox stocks at the CDC and in Russia has been planned by the WHO. However, it has been postponed several times because of the controversy over the value of keeping smallpox stocks (MicroFocus 13.8).

MONKEYPOX IS A NEW DISEASE TO THE WESTERN HEMISPHERE

In May 2003, an outbreak of monkeypox was reported in the Midwest (Wisconsin, Illinois, Indiana, Missouri, Kansas, and Ohio) among dozens of people who became sick after having contact with wild or exotic mammalian pets (primarily prairie dogs or Gambian giant rats). Pet prairie dogs were infected at a pet shop through contact with Gambian giant rats and dormice that originated in Ghana. This is the first time that there has been an outbreak of monkeypox in the Western Hemisphere (FIGURE 13.19).

Monkeypox is a rare viral disease that occurs mostly in central and western Africa. It is called "monkeypox" because it was first found in 1958 in laboratory primates. Blood tests of animals in Africa later found that other types of animals probably had monkeypox. Scientists also recovered the virus that causes monkeypox from an African squirrel, which might be the common host for the disease. Rats, mice, and rabbits can get monkeypox also. In 1970, the first case of monkeypox was reported in humans.

The monkeypox virus is a double-stranded DNA virus in the same group of *Poxviridae* as the smallpox and cowpox viruses. People get monkeypox from an

MicroFocus 13.8

"SHOULD WE OR SHOULDN'T WE?"

One of the liveliest debates in microbiology is whether the last remaining stocks of smallpox viruses should be destroyed. Here are some of the arguments.

For Destruction:

■ People are no longer vaccinated, so if the virus should escape the laboratory, a deadly epidemic could ensue.

■ The DNA of the virus has been sequenced, and many clones of fragments are available for performing research experiments; therefore, the whole virus is no longer necessary.

■ Eradicating the disease means eradicating the remaining stocks of laboratory virus, and the stocks must be destroyed to complete the project.

■ If the United States and Russia destroy their smallpox stocks, it will send a message that biological warfare cannot be tolerated.

Against Destruction:

■ Future studies of the virus are impossible without the whole virus. Indeed, certain sequences of the viral genome defy deciphering by current laboratory means.

■ Studying the genome of the virus without the whole virus will not provide insights into how the virus causes disease.

■ Mutated viruses could cause smallpox-like diseases, so continued research on smallpox is necessary in order to be prepared.

■ Smallpox viruses may be secretly retained in other labs in the world for bioterrorism purposes, so destroying the stocks may create a vulnerability.

Smallpox viruses also may remain active in buried corpses.

■ Destroying the virus impairs the scientist's right to perform research, and the motivation for destruction is political, not scientific.

Now it's your turn. Can you add any insights to either list? Which argument do you prefer?

P.S. In April 2002, the World Health Assembly of the World Health Organization recommended postponing the destruction of all remaining smallpox stocks until all research and drug development is concluded. This will allow time to prepare for a natural outbreak or potentially deliberate (bioterrorist) release of variola virus.

---

infected animal through bites or contact with the animal's blood, body fluids, or its rash. The disease also can spread from person to person through large respiratory droplets during long periods of face-to-face contact, or by touching body fluids of a sick person or objects such as bedding or clothing contaminated with the virus. After several generations of transmission, the disease apparently stops spreading.

In humans, the signs and symptoms of monkeypox are similar to smallpox, but usually much milder and less contagious. After a 12-day incubation period, people infected with the virus will experience fever, headache, muscle aches, and backache; their lymph nodes may swell and many develop a cough. One to three days (or longer) after the fever starts, a rash develops that progresses into raised bumps filled with fluid. It often starts on the face and spreads across the body. The bumps go through several stages before they get crusty, scab over, and fall off. The illness usually lasts for two to four weeks.

In Africa, monkeypox has killed between 1 percent and 10 percent of people who were infected. However, this risk probably would be lower in the United States, where nutrition and access to medical care are better. Of the 72 cases reported to the CDC by July 2003, there were no deaths although 18 were hospitalized.

There is no specific treatment for monkeypox. However, since the monkeypox virus is closely related to smallpox, the smallpox vaccine may protect people from getting monkeypox. In Africa, for example, people who received the smallpox vac-

FIGURE **13.19**

The First Case of Monkeypox

This outbreak occurred in the Midwest in June 2003. It represents the first outbreak of monkeypox in the Western Hemisphere and demonstrates the danger associated with the import of exotic animals.

TEXTBOOK CASES

1. On April 9, 2003, a Texas animal distributor received a shipment of some 800 small mammals from Accra, Ghana. This included rope squirrels, Gambian giant rats, and dormice.

2. Twelve days later, an Illinois distributor received Gambian rats and dormice from the Texas distributor. A Gambian rat was housed with prairie dogs that the distributor had before the prairie dogs were sold to distributors in six states. Additional prairie dogs were sold at animal swap meets.

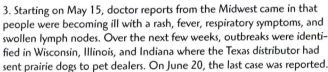

3. Starting on May 15, doctor reports from the Midwest came in that people were becoming ill with a rash, fever, respiratory symptoms, and swollen lymph nodes. Over the next few weeks, outbreaks were identified in Wisconsin, Illinois, and Indiana where the Texas distributor had sent prairie dogs to pet dealers. On June 20, the last case was reported.

4. In the June 30, 2003, final report, 72 cases of monkeypox were reported to the CDC from Wisconsin, Illinois, Indiana, Missouri, Kansas, and Ohio. Of these, 37 were laboratory confirmed and 35 were suspect and probable cases. Confirmation of monkeypox was determined by DNA analysis, immunological testing, and electron microscopy.

cine in the past had a lower risk of contracting monkeypox. Besides the smallpox vaccine being effective at protecting people against monkeypox when it is given before they are exposed to monkeypox, experts also believe that vaccination after exposure may help prevent the disease or make it less severe. People with life-threatening allergies to latex or to the smallpox vaccine or any of its ingredients (polymyxin B, streptomycin, chlortetracycline, neomycin) should not get the smallpox vaccine.

MOLLUSCUM CONTAGIOSUM FORMS MILDLY CONTAGIOUS SKIN LESIONS

Molluscum contagiosum is a viral disease accompanied by wart-like skin lesions. The lesions are firm, waxy, and elevated with a depressed center. When pressed, they yield a milky, curd-like substance. Although usually flesh toned, the lesions may appear white or pink. Possible areas involved include the facial skin and eyelids in children, and the external genitals in adults. The lesions may be removed by excising them (cutting them out) and pose no public health threat.

The virus of molluscum contagiosum is an enveloped double-stranded DNA virion of the *Poxviridae* family. Transmission is generally by contact, such as by sexual contact. A characteristic feature of the disease is the presence of large cytoplasmic bodies called **molluscum bodies** in infected cells from the base of the lesion.

WARTS OR CERVICAL CANCER CAN BE PRODUCED BY PAPILLOMA VIRUSES

Warts are small, usually benign skin growths that are commonly due to viruses. **Plantar warts** occur on the soles of the feet. **Genital warts** are often transmitted in sexual contact. These warts are sometimes called **condylomata**, from the Greek *kondyloma*, meaning "knob," a reference to the fig-like appearance of the warts. They are usually moist and pink.

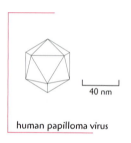

40 nm

human papilloma virus

One of the primary causes of warts are the **human papilloma viruses (HPV)**, a collection of over two dozen types of icosahedral, naked double-stranded DNA virions of the *Papovaviridae* family. In most cases, the skin warts produced by certain strains cause a minor problem. However, evidence suggests that other strains of HPV may be associated with **cervical cancer**. Indeed, in 1994 researchers reported the identification of DNA from 25 types of papilloma viruses in the tumor cells of 95 percent of patients having cervical cancer.

Genital warts caused by papilloma viruses may be transferred during sexual intercourse, and some virologists suggest that this condition may be more prevalent than genital herpes. An estimated 4 million Americans are believed to be infected, and many harbor and transmit the virus without experiencing symptoms. Malignancies of the cervix, vagina, penis, and anus also have yielded papilloma viruses. Moreover, a pregnant woman with genital warts may transmit the viruses during the birth process, and studies show that in the newborn the viruses may lodge in the larynx, trachea, and the lungs. A controversial study reported in 1999 has linked the viruses to prostate cancer.

TABLE 13.3 presents a summary of the dermotropic viral diseases covered in this chapter.

KAWASAKI DISEASE IS THE LEADING CAUSE OF CHILDHOOD HEART DISEASE

Although Kawasaki disease has not yet been identified as a viral disorder, the course of the disease suggests that an infectious agent is involved. We therefore shall consider it here.

Kawasaki disease (KD) takes its name from the Japanese pediatrician Tomisaku Kawasaki, who described its symptoms in 1967. Children one to two years old account for most patients. The illness is characterized by a high fever and sore throat; then, **red spots** appear, first on the extremities and next the body trunk. As the disease progresses, there is a characteristic peeling of the skin about the fingers,

TABLE 13.3

A Summary of Dermotropic Viral Diseases

DISEASE	CLASSIFICATION OF VIRUS	TRANSMISSION	ORGANS AFFECTED	VACCINE	SPECIAL FEATURES	COMPLICATIONS
Herpes simplex	Herpesviridae	Contact	Skin Pharynx Genital organs	Not available	Characteristic lesions Lipschütz bodies Acyclovir treatment	Encephalitis Congenital infections Neonatal herpes
Chickenpox (varicella)	Herpesviridae	Droplets Contact	Skin Nervous system	Attenuated viruses	Characteristic lesions Crops Acyclovir treatment	Herpes zoster (shingles) Reye's syndrome Pneumonia
Measles (rubeola)	Paramyxoviridae	Droplets Contact	Respiratory tract Skin, blood	Attenuated viruses	Koplik spots Progress of rash Hemagglutination inhibition	SSPE Pneumonia Encephalitis
Rubella (German measles)	Togaviridae	Droplets Contact	Skin, blood	Attenuated viruses	Skin rash Mild cold symptoms	Congenital rubella syndrome
Fifth disease (Erythema infectiosum)	Parvoviridae	Droplets (?)	Skin, blood	Not available	"Slapped-cheek" appearance Lacy skin patterns	None established
Mumps	Paramyxoviridae	Droplets	Salivary glands Blood	Attenuated viruses	Swollen glands Hemagglutination inhibition	Orchitis Encephalitis Meningitis
Smallpox (variola)	Poxviridae	Contact Droplets Fomites	Skin Blood	Cowpox viruses	Characteristic lesions Macules, papules, vesicles, pox	Permanent scarring
Monkeypox	Poxviridae	Contact	Skin	Smallpox vaccine	Characteristic rash	Not known
Molluscum contagiosum	Poxviridae	Contact	Skin	Not available	Characteristic lesions	None
Warts	Papovaviridae	Contact	Skin	Not available	Characteristic lesions	None
Kawasaki disease	(Unknown)	(Unknown)	Skin, blood	Not available	Skin rash Desquamation	Heart involvement

a phenomenon called **desquamation**. Complications of KD usually involve the cardiovascular system, and internal bleeding may occur. Treatment is generally directed at minimizing these possibilities. Intravenous gamma globulin is effective if given at onset of fever.

KD was first detected in the United States in 1971. Since then, it has been reported with increasing frequency, and about 2,500 cases were reported through the 1980s. A notable outbreak in 1985 in the Denver, Colorado, area involved 61 people.

There is some laboratory evidence that a staphylococcus or a streptococcus may be involved in KD. The possibility also exists that the disease is an immune-related syndrome complicating some unknown viral disease in the same manner that Reye's syndrome follows influenza and chickenpox. Some physicians therefore refer to the illness as **Kawasaki syndrome**. Indeed, the CDC has made no mention of the disease in its reports since 1989, and most other microbiology textbooks do not discuss it. Perhaps KD also will disappear from this textbook one day, but for the time being, it is important to be aware of its existence.

Note to the Student

In this chapter, we have had an opportunity to note the changing complexion of microbiology. We have seen, for example, how measles viruses are increasingly associated with neurological problems, how Reye's syndrome has become linked to influenza and chickenpox, how an association is growing between cervical tumors and herpes simplex viruses, and how chickenpox and herpes zoster, once thought to be separate diseases, are now linked to the same virus. Moreover, we note that despite modern detection devices, there is still no identifiable agent for Kawasaki disease. Most recently, diseases caused by the SARS-associated coronavirus and monkeypox virus point to the continuing threat of emerging infectious diseases.

The point is that microbiology is a dynamic and ever-changing science. This dynamism implies that microbiologists and physicians must continually adjust to new truths as they emerge. A friend once told me that if you dip a tennis ball into the ocean, the water dripping from the ball represents all that is known; the ocean represents all that is waiting to be discovered.

Summary of Key Concepts

13.1 PNEUMOTROPIC VIRAL DISEASES

■ **Influenza Is a Highly Communicable Disease**. Influenza is caused by three different orthomyxoviruses, types A, B, and C. The spike proteins neuraminidase and hemagglutinin are necessary for viral entry and exit during an infection. Antigenic drift and antigenic shift account for the yearly differences in flu strains and for flu pandemics. Symptoms include sudden chills, headache, fatigue, and chest pain. Influenza is best prevented with yearly vaccination, although new antiviral drugs can shorten the duration of symptoms.

■ **Adenovirus Infections Often Are Responsible for Colds.** Adenoviruses cause some types of colds. These DNA viruses also cause keratoconjunctivitis. They have been used in gene therapy as the vector to carry therapeutic genes.

■ **Respiratory Syncytial Disease Affects the Lower Respiratory Tract in Young Children.** The RS virus is a paramyxovirus. After infecting cells, it causes the cells to fuse into syncytia. RS disease often causes a type of viral pneumonia in children and an upper respiratory disease in adults. Immune globulin and antivirals have been used to lessen the symptoms.

■ **Parainfluenza Accounts for 40 Percent of Acute Respiratory Infections in Children.** Parainfluenza is another paramyxovirus that produces milder symptoms than influenza. No specific therapy exists.

■ **Rhinovirus Infections Produce Inflammation in the Upper Respiratory Tract.** Over 100 rhinoviruses are members of the picornaviruses. They are transmitted through airborne droplets or by contaminated objects. Symptoms produce a typical head cold.

■ **Severe Acute Respiratory Syndrome Spreads through Close Person-to-Person Contact.** SARS represents a newly emerging viral disease caused by a coronavirus. It is spread by person-to-person contact. Symptoms include fever, headache, feeling of discomfort, and body aches. A dry cough and difficulty breathing often occur. In severe illness, insufficient oxygen reaches the blood and mechanical ventilation is required.

13.2 DERMOTROPIC VIRAL DISEASES

- **Herpes Simplex Are Widespread, Often Recurrent Infections.** Herpes simplex describes a wide spectrum of viral diseases commonly found in the environment. Among the herpesviruses are ones that cause cold sores (type I) and those that cause genital herpes (type II). Several antiviral drugs have been developed to treat herpes simplex infections.

- **Other Herpesvirus Infections Have Been Detected Recently.** Herpesviruses cause other infections. Human herpesvirus 6 causes roseola and has been implicated in multiple sclerosis. Human herpesvirus 8 causes Kaposi's sarcoma, which is very prevalent in AIDS patients.

- **Chickenpox (Varicella) Is a Common, Highly Contagious Disease.** Another member of the herpesviruses is varicella-zoster. This virus causes chickenpox, which is one of the most highly contagious diseases. The same virus causes shingles in adults, which can be a painfully debilitating disease. Acyclovir has been successful at lessening the symptoms.

- **Measles (Rubeola) Is Another Highly Communicable Disease.** The measles virus is a member of the paramyxoviruses. Symptoms of measles include a hacking cough, sneezing, eye redness, sensitivity to light, and a high fever. Koplik spots are common along the gum line. The characteristic red rash appears a few days after the Koplik spots appear.

- **Rubella (German Measles) Is an Acute but Mildly Infectious Disease.** Rubella is a member of the togaviruses, a group of single-stranded (+ strand) RNA viruses. Infection leads to a fever and a pale-pink rash beginning on the face and spreading over the body and extremities. Congenital rubella syndrome can develop if a pregnant woman has rubella.

- **Fifth Disease (Erythema Infectiosum) Produces a Mild Rash.** Fifth disease is caused by the B19 virus, which is a member of the parvoviruses. These single-stranded DNA viruses are spread by the respiratory route. Infection produces a red rash that resembles a slapped cheek. In adults, an infection can produce painful joints.

- **Mumps Is Spread by Respiratory Droplets.** Mumps is produced by a viral infection from a member of the paramyxoviruses. The disease produces swollen parotid (salivary) glands. The swelling of the testes (orchitis) may be a complication of the disease.

13.3 OTHER DERMOTROPIC VIRAL DISEASES

- **Smallpox (Variola) Is a Contagious and Sometimes Fatal Disease.** The most dangerous virus in the groups described in this chapter is variola, a membrane of the poxviruses. Smallpox symptoms start with a fever and body weakness followed by the development of pustules on the face and body. Survival leaves pockmarks where pustules had formed and scabbed. Smallpox was the first disease purposely eradicated from planet Earth, which has left the possibility of the virus being used by bioterrorists as a biological weapon.

- **Monkeypox Is a New Disease to the Western Hemisphere.** Monkeypox is a new disease to the Western Hemisphere. This poxvirus causes mild symptoms. In the recent outbreak in the United States, most infections came from animal bites from wild prairie dogs that had been infected from an imported Gambian giant rat. The disease produces a rash that forms raised bumps filled with fluid before they crust and scab over.

- **Molluscum Contagiosum Forms Mildly Contagious Skin Lesions.** The disease is caused by another poxvirus but not closely related to smallpox or monkeypox. The virus produces mild wart-like skin lesions that pose no public health threat.

- **Warts or Cervical Cancer Can Be Produced by Papilloma Viruses.** Different strains of the human papillomaviruses (HPV) produce warts or cervical cancer. Many of the strains of HPV produce benign viral tumors, such as plantar warts, on the skin. Genital warts are caused by other HPV strains and are transmitted by sexual intercourse. Malignancies of the cervix, vagina, and penis have been associated with such HPV infections.

- **Kawasaki Disease Is the Leading Cause of Childhood Heart Disease.** Most cases of childhood heart disease are due to Kawasaki disease. It is not known if this disease is caused by a virus or other microbe. Most infections occur in children one to two years old. The symptoms include high fever and a sore throat. Red spots form on the extremities before spreading over the body. Complications can lead to heart disease and internal bleeding.

Questions for Thought and Discussion

Answers to selected questions can be found in Appendix C.

1. In the mid-1980s, the nation's colleges for the deaf reported an unprecedented demand for admission. For example, at the National Technical Institute for the Deaf at Rochester Institute of Technology, the student body swelled from 750 students to 1,250 students. How was this related to the events of a previous generation involving rubella?

2. In February 1992, the CDC reported an outbreak of measles at an international gymnastics competition in Indianapolis, Indiana. A total of 700 athletes and numerous coaches and managers from 51 countries were involved. Although the potential for a disastrous international epidemic was high, it never materialized. What steps do you think the local health agencies took to quell the spread of the disease?

3. A little girl experiences frequent and severe vomiting for a period of hours. She soon becomes sleepy and glassy eyed. When disturbed, she quickly becomes irritated and combative. One week before, she had recovered from chickenpox. What is she experiencing, and what course of action must be taken?

4. Thomas Sydenham, the "English Hippocrates," was a London physician in the seventeenth century. In 1661, he differentiated measles from scarlet fever, smallpox, and other fevers, and set down the foundations for studying these diseases. How would a modern Thomas Sydenham go about distinguishing the variety of look-alike skin diseases discussed in this chapter?

5. In the United Kingdom, the approach to rubella control is to concentrate vaccination programs on young girls just before they enter the childbearing years. In the United States, the approach is to immunize all children at the age of 15 months. Which approach do you believe is preferable? Why?

6. One way of avoiding the viruses that cause common colds is to adopt the motto, "Let us spray." This motto refers to using a disinfectant spray to destroy viruses on environmental surfaces. How many *other* ways can you name for avoiding cold viruses?

7. In 1994, the fitness file of a local newspaper carried a story on "The New Herpes." The reference was to human papilloma viruses and genital warts as the herpes simplex of the 1990s. How many similarities can you find between these two diseases? Would you agree with the comparison?

8. Despite its availability and effectiveness, many physicians do not recommend the chickenpox vaccine to parents and their children. Indeed, the CDC reported that fewer than 20 percent of candidate children were immunized during 1995 and 1996. Why do you think some physicians are reluctant to use the vaccine? Do you believe their skepticism is warranted? What might you recommend to justify or to overcome the reluctance to use the vaccine?

9. Most physicians agree that there would be great demand for a genital herpes vaccine. However, there is much opposition to marketing a vaccine that contains attenuated herpes simplex viruses. Why is this so? What alternatives are there for a useful vaccine against genital herpes?

10. It is not uncommon for a person with respiratory disease to visit the doctor and be told, "Don't worry about it. It's just a touch of the flu. I'll give you a shot of penicillin before you leave." Suppose the person wanted to know if it really was influenza. What diagnostic tests would have to be performed? Also, why is the doctor inclined to give a penicillin injection?

11. A child experiences "red bumps" on her face, scalp, and back. Within 24 hours, they have turned to tiny blisters and become cloudy, some developing into sores. Finally, all become brown scabs. New "bumps" keep appearing for several days, and her fever reaches 39°C by the fourth day. Then the blisters stop coming and the fever drops. What disease has she had?

12. The great seventeenth-century physician William Harvey, who discovered how blood circulates, was a great fan of garlic therapy to treat disease. In one of his writings, Harvey recommended putting a clove of garlic inside your shoe when you have a respiratory illness. What do you think of Harvey's recommendation?

13. One day in March 1977, a Boeing 737 bound for Kodiak, Alaska, developed engine trouble and was forced to land. While the company rounded up another aircraft, the passengers sat waiting for 4 hours in the unventilated cabin. One passenger, it seemed, was in the early stages of influenza and was coughing heavily. By the week's end, 38 of the 54

passengers on the plane had developed influenza. What lessons does this incident teach?

14. Although smallpox viruses are considered to be gone from the environment, there remains a closely related virus that causes monkeypox in nature. By what genetic mechanisms could this virus conceivably become a smallpox virus?

15. A man experiences an attack of shingles and is warned by his doctor to stay away from children as much as possible. Why is this advice given? Is it justified?

Review

On completing your study of pneumotropic and dermotropic viral diseases, test your comprehension of the chapter contents by circling the choices that best complete each of the following statements. The answers are listed in Appendix D.

1. Rhinoviruses are a collection of (RNA, DNA) viruses having (helical, icosahedral) symmetry and the ability to infect the (air sacs, nose), causing (mild, serious) respiratory symptoms.

2. Herpes simplex is a viral disease that can be transmitted by (breathing contaminated air, contact) and is characterized by thin-walled (blisters, ulcers) that often appear during periods of (emotional stress, exercising), but can be treated with a drug called (deoxycyclovir, acyclovir).

3. In children, the skin lesions of chickenpox occur (all at once, in crops) and resemble (teardrops, pitted scars), but in adults the lesions are known as (shingles, erythemas) and resemble blotchy patches of (blue, red) that are very (itchy, painful).

4. For generations, rubella was thought to be a mild form of (chickenpox, measles) because it also was accompanied by (a skin rash, brain lesions) and was transmitted by (contaminated water, airborne droplets).

5. The complications of influenza include (Reye's, Koplik) syndrome; for mumps, the complication is a disease of the (testes, pancreas) called (colitis, orchitis), and for measles, it is a disease of the (brain, liver) known as (SSPE, GBS).

6. One of the early signs of (smallpox, measles) is a series of (Koplik spots, Lipschütz bodies) occurring in the (lungs, mouth) and signaling that a (red rash, blue-green rash) is forthcoming.

7. Although the agent of (fifth, sixth) disease has not been identified with certainty, the leading candidate is the (B29, B19) strain of (picornavirus, parvovirus), a small (DNA, RNA) virus.

8. After transmission by (mosquitoes, airborne droplets), the virus of (mumps, Kawasaki disease) spreads by the blood to the (salivary, sweat) glands, where it interferes with fluid secretion.

9. Although now eradicated, (mumps, smallpox) can be prevented by immunizations with (fowlpox, cowpox) virus in a method first devised in 1798 by Edward (Jennings, Jenner).

10. Antigenic variation among (mumps, influenza) viruses seriously hampers the development of a highly effective (vaccine, treatment), and a life-threatening situation can occur if secondary infection due to (fungi, bacteria) complicates the primary infection.

11. Respiratory syncytial disease is caused by a (DNA, RNA) virus that infects the (lungs, intestines) of (adults, children) and induces cells to (clump together, move apart) and form giant cells called (syncytia, tumors).

12. Severe acute respiratory syndrome is caused by a (coronavirus, orthomyxovirus), a (naked, enveloped) virus that is spread by (sexual, person-to-person) contact.

13. Adenoviruses include a collection of (DNA, RNA) viruses that induce the formation of (granules, inclusions) and are responsible for (yellow fever, common colds), as well as infections of the (eye, ear).

14. Genital herpes is caused by a (helical, icosahedral) virus that is believed to affect 10 to 20 (thousand, million) Americans each year, causing blisters with (thick, thin) walls that disappear in about three (days, weeks), only to reappear when (stress, physical injury) occurs.

15. The TORCH diseases are a set of (infectious, physio-logical) diseases transmitted by (airborne droplets, transplacental passage), occurring in (the elderly, newborns), and including (rubeola, rubella) and (herpes simplex, humoral disease).

16. The MMR vaccine contains (inactivated, attenuated) viruses and is used primarily in (children, older adults) to provide (long-term, short-term) immunity to such diseases as (measles, molluscum contagiosum), (mononucleosis, mumps), and (German measles, influenza).

http://microbiology.jbpub.com

The site features **eLearning,** an on-line review area that provides quizzes and other tools to help you study for your class. You can also follow useful links for in-depth information, or just find out the latest micro-biology news.

Viscerotropic and Neurotropic Viral Diseases

I don't think we are losing the war, but we're certainly not finished with the war.

—Ronald Valdiserri, CDC Deputy Director, speaking on the upsurge of HIV cases in the United States

The year 2001 marked 20 years since the acquired immune deficiency syndrome (AIDS) epidemic started. Now 23 years into the epidemic, there still are only somber numbers to announce. Worldwide, an estimated 42 million people are infected with the human immunodeficiency virus (HIV), 20 million have died from AIDS, and 750,000 babies are born each year with HIV infection (**FIGURE 14.1**). The Joint United Nations Programme on HIV/AIDS (UNAIDS) has projected that 70 million people will die from AIDS in the next 20 years—if a cure is not found. Indeed, there still is no cure or vaccine, and antiretroviral therapies can be toxic and quickly become less useful as HIV develops resistance. HIV certainly is one of the most complex and deadly viruses to ever strike humanity.

The epidemic is worst in sub-Saharan Africa where almost 30 million people have HIV infection. In some African cities, a staggering 30 to 50 percent—almost half the population—is HIV-positive. Without a cure, by 2020 more than 25 percent of the workforce in some African cities will be lost because of AIDS. In July 2003, President Bush pledged $15 billion to fight AIDS in Africa and the Caribbean where almost half of the world's HIV infections are located. The plan to combat AIDS calls for antiviral treatment for 2 million HIV-infected people who cannot afford the costly cocktail of drugs that can prolong and improve their lives. The initiative also provides hospice care for the dying, helps some of the 13 million children who have

■ **14.1 Viscerotropic Viral Diseases**
Yellow Fever Is Transmitted by Mosquitoes
Dengue Fever Can Progress to a Hemorrhagic Fever
Infectious Mononucleosis Is Most Common in Children and
  Young Adults
Hepatitis A Is Transmitted Most Often by the Fecal-Oral Route
Hepatitis B Can Produce a Chronic Liver Infection
Hepatitis C Often Causes Cirrhosis and Liver Failure
Other Viruses Also Cause Hepatitis

■ **14.2 Other Viscerotropic Viral Diseases**
Viral Gastroenteritis Is Caused by Any of Several Viruses
Viral Fevers Can Be Caused by a Variety of Viruses
Hantavirus Pulmonary Syndrome Produces Sudden Respiratory Failure
Cytomegalovirus Disease Can Produce Serious Birth Defects
Acquired Immune Deficiency Syndrome Results from Immune System
  Dysfunction

■ **14.3 Neurotropic Viral Diseases**
Rabies Is a Highly Fatal Infectious Disease
Polio May Be the Next Infectious Disease Eradicated
Arboviral Encephalitis Is Caused by Blood-Sucking Arthropods
West Nile Fever Is an Emerging Disease in the Western Hemisphere
Lymphocytic Choriomeningitis Follows Exposure to Rodent Feces

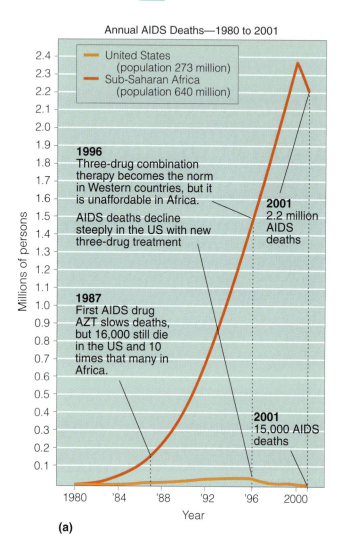

Annual AIDS Deaths—1980 to 2001

1996
Three-drug combination therapy becomes the norm in Western countries, but it is unaffordable in Africa.

AIDS deaths decline steeply in the US with new three-drug treatment

2001
2.2 million AIDS deaths

1987
First AIDS drug AZT slows deaths, but 16,000 still die in the US and 10 times that many in Africa.

2001
15,000 AIDS deaths

United States (population 273 million)
Sub-Saharan Africa (population 640 million)

Millions of persons

Year

(a)

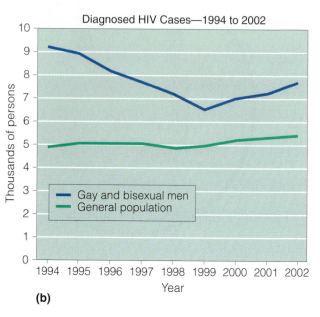

Diagnosed HIV Cases—1994 to 2002

Thousands of persons

Gay and bisexual men
General population

Year

(b)

FIGURE 14.1

Annual AIDS Deaths and Diagnosed Cases

(a) Deaths from AIDS in the United States and Sub-Saharan Africa between 1980 and 2001. (b) In the United States, a study of 25 states indicates that there was a 7.1 percent increase in newly diagnosed cases in gay and bisexual men from 2001 to 2002. This was the third straight year the number has increased. During the same time interval, the general population showed a 2.2 percent increase in AIDS diagnoses worldwide. (©2001, *The Washington Post*. Reprinted with permission. Inset: data from the CDC.)

---

lost one or both parents, and intensifies prevention programs through strategies like sexual abstinence, education, and promotion of condom use.

Africa is not the only worry. In China, in 2001 the number of reported HIV infections rose by 67 percent and in some provinces UNAIDS reports the HIV infection rate among drug users is greater than 70 percent. In Eastern Europe, sexually transmitted diseases have been doubling every year since 1998, which means HIV infections also are increasing at alarming rates.

Compared to Africa, Asia, and Eastern Europe, in most developed nations HIV prevalence hardly has changed, in part because of the number of people on anti-retroviral drugs. News in the United States reflects this where the number of new HIV cases has remained level in the heterosexual community—that was until 2002.

In the summer of 2003, federal health authorities announced that the number of diagnosed cases of HIV rose in the United States in 2002 for the first time in a decade. The CDC reported 42,136 AIDS diagnoses in 2002, which represents a 2.2 percent increase from the previous year and the first rise since 1993. In 2002, HIV diagnoses among gay and bisexual men rose 7.1 percent, an increase of nearly 18 percent since 1999.

American health officials attribute this disturbing upswing to a growing complacency about the dangers of AIDS. In addition, the younger generation does not remember the devastation of the AIDS epidemic and perhaps feels safer with the advent of life-extending antiretroviral drugs available.

New cases continue to rise because people are not being diagnosed early enough to prevent them passing the infection to others. American officials also believe people are finding it difficult to adhere to the often complex drug regimens. In addition, complacency comes from the fact that there was a 5.9 percent decline in deaths in 2002 (16,371 AIDS deaths) from 2001.

A major focus of the chapter will be on AIDS. We also shall survey Ebola fever, hepatitis, polio, and other diseases that are discussed in the news media regularly.

Overall, the diseases addressed in this chapter fall into two general categories. Some illnesses, such as yellow fever and mononucleosis, are regarded as **viscerotropic** diseases because they affect the blood and visceral organs. The second category of illnesses are the **neurotropic** diseases, such as polio, rabies, and West Nile encephalitis, which affect the central nervous system. As in Chapter 13, each disease is presented as a separate essay, and you may select the order of study most suitable to your needs.

14.1

Viscerotropic Viral Diseases

The viscerotropic viral diseases affect such organs as the blood, liver, spleen, and the small and large intestines. To reach these organs, the viruses are generally introduced to the body tissues by arthropods or by contaminated food and drink, as we shall see in the discussions that follow.

YELLOW FEVER IS TRANSMITTED BY MOSQUITOES

The earliest known outbreak of **yellow fever** in the Western Hemisphere took place in Central America in 1596. The disease spread rapidly, and it soon rendered large regions of the Caribbean and tropical Americas almost uninhabitable. In time, natives developed immunity or suffered only mild cases, but the mortality rate in outsiders remained high. Finally, in 1901, a group led by Walter Reed identified mosquitoes as the agents of transmission (MicroFocus 14.1). With widespread vector control, the incidence rate of the disease gradually declined. Today, yellow fever is endemic in 33 countries in South America and Africa.

Yellow fever was the first human disease associated with a virus. The causative agent is a single-stranded (+ strand) RNA virion of the *Flaviviridae* family with icosahedral symmetry and an envelope. It is an example of an **arbovirus** because it is *ar*thropod-*bo*rne.

Yellow fever occurs in nature in monkeys and other jungle animals, where the virus is transmitted by various mosquitoes, including species of *Haemogogus*. Epidemics occur when a person is bitten in the forest and then travels to urban areas where the virus is passed via a blood meal by a different mosquito, *Aedes aegypti*, which then transmits the virus among humans. (The tiger mosquito *Aedes albopictus* also has been implicated in transmission.) *Aedes aegypti* is common in the Caribbean region and in the southern and eastern United States, and yellow fever was a problem in these areas for many generations. In 1803, for example, emperor Napoleon of France sent troops to quell an uprising in Haiti, but yellow fever killed

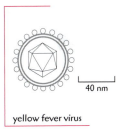

yellow fever virus

40 nm

MicroFocus 14.1

"FOR THE CAUSE OF HUMANITY . . ."

During the Spanish American War, the US government became disease conscious because more soldiers were dying from disease than from bullet wounds. Yellow fever was particularly bad in Cuba, where it exacted a heavy toll. When the war was over, Cuba remained under US control, and the Surgeon General sent a commission of four men to study the disease. Led by Major Walter Reed, the group included three assistant surgeons: James Carroll, Jesse W. Lazear, and Aristides Agaramonte. On June 25, 1900, the four men assembled in Cuba and began their work.

At first the commission devoted its energy to isolating a bacterium, but none could be found. As the weeks wore on, the investigators were impressed with the peculiar way the disease jumped from house to house, even when there was no contact with infected persons or contaminated objects. They also visited Carlos J. Finlay, a physician from Havana who insisted that mosquitoes were involved in transmission. If his theory was true, then the disease could be interrupted by simply killing the mosquitoes.

By now it was August, and Reed had been called back to Washington. Carroll, Lazear, and Agaramonte pushed forward and bred mosquitoes from eggs given them by Finlay. They allowed the mosquitoes to feed on patients with established cases of yellow fever, and then they applied the insects to the skin of volunteers, including themselves. The results were inconclusive: Some volunteers got yellow fever, but others did not. Two accidents then saved the research. One was fortunate, the other tragic.

One day in late August, Carroll decided to feed an "old" mosquito some of his own blood, lest it die. Three days later, Carroll was ill with the fever.

Lazear's notebook recorded that the insect had fed "*twelve days* before on a yellow fever patient, who was then in his *second day* of the disease." This, they discovered, was the proper combination of two factors necessary for a successful transmission. The disease could be reproduced over and over again, if this procedure was followed.

Then came tragedy. Lazear was working at the bedside of a yellow fever patient when a stray mosquito settled on his wrist. For reasons not clear, Lazear let the insect drink its fill. Five days later, he developed yellow fever; on the seventh day of his illness, he died. Lazear had been bitten previously, but apparently by uninfected mosquitoes. This time, the mosquito was infected.

Reed returned to Cuba in October, and a new set of experiments was planned to prove once and for all that clothing and other objects could not transmit yellow fever. The experiments were gruesome: Some volunteers slept in blood-soaked and vomit-stained garments of disease victims; others allowed themselves to be mercilessly bitten by mosquitoes; still others came forward to be injected with blood from yellow fever victims. By late 1900, there was no doubt that mosquitoes were the carriers of yellow fever.

It has been said that the experiments performed in Cuba are among the noblest in the history of medicine. Two volunteers, Private John R. Kissinger and clerk John J. Moran, were asked why they were agreeing to such life-threatening experiments. "We volunteer," they replied, "solely for the cause of humanity and in the interest of science."

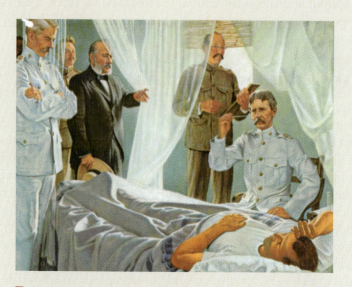

■ *A painting by Robert Thom depicting members of the yellow fever commission at the bedside of Private John Kissinger, after he was bitten by infected mosquitoes. Left to right: Major W. C. Gorgas (Havana sanitation officer), Aristides Agaramonte, Carlos J. Finlay, James Carroll, and Walter Reed. The experiments were conducted at Camp Lazear, named for Jesse W. Lazear, who had died of yellow fever the previous summer.*

thousands of his men. Soon thereafter, Napoleon came to think of the Americas as a fever-ridden land, and when President Thomas Jefferson sent emissaries to negotiate the purchase of the French New Orleans region, Napoleon offered the entire Louisiana territory at a bargain price.

Yellow fever can be a fatal disease. WHO estimates there are over 200,000 cases and 30,000 deaths annually. Mosquitoes inject the viruses into the bloodstream, and fever mounts within days. Infection of the liver causes an overflow of bile pigments into the blood, a condition called **jaundice**, and the complexion becomes yellow (the disease often is called "yellow jack"). The gums bleed, the stools turn bloody, and the delirious patient often vomits blood. Patients die of internal bleeding, and mortality rates are very high, as in a notable Philadelphia epidemic of 1793 (**MicroFocus 14.2**).

Except for supportive therapy, no treatment exists for yellow fever. However, the disease can be prevented by immunization with either of two vaccines. The more widely used vaccine contains the 17D strain of yellow fever virus cultivated in chicken eggs.

DENGUE FEVER CAN PROGRESS TO A HEMORRHAGIC FEVER

Dengue fever has been known since David Bylon, a physician in the Dutch East Indies, described an outbreak in 1779. The disease takes its name from the Swahili word *dinga*, meaning "cramp-like attack," a reference to the symptoms. Like yellow fever, dengue fever is caused by an RNA virion of the *Flaviviridae* family that multiplies in white blood cells and platelets. The virus is closely related to the yellow fever virus, except that four strains of dengue fever virus are known to exist. Transmission is by the *Aedes aegypti* mosquito and by the tiger mosquito *Aedes albopictus* (**FIGURE 14.2**).

High fever and prostration are early signs of dengue fever. These are followed by sharp pain in the muscles and joints, and patients often report sensations that their bones are breaking. The disease therefore has been called **breakbone fever**. Another name, **saddleback fever**, is used because of temperature fluctuations. After about a week, the symptoms fade. Death is uncommon, but if one of the other strains of

MicroFocus 14.2

THE PHILADELPHIA STORY

Yellow fever was one of the most dramatic diseases ever to strike the United States. During the 1700s, historians chronicled 35 separate outbreaks as the disease ravaged the country. Nowhere did yellow fever strike harder than in Philadelphia.

In 1793, Philadelphia was the capital of the United States, and its largest city, with a population of 40,000. When yellow fever broke out, the panic rivaled that in Europe during the plague years. In patients, the eyes glazed, the flesh

yellowed, and delirium developed. People died, not here and there, but in clusters and in alarming patterns. Friends recoiled from one another. If they met by chance, they did not shake hands but nodded distantly and hurried on. The air felt diseased, and people dodged to the windward of those they passed. The deaths went on, great ugly scythings of humanity.

At the height of the epidemic, thousands fled Philadelphia, and officials posted warning notices on all homes

where people were infected. Those who could not get away, including most of the city's poor, sought protection by breathing through cloth masks soaked in garlic juice, vinegar, or camphor. Benjamin Rush, a noted American physician (and signer of the Declaration of Independence), prescribed a frightening course of purges, bloodlettings, vomiting, and immersion in icewater to reduce fever. Nearly all the 24,000 people who remained in Philadelphia were afflicted. Almost 5,000 died.

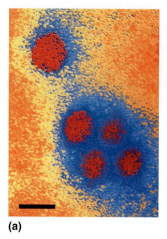
(a)

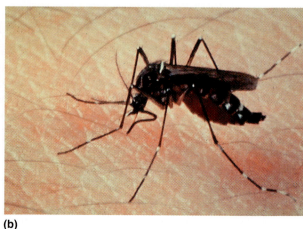
(b)

Hemorrhagic:
referring to bleeding and accumulating blood.

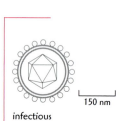
infectious
mononucleosis virus

150 nm

dengue virus later enters the body, a condition called **dengue hemorrhagic fever** may occur. In this condition, a rash from skin hemorrhages appears on the face and extremities, and severe vomiting and shock ensue as the blood pressure decreases dramatically.

Dengue fever traditionally has been confined to Southeast Asia. However, in 1963, the disease broke out in Central America, and it has occurred sporadically in the Americas since then. In 2003, globally, the dengue fever virus infected about 50 million people and killed some 12,000. Also, developing a vaccine may be more difficult than once thought (MicroFocus 14.3).

INFECTIOUS MONONUCLEOSIS IS MOST COMMON IN CHILDREN AND YOUNG ADULTS

The name **infectious mononucleosis** (or "mono" in the vernacular) is familiar to young adults because the disease is common in this age group. It is sometimes called the "kissing disease" because it is spread by contact with saliva. Droplets and contaminated objects, such as table utensils and drinking glasses, also may carry the virus.

Infectious mononucleosis is a blood disease, especially of antibody-producing B lymphocytes of the lymph nodes and spleen. Enlargement of the lymph nodes ("swollen glands") is accompanied by a sore throat, fever, and a high count of damaged **B lymphocytes**, a type of mononuclear white blood cell (hence the disease's name, mononucleosis). Mononucleosis usually runs its course in three to four weeks. Among the most dangerous complications are defects of the heart, paralysis of the face, and rupture of the spleen. The liver may be involved and jaundice may occur, a condition some physicians refer to as hepatitis. Those who recover usually become carriers for several months and shed the viruses into their saliva.

The diagnostic procedures for mononucleosis include detection of an elevated lymphocyte count and the observation of **Downey cells**, the damaged B lymphocytes with vacuolated and granulated cytoplasm. The patient also experiences an elevation of **heterophile antibodies** (antibodies reacting with antigens from unrelated species). Such antibodies can be detected by the **Paul-Bunnell test**, performed by mixing samples of the patient's serum with sheep or horse erythrocytes and observing the cells for agglutination. FIGURE 14.3 describes an adaptation of this test. The disease strikes an estimated 100,000 people annually in the United States.

MicroFocus 14.3

STAYING ONE STEP AHEAD

The dengue fever viruses are found in the tropical areas of the world where 40 percent of the world's population lives. Consequently, a large number of people are at risk of contracting dengue fever or dengue hemorrhagic fever (DHF). In fact, over 50 million individuals become infected each year and more than 12,000 die. So, producing a vaccine would seem to be a high priority.

At first, producing a vaccine against dengue fever doesn't seem that difficult. Grow the virus and genetically engineer a safe vaccine that incorporates a surface protein. However, once one realizes that there are four different dengue viruses (dengue-1, -2, -3, and -4), the quest for a vaccine becomes a bit more challenging because a vaccine against one or two of the dengue viruses actually could increase the risk of an infected person contracting the more serious DHF. Still, isn't it possible to develop a vaccine against all four dengue viruses? In fact, progress has been made in the development of recombinant vaccines that may protect against all four dengue viruses. It has been difficult, but hopes are that a vaccine will be ready for public health use within the first decade of this century.

But, now there may be a new hurdle. John Aaskov and colleagues at Australia's Queensland University of Technology have reported that the dengue fever viruses appear to mutate quickly. In 2003, they reported that two of the dengue-1 strains that caused 95 percent of the dengue fever cases in Myanmar in 2001 were mutated strains. More surprising is that the mutations had occurred locally within one year. The team also reported that two different strains of dengue-2 had recombined to form a new third strain. This is all on top of their 2002 report that in Thailand two new strains of dengue-3 had evolved within one year and replaced the dominant local strain. This resulted in the worst outbreak of dengue fever ever recorded in Thailand.

The results from Southeast Asia suggest that the dengue fever viruses can mutate quite rapidly and trigger new epidemics. If this isn't bad enough, the reports also spell trouble for potential vaccines. If the viruses can mutate at rapid rates, a dengue virus vaccine probably will not be effective against a newly mutated strain. The virus would stay a step ahead of an effective vaccine.

If the rapid rate of evolution of dengue fever viruses is verified, Aaskov has a possible solution for a vaccine. If dengue recombinant vaccines can be produced efficiently and quickly, perhaps a vaccine from last year could be reformulated to meet the "current strain" that is spreading this year—it would be similar to what is done for the influenza vaccine each year. It will be interesting to see how vaccine development proceeds.

The virus of infectious mononucleosis is the **Epstein-Barr virus** (**EBV**), a double-stranded DNA virus of the *Herpesviridae*. Up to 95 percent of the American population between 35 and 40 years of age has been infected with EBV. Infants usually are susceptible to EBV as soon as their maternal antibodies (present at birth) disappear. Many children who become infected with EBV show no symptoms or the symptoms are indistinguishable from other typical childhood illnesses. If a person is not infected as an infant or young child, an infection with EBV during adolescence or young adulthood runs a 35 to 50 percent chance of causing infectious mononucleosis. **EBV disease**, involving infection of B lymphocytes, is assumed to be a precursor to mononucleosis.

EBV has been detected in patients who have **Burkitt's lymphoma**, a tumor of the connective tissues of the jaw that is prevalent in areas of Africa. First isolated in the early 1960s by British virologists M. Anthony Epstein and Yvonne M. Barr, the Epstein-Barr virus was a surprising revelation and an important breakthrough in medicine because it demonstrated the link between viruses and cancer.

Contemporary virologists continue to search for reasons why EBV is associated with tumors on one continent and infectious mononucleosis on another. Some cancer specialists theorize that the malaria parasite, common in central Africa, acts as an irritant of the lymph gland tissue, thereby stimulating tumor development. All Burkitt's lymphoma cells have a chromosomal translocation that activates an oncogene. Therefore, EBV infection may stimulate or increase the frequency of the translocation.

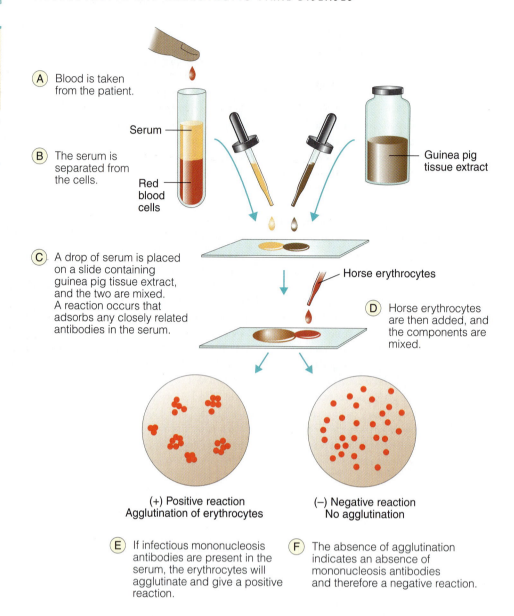

A Blood is taken from the patient.

Serum

B The serum is separated from the cells.

Red blood cells

Guinea pig tissue extract

C A drop of serum is placed on a slide containing guinea pig tissue extract, and the two are mixed. A reaction occurs that adsorbs any closely related antibodies in the serum.

Horse erythrocytes

D Horse erythrocytes are then added, and the components are mixed.

(+) Positive reaction
Agglutination of erythrocytes

(−) Negative reaction
No agglutination

E If infectious mononucleosis antibodies are present in the serum, the erythrocytes will agglutinate and give a positive reaction.

F The absence of agglutination indicates an absence of mononucleosis antibodies and therefore a negative reaction.

EBV also has been linked to **chronic fatigue syndrome (CFS)**. The symptoms include sore throat, aching muscles, sleep disturbances, swollen lymph nodes, and prolonged, overwhelming fatigue (patients say they feel as limp as Raggedy Ann dolls). Evidence of EB virus involvement is based on the presence of these viruses and their antibodies in some affected individuals. The relationship is not conclusive, however. Rather, the possibility exists that CFS may have multiple causes including viruses, immunological dysfunction, and nutritional deficiencies.

HEPATITIS A IS TRANSMITTED MOST OFTEN BY THE FECAL-ORAL ROUTE

Some years ago, the members of a university football team paused during practice and drank water taken from a local well. But this was no ordinary water. During the previous week, the water had been contaminated by viruses seeping into the well from a cesspool high above. Within days, all the players began to feel ill, and soon the unmistakable signs of hepatitis appeared. MicroFocus 14.4 recounts another such incident.

hepatitis A virus

Hepatitis is an acute inflammatory disease of the liver caused by several viruses. **Hepatitis A (infectious hepatitis)** is the form most commonly transmitted by food or water contaminated by the feces of an infected individual. An infected food handler often is involved, and outbreaks have been traced also to day-care centers where workers contact contaminated feces. In addition, the disease may be transmitted by raw shellfish such as clams and oysters, since these animals filter and concentrate the viruses from contaminated seawater.

Hepatitis A is caused by a small, single-stranded (+ strand) RNA virion belonging to the *Picornaviridae* family. The virion lacks an envelope and has icosahedron symmetry (FIGURE 14.4). Hepatitis A viruses are very resistant to chemical and physical agents, and several minutes of exposure to boiling water may be necessary to inactivate them.

The incubation period for hepatitis A is usually between two and four weeks. Therefore, hepatitis A is sometimes called **short-incubation hepatitis** (relative to hepatitis B, or long-incubation hepatitis). Initial symptoms include anorexia, nausea, vomiting, and low-grade fever. Discomfort in the upper-right quadrant of the abdomen follows as the liver enlarges. Considerable jaundice usually follows the onset of symptoms by one or two weeks (the urine darkens, as well), but many cases are without jaundice. The symptoms may last for several weeks, and relapses are common. A long period of convalescence generally is required, during which alcohol and other liver irritants are excluded from the diet. Recovery brings life-long immunity.

Diagnostic procedures for hepatitis A are based on liver function tests, observation of characteristic symptoms, and the demonstration of hepatitis A antibodies in the serum. The virus is excreted in large numbers in the stools about two weeks before symptoms appear. In one recent incident, for example, a restaurant worker in New Jersey became infected on May 9 but showed no symptoms, even though he was shedding hepatitis viruses. Hepatitis symptoms developed at the end of May, and during the first three weeks of June, 56 cases of hepatitis broke out among patrons of the restaurant.

MicroFocus 14.4

THIRTY-TWO AND COUNTING

For some, the number 13 is unlucky, but for the town of Peter's Creek, Alaska, the unlucky number was 32. It was late spring 1988, and the weather was unusually hot for that time of year. Between May 23 and June 10, 32 unfortunate people contracted hepatitis A, and things went downhill fast.

The outbreak of hepatitis was traced to a local convenience market, and the culprit was the ice slush that so many people enjoy on a hot day. The slush was contaminated, possibly by a certain store employee. Although the employee refused to be tested, his sister had had hepatitis A recently, and he had looked somewhat jaundiced at the time. He was one of two store employees responsible for preparing the slush each day. (Later, it was learned that he used water from the bathroom sink to make the slush.)

For the unlucky patients, there were many days of abdominal pain, fever, jaundice, and a serious liver disease. They would have to avoid fats and oils (no fried foods, mayonnaise, or oily salad dressings), and they could not have alcohol of any type. For many, there was the added burden of knowing they had infected others, because 23 additional cases soon developed. It was not a summer to remember fondly.

FIGURE 14.4

Hepatitis A Viruses

A transmission electron micrograph of hepatitis A viruses. The particles were coated with antibodies to assist staining. The coating accounts for the halo around the viruses. (Bar = 60 nm.)

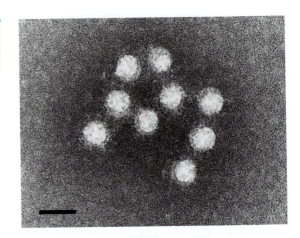

There is no treatment for hepatitis A except for prolonged rest and relieving symptoms. In those exposed to the virus, it is possible to prevent development of the disease by administering **hepatitis A immune globulin** within two weeks of infection. This preparation consists of antiviral antibodies obtained from blood donors. Blood is routinely screened for hepatitis antibodies, and if large amounts are found, the blood serum is used for immune globulin. In the New Jersey outbreak cited previously, 1,430 people were given injections during a two-day clinic held on June 19 and 20. And over 30,000 individuals received injections during a 1997 outbreak related to strawberries (MicroFocus 14.5).

MicroFocus 14.5

OUTFLOW

During March 1997, a fourth-grade student at Madison Elementary School in Michigan became violently ill with fever, vomiting, terrible abdominal pains, and urine the color of dark tea. When her parents took her to the hospital emergency room in Marshall, they learned that other children from her school were equally ill. An outbreak of hepatitis A had begun.

By the end of March, over 100 children and staff members had hepatitis A, and three school districts from two separate counties were involved. Investigators began a hunt for the epidemic's source and narrowed the list of suspected foods to frozen sliced strawberries. By mid-April, 264 children, parents, staff members, and visitors to the three schools were sick with hepatitis A. The outbreak was one of the largest in public health history.

The search for the source of the contaminated strawberries was exhaustive but fruitless. By June 1997, CDC officials confirmed their failure to locate an origin. In the interim period, however, much had happened. For example, the strawberries were traced to farms in northern Mexico, and relations between the United States and Mexico became strained as diplomats traded insinuations and insults. Moreover, the strawberry industry, already reeling from a 1996 erroneous involvement with *Cyclospora*, was crashing further.

In San Diego, the company supplying the strawberries was also in trouble. Federal law requires that US farmers supply all produce for federally supported school lunch programs (as in Madison); the company had purchased its strawberries in Mexico. A federal indictment of the company followed.

There was also a rush on immune globulin, the antibody preparation used to prevent the development of hepatitis A. Because strawberries from the implicated batch were traced to California school freezers, 9,000 children from Los Angeles received injections of immune globulin; in Georgia, 10,000 children lined up for shots; and in Tennessee, 8,600 children rolled up their sleeves. Fortunately, no cases developed outside of Michigan.

A final effect of the epidemic was a review of questionable sanitary conditions associated with strawberry picking, preparation, freezing, and distribution. New laws and regulations were in the offing, as public health agencies sought some benefit amid the wrongful suffering of the Michigan children. Next time, they assured, things would be different.

Maintaining high standards of personal and environmental hygiene, and removing the source of contamination, are essential to interrupting the spread of hepatitis A. Moreover, in 1995, the Food and Drug Administration (FDA) licensed a vaccine composed of formalin-inactivated viruses. Known commercially as Havrix, the vaccine is administered into the shoulder muscle in two doses to those between ages 2 and 18 (pediatric formulation) and in three doses to those over age 18 (adult formulation). A second vaccine (Vaqta) requiring only one dose was licensed in 1996. In 2001, 10,609 Americans contracted hepatitis A, the lowest number of cases ever reported, due primarily to recommended childhood vaccinations.

HEPATITIS B CAN PRODUCE A CHRONIC LIVER INFECTION

Hepatitis B (serum hepatitis) is the second major type of hepatitis. It is caused by a double-stranded DNA virus known as a **hepadnavirus** (*hepatitis-DNA-virus*) of the family *Hepadnaviridae*. The hepadnavirus can appear in three forms. In its most frequently observed form, the virus is seen as spherical particles measuring about 22 nm in diameter. These small particles appear to be composed exclusively of an antigenic protein substance called **hepatitis B surface antigen**, or **HBsAg**. A second form is elongated particles also composed of HBsAg in tubular or filamentous rods up to 200 nm long. The third form is the more traditional virus, a virion containing HBsAg as an outer envelope surrounding an inner nucleocapsid of double-stranded DNA enclosed in a shell composed of **hepatitis B core antigen**, or **HBcAg**. This third form has been called the **Dane particle** since it was first reported by D. S. Dane in 1970. Apparently, the overproduction of the surface HBsAg during viral replication yields this protein in large amounts. The antigen has remained an important diagnostic sign for hepatitis B since first reported by Nobel laureate Baruch Blumberg in the early 1970s.

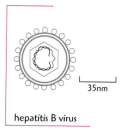

hepatitis B virus

Transmission of hepatitis B usually involves direct or indirect contact with an infected body fluid such as blood or semen. For example, transmission may occur by contact with blood-contaminated needles used in hypodermic syringes or for tattooing, acupuncture, or ear piercing. Blood-contaminated objects such as fiber-optic **endoscopes**, instruments, and renal dialysis tubing are implicated also. Moreover, transmission may take place by contact with saliva, including contact made in the dental office (**FIGURE 14.5**). Hepatitis B is also an important **sexually transmitted disease**, particularly when anal intercourse takes place. This is because bleeding often occurs during anal intercourse, and viruses can enter the bloodstream of the receptive partner from the semen of the infected individual. (A similar situation holds for AIDS.)

Endoscope:
an instrument consisting of a fiber-like strand that is inserted into tissue to observe it.

The clinical course of hepatitis B is basically the same as for hepatitis A, but more severe illness is generally associated with hepatitis B. The disease has an incubation period of four weeks to six months and is therefore known as **long-incubation hepatitis** (relative to hepatitis A). Among adults, the common findings during the early stage are fatigue, anorexia, and taste changes. (Smokers experience a notable distaste for cigarettes.) Dark urine and clay-colored stools are present several days before jaundice appears. An uncomfortable sense of fullness and tenderness is felt in the upper-right quadrant of the abdomen. Recovery usually occurs about three to four months after the onset of jaundice, and about 10 percent of patients remain carriers for several months. In rare cases, extensive liver damage may occur, including liver cancer. The latter is called **hepatocarcinoma**.

Injections of alpha-interferon (Intron A) can influence the course of hepatitis B. Lamivudine (3TC, Epivir), a base analog, also has been FDA approved as a treatment.

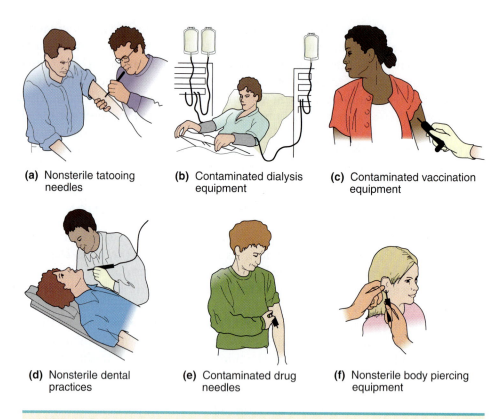

(a) Nonsterile tatooing needles

(b) Contaminated dialysis equipment

(c) Contaminated vaccination equipment

(d) Nonsterile dental practices

(e) Contaminated drug needles

(f) Nonsterile body piercing equipment

FIGURE 14.5

Some Methods for the Transmission of Hepatitis B

Moreover, prophylactic therapy may be rendered by injections of **hepatitis B immune globulin**. This preparation consists of antibodies concentrated from the serum of blood donors. In addition, the disease can be prevented by immunization with the hepatitis B vaccine. Since 1987, this vaccine has consisted of hepatitis B surface antigens (HBsAg) produced by genetically engineered yeast cells. The vaccine is known commercially as Recombivax HB or Engerix-B (depending on the company that produces it). Recommended for all age groups (including infants), it is particularly valuable for health care workers who might be exposed to blood from patients. For infant use, it is combined with the Hib vaccine as Comvax. As a result of child and adolescent vaccinations, there has been a 60 percent decrease in reported cases of hepatitis B in the United States (7,843 cases in 2001).

HEPATITIS C OFTEN CAUSES CIRRHOSIS AND LIVER FAILURE

Hepatitis C is caused by a single-stranded (+ strand) RNA virus of the *Flaviviridae* family, a virus that has not yet been cultivated in laboratory animals or human cell cultures. There are few early symptoms associated with hepatitis C, and liver damage develops slowly, but insidiously. Often, the liver cirrhosis is beyond repair when the disease is finally recognized. Indeed, damage from hepatitis C is the primary reason for liver transplants in the United States.

Cirrhosis:
progressive deterioration of the liver.

Hepatitis C was formerly a serious problem in blood transfusion recipients, but a laboratory screening test for the virus has reduced the risk to 1 in 100,000 units of blood transfused. (A home screening test called Hepatitis C Check also has received FDA approval.) The disease is observed most often in injection drug users and sometimes is diagnosed in individuals who use cocaine or have tattoos or body piercings. However, it also has been spread inadvertently through medical intervention as MicroFocus 14.6 vividly demonstrates. Alpha-interferon and ribivirin are approved therapies for the disease.

The screening test for hepatitis C followed identification of the virus by Michael Houghton of the biotechnology company Chiron. In addition to its diagnostic purpose, the test has been used to hunt for antibodies against hepatitis C in preserved blood. One such search in 1999 uncovered antiviral antibodies in frozen blood samples taken from servicemen in 1948 during a serious epidemic of strep infection. The finding confirmed that hepatitis C has been in the United States at least 50 years and is not a new disease, as some epidemiologists suspected. Indeed, the late blooming of hepatitis C suggests that this disease may be a major factor during the twenty-first century. Almost 4 million Americans are infected with the hepatitis C virus and 2.7 million of those have chronic infections. Still, new infections have declined from 240,000 in 1980 to 25,000 in 2001. TABLE 14.1 compares hepatitis A, B, and C.

MicroFocus 14.6

WHAT'S WORSE—THE DISEASE OR MEDICAL INTERVENTION?

Egypt has a population of 62 million and contains the highest prevalence of hepatitis C in the world. The Egyptian Ministry of Health estimates a national prevalence rate of at least 12 percent (or 7.2 million people). Chronic hepatitis C is the main cause of liver cirrhosis and liver cancer in Egypt, and indeed, one of the top five leading causes of death. The highest concentration of the hepatitis C virus (HCV) appears in farming people living in the Nile delta and rural areas. So, why does Egypt have such a high prevalence rate of HCV?

Several recent studies suggest that HCV was transmitted through the contamination of reusable needles and syringes used in the treatment of schistosomiasis, a disease caused by a blood parasite (see Chapter 17). Schistosomiasis is a common parasitic disease in Egypt and can cause urinary or liver damage over many years. Farmers and rural populations are at greatest risk of acquiring the disease through swimming or wading in contaminated irrigation channels or standing water.

Prior to 1984, the treatment for schistosomiasis was intravenous tartar emetic. Between the 1950s and the 1980s, hundreds of thousands of Egyptians received this standard treatment, called parenteral antischistosomal therapy (PAT). Today, drugs for schistosomiasis are administered in pill form.

Evidence suggests that inadequately sterilized needles used in the PAT campaign contributed to the transmission of HCV. Needles were routinely recycled and not properly sterilized at that time due to cost and limited resources. Overall, despite improvement in schistosomiasis-induced morbidity, the PAT campaign set the stage for the world's largest transmission of blood-borne pathogens resulting from medical intervention. The treatment campaign was conducted with the best of intentions, using accepted sterilization techniques of the time. However, much of the PAT campaign was carried out before disposable syringes and needles were available. In addition, no one was aware of HCV prior to the 1980s or the dangers associated with blood exposure.

Further evidence for the correlation between the PAT campaign and hepatitis C was the drop in the hepatitis C rate when PAT injections were replaced with oral medications. Sadly, in part because of the high number of people who were infected, the risk of transmission remains high in the Egyptian population today.

TABLE 14.1

A Comparison of Three Types of Hepatitis

CHARACTERISTIC	HEPATITIS A	HEPATITIS B	HEPATITIS C
Alternate names	Infectious hepatitis	Serum hepatitis	Posttransfusion hepatitis NANB hepatitis
Virus	RNA virus *Picornaviridae*	DNA virus Dane particle *Hepadnaviridae*	RNA virus *Flaviviridae*
Incubation period	2–4 weeks	4 weeks to 6 months	2 weeks to 6 months
Major transmission	Food and water Saliva contact Sexual contact	Body fluids Blood Sexual contact	Blood
Symptoms	Jaundice Abdominal pain	Jaundice Abdominal pain	Jaundice Abdominal pain
Illness severity	Moderate	High	High
Diagnosis	Liver function tests Symptoms Antibodies in serum	Liver function tests Symptoms HBsAg in serum	Liver function tests Symptoms Antibodies in serum
Carrier state	Rarely develops	Develops	Develops
Nosocomial problem	No	Yes	Yes
Prevention	Vaccine	Vaccine	None
Liver cancer	Not likely	Possible	Not established

OTHER VIRUSES ALSO CAUSE HEPATITIS

Delta hepatitis appears to be caused by two viruses: the hepatitis B virus and the hepatitis D (delta) virus (delta is the fourth letter of the Greek alphabet, equivalent to D). Delta viruses were discovered by Italian investigators in 1977. They consist of a protein fragment called the **delta antigen** and a segment of RNA. Apparently, the viruses can only cause liver damage when hepatitis B virus also is present (the press labeled them the "piggyback viruses"). Injection drug users were a major focus of an outbreak in Massachusetts in 1984. Chronic liver disease with cirrhosis is two to six times more likely in a co-infection.

Hepatitis E appears to be caused by a single-stranded (+ strand) RNA virus that does not have an envelope and has not yet been identified to a specific family. The virus is apparently transmitted to humans via fecally contaminated drinking water. Pregnant women seem to be particularly susceptible to illness, especially in the later stages of pregnancy. No evidence of chronic infection has been noted. **Hepatitis G** involves a chronic liver illness. It is a single-stranded (+ strand) RNA virus of the *Flaviviridae* with an envelope and icosahedral symmetry. Like Hepatitis B and C, the virus is transmitted by blood, blood products, or sexual intercourse. It appears to cause persistent infections in 15 to 30 percent of infected adults.

To this point . . .

We have begun a study of the viscerotropic viral diseases by focusing on several diseases of the blood and visceral organs. The first two diseases, yellow fever and dengue fever, are caused by viruses injected directly into the bloodstream by mosquitoes. Yellow fever has great historical interest because epidemics were widespread during past generations, but vaccines and arthropod control now limit its spread in many parts of the world. Dengue fever remains a threat in many countries. High fever reflects viral invasion of the blood in both cases.

We then turned to infectious mononucleosis and hepatitis. Ongoing research into these diseases reveals new information, while questions continue to surface. For example, the role of the Epstein-Barr virus in infectious mononucleosis is still uncertain, and the relationship of the disease to Burkitt's lymphoma is being studied. In hepatitis, we saw how numerous different forms have emerged since the 1970s, each caused by a different virus. Infection of the liver is the common element in all types of hepatitis, and convalescence is generally long because damage to the liver is not easily repaired. Liver disease is also a consequence of yellow fever, and often it occurs in infectious mononucleosis.

We shall now focus attention on a series of viral diseases of the gastrointestinal tract as we study viral gastroenteritis, and we shall study a number of viral fevers in which the pathogens affect the bloodstream. In addition, we will consider acquired immune deficiency syndrome (AIDS). Many of the viscerotropic diseases in this section occur sporadically and are not well understood. However, the potential for epidemics is great.

14.2

Other Viscerotropic Viral Diseases

We continue with the viscerotropic diseases by examining additional diseases of the gastrointestinal tract and blood. Many emerging viral diseases are in this section, and we include a discussion of acquired immune deficiency syndrome (AIDS). Recall that our classification system is arbitrary, and that many of the diseases also affect organs other than the visceral organs. An example is seen in some of the agents of viral gastroenteritis that we examine first.

VIRAL GASTROENTERITIS IS CAUSED BY ANY OF SEVERAL VIRUSES

Viral gastroenteritis is a general name for a common illness occurring in both epidemic and endemic forms. It affects all age groups worldwide and may include some of the frequently encountered traveler's diarrheas. Public health officials believe that the disease is second in frequency to the common cold, among infectious illnesses affecting people in the United States. (In developing nations, gastroenteritis is estimated to be the second leading killer of children under the age of 5, accounting for 23 percent of all deaths in this age group.) Clinically the disease varies, but usually it has an explosive onset with varying combinations of diarrhea, nausea, vomiting, low-grade fever, cramps, headache, and malaise. It can be severe in infants, the elderly, and patients whose immune systems are compromised by other illnesses. Some people mistakenly call it "stomach flu."

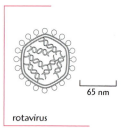

rotavirus

65 nm

One cause of viral gastroenteritis is the human **rotavirus**, a virus first described in 1973 in Australia. The virion contains eleven segments of double-stranded RNA as well as inner and outer capsids, but no envelope. A member of the *Reoviridae* family, it is named for its circular appearance (*rota* is Latin for "wheel"). FIGURE 14.6a shows this virus. Transmission occurs by ingestion of contaminated food or water or from contaminated surfaces. A possible mechanism for its action on the body is summarized in MicroFocus 14.7. The CDC considers rotaviruses the single most important cause of diarrhea in infants and young children admitted to hospitals. Many cases involve severe dehydration and death. Some 55,000 hospitalizations result from rotavirus infections each year in the United States and over 600,000 children die each year worldwide.

In 1998, officials at the FDA approved RotaShield, a vaccine to protect against rotavirus infection. However, in July 1999, the Advisory Committee on Immunization Practices of the FDA removed the RotaShield vaccine after some infants developed bowel obstructions following vaccination.

A second cause of viral gastroenteritis is the noroviruses (FIGURE 14.6b). Noroviruses (formerly called the Norwalk-like viruses) are transmitted primarily through the fecal-oral route, either by consumption of contaminated food or water, or by direct person-to-person spread (FIGURE 14.7). Contamination of surfaces also may act as a source of infection. The CDC estimates that some 23 million cases of viral gastroenteritis and at least 50 percent of all foodborne outbreaks of viral gastroenteritis are caused by norovirus infections.

The noroviruses belong to the family *Caliciviridae*. These viruses contain single-stranded (+ strand) RNA and they have icosahedral symmetry. Currently, there are

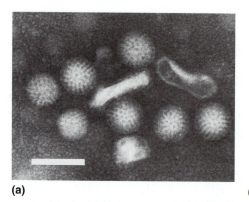

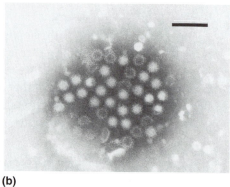

(a)   (b)

FIGURE 14.6

Two Viruses that Cause Gastroenteritis

(a) Rotaviruses observed in the diarrheal stool of an infant with gastroenteritis. Note the wheel-like circular appearance from which the rotavirus takes its name. (Bar = 100 nm.) (b) Noroviruses from stool specimens of a patient with gastroenteritis. The viruses appear to have icosahedral symmetry. (Bar = 100 nm.)

MicroFocus 14.7

SMELLY INSIGHTS

The cage was not supposed to smell of diarrhea. Mouse number 3 had received the control protein, not the intact rotavirus. The protein should have been harmless. Where did all this mess come from?

Then it hit—could the protein be causing the diarrhea? And if so, how? Is this one of those moments every researcher dreams about? Quick, check the data. Whom do I tell? Can I repeat the experiment? Where do we go from here? Wow!

It happened to Judith M. Ball, virus researcher at Baylor College of Medicine. She and her colleague Mary Estes were studying rotaviruses, a cause of

often-fatal diarrhea in millions of children. They had isolated a protein called nonstructural glycoprotein number 4 (NSP4), so named because it was not involved in the structure of the rotaviral capsid (*nonstructural*), it was a *protein*, and it was number 4 in the series of proteins studied. Now they were testing NSP4 to confirm that it was simply another viral protein of little pathological consequence. It was not supposed to cause diarrhea.

But it did. And Bell's insight opened the door to a new biochemical process and a novel concept of viral pathology: Experiments showed that biochemically, NSP4 protrudes through the capsid and

frees the virus from surrounding cellular membranes so it can replicate. For the viral pathologist, further studies indicated that in the body, the protein induces diarrhea by mobilizing intracellular calcium; this leads to increased calcium secretion and an outpouring of water characteristic of diarrhea. The protein is a toxin, the first digestive tract toxin of viral origin ever detected.

As she prepared her research paper for *Science* magazine in 1996, Ball recalled how it began with a messy cage and a glorious moment in time. Ball contends she was lucky. Her colleagues prefer to point to the dictum of Louis Pasteur: "Chance favors the prepared mind."

---

at least four norovirus groups. Noroviruses are highly contagious, and as few as 10 virions can cause illness in an individual.

The incubation period for norovirus gastroenteritis is about 36 hours. Symptoms include acute-onset vomiting (more common in children), watery non-bloody diarrhea with abdominal cramps, and nausea. Low-grade fever occasionally occurs. Symptoms usually last 24 to 60 hours and recovery is complete. Dehydration is the most common complication. There is no treatment for norovirus gastroenteritis. Fluid replacement and correcting electrolyte imbalances are required.

Viral gastroenteritis also can be caused by either of two **enteroviruses**. Enteroviruses are small, icosahedral single-stranded (+ strand) RNA virions of the *Picornaviridae* family. They are currently considered a less frequent cause of gastroenteritis than other viruses and are notable for the variety of infections they may cause. An antiviral drug called pleconaril appears to limit their replication.

One enterovirus is the **Coxsackie virus**, first isolated in 1948 by Gilbert Dalldorf and Grace Mary Sickles from the stool of a patient residing in Coxsackie, New York. The viruses occur in many strains within two groups, A and B. Strains B4 and B5 Coxsackie viruses are associated most commonly with gastroenteritis. Group B viruses also are implicated in **pleurodynia** (or Bornholm's disease), a painful disease of the rib muscles; and **myocarditis**, a serious disease of the heart muscle and valves, sometimes resulting in a heart attack or the need for a heart transplant. In addition, group B Coxsackie viruses are among the most frequent causes of **aseptic meningitis**. Group A viruses have been isolated from cases of respiratory infections, conjunctivitis, and **herpangina**, a disease of children with abrupt fever onset and punched-out vesicles on the soft palate, tongue, tonsils, and hands ("hand, foot, and mouth disease"). Enterovirus 71 is involved.

Some virologists believe that Coxsackie viruses are the so-called **24-hour viruses** responsible for brief bouts of diarrhea. And an intriguing 1994 report in the *New*

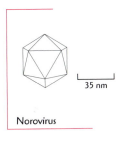

35 nm

Norovirus

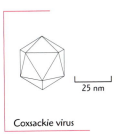

25 nm

Coxsackie virus

FIGURE 14.7

Outbreaks of Norovirus on Cruise Ships

Between January 1 and December 2, 2002, CDC's Vessel Sanitation Program recieved reports of 21 outbreaks of acute gastroenteritis on 17 cruise ships. Nine reports were confirmed by laboratory analysis to be associated with noroviruses. Such outbreaks demonstrate how easily noroviruses can be transmitted from person to person in a closed environment.

TEXTBOOK CASES

1. On July 18, a cruise ship embarked for a seven-day cruise from Vancouver to Alaska with 1,318 passengers and 564 crewmembers.

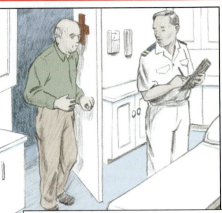

2. The next day, five passengers reported to the ship's infirmary with symptoms of acute gastroenteritis.

3. On disembarking, a total of 167 passengers and nine crewmembers had reported illness with symptoms of vomiting and diarrhea. Five of 10 stool specimens from ill passengers were positive for norovirus. After passengers disembarked, the ship was disinfected in accordance with CDC recommendations.

4. On the same day, a new group of passengers embarked for another seven-day cruise. During the cruise, 189 of 1,336 passengers and 30 of 571 crew members had acute gastroenteritis with diarrhea and vomiting. An environmental health inspection conducted by CDC revealed no sanitary deficiencies. The cruise line owner voluntarily took the ship out of service for one week for aggressive cleaning and sanitizing. No outbreaks were reported on subsequent cruises.

Diabetes:
a disease in which the passage of glucose into body cells is interrupted.

England Journal of Medicine indicated that Coxsackie viruses might trigger the development of insulin-dependent **diabetes** in genetically susceptible individuals. According to the report, the viruses stimulate the immune system to produce antibodies that attack the pancreatic cells responsible for producing insulin.

The second enterovirus that causes viral gastroenteritis is the **echovirus**. Echoviruses were discovered in the early 1950s. They take their name from the acronym ECHO, for enteric (intestinal), cytopathogenic (pathogenic to cells), human (human host), and orphan (a virus without a famous disease). Echoviruses occur in many strains and cause gastroenteritis as well as aseptic meningitis. The meningitis, however, is usually less severe than bacterial meningitis. Echoviruses are also a cause of respiratory infections and rapidly developing maculopapular skin rashes called **exanthems**. A Massachusetts outbreak called the Boston exanthem attracted attention in 1954.

VIRAL FEVERS CAN BE CAUSED BY A VARIETY OF VIRUSES

A series of viruses may cause human illnesses characterized by high fever. These diseases occur sporadically and are usually rare in the United States.

An example of a viral fever is **Colorado tick fever**. This disease is caused by an RNA virus of the *Reoviridae* family. It is transmitted by the tick *Dermacentor andersoni* and accounts for about 200 cases of disease per year, chiefly in the state of Colorado. Characteristic symptoms include alternating periods of fever and relief (saddleback fever), with pain in the muscles, joints, and eyes. **Leukopenia**, and the presence of viruses inside the red blood cells, are other factors that mark the disease.

Leukopenia: a reduced number of white blood cells.

Sandfly fever is another example of a viral fever. It is caused by a virus in the *Bunyaviridae* family. The disease is prevalent where sandflies of the genus *Phlebotomus* abound. It occasionally breaks out in Mediterranean regions, Southeast Asia, and parts of Central America. Patients suffer recurrent high fever and joint and bone pains resembling those in dengue fever.

Rift Valley fever is named for a region in eastern Africa called the Rift Valley. This is an immense earthquake-prone area. In addition to affecting humans, the disease affects animals and causes extensive losses of sheep and cattle. Indeed, the virus, which is in the family *Bunyaviridae*, is so dangerous to animals that federal law prohibits their transport to the mainland of the United States. Transmission is by several genera of mosquitoes, and dengue-like pain in the bones and joints accompanies the fever, which lasts for about a week.

Certain viral fevers are accompanied by severe hemorrhagic lesions of the tissues. These diseases are classified as **viral hemorrhagic fevers**. One example, **Lassa fever**, is so named because it was first reported in the town of Lassa, Nigeria, in 1969 (**FIGURE 14.8**). The disease is caused by a single-stranded (− strand) RNA virus of the *Arenaviridae* family. Infection is accompanied by severe fever, prostration, and patchy blood-filled hemorrhagic lesions of the throat. The fever persists for weeks, the pharyngeal lesions bleed freely, and profuse internal hemorrhaging is common. At least four epidemics have been identified in Africa since 1969, and the disease is occasionally seen in the United States. Lassa fever is responsible for about 5,000 deaths per year in West Africa. It can be successfully treated with ribivirin. John G. Fuller has written a lucid and vivid account of the discovery of Lassa fever in his book *Fever! The Hunt for a New Killer Virus*. The book should be read by anyone who believes there are no remaining frontiers in medicine.

A rare but severe viral hemorrhagic fever is **Marburg hemorrhagic fever**, named for Marburg, West Germany, where an outbreak occurred in 1967. Virologists identified the virus in tissues of green vervet monkeys imported from Africa. The patient experiences alternating periods of fever and relief, bleeding from the gums and throat, and gastrointestinal hemorrhaging; mortality rates during epidemics tend to be high. The viral agent is a filovirus in the *Filoviridae* consisting of single-stranded (− strand) RNA viruses that exist as long thread-like viruses (*filum* is Latin for "thread") that often take the shape of a fishhook or U. Exactly how humans become infected is not known, although body fluids or blood from an infected individual might be a source. No treatment is available and the fatality rate is up to 25 percent.

Another filovirus, the Ebola virus, captured headlines in 1995. During that summer, an outbreak of **Ebola hemorrhagic fever** occurred in Zaire and Sudan,

(A) The bush rat is a staple in the diet of certain African natives. The traditional rat hunt begins with a fire in the grasslands that drives the rats into the open, where they are clubbed.

(B) Occasionally the rats will run into local houses for shelter.

(C) Lassa fever viruses apparently are transmitted by arthropods in rat fur or rat excretions to susceptible people visiting the area.

(D) When infected rats are consumed as food, infection can occur.

FIGURE 14.8

Transmission of Lassa Fever Virus

There are several ways Lassa fever virus can be transmitted from nature.

and an astonished world watched as newspaper reports spoke of blood spouting from patients' eyes, nose, ears, and gums; organs turning to liquid; and a horror story similar to the one recounted in *The Hot Zone* by Richard Preston. (The salient difference was that the Ebola viruses in Preston's book were loose near Washington, D.C.) When it was over, hundreds of Africans had died and the media were alerting readers to an unknown microbial world lurking in the wilderness. The phrase "emerging viruses" was being mentioned with increased frequency on the late news. Through 2002, there have been 14 confirmed outbreaks of Ebola hemorrhagic fever in Africa. Over 1,600 cases have been reported.

How Ebola virus causes massive internal bleeding and hemorrhaging was a subject of speculation until 1998, when researchers at the University of Michigan discovered that the virus encodes at least two glycoproteins. One attaches to endothelial cells lining the veins and arteries, where it encourages viral entry. Viral replication and damage to the cells weaken the blood vessels, causing them to leak; catastrophic bleeding follows. The second glycoprotein apparently attaches to neutrophils (types of white blood cells) and thereby limits phagocytosis and the immune response. Thus, the patient bleeds to death internally before a reasonable immunological defense can be mounted. The Ebola virus is shown in FIGURE 14.9a.

Other viral hemorrhagic fevers, all caused by *Bunyaviridae* or *Arenaviridae* are Congo-Crimea hemorrhagic fever, which occurs worldwide; Oropouche fever, which affects regions of Brazil; and Junin and Machupo, the hemorrhagic fevers of Argentina and Bolivia, respectively. An arenavirus called the Sabia virus has caused hemorrhagic illnesses in Brazil, and in 1994, it caused disease in a Yale University researcher, who survived the ordeal. A virus called the Guanarito virus is associated with Venezuelan hemorrhagic fever.

HANTAVIRUS PULMONARY SYNDROME PRODUCES SUDDEN RESPIRATORY FAILURE

Still another hemorrhagic fever is **Korean hemorrhagic fever,** caused by the single-stranded (– strand) RNA-containing Hantaan virus (*Bunyaviridae*), named for the Hantaan River in Korea. American servicemen experienced the disease during the Korean War, and an outbreak occurred among Marines in Korea in 1986. Then, in the summer of 1993, a brief epidemic occurred among residents of the southwestern United States. Its diagnosis is described in MicroFocus 14.8. Symptoms of the unknown disease included a rapidly developing flu-like illness, with blood hemorrhaging and acute renal and respiratory failure. The disease was given the technical name **hantavirus pulmonary syndrome (HPS),** and the strain of virus was named Sin Nombre virus (Spanish for "no name") (FIGURE 14.9b). CDC investigators identified airborne viral particles from the dried urine and feces of rodents (especially deer mice) as the vector for the virus. By the end of 1993, 91 cases were confirmed in 20 states. Forty-eight patients died. As of June 2002,

---

FIGURE 14.9

The Ebola Virus and Hantavirus

(a) Transmission electron micrograph of Ebola viruses, the cause of Ebola hemorrhagic fever. It is one of the group of filoviruses, so-called for its thin and long shape. Here the viral filament is seen looping back on itself. (Bar = 140 nm.) (b) A false color transmission electron micrograph of the hantavirus, the cause of hantavirus pulmonary syndrome. This enveloped virus with a helical nucleocapsid was first isolated in 1993 following an outbreak of hantavirus disease in the southwestern United States. (Bar = 200 nm.)

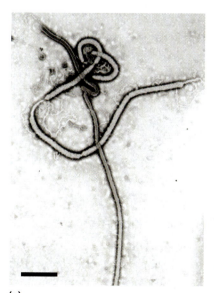

(a)

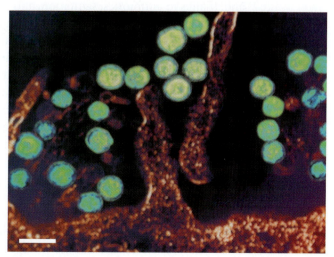

(b)

MicroFocus 14.8

"WE CAN'T FIND IT BUT WE KNOW IT'S THERE!"

A microbial identification of an unusual sort was made in 1993. That year, a strange epidemic of viral disease broke out in the southwestern United States at the border area of four states (Arizona, Colorado, New Mexico, and Utah). The disease temporarily came to be known as the "four corners disease." It spread quickly among area residents, causing severe hemorrhaging, pneumonia, and kidney disease, and it resulted in 35 deaths among 52 patients. Scientists performed a battery of tests, hunting for myriad viruses and their antibodies, but they came up empty each time.

Then they hit on a novel idea. Instead of looking for the virus, why not look for the viral genome—the DNA of the virus.

Scientists obtained DNA from numerous different viruses in culture collections and replicated the DNA fragments to produce a variety of "gene probes." A gene probe is a small, single-stranded molecule of DNA that will hunt down and unite with its complementary single-stranded DNA molecule (Chapter 7). Acting like a right hand searching for a left hand, the gene probe emits a radioactive signal when the union has taken place (if the complementary strand cannot be found, no signal is sent).

Now the scientists were ready to give their theory a try. They began by extracting DNA from the tissues of disease patients. Then they combined the DNA with gene probes from a constellation of viruses and held their breath to see

which would send a signal. Their answer came a few short minutes later when a sooty band of radioactivity appeared on their instruments; the DNA from the patient was uniting with the gene probe from the hantavirus, a very rare virus named for the Hantaan River in Korea. The infecting virus must be a hantavirus.

Not only was the mystery solved, but scientists now had a useful tool for diagnosing and tracking the disease. And they could work to interrupt its spread because they knew how to locate the disease and what was causing it. They also could point out the face of the enemy—well, not really, because they had not seen the virus, only its footprint.

there have been a total of 318 cases of HPS reported from 31 states. The fatality rate is over 35 percent.

Hantavirus pulmonary syndrome has now become a "regular" in the lexicon of medicine not only in the United States but throughout much of the Americas (FIGURE 14.10).

Researchers are attempting to produce a hantavirus vaccine by inserting hantaviral genes into a cowpox (vaccinia) virus. This type of genetically engineered vaccine has been produced for other diseases as well.

In the 1960s and 1970s, public health officials believed they had triumphed over infectious disease. Smallpox was on the way to extinction; polio was under control; and, thanks to antibiotics, sanitation, and pesticides, such maladies as tuberculosis, cholera, and malaria were disappearing. But then in one wave after another, nature counterattacked with Marburg disease, Lassa fever, Ebola fever, and hantavirus disease, as well as Legionnaires' disease, hepatitis C, *E. coli* O157:H7, Lyme disease, and AIDS. In all, at least 30 newly identified pathogens emerged. By the end of the century, health officials were concluding that new pathogens would continue to threaten human existence for all time. The trick, researchers suggest, is to adopt a guerrilla strategy in which reliable intelligence and rapid reaction are the keys to survival. The viral fevers taught science a valuable lesson for the new century.

CYTOMEGALOVIRUS DISEASE CAN PRODUCE SERIOUS BIRTH DEFECTS

The cytomegalovirus (CMV) is an icosahedral DNA virion of the *Herpesviridae*. The virus takes its name from the enlarged cells ("cyto-megalo") found in infected tissues. Usually, these are cells of the salivary glands, epithelium, or liver.

FIGURE 14.10

Hantavirus Pulmonary Syndrome Cases, the Americas 1993–2001

Cases of HPS have been reported by the Pan American Health Organization in North, Central, and South America. The number of reported cases are indicated.

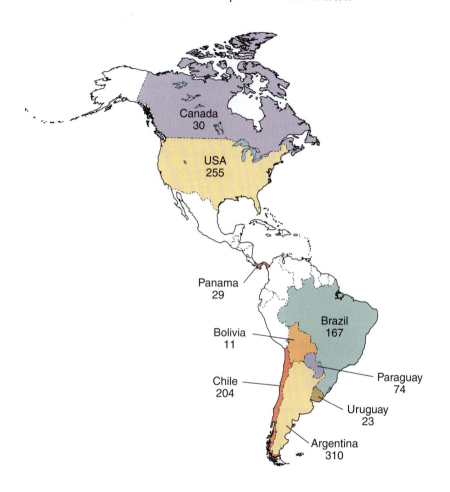

Cytomegalovirus disease may be among the most common diseases in American communities. Fever, malaise, and, in some cases, an enlarged spleen, develop, but few other signs of disease are observed. Most patients recover uneventfully. However, if a woman is pregnant, a serious congenital disease may ensue as the viruses pass into the fetal bloodstream and damage the fetal tissues. Mental impairment is sometimes observed in young patients. The C in the **TORCH** group of diseases refers to cytomegalovirus disease. The other letters stand for toxoplasmosis (T), rubella (R), and herpes simplex (H). The O is for other diseases, such as syphilis.

The cytomegalovirus also has demonstrated its invasive tendency in patients who have acquired immune deficiency syndrome (AIDS). Up to one-third of AIDS patients experience **CMV-induced retinitis**, a serious infection of the retina that can lead to blindness. Ganciclovir often is used to treat the disease, with foscarnet as an alternative drug. In immunocompromised individuals, cytomegaloviruses also can infect the lungs, liver, brain, and kidneys and cause death. Patients undergoing cancer therapy or receiving organ transplants may be susceptible to CMV disease because immunosuppressive drugs often are administered to these patients. The virus also has been implicated in cardiovascular disease (Chapter 8).

ACQUIRED IMMUNE DEFICIENCY SYNDROME RESULTS FROM IMMUNE SYSTEM DYSFUNCTION

In 1981, physicians described a syndrome involving a deficiency of the immune system. This clinical entity included the development of certain opportunistic infections, as well as an unusual type of skin cancer called Kaposi's sarcoma. The most plausible factor was a virus, but a definitive agent was not identified until 1984, when a French group led by Luc Montagnier isolated the infectious agent. In 1986, the virus was given its current name of **human immunodeficiency virus (HIV)**. By that time, the disease was well known as **acquired immune deficiency syndrome (AIDS)**.

HIV is a single-stranded (+ strand) RNA virus with icosahedral symmetry and an envelope that contains spikes, as **FIGURE 14.11** illustrates. Within its genome it contains molecules of an enzyme called **reverse transcriptase**. When the RNA is released in the cytoplasm of a host cell, the enzyme synthesizes a molecule of DNA using the genetic message in the RNA as a template. (This reversal of the usual mode of genetic information transfer [transcription] gives the virus its name, retrovirus, and the enzyme its name, reverse transcriptase.) The DNA molecule, now integrates into the host's DNA as a provirus, and from that location, it transcribes its genetic message into new particles of HIV. The whole viruses then "bud" from the host cell to infect other cells, as **FIGURE 14.12** indicates. This individual now has **HIV infection**.

The normal host cell of HIV is a cell of the immune system called a **T lymphocyte (T cell)**. These cells participate in cell-mediated immunity. In this process, some T lymphocytes mature and proliferate to form a colony of cytotoxic T cells that respond to the presence of protozoa, fungi, and infected cells. (These processes are

FIGURE 14.11

A Diagram of the Human Immunodeficiency Virus (HIV)

The virus consists of two molecules of RNA and molecules of reverse transcriptase. A protein capsid surrounds the genome, and an envelope with spikes of protein lies outside the capsid and matrix.

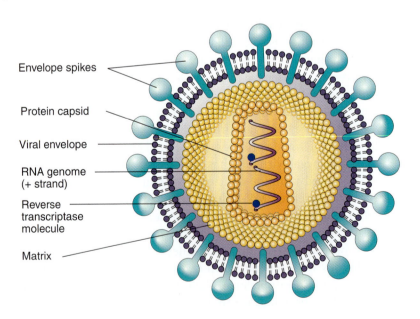

Envelope spikes

Protein capsid

Viral envelope

RNA genome (+ strand)

Reverse transcriptase molecule

Matrix

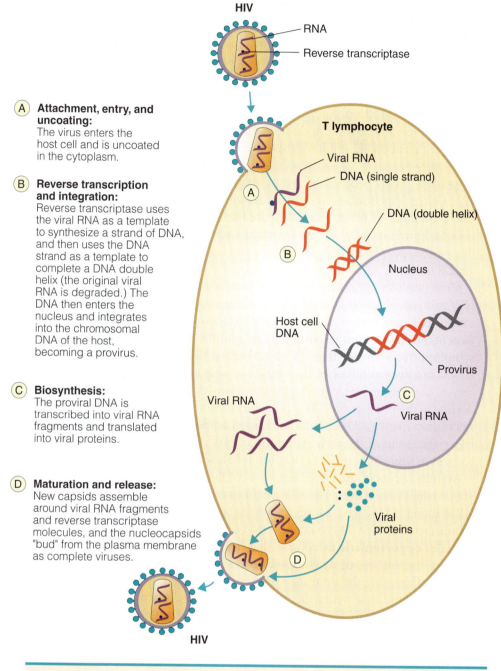

HIV

RNA

Reverse transcriptase

T lymphocyte

Viral RNA

DNA (single strand)

DNA (double helix)

Nucleus

Host cell DNA

Provirus

Viral RNA

Viral RNA

Viral proteins

HIV

(A) **Attachment, entry, and uncoating:**
The virus enters the host cell and is uncoated in the cytoplasm.

(B) **Reverse transcription and integration:**
Reverse transcriptase uses the viral RNA as a template to synthesize a strand of DNA, and then uses the DNA strand as a template to complete a DNA double helix (the original viral RNA is degraded.) The DNA then enters the nucleus and integrates into the chromosomal DNA of the host, becoming a provirus.

(C) **Biosynthesis:**
The proviral DNA is transcribed into viral RNA fragments and translated into viral proteins.

(D) **Maturation and release:**
New capsids assemble around viral RNA fragments and reverse transcriptase molecules, and the nucleocapsids "bud" from the plasma membrane as complete viruses.

FIGURE 14.12

The Replication Cycle of the Human Immunodeficiency Virus (HIV)

explored in more detail in Chapter 19.) Infection by HIV leads to the failure of this immune response and places the host at risk for opportunistic pathogens and cancers, similar to those in patients with other immunodeficiencies.

The major T cell infected by HIV is the helper T lymphocyte that bears the **CD4 receptor site**. (The helper T cell also is called a CD4 cell because of the site.) A second coreceptor site also is involved. Other cells infected may include those of the central nervous system and macrophages in the blood and tissue. For a period of time, the body keeps up with the virus and replaces T cells as they are destroyed. Eventually, however, the infected individual suffers a severe decline in CD4 cells in the blood. Without these cells, the cytotoxic T cells cannot mature.

At the end of 2001, a total of over 816,000 cases of AIDS were reported to the CDC since the epidemic's beginning. Close to 500,000 deaths have occurred from complications such as opportunistic illnesses. It is important to note that AIDS is the end result of HIV infection. Approximately 1 million Americans are believed to have HIV infection and 25 percent do not know they are infected. Worldwide, scientists estimate that about 42 million people are infected with HIV.

■ **SYMPTOMS.** Many people are asymptomatic when they first become infected with HIV. Others experience a flu-like illness within a month or two of HIV exposure. The symptoms of fever, headache, tiredness, and enlarged lymph nodes in the neck and groin usually disappear within a week to a month. Importantly, during this period, people are very infectious, and large amounts of virus are present in genital fluids.

More persistent or severe symptoms may not appear for 10 years or more after HIV first enters the body in adults, or within two years in children born with HIV infection. This period of "asymptomatic" infection varies between infected individuals. Some have symptoms within a few months of infection, while others can be symptom free for more than 10 years. During this asymptomatic period though, HIV is not dormant. Rather, the virus is actively multiplying, infecting, and killing cells of the immune system. The most obvious effect of HIV infection is a gradual decline in the number of CD4 T cells, although some may have an abrupt and dramatic drop in their CD4 counts. In addition, a person with CD4 cells above 200 may experience some of the early symptoms of HIV disease. Others may have no symptoms even though their CD4 cell count is below 200.

An infected individual does not have AIDS until they reach the most advanced stages of HIV infection. The CDC defines an HIV-infected individual as having AIDS if their CD4 T cell count is below 200 per cubic millimeter of blood. Healthy adults usually have CD4 cell counts of 1,000 or more. In addition, the definition includes **opportunistic infections** that healthy people normally would not have (**FIGURE 14.13**). In people with AIDS, these infections often are severe and eventually fatal because the immune system is so damaged by HIV that the body cannot fight off multiple pathogen infections. The more noteworthy of the fungal and protozoan diseases will be discussed in Chapter 15 and 16. People with AIDS also are likely to develop various cancers, especially those caused by viruses, such as **Kaposi's sarcoma** and **cervical cancer**, or cancers of the immune system known as **lymphomas**.

Scientists have been studying those few individuals who were infected with HIV 10 or more years ago and yet have not developed symptoms of AIDS. The scientists are trying to discover what factors make these people "non-progressors." Is there something different about their immune systems, are they infected with a less aggressive HIV strain, or are their genes in some way protecting them from the effects of HIV? It is hoped that by understanding the body's natural method of control, ideas will surface for protective HIV vaccines and other vaccines to prevent the disease from progressing.

■ **TRANSMISSION.** HIV is transmitted primarily by contamination with infected blood or semen. Intimate, unprotected sexual contact, including anal intercourse, is a common method of transmission. Rectal tissues bleed and give access to the virus. Unpro-

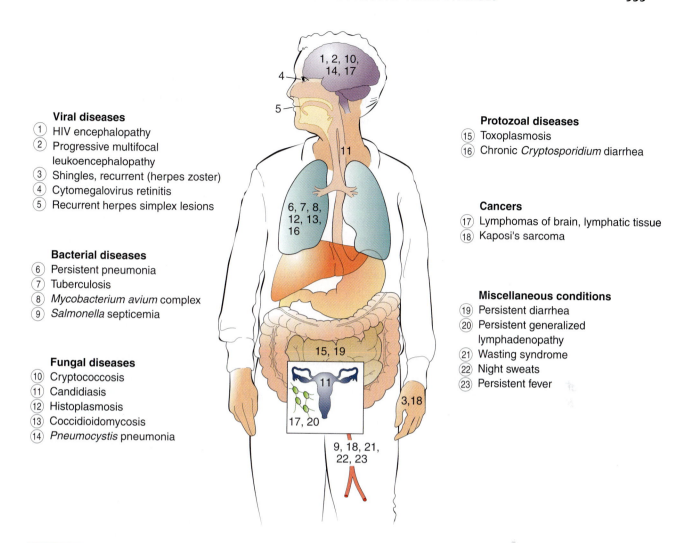

Viral diseases
1. HIV encephalopathy
2. Progressive multifocal leukoencephalopathy
3. Shingles, recurrent (herpes zoster)
4. Cytomegalovirus retinitis
5. Recurrent herpes simplex lesions

Bacterial diseases
6. Persistent pneumonia
7. Tuberculosis
8. *Mycobacterium avium* complex
9. *Salmonella* septicemia

Fungal diseases
10. Cryptococcosis
11. Candidiasis
12. Histoplasmosis
13. Coccidioidomycosis
14. *Pneumocystis* pneumonia

Protozoal diseases
15. Toxoplasmosis
16. Chronic *Cryptosporidium* diarrhea

Cancers
17. Lymphomas of brain, lymphatic tissue
18. Kaposi's sarcoma

Miscellaneous conditions
19. Persistent diarrhea
20. Persistent generalized lymphadenopathy
21. Wasting syndrome
22. Night sweats
23. Persistent fever

FIGURE 14.13

Opportunistic Illnesses in AIDS Patients

This figure illustrates the variety of opportunistic illnesses that affect the body when its immune system has been compromised as a result of infection with HIV. Note the various systems that are affected and the myriad organisms involved.

tected vaginal intercourse is also a high-risk sexual activity, especially if lesions, cuts, or abrasions of the vaginal tract exist. The use of condoms has been shown to decrease the transmission of HIV significantly. The sharing of blood-contaminated needles by injection drug users also transmits HIV. Needle-exchange programs now are being used as a way of interrupting this transmission method. Moreover, transplacental transfer from mother to child is an acknowledged method of HIV transmission.

Viral transmission also can occur through blood products used for medical purposes. Packed red blood cells and blood factor concentrates may contain the virus, but extensive tests are now performed to preclude their presence. Health care workers are at risk of acquiring HIV during their professional activities, such as through an accidental needle stick. Health care workers always should practice established infection-control procedures (standard precautions).

■ **DIAGNOSIS.** A number of diagnostic tests can be used to determine whether an exposed person is HIV infected. HIV antibodies are detected by the **ELISA test** (discussed in Chapter 20). The **Western blot** analysis also is used to confirm the diagnosis. In 2003, the FDA approved the OraQuick Rapid HIV-1 test that can detect HIV-1 antibodies in blood samples in 20 minutes. The **viral load test** detects the RNA of HIV and is available to assess the extent of infection. Using this test, almost 100 percent of infected individuals can be detected. Gene probes and PCR amplification methods also can be used, as Chapter 20 explains.

■ **TREATMENT.** Protocols for HIV infection and AIDS have been researched for many years. The drug used since 1987 is **azidothymidine**, commonly known as **AZT**. AZT interferes with reverse transcriptase activity and acts as a chain terminator as it inhibits DNA synthesis (**FIGURE 14.14**). Other anti-HIV agents approved for use include **dideoxyinosine (ddI), dideoxycytidine (ddC),** and **lamivudine (3TC).** These **reverse transcriptase inhibitors** inhibit DNA synthesis. This decreases the viral load in patients and increases the survival of patients by delaying the start of opportunistic infections.

Still another group of anti-HIV agents are the **protease inhibitors.** These drugs interfere with the processing step of capsid production in the synthesis of HIV particles. The drugs inhibit the enzyme protease, which is responsible for sectioning molecules of large protein for use in the capsid. Several protease inhibitors are now in use, including **saquinavir (Invirase), indinavir (Crixivan),** and **ritonavir (Norvir).** HIV can become resistant to any of these drugs. Therefore, more effective treatment requires a combination of inhibitors. When inhibitors such as AZT and 3TC are used in combination, such treatment is referred to as **highly active antiretroviral therapy (HAART).** Although HAART is not a cure, it has been significant in reducing the number of deaths from AIDS.

A third and new group of antiretroviral agents are HIV **fusion inhibitors.** One, called **enfuvirtide** (Fuzeon), was approved by the FDA in 2003 for use in combination with other antiretroviral drugs. The fusion inhibitors work by blocking receptor recognition to the surface of CD4 cells, preventing the fusion of the viral envelope with the cell's plasma membrane. Such inhibition blocks viral entry into the CD4 cells.

■ **PREVENTION.** The only way to prevent HIV infection is to avoid behaviors that put a person at risk of infection, such as sharing needles and having unprotected sex. A successful vaccine for HIV and AIDS has yet to be developed. Two types of vaccines are being considered. **Preventive vaccines** are for HIV-negative individuals; given before exposure to HIV, the vaccines would prevent infection from occurring. **Therapeutic vaccines** are for HIV-positive individuals; they would help the individual's immune system control HIV so the disease could not progress or be transmitted to others.

There are two major reasons why a successful vaccine has been lacking. First, HIV, like dengue fever virus, continually mutates and recombines. Such behavior by the virus means a vaccine has to protect individuals against many strains of the virus. In addition, since HIV infects CD4 cells, the vaccine needs to activate the very cells that are infected by the virus. Although at times it appears that vaccine research is an uphill battle, scientists are optimistic that a safe and effective HIV vaccine can be produced. Importantly, progress in basic and clinical research is moving forward and scientists are inching closer to identifying products suitable for a successful HIV vaccine.

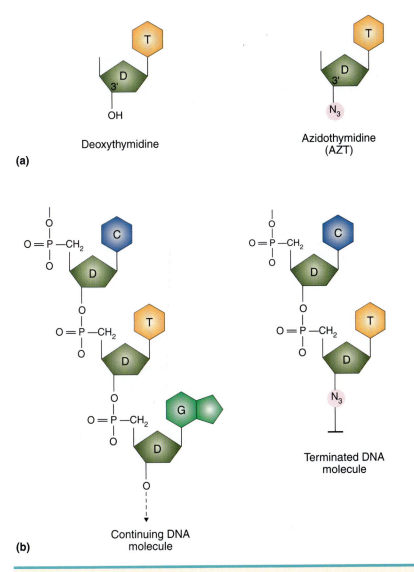

(a)

Deoxythymidine

Azidothymidine (AZT)

(b)

Continuing DNA molecule

Terminated DNA molecule

FIGURE 14.14

How AZT Works

(a) Azidothymidine (AZT) is similar to the nucleotide deoxythymidine, except where deoxythymidine has a –OH group AZT has an –N$_3$ group. The –OH is used to hook onto the next nucleotide in the growing DNA chain, but the –N$_3$ cannot perform this function, and nothing will attach to it. (b) When AZT is present, the DNA polymerase sometimes picks it up instead of deoxythymidine because of the structural similarity between the two molecules. After the AZT molecule is placed in the chain, however, another nucleotide cannot attach, and the DNA chain stops growing. With DNA synthesis halted, the virus cannot be constructed. AZT is said to be a chain terminator.

Although a vaccine for HIV infection and AIDS has not yet been found, evidence indicates that inactivated whole viruses may protect susceptible individuals. However, there is reluctance to use such vaccines because of the possibility that active viruses may be present. Other vaccines use synthetic viral fragments, such as the gp41 and gp120 proteins found in the envelope of HIV. More information about a possible AIDS vaccine is presented in Chapter 20. While the vaccine is being developed, emphasis continues to be placed on preventing the infection, because vaccine development will probably require several more years. TABLE 14.2 summarizes AIDS and other diseases in this section.

TABLE 14.2

A Summary of Viscerotropic Viral Diseases

DISEASE	CLASSIFICATION OF VIRUS	TRANSMISSION	ORGANS AFFECTED	VACCINE	SPECIAL FEATURES	COMPLICATIONS
Yellow fever	*Flaviviridae*	*Aedes aegypti* (mosquito)	Liver Blood	Inactivated viruses	Jaundice	Hemorrhaging
Dengue fever	*Flaviviridae*	*Aedes aegypti* (mosquito)	Blood Muscles	Not available	Breakbone fever	Hemorrhagic fever
Infectious mononucleosis	*Herpesviridae*	Saliva Contact Droplets	Blood Lymph nodes Spleen	Not available	Downey cells Paul-Bunnell test	Splenic rupture Jaundice
Hepatitis A	*Picornaviridae*	Food Water Contact	Liver	Inactivated virus	Jaundice	Liver damage
Hepatitis B	*Hepadnaviridae*	Contact with body fluids	Liver	Synthetic proteins	Jaundice	Liver cancer
Hepatitis C	*Flaviviridae*	Contact with body fluids	Liver	Not available	Jaundice	Liver damage
Viral gastroenteritis	Many viruses	Food Water	Intestine	Not available	Diarrhea	Dehydration Meningitis
Viral fevers	Many viruses	Contact Arthropods Animals	Blood	Not available	Fever Joint pain	Hemorrhaging
Cytomegalo-virus disease	*Herpesviridae*	Contact Congenital transfer	Blood Lung	Not available	Enlarged cells	Fetal damage
Acquired immune deficiency syndrome	*Retroviridae*	Contact with body fluids and blood products	Blood Lymph nodes Brain	Not available	Immune deficiency	Opportunistic illnesses

To this point . . .

We have discussed additional viscerotropic viral diseases, beginning with a series of diseases of the gastrointestinal tract. Many viruses are involved in this disorder, including the rotavirus, the noroviruses, and two types of enteroviruses. We noted that infections are accompanied by varying combinations of diarrhea and malaise, and that the diseases can be severe in certain individuals.

We then surveyed a number of viral fevers, three of which are transmitted by arthropods. The viral fevers are rare in the United States, but they commonly occur in other parts of the world, such as South America and Africa, and represent a potential source of epidemics if they break out in susceptible populations. We also saw how cytomegalovirus is widespread in Americans but of little consequence except in pregnant women and those with suppressed immune systems, such as people infected with HIV.

The section closed with a discussion of HIV infection and AIDS, a disease that is currently considered a major health problem in the United States. The disease is caused by the human immunodeficiency virus (HIV). It involves the immune system and paves the way for opportunistic illnesses.

In the final section of this chapter, we shall survey several neurotropic viral diseases. These diseases have substantial importance because they affect the nervous system and often result in death or permanent paralysis. Two of the diseases, rabies and polio, can be prevented with immunization. Another neurotropic disease, West Nile fever, represents a newly emerging disease in the Western Hemisphere.

14.3 Neurotropic Viral Diseases

Neurotropic viral diseases affect the human nervous system. This fragile system suffers substantial damage when viruses replicate in the tissue. Rabies, a highly fatal and well known disease, is symbolic, as we shall see in the paragraphs ahead. Other diseases and their viruses are less well known. For example, in 1999, an episode of brain illness was linked to a newly recognized paramyxovirus called the **Nipah virus**. Over 250 cases of illness occurred in Malaysia and Singapore, and exposure to pigs seemed to be the primary mode of transmission.

RABIES IS A HIGHLY FATAL INFECTIOUS DISEASE

Rabies is notable for having the highest mortality rate of any human disease, once the symptoms have fully materialized. Few people in history have survived rabies, and in those who did, it is uncertain whether the symptoms were due to the disease or the therapy.

Rabies can occur in most warm-blooded animals, including dogs and cats, horses and rats, and skunks and bats (**FIGURE 14.15**). The disease also has been identified in Alaskan caribou, Russian wolves, and American prairie dogs.

The rabies virus is a single-stranded (– strand) RNA virion of the *Rhabdoviridae* family with a meager five genes in its genome. It is rounded on one end, flattened on the other, and looks like a bullet (**FIGURE 14.16**). The virus enters the tissue through a skin wound contaminated with the saliva, urine, blood, or other fluid from an infected animal. The air in a cave inhabited by diseased bats can transmit the virus also. Indeed, rabid bats are a primary source of human infection in the United States.

The incubation period for rabies varies according to the amount of virus entering the tissue and the wound's proximity to the central nervous system. As few as six days or as long as a year may elapse before symptoms appear. A bite from a rabid animal does not ensure transmission, however, because experience shows that only 5 to 15 percent of inoculated individuals develop the disease.

Early signs of rabies are abnormal sensations such as tingling, burning, or coldness at the site of the bite. Fever, headache, and increased muscle tension develop, and the patient becomes alert and aggressive. Soon there is paralysis, especially in

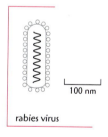

100 nm

rabies virus

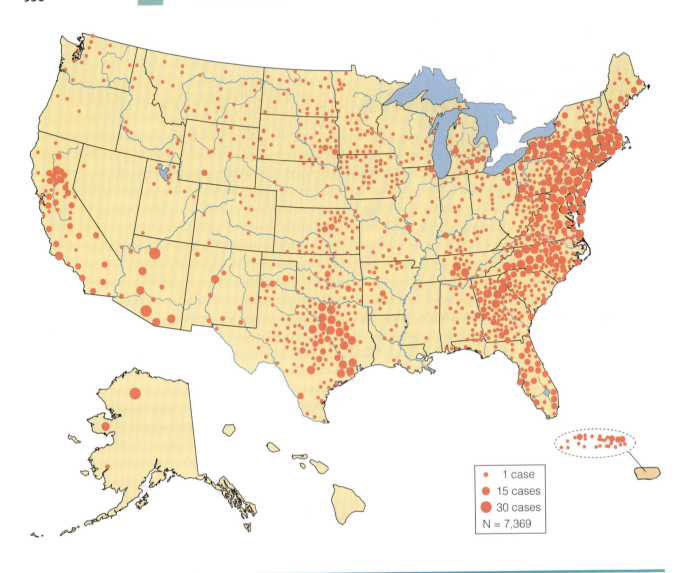

FIGURE 14.15

Reported Cases of Animal Rabies, 2001

There were 7,437 cases of rabies in animals reported in 49 states, the District of Columbia, and Puerto Rico in 2001. No cases in humans were reported.

the swallowing muscles of the pharynx, and saliva drips from the mouth. Brain degeneration, together with an inability to swallow, increases the violent reaction to the sight, sound, or thought of water (the word "rabies" comes from the Latin *rabere* for "rage"). The disease therefore has been called **hydrophobia**—literally, the fear of water. Death usually comes within days from respiratory paralysis.

A person who is bitten by an animal, particularly a wild animal, should be treated as if the animal were rabid. Before 1980, this meant up to two-dozen injections of duck embryo vaccine given at a 45-degree angle in the abdominal fat. Since 1980, however, physicians have used a vaccine composed of inactivated viruses cultivated in human embryonic lung cells. (A less expensive vaccine prepared in chick embryo cells is now being tested.) Because human tissue is used, allergic responses

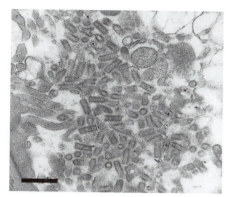

FIGURE 14.16

The Rabies Virus

An accumulation of rabies viruses in the salivary gland tissue of a canine. The viruses are elongated particles, some having the tapered shape of bullets. This configuration is characteristic of the rhabdoviruses. (Bar = 100 nm.)

are much reduced. For people suffering from animal bites, five injections are given in the shoulder muscle of the arm. These injections are preceded by thorough cleansing and one dose of rabies immune globulin to provide immediate antibodies at the site of the bite. This **postexposure immunization** usually is accompanied by a tetanus booster, but the latter is omitted if exposure is not certain, as occurred in the incident described in FIGURE 14.17. For high-risk individuals such as veterinarians, trappers, and zoo workers, a preventive immunization of three injections may be given.

Rabies historically has been a major threat to animals. One form, called **furious rabies**, is accompanied by violent symptoms as the animal becomes wide eyed, drools, and attacks anything in sight. In the second form, **dumb rabies**, the animal is docile and lethargic, with few other symptoms. In 2001, more than 7,400 wildlife cases were reported throughout the United States (Figure 14.15), with raccoons and skunks accounting for 67 percent of the cases. There were no human cases in 2001, due in part to postexposure immunizations, of which over 25,000 are given annually. (However, over 50,000 die of rabies worldwide.) Health departments now are conducting a novel campaign to immunize wild animals using vaccine air-dropped within biscuit-sized baits of dog food and fish meal.

POLIO MAY BE THE NEXT INFECTIOUS DISEASE ERADICATED

The name **polio** is a shortened form of **poliomyelitis**, a word derived from the Greek *polios* for "gray" and *myelon* for "matter." The "gray matter" is the nerve tissue of the spinal cord and brain, which are affected in the disease. Viruses that cause polio are among the smallest virions, measuring 27 nm in diameter. They are composed of single-stranded (+ strand) RNA and are icosahedral virions of the *Picornaviridae* family (FIGURE 14.18).

25 nm

poliovirus

Polioviruses usually enter the body by contaminated water and food. They multiply first in the tonsils and then in lymphoid tissues of the gastrointestinal tract, causing nausea, vomiting, and cramps. In many cases, this is the extent of the problem. Sometimes, however, the viruses pass through the bloodstream and localize on the meninges, where they cause **meningitis**. Paralysis of the arms, legs, and body trunk may result. In the most severe form of polio, the viruses infect the medulla of the brain, causing **bulbar polio** (the medulla is bulb-like). Nerves serving the upper body torso are affected. Swallowing is difficult, and paralysis develops in the tongue, facial muscles, and neck. Paralysis of the diaphragm muscle causes labored breathing and may lead to death.

FIGURE 14.17

An Unusual Case of Rabies

This case was unusual because rabies is not usually associated with calves. Transmission had probably occurred during contact with a wild animal.

TEXTBOOK CASES

1. Twenty-seven guests attended a wedding reception in rural Columbia County, New York, on November 13, 1994. The bride and groom sparkled on their wedding day.

2. During the reception, the guests were invited to stroll around the farm. A young calf was the center of attention, and most of the guests had an opportunity to pet the animal.

3. During the next 2 weeks, the calf behaved erratically, with twitching movements and paralysis developing first in its limbs, then in its breathing muscles. The veterinarian could do little to help. On November 27, the calf died.

4. Because there were numerous reports of rabies in raccoons in New York, the veterinarian suggested that the calf's head be sent to the lab for analysis. The lab report was positive for rabies.

5. The family members and all 27 guests were contacted, and all received postexposure immunizations to prevent their getting the disease.

Virologists have identified three types of poliovirus: type I, the **Brunhilde** strain, causes a major number of epidemics and is sometimes a cause of paralysis; type II, the **Lansing** strain, occurs sporadically but invariably causes paralysis; and type III, the **Leon** strain, usually remains in the intestinal tract. Once the method of laboratory cultivation of polioviruses was established by Enders, Weller, and Robbins, a team led by Jonas Salk grew large quantities of the viruses and inactivated them with formaldehyde to produce the first polio vaccine in 1955. Albert Sabin's group subsequently developed a vaccine containing attenuated (weakened) polioviruses. This vaccine was in widespread use by 1961 and could be taken orally as compared with Salk's vaccine, which had to be injected. Both vaccines are referred to as **trivalent** because they con-

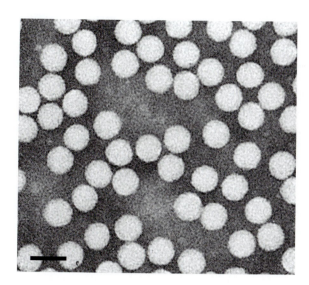

FIGURE 14.18

Polioviruses

A transmission electron micrograph of the viruses that cause polio. Although the particles appear to be circular, their symmetry has been found to be icosahedral. With a diameter of about 27 nm, these are among the smallest viruses that cause human disease. (Bar = 40 nm.)

tain all three strains of virus. The vaccines have contributed substantially to the reduction of polio, as FIGURE 14.19 shows.

In 1988, the forty-first World Health Assembly, with funds raised by Rotary International, launched a global initiative to eradicate polio by the end of the year 2000. To meet this goal, progress was made in the 1980s to eliminate the poliovirus from the Americas, and to protect all children from the disease (MicroFocus 14.9). In the 15 years since the Global Polio Eradication Initiative was launched, the number of cases of polio has fallen by over 99 percent, from more than 350,000 cases in 1988 to 1,919 reported cases in 2002. In the same period, the number of polio-infected countries was reduced from 125 to 7.

In 1994, the WHO certified that the 36 countries in the Americas were polio free. In 2000, the WHO declared the Western Pacific Region (37 countries and areas including China) polio free, and in 2002, the European Region (51 countries) was certified polio free. Widely endemic on five continents in 1988, polio now is limited to parts of Africa and south Asia. Obviously, the goal to eradicate polio by year 2000 has not been met. War and other social issues around the globe have made goal achievement more difficult than first thought. In July 2003, Dr. David Heymann, a top WHO expert, announced polio could be eradicated by 2005 if governments in India, Nigeria, Pakistan, and Egypt fully backed extensive immunization campaigns. These four countries account for 99 percent of all new cases. As Figure 14.19 shows, polio also must be eradicated from Afghanistan, Niger, and Somalia.

A discouraging legacy of the polio epidemics is **postpolio syndrome (PPS)**. What appears to be occurring is that many people who had polio 10 to 40 years ago now are experiencing the initial ailments they had with polio. This includes muscle weakness and atrophy, general fatigue and exhaustion, muscle and joint pain, and breathing or swallowing problems. The National Institute of Neurological Disorders and Stroke (NINDS) estimates that of the 300,000 polio survivors in the United States, PPS affects 25 to 50 percent or more of these individuals.

Several theories have been put forward to explain the cause of PPS and the progressive muscle weakness. One theory suggests that the poliovirus remained dormant in the body after recovery and PPS represents the reactivated virus. Another theory proposes that the initial polio infection generated an autoimmune reaction (Chapter 21), which over the years has caused the body's immune system to attack

FIGURE 14.19

The Race to Eradicate Polio

The World Health Organization and other health agencies are attempting to eradicate polio from the face of the Earth by early in the twenty-first century. Progress of the campaign is illustrated by the global incidence rate of polio in 1988, 1995, and 2002.

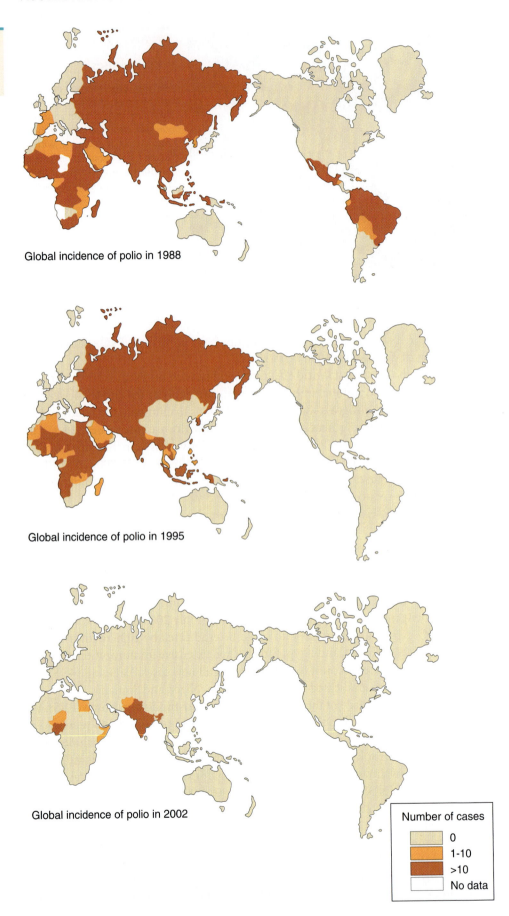

Global incidence of polio in 1988

Global incidence of polio in 1995

Global incidence of polio in 2002

Number of cases

0

1-10

>10

No data

MicroFocus 14.9

BLITZ

One ordinary morning at the end of December 1997, two million people set out to eradicate polio in India. Coming from every conceivable corner of a vast country, they arrived at 650,000 Indian villages, where they set up immunization posts. Children came to them by the thousands, then the hundreds of thousands, then the millions. By the end of the day, 127 million children had received polio immunizations.

This was the National Immunization Day for India, one in a series of such events across the world. The campaigns are coordinated by the World Health Organization, with help from UNICEF (which provides the oral polio vaccine) and Rotary International (which helps recruit volunteers). Their success is illustrated by the 1996 achievement—over

420 million children under age 5 vaccinated, nearly two-thirds of the world's population of children in that age group.

Then, there are the Days of Tranquility. These are pauses in wars and civil strife to allow children to be immunized. In El Salvador, a day is designated each year, and health care workers cross enemy lines to vaccinate the local children. In Sri Lanka, polio vaccine was passed across front lines during Days of Tranquility in 1995 and 1996. Cease-fires have been called in the Sudan to pass out polio vaccine (the days also give warring armies a glimpse of peace).

Polio is a particularly desirable target for eradication because, like smallpox, the disease occurs only in humans. Moreover, the virus exists in the body for only a short period of time, and the

vaccine provides effective intervention for the disease. These reasons, plus the involvement of the world's nations, have generated high optimism that one day, polio may be nothing more than a memory.

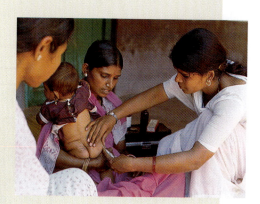

motor neurons as if they were foreign substances. Currently, the most accepted theory is that the loss of motor neurons that occurs during polio was compensated by the remaining healthy neurons. Eventually, the heavy burden and stress placed on these nerve cells causes them to gradually deteriorate and lead to the muscle weakness characteristic of PPS.

ARBOVIRAL ENCEPHALITIS IS CAUSED BY BLOOD-SUCKING ARTHROPODS

The term **encephalitis** means acute inflammation of the brain. Used in the general sense, encephalitis may refer to any brain disorder, much as pneumonia refers to a lung disorder. In this section, we shall use encephalitis to mean a number of viral disorders that are *ar*thropod*bo*rne (hence arboviral).

Arboviral encephalitis may be caused by a series of RNA viruses, usually of the *Togaviridae* or *Bunyaviridae* families. In humans, viral encephalitis is characterized by sudden, very high fever and a severe headache. Normally, the patient experiences pain and stiffness in the neck, with general disorientation. Patients become drowsy and stuporous, and may experience a number of convulsions before lapsing into a coma. Paralysis and mental disorders may afflict those who recover. Mortality rates are generally high.

There are many forms of arboviral encephalitis and various vectors of the disease. One form is **St. Louis encephalitis (SLE)**, named for the city where it was first identified in 1933. This disease, transmitted by mosquitoes, resurfaced in 1975 with 1,300 cases nationwide. Other forms are California encephalitis, La Crosse encephalitis, and

Japanese B encephalitis, all transmitted by mosquitoes. Russian encephalitis and Louping ill encephalitis are forms transmitted by ticks.

Arboviral encephalitis is also a serious problem in horses, causing erratic behavior, loss of coordination, and fever (**FIGURE 14.20**). In addition to the economic loss sustained by the death of the horses, the disease is transmissible from horses to humans by ticks, mosquitoes, and other arthropods. Important forms are Eastern equine encephalitis (EEE), Western equine encephalitis (WEE), and Venezuelan Eastern equine encephalitis (VEEE). The diseases occur in many animals in nature and are a particular problem in birds because birds are natural reservoirs of the viruses and spread them during annual migrations.

FIGURE 14.20

An Outbreak of Eastern Equine Encephalitis

This outbreak occurred in Florida during 1991. Horses in transit or flocks of birds may have been responsible for transporting the viruses from central to northern Florida.

1. During the summer of 1991, a group of horses at a northern Florida stable were observed displaying erratic behaviors. Many horses were unsteady on their feet, and some walked in circles.

2. As the days passed, the symptoms worsened. Many of the horses held their heads low, as if asleep. Worried veterinarians suspected Eastern equine encephalitis (EEE).

3. Shortly thereafter, a number of the stable's trainers, jockeys, and stable hands developed piercing headaches, occasional numbness in the arms and legs, and stiff necks.

4. Some of them were hospitalized. Blood tests confirmed that they harbored the EEE virus. To interrupt the epidemic, investigators began a search for the agent of transmission.

5. Heavy mosquito swarms were reported at a tire depot in central Florida. Suspecting arthropod involvement, researchers collected *Aedes albopictus* mosquitoes from pools of water in the tires.

6. Public health officials successfully isolated the EEE virus from the mosquitoes and concluded they were the epidemic's source. This was the first isolation of the EEE virus from *A. albopictus*.

WEST NILE FEVER IS AN EMERGING DISEASE IN THE WESTERN HEMISPHERE

Since the first outbreak in 1999 in New York City, each year West Nile virus (WNV) has moved farther west across the United States (FIGURE 14.21). In 2003, there were over 8,500 cases reported in 45 states and 200 deaths from encephalitis. Now, experts believe the virus is here to stay and will cause seasonal epidemics that flare up in the summer and continue into the fall.

WNV is a member of the *Flaviviridae* and consists of a single-stranded (+ strand) RNA genome with an envelope and icosahedral symmetry (FIGURE 14.22). Before the outbreak in the United States in 1999, WNV was established in Africa, West Asia, and the Middle East. It is closely related to St. Louis encephalitis virus. The virus has a somewhat broad host range and can infect humans, birds, mosquitoes, horses, and some other mammals.

The virus is spread by the bite of an infected mosquito, which itself becomes infected when it feeds on infected birds (FIGURE 14.23). Infected mosquitoes then spread WNV to humans and other animals when they bite. In 2002, health officials reported a very small number of cases where WNV was transmitted through blood transfusions, organ transplants, breast-feeding, and even during pregnancy from mother to baby. The virus is not spread through casual contact such as touching or kissing a person who is infected.

FIGURE 14.21

West Nile Virus in the United States, 1999–2002

The map shows the spread of the West Nile Virus across the United States from 1999 through 2002.

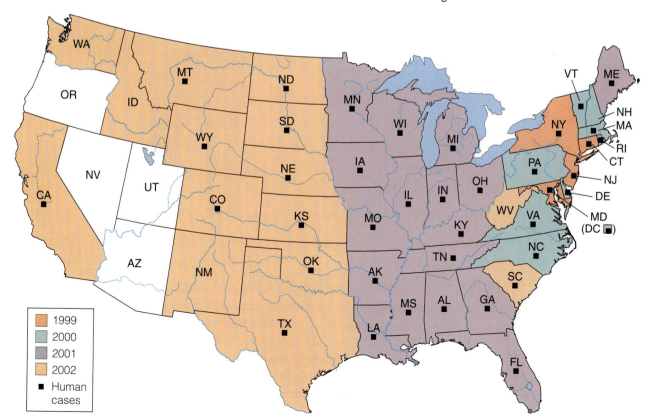

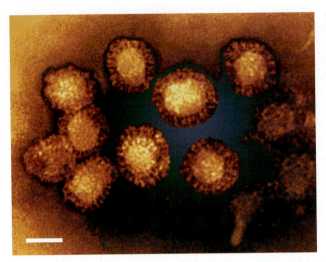

FIGURE 14.22

West Nile Virus

False color transmission electron micrograph of the West Nile virus (WNV). WNV is transmitted by mosquitoes and is known to infect both humans and animals (such as birds). Note the spikes projecting from the viral envelope. (Bar = 40 nm.)

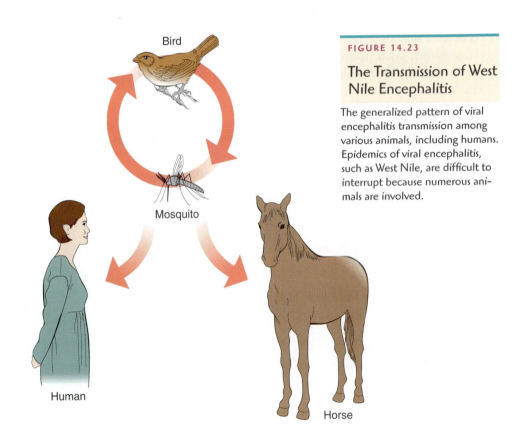

Bird

Mosquito

Human

Horse

FIGURE 14.23

The Transmission of West Nile Encephalitis

The generalized pattern of viral encephalitis transmission among various animals, including humans. Epidemics of viral encephalitis, such as West Nile, are difficult to interrupt because numerous animals are involved.

WNV causes a potentially serious illness. People typically develop symptoms between 3 and 14 days after they are bitten by the infected mosquito. Approximately 80 percent of people who are infected remain asymptomatic. Most of the remaining infected individuals display a mild **West Nile fever**, including fever, headache, body aches, nausea, vomiting, and sometimes swollen lymph glands. They also may develop a skin rash on the chest, stomach, and back. Symptoms typically last a few days. About one in 150 people infected with WNV will develop **encephalitis** or **meningitis** that affects the central nervous system. Symptoms can include high fever, headache, stiff neck, stupor, disorientation, coma, convulsions, muscle weakness, vision loss, numbness, and paralysis. These symptoms may last several weeks, and the neurological effects may be permanent. Death can result.

There is no specific treatment for WNV infection. In cases of encephalitis and meningitis, people may need to be hospitalized so they can receive supportive treatment including intravenous fluids, help with breathing, and nursing care. People who spend a lot of time outdoors are more likely to be bitten by an infected mosquito. These individuals should take special care to avoid mosquito bites. Also, people over the age of 50 are more likely to develop serious symptoms from a WNV infection if they do get sick, so they too should take special care to avoid mosquito bites.

LYMPHOCYTIC CHORIOMENINGITIS FOLLOWS EXPOSURE TO RODENT FECES

Lymphocytic choriomeningitis (LCM) is usually found in mice, hamsters, and other rodents, where it is transmitted by feces and dustborne viruses from urine. Several outbreaks in humans have been recorded since 1960, most related to pet hamsters or hamster colonies in laboratories. In 1984, a case was traced to mice in the patient's home. The lymphocytic choriomeningitis virus is a single-stranded (– strand) RNA virus of the *Arenaviridae* family.

LCM is a mild disorder, often with influenza-like symptoms. Fever and malaise precede headache, drowsiness, and stupor, and the meninges of the brain are infiltrated with large numbers of lymphocytes (hence the disease's name). The symptoms subside within a week, and the mortality rate is very low. Aseptic meningitis may be the only symptom of disease.

TABLE 14.3 summarizes the neurotropic viral diseases. MicroInquiry 14 presents several case studies concerning some of the viral diseases discussed in this chapter.

TABLE 14.3

A Summary of Neurotropic Viral Diseases

DISEASE	CLASSIFICATION OF VIRUS	TRANSMISSION	ORGANS AFFECTED	VACCINE	SPECIAL FEATURES	COMPLICATIONS
Rabies	*Rhabdoviridae*	Contact with body fluids	Brain Spinal cord	Inactivated viruses	Paralysis Hydrophobia	Death
Polio	*Picornaviridae*	Food Water Contact with human feces	Intestine Spinal cord Brain	Inactivated viruses Attenuated viruses	Infection of medulla	Permanent paralysis
Arboviral encephalitis	Many viruses	Arthropods	Brain	Not available	Encephalitis	Coma Seizures
West Nile fever encephalitis meningitis	*Flaviviridae*	Mosquitoes	Brain	Not available	Neurological effects can be permanent	Death
Lymphocytic choriomeningitis	*Arenaviridae*	Dust Contact	Brain Meninges	Not available	Lymphocytes in brain	Paralysis

MicroInquiry 14

VISCEROTROPIC AND NEUROTROPIC VIRAL DISEASE IDENTIFICATION

Below are descriptions of several viscerotropic and neurotropic viral diseases based on material presented in this chapter. Read the case history and then answer the questions posed. Answers can be found in Appendix E.

Case 1: A 27-year-old male with a history of intravenous and over-the-counter drug abuse comes to the emergency room complaining of nausea, vomiting, headache, and abdominal pain. He indicates he has had these symptoms for two days. His vital signs are normal and he presents with slight jaundice. Physical examination indicates discomfort in the upper right quadrant. Serological tests for HBsAg and serum antibodies to HAV are negative.

14.1a. Based on these symptoms and clinical findings, what disease does the patient have and what virus is the cause?

14.1b. What factors lead you to this conclusion?

14.1c. What is the significance of discomfort in the upper right quadrant?

14.1d. What is the probable source of the infection?

14.1e. To what other diseases is the patient at risk?

Case 2: A 19-year-old female comes to a neighborhood clinic complaining of difficulty swallowing. She tells the physician she has had a sore throat and fever for about a week and has been extremely tired. Examination indicates she has enlarged tonsils and a high count of B lymphocytes. Her airways are clear and she has a clean chest X ray. Throat cultures for a bacterial infection are negative but a Monospot test is positive.

14.2a. What viral disease would be consistent with the patient's symptoms and clinical tests? What virus is the causative agent?

14.2b. Why was a bacterial infection considered?

14.2c. What is the significance of the elevated B lymphocyte count?

14.2d. What dangerous complications can occur from this disease?

14.2e. In developing nations, what tumor often is associated with an infection by the same virus? How might the virus influence tumor formation?

Note to the Student

In the remote tropics of South America and Africa there lurks a coterie of viruses that infect animals, but seldom bother humans. Occasionally, however, humans stumble into their paths, with results horrifying enough to mark the annals of medicine. The hemorrhagic fever viruses, for example, make the internal organs bleed and rot, and patients ooze contagious, virus-laden blood from their eyes, ears, nose, and other orifices. Marburg viruses appeared in 1967, then Lassa fever viruses in the 1970s, followed by Ebola viruses and hantaviruses in the years thereafter. In 1994, another hemorrhagic fever virus, the Sabia virus, escaped from a high-security laboratory at Yale University and roamed the streets of New Haven and Boston before felling its human host with symptoms (the patient recovered, and none of his 80 contacts developed the disease).

Where have these viruses come from and what do they portend? The strange new pathogens erupting on American soil are believed to be ancient organisms that lacked the opportunity to attack until humans blundered into their habitat. Lassa virus, Marburg virus, Ebola virus, hantavirus—all were probably confined for millennia to isolated human groups or animals; then civilization's encroachment on the forest let them broaden their range. Now it was up to nature to trigger an expansion of the host organism. Previous to the 1993 hantavirus epidemic, for instance, heavy rains had disturbed the desert and mountain ecology of the southwestern United States, leading to an abundance of piñon nuts and grasshoppers. Suddenly, the deer mouse population exploded, and hantaviruses had a ready

Case 3: A 59-year-old man visits the local hospital emergency room complaining of nausea, abdominal discomfort with vomiting, and fever. On physical examination, he has a temperature of 38°C and appears jaundiced. He indicates the symptoms appeared abruptly two weeks ago after he had returned from a tour to Egypt. While there, the patient indicated he lived like the locals, ate in local restaurants, and often drank the local water. He did not have any vaccinations before leaving on his trip.

14.3a. Name two viral diseases discussed in this chapter that the patient might have.

14.3b. What most likely is the disease the patient has? Why would you make that conclusion?

14.3c. Identify a possible transmission route for this virus.

14.3d. What serological tests could be done to pinpoint the illness?

14.3e. If the patient had sought medical intervention earlier in the illness, what treatment could have been prescribed?

Case 4: A 27-year-old woman comes to a neighborhood clinic with a fever, backache, headache, and bone and joint pain. She indicates these symptoms appeared about two days ago. She also complains of eye pain. Questioning the woman reveals that she has just returned from a trip to Bangladesh where she was doing ecological research in the tropical forests. Skin examination shows remnants of several mosquito bites, which the woman corroborates. She indicates this was her first trip to Southeast Asia. She also reports that she had been taking some antibiotics that she had been given.

14.4a. Based on the woman's symptoms, what viral disease does she most likely have?

14.4b. What clues lead you to this diagnosis?

14.4c. What is unique about this virus?

14.4d. Explain why the antibiotics did not help the woman's condition.

14.4e. Why might the woman want to seriously consider not returning to Southeast Asia to continue her ecological work?

---

mode of transport. In other cases, humans luck out before the viruses can find an alternate host: Ebola virus kills its victims so quickly that it does not have time to spread. And it does not have an animal host—not yet, at least.

As the world shrinks to a global village, a great biological soup is emerging. It is a soup into which dangerous microorganisms of long-isolated ecologies are being mixed. Air travel provides an excellent opportunity for an ill person to carry viruses across the ocean in a matter of hours. Military bases in far corners of the globe and increased trade with new nations promote the exposure of foreigners to exotic viruses. Europeans once encountered dangerous new microorganisms in the New World. Now their descendants are confronting novel pathogens in their own new worlds.

To many scientists, these are exciting times because their search for disease agents is reminiscent of searches made by Pasteur and Koch over 100 years ago. To others, the times are foreboding because they must deal with immediate threats to life. They ask, what else is out there?

Contrary to what some people believe, science does not have an answer for everything. If you find yourself becoming complacent about disease and its control, I urge you to read Fuller's book about Lassa fever (*Fever*) or Preston's book about Ebola fever (*The Hot Zone*). Better yet, read Laurie Garrett's *The Coming Plague*. It will help you realize that the wave of emerging diseases is far from spent.

Summary of Key Concepts

14.1 VISCEROTROPIC VIRAL DISEASES

■ **Yellow Fever Is Transmitted by Mosquitoes.** Yellow fever is a hemorrhagic fever caused by an arbovirus in the *Flaviviridae* family. Symptoms include fever, jaundice, bleeding gums, a bloody stool, internal bleeding, and delirium. Mortality rates can be very high. No treatment exists, although a yellow fever vaccine is available for prevention.

■ **Dengue Fever Can Progress to a Hemorrhagic Fever.** The dengue fever virus is a member of the *Flaviviridae*. Dengue fever results from a mosquito bite and causes fever, and muscle and joint pain. Recovery usually is complete; however, infection by another strain of the virus can cause dengue hemorrhagic fever. A rash due to skin hemorrhaging appears on the face and extremities. Vascular shock can lead to death.

■ **Infectious Mononucleosis Is Most Common in Children and Young Adults.** The Epstein-Barr virus (EBV), a member of the *Herpesviridae*, is responsible for mononucleosis in many young adults in developed nations. Fever, sore throat, and lymph node swelling occur. EBV also is responsible for Burkitt's lymphoma typically found in developing nations.

■ **Hepatitis A Is Transmitted Most Often by the Fecal-Oral Route.** Hepatitis A is caused by a picornavirus. Infection follows the fecal-oral route. Symptoms of infection include anorexia, nausea, vomiting, and fever, followed by jaundice. Recovery confers life-long immunity.

■ **Hepatitis B Can Produce a Chronic Liver Infection.** Hepatitis B causes a more serious inflammation of the liver. A member of the *Hepadnaviridae*, the virus must breach the skin. Blood transfusions, contaminated needles, and sexual intercourse are typical means of transmission. The virus infects the liver, causing inflammation. Liver damage can lead to death.

■ **Hepatitis C Often Causes Cirrhosis and Liver Failure.** The hepatitis C virus is a member of the *Flaviviridae*. Infection leads to a slow and damaging liver disease, causing cirrhosis and potential liver failure.

■ **Other Viruses Also Cause Hepatitis.** Besides the viruses responsible for hepatitis A, B, and C, there are additional hepatitis viruses. Hepatitis D (delta hepatitis) virus only causes liver infection in the presence of hepatitis B virus. A co-infection increases the risk of liver cirrhosis. Hepatitis G virus also is transmitted by contaminated blood products or sexual intercourse. It causes persistent infections in up to 30 percent of infected individuals.

14.2 OTHER VISCEROTROPIC VIRAL DISEASES

■ **Viral Gastroenteritis Is Caused by Any of Several Viruses.** Viral gastroenteritis can be caused by viruses in the *Reoviridae*. Transmission typically follows the fecal-oral route and severe dehydration or death can occur in children. Noroviruses, in the family *Caliciviridae* also are transmitted along the fecal-oral route. Symptoms include acute-onset vomiting (more common in children), watery non-bloody diarrhea with abdominal cramps, and nausea. Low-grade fever occasionally occurs. The Coxsackie virus and echovirus, both in the *Picornaviridae*, also cause viral gastroenteritis.

■ **Viral Fevers Can Be Caused by a Variety of Viruses.** Viral fevers include a variety of viruses responsible for Colorado tick fever, sandfly fever, and Rift Valley fever. Viral hemorrhagic fevers include Lassa fever, Marburg hemorrhagic fever, and Ebola hemorrhagic fever. It is not exactly known how the latter two viruses, which make up the *Filoviridae*, are transmitted, but both have caused substantial numbers of deaths.

■ **Hantavirus Pulmonary Syndrome Produces Sudden Respiratory Failure.** Hantavirus is a member of the *Bunyaviridae* and is the cause of hantavirus pulmonary syndrome. A flu-like illness quickly develops into blood hemorrhaging, and renal and respiratory failure.

■ **Cytomegalovirus Disease Can Produce Serious Birth Defects.** Cytomegalovirus can cause a severe disease in newborns and is a major cause of birth defects.

■ **Acquired Immune Deficiency Syndrome Results from Immune System Dysfunction.** AIDS is the final stage of an infection of lymphocytes caused by the human immunodeficiency virus (HIV). HIV is a member of the *Retroviridae* and forms a provirus in host cells. Infection and destruction of CD4 T cells eventually lead to an inability of the immune system to fend off opportunistic diseases. The presence of one or more of these diseases along with a drop in CD4 cells signals the individual has AIDS. Transmission is though blood, blood products, contaminated needles, or unprotected sexual intercourse. Many antiretroviral drugs are available to slow the progression of disease. However, a successful vaccine has eluded vaccine developers.

14.3 NEUROTROPIC VIRAL DISEASES

■ **Rabies Is a Highly Fatal Infectious Disease.** Rabies typically is transmitted by the bite of a rabid animal. Infections by this rhabdovirus lead to fever, headache, aggressive behavior, and sensitivity to light. Without treatment, infection is almost 100 percent fatal.

■ **Polio May Be the Next Infectious Disease Eradicated.** The poliovirus is a member of the *Picornaviridae*, which affects nervous tissue, causing permanent paralysis and muscle atrophy. A polio vaccine has eliminated the disease from most of the world. Total eradication is projected to be accomplished by 2005.

■ **Arboviral Encephalitis Is Caused by Blood-Sucking Arthropods.** A group of RNA viruses is responsible for arboviral encephalitis. This includes St. Louis encephalitis in humans and several diseases that cause serious problems in horses. These include Eastern equine encephalitis, Western equine encephalitis, and Venezuelan Eastern equine encephalitis.

■ **West Nile Fever Is an Emerging Disease in the Western Hemisphere.** Recently, the most noteworthy of the arboviral encephalitis viruses is West Nile virus. This flavivirus is transmitted by mosquitoes. Mild infections cause fever, headache, body aches, nausea, and vomiting. A skin rash on the chest, stomach, and back may develop. About 1 in 150 people infected develop West Nile encephalitis or meningitis. Symptoms include high fever, headache, stiff neck, stupor, disorientation, coma, convulsions, muscle weakness, vision loss, numbness, and paralysis. These symptoms may last several weeks and death can result.

■ **Lymphocytic Choriomeningitis Follows Exposure to Rodent Feces.** Lymphocytic choriomeningitis comes from exposure to contaminated rodent feces. A member of the *Arenaviridae*, the virus produces a mild illness, and the symptoms of fever, headache, drowsiness, and stupor subside in about a week.

Questions for Thought and Discussion

Answers to selected questions can be found in Appendix C.

1. In 1995, the New York newspaper *Newsday* reported that during the previous year, 1,700 cases of rabies had occurred in New York State. Public health officials report, to the contrary, that rabies cases in the entire United States rarely exceed single digits. What might have been the source of *Newsday*'s error?

2. In 1996, a medical school student at the University of Maryland was assigned a case for a seminar on undetermined diagnoses in difficult cases. His subject was identified only as E. P., a gentleman who lived in the mid-1800s and was a well-known poet and animal lover. The man was taken to a hospital with delirium and tremors. He was confused and combative, and he refused to drink any water or other liquid. He died on October 7, 1849. The traditional diagnosis had been alcoholism, but the student made a different diagnosis. What do you think it was? When told the man's name, the student thought "Nevermore!" Who was the man?

3. A *New York Times* crossword puzzle once contained a space in the Across column for a term containing 14 letters. The only clue given was "dengue." By using the Down column, you could see that the second-to-last letter was an *e* and that the fifth-to-last letter was an *f*. What was the answer?

4. In 1996, the CDC's Advisory Committee on Immunization Practices (ACIP) recommended that the polio vaccine be used on a sequential schedule: two doses of injectable, inactivated (Salk) vaccine at 2 and 4 months of age, followed by two doses of oral, attenuated (Sabin) vaccine at 12 to 18 months and 4 to 6 years. This was a departure from a policy established in 1987, when the ACIP first recommended exclusive use of the oral Sabin vaccine. Why do you

believe it changed its recommendations? And how well do you think the public and medical community will react to this change? (Chapter 20 details still more modifications to the policy.)

5. Health authorities shifted into adrenaline overdrive when an outbreak of Ebola hemorrhagic disease occurred in Reston, Virginia, in 1989. What sparks such a dramatic response when a disease like Ebola fever breaks out?

6. Walt Disney World uses a series of sentinel chickens strategically placed on the grounds to detect any signs of viral encephalitis. Why do you suppose they use chickens? Why is Disney World particularly susceptible to outbreaks of viral encephalitis? And what recommendations might be offered to tourists if the disease broke out?

7. During 1997, El Niño brought abundant rain and a mild winter to the southwestern United States. The conditions encouraged a burgeoning rodent population. (For example, deer mice that usually reproduce twice a year were able to turn out three litters.) Which viral disease did public health officials anticipate for 1998? What precautions did epidemiologists give residents?

8. Written on some blood donor cards is the notation "CMV(+)." What do you think the letters mean, and why are they placed there?

9. A student in a biology laboratory uses a sterile lancet from a package to pierce the skin and obtain blood for a blood-typing exercise. The student places the lancet down on the desktop, whereupon a nearby student picks it up and uses it again to pierce the skin. What is the danger?

10. At a college campus some years ago, a group of students were passing around a wine bottle. Several days later, one member of the group developed hepatitis A and the others requested that the college infirmary distribute immune globulin shots. When the infirmary refused, the students demonstrated, causing a stir. A reporter from a local newspaper wrote about the incident and recounted it accurately except for the last line of the article, which read: "Hepatitis A is normally transmitted by contaminated syringes." What is microbiologically wrong with this statement, and what are its implications?

11. Since 1990, the makers of one of the recombinant hepatitis B vaccines (Engerix-B) have offered free immunizations to emergency medical technicians (EMTs) throughout the United States. If you were an EMT, would you accept the offer?

12. Sicilian barbers are renowned for their skill and dexterity with razors (and sometimes their singing voices). In 1995, French researchers studied a group of 37 Sicilian barbers and found that 14 had antibodies against hepatitis C, despite never having been sick with the disease. By comparison, when a random group of 50 blood donors was studied, none had the antibodies. What might account for the high incidence of exposure to hepatitis C among the barbers?

13. In many diseases, the immune system overcomes the infectious agent, and the person recovers. In certain diseases, the infectious agent overcomes the immune system, and death follows. Compare this broad overview of disease and resistance to what is taking place with AIDS, and explain why AIDS is probably unlike any other disease encountered in medicine.

14. The restaurant industry is currently weighing a requirement that all restaurant workers be immunized with hepatitis A vaccine. Would you favor or oppose this requirement?

15. A diagnostic test has been developed to detect hepatitis C in blood intended for transfusion purposes. Obviously, if the test is positive, the blood is not used. However, there is a lively controversy as to whether the blood donor should be informed of the positive result. What is your opinion? Why?

16. An epidemiologist notes that India has a high rate of dengue fever but a very low rate of yellow fever. What might be the cause of this anomaly?

Review

On completing your study of these pages, test your understanding of their contents by deciding whether the following statements are true or false. If the statement is true, write "True" in the space. If false, substitute a word for the underlined word to make the statement true. The answers are listed in Appendix D.

_____ 1. Both yellow fever and dengue fever are caused by a <u>DNA</u> virus transmitted by the mosquito.

_____ 2. 80 percent of people infected by West Nile virus experience <u>mild symptoms</u>.

_____ 3. The Coxsackie virus is a well-known cause of <u>gastroenteritis</u> in humans.

_____ 4. Downey cells are a characteristic sign of the viral disease <u>infectious mononucleosis</u>.

_____ 5. The echovirus is known to cause disease of the <u>spleen</u> in humans who acquire the virus.

_____ 6. Polio may be caused by any of <u>three</u> strains of poliovirus.

_____ 7. The term *hydrophobia* means "fear of water," and it is commonly associated with patients who have <u>encephalitis</u>.

_____ 8. Hepatitis is primarily a disease of the <u>liver</u>.

_____ 9. The cell most often affected by HIV, the AIDS virus, is the human <u>monocyte</u>.

_____ 10. Because of the characteristic symptoms, dengue fever is sometimes called <u>breakbone fever</u>.

_____ 11. A small RNA virus is regarded as the cause of hepatitis <u>A</u>.

_____ 12. Norovirus and rotavirus are both considered to be agents of viral <u>encephalitis</u>.

_____ 13. The cytomegalovirus can cause serious disease of the <u>lungs</u> in AIDS patients.

_____ 14. The <u>Epstein-Barr</u> virus is most probably the cause of infectious mononucleosis.

_____ 15. A vaccine is available to prevent <u>yellow fever</u> but not to prevent AIDS.

_____ 16. The first outbreak of West Nile virus in the Western Hemisphere occurred in <u>Miami</u>.

_____ 17. The Salk and Sabin vaccines are used for immunizations against <u>hepatitis</u>.

_____ 18. HIV is a <u>reovirus</u> in which the RNA of the genome is used as a template to synthesize DNA.

_____ 19. Pleurodynia and myocarditis are both related to infection by <u>Norovirus</u>.

_____ 20. Hepatitis B is most commonly transmitted by contact with infected semen or infected <u>blood</u>.

_____ 21. One of the organs that suffers damage during infections with yellow fever virus is the <u>kidney</u>.

_____ 22. The vaccine currently in use to protect against hepatitis A consists of <u>genetically engineered proteins</u>.

_____ 23. One of the most important causes of diarrhea in infants and young children admitted to hospitals is the <u>Epstein-Barr virus</u>.

_____ 24. Filoviruses are long, thread-like viruses that cause hemorrhagic fevers and include the <u>Marburg virus</u>.

_____ 25. Both *Aedes aegypti* and *Aedes albopictus* may be capable of transmitting the virus of <u>dengue fever</u>.

_____ 26. Enlargement of the lymph nodes, sore throat, mild fever, and a high count of B-lymphocytes are characteristic symptoms in people who have <u>polio</u>.

_____27. The resistance of hepatitis A viruses to chemical and physical changes in the environment is generally considered to be <u>low</u>.

_____28. The "four corners disease" that broke out in the United States in 1993 was eventually related to <u>Lassa fever viruses</u>, which were possibly transmitted among individuals by the deer mouse.

_____29. The C in the TORCH group of diseases stands for the <u>cephalovirus</u>, which can be transmitted from a pregnant woman to her unborn child.

_____30. The current protocol for rabies immunizations is to give the injections into the <u>stomach</u>.

http://microbiology.jbpub.com

The site features **eLearning,** an on-line review area that provides quizzes and other tools to help you study for your class. You can also follow useful links for in-depth information, or just find out the latest microbiology news.

The Fungi

Mushrooms always grow in damp places, which is why they look like umbrellas.

—A midwestern eighth-grader sharing his insightful observation of a typical fungus

IRELAND OF THE 1840s was an impoverished country of 8 million people. Most were tenant farmers paying rent to landlords who were responsible, in turn, to the English property owners. The sole crop of Irish farmers was potatoes, grown season after season on small tracts of land. What little corn was available was usually fed to the cows and pigs.

Early in the 1840s, heavy rains and dampness portended calamity. Then, on August 23, 1845, *The Gardener's Chronicle and Agricultural Gazette* reported: "A fatal malady has broken out amongst the potato crop. On all sides we hear of the destruction. In Belgium, the fields are said to have been completely desolated."

The potatoes had suffered before. There had been scab, drought, "curl," and too much rain, but nothing was quite like this new disease. It struck down the plants like frost in the summer. Beginning as black spots, it decayed the leaves and stems, and left the potatoes a rotten, mushy mass with a peculiar and offensive odor. Even the harvested potatoes rotted.

The winter of 1845 to 1846 was a disaster for Ireland. Farmers left the rotten potatoes in the fields, and the disease spread. The farmers first ate the animal feed and then the animals. They also devoured the seed potatoes, leaving nothing for spring planting. As starvation spread, the English government attempted to help by importing corn and establishing relief centers. In England, the potato disease had few repercussions because English agriculture included various grains. In Ireland, however, famine spread quickly.

After two years, the potato rot seemed to slacken, but in 1847 ("Black '47") it returned with a vengeance. Despite relief efforts by the English, over two million Irish died from starvation. Eventually, about 900,000 survivors set off for Canada and the United States. Those who stayed had to deal with economic and political upheaval as well as misery and death.

The potato blight faded in 1848, but it did not vanish. Instead, it emerged again during wet seasons and blossomed anew. In the end, hundreds of thousands of Irish left the land and moved to cities or foreign countries. During the 1860s, great waves of Irish immigrants came to the United States—all resulting from *Phytophthora infestans* that caused late blight of potatoes.

Such are the historic, political, economic, and sociological effects of one species of fungus. Other fungal diseases of fruits, grains, and vegetables can be equally devastating, and we shall see numerous examples in this chapter as we survey the fungi. In addition, we shall take note of several widespread human and animal diseases caused by fungi, and we shall encounter many beneficial fungi such as those used to make antibiotics, breads, foods, and insecticides. Our study will begin with a focus on the unique structures, growth patterns, and life cycles of fungi.

15.1 Characteristics of Fungi

The fungi (sing., fungus) are a diverse group of eukaryotic microorganisms, with over 100,000 identifiable species. For many decades, fungi were classified as plants, but laboratory studies have revealed at least four properties that distinguish fungi from plants: Fungi lack chlorophyll, while plants have this pigment; the cell walls of fungal cells contain a carbohydrate called **chitin**, not found in plant cell walls; though generally filamentous, fungi are not truly multicellular like plants, because the cytoplasm of one fungal cell can mingle through pores with the cytoplasm of adjacent cells; and fungi are heterotrophic eukaryotes, while plants are autotrophic eukaryotes. Mainly for these reasons, fungi are placed in their own kingdom **Fungi**, in the Whittaker classification of organisms. The study of fungi is called **mycology**, and a person who studies fungi is a mycologist. Invariably, the prefix *myco-* will be part of a word referring to fungi, since *mykes* is Greek for "fungus."

Fungi generally have life cycles involving two phases: a growth phase and a reproductive phase. A major group of fungi, the **molds**, grow as long, tangled strands of cells that give rise to visible colonies (**FIGURE 15.1**). Another group, the **yeasts**, are unicellular organisms whose colonies visually resemble bacterial colonies.

FUNGI SHARE A COMBINATION OF CHARACTERISTICS

With the notable exception of yeasts, fungi consist of masses of intertwined filaments of cells called **hyphae** (sing., hypha). The hypha is the morphological unit of the fungus and is visible only with the aid of a microscope (**FIGURE 15.2**). Hyphae have a broad diversity of forms (as photographs in this chapter illustrate), and many hyphae are highly branched with reproductive structures. A thick mass of hyphae is called a **mycelium** (pl., mycelia). This mass is usually large enough to be seen with the unaided eye, and generally it has a rough, cottony texture.

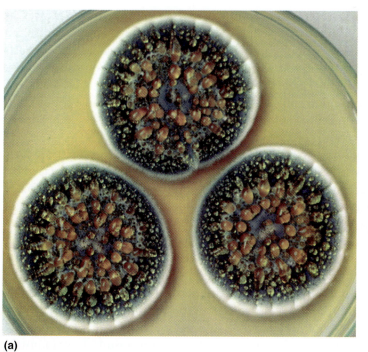

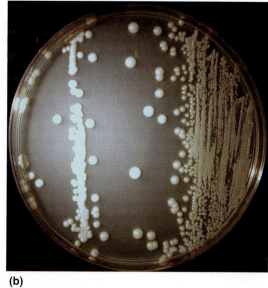

(a)

(b)

FIGURE 15.1

Fungal Colonies

(a) On growth media, molds, such as *Penicillium chrysogenum*, grow as fuzzy colonies visible to the naked eye. The spores are the darker green regions of the colonies. (b) Petri dish culture showing colonies of the yeast-like fungus *Torulopsis glabrata*. A food spoilage organism, this species obtains its nutrional requirements from the breakdown of dead or decaying material.

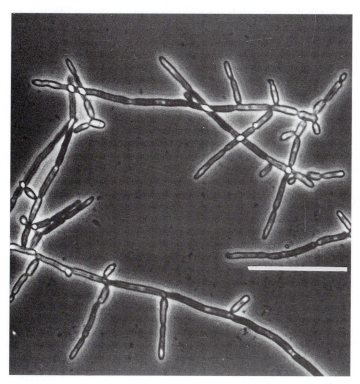

FIGURE 15.2

The Hypha

A phase–contrast photomicrograph of *Candida albicans* in its hyphal form. In the human body, phagocytes have a difficult time coping with these filamentous structures, a factor that may contribute to the pathogenicity of the fungus. (Bar = 40 μm.)

Each cell of a single hypha is eukaryotic, with one or more nuclei, as well as organelles typical of eukaryotes. Thus, fungal cells contain mitochondria, an endomembrane system, ribosomes, and a cytoskeleton. Fungal cells lack chloroplasts and the ability to carry out photosynthesis. The cell wall is composed of large amounts of chitin. **Chitin** is a polymer of acetylglucosamine units, that is, glucose molecules containing amino and acetyl groups. The cell wall provides rigidity and strength, which, like the cell wall of bacteria, allows the cells to resist bursting due to high internal water pressure.

In many species of fungi, cross-walls, or **septa** (sing., septum), within the hyphae divide the cytoplasm into separate cells (**FIGURE 15.3**). Such fungi are described as **septate**. The septa are incomplete, however, and pores allow adjacent cytoplasms to mix. In other fungal species, the cells have no septa (**nonseptate**), and the cytosol and organelles of neighboring cells mingle freely. The common bread mold *Rhizopus stolonifer* is nonseptate, while the blue-green mold that produces penicillin, *Penicillium notatum*, is septate. Both types of fungi are considered to be **coenocytic**, that is, each cell in a hypha contains two or more nuclei.

Since fungi absorb preformed organic matter, they are described as **heterotrophic** organisms. Most are saprobes, feeding on dead organic matter. Together with the bacteria, these fungi decompose vast quantities of dead organic matter that would otherwise accumulate and make the Earth uninhabitable (**MicroFocus 15.1**). In industrial settings, these decompositions can be profitable. The fungus *Trichoderma*, for example, produces enzymes that degrade cellulose and give jeans a "stone-washed" appearance. Others are pathogens, living on plants or animals, often causing disease.

FUNGAL GROWTH IS INFLUENCED BY SEVERAL FACTORS

Fungi acquire the nutrients they need for growth through **absorption**. The molds and unicellular yeasts secrete enzymes into the surrounding environment that break down (hydrolyze) complex organic compounds into simpler ones. As a result of this extracellular digestion, simpler compounds, like glucose and amino acids, can be absorbed. Since the molds are composed of numerous hyphae, the mycelium forms

---

FIGURE 15.3

Hypha Structure

Molds can have hyphae that are septate or nonseptate. (a) Septa compartmentalize hyphae into separate cells. The septa have a pore through which cytoplasm and nuclei can move. (b) Other fungi have nonseptate hyphae. In both types, the cells are coenocytic (multinucleate).

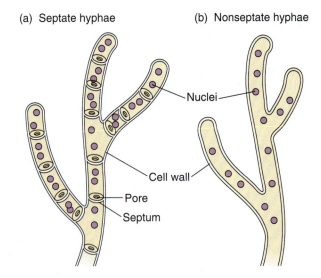

(a) Septate hyphae   (b) Nonseptate hyphae

Nuclei

Cell wall

Pore

Septum

MicroFocus 15.1

WHEN FUNGI RULED THE EARTH

About 250 million years ago, at the close of the Permian period, a catastrophe of epic proportions visited the Earth. Apparently, over 90 percent of animal species in the seas vanished. The great Permian extinction, as it is called, also wreaked havoc on land animals and cleared the way for dinosaurs to inherit the planet.

But land plants managed to survive, and before the dinosaurs came, they spread and enveloped the world. At least, that is what paleobiologists tradi-

tionally believed. Now, however, they are revising their outlook and finding a significant place for the fungi. In 1996, Dutch scientists from Utrecht University presented evidence that land plants were decimated by the extinction and that for a brief geologic span, dead wood covered the planet. During this period, they suggest, the fungi emerged, and wood-rotting species experienced a powerful spike in their populations. Support for their theory is offered by numerous findings of fossil fungi from the

post-Permian period. The fossils are bountiful, and they come from all corners of the globe. Significantly, they contain fungal hyphae, the active feeding forms rather than the dormant spores.

And so it was that fungi proliferated wildly and entered a period of feeding frenzy where they were the dominant form of life on Earth. It's something worth considering next time you turn up your nose at a lowly fungus contaminating a cup of yogurt.

a tremendously large surface area for nutrient absorption. It is the "feeding network" for these organisms (MicroFocus 15.2).

In the yeasts, growth occurs by cell division. Yeast cells undergo many rounds of mitosis and cytokinesis for form a large population of cells. In the molds, mitosis also occurs within the hyphae. Growth of these mycelial fungi involves lengthening and branching of the hyphae from the hyphal tips. This is correlated with a simultaneous increase in cell cytoplasm and flow to the hyphal tips.

Fungal growth is influenced by many factors in the environment. Besides the availability of nutrients, oxygen, temperature, and pH are important. The majority

MicroFocus 15.2

A HUMONGOUS AND ANCIENT INDIVIDUAL

What's bigger than a blue whale and older than a California redwood? A fungus, of course! In the Malheur National Forest of Oregon's Blue Mountains lies *Armillaria*, a plant pathogen that infects evergreen trees. The visible part of *Armillaria* is the golden mushrooms it produces. However, underground in the soil lies an invisible mycelium that invades the roots of the evergreen tree, absorbing its water and carbohydrates. This often spells the demise of the infected tree.

In 2001, scientists studying in the Malheur National Forest discovered a single *Armillaria* that spreads through 2,200 acres of forest soil. It measures over nine

square kilometers—about the area of 1,600 football fields! To be of this size, scientists believe the fungus germinated from a single spore somewhere between 2,000 and 8,500 years ago. Now that's old and humongous, considering the mycelium of most *Armillaria* species covers only 20 or 30 acres.

However, this was not the first sighting of large *Armillaria* mycelia. In 1992, another species of *Armillaria* was discovered in a Michigan hardwood forest that measured about 37 acres in size. In eastern Washington, an *Armillaria* mycelium measuring about 1,500 acres in size has been discovered. None have topped the Oregon fungus.

The world's biggest fungus is challenging the concept of what constitutes an individual. Is the Oregon *Armillaria* a single individual organism? Researchers had thought that an individual fungus organism grew in a confined cluster in the forest, identified by the ring-shaped patches of dead trees that they spotted from the air. No one expected to find that the well-separated clusters represented one contiguous organism. "If you could take away the soil and look at it, it's just one big heap of fungus with all of these filaments that go out under the surface," says Dr. Catherine Parks, who was one of the discoverers of the giant fungus.

of fungi are aerobic organisms, with the notable exception of the facultative yeasts that multiply in the presence of oxygen, or under fermentation conditions.

Temperature is very important in determining the amount and rate of fungal growth. Most fungi grow best at about 25°C, a temperature close to normal room temperature. Notable exceptions are the pathogenic fungi, which thrive at 37°C, body temperature. Usually these fungi also grow on nutrient media at 25°C. Such fungi are described as **dimorphic** (two forms), having a yeast-like stage at 37°C and a mycelial stage at 25°C. Psychrophilic fungi grow at still lower temperatures, such as the 5°C found in a normal refrigerator.

Many fungi thrive under mildly acidic conditions at a pH from 5 to 6. Acidic soil therefore may favor fungal turf diseases, in which case lime (calcium carbonate) should be used to neutralize the soil. Mold contamination is common also in acidic foods such as sour cream, applesauce, citrus fruits, yogurt, and most vegetables. Moreover, the acidity in breads and cheese encourages fungal growth. Blue cheese, for example, consists of milk curds in which the mold *Penicillium roqueforti* is growing.

In the laboratory, fungi are grown like bacteria in tubes or culture plates (figure 15.3). Normally, a high concentration of sugar is conducive to growth, and laboratory media for fungi usually contain extra glucose in addition to an acidic environment. Examples of such media are **Sabouraud dextrose agar** and **potato dextrose agar**. It should be noted that the chemical composition of the medium may influence the appearance of the fungus, as FIGURE 15.4 shows.

In nature, the fungi are important links in ecological cycles because they, along with bacteria, rapidly decompose animal and plant matter. Working in immense numbers, fungi release carbon and minerals back to the environment and make them available for recycling. However, fungi may be a liability for industries because they often contaminate leather, hair products, lumber, wax, cork, and polyvinyl plastics.

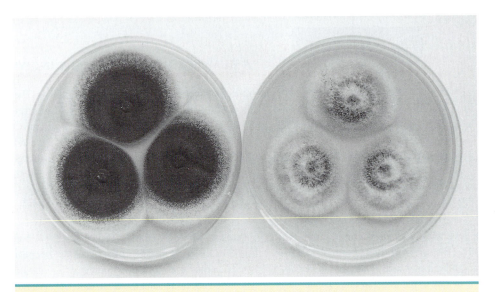

FIGURE 15.4

Variations in a Mold

The mold *Metarrhizium* growing on two different types of fungal medium and displaying two different growth patterns. In microbiology, the nature of the growth environment often influences the appearance of the microorganism.

Many fungi live in a mutually beneficial relationship with other species in nature, an association called **mutualism**. In the southwestern Rocky Mountains, for instance, a fungus of the genus *Acremonium* thrives on the blades of a species of grass called *Stipa robusta* ("robust grass"). The fungus produces a powerful poison that can put horses and other animals to sleep for about a week (the grass is called "sleepy grass" by the locals). Thus, the grass survives where others are nibbled to the ground, reflecting the mutually beneficial interaction between plant and fungus.

Other fungi called **mycorrhizal fungi** also live harmoniously with plants. The hyphae of these fungi envelop or invade the roots of plants (and sometimes their stems). Though poised to suck the plants dry, the fungi are gentle neighbors. Mycorrhizal fungi consume some of the carbohydrates produced by the plants, but in return they contribute minerals and fluids to the plant's metabolism. Mycorrhizal fungi have been found in plants from salt marshes, deserts, and pine forests. Indeed, in 1995, researchers from the University of Dayton reported that over 50 percent of the plants growing in the large watershed area of southwestern Ohio contain mycorrhizal fungi.

REPRODUCTION IN FUNGI INVOLVES SPORE FORMATION

Reproduction in fungi may take place by asexual as well as by sexual processes. The principal structure of asexual reproduction is the **fruiting body**. In the asexual process, thousands of **spores** are produced, all resulting from the mitotic divisions of a single cell and all genetically identical. Each spore has the capability of germinating to reproduce a new hypha that will become a mycelium (**FIGURE 15.5**).

Sporulation usually occurs in structures developing from the hyphae. Certain asexual spores develop within a sac called a **sporangium**. Appropriately, these spores are called **sporangiospores**. Other spores develop on supportive structures called **conidiophores**. These unprotected spores are known as **conidia** (sing., conidium), from the Greek *conidios*, meaning "dust." The bread mold *Rhizopus* produces sporangiospores, while the blue-green mold *Penicillium* as well as other fungi produce conidia (**FIGURE 15.6**). Fungal spores are extremely light and are blown about in huge numbers by wind currents. Many people have an allergic reaction when they inhale spores, and local newspapers may report the daily mold spore count to alert sufferers.

Some asexual modes of reproduction do not involve a spore-containing structure. For example, spores may form by fragmentation of the hypha. This process yields **arthrospores**, from the Greek *arthro-* for "joint." The fungi that cause athlete's foot multiply in this manner. Another asexual process is called **budding**. In

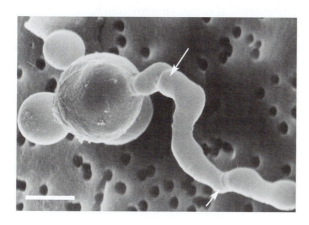

FIGURE 15.5

A Germinating Spore

A scanning electron micrograph of a spore of the fungus *Cephalosporium* germinating to form a hypha. Note the septa (arrows) between cells of the hypha. (Bar = 2 μm.)

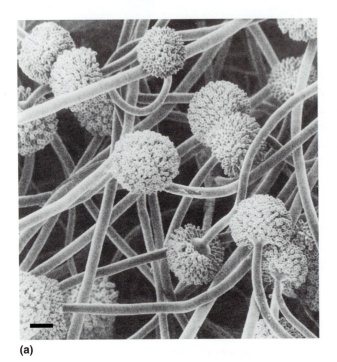

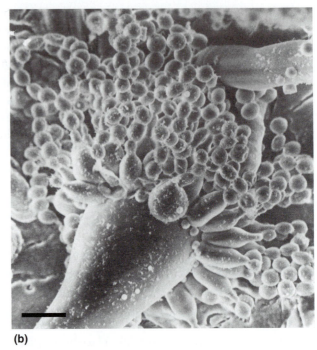

(a)

(b)

FIGURE 15.6

The Fungus *Aspergillus niger*

(a) A scanning electron micrograph showing the mold-like phase of *Aspergillus niger*. Many conidiophores are present within the mycelium. Conidiophores contain conidia, the unprotected spores of certain fungi. (Bar = 10 µm.) (b) A close-up view of one conidiophore with a mass of conidia. (Bar = 3 µm.)

Mating type:
a factor or factors that determine if one fungal strain can "mate" with another strain.

budding, the cell becomes swollen at one edge, and a new cell called a **blastospore**, or **bud**, develops from the parent cell and breaks free to live independently. Yeasts multiply in this way.

Many fungi also produce spores by sexual reproduction. In this process, the cells, or hyphae, of opposite **mating types** come together and fuse. Since the nuclei are genetically different in each mating type, the fusion cell represents a **heterokaryon**; that is, a cell with genetically dissimilar nuclei. The fusion of nuclei follows, and the mixing of chromosomes temporarily forms a double set of chromosomes, a condition known as **diploid** (from the Greek *diploos*, meaning "twofold"). Eventually the chromosome number is halved by meiosis, and the cell returns to having a single set of chromosomes, the so-called **haploid** condition (from the Greek *haploos* for "single"). Spores develop from cells in the haploid condition. A visible fruiting body often results during sexual reproduction, and it is the location of the spores. A mushroom is a fruiting body. FIGURE 15.7 summarizes the life cycle of a typical fungus.

Sexual reproduction is advantageous because it provides an opportunity for the evolution of new genetic forms better adapted to the environment than the parent forms. For example, a fungus may become resistant to fungicides as a result of chromosomal changes during sexual reproduction. Separate mycelia of the same fungus may be involved in sexual reproduction, or the process may take place between separate hyphae of the same mycelium.

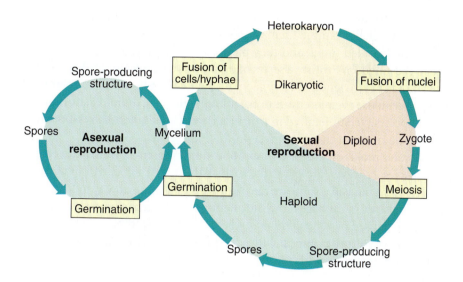

FIGURE 15.7

A Typical Fungal Life Cycle

Most fungi have both sexual and asexual reproduction characterized by spore formation. However, some fungi appear to lack a sexual phase.

To this point . . .

We have discussed aspects of the structures, growth, and reproductive patterns in fungi. We began by exploring some details of the hyphae and mycelium, and then we examined the temperature, pH, and oxygen requirements for growth. Most fungi grow at temperatures close to room temperature and under conditions that are slightly acidic.

The discussion then turned to the two processes of reproduction that take place in the fungi. All fungi reproduce by an asexual process during which spores are formed. Most species also reproduce by a sexual process in which sexually different mating types fuse and eventually result in the formation of sexual spores. Genetic variation is an advantage of such a process.

In the next section, we will examine the four phyla in the kingdom Fungi, which are based on sexual reproduction and spore formation.

15.2

The Classification of Fungi

The classification of fungi is currently in a state of flux, owing in part to new biochemical analyses and DNA studies. Originally, fungi were considered part of the plant kingdom (they are still discussed in many botany courses). Then in 1968, they received their own kingdom status with Whittaker's classification scheme. Currently, they are still considered a kingdom, but under the domain Eukarya in Woese's three-domain system. Therefore, we continue to speak of the kingdom Fungi. Within the kingdom, the subcategories traditionally have been called divisions, as they are in botany. However, mycologists are accepting the term *phylum* (pl., phyla) to further distance the fungi from the plants.

The fungi we discuss in this chapter are classified in four phyla in the kingdom Fungi, and we shall examine them momentarily. However, several other organisms were formerly called fungi, and they merit brief attention. For example, biologists place in the kingdom Protista a series of organisms called **water molds** (Oomycota). Water molds have coenocytic hyphae, but their cell walls are composed predominantly of cellulose (versus chitin in true fungi); also, the diploid condition prevails in most species (versus haploid in true fungi); and finally, flagellated spores occur in the life cycle (none of the four phyla produce motile cells).

Most water molds are important saprobic decomposers in freshwater ecosystems. Some species, however, are parasitic, such as those that infect fish in aquaria. Also included in the parasite group are the organisms that cause downy mildew in grapes, white rust disease in cabbages, and the infamous late blight in potatoes. The effects of this disease, caused by *Phytophthora infestans*, are discussed in the chapter opening.

The kingdom Protista also includes the **plasmodial slime molds** (Myxomycota) and the **cellular slime molds** (Acrasiomycota). Although these organisms resemble amoebas in part of their life cycle, they also produce spores that are the basis for the "mold" connotation.

Another "basal group" of fungi are the **chytrids**. Chytrids give us clues about the origin of fungi. First, chytrids are predominantly aquatic, and not terrestrial. This means that fungi probably got their start in the water, as did plants and animals. Secondly, chytrids have flagellated reproductive cells. No other fungi have flagella, suggesting that the other fungi lost this trait at some point in their evolutionary history. Finally, like other fungi, chytrids have chitin strengthening their cell walls. Until recently, few chytrids have had any noticeable impact. A few species infect algae and another species, *Synchytrium endobioticum*, causes potato wart disease, which can be controlled by the development of resistant varieties. The same cannot be said for many frog species (MicroFocus 15.3).

Within the kingdom Fungi, mycologists recognize three phyla based on the format of sexual reproduction, and they delineate one phylum where sexual stages have not yet been identified for the members. Generally, fungal distinctions are made on the basis of structural differences or physiological or biochemical patterns. However, DNA analyses are becoming an important tool for drawing relationships among various fungi. Next, we shall briefly examine each of the four groups.

THE ZYGOMYCOTA HAVE NONSEPTATE HYPHAE

The phylum **Zygomycota** consists of a group of fungi (zygomycetes) that inhabit terrestrial environments. The zygomycetes have coenocytic hyphae, but form complete septa where the reproductive cells are located. During sexual reproduction, sexually opposite mating types fuse, forming a heterokaryotic, diploid **zygospore** (FIGURE 15.8). Following nuclear fusion, a mature zygospore forms in which meiosis occurs. After a period of dormancy, the zygospore germinates and releases sporangiospores from a sporangium. Elsewhere in the mycelium, thousands of asexually produced sporangiospores are being produced within sporangia. Both sexually produced and asexually produced spores are dispersed on wind currents.

The well-known member of the Zygomycota is the common bread mold, *Rhizopus stolonifer*. The hyphae of this fungus form a white or gray mycelium, with upright sporangiophores, each bearing globular sporangia. Thousands of sporangiospores are formed in each sporangium. Occasional contamination of bread is compensated by the beneficial roles *Rhizopus* plays in industry. One species, for example, ferments rice to sake, the rice wine of Japan; another species is used in the

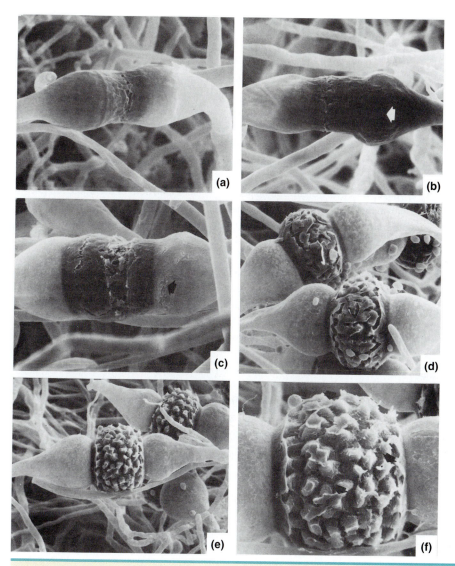

FIGURE 15.8

Sexual Reproduction in the Zygomycota

A sequence of scanning electron micrographs showing zygospore formation in the mold *Rhizopus*. (a) Sexually opposite hyphae fuse and form a fusion septum. (b) Cells at the septum begin to swell (arrow) and show early signs of a zygospore. (c) The outer primary wall begins to rupture. (d) The rupturing continues and (e) the zygospore is revealed. The zygospore continues to mature. (f) A magnified view of the zygospore, showing its surface characteristics and the remnants of the primary wall. When the zygospore later germinates, it will produce a sporangium that releases haploid sporangiospores to propagate the fungus.

production of cortisone, a drug that reduces inflammation in body tissues. These processes are explored further in Chapter 27.

THE ASCOMYCOTA ARE THE SAC FUNGI

Members of phylum **Ascomycota** usually are called ascomycetes. They are very diverse, varying from unicellular yeasts to powdery mildews, cottony molds, and large and complex "cup fungi" (**FIGURE 15.9**). The latter form a cup-shaped fruiting

THE DAY THE FROGS DIED

In the early 1990s, researchers in Australia and Panama started reporting massive declines in the number of amphibians in ecologically pristine areas. As the decade progressed, massive die-offs occurred in dozens of frog species and a few species even became extinct. Once filled with frog song, the forests were quiet. "They're just gone," said one researcher.

Hypotheses to explain these declines included, among others, habitat modification, introduction of new predators, increased ultraviolet radiation, adverse weather changes, and infectious disease. By 1998, infectious disease was identified as one of the most likely reasons for the decline. Investigations in Australia and throughout the world identified a fungal disease as causing much of the global amphibian deaths. The disease is chytridiomycosis and it is caused by a chytrid called *Batrachochytrium dendrobatidis*.

Chytrids are a group of aquatic fungi that produce flagellated spores. Normally, the presence of flagellated cells would exclude an organism from the kingdom Fungi, but molecular systematists have discovered that chytrids have cell walls of chitin and coenocytic hyphae, as well as proteins and nucleic acids more like the fungi than any other group. The phylum Chytridiomycota has been established for these organisms and *B. dendrobatidis* is the only known chytrid that infects vertebrates.

As of 2002, more than 94 amphibian species in Australia, New Zealand, South and Central America, North America, Europe, and Africa had been discovered that were infected with the chytrid. Chytridiomycosis is highly infectious and may be spread by movement (export/import) of infected animals. The fungus uses the frog's keratinized skin as a nutrient source and epidermal sloughing of the frog's skin is one sign of the

disease. Other signs include abnormal posturing, lethargy, and loss of the "right side up" reflex. The disease appears to be most severe in frog species confined to mountain rain forests.

Chytridiomycosis is an emerging disease, this time threatening the amphibian populations of the world. It is just one example of the dangers infectious agents pose for wildlife. In the case of the frog declines, there can be ripple effects. Without frog larvae, mountain streams may become overgrown with algae. In addition, there are certain species of snakes that survive only on amphibians and so their populations could decline.

Are frog deaths from chytridiomycosis a sign of a yet unseen shift in the ecosystem, much like a canary in a coal mine? Some believe this type of "pathogen pollution" may be as serious as chemical pollution.

(a)

(b)

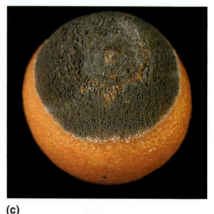

(c)

FIGURE 15.9

Three Common Ascomycetes

(a) The cup fungus *Cookeina tricholoma*, an ascomycete whose ascocarp resembles a cup on a short stalk. This fungus is often found on rotting wood. (b) The edible morel, an ascomycete prized for its delicate taste. (c) A *Penicillium* species growing on a piece of rotten citrus fruit. This is the ascomycete that produces the antibiotic penicillin.

body composed of hyphae tightly woven and packed together. The structure is called an **ascocarp**. The hyphae of an ascomycete are septate, with large pores allowing a continuous flow of cytoplasm.

Though their mycelia vary considerably, all ascomycetes form in the ascocarp a reproductive structure called an **ascus** during sexual reproduction. An ascus is a sac (ascomycetes are "sac fungi"), within which up to eight haploid **ascospores** form (FIGURE 15.10). Most of the ascomycetes also reproduce asexually by means of conidia, produced in chains at the end of a conidiophore.

Certain members of the ascomycetes are extremely beneficial. One example is the yeast *Saccharomyces*, used in brewing and baking (to be discussed shortly). Another example is *Aspergillus*, which produces such products as citric acid, soy sauce, and vinegar and is used in genetics research (FIGURE 15.11). A third is *Penicillium*, various species of which produce the antibiotic penicillin, as well as such cheeses as Roquefort and Camembert (Chapter 25). The edible morels and truffles also are classified in this group.

Ascomycetes are the most frequent fungal partner in **lichens**. A lichen is a mutualistic association between a fungus and a photosynthetic organism such as a

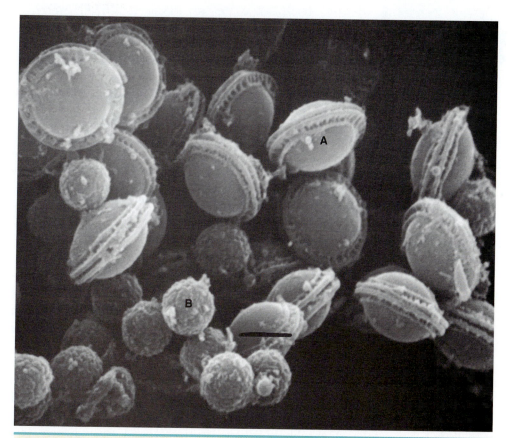

FIGURE 15.10

Two Types of Ascomycete Spores

Ascospores (A) and conidia (B) of the fungus *Aspergillus quadrilineatus*. This fungus was cultivated from the nasal sinuses of an ill patient who recently had received a bone marrow transplant. The sexually produced ascospores display a series of so-called equatorial crests at their midlines. The asexually produced conidia are round spores. (Bar = 1 μm.)

(a)

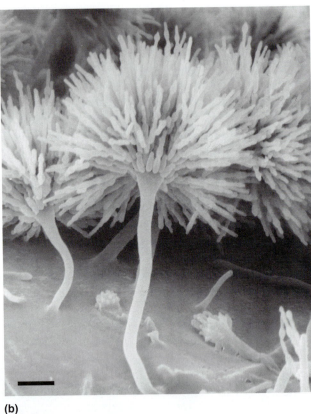

(b)

FIGURE 15.11

Normal and Mutant Fungi

Scanning electron micrographs of the conidiophores of *Aspergillus nidulans*. (a) The normal or "wild type" form of the fungus is depicted. At the tip of the hypha, the conidiophores contain hundreds of asexually produced spores (conidia), any of which can germinate to reproduce the fungus. (Bar = 3 μm.) (b) A mutated form of *A. nidulans*. This organism was produced by mutating the regulatory genes of the fungus. As a result of the molecular manipulations, distinctive structural variations have occurred in the fungus, and the production of spores has been interrupted. (Bar = 10 μm.)

cyanobacterium or green alga (**FIGURE 15.12a**). Most of the visible body of a lichen is the fungus. Its hyphae penetrate the cells of the photosynthetic partner and receive carbohydrate nutrients. The photosynthetic organism receives fluid from the water-husbanding fungus. Together, the organisms form a composite that readily grows in environments where neither organism could survive by itself (e.g., rock surfaces). Indeed, in some harsh environments, lichens support entire food chains. In the Arctic tundra, for example, reindeer graze on carpets of reindeer moss, actually a type of lichen. Lichens often are grouped by appearance into leafy lichens (foliose), shrubbery lichens (fruticose), and crusty lichens (crustose), as **FIGURE 15.12b** shows.

On the deficit side, some ascomycetes attack valuable plants. For instance, one member of the group parasitizes crops and ornamental plants, causing powdery mildew. Another species has almost entirely eliminated the chestnut tree from the American landscape. Still another ascomycete presently is attacking elm trees in the United States (Dutch elm disease) and is threatening the extinction of this plant.

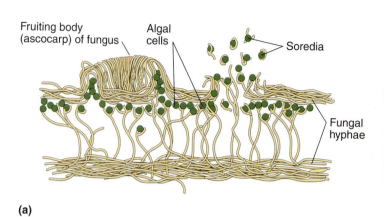

Labels: Fruiting body (ascocarp) of fungus, Algal cells, Soredia, Fungal hyphae

(a)

(b)

FIGURE 15.12

Lichens

(a) A cross section of a lichen, showing the upper and lower surfaces where tightly coiled fungal hyphae enclose photosynthetic algal cells. On the upper surface, a fruiting body, or ascocarp, has formed. Airborne clumps of algae and fungus called soredia are dispersed from the ascocarp to propagate the lichen. Loosely woven fungi at the center of the lichen permit the passage of nutrients, fluids, and gases. (b) A typical "crusty" lichen growing on the surface of a rock. Lichens are rugged organisms that can tolerate environments where there are few nutrients and extreme conditions. Their organic matter often forms the foundation of a local food chain.

Two other ascomycete pathogens are *Claviceps purpurea*, which causes ergot disease of rye plants, and *Aspergillus flavus*, which attacks a variety of foods and grains (Chapter 25).

THE BASIDIOMYCOTA ARE THE CLUB FUNGI

Members of the phylum **Basidiomycota**, commonly known as basidiomycetes, are club fungi. They include the mushrooms, as well as the shelf fungi, puffballs, and other fleshy fungi, plus the parasitic rust and smut fungi. The name basidiomycete refers to the reproductive structure on which sexual spores are produced after hyphal fusion and meiosis have taken place. The structure, resembling a club, is called a **basidium** (pl., basidia), which is Latin for "small pedestal." Its sexually produced spores are known as **basidiospores**.

Perhaps the most familiar member of the group is the edible mushroom. Indeed, the Italian word *fungi* means "mushroom." Its mycelium forms below the ground, and after sexual fusion has taken place, the tightly compacted hyphae force their way to the surface and grow into a fruiting body called a **basidiocarp**, which is the mushroom and its cap (**FIGURE 15.13**). Basidia develop on the underside of the cap along the gills, and each basidium may have up to eight basidiospores. Edible mushrooms belong to the genus *Agaricus*, but one of the most potent toxins known to science is produced by another species of a visually similar genus, *Amanita*. Every year outbreaks of mushroom poisoning, most related to this genus, are reported to the CDC, including the episode described in **FIGURE 15.14**. The puffballs produce a powdery mass of spores that have been known to cause serious respiratory illness when the spores are inhaled.

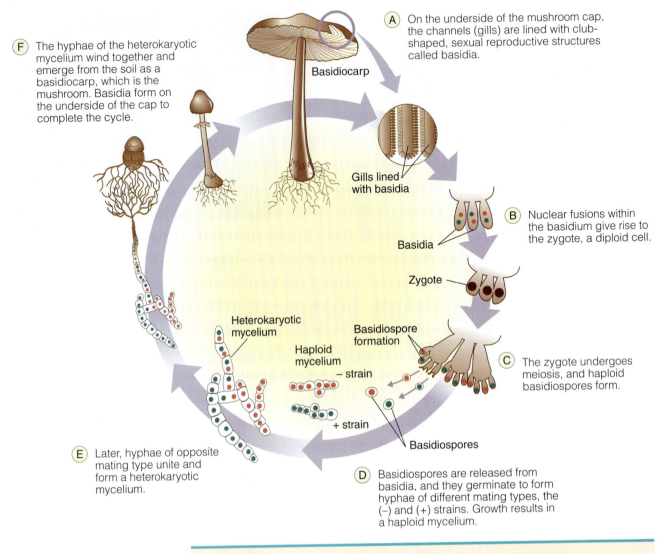

F The hyphae of the heterokaryotic mycelium wind together and emerge from the soil as a basidiocarp, which is the mushroom. Basidia form on the underside of the cap to complete the cycle.

Basidiocarp

A On the underside of the mushroom cap, the channels (gills) are lined with club-shaped, sexual reproductive structures called basidia.

Gills lined with basidia

Basidia

B Nuclear fusions within the basidium give rise to the zygote, a diploid cell.

Zygote

Heterokaryotic mycelium

Haploid mycelium

Basidiospore formation

– strain

C The zygote undergoes meiosis, and haploid basidiospores form.

+ strain

Basidiospores

E Later, hyphae of opposite mating type unite and form a heterokaryotic mycelium.

D Basidiospores are released from basidia, and they germinate to form hyphae of different mating types, the (–) and (+) strains. Growth results in a haploid mycelium.

FIGURE 15.13

The Life Cycle of a Typical Basidiomycete

Agricultural losses due to the basidiomycetes of rust and smut diseases are considerable. **Rust diseases** are so named because of the orange-red color of the infected plant. The diseases strike wheat, oats, and rye, as well as trees used for lumber, such as white pines. Many rust fungi require alternate hosts to complete their life cycles, and local laws often prohibit the cultivation of certain crops near rust-sensitive plants. For example, it may be illegal to raise gooseberries near white pine trees. **Smut diseases** give a black, sooty appearance to plants. They affect corn, blackberries, and a number of grains, and cause untold millions of dollars' worth of damage yearly.

THE DEUTEROMYCOTA LACK A SEXUAL REPRODUCTION PHASE

Certain fungi lack a known sexual cycle of reproduction and consequently are labeled with the botanical term "imperfect." These imperfect fungi are placed in the fourth phylum **Deuteromycota**, where reproduction occurs only by an asexual

FIGURE 15.14

A Fatal Case of *Amanita* Food Poisoning

This case was related to wild mushrooms. Warm, heavy rainfall in the preceding weeks may have contributed to their unanticipated appearance in the environment.

TEXTBOOK CASES

1. In early January 1997, a man decided to play a round of golf at a course near his home in northern California. At the ninth hole, the man drove his ball into the woods. As he searched for the ball, he noticed a cluster of wild mushrooms and decided to pick some for dinner.

2. He cooked the mushrooms that evening, along with a steak. As he lived alone, he did not share the mushrooms with anyone else.

3. Three days later, the man experienced severe diarrhea and became very weak. He went to the emergency room, where liver tests revealed abnormal liver function.

4. His symptoms worsened, and within two days, his liver and kidneys were deteriorating. Doctors placed him on a kidney dialysis machine but it was too late. He died soon thereafter.

5. Investigators retraced the man's recent activities. At the golf course they found *Amanita phalloides*, an extremely poisonous mushroom. Lab tests confirmed the man's death from mushroom poisoning.

method. It should be noted that a sexual cycle probably exists for these "deuteromycetes," but it has not been seen by mycologists.

When the sexual cycle is discovered, the deuteromycete is reclassified into one of the other three divisions. A case in point is the fungus known as *Histoplasma capsulatum*. This fungus causes histoplasmosis, a disease of the human lungs and other internal organs. When the organism was found to produce ascospores, it was reclassified with the ascomycetes and given the new name *Emmonsiella capsulata*.

However, some traditions die slowly, and certain mycologists insisted on retaining the old name because it was familiar in clinical medicine. Thus, mycologists decided to use two names for the fungus: the new name, *Emmonsiella capsulata*, for the sexual stage; and the old name, *Histoplasma capsulatum*, for the asexual stage.

Many fungi pathogenic for humans are deuteromycetes. These fungi usually reproduce by budding or fragmentation, and segments of hyphae are commonly blown about with dust or deposited on environmental surfaces. For example, fragments of the athlete's foot fungus are sometimes left on towels and shower room floors. Others reproduce by spores, as **MicroFocus 15.4** describes. Nonpathogenic fungi are also classified here, as exemplified by species of *Pseudomassaria*. In 1999, this fungus was found to produce a compound that mimics insulin by assisting glucose passage into human cells; it could possibly be used by diabetics one day. (The compound has an advantage over insulin because it can be taken orally.) Other fungi in the Deuteromycota also have medical potential (**MicroFocus 15.5**).

The four phyla of fungi are compared in **TABLE 15.1**.

MicroFocus 15.4

MOLD IN YOUR HOME

Nobody likes mold growing on a piece of cheese in their refrigerator. But, what about molds that are actually growing in your home? The most common indoor molds are *Cladosporium*, *Penicillium*, *Aspergillus*, and *Alternaria*. Apart from triggering allergic reactions in sensitive people, these molds usually are not a problem indoors. Under certain circumstances, however, a mold might produce potentially toxic substances (mycotoxins). Take *Stachybotrys chartarum* as a case in point.

S. chartarum (also known as black mold or toxic mold) is a slimy, greenish-black mold in the phylum Deuteromycota. It likes to grow on such things as fiberboard and gypsum board that most of us probably have in the walls of our home. Growth will occur on these materials if they become wet from water damage or flooding, excessive humidity, water leaks, or condensation.

From January 1993 to November 1994, a group of infants in Cleveland died from an unexplained hemorrhagic lung disease that was subsequently given the label acute idiopathic pulmonary hemosiderosis (AIPH). Results from an epidemiological study found that most of the affected infants lived in water-damaged homes, suggesting a possible fungal disease. An initial microbiological investigation concluded that these infants probably were exposed in their homes to mycotoxins from *S. chartarum*. *S. chartarum* produces trichothecene mycotoxins, which are potent inhibitors of DNA replication, transcription, and translation. Direct inhalation of spores by animals can lead to an often-fatal disease called stachybotrytoxicosis, which is characterized by leucopenia (reduced white blood cell count) and hemorrhage (bleeding).

That wasn't the end of the story. A review of the case revealed some important flaws in the investigation. No link between *S. chartarum* and the cases of AIPH could be made after all. While it's clear that *S. chartarum* produces mycotoxins and that some strains cause disease in animals, the human health risks from exposure to *S. chartarum* are uncertain. Despite this lack of epidemiological evidence, recent stories have reported cases of severe health problems from *S. chartarum*, including memory loss and severe lung problems in infants and the elderly.

Mold growing in a home, whether it is *S. chartarum* or other molds, indicates that there is a problem with water or moisture. Once mold starts to grow in insulation or gypsum board, the only way to deal with the problem is by removal and replacement. In some cases, the mold growth was so extensive in the home that people have burned down their houses and everything inside to get rid of the mold. The Environmental Protection Agency (EPA) says that one "does not need to take any different precautions with *S. chartarum* than with other molds." (The people who burned down their houses might disagree.) If there is water damage, prompt cleaning of walls and other flood-damaged items with water mixed with chlorine bleach, diluted 10 parts water to 1 part bleach, is necessary to prevent mold growth—and moldy items do need to be removed—one way or another.

MicroFocus 15.5

NOT ALL FUNGI ARE BAD

In 1989, researchers at Johns Hopkins University discovered that taxol, a chemical derived from yew trees, could greatly reduce the size of tumors in women suffering from ovarian cancer. Two years later, in January 1993, the Food and Drug Administration approved taxol for ovarian cancer, while noting that the drug might be useful for breast, head, and neck tumors. Unfortunately, the exhilaration that accompanied approval of the new treatment was counterbalanced by the cost of the drug (about $1,000 per treatment cycle) and the fear that the yew tree might be overfarmed to provide bark for the drug.

Then, in 1993, a new twist was added to the taxol story. In April, Montana researchers discovered growing within the bark of the yew a fungus that produces taxol on its own. Plant pathologist Gary Strobel and chemist Andrea Stierle, both from Montana State University, led the research. Under Strobel's intuitive direction, Stierle searched the Montana woods for local yews (*Taxus pacifica*) that would yield taxol. Finding one such yew, they went a step further and isolated a fungus from within the folds of the yew's bark. The fungus continued to produce taxol even after removal from its host plant.

They named the fungus *Taxomyces andreanae* (Andrea's taxus–fungus). *T. andreanae*'s yield of taxol is low, which has sent Strobel on the quest for better taxol-producing species. He has discovered a fungus associated with an Asian yew tree that produces 1,000-times more taxol than the Montana species. This makes commercial taxol production much more likely. For example, enormous fermentation tanks can be used to produce enormous amounts of the fungus and much larger amounts of the drug. Moreover, genetic engineering techniques can be used to pinpoint and clone the taxol genes, then transfer them to high-yield vector organisms such as bacteria. Strobel and the the drug companies believe that these and other approaches can work, making the fungus's future appear bright.

 (a)

 (b)

■ *(a) The fungus* Taxomyces andreanae *showing its hyphal strands and fruiting bodies with spores. (Bar = 5 μm.) (b) Strobel and Stierle (left) with the yews from which the fungus was obtained.*

TABLE 15.1

Comparisons of the Phyla of Fungi

PHYLUM	COMMON NAME	CROSS WALLS (SEPTA)	SEXUAL STRUCTURE	SEXUAL SPORE	ASEXUAL SPORE	REPRESENTATIVES
Zygomycota	Zygomycetes	No	Zygospore	Zygospore	Sporangiospore	*Rhizopus*
Ascomycota	Ascomycetes, sac fungi	Yes	Ascus	Ascospore	Conidia	*Saccharomyces Aspergillus Penicillium*
Basidiomycota	Basidiomycetes, club fungi	Yes	Basidium	Basidiospore	Conidia fragments	*Agaricus Amanita*
Deuteromycota	Deuteromycetes, imperfect fungi	Yes	Unknown	Unknown	Conidia fragments	*Candida Trichophyton*

THE YEASTS ARE MICROSCOPIC, UNICELLULAR FUNGI

The word *yeast* refers to a large variety of unicellular fungi (as well as the single-cell stage of any fungus). Included in the group are nonspore-forming yeasts of the Deuteromycota, as well as certain yeasts that form basidiospores or ascospores and thus belong to the Basidiomycota or Ascomycota. The yeasts we shall consider here are the species of *Saccharomyces* used extensively in brewing, baking, and as a food supplement. Pathogenic yeasts will be discussed shortly.

Saccharomyces literally means "sugar-fungus," a reference to the ability of the organism to ferment sugars. The most commonly used species of *Saccharomyces* are *S. cerevisiae* and *S. ellipsoideus*, the former used for bread baking and alcohol production, the latter for alcohol production. Yeast cells are about 8 µm long and about 5 µm in diameter. They reproduce chiefly by budding (**FIGURE 15.15**), but a sexual cycle also exists in which cells fuse and form an enlarged cell (an ascus) containing smaller cells (ascospores). The organism is therefore an ascomycete.

The cytoplasm of *Saccharomyces* is rich in B vitamins, a factor that makes yeast tablets valuable nutritional supplements. One pharmaceutical company adds iron to the yeast and markets its product as Ironized Yeast, recommended for people with iron-poor blood.

The baking industry relies heavily upon *S. cerevisiae* to supply the texture in breads. Flour, sugar, and other ingredients are mixed with yeast, and the dough is set aside to rise. During this time, the yeasts break down glucose and other carbohydrates, and produce carbon dioxide through the chemistry of glycolysis and the Krebs cycle (Chapter 5). The carbon dioxide expands the dough, causing it to rise. Protein-digesting enzymes, also from the yeast, partially digest the gluten protein of the flour to give bread its spongy texture (**MicroFocus 15.6**). To make bagels, the dough is boiled before baking; for sourdough bread, *Lactobacillus* species are added to give an acidic flavor to the bread; for rye bread, rye flour is substituted. In all these modifications, yeast remains an essential ingredient.

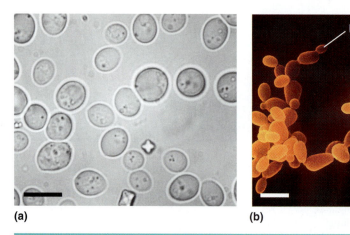

(a)   (b)

FIGURE 15.15

Yeasts

Two views of *Saccharomyces*, the common baking and brewing yeast. (a) A photomicrograph of yeast cells used in beer fermentation. (Bar = 10 µm.) (b) A false color scanning electron micrograph of *Saccharomyces cerevisiae*. Several cells are budding. (Bar = 10 µm.)

MicroFocus 15.6

THE DRIER THE BETTER

Need a quick microbiology laboratory? Simply go to the grocery store and purchase a package of "active dry yeast." Open the package, pour the contents into a bit of warm water, and you're on your way—instant microorganisms. You can study the yeasts with the microscope, investigate their physiology, and if you're really with it, rearrange their genes. All this from a package.

It wasn't always that way, though. Scientists could never figure out how to keep yeasts alive and in a dry state at the same time. If someone wanted to make bread, they had to get a starter from the "mother" dough where the yeast was growing; if the objective was wine fermentation, a trip to the culture supply or the old wine was necessary.

Then, during World War II, German prisoners were found in possession of a curious brown powder—it was the elusive dry yeast. They had the yeast for use as a nutritious food, or for bread when added to a bit of dough (or "Battlefield Red," if added to crushed grapes). Of course, the prisoners were not about to reveal the secret of the dried yeast, probably because they did not know what it was.

The postwar period was a different matter. In the euphoria of sharing, the Western world learned the secret of trehalose. Trehalose is a disaccharide. Added to yeast cells, it stabilizes the cell membrane, prevents cell damage due to drying, and keeps the yeast alive and active. Now everyone knew the answer, including an entrepreneur named Arthur Fleischmann—millionaire founder and owner of Fleischmann's Active Dry Yeast.

Yeasts are plentiful where there are orchards or fruits (the haze on an apple is a layer of yeasts). In natural alcohol fermentations, wild yeasts of various *Saccharomyces* species are crushed with the fruit; in controlled fermentations, *S. ellipsoideus* is added to the prepared fruit juice. Now the chemistry is identical with that in dough: The fruit juice bubbles profusely as carbon dioxide evolves through the reactions of glycolysis and the Krebs cycle. When the oxygen is depleted, the yeast metabolism shifts to fermentation, and the pyruvate from glycolysis changes to consumable ethyl alcohol (Chapter 5).

The products of yeast fermentation depend on the starting material. For example, when yeasts ferment barley grains, the product is beer; if grape juice is fermented, the product is wine. Sweet wines contain leftover sugar, but dry wines have little sugar. Sparkling wines such as champagne continue to ferment in thick bottles as yeast metabolism produces additional carbon dioxide. For spirits such as whiskey, rye, or scotch, some type of grain is fermented and the alcohol is distilled off. Liqueurs are made when yeasts ferment fruits such as oranges, cherries, or melons. Virtually anything that contains simple carbohydrates can be fermented by *Saccharomyces*. The huge share of the US economy taken up by the wine and spirits industries is testament to the significance of the fermentation yeasts. A fuller discussion of fermentation processes is presented in Chapter 27.

MicroInquiry 15 looks at the relationship of the fungi to other eukaryotic kingdoms.

EVOLUTION OF FUNGI

For quite some time fungi were classified as plants and were thought to share an evolutionary lineage with those multicellular eukaryotes. The fossil record provides evidence that fungi and plants "hit the land" about the same time. There are fossilized fungi that are over 450 million years old, which is about the time that plants started to colonize the land. In fact, some fossilized plants that are thought to represent the first to colonize land, appear to have mycorrhizal associations.

Taxonomists believe that the fungi represent a monophyletic line; that is, a single ancestral species gave rise to the four fungal phyla (Chytridiomycota, Zygomycota, Ascomycota, and Basidiomycota). We are ignoring the Deuteromycota since most have relationships with the Zygomycota, Ascomycota, or Basidiomycota if a sexual stage is observed. Only the Chytridiomycota (chytrids) have flagella, suggesting they are the oldest line and that fungal ancestors were flagellated and aquatic (FIG-URE A). The chytrids may have evolved from a protistan ancestor and, as fungi colonized the land, they evolved different reproductive styles, which systematists separated into three nonflagellated phyla described in this chapter.

So, what evolutionary relationships do the fungi have with the eukaryotic kingdoms Plantae and Animalia?

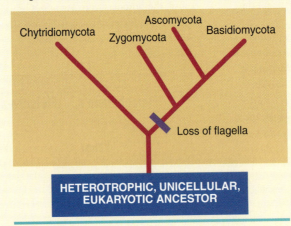

FIGURE A

Taxonomic Scheme for the Relationships between Fungal Phyla

A tremendous amount of research has been carried out to look at that very question. Table A has some data that you need to analyze and from which you need to make a conclusion as to what relationships are best supported by the evidence. You do not need to understand the function or role for every characteristic listed. The answers can be found in Appendix E.

15.1. Based on the data presented, which groups most likely evolved from a *common* protistan ancestor?

15.2. Draw a taxonomic scheme showing the relationships between Plantae, Animalia, and Fungi kingdoms.

To this point . . .

In this section we discovered that the sexual cycle is the basis for the classification of fungi into four phyla. Members of the Zygomycota form zygospores. In fungi of the Ascomycota, a sac forms containing up to eight ascospores. Members of the Basidiomycota form a club-like supportive structure, the basidium, on which basidiospores develop. Deuteromycota members have no known sexual cycle. The section closed with a discussion of the economically important baking and fermentation yeasts.

We now focus our attention on the fungal diseases of humans. Because these diseases generally do not occur in widespread epidemics, their names may be unfamiliar. However, some diseases may endanger human life, especially if the immune system has been compromised. For this reason, fungal diseases often occur as complications of other diseases, or in situations where a patient is undergoing treatment for an unrelated problem.

TABLE A

Comparisons of Organismal, Cellular, and Biochemical Characteristics

CHARACTERISTIC	FUNGI	PLANTS	ANIMALS
Have chloroplasts		✓	
Carry out photosynthesis		✓	
Have mitochondria	✓	✓	✓
Protein sequence similarities	✓		✓
Elongation factor 3 protein (translation) similarities	✓		✓
Types of polyunsaturated fatty acids			
Alpha linolenic		✓	
Gamma linolenic	✓		✓
Cytochrome system similarities	✓		✓
Mitochondrial UGA codes for tryptophan	✓		✓
Plate-like mitochondrial cristae	✓		✓
Presence of chitin	✓		some
Type of glycoprotein bonding			
O-linked	✓		✓
N-linked	✓		✓
Absorptive nutrition	✓		some
Lipid reserves	✓	✓	✓
Glycogen reserves	✓		✓
Pathway for lysine biosynthesis			
Aminoadipic acid pathway	✓		
Diaminopimelic pathway		✓	
Presence of microtubules	✓	✓	✓
Microtubule chemical composition similarities	✓		✓
Type of sterol intermediate			
Lanosterol	✓		✓
Cycloartenol		✓	
Mitochondrial ribosomes 5S RNA present		✓	
Presence of endomembrane system	✓	✓	✓
Presence of lysosomes	✓		✓

Fungal Diseases of Humans

In humans, fungal diseases, called **mycoses**, often affect many body regions. For example, several diseases, including ringworm and athlete's foot, involve the skin areas, while others, such as cryptococcosis and histoplasmosis, occur in the lungs before spreading to other body areas. One disease, candidiasis, may take place in the oral cavity, intestinal tract, skin, vaginal tract, and other body locations depending on the conditions that stimulated its development. In a great many fungal diseases, a weakened immune system contributes substantially to the occurrence of the infection, as we often will see in this section.

CRYPTOCOCCOSIS USUALLY OCCURS IN IMMUNOCOMPROMISED INDIVIDUALS

Cryptococcosis is among the most dangerous fungal diseases in humans. It affects the lungs and the meninges (the coverings of the brain and spinal cord) and is estimated to account for over 25 percent of all deaths from fungal disease.

Cryptococcosis is caused by a yeast known as *Cryptococcus neoformans*. The organism is found in the soil of urban environments and grows actively in the droppings of pigeons, but not within the pigeon tissues. Cryptococci may become airborne with gusts of wind, and the organisms subsequently enter the respiratory passageways of humans. Air conditioner filters are hazardous because they trap large numbers of cryptococci.

C. neoformans cells, having a diameter of about 5 to 6 μm, are embedded in a gelatinous capsule that provides resistance to phagocytosis (**FIGURE 15.16**). The cells penetrate to the air sacs of the lungs, but symptoms of infection are generally rare. However, if the cryptococci pass into the bloodstream and localize in the meninges and brain, the patient experiences piercing headaches, stiffness in the neck, and paralysis. Diagnosis is aided by the observation of encapsulated yeasts in respiratory secretions or cerebrospinal fluid (CSF) obtained by a spinal tap.

Untreated cryptococcal meningitis may be fatal. However, intravenous treatment with the antifungal drug **amphotericin B** is usually successful, even in severe cases. Because this drug has toxic side effects such as kidney damage and anemia, the patient should be monitored continually.

Resistance to cryptococcal meningitis appears to depend upon the proper functioning of a branch of the immune system governed by T lymphocytes (T cells). When these cells are absent in sufficient quantities, the immune system becomes severely compromised, and cryptococci can invade the tissues as opportunists. In patients with **acquired immune deficiency syndrome** (AIDS), one of the causes of death is cryptococcosis.

During the early 1980s, mycologists identified a sexual stage for *C. neoformans*. The stage is related to the smut fungi of the Basidiomycota and is called *Filobasidiella neoformans*.

CANDIDIASIS OFTEN IS A MILD, SUPERFICIAL INFECTION

Candida albicans often is present in the skin, mouth, vagina, and intestinal tract of healthy humans and other animals, where it lives without causing disease (**FIGURE**

15 μm

Cryptococcus neoformans

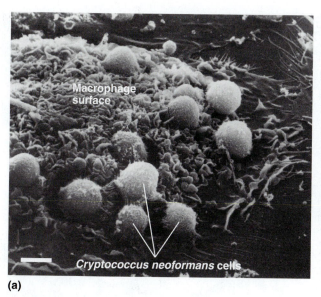

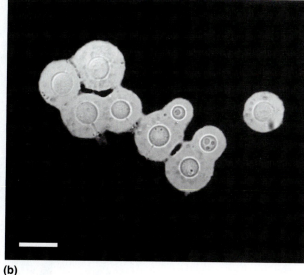

(a)   (b)

FIGURE 15.16

Cryptococcus neoformans

(a) A scanning electron micrograph of *C. neoformans* cells clinging to the surface of a human macrophage (a type of phagocytic white blood cell). Internalization of the fungal cells is retarded by the capsules they possess. In this view, the cells have been attached to the macrophage surface for a long 30 minutes, and still they have not been internalized. (Bar = 5 μm.) (b) A negatively stained photomicrograph of *C. neoformans*. A distinct capsule surrounds each oval, yeast-like cell. This capsule provides resistance to phagocytosis and enhances the pathogenic tendency of the fungus. (Bar = 10 μm.)

15.17). The organism is a small deuteromycete yeast that forms filaments called pseudohyphae when cultivated in laboratory media. When immune system defenses are compromised, or when changes occur in the normal microbial population in the body, *C. albicans* flourishes and causes numerous forms of **candidiasis**.

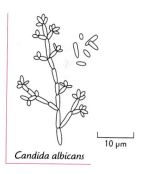

Candida albicans

One form of candidiasis occurs in the vagina and is often referred to as **vulvovaginitis**, or a "**yeast infection**." Symptoms include itching sensations (pruritis), burning internal pain, and a white "cheesy" discharge. Reddening (erythema) and swelling of the vaginal tissues also occur. Diagnosis is performed by observing *C. albicans* in a sample of vaginal discharge or vaginal smear, and by cultivating the organisms on laboratory media. Treatment is usually successful with **nystatin** (Mycostatin) applied as a topical ointment or suppository. **Miconazole, clotrimazole**, and **ketoconazole** are useful alternatives. There are some 20 million cases reported every year in the United States.

Vulvovaginitis is considered a sexually transmitted disease (but the disease is usually much milder in men than in women). In addition, studies have shown that excessive antibiotic use may encourage loss of the rod-shaped lactobacilli that are normally present in the vaginal environment. Without lactobacilli as competitors, *C. albicans* flourishes. Other predisposing factors are the contraceptive intrauterine device (IUD), corticosteroid treatment, pregnancy, diabetes, and tight-fitting garments, which increase the local temperature and humidity.

Oral candidiasis is known as **thrush**. This disease is accompanied by small, white flecks that appear on the mucous membranes of the oral cavity and then grow together to form soft, crumbly, milk-like curds. When scraped off, a red, inflamed base is

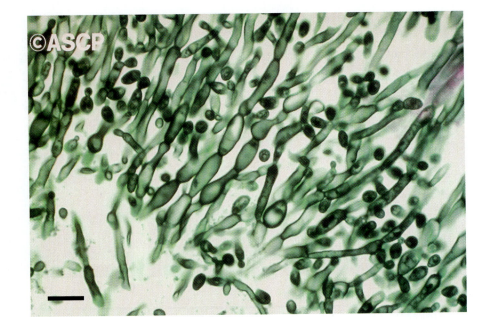

FIGURE 15.17

The Agent of Candidiasis

A photomicrograph of *Candida albicans* stained with methenamine. Oval yeast cells and psuedohyphae are apparent. (Bar = 40 μm.)

revealed. Oral suspensions of gentian violet and nystatin ("swish and spit") are effective for therapy. The disease is common in newborns, who acquire it during passage through the vagina (birth canal) of infected mothers. Children also may contract thrush from nursery utensils, toys, or the handles of shopping carts. Candidiasis may be related to a suppressed immune system. Indeed, thrush may be an early sign of AIDS in an adult patient.

Candidiasis in the intestinal tract is closely tied to the use of antibiotics. Certain drugs destroy bacteria normally found here and allow *C. albicans* to flourish. In the 1950s, yogurt became popular as a way of replacing the bacteria. Today when intestinal surgery is anticipated, the physician often uses antifungal agents to curb *Candida* overgrowth. Moreover, people whose hands are in constant contact with water may develop a hardening, browning, and distortion of the fingernails called **onychia**, also caused by *C. albicans*.

DERMATOPHYTOSIS IS AN INFECTION OF THE BODY SURFACE

Dermatophytosis is a general name for a fungal disease of the hair, skin, and nails caused by a wide variety of fungi. The diseases are commonly known as **tinea infections**, from the Latin *tinea* for "worm," because in ancient times, worms were thought to be the cause. The tinea infections include tinea pedis, athlete's foot; tinea capitis, ringworm of the head; tinea corporis, ringworm of the body; tinea cruris, ringworm of the groin, or "jock itch"; and tinea unguium, ringworm of the nails.

The causes of dermatophytosis are a series of fungi called **dermatophytes**. One example is species of *Trichophyton*, an ascomycete whose sexual stage is named *Arthroderma*. Another example is certain species of *Microsporum*, also an ascomycete, whose sexual stage is named *Nannizzia*. A third cause is species of *Epidermophyton*, currently considered a deuteromycete.

Dermatophytosis is commonly accompanied by blister-like lesions appearing on the skin, along the nail plate, or in the webs of the toes or fingers. Often a thin, fluid discharge exudes when the blisters are scratched or irritated. As the blisters dry, they

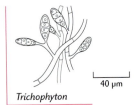

40 μm

Trichophyton

leave a scaly ring (FIGURE 15.18). Centuries ago, people believed that worms inhabited the scaly ring, hence the name ringworm. The symptoms of dermatophytosis vary considerably and may include loss of hair, change of hair color, and local inflammatory reactions.

The majority of dermatophytes grow readily on Sabouraud dextrose agar, and trained mycologists usually can diagnose the disease by observing the type of hypha and arthrospore present. Moreover, infected hairs and fungal cultures fluoresce in ultraviolet light. If protected from dryness, the dermatophytes live for weeks on wooden floors of shower rooms or on mats. People transmit the fungi by contact (FIGURE 15.19) and on towels, combs, hats, and numerous other types of fomites (inanimate objects). They also acquire the fungi by contact with household pets, because tinea diseases affect cats and dogs.

Treatment of dermatophytosis often is directed at changing the conditions of the skin environment. Commercial powders dry the diseased area, while ointments change the pH to make the area inhospitable for the organism. Certain acids such as undecylenic acid (Desenex) and mixtures of acetic acid and benzoic acid (Whitfield's ointment) are active against the fungi. Also, **tolnaftate** (Tinactin) and **miconazole** (Micatin) are useful as topical agents for infections not involving the nails and hair. **Griseofulvin**, administered orally, is a highly effective chemotherapeutic agent for severe dermatophytosis. This drug causes shriveling of the hyphae, possibly by interfering with nucleic acid synthesis. FIGURE 15.20 shows a case of nail ringworm treated with **itraconazole**.

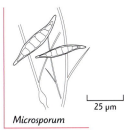

Microsporum

25 µm

HISTOPLASMOSIS CAN PRODUCE A SYSTEMIC DISEASE

On January 4, 1988, a group of 17 students from an American university crawled into a cave in a national park in Costa Rica to observe the numerous bats whose droppings covered the floor. Within three weeks, 15 students developed fever, headache, cough, and severe chest pains. Twelve patients tested positive for *Histoplasma capsulatum*, and all were treated for histoplasmosis.

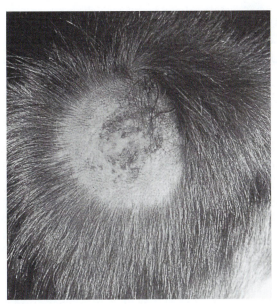

(a)

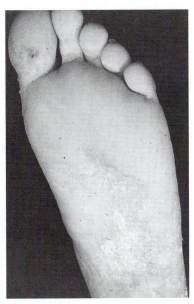

(b)

FIGURE 15.18

Two Views of Dermatophytosis

(a) Ringworm of the scalp, due to *Trichophyton mentagrophytes*. The lesions have crusted to form scaly blisters in this view. (b) Athlete's foot, caused by *Trichophyton rubrum*. The scaly blisters can be seen on the soles of the feet and in the webs between the toes.

FIGURE 15.19

An Outbreak of Ringworm

This outbreak occurred among participants at an international wrestling meet. The incident happened in Schaumberg, Illinois, in 1992. It was believed to be one of the first epidemics of transmissible ringworm reported in the United States.

1. In the fall of 1992, a number of wrestlers from the United States and abroad attended a meet in Schaumberg, Illinois.

2. The wrestlers competed in several divisions until the divisional champions were decided. Skin contact between the wrestlers is routine during the bouts.

3. On returning home, a number of participants noticed scaly, pink blotches on their shoulders, neck, or face.

4. Mycologists took skin samples and scrapings from those affected and cultivated *Trichophyton tonsurans*. The diagnosis was ringworm (tinea corporis).

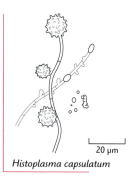

Histoplasma capsulatum

20 μm

Histoplasmosis is a lung disease prevalent in the Ohio and the Mississippi River valleys. The causative agent is *Histoplasma capsulatum*, an ascomycete whose sexual phase is named *Emmonsiella capsulata*. Infection usually occurs from the inhalation of spores in dry, dusty soil, and the disease often is called summer flu. Most people recover without treatment. However, a small percentage of people develop a disseminated form of histoplasmosis with tuberculosis-like lesions of the lungs and other visceral organs. (The singer Bob Dylan suffered from lung and pericardial disease in 1997, as a result of histoplasmosis.) AIDS patients are vulnerable to this condition. Amphotericin B or ketoconazole may be used in treatment.

The fungus of histoplasmosis often is found in the air of chicken coops and bat caves. Although *Histoplasma* does not affect birds or bats, it grows in the droppings

FIGURE 15.20

A Pathogen and the Disease

(a) A phase-contrast photomicrograph of *Microsporum racemosum*, an ascomycete and the cause of fungal disease of the nails. The long oval bodies are the conidia of the fungus. (Bar = 20 µm.)
(b) The nail of a 60-year-old female patient infected with *M. racemosum*. This incident began with a puncture wound by a fish bone. The fungus was identified by genetic analysis of its ribosomal RNA, and the disease was successfully resolved by treatment with itraconazole for 12 weeks.

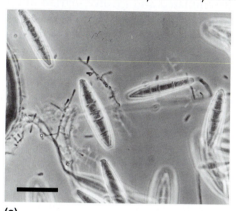
(a)

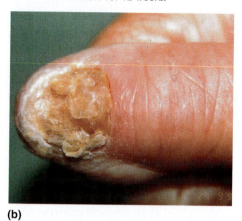
(b)

of these animals, as the outbreak in Costa Rica illustrates. Prolonged exposure to the air therefore may be hazardous. The disease is sometimes called Darling's disease after Samuel Darling, who described the cause in 1915.

BLASTOMYCOSIS IS ENDEMIC IN THE SOUTHEASTERN UNITED STATES

Blastomycosis occurs principally in Canada, the Great Lakes region, and areas of the United States from the Mississippi River to the Carolinas. The pathogen is *Blastomyces dermatitidis*, an ascomycete whose sexual phase is named *Ajellomyces dermatitidis*. The fungus is dimorphic, appearing in the human as a yeast with a figure-8 appearance. Blastomycosis also is referred to as Gilchrist's disease for Thomas C. Gilchrist, the American dermatologist who first described it in 1896.

Blastomycosis is associated with dusty soil and bird droppings, particularly in and near barns and sheds. Entry to the body may occur through cuts and abrasions, and raised wart-like lesions often are observed on the face, hands, and legs. Inhalation leads to lung lesions with persistent cough and chest pains. Healing is generally spontaneous.

The disseminated form of blastomycosis may involve many internal organs and may prove fatal. Amphotericin B used in therapy is thought to change the permeability of the fungal cell membrane and induce a leakage of cytoplasm.

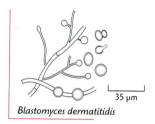
Blastomyces dermatitidis

COCCIDIOIDOMYCOSIS PRIMARILY IS A RESPIRATORY INFECTION

Travelers to the San Joaquin Valley of California and dry regions of the southwestern United States may be exposed to a fungal disease known as **coccidioidomycosis**, or "valley fever." Its cause is *Coccidioides immitis*, a protozoan-like fungus of the Deuteromycota. The organism produces arthrospores by a unique process of endospore and **spherule** formation, shown in **FIGURE 15.21**. When inhaled into the human lungs, *C. immitis* induces an influenza-like disease, with a dry, hacking cough, chest pains, and high fever. During most of the 1980s, about 450

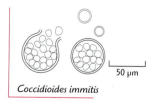
Coccidioides immitis

annual cases of coccidioidomycosis were reported to the CDC. In 1991, however, that number jumped to over 1,200 cases, and in 2001, the number of reports exceeded 3,900, 97 percent reported from California and Arizona.

Coccidioidomycosis is usually transmitted by dust particles laden with fungal spores. Cattle, sheep, and other animals deposit the spores in soil, and they become airborne with gusts of wind. Indeed, cases have been traced to standing on a railroad platform in the southwestern United States "to get a breath of fresh air." Many cases are self-limiting, but others are disseminated and involve myriad internal organs and structures, including the meninges of the spinal cord. Spherules can be located in the sputum and biopsied tissue of patients. Amphotericin B is prescribed for severe cases.

PNEUMOCYSTIS PNEUMONIA CAUSES ILLNESS AND DEATH WITHIN IMMUNOSUPPRESSED INDIVIDUALS

Pneumocystis **pneumonia (PCP)** currently is the most common cause of nonbacterial pneumonia in Americans with suppressed immune systems. The causative organism, *Pneumocystis jiroveci* (previously called *Pneumocystis carinii*) was first observed in 1910 by John Carini in his studies with rats. The disease, originally described as an atypical pneumonia, remained in relative obscurity until the 1980s, when it was recognized as the cause of death in over 50 percent of patients dying from the effects of AIDS. It should be noted that although the organism's structure and behavior are more typical of a protozoan (Chapter 16), recent evidence supports the assignment of *P. jiroveci* as an ascomycete. Analysis of its ribosomal RNA, for example, shows a closer relationship to fungal RNA than to protozoal RNA.

P. jiroveci has a complex life cycle that takes place entirely in the alveoli of the lung. A feeding stage, called the **trophozoite**, swells to become a precyst stage, in which up to eight sporozoites develop. When the cyst is mature, it opens and liberates the sporozoites, which enlarge and undergo further reproduction and maturation to trophozoites.

Present evidence indicates that *P. jiroveci* is transmitted by droplets from the respiratory tract. A wide cross section of individuals harbors the organism without symptoms, mainly because of the control imposed by T lymphocytes. However, when the immune system is suppressed, as in AIDS patients, *Pneumocystis* trophozoites and cysts fill the alveoli and occupy all the air spaces. A nonproductive cough develops, with high fever and difficult breathing. Progressive deterioration leads to consolidation of the lungs and, eventually, death.

The current treatment of choice for PCP is trimethoprim-sulfamethoxazole (cotrimoxazole). Another drug, pentamidine, is used for patients who do not respond to co-

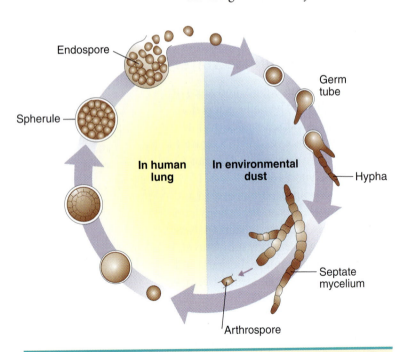

Endospore

Spherule

In human lung

In environmental dust

Germ tube

Hypha

Septate mycelium

Arthrospore

FIGURE 15.21

The Developmental Cycle of *Coccidioides immitis*

Outside the body, the organism exists as a septate mycelium. It segments to form airborne arthrospores, which are inhaled. In the respiratory tract, the arthrospores swell to yield a large body, the spherule, that segments and breaks down to release endospores. When released to the environment, the endospores form germ tubes and then the mycelium.

trimoxazole therapy. These drugs, however, have limited value in immunosuppressed individuals, as evidenced by the high death rate in AIDS patients.

OTHER FUNGI ALSO CAN CAUSE MYCOSES

A number of other fungal diseases deserve brief mention because they are important in certain parts of the United States, or they affect individuals in certain professions. Generally the diseases are mild, although complications may lead to serious tissue damage.

Aspergillosis is a unique disease because the fungus enters the body as conidia and then grows as a mycelium. Disease usually occurs in an immunosuppressed host or where an overwhelming number of conidia has entered the tissue. The most common cause is *Aspergillus fumigatus*, an ascomycete. Infection of the lung may yield a round ball of mycelium called an **aspergilloma**, requiring surgery for removal. Conidia in the earwax lead to a painful ear disease known as **otomycosis**. Disseminated *Aspergillus* causes blockage of blood vessels, inflammation of the inner lining of the heart, or clots in the heart vessels. Amphotericin B therapy is usually necessary. It should be noted that one species of *Aspergillus* is very helpful, as MicroFocus 15.7 discusses.

A closely related fungus, *Aspergillus flavus*, produces toxic compounds called **aflatoxins**. The mold is found primarily in warm, humid climates, where it contaminates agricultural products such as peanuts, grains, cereals, sweet potatoes, corn, rice, and animal feed. Aflatoxins are deposited in these foods and ingested by humans where they are thought to be carcinogenic, especially in the liver. Contaminated meat and dairy products are also sources of the toxins. Half of the cancers in sub-Saharan Africa are liver cancers and 40 percent of analyzed foods contain aflatoxins, highlighting the threat. Fungal toxins are called **mycotoxins**.

Another fungus that produces a powerful toxin is *Claviceps purpurea*. This member of the ascomycetes grows as hyphae on kernels of rye, wheat, and barley. As hyphae penetrate the plant, the fungal cells gradually consume the substance of the grain,

---

MicroFocus 15·7

"NOT WITHOUT MY BEANO!"

Some people would not dare sit down to a meal of corned beef and cabbage without a knife, fork, soda bread—and, of course, their Beano. Neither would they have Brussels sprouts with their steak, or broccoli with their fried chicken unless they were sure their Beano was nearby. Same for *pasta y fagiola* ("pasta fahzoole")—no Beano? No thanks!

To the scientist, there is really no mystery: Beano is the trade name for an enzyme preparation from the mold *Aspergillus niger*. The enzyme breaks down galactose, a disaccharide in beans,

cabbage, broccoli, Brussels sprouts, and other "strong vegetables" and high-fiber foods. Normally, galactose is broken down by the body's natural enzyme (alpha-galactosidase). But in the absence of the enzyme, bacteria in the large intestine will break down the galactose. Unfortunately, they do so at a heavy price: gas (flatulence), bloating, embarrassment—and an unwillingness to go back for seconds.

Enter Beano. All that's necessary is a couple of tablets or drops of liquid with the first bites of food (it tastes somewhat like soy sauce). Then, the fungal enzyme

takes over and breaks down the galactose, leaving none for the bacteria. And leaving a happy memory of the meal—or so says the manufacturer.

and the dense tissue hardens into a purple body called a **sclerotium**. A group of peptide derivatives called alkaloids are produced by the sclerotium and deposited in the grain as a substance called **ergot**. Products such as bread made from rye grain may cause ergot rye disease, or **ergotism** (MicroFocus 15.8). Symptoms may include numbness, hot and cold sensations, convulsions with epileptic-type seizures, and paralysis of the nerve endings. Lysergic acid diethylamide (LSD) is a derivative of an alkaloid in ergot. Commercial derivatives of these alkaloids are used to cause contractions of the smooth muscles, such as to induce labor or relieve migraine headaches.

Sporotrichosis is an occupational hazard of those who work with wood, wood products, or the soil. The disease can be contracted by handling sphagnum (peat) moss used to pack tree seedlings. It also is transmitted by punctures with rose thorns and is often referred to as **rose thorn disease**. The causative agent is *Sporothrix schenkii* (sometimes called *Sporotrichum schenkii*), which is a dimorphic fungus (FIGURE 15.22). Pus-filled purplish lesions form at the site of entry, and "knots" may be felt under the skin. Dissemination, though rare, may occur to the bloodstream, where blockages may cause swelling of the tissues (edema). In 1988, an outbreak of 84 cases of cutaneous sporotrichosis occurred in people who handled conifer seedlings packed with sphagnum moss from Pennsylvania (FIGURE 15.23). Cutaneous infections are controlled with potassium iodide, but systemic infections require amphotericin B therapy.

TABLE 15.2 summarizes the fungal diseases of humans.

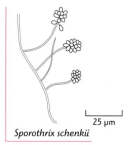

25 μm

Sporothrix schenkii

MicroFocus 15.8   DOUBLE, DOUBLE TOIL AND TROUBLE; FIRE BURN, AND CAULDRON BUBBLE

As an undergraduate, Linnda Caporael was missing a critical history course for graduation. Little did she know that through this course she was about to provide a possible answer for one of the biggest mysteries of early American history—the cause of the Salem Witch Trials. These trials in 1692 led to the execution of 20 people who had been accused of being witches in Salem, Massachusetts.

Linnda Caporael, now a behavioral psychologist at New York's Rensselaer Polytechnic Institute, in preparation of a paper for her history course had read a book where the author could not explain the hallucinations among the people in Salem during the witchcraft trials. Caporael made a connection between the "Salem witches" and a French story of ergot poisoning in 1951. In Pont-Saint-Esprit, a small village in the south of France, more than 50 villagers went insane for a month after eating ergotized rye flour. Some people had fits

of insomnia, others had hallucinogenic visions, and still others were reported to have fits of hysterical laughing or crying. In the end, three people died.

Caporael noticed a link between these bizarre symptoms, those of Salem witches, and the hallucinogenic effects of drugs like LSD, which is a derivative of ergot. Could ergot possibly have been the perpetrator? Caporael's research would not be the only one to propose that ergot had affected historical events.

During the Dark Ages, Europe's poor lived almost entirely on rye bread. Between 1250 and 1750, ergotism, then called "St. Anthony's fire," led to miscarriages, chronic illnesses in people who survived, and mental illness. Ergotism even has been proposed as precipitating some of the events leading to the French Revolution. Importantly, hallucinations were considered "work of the devil."

Toxicologists know that eating ergotized food can cause violent muscle spasms, delusions, hallucinations, crawl-

ing sensations on the skin, and a host of other symptoms—all of which, Linnda Caporael found in the records of the Salem witchcraft trials. Ergot thrives in warm, damp, rainy springs and summers, which were the exact conditions Caporael says existed in Salem in 1691. Add to this that parts of Salem village consisted of swampy meadows that would be the perfect environment for fungal growth and that rye was the staple grain of Salem—and it is not a stretch to suggest that the rye crop consumed in the winter of 1691–1692 could have been contaminated by large quantities of ergot.

Caporael concedes that, as with the French Revolution suggestion, ergot poisoning can't explain all of the events at Salem. Some of the behaviors exhibited by the villagers probably represent instances of mass hysteria. Still, as people reexamine events of history, it seems that just maybe ergot poisoning did play some role—and, hey, not bad for an undergraduate history paper!

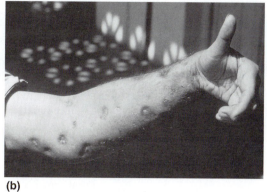

FIGURE 15.22

Sporothrix schenkii

(a) A false color scanning electron micrograph of hyphae (orange) and conidia (purple) formed on conidiophores. (Bar = 8 μm.) (b) A patient showing the lesions of sporotrichosis on an infected arm. Characteristic "knots" can be felt under the skin.

(a)

(b)

FIGURE 15.23

An Outbreak of Sporotrichosis

In this outbreak, *Sporothrix schenkii* was isolated from the packing moss used at the Pennsylvania nursery, which supplied the trees and seedlings. Eighty-four cases of sporotrichosis in 15 states were identified in the outbreak.

TEXTBOOK CASES

1. In May 1988, a man visited an Illinois physician complaining of a swollen right hand and forearm. He had almond-sized "knots" under the skin of his right arm. The physician diagnosed sporotrichosis and placed the man on potassium iodide therapy.

2. On the next visit, the man brought along his neighbor, who had similar symptoms. Once again, the physician diagnosed sporotrichosis. The neighbor mentioned that the two men work together to raise and sell Christmas trees: to make a little money on the side.

3. In July, the physician examined a third patient. On questioning, the patient explained that he had recently participated in a sale of Colorado blue spruce seedlings to children. The source of the seedlings was the same Pennsylvania nursery that supplied Christmas tree seedlings to the first two men.

4. The physician reported his observations to the Illinois State Health Department. He learned that his patients were part of a growing list of cases of sporotrichosis being reported nationally. All the cases were related to fungus-contaminated moss used to pack trees and seedlings at the same nursery in Pennsylvania.

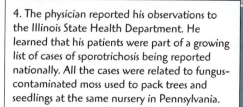

TABLE 15.2

A Summary of Fungal Diseases of Humans

ORGANISM	DIVISION (PHYLUM)	DISEASE	TRANSMISSION
Cryptococcus neoformans	Basidiomycota	Cryptococcosis	Airborne cells
Candida albicans	Deuteromycota	Candidiasis Vaginitis Thrush, onychia	Airborne Sexual contact Skin contact
Trichophyton species *Microsporum* species *Epidermophyton* species	Ascomycota Ascomycota Deuteromycota	Tinea pedis Tinea capitis Tinea corporis	Contact with hyphal fragments
Histoplasma capsulatum	Ascomycota	Histoplasmosis	Airborne spores
Blastomyces dermatitidis	Ascomycota	Blastomycosis	Airborne spores Open wound
Coccidioides immitis	Deuteromycota	Coccidioidomycosis	Airborne spores
Pneumocystis jiroveci	Ascomycota?	*Pneumoncystis* pneumonia	Droplets
Aspergillus fumigatus	Ascomycota	Aspergillosis Otomycosis	Airborne spores
Sporothrix schenkii	Deuteromycota	Sporotrichosis	Spores Puncture wound

Note to the Student

During the 1980s, an epidemiologist at the University of Virginia initiated a study of infectious diseases acquired in the hospital, fully expecting most diseases to be caused by viruses and bacteria. To his surprise, almost 40 percent of the diseases were due to fungi.

The study was one of the first pieces of evidence pointing to an emerging threat of fungal disease. In most cases, an impaired immune system is involved—a system overwhelmed with human immunodeficiency virus, or a system depressed by cancer chemotherapy or antirejection drugs following a transplant. But, there is also the problem of a meager arsenal of antifungal drugs, and the fungal pathogens are developing resistances to drugs of long-standing use. Moreover, clinical research in mycology has not kept up with that in clinical virology and bacteriology, in part due to the lack of established cultivation methods for fungal pathogens.

ORGAN AFFECTED	DIAGNOSIS	TREATMENT	COMMENT
Lungs Spinal cord Meninges	Examination of spinal fluid	Amphotericin B	Associated with pigeon droppings
Intestine Vagina Skin, mouth	Urine examination Vaginal smears Lab cultivation	Nystatin Miconazole Ketoconazole	Normally in human intestine
Skin	Lab cultivation Tissue examination	Undecylenic acid Griseofulvin Miconazole Itraconazole	Widely encountered skin diseases
Lungs Various organs	Lab cultivation Tissue examination	Amphotericin B	Associated with birds and bats
Lungs Various organs	Lab cultivation Tissue examination	Amphotericin B	Associated with bird droppings
Lungs	Lab cultivation Tissue examination	Amphotericin B	Common in southwestern US
Lungs	Lung examination	Trimethoprim- sulfamethoxazole	Associated with AIDS
Lungs Ears	Lab cultivation Tissue examination	Amphotericin B	Hyphae grow in body
Skin Lymph vessels	Lab cultivation Tissue examination	Amphotericin B Potassium iodide	Associated with rotten wood

The list of fungal pathogens also appears to be expanding. In addition to the well-known pathogens (*Candida, Histoplasma, Cryptococcus,* and others), mycologists are proposing that up to 150 species of fungi may be pathogenic. *Aspergillus* species, for example, have increasingly been related to pneumonia; the plant pathogen *Fusarium* has been isolated from human blood disease; the mushroom *Coprinus* riddled a prosthetic heart valve in a recent case of endocarditis; and even the yeast *Saccharomyces* was found to cause infection in one person's burned tissue.

As of mid-2003, the CDC requests physicians to report 49 infectious diseases, but only one is of fungal origin (coccidioidomycosis). However, as more and more fungi are being recognized for their pathogenic potential, the numbers of reported fungal disease may increase. Once considered a nuisance, fungi are now coming out of the shadows as serious threats to human health.

Summary of Key Concepts

15.1 CHARACTERISTICS OF FUNGI

- **Fungi Share a Combination of Characteristics.** Fungi are a group of eukaryotic microorganisms distinguished from plants by their lack of chlorophyll, by differences in their cell walls, and by the fact that fungi are not truly multicellular. Moreover, fungi are heterotrophic, while plants are autotrophic. Most fungi consist of masses of hyphae that form a mycelium. Cross-walls separate the cells of hyphae in many fungal species.

- **Fungal Growth Is Influenced by Several Factors.** Fungi secrete enzymes into the surrounding environment and absorb the break-down products. Tremendous absorption can occur when there is a large mycelium surface. Yeasts grow in number by mitosis while mycelial fungi grow in size through lengthening of the hyphae at the hyphal tips. Most fungi are aerobic, grow best around 25°C, and prefer slightly acidic conditions.

- **Reproduction in Fungi Involves Spore Formation.** Reproductive structures generally occur at the tips of hyphae. Masses of asexually produced spores within or at the tip of the hypha provide the mechanisms for propagating the fungi. Spores also can be produced by sexual reproduction, in which case the format of the reproductive process provides a basis for separating fungi into three phyla. In a fourth group, no sexual reproduction phase has been observed, and reproduction occurs solely asexually.

15.2 THE CLASSIFICATION OF FUNGI

- **The Zygomycota Have Nonseptate Hyphae.** The phylum Zygomycota has nonseptate fungi. The sexual phase is characterized by the formation of a zygospore that releases haploid spores that germinate into a new mycelium. The best-known member is *Rhizopus stolonifer*, a common bread mold.

- **The Ascomycota Are the Sac Fungi.** The mycelial fungi in the phylum Ascomycota have septate hyphae. The phylum includes the unicellular yeasts and mildews. It also contains the cup fungi that have a cup-shaped fruiting body. Within this body, ascospores are produced that germinate to form a new haploid mycelium. Asexual reproduction is through the dissemination of conidia. Several ascomycetes have industrial applications. Lichens often are a mutualistic association between an ascomycete and either a cyanobacterium or a green alga.

- **The Basidiomycota Are the Club Fungi.** The phylum Basidiomycota includes the mushrooms. Within these fruiting bodies, basidiospores are produced. On germination, they produce a new haploid mycelium. There are numerous edible and very poisonous mushrooms in the phylum. Rusts and smuts that cause many plant diseases are additional members of the phylum.

- **The Deuteromycota Lack a Sexual Reproduction Phase.** Those fungi in which mycologists have not observed a sexual phase are said to be "imperfect" and placed in the phylum Deuteromycota. Many human fungal diseases involve fungi in this phylum.

- **The Yeasts Are Microscopic, Unicellular Fungi.** The best-known yeasts are species of *Saccharomyces* that are involved in baking and brewing. *Saccharomyces* is one of the unicellular ascomycetes.

15.3 FUNGAL DISEASES OF HUMANS

- **Cryptococcosis Usually Occurs in Immunocompromised Individuals.** Cryptococcosis is caused by *Cryptococcus neoformans*. It often infects immunosuppressed individuals, such as AIDS patients, and therefore represents an opportunistic disease. Cryptococcosis affects the lungs and spinal cord. Amphotericin B is available to alleviate the symptoms of the disease.

- **Candidiasis Often Is a Mild, Superficial Infection.** Candidiasis, caused by *Candida albicans*, is an opportunistic disease that, like *C. neoformans*, most often occurs in immunosuppressed individuals, such as AIDS patients. Candidiasis can occur in numerous organs such as the skin, intestines, vaginal tract (vulvovaginitis), and oral cavity (thrush). Various antifungal drugs are available for treatment.

- **Dermatophytosis Is an Infection of the Body Surface.** Fungal dermatophytes cause a number of skin diseases, including athlete's foot (tinea pedis), ringworm of the head (tinea capitis), ringworm of the body (tinea corporis), "jock itch" (tinea cruris), and ringworm of the nails (tinea unguium).

- **Histoplasmosis Can Produce a Systemic Disease.** Histoplasmosis is endemic to the Ohio and Mississippi River valleys. Inhalation of spores of *Histoplasma capsulatum* produces an infection that usually does not require treatment. However, if the disease becomes disseminated, it produces tuberculosis-like lesions that can be fatal.

■ **Blastomycosis Is Endemic in the Southeastern United States.** Blastomycosis is endemic in the southeastern United States in dusty soil and bird droppings. The causative agent, *Blastomyces dermatitidis*, is a dimorphic fungus that grows as a yeast form in humans and produces wart-like lesions on the face, hands, and legs. Amphotericin B can be used for treatment.

■ **Coccidioidomycosis Primarily Is a Respiratory Infection.** *Coccidioides immitis* is endemic in the valleys of California and deserts of the southwest, especially Arizona. The disease, coccidioidomycosis, is a respiratory infection that develops from breathing *C. immitis* spores. For most infected individuals, it produces flu-like symptoms. In a few people, the disease becomes disseminated throughout several body organs. Severe cases can be fatal. Treatment involves amphotericin B therapy.

■ ***Pneumocystis* Pneumonia Causes Illness and Death within Immunosuppressed Individuals.** *Pneumocystis jiroveci*, formerly called *P. carinii*, is an opportunistic fungus in individuals with an impaired immune system. The fungus infects the alveoli of the lungs leading to a nonproductive cough and fever. Progressive deterioration is called *Pneumocystis* pneumonia (PCP). Without treatment, infection often leads to death.

■ **Other Fungi Also Can Cause Mycoses.** A few other fungi also cause human illness and disease. *Aspergillus fumigatus* causes aspergillosis, which is a lung infection. *Aspergillus flavus* produces a mycotoxin, called aflatoxin, which is carcinogenic. *Claviceps purpurea* produces alkaloid derivatives in grain called ergot. Eating ergotized breads can cause ergotism, which produces a variety of symptoms including convulsions with epileptic-type seizures and paralysis of nerve endings. *Sporothrix schenkii* causes sporotrichosis, which is transmitted by skin punctures. Pus-filled lesions form, generating "knots" under the skin. Potassium iodide is used for such skin infections.

Questions for Thought and Discussion

Answers to selected questions can be found in Appendix C.

1. In the 1980s in a suburban community, a group of residents obtained a court order preventing another resident from feeding the flocks of pigeons that regularly visited the area. Microbiologically, was this action justified? Why?

2. A homemaker decides to make bread. She lets the dough rise overnight in a warm corner of the room. The next morning she notices a distinct beer-like aroma in the air. What is she smelling, and where did the aroma come from?

3. Fungi are extremely prevalent in the soil, yet we rarely contract fungal disease by consuming fruits and vegetables. Why do you think this is so?

4. In 1991, the US Food and Drug Administration approved for over-the-counter sales a number of antifungal drugs such as clotrimazole and miconazole (Gyne-Lotrimin and Monistat, respectively). It thus became possible for a woman to diagnose and treat herself for a vaginal yeast infection. Should she do it?

5. Why is it a good idea to occasionally empty a package of yeast into the drain leading to a cesspool or a septic tank? Why are yeasts accused of having "metabolic schizophrenia"?

6. A woman has a continuing problem of ringworm, especially of the lower legs in the area around the shins. Questioning reveals that she has five very affectionate cats at home. Is there any connection between these facts?

7. A student of microbiology proposes a scheme to develop a strain of bacteria that could be used as a fungicide. Her idea is to collect the chitin-containing shells of lobsters and shrimp, grind them up, and add them to the soil. This, she suggests, will build up the level of chitin-digesting bacteria. The bacteria would then be isolated and used to kill fungi by digesting the chitin in fungal cell walls. Do you think her scheme will work? Why?

8. Mr. A and Mr. B live in an area of town where the soil is acidic. Oak trees are common, and azaleas and rhododendrons thrive in the soil. In the spring, Mr. A spreads lime on his lawn, but Mr. B prefers to save the money. Both use fertilizer, and both have magnificent lawns. Come June, however, Mr. B notices that mushrooms are popping up in his lawn and that brown spots are beginning to appear. By July, his lawn has virtually disappeared. What is happening in Mr. B's lawn, and what can Mr. B learn from Mr. A?

9. On June 27, 1995, a crew of five workers began a partial demolition of an abandoned city hall building in a Kentucky community. Three weeks later, all five required treatment for acute respiratory illness, and three were hospitalized. Cells obtained from the patients by lung biopsy revealed oval bodies. When the construction site was inspected, epidemiologists found an accumulation of bat droppings, and neighbors said they had seen bats in the area in recent weeks. From the information, can you surmise the nature of the disease in the demolition crew?

10. In 1992, residents of a New York community, unhappy about the smells from a nearby composting facility and concerned about the health hazard posed by such a facility, had the air at a local school tested for the presence of fungal spores. Investigators from the testing laboratory found abnormally high levels of *Aspergillus* spores on many inside building surfaces. Is there any connection between the high spore count and the composting facility? Is there any health hazard involved?

11. On January 17, 1994 a serious earthquake struck the Northridge section of Los Angeles County in California. From that date through March 15, 170 cases of coccidioidomycosis were identified in adjacent Ventura County. This number was almost four times the previous year's number of cases. Can you guess the connection between the two events?

12. In Arizona, during the five-year period of 1990 to 1995, the incidence of cases of coccidioidomycosis increased 144 percent. How many reasons can you postulate for this extremely high increase?

http://microbiology.jbpub.com

The site features **eLearning,** an on-line review area that provides quizzes and other tools to help you study for your class. You can also follow useful links for in-depth information, or just find out the latest microbiology news.

Review

The significance of the fungi is broad and diverse, as this chapter has demonstrated. To test your knowledge of the important fungi, match the statement on the left to the organism on the right by placing the correct letter in the available space. A letter may be used once, more than once, or not at all. Answers are listed in Appendix D.

_____ 1. Causes late blight of potatoes.

_____ 2. Produces a widely used antibiotic.

_____ 3. Used for bread baking.

_____ 4. Growth can be interrupted with griseofulvin.

_____ 5. Causes "valley fever" in the southwestern US.

_____ 6. Common white or gray bread mold.

_____ 7. Poisonous mushroom.

_____ 8. Sexual phase known as *Emmonsiella*.

_____ 9. Agent of rose thorn disease.

_____ 10. Edible mushroom.

_____ 11. Can overgrow the intestine when antibiotic consumed.

_____ 12. Known to cause Darling's disease.

_____ 13. Has nonseptate hyphae.

_____ 14. Produces citric acid, soy sauce, and vinegar.

_____ 15. Associated with the droppings of pigeons.

_____ 16. Agent of ergot disease in rye plants.

_____ 17. Cause of vaginal yeast infections in women.

_____ 18. Used to produce wine from grape juice.

_____ 19. One of the causes of athlete's foot.

_____ 20. Often found in chicken coops and bat caves.

_____ 21. Produces a toxic aflatoxin.

_____ 22. Agent of dermatophytosis.

_____ 23. Reproduction includes a spherule.

_____ 24. Nystatin and miconazole to inhibit.

_____ 25. Involves a trophozoite stage.

A. *Amanita phalloides*

B. *Claviceps purpurea*

C. *Sporothrix schenkii*

D. *Blastomyces dermatitidis*

E. *Agaricus* species

F. *Acremonium* species

G. *Aspergillus* species

H. *Saccharomyces ellipsoideus*

I. *Phytophthora infestans*

J. *Metarrhizium* species

K. *Coccidioides immitis*

L. *Stipa robusta*

M. *Epidermophyton* species

N. *Pneumocystis jiroveci*

O. *Cryptococcus neoformans*

P. *Candida albicans*

Q. *Penicillium notatum*

R. *Synchytrium endobioticum*

S. *Histoplasma capsulatum*

T. *Batrachochytrium dendrobatidis*

U. *Rhizopus stolonifer*

V. *Cookeina tricholoma*

W. *Saccharomyces cerevisiae*

X. *Stachybotrys chartarum*

Y. *Taxomyces andreanae*

Z. *Aspergillus flavus*

16

The Protozoa

It races through the bloodstream, hunkers down in the liver, then rampages through red blood cells before being sucked up by its flying, buzzing host to mate, mature, and ready itself for another wild ride through a two-legged motel.

—The editor of *Discover* magazine describing, in flowery terms, the life cycle of the protozoan that causes malaria

PRIL 12, 1993, SHOULD HAVE BEEN a festive day in Milwaukee, Wisconsin. The baseball home opener was scheduled, and fans were eager to see the Brewers play the California Angels. But the scoreboard contained an ominous message: "For your safety, no city of Milwaukee water is being used in any concession item." The city was in the throes of an epidemic, and a protozoan was to blame.

The protozoan was *Cryptosporidium parvum*, an intestinal parasite that causes mild to serious diarrhea, especially in infants and the elderly. As the protozoa attach themselves to the intestinal lining, they mature, reproduce, and encourage the body to release large volumes of fluid. The infection is accompanied by abdominal cramps, extensive water loss, and in many cases, vomiting and fever.

Even as the first ball was being thrown out at the stadium, health inspectors were analyzing Milwaukee's water purification plants to see how a protozoan could have reached the city's water supply. *Cryptosporidium* is a waterborne parasite commonly found in the intestines of cows and other animals. Perhaps, they guessed, the heavy rain and spring thaw had washed the protozoan from farm pastures and barns into the Milwaukee River. The river might have brought *Cryptosporidium* into Lake Michigan from which the city drew its water. Indeed, the mouth of the river was very close to the intake pipe from the lake. Moreover, they added, *Cryptosporidium* can resist the chlorine treatment used to control bacteria in water; and the tests to detect bacterial contamination do not detect protozoa, such as *Cryptosporidium*.

As researchers worked to unravel the mystery, the game went on. Soda was available, but only from bottles. Drinking fountains were turned off. Two huge US Army tanks stood by to provide reserve water for the 50,000 fans in attendance. In the city, tens of thousands of Milwaukeeans made the mildly embarrassing trip to the drugstore to stock up on toilet paper and antidiarrheal medications. (A large window sign at a local Walgreen's proudly proclaimed: "We have Imodium A-D.") Back at the ballgame, things were not going much better—the Brewers lost to the Angels 12 to 5.

Cryptosporidium parvum (**FIGURE 16.1**) will be one of the protozoa we study in this chapter. We shall encounter other protozoa that infect the human intestine as well as several protozoa that live primarily in the blood and other organs of the body. Many of the diseases we encounter (for example, malaria) will have familiar names, but others, such as *Cryptosporidium* infections, are emerging diseases in our society (indeed, *Cryptosporidium* was not known to infect humans before 1976). Our study will begin with a focus on the characteristics of protozoa.

16.1
Characteristics of Protozoa

Protozoa are a group of about 65,000 species of single-celled organisms. They take their name from the Greek words *protos* and *zoon*, literally meaning "first animal." This name refers to the position many biologists formerly believed protozoa occupy in the evolution of living things. Though often studied by zoologists, protozoa also interest microbiologists because they are unicellular, have a microscopic size, and are involved in disease. The discipline of **parasitology** is generally concerned with the medically related protozoa and the multicellular parasites (Chapter 17).

PROTOZOA CONTAIN TYPICAL EUKARYOTIC ORGANELLES

Protozoa are among the largest organisms encountered in microbiology, some forms reaching the size of the period at the end of this sentence. With only a few

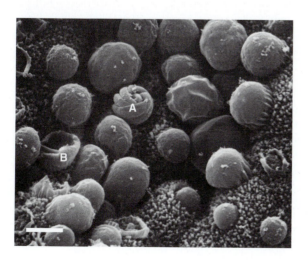

FIGURE 16.1

A Waterborne Protozoan

A scanning electron micrograph of *Cryptosporidium parvum* at the surface of intestinal tissue. The parasites are the globular sac-like bodies. Within the globes are numerous long, thin forms of the parasite called merozoites. After release from the globes, the merozoites will infect nearby cells. One parasite (A) has lost its membrane and the merozoites can be seen crowded together. The crater-like structures (B) are parasites from which the merozoites already have been released. (Bar = 4 μm.)

exceptions, protozoa have no chloroplasts and thus cannot produce carbohydrates by photosynthesis. Although each protozoan is composed of a single cell, the functions of that cell bear a resemblance to the functions of multicellular animals rather than to those of an isolated cell from that animal.

Most protozoa are free-living and thrive where there is water. They may be located in damp soil and mud; in drainage ditches and puddles; and in ponds, rivers, and oceans. Some species of protozoa remain attached to aquatic plants or rocks, while other species swim about. The film of water on an ordinary dirt particle often contains protozoa. **FIGURE 16.2** illustrates some of the diversity that exists within the protozoal groups.

FIGURE 16.2

Diversity Among Protozoa

The four groups are represented, with the structures for movement (pseudopods, flagella, cilia) shown for members of groups I–III. Group IV lacks such structures.

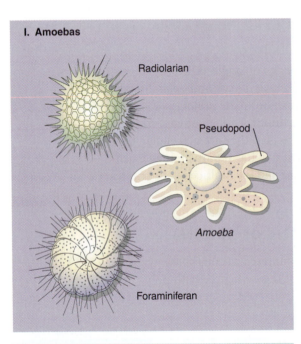

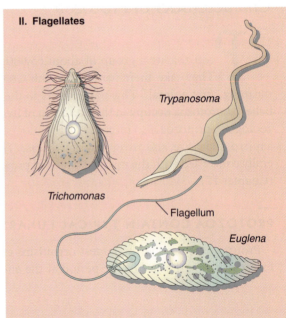

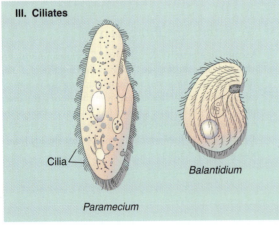

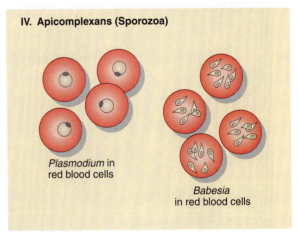

Protozoal cells are surrounded by a plasma membrane. In some species, the membrane forms part of a complex outer covering called a **pellicle**. The cytoplasm contains the typical eukaryotic organelles and each cell has a nucleus. In addition, freshwater protozoa continually take in water by the process of osmosis and eliminate the excess water via organelles called **contractile vacuoles**. These vacuoles expand with water drawn from the cytoplasm and then appear to "contract" as they release water through a temporary opening in the cell membrane. Many protozoa also contain flagella or cilia, which permit cell motility.

MOST PROTOZOA ARE HETEROTROPHS

The majority of protozoa are heterotrophs. They obtain their nutrients by engulfing food particles by phagocytosis or through special organs of ingestion (**FIGURE 16.3**). A membrane then encloses the particles to form an organelle called a **food vacuole**. The vacuole joins with **lysosomes**, and digestive enzymes from the lysosomes proceed to break down the ingested particles. Nutrients are absorbed from the vacuole, and any undigested residue is eliminated from the cell. Some protozoa have the ability to digest the cellulose in wood.

Except for the parasitic organisms of disease and the species that feed on bacteria, protozoa are saprobic. Most protozoa are aerobic, obtaining their oxygen by diffusion through the cell membrane. The feeding form of a protozoan is commonly known as the **trophozoite** (*troph-* is the Greek stem for "food"). Another form, the **cyst**, is a dormant, highly resistant stage that develops in some protozoa when the organism secretes a thick case around itself during times of environmental stress. Reproduction in protozoa usually occurs by the asexual process of binary fission (**FIGURE 16.4**). Many protozoa also have a sexual stage.

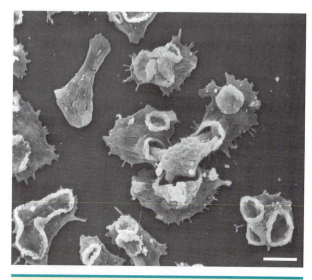

FIGURE 16.3

Feeding Behavior in Protozoa

This scanning electron photograph shows three amoebas attacking and devouring a fourth, presumably dead, amoeba. The amoebas are *Naegleria fowleri*, a cause of meningoencephalitis in humans. Each amoeba possesses a series of sucker-like structures that stick to the prey and serve as portals for ingestion. (Bar = 10 µm.)

To this point . . .

We have introduced the protozoa and have noted that they are a group of heterotrophic, unicellular, microscopic organisms often involved in human disease. Most protozoa are found in aquatic environments, and the majority are free-living. Nutrients are commonly obtained through phagocytosis by the trophozoite form of the organism. A cyst may be formed under conditions of environmental stress.

In the next section, we will briefly examine the four groups of protozoa.

FIGURE 16.4

Binary Fission in *Amoeba proteus*

In this sequence of photomi-crographs, the cytoplasm is seen separating to form two new individuals. Few visible changes are apparent for the first 15 minutes, but once cell division begins, separation takes place rapidly. (Bar = 5 μm.)

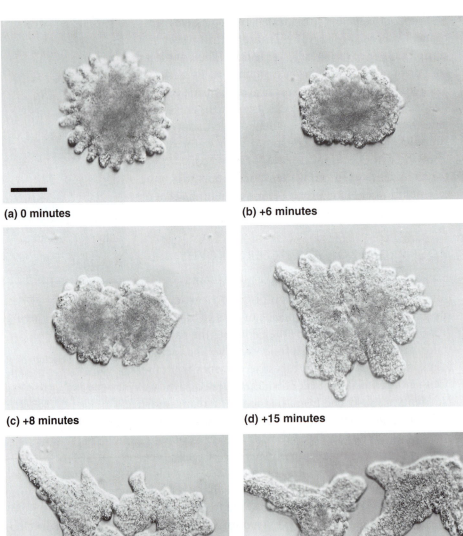

(a) 0 minutes

(b) +6 minutes

(c) +8 minutes

(d) +15 minutes

(e) +18 minutes

(f) +21 minutes

16.2

The Classification of Protozoa

When Robert Whittaker assigned protozoa to the kingdom **Protista** in 1969, he did so as a matter of convenience, rather than on the basis of evolutionary relationships. During the 1970s and 1980s, the kingdom's boundaries were extended to include some multicellular organisms (e.g., certain seaweeds), as well as some fungus-like organisms (e.g., slime molds and water molds), and a pot-pourri of other organisms that needed a taxonomic home. The tendency was to place in the kingdom Protista any eukaryotes that did not comfortably fit into the definitions of plants, animals, or fungi.

The kingdom Protista probably is a product of ignorance that reflects how many different types are related to one another. However, excellent new research in the area is shedding light on the taxonomic relationships, and in the years ahead, the kingdom Protista may be replaced by several new kingdoms. Molecular variations in ribosomal subunits, for example, are the basis for one new classification scheme, the three-domain system.

However, that does not resolve the immediate problem of what to do with the protozoa. While the classification system is in a state of flux, we shall use type of motion as a criterion and discuss four groups of protozoa using the terms familiar to most students of biology: **amoebas** (protozoa that move by pseudopodis); **flagellates** (protozoa moving by flagella); **ciliates** (organisms with cilia); and **apicomplexans**, or **sporozoa** (protozoa exhibiting no motion in the adult form). This organization will give us considerable latitude when considering the protozoa of medical significance, while permitting us to introduce some protozoa of ecological and industrial importance. However, it should be remembered that the groupings are artificial, informal, and temporary.

AMOEBAS FORM PSEUDOPODS

Amoebas have been placed in the group **Sarcodina** (sarcodines) by many biologists and in the group **Rhizopoda** (rhizopods) by others. These organisms move as their cytoplasm flows into temporary formless projections called **pseudopods** ("false-feet"). The amoeba is the classic example of the group, and thus the motion is called **amoeboid motion** (FIGURE 16.5). Pseudopods also capture small algae and other protozoa in the process of phagocytosis. It is interesting to note that **actin** filaments and **myosin**, the well-known proteins of muscle tissue, are associated with movement of the pseudopods.

An amoeba may be as large as 1 millimeter in diameter. It usually lives in freshwater and reproduces by binary fission. Amoebas may be found in home humidifiers, where they have been known to cause an allergic reaction called **humidifier fever**. Far more serious are the parasitic amoebas that cause amoebiasis and a form of encephalitis.

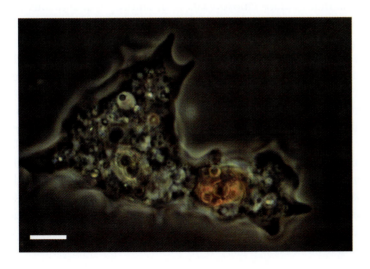

FIGURE 16.5

Freshwater Amoeba

A phase-contrast photomicrograph of a freshwater amoeba. Pseudopodia can be seen extending from the edges of the cell (top center). Several food vacuoles are present within the cell. (Bar = 10 μm.)

Two large groups of marine amoebas have ecological significance. The first group, the **radiolarians**, are abundant in the Indian and Pacific oceans. These amoebas have spherical shells with highly sculptured glassy skeletons from which pseudopods extend, reminiscent of vintage Christmas ornaments. When the protozoa die, their skeletal remains litter the ocean floor with deposits called radiolarian ooze. The second group, the **foraminiferans**, have chalky skeletons, often in the shape of snail shells with openings between sections (the name means "little window"). Foraminiferans flourished during the Paleozoic era, about 225 million years ago. Their shells in ocean sediments therefore serve as depth markers for oil-drilling rigs and as estimates of the age of the rock. Geologic upthrust has brought the sediments to the surface in several places around the world, such as the White Cliffs of Dover (**FIGURE 16.6**).

Reports first published in 1987 indicated that amoebas in the genus *Acanthamoeba* (e.g., *A. castellani*) can cause corneal infection in people who wear contact lenses. They were reminded to adhere to recommended care and use procedures, and ophthalmologists and optometrists were advised to increase patient education. Recent research indicates that bacterial infection can be a cofactor in *Acanthamoeba* infection of the eye. Many cases are misdiagnosed as herpes simplex infections.

FLAGELLATES MOVE BY ONE OR MORE FLAGELLA

Flagellates traditionally have been placed in the group **Mastigophora**. They often have the shape of a vase, and all move by means of one or more whip-like, undulating **flagella** (*mastig-* is Greek for "whip"). The flagellum can either push or pull the organism, depending on the species. Flagella occur singly, in pairs, or in large numbers. Each flagellum has the characteristic 9 + 2 arrangement of microtubules found in all eukaryotic flagella. Undulations sweep down the flagella to the tip, and the lashing motion forces water outward to provide locomotion. The movement resembles the activity of a fish sculling in water. In Chapter 4, we described bacterial flagella whose structure, size, and type of movement differ from the flagellates.

Almost half the known species of protozoa are flagellates. An example is the green flagellate *Euglena* often found in freshwater ponds. This organism is unique because it is one of the few genera of protozoa that contain chloroplasts with chlorophyll, and thus it is capable of photosynthesis. Some botanists claim it to be a plant, but zoologists point to its ability to move using flagella and suggest that it is more animal-like. Still other biologists point out that it may be the basic stock of evolution from which both animal and plant forms once arose.

Some species of flagellated protozoa are free-living, but most live together with plants or animals. Several species, for example, are found in the gut of the termite, where they participate in a mutualistic relationship. Other species are parasitic in humans and cause disease of the nervous, urogenital, or gastrointestinal systems. Still others include the dinoflagellates that cause the infamous **red tides**. Chapter 26 discusses these species in more detail. One dinoflagellate, *Pfiesteria piscicida*, has been linked to extensive fish kills in waters from Alabama to Delaware (**MicroFocus 16.1**).

FIGURE 16.6

The White Cliffs of Dover, England

These cliffs are composed of the remains of foraminiferans that thrived in the oceans millions of years ago.

MicroFocus 16.1

HOW TO SKIN A FISH—PROTOZOAN STYLE

In 1997, *Newsweek* magazine labeled it the "cell from hell." It was a flagellated protozoan known as *Pfiesteria piscicada*, and it was destroying the fish population along the southeastern coast of the United States. Silvery menhaden (a type of herring) were turning up with quarter-sized lesions on their bodies, and a *Pfiesteria*-hysteria was setting in. The seminal research had been done by JoAnn Burkholder and other scientists from North Carolina State University. Their research was pointing to one or more toxins produced by the protozoan. During August of that year, tens of thousands of fish died in a 4.5-mile stretch of Maryland's Pocomoke River. In 2000, a *Pfiesteria* outbreak killed about five million menhaden along the Delaware coast.

More recently, scientists are starting to better understand the organism and the disease. *P. piscicada* normally is non-toxic and feeds on algae and bacteria in the water. However, when fish like the menhaden school, the behavior triggers the protozoan cells to produce the toxins. Some toxins stun the fish, making them lethargic. While the fish are inca-pacitated, the *P. piscicada* cells attach to the fish and feed on their tissues and blood by secreting other toxins that break down fish skin tissue, opening bleeding sores or lesions. Given that enough protozoan cells are present, they could actually skin the fish alive. So, fish are killed by the toxins released by *P. piscicada*, or by secondary infections that attack the fish once the toxins have caused lesions to develop.

The *Pfiesteria* toxins in the water break down within a few hours and the toxic outbreak comes to an end as the protozoan cells change back into non-toxic forms. However, once fish are weakened by the toxins, *Pfiesteria*-related fish lesions or fish kills may persist for days or possibly weeks.

If there were any human health problems, it would come from the toxins present in the river or estuary. Some initial studies suggest that the *Pfiesteria* toxins may cause memory loss and confusion, along with skin, respiratory, and gastrointestinal problems.

Other factors also may influence toxic *Pfiesteria* outbreaks. Warm, brackish, slow moving waters and high levels of nutrients, such as nitrogen and phosphorus from farm runoff, could encourage the growth of *Pfiesteria* by stimulating the growth of algae on which the nontoxic forms of *Pfiesteria* feed. Evidence also suggests that nutrients may directly stimulate the growth of *Pfiesteria*. As of this writing, more research needs to be done to understand what causes the "*Pfiesteria* blooms" and exactly how serious the toxins are to human health.

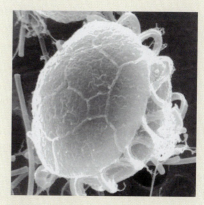

■ *Pfiesteria shumwayae, a member of the toxic* Pfiesteria *complex responsible for killing millions of fish.*

CILIATES HAVE TWO TYPES OF NUCLEI

Ciliates, which have been classified traditionally as **Ciliophora**, are among the most complex cells on Earth. They range in size from a microscopic 10 μm to a huge 3 mm (about the same relative difference between a football and a football field). All species are covered with hair-like **cilia** (sing., cilium) in longitudinal or spiral rows. The movement of the cilia is coordinated by a network of fibers running beneath the surface of the cell. Cilia beat in a synchronized pattern, much like a field of wheat bending in the breeze or the teeth on a comb bending if you pass your thumb across the row. The organized "rowing" action that results speeds the ciliate along in one direction. By contrast, flagellar motion tends to be jerky and much slower.

The complexity of ciliates is illustrated by the slipper-shaped *Paramecium*. This organism has a primitive gullet, as well as a "mouth" into which food particles are swept. The cells have a single large macronucleus that controls cell metabolism, and one or more micronuclei that are used for genetic recombination. During **sexual**

recombination, called **conjugation**, two cells make contact, and a cytoplasmic bridge forms between them. A micronucleus from each cell undergoes two divisions to form four micronuclei, of which three disintegrate and only one remains to undergo mitosis. Now a "swapping" of micronuclei takes place, followed by a union to re-form the normal micronucleus (**FIGURE 16.7**). This genetic recombination is somewhat analogous to what occurs in bacteria. It is observed during periods of environmental stress, a factor that suggests the formation of a genetically different and, perhaps, better-adapted organism. Reproduction at other times is by asexual binary fission.

Another feature of *Paramecium* is the **kappa factors**. These nucleic acid particles are apparently responsible for the synthesis of toxins that destroy ciliates lacking the factors. *Paramecium* species also possess **trichocysts**, organelles that discharge filaments to defend against predators. A third feature is the **contractile vacuole** used to "bail out" excess water from the cytoplasm. These organelles are present in freshwater ciliates but not in saltwater species because little excess water exists in the cells.

Ciliates have been the subject of biological investigation for many decades. They are found readily in almost any pond or gutter water; they have a variety of shapes; they exist in several colors, including light blue and pink; they exhibit elaborate and

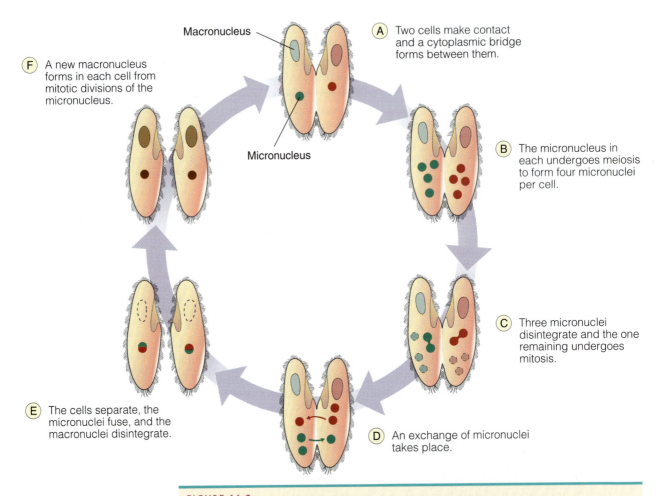

Macronucleus

Micronucleus

A Two cells make contact and a cytoplasmic bridge forms between them.

B The micronucleus in each undergoes meiosis to form four micronuclei per cell.

C Three micronuclei disintegrate and the one remaining undergoes mitosis.

D An exchange of micronuclei takes place.

E The cells separate, the micronuclei fuse, and the macronuclei disintegrate.

F A new macronucleus forms in each cell from mitotic divisions of the micronucleus.

FIGURE 16.7

Sexual Recombination in *Paramecium*

controlled behavior patterns; and they have simple nutritional requirements, which makes cultivation easy.

APICOMPLEXANS (SPOROZOA) ARE OBLIGATE PARASITES

Apicomplexans are so named because one end of the cell (the *api*cal end) contains a *complex* of organelles used for penetrating host cells. The group name is **Apicomplexa**. Another name for the group is **Sporozoa**, because these protozoa once were believed to form spores. Another reason is because one stage in the life cycle is an infectious form known as the sporozoite.

Apicomplexans are parasitic protozoa with complex life cycles that include alternating sexual and asexual reproductive phases. These phases often occur in different hosts.

Protozoa in this group are notable for the absence of cilia or flagella in the adult form. Two species, the organisms of malaria and toxoplasmosis, are of special significance, the first because it is one of the most prolific killers of humans, the second because of its association with the disease AIDS. Other notable species include *Isospora belli*, a cause of the human intestinal disease **coccidiosis**; *Sarcocystis* species, which live in the intestines as well as the muscle tissue of humans and animals; and *Encephalitozoon* species, which cause disseminated illness in AIDS patients (**FIGURE 16.8**).

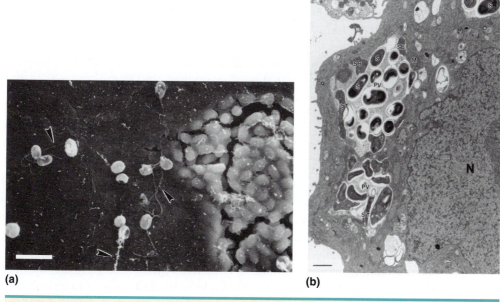

(a)   **(b)**

FIGURE 16.8

Encephalitozoon species

(a) A scanning electron micrograph of an *Encephalitozoon* species from cell culture. Note the delicate thread-like tubules extending from the ends of the cells (arrowheads). The significance of these tubules is unknown. (Bar = 200 μm.) (b) A transmission electron micrograph of a cell infected with *Encephalitozoon*. The parasite is multiplying within a vacuole termed the parasitophorus vacuole (PV). Within the vacuole, the *Encephalitozoon* cells have formed young, spore-like bodies called sporoblasts (SB), spore-like bodies (S), and developmental forms called the meront (M) and sporont (ST). The nucleus (N) of the cell can be observed. (Bar = 200 μm.)

To this point . . .

We outlined the salient features of the four groups of protozoa. Members of the first group, the amoebas, move by means of amoeboid motion using their pseudopodis. Radiolarians and foraminiferans are important members of this group. The second group includes flagellated protozoa, and Euglena *is an often-studied, photosynthetic member of the group. Ciliates are extremely complex, with many features of multicellular animals. The genetic recombination mechanism exhibited by* Paramecium *illustrates the complexity. The final group, apicomplexans (sporozoa), contains protozoa whose life cycles are complex. Motion is not observed in the adult forms of these organisms.*

We now will examine several human diseases caused by protozoal parasites and studied by parasitologists. The survey will be organized according to the groups of protozoa, beginning with amoebas and concluding with the flagellates. Protozoal diseases occur worldwide, and public health agencies consider them to be a global health problem. The diseases are particularly prevalent in tropical and subtropical regions.

16.3 Protozoal Diseases Caused by Amoebas and Flagellates

Diseases from amoebas and flagellates occur in a variety of systems of the human body. For example, some diseases, such as amoebiasis and giardiasis, take place in the digestive system, while others, such as sleeping sickness and leishmaniasis, occur in the blood. Still others, such as trichomoniasis, develop in the urogenital tract. Associated with these diseases is an equally diverse series of modes of transmission, as we shall observe in the discussions ahead.

AMOEBIASIS IS THE SECOND LEADING CAUSE OF DEATH FROM PARASITIC DISEASE

Amoebiasis occurs throughout all areas of the world, from tropical to subpolar regions. The disease primarily affects people (both children and adults) who are undernourished and living in unsanitary conditions. Although an intestinal illness at first, it can spread to various organ systems. Some 40,000 to 100,000 people die each year from amoebiasis.

The causative agent of amoebiasis is *Entamoeba histolytica*. In nature, the organism exists in the **cyst** form. It enters the body by food or water contaminated with human or animal feces, or by direct contact with feces. Contact with soiled diapers, such as in a day-care center, thus may be hazardous. The organisms pass through the stomach as cysts, and the **trophozoite** amoebas emerge in the distant portion of the small intestine and in the large intestine (**FIGURE 16.9**).

Entamoeba histolytica has the ability to destroy tissue (*histolytica* means "tissue-lysing"). Using their protein-digesting enzymes, the amoebas penetrate the wall of the large intestine, causing lesions and deep ulcers. Patients experience sharp pain similar to that in appendicitis, but relatively little diarrhea or dysentery because the

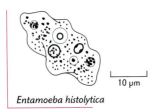

Entamoeba histolytica

10 μm

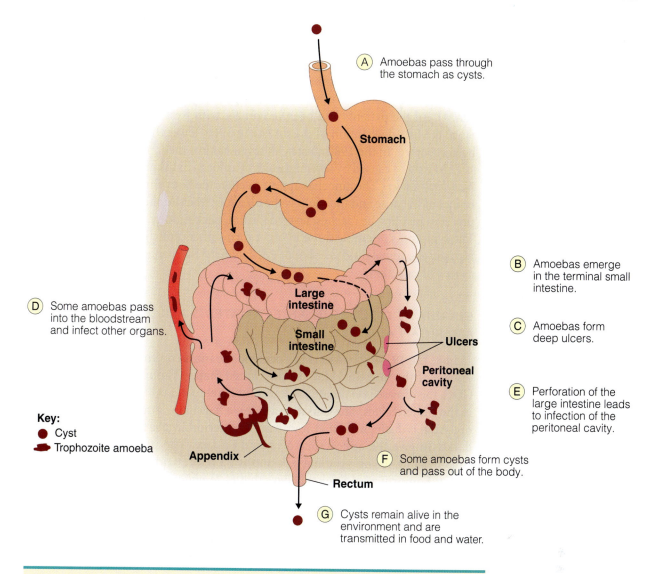

A Amoebas pass through the stomach as cysts.

B Amoebas emerge in the terminal small intestine.

C Amoebas form deep ulcers.

D Some amoebas pass into the bloodstream and infect other organs.

E Perforation of the large intestine leads to infection of the peritoneal cavity.

F Some amoebas form cysts and pass out of the body.

G Cysts remain alive in the environment and are transmitted in food and water.

Key:
● Cyst
◗ Trophozoite amoeba

Stomach

Large intestine

Small intestine

Ulcers

Peritoneal cavity

Appendix

Rectum

FIGURE 16.9

The Course of Amoebiasis Due to *Entamoeba histolytica*

ulcers are separated and do not drastically affect water absorption (the older term **amoebic dysentery** has thus been replaced by amoebiasis). In severe cases, tissue invasion will extend to blood vessels of the intestinal wall, and bloody stools will follow. Ulcer perforation into the peritoneum, the cavity outside the visceral organs, is a possibility. The amoebas also invade the blood and may spread to the liver or lung, where fatal abscesses may develop.

Metronidazole and paromomycin commonly are used to treat amoebiasis, but the drugs do not affect the cysts, and repeated attacks of amoebiasis may occur for months or years. The patient often continues to shed cysts in the feces to infect other people. Amoebiasis was originally a disease of the tropics, but soldiers returning after World War II brought it to the United States, and about 5 percent of Americans are believed to be infected. In recent years, waves of immigrants from Mexico and Caribbean nations have added to the incidence of the disease.

PRIMARY AMOEBIC MENINGOENCEPHALITIS IS A SUDDEN AND DEADLY INFECTION

Naegleria fowleri

10 μm

In the summer of 1980, the Centers for Disease Control and Prevention (CDC) noted an unusual cluster of seven cases of **primary amoebic meningoencephalitis (PAM)**. All the patients had been in contact with freshwater in a tropical setting. Since the first description of the disease in 1965, fewer than three dozen cases had been reported. In the 1980 cluster, all seven victims died.

PAM may be caused by several species of amoeba in the genus *Naegleria*, especially *Naegleria fowleri*. The amoebas appear to enter the body through the mucous membranes of the nose and then follow the olfactory tracts to the brain. Nasal congestion precedes piercing headaches, fever, delirium, neck rigidity, and occasional seizures. The symptoms resemble those in other forms of encephalitis and meningitis. *Naegleria* in the spinal fluid is a sign of PAM. Ninety-five percent of patients die within a week of infection.

GIARDIASIS OCCURS IN AN ESTIMATED 5 PERCENT OF AMERICAN ADULTS

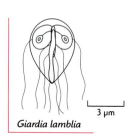

Giardia lamblia

3 μm

Since the 1970s, **giardiasis** has become the most commonly detected protozoal disease of the intestinal tract in the United States. Though not reportable to the CDC, the disease is estimated to cause 100,000 to 2.5 million infections each year in the United States. The causative agent is a flagellate named *Giardia lamblia*. This organism is distinguished by four pairs of anterior flagella and two nuclei that stain darkly to give the appearance of eyes on a face. The protozoan can be divided equally along its longitudinal axis and is therefore said to display **bilateral symmetry**. Some microbiologists believe that *G. lamblia* was described as early as 1681 by Anton van Leeuwenhoek in samples of his stool.

Giardiasis is commonly transmitted by water that contains *Giardia* cysts stemming from cross-contamination of drinking water with sewage. In the 1970s, for example, 38 cases in Aspen, Colorado, broke out after a sewer line was obstructed and sewage leaked into the town's water supply. Recent years also have witnessed outbreaks in day-care centers and schools resulting from contact with feces. In addition, the disease has spread to wild animals (especially beavers), where it is now very common and from which it can be obtained via contaminated water.

Giardia lamblia passes through the stomach as cysts, and the trophozoites emerge as flagellates in the duodenum. They adhere to the intestinal lining using sucker devices and multiply rapidly. The patient feels nauseous, experiences gastric cramps and flatulence, and emits a foul-smelling watery diarrhea that may last for weeks. Some microbiologists maintain that the diarrhea arises from overgrowth of the intestinal wall with parasites, while others suggest injury to the tissue (**FIGURE 16.10**).

Diagnosis of giardiasis depends on the microscopic identification of trophozoites or cysts in freshly passed fecal material. A reliable alternative to stool examination is the use of the **Enterotest capsule**. In this procedure, the patient swallows a weighted gelatin capsule attached to a string. The free end of the string is then taped to the mouth. After four hours the capsule is withdrawn, and the bile-stained mucus is scraped from the capsule and examined for trophozoites. The organisms exhibit an erratic turning motion, similar to a falling leaf.

Treatment of giardiasis may be administered with drugs such as quinacrine (Atabrine), and furazolidine (Furoxone), or metronidazole. However, these drugs

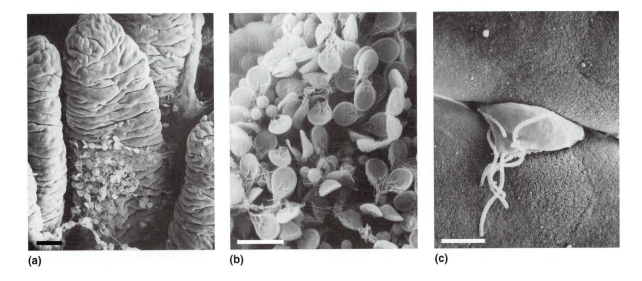

(a)   (b)   (c)

FIGURE 16.10

Giardia lamblia Infection of the Intestine

(a) A mass of trophozoites adhering to the base of a villus in the intestine of an animal. (Bar = 5 µm.)
(b) A view of the *Giardia* population at the base. Note the flat shape of the cells with numerous flagella and a disk-like sucker device for attaching to the tissue. (Bar = 5 µm.) (c) A single *Giardia* trophozoite wedged into the wall of the villus. Three pairs of flagella protrude posteriorly. (Bar = 1 µm.)

have side effects that the physician may wish to avoid by letting the disease run its course without treatment. Those who recover often become **carriers** and excrete the cysts for years. Giardiasis is sometimes mistaken for viral gastroenteritis and is considered a type of **traveler's diarrhea**. MicroInquiry 16 gives you a chance to be a disease detective and solve an outbreak of giardiasis.

TRICHOMONIASIS AFFECTS OVER 10 PERCENT OF SEXUALLY ACTIVE INDIVIDUALS

Trichomoniasis is among the most common diseases in the United States, with an estimated 2.5 million people affected annually. The disease is transmitted primarily by sexual contact and is considered a **sexually transmitted disease**. Fomites (inanimate objects) such as towels and clothing have been implicated in transmission also.

Trichomonas vaginalis, the causative agent, is a pear-shaped protozoan with two pairs of anterior flagella and one posterior flagellum. It thrives in the slightly acidic environment of the human vagina. Establishment may be encouraged by physical or chemical trauma, including poor hygiene, drug therapy, diabetes, or mechanical contraceptive devices such as the intrauterine device (IUD). The organism has no cyst stage.

In females, trichomoniasis is accompanied by intense itching (pruritis), and burning pain during urination. Usually, a creamy white, frothy discharge also is present. The symptoms are frequently worse during menstruation, and erosion of the cervix may occur. In males, the disease occurs primarily in the urethra, with

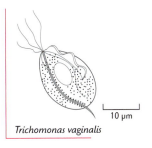

10 µm

Trichomonas vaginalis

MicroInquiry 16

BEING A DISEASE DETECTIVE

An exciting and challenging responsibility of an epidemiologist—or "disease detective"—is to investigate an infectious disease outbreak. Frequently, the cause and source of the outbreak are unknown. The following is an actual case investigated by epidemiologists at the Centers for Disease Control and Prevention (CDC).[1] In investigating an outbreak, speed is important, but getting the right answer is essential, too. To satisfy both requirements, epidemiologists approach investigations systematically by using a somewhat structured set of steps that we will explore more in Chapter 18. For now, see how well you do at figuring out the source for this outbreak of giardiasis. Answers can be found in Appendix E.

In April 1988, the Albuquerque Environmental Health Department and the New Mexico Health and Environment Department received reports of giardiasis that had occurred in members of a church youth group in Albuquerque. The reports indicated the first two members had onset of diarrhea on March 3 and 4,

respectively; stool specimens from both individuals were positive for *Giardia lamblia* cysts. The two individuals had only church youth group activities in common.

The health department investigation found out that the youth group had dinner once a week at the church, where parents prepared the food. Since no record of who attended each meal was kept, a survey of all families attending the church was conducted to identify any family members who had eaten at any youth group dinners in March and any who had experienced a case of diarrhea since February 1, 1988. Of the 148 persons who attended at least one dinner in March, 42 reported diarrheal illness.

Interviews indicated onset of illness occurred from March 3 to March 30, and illness lasted 1 to 32 days (median: 20 days). Importantly, 21 of 108 persons who ate the youth group dinner on March 2 developed an illness meeting the case definition for giardiasis, compared with only one of 40 who did not eat that meal.

The 21 persons who developed illness reported that the most frequent symptoms were fatigue, diarrhea, abdominal cramps, bloating, and weight loss. Patients ranged in age from 11 to 58 years; 14 were female; 15 sought care from a physician. Fourteen patients submitted stool specimens for ova and parasite examination; 10 specimens were positive for *Giardia* cysts. Stool specimens tested for *Shigella, Salmonella, Campylobacter,* and *Yersinia* were negative. The foods served at the dinner on March 2 included tacos (with meat, onions, tomatoes, lettuce, cheese, salsa, sour cream, and tortillas), corn, peaches, cupcakes, soft drinks, coffee, and tea. By the time of the investigation, no food samples remained for microbiologic testing. Persons who became ill reported eating lettuce, salsa, onions, and tomatoes, or drinking tea/coffee. Water consumption was not associated with illness.

All the implicated foods, except for the commercially prepared salsa (pasteurized and canned), were prepared in the church kitchen. The lettuce and

---

pain on urination and a thin, mucoid discharge. The disease can occur concurrently with gonorrhea (**FIGURE 16.11**).

Direct microscopic examination of clinical specimens is the most rapid and least expensive technique for identifying *T. vaginalis*. This is accomplished by making a wet mount preparation of the discharge and observing the quick, jerky motion of the protozoa. The drug of choice for treatment is orally administered metronidazole (Flagyl). Both the patient and the sexual partner should be treated concurrently to prevent transmission or reinfection.

TRYPANOSOMIASIS IS TWO DIFFERENT DISEASES

Trypanosomiasis is a general name for two diseases caused by species of *Trypanosoma*. Protozoa in this genus are elongated flagellates having a characteristic undulating membrane that waves as the organism moves (**FIGURE 16.12**). The two diseases caused by trypanosomes are traditionally known as African sleeping sickness and Chagas's disease.

African sleeping sickness cycles between humans and the **tsetse fly** *Glossina palpalis.* The insect bites an infected patient or animal, and the trypanosomes localize

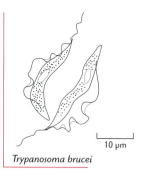

10 μm

Trypanosoma brucei

tomatoes were rinsed at the kitchen's main sink; the outer leaves of the lettuce were removed; and the lettuce, tomatoes, and onions were chopped on the same cutting board, which was not washed between items. The dinner of March 2 was prepared by eight women whose children were in the youth group; all ate the meal. Although the woman who prepared the lettuce and tomatoes taught preschool and had a child in preschool, neither she nor her child was ill when the meal was prepared. None of the eight food preparers reported symptoms at the time of meal preparation;

however, five became ill with diarrhea after March 8. Three had stool specimens positive for *Giardia* cysts.

The church is on the municipal water system. An examination of possible connections between the church's water system and the sanitary sewer system identified five potential sites. However, water samples taken at the time of the examination had adequate chlorine levels and were negative for bacterial contamination. On April 4, after the health department investigation had started, the church stopped using municipal water and began catering meals. After

the examination of the water/sewer systems was conducted, the church flushed every outlet simultaneously for three hours. No new cases occurred after these measures were completed.

16.1 From the information given, determine the most likely vehicle of transmission of *Giardia lamblia* in this outbreak of giardiasis. Explain your reasoning.

[1] Adapted from the Centers for Disease Control *Morbidity and Mortality Weekly Report.* June 16, 1989; 38(23): 405–407.

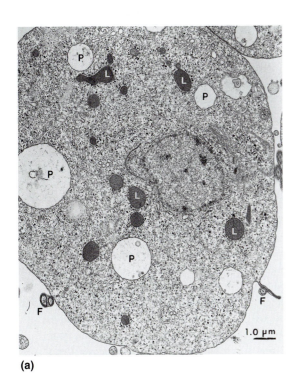

(a)

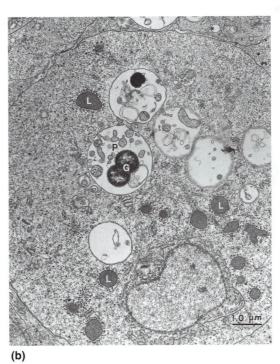

(b)

FIGURE 16.11

Trichomoniasis and Gonorrhea

Transmission electron micrographs illustrating that *Trichomonas* cells have the capacity to phagocytize *Neisseria gonorrhoeae*, the bacterial agent of gonorrhea, when the two organisms are incubated together. (a) Prior to experimental infection with *N. gonorrhoeae*, the protozoan contains many lysosomes (L) as well as relatively empty food vacuoles, or phagosomes (P). (b) After 5 minutes of incubation with *N. gonorrhoeae*, an intact diplococcus (G) is observed within a phagosome, and other phagosomes show evidence of bacterial debris. When trichomoniasis and gonorrhea occur simultaneously, the bacteria apparently serve as a food source for the protozoa.

in the insect's salivary gland. After a two-week development in the fly, transmission occurs during a bite. The point of entry becomes painful and swollen in several days, and a chancre similar to that in syphilis is observed. Invasion of, and multiplication in, the bloodstream follows soon thereafter.

Two types of African sleeping sickness exist. One form, common in Central Africa, is caused by *Trypanosoma brucei* variety *gambiense*. It is accompanied by chronic bouts of fever, as well as severe headaches, changes in sleep patterns and behavior, and a general wasting away. As the trypanosomes invade the brain, the patient slips into a coma (hence the name, "sleeping sickness"). The second form, common in East Africa, is due to *Trypanosoma brucei* variety *rhodesiense*. This disease is more acute, with higher fever and rapid coma and death.

African sleeping sickness exists wherever the tsetse fly is found, and Africa provides the right combination of temperature and moisture for this insect. In 1898, David Bruce identified the trypanosomes in tsetse flies and recommended insect control (MicroFocus 16.2). Bruce's discovery opened the door to the British colonization of Africa. Today the disease is checked by clearing brushlands and treating areas where the insects breed. Patients are treated with a drug known as pentamidine isethionate. Alternative drugs are arsenic-based melarsoprol and difluoromethylornithine, both of which are extremely expensive and toxic.

American trypanosomiasis is caused by *Trypanosoma cruzi*, a trypanosome discovered by Carlos Juan Chagas. In his honor, the disease is called **Chagas' disease**. Triatomid bugs of the genera *Triatoma* and *Rhodnius* are essential to transmission of the trypanosomes. (The insects also are known as reduviid bugs.) The insects are found in the cracked walls of mud and adobe houses. They feed at night and bite where the skin is thin, such as on the lips, face, or forearms. For this reason, they are called "kissing bugs." The insects deliver nitric oxide to the wound via their saliva to keep the blood vessel open while they complete their blood meal.

MicroFocus 16.2

CLOSE, BUT NO CIGAR

In the late 1800s, new trade routes opened in Africa, and travel increased substantially into the hitherto unknown equatorial belt. Among the most adventurous explorers were the British, but they regularly fell ill with the "sleeping sickness." Therefore, in 1894, the British government sent a medical team headed by David Bruce to investigate the disease and find its cause. In several types of animals, the researchers located a new parasite (later named *Trypanosoma brucei* for the team leader), but they could not find it in human victims. Interestingly, they also found the parasite in tsetse flies.

As it turned out, the British were not the only ones interested in sleeping sickness. The Italians also had a medical team in Africa, and a group headed by Aldo Castellani found Bruce's parasite in the nervous system of infected humans. Bruce read Castellani's report and began a search for parasites in human blood, for if they could reach the nervous system, the blood would be the logical route. Sure enough, they were in the blood—in scores of thousands. But, how did they get there? The tsetse flies, of course! All the pieces of the puzzle seemed to fit: The sickness is transmitted among animals by tsetse flies, which also bite humans and inject the parasites into the blood. Passage to the brain follows.

The solution was obvious: Stop the tsetse flies and thereby stop the sleeping sickness. After all, that's what Gorgas and Reed were doing with mosquitoes and yellow fever in the Caribbean islands and Central America. But it was not to be. Yellow fever is primarily a human disease, but sleeping sickness affects numerous wild animals, and the animals could not be kept away from human populations. Moreover, tsetse flies breed everywhere, from waterlogged river banks to arid deserts to savannas and grasslands (mosquitoes breed only where water collects). The effort to stop the disease was valiant, but it was doomed to failure. To this day, sleeping sickness remains a threat to human life in Africa.

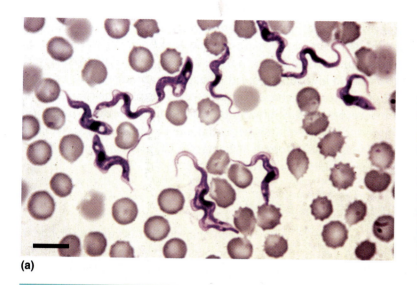

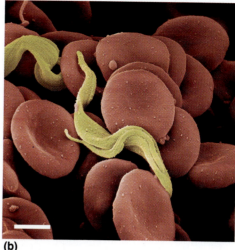

(a)

(b)

FIGURE 16.12

Trypanosoma brucei

(a) Light micrograph of the parasitic protozoan *Trypanosoma brucei*, which causes African sleeping sickness. The stained protozoan cells are seen in a human blood sample. The undulating membrane and flagellum are evident. (Bar = 7 μm.) (b) A false color scanning electron micrograph of *Trypanosoma brucei* (yellow) among red blood cells. The cells of *T. cruzi*, which causes American trypanosomiasis look identical to *T. brucei* cells. (Bar = 5 μm.)

In humans, Chagas' disease is characterized by fever and widespread tissue damage, especially in the heart. The trypanosomes destroy the cardiac nerves so thoroughly that the victim experiences sudden heart failure. Organisms may reach the brain also, where they induce coma and death. A recent estimate put the number of cases in South and Central America at 12 million. Although there is no completely effective treatment for the disease, some relief is experienced with nifurtimox (also known as Bayer 2502). MicroFocus 16.3 explores an avenue to prevention.

MicroFocus 16.3

GENE ENGINEERS TO THE RESCUE

Dealing with Chagas' disease in Latin and South America has always been difficult: Health care workers have plastered the walls of native homes to try and eliminate the triatomid (reduviid) bugs that spread the disease. They have used insecticides at various formulations to kill the intermediary insect. Still, the epidemic continues to rage.

Now, gene engineers have entered the picture. Yale University researchers have isolated *Rhodococcus rhodnii*, a bacterium living harmlessly in the gut of the insect. Within the triatomid bug (genus *Rhodnius*), the bacterial rod lives a peaceful, unassuming existence, deriving nutrients from the insect and contributing a natural defense against invading bacteria. The relationship is an example of mutualism in nature.

Researchers have cultivated *R. rhodnii* and altered its genome so it produces a powerful drug called cecropin A. This peptide compound is extremely active against *Trypanosoma cruzi*, the agent of Chagas' disease. When produced within the insect's gut, the drug kills the parasite and prevents its transmission to the next individual. Researchers next hope to introduce to the environment a number of triatomid bugs with their built-in medical arsenal. In the best-case scenario, the new insects will replace the traditional ones and bring the drug to where it is needed most—the insect gut. The insect will survive, the human will survive, and the most dangerous member of the triumvirate, the protozoal parasite, will be eliminated.

An interesting theory emerged in the 1980s to explain why victims of trypanoso-miasis suffer waves of blood invasion of parasites and accompanying waves of fever. Researchers found that proteins on the membrane surface of the trypanosome were different as each new blood invasion occurred. Thus, the antibodies formed against the preceding parasites were ineffective against the new variants. The change in the trypanosome's surface proteins is apparently due to chromosomal **insertion sequences**, which produce copies of themselves and move along the DNA, changing the genetic codes for membrane proteins (Chapter 7). Charles Darwin was one of the prominent sufferers of the disease (MicroFocus 16.4).

LEISHMANIASIS IS OF TWO FORMS

10 μm

Leishmania donovani

Leishmaniasis is a rare disease in the United States, but it occurs worldwide in large-scale epidemics. (An estimated 12 million people in the tropics and subtrop-ics suffer from the disease.) The responsible protozoa include several species of *Leishmania*, such as *Leishmania donovani* and *L. tropica* (FIGURE 16.13). Transmis-sion is by the **sandfly** of the genus *Phlebotomus*.

One form of leishmaniasis is a **visceral disease** called **kala-azar**, meaning "black fever." This disease is characterized by infection of the body's white blood cells and is accompanied by irregular bouts of fever, swollen spleen and liver, progressive anemia, and emaciation. About 90 percent of cases are fatal, if not treated.

MicroFocus 16.4

CHAGAS AND DARWIN

In 1909, Carlos Chagas was a young Brazilian doctor of 29 when he arrived in a small town north of Rio de Janeiro. Chagas was there to study malaria, but another problem caught his attention: Many of the local people were suffering from lethargy, shortness of breath, and irregular heartbeat. And no one knew what was the cause.

Chagas set aside his interest in malaria and began a search for the parasite of this strange new disease. He quickly tracked down the vector, a cricket-like triatomid bug that lived in the walls of thatched houses and sucked the blood of sleeping inhabitants. From the bug he extracted a whip-like protozoan similar to the trypanosome of African sleeping sickness. In rapid succession, Chagas proved that the trypanosome could infect monkeys; he found it in a cat in a bug-infested house, and he isolated it from the blood of a young girl displaying the symptoms of the disease (now rec-ognized as Chagas' disease).

To be sure, the disease had not been described previously, but neither was it a new disease. Unbeknown to Chagas, it had probably claimed the life of Charles Darwin many years before. Historians record that Darwin's health declined perceptibly on his return to England from South America (during his famous voyage aboard HMS *Beagle*). Some writ-ers maintain that the illness was psycho-somatic—Darwin took a public battering on publication of his theory of natural selection—but Saul Adler, a tropical medicine researcher, believes otherwise. Adler believes that Darwin suffered from Chagas' disease. Indeed, while in Argentina, Darwin wrote: ". . . at night I experienced an attack (for it deserves no less a name) of *Reduvius*, the giant black bug of the Pampas. It is most dis-gusting to feel soft, wingless insects about an inch long, crawling over one's body. Before sucking they are quite thin, but afterwards they become round and bloated with blood."

The bug that Darwin describes is probably *Triatoma infestans*, the vector that Chagas would identify generations later. Furthermore, Darwin's symptoms matched those in Chagas' patients. The disease would linger in Darwin's tissues for 40 years and reduce the vigorous adventurer to a shell of his former self.

■ *Charles Darwin*

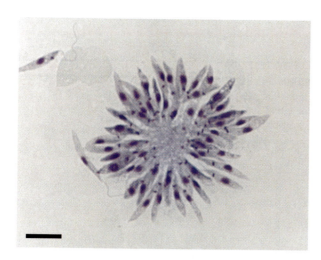

FIGURE 16.13

A Photomicrograph of *Leishmania tropica*

L. tropica is shown in the rosette patterns, in which long, thin flagellates are clustered, giving the appearance of petals on a flower. The rosette pattern is found in older culture media. Some researchers believe that the organisms are in a feeding frenzy at this point, but there is no evidence to support this theory. (Bar = 25 μm.)

A second form of leishmaniasis is a disfiguring, **cutaneous disease**, less severe because the parasites remain in the skin, as shown in **FIGURE 16.14**. This disease often is referred to as an **oriental sore** ("Oriental" because epidemics have been documented in China). It is known also in the Middle East as the "rose of Jericho" disease from the flowery appearance of the skin lesion. Hundreds of American soldiers have been infected during the Iraq conflict with the cutaneous form, which they call the "Bhagdad boil."

Control of the sandfly remains the most important method for preventing outbreaks of leishmaniasis. The antimony compound, stibogluconate, is used to treat established cases of leishmaniasis. At least 7 cases of visceral leishmaniasis and 16 cases of cutaneous leishmaniasis occurred in military personnel associated with Operation Desert Storm, in 1991 (**FIGURE 16.15**). In 1993, visceral leishmaniasis broke out in epidemic proportions in civil war–torn regions of Sudan. In 1997, investigators identified a transposon in the genome of *Leishmania* (appropriately called *mariner*) that moves about, changing the organism's genetic character, and thereby rendering it unaffected by the immune response.

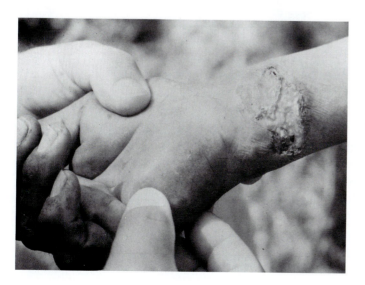

FIGURE 16.14

The Leishmaniasis Sore

A cutaneous sore on the wrist of a man with leishmaniasis. Note the shallow, circular nature of the ulcer.

FIGURE 16.15

An Outbreak of Leishmaniasis

This outbreak occurred among military personnel who fought in Operation Desert Storm during 1991.

1. During 1991, approximately 500,000 military personnel took part in Operation Desert Storm in Saudi Arabia, Kuwait, and other countries of the Persian Gulf region.

2. While stationed in the Middle East during and after the fighting, many individuals were subjected to the bites of sandflies, the arthropods that transfer the protozoan *Leishmania tropica*.

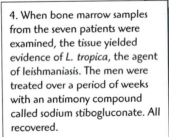

3. On returning to the United States after several months' duty, seven men displayed the symptoms of leishmaniasis, including high fever, chills, malaise, liver and spleen involvement, and gastrointestinal distress. Several had low-volume watery stools and abdominal pain.

4. When bone marrow samples from the seven patients were examined, the tissue yielded evidence of *L. tropica*, the agent of leishmaniasis. The men were treated over a period of weeks with an antimony compound called sodium stibogluconate. All recovered.

To this point . . .

We have considered seven protozoal diseases caused by amoebas or flagellates. Amoebiasis is a disease of the intestinal tract that has become fairly widespread in the United States since it was introduced from tropical areas. The cyst form may be contracted from contaminated food or water or by contact with infected individuals. A similar situation holds for giardiasis. Sewage leaks are another community problem that may lead to giardiasis.

We also studied trichomoniasis, a sexually transmitted disease that is among the most common illnesses in Americans. Women are particularly susceptible to this disease, but *men can be carriers. The discussion then moved to arthropod-borne protozoal diseases, in which species of* Trypanosoma *and* Leishmania *rely on insects to complete their life cycles. The arthropod represents the "weak link" in the transmission of these diseases, and control can be effected by eliminating the arthropod. This is one reason leishmaniasis and sleeping sickness are not prevalent where sanitation methods are established. The diseases are uncommon in the United States also because the necessary arthropods are not usually found here.*

We now will focus our attention on a disease caused by a ciliate and several diseases due to apicomplexans (sporozoa). In this group we shall find rapidly emerging problems in medical microbiology, as well as one illness, malaria, that is the most widespread human disease recognized today.

16.4
Protozoal Diseases Caused by Ciliates & Apicomplexans (Sporozoa)

Only one serious human disease is associated with a ciliated protozoan, but numerous diseases are related to the second group in this section. Moreover, the roster of diseases caused by apicomplexans (sporozoa) continues to grow, partly because of their association with a weakened immune system. Compromised immunity can be related to a disease such as AIDS, as well as to immunosuppressant cancer therapy or treatment following an organ transplant. In these cases, the protozoal disease may present a serious challenge to the patient.

BALANTIDIASIS IS CONTRACTED FROM WATER OR FOOD CONTAMINATED WITH PIG FECES

Balantidiasis is an intestinal disease caused by *Balantidium coli.* The protozoan is among the largest organisms to infect humans. It measures up to 100 μm in length by 70 μm in width, with cilia over its entire surface, and has a large kidney-shaped nucleus as well as a small micronucleus. Trophozoite and cyst stages exist.

Balantidiasis is a rare disease in temperate climates. It is spread by contaminated water or food, especially pork. Cysts pass through the stomach, and the ciliated trophozoites emerge from cysts in the intestines, where they cause mild ulceration. Profuse diarrhea, nausea, and rapid weight loss are characteristic signs of disease. Often the disease is chronic, since cysts remain in the intestinal wall. Metronidazole or paromomycin are used for treatment. Though cases are uncommon in the United States, some concern has been voiced about symptomless carriers returning from tropical regions of the world.

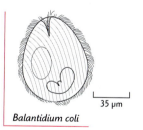

35 μm

Balantidium coli

TOXOPLASMOSIS IS ONE OF THE MOST COMMON INFECTIOUS DISEASES

Toxoplasmosis was first recognized as a clinical disease in 1909 by Charles Nicolle, the French investigator who associated lice with epidemic typhus. Nicolle assumed that toxins were a factor in the disease and named it accordingly, but no evidence

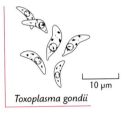

Toxoplasma gondii

exists for this involvement. Instead, the symptoms arise from tissue damage caused by the growth of protozoa.

The cause of toxoplasmosis is *Toxoplasma gondii*. *T. gondii* exists in three forms: the trophozoite, the cyst, and the oocyst. Trophozoites are crescent-shaped or oval organisms without evidence of locomotor organelles (FIGURE 16.16a). Located in tissue during the acute stage of disease, they force their way into all mammalian cells (FIGURE 16.16b), with the notable exception of erythrocytes. To enter cells, the parasites form a ring-shaped structure of the host cell membrane, then pull the membrane over themselves, much like pulling a sock over the foot. Cysts develop from the trophozoites within host cells and may be the source of repeated infections. Muscle and nerve tissue are common sites of cysts. **Oocysts** are oval bodies that develop from the cysts by a complex series of asexual and sexual reproductive processes.

Toxoplasma gondii exists in nature in the cyst and oocyst forms. Grazing animals acquire these forms from the soil and pass them to humans via contaminated beef, pork, or lamb (FIGURE 16.17). Rare hamburger meat is a possible source. Domestic cats acquire the cysts from the soil or from infected birds or rodents. Oocysts then form in the cat. Humans are exposed to the oocysts when they forget to wash their hands after contacting cat feces while changing the cat litter or working in the garden. Touching the cat also can bring oocysts to the hands, and contaminated utensils, towels, or clothing can contact the mouth and transfer oocysts.

Toxoplasmosis develops after trophozoites are released from the cysts or oocysts in the host's gastrointestinal tract. *T. gondii* invades the intestinal lining and spreads throughout the body via the blood. Patients develop fever, malaise, sore throat, and swelling of the spleen, liver, and lymph nodes. In these respects the disease resembles

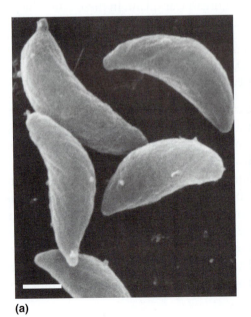

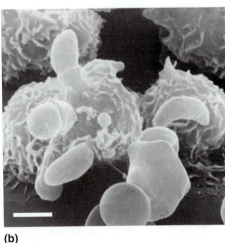

(a)   (b)

FIGURE 16.16

Toxoplasma gondii, the Cause of Toxoplasmosis

(a) A scanning electron micrograph of numerous parasites in the crescent-shaped trophozoite stage. (Bar = 2 μm.) (b) A white blood cell simultaneously attacked by several trophozoites. There are at least three sites where invasion of the white blood cells is taking place. (Bar = 5 μm.)

infectious mononucleosis. Lesions also may occur on the retina of the eye, and virtually any organ may be involved. Complications, however, are rare. Diagnosis may be made by isolating trophozoites from the blood or other body fluids. Sulfonamide drugs are commonly used in therapy.

Pregnant women are at risk of developing toxoplasmosis because the protozoa may cross the placenta and infect the fetal tissues, as Figure 16.17 illustrates. Neurological

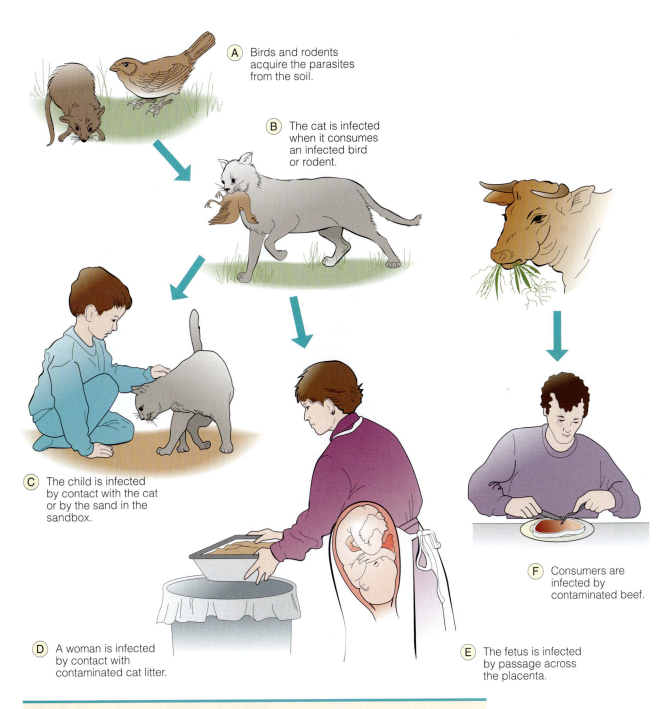

(A) Birds and rodents acquire the parasites from the soil.

(B) The cat is infected when it consumes an infected bird or rodent.

(C) The child is infected by contact with the cat or by the sand in the sandbox.

(D) A woman is infected by contact with contaminated cat litter.

(E) The fetus is infected by passage across the placenta.

(F) Consumers are infected by contaminated beef.

FIGURE 16.17

The Cycle of Toxoplasmosis in Nature

damage, lesions of the fetal visceral organs, or spontaneous abortion may result. **Congenital infection** is least likely during the first trimester, but damage may be substantial when it occurs. By contrast, congenital infection is more common if the woman is infected in the third trimester, but fetal damage is less severe. Lesions of the retina are the most widely documented complication in congenital infections. The T in the **TORCH** group of diseases refers to toxoplasmosis (the others are rubella, cytomegalovirus, and herpes simplex; O is for other diseases, such as syphilis).

Toxoplasma gondii also is known to cause severe disease in immunosuppressed individuals. For example, in patients with AIDS, the normal immune defenses that prevent the spread of disease have been destroyed, and *T. gondii* attacks the brain tissue. The infected tissue attracts immune system components, and the resulting inflammation and swelling often result in cerebral lesions, seizures, and death. Often the disease stems from opportunistic parasites already in the body. Patients receiving immunosuppressant therapy for cancer, or to prevent organ transplant rejection, are similarly at risk.

Toxoplasma gondii is regarded as a universal parasite. Some researchers suggest that it is the most common parasite of humans and other vertebrates. It is important to ranchers, dairy product producers, pet breeders, and anyone who comes in contact with a domestic cat. Indeed, evidence indicates that where there are no cats, toxoplasmosis is rare. Hundreds of millions of humans worldwide are estimated to be infected.

MALARIA IS CAUSED BY FOUR DIFFERENT PROTOZOAL SPECIES

Though the disease malaria has been known to exist since at least 1000 b.c., the word *malaria* has been used only since the 1700s. Before that time, the disease was known as ague, from the French *aigu*, meaning "sharp" (a reference to the sharp fever that accompanies the disease). During the 1700s, however, Europeans began using the Italian word for "bad air" (*mal-aria*), reflecting the theory that the disease was somehow related to an unknown atmospheric influence called miasma. Wave after wave of malaria swept over the world during that century, and few regions were left untouched. American pioneers settling in the Mississippi and Ohio valleys suffered great losses from the disease.

Between 300 and 500 million of the world's population now suffer the chills, fever, and life-threatening effects of **malaria**. The disease exacts its greatest toll in Africa, where the WHO estimates that over 1 million children under the age of 5 die from malaria annually; that is equivalent to one child dying every 30 seconds! No infectious disease of contemporary times can claim such a dubious distinction. Though progress has been made in the control of malaria, the figures remain appallingly high. (Even the United States is involved in the malaria pandemic—over 1,000 cases are diagnosed annually.) MicroFocus 16.5 looks at a program to reduce the malaria burden.

Malaria is caused by four species of *Plasmodium: P. vivax, P. ovale, P. malariae,* and *P. falciparum.* All are transmitted by the female *Anopheles* mosquito, which consumes human blood to provide chemical components for her eggs. The life cycle of the parasites has three important stages: the sporozoite, the merozoite, and the gametocyte. Each is a factor in malaria.

The mosquito sucks human blood from a person with malaria and acquires **gametocytes**, the form of the protozoan found in red blood cells (**FIGURE 16.18**). Within the insect a transition to sporozoites takes place, and the **sporozoites** then migrate to the salivary gland. When the mosquito bites another human, several hundred sporozoites enter the person's bloodstream and quickly migrate to the liver. After several hours, the transformation of one sporozoite to 25,000 **merozoites** has

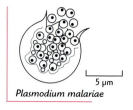

5 μm

Plasmodium malariae

Merozoite:
an intermediate stage in the life cycle of the malaria parasite *Plasmodium.*

MicroFocus 16.5

ROLL BACK MALARIA

The World Health Organization (WHO) estimates there are over 300 million cases of malaria each year, resulting in more than a million deaths. Over 90 percent of these deaths occur in sub-Saharan Africa where most victims are under five years of age.

The United Nations General Assembly has designated 2001–2010 the decade to "Roll Back Malaria" in developing nations, especially in Africa. Roll Back Malaria is a global partnership designed to cut by 50 percent the world's malaria burden by 2010. The program will enable countries to take effective and sustained action against malaria by providing their citizens with rapid and effective treatment and attempting to prevent and control malaria in pregnant females.

Since over 70 percent of all malarial deaths occur in children under five years of age, a child's most vulnerable period for contracting malaria starts at six months, when the mother's protective immunity wears off and before the infant has established its own fully functional immune system. During this window of vulnerability, a child's condition can deteriorate quickly and the child can die within 48 hours after the first symptoms appear. Therefore, a key part of Roll Back Malaria is aimed at children and a series of interim goals have been put forward. They propose that by 2005:

■ At least 60 percent of those suffering from malaria should have access to inexpensive and correct treatment within 24 hours of the onset of symptoms.

■ At least 60 percent of those at risk of malaria, particularly pregnant women and children less than five years old, should have access to suitable personal and community protective measures, such as insecticide-treated mosquito nets.

■ At least 60 percent of all pregnant women who are at risk of malaria, especially those in their first pregnancies, should receive intermittent preventive treatment.

Through worldwide partnerships, the Roll Back Malaria program will allow countries that experience a high malaria burden to take effective and sustainable action to halve the malaria burden by 2010.

been completed, and the merozoites emerge from the liver to invade the red blood cells. While in the red blood cells (RBCs), the merozoites can synthesize any of about 150 proteins that attach to RBC membranes and anchor the RBCs to the blood vessels. By constantly switching among 150 genes (for 150 proteins), the malarial parasite avoids detection by the body's immune system.

Within the human red blood cells, the merozoites undergo another series of transformations that result in several gametocytes and thousands of new merozoites. In response to a biochemical signal, thousands of RBCs rupture simultaneously, thereby releasing the parasites and their toxins. Now the excruciating **malaria attack** begins. First, there is intense cold, with shivers and chattering teeth. The temperature then rises rapidly to 40°C, and the sufferer develops intense fever, headache, and delirium. After two or three hours, massive perspiration ends the hot stage, and the patient often falls asleep, exhausted.

During this quiet period, the merozoites enter a new set of red blood cells and repeat the cycle of transformations. *P. vivax* and *P. ovale* spend about 48 hours in the red blood cells, so that 48 hours pass between malaria attacks. This is **tertian malaria** (*tertian* is from the Latin for "three-day," based on the Roman custom of calling the day the event happened the first day of the cycle). For *P. malariae*, the cycle takes 72 hours, and the disease is called **quartan malaria** ("four-day" malaria). The cycle of *P. falciparum* is not defined, and attacks may occur at widely scattered intervals. This type of malaria, the most lethal, is known as **estivo-autumnal malaria**, referring to the summer-fall periods when mosquitoes breed heavily.

Death from malaria may be due to a number of factors related to the loss of red blood cells. Substantial anemia develops, and the hemoglobin from ruptured blood cells enters the urine; malaria is, therefore, sometimes called **blackwater fever**. Cell

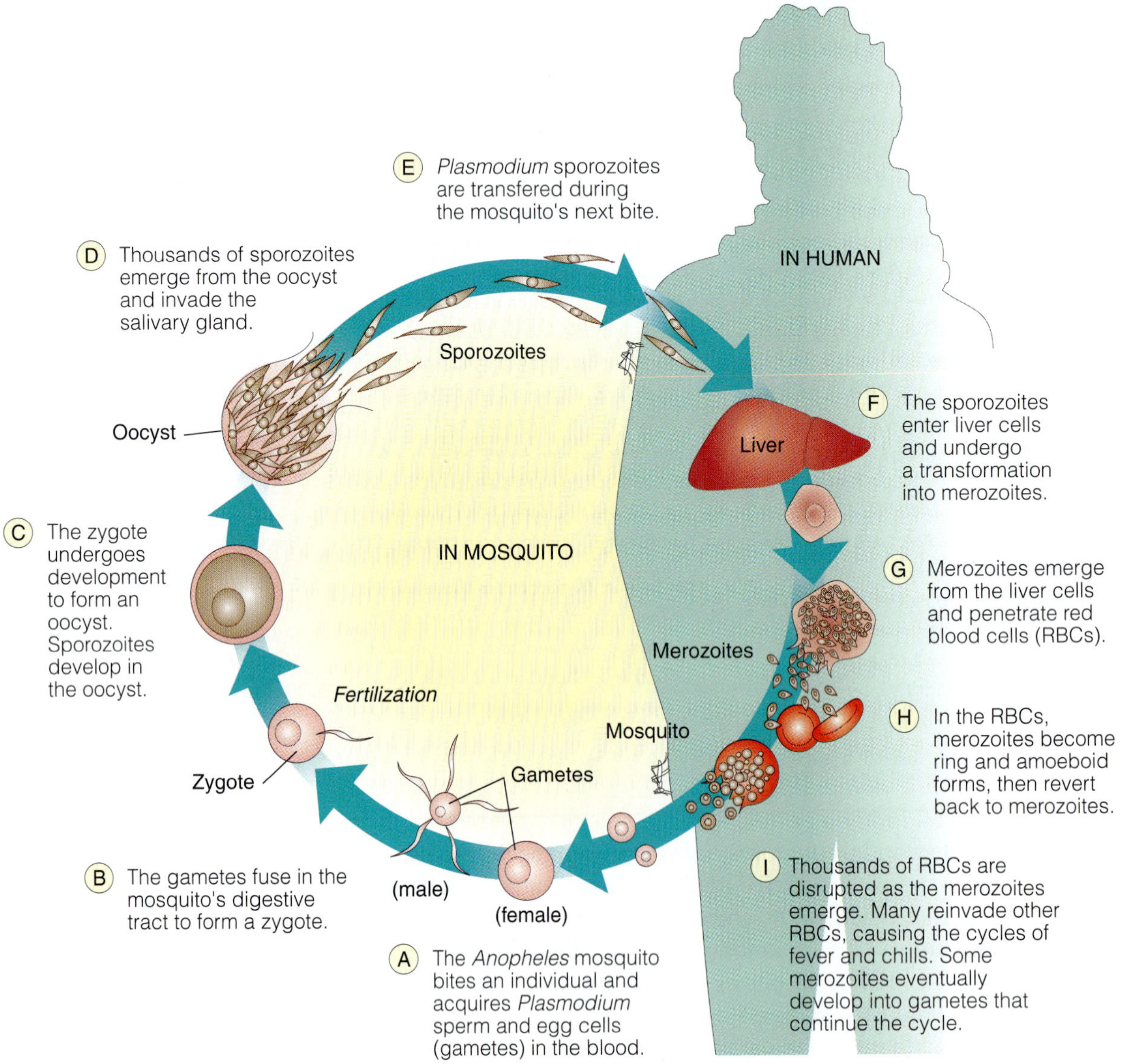

E   *Plasmodium* sporozoites are transfered during the mosquito's next bite.

D   Thousands of sporozoites emerge from the oocyst and invade the salivary gland.

Sporozoites

IN HUMAN

Oocyst

Liver

F   The sporozoites enter liver cells and undergo a transformation into merozoites.

C   The zygote undergoes development to form an oocyst. Sporozoites develop in the oocyst.

IN MOSQUITO

G   Merozoites emerge from the liver cells and penetrate red blood cells (RBCs).

Merozoites

Fertilization

Mosquito

H   In the RBCs, merozoites become ring and amoeboid forms, then revert back to merozoites.

Zygote

Gametes

B   The gametes fuse in the mosquito's digestive tract to form a zygote.

(male)

(female)

I   Thousands of RBCs are disrupted as the merozoites emerge. Many reinvade other RBCs, causing the cycles of fever and chills. Some merozoites eventually develop into gametes that continue the cycle.

A   The *Anopheles* mosquito bites an individual and acquires *Plasmodium* sperm and egg cells (gametes) in the blood.

FIGURE 16.18

The Malaria Cycle

Malaria continues to be among the most widespread infectious diseases in the world.

fragments accumulate in the small vessels of the brain, kidneys, heart, liver, and other vital organs and cause clots to form. Heart attacks, cerebral hemorrhages, and kidney failure are common. The intense fever leads to convulsions.

Since its discovery about 1640, **quinine** has been the mainstay for treating malaria. When the trees used as a source of this drug fell into the hands of the Japanese in World War II, American researchers developed chloroquine for the active stage of malaria and primaquine for the dormant stage. Chloroquine remained an important mode of therapy until recent years, when drug resistance began emerging in *Plasmodium* species. Since 1989, an alternative drug called **mefloquine** has been recommended for individuals entering malaria regions of the world. However, serious medical side effects, including cognitive functioning problems, have been associated with some people taking the drug. Another drug called proguanil is recommended if the person cannot tolerate mefloquine. Since 1991, the CDC also has recommended using quinidine gluconate, a quinine derivative, for treating complicated *P. falciparum* infections. In clinical trials is another drug, Artemether (Artenam), which appears to be as effective as quinine for treating malaria. Prospects for a malaria vaccine emerged in the 1990s (**MicroFocus 16.6**).

BABESIOSIS IS FOUND IN THE NORTHEASTERN UNITED STATES

Babesiosis is a malaria-like disease caused by *Babesia microti*. The protozoa live in **ticks** of the genus *Ixodes*, and are transmitted when these arthropods feed in human skin. Areas of coastal Massachusetts, Connecticut, and Long Island, New York, have experienced outbreaks in recent years.

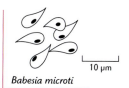

Babesia microti

10 µm

MicroFocus 16.6

MALARIA VACCINE CANDIDATES

Early in the 1960s, international health authorities thought they had malaria licked. Drugs such as quinine and chloroquine, and insecticides such as DDT, had reduced dramatically the incidence of malaria, and some officials were bold enough to speak of eradication. But then reality dawned: *Plasmodium falciparum*, the most lethal of the malaria species, showed signs of resistance to chloroquine; mosquitoes with DDT resistance began emerging; developing nations became complacent about malaria and relaxed their vigilance; and the campaign to control the disease in remote parts of the globe failed.

In 1967, Ruth Nussenzweig and her coworkers at New York University showed that irradiated sporozoites injected into mice provoke an immune response. A vaccine with whole sporozoites was not feasible, though, because sporozoites were impossible to obtain in the massive quantities needed for a vaccine. They needed the immune-stimulating antigen on the sporozoite cell surface. By 1984, they located the antigen and in 1985, the gene that codes for the antigen was identified.

Meanwhile, in Colombia a research group led by Manuel Elkin Patarroyo developed a vaccine containing four different and synthetic protein antigens derived from *P. falciparum*. In March 1993, the Colombian researchers announced that the vaccine brought a 39 percent reduction of malaria cases. Later that year, a second trial resulted in 68 percent fewer cases among vaccine recipients. The scientific community noted that reductions of 39 percent and 68 percent are not extraordinary for a vaccine; nevertheless, when over 1 million people die of malaria annually, these reductions constitute significant numbers. The vaccine's true efficacy remains to be determined.

Research into a malaria vaccine continues today. There are three general categories of malaria vaccine candidates being tested, each targeting a different stage of the plasmodium parasite. Sporozoite vaccines are directed against the sporozoite and liver stages of the parasite infection. The sporozoites are the form transferred into humans by the bite of an infected mosquito. An effective sporozoite vaccine would produce antibodies that prevent infection by blocking invasion of liver cells or activate populations of T cells that would destroy infected liver cells, preventing release of merzoites into the bloodstream.

Merozoite-specific vaccines would interfere with the blood stage of infection when merozoites invade and replicate in red blood cells. Such a vaccine would reduce both the severity and duration of the disease by reducing the number of merozoites released into the blood.

A third group of vaccines would produce human antibodies against the gametes stage of the parasite that mates in the mosquito gut. The idea is that antibodies produced against the gametes, which develop in the human, would be taken up into the mosquito during a blood meal. In the mosquito, the antibodies would block gamete fertilization and subsequent parasite development in the mosquito would not occur.

Most investigators now acknowledge that a combination vaccine (multiantigen, multistage) probably will be the best approach. The various malaria vaccine candidate antigens are being produced in different ways: as recombinant proteins, as DNA vaccines, and as protein spikes on viruses.

Also of note is that the *P. falciparum* genome was sequenced in 2002, so functional genomics may discover an "Achilles heel" that would be a target for other types of vaccines.

Babesia microti penetrates human red blood cells. As the cells disintegrate, a mild anemia develops. Piercing headaches accompany the disease and, occasionally, meningitis occurs. A suppressed immune system appears to favor establishment of the disease. However, babesiosis is rarely fatal, and drug therapy is not recommended. Carrier conditions may develop in recoverers, and spread by blood transfusion is possible. Travelers returning from areas of high incidence therefore are advised to wait several weeks before donating blood to blood banks. Tick control is considered the best method of prevention.

Babesia has a significant place in the history of American microbiology because in the late 1800s, Theobald Smith located *B. bigemina* in the blood of cattle suffering from Texas fever. His report was one of the first linking protozoa to disease, and, in part, it necessitated that the then-prevalent "bacterial" theory of disease be modified to the "germ" theory of disease (Chapter 1).

CRYPTOSPORIDIOSIS IS CAUSED BY A HIGHLY INFECTIOUS PROTOZOAN

Before 1976, **cryptosporidiosis** was recognized as a cause of diarrhea in animals but was unknown in humans. Since that year, however, microbiologists have observed the disease in humans, and the number of cases has risen markedly.

Cryptosporidiosis is caused by *Cryptosporidium parvum* and other species of *Cryptosporidium* such as *C. coccidi*. The organism is similar to *Toxoplasma gondii* in that it has a complex life cycle involving trophozoite, sexual, and oocyst stages (**FIGURE 16.19**). Patients with competent immune systems appear to suffer limited diarrhea that lasts one or two weeks and does not require hospitalization. However, individuals with suppressed immune systems, such as AIDS patients, experience cholera-like, profuse diarrhea that is severe and often irreversible. These patients undergo dehydration and emaciation, and often die of the disease.

Cryptosporidiosis has an incubation period of about one week, which explains why few cases are diagnosed properly because most people relate their nausea and diarrhea to something they ate a day or two before. In healthy adults, the infection usually lasts about 10 days, but the diarrhea may remain for up to two months. A fecal-oral route is the major mode of transmission (as noted in the chapter opening), but physical contact also can transmit *Cryptosporidium* (children in day-care centers are at risk). Experiments indicate that the **dose level** to establish infection is very low (as low as 30

Dose:
the number of parasites needed to establish an infection in the body.

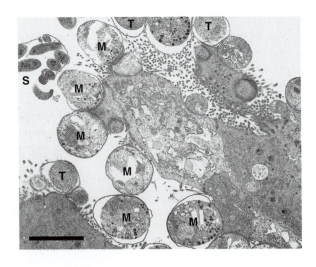

FIGURE 16.19

Cryptosporidium coccidi

A transmission electron micrograph of the intestinal lining of an animal infected with *Cryptosporidium*. Many aspects of the complex life cycle are visible, including the trophozoite stage (T), and the schizont stage (S) releasing eight merozoites. The macrogametes (M) are large cells that fuse in the sexual reproductive cycle of the organism. (Bar = 2 μm.)

organisms), and no drug is widely accepted to treat the disease at this writing, although nitazoxanide has shown promise. High-level infections can be detected in patients by performing the **acid-fast test** (Chapter 3) on stool specimens and noting the presence of acid-fast oocysts.

CYCLOSPORIASIS DEVELOPS AFTER EATING CONTAMINATED FRESH PRODUCE

Beginning in 1996 and continuing through 1997 and 1998, public health officials noted a series of clusters of intestinal disease related to raspberries imported from Guatemala. In all cases, the outbreaks were related to the protozoan *Cyclospora cayetanensis*, shown in **FIGURE 16.20**.

In Americans, *C. cayetanensis* was first observed in 1986 in travelers returning from Mexico and Haiti. The organism is a coccidian parasite, previously found only in reptiles and limited species of mammals. The organisms produce oocysts, each containing two sporocysts; they appear microscopically as spheres 8 to 10 μm in diameter, with a cluster of membrane-enclosed globules. In this regard, they are similar to, but larger than, the oocysts of another coccidian parasite, *Cryptosporidium*. Differential diagnosis is important because *C. cayetanensis* responds to the drug combination of trimethoprim-sulfamethoxazole, whereas *Cryptosporidium* does not.

Cyclosporiasis has an unusually long incubation period of one week. This lengthy period can be a clue to identifying the disease (but it also impairs the ability of people to recall how they were exposed, and the responsible food may have been discarded so it cannot be tested). Symptoms of the disease include watery diarrhea, nausea, abdominal cramping, bloating, and vomiting. Treatment is successful with the drugs noted above, but the symptoms often return. Moreover, the symptoms often remain for over one month during the first illness.

FIGURE 16.20

The Source and Cause of Cyclosporiasis

(a) During 1996, 1997, and 1998, raspberries imported from Guatemala were identified as a source of cyclosporiasis, an intestinal disease caused by an apicomplexan protozoan. (b) The agent of cyclosporiasis is *Cyclospora cayetanensis*, shown in the photomicrograph in the oocyst stage after acid-fast staining. (Bar = 10 μm.)

(a)

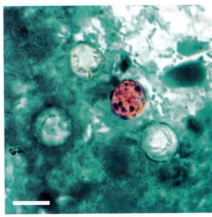

(b)

TABLE 16.1

A Summary of Protozoal Diseases in Humans

ORGANISM	TYPE	ORGANS OF MOTION	DISEASE	TRANSMISSION
Entamoeba histolytica	Amoeba	Pseudopodia	Amoebiasis	Water Food
Naegleria fowleri	Amoeba	Pseudopodia	Primary amoebic meningoencephalitis	Water
Giardia lamblia	Flagellate	Flagella	Giardiasis	Water Contact
Trichomonas vaginalis	Flagellate	Flagella	Trichomoniasis	Sexual contact
Trypanosoma brucei	Flagellate	Flagella	African sleeping sickness	Tsetse fly (Glossina)
Trypanosoma cruzi	Flagellate	Flagella	South American sleeping sickness	Triatomid bug (Triatoma)
Leishmania donovani	Flagellate	Flagella	Leishmaniasis (Kala-azar)	Sandfly (Phlebotomus)
Balantidium coli	Ciliate	Cilia	Balantidiasis	Food Water
Toxoplasma gondii	Apicomplexan	None in adult	Toxoplasmosis	Domestic cats Food
Plasmodium species	Apicomplexan	None in adult	Malaria	Mosquito (Anopheles)
Babesia microti	Apicomplexan	None in adult	Babesiosis	Tick (Ixodes)
Cryptosporidium parvum	Apicomplexan	None in adult	Cryptosporidiosis	Water, food
Cyclospora cayetanensis	Apicomplexan	None in adult	Cyclosporiasis	Food

How raspberries become contaminated during outbreaks is not clear. Public health officials believe that tainted water used for washing the berries may be the source, the berries may be handled by someone whose hands are contaminated, or they may be contaminated during shipping. Regardless of the source, the CDC and FDA have insisted that Guatemalan growers and exporters of raspberries introduce control measures focusing on improved water quality and better sanitation methods on local farms. Transmission from person to person is unlikely. An earlier link of cyclosporiasis to strawberries (MicroFocus 16.7) proved erroneous.

TABLE 16.1 summarizes the protozoal diseases of humans.

ORGAN AFFECTED	DIAGNOSIS	TREATMENT	COMMENT
Intestine Liver Lungs	Stool examination	Paromomycin Metronidazole	Deep intestinal ulcers
Brain	Spinal fluid examination	None effective	Uncommon in US
Intestine	Stool examination Gelatin capsule	Quinacrine Furazolidine	Incidence increasing in US
Urogenital organs	Urine or swab examination	Metronidazole Tinidazole	May result in sterility
Blood Brain	Blood smear	Various drugs	Two types, depending on region
Blood Brain Heart	Blood smear	Various drugs	Common in South America
White blood cells Skin Intestine	Tissue examination	Antimony	Ulcers yield skin disfiguration
Intestine	Stool examination	Paromomycin Metronidazole	Symptomless carriers common
Blood Eyes Tissue cells	Blood examination	Sulfonamide drugs	Congenital damage possible Associated with AIDS
Liver Red blood cells	Blood examination	Quinine Mefloquine Proguanil	World's most urgent public health problem
Red blood cells	Blood examination	None recommended	Carrier state possible
Intestine	Stool examination	None effective	Associated with AIDS
Intestine	Stool examination	Trimethoprim Sulfamethoxazole	Long incubation period

MicroFocus 16.7

THE BEST NEWS SINCE SHORTCAKE

Strawberry growers and distributors were about to take a hit. A serious outbreak of diarrheal illness was being blamed on their livelihood, and they were shaking in their boots. The outbreak was developing in Houston, Texas, where 62 cases were confirmed. All indications pointed to the strawberries. Already the government memos were being prepared: People should avoid California strawberries. Clearly, 1996 was not shaping up as a banner year.

Then news of another outbreak reached them, this time in Charleston, South Carolina. Intestinal illness had developed after a luncheon attended by 64 people; 37 of the 64 were sick with abdominal cramping, vomiting, and diarrhea. For dessert, the management had served strawberries.

However, they learned that the management also had served raspberries with the same dessert. And there was another luncheon at the same restaurant that same day. Ninety-five people had attended that second luncheon, and nobody got sick! The only fruit the management served was strawberries.

Hmmm. Maybe it wasn't the strawberries after all. Maybe it was the raspberries? The ensuing weeks would be considerably brighter for the strawberry growers, but nightmarish for the raspberry producers.

Cyclospora-related illness broke out in New York City, New Jersey, Toronto, and a host of other locales. In each case, raspberries were implicated, but strawberries were nowhere to be seen. The strawberry industry had dodged the bullet. Whew! Bring on the shortcake.

On an October day in 1983, a local medical laboratory notified the Los Angeles Department of Health Services of a large number of stool samples containing *Entamoeba histolytica*, the protozoan of intestinal amoebiasis. The laboratory also had reported 38 cases of amoebiasis during the previous two months. Department officials were mystified because there had been no increase in the number of specimens the laboratory examined, no clustering of cases of amoebiasis, few patients in high-risk categories (tourists, immigrants, or institutionalized patients), and no instances of increased reporting from other laboratories. When Health Department investigators reexamined 71 slides of fecal material from the 38 cases, they found only 4 with *Entamoeba histolytica*. Officials concluded that the laboratory was in error, and that the bodies thought to be amoebas were in fact white blood cells.

Diagnostic methods for protozoal diseases have not changed fundamentally for many generations, and the sophisticated devices used for bacterial and viral detection have not yet reached the parasitology laboratory. Cultivation methods for pathogenic protozoa are difficult, and a correct diagnosis often depends on direct observation of tissue specimens, together with a sense of intuition and understanding of the patterns of disease. In the Los Angeles case described above, these were apparently lacking.

With the increasing prevalence of protozoal diseases in recent years, parasitologists have strengthened their role in the health care delivery system. Giardiasis has become a well-known intestinal disease, and trichomoniasis remains among the most common sexually transmitted diseases. Diseases such as cryptosporidiosis and toxoplasmosis were relatively new in the 1990s, and increasing coverage in media reports is testimony to their importance. Even malaria is still encountered in travelers and immigrant groups.

It takes years of training and experience to be a good parasitologist. A keen and discriminating eye is essential, and the right questions must be asked to distinguish between organisms and tissue debris. If your future goals are not yet established, you might wish to consider parasitology as a career.

Summary of Key Concepts

16.1 CHARACTERISTICS OF PROTOZOA

- **Protozoa Contain Typical Eukaryotic Organelles.** Protozoa are of interest to microbiologists because they are unicellular, have a microscopic size, and cause infectious disease. They usually are much larger than prokaryotic organisms, and the functions of their cells bear a resemblance to the functions of multicellular organisms.

- **Most Protozoa Are Heterotrophs.** The majority of protozoa are heterotrophic. Some exist in the trophozoite and cyst forms, the latter a very resistant form. Protozoa usually reproduce asexually by binary fission, although some also have a sexual stage.

16.2 THE CLASSIFICATION OF PROTOZOA

- **Amoebas Form Pseudopods.** Protozoa such as amoebas move by means of pseudopods. They usually live in freshwater and reproduce by binary fission. The radiolarians and foraminiferans also are marine amoebas with hard glassy or chalky skeletons.

- **Flagellates Move by One or More Flagella.** This group of protozoa moves by means of one or more flagella that are composed of microtubules. The flagellate *Euglena* is one of the few protozoa that contain chloroplasts.

- **Ciliates Have Two Types of Nuclei.** Ciliates, like *Paramecium*, have their cell surfaces covered by cilia. The cells also contain two different types of nuclei—macronuclei, which control metabolic events and micronuclei, which play a critical role in genetic recombination through conjugation.

- **Apicomplexans (Sporozoans) Are Obligate Parasites.** Nonmotile protozoa are termed apicomplexans, or sporozoa. All members of the group are obligate parasites.

16.3 PROTOZOAL DISEASES CAUSED BY AMOEBAS AND FLAGELLATES

- **Amoebiasis Is the Second Leading Cause of Death from Parasitic Disease.** The most serious amoeba-related disease is amoebiasis, due to *Entamoeba histolytica*. Intestinal ulcers and sharp, appendicitis-like pain accompany the disease.

- **Primary Amoebic Meningoencephalitis Is a Sudden and Deadly Infection.** The amoeba *Naegleria fowleri* enters the nose and follows the olfactory tracts to the brain. Most patients die within seven days on the infection.

- **Giardiasis Occurs in an Estimated 5 Percent of American Adults.** Giardiasis is an intestinal disease, but the protozoa do not penetrate the tissue. The agent, a flagellate called *Giardia lamblia*, is acquired from contaminated food and water.

- **Trichomoniasis Affects Over 10 Percent of Sexually Active Individuals.** Trichomoniasis, a sexually transmitted disease caused by the flagellate *Trichomonas vaginalis*, is among the most common diseases in the United States.

- **Trypanosomiasis Is Two Different Diseases.** Arthropod vectors play a role in this protozoal disease. African sleeping sickness (African trypanosomiasis) is transmitted by tsetse flies. *Trypanosoma brucei* variety *gambiense* is prevalent in Central Africa while *Trypanosoma brucei* variety *rhodesiense* is common in East Africa. Blood invasion and tissue involvement characterize the disease. Chagas' disease (American trypanosomiasis) is characterized by fever and heart tissue damage. The protozoan, *Trypanosoma cruzi*, is transmitted by the triatomid bugs.

- **Leishmaniasis Is of Two Forms.** Leishmaniasis is transmitted by sandflies and involves blood (visceral leishmaniasis; kala-azar) and tissue (cutaneous leishmaniasis) invasion. About 90 percent of cases involving the visceral form are fatal if not treated.

16.4 PROTOZOAL DISEASES CAUSED BY CILIATES AND APICOMPLEXANS (SPOROZOA)

- **Balantidiasis Is Contracted from Water or Food Contaminated with Pig Feces.** *Balantidium coli* is known as an intestinal parasite acquired by a fecal-oral route. Infections usually occur in tropical regions of the world.

- **Toxoplasmosis Is One of the Most Common Infectious Diseases.** An apicomplexan, *Toxoplasma gondii*, causes serious disease in AIDS patients. *Toxoplasma* affects various organs, especially those of the nervous system. Infections in pregnant women can lead to congenital infections of the fetus.

- **Malaria Is Caused by Four Different Protozoal Species.** Another nonmotile protozoan, *Plasmodium*, is the cause of malaria, one of the most serious global health problems. Transmitted by mosquitoes, *Plasmodium* species destroy the red blood cells and often bring on death.

- **Babesiosis Is Found in the Northeastern United States.** A similar disease to malaria is called babesiosis. The protozoan, *Babesia microti*, is transmitted by ticks. The disease is rarely fatal and remains more localized in occurrence.

- **Cryptosporidiosis Is Caused by a Highly Infectious Protozoan.** *Cryptosporidium* causes a cholera-like diarrhea. The protozoan, *Cryptosporidium parvum*, causes a more serious disease in people with suppressed immune systems.

- **Cyclosporiasis Develops after Eating Contaminated Fresh Produce.** Cyclosporiasis is an intestinal disease contracted from eating fresh produce contaminated with *Cyclospora cayetanensis*. Symptoms include watery diarrhea, nausea, abdominal cramps, bloating, and vomiting.

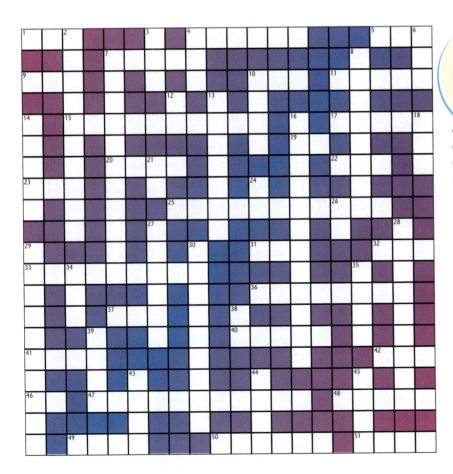

Review

A major theme of this chapter has been the protozoal diseases that affect humans. To review the chapter contents, solve the following crossword puzzle. The answers to the puzzle are in Appendix D.

■ ACROSS

1. Pet that is a reservoir for *Toxoplasma gondii*.
4. One of the species that causes malaria (abbr).
5. *Trichomonas* has _____ posterior flagellum.
7. Discovered the agent of African sleeping sickness.
9. Pregnant women should not clean an animal's litter _____.
10. Agency that summarizes protozoal disease statistics.
11. Specimen examined to detect *Trichomonas* infection.
15. Genus of apicomplexans involved in malaria.
17. Possible mode of transmission of *Giardia lamblia*.

Questions for Thought and Discussion

Answers to selected questions can be found in Appendix C.

1. A newspaper article written in the 1980s asserted that parasitology is a "subject of low priority in medical schools, largely because the diseases are considered as exotic infections that occur in remote parts of the world." Do you agree with the contention that they are "exotic infections that occur in remote parts of the world"? Would you favor increased attention to protozoal diseases and more study of the general field of parasitology in the medical school curriculum? Why?

2. In the early part of this century, quinine syrup was taken to prevent malaria during visits to tropical regions of the world. To most people, the taste of quinine is very bitter, but the British found a way to make it less objectionable by mixing it with a type of alcoholic liquor. What do you suppose was that liquor, and how did the drink come to be known?

3. On returning from the Persian Gulf War, American servicemen were not accepted as blood donors because there was concern about their having been exposed to *Leishmania* species. What is the connection?

19. Amoeba (initials) that may cause nervous system illness.
20. Arthropod that transmits *Babesia microti*.
22. Climate where malaria is commonly found.
23. Emitted from intestine during cases of giardiasis.
25. Sexually transmitted flagellated protozoan.
31. Type of cell (abbr) infected by *Toxoplasma gondii*.
32. In _____, trichomoniasis produces a thin, mucoid discharge.
33. Induces intestinal ulcers and sharp pains after water transmission.
36. Arthropod that transmits *Plasmodium* species.
37. System (abbr) affected during cases of trypanosomiasis.
40. Mode of locomotion of ciliates.
41. How a protozoan is able to _____ determines its group designation.
42. Once believed to transmit malaria.
46. There are _____ flagella in *Balantidium coli*.
47. Genus of protozoa involved in sleeping sickness.
48. Extraordinarily high during cases of malaria.

49. Resistant form present in *Blantidium* and *Toxoplasma*.
50. Insect fluid that contains malarial parasites.
51. Number of groups of protozoa.

■ DOWN

2. Causes a disease that resembles infectious mononucleosis.
3. *Triatoma* species live in the _____ walls of South American huts.
4. Can be infected by *Toxoplasma*.
6. Body organ infected by *Acanthamoeba* species.
7. Ciliated protozoan that causes intestinal illness.
8. Genus of arthropod that transmits *T. cruzi*.
10. Possible complication of trypanosomiasis.
12. Sporozoa display _____ evidence of locomotion in the adult form.
13. Mastigophora species with two nuclei and eight flagella.
14. Many protozoal diseases can be treated with _____.
16. A fungus, formerly classified as a protozoan, which causes a lung disease common in AIDS patients.

18. Cell (abbr) in which *Plasmodium* undergoes its life cycle.
21. Apicomplexan (initials) that induces cholera-like diarrhea.
24. Trichomoniasis is associated with an elevated vaginal _____.
26. State (abbr) where babesiosis is known to occur.
27. Fly that transmits protozoan that causes trypanosomiasis.
28. Stage in the life cycle of *Plasmodium*.
29. Flagellated protozoan that causes kala-azar and oriental sore.
30. Group to which amoebas belong.
34. Number of recognized forms of trypanosomiasis.
35. Serious human disease complicated by protozoal infection.
38. Protozoan (initials) that can contaminate fresh produce.
39. Rare hamburger _____ can be a mode of transmission for *Toxoplasma*.
43. Resistant form present in *Entamoeba* and *Giardia*.
44. Environment from which amoebic cysts can be obtained.
45. The motion of *Giardia* resembles that of a falling _____.

4. The incidence of malaria is said to be skyrocketing in Africa, in part because of the antidrought measures being used in certain regions and the agricultural improvements being developed in other areas. Why is the malaria increase related to these measures?

5. The term *irritable bowel syndrome* often is used when a physician cannot explain why a patient is suffering long-lasting diarrhea, abdominal pain, and fatigue. (Such a diagnosis is euphemistically called a wastebasket diagnosis.) What candidates for this syndrome might be found in this chapter, and what is the description of each organism sought out in the cases you mention?

6. Recent outbreaks of waterborne protozoal disorders of the intestinal tract have raised the possibility that municipalities may have to begin filtering their water instead of just chlorinating it. Cost estimates for the filtering technology could reach into the millions or billions of dollars. Do you think the average taxpayer will consider the money well spent?

7. An anonymous CDC epidemiologist once said in an interview: "Day-care centers are the open sewers of the twentieth century." Would you agree? Why or why not?

8. You and a friend who is three months pregnant stop at a hamburger stand for lunch. Based on your knowledge of toxoplasmosis, what helpful advice can you give your friend? On returning home, you notice that she has two cats. What additional information might you be inclined to share with her?

9. Malaria is widely accepted as a contributing factor to the downfall of Rome. Over the decades, the incidence of malaria seemed to increase with expansion of the Roman Empire, which stretched from the Sahara desert to the borders of Scotland, and from the Persian Gulf to the western shores of Portugal. How do you suspect the disease and the expansion are connected?

10. The flagellated protozoan *Giardia lamblia* is named for Alfred Giard, French biologist of the late 1800s, and Vilem Dusan Lambl, Bohemian physician of the same period. Unfortunately, this information does not tell us much about the organism (except who did much of the descriptive work). Quite to the contrary, the names of other protozoa in this chapter tell us much. What are some examples of the more informative names?

11. It has been said that until recent times, many victims of a particular protozoal disease were buried alive because their life processes had slowed to the point where they could not be detected with the primitive technology available. Which disease was probably present?

12. In 1989, giardiasis was the most reported disease in the state of Vermont. Assuming that Vermont residents have reasonably good standards of hygiene, how is it that a disease like giardiasis can top the list of reported diseases in that state?

13. The outbreak of infections due to *Cryptosporidium* has prompted a reevaluation of drinking water purification systems in the United States. Of particular concern is the observation that *Cryptosporidium* species are unaffected by chlorine, the major disinfectant used in drinking water. Indeed, in a 1995 interview, the head of a research team studying the chlorine resistance of *Cryptosporidium* said, "You can wash these things in Clorox and they will smile right back at you." And an earlier CDC study showed that *Cryptosporidium* could actually grow on Clorox powder. In view of these findings, what alternatives might be available for safeguarding the nation from *Cryptosporidium*-contaminated water?

14. African herdsmen know that to prevent trypanosomiasis, cattle should be driven across the "fly belt" from one grazing ground to another only at night. What does this bit of folk knowledge tell you about tsetse flies?

15. In 1991, cardiologists at a Los Angeles hospital hypothesized that a few patients with a certain protozoal disease easily could be lost among the far larger population of heart disease sufferers patronizing county clinics. They proved their theory by finding 25 patients with this protozoal disease among patients previously diagnosed as having coronary heart disease. Which protozoal disease was involved?

16. In the 1950s, this African country was called the Belgian Congo. Then in the 1960s, it became Zaire. Currently it is the Democratic Republic of the Congo. Through all these political changes and episodes of civil strife, the epidemic of African sleeping sickness has remained. Why?

http://microbiology.jbpub.com

The site features **eLearning**, an on-line review area that provides quizzes and other tools to help you study for your class. You can also follow useful links for in-depth information, or just find out the latest microbiology news.

The Multicellular Parasites

Pass the Sushi, Carefully

—Headline to an article alerting consumers to the possible parasites in raw fish

O N FEBRUARY 13, 1984, A GROUP of 18 students from the United States arrived in Kenya for various purposes of study. The weather was hot and humid, so the students decided to improvise a small swimming pool. They carefully placed rocks and branches across a small stream and within a few hours, they had their pool. The water was cool and refreshing, and they congratulated themselves on their ingenuity.

But soon a problem developed. Several students broke out with an itchy rash, and subsequently, 14 of the group became acutely ill with fever, diarrhea, malaise, and weight loss. Two of the students developed paralysis of the lower extremities and had to return home for treatment. Stool examinations revealed the eggs of *Schistosoma mansoni*, a parasitic worm. Apparently the water had been contaminated. What had begun as a pleasant day turned into a nightmare.

Schistosomiasis is one of the diseases caused by the multicellular parasites we shall study in this chapter. Multicellular parasites, including the flatworms and roundworms, probably infect more people worldwide than any other group of organisms. They range in size from the tiny flukes, which must be studied with a microscope, to the tapeworms, which sometimes reach 45 meters in length. In the strict sense, flatworms and roundworms are animals, but they are studied also in microbiology because of their small size and ability to cause disease. Together with the protozoa discussed in Chapter 16, they are the subject of study of the biological discipline known as **parasitology**.

Our review of the multicellular parasites will be a brief one. Often simply referred to as **helminths** (Greek for "worm"), we shall include descriptions of these parasites, their life cycles, and the types of organisms and tissues they infect. You may note that the diseases in this chapter have few unique symptoms and are characterized by an abundance of parasites in a particular area of the body. Often the body will tolerate the parasites until the worm burden becomes immense. At that point, interference with an organ's function develops, and disease symptoms follow.

17.1

Flatworms

Flatworms belong to the animal phylum **Platyhelminthes**, a name derived from the Greek *platy-* for "flat" and *helminth* for "worm." All the parasites in this phylum have flattened bodies that are slender and broadly leaf-like, or long and ribbon-like (**FIGURE 17.1**). The animals exhibit **bilateral symmetry**, meaning that when cut in the longitudinal plane, the body yields identical halves.

As multicellular animals, flatworms have tissues functioning as organs in organ systems. They have no specialized respiratory or circulatory structures, and they lack a digestive tract. The gut simply consists of a sac with a single opening. Complex reproductive systems are found in many animals of the group, and a large number of species have both male and female reproductive organs. These organisms are termed **hermaphroditic**.

Flatworms are divided into three classes. The first class, Turbellaria, includes free-living flatworms that do not cause disease and therefore will not be covered here. The common planarian studied in general biology classes is a typical member of the

---

FIGURE 17.1

Two Examples of Flatworms

(a) A scanning electron micrograph of the fluke *Fasciola hepatica*. Note the flattened, broad, and leaf-like shape of this parasite. *F. hepatica* thrives in the liver of humans. The ventral and oral suckers used for attachment to the tissue can be seen. (Bar = 6 mm.) (b) An unmagnified view of the tapeworm *Taenia saginata*. This flatworm is long and ribbon-like, and consists of hundreds of visible segments. *T. saginata* infects the human intestinal tract.

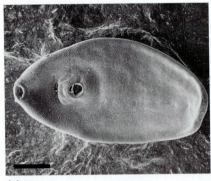
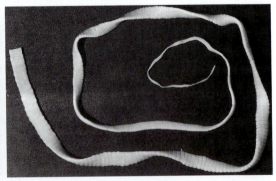

(a)   (b)

Turbellaria. The second class, Trematoda, includes the flukes. We shall begin our coverage with this group of parasites. The third class, Cestoda, consists of tapeworms. These parasites will be discussed after the flukes.

FLUKES USE SNAILS AS AN INTERMEDIATE HOST

Flukes are leaf-like parasitic worms of the class **Trematoda**. Generally, flukes have complex life cycles that may include encysted egg stages and temporary larval forms. Sucker devices are commonly present to enable the parasite to hold fast to its host. In many cases, two hosts exist: an **intermediate host**, which harbors the larval form, and a **definitive host**, or final host, which harbors the sexually mature adult form. In this chapter, we shall be concerned with parasites whose definitive host is a human.

The life cycle of a fluke often contains several phases. In the human host, the parasite produces fertilized eggs generally released in the feces. When the eggs reach water, they hatch and develop into tiny ciliated larvae called **miracidia** (sing., miracidium). The miracidia penetrate snails (the intermediate host) and go through a series of asexual reproductive stages, often including **sporocyst** and **redia** stages. Rediae become tadpole-like larvae called **cercariae**, which are released to the water. Now the cercariae develop into encysted forms called **metacercariae**, which make their way back to humans. We shall see variations on this basic life cycle in the discussions of fluke diseases that follow.

SCHISTOSOMIASIS IS VERY COMMON WORLDWIDE

Three important species of flukes invade the bloodstream in humans: *Schistosoma mansoni*, *S. japonicum*, and *S. haematobium*. The first is distributed throughout Africa and South America, the second in the Far East, and the third mainly in Africa and India. The WHO estimates that 250 million people worldwide are infected with *Schistosoma*, including about 400,000 individuals in the United States. The disease is called **schistosomiasis**. It ranks second only to malaria in causing disease in the tropics. In some regions, the term **bilharziasis** is still used; it comes from the older name for the genus, *Bilharzia*. MicroFocus 14.6 discussed how treatment for schistosomiasis spread hepatitis C through much of Egypt.

Species of *Schistosoma* measure about 10 mm in length. Male and female species mate in the human liver and produce eggs that are released in the feces (**FIGURE 17.2**). The eggs hatch to miracidia in freshwater, and the miracidia make their way to snails, where conversions to sporocysts and cercariae take place. The cercariae escape from the snails and attach themselves to the bare skin of humans. Cercariae become young schistosomes, which infect the blood and cause fever and chills. The major effects of disease are due to eggs carried by the bloodstream to the liver. Liver damage is usually substantial. Eggs also gather in the intestinal wall causing ulceration, diarrhea, and abdominal pain. Bladder infection is signaled by bloody urine and pain on urination. The antihelminthic drug praziquantel is used for treatment. Outbreaks may have substantial consequences, as **MicroFocus 17.1** explains.

Certain species of *Schistosoma* penetrate no farther than the skin because the definitive hosts are birds instead of humans. The cercariae of these schistosomes cause dermatitis in the skin and a condition commonly known as **swimmer's itch**. After penetration, the cercariae are attacked and destroyed by the body's immune system, but they release allergenic substances that cause the itching and body rash (**FIGURE 17.3**). The condition, not a serious threat to health, is common in northern lakes in the United States.

Schistosoma mansoni

The Life Cycle of the Blood Fluke, *Schistosoma mansoni*

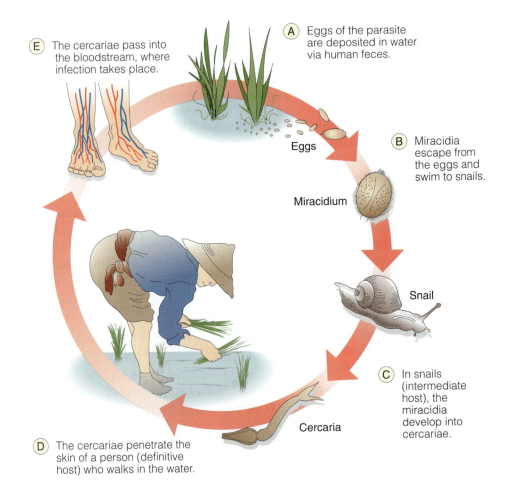

Ⓐ Eggs of the parasite are deposited in water via human feces.

Ⓔ The cercariae pass into the bloodstream, where infection takes place.

Ⓑ Miracidia escape from the eggs and swim to snails.

Eggs

Miracidium

Snail

Ⓒ In snails (intermediate host), the miracidia develop into cercariae.

Cercaria

Ⓓ The cercariae penetrate the skin of a person (definitive host) who walks in the water.

MicroFocus 17.1

INVASION

In the late 1940s, clashes between the nationalist Chinese, headed by Chiang Kai-Shek, and the communist Chinese, led by Mao Tse Tung, heated up to the point of open warfare. After months of fighting, the communist forces drove Chiang's troops off the Chinese mainland and onto the island of Taiwan (then called Formosa). It soon became apparent to world leaders that a communist invasion of Taiwan was imminent.

However, invading an island takes a certain type of training. Therefore, communist generals gathered their troops by the hundreds of thousands in a swampy area near the southern coast of mainland China to learn the methods of amphibious warfare. Unbeknown to them, however, the area was infested with parasites and snails, and within days, over 30,000 men were infected with *Schistosoma*. The troops suffered severe fever, liver damage, and intestinal distress, and Chinese leaders were forced to postpone their assault on Taiwan for several months.

The delay was critical because it gave the nationalist Chinese precious extra time to prepare for invasion. In addition, intelligence reports of the impending invasion reached US President Harry Truman, and he sent ships from the Seventh Naval Fleet to the waters off Taiwan to preserve the neutrality of Chiang's forces. Eventually, these and other factors persuaded the communist Chinese to abandon their assault, and the republic of Taiwan was left intact.

We shall never know whether the invasion would have succeeded or what the political ramifications might have been. Indeed, some historians point out that the United States was preparing to back the nationalist Chinese, while the former Soviet Union was readying itself to support the communists. One thing does appear certain: A worm played a significant role in the turn of world events.

FIGURE 17.3

An Outbreak of Schistosomiasis in Delaware

1. On October 19, 1991, a group of 37 students from a local high school biology class visited a shellfishing area at Cape Henlopen State Park in Delaware. The tide was low, the water was calm, and the weather was sunny and unseasonably warm. Several ducks and geese were nearby.

2. The students and their teacher spent about 2 hours wading in the water collecting specimens. There were many clams, oysters, and snails in the water, as well as other marine specimens of interest.

3. Within 10 days of the trip, 29 of the 37 students developed pruritic dermatitis, an itching skin condition. Eleven of the affected students visited their doctors for treatment. Eventually, 36 of the 37 individuals developed pruritis; the only unaffected student was one who did not go into the water.

4. Public health investigators examined snails from the seawater. Schistosomes normally associated with ducks, geese, and other birds were found in a small percentage of snails examined. Also, it is known that thousands of schistosomes can be released from a single snail, and the release is favored by weather conditions seen that day.

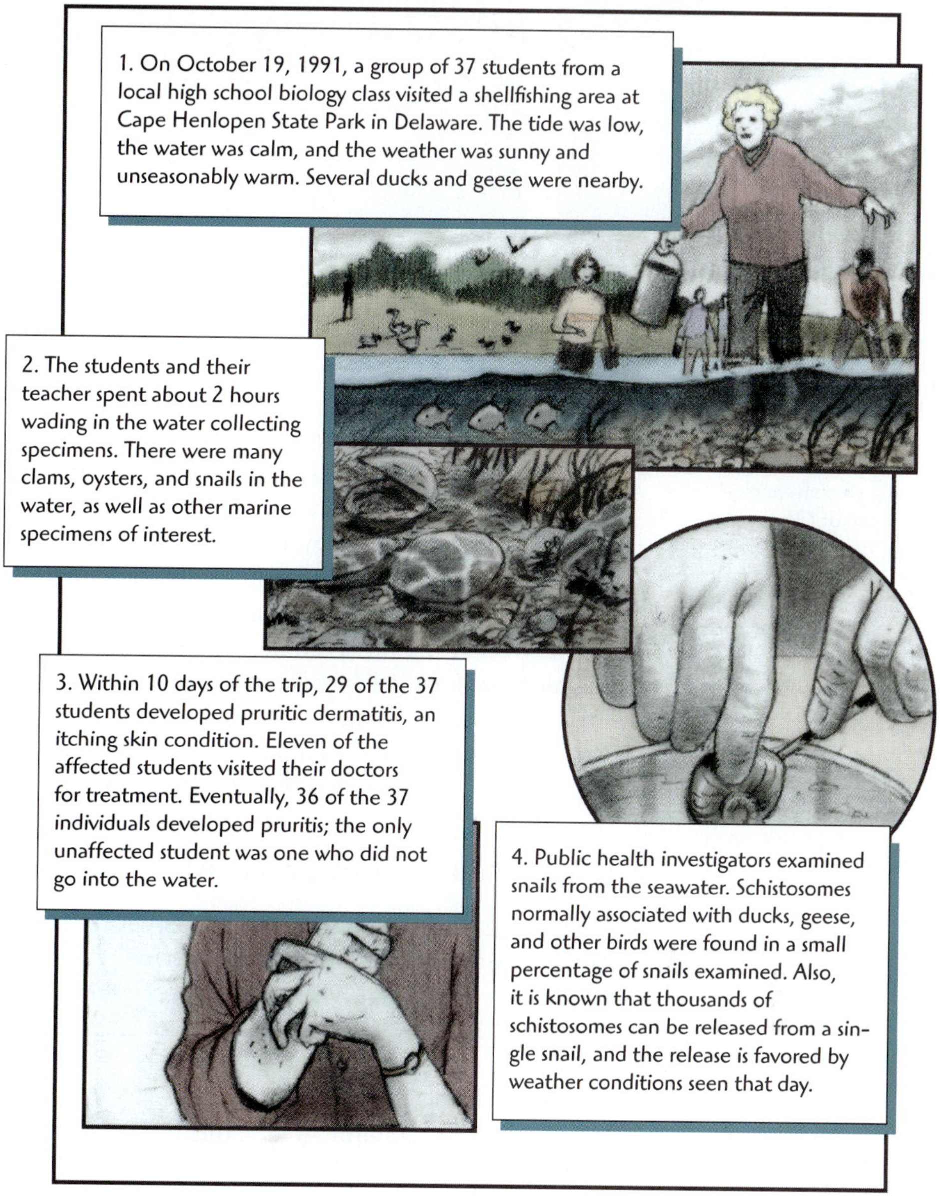

CHINESE LIVER FLUKE DISEASE INVOLVES TWO INTERMEDIATE HOSTS

The **Chinese liver fluke** is so named because it infects the liver and is common in many regions of Asia, especially China and Japan. The organism is named *Clonorchis sinensis*. It has oral and ventral sucker devices (**FIGURE 17.4**) and is hermaphroditic, with a complex reproductive system.

Eggs of *C. sinensis* contain complete miracidia, which emerge after the eggs enter water. The miracidia penetrate snails and change into sporocysts, which then produce a generation of elongated rediae. These rediae become cercariae, which escape

Clonorchis sinensis

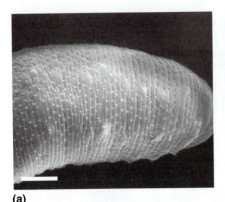

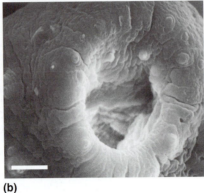

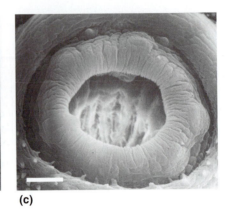

(a)   (b)   (c)

FIGURE 17.4

The Chinese Liver Fluke

Scanning electron microscope views of *Clonorchis sinensis*, the Chinese liver fluke. (a) A whole mount showing the anterior aspect of the parasite. (Bar = 2 mm.) (b) The oral sucker device. (Bar = 1 mm.) (c) The ventral sucker device. (Bar = 1 mm.)

from the snail and bore into the muscles of fish, the second intermediary host. Now the cercariae develop into encysted metacercariae, and humans acquire the metacercariae by consuming raw or poorly cooked fish. Public health officials recommend heating fish to a minimum of 50°C for 15 minutes to destroy the cysts. In humans, the metacercariae become adults and migrate from the small intestine up the bile duct to the gall bladder and liver, where infection takes place.

The effect of *C. sinensis* on humans depends on the extent of infection. Substantial infection in the gall bladder may lead to duct blockage and poor digestion of fats. There also may be damage to the liver as eggs accumulate in the tissues of this organ (**FIGURE 17.5**). Often the patient is without symptoms because of the low number of parasites. Cats, dogs, and pigs may carry the metacercaria cysts also.

OTHER FLUKES ALSO CAUSE HUMAN DISEASE

The **intestinal fluke** of humans is known as *Fasciolopsis buski*, a large fluke that can be as long as 8 cm. The parasite lives in the duodenum, where it causes diarrhea and intestinal blockage. From the feces, eggs enter water in such places as rice paddies and drainage ditches, and miracidia emerge. Snails are the next host for the miracidia, and cercariae escape from the snail and swim to blades of grass and vegetation where metacercariae form. Such plants as water chestnuts and water bamboo are likely to be contaminated. Consumption of these vegetables raw or poorly cooked leads to human infection.

The human **lung fluke** is *Paragonimus westermani*. This parasite is common in Asia and the South Pacific. Its eggs are coughed up from the lung, swallowed, and then excreted in the feces. Cercariae develop in snails, and metacercariae later form in crabs. Infection follows when people eat the poorly cooked crabmeat, and the flukes pass from the intestine to the blood to the lungs. Difficult breathing and chronic cough develop as the parasites accumulate in the lungs. Fatalities are possible.

The human **liver fluke** is *Fasciola hepatica*, a leaf-like flatworm common in sheep and cattle. Eggs from the animal reach the soil in feces, and if snails are present, the conversions to cercariae and encysted metacercariae follow. Parasite cysts gather on vegetation such as watercress, and ingestion by humans follows. The parasites penetrate the intestinal wall and migrate to the liver, where tissue damage may be substantial, especially if fluke numbers are high.

MicroInquiry 17 examines how such parasites can manipulate their host.

Fasciolopsis buski

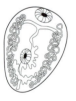

Paragonimus westermani

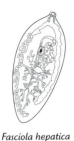

Fasciola hepatica

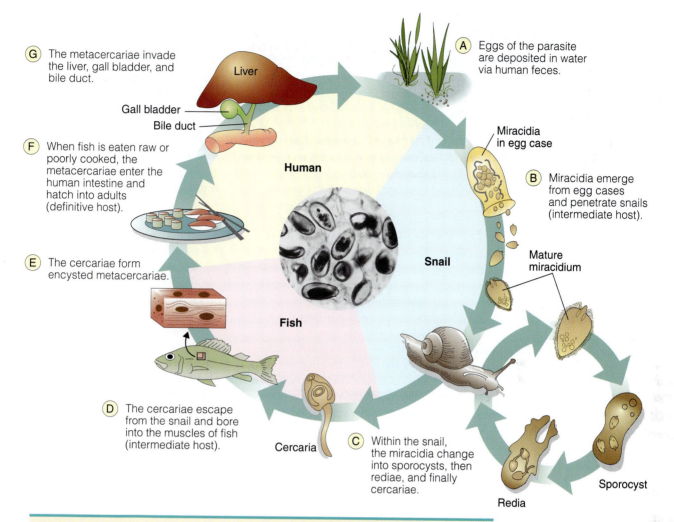

G The metacercariae invade the liver, gall bladder, and bile duct.

A Eggs of the parasite are deposited in water via human feces.

Liver

Gall bladder

Bile duct

Miracidia in egg case

Human

B Miracidia emerge from egg cases and penetrate snails (intermediate host).

F When fish is eaten raw or poorly cooked, the metacercariae enter the human intestine and hatch into adults (definitive host).

Mature miracidium

E The cercariae form encysted metacercariae.

Snail

Fish

D The cercariae escape from the snail and bore into the muscles of fish (intermediate host).

Cercaria

C Within the snail, the miracidia change into sporocysts, then rediae, and finally cercariae.

Sporocyst

Redia

FIGURE 17.5

The Life Cycle of the Chinese Liver Fluke, *Clonorchis sinensis*

The photograph shows eggs of *C. sinensis* in human tissue. The thick protective covering of the eggs is evident.

TAPEWORMS ARE FLATWORMS THAT LIVE IN THE HOST INTESTINES

Tapeworms belong to the third class of flatworms, the **Cestoda**. These worms have long, flat bodies consisting of a head region and a ribbon-like series of segments called **proglottids**. The head region, called the **scolex**, contains hooks or sucker-like devices that enable the worm to hold fast to host tissue. Behind the scolex is a neck region in which new proglottids are formed. These new proglottids constantly push the mature proglottids to the rear. The most distant ones, called **gravid proglottids**, are filled with fertilized eggs. As they break free, they spread the eggs of the tapeworm.

Tapeworms generally live in the intestines of a host organism. In this environment they are constantly bathed by nutrient-rich fluid, from which they absorb food already digested by the host. Tapeworms have adapted to a parasitic existence and

PARASITES AS MANIPULATORS

Over the last several chapters, we have been discussing viral, bacterial, and fungal pathogens, and protozoan and multicellular parasites. You might conclude that these microbes make their living "sponging off" their hosts with the ultimate aim to reproduce. In fact, for a long time scientists thought that these organisms simply evolved so they could take advantage of what their hosts had to offer. In reality, new research suggests that these parasites may be quite a bit more sophisticated than previously thought. Two cases involving multicellular parasites are submitted for your inquiry.

Case 1. The lancet fluke, *Dicrocoelium dendriticum*, infects cows. The cows eat grass on which the fluke cysts exist—well, sort of exist as we will see below. Here is the scenario. Since the cow is the definitive host, the fluke needs to make sure that reproduction in an intermediate host will produce cysts that will be eaten by other cows as they graze on the grass.

17.1a. As a parasite, how do you ensure cysts will be on grass blades that cows eat?

Infected cows produce dung that contains fluke eggs. Snails (intermediate host #1) forage on the dung and in the process ingest the fluke eggs. The eggs hatch and bore their way from the snail gut to the digestive gland where fluke larvae are produced. In an attempt to fight off the infection, the snail smothers the parasites in balls of slime that it coughs up into the grass.

Ants (intermediate host #2) come along and swallow these nutrient-rich slime balls. Now, here is the really interesting part, because the lancet fluke has yet to get cysts onto grass blades so they will be eaten by more cows. In infected ants, the larvae travel to the ant's head and specifically the nerves that control the ant's mandibles. While most of the larvae then return to the abdomen where they form cysts, a few flukes remain in the head and control ant behavior. As evening comes and the temperature cools, the head-infected ants with a belly full of cysts leave their colony, climb to the tip of grass blades, and hold on tight with their mandibles. The lancet fluke has taken control of the ants and placed them in a position most favorable to be eaten by a grazing cow.

17.1b. What happens if no cow comes grazing that evening?

If a grazing cow does not come by, the next morning the flukes release their influence on the ants, which return to their "normal duties." But, the next evening, the flukes again take control and the ants march back up to the tips of grass blades. Should a cow eat the

---

have lost their intestines, but they still retain well-developed muscular, excretory, and nervous systems.

Tapeworms are widespread parasites that infect practically all mammals, as well as many other vertebrates. Since they are more dependent on their hosts than flukes, tapeworms have precarious life cycles. Tapeworms have a limited range of hosts, and the chances for completing the cycle are often slim. With rare exceptions, tapeworms require at least two hosts, as we shall see in the following examples of tapeworm diseases. All tapeworm infections can be treated with praziquantel.

BEEF AND PORK TAPEWORM DISEASES USUALLY ARE CHRONIC BUT BENIGN

Humans are the definitive hosts for both the **beef tapeworm** *Taenia saginata* and the **pork tapeworm** *Taenia solium*. People infected with one of these tapeworms expel numerous gravid proglottids daily. The proglottids accumulate in the soil and are consumed by cattle or pigs. Embryos from the eggs travel to the animal's muscle, where they encyst. Humans then acquire the cysts in poorly cooked beef or pork.

The beef tapeworm may reach 10 meters in length, while the pork tapeworm length is 2 to 8 meters. Each tapeworm may have up to 2,000 proglottids. Attachment via the scolex occurs in the small intestine, and obstruction of this organ may result. In most cases, however, there are few symptoms other than mild diarrhea, and

Taenia saginata

grass with the ants, the fluke cysts in the ant abdomen quickly hatch and another cycle of reproduction begins in the cow. Now, that is quite a behavioral driving force to control your destiny; in this case, to ensure fluke cysts are positioned so they will be eaten by the cow.

Case 2. (This case is based on research carried out in a coastal salt marsh by Kevin Lafferty's group at the University of California at Santa Barbara.) Another fluke, *Euhaplorchis californiensis*, uses shorebirds as its definitive host. Infected birds drop feces loaded with fluke eggs into the marsh. Horn snails (intermediate host) eat the droppings containing the eggs. The eggs hatch and castrate the snails. Larvae are produced in the water and latch onto the gills of their second intermediate host, the California killifish. The larvae move along the blood vessels to a nerve that carries them to the brain where the larvae form a thin layer on top of the brain. There they stay, waiting for the fish to be eaten by a shorebird. Once in the bird's gut, the adults develop and produce another round of fertilized eggs. Here, we have two questions.

17.2a. What is the purpose for the larvae castrating the snails, which would appear to be at a disadvantage reproductively versus uninfected snails?

An experiment was set up to answer this question. Throughout the salt marsh, Lafferty set up cages with uninfected snails and other cages with infected ones. The results were as expected—the uninfected snails produced many more offspring than did the infected snails. Lafferty believes a marsh full of uninfected snails would reproduce so many offspring as to deplete the algae and increase the population of crabs that feed on the snails. So, the parasite actually is controlling the snail population and keeping the salt marsh ecosystem balanced—again for its benefit.

17.2b. Why do the larvae in the killifish take up residence on top of the fish's brain?

Through a series of experiments, Lafferty's group discovered that infected fish underwent a swimming behavior not observed in the uninfected fish. The infected killifish darted near the water surface, a very risky behavior that makes it more likely the fish would be caught by shorebirds. In fact, experiments showed infected fish were 30 times more likely to be caught by shorebirds than uninfected fish. Now, from the parasite's perspective, they want to be caught so another round of reproduction can start in the shorebird. So again, the fluke has manipulated the situation to its benefit.

a mutual tolerance may develop between parasite and host. The notion that a tapeworm causes severe emaciation is largely unfounded.

FISH TAPEWORM DISEASE RESULTS FROM EATING UNDERCOOKED, INFECTED FRESHWATER FISH

The **fish tapeworm** is the longest human parasite, some species measuring 45 meters in length. The parasite is named *Diphyllobothrium latum*. Its life cycle is complex and includes two intermediate hosts: a small shrimp-like crustacean called a **copepod** and a fish (**FIGURE 17.6**). The copepod acquires tapeworm embryos from proglottids excreted into water by humans. Next, the copepod is eaten by fish such as minnows, and the embryo finds its way to the fish muscle. Minnows may be eaten by larger fish such as trout, perch, and pike, and the embryos are passed along. Consumption of raw or poorly cooked fish by humans completes the cycle.

Diphyllobothrium latum

Infections with fish tapeworms occur throughout the world. In the United States, infections are most common in the Great Lakes region. Obstruction of the small intestine may occur, and anemia may develop in the patient, possibly due to the parasite's consumption of vitamin B_{12} needed for red blood cell formation. Cooking at 50°C for 15 minutes is generally sufficient to destroy the cysts, but the recent popularity of raw fish dishes has contributed to a rise in the incidence rates of the disease.

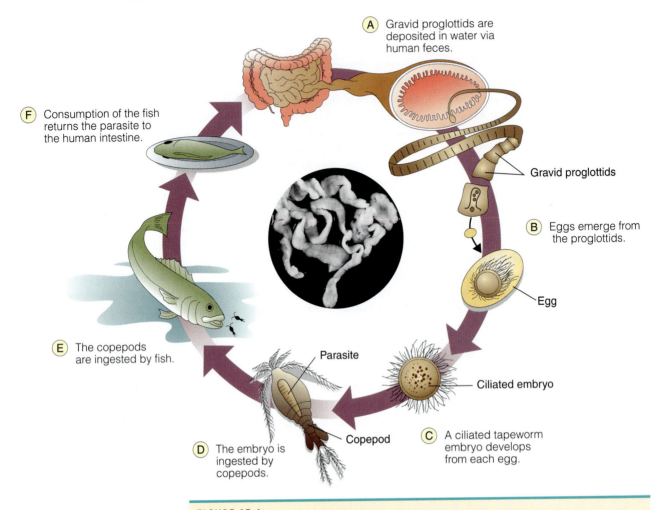

Ⓐ Gravid proglottids are deposited in water via human feces.

Ⓕ Consumption of the fish returns the parasite to the human intestine.

Gravid proglottids

Ⓑ Eggs emerge from the proglottids.

Egg

Ⓔ The copepods are ingested by fish.

Parasite

Ciliated embryo

Copepod

Ⓓ The embryo is ingested by copepods.

Ⓒ A ciliated tapeworm embryo develops from each egg.

FIGURE 17.6

The Life Cycle of the Fish Tapeworm, *Diphyllobothrium latum*

The photograph shows *D. latum* isolated from a human patient.

A FEW OTHER TAPEWORMS ALSO CAUSE DISEASE

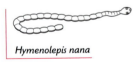

Hymenolepis nana

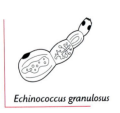

Echinococcus granulosus

The **dwarf tapeworm** *Hymenolepis nana* is so named because it is only 15 to 50 mm long. The most common tapeworm in humans worldwide, *H. nana* lives in the small intestine, where it holds fast with four sucker devices and a row of small hooks. Eggs released in the feces spread among humans by contaminated food or contact with objects or an infected individual. People living in the southeastern United States often experience infections.

Dogs and other canines such as wolves, foxes, and coyotes are the definitive hosts for the **dog tapeworm** called *Echinococcus granulosus* (**FIGURE 17.7**). Eggs reach the soil in feces and spread to numerous intermediary hosts, one of which is humans. Contact with a dog also may account for transmission. In humans, the parasites travel by the blood to the liver, where they form thick-walled **hydatid cysts**. Surgery may be necessary for their removal. Completion of the life cycle takes place when a dog consumes infected animal liver, such as after a kill. Dog food is another possible source of cysts. The adult tapeworm then emerges to parasitize the dog.

The flatworm diseases of humans are summarized in **TABLE 17.1**.

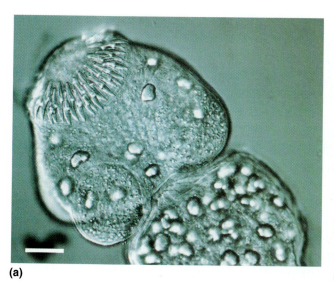

(a)

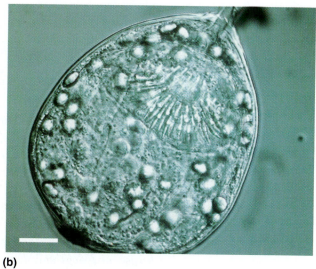

(b)

FIGURE 17.7

The Dog Tapeworm

Two views of the dog tapeworm *Echinococcus granulosus*, as seen by phase-contrast microscopy. (Bars = 20 μm.) (a) The head region, or scolex, of *E. granulosus*. The hooks at the end of the scolex are used to attach to the infected tissue; the indentations at the base of the scolex emphasize its presence. (b) The hydatid cyst of *E. granulosus* isolated from a lung section of a 31-year-old woman immigrant from the Sudan.

TABLE 17.1

A Summary of Flatworm Parasitic Diseases in Humans

ORGANISM	DISEASE	TRANSMISSION	ORGANS AFFECTED	CHARACTERISTIC SIGN	ANIMAL HOST
Schistosoma mansoni *S. japonicum* *S. haematobium*	Blood fluke disease	Water contact	Skin Liver Blood	Rash Liver damage Fever	Snail
Clonorchis sinensis	Chinese liver fluke disease	Fish consumption	Gall bladder Liver	Liver damage Poor fat digestion	Snail Fish
Fasciolopsis buski	Intestinal fluke disease	Consumption of water plants	Intestine	Diarrhea	Snail
Paragonimus westermani	Lung fluke disease	Consumption of crabs	Lungs	Cough Poor breathing	Snail Crab
Fasciola hepatica	Liver fluke disease	Consumption of water plants	Liver	Liver damage	Snail Cattle
Taenia saginata	Beef tapeworm disease	Beef consumption	Intestine	Diarrhea	Cattle
Taenia solium	Pork tapeworm disease	Pork consumption	Intestine	Diarrhea	Pig
Diphyllobothrium latum	Fish tapeworm disease	Fish consumption	Intestine	Diarrhea Anemia	Copepod Fish
Hymenolepis nana	Dwarf tapeworm disease	Food Contact	Intestine	Diarrhea	None significant
Echinococcus granulosus	Dog tapeworm disease	Contact	Liver	Liver damage	Dog, other canines

To this point . . .

We have surveyed many of the flatworm parasites that infect humans. The flat-worms include flukes, which are slender and broadly leaf-like, and tapeworms, which are long and ribbon-like. Flukes belong to the class Trematoda. They have complex life cycles, with snails serving as an intermediate host. Stages in the life cycle include the miracidium, cercaria, and metacercaria, among others. The blood fluke is acquired by contact with water, while the Chinese liver fluke is ingested in raw or poorly cooked fish. Intestinal flukes and lung flukes also are acquired in foods.

We then discussed the tapeworms of the class Cestoda. These parasites generally live in the human intestinal tract. Proglottids carry tapeworm eggs to the soil or water, from which they are consumed by animals or fish. Human infection is reestablished when contaminated meat or fish is eaten. Beef, pork, and fish tapeworm diseases are passed along in this way. Contaminated food or objects may be the source of dwarf tapeworm disease, and dogs may pass dog tapeworms to humans. In dog tapeworm disease, humans are an intermediary host.

We now will focus on the roundworms. Roundworms are anatomically more complex than flatworms and are classified in a completely different phylum of animals. We shall see how the life cycles of roundworms are considerably more simple than the cycles of flatworms, and how infection is spread much more easily. Among the roundworms are those that cause pinworm disease and trichinosis, two diseases that are common in the United States.

17.2

Roundworms

Roundworms occupy every imaginable habitat on Earth. They live in the sea, in freshwater, and in soil from polar regions to the tropics. Good top-soil, for example, may contain billions of roundworms per acre. They parasitize every conceivable type of animal and plant, causing both economic damage and serious disease.

The roundworms are a subgroup of the phylum **Aschelminthes**, from the Greek *asc-* for "sac" and *helminth* for "worm." The sac refers to a digestive tract set apart from the internal muscles in a sac-like or pouch-like arrangement, a feature not found in flatworms. In reality, the sac is a tubular intestine open at the mouth and anus. Food thus can move in one direction, a substantial improvement over the blind sac arrangement in Platyhelminthes. This is one reason the Aschelminthes are considered to be more evolutionarily advanced than the Platyhelminthes.

Roundworms have separate sexes. Following fertilization of the female by the male, the eggs hatch to larvae that resemble miniature adults. Growth then occurs by cellular enlargement and mitosis. Damage in hosts is generally caused by large worm burdens in the blood vessels, lymphatic vessels, or intestines (**FIGURE 17.8**). Also, the infestation may result in nutritional deficiency or damage to the muscles.

Roundworms traditionally have been known as **nematodes** because they are thread-like (*nema* is Latin for "thread"). Indeed, in some texts the phylum of round-worms is called **Nematoda**, and in other books it is referred to as **Nemathelminthes**.

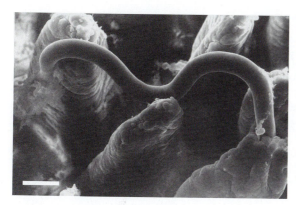

FIGURE 17.8

The Roundworm *Trichinella spiralis*

A scanning electron micrograph of the roundworm *Trichinella spiralis* in human intestinal tissue. This parasite is the cause of trichinosis. In the photograph, the worm is emerging from one intestinal villus and entering another villus. (Bar = 0.4 mm.)

PINWORM DISEASE IS THE MOST PREVALENT HELMINTHIC INFECTION IN THE UNITED STATES

The widely encountered **pinworm** is a roundworm called *Enterobius vermicularis*. This worm is the most common helminthic parasite in the United States, with an estimated 30 percent of children and 16 percent of adults serving as hosts. The male and female worms live in the distant part of the small intestine and in the large intestine, where the symptoms of infection include diarrhea and itching in the anal region. The female worm is about 10 mm long, and the male is about half that size.

The life cycle of the pinworm is relatively simple. Females migrate to the anal region at night and lay a considerable number of eggs. The area itches intensely, and scratching contaminates the hands and bed linens with eggs. Reinfection may then take place if the hands are brought to the mouth or if eggs are deposited in foods by the hands. The eggs are swallowed, whereupon they hatch in the duodenum and mature in the regions beyond.

Diagnosis of pinworm disease may be made accurately by applying the sticky side of cellophane tape to the area about the anus and examining the tape microscopically for pinworm eggs (**FIGURE 17.9**). Mebendazole or pyrantel pamoate are effective for controlling the disease, and all members of an infected person's family should be treated because transfer of the parasite probably has taken place. Even without medication, however, the worms will die in a few weeks, and the infection will disappear as long as reinfection is prevented.

Enterobius vermicularis

WHIPWORM DISEASE COMES FROM EATING FOOD CONTAMINATED WITH PARASITE EGGS

The **whipworm** *Trichuris trichiura* acquired its name from the observation that its anterior end is long and slender like a buggy whip. Infection takes place in the human intestine, especially near the junction of the small and large intestines, as shown in **FIGURE 17.10**. Damage to the intestinal lining may be severe, and appendicitis-like pain is sometimes experienced, with some anemia resulting from ingestion of blood by the parasite.

The female whipworm is approximately 40 mm long; the male is shorter with a characteristically curled tail. Eggs eliminated in the human feces hatch to larvae after about two weeks in the soil. Transmission occurs by soil-contaminated food and water as well as by contact with soiled hands. Whipworm disease is encountered where the environment is hot and moist (such as in the tropics), and where poor

Trichuris trichiura

FIGURE 17.9

Diagnosing Pinworm Disease

The transparent tape technique used in the diagnosis of pinworm disease.

(A) Clear plastic tape is pulled back over the end of the slide to expose the gummed surface.

(B) The tape, still attached to the slide, is looped over a wooden stick.

(C) The gummed surface of the tape is touched several times to the anal region.

(D) The tape is replaced on the slide.

(E) The slide is smoothed down with cotton or gauze. It is then examined under a microscope for pinworm eggs.

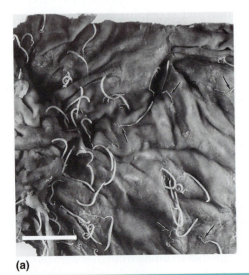

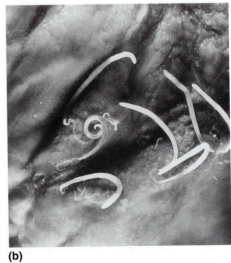

(a)   (b)

FIGURE 17.10

The Whipworm *Trichuris trichiura*

(a) A view of the whipworm *Trichuris trichiura* in human intestinal tissue. The typical "buggy whip" is evident (arrows). (Bar = 40 mm.) (b) Several female whipworms and a male whipworm (curled tail) are seen in their actual size.

sanitary facilities exist. Patients often have concurrent infections with other parasites. Diagnosis depends on the identification of eggs in the feces. Patients are treated with mebendazole.

ROUNDWORM DISEASE IS PREVALENT IN THE AMERICAN SOUTH

Infection with "roundworms" usually implies infection with *Ascaris lumbricoides*. One of the largest intestinal nematodes, the female *A. lumbricoides*, may be up to 30 cm long, and the male, 20 cm long. The parasite resembles an earthworm and is the most worm-like of the helminthic parasites.

Ascaris lumbricoides

A female *Ascaris* is a prolific producer of eggs, sometimes generating over 200,000 per day. The eggs are fertilized and passed to the soil in the feces, where they hatch to larvae. The larvae then attach to plants and are ingested. In many parts of the world, human feces, or nightsoil, is used as fertilizer for crops. This adds to the spread of the parasite. Contact with contaminated fingers and consumption of water containing soil runoff are other possible modes of transmission.

After the larvae have been consumed, the tiny worms grow in the small intestine. Abdominal symptoms develop as the worms reach maturity in about two months. Intestinal blockage may be a consequence when tightly compacted masses of worms accumulate, and perforation of the small intestine is possible. In addition, roundworm larvae may pass to the blood and infect the lungs, causing pneumonia. If the larvae are coughed up and then swallowed, intestinal reinfection occurs.

Except for pinworms, *A. lumbricoides* is the most prevalent multicellular parasite in the United States. The WHO estimates that hundreds of millions of people are infected worldwide. Tropical and subtropical regions are the primary foci of disease, but areas of the southwestern United States are heavily infested because eggs remain viable in the moist clay soil of this region. Again, mebendazole is the drug for treatment.

MicroFocus 17.2 looks at roundworms and raw fish (sushi) consumption.

TRICHINOSIS OCCURS WORLDWIDE

Trichinosis is a term familiar to anyone who enjoys pork and pork products, because packages of pork usually contain warnings to cook the meat thoroughly to avoid this disease. Ironically, trichinosis is common where living standards are high enough for pork to be eaten routinely, but the disease is rare where pork is a luxury.

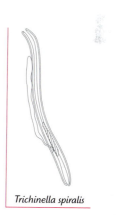

Trichinella spiralis

Trichinosis is caused by the small roundworm *Trichinella spiralis*. The worm lives in the intestines of pigs and several other mammals. Larvae of the worm migrate through the blood and penetrate the pig's skeletal muscles, where they remain in cysts. When raw or poorly cooked pork is consumed, the cysts pass into the human intestines and the worms emerge. Intestinal pain, vomiting, nausea, and constipation are common symptoms.

Complications of trichinosis occur when *T. spiralis* larvae migrate to the muscles and form cysts. The patient commonly experiences pain in the breathing muscles of the ribs, loss of eye movement due to cyst formation in the eye muscles, swelling of the face, and hemorrhaging in various body tissues. Some sufferers also develop a cough and skin lesions, and the larvae may invade the brain, where they cause paralysis. Paralysis and death occurred in a man in New York in 1985 after he contracted trichinosis by consuming pork inadvertently ground with beef. The man had a habit of nibbling the meat while preparing hamburgers.

The cycle of trichinosis is completed as cysts are transmitted back to nature in the human feces (FIGURE 17.11). Consumption of human waste and garbage

MicroFocus 17.2

HOLD THE ANISAKIASIS ON MY SUSHI—PLEASE!

With the increasing popularity of sushi and sashimi bars, patrons should be sure that the establishment receives only the best and freshest raw fish. Although fewer than 10 cases of illness from eating parasite-infected raw or undercooked fish are diagnosed in the United States each year, health experts suspect many other cases go undetected.

In North America, anisakiasis refers to disease caused by the accidental ingestion of larvae of the roundworms *Anisakis simplex* (herring worm) or *Pseudoterranova decipiens* (cod or seal worm). Individuals having anisakiasis often realize they are infected when they sense a tingling or tickling sensation in the throat and then cough up or manually remove a worm. One nematode is the usual number recovered from a patient. In more severe cases, the larvae pass into the bowel and cause acute abdominal pain and a nauseous feeling similar to appendicitis. Symptoms typically start within six hours after consumption of raw or undercooked

seafood. Most cases resolve themselves but severe cases may require surgical or endoscopic removal of the parasite.

Seafoods are the principal source of human infections from the larval worms. The adults of *A. simplex*, for example, are found in the stomachs of the definitive hosts, such as whales and dolphins. Fertilized eggs from the female parasite pass out of the host with the host's feces. The eggs develop into larvae that hatch in seawater. These larvae infect copepods (intermediate hosts) or other small invertebrates. The larvae grow in the invertebrate and become infective for the next intermediate host, a fish (cod, haddock, fluke, pacific salmon, herring, flounder, or monkfish) or larger invertebrate host such as a squid. There, the larvae may penetrate through the digestive tract into the muscle of the host. When fish or squid containing the larvae are ingested by marine mammals (whales, dolphins), the reproduction cycle is complete after the larvae molt twice and develop into adult worms.

Humans are considered accidental hosts because of eating raw or undercooked infected marine fish.

Examining fish on a light table, called candling, has been used by commercial processors to limit the number of roundworms in certain white-flesh fish that typically are infected. This method is not completely effective and will not adequately remove even the majority of roundworms from fish with pigmented flesh. Worldwide, the highest number of reported cases of anisakiasis occurs in Japan, the Pacific coast of South America, and Scandinavia because of the large volume of raw fish people regularly eat.

The US Food and Drug Administration (FDA) recommends that all fish and shellfish intended for raw (or semi-raw, such as marinated or partly cooked) consumption be blast frozen to −35°C (−31°F) or below for 15 hours, or be regularly frozen to −20°C (−4°F) or below for seven days to ensure killing of any parasites.

---

then brings the cysts to the pig. Modern methods of agriculture provide standardized feed for pigs, but many pigs are still exposed to the cysts. Consumers should be aware that poorly cooked pork is the principal source of the approximately 150 cases of trichinosis reported in the United States annually. Pickling, smoking, and heavy seasoning are not adequate substitutes for thorough heating, but freezing greatly reduces larval viability. Routine inspection of pigs for *T. spiralis* cysts is not common practice in slaughterhouses, and thus the burden for the prevention of trichinosis falls to the consumer. In 1986, the US government approved the use of low-dose irradiation of fresh pork to combat trichinosis. Drugs have little effect on cysts.

HOOKWORM DISEASE IS AN INFECTION OF THE UPPER INTESTINE

Necator americanus

Hookworms are roundworms that have a set of hooks or sucker devices for firm attachment to tissues of the host. Two hookworms, both about 10 mm in length, may be involved in human disease. The first is the **Old World hookworm**, *Ancylostoma duodenale*, which is found in Europe, Asia, and the United States; the second is the **New World hookworm**, *Necator americanus*, which is prevalent in the Caribbean islands (where it may have been brought by slaves from Africa).

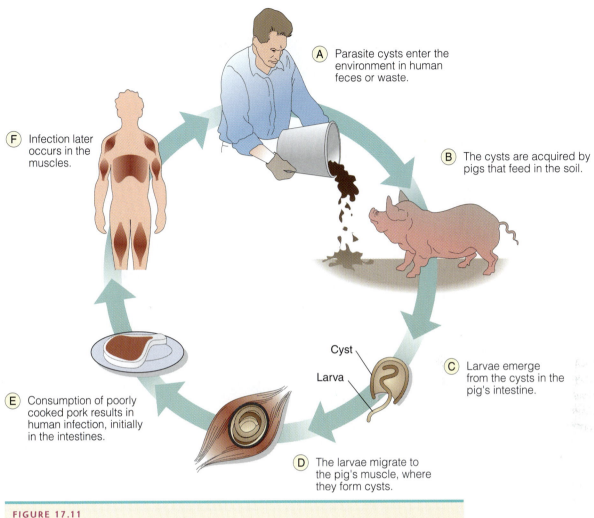

A Parasite cysts enter the environment in human feces or waste.

B The cysts are acquired by pigs that feed in the soil.

C Larvae emerge from the cysts in the pig's intestine.

D The larvae migrate to the pig's muscle, where they form cysts.

E Consumption of poorly cooked pork results in human infection, initially in the intestines.

F Infection later occurs in the muscles.

Cyst

Larva

FIGURE 17.11

The Life Cycle of *Trichinella spiralis*

Hundreds of millions of people around the globe are believed to be infected by hookworms. These parasites live in the human intestine, where they suck blood from the tissues. Hookworm disease therefore is accompanied by blood loss and is generally manifested by anemia. Cysts also may become lodged in the intestinal wall, and ulcer-like symptoms may develop.

The life cycle of a hookworm involves only a single host, the human (FIGURE 17.12). Hookworm eggs are excreted to the soil, where the larvae emerge as long, rod-like **rhabditiform** larvae. These later become thread-like **filariform** larvae that attach themselves to vegetation in the soil. When contact with bare feet is made, the filariform larvae penetrate the skin layers and enter the bloodstream. Soon, they localize in the lungs and are carried up to the pharynx in secretions, and then swallowed into the intestines.

Hookworms are common where the soil is warm, wet, and contaminated with human feces. The disease is prevalent where people go barefoot. Mebendazole may be used to reduce the worm burden, and the diet may be supplemented with iron to replace that lost in the loss of blood. It should be noted that dogs and cats also harbor hookworm eggs and pass them in the feces.

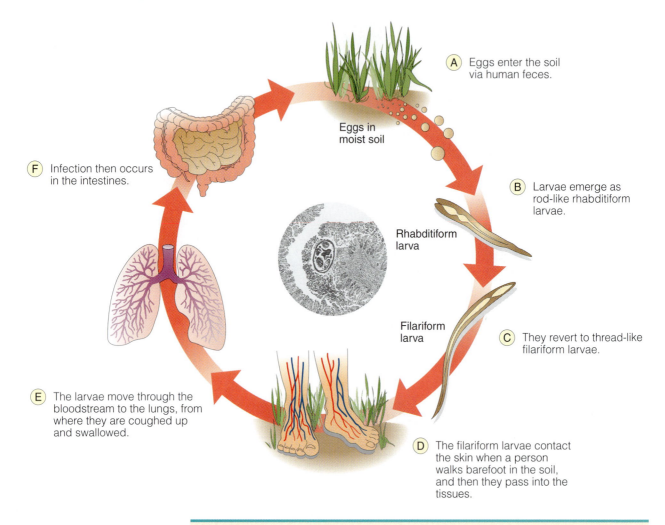

A Eggs enter the soil via human feces.

Eggs in moist soil

F Infection then occurs in the intestines.

B Larvae emerge as rod-like rhabditiform larvae.

Rhabditiform larva

Filariform larva

C They revert to thread-like filariform larvae.

E The larvae move through the bloodstream to the lungs, from where they are coughed up and swallowed.

D The filariform larvae contact the skin when a person walks barefoot in the soil, and then they pass into the tissues.

FIGURE 17.12

The Life Cycle of the Hookworms *Ancylostoma duodenale* and *Necator americanus*

The photograph shows an egg lodged in intestinal tissue.

Strongyloides stercoralis

STRONGYLOIDIASIS IS A PARASITIC INTESTINAL INFECTION

Strongyloidiasis is caused by *Strongyloides stercoralis*, a parasite that resembles hookworms in appearance, distribution, and life cycle. Adult worms inhabit the small intestine, especially the duodenum, where they cause abdominal pain, nausea, vomiting, and diarrhea alternating with constipation. Pulmonary symptoms mimic the pneumonia induced by hookworms. A drug called thiabendazole provides effective therapy.

Strongyloidiasis is of importance to Americans because many Vietnam veterans were exposed to the parasites in Southeast Asia. In a 1981 study, for example, doctors tested 530 veterans of Pacific wars for the parasite and found that 43 harbored it in their stools. In some cases, this was almost 40 years after the initial infection. Many reported regular five-day episodes of itchy skin rash, another sign of the disease.

FILARIASIS IS A LYMPHATIC SYSTEM INFECTION

Filariasis is a parasitic disease caused by a roundworm named *Wuchereria bancrofti*. The worm breeds in the tissues of the human lymphatic system and causes extensive inflammation and damage to the lymphatic vessels and lymph glands. After years of infestation, the arms, legs, and scrotum swell enormously and become distorted with fluid. This condition is known as **elephantiasis** because of the gross swelling of lymphatic tissues and the resemblance of the skin to elephant hide (FIG-URE 17.13).

The female form of *Wuchereria bancrofti* is about 100 mm long. Its fertilized eggs give rise to tiny eel-like microfilariae, which enter the human bloodstream. The microfilariae then are ingested by mosquitoes during a blood meal, where they develop into infective larvae passed along to another human during the next blood meal. The larvae subsequently grow to adults and infect the lymphatic system. Infection is limited to the number of larvae injected by the mosquito, since microfilariae cannot grow to adulthood without first passing through the mosquito.

Filariasis is prevalent where mosquitoes are plentiful, such as in the hot, humid climates of Central and South America and the Caribbean islands. Missionaries, emigrants, and visitors from these areas often carry the worms in their tissues.

Wuchereria bancrofti

GUINEA WORM DISEASE CAUSES SKIN ULCERS

The **Guinea worm** is a roundworm, thought to have been introduced to Africa, India, and the Middle East by navigators who plied the sea-lanes between these areas and South Pacific islands such as New Guinea. The worm's scientific name is *Dracunculus medinensis*.

The male Guinea worm is small, but the female may be up to 100 cm in length. Often the worm lies just below the skin of humans and causes swellings that resemble varicose veins. It causes a skin ulcer through which larvae are discharged into water. A small shrimp-like crustacean called a copepod picks up the larvae and transmits them to another human when the copepod is consumed in water. Alternately, the copepod is eaten by fish, which then transmit the larvae (MicroFocus 17.3). Once in humans, the larvae migrate to the human skin and grow to adults.

The Guinea worm exposes itself through the skin ulcer and thus can be removed by careful winding on a stick. This primitive but time-honored method must be performed cautiously and slowly. The site of the ulcer burns intensely, and the

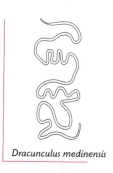

Dracunculus medinensis

FIGURE 17.13

The Effect of Filariasis

Elephantiasis of the leg caused by the parasite *Wuchereria bancrofti*. The worm breeds in the tissues of the lymphatic vessels and damages them. As fluid accumulates, the legs swell and become distorted.

WELCOME TO NEW YORK CITY!

The Guinea worm had long been a parasite among Scandinavian fishermen and their families. Individuals acquired the worm by drinking water that contained infected crustaceans, or by eating fish that had previously eaten the crustaceans. For the Guinea worm, New York City appeared to be a long way off, but a somewhat circuitous route made the transition possible during the 1800s.

Scandinavians immigrated to the United States in substantial numbers during the nineteenth century. They settled along the shores of the Great Lakes, and, as their feces made their way into the lakes, the Guinea worm followed along. Eventually, fish from the Great Lakes became infected with larvae of the worm.

As the years passed, substantial commerce in fish developed between the Midwest and New York City. Jewish homemakers were particularly fond of pike, pickerel, and carp from the Great Lakes, and they used the fish to make a delicacy called gefilte fish. Carefully they pressed minced fish, eggs, and seasonings into balls and boiled the mixtures. During cooking, however, they often tasted the gefilte fish to see if it was done. While sampling the uncooked fish, they unwittingly acquired roundworm larvae and became new hosts for the Guinea worms. The transition was complete.

Guinea worm infection is now rare in New York City or elsewhere in the United States. Sanitary practices, fish inspection, and the commercial production of gefilte fish have limited the spread of the worm. Gefilte fish is still popular among Jewish people, but the unwanted hitchhikers largely have been eliminated.

---

fever and local burning are often described as "fiery." Partly because of this perception, it is believed that the "fiery serpents" mentioned in the Bible's Book of Numbers may have been Guinea worms. Also, the stick with worms wound around it may be the source of the serpent on a staff that is the symbol of healing used by the medical profession.

As of 1991 over 420,000 cases of Guinea worm disease were reported in 17 African countries and parts of India and Pakistan. By that time, eradication programs had been established in Ghana and Nigeria to interrupt the spread of the disease by providing safe sources of drinking water, by teaching populations at risk to boil or filter contaminated water, and by treating drinking water with chemicals. By 2000, the global incidence was about 75,000 cases. The disease still is found in sub-Saharan Africa, especially in Sudan and West Africa.

EYEWORM DISEASE OCCURS IN THE RAIN AND SWAMP FORESTS OF AFRICA

Loa loa

The **eyeworm** is a type of roundworm called *Loa loa*. This parasite, native to West and Central Africa, is about 60 mm long. It is often found in insects such as deerflies and horseflies and is injected into the subcutaneous tissues of humans during an insect bite. The parasite may remain in the connective tissues for many months.

Loa loa is called the eyeworm because it is often attracted to the surface of the eye by warm temperatures. Commonly, it appears on the cornea (FIGURE 17.14). Here it causes conjunctivitis and painful irritation of the eye muscles. Swellings of the extremities are observed also. The parasite returns to the subcutaneous tissues as the skin temperature cools.

TABLE 17.2 summarizes the human diseases caused by roundworms.

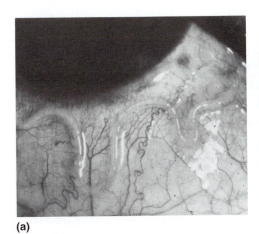
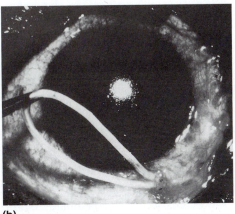

(a) (b)

FIGURE 17.14

The Eyeworm *Loa loa*

Two views of the human eye showing infection with *Loa loa*. (a) The worm is seen below the conjunctiva along the white of the eye. (b) The worm is being removed from the eye.

TABLE 17.2

A Summary of Roundworm Parasitic Diseases in Humans

ORGANISM	DISEASE	TRANSMISSION	ORGANS AFFECTED	CHARACTERISTIC SIGN	ANIMAL HOST
Enterobius vermicularis	Pinworm disease	Contact Food, clothing	Intestine	Anal itching	None significant
Trichuris trichiura	Whipworm disease	Food Water	Intestine	Abdominal pain	None significant
Ascaris lumbricoides	Roundworm disease	Food, water Contact	Intestine Lungs	Emaciation Pneumonia	None significant
Trichinella spiralis	Trichinosis	Pork consumption	Intestine Muscles Eyes	Diarrhea Muscle pain Loss of eye movement	Pig
Ancylostoma duodenale Necator americanus	Hookworm disease	Contact with moist vegetation	Intestine Lungs Lymph	Anemia Abdominal pain	None significant
Strongyloides stercoralis	Strongyloidiasis	Contact with moist vegetation	Intestine Lungs	Anemia Abdominal pain	None significant
Wuchereria bancrofti	Filariasis	Mosquito	Lymph vessels	Edema Elephantiasis	Mosquito
Dracunculus medinensis	Guinea worm disease	Food Water	Skin	Skin ulcer	Copepod Fish
Loa loa	Eyeworm disease	Deerflies Horseflies	Eye Connective tissues	Conjunctivitis	Insects

Note to the Student

Commensalism refers to an association between two populations of organisms in which only one population benefits while causing no apparent harm to the other. Certain bacteria that inhabit the human skin are one example.

Distasteful as it may be, we must now broaden our definition of commensalism to include certain multicellular parasites, for indeed, many of us harbor these worms in our organ menageries. As we have seen in this chapter, infection by multicellular parasites need not be fatal. Moreover, we should realize how widespread the parasites are. According to estimates by the World Health Organization, more people are infected with multicellular parasites than anyone previously imagined. With the emergence of developing nations and travels by Americans to remote parts of the world, the parasites will continue to enter our society in ever-increasing numbers. They may represent an area of public health concern in the years ahead.

Summary of Key Concepts

17.1 FLATWORMS

■ **Flukes Use Snails as an Intermediate Host.** Snails are intermediate hosts for the larval stage of fluke reproduction, while humans are the definitive host for the sexually mature adults. The life cycles include several phases, including miracidium, sporocyst, redia, cercaria, and metacercaria.

■ **Schistosomiasis Is Very Common Worldwide.** One of the major fluke parasites of humans is the blood fluke (*Schistosoma*) that causes schistosomiasis. Symptoms include fever and chills. If eggs are carried by the blood to the liver, liver damage results. Bladder infections produce bloody urine.

■ **Chinese Liver Fluke Disease Involves Two Intermediate Hosts.** The Chinese liver fluke (*Clonorchis*) has two intermediate hosts, the snail and fish. Symptoms of a human infection may be duct blockage of the gall bladder and poor digestion of fats. Liver damage also can occur.

■ **Other Flukes Also Cause Human Disease.** The lung fluke (*Paragonimus*) is common in Asia and the South Pacific when people eat undercooked, infected crabmeat. The flukes then pass to the lungs and cause breathing difficulties. The intestinal fluke (*Fasciolopsis*) penetrates the human intestinal wall and then migrates to the liver, causing substantial liver damage.

■ **Tapeworms Are Flatworms that Live in the Host Intestines.** Tapeworms are long, segmented worms belonging to the class Cestoda in the animal phylum Platyhelminthes. The worms consist of segments called proglottids and a head region called the scolex, often with hooks or sucker devices for attachment to the host tissue. Tapeworms can infect all mammals and generally are transmitted to humans in foods.

■ **Beef and Pork Tapeworm Diseases Usually Are Chronic but Benign.** Examples of foodborne tapeworms are the beef tapeworm (*Taenia saginata*) and the pork tapeworm (*T. solium*). Living in the intestines, the tapeworms expel many gravid proglottids daily. Humans acquire the parasite as cysts in poorly cooked beef or pork. Symptoms are mild.

■ **Fish Tapeworm Disease Results from Eating Undercooked Infected Freshwater Fish.** The fish tapeworm (*Diphyllobothrium*) includes two intermediate hosts, copepods and fish. Infection may obstruct the intestine and cause anemia.

■ **A Few Other Tapeworms Also Cause Disease.** Other tapeworms are acquired from the soil, including the dwarf tapeworm, which is the most common tapeworm in humans. The dog tapeworm (*Echinococcus granulosus*) can be spread to humans, who are an intermediate host.

17.2 ROUNDWORMS

■ **Pinworm Disease Is the Most Prevalent Helminthic Infection in the United States.** The pinworm (*Enterobius*) is acquired as eggs from the soil. Symptoms include diarrhea and anal itching.

■ **Whipworm Disease Comes from Eating Food Contaminated with Parasite Eggs.** The whipworm (*Trichuris*) causes damage to the intestinal lining and appendicitis-like pain may be experienced.

■ **Roundworm Disease Is Prevalent in the American South.** The roundworm (*Ascaris*) larvae grow in the small intestine and reach maturity after two months. Intestinal blockage may occur and the small intestine may be perforated. Spread to the blood and lungs can cause pneumonia.

■ **Trichinosis Occurs Worldwide.** An important human parasite is the pork roundworm (*Trichinella*) acquired in undercooked pork. Symptoms include intestinal pain, nausea, vomiting, and constipation.

Complications occur when the parasite larvae move to the muscles and form cysts. Larvae also can invade the brain and cause paralysis. Death can occur.

■ **Hookworm Disease Is an Infection of the Upper Intestine.** Hookworms (*Necator* and *Ancylostoma*) are soilborne. On infection, they suck blood from the human intestinal wall, causing anemia. Cysts can form in the intestinal wall.

■ **Strongyloidiasis Is a Parasitic Intestinal Infection.** This intestinal infection is caused by *Strongyloides stercoralis.* Infection causes abdominal pain, nausea, vomiting, and diarrhea that alternate with constipation. Pneumonia-like symptoms can be caused by lung infections.

■ **Filariasis Is a Lymphatic System Infection.** The filarial worm *(Wuchereria)* that causes elephantiasis is mosquitoborne. Blockage of the lymphatic system produces swelling as fluid accumulates in the legs.

■ **Guinea Worm Disease Causes Skin Ulcers.** Skin ulcers are produced by the Guinea worm, *Dracunculus medinensis.*

■ **Eyeworm Disease Occurs in the Rain and Swamp Forests of Africa.** Arthropods transmit the eyeworm (*Loa loa*). The worm can remain in the connective tissue of the eye for several months.

Questions for Thought and Discussion

Answers to selected questions can be found in Appendix C.

1. A diplomat visits a foreign country with which relations have recently been established. She observes that the people eat raw fish, wear no shoes, consume snails as a regular part of their diet, enjoy watercress and water chestnuts in their salads, and do not believe in pesticides. What report might she make on her return to the United States?

2. As of 2003, the World Health Organization reported that malaria was the most prevalent tropical disease (300 million cases per year). The next two were schistosomiasis and filariasis (200 million and 90 million annual cases, respectively). How do you believe the incidence of these diseases can be reduced on a global scale?

3. Because tapeworms have no intestines, they must obtain their nutrients by absorbing organic matter from the external environment in the intestines of humans or animals. How does this observation dispel the notion that evolution always yields animals more complex than their predecessors?

4. Federal law now stipulates that food scraps fed to pigs must be cooked to kill any parasites present. It also is known that feedlots for swine are generally more sanitary than they have been in the past. As a result of these and other measures, the incidence of trichinosis in the United States has declined, and the acceptance of "pink pork" has increased. Do you think this is a dangerous situation? Why?

5. In the 1960s, the Aswan High Dam was built in Egypt to retain the water of the Nile River for irrigation.

The agricultural productivity of the region increased significantly but was accompanied by a dramatic increase in the snail population in the water. Before long, the number of cases of schistosomiasis in farmers had doubled. How are all these events related?

6. The Greek stems *di-* and *phyllo-* infer "double-thin" (filo dough is thin dough used in Greek pastry). *Bothros* is also a Greek word, meaning "dirty water." *Latum* is the Latin word for "broad" or "wide." How does this knowledge of classical languages help in remembering the description of the parasite *Diphyllobothrium latum*?

7. In Japan, eating raw fish is a traditional and honored custom. Are there any hazards involved in this practice?

8. Why are some veterinarians inclined to recommend that dog and cat owners avoid buying pet food that contains liver?

9. A person finds that he has difficulty digesting fats. Mayonnaise makes him ill, fried foods cause diarrhea, and he cannot eat salads made with oily salad dressings. However, he does enjoy raw fish dishes, such as sushi and sashimi. What might be his problem?

10. A group of campers was sitting around a fire on a chilly summer evening, when one noticed a threadlike body in the eye of another camper. What might the body have been, and what conditions might have led to its appearance at that time?

11. A newspaper report once indicated that for every case of cancer, the National Institutes of Health spends $209 annually on research. By contrast, for every case of blood fluke disease (schistosomiasis), the amout is $0.04. What factors might account for this sharp distinction in spending? Would you support increased expenditures for schistosomiasis research?

12. Trichinosis may occur in Italian, Polish, and German communities, but it is extremely rare in Jewish communities. Why is this so? How can the local butcher play a key role in preventing the spread of trichinosis among his customers?

13. It has been suggested that with fewer and fewer of the world's peoples walking barefoot, the possibility of infection by skin-penetrating parasites will continue to decline. How many other instances can you name where a change in living habits will bring about a decline in parasitic diseases? Now consider the reverse. How many instances can you name where changes in peoples' habits will bring an increasing incidence of parasitic disease?

14. Certain restaurants offer a menu item called steak tartare. Aficionados know that the beef in this dish is served raw. What hazard might this meal present to the restaurant patron?

15. A biologist, writing in a 1984 publication, stated: "Perhaps the most important reason for discussing parasitical diseases is that they highlight just how enmeshed we are in the web of life. . . ." How many examples can you find in this chapter to support this concept?

http://microbiology.jbpub.com

The site features **eLearning,** an on-line review area that provides quizzes and other tools to help you study for your class. You can also follow useful links for in-depth information, or just find out the latest microbiology news.

Review

Consider the characteristic on the left and the three possible choices on the right. Select the disease(s) or parasite name(s) that best apply to the characteristic, and place the letter(s) next to the characteristic. The answers are listed in Appendix D.

_____ 1. Transmitted by an arthropod.
   a. Filariasis
   b. Trichinosis
   c. Hookworm disease

_____ 2. Beef tapeworm.
   a. *Echinococcus granulosus*
   b. *Fasciola hepatica*
   c. *Taenia saginata*

_____ 3. Type of fluke.
   a. *Schistosoma*
   b. *Necator*
   c. *Fasciola*

_____ 4. Causes a lung infection.
   a. Guinea worm disease
   b. Trichinosis
   c. Strongyloidiasis

_____ 5. Infects the human intestines.
   a. *Trichinella spiralis*
   b. *Ascaris lumbricoides*
   c. *Fasciolopsis buski*

_____ 6. Type of tapeworm.
   a. *Paragonimus*
   b. *Hymenolepis*
   c. *Fasciola*

_____ 7. Snail is the intermediate host.
   a. Blood fluke
   b. Dog tapeworm
   c. Intestinal fluke

_____ 8. Attaches to host tissue by hooks.
   a. *Necator*
   b. *Diphyllobothrium*
   c. *Loa*

_____ 9. Affects pigs as well as humans.
   a. *Loa loa*
   b. *Trichinella spiralis*
   c. *Clonorchis sinensis*

_____ 10. Life cycle includes miracidium and cercaria.
   a. *Schistosoma*
   b. *Fasciola*
   c. *Paragonimus*

_____ 11. Acquired by consuming contaminated pork.
   a. *Taenia soleum*
   b. *Loa loa*
   c. *Necator americanus*

_____ 12. Classified in the phylum Aschelminthes.
   a. *Dracunculus*
   b. *Wuchereria*
   c. *Schistosoma*

_____ 13. Infects tissues of the human eye.
   a. *Schistosoma*
   b. *Clonorchis*
   c. *Loa*

_____ 14. Male and female forms exist.
   a. *Hymenolepis*
   b. *Ascaris*
   c. *Trichuris*

_____ 15. Forms hydatid cysts.
   a. Fish tapeworm
   b. Guinea worm
   c. Dog tapeworm

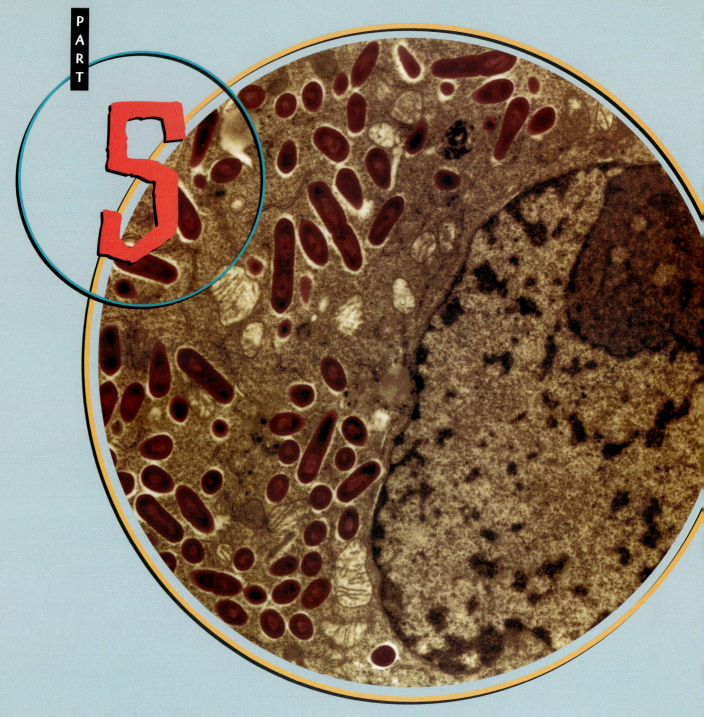

Disease and Resistance

I n past centuries, the spread of disease appeared to be willfully erratic. Illnesses would attack some members of a population while leaving others untouched. A disease that for many generations had taken small, steady tolls would suddenly flare up in epidemic proportions. And strange, horrifying plagues descended unexpectedly on whole nations.

Scientists now know that humans live in a precarious equilibrium with the microorganisms that surround them. Generally the relationship is harmonious, because humans can come in contact with most microorganisms and develop resistance to them. However, when the natural resistance is unable to overcome the aggressiveness of microorganisms, disease sets in. In other instances, the resistance is diminished by a pattern of human life that gives microorganisms the edge. For example, during the Industrial Revolution of the 1800s, many thousands of Europeans moved from rural areas to the cities. They sought new jobs, adventure, and prosperity. Instead, they found endless labor, unventilated factories, and wretched living conditions—and they found disease.

In Part 5 of this text, we shall explore the infectious disease process and the mechanisms by which the body responds to disease. Chapter 18 will open with an overview of the host-parasite relationship and the factors that contribute to the establishment of disease. In Chapter 19, the discussion turns to nonspecific and specific methods by which body resistance develops, with emphasis on the immune system. Various types of immunity are explored in Chapter 20, together with a survey of laboratory methods that use the immune reaction in the diagnosis of disease. In Chapter 21 the discussion centers on immune disorders that lead to serious problems in humans. In these chapters, we shall ferret out the roots of infectious disease and resistance and come to understand them at the fundamental level.

■ Legionella pneumophila *infecting a macrophage*

HEALTH AND MEDICINE

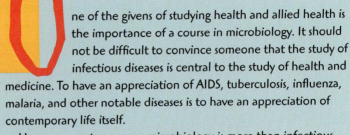

One of the givens of studying health and allied health is the importance of a course in microbiology. It should not be difficult to convince someone that the study of infectious diseases is central to the study of health and medicine. To have an appreciation of AIDS, tuberculosis, influenza, malaria, and other notable diseases is to have an appreciation of contemporary life itself.

However, contemporary microbiology is more than infectious diseases. Over the past 25 years or so, the discipline of immunology has ingrained itself into modern health care. For instance, dealing with allergy is dealing with an immunological problem. Diseases such as lupus and rheumatoid arthritis are the domain of immunology, as is Rh disease of the newborn. Geriatric medicine is related to immunology because the immune systems of the elderly function less efficiently and therefore leave individuals more susceptible to infectious disease.

Many microbiological tests performed in today's laboratories are based in immunology. Where older tests attempted to detect the presence of microorganisms, the newer tests detect antibodies produced in response to these organisms. Antibody tests also are being used to detect hormones in blood and urine (such as the hormone test for pregnancy). Before a transplant is performed, immunological tests are required to determine the compatibility of recipient and donor tissues. Antibody tests are used even in forensic medicine to identify blood types and other secretions.

As a subdiscipline of microbiology, immunology has emerged to become an important facet of the health care system. Many health science curricula offer separate courses in immunology, but for many undergraduates, the exposure to immunology begins and ends with the microbiology course. For this reason, it is important to understand not only the philosophical significance of immunology, but also the practical benefit of studying the immune system.

Infection and Disease

Tastes like swamp water.

—Australian physician Barry J. Marshall remarking to a colleague, as he secretly drank a culture of *Helicobacter pylori* to prove it causes peptic ulcers

CHOLERA BROKE OUT IN EUROPE in the 1840s and reached London in June 1849. One of the most severely affected areas was the district around Golden Square, where in a 10-day period in August, over 500 people died of the disease.

A physician named John Snow lived close to Golden Square. Snow had long been interested in cholera, and the outbreak provided an opportunity to continue and concentrate his study. The prevailing wisdom was that bad air and direct contact spread the disease, but Snow believed they played negligible roles. His beliefs would be strengthened by his observations.

Snow noted that most cholera patients in the Golden Square district drew their water from a well on Broad Street. The water was obtained by a hand-operated pump accessible to all. Snow discovered that the well was contaminated by the cesspool overflow from a tenement in which a cholera patient lived, and he concluded that water was the source of the disease. On September 7, 1849, he presented his findings to the local community council. Snow's study impressed the council members, and they inquired how he intended to stop the epidemic. Snow thought for a moment and replied with the now-classic solution: "Take the handle off the Broad Street pump." By the following day, the handle was gone; shortly thereafter, the epidemic subsided.

Unfortunately, not all epidemics are quite as easy to interrupt. In this chapter, we shall discuss the complex mechanisms that underlie the spread and development of infectious disease. Individual diseases are considered in many other chapters of this text, and our purpose here is to bring together many concepts of disease and synthesize an overview of the host-parasite relationship. We shall summarize much of the important terminology used in medical microbiology and focus on the methods used by microorganisms to establish themselves in the tissues. This study will prepare us for a detailed discussion of resistance mechanisms in the following chapters.

18.1 The Host–Parasite Relationship

The word **infection** is derived from Latin origins meaning "to mix with" or "to corrupt." The term refers to the relationship between two organisms, the host and the parasite, and the competition for supremacy that takes place between them. A host whose resistance is strong remains healthy, and the parasite is either driven from the host or assumes a benign relationship with the host. By contrast, if the host loses the competition, disease develops. The term **disease** appears to have originated from Latin stems that mean "living apart," a reference to the separation of ill individuals from the general population. Disease may be conceptualized as any change from the general state of good health. It is important to note that disease and infection are not synonymous; a person may be infected without becoming diseased.

THE HUMAN BODY IS HOST TO ITS OWN MICROBIOTA

The human body harbors more than 10 trillion microbes that represent its **microbiota** or **normal flora** (FIGURE 18.1). The human microbiota is a population of microorganisms that reside in the body without causing disease. Some organisms in the population establish a permanent relationship with the body, while others are more transient and are present for limited periods of time. In the large intestine of

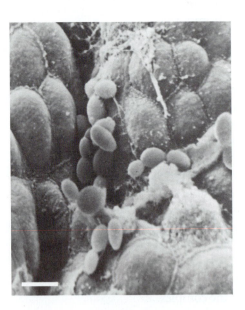

FIGURE 18.1

The Host Microbiota

A scanning electron micrograph of the intestinal lining showing *Candida albicans* attached to the surface. The yeast form of the cells is apparent. (Bar = 4 μm.)

humans, for example, *Escherichia coli* and *Candida albicans* are almost always found, but streptococci are transient.

The relationship between the body and its microbiota is an example of a **symbiosis**, or living together. In some cases, the symbiosis is beneficial to both the body and the microorganisms. This relationship is called **mutualism**. For example, species of *Lactobacillus* live in the human vagina and derive nutrients from the environment while producing acid to prevent the overgrowth of other organisms. In other cases, the symbiosis is beneficial only to the microorganisms, in which case the symbiosis is called **commensalism**. *Escherichia coli* is generally presumed to be a commensal in the human intestine, although some evidence exists for mutualism because the bacteria produce certain amounts of vitamins B and K. In addition, the microbiota usually will out-compete invading pathogens, thereby protecting the body from dangerous infections.

Microbiota may be found in several body tissues. On the skin, for instance, there are various forms of viruses, fungi, and bacteria, particularly staphylococci and *Propionibacterium acnes*. The oral cavity (tongue, teeth, and gums) commonly contains members of the genera *Neisseria*, *Leptotrichia*, and *Bacteroides*, as well as many diphtheria-like bacilli (diphtheroids), fungal spores, and streptococci. Many of the bacteria are related to bad breath, as MicroFocus 18.1 explores. The upper respiratory tract is the site of all these organisms, as well as pneumococci and species of *Haemophilus* and *Mycoplasma*. These organisms may cause respiratory disease if the body defenses are compromised.

MicroFocus 18.1

BIG BUSINESS

The newest wrinkle in the franchise business is the bad-breath clinic, the "halitosis heaven." Advertisements for these clinics promise cure rates of 98 percent, and, for those unable to visit in person, home treatment programs are available. People with a breath problem can even use the Internet to "talk" with halitosis experts and receive treatment—as long as a credit card is available. Even NBC's *Today Show* and ABC's *20/20* have featured segments on the perils of dragon mouth.

But, is a breath makeover really necessary? Generally not, if you realize the cause. High on the list are the pungent foods we eat—for example, garlic seeps out of our digestive tract and comes back to haunt us through the lungs and skin. And there are numerous disease-associated bad breaths: the fruity odors of diabetes, the fishy odors from kidney problems, and the cheesy smells from tonsil inflammations.

Aside from these, 90 percent of halitosis cases are related to anaerobic bacteria breaking down the tiny particles of food between the teeth, under the gums, and beneath braces or dentures. As they multiply merrily in their myriad morsels of munchies, bacteria produce sulfur gases not unlike those in a swamp. (Indeed, a rotten egg and a rotten mouth find common ground in the same sulfur gases.) The area at the back of the tongue is a notoriously good hangout for bacteria, especially since everything we eat passes that way.

So, what's a clean breath maven to do? First off, brush and floss vigorously and often to get rid of the bacteria and their meals. (It's a good idea to gently brush your tongue, especially the back part, as well.) Chew gum, preferably sugar-free, to increase saliva flow and keep saliva's natural antibacterial agents flowing. Snack on apples, carrots, and other fiber-rich foods that scrub the teeth between meals. And drink lots of water to reduce stagnation in the mouth and prevent bacteria from gaining a foothold.

The over-the-counter cures for bad breath (mints, sprays, and mouthwashes) have become a billion-dollar industry. But, we do not need to spend a penny if we use a bit of common sense. And a bit of microbiology.

The stomach in humans is generally without a normal flora, mainly because of the low pH of its contents. However, the latter part of the small intestine and the large intestine abound with microorganisms. *Bacteroides* species are numerous, together with *Clostridium* spores, various streptococci, and a number of gram-negative rods, including species of *Enterobacter, Klebsiella, Proteus,* and *Pseudomonas.* MicroFocus 18.2 describes some benefits of the gut bacteria. In females, *Lactobacillus* is a notable component of the vagina; other organisms may be located near the urogenital orifices in both males and females. The blood, urine, and internal organs are usually **sterile** (devoid of microorganisms) unless disease is in progress.

Organisms of the microbiota are introduced when the newborn passes through the birth canal. Additional organisms enter when breathing begins and upon first feeding, as FIGURE 18.2 illustrates. Within two to three days, most common micro-

MicroFocus 18.2

A WARM AND FUZZY GUT FEELING

There is a tendency to think of all bacteria and other microorganisms as the "bad guys" because just a few may cause disease. In reality, many microbes actually help us evade and eliminate the "bad guys" as well as help us in other ways. Take your gut (gastrointestinal tract) as an example.

Adults contain somewhere between 500 and 1,000 different species of bacteria in their gut. If you add up all these bacteria (part of the total normal human microbiota), they outnumber all the cells in your body at least tenfold. What do all these bacteria do? We know that the well-known residents, like *Escherichia coli*, help with water reabsorption and produce some of the vitamin K we need in our diet. In return, we provide a very hospitable environment for the bacteria and they can survive on some of the digested food we process. But, what about all the other bacteria that have been identified in the gut? What role do they play—if any?

Since the 1950s, investigators have manipulated and engineered special strains of mice that have germ-free guts; that is, their guts are without any microorganisms; not even the microbiota are present. Interestingly, such mice need to consume about 30 percent *more* food calories to maintain their weight. That might seem surpris-

ing. Less surprising is that the mice also are very susceptible to infections and disease. Without the microbiota, the classic view states there is nothing to compete with invading pathogens, so they have an easier time establishing an infection.

So, one way to examine what the various members of the normal microbiota do is to introduce each species separately into germ-free mice and see what happens. That is exactly what Jeffrey Gordon and colleagues at Washington University School of Medicine in St. Louis did.

Bacteroides thetaiotaomicron, a gram-negative anaerobe, is less well known than bacteria like *E. coli.* However, *B. thetaiotaomicron* exists in the human gut at concentrations 1,000 times greater than *E. coli,* so it must be doing something. What Gordon and the other investigators did was introduce *B. thetaiotaomicron* in the guts of germ-free mice and monitor what happened. They quickly discovered that the mice synthesize fucose, a monosaccharide sugar, onto the surface of the intestinal cells. Germ-free mice do not. Apparently, *B. thetaiotaomicron* has provided a signal or stimulus to the intestinal cells telling them to turn on the genes for fucose synthesis. Why fucose? This is the sugar that *B. thetaiotaomicron* uses for energy and metabolism.

Using DNA techniques that allow large numbers of genes to be monitored all at once, the Washington University group discovered that *B. thetaiotaomicron* actually triggers the intestinal cells in germ-free mice to turn on or turn off some 100 genes of the approximately 25,000 genes in the mouse genome. Some of these genes help the mice absorb and metabolize sugars and fats. Other genes activated by the bacteria produce products that help protect the intestinal wall from being penetrated by other normal microbiota and pathogens. So, maybe the normal microbiota does something more active than simply competing with invading pathogens. Other genes activated stimulate the growth of new blood vessels. This explains why germ-free mice had to eat 30 percent more calories to maintain body weight. Germ-free mice have a less well developed blood vessel system and thus were inefficient at absorbing nutrients.

The result is that the physical development of the normal gut in mice (and would assume in humans too) is dependent on the normal microbiota. It might be scary to think about, but much of our gut physiology might be controlled by the microbial genomes of our microbiota! Scary, yes, but I have a warm and fuzzy gut feeling knowing they are there.

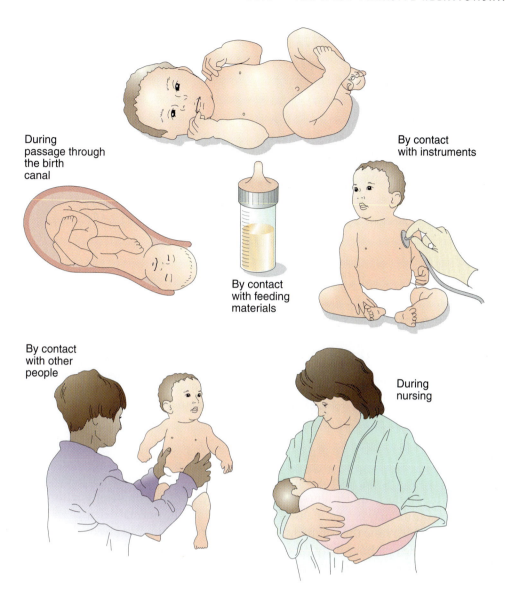

FIGURE 18.2

Five Possible Origins
of the Microbiota
in a Newborn

During
passage through
the birth
canal

By contact
with instruments

By contact
with feeding
materials

By contact
with other
people

During
nursing

biota organisms have appeared. During the next few weeks, contact with the mother and other individuals will expose the child to additional microorganisms. The human microbiota remains throughout life, undergoing changes in response to the internal and external environment of the individual.

PATHOGENS DIFFER IN THEIR ABILITY TO CAUSE DISEASE

Pathogenicity refers to the ability of a microorganism to gain entry to the host's tissues and bring about a physiological or anatomical change, resulting in altered health and leading to disease. The word is derived from the Greek term *pathos*, meaning "suffering." The term **pathogen** has the same root and refers to an organism having the ability to cause disease. The symbiotic relationship between host and parasite is called **parasitism.** FIGURE 18.3 compares some of the most prolific pathogens.

Microorganisms vary greatly in their pathogenicity. Certain pathogens, such as the cholera, plague, and typhoid bacilli, are well known for their ability to cause serious

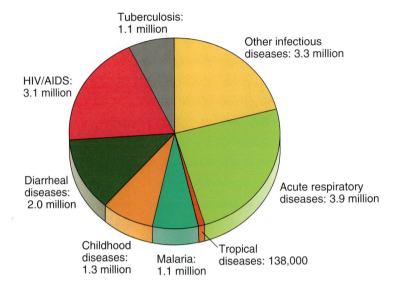

Tuberculosis: 1.1 million

Other infectious diseases: 3.3 million

HIV/AIDS: 3.1 million

Diarrheal diseases: 2.0 million

Acute respiratory diseases: 3.9 million

Childhood diseases: 1.3 million

Malaria: 1.1 million

Tropical diseases: 138,000

FIGURE 18.3

Infectious Disease Deaths Worldwide—2001

This pie chart depicts the leading causes of infectious disease and the number of worldwide deaths they caused in 2001, as reported by the World Health Organization. Tropical diseases include trypanosomiasis, Chagas' disease, schistosomiasis, leishmaniasis, and filariasis.

human disease. Others, such as common cold viruses, are considered less pathogenic because they induce milder illnesses. Still other microorganisms are **opportunistic**. These organisms may be commensals in the body until the normal defenses are suppressed, at which time the commensals seize the "opportunity" to invade the tissues and express their pathogenicity. An example is observed in individuals with acquired immune deficiency syndrome (AIDS). These patients are highly susceptible to opportunistic organisms such as *Pneumocystis jiroveci* (formerly called *P. carinii*) and *Toxoplasma gondii*. Other new pathogens and diseases are depicted in **FIGURE 18.4**. **MicroFocus 18.3** illustrates that host-parasite relationships go beyond humans.

Diseases caused by opportunistic organisms illustrate how a shift in the body's delicate balance of controls may convert "infection" to "disease." Another example happened in the 1950s when antibiotics came into widespread use. As bacteria in the intestinal microbiota disappeared, the biological controls exerted on *Candida albicans* vanished and candidiasis became common. Thus, a chance upset in resistance mechanisms or control shifts may enhance the ability of organisms to establish disease.

A sobering concept has emerged in recent years. Microbiologists once believed that organisms were either pathogenic or nonpathogenic. This distinction has been blurred by the realization that otherwise benign organisms may become pathogenic when body defenses weaken or fail. The AIDS epidemic points up this concept well. Transplant and cancer patients treated with radiation and immunosuppressant drugs also are affected. It now is recognized that pathogenicity is a function of the aggressive nature of the parasite, as well as the level of resistance in the host.

The word **virulence** is used to express the degree of pathogenicity of a parasite. This term is derived from the Latin *virulentus*, meaning "full of poison." An organism that invariably causes disease, such as the typhoid bacillus, is said to be "highly virulent." By comparison, an organism that sometimes causes disease, such as *Candida albicans*, is labeled "moderately virulent." Certain organisms, described as **avirulent**, are not regarded as disease agents. The lactobacilli and streptococci found in yogurt are examples. However, it should be noted that any microorganism has the ability to change genetically and become virulent.

In recent years, a new term **pathogenicity islands** has been used to refer to the clusters of genes responsible for virulence. The genes encode many of the virulence factors (e.g., enzymes and toxins) discussed later in this chapter. The pathogenicity islands are fairly large segments of the genome of pathogenic strains that are absent in nonpathogenic strains. These islands have many of the properties of intervening sequences, suggesting they move by lateral gene transfer (Chapter 7). These blocs of genetic information could move from a pathogenic strain into an avirulent organism, converting it to a pathogen. Such lateral transfer processes show how the evolution of pathogenicity can make quick jumps.

MicroFocus 18.3

WE ARE NOT ALONE

When we speak of host-parasite relationships, we tend to be human-centric. We understandably want to know how we as hosts are affected by microbial pathogens. There is nothing wrong with this view. It's just that today there are other host-parasite relationships of equal importance but with possibly more tragic results. Specifically, the hosts are the great apes (chimpanzees and gorillas) of western equatorial Africa and the parasite is the Ebola virus.

The Ebola virus has been one of the most feared parasites to emerge recently. It has caused Ebola hemorrhagic fever (EHF) in humans during several outbreaks in Africa since 1976, producing fatality rates in humans as high as 80 percent. In 2003, we still do not know how the virus is transmitted or what the reservoir is for this virus.

In western equatorial Africa (Cameroon, Equatorial Guinea, the Central African Republic, the Republic of Congo, and Gabon), conservationists have witnessed hundreds of endangered western gorillas and common chimpanzees die. Since 1983, the numbers of great apes have declined by as much as

56 percent in some areas. Previously, habitat loss due to logging and expanding agriculture posed the greatest threat to the great apes. Now death is primarily due to two factors. One is the slaughter of apes for what is called "bush meat." Commercial hunters kill the great apes to supply logging camps and cities with smoked bush meat, despite efforts by governments and other groups to outlaw the practice. Interestingly, the 2001–02 Ebola outbreak was in the Gabon/Republic of Congo area. Whether there is a relationship between bush meat and transmission of Ebola virus remains to be discovered.

The second factor, which is even more ominous, is the death from the Ebola virus. EHF quickly and silently is eliminating the great ape populations. In Gabon, which contains 80 percent of the world's western gorillas and most common chimpanzees, thousands of these animals have died from the Ebola virus. In the Minkébe forest of northern Gabon alone, ape densities have declined 99 percent in the last 10 years.

Spreading some 40 kilometers a year, the latest EHF epidemic is in the Republic of Congo and approaching the Lossi

sanctuary and the Odzala National Park, which contains one of the highest densities of chimpanzees and gorillas in the world.

How Ebola was introduced into the great ape population is uncertain. Many believe the destruction of ape habitats by humans forced the apes to encounter an unknown reservoir species that transmitted the disease. Also, there are regions where the ape densities are so abnormally high (in some places eight times normal levels) that disease is easily spread from animal to animal. If densities were much thinner, a natural outbreak of EHF could not spread easily and would "burn itself out."

So, authorities and conservationists are puzzled what to do. Some say EHF will burn itself out, *hopefully*. Others want to develop vaccination programs or relocate healthy animals to disease-free areas. It would seem that without some form of intervention and a reduction in ape densities, the outlook is bleak. A 2003 survey echoes these thoughts—in 30 years, the apes of Africa could be extinct.

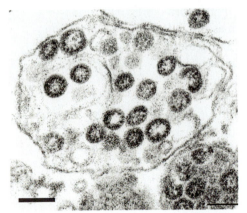

(a)

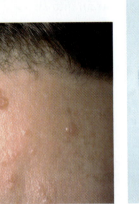

(b)

Newly Emerging Pathogens

(a) Transmission electron micrograph of the SARS coronavirus in bronchialveolar-lavage fluid from a patient with severe acute respiratory syndrome (SARS). Numerous intracellular virions can be seen. (Bar = 170 nm.) (b) Human monkeypox rash on an infected patient. Monkeypox is a severe smallpox-like illness caused by monkeypox virus. In June 2003, an outbreak of monkeypox was reported for the first time in the United States. (c) False-color transmission electron micrograph of the Ebola virus. Ebola hemorrhagic fever is a severe, often fatal disease in humans, monkeys, gorillas, and chimpanzees. The virus has caused sporadic outbreaks since 1976. (Bar = 200 nm.)

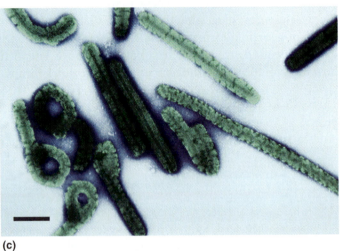

(c)

DISEASES PROGRESS THROUGH A SERIES OF STAGES

Disease is a dynamic series of events expressing the competition between parasite and host (although certain diseases, such as botulism, follow an ingestion of toxins). In most instances, there is a recognizable pattern in the progress of the disease following the entry of the pathogen into the host. Certain periods may be distinguished (FIGURE 18.5).

The episode of disease begins with an **incubation period**, reflecting the time that elapses between the entry of the microbe into the host and the appearance of the first symptoms. Incubation periods may be a short one to three days, as in cholera; a moderate one to two weeks, as in measles; or a long three to six years, as in leprosy. Such factors as the number of organisms, their generation time and virulence, and the level of host resistance determine the incubation period's length. The location of entry also may be a determining factor. For instance, the incubation period for rabies may be as short as several days or as long as a year, depending on how close to the central nervous system the viruses enter the body.

The next phase in disease is a time of mild **signs** or **symptoms**, called the **prodromal** phase. For many diseases, this period is characterized by general symptoms such as nausea, fever, headache, and malaise, which indicate that the competition for supremacy has begun. Although the patient feels miserable, there is some evidence that the symptoms are beneficial (MicroFocus 18.4).

The **acme period** or **climax** follows. This is the acute stage of the disease, when specific symptoms appear. Examples are the skin rash in scarlet fever, jaundice in

Sign:
a change in body function discovered on examination of the patient (such as a fever).

Symptom:
a change in body function experienced by the patient (such as headache or sore throat).

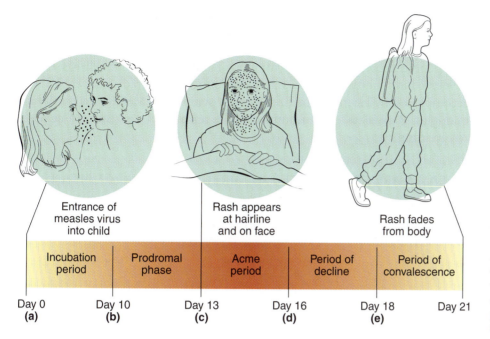

Entrance of measles virus into child

Rash appears at hairline and on face

Rash fades from body

| Incubation period | Prodromal phase | Acme period | Period of decline | Period of convalescence |

Day 0 **(a)** Day 10 **(b)** Day 13 **(c)** Day 16 **(d)** Day 18 **(e)** Day 21

FIGURE 18.5

The Course of Disease, as Typified by Measles

(a) A child is exposed to measles viruses in respiratory droplets, and the incubation period begins. (b) At the end of this period, the child experiences fever, respiratory distress, and general weakness as the prodromal phase ensues. (c) The acme period begins with the appearance of specific measles symptoms, such as the body rash. The period reaches a peak as the rash covers the body. (d) The rash fades first from the face and then the body trunk as the period of decline takes place. (e) With the period of convalescence, the body returns to normal.

hepatitis, Koplik spots and rash in measles, and swollen lymph nodes in infectious mononucleosis. Often, patients suffer high fever and chills, the latter reflecting differences in temperature between the superficial and deep areas of the body. Dry skin and a pale complexion may result from constriction of the skin's blood vessels to conserve heat.

As the symptoms subside, the host enters a **period of decline**. This period may be preceded by a crisis period, after which recovery is often rapid. In other cases, the period of decline may last a long time. Sweating is common as the body releases excessive amounts of heat, and the normal skin color soon returns as the blood vessels dilate. The sequence comes to a conclusion after the body passes through a **period of convalescence**. During this time, the body's systems return to normal.

MicroFocus 18.4

ILLNESS MAY BE GOOD FOR YOU

For most of the twentieth century, medicine's approach to infectious disease was relatively straightforward: Note the symptoms and fix them. But, that may change in the future, as Darwinian medicine gains a stronger foothold. Proponents of Darwinian medicine ask *why* the body has evolved its symptoms, and question whether relieving the symptoms may leave the body at greater risk.

Consider coughing, for example. In the rush to stop a cough, we may be neutralizing the body's mechanism for clearing pathogens from the respiratory tract. Nor may it be in our best interest to stifle a fever, since fever enhances the immune response to disease. Many physicians view iron insufficiency in the blood as a symptom of disease, yet many bacterial species (e.g., tubercle bacilli) require this mineral, and while iron is sequestered out of the blood in the liver, they cannot grow well. Even diarrhea can be useful—it helps propel pathogens from the intestine and assists the elimination of enterotoxins.

Darwinian biologists point out that disease symptoms have evolved over the vast expanse of time and probably have benefits that are waiting to be understood. They are not suggesting a major change in how doctors treat their patients, but they are pushing for more studies on whether symptoms are part of the body's natural defenses. It's not quite time to throw out the Nyquil, Tylenol, or Imodium. Not yet, at least.

During disease progression, diseases may be described as clinical or subclinical. A **clinical** disease is one in which the symptoms are apparent, while a **subclinical** disease is accompanied by few obvious symptoms. Many people, for example, have experienced subclinical cases of mumps or infectious mononucleosis and have developed immunity to future attacks. By contrast, certain diseases are invariably accompanied by clearly recognized clinical symptoms. Measles and malaria are examples. MicroFocus 18.5 recounts how the remains of a disease can be used to resolve conflicts.

DISEASES ARE TRANSMITTED BY DIRECT OR INDIRECT MEANS

The agents of disease may be transmitted by a broad variety of methods conveniently divided into two general categories: direct methods and indirect methods, as illustrated in FIGURE 18.6.

FIGURE 18.6

Methods of Transmitting Disease

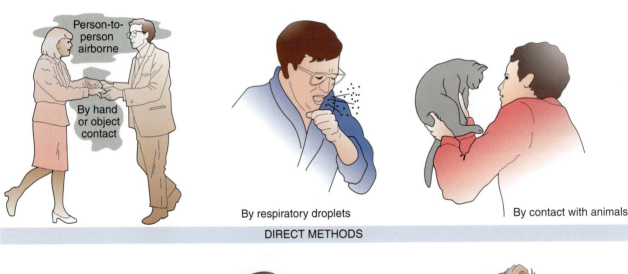

Person-to-person airborne

By hand or object contact

By respiratory droplets

By contact with animals

DIRECT METHODS

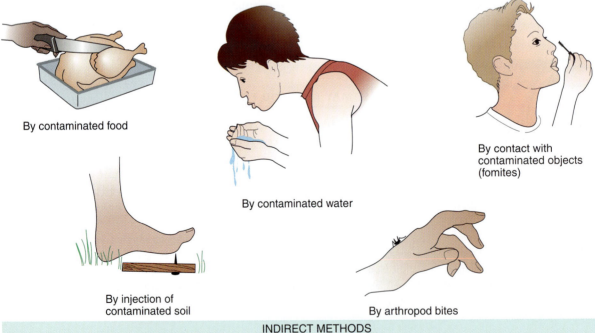

By contaminated food

By contaminated water

By contact with contaminated objects (fomites)

By injection of contaminated soil

By arthropod bites

INDIRECT METHODS

MicroFocus 18.5

THE TRUTH IN COAL DUST

The story is told of a rich man who died and left his considerable estate to his only son, whom he had not seen in many years. Some time later, two men showed up, each with credible documents and each claiming to be the man's son and heir.

The boy's pediatrician was called in to consult. He didn't recognize either man, but after reviewing the son's medical records, he had an idea. He sent both men down to the cellar to shovel coal, telling them not to wipe away the coal dust. When they returned, one of the men had tiny bits of white showing through the coal dust on his face. This man, the pediatrician said, was the true son. The answer was in the medical records: The son had had smallpox as a boy, and even though the scars disappeared, the doctor knew that dust would not stick to places on the skin where the smallpox scars had been. The impostor was exposed, and the son collected his fortune.

Direct methods of transmission imply close or personal contact with one who has the disease. Hand-shaking, kissing, sexual intercourse, and contact with feces are examples. Such diseases as gonorrhea and genital herpes are spread by direct contact. Direct contact also may mean exposure to **respiratory droplets**, the tiny particles of mucus and saliva expelled from the respiratory tract during a cough or sneeze. Diseases spread by this method include influenza, measles, pertussis (whooping cough), and streptococcal sore throat. For some diseases, direct contact with an animal is necessary. Rabies, leptospirosis, and toxoplasmosis are typical.

Indirect methods of disease transmission include the consumption of contaminated food or water (MicroFocus 18.6), and contact with fomites. Foods are contaminated during processing or handling, or they may be dangerous when made from diseased animals. Poultry products, for example, are often a source of salmonellosis because *Salmonella* species frequently infect chickens; and pork may spread trichinosis because *Trichinella* parasites may live in muscles of the pig. **Fomites** are inanimate objects that carry disease organisms. For instance, bed linens may be contaminated with pinworm eggs, and contaminated syringes and needles may transport the viruses of hepatitis B and AIDS.

Arthropods represent another indirect method of transmission. Living organisms that carry disease agents from one host to another, such as arthropods, are called

MicroFocus 18.6

DINNER AT LA CASA CUCARACHA

"We're having a wonderful time in this tropical wonderland. Last night, we went to a great restaurant. We walked down Cabeza de Vaca Boulevard, just as they told us, and there on the corner of Ponce de Leon square, we saw it—La Casa Cucaracha.

"The shrimp were so icy cold, we couldn't eat them fast enough. [Uh oh. Should never have cold seafood in that part of the world.] The salad was crisp and delicious, especially the lettuce. [Shouldn't have salad either; the fixins' were probably washed in the local water.] And the waiter was so helpful— he peeled the orange for me right there at the table. [Another no-no; his hands probably touched the peeled fruit.]

"Dinner was superb. We had crab meat *au gratin* just warm enough to bring out the flavor. [It should have been steaming hot.] The local vegetables were good enough to eat raw [another mistake], but the potato was a run-of-the-mill baked potato. [Probably the only safe thing you ate.] For dessert we had local fruits on a lovely bed of crushed ice. [The water, again.]

"How am I feeling today? OK, I guess. On second thought, I do have this little pain in my stomach. . . ."

vectors. In some cases, the arthropod may be a **mechanical vector** of disease because it passively transports microorganisms on its legs and other body parts. House flies can carry diseases picked up on their feet. In other cases, the arthropod itself is diseased and serves as a **biological vector.** In malaria and yellow fever, for instance, the protozoan or virus infects the mosquito and accumulates in its salivary gland, from which they are injected during the next bite.

FIGURE 18.7 shows how diseases can spread in a hospital setting.

RESERVOIRS REPRESENT SOURCES OF DISEASE ORGANISMS

For a disease to perpetuate itself, a continuing source of disease organisms in nature is necessary. These sources are called **reservoirs** (FIGURE 18.8). In smallpox, the sole reservoir of viruses is humans, and the World Health Organization was able to limit the spread of the virus and eradicate smallpox from the world by locating all human reservoirs (Chapter 13). A **carrier** is a special type of reservoir. Generally a carrier is one who has recovered from the disease but continues to shed the disease agents. For instance, people who have recovered from typhoid fever or amoebiasis

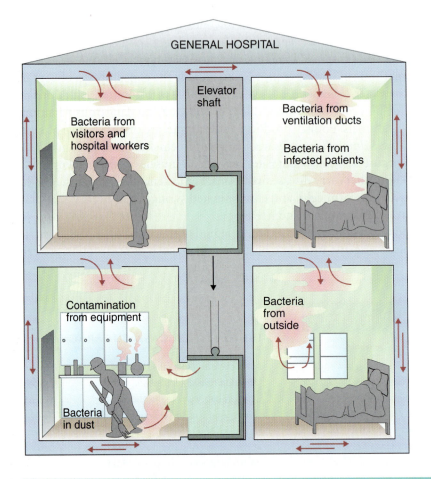

FIGURE 18.7

Examples of Microorganism Transmission in a Hospital

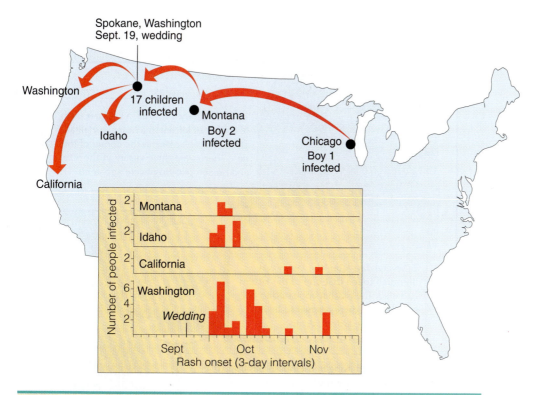

FIGURE 18.8

How a Reservoir May Introduce Disease to a Population

Measles broke out in a Chicago neighborhood late in August 1983. A 2-year-old boy contracted the virus, and during the incubation period, he returned home to Billings, Montana. On September 4, he broke out in a rash and infected boy number 2. The second boy developed a rash on September 19 while at a wedding in Spokane, Washington. The wedding was attended by about 375 people from Idaho, Montana, Oregon, and Washington. Seventeen children at the wedding were infected. They eventually spread measles to numerous other individuals, as well as to health care workers and families of health care workers, as represented in the diagram. In all, 42 people contracted the disease from the initial human reservoir (carrier).

become carriers for many weeks after the symptoms of disease have left. Their feces may spread the disease to others via contaminated food or water.

Animals are a second reservoir of disease. Domestic house cats usually show no symptoms of toxoplasmosis but are able to transmit the protozoan *Toxoplasma gondii* to humans, where the disease manifests itself. Cats also can transmit Q fever, as **FIGURE 18.9** indicates. Water and soil make up a third group of reservoirs since they often are contaminated with disease agents.

DISEASES CAN BE DESCRIBED BY THEIR BEHAVIOR AND OCCURRENCE IN A POPULATION

Most diseases studied in this text are **communicable diseases**, that is, they are transmissible directly or indirectly among hosts in a population. Certain communicable diseases are described as being **contagious** because they pass with particular ease among hosts. Chickenpox, measles, and genital herpes fall into this category. One reason for easy transmission is that the focus of disease often is on or close to

TEXTBOOK CASES

FIGURE 18.9

An Outbreak of Q Fever Associated with a Cat

This incident happened in Halifax, Nova Scotia, during the winter months when windows in the home are usually closed and there is little circulation of outside air through the house.

1. On February 14, 1987, a cat gave birth to a litter of kittens in a home in Halifax. The birth took place in the corner of the den. There were three kittens in the litter, and no complications were apparent.

2. Several days later, the family was distressed to see that one of the kittens had died. The mother appeared to be healthy, and there seemed to be no reason for the kitten's death.

3. The man living in the house was host to a regular poker game. The group gathered in the den at least once a week. That week they played poker the day after the kitten died.

4. Several weeks later, all six of the poker players began coughing, sneezing, and displaying respiratory distress. One man with a history of heart disease became seriously ill and died.

5. Physicians detected pneumonia in all the men, and health researchers began a search to discover the cause. They examined the cat and found evidence of *Coxiella burnetii* in its uterine tissue. Evidence of the same organism was found in blood samples from the five remaining men. The disease was identified as Q fever. All five men received doses of antibiotics and recovered.

the body surface. **Noncommunicable diseases** are singular events in which the agent is acquired directly from the environment and is not easily transmitted to the next host. In tetanus, for example, penetration of soil containing *Clostridium tetani* spores to the anaerobic tissue of a wound must occur before this disease develops.

Diseases also may be described as endemic, epidemic, or pandemic. An **endemic disease** is one that occurs at a low level in a certain geographic area. Plague in the American Southwest is an example. By comparison, an **epidemic disease** (or epidemic) breaks out in explosive proportions within a population. Influenza often causes epidemics. This should be contrasted with an **outbreak**, which is a more contained epidemic. An abnormally high number of measles cases in one US city would be classified as an outbreak. A **pandemic disease** (or pandemic) occurs worldwide. The most obvious example here would be AIDS.

Health organizations, such as the Centers for Disease Control and Prevention (CDC) and the WHO, learn a lot about diseases by analyzing data associated with a disease. MicroInquiry 18 explores such epidemiological data.

DISEASES CAN BE CLASSIFIED BY SEVERITY AND HOST INVOLVEMENT

Besides diseases being classified as "communicable" or "noncommunicable" in their ability to be transmitted, or "endemic" or "epidemic" indicating how widespread the disease is, diseases also can be classified in two other ways.

The words "acute" and "chronic" are applied to diseases as relative measurements of their severity. An **acute disease** develops rapidly, is usually accompanied by severe symptoms, comes to a climax, and then fades rather quickly. Cholera, epidemic typhus, and SARS are examples of acute diseases. A **chronic disease**, by contrast, often lingers for long periods of time. The symptoms are slower to develop, a climax is rarely reached, and convalescence may continue for several months. Hepatitis A, trichomoniasis, and infectious mononucleosis are examples of chronic diseases. Sometimes an acute disease may become chronic when the body is unable to rid itself completely of the parasite. For example, one who has giardiasis or amoebiasis may experience sporadic symptoms for many years.

Infections may develop in one of two ways. A **primary infection** occurs in an otherwise healthy body, while a **secondary infection** develops in a weakened individual. In the influenza pandemic of 1918 and 1919, hundreds of millions of individuals contracted influenza as a primary infection and many developed pneumonia as a secondary infection. Numerous deaths in the pandemic were due to pneumonia's complications.

As the names imply, **local diseases** are restricted to a single area of the body, while **systemic diseases** are those disseminating to the deeper organs and systems. Thus, a staphylococcal skin boil beginning as a localized skin lesion may become more serious when staphylococci spread and cause systemic disease of the bones, meninges, or heart tissue.

The word commonly used for dissemination of living bacteria through the bloodstream is **bacteremia**. **Septicemia** refers to the multiplication of bacteria in the blood. This used to be called "blood poisoning" in older textbooks but is now regarded as streptococcal or staphylococcal blood disease. **Fungemia** refers to the spread of fungi, **viremia** to the spread of viruses, and **parasitemia** to the spread of protozoa and multicellular worms through the blood.

A word of caution might be appropriate at this juncture. We have used the word "disease" loosely to define a clinical condition initiated by an infectious microorganism. It should not be inferred, however, that this description is true of all diseases. For example, physiological diseases, such as diabetes mellitus, are due to a malfunction of a body organ or system; nutritional diseases, such as scurvy and beri-beri, are caused by a dietary insufficiency; and genetic diseases, such as sickle-cell anemia and cystic fibrosis, are traced to defects in human genes. In these cases, infectious agents or parasites are not involved. Therefore, as we use the word "disease," remember that we are referring to "infectious disease."

MicroInquiry 18

APPLICATIONS OF MEDICAL EPIDEMIOLOGY

Epidemiology, as applied to infectious disease, is a scientific study from which health problems are identified. In this inquiry, we are going to look at just a few of the applications of medical epidemiology for analyzing the patterns of illness. Answers can be found in Appendix E.

One of the important measures is to assess disease occurrence. The **incidence** of a disease is the number of reported cases in a given time frame. **FIGURE A** is a line graph showing the number of new cases of AIDS per year in the United States.

The **prevalence** of a disease refers to the percentage of the population that is affected at a given time.

18.1a. What was the incidence of AIDS in 1993 and 2001?

18.1b. Assuming that 264,000 were living with AIDS in 1993 and 370,000 in 2001, how has the prevalence of AIDS changed between 1993 and 2001? (Assume the population of the United States has remained at 290 million).

Descriptive epidemiology describes activities (time, place, people) regarding the distribution of diseases within a population. Once some data have been collected on a disease, epidemiologists can analyze these data to characterize disease occurrence. Often a comprehensive description can be provided by showing the disease trend over time, its geographic extent (place), and the populations (people) affected by the disease.

CHARACTERIZING BY TIME.

Traditionally, drawing a graph of the number of cases by the date of onset shows the time course of an epidemic. **FIGURE B** is an **epidemic curve**, or "epi curve," a histogram providing a visual display of the magnitude and time trend of an Ebola outbreak in Africa.

Look at the epi curve to the right. One important aspect of a bar graph is to consider its overall shape. An epi

curve with a single peak indicates a **single source** (or "**point source**") **epidemic** in which people are exposed to the same source over a relatively short time. If the duration of exposure is prolonged, the epidemic is called a **continuous common source epidemic**, and the epi curve will have a plateau instead of a peak. Person-to-person transmission is likely and its spread may have a series of plateaus one incubation period apart.

18.2a. Identify the type of epi curve drawn in Figure B and explain what the onset says about the nature of disease spread.

18.2b. Is there more than one plateau? Explain the significance

---

FIGURE A

Acquired Immunodeficiency Syndrome (AIDS)

Number of cases reported by year in the United States and US territories for the years 1981 to 2001. Source: CDC.

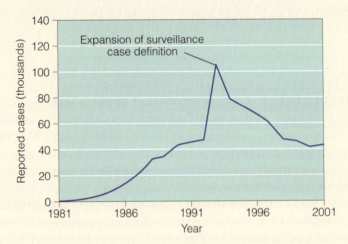

FIGURE B

Ebola Hemorrhagic Fever (Congo and Gabon)

Number of cases of Ebola hemorrhagic fever by week from October 2001 to March 2002. Source: *Weekly Epidemiological Record*, No. 26, June 27, 2003.

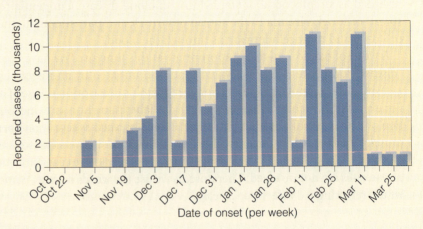

that multiple plateaus might have in interpreting the spread of the Ebola hemorrhagic fever outbreak.

CHARACTERIZING BY PLACE.

Analysis of a disease or outbreak by place provides information on the geographic extent of a problem and may show clusters or patterns that provide clues to the identity and origins of the problem. It is a simple and useful technique to look for geographic patterns where the affected people live, work, or may have been exposed.

A geographic distribution for Lyme disease in shown in **FIGURE C**. It is similar to a spot map, where each case of a disease in a community or state may be shown to reflect clusters or patterns of disease. Figure C identifies the numbers of cases of Lyme disease by county in the Unites States in 2001.

18.3a. From this county map, what inferences can you draw with regard to the reported cases of Lyme disease?

18.3b. What suggestions could you make for the high incidence in the one county in Texas that had greater than 15 reported cases?

CHARACTERIZING BY PERSON.

Populations at risk for a disease can be determined by characterizing a disease or outbreak by person. Persons also refer to populations identified by personal characteristics (e.g., age, race, gender) or by exposures (e.g., occupation, leisure activities, drug intake). These factors are important because they may be related to disease susceptibility and to opportunities for exposure.

Age and gender often are the characteristics most strongly related to exposure and to the risk of disease. For example, **FIGURE D** is a histogram showing the incidence of pertussis (whooping cough) in the United States in 2001.

18.4a. Look at the histogram and describe what important information is conveyed in terms of the majority of cases and relative incidence in 2001.

18.4b. As a health care provider, what role do you see for vaccinations and booster shots with regard to this disease?

FIGURE C

Lyme Disease

Counties in the mainland United States showing the incidence of Lyme disease in 2001. Note that the counties in orange represent 90 percent of the total number of cases reported in 2001. Source: CDC.

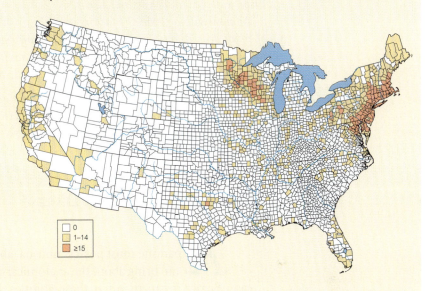

- □ 0
- □ 1–14
- ▨ ≥15

FIGURE D

Pertussis

The reported number of cases of pertussis by age group in the United States in 2001. Source: CDC.

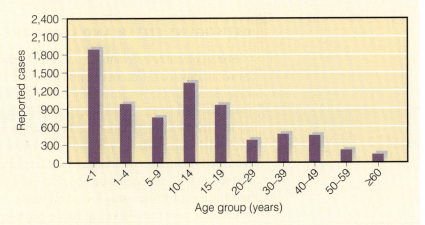

To this point . . .

We have begun our study of infection and disease by studying the host-parasite relationship. Infection refers to the relationship between the host and parasite, while disease results from the change in the state of the host's health brought about by damage caused by the parasite. We noted that the body is inhabited by a normal flora of organisms, some of which are commensals and opportunists. The term pathogenicity was explored in the general sense, and we noted how a shift in the delicate balance of body control can lead to disease. Virulence refers to the degree of pathogenicity of a parasite.

Next, we surveyed the periods during the progress of disease and used many examples from various diseases to illustrate the basic concepts. Clinical and subclinical diseases vary on the basis of clinical symptoms, and the transmission of disease occurs by direct and indirect methods, including reservoirs and carriers. We defined communicable and noncommunicable diseases, acute and chronic diseases, primary and secondary diseases, and local and systemic diseases.

We now will focus on the factors contributing to the establishment of disease. Portal of entry, dose, and tissue penetration are examples of these factors. We shall then survey some of the virulence factors (enzymes and toxins) used by the parasite to overcome body defenses and interfere with vital metabolic processes.

18.2

The Establishment of Disease

A parasite must possess unusual abilities if it is to overcome host defenses and bring about the anatomical or physiological changes leading to disease. Before it can manifest these abilities, however, the parasite must first gain entry to the host in sufficient numbers to establish a population (FIGURE 18.10). Next, it must be able to penetrate the tissues and grow at that location. Disease is therefore a complex series of interactions between parasite and host. In this section, we shall examine some of the factors that determine whether disease can occur, with a focus on the microorganism. The events are summarized in FIGURE 18.11.

PATHOGENS ENTER AND LEAVE THE HOST THROUGH PORTALS

Portal of entry refers to the site at which the pathogen enters the host. It varies considerably for different organisms and is a key factor in the establishment of disease. For example, tetanus may occur if *Clostridium tetani* spores are introduced from the soil to the anaerobic tissue of a wound, but tetanus will not develop if spores are consumed with food because the spores do not germinate in the human intestinal tract. This is why one can eat a freshly picked radish without fear of tetanus. One reason offered for specific portals of entry is the presence of adhesive factors on the surface of parasites. For example, gonococci attach by means of pili to specific receptor sites only found on tissues of the urogenital system.

Certain parasites have multiple portals of entry. The tubercle bacillus, for instance, may enter the body in respiratory droplets, contaminated food or milk,

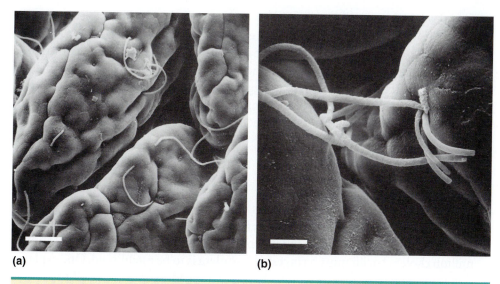

(a)   (b)

FIGURE 18.10

A Population of Intestinal Bacteria

Two scanning electron micrographs of the villi of the intestine showing filamentous bacteria. (a) In this view, the bacteria appear to emerge from clefts in the tissue and extend out onto the surface of the villus. (Bar = 25 μm.) (b) At higher magnification, the attachment in the cleft of tissue can be seen more clearly. (Bar = 6 μm.)

or skin wounds. The bacterium causing Q fever enters by all these methods, as well as by an arthropod bite. The tularemia bacillus may enter the eye by contact, the skin by an abrasion, the respiratory tract by droplets, the intestines by contaminated meat, or the blood by an arthropod bite.

It should be noted that upon entry to a host, a pathogen is confronted with a profoundly different environment than it is used to. Adaptation to this environment requires the genetic machinery that enables the pathogen to grow and multiply. Indeed, the expression of pathogenicity may be delayed until such time that the microorganism finds its favorite niche in the body.

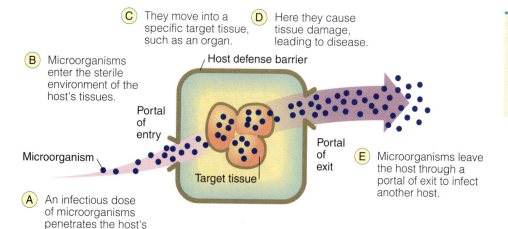

C They move into a specific target tissue, such as an organ.

D Here they cause tissue damage, leading to disease.

Host defense barrier

B Microorganisms enter the sterile environment of the host's tissues.

Portal of entry

Microorganism

Target tissue

Portal of exit

E Microorganisms leave the host through a portal of exit to infect another host.

A An infectious dose of microorganisms penetrates the host's defensive barrier.

FIGURE 18.11

The Generalized Events in the Establishment of Disease

At the conclusion of its pathogenicity cycle, a pathogen must be able to leave the body through some suitable **portal of exit**, as shown in FIGURE 18.12. This is of more than passing importance because easy transmission permits the pathogen to continue its pathogenic existence in the world. Modern microbiologists, such as Amherst College's Paul Ewald, have suggested that if a microorganism cannot find a suitable portal of exit or mode of transmission, it may be replaced by less pathogenic species. Ewald's theory is described in MicroFocus 18.7.

DISEASE ONSET IS INFLUENCED BY THE NUMBER OF INVADING PATHOGENS

Dose refers to the number of parasites that must be taken into the body for disease to be established. Experiments indicate, for example, that the consumption of a few hundred thousand typhoid bacilli will probably lead to disease. By contrast, many million cholera bacilli must be ingested if cholera is to be established. One explanation is the high resistance of typhoid bacilli to the acidic conditions in the stomach, in contrast to the low resistance of cholera bacilli. Also, it may be safe to eat fish when the water contains hepatitis A viruses, but eating raw clams from that same water can be dangerous because clams are filter-feeders, and the concentration (or dose) of hepatitis A viruses is much higher in these animals.

Often the host is exposed to low doses of a pathogen and, as a result, develops immunity. For instance, many people can tolerate low numbers of mumps viruses

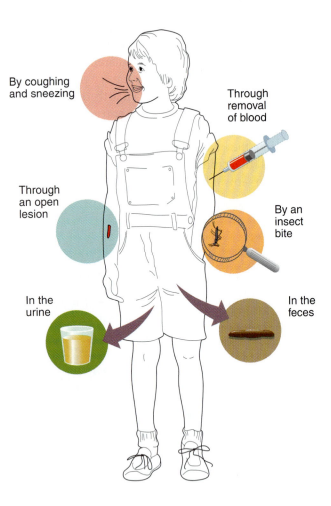

FIGURE 18.12

Six Different Portals of Exit from the Body

By coughing and sneezing

Through removal of blood

Through an open lesion

By an insect bite

In the urine

In the feces

MicroFocus 18.7

CONTROLLING THE PATHOGEN

What if we could control the destiny of a pathogen? What if we could make a virulent pathogen less virulent, or even induce it to become harmless? Suppose, for example, we could "tame" the human immunodeficiency virus (HIV), and thereby put an end to the AIDS epidemic?

It may be possible, says Paul Ewald, of Amherst College in Massachusetts. Ewald maintains that cutting off HIV from access to new hosts would induce milder strains, possibly even harmless strains, to evolve. Following this reasoning, it would make sense to take funds from anti-HIV drug programs and put them into transmission-prevention programs, as he suggests.

Ewald's novel approach to medicine is presented in his seminal book, *Evolution of Infectious Disease*. His theory is based on the observation that when a killer pathogen cannot spread from person to person, it cannot infect enough people to survive, and it will soon die out. But, it will not leave a total void. Rather, its niche will be filled by a milder strain that will not make people sick enough to die. This milder strain can be passed among people because they leave their homes

and make contact with one another. Thus, it is in the interest of the pathogen to be mild, especially when it depends on the host to transmit it.

Sometimes the pathogen does not need its host for transmission, so it doesn't have to evolve to a mild form. (Consider, for instance, diseases such as malaria, sleeping sickness, yellow fever, and plague.) The common thread among these diseases is transmission by an arthropod. The pathogen can remain a killer because transmission is relatively easy, as long as the correct arthropod is available. In these cases, Ewald's transmission-prevention pro-

gram (arthropod control) would be extremely beneficial in encouraging the pathogens to evolve to mild forms.

Much of Ewald's work has been with cholera. His research has shown that water heavily polluted with feces contains virulent *Vibrio cholerae*, but that milder strains inhabit clean water. The implication is that cholera bacilli evolve to milder forms when they cannot use water as a transfer vehicle. Once again, transmission prevention would appear not only to interrupt the epidemic but, more importantly, to force the pathogen into a more benign mode of existence.

Ewald's boldness and originality are hailed by some scientists, but viewed with skepticism by others who see a need for independent corroboration and much deeper study. Trying to understand a microorganism's evolution in response to its environment is difficult, the skeptics say, but trying to understand the evolution when another species (humans) is involved can complicate this issue much further. Still, Ewald's views have placed transmission-prevention programs in a new light, while encouraging epidemiologists to continue fighting the good fight.

---

without exhibiting disease. They may be surprised to find that they are immune to mumps when it breaks out in their family at some later date. Other individuals have developed immunity to fungal pathogens after several low-grade exposures. Cases of histoplasmosis, for example, are regarded as a mild "summer flu" in the Ohio and Mississippi valleys, but histoplasmosis can be serious in an individual not previously exposed to the disease. Blastomycosis can be equally dangerous, as FIGURE 18.13 illustrates. It should be noted that immune people remain healthy because they enjoy strong resistance, again pointing to the notion that resistance and disease are fundamentally inseparable.

MOST PATHOGENS MUST PENETRATE HOST TISSUES

Most knowledge of tissue penetration (FIGURE 18.14) has been developed from studies on histological preparations, tissue cultures, and animals, and an understanding of the penetration process is generally incomplete. Experiments reported by

FIGURE 18.13

An Outbreak of Blastomycosis

This outbreak occurred in Boulder, Colorado, during the summer of 1998. Growth of the responsible fungus may have been encouraged by the animal feces present, and unusually heavy rainfall may have contributed to the high humidity needed for fungal growth. Epidemiologists postulated that spores in the dust were inhaled by the men infected.

TEXTBOOK CASES

1. During June 1998, officials from the City of Boulder (Colorado) Open Space program decided to move ahead on a planned project to relocate a colony of prairie dogs from an area at the outskirts of the city to make way for a park.

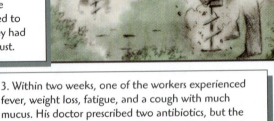

2. By early August, the animals had been relocated, and 15 workers were busy excavating the site, especially the abandoned prairie dog tunnels and burrows. Their gasoline-powered augers created a lot of dust. No protective clothing or facemasks were provided to the men at the site. As a result, they had an opportunity to breathe much dust.

3. Within two weeks, one of the workers experienced fever, weight loss, fatigue, and a cough with much mucus. His doctor prescribed two antibiotics, but the symptoms worsened. He was hospitalized by September.

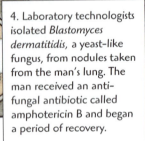

4. Laboratory technologists isolated *Blastomyces dermatitidis*, a yeast-like fungus, from nodules taken from the man's lung. The man received an antifungal antibiotic called amphotericin B and began a period of recovery.

5. Days later, another man from the project reported to the same hospital with identical symptoms. An alert doctor saw the similarities and tested the man for *B. dermatitidis*. When the test was positive, the doctor called epidemiologists.

Stanley Falkow and his coworkers at Stanford University indicate that genes for cell penetration exist on the chromosome of certain bacteria. These genes appear to code for surface proteins that assist penetration. In the late 1980s, Falkow's group successfully isolated the penetration genes from *Yersinia pseudotuberculosis* and inserted them into *E. coli*, which then displayed penetration. Falkow has written that pathogens use their invasive tendencies to gain access to privileged niches denied to nonpathogens. For example, pathogenic *E. coli* O157:H7 penetrates to sites in the small intestine tissue and urinary system that are not available to commensal *E. coli* strains remaining in the lumen of the small intestine. Other researchers have reported genes for penetration on the plasmids of *Shigella flexneri*.

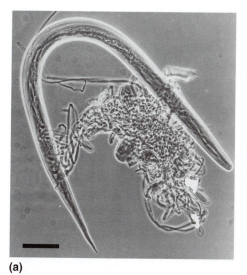

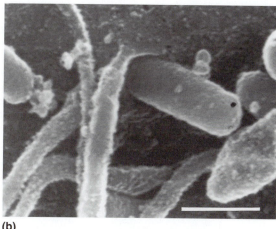

(a)   (b)

FIGURE 18.14

Two Examples of Tissue Penetration by Parasites

(a) The fungus *Dactylaria* attacking the roundworm *Panagrelus* in two places. Fungal penetration has been studied by observing how the fungus penetrates the outer membranes of the worm and parasitizes its tissues. (Bar = 25 μm.) (b) An *Escherichia coli* cell experimentally reengineered to produce a surface protein that permits invasion of tissue cells. In this scanning electron micrograph, *E. coli* is invading a human lung cell. The finger-like protrusions are part of the normal outer membrane of the lung cell. (Bar = 1 μm.)

In some cases, penetration may not be critical to disease. The pertussis bacillus, for example, remains on the surface layers of the respiratory tract while producing the toxins that lead to disease. The cholera bacillus does likewise in the intestine. The ability of a pathogen to penetrate tissues and cause structural damage is called **invasiveness**. The bacilli of typhoid fever and the protozoa that cause amoebiasis are well known for their invasiveness. A species of bacteria related to the typhoid bacillus is shown invading tissue in **FIGURE 18.15**. By penetrating the tissue of the gastrointestinal tract, these organisms cause ulcers and sharp, appendicitis-like pain characteristic of the respective diseases.

Successful penetration and invasiveness require that pathogens have one or more **virulence factors** that facilitate entry into, and disease in, the host. These factors include adhesins, enzymes, and toxins.

ADHESINS ARE IMPORTANT VIRULENCE FACTORS

Although penetration may not be essential to pathogenicity, **adhesion** is often a key factor. Adhesion localizes pathogens at appropriate tissue sites and may result in internalization by phagocytosis or bacterial-induced endocytosis. A variety of molecules and structures collectively known as **adhesins** are involved. Adhesins often are associated with capsules, flagella, or pili (Chapter 4), as shown in **FIGURE 18.16**, or they may act independently of these structures. The M protein of *Streptococcus* species is an example of the latter (Chapter 8). The host cell is often an active partner in the adhesion because the pathogen triggers it to express target receptor sites for adhesin binding. Also, many viruses have spikes on the capsid or envelope that allow for attachment.

FIGURE 18.15

Tissue Invasion

A series of transmission electron micrographs displaying the interaction between invasive *Salmonella typhimurium* and epithelial cells of the human intestine. (Bars = 1 μm.) (a) At 30 minutes postinfection, bacteria are seen adhering to the tips of intestinal microvilli. (b) A bacterium is being engulfed by an epithelial cell. (c) At 1 hour postinfection, salmonellae can be observed within vacuoles in the cells. (d) At 12 hours postinfection, a number of vacuoles containing bacteria unite with one another, and bacteria multiply within this large vacuole. (e) At 24 hours postinfection, the epithelial cell is filled with salmonellae and is breaking down to release the bacteria.

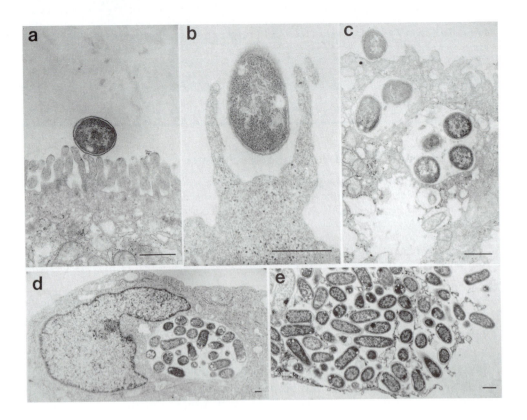

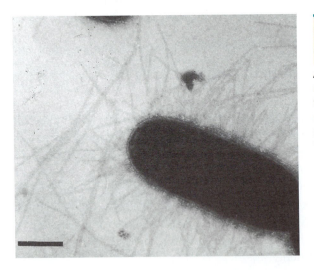

FIGURE 18.16

Pili in *E. coli*

A transmission electron micrograph of *Escherichia coli* displaying the pili used for adhesion to the tissue. Adhesins in pili such as these increase the pathogenicity of the organism by allowing it to localize at its appropriate tissue site. (Bar = 0.5 μm.)

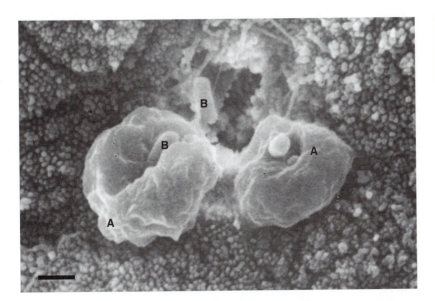

FIGURE 18.17

Salmonella Penetration of Intestinal Cells

Researchers have discovered that microvilli of the intestinal epithelium undergo dramatic changes when *Salmonella* cells come into close contact. The microvilli themselves disappear, and tiny membrane blebs (or "ruffles") spring up in their place, engulfing the salmonellae. Some two hours later the microvilli reappear, but the bacteria are now within the cells. In this scanning electron micrograph, the membrane blebs (A) can be seen as puffy structures. A bacterium (B) is within the cavity of one bleb and another is adjacent to it. (Bar = 3 μm.)

Cell internalization by endocytosis or phagocytosis is well researched (Chapter 19). It has become apparent that some bacterial pathogens also can induce nonphagocytic cells to take up the pathogens. This process allows the pathogen to enter a protective niche or to pass through otherwise impenetrable barriers, such as the blood-brain barrier. Pathogens, such as *Listeria*, have cell membrane adhesive proteins that form a zipper-like binding of pathogen to host cell. This triggers the production of host plasma membrane "ruffles" (like a flamenco dancer's skirt). As a result of this molecular adhesion and cross-talk, phagocytosis occurs and the pathogen enters the host cell cytoplasm. **FIGURE 18.17** shows these changes.

Researchers have found that certain genera of bacteria (e.g., *Listeria* and *Shigella*) can use the host cell's actin for cell-to-cell invasion. These pathogens synthesize an actin tail that propels the organism through the cell's cytoplasm (**MicroFocus 18.8**).

MicroFocus 18.8

TRAVELING LIGHT

Rather than bringing along a personal source of intercellular motility, a *Shigella* bacterium uses what is available and designs it to its own specifications. The organism possesses a gene that allows it to take active protein from the scaffolding of a host cell and jerry-rig the protein to form a comet-like tail. Then the bacterium moves from cell to cell, avoiding body defenses and causing intestinal illness.

The gene for this opponent retooling has been identified by researchers from the Whitehead Institute in Massachu-setts. A team led by Julie A. Therist isolated the gene and transferred it to nonmotile *E. coli* cells. When placed in a culture of tissue cells, the *E. coli* quickly grew *Shigella*-like tails and began thrusting about. What was first believed to be a very complicated system turned out to be quite simple.

Students of anatomy and physiology know actin as the major protein of thin filaments in the sliding-filament model of muscle contraction. (Can you name the protein of the thick filaments?) Students of general biology will recognize actin as a component of the cytoskeleton. And now "actin" has come to microbiology. Welcome, actin.

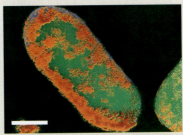

■ *A false color transmission electron micrograph of a* Shigella *bacterium. (Bar = 0.5 μm.)*

When the bacterium thuds against the host's plasma membrane, it distorts and indents the adjacent cell. Using a molecule called **cadherin**, the pathogen bridges the junction between the two cells and enters the next cell (somewhat like moving from train car to train car through connecting doors). A remarkable photograph of *Listeria* moving from cell to cell is presented in Chapter 9, and another series is shown in FIGURE 18.18. The system allows bacterial invasion to occur without bacteria leaving the cellular environment.

The genes that control such activities as these appear to cluster in bacterial cells as a pathogenicity island. A so-called invasion complex has been identified in *Listeria* and *Shigella* species. While fascinating in itself, the study of microbial pathology is doubly important because it provides a glimpse of the biology of humans as well.

ENZYMES CAN INCREASE THE VIRULENCE OF A PATHOGEN

The virulence of a parasite depends to some degree on its ability to produce a series of extracellular enzymes that help the pathogen resist body defenses. The enzymes act on host cells and interfere with certain functions or barriers meant to retard invasion. TABLE 18.1 summarizes the activities of these enzymes.

An example of a bacterial enzyme is the **coagulase** produced by virulent staphylococci (FIGURE 18.19a). Coagulase catalyzes the formation of a blood clot from fibrinogen proteins in human blood. The clot sticks to staphylococci, protecting them from phagocytosis. Part of the walling-off process observed in a staphylococcal skin boil is due to the clot formation. Coagulase-positive staphylococci may be identified in the laboratory by combining staphylococci with human or rabbit plasma. The formation of a clot in the plasma indicates coagulase activity.

Many streptococci have the ability to produce the enzyme **streptokinase**. This substance dissolves fibrin clots used by the body to restrict and isolate an infected area. Streptokinase thus overcomes an important host defense and allows further tissue invasion by the bacteria.

TABLE 18.1

A Summary of Enzymes that Contribute to Virulence

ENZYME	SOURCE	ACTION	EFFECT
Coagulase	Staphylococci	Forms a fibrin clot	Allows resistance to phagocytosis
Streptokinase	Streptococci	Dissolves a fibrin clot	Prevents isolation of infection
Hyaluronidase	Pneumococci Streptococci Staphylococci	Digests hyaluronic acid	Allows tissue penetration
Leukocidin	Staphylococci Streptococci Certain rods	Disintegrates phagocytes	Limits phagocytosis
Hemolysins	Clostridia Staphylococci	Dissolves red blood cells	Induces anemia and limits oxygen delivery

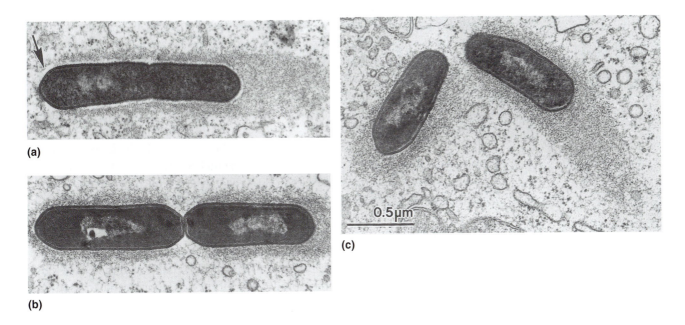

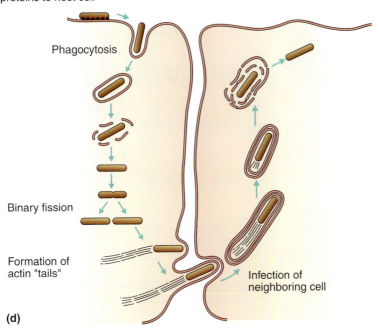

FIGURE 18.18

Movement in *Listeria monocytogenes*

Listeria monocytogenes is a gram-positive rod responsible for outbreaks of a foodborne disease known as listeriosis (Chapter 9). After entering a host cell, the bacterium moves about by forming a polar tail composed of filaments of a protein called actin. (a) *Listeria* is seen in the cytoplasm of a host macrophage. A short tail is observed at the right pole but not the left pole (arrow). (b) A *Listeria* cell is dividing in the host cell cytoplasm, and actin filaments are seen at all surfaces except where division is occurring. (c) The two new rods are each producing their polar tails. (d) The invasive cycle. After internalization and release into the macrophage cytoplasm, the cells divide to form a larger population. Actin tail formation facilitates spread to adjacent cells.

Hyaluronidase is sometimes called the spreading factor because it enhances penetration of a pathogen through the tissues. The enzyme digests hyaluronic acid, a polysaccharide that binds cells together in a tissue (**FIGURE 18.19b**). The term *tissue cement* is occasionally applied to this polysaccharide. Hyaluronidase is an important virulence factor in pneumonococci and certain species of streptococci and staphylococci. In addition, gas gangrene bacilli use it to facilitate spread through the muscle tissues.

Leukocidins and hemolysins are enzymes that destroy blood cells. **Leukocidins** are products of staphylococci, streptococci, and certain bacterial rods. The enzymes disintegrate circulating neutrophils and tissue macrophages, both of which are active phagocytes. Usually disintegration occurs before phagocytosis has taken place, but occasionally it occurs after the phagocyte has engulfed the parasite. The enzymes attach to the phagocyte's membrane and trigger changes leading to the release of lysosomal enzymes in the cytoplasm. The phagocyte quickly disintegrates.

Hemolysins are a group of enzymes that dissolve red blood cells. Studies show that hemolysins combine with the membranes of erythrocytes and cause the formation of pores, after which lysis takes place as cytoplasmic contents spill out. During cases of gas gangrene, hemolysins lead to substantial anemia. Staphylococci and streptococci also

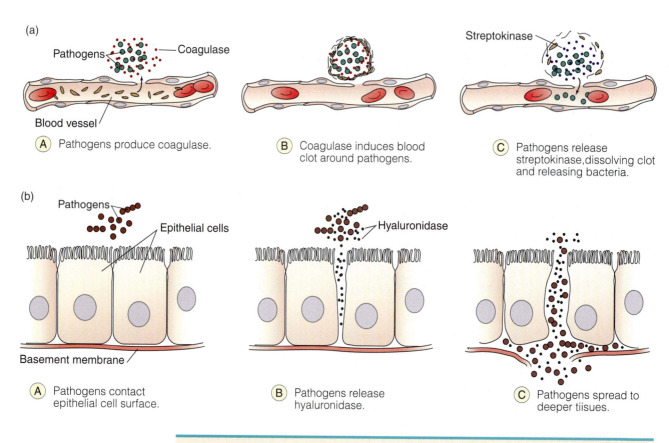

(a)

Pathogens — Coagulase

Blood vessel

A Pathogens produce coagulase.

B Coagulase induces blood clot around pathogens.

Streptokinase

C Pathogens release streptokinase, dissolving clot and releasing bacteria.

(b)

Pathogens

Epithelial cells

Basement membrane

A Pathogens contact epithelial cell surface.

Hyaluronidase

B Pathogens release hyaluronidase.

C Pathogens spread to deeper tiisues.

FIGURE 18.19

Enzyme Virulence Factors

(a) Some bacteria produce the enzyme coagulase, which triggers clotting of blood plasma. Bacteria within the clot can break free by producing streptokinase. (b) Some invasive bacteria produce the enzyme hyaluronase, which degrades the cementing polymer holding cells of the intestinal lining together.

are known to produce these virulence factors. The reason for lysing the red blood cells is to get at the iron in hemoglobin. Bacteria need iron for metabolism and the only source for iron in humans is the red blood cells. In the laboratory, hemolysin producers can be detected by hemolysis, a destruction of blood cells in a blood agar medium (Chapter 8).

Furthermore, if a pathogen exists in a biofilm, its virulence can be enhanced because here it can resist body defenses and drugs. A **biofilm** is a sticky layer of extracellular polysaccharides and proteins enclosing a colony of bacteria at the tissue surface (Chapter 4). Phagocytes and antibodies have difficulty reaching the microorganisms in this slimy conglomeration of armor-like material (FIGURE 18.20). Moreover, microorganisms often survive without dividing in a biofilm. This makes them impervious to the antibiotics that attack dividing cells. (Indeed, the antibiotics do not penetrate the biofilm easily.) The target of the antibiotic may be hidden by a chemical reaction between target molecule and biofilm material. CDC officials have estimated that fully 65 percent of human infections involve biofilms.

TOXINS ARE POISONOUS VIRULENCE FACTORS

Toxins are microbial poisons that profoundly affect the establishment and course of disease because a single toxin can make an organism virulent. The ability of pathogens to produce toxins is referred to as **toxigenicity**. Toxins present in the blood is called **toxemia**. Two types of toxins are recognized: exotoxins and endotoxins (TABLE 18.2).

Exotoxins are produced by gram-positive and gram-negative bacteria. They are protein molecules, manufactured during the metabolism of bacteria. Exotoxins are released into the surrounding environment of the tissue as they are produced (*exo-* is Greek for "outside"). They dissolve in the blood fluid and circulate to their site of activity. The symptoms of disease soon develop.

The exotoxin produced by the botulism bacillus *Clostridium botulinum* is among the most lethal toxins known. One pint of the pure toxin is believed sufficient to destroy the world's population. In humans, the toxin inhibits the release of acetylcholine at the synaptic junction, a process that leads to the paralysis seen in botulism (Chapter 9). Another exotoxin is produced by tetanus bacilli, *C. tetani*. In this case, the exotoxin blocks the relaxation pathway that follows muscle contraction, thereby permitting volleys of spontaneous nerve impulses and uncontrolled muscular contractions.

Another exotoxin is produced by *Corynebacterium diphtheriae*, the diphtheria bacillus. The exotoxin interferes with the protein synthesis in the cytoplasm of epithelial cells of the upper respiratory tract. Disintegrated cells then accumulate with mucus, bacteria, fibrous material, and white blood cells, resulting in life-threatening respiratory blockages. Other exotoxins are formed by the bacteria that cause scarlet fever, staphylococcal food poisoning, pertussis, and cholera.

When toxins function in a particular organ system, they are given more clearly defined names. For example, the botulism toxin is called a **neurotoxin** because of its activity in the nervous system, while the staphylococcal toxin is called an **enterotoxin** since it functions in the gastrointestinal tract.

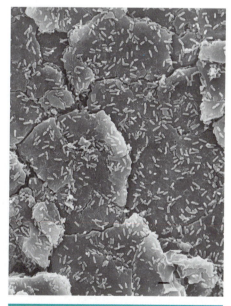

FIGURE 18.20

Bacteria in a Biofilm

A scanning electron micrograph of *Pseudomonas aeruginosa* found in a biofilm in a latex catheter in a hospital setting. Within biofilms like these, the bacteria can resist disinfectant treatment of the equipment. In the human body, biofilms add to the virulence of a pathogen by helping it resist drugs and natural body defenses. (Bar = 5 μm.)

TABLE 18.2

A Comparison of Exotoxins and Endotoxins

CHARACTERISTIC	EXOTOXINS	ENDOTOXINS
Source	Gram-positive and gram-negative bacteria	Gram-negative bacteria
Location in bacterium	Cytoplasm	Cell wall
Chemical composition	Protein	Lipid-polysaccharide-peptide
Antibodies elicited	Yes	No
Conversion to toxoid	Possible	Not possible
Liberation of toxin	On production by the cell	On disintegration of the cell
Representative effects	Interfere with synaptic activity Interrupt protein synthesis Increase capillary permeability Increase water elimination	Increase body temperature Increase hemorrhaging Increase swelling in tissues Induce vomiting, diarrhea

The body responds to exotoxins by producing antibodies called **antitoxins**. When toxin and antitoxin molecules combine with each other, the toxin is neutralized (Chapter 19). This process represents an important defensive measure in the body. Therapy for people who have botulism, tetanus, or diphtheria often includes injections of antitoxins (immune globulin) to neutralize the toxins.

Because exotoxins are proteins, they are susceptible to the heat and chemicals that normally denature proteins. A chemical such as formaldehyde may be used to alter the toxin and destroy its toxicity without hindering its ability to elicit an immune response in the body. The result is a **toxoid**. When the toxoid is injected into the body, the immune system responds with antitoxins that circulate and provide a measure of defense against disease. Toxoids are used for diphtheria and tetanus immunizations in the diphtheria-tetanus-acellular pertussis (DTaP) vaccine.

Endotoxins are part of the cell wall of bacteria, and as such, they are released only upon disintegration of the cell. They are present in the outer membrane in many gram-negative bacilli and are composed of lipid-polysaccharide complexes (Chapter 4). The lipid portion of the lipopolysaccharide (LPS) of the outer membrane is the toxic portion. Endotoxins do not stimulate an immune response in the body, nor can they be altered to prepare toxoids. They appear to function by activating a blood-clotting factor to initiate blood coagulation and by influencing the complement system (Chapter 19). The toxins of plague bacilli are especially powerful.

Endotoxins manifest their presence by certain signs and symptoms. Usually an individual experiences an increase in body temperature, substantial body weakness and aches, and general malaise. Damage to the circulatory system and shock may occur also. In this case, the permeability of the blood vessels changes and blood leaks into the intercellular spaces, where it is useless. The tissues swell, the blood pressure drops, and the patient may lapse into a coma. This condition, commonly called **endotoxin shock**, may accompany antibiotic treatment of diseases due to gram-negative bacilli because endotoxins are released as the bacilli are killed by the antibiotic.

Endotoxins usually play a contributing rather than a primary role in the disease process. Certain endotoxins reduce platelet counts in the host and thereby hinder clot formation. Other endotoxins are known to increase hemorrhaging. Like exotoxins, endotoxins add to the virulence of a parasite and enhance its ability to establish disease.

To this point . . .

In our continued study of infection and disease, we examined the factors that are involved in establishing a disease. Not only do pathogens need the correct portal of entry to initiate disease, they need a suitable portal of exit that will ensure the transmission of the pathogens to additional hosts. Besides entry, an essential number of pathogens (dose) must enter the body to establish a disease. Usually entry requires that pathogens penetrate host tissues. Thus, a pathogen's invasiveness also is important in establishing a disease.

Next, we studied the virulence factors that pathogens might have that increase their ability to cause disease. First, we learned that many microorganisms have specific proteins, called adhesins that allow the microbe to adhere to appropriate tissues. Adhesins may be present on bacterial capsules, flagella, or pili. Viruses attach by spikes on the capsid or envelope. Such adhesion mechanisms usually result in the internalization of the pathogen into the host cell by endocytosis or phagocytosis. Many bacteria use enzymes to maximize their pathogenicity. Secretion of coagulases produces bacteria-filled blood clots that protect the clusters of pathogens from being phagocytized and streptokinase enzymes allow the bacteria to escape from host-stimulated clots. Bacteria that can produce hyaluronidases can drill through cell layers to infect deeper tissues. Another set of enzymes, the hemolysins, lyse red blood cells to obtain the iron in the hemoglobin molecules. Again, pathogenicity is maximized by getting the iron needed for metabolism.

The last group of virulence factors we examined was the exotoxins and endotoxins. Exotoxins are released by gram-positive and gram-negative bacteria and have specific disease effects depending on the specific toxin. Antitoxins are produced against exotoxins and inactivated toxins (toxoids) can be used to prepare vaccines. Endotoxins represent the lipopolysaccharide in the outer membrane of gram-negative bacteria. These toxins cause more generalized effects, such as fever, tissue swelling, and vomiting and diarrhea.

Before ending this chapter, we will look at the blood and blood components that make up the human circulatory system. We also will see the importance of the lymphatic system. Both are crucial to immune system function, the topic of the next several chapters.

18.3

The Human Circulatory and Lymphatic Systems

The circulatory and lymphatic systems of the human host serve as the principal vehicle for the dissemination of parasites and their toxins. Many important factors in the host defense systems also operate within these systems. These two concepts add to the importance of a basic understanding of circulatory and lymphatic system components and their capabilities. Our outline of the systems also will serve as a bridge between the disease process discussed in this chapter and the resistance process surveyed in the next chapter. The emphasis will be on the blood and its components.

BLOOD CONSISTS OF THREE MAJOR COMPONENTS

Blood is the fluid and formed elements by which oxygen is carried to the tissues. It consists of three major components: the fluid, the clotting agents, and the cells. The fluid portion, called **serum**, is an aqueous solution of minerals, salts, proteins, and other organic substances. When clotting agents, such as fibrinogen and prothrombin, are present, the fluid is referred to as **plasma**. The pH of arterial blood is about 7.35 to 7.45.

Three types of cells circulate in the blood: the red blood cells, or erythrocytes; the white blood cells, or leukocytes; and the platelets, or thrombocytes. **Erythrocytes** arise in the bone marrow and carry oxygen to the tissues loosely bound to the red pigment hemoglobin. A normal adult has about 5 million erythrocytes per microliter (μl) of blood. After circulating for about 120 days, erythrocytes disintegrate in the spleen, liver, and bone marrow. The hemoglobin then is converted to bilirubin, a pigment that gives bile a deep yellow color. Normally, bilirubin is carried to the liver for degradation in a protein-bound form. However, if the liver is damaged by a disease, or if too many erythrocytes disintegrate, excess bilirubin and bile pigments may enter the bloodstream. This condition, called **jaundice**, is responsible for the yellow color of the complexion during cases of hepatitis and yellow fever (Chapter 14).

The white blood cells, or **leukocytes**, have no pigment in their cytoplasm and therefore appear gray when unstained. These cells also are produced in the bone marrow. They number about 4,000 to 12,000 per μl of blood and have different lifespans, depending on the type of cell. **TABLE 18.3** shows the different types of white blood cells, along with the other components of blood.

Neutrophils have a multilobed nucleus and therefore are referred to as **polymorphonuclear leukocytes**, or **PMNs** . In their cytoplasm, there are many lysosomes that contain digestive enzymes. Neutrophils (**FIGURE 18.21**) function chiefly as **phagocytes**, cells that carry out phagocytosis. They pass out of the circulation through pores in the vessels and squeeze into narrow passageways among the cells to engulf particles. Approximately 50 to 70 percent of the leukocytes are neutrophils. Their lifespan is about 12 hours.

FIGURE 18.21

A View of Neutrophils

A scanning electron micrograph of neutrophils. Note the irregular shapes of the cells, demonstrating the amoeba-like projections used in phagocytosis and motility. (Bar = 10 μm.)

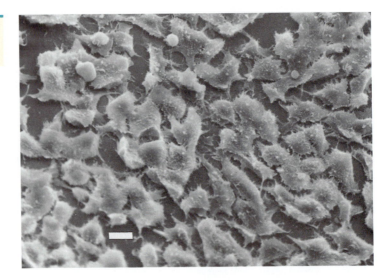

TABLE 18.3

The Cellular Composition of Human Blood

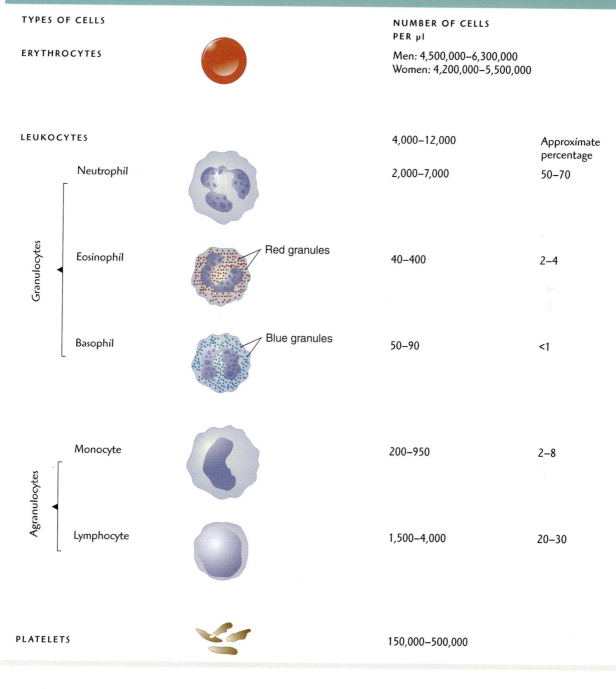

TYPES OF CELLS			NUMBER OF CELLS PER µl	
ERYTHROCYTES			Men: 4,500,000–6,300,000 Women: 4,200,000–5,500,000	
LEUKOCYTES			4,000–12,000	Approximate percentage
Granulocytes	Neutrophil		2,000–7,000	50–70
	Eosinophil	Red granules	40–400	2–4
	Basophil	Blue granules	50–90	<1
Agranulocytes	Monocyte		200–950	2–8
	Lymphocyte		1,500–4,000	20–30
PLATELETS			150,000–500,000	

Basophils are leukocytes whose granules stain blue with basic dyes such as hematoxylin. The basophils number about 50 to 90 per μl of blood and represent less than 1 percent of the total number of leukocytes. They function in allergic reactions, as their granules release physiologically active substances such as histamine. This process is discussed in detail in Chapter 21.

Eosinophils are the third type of leukocyte, representing about 2 to 4 percent of the total number. These cells exhibit red cytoplasmic granules when an acidic dye such as eosin is applied. Substances in the granules contain toxic compounds to defend against multicellular parasites, such as flukes and tapeworms. Eosinophils have phagocytic activity, although it is less significant than that of neutrophils. Eosinophils, basophils, and neutrophils are often called *granulocytes.*

Another major phagocyte of the circulatory system is the **monocyte**. This cell has a single, bean-shaped nucleus that takes up most of the area of the cytoplasm. Monocytes lack granules and account for approximately 2 to 8 percent of the leukocytes. In the tissues, monocytes mature into a type of phagocyte called the **macrophage**. In contrast to the two-week lifespan of monocytes, macrophages may live for several months.

Another type of leukocyte is the **lymphocyte**. This cell arises in the bone marrow and migrates to the lymph nodes after modification. It has a single, large nucleus and no granules. Monocytes and lymphocytes often are called *agranulocytes.* Lymphocytes make up about 20 to 30 percent of the white blood cells in the human body. They function in the immune system as B lymphocytes and T lymphocytes (Chapter 19). Their numbers increase dramatically during the course of certain diseases, such as infectious mononucleosis.

Blood platelets represent the third type of cell in the circulatory system. They are small, disk-shaped ("plate-let") cells that originate from cells in the bone marrow. Platelets have no nucleus and function chiefly in the blood-clotting mechanism.

THE LYMPHATIC SYSTEM HAS THREE FUNCTIONS

The fluid that surrounds the tissue cells and fills the intercellular spaces is called tissue fluid, or **lymph**. Lymph is similar to serum except that lymph has fewer proteins. Lymph bathes the body cells, supplying oxygen and nutrients while collecting wastes. It is pumped along in tiny vessels by the contractions of skeletal muscle cells. Eventually the tiny lymph vessels unite to form large vessels that compose a lymphatic system. On the right side, the system empties into a large vein just before the heart. FIGURE 18.22 illustrates the interrelationship of the lymphatic system and the circulatory system in humans.

Pockets of lymphatic tissue located along the lymph vessels are known as **lymph nodes**. Lymph nodes are prevalent in the neck, armpits, and groin. They are bean-shaped organs containing phagocytes, which engulf particles in the lymph, and lymphocytes, which respond specifically to substances in the circulation. Since resistance mechanisms are closely associated with the lymph nodes, it is not surprising that they become enlarged during periods of disease (sometimes they are called "swollen glands"). The tonsils, adenoids, spleen, Peyer's patches of the small intestine, and appendix are specialized types of lymph nodes. Their locations are noted in FIGURE 18.23.

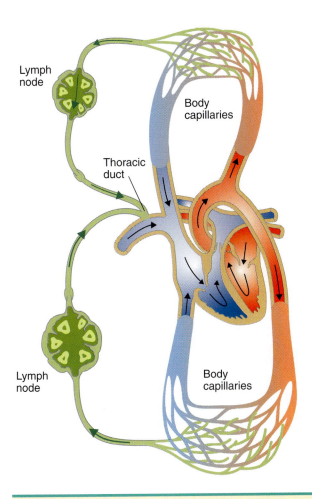

Lymph node

Body capillaries

Thoracic duct

Lymph node

Body capillaries

FIGURE 18.22

The Interrelationship of the Lymphatic System and the Circulatory System

Blood fluid passes out of the arteries in the upper and lower parts of the body. It enters a system of lymphatic ducts that arise in the tissues. The fluid, called lymph, passes through lymph nodes and on the right side makes its way back to the general circulation via the thoracic duct. The thoracic duct enters a main vein just before the vein enters the heart. A similar system exists on the left side.

The lymphatic system has three major functions: (1) return fluids and solutes to the blood; (2) transport lipids to the circulatory system; and (3) produce, maintain, and distribute lymphocytes necessary for defense against pathogens.

Pathogens possess a wealth of virulence factors that add to pathogenicity, but the lymphatic system of the host has an equally formidable array of defensive capabilities. For example, in the 1300s an estimated one-third of the population of Europe fell to bubonic plague. However, two-thirds survived, and without the benefit of antibiotics, antibody injections, or other treatments now used for disease. The nature and function of the defensive mechanisms that led to this survival and the resistance modes that operate in all humans are the major topics of Chapter 19.

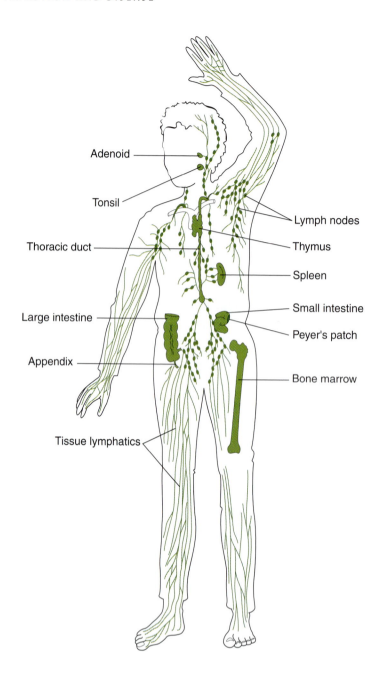

Adenoid

Tonsil

Thoracic duct

Large intestine

Appendix

Tissue lymphatics

Lymph nodes

Thymus

Spleen

Small intestine

Peyer's patch

Bone marrow

FIGURE 18.23

The Human Lymphatic System

The human lymphatic system consists of lymphocytes, lymphatic organs, lymph vessels, and lymph nodes located along the vessels. The lymphatic organs are illustrated, and the preponderance of lymph nodes in the neck, armpits, and groin is apparent.

Note to the Student

You may have noted that infection, disease, and host death are widely separated phenomena with many intermediary steps. Consider, for example, what happens when cholera strikes. The bacillus, *Vibrio cholerae*, multiplies along the walls of the intestine and produces toxins. The toxins cause such massive diarrhea that up to six quarts of fluid may be lost over the next several hours. The blood thickens, urine production ceases, the skin becomes dry, wrinkled, and cold, and the sluggish flow of blood to the brain leads to shock, coma, and death.

Now consider where this leaves the cholera bacilli. After the host dies, the bacilli are pushed to the side by the enormous number and variety of other microorganisms normally found in the intestine. For the cholera bacilli, their source of nourishment is gone, and the environment is now filled with hostile predators. The situation has gotten thoroughly out of hand, and soon the cholera bacilli will perish just as the host perished.

It is clear that, in the long run, host death is rarely beneficial to the parasite. Is it possible that the death of the host is, in fact, a biological accident?

Summary of Key Concepts

18.1 THE HOST-PARASITE RELATIONSHIP

Concepts about infection and disease, and the factors leading to the establishment of disease, were the two principal focal points of this chapter. Infection refers to the relationship between two organisms, the host and the parasite (pathogen), and the competition taking place between them. Disease may be considered a change from a general condition of good health arising from the parasite's victory in the competition.

- **The Human Body Is Host to Its Own Microbiota.** The human body contains large resident populations of microorganisms that are not harmful and are called the human body microbiota. Any part of the body that has exposure to the outside environment contains populations of human microbiota. One of the advantages of the microbiota is that they usually out-compete invading pathogens and so help protect the body against potentially dangerous infections.

- **Pathogens Differ in Their Ability to Cause Disease.** Pathogenicity, the ability of the parasite to cause disease, is fundamentally inseparable from the host's ability to resist disease. Thus, disease and resistance go hand-in-hand. Virulence refers to the degree of pathogenicity that a parasite displays.

- **Diseases Progress through a Series of Stages.** The progress of disease follows a regular pattern that includes an incubation period, prodrome phase, and periods of acme, decline, and convalescence.

- **Diseases Are Transmitted by Direct or Indirect Means.** Diseases may be transmitted by direct or indirect methods. Indirect methods include consumption of contaminated food or water, contaminated inanimate objects (fomites), and arthropods. Arthropods can be mechanical or biological vectors for the transmission of disease.

- **Reservoirs Represent Sources of Disease Organisms.** Reservoirs include humans, who represent carriers of a disease, arthropods, and food and water in which some parasites survive.

- **Diseases Can Be Described by Their Behavior and Occurrence in a Population.** A disease may be communicable, such as measles, or noncommunicable, such as tetanus. The occurrence of diseases falls into three categories: endemic, epidemic, and pandemic. An "outbreak" is essentially the same as an epidemic, although usually an outbreak is more confined in terms of disease spread.

■ **Diseases Can Be Classified by Severity and Host Involvement.** A disease also may be classified as acute (rapid and short duration) or chronic (lingering and long duration). A primary infection is an illness caused by a pathogen in an otherwise healthy host, while a secondary infection involves the development of other diseases as a result of the primary infection lowering host resistance. Local diseases are restricted to a specific part of the body while systemic diseases spread to several parts of the body and deeper tissues.

18.2 THE ESTABLISHMENT OF DISEASE

A parasite must have unusual abilities in order to bring about disease, and certain conditions must exist if disease is to be established.

■ **Pathogens Enter and Leave the Host through Portals.** To cause disease, most pathogens must enter the body through a correct portal of entry. To efficiently spread the disease to other hosts, the disease organism also must leave the body through an appropriate portal of exit.

■ **Disease Onset Is Influenced by the Number of Invading Pathogens.** The dose represents the number of pathogens taken into the body. In some cases, a small dose will not cause disease while for more virulent pathogens, a low dose may cause disease. High doses also do not necessarily mean a disease will occur.

■ **Most Pathogens Must Penetrate Host Tissues.** The possibility of disease is enhanced if a parasite can penetrate host tissues. The pathogen's ability to penetrate tissues and cause damage is referred to as invasiveness. Virulence and invasiveness are strongly dependent on the spectrum of virulence factors a pathogen possesses.

■ **Adhesins Are Important Virulence Factors.** Virulence factors, called adhesins, allow bacteria to adhere to specific cells. Adhesins can be found on capsules, flagella, or pili of some bacteria. Viruses contain spikes attached to the envelope or capsid, which allows for attachment to host cells. Adhesion or attachment usually leads to endocytosis or phagocytosis of the pathogens.

■ **Enzymes Can Increase the Virulence of a Pathogen.** Many bacteria produce enzymes to overcome the body's defenses. These include coagulase, streptokinase, and hyaluronidase enzymes. In addition, some bacteria produce lytic enzymes, such as leukocidins and hemolysins. The latter lyse red blood cells, providing bacteria with access to the iron-containing hemoglobin.

■ **Toxins Are Poisonous Virulence Factors.** Many bacteria synthesize toxins that interfere with body processes. These toxins can be classified as exotoxins or endotoxins, depending on their physiological effects and activity. Exotoxins are proteins released by gram-positive and gram-negative cells. Their effects depend on the enzyme produced and host system affected. Exotoxins are the lipopolysaccharides released from dead gram-negative cells. Their effects on the host are more universal.

18.3 THE HUMAN CIRCULATORY AND LYMPHATIC SYSTEMS

An understanding of the circulatory and lymphatic systems is important because the disease process often occurs here, and because resistance arises from factors in these systems.

■ **Blood Consists of Three Major Components.** The blood components are the serum (or plasma), clotting agents (blood platelets), and cells. In terms of infection, the most important cells are the leukocytes. Leukocytes consist of the granulocytes (neutrophils, eosinophils, and basophils). The agranulocytes include the monocytes that mature into macrophages and the lymphocytes (B lymphocytes and T lymphocytes).

■ **The Lymphatic System Has Three Functions.** The lymphatic system functions to (1) return tissue fluids (lymph) and solutes back to the circulatory system; (2) transport lipids to the circulatory system; and (3) produce, maintain, and distribute lymphocytes necessary for defense against invasive pathogens.

http://microbiology.jbpub.com

The site features **eLearning,** an on-line review area that provides quizzes and other tools to help you study for your class. You can also follow useful links for in-depth information, or just find out the latest microbiology news.

Questions for Thought and Discussion

Answers to selected questions can be found in Appendix C.

1. In 1840, Great Britain introduced penny postage and issued the first adhesive stamps. However, politicians did not like the idea because it deprived them of the free postage they were used to. Soon, a rumor campaign was started, saying that these gummed labels could spread disease among the population. Can you see any wisdom in their contention? Would their concern "apply" today?

2. A woman takes an antibiotic to relieve a urinary tract infection caused by *Escherichia coli*. The infection resolves, but in two weeks, she develops a *Candida albicans* infection of the vaginal tract. What conditions may have caused this to happen? What nonantibiotic course of treatment might a doctor prescribe to solve this problem?

3. In his classic book *The Mirage of Health* (1959), the French scientist René Dubos develops the idea that health is a balance of physiological processes, a balance that takes into account such things as better nutrition and better living conditions. (Such a view opposes the more short-sighted approach of locating an infectious agent and developing a cure.) From your experience, describe several other things that Dubos might add to his list of "balancing agents."

4. The transparent windows placed over salad bars are commonly called "sneeze bars" because they help prevent nasal droplets from reaching the salad items. What other suggestions might you make to prevent disease transmission via the salad bar?

5. In 1892, a critic of the germ theory of disease named Max von Pettenkofer sought to discredit Robert Koch's work by drinking a culture of cholera bacilli diluted in water. Von Pettenkofer suffered nothing more than mild diarrhea. What factors may have contributed to the failure of the bacilli to cause cholera in his body?

6. Between 1982 and 1992, the number of Americans dying of infectious diseases rose 58 percent. That rise continues to this day. Population shifts, modern travel patterns, and microbial evolution are three of the many reasons given for the emergence and reemergence of infectious diseases. How many other reasons can you name?

7. While slicing a piece of garden hose, a gardener cut himself with a sharp knife. The wound was deep, but it closed quickly. Shortly thereafter, he reported to the emergency room of the community hospital, where he received a tetanus shot. What did the tetanus shot contain, and why was it necessary?

8. In the early 1970s, the Egyptian demographer Abdul Waheed Omran observed that in many modern industrial nations, the major killers no longer were infectious diseases. As people lived longer, he maintained, they succumbed to "diseases of civilization," such as cancer, heart disease, diabetes, obesity, and osteoporosis. Omran was among the first to recognize what he called an "epidemiological transition." Scientists now recognize that a new epidemiological transition is taking place in our era. What is this new transition, and what evidence do you think the scientists cite for the transition?

9. It has been estimated that in a sneeze, 4,600 droplets are shot forth with a muzzle velocity of 152 feet per second. Scientists believe that the droplets from a single sneeze hang suspended in the air for up to 30 minutes, and collectively may contain over 35 million viruses. How many diseases can you name that are transmitted by droplets such as these?

10. An environmental microbiologist has created a stir by maintaining that a plume of water is aerosolized when a toilet is flushed, and that the plume carries bacteria to other items in the bathroom, such as toothbrushes. Assuming this is true, what might be two good practices to follow in the bathroom?

11. You are a microorganism hunter assigned to make a list of the ten worst "hot zones" in your home. The title of your top-ten list will be "Germs, Germs Everywhere." What places will make your list, and why?

12. A man takes a roll of dollar bills out of his pocket and "peels" off a few to pay the restaurant tab. Each time he peels, he wets his thumb with saliva. What is the hazard involved?

13. In your opinion, would an epidemic disease or an endemic disease pose a greater threat to public health in the community? Why? Given the choice, would you rather experience a chronic disease or an acute disease?

14. While touring a day-care center, a visitor notices that a nail brush is next to each sink. When asked why, the center director replies that all personnel are requested to use the brush after they wash their hands, especially after changing a child's diaper. Why is this a good idea?

15. When Ebola fever broke out in Africa in 1995, public health epidemiologists noted how quickly the responsible virus killed its victims and guessed that the epidemic would end shortly. Sure enough, within three weeks it was over. What was the basis for their prediction? What other conditions had to apply for them to be accurate in their guesswork?

Review

Test your knowledge of this chapter's contents by determining whether the following statements are true or false. If the statement is true, write "True" in the space. If false, substitute a word for the underlined word to make the statement true. The answers are listed in Appendix D.

_____ 1. An <u>epidemic</u> disease is one that occurs at a low level in a certain geographic area.

_____ 2. A parasite that is <u>invasive</u> has the ability to penetrate tissues and cause structural damage.

_____ 3. Among the microbial enzymes that are able to destroy blood cells are <u>hemolysins</u> and leukocidins.

_____ 4. The term <u>disease</u> refers to the living together of two organisms and the competition that takes place between them for supremacy.

_____ 5. Organs of the human body that do not have a normal flora include the blood and the <u>small intestine</u>.

_____ 6. A <u>commensalism</u> is a form of symbiosis in which only one organism benefits but no damage occurs to the other.

_____ 7. An organism with <u>high</u> virulence is generally unable to cause disease in the body.

_____ 8. A <u>sign</u> is a change in body function experienced by the patient.

_____ 9. A <u>biological</u> vector is an arthropod that carries pathogenic microorganisms on its feet and body parts.

_____ 10. Those organisms that cause disease when the immune system is depressed are known as <u>opportunistic</u> organisms.

_____ 11. A <u>reservoir</u> is one who has recovered from a disease but continues to shed the disease agents.

_____ 12. The human body responds to the presence of toxins by producing <u>endotoxins</u>.

_____ 13. The term <u>bacteremia</u> refers to the spread of bacteria through the bloodstream.

_____ 14. A pathogenic staphylococcus is able to form a fibrin clot through its production of <u>hyaluronidase</u>.

_____ 15. A <u>toxoid</u> is an immunizing agent prepared from an exotoxin.

_____ 16. Few symptoms are exhibited by a person who has a <u>subclinical disease</u>.

_____ 17. Among the <u>indirect</u> methods of disease transmission are kissing, hand-shaking, and contact with feces.

_____ 18. Certain parasites, such as the bacterium that causes <u>tuberculosis</u>, have multiple portals of entry.

_____ 19. A <u>chronic</u> disease is one that develops rapidly, is usually accompanied by severe symptoms, and comes to a climax.

_____ 20. The <u>acme</u> period is the time between the entry of the parasite to the host and the appearance of symptoms.

19 Resistance and the Immune System

Interlopers are vigorously attacked and their molecular signatures are memorized so that next time they can be carded and stopped at the door.

—*New York Times* reporter George Johnson describing the immune system's activity

Think of the human body as a doughnut. Just as the hole passes through the center of the doughnut, so too the gastrointestinal (GI) tract passes through the center of the human body. The body secretes enzymes into the hole, digests the food we eat, absorbs what it needs, and lets the remainder pass out of the hole. There is no natural opening to the internal tissues in the gastrointestinal tract, and the fact that something is in the GI tract does not mean that it is in the body.

The same is true for the respiratory and urinary tracts. The respiratory tract leads to dead-end pouches of the alveoli, while the urinary tract leads first to the urinary bladder and then to the kidneys, where it terminates at cup-shaped structures called Bowman's capsules. In both instances, a layer of cells shields the blood and the body's internal organs from the outside environment. Close examination of other parts of the body reveals similar dead ends, and it becomes clear that the body, like the doughnut, is a closed container. Only after the walls of this container are penetrated can most diseases be established.

The skin and its extensions into the gastrointestinal, respiratory, and urinary tracts represent a major form of resistance to infection and disease. This form is **nonspecific**, because it exists in all humans and is present from birth. Also, it protects against all parasites. (The immunity is said to be "innate.") Other forms of resistance are **specific**, because they come about in response to a particular parasite and are directed solely at that parasite (FIGURE 19.1).

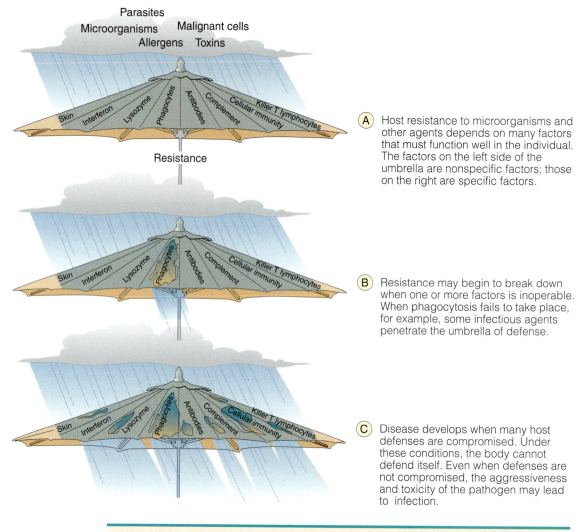

Parasites
Microorganisms   Malignant cells
Allergens   Toxins

Resistance

(A) Host resistance to microorganisms and other agents depends on many factors that must function well in the individual. The factors on the left side of the umbrella are nonspecific factors; those on the right are specific factors.

(B) Resistance may begin to break down when one or more factors is inoperable. When phagocytosis fails to take place, for example, some infectious agents penetrate the umbrella of defense.

(C) Disease develops when many host defenses are compromised. Under these conditions, the body cannot defend itself. Even when defenses are not compromised, the aggressiveness and toxicity of the pathogen may lead to infection.

FIGURE 19.1

The Relationship Between Host Resistance and Disease

The focus of this chapter will be to examine both forms of resistance and to show how good health depends upon their proper functioning. The discussion is related to the previous chapter's study of disease, except that the emphasis switches from the parasite to the host.

19.1

Nonspecific Resistance

Nonspecific resistance to disease involves a broad group of factors, many of which are still not defined. It depends upon the general well-being of the individual and proper functioning of the body's systems. Accordingly, it takes into account such determinants as nutrition, fatigue, age, gender, and climate. Specific examples of these factors are highlighted in discussions of individual diseases, and therefore we shall not pause to delineate them here.

THERE ARE BROAD TYPES OF NONSPECIFIC RESISTANCE

One form of nonspecific resistance is called **species immunity**. This genetically determined resistance implies that diseases affecting one species will not affect another. For example, humans do not contract hog cholera, while hogs do not contract AIDS (their cells lack the necessary receptor sites for HIV). Similarly, cattle plague is unknown in humans, while gonorrhea does not occur in cattle. Immunities such as these are probably based on physiological, anatomical, and biochemical differences. MicroFocus 19.1 presents an interesting offshoot of species immunity.

Behavioral immunities exist among various races and peoples of the world. Many of these immunities are due to nonspecific factors related to a people's way of life. For example, in the 1700s, Tamil laborers were brought from southern India to work on plantations in Malaya. The laborers continued the custom of bringing water into their houses only once a day and not storing it between times. This deprived mosquitoes of indoor breeding places and reduced the incidence of malaria among the laborers.

Racial immunities reflect the evolution of resistant humans. For instance, black Africans affected by the genetic disease sickle-cell anemia do not contract malaria, presumably because the parasite cannot penetrate distorted red blood cells. Some investigators hold to the theory that **population immunities** exist because parasites have adapted to the body's environment. Americans, for example, generally view measles as a mild disorder, but when the disease was introduced to Greenland in the early 1960s, it exacted a heavy toll of lives in a population that had no previous

MicroFocus 19.1

"THANKS, BUT I'LL PASS!"

There are many reasons for avoiding cannibalism, but research reported in 1998 offers another disincentive: Animals that practice cannibalism probably will ingest lethal microorganisms.

The research was performed at the University of North Carolina by David Pfennig and his colleagues. Pfennig's group raised four groups of naturally cannibalistic salamanders, giving each group one of the following menus: healthy salamanders of another species, ill salamanders of another species (ill because they were exposed to bacteria-contaminated water), healthy salamanders of their own species, or ill salamanders of their own species. Of the four groups, three did very well; the one that did poorly were the salamanders that ate ill members of

their own species. Nearly half died of bacterial infection.

The results point to the concept that bacteria, as well as most microbes, specialize in particular species. If cannibalism takes place among members of the same species, the bacteria are easily transferred, and the salamanders die. But, bacteria infecting a different species are not necessarily able to infect the cannibalistic species. Thus, the salamanders eating the sick animals of another species do not themselves become ill.

In the cold calculus of biology, it can make sense for an animal to reabsorb crucial energy when it consumes its young during periods of starvation ("crisis cannibalism"). Evolutionary history, however, seems to show that animals rejecting cannibalism do better in the

long run. Perhaps they avoid self-injury by refusing to attack one another. Then again, perhaps they are really avoiding species-specific microorganisms. After all, if microbes target the dinner, why shouldn't they target the diner as well?

■ *A tiger salamander larva eating a member of its species.*

exposure to the virus. A similar event took place when the Spanish conquistadors arrived in the New World in the 1500s (MicroFocus 19.2).

In addition to species and other immunities, nonspecific resistance often is referred to as **innate immunity** because genetically encoded receptors, present in the body from birth, recognize microbial features common to many pathogens. These defense mechanisms operate against all foreign substances and organisms. We shall review these processes next.

MECHANICAL AND CHEMICAL BARRIERS ARE THE FIRST DEFENSES AGAINST INFECTION

Mucous membrane:
the moist epithelium lining of the digestive and respiratory tracts.

Sebum:
a mixture of fatty acids, proteins, and salts secreted by oil glands.

As noted in Chapter 18, there are several routes by which pathogens can enter the body. Therefore, the intact skin and the **mucous membranes** that extend into the body cavities are among the most important resistance factors at these portals of entry. The skin's epidermis is constantly being sloughed off, and with it go microorganisms. The keratin in skin is a poor source of carbon for microorganisms; the fatty acids in **sebum** and sweat are antimicrobic; and, from the viewpoint of a microorganism, the low water content of the skin presents a veritable desert. Toxins notwithstanding, unless penetration of these barriers occurs, disease is rare. However, sometimes direct contact with certain dermatophytes can lead to diseases such as athlete's foot.

Penetration of the skin barrier is a fact of everyday life. A cut or abrasion, for example, allows staphylococci to enter the blood, and the bite of an infected arthropod acts as a hypodermic needle permitting the pathogen to enter. Yellow fever viruses, malaria parasites, most species of rickettsiae, and plague bacilli are but a few

MicroFocus 19.2

CONQUEST BY DISEASE

History books teach that Spain's conquest of the Aztec nation of Central America was due to horses, gunpowder, and the superior force of Spanish arms. Conquest meant overcoming millions of people and overturning long-standing traditions of religion and culture, and there were only 800 men with Hernando Cortez the day he landed in Mexico in 1518. Cortez and the Spanish eventually toppled the Aztec nation, but their strongest ally was not gunpowder and arms; it was disease.

Mexico was totally unprepared for smallpox. The disease was new to the country, and the Aztec population was without any trace of immunity. The Spanish, on the other hand, had contended with European outbreaks of smallpox for generations, and they were

relatively immune. Little thought was given to the consequences when slaves sick with smallpox arrived with the Spaniards. Soon the Aztec community was infected.

By April 1521, Cortez had established a colony on Mexican soil and marched inland to attack Tenochtitlan, the stronghold of the Aztec nation. The siege continued for four months, until the city finally fell on August 13, 1521. Expecting to plunder the city, the Spanish rushed in, but were shocked to find the houses filled with the dead. A smallpox epidemic was raging. Half the population had succumbed.

Nor did it end here. Eleven years later, Spanish invaders introduced another epidemic of disease, believed to be measles. Thousands of Aztecs died. In 1545, disease broke out again. Contemporary

writings describing the symptoms indicate that it was either epidemic typhus or typhoid fever. In one province 150,000 people are estimated to have died. Influenza raged in 1558 and 1559, and mumps broke out with fatal consequences in 1576.

By the end of the century, the Aztec nation was battered into submission. Native authority figures succumbed, and the surviving Aztecs dutifully obeyed the commands of Spanish landowners, tax collectors, and missionaries. To both the conqueror and the conquered, the divine and natural orders had spoken out loudly against the native beliefs; the Aztecs would offer no further resistance. By one historian's account, almost 19 million of the original population of 25 million died of disease by 1595.

examples. Other means of penetrating this mechanical barrier include splinters, tooth extractions, burns, shaving nicks, war wounds, and injections.

Certain features of the mucous membranes provide resistance to parasites. For instance, cells of the mucous membranes along the lining of the respiratory passageways secrete **mucus**, which traps heavy particles and microorganisms. The cilia of other cells then move the particles along the membranes up to the throat, where the particles are swallowed. Stomach acid usually destroys any microorganisms.

Mucus:
a sticky secretion of glycoproteins, leukocytes, and salts.

Resistance in the vaginal tract is enhanced by the low pH. This develops when *Lactobacillus* species in the microbiota break down glycogen to various acids. Many researchers believe that the disappearance of lactobacilli during antibiotic treatment encourages diseases such as candidiasis and trichomoniasis to develop. In the urinary tract, the slightly acidic pH of the urine promotes resistance to parasites, and the flow of urine flushes microorganisms away.

A natural barrier to the gastrointestinal tract is provided by stomach acid, which has a pH of approximately 2.0. (A cotton handkerchief placed in stomach acid would dissolve in a few short moments.) Most organisms are destroyed in this environment. Notable exceptions include typhoid and tubercle bacilli, protozoal cysts, and polio and hepatitis A viruses, as well as *Helicobacter pylori*, a major cause of peptic ulcers. **Bile** from the gallbladder enters the system at the duodenum and serves as an inhibitory substance. In addition, duodenal enzymes digest the proteins, carbohydrates, fats, and other large molecules of microorganisms, while antimicrobial peptides damage or kill pathogens.

Bile:
the fluid secreted by the liver to aid in fat digestion.

Another chemical inhibitor of a nonspecific nature is the enzyme **lysozyme**. This protein was described in the early 1920s by Alexander Fleming, who later gained recognition for the discovery of penicillin. Lysozyme is found in human tears, sweat, and saliva. It disrupts the cell walls of gram-positive bacteria by digesting peptidoglycan. Another inhibitor is **interferon**. Interferon is actually a group of substances (interferons) produced by body cells in response to invasion by viruses. Interferons trigger the production of inhibitory substances that "interfere" with viral reproduction. A thorough account of the interferons is presented in Chapter 12.

Lastly, but importantly, are the normal microbiota of the body surfaces that usually outcompete pathogens for nutrients and attachment sites. The microbiota also produce antimicrobial proteins.

TABLE 19.1 summarizes many of the nonspecific mechanical and chemical resistance mechanisms.

PHAGOCYTOSIS REPRESENTS PART OF A SECOND LINE OF NONSPECIFIC DEFENSES

Shortly after the germ theory of disease was verified by Koch, Elie Metchnikoff made a chance discovery that clarified how living cells could protect themselves against microorganisms. Metchnikoff noted that motile cells in the larva of a starfish gathered around a wooden splinter placed within the cell mass. He suggested that the cells actively sought out and engulfed foreign particles in the environment to provide resistance. Metchnikoff's theory of phagocytosis, published in 1884, was received with skepticism, because it appeared to conflict with the antitoxin theory then in vogue. Many investigators believed that antitoxins produced by the body were the sole basis for resistance, but in succeeding years they came to appreciate phagocytosis as equally important.

In contemporary microbiology, **phagocytosis** ("cell-eating") is viewed as a major form of nonspecific defense in the body. If a pathogen breaches the first line of

TABLE 19.1

Nonspecific Mechanical and Chemical Barriers to Disease

RESISTANCE MECHANISM	ACTIVITY
Skin layers	Provide a protective covering to all body tissues
Mucous membranes of body cavities	Trap airborne particles in mucus Sweep particles along by cilia
Acidity in the vagina and stomach	Acidic pH toxic to microorganisms
Bile	Inhibitory to most microorganisms
Duodenal enzymes	Digest structural and metabolic chemical components of microorganisms
Lysozyme in tears, saliva, secretions	Digests cell walls of gram-positive bacteria
Interferons	Inhibit replication of viruses
Normal microbiota	Compete for nutrients and attachment sites Produce antimicrobial substances

mechanical or chemical defenses, the intruder usually is recognized and engulfed by cells called **phagocytes** (FIGURE 19.2). There are two groups of phagocytes, the macrophages and the neutrophils. **Macrophages** arise from circulating monocytes (Chapter 18) and reside in many tissues of the body, including the spleen, lymph nodes, gastrointestinal tract, lungs, and connective tissues. In other words, many of the macrophages are located just where a pathogen might cross the skin or mucous membranes.

The **neutrophils** have a short life and are abundant in the blood. Unlike macrophages, they are not found in healthy tissue. Macrophages will be the first to interact with an invading pathogen, but they are joined by large numbers of neutrophils to eliminate or limit the spread of the pathogen. Neutrophils have a greater variety of antimicrobial substances and are more likely to kill pathogens.

Phagocytosis begins with an invagination, or folding in, of the cell membrane to form a phagocytic vesicle, or **phagosome** (FIGURE 19.3). The phagosome then pinches off and fuses with several lysosomes, organelles that contribute digestive enzymes, proteins, and peptides to the digestion process. Within the **phagolysosome**, lysosome products contribute in two ways to the destruction of engulfed pathogens. Lysosome enzymes, such as **lysozyme** digest bacterial cell walls while **acid hydrolases** digest bacterial contents. In addition, other products generated by the lysosome or in the cytoplasm also help in the killing. These products include **hydrogen peroxide (H_2O_2)**, **nitric oxide (NO)**, and **superoxide anions (O_2^-)**. All three are toxic to bacteria. The process is completed as waste materials are eliminated from the phagocyte.

Neutrophils are short lived and usually die after a round of phagocytosis. Macrophages survive many rounds of phagocytosis and, through endomembrane activity, generate new lysosomes to continue the digestive process. If the host has the

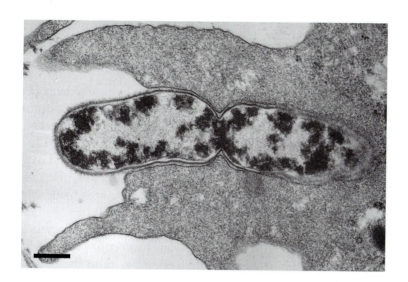

FIGURE 19.2

Phagocyte Ingestion

A transmission electron micrograph of a dividing bacillus being engulfed by a neutrophil. The extensions of the cell surround the bacilli, and membranes of the phagocytic vesicle follow the contours of the organism being ingested. (Bar = 0.5 μm.)

FIGURE 19.3

The Mechanism of Phagocytosis

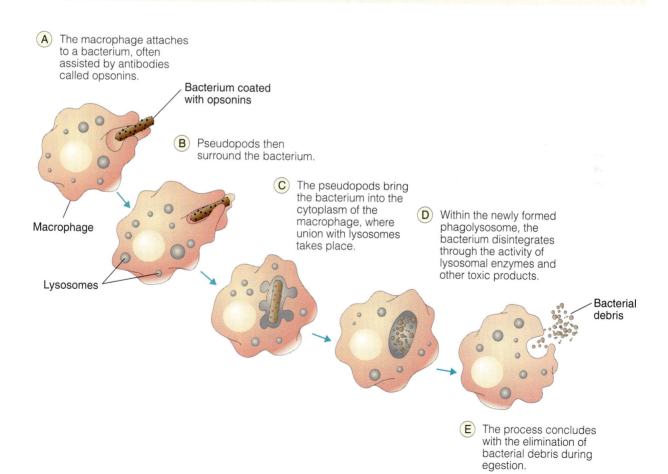

(A) The macrophage attaches to a bacterium, often assisted by antibodies called opsonins.

Bacterium coated with opsonins

(B) Pseudopods then surround the bacterium.

(C) The pseudopods bring the bacterium into the cytoplasm of the macrophage, where union with lysosomes takes place.

(D) Within the newly formed phagolysosome, the bacterium disintegrates through the activity of lysosomal enzymes and other toxic products.

Macrophage

Lysosomes

Bacterial debris

(E) The process concludes with the elimination of bacterial debris during egestion.

advantage, the pathogens will be cleared and the infection ends. However, the host-parasite relationship is always evolving to give either the host or the parasite the advantage. Sometimes the pathogen gets the upper hand, as MicroFocus 19.3 explains.

Although phagocytes are effective at killing pathogens and clearing many infections, collateral damage often occurs. During the fierce phagocytic activity occurring at the infection site, lysosome enzymes often are released inadvertently into the surrounding tissue. In addition, neutrophils release lysosome enzymes while dying. The released lysosome products have damaging effects on surrounding tissue and often are the major reason for tissue damage from an infection. Indeed, **pus**, often a sign of infection, is composed of dead and dying neutrophils and damaged tissue arising from lysosome damage.

To limit host damage, the body tries to regulate the numbers of phagocytes and antimicrobial products involved in the defense against infection. For example, during a bacterial infection, the bacteria themselves release short peptides that attract phagocytes to the site of infection. In addition, binding of pathogens to macrophages during phagocytosis stimulates the phagocytes to release chemicals called **chemokines** that attract neutrophils and additional phagocyte cells. Therefore, the chemokine response must be managed so there is not excessive damage at the infection site.

The interaction between parasite and phagocyte also is enhanced by the presence of antibodies. These protein molecules attach to parasites and increase adherence to the phagocytic cell at specific receptor sites. In other situations, components of the complement system (discussed shortly) bind the parasite to the phagocyte. Enhanced phagocytosis is called **opsonization**, and the antibodies or complement components that encourage it are termed **opsonins**, from Latin stems meaning "to prepare for."

MicroFocus 19.3

AVOIDING THE "BLACK HOLE" OF THE PHAGOCYTE

Lysosomes contain quite a battery of enzymes, proteins, and other substances that are very efficient at digesting or being toxic to almost anything with which they come in contact. However, not all bacteria are automatically engulfed by phagocytes. Some bacteria have a thick capsule (Chapter 4) that is not easily internalized by phagocytes while other pathogens simply invade in such large numbers that they overwhelm the host defenses. More interesting though are pathogens that have evolved strategies to evade the lysosome—the "black hole" of the phagocyte.

Listeria monocytogenes is internalized by phagocytes. However, once in a phagosome, the bacterium releases a pore-forming toxin that lyses the phagosome membrane. This allows the *L. monocytogenes* cells to enter and reproduce in the phagocyte cytoplasm before lysosomes can fuse with the phagosome membrane. The bacteria then spread to adjacent cells by the formation of actin tails described in Chapter 18.

Other pathogens prevent lysosome fusion with the phagocytic vesicle in which they are contained. The causative agent of tuberculosis, *Mycobacterium tuberculosis*, is internalized by macro-phages as normal. However, once in a phagosome, the bacterial cells prevent fusion of lysosomes with the phagosome. In fact, the tubercle bacteria actually reproduce within the phagosome. A similar strategy is used by *Legionella pneumophila*. *Toxoplasma gondii*, the protozoan responsible for toxoplasmosis (Chapter 16) evades the lysosomal killing machine of macrophages by enclosing itself in its own membrane vesicle that does not fuse with lysosomes.

Avoiding digestion by lysosomes is key to survival for these pathogens.

MicroFocus 19.4

BUT IT DOESN'T ITCH!

How many of us at some time have had a small red spot or welt on our skin that looks like an insect bite—but it doesn't itch? When most of us are bitten by an insect, such as an ant, bee, or mosquito, the site of the bite starts itching almost immediately. We quickly reach for the anti–itch cream to bring us some relief. Yet, there are cases where bites do not itch, such as a tick bite. So, what's the difference?

Most insect bites trigger a typical inflammatory reaction. Take a mosquito as an example. When it bites and takes a blood meal, saliva is injected into the skin that contains substances so the blood does not coagulate. The bite causes tissue damage and the activation of chemical mediators that trigger the typical characteristics of an inflammatory response: redness, warmth, swelling, and pain at the injured site. One of the major chemical mediators is histamine, released by the damaged tissue and immune cells in the tissue. An itch actually is a form of pain and arises when his-

tamine affects nearby nerves. Itching is the result.

In a tick bite, the scenario is slightly different. The bite triggers the same set of inflammatory reactions and histamine is produced. However, in the tick saliva is another molecule called histamine-binding protein (HBP). HBP binds to the histamine, preventing the mediator from affecting the nearby nerves and no itching occurs.

Now, what is the advantage to the tick of producing HBP? Hmmm.

INFLAMMATION PLAYS AN IMPORTANT ROLE IN FIGHTING INFECTION

Inflammation is a nonspecific defensive response by the body to an injury in the tissue. It develops after a mechanical injury, such as an injury or blow to the skin, or from exposure to a chemical agent, such as acid or bee venom. An inflammatory response also may be due to an infection by a living organism, such as a parasite.

In the case of an infection, the pathogen's presence and tissue injury sets into motion a process that limits the spread of the infection (FIGURE 19.4). At the site of tissue damage, chemical mediators, such as **histamine**, are produced that trigger a local dilation of the blood vessels (vasodilation) and increase capillary permeability. This allows the flow of plasma into the injured tissue and fluid accumulation (edema) at the site of infection. Capillary permeability and chemokine production also bring phagocytes into the injured tissue. Neutrophils adhere to the blood vessels close to the injury and then migrate between capillary cells (diapedesis) into the tissue. There the neutrophils augment phagocytosis of the pathogens by front-line macrophages. Soon, additional macrophages migrate into the area as neutrophils die.

The inflamed area exhibits four characteristic signs: *rubor* (redness from blood accumulation); *calor* (heat from the warmth of the blood); *tumor* (swelling from the accumulation of fluid), and *dolor* (pain from injury to local nerves). Sometimes itching occurs, as MicroFocus 19.4 describes.

Again, pus is a product of phagocytosis during inflammation. When pus becomes enclosed in a wall of fibrin through activation of the clotting mechanism, a sac may form. This sac is an **abscess**, or **boil**, often due to staphylococci. When several abscesses accumulate, an enlarged structure called a **carbuncle** results (Chapter 11).

Inflammation and phagocytosis are thus interrelated. Their purpose is to confine the pathogen to the site of entry and to repair or replace the tissue that has been injured.

FIGURE 19.4

The Process of Inflammation

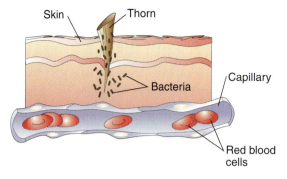

(A) A thorn pierces the skin, causing a mechanical injury and bringing bacteria into the tissue.

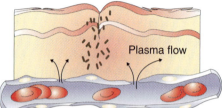

(B) Capillary walls open (dilate), and plasma flows to the site of injury, making it red, swollen, and warm.

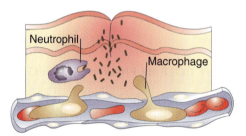

(C) Phagocytes (neutrophils and macrophages) arrive and begin phagocytosis.

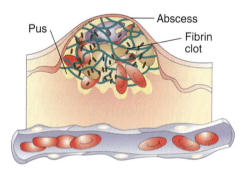

(D) A fibrin wall accumulates around the mixture of plasma, leukocytes, bacteria, and tissue cells, collectively called pus. An abscess becomes apparent.

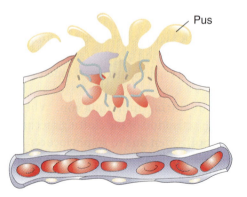

(E) Lancing of the abscess releases the pus.

MODERATE FEVER IS BENEFICIAL TO HOST DEFENSES

Fever is an abnormally high body temperature that may provide a nonspecific mechanism of defense against disease. Fever-producing substances, called **pyrogens**, are produced by activated macrophages as well as bacteria, viruses, and other microorganisms. These chemicals affect a region at the base of the brain called the hypothalamus and stimulate it to raise the body temperature several degrees. As this takes place, cell metabolism increases and blood vessels constrict, thus denying blood to the skin and keeping its heat within the body. Patients thus experience cold skin and chills along with the fever.

A moderate fever in adults may be beneficial because it appears to inhibit the growth of certain organisms. The increased metabolism in cells also encourages rapid tissue repair and raises the level of phagocytosis, while reducing the amount of bloodborne iron available to parasites. However, if the temperature rises above 40.6°C (105°F), convulsions and death may result. Also, infants with a fever above 38°C (100°F) or older children with a fever of 39°C (102°F) need medical attention.

NATURAL KILLER CELLS RECOGNIZE AND KILL ABNORMAL CELLS

Natural killer (NK) cells are a unique group of defensive lymphocytes that roam the body in blood and lymph and kill cancer cells and virus-infected cells as part of innate immunity. NK cells act spontaneously and without any previous immune activation. The name "natural" killer cells reflects the nonspecific nature of the killing activity.

NK cells are not phagocytic; rather, they contain on their surfaces a set of special receptor sites that must form a complex with the target cells before a cell-to-cell interaction takes place. The receptor sites match with a group of class I MHC proteins (to be discussed presently), which are found on all body cells. When the normal class I MHC proteins are present, the NK cell recognizes the target cell as one of the body's own, and spares it. However, when these MHC proteins are absent or in reduced amounts (as on a cancer cell, organ transplants, or virus-infected cells), then the NK cell binds to the target cell, damages its cell membrane, and induces lysis.

COMPLEMENT MARKS PATHOGENS FOR DESTRUCTION

Complement (also called the complement system) is a series of nearly 30 proteins that circulate in the bloodstream. Though complement is better known for its relationship to the immune system (it "complements" antigen–antibody activity), the protein series is also a nonspecific deterrent to disease. If a parasite causes an infection, complement proteins become active through a cascade of steps that assist in the inflammatory response and phagocytosis. They also bring about the destruction of the invading parasites through cell lysis. Two of the complementary pathways for complement activation are described below (FIGURE 19.5).

One pathway, called the **classical pathway**, involves antigen-antibody complexes (discussed presently) activating a cascade of complement proteins that culminate in the activation of complement protein C3. The other pathway, called the **alternative pathway**, involves the binding of C3 molecules directly to the pathogen cell surface. So, both pathways converge by the activation of the C3 complement protein. C3 then is hydrolyzed into a C3a and C3b fragment. C3b

FIGURE 19.5

Overview of Complement Function

The early events of both the classical and alternative pathways of complement activation culminate in the activation of complement protein 3 (C3) and the effector functions of complement. Complement fragment C3b binds to the membrane and opsonizes bacteria, allowing phagocytes to internalize them (phagocytosis). The small fragments, C5a and C3a, are mediators of local inflammation. C3b also activates another complement cascade, whereby the terminal components of complement assemble into a membrane–attack complex (MAC) that can damage the membrane of many pathogens.

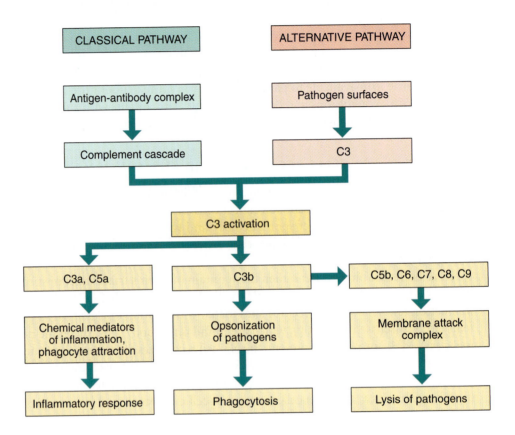

activates another complement protein C5, which also is hydrolyzed into C5a and C5b fragments.

Complement fragments C3a and C5a help activate an inflammatory response through vasodilation and increased capillary permeability. C3b also acts as an opsonin by binding to the pathogen surface (opsonization), enhancing phagocytosis by phagocytes.

C3b plays one more important role. The complement fragment triggers another complement cascade that assembles into what are called **membrane attack complexes** (**MACs**). These complexes form large holes in the membranes of many microorganisms, especially gram-negative bacteria and enveloped viruses. The MACs are so prevalent that they lyse the cell, killing the pathogen. This teamwork within the complement system and inflammation provides a formidable obstacle to pathogen invasion.

To this point . . .

We have begun the study of resistance to disease by outlining some nonspecific mechanisms that defend against all parasites from the time of birth. Species and behavioral immunities are examples. Other examples are mechanical and chemical barriers, including the barriers presented by the skin and mucous membranes, the acidity in the stomach and vagina, and the antimicrobial substances lysozyme and interferon.

We examined phagocytosis in depth since this process is a major form of nonspecific defense. The process may involve particles as well as dissolved materials. We also noted the value of inflammation, fever, and natural killer cells in nonspecific resistance. The complement system was formerly believed to be solely a part of the specific mechanism of resistance, but scientists now realize its nonspecific role.

We now will focus on specific resistance as reflected by the immune system. After a brief discussion of the beginnings of immunology, we shall survey antigens, the substances that stimulate the immune response. From there, we shall explore the origin of the immune system and see how it is established in the body. The system is actually dual in nature, although the processes are interlocking. The discussion therefore will diverge, first to study the system of cell-mediated immunity and later to examine antibody-mediated immunity. In a sense, this is the most important section in the entire textbook because it outlines the processes by which specific resistance to specific parasites takes place. How well the resistance operates largely dictates whether a person will remain free of infectious disease or will recover from disease.

19.2

Specific Resistance and the Immune System

In the late 1800s, the mechanisms of specific resistance to infectious disease were largely obscure because no one was really sure how the body responded when infected. However, medical bacteriologists were aware that certain proteins of the blood unite specifically with chemical compounds of microorganisms. The blood proteins were named *Bence Jones proteins* after Henry Bence Jones, a British physician who identified them in the urine of a patient. In 1922, researchers from Johns Hopkins University showed that Bence Jones proteins (later found to be parts of antibodies) were unlike normal serum proteins, but the nature of the proteins remained a matter of conjecture.

Until the 1950s, specific resistance to disease was virtually synonymous with immunity. By that time, vaccines were available for numerous diseases, and immunologists saw themselves as specialists in disease prevention. The explosion of interest in the biological sciences after World War II spilled over to immunology, and soon it became apparent that specific resistance is a phenomenon with broader implications, including organ transplantation, allergic reactions, and resistance to cancer. In addition, the groundwork was laid for deciphering the nature and function of the Bence Jones proteins. This work would lead to the elucidation of antibody structure in the 1960s, and the maturing of immunology to one of the key scientific disciplines of our times.

In this section, we shall study the immune system as it relates to specific defense mechanisms to disease. It should be noted that the immune system also relates to the cancer patient, transplant recipient, laboratory diagnostician, and hay fever sufferer. Some of these topics are explored in detail in Chapters 20 and 21. Research in immunology provides an important window to the disease process, as well as many other processes of life. Our study will begin with a survey of the substances that stimulate the immune response.

ANTIGENS STIMULATE AN IMMUNE RESPONSE

Antigens are chemical substances capable of mobilizing the immune system and provoking an immune response. They enter the body through the typical portals of entry, such as the mucous membranes of the respiratory tract, a wound, an arthropod bite through the skin, or an injury to the gastrointestinal tract. Most antigens are large, complex molecules (macromolecules), that are not normally found in the body and are consequently referred to as "nonself." Antigens exhibit two important properties: **immunogenicity**, the ability to stimulate cells of the immune system; and **reactivity**, the ability to react with products of immune system cells (antibodies), or with the cells themselves. Antigens often are called **immunogens**.

The list of antigens is enormously diverse. It includes milk proteins, substances in bee venom, hemoglobin molecules, viruses, bacterial toxins, and chemical substances found in bacterial flagella, pili, and capsules (**FIGURE 19.6**). The most common antigens are proteins, polysaccharides, and the chemical complexes formed between these substances and lipids or nucleic acids. Proteins are the most potent antigens because their amino acids have the greatest array of building blocks, an array leading to a huge variety of combinations and hence, diversity in three-dimensional structures. Polysaccharides are less potent antigens than proteins because they lack chemical diversity and rapidly break down in the body. Lipids also can be antigenic, as exemplified by the cell wall lipids (mycolic acid) of tubercle bacilli.

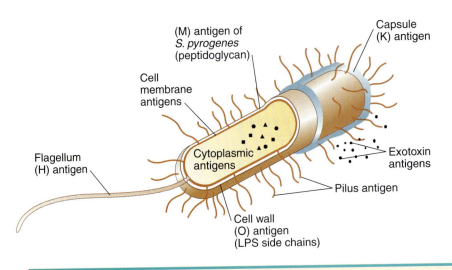

FIGURE 19.6

The Various Antigens Possible on a Bacterial Cell

Each different antigen in this idealized bacterium is capable of stimulating the immune system.

Antigens usually have a molecular weight of over 10,000 daltons. Because of this large size, the antigen molecule easily is phagocytized by macrophages, the necessary first step in the immune process. Antigens do not stimulate the immune system directly. Rather, the immune system recognizes a small part of the antigen molecule called the **antigenic determinant**, or **epitope**. An antigenic determinant (epitope) contains about six to eight amino acid molecules or monosaccharide units. Each antigenic determinant has a characteristic three-dimensional shape and a molecular weight of about 350 daltons. An antigen may have numerous antigenic determinants, and a structure such as a bacterial flagellum may have hundreds of these molecules. Antigenic determinants are unique microbial fingerprints to which the immune system responds (**FIGURE 19.7**).

Small nucleotides, hormones, peptides, and other molecules usually are not capable of stimulating the immune system by themselves; they are not immunogenic. However, when they link to proteins in the body, the immune system may recognize the combination as foreign and respond to it. Allergy reactions (Chapter 21) are examples of immune responses to these combinations. In the combination, the small molecule that by itself lacks immunogenicity is known as a **hapten** (*haptein* is Greek for "grasp," a reference to the interactions between the hapten molecule and

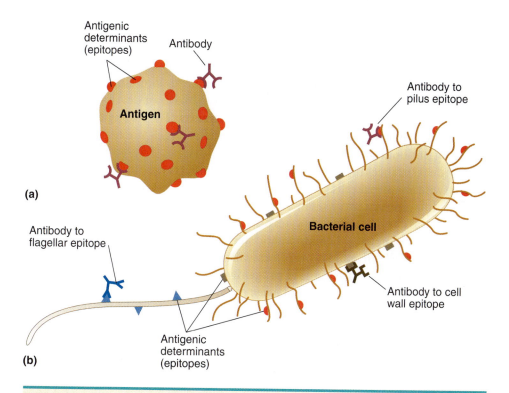

FIGURE 19.7

Antigens and Antigenic Determinants

(a) A large antigen molecule showing a number of antigenic determinants, also called epitopes. These are the sites where products of the immune system, such as antibodies, will interact. The reaction between antigenic determinants and antibodies neutralizes the antigen. (b) A bacterium with various sources of antigens and the antibodies that react with the antigenic determinants on these antigens. Note the variations in structure of the antigenic determinants and the corresponding antibodies that unite specifically with them.

the immune system's antibody). Examples of haptens are penicillin molecules, molecules in poison ivy plants, and molecules in certain cosmetics and dyes. When chemically linked to a protein carrier, they become immunogenic.

Under normal circumstances, one's own chemical substances do not stimulate an immune response. This failure to stimulate the immune system occurs because the substances are interpreted as "self." Immunologists believe that prior to birth, the body's proteins and polysaccharides destroy the immune system cells that otherwise might respond to "self antigens." The individual thereby develops a tolerance of "self" and remains able to respond only to "nonself." This theory, known as **specific immunologic tolerance**, was advanced in the late 1950s by Frank MacFarlane Burnet (a 1960 Nobel laureate) and David Talmage. Cells responding to self also are destroyed in the thymus, as we shall see shortly. **Autoantigens** are a person's own chemical substances that stimulate an immune response when self-tolerance breaks down (as in lupus erythematosus, Chapter 21).

THE IMMUNE SYSTEM ORIGINATES FROM GROUPS OF STEM CELLS

The **immune system** is a general term for the complex series of cells, factors, and processes providing an adaptive and specific response to antigens associated with microorganisms, or with potentially harmful molecules such as microbial toxins. As such, the system lends specific resistance against infection and disease.

The cornerstones of the immune system are a set of body cells known as **lymphocytes**. These cells are distributed throughout the body, where they comprise part of the lymphoid system (Chapter 18). Lymphocytes are small cells, about 10 to 20 μm in diameter, each with a large nucleus taking up almost the entire space of the cytoplasm. Under the microscope, all lymphocytes look similar. However, two types of lymphocytes can be distinguished on the basis of developmental history, cellular function, and unique biochemical differences. The two types are B lymphocytes and T lymphocytes. **B lymphocytes** (**B cells**) largely are responsible for producing antibodies that provide resistance to disease. **T lymphocytes** (**T cells**) provide resistance through direct cell-to-cell contact with and lysis of infected or otherwise abnormal cells.

The immune system arises in the fetus about two months after conception (FIGURE 19.8). At this time, lymphocytes originate from primitive cells in the yolk sac and bone marrow known as **hematopoietic stem cells**. These cells then differentiate into two types of cells: **myeloid progenitors**, which become red blood cells or most of the white blood cells; and **lymphoid progenitors**, which become lymphocytes of the immune system. (The Greek word *poien* means "to make"; thus, hematopoietic cells are "blood-making cells.") We now shall follow the fate of the lymphopoietic cells.

Lymphoid progenitors take either of two courses. From the bone marrow, some of the cells proceed to an organ of the thoracic cavity called the **thymus**. This flat, bilobed organ lies below the thyroid gland near the top of the heart. The thymus is large in size at birth and increases in size until the age of puberty, when it begins to shrink. Within the thymus, the progenitor cells mature over a two- or three-day period and are modified by the addition of surface receptor proteins. They emerge from the organ as T lymphocytes (T for thymus). Mature T cells are ready to engage in cell-mediated immunity and are said to be **immunocompetent**. The T lymphocytes colonize the lymph nodes, spleen, tonsils, and other lymphoid organs, and they become a major portion of the lymphoid system.

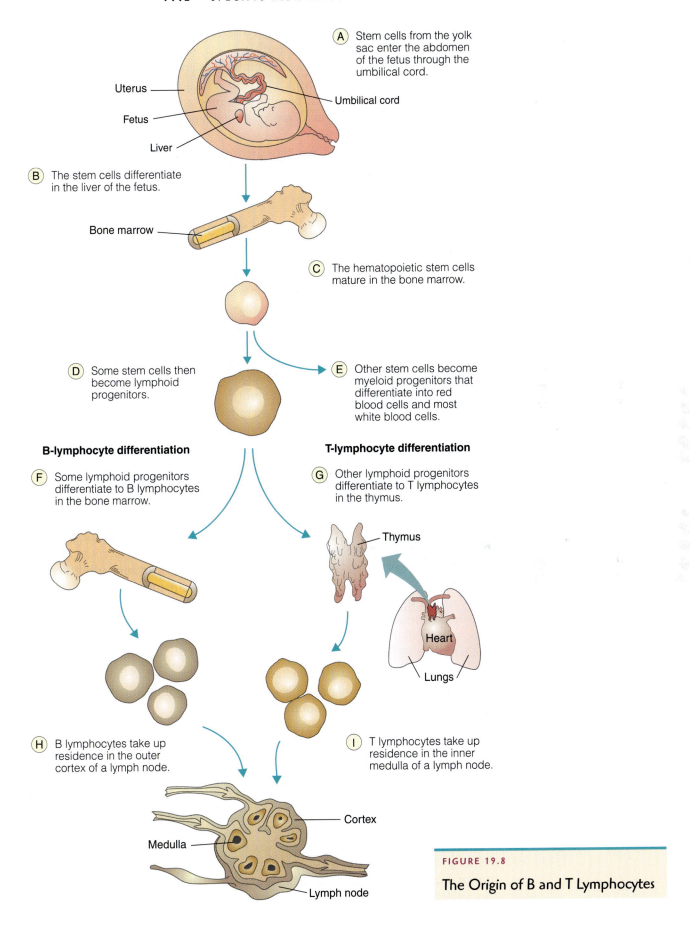

(A) Stem cells from the yolk sac enter the abdomen of the fetus through the umbilical cord.

Uterus

Fetus

Liver

Umbilical cord

(B) The stem cells differentiate in the liver of the fetus.

Bone marrow

(C) The hematopoietic stem cells mature in the bone marrow.

(D) Some stem cells then become lymphoid progenitors.

(E) Other stem cells become myeloid progenitors that differentiate into red blood cells and most white blood cells.

B-lymphocyte differentiation

(F) Some lymphoid progenitors differentiate to B lymphocytes in the bone marrow.

T-lymphocyte differentiation

(G) Other lymphoid progenitors differentiate to T lymphocytes in the thymus.

Thymus

Heart

Lungs

(H) B lymphocytes take up residence in the outer cortex of a lymph node.

(I) T lymphocytes take up residence in the inner medulla of a lymph node.

Cortex

Medulla

Lymph node

FIGURE 19.8

The Origin of B and T Lymphocytes

It should be noted that a large percentage of developing T cells are destroyed in the thymus (a type of "programmed cell death"). These cells would normally react with self antigens. Therefore, the mature T lymphocytes emerging are the cells able to interact with nonself (foreign) antigens.

The B lymphocytes mature and become immunocompetent in the bone marrow. For this reason, the lymphocyte is known as the B lymphocyte. Like the T lymphocytes, B lymphocytes mature with surface receptor proteins on their membranes and become immunocompetent. Once they are immunocompetent, the B cells move through the circulation to colonize organs of the lymphoid system, where they join the T cells.

T AND B CELLS HAVE RECEPTORS TO RECOGNIZE ANTIGENS

As noted, the maturation of both T cells and B cells is accompanied by the insertion of surface **receptor proteins** spanning the width of the cell membrane. The receptor protein enables the lymphocyte to recognize a specific antigenic determinant (epitope) and bind to it. After this recognition and binding has occurred, the lymphocyte is said to be **committed**. In T cells, the receptor protein at the cell surface is composed of two chains of glycoproteins linked to one another by disulfide bonds. In B cells, the receptor protein is an antibody molecule. About 100,000 surface receptors are found on each T or B cell.

The location of highly specific receptor proteins on the lymphocyte surface is one of the more remarkable discoveries of contemporary immunology. The implication is that even before an antigen enters the body, an immunocompetent cell is already waiting for it. Moreover, the genetic code for synthesizing the surface receptor is present in an individual even though that individual has not ever been exposed to an antigen. We may never experience malaria, for example, yet we already have surface receptor proteins for recognizing and binding to the antigens of malaria parasites. This concept is engaging and thought provoking.

B cells can recognize antigenic determinants on an antigen. However, the recognition between antigenic determinants and T cells depends on surface receptor proteins plus another set of glycoprotein molecules called the **major histocompatibility complex proteins**, or **MHC proteins**. (These proteins are sometimes called MHC antigens because they stimulate an antibody response in other individuals.) MHC proteins are embedded in the membranes of all cells of the body. At least 20 different genes encode MHC proteins, and at least 50 different forms of the genes exist. Thus, the variety of MHC proteins existing in the human population is enormous, and the chance of two individuals having the same MHC proteins is incredibly small. (The notable exception is identical twins.) The MHC proteins define the uniqueness of the individual and play a role in the immune response.

There are two important classes of MHC proteins, and both help define the individual as self. **Class I MHC proteins** are found on virtually all nucleated cells of the body, but **class II MHC proteins** can be found only on B lymphocytes, macrophages, and a few other cell types. These class II proteins also function in the immune response. There is evidence that certain human diseases are associated with the MHC proteins (Chapter 21).

The actual immune response originates with the entry of antigens into the body and their penetration into the lymphatic or cardiovascular system. Here the antigens are phagocytized by macrophages and other phagocytic cells, and the antigens are broken down (**FIGURE 19.9**). The macrophages then display peptides derived from the antigens as MHC-peptide complexes on their surface and are referred to as

MHC protein:
proteins on the surface of body cells that define the uniqueness of an individual.

FIGURE 19.9

Phagocytosis

This false color scanning electron micrograph shows a macrophage (gray) engulfing a species of *Leishmania* (purple), the protozoan that causes leishmaniasis (Chapter 16). Note the highly irregular fluid nature of the surface of the macrophage. (Bar = 10 μm.)

antigen-presenting cells (**APCs**). Macrophages move to the lymphoid organs, where T cells and B cells are waiting. Another important APC is the **dendritic cell**, a cell with long, finger-like extensions that form lacy networks in virtually all tissues and phagocytize infected cells nearby. The phagocytosis and transport are extremely important because research evidence indicates that unprocessed antigens stimulate the immune system poorly.

Within the tissues of the lymphoid organs (spleen, lymph nodes, tonsils, and so on), the T cells and B cells are waiting to implement either of the two arms of the immune system. *Antibody-mediated immunity* will result in activated B cells and antibodies that react with microorganisms (i.e., free bacteria and viruses), soluble antigens, and non-self cells within the body's environment. *Cell-mediated immunity* will result in activated T cells for direct interaction with eukaryotic pathogens, as well as antigen-marked cells (such as virus-infected and transplanted cells). In the next section, we shall discuss both types of immunity, beginning with cell-mediated immunity.

MicroFocus 19.5 addresses an alternative theory of how the immune system operates.

CELL-MEDIATED IMMUNITY IS A RESPONSE BY ANTIGEN-SPECIFIC T LYMPHOCYTES

The body's defense against microorganisms infecting its cells is centered in **cell-mediated immunity** (**CMI**), sometimes called cellular immunity. Cell-mediated immunity responds to cells that have been infected with pathogens such as viruses, rickettsiae, and certain bacteria, including *Mycobacterium tuberculosis*. During infection, pathogens alter the infected cells and change the molecular architecture of the cell's surface. The infected cells therefore act as signposts and become objects of an attack by cells of CMI. Protozoa and fungi also stimulate CMI, as do cancer cells and the cells of transplanted tissue.

The T cells participating in CMI make up two well-defined subpopulations: **cytotoxic T cells** and **helper T cells**. The two cells differ in the proteins on their cell surface. Cytotoxic T cells have CD8 proteins while helper T cells have CD4 proteins (**FIGURE 19.10**). Such proteins act as coreceptors.

The process of cell-mediated immunity is shown in **FIGURE 19.11**. It begins this way: In the body where infected tissue exists, a series of macrophages (as well as dendritic cells, and other phagocytic cells) have engulfed viruses or bacteria. In the

FIGURE 19.10

Distinctive Receptors on T Lymphocytes

(a) Helper T cells have a T-cell receptor and a CD4 coreceptor that recognize antigen peptides associated with class II MHC molecules on antigen-presenting cells. b) Cytotoxic T cells have a T-cell receptor and a CD8 coreceptor that recognize antigen peptides associated with class I MHC molecules on virus-infected cells, tumor cells, and foreign tissue transplants.

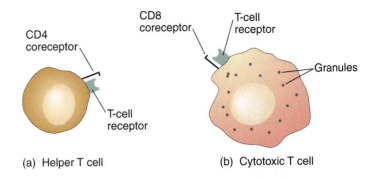

(a) Helper T cell   (b) Cytotoxic T cell

macrophage's cytoplasm, the antigens have been broken down into antigen peptides, and the latter have been captured and returned to the macrophage's surface, nestled within class II MHC proteins. The antigen-presenting macrophage now has a series of "red flags" on its surface. The APC travels to an organ of the lymphoid system, and CMI is ready to continue.

When the APC enters the lymphoid organ, a hunt begins. With its exposed class II MHC/peptide complex, the macrophage mingles among the myriad groups of inactive helper T cells, searching for the cluster having the surface receptors that recognize the MHC/peptide. This process requires considerable time and energy because only one cluster of T lymphocytes will have receptors that match. Soon the correct cluster is found.

MicroFocus 19.5

WHO GOES THERE? HARMLESS OR HARMFUL?

The traditional role of the immune system, and the way it is described in this text, is to defend the body against anything that is foreign; that is, the immune system discriminates between self and nonself by tolerating self and rejecting or attacking nonself. If this notion is correct, how do you explain the following? Immature immune systems of mice can produce an immune response when confronted with viruses and the immune systems in adult mice can be "taught" to tolerate cells they never before encountered. Traditional thinking says immature systems cannot respond because they are busy "learning" how to distinguish self from nonself, and in adults, tolerance should not be allowed because self from nonself recognition is established.

As far back as 1994, Holly Matzinger saw the immune system differently. Matzinger, a research scientist at the National Institute of Allergy and Infectious Diseases in Bethesda, Maryland, advocates the so-called danger model of immunity, which suggests the immune system is more concerned with damage and danger to self than simply foreignness. In other words, the immune system wants to know: "Who goes there? Harmless or harmful?"

If the immune system ignores harmless materials while fighting dangerous ones, then the system would not react to the foods we eat (allergies notwithstanding) but would react to invading parasites. While the self/nonself model requires the immune system to educate itself as to what's dangerous and what's

not, the danger model maintains that the immune system will spring into action only when something is associated with potentially doing harm. Matzinger says antigen-presenting cells are activated by danger/alarm signals from injured tissues or distressed cells, and not from recognition of foreign material at the injured site.

At this writing, the danger model has been supported by some recent discoveries, although Matzinger admits the model does not address the mechanism of allergies. So, perhaps there is common ground for both the traditional and contemporary models. As with so much of science, there is still much to learn and understand.

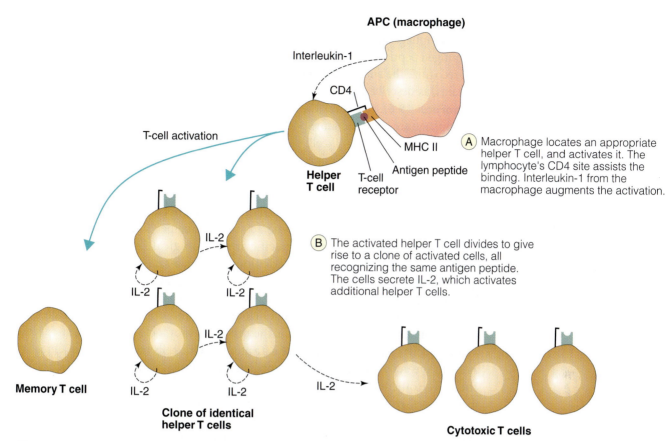

APC (macrophage)

Interleukin-1

CD4

T-cell activation

MHC II

Helper T cell

T-cell receptor

Antigen peptide

(A) Macrophage locates an appropriate helper T cell, and activates it. The lymphocyte's CD4 site assists the binding. Interleukin-1 from the macrophage augments the activation.

IL-2

IL-2   IL-2

IL-2

IL-2   IL-2

(B) The activated helper T cell divides to give rise to a clone of activated cells, all recognizing the same antigen peptide. The cells secrete IL-2, which activates additional helper T cells.

Memory T cell

Clone of identical helper T cells

IL-2

Cytotoxic T cells

(C) Some memory T cells form from the division of the activated T cells. They will be distributed to the tissues for long-term immunity.

(D) Cytotoxic T cells are activated by infected target cell and IL-2 from helper T cells. The cytotoxic T cells divide and form a clone of cytotoxic T cells. These cells leave the lymphoid tissues and travel to the infection site.

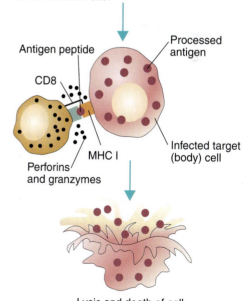

Antigen peptide

Processed antigen

CD8

MHC I

Infected target (body) cell

Perforins and granzymes

Lysis and death of cell

(E) At the infected tissues, the cytotoxic T cells unite with infected cells at antigen peptides enclosed within class I MHC proteins. The cytotoxic cells release perforin and granzymes, which enter the infected cells and destroy them. Specific defense is thus rendered.

FIGURE 19.11

The Process of Cell-Mediated Immunity

It is important to realize that recognition is between the class II MHC proteins on the macrophage surface and receptors on the T-lymphocyte surface. (Remember that the antigen peptides are nestled within the MHC proteins.) Indeed, the T lymphocyte will recognize the antigenic determinants only within the spatial context of the MHC proteins, as described by 1996 Nobel laureates Peter Doherty and Rolf Zinkernagel (MicroFocus 19.6). It is as if the T lymphocyte must first ascertain that it and the macrophage are from the same body before it will respond.

As mentioned, helper T cells also have another surface protein, called the **CD4 receptor**. This coreceptor enhances binding of the helper T cell to the macrophage because the CD4 protein recognizes the peptide associated with the class II MHC protein. (Incidentally, the CD4 receptor is where HIV binds to T cells.) Once the recognition has been made, the inactive T cell becomes an **effector T cell**.

Now the APC (the macrophage) interacts with the helper T cell, and things become livelier with the secretion of specific immunological growth factors called **cytokines**. One cytokine, called **interleukin-1 (IL-1)**, is secreted by the macrophage when it binds with a helper T cell. The cytokine stimulates T-cell activation. Interleukin-1 causes the helper T cell to secrete another cytokine, **interleukin-2 (IL-2)**, which stimulates cell division of that helper T cell and others that have been activated by macrophages. The result is the production of a **clone** of antigen-specific helper T cells (Figure 19.11). These activated effector cells have two roles. One is to assist in the activation of antibody-mediated immunity, which we will discuss just ahead. The other role is to activate the cytotoxic T lymphocytes, which are key participants in cell-mediated immunity.

After the cytotoxic T cells bind to the APC, under the influence of helper T-cell cytokines, cytotoxic T cells proliferate to form a clone of cytotoxic T cells capable of killing infected cells. The cytotoxic T cells leave the lymphoid tissue and enter the

MicroFocus 19.6

COMPELLING

They were assigned to the same cramped lab because there was no room elsewhere: Peter Doherty, the native Australian, and Rolf Zinkernagel, the visiting researcher from Switzerland. It was the early 1970s, and research space at the John Curtin School of Medical Research in Canberra was limited. But they would make do.

Both researchers were interested in how the brain is affected by the virus of lymphocytic choriomeningitis (LCM). The duo infected a mouse with LCM virus and collected its activated cytotoxic T lymphocytes. As they anticipated, the latter killed the virus-infected brain cells when the cells were combined in a test tube. What happened next, however, perplexed them: The same activated T lymphocytes failed to kill virus-infected cells from another strain of mice.

Doherty and Zinkernagel scratched their heads. Then, they remembered a report by Harvard immunologist Hugh McDevitt linking immune activities to the major histocompatibility complex (MHC) proteins at body cell surfaces. The MHC proteins are a sort of "molecular identification card" unique to the cells of a particular individual. At that time, they were known to be involved in the rejection of organs transplanted between unrelated people. The researchers wondered whether they might be involved also in T lymphocyte activity.

Then followed many years of work with T lymphocytes and infected cells bearing various combinations of MHC proteins. Time after time, the T lymphocyte attacked only the cell with matching MHC proteins. The theory was gradually emerging: An attacking T lymphocyte must recognize two signals, a signal from the virus and a signal from the infected cell (i.e., the antigen peptide and the MHC protein).

Doherty and Zinkernagel proposed the model that turned out to be correct. Their theory was compelling, and so was their prize: the 1996 Nobel Prize in Physiology or Medicine.

lymph and blood vessels. They circulate until they come upon their **target cells**, the infected cells displaying telltale antigenic peptides (the "red flags") on their surface. What has happened in the interim is this: A cell infected by an indwelling microorganism has been producing microbial proteins. Class I MHC proteins have picked up some of these peptides and transported them to the cell surface, displaying them within the class I MHC proteins. These are the "red flags" detected by the cytotoxic T cells as they arrive on the scene; they are the same antigen peptides originally recognized by the helper T cells attached to APCs and with which the cytotoxic T cells will react.

Now the cell-to-cell interaction begins. The cytotoxic T lymphocyte joins its receptor proteins with the antigen peptides and uses one of its surface receptors called **CD8** to bind to class I MHC/peptide on the infected cell surface (Figure 19.11). Then, the cytotoxic T cell releases a number of active substances, including a toxic protein called **perforin**. Perforin inserts into the membrane of the infected cell, forming cylindrical pores in the membrane. Ions, fluids, and cell structures escape, thereby bringing about lysis and the so-called "lethal hit." In addition, cytotoxic T cells release **granzymes** that enter the target cell and trigger DNA fragmentation. Cell death not only deprives the parasite of a place to live, but it also exposes the parasite to antibodies in the extracellular fluid. A similar type of action is brought about by natural killer (NK) cells, but there is no response to antigen peptides involved in this nonspecific defense. Rather, NK cells react to body cells lacking class I MHC molecules.

Cytotoxic T cells also are active against tumor cells because these cells display distinctive molecules on their surfaces (**FIGURE 19.12**). The molecules are not present in other body cells, so they are viewed as antigen peptides. Harbored within class I MHC proteins at the cell surface, the antigen peptides react with receptors on cytotoxic T cells and the tumor cells are subsequently killed. However those cancer agents that reduce the level of class I MHC proteins at the cell surface reduce the reactivity of the cell and encourage immunologic escape and survival.

When T cells start dividing after activation by APCs, a special clone of T cells forms, one that will provide resistance in the event the antigen reenters the body anytime in the future. These lymphocytes are called **memory T lymphocytes** (Figure 19.11). They distribute themselves to virtually all parts of the body and remain in the tissues to provide a type of long-term immunity. Should the antigens be detected once again in the tissues, the memory T cells will multiply rapidly, interact with the infected cells quickly, secrete cytokines without delay, and set into motion the process of providing instantaneous CMI. This is one reason we enjoy long-term immunity to a given disease after having contracted and recovered from that disease.

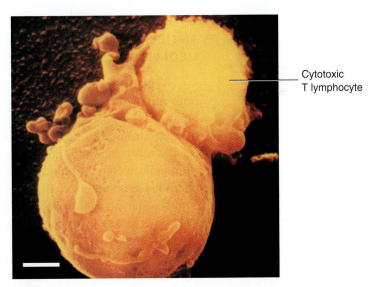

Cytotoxic
T lymphocyte

FIGURE 19.12

A "Lethal Hit"

A scanning electron micrograph of a cytotoxic T lymphocyte attacking a tumor cell. The activity of T lymphocytes represents a key defensive mechanism by the immune system. (Bar = 2 μm.)

To this point . . .

We have delved into specific resistance by describing the antigens that stimulate the immune system. The list of possible antigens is enormously diverse, but most are polysaccharides and proteins with a series of antigenic determinants. Haptens may function as antigenic determinants, but first they must complex with protein or polysaccharide. We saw why the immune system will not usually respond to the body's own substances (self) in the theory of specific immunologic tolerance.

The discussion then moved on to the immune system. We explored the origin of the system in terms of B lymphocytes and T lymphocytes, and noted how each colonizes the lymphoid tissues to form the immune system. The process of immunity generally begins with the phagocytosis of antigens and delivery of antigen peptides to the lymphoid tissues. The immune process then diverges to either of two systems. The focus next turned to cell-mediated immunity, the complex process in which helper T cells become activated through binding with antigen-presenting cells. Through cytokine secretion, helper T cells activate cytotoxic T cells that migrate to the antigen site and interact with abnormal cells, causing cell death through cell lysis and DNA fragmentation.

We now will explore the second arm of immunity, called antibody-mediated immunity. Here the B lymphocytes play a central role, and antibodies function in the immune response. We shall examine the structures and types of antibodies, how they are related to the genes of B cells, and the processes by which they interact with antigens and provide specific resistance. Together with cell-mediated immunity, antibody-mediated immunity represents the key to continued good health.

ANTIBODY-MEDIATED (HUMORAL) IMMUNITY IS A RESPONSE MEDIATED BY ANTIGEN-SPECIFIC B LYMPHOCYTES

Cell-mediated immunity is provided by attacking cells, but **antibody-mediated immunity (AMI)** is centered in antibodies, a series of protein molecules circulating in the body's fluids. (A body fluid is known also as a "humor," and antibody-mediated immunity has traditionally been called **humoral immunity**, but we shall use the more contemporary name.) In AMI, antibodies react with such things as toxin molecules in the bloodstream, as well as with antigens on microbial surfaces or microbial structures (e.g., flagella, pili, capsules), and with viruses in the extracellular fluid. The interaction of antibody and antigen leads to elimination of the antigen by various means, as we shall discuss in this section.

Antibody-mediated immunity, the second arm of the immune system, is well known because its origins trace back to the 1890s. In that period, Emil von Behring and Shibasaburo Kitasato determined that animals surviving a serious bacterial infection have "protective factors" in their blood to increase resistance to future attacks of the same infection. Today, these protective factors are known to be **antibodies**. Antibody formation and activity are the major topics of this section.

AMI begins with an encounter between pathogen or toxin and cells of the immune system. In this case, the major participating cell is the B lymphocyte. Antigens stimulating the process are usually derived from the bloodstream, such as those associated with bacteria, viruses, and certain organic substances as we have noted. The antigen epitopes interact with B cells, as shown in FIGURE 19.13.

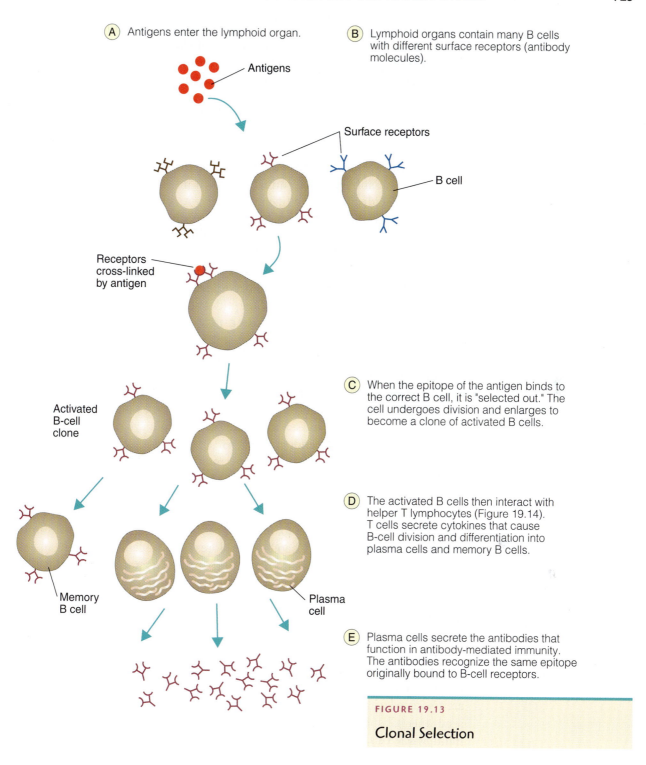

(A) Antigens enter the lymphoid organ.

(B) Lymphoid organs contain many B cells with different surface receptors (antibody molecules).

Antigens

Surface receptors

B cell

Receptors cross-linked by antigen

(C) When the epitope of the antigen binds to the correct B cell, it is "selected out." The cell undergoes division and enlarges to become a clone of activated B cells.

Activated B-cell clone

(D) The activated B cells then interact with helper T lymphocytes (Figure 19.14). T cells secrete cytokines that cause B-cell division and differentiation into plasma cells and memory B cells.

Memory B cell

Plasma cell

(E) Plasma cells secrete the antibodies that function in antibody-mediated immunity. The antibodies recognize the same epitope originally bound to B-cell receptors.

FIGURE 19.13

Clonal Selection

Activation of the B cells begins when antigens enter a lymphoid organ and bring the antigenic determinants close to the appropriate B cells that can respond. As with CMI, antigenic determinants must match with surface receptor proteins on the B lymphocytes, as Figure 19.13 illustrates. The surface proteins on B cells are antibodies. Activation of the B cells requires that epitopes cross-link antibody molecules. Haptens may be poor stimulators of B cells because of their inability to perform cross-linking.

The selection of a single cluster of B lymphocytes by a particular antigenic determinant is the underlying basis for the **theory of clonal selection**, shown in Figure 19.13. First proposed in the 1950s by Nobel laureates Frank MacFarlane Burnet and Peter Medawar, the updated theory of clonal selection explains how antigenic determinants and specific B lymphocytes will function in the immune process, considering the enormous variety of possible B cells available in the lymphoid tissue. A detailed discussion of the contemporary theory follows.

Once the antigenic determinant has bound with its corresponding surface receptor protein, and after cross-linking has occurred, the combination of receptor protein and antigenic determinant is taken into the cytoplasm of the B cell. Then, the B cell displays the antigen peptide on its surface within a class II MHC molecule. While these processes have been going on, macrophages and other phagocytes bearing the antigenic determinants have contacted helper T cells. Now the helper T cell is activated, and it forms a clone of helper T cells keyed to the particular antigen peptide. Cytokines such as interleukin-2 are produced by the helper T cells to stimulate the cells to become more active. These activated helper T cells then recognize the same MHC/peptide molecule complex on the B-cell surface, and they bind to the B cells (**FIGURE 19.14**). This immunologic cooperation between the B lymphocyte and the helper T lymphocyte continues the immune response. Interleukin-2, along with other cytokines, assists B-cell activation.

At this point, antibody-mediated immunity has been initiated, and B lymphocytes bearing certain receptors have been selected by their reaction with specific antigenic determinants. Antigens that evoke this sort of response are called **T-dependent antigens** because they require the services of helper T lymphocytes. We should note that some antigens are **T-independent antigens**. These substances (such as in bacterial capsules and flagella) do not require the intervention of helper T lymphocytes. Instead, they bind directly with the receptor proteins on the B-lymphocyte surface and stimulate the cells. However, the immune response is generally weaker, and no memory cells are produced. Indeed, when T-independent antigens are used in vaccines, scientists must find ways to bind the antigens to substances that will stimulate T-lymphocyte intervention. Chapter 20 explores this problem in more detail.

Another group of antigens worth mentioning are the **superantigens**. "Regular" antigens must be broken down and processed to antigen peptides before they are ushered by the macrophage's MHC proteins to the T-lymphocyte surface, but superantigens bind directly to the MHC proteins and to the T-cell receptor without any internal processing. The cross-linking of MHC proteins on macrophages with T-cell receptors stimulates massive numbers of activated helper T lymphocytes to form, with an unusually high secretion of cytokines. The result is an extremely vigorous immune response that can lead to shock and death. Superantigens and massive cytokine release are associated with the staphylococcal toxins of toxic shock syndrome and scalded skin syndrome (Chapter 11), and possibly with immune system diseases such as arthritis, multiple sclerosis, and psoriasis. Having digressed for these exceptions, we now return to our survey of antibody-mediated immunity.

Once activated by helper T lymphocytes and cytokines, the B cells multiply and mature into a clone of **plasma cells**. Plasma cells are large, complex cells having no surface protein receptors. Their sole purpose is to produce antibodies. A plasma cell lives about four to five days, during which time it produces an incredible 2,000 antibody molecules per second. Each antibody molecule has the same antigen-binding property as the receptor on the original B lymphocyte, and all the antibody molecules eventually reach the circulation. In a short time, the body becomes saturated with antibodies. Antibody molecules fill the blood, lymph, saliva, sweat, and all other body

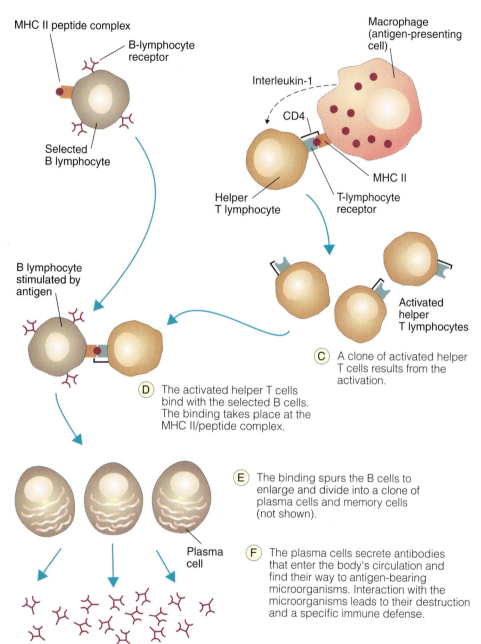

A The antigen-receptor complex has been taken into the selected B lymphocyte, and the antigen peptide is now displayed on the B cell's surface within the class II MHC protein.

B Meanwhile, an antigen-presenting macrophage has activated a helper T lymphocyte. Interleukin-1 from the macrophage assists the activation.

FIGURE 19.14

The Process of Antibody-Mediated Immunity

MHC II peptide complex

B-lymphocyte receptor

Selected B lymphocyte

Macrophage (antigen-presenting cell)

Interleukin-1

CD4

MHC II

Helper T lymphocyte

T-lymphocyte receptor

B lymphocyte stimulated by antigen

Activated helper T lymphocytes

C A clone of activated helper T cells results from the activation.

D The activated helper T cells bind with the selected B cells. The binding takes place at the MHC II/peptide complex.

E The binding spurs the B cells to enlarge and divide into a clone of plasma cells and memory cells (not shown).

Plasma cell

F The plasma cells secrete antibodies that enter the body's circulation and find their way to antigen-bearing microorganisms. Interaction with the microorganisms leads to their destruction and a specific immune defense.

secretions. By sheer force of numbers they come upon the original site of antigens, bind to the antigens, and mark them for destruction. (In many cases, however, they are too late to prevent the initial surge of infection, and symptoms already may be present when they arrive to provide their specific defense.)

At any one time, antibody molecules represent about 17 percent of the total protein in a person's blood fluid. Antibodies are extremely diverse, and the cells of a single species of bacterium may elicit the formation of hundreds of different kinds of antibodies corresponding to hundreds of different kinds of bacterial epitopes. Each unique antibody molecule is then capable of uniting with the unique antigen molecule. The term **immunoglobulin (Ig)** is used interchangeably with *antibody* because antibodies exhibit the properties of *globulin* proteins and are used in the *immune* response. In this text, we shall use the terms antibody and immunoglobulin interchangeably.

We shall study the interaction of antibodies and antigens shortly, but at this point it is important to note that certain B cells do not become plasma cells. Instead, they mature into a clone of **memory B cells**. Memory B lymphocytes remain in the lymphoid tissues for many years, in some cases for a person's lifetime. Should the parasite and its antigens reenter the body, the memory cells will divide and mature into plasma cells and produce antibodies without delay. The antibodies rapidly flood the bloodstream and bring about an immediate neutralization of the antigens. The symptoms of infection rarely occur this time.

ANTIBODIES RECOGNIZE SPECIFIC EPITOPES

Details about the nature of antibodies began to emerge in the late 1950s and early 1960s, with the union of immunology and molecular biology. Research focused on the Bence Jones proteins, identified a century earlier by Henry Bence Jones. A substantial body of information and numerous theories existed about the functions and origins of these proteins, but little was known about their detailed chemical composition. Investigators expected that an understanding of the structure of the proteins (now known to be antibodies) would provide an understanding of their unique specificity for antigens. The answers were provided by Gerald M. Edelman and Rodney M. Porter, who described the chemical composition and structure of the proteins and received the 1972 Nobel Prize in Physiology or Medicine for their work.

The basic antibody molecule consists of four polypeptide chains: two identical **heavy (H) chains** and two identical **light (L) chains**. These chains are joined together by disulfide linkages to form a Y-shaped structure, illustrated in FIGURE 19.15. Each heavy chain consists of about 400 amino acids, while each light chain has about 200 amino acids. The antibody molecule formed is called a **monomer**. It has two identical halves, each half consisting of a heavy chain and a light chain.

Constant and variable regions exist within each polypeptide chain. The amino acids in the **constant regions** of both light and heavy chains are virtually identical among different types of antibodies. However, the amino acids of the **variable region** differ across the hundreds of thousands of antibody types. Thus, the variable regions of a light and heavy chain combine to form a highly specific, three-dimensional structure, called the **antigen binding site**, which is somewhat analogous to the active site of an enzyme. This portion of the antibody molecule is uniquely shaped to "fit" a specific epitope or antigenic determinant. Moreover, the "arms" of the antibody are identical so that a single antibody molecule may combine with two identical antigen molecules. As we will see, these combinations may lead to a complex of antibody and antigen molecules.

When an antibody molecule is treated with papain, an enzyme from the papaya fruit, the antibody separates at the hinge joint, and two functionally different segments are isolated (Figure 19.15). One segment (actually two identical segments), called the **Fab fragment**, for *f*ragment-*a*ntigen-*b*inding, is the portion that will combine with the antigenic determinant. The second segment, called the **Fc fragment**, for *f*ragment able to be *c*rystallized, performs various functions: it is the part of the antibody molecule that combines with phagocytes in opsonization; it appears

FIGURE 19.15

Details of an
Antibody Molecule

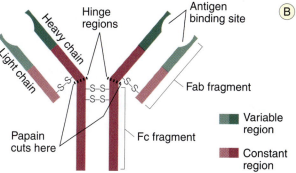

(A) The antibody molecule is a protein consisting of four polypeptide chains: two light chains and two heavy chains connected by disulfide (S-S) linkages. The heavy chains bend at a hinge point.

(B) The variable region is where the amino acid compositions of various antibodies differ. In the constant region, the amino acid compositions are similar in different antibodies.

(C) On treatment with papain enzyme, cleavage occurs at the hinge point, and three fragments result: two identical Fab fragments and one Fc fragment.

to neutralize viral receptor sites; it attaches to certain cells in allergic reactions; and it activates the complement system in resistance mechanisms.

THERE ARE FIVE IMMUNOGLOBULIN CLASSES

Five classes of antibodies have been identified, based on differences in the heavy chains of the constant region. Using the abbreviation Ig (immunoglobulin), the five classes are designated IgM, IgG, IgA, IgE, and IgD. The structures of these antibodies are illustrated in **FIGURE 19.16**. IgM consists of a **pentamer** (five monomers), whose tail segments are connected by a glycoprotein called a joining (J) chain. IgA is a **dimer** (two monomers), with the segments also connected by a **J chain**. In addition, a secretory component of IgA enables the molecule to leave secretory epithelial cells (**MicroFocus 19.7**). The remaining three antibodies are monomers.

IgM, the largest antibody molecule (M stands for macroglobulin), is the first immunoglobulin to appear in the circulation after the stimulation of B lymphocytes. It is the principal component of the **primary antibody response** (**FIGURE 19.17**), meaning that the presence of IgM in the serum of a patient indicates a very recent infection. Because of its size, most IgM remains in the circulation. Research indicates that IgM is formed during fetal infections with rubella or toxoplasmosis, indicating that a certain immunological competence exists in the fetus. About 5 to 10 percent of the antibody components of normal serum consists of this antibody. IgM is one of the antibodies bound to the B lymphocyte surface as a receptor protein.

IgG is the classical **gamma globulin**. This antibody is the major circulating antibody, comprising about 80 percent of the total antibody content in normal serum. IgG appears about 24 to 48 hours after antigenic stimulation and continues the antigen-antibody interaction begun by IgM. It is thus the principal antibody of the **secondary antibody response**. In addition, it provides long-term resistance to disease as a product of the memory B cells. Booster injections of a vaccine raise the level of this antibody considerably in the serum. IgG is also the **maternal antibody** that

J chain:
a glycoprotein in IgM and IgA molecules that connects the subunits of the antibody.

Primary antibody response:
the first antibody reaction of the immune system to the presence of antigens.

Secondary antibody response:
a second or ensuing response by the immune system to an antigen.

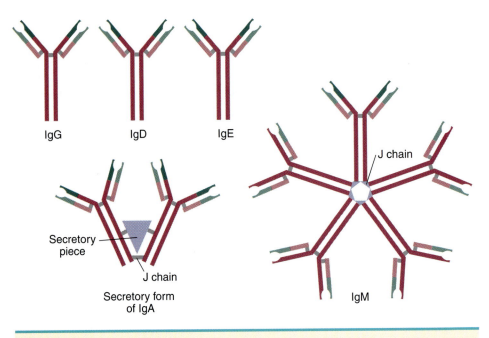

FIGURE 19.16

The Structures of the Five Antibody Classes

Note the complex structures of IgM (a pentamer) and IgA (a dimer). IgG, IgE, and IgD consist of monomers, each composed of two heavy chains and two light chains of amino acids.

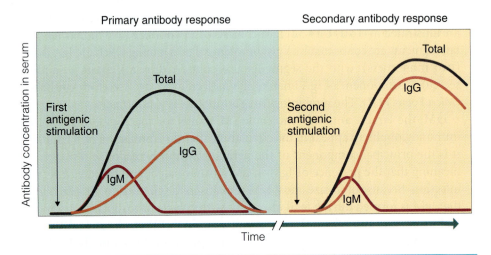

FIGURE 19.17

The Primary and Secondary Antibody Responses

After the initial antigenic stimulation, IgM is the first antibody to appear in the circulation. Initially, it is the principal component of the primary antibody response. Later, IgM is supplemented by IgG. On a second or ensuing exposure to the same antigen, the production of IgG is more rapid, and the concentration in the serum reaches a higher level than previously. Thus, the IgG antibody is more concentrated in the secondary antibody response.

MicroFocus 19.7

THE OTHER ARMY

Most people are familiar with the body's army, headquartered downtown—the antibodies and T lymphocytes that constitute the blood-centered system of immunity. Few, however, know much about the army of the suburbs—the specialized cells and antibodies that protect the body at its vulnerable entryways, in locations such as its respiratory, gastrointestinal, and urinary tracts. The cells and antibodies in the fragile mucous membranes lining these systems constitute a network now known as the mucosal immune system.

The extent of communication between the blood and mucosal systems is uncertain at this writing. Injected vaccines, for example, rarely evoke IgA production for mucosal immunity. However, stimulating one region of mucosal tissue to elicit IgA appears to produce an immune response at a distant mucosal surface. Thus, a vaccine in a nasal spray can elicit an antibody response along the gastrointestinal tract. This interrelatedness has led immunologists to think in terms of a single mucosal system.

Investigations of mucosal immunity have lagged behind those of blood-borne immunity in part because tissue samples from the mucosa are difficult to study. Studies are being encouraged though because stopping infection at its point of entry is highly desirable. Also, vaccine research would benefit considerably because oral vaccines are preferred to injectable types, and developing an oral vaccine requires a basic understanding of mucosal immunity. Indeed, the output of the mucosal system appears to outdistance that of the bloodborne system. It has been estimated, for instance, that over 3 grams of IgA are secreted as compared to only 1 gram of IgA released into the bloodstream. On this basis alone, the "suburban army" merits a greater share of attention.

crosses the placenta and renders immunity to the fetus until the child can make antibodies at about six months of age.

Approximately 10 percent of the total antibody in normal serum is **IgA**. One form of this antibody, called **serum IgA**, exists in the serum and is similar to IgG. A second form accumulates in body secretions and is referred to as **secretory IgA**. This antibody provides resistance in the respiratory and gastrointestinal tracts, possibly by inhibiting the attachment of pathogens to the tissues. It also is located in tears and saliva, and in the colostrum, the first milk secreted by a nursing mother. When consumed by a child, the antibodies provide resistance to potential gastrointestinal disorders. The secretory component comes from epithelial cells and helps move the antibody into the secretions.

IgE plays a major role in allergic reactions by sensitizing cells to certain antigens. This process is discussed in Chapter 21. **IgD** is a cell surface receptor on the B lymphocytes together with IgM. For this reason, it is called a membrane antibody and is thought to function in activating B cells.

TABLE 19.2 summarizes the characteristics of the five types of antibodies.

ANTIBODY DIVERSITY IS A RESULT OF GENE REARRANGEMENT

For decades, immunologists were puzzled as to how an enormous variety of antibodies (perhaps a million or more different types) could be encoded by the limited number of genes associated with the immune system. Because antibodies, like all proteins, are specified by genes, it would be reasonable to assume that an individual must have a million or more genes for antibodies. However, geneticists point out that human cells have only about 35,000 genes for all their functions.

The antibody diversity problem has been resolved by showing that embryonic cells contain about 300 genetic segments that can be shuffled (like transposons) and combined in each B cell as it matures. The process, known as **somatic recombination**, is a random mixing and matching of gene segments to form unique antibody

TABLE 19.2

Properties of Five Types of Antibodies

DESIG-NATION	PERCENTAGE OF ANTIBODY IN SERUM	LOCATION IN BODY	MOLECULAR WEIGHT (DALTONS)	NUMBER OF MONOMERS	COMPLEMENT BINDING	CROSSES PLACENTA	CHARACTERISTICS
IgM	5–10	Blood Lymph	900,000	5	Yes	No	First antibody formed in primary response
IgG	80	Blood Lymph	150,000	1	Yes	Yes	Principal component of secondary response
IgA	10	Secretions Body cavities	350,000	2	No	No	Protection in body cavities
IgE	<1	Blood Lymph	200,000	1	No	No	Role in allergic reactions
IgD	0.05	Blood Lymph	180,000	1	No	No	Receptor sites on B lymphocytes

genes. Information encoded by these genes then is expressed in the surface receptor proteins of B lymphocytes and in the antibodies later expressed by the stimulated clone of plasma cells.

The process of somatic recombination was first postulated in the late 1970s by Susumu Tonegawa, winner of the 1987 Nobel Prize in Medicine or Physiology. According to the process, the gene segments coding for the light and heavy chains of an antibody are located on different chromosomes. The light and heavy chains are synthesized separately, then joined to form the antibody. One of 8 constant genes, one of 4 joiner genes, one of 50 diversity genes, and one of up to 300 variable genes can be used to form a heavy chain. One of 300 variable genes is selected and combined with one of 5 joiner genes and a constant gene to form the active light-chain gene. After deletion of intervening genes, the new gene can function in protein synthesis, as shown in FIGURE 19.18.

Tonegawa's discovery was revolutionary because it questioned two dogmas of biology: that the DNA for a protein needs to be one continuous piece (for antibody synthesis, the gene segments are separated); and that every body cell has exactly the same DNA (the antibody genes for different B lymphocytes differ). Current evidence suggests that more than 600 different antibody gene segments exist per cell. Additional diversity is generated through imprecise recombination and somatic mutation. Therefore, the total antibody diversity produced by the B cells is in the range of 10^8 to 10^{11} immunoglobulin possibilities.

ANTIBODY INTERACTIONS MEDIATE THE DISPOSAL OF ANTIGENS (PATHOGENS)

In order for specific resistance to develop, antibodies must interact with antigens in such a way that the antigen is altered. The alteration may result in death to the microorganism that possesses the antigen, inactivation of the antigen, or increased susceptibility of the antigen to other body defenses.

Certain antibodies, called **neutralizing antibodies**, react with viral capsids and prevent viruses from entering their host cells. Influenza viruses are inhibited by neu-

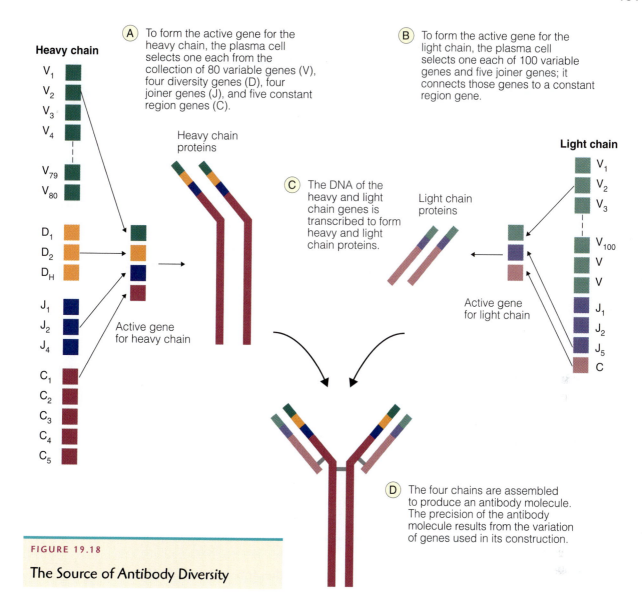

Heavy chain

V_1
V_2
V_3
V_4
V_{79}
V_{80}

D_1
D_2
D_H

J_1
J_2
J_4

C_1
C_2
C_3
C_4
C_5

(A) To form the active gene for the heavy chain, the plasma cell selects one each from the collection of 80 variable genes (V), four diversity genes (D), four joiner genes (J), and five constant region genes (C).

Heavy chain proteins

Active gene for heavy chain

(B) To form the active gene for the light chain, the plasma cell selects one each of 100 variable genes and five joiner genes; it connects those genes to a constant region gene.

Light chain

V_1
V_2
V_3
V_{100}
V
V
J_1
J_2
J_5
C

(C) The DNA of the heavy and light chain genes is transcribed to form heavy and light chain proteins.

Light chain proteins

Active gene for light chain

(D) The four chains are assembled to produce an antibody molecule. The precision of the antibody molecule results from the variation of genes used in its construction.

FIGURE 19.18

The Source of Antibody Diversity

raminidase antibodies in this way (**FIGURE 19.19**). **Neutralization** also represents a vital mechanism of defense against toxins. The antibodies, known as **antitoxins**, alter the toxin molecules near their active sites and mask their toxicity. Neutralization of toxin molecules also increases their size, thus encouraging phagocytosis while lessening their ability to diffuse through the tissues. Antibodies also coat bacteria (**opsonization**), preventing bacterial attachment to cells.

Antibodies called **agglutinins** react with antigens on the surface of organisms such as bacteria. This action causes clumping, or **agglutination**, of the organisms. Movement is inhibited if antibodies react with antigens on the flagella of microorganisms, and the organisms may even be clumped together by their flagella. The reaction of antibodies with pilus antigens prohibits attachment of an organism to the tissues while agglutinating them and increasing their susceptibility to phagocytosis.

Another form of antibodies is the **precipitins**. These antibodies react with dissolved antigens and convert them to solid precipitates (**precipitation**). In this form, antigens usually are inactive and more easily phagocytized. In all these interactions,

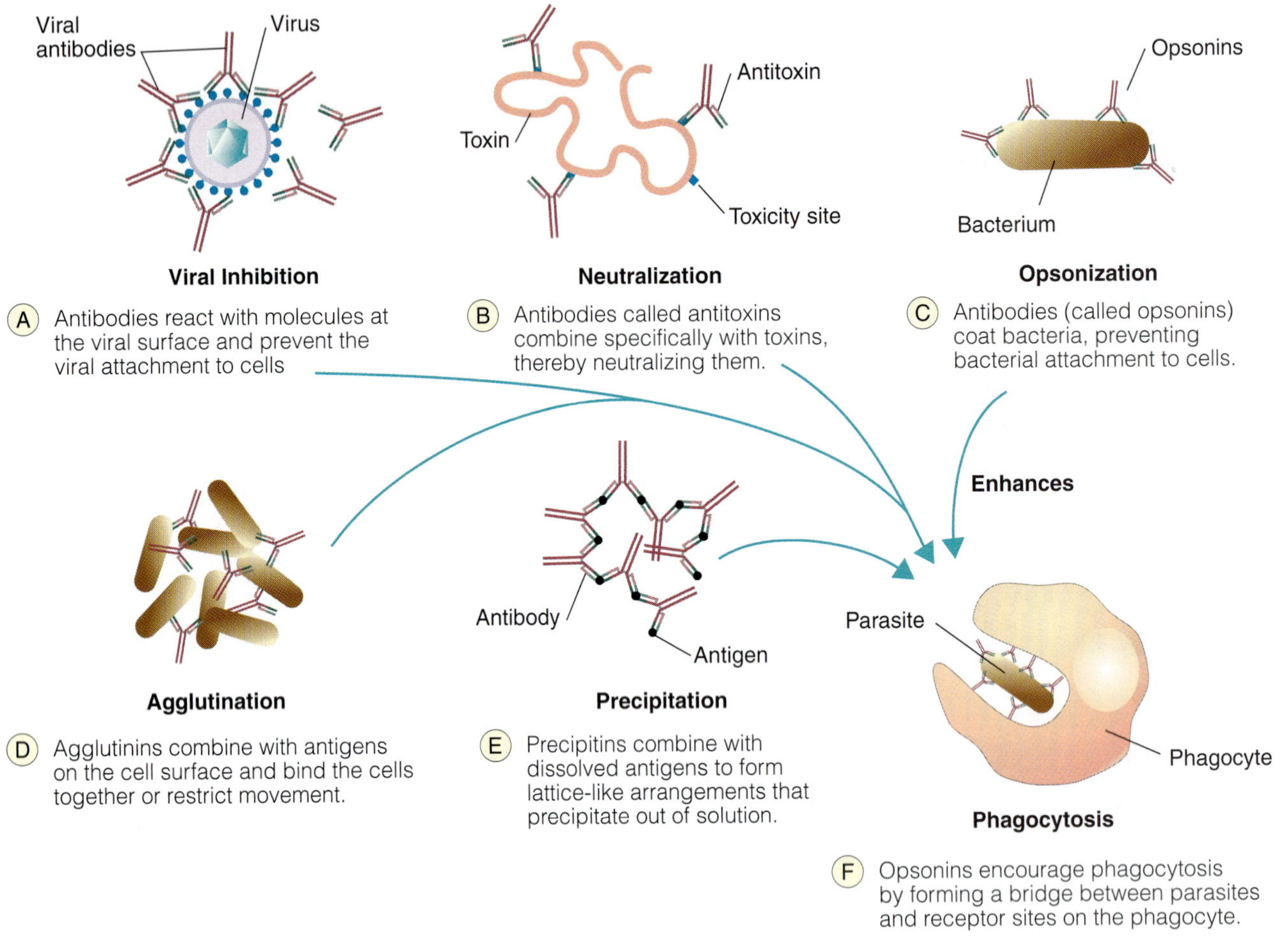

Viral
antibodies — Virus

Viral Inhibition

(A) Antibodies react with molecules at
the viral surface and prevent the
viral attachment to cells

Antitoxin

Toxin —

Toxicity site

Neutralization

(B) Antibodies called antitoxins
combine specifically with toxins,
thereby neutralizing them.

Opsonins

Bacterium

Opsonization

(C) Antibodies (called opsonins)
coat bacteria, preventing
bacterial attachment to cells.

Enhances

Agglutination

(D) Agglutinins combine with antigens
on the cell surface and bind the cells
together or restrict movement.

Antibody

Antigen

Precipitation

(E) Precipitins combine with
dissolved antigens to form
lattice-like arrangements that
precipitate out of solution.

Parasite

Phagocyte

Phagocytosis

(F) Opsonins encourage phagocytosis
by forming a bridge between parasites
and receptor sites on the phagocyte.

FIGURE 19.19

Mechanisms by Which Antibodies Interact with Antigens

In all cases, the interaction facilitates phagocytosis by phagocytes.

the antigen-antibody complexes formed enhance phagocytosis and thus the removal of the infecting pathogen or the toxic chemical.

A final example of antigen-antibody interaction involves the complement system (complement), introduced earlier in the chapter. The complement system is a series of nearly 30 proteins that functions in a cascading set of reactions. One set of reactions leads to lysis of bacterial cells. In the **classical pathway** for complement activation, a complement cascade at the cell surface results in the formation of a **membrane attack complex** (FIGURE 19.20). This complex forms pores in the membrane, increasing cell membrane permeability and inducing the cell to undergo lysis through the leakage of water from its cytoplasm (in a process reminiscent of cytotoxic T-lymphocyte activity with perforin).

FIGURE 19.21 summarizes the antibody- and cell-mediated immune responses. MicroInquiry 19 investigates some alternative immune system outcomes.

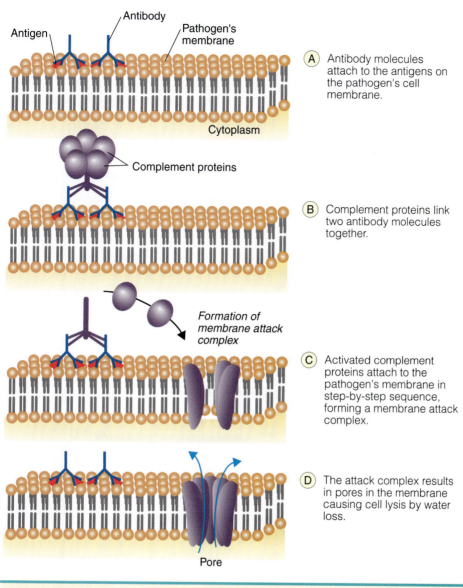

A Antibody molecules attach to the antigens on the pathogen's cell membrane.

B Complement proteins link two antibody molecules together.

C Activated complement proteins attach to the pathogen's membrane in step-by-step sequence, forming a membrane attack complex.

D The attack complex results in pores in the membrane causing cell lysis by water loss.

FIGURE 19.20

The Classical Pathway of Complement Activity

CAN THINKING "WELL" KEEP YOU HEALTHY?

The idea that mental states can influence the body's susceptibility to and recovery from disease has a long history. The Greek physician Galen asserted that cancer struck more frequently in melancholy women than in cheerful women. During the twentieth century, the concept of mental state and disease was researched more thoroughly, and a firm foundation was established linking the nervous system and the immune system (MicroFocus 19.8). A new field called *psychoneuroimmunology* has emerged.

One such link exists between the hypothalamus and the T lymphocytes. The **hypothalamus** is a portion of the brain located beneath the cerebrum. It produces

Summary of the Immune Response

Antibody-mediated (humoral) immunity produces antibodies that respond and bind to antigens. Antibody-antigen complexes facilitate complement binding and phagocytosis. Cell-mediated immunity stimulates helper T cells that activate B cells and cytotoxic T cells. Cytotoxic T cells bind to and kill infected cells or other abnormal cells. Memory B and T cells are important in a second or ensuing exposure to the same antigen.

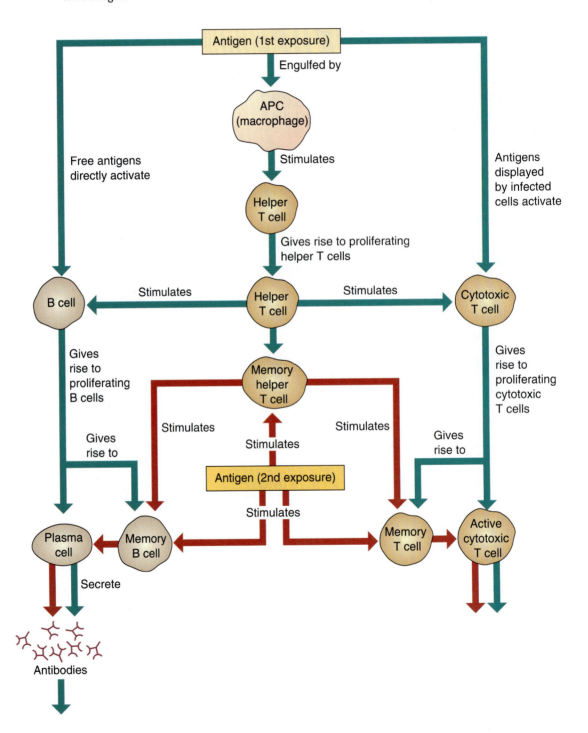

MicroFocus 19.8

SICK OVER EXAMS

At the end of every semester, medical students at Ohio State University suffer through the Day of the Big Bleed. First, they "bleed" over their final exams; then they bleed, quite literally, for psychologist Janice Kiecolt-Glaser and immunologist Ronald Glaser, her husband. The Glasers are attempting to learn whether the stress of final exams diminishes the activity of the immune system.

The "thinking well" phenomenon has been known for years—a positive attitude can help the immune system's activities, while stress can place a burden on the system. Providing proof of this phenomenon has been difficult, but the Glasers are among those determined to show that a correlation exists between stress and reduced immune activity. Already, they have demonstrated reduced activity in natural killer cells in blood taken from students during exam week. Also, they have shown that in herpes-infected students, the virus is more active when exams are going on (a reflection of reduced body defense).

In a recent study, 57 Ohio State medical students received blister wounds on their arms. Split into two groups, group 1 was allowed to contend with the stress of medical exams. Group 2 received standard relaxation therapy prior to the exams. Group 1 showed delayed wound healing, indicative of reduced immune system function. Group 2 had much faster wound healing corresponding to normal immune system function. The lessons appear straightforward: Prepare thoroughly for exams, think of clear mountain streams, and stay healthy.

a chemical-releasing factor that induces the pituitary gland, positioned just below the hypothalamus, to secrete the hormone ACTH, which targets the adrenal glands. The adrenal glands, in turn, secrete steroid hormones (glucocorticoids) that influence the activity of T cells in the thymus gland.

Another link is established by branches of the **autonomic nervous system** that extend into the lymph node and spleen tissues. The autonomic nervous system automatically regulates the functioning of such organs as the heart, stomach, and lungs through myriad nerve fibers. The direct anatomical link between the immune and nervous systems allows a direct two-way communication.

A third link between the nervous and immune systems starts with **thymosins**, a family of substances originating in the thymus gland. When experimentally injected into brain tissue, thymosins stimulate the pituitary gland via the hypothalamus to release hormones, including the one that stimulates the adrenal gland. Although the precise functions of thymosins are yet to be determined, it appears that they serve as specific molecular signals between the thymus and the pituitary gland. A circuit is apparently present that stimulates the brain to adjust immune responses and the immune system to alter nerve cell activity.

The outcome of these discoveries is the emergence of a strong correlation between a patient's mental attitude and the progress of disease. Rigorously controlled studies conducted in recent years have shown that the aggressive determination to conquer a disease can increase one's lifespan. Therapies can consist of relaxation techniques, as well as mental imagery that disease organisms are being crushed by the body's stalwart defenses. Behavioral therapies of this nature can amplify the body's response to disease and accelerate the mobilization of its defenses.

Few reputable practitioners of behavioral therapies believe that such therapies should replace drug therapy. However, the psychological devastation associated with many diseases such as AIDS cannot be denied, and it is this intense stress that the "thinking well" movement attempts to address. Very often, for instance, a person learning of a positive HIV test goes into severe depression, and because depression can adversely affect the immune system, a double dose of immune suppression

MicroInquiry 19

IMMUNE SYSTEM OUTCOMES

Immune system function is a very complicated and complex system of checks and balances. As potent as the immune system can be, some pathogens can evade immune system attack due to some microbial component. In addition, genetic deficiencies within the host can make the immune system less efficient, and in some cases can have overt consequences in attempting to fight infection and disease caused by pathogens. These tend to be most obvious in children since the deficiencies are congenital.

Let's look at several cases. Based on your understanding of the immune system described in this chapter, see if you can predict the consequences of or reasons for the immune system dysfunction within the host. Answers can be found in Appendix E.

Case 1. A 35-year-old female has a normally functioning immune system. She has had the typical flus and colds in her life and an occasional bacterial infection. In all cases, she has recovered without medical intervention. In one of the bacterial infections from which she recovered, she was infected with two species of gram-negative bacteria. One species contained a capsule and the other did not have a capsule.

19.1a. Describe how the complement system will respond to these two bacterial species in terms of membrane-attack complexes.

19.1b. Explain how the complement system will facilitate the disposal of these two species.

Case 2. A 7-year-old girl suffers from DiGeorge's syndrome. This is a congenital birth defect in which the thymus fails to develop during embryogenesis. She has been susceptible to several bacterial and some viral infections.

19.2a. Suppose she is infected by a bacterium that invades the bloodstream but does not penetrate host cells. How will her immune system respond to this infection?

19.2b. Explain the consequences if she is infected by the hepatitis A virus, a non-enveloped RNA virus.

Case 3. An 8-year-old boy suffers from frequent and severe bacterial infections. The family physician thinks that he suffers from agammaglobulinemia (absence of immunoglobulins). The tests performed indicate he has normal immunoglobulin levels. However, in

ensues. Perhaps by relieving the psychological trauma, the remaining body defenses can adequately handle the virus.

As with any emerging treatment method, there are numerous opponents of behavioral therapies. Some opponents argue that naive patients might abandon conventional therapy; another argument is that therapists might cause enormous guilt to develop in patients whose will to live cannot overcome failing health. Proponents counter with the growing body of evidence showing that patients with strong commitments and a willingness to face challenges—signs of psychological hardiness—have relatively greater numbers of T cells than passive, nonexpressive patients. To date, no study has proven that mood or personality has a life-prolonging effect on immunity. Still, doctors and patients are generally inspired by the possibility that one can use his or her mind to help stave off the effects of infectious disease. Though unsure of what it is, they generally agree that *something* is going on.

doing the blood work, a deficiency in complement protein C3 was discovered.

19.3a. What is the role of C3 in the normal complement cascade?

19.3b. Explain why a deficiency in C3 causes the frequent bacterial infections experienced by the patient.

Case 4. A 7-year-old boy has been suffering since birth from repeated bacterial infections and impaired healing of wounds. In skin infections, analysis shows a much reduced level of granulocytes at the lesions even though these leukocytes are found in the blood. The patient has apparently effective cellular- and antibody-mediated immunity.

19.4a. Provide two explanations for the lack of granulocytes at the lesions.

19.4b. With effective cellular- and antibody-mediated immunity, why does the patient still show susceptibility to bacterial infections?

Case 5. A woman gives birth to a newborn that dies five days after birth. At autopsy, the newborn is found to contain no lymploid progenitor cells. Her physician tells her the newborn died of a very rare congenital disease called reticular dysgenesis. In fact, most cases of the disease have been stillbirths.

19.5a. Explain the role of lymploid progenitor cells in the generation of the immune system.

19.5b. Why are the few cases of reticular dysgenesis usually stillbirths?

Case 6. A 5-month-old girl has suffered recurrent infections of prolonged diarrhea due to rotavirus infections. Clinical analysis indicates the infant has lowered levels of helper T cells. Genetic analysis indicates the infant lacks normal interleukin-2 (IL-2) receptors.

19.6a. Explain how the lack of normal IL-2 receptors could bring about the lowered T-cell count.

19.6b. Why is this infant susceptible to this virus as well as other viral, bacterial, and fungal infections?

Note to the Student

It may have occurred to you that an episode of disease is much like a war. First, the invading microorganisms must penetrate the natural barriers of the body; then they must escape the phagocytes and other chemicals that constantly patrol the body's waterways and tissues. Finally, they must elude the antibodies or T lymphocytes that the body sends out to combat them. How well we do in this battle will determine whether we survive the disease.

Obviously, we have done very well, because most of us are in good health today. As it turns out, most infectious organisms are stopped at their point of entry to the body. Cold viruses, for example, get no farther than the upper respiratory tract. Many parasites come into the body by food or water, but they rarely penetrate beyond the intestinal tract. The staphylococci in a wound may cause inflammation at the infection site, but this is usually the extent of the problem.

To be sure, drugs and medicines help in those cases where diseases pose life-threatening situations. In addition, sanitation practices, vector control, care in the preparation of food, and other public health measures prevent microorganisms from reaching the body in the first place. However, in the final analysis, body defense represents the bottom line in protection against disease, and, as history has shown, body defenses work very well. Indeed, as Lewis Thomas has suggested in *The Lives of a Cell*, a microorganism that catches a human is in considerably more danger than a human who has caught a microorganism.

Summary of Key Concepts

19.1 NONSPECIFIC RESISTANCE

■ **There Are Broad Types of Nonspecific Resistance.** Species immunity, behavioral immunity, racial immunity, and population immunity represent four forms in which nonspecific resistance occurs. In addition, individuals possess innate immunity, a nonspecific defense against parasites.

■ **Mechanical and Chemical Barriers Are the First Defenses against Infection.** The skin and mucous membranes represent mechanical barriers against infection. Mucus traps pathogens, and cilia in the respiratory tract remove particles and pathogens. Chemical defenses include an acidic urinary tract pH, and stomach acid and bile in the gastrointestinal tract. Lysozyme and interferon help destroy bacteria and block virus replication, respectively. The normal microbiota on body surfaces also limits pathogen spread and infection.

■ **Phagocytosis Represents Part of a Second Line of Nonspecific Defenses.** Phagocytosis is the internalization of particle material and pathogens by phagocytes (neutrophils and macrophages). Pathogens are internalized in a phagosome that then fuses with lysosomes that contain digestive enzymes and toxic chemicals that usually destroy the pathogen. Antibody-coated (opsonized) parasites enhance the efficiency of phagocytosis.

■ **Inflammation Plays an Important Role in Fighting Infection.** Tissue injury by objects or pathogens triggers the release of chemical mediators, such as histamine. These mediators trigger vasodilation, and increase capillary permeability and edema at the site of infection. The inflamed area exhibits redness, warmth, swelling, and pain.

■ **Moderate Fever Is Beneficial to Host Defenses.** Moderate fever inhibits the growth of pathogens while increasing metabolism for tissue repair and enhancing phagocytosis.

■ **Natural Killer Cells Recognize and Kill Abnormal Cells.** Natural killer (NK) cells are defensive lymphocytes that act spontaneously on virus-infected cells, cancer cells, or organ transplants. NK cells release cell-destroying chemicals.

■ **Complement Marks Pathogens for Destruction.** Complement is a set of more than 20 serum proteins that enhance phagocytosis through inflammation, opsonize pathogens for phagocytosis, and lyse cells through the formation of membrane attack complexes.

19.2 SPECIFIC RESISTANCE AND THE IMMUNE SYSTEM

■ **Antigens Stimulate an Immune Response.** Antigens are chemical substances that stimulate T- and B-cell activation (immunogenicity) and the synthesis of antibodies to which they react (reactivity). That part of the antigen that stimulates immune system activity is the antigenic determinant or epitope. Haptens are small molecules that by themselves do not stimulate an immune system reaction unless they are linked to a larger protein carrier.

■ **The Immune System Originates from Groups of Stem Cells.** In humans, immune system cells arise from stem cells in the bone marrow. One group of cells derived from these stem cells are the lymphopoid progenitor cells, which give rise to T lymphocytes and B lymphocytes. B lymphocytes (B cells) mature in the bone marrow while T lymphocytes (T cells) leave the bone marrow and migrate to the thymus where they mature. Immunocompetent B and T cells colonize the lymphoid organs.

■ **T and B Cells Have Receptors to Recognize Antigens.** T cells have T-cell receptors on the plasma membrane that recognize peptides contained within class II major histocompatibility complex (MHC) proteins (found on macrophages and B cells). B cells have antibody molecules anchored to the plasma membrane that act as receptors to bind antigen (epitope). Once the epitope is bound to T and B cells, those cells are committed.

■ **Cell-Mediated Immunity Is a Response by Antigen-Specific T Lymphocytes.** Cell-mediated immunity is triggered by T cells. The helper T cell subgroup that recognizes an MHC/peptide complex on macrophages (antigen-presenting cells) binds to that complex. Binding stimulates the release of interleukin-1 by the macrophage that activates the helper T cells, which divide into a clone of identical helper T cells and memory T cells. Helper T cells then can activate B cells (antibody-mediate immunity) or activate cytotoxic T cells. Both activations occur through cytokine interactions. Activated cytotoxic T cells recognize abnormal cells (virus-infected cells, tumors, tissue transplants) that present class I MHC proteins with bound antigen peptide. Binding of cytotoxic T cells triggers the release of perforins and granzymes that lyse and kill the abnormal cells.

- **Antibody-Mediated (Humoral) Immunity Is a Response Mediated by Antigen-Specific B Lymphocytes.** B cells bind antigen directly to anchored antibody receptors and then process antigen as peptides attached to class II MHC proteins. Activated helper T cells bind to the presenting B cells. Through cytokine stimulation, B cells divide into a clone of identical B cells and then differentiate into plasma cells and B memory cells. Plasma cells secrete antibody that recognizes the same epitope as that recognized by the B-cell receptor.

- **Antibodies Recognize Specific Epitopes.** The structure of an antibody is designed to recognize a specific epitope. Each antibody monomer consists of four polypeptide chains; two identical shorter light chains and two identical longer heavy chains. The variable regions of the light and heavy chains on each arm form identical antigen-binding sites where the epitope (antigen) binds.

- **There Are Five Immunoglobulin Classes.** Five classes of antibody molecule exist. IgM is a pentamer found in the blood and lymph. It activates complement and is the first antibody produced in a primary antibody response. IgG makes up the vast majority of antibody in blood and lymph, activates comple-

ment, and crosses the placenta. Secretory IgA is found in the body cavities where it binds bacteria and viruses before they can infect tissues. Although IgE makes up a very small percentage of antibodies in blood and lymph, it plays a major role in allergic reactions. IgD normally is bound to B cells as a cell surface receptor.

- **Antibody Diversity Is a Result of Gene Rearrangement.** Antibodies can be produced that recognize almost any antigen. To do this, somatic recombination between a large number of genes (variable, diversity, joiner, constant) for the heavy and light chains occurs.

- **Antibody Interactions Mediate the Disposal of Antigens (Pathogens).** The actual disposal or removal of antigen is facilitated by antibody-antigen interactions (inhibition, neutralization, opsonization, agglutination, precipitation) that result in phagocytosis by phagocytes. In addition, the formation of membrane-attack complexes by complement directly lyses and kills pathogens.

- **Can Thinking "Well" Keep You Healthy?** A new science of psychoneuroimmunology has developed based on research linking the nervous system with immune system function. Under stress conditions, the nervous system can depress the immune system's reactions to infection and, in part, be responsible for an unhealthy physiology.

Questions for Thought and Discussion

Answers to selected questions can be found in Appendix C.

1. The opening of this chapter suggests that for many diseases, a penetration of the walls surrounding the human body must take place. Can you think of any diseases where penetration is not a prerequisite to illness?

2. It has been said that no other system in the human body depends and relies on signals as greatly as the immune system. What evidence can you offer to support or reject this concept?

3. Lysozyme is a valuable natural enzyme for the control and destruction of gram-positive bacteria. Can you postulate why this substance is not a commercially available pharmaceutical product?

4. The ancestors of modern humans lived in a sparsely settled world where communicable diseases were probably very rare. Suppose that by using some magical scientific invention, one of those individuals was thrust into the contemporary world. How do you suppose he or she would fare in relation to infectious disease? What is the immunological basis for your answer?

5. In the book and movie *Fantastic Voyage*, a group of scientists is miniaturized in a submarine (the *Proteus*) and sent into the human body to dissolve a blood clot. The odyssey begins when the miniature submarine carrying the scientists is injected into the bloodstream. Can you think of any natural opening to the body interior they could have used instead of the injection?

6. One of the most intriguing aspects of dental research is that a vaccine may eventually be produced to protect against tooth decay. A study in the 1980s, for example, showed that reduced incidence of decay correlated with high levels of salivary IgA specific for *Streptococcus mutans*. What problems do you foresee in the development of such a vaccine, and what advantages and disadvantages might there be in its use?

7. A University of Cincinnati immunologist has demonstrated that cockroaches injected with small doses of honeybee venom can develop resistance to future injections of venom that would ordinarily be lethal. Does this finding imply that cockroaches have an immune system? Which might be the next steps for the research to take? What does this research tell you about the cockroach's ability to survive for three or four years, far longer than most other insects?

8. Cattle are known to have the alcoholic carbohydrate erythritol in their placentas. When brucellosis occurs in pregnant cattle, the causative agent *Brucella abortus* metabolizes the erythritol and grows in the placenta, causing abortion of the fetal calf. Humans, by contrast, have no erythritol in the placenta and therefore do not suffer similar effects when they contract brucellosis. What type of immunity does this exemplify?

9. The botulism toxin is the most powerful toxin known to science; the tetanus toxin is the second most powerful. Despite this, the botulism toxin may be less dangerous than the tetanus toxin. Why is this so?

10. Why is it a good idea to cry when something gets into your eye?

11. Phagocytes have been described as "bloodhounds searching for a scent" as they browse along the walls

Review

A major theme of this chapter is antigens and antibodies, two key aspects of the immune system. To test your knowledge of antigens and antibodies, solve the following puzzle. The answers are given in Appendix D.

■ ACROSS

1. Does not induce an immune response.
4. Type of immunity (abbr) where T lymphocytes function.
5. Antigenic on rare occasions.
6. Immunity related to antibodies also is said to be hum-_____.
10. Small molecules may act as antigens called _____-tens.
12. Alternate name for antibody.
14. Protein group; functions in antigen- antibody reaction.

of blood vessels. The scent they usually seek is a chemotactic factor, a peptide released by a bacterium. Does it strike you as unusual that a bacterium should release a substance to attract the "bloodhound" that will eventually lead to the bacterium's demise?

12. One of the earliest symptoms of disease in the body is swollen lymph nodes (erroneously referred to as "swollen glands"). Why?

13. A microbiology professor has suggested that an antigenic determinant arriving in the lymphoid tissue is like a parent searching for the face of a lost child in a crowd of a million children. Would you agree with this analogy? Why or why not?

14. A 1989 high school biology text contains the following statement: "T cells do not produce circulating antibodies. They carry cellular antibodies on their surface." What is fundamentally incorrect about this statement, and what would you say in your letter to the publisher after reading the statement?

15. Some years ago, a novel entitled *Through the Alimentary Canal with Gun and Camera* appeared in bookstores. The book described a fictitious account of travels through the human body. What perils, microbiologically speaking, would you encounter if you were to take such an adventure?

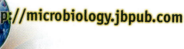

The site features **eLearning,** an on-line review area that provides quizzes and other tools to help you study for your class. You can also follow useful links for in-depth information, or just find out the latest microbiology news.

16. End of antibody molecule where antigen reacts.
19. Antigenic materials are classified as _____-self.
20. Cells that produce antibody molecules.
21. Number of monomeric units in IgM.
23. Antibody response in which IgM is the principal component.
26. Largest antibody.
29. Number of heavy chains in a monomeric antibody molecule.
30. A person's own chemical substance is an _____-antigen.
33. Alternate name for an antigenic determinant.
34. Reaction in which IgE functions.
35. Cell that phagocytizes microorganisms and begins immune response.
36. Number of amino acid molecules possible in an antigenic determinant.
37. White blood cells involved in immune response.

■ **DOWN**
1. Medium for transport of antibodies.
2. Type of acid in antibody molecule.
3. Antigens generally have a _____ molecular weight.
7. Number of polypeptide chains in a monomeric antibody molecule.
8. Nobel Prize for study of antibody diversity.
9. Structure of IgA.
11. Sole organic material in antibody molecule.
12. Body system depending on antibody activity.
13. Secretory antibody.
14. Region of identity among antibody molecules.
15. Fragment of antibody that combines with antigenic determinant.
17. Sensitizes cells to certain antigens.
18. Proposed theory of clonal selection.
22. Major antibody of the secondary antibody response.
24. Antibody molecule with two identical halves.

25. Type of blood cell (abbr) derived from myeloid progenitor cells.
26. Antibody found only at the surface of B lymphocytes.
27. An antibody molecule reacts with an epi-_____.
28. Type of immunity (abbr.) based on activity of B lymphocytes.
31. Nodes where immune system cells gather.
32. Molecules functioning in recognition reactions between cells.
34. Nucleic _____ usually are poor antigens.

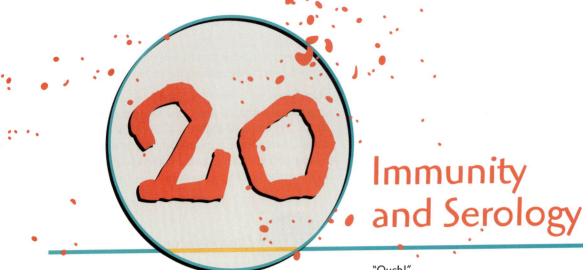

20

Immunity and Serology

"Ouch!"

—The plaintive cry of a child receiving yet another vaccination

T RADITION TELLS US THAT as early as the eleventh century, Chinese doctors ground up smallpox scabs and blew the powder into the noses of healthy people to protect them against the ravages of smallpox. Modern vaccines work less haphazardly. They are composed of chemically altered toxins, treated microorganisms, or chemical parts of microorganisms. Such vaccines work by exploiting the immune system's ability to recognize antigens and respond with antibodies and other factors that remain in the body and later attack the toxins or microorganisms should they appear.

The technology continues to improve. Using the techniques of genetic engineering, researchers are now striving to eliminate all but the absolutely necessary material required to stimulate the immune system. Scientists identify the antigen that triggers the immune response, clone the genes that encode the antigen's production, and insert the genes into organisms such as yeasts and *Escherichia coli*. These biological factories promptly churn out millions of copies of the antigen. The antigens then are extracted from the culture media, purified, and concentrated to form the new vaccine. A synthetic vaccine for hepatitis B is available now, and laboratories are working on synthetic vaccines for such diseases as genital herpes, rabies, AIDS, typhoid fever, diphtheria, malaria, and cholera. They also are exploring new methods for administration, as MicroFocus 20.1 explores. In the view of many scientists, the current era represents a golden age of vaccine technology.

MicroFocus 20.1

PINCUSHION PARANOIA

The crying starts as you round the corner and little Michael realizes that he's on the way to the doctor's office. At two months of age, he got his hepatitis B shot, then his DTaP, then Hib, then polio, then another and another. At four months, they started all over—DTaP, polio, Hib, hepatitis B—and they even threw in a rotavirus shot. Back to the doctor's at six months for another series. And now he is a year old, and he has been hearing new words: varicella and MMR. It was just no fun being a kid!

It may be too late for Michael, but the future looks brighter for Michael Junior. Researchers are experimenting with diphtheria and tetanus immunizations using a skin patch carrying toxoids. The patch contains 400 microscopic needles that introduce vaccines just beneath the skin layers but above the nerve endings, so there is no pain.

Moreover, in 2003 a California company received approval from the FDA for a nasal spray (FluMist) that delivers active influenza viruses to the mucosal surfaces of the nose. Approved for

healthy individuals 5 to 49 years of age, the device looks like a syringe, but instead of a needle it has a sprayer that squirts an aerosol mist into each nostril. By encouraging antibody production in the nasal passageways, the vaccine encourages defense at the portal of entry for flu viruses. The principal antibody stimulated is IgA.

If things go well, Michael Junior may even be offered a banana for his hepatitis B immunization. Researchers at New York's Boyce Thompson Institute already have spliced *E. coli* genes into potato plants and produced potatoes containing *E. coli* antigens. Although the "veggie vaccine" is not especially tasty (it must be eaten raw to avoid destroying the antigens), the process has been deemed a success, and experiments are continuing to introduce the same genes to banana plants. Bananas are attractive because children like them (and children are the main recipients of vaccines), the fruit grows in tropical countries (where vac-

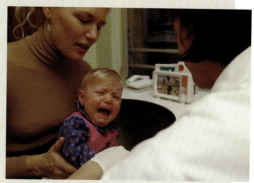

cines are widely used and needed to prevent childhood diseases), and the fruit does not have to be cooked. On the downside, there is the matter of transporting the antigens through the stomach's hydrochloric acid and past the duodenum's enzymes. Unfortunately, several years of research lie ahead before the banana replaces the needle.

So, for the time being, Michael must bear it as best he can. But someday he'll be off to college, and they will send his tuition bill home. Ah, revenge is sweet.

In this chapter, we shall study the role of vaccines in the immune response while examining the four major mechanisms by which immunity comes about. Antibodies occupy a central position in each of these mechanisms. We also shall study how antibodies may be detected in a disease patient by a variety of laboratory tests. These diagnostic procedures help the physician understand the disease and prescribe a course of treatment. Immune mechanisms generally protect against disease, but when they fail, the laboratory tests provide a clue about what is taking place in the patient.

20.1
Immunity to Disease

The word *immune* is derived from the Latin *immuno*, meaning "safe" or "free from." In its most general sense, the term implies a condition under which an individual is protected from disease. This does not mean, however, that one is immune to all diseases, but rather to a specific disease or group of diseases.

Two general types of immunity are recognized: innate immunity and acquired immunity. **Innate immunity** is an inborn capacity for resisting disease. It begins at birth and depends on genetic factors expressed as physiological, anatomical, and biochemical differences among living things. Examples of innate immunity are the lysozyme found in tears, saliva, and other body secretions; the acidic pH of the gastrointestinal and vaginal tracts; and the interferon produced by body cells to protect against viruses. Innate immunity is synonymous with nonspecific resistance, discussed in Chapter 19.

Acquired immunity, by contrast, begins after birth. It depends on the presence of T lymphocytes, antibodies, and other factors originating in the immune system. It is the cell-mediated and antibody-mediated immunity described in Chapter 19. Four types of acquired immunity are generally recognized, as the following discussions will explore (**FIGURE 20.1**). Although the emphasis will be on antibodies and antibody-mediated immunity, it should be remembered that cell-mediated immunity is also an important consideration in the total spectrum of resistance to infectious disease.

FIGURE 20.1

The Four Types of Acquired Immunity

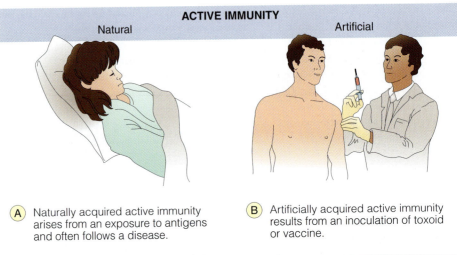

ACTIVE IMMUNITY

Natural   Artificial

(A) Naturally acquired active immunity arises from an exposure to antigens and often follows a disease.

(B) Artificially acquired active immunity results from an inoculation of toxoid or vaccine.

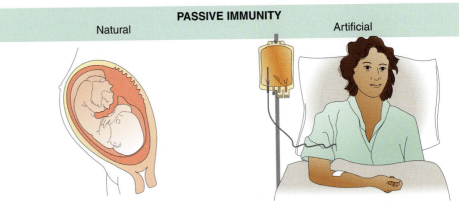

PASSIVE IMMUNITY

Natural   Artificial

(C) Naturally acquired passive immunity stems from the passage of IgG across the placenta from the maternal to the fetal circulation.

(D) Artificially acquired passive immunity is induced by an injection of antibodies taken from the circulation of an animal or another person.

NATURALLY ACQUIRED ACTIVE IMMUNITY DEVELOPS FROM EXPOSURE TO AN INFECTIOUS AGENT

Active immunity develops after antigens enter the body and the individual's immune system actively responds with antibodies and specific lymphocytes. The exposure to antigens may be unintentional or intentional. When it is unintentional, the immunity that develops is called naturally acquired active immunity.

Naturally acquired active immunity usually follows a bout of illness and occurs in the "natural" scheme of events. However, this is not always the case, because subclinical diseases also may bring on the immunity. For example, many people have acquired immunity from subclinical cases of mumps or from subclinical fungal diseases such as cryptococcosis. This is the **primary antibody response** (Chapter 19).

Memory cells residing in the lymphoid tissues are responsible for the production of antibodies that yield naturally acquired active immunity. The cells remain active for many years and produce IgG immediately upon later entry of the pathogen to the host. Such an antibody response is called the **secondary antibody response**. It also is sometimes called the **secondary anamnestic response**, from the Greek *anamnesis*, for "recollection."

ARTIFICIALLY ACQUIRED ACTIVE IMMUNITY IS ESTABLISHED BY VACCINATION

Artificially acquired active immunity develops after the immune system produces antibodies following an intentional exposure to antigens. The antigens are usually contained in an immunizing agent such as a vaccine or toxoid, and the exposure to antigens is "artificial."

Viral vaccines consist of either **inactivated viruses**, incapable of multiplying in the body, or **attenuated viruses**, which multiply at low rates in the body but fail to cause symptoms of disease. The Salk polio vaccine typifies the former, while the Sabin oral polio vaccine represents the latter. Bacterial vaccines fall into similar categories. The formerly used pertussis (whooping cough) vaccine consisted of dead cells, while the BCG tuberculosis vaccine (Chapter 8) is composed of attenuated bacteria. Whole-microorganism viral and bacterial vaccines are commonly called **first-generation vaccines**. MicroFocus 20.2 describes how the components of the flu vaccine are decided each year.

One advantage of vaccines made with attenuated organisms is that organisms multiply for a period of time within the body, thus increasing the dose of antigen administered. This higher dose results in a higher level of immune response than that obtained with the single dose of inactivated organisms. Also, attenuated organisms can spread to other people and reimmunize them, or immunize them for the first time. However, attenuated organisms may be a health hazard because of this same ability to continue multiplying. In 1984, for example, a recently immunized soldier spread vaccinia (cowpox) viruses to his daughter. She, in turn, infected seven young friends at a slumber party.

Currently, there are no widely used bacterial vaccines made with whole-organisms and used for long-term protection. The older pertussis vaccine was replaced by the acellular pertussis vaccine composed of *Bordetella pertussis* extracts. Some whole organism vaccines are used for temporary protection. For instance, bubonic plague and cholera vaccines are available to limit an epidemic. In these cases, the immunity lasts only for several months because the material in the vaccine is weakly antigenic.

MicroFocus 20.2

DESIGNER FLU VACCINES

Each flu season approximately 10 to 20 percent of Americans get the flu. Of this number, more than 114,000 are hospitalized and some 36,000 die per year from the complications of flu. Many of these hospitalizations and deaths could be prevented with a yearly flu shot, especially for those people at increased risk (people over 50, immunocompromised individuals, and health care workers in close contact with flu patients). Since influenza viruses change often (Chapter 13), the influenza vaccine is updated each year to make sure it is as effective as possible. How is each year's vaccine designed?

Each flu season, information on circulating influenza strains and epidemiological trends is gathered by the World Health Organization (WHO) Global Influenza Surveillance Network. The network consists of 112 national influenza centers in 83 countries and four WHO Collaborating Centers for Reference and Research on Influenza located in Atlanta, United States; London, United Kingdom; Melbourne, Australia; and Tokyo, Japan. The national influenza centers sample patients with influenza-like illness and submit representative isolates to WHO Collaborating Centres for immediate strain identification.

Twice a year (February: northern hemisphere; September: southern hemisphere), WHO meets with the Directors of the Collaborating Centres and representatives of key national laboratories to review the results of their strain identifications and to recommend the composition of the influenza vaccine for the next flu season. Since 1972, WHO has recommended 39 changes in the influenza vaccine formulation.

In the United States, the Food and Drug Administration or the Centers for Disease Control and Prevention provide the viral strains to American vaccine manufacturers in February. Each virus strain is grown separately in chicken eggs. After it has replicated many times, the fluid containing the viruses is removed, and the viruses purified and inactivated. Then, the appropriate strains are mixed together with a carrier fluid. For the 2003–04 season, the American trivalent vaccine consisted of influenza A/New Caledonia/20/99-like (H1N1), influenza A/Panama/2007/99 (H3N2), and either influenza B/Hong Kong/330/01 or B/Hong Kong/1434/02 viruses. Production usually is completed in August and ready for shipment in October.

Weakly antigenic vaccines are available also for laboratory workers who deal with rickettsial diseases such as Rocky Mountain spotted fever, Q fever, and typhus. The danger in these vaccines is that the residual egg protein in the cultivation medium for rickettsiae may cause allergic reactions in recipients. TABLE 20.1 presents a summary of currently available bacterial and viral vaccines.

Immunizing agents that stimulate immunity to toxins are known as **toxoids**. These agents are currently available for protection against diphtheria and tetanus, two diseases whose major effects are due to exotoxins. Toxoids are prepared by incubating toxins with a chemical, such as formaldehyde, until the toxicity is lost.

To avoid multiple injections of immunizing agents, it is advantageous to combine vaccines into a **single-dose vaccine**. Experience has shown this is possible for the diphtheria-pertussis-tetanus vaccine (DPT), the newer diphtheria-tetanus-acellular pertussis vaccine (DTaP), the measles-mumps-rubella vaccine (MMR), and the trivalent oral polio vaccine (TOP). There is even a vaccine that will immunize against four diseases simultaneously: In 1993, the FDA approved a combined vaccine that includes diphtheria and tetanus toxoids, pertussis vaccine, and *Haemophilus influenzae* b (Hib) vaccine. Marketed as Tetramune, the quadruple vaccine is used in children age two months to five years to protect against the DPT diseases, as well as *Haemophilus* meningitis. For other vaccines, however, a combination may not be useful because the antibody response is lower for the combination than for each vaccine taken separately. Immunologists believe that poor phagocytosis by macrophages is one reason. FIGURE 20.2 outlines the recommendations for these vaccines in healthy infants and children.

TABLE 20.1

The Principal Bacterial and Viral Vaccines Currently in Use

DISEASE	ROUTE OF ADMINISTRATION	RECOMMENDED USAGE/COMMENTS
Contain Killed Whole Bacteria		
Cholera	Subcutaneous (SQ) injection	For travelers; short-term protection
Typhoid	SQ and intramuscular (IM)	For travelers only; variable protection
Plague	SQ	For exposed individuals and animal workers; variable protection
Contain Live, Attenuated Bacteria		
Tuberculosis (BCG)	Intradermal (ID) injection	For high-risk occupations only; protection variable
Subunit Bacterial Vaccines (Capsular Polysaccharides)		
Meningitis (meningococcal)	SQ	For protection in high-risk individuals, such as military recruits; short-term protection
Meningitis (H. influenzae)	IM	For infants and children; may be administered with DTaP
Pneumococcal pneumonia	IM or SQ	Important for people at high risk: the young, elderly, and immuno-compromised; moderate protection
Pertussis	IM	For newborns and children
Subunit Bacterial Vaccine (Protective Antigen)		
Anthrax	SQ	For lab workers and military personnel
Toxoids (Formaldehyde-Inactivated Bacterial Exotoxins)		
Diphtheria	IM	A routine childhood vaccination; highly effective in systemic protection
Tetanus	IM	A routine childhood vaccination; highly effective
Botulism	IM	For high-risk individuals, such as laboratory workers
Contain Inactivated Whole Viruses		
Polio (Salk)	IM	Routine childhood vaccine; highly effective; safer than Sabin vaccine
Rabies	IM	For individuals sustaining animal bites or otherwise exposed; highly effective
Influenza	IM	For adults over 50; requires constant updating for new strains
Hepatitis A	IM	Protection for travelers and anyone at risk; effectiveness not established
Contain Attenuated Viruses		
Adenovirus infection	Oral	For immunizing military recruits
Measles (rubeola)	SQ	Routine childhood vaccine; highly effective
Mumps (parotitis)	SQ	Routine childhood vaccine; highly effective
Polio (Sabin)	Oral	Routine childhood vaccine; highly effective; possible vaccine-induced polio
Smallpox (vaccinia)	Pierce outer layers of skin	For lab workers, military personnel, health care workers
Rubella	SQ	Routine childhood vaccine; highly effective
Chickenpox (varicella)	SQ	Routine childhood vaccine; immunity can diminish over time; effectiveness not yet established
Yellow fever	SQ	For travelers, military personnel in endemic areas
Recombinant Viral Vaccine		
Hepatitis B	IM	Medical, dental, laboratory personnel; newborns, others at risk; highly effective

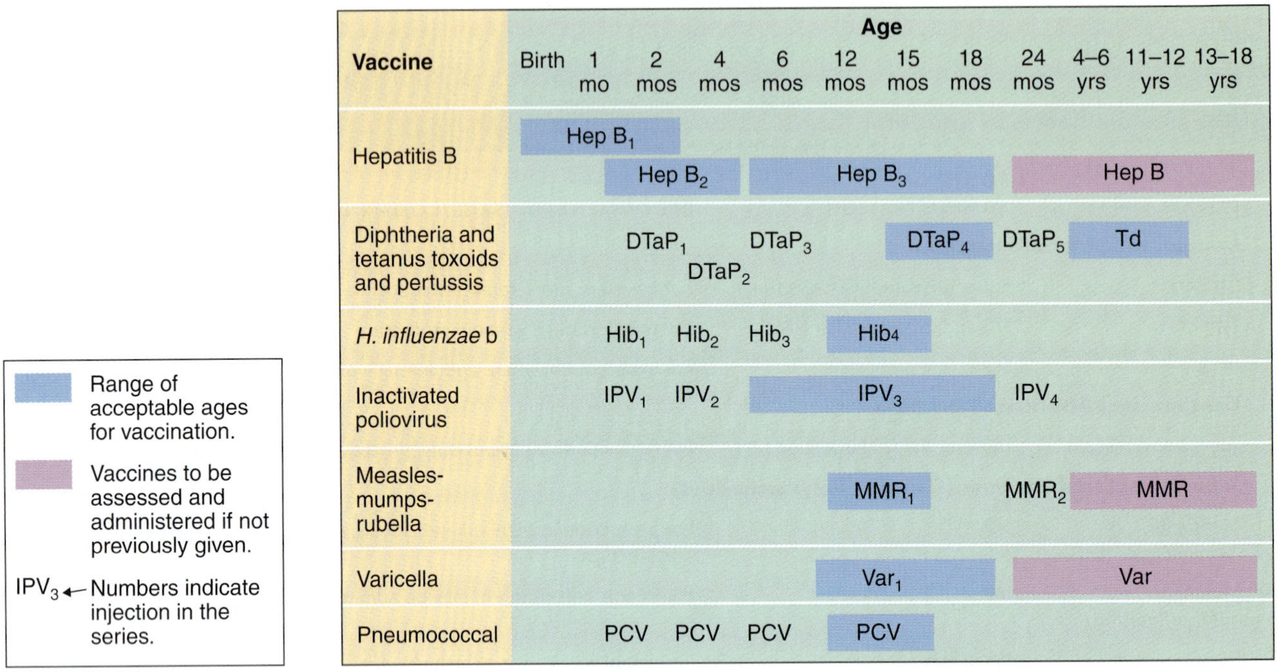

Vaccine	Birth	1 mo	2 mos	4 mos	6 mos	12 mos	15 mos	18 mos	24 mos	4–6 yrs	11–12 yrs	13–18 yrs
Age												
Hepatitis B	Hep B₁	Hep B₁	Hep B₂	Hep B₂	Hep B₃	Hep B₃	Hep B₃				Hep B	Hep B
Diphtheria and tetanus toxoids and pertussis			DTaP₁	DTaP₂	DTaP₃		DTaP₄	DTaP₄	DTaP₅	DTaP₅	Td	
H. influenzae b			Hib₁	Hib₂	Hib₃	Hib₄	Hib₄					
Inactivated poliovirus			IPV₁	IPV₂	IPV₃	IPV₃	IPV₃	IPV₃	IPV₄	IPV₄		
Measles-mumps-rubella						MMR₁	MMR₁			MMR₂	MMR	MMR
Varicella						Var₁	Var₁	Var₁			Var	Var
Pneumococcal			PCV	PCV	PCV	PCV	PCV					

Legend:
- Range of acceptable ages for vaccination.
- Vaccines to be assessed and administered if not previously given.
- IPV₃ ← Numbers indicate injection in the series.

FIGURE 20.2

Childhood and Adolescent Immunization Schedule—2003

Each year the Advisory Committee on Immunization Practices of the CDC reviews the childhood and adolescent immunization schedule. This schedule indicates the recommended ages for routine administration of licensed vaccines. "Catch-up" vaccination may be administered if a visit to the doctor is missed. Additional recommendations concerning the schedule are available from the CDC at www.cdc.gov.

Immunologists foresee the day when preparations called **subunit vaccines**, or **second-generation vaccines**, which contain only parts or subunits of microorganisms, will be used exclusively. For example, pili from bacteria may be extracted and purified for use in a vaccine to stimulate antipili antibodies (**FIGURE 20.3**). These would inhibit the attachment of bacteria to tissues and facilitate phagocytosis. Another example is the subunit vaccine for pneumococcal pneumonia, licensed for use in 1983. The vaccine contains 23 different polysaccharides from the capsules of 23 strains of *Streptococcus pneumoniae*. An example of a **conjugate vaccine** is the Hib vaccine against *Haemophilus influenzae* b, the agent of *Haemophilus* meningitis (Chapter 8). Composed of capsular polysaccharides linked to tetanus or diphtheria toxoid, the vaccine has been available since 1990 and has been a critical factor in reducing the incidence of *Haemophilus* meningitis, from 18,000 cases annually (1986) to a few dozen cases in recent years (2001).

Another form of vaccine is the **synthetic vaccine**, or **third-generation vaccine**. This preparation represents a sophisticated and practical application of recombinant DNA technology. To produce the vaccine, three major technical problems must be solved: The immune-stimulating antigen must be identified, living cells must be reengineered to produce the antigens, and the size of the antigens must be increased to promote phagocytosis and the immune response.

The genetic engineering process has worked for a recombinant subunit vaccine for hepatitis B. The vaccine is marketed by different companies as Recombivax HB or Engerix-B. Because the vaccine is not made from blood fragments (as was the previous hepatitis B vaccine), it relieves the fear of contracting human immunodeficiency virus (HIV) from contaminated blood. Many immunologists believe that the synthetic agents will usher in a renaissance of vaccines. In 1993, for example, biotechnologists announced the development of an experimental cholera vaccine containing *Vibrio cholerae* whose genes for toxin production were experimentally removed. An AIDS vaccine is also on the horizon (MicroFocus 20.3).

Part of the renaissance is a group of vaccines called **DNA vaccines**. These vaccines consist of plasmids engineered to contain a protein-encoding gene. Unlike replicating viruses or live bacteria, plasmids are not infectious or replicative, nor do they encode any proteins other than those specified by the plasmid genes, so they have a measure of safety. However, the vaccine is difficult to make because a plasmid must contain a promoter site, a convenient cloning site for the gene of interest, a polyadenine tail used as a termination sequence, an origin of replication, and a selectable marker sequence such as an ampicillin-resistance gene (in addition, of course, to the protein-encoding gene or genes).

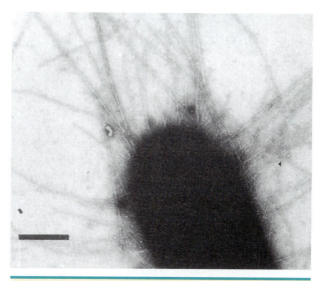

FIGURE 20.3

A Possible Approach to Antibody Use

A transmission electron micrograph of a cell of *Escherichia coli* showing the numerous pili at the pole of the cell. Pili are used to help the bacterium adhere to the infected tissue. A vaccine containing pilus antigens could be used to stimulate antipili antibodies, which bind to the pili and prevent the bacterium from adhering. (Bar = 0.5 μm.)

One highlight of the DNA vaccine is that no special formulation is necessary. Animals can be immunized against disease merely by injecting plasmids suspended in a saline (salt) solution. And delivery of the plasmids to target cells can be accomplished by nasal spray; injection into the muscle or vein, and under the skin; and by a so-called gene gun. A gene gun is a propulsion device that shoots DNA-coated gold beads into the skin. Furthermore, deploying plasmids as DNA vaccines has the added advantage of stimulating both antibody-mediated immunity (AMI) and cell-mediated immunity (CMI), because DNA vaccines encode proteins that are released from the cell (for AMI) and proteins that fix themselves to the target cell surface (for CMI). DNA vaccines also appear to elicit a strong immune response possibly due to a stimulus by other genes in the plasmid. The vaccines also are more stable than conventional vaccines at low and high temperatures (which makes shipping easier). At this writing, DNA vaccines have been used experimentally to protect against influenza, *Salmonella typhi*, HIV infection, herpes simplex, and hepatitis B. MicroFocus 20.4 explores their discovery.

Vaccines for immunizations may be administered by injection, oral consumption, or nasal spray, as currently used for some respiratory viral diseases. **Booster immunizations** commonly follow as a way of raising the antibody level by stimulating the memory cells to induce the secondary anitbody response. This is why a tetanus booster is given to anyone who sustains a deep puncture wound by a soil-contaminated object. TABLE 20.2 summarizes the vaccines universally recommended by the CDC as of 2003.

Substances called **adjuvants** increase the efficiency of a vaccine or toxoid by increasing the availability of the antigen in the lymphatic system. Common adjuvants

AN AIDS VACCINE—WHY ISN'T THERE ONE AFTER ALL THESE YEARS?

Producing an AIDS vaccine might appear rather straightforward: Cultivate a huge batch of human immunodeficiency virus (HIV), inactivate it with chemicals, purify it, and prepare it for marketing. In fact, this was the mentality and approach in the mid-1980s. Unfortunately, things are not quite so simple when HIV is involved. Here are the major reasons why a vaccine remains elusive after all these years.

The effects of a bad batch of weakened or inactivated vaccine would be catastrophic; and people generally are reluctant to be immunized with whole HIV particles, no matter how reassuring the scientists. In addition, how the body actually protects itself from pathogens such as HIV remains to be clearly understood. Without that understanding, an effective vaccine cannot be developed.

Vaccine development also has been slow because of the high mutation rate of the virus. Some HIV strains around the world vary by as much as 35 percent in the capsid and envelope proteins they possess. Thus, HIV is more mutable than the influenza viruses, and no vaccine has been developed yet to make one completely immune to influenza. It is hard to make a vaccine targeted at one strain when the virus keeps changing its coat.

Additional reasons why a vaccine has not been developed include factors such as a short-lived vaccine would protect an individual only for a short time, necessitating an endless series of scheduled booster shots to which few would adhere. A vaccine also might act like dengue fever, where vaccination could actually make someone more at risk if they actually were to contract the disease. In 2002, an HIV patient who was holding the virus in check became infected with another strain of HIV through unprotected sex. This patient became superinfected. Although his immune system kept the original strain in check, it was powerless to control the new strain. So, could a vaccine produce the same result if the individual was infected with another strain of HIV? Almost every AIDS vaccine researcher around the world is concerned about the unknown factors concerning an AIDS vaccine.

On the research side, as of 2003, no vaccine trials have been shown to stimulate cellular immunity to a level necessary to destroy HIV. The number of cytotoxic T cells and memory T cells produced is not up to the job of combating HIV. Almost since the inception of HIV vaccine research, there has been a lack of funds and leadership needed for rapid progress. In March 2002, Anthony Fauci, Director of the National Institute for Allergy and Infectious Diseases, reported to the Presidential Advisory Council on HIV and AIDS that a "broadly effective AIDS vaccine could be a decade or more away." Let's keep our fingers crossed and hope some breakthroughs are made soon.

A HAPPENING

It was another of those remarkable moments in science, an unexpected observation that opened the door to a whole new type of vaccine. It happened in 1989 at Vical Inc., a California biotechnology company. Biochemist Philip Felgner and his research group were experimenting with plasmids, the ultramicroscopic ringlets of cytoplasmic DNA that carry many nonessential genes in bacteria. Felgner wished to learn whether the plasmids could carry genes into a mouse when wrapped in lipid-containing bodies called liposomes. He could hardly imagine what he was about to discover.

Felgner's protocol was simple. Some plasmids were packaged in the tiny, spherical liposomes, then injected into the muscle of a mouse. As a control, some unpackaged plasmids were injected into another mouse. The latter plasmids, by all expectations, should remain inert or be destroyed.

But they were not destroyed; nor did they remain inert. Instead, the cells receiving them began synthesizing the proteins encoded by the naked plasmids. Somehow, the plasmids had remained intact, found the necessary biochemical machinery, and induced the cell to begin sputtering protein. And, adding to the wild results, the proteins were stimulating the mouse's immune system to produce antibodies. The realization slowly dawned on Felgner and his group: They had discovered a new way to immunize an animal; they had produced a DNA vaccine.

Scientists are constantly taught never to anticipate their results when performing an experiment. The trick is to formulate a hypothesis, devise a reasonable set of experimental conditions, turn on the juice, and sit back while Mother Nature reveals her truths. Usually the process is slow and plodding. But every now and then, nature astounds us with a bit of knowledge that makes all the dreary days worthwhile. It has happened innumerable times in science, and it happened once again in 1989 in San Diego.

Oh, by the way, the plasmids packaged in liposomes also encoded protein.

TABLE 20.2

Universally Recommended Vaccinations for Children, Adolescents, and Adults

POPULATION	VACCINATION	DOSAGE
All young children	Measles, mumps, and rubella	2 doses
	Diphtheria-tetanus toxoid and pertussis vaccine	5 doses
	Polio	4 doses
	Haemophilus influenzae b	3–4 doses
	Hepatitis B	3 doses
	Varicella	1 dose
Previously unvaccinated or partially vaccinated adolescents	Hepatitis B	3 doses, total
	Varicella	If no previous history of varicella, 1 dose for children <12 years, 2 doses for children ≥12 years
	Measles, mumps, and rubella	2 doses, total
	Tetanus–diphtheria toxoid	If not vaccinated during previous 5 years, 1 combined booster during ages 11–16
All adults (19–49)	Varicella	2 doses for susceptible
	Tetanus–diphtheria (Td) toxoid	1 dose administered every 10 years
	Hepatitis A	2 doses total
	Hepatitis B	3 doses total
All adults >50 years	Influenza	1 dose administered annually
All adults >65 years	Pneumococcal	1 dose

include aluminum sulfate ("alum") and aluminum hydroxide in toxoid preparations, as well as mineral oil or peanut oil in viral vaccines. The particles of adjuvant linked to antigen are taken up by macrophages and presented to lymphocytes more efficiently than dissolved antigens. Experiments also suggest that adjuvants may stimulate the macrophage to produce interleukin-1, a lymphocyte-activating factor, and thereby reduce the necessity for helper T-lymphocyte activity. Moreover, adjuvants provide slow release of the antigen from the site of entry and provoke a more sustained immune response. MicroFocus 20.5 describes a new category of antibodies.

NATURALLY ACQUIRED PASSIVE IMMUNITY COMES FROM ACQUIRING MATERNAL ANTIBODIES

Passive immunity develops when antibodies enter the body from an outside source (in contrast to active immunity, in which individuals synthesize their own antibodies). The infusion of antibodies may be unintentional or intentional, and thus, natural or artificial. When unintentional, the immunity that develops is called naturally acquired passive immunity.

Naturally acquired passive immunity also is called **congenital immunity**. It develops when antibodies pass into the fetal circulation from the mother's bloodstream via the placenta and umbilical cord. These antibodies, called **maternal antibodies**, remain with the child for approximately three to six months after birth and fade as the child's immune system becomes fully functional. Certain antibodies, such

MicroFocus 20.5

BIOPHARMING

Just when you thought you had seen it all, something new pops onto the radar screen. This time it's plantibodies. That's right, plantibodies—corn plants that produce human antibodies.

The proud parents are scientists from Agracetus Inc., a biotechnology company in Wisconsin. Using a secret gene-delivery system, the researchers force genes that encode human antibodies into the genome of corn seeds. Then they plant the latter in an unpretentious cornfield. When harvested by conven-

tional agricultural methods, the corn kernels yield human antibodies that can be used to ferry radioisotopes to human tumor cells and destroy them.

As of mid-2003, several companies were testing pharmaceutical products grown in genetically engineered plants. Already planted in the fields are soybeans that produce herpes simplex antibodies (for treating genital herpes), and tobacco plants that manufacture streptococcal antibodies. The latter might be added to a mouthwash to

fight tooth decay associated with *Streptococcus mutans*.

Many people see these innovative and imaginative processes, called biopharming, as the culmination of decades of biotechnology research. Others see them as the first offspring of a new family of biotechnology products. Indeed, researchers point to Winston Churchill's immortal words: "This is not the end; this is not even the beginning of the end; it is, perhaps, the end of the beginning."

as measles antibodies, remain for 12 to 15 months. The process occurs in the "natural" scheme of events.

Maternal antibodies play an important role during the first few months of life by providing resistance to diseases such as pertussis, staphylococcal infections, and viral respiratory diseases. Because the antibodies are of human origin and are contained in human serum, they will be accepted without problem. The only antibody in the serum is IgG.

Maternal antibodies also pass to the newborn through the first milk, or **colostrum**, of a nursing mother, as well as during future breast-feedings. In this instance, IgA is the predominant antibody, although IgG and IgM also have been found in the milk. The antibodies accumulate in the respiratory and gastrointestinal tracts of the child and apparently lend increased disease resistance (Chapter 19).

ARTIFICIALLY ACQUIRED PASSIVE IMMUNITY IS PRODUCED FROM THE INJECTION OF ANTISERA

Artificially acquired passive immunity arises from the intentional injection of antibody-rich serum into the patient's circulation. The exposure to antibodies is thus "artificial." In the decades before the development of antibiotics, such an injection was an important therapeutic tool for the treatment of disease. The practice still is used for viral diseases such as Lassa fever and arthropodborne encephalitis, and for bacterial diseases in which a toxin is involved. For example, established cases of botulism, diphtheria, and tetanus are treated with serum containing the respective antitoxins.

Various terms are used for the serum that renders artificially acquired passive immunity. **Antiserum** is one such term. Another is **hyperimmune serum**, which indicates that the serum has a higher-than-normal level of a particular antibody. If the serum is used to protect against a disease such as hepatitis A, it is called **prophylactic serum**. When the serum is used in the therapy of an established disease, it is called **therapeutic serum**. When serum is taken from the blood of a convalescing patient, physicians refer to it as **convalescent serum**. Another common term,

Antiserum:
serum that is rich in a particular antibody.

gamma globulin, takes its name from the fraction of blood protein in which most antibodies are found. Gamma globulin usually consists of a pool of sera from different human donors, and thus it contains a mixture of antibodies, including those for the disease to be treated.

Passive immunity must be used with caution because in many individuals, the immune system recognizes foreign serum proteins as antigens and synthesizes antibodies against them in an allergic reaction. When antibodies interact with the proteins, a series of chemical molecules called immune complexes may form (Chapter 21), and with the activation of complement, the person develops a disease called **serum sickness**. This often is characterized by a hive-like rash at the injection site, accompanied by labored breathing and swollen joints. To avoid the disease, it is imperative that the patient be tested for allergies before serum therapy is instituted. If an allergy exists, minuscule doses should be given to eliminate the allergic state, and then a large therapeutic dose can be administered.

Serum sickness: a type of allergic reaction in which the immune system forms antibodies against protein in antiserum.

Artificially acquired passive immunity provides substantial and immediate protection against disease, but it is only a temporary measure. The immunity that develops from antibody-rich serum usually wears off within days or weeks. Among the serum preparations currently in use are those for hepatitis A and chickenpox. Both are made from the serum of blood donors routinely screened for hepatitis A and chickenpox. The four types of immunity are summarized in TABLE 20.3.

HERD IMMUNITY RESULTS FROM EFFECTIVE VACCINATION PROGRAMS

A population without a vaccination program is vulnerable to disease epidemics. Many people will suffer from the disease; some may die, while others could be left with a permanent disability. Even with a vaccination program, if insufficient numbers of citizens get the vaccination, the pathogen still can infect those who are not protected (FIGURE 20.4).

Vaccinations are never meant to reach 100 percent of the population. This is partly due to some people simply not being vaccinated and because some individuals simply respond poorly to a vaccine. They remain susceptible and unprotected. Still, the lack of 100 percent vaccination is not seen as a problem as long as most of the population is immune to an infectious disease. This makes it unlikely that a susceptible person will come in contact with an infected individual. This phenomenon, called **herd immunity**, implies that if enough people in a population are immunized against certain diseases, then it is very difficult for those diseases to spread.

Microbiologists and epidemiologists have found that when approximately 90 percent of the population is vaccinated, the spread of the disease is effectively stopped. The fact that the other 10 percent of the "herd" or population remain susceptible is allowable because it is very hard for pathogens "to find someone" who isn't vaccinated. Susceptible individuals are protected from catching the disease and if the person should catch the disease, there are so many vaccinated people that it is unlikely the person would be able to spread it.

Herd immunity can be affected by several factors. One is the environment. People living in crowded cities are more likely to catch a disease if they are not vaccinated than non-vaccinated people living in rural areas because of the constant close contact with other people in the city. Another factor is the strength of an individual's immune system. People whose immune systems are compromised, either because they have a disease or because of medical treatment (anticancer or antirejection

TABLE 20.3

Characteristics of the Four Types of Immunity

TYPE OF IMMUNITY	IMMUNIZING AGENT	EXPOSURE TO IMMUNIZING AGENT	EFFECTIVE DOSE REQUIRED	RELATIVE TIME UNTIL IMMUNITY	RELATIVE DURATION OF IMMUNITY
Naturally acquired active	Antigens	Unintentional	Small	Long	Long (lifetime)
Artificially acquired active	Antigens	Intentional	Small	Long	Long (months to years)
Naturally acquired passive	Antibodies	Unintentional	Large	Short	Short (4–6 months)
Artificially acquired passive	Antibodies	Intentional	Large	Short	Short (up to 6 weeks)

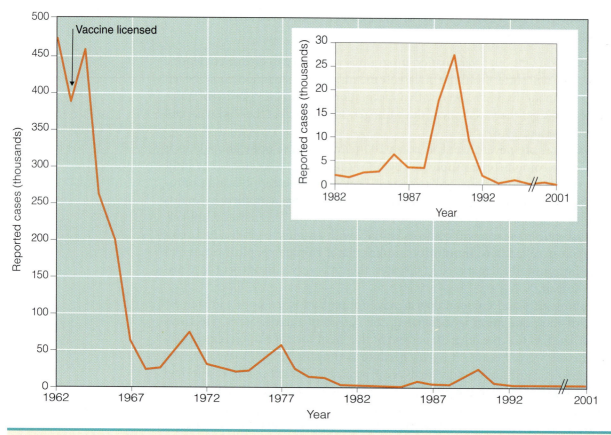

FIGURE 20.4

Measles Epidemics and Herd Immunity

With the introduction of the measles vaccine in 1962, vaccinations quickly built up a high herd immunity and few measles epidemics occurred. However, between 1989 and 1991 (inset), a measles epidemic occurred among inadequately vaccinated college students and unvaccinated preschool children. The epidemics were the result of decreased vaccinations that lowered the herd immunity.

USUAL ROUTE OF INTRODUCTION	SOURCE OF ANTIBODIES	FUNCTION	EFFECTIVENESS IN NEWBORN	EFFECTIVENESS IN ADULT	ORIGIN
Various tissues	Self	Therapeutic Prophylactic	Low	High	Clinical or subclinical disease
Intramuscular or intradermal	Self	Prophylactic	Low	High	Toxoid or vaccine
Intravenous	Other than self	Prophylactic	High	Low	Transplacental passage of antibodies
Intravenous	Other than self	Prophylactic Therapeutic	High	Moderate	Serum that contains antibodies

drugs), may not be able to be immunized. They stand at a greater risk of catching the disease to which others are immunized.

DO VACCINES HAVE DANGEROUS SIDE EFFECTS?

If you look back at Figure 20.2, you will notice that the current immunization schedule in the United States for young children consists of up to 12 shots of five vaccines by six months of age and an additional four shots of six more vaccines by 18 months. The vaccination records show that vaccinations have been very successful at eliminating or greatly reducing the incidence of childhood diseases in the United States. Still, many parents wonder if this number of vaccines can have adverse effects on a developing young child, and if so, should they avoid taking their child to the doctor for routine vaccinations.

Although established vaccines are very safe, some vaccines do have side effects in certain people. Some people suffer mild fever, soreness at the injection site, or malaise from a vaccination. People who are allergic to eggs should not be vaccinated against the flu virus because the virus is replicated in chicken eggs. A recent example of risk assessment is the smallpox vaccine. In 2002–03, there was much discussion weighing the risks of the vaccine's potential side effects (mostly mild but potentially deadly for a few) versus the risk of a smallpox bioterrorist attack. The smallpox vaccine certainly is not recommended for young children.

Perhaps the vaccine of most concern regarding standard childhood vaccines is MMR. In 1998, a paper was published suggesting that there might be a link between the MMR vaccine and autism in children. The thought was that perhaps the sheer number of vaccinations could overwhelm a child's immune system and cause neurological damage. To make a long story short, as of 2003, studies carried out by other scientists and researchers have found no evidence to support the autism claim. Micro-Focus 20.6 looks at the consequences of declining vaccinations and herd immunity.

However, another substance has been of concern. This is **thimerosal**, a substance put in some vaccines (not MMR) as a preservative. Thimerosal contains

MicroFocus 20.6

THE RISK OF NOT VACCINATING

Data from epidemiological studies clearly show that vaccinations essentially eliminate outbreaks of infectious disease. Yet, about half the states allow parents to apply for vaccination waivers to enter their children in school. In mid-2003, the Centers for Disease Control and Prevention (CDC) reported that only about 75 percent of American children have been vaccinated on schedule against the nine diseases mandated by federal law. The CDC said that although coverage varied widely among states and major cities, there were areas in the country where too many children are not up-to-date on their shots. In addition, only a small percentage of adults receive booster shots for diseases like tetanus and diphtheria. Two examples point out the risk of not vaccinating.

In 1998, Colorado allowed vaccination exemptions for 2 percent of eligible children. This was double the exemption rate nationally. The analysis of Colorado and national health records showed that between 1987 and 1998, Colorado children three to eight years of age had a 22-times higher risk of contracting measles and nearly six times the risk of contracting pertussis (whooping cough) than did children nationally. One assumes the large number of unvaccinated children (lower herd immunity) was the main reason for the higher number of disease cases.

In 2003, a report indicated that measles outbreaks were becoming more prevalent in the United Kingdom. The reason was that more parents were not having their children vaccinated with the MMR vaccine because of its alleged side effects. In 1998, vaccinations dropped as there were reports that the MMR vaccine could cause autism. Even though such reports have been shown to be incorrect, vaccinations with the MMR vaccine have continued to drop. Herd immunity is losing its prescribed effects and, unless reversed, health officials fear that a continued drop could lead to the reestablishment of measles as an endemic disease in the United Kingdom.

high concentrations of mercury, which is a neurotoxin. Although considered a low-level exposure, the American Academy of Pediatrics stated in 1999 that the levels of mercury in the vaccines could be dangerous to young children. Thimerosal has been removed from all childhood vaccines. Whether it could have caused damage to or side effects in children prior to its removal has not been decided.

Realize that few of the millions of people that are vaccinated each year suffer any serious consequence. The risks of contracting a childhood disease from not being vaccinated are thousands of times greater than the risks associated with any vaccine.

To this point . . .

We have discussed four different types of acquired immunity with examples of each. Naturally acquired active immunity develops after a bout of illness or following a subclinical disease. The exposure to antigens is unintentional, and the immune system produces antibodies. By contrast, artificially acquired active immunity requires an intentional exposure to antigens, such as when one receives a vaccine or toxoid. Subunit vaccines, synthetic vaccines, and DNA vaccines contain no organisms and thus are an improvement over whole-organism vaccines. As before, immunity comes about when the immune system produces antibodies.

Passive immunity, be it naturally or artificially acquired, involves the patient receiving antibodies that were produced in another animal or human.

We concluded the section by discussing the importance of herd immunity. If herd immunity remains high in a population, disease epidemics are rare. Also, although a few vaccines may have side effects in sensitive individuals, vaccinations far outweigh the potential disease consequences of not being vaccinated.

In the next section, we shall shift our attention to the activity of antibodies in the laboratory. We will focus on serological reactions, in which antigen-antibody interactions are used for diagnostic purposes. We shall survey multiple forms of these tests and note the advances in technology that have made the serology laboratory a highly sophisticated environment for conducting practical immunology.

20.2

Serological Reactions

Antigen-antibody interactions studied under laboratory conditions are known as **serological reactions** because they commonly involve serum from a patient. In the late 1800s, serological reactions first were adapted to laboratory tests used in the diagnosis of disease. The principle was simple and straightforward: If the patient had an abnormal level of a specific antibody in the serum, a suspected disease agent probably was present. Today, serological reactions have diagnostic significance, as well as more broad-ranging applications. For example, they are used to confirm identifications made by other procedures and detect organisms in body tissues. In addition, they help the physician follow the course of disease and determine the immune status, and they aid in determining taxonomic groupings below the species level.

SEROLOGICAL REACTIONS HAVE CERTAIN CHARACTERISTICS

The reactants in a serological reaction generally consist of an antigen and a **serum** sample. The nature of either must be known; the object is to determine the nature of the other. In some cases, the unknown can be determined merely by placing the reactants on a slide and observing the presence or absence of a reaction. However, serological tests are not always quite so direct. For example, an antigen may have only one antigenic determinant, and the combination with an antibody molecule on a one-to-one basis may be invisible. To solve this dilemma, a second-stage reaction using an indicator system may be required, or a contact signal may be necessary. We shall see how this works in several tests.

Another possible drawback to a successful serological reaction is that antigen or antibody solutions may require considerable dilution to reach a concentration at which a reaction will be most favorable. This may be used to the physician's advantage, however, because the dilution series is a valuable way of determining the titer of antibodies, as shown in **FIGURE 20.5**. The **titer** is the most dilute concentration of serum antibody that yields a detectable reaction with its specific antigen. This number is expressed as a ratio of antibody to total fluid (e.g., 1:50) and is used to indicate the amount of antibodies in a patient's serum. For instance, the titer of influenza antibodies may rise from 1:20 to 1:320 as an episode of influenza progresses, and then continue upward, stabilizing at 1:1,280 as the disease reaches its peak. A rise in the titer of a certain antibody also indicates that an individual has a case of that disease, an important factor in diagnosis.

Serum:
the clear, cell-free fluid portion of the blood that contains no clotting agents.

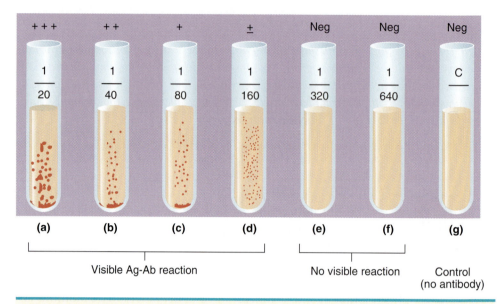

+++	++	+	±	Neg	Neg	Neg
$\frac{1}{20}$	$\frac{1}{40}$	$\frac{1}{80}$	$\frac{1}{160}$	$\frac{1}{320}$	$\frac{1}{640}$	C
(a)	(b)	(c)	(d)	(e)	(f)	(g)

Visible Ag-Ab reaction   No visible reaction   Control
                                                        (no antibody)

FIGURE 20.5

The Determination of Titer

A sample of antibody-containing serum was diluted in saline solution to yield the dilutions shown. An equal amount of antigen was then added to each tube, and the tubes were incubated. An antigen-antibody interaction may be seen in tubes (a) through (d), but not in tubes (e) or (f) or the control tube, (g). The titer of antibody is the highest dilution of serum antibody in which a reaction is visible—in this case, tube (d). The titer, therefore, is expressed as 1:160.

Haptens may pose a problem in a serological reaction because, as partial antigens, they usually have only a single antigenic determinant. Technologists have circumvented this problem by conjugating the haptens to carrier particles, such as polystyrene beads. When the hapten unites with an antibody, the entire bead is involved in the complex, and a visible reaction occurs.

Serology has become a highly sophisticated and often automated branch of immunology. As the following tests illustrate, the serological reactions have direct application to the clinical laboratory—not only in microbiology, but in other fields as well.

Serology:
a branch of immunology that studies serological reactions.

NEUTRALIZATION INVOLVES ANTIGEN-ANTIBODY REACTIONS

Neutralization is a serological reaction in which antigens and antibodies neutralize each other. The reaction is used to identify toxins and antitoxins, as well as viruses and viral antibodies. Normally, little or no visible evidence of a neutralization reaction is present, and the test mixture therefore must be injected into a laboratory animal to determine whether neutralization has taken place.

An example of a neutralization test is the one used to detect botulism toxin in food. Normally the botulism toxin is lethal to a laboratory animal, and if a sample of the food contains the toxin, the animal will succumb after an injection. However, if the food is first mixed with botulism antitoxins, the antitoxin molecules neutralize the toxin molecules, and the mixture has no effect on the animal. Conversely, if the toxin was produced by some other organism, no neutralization will occur, and the mixture will still be lethal to the animal. A similar test for diphtheria is the **Schick test**. In this test, a person's immunity to diphtheria can be determined by injecting diphtheria toxin intradermally. No skin reaction will occur if the person has neutralizing antibodies. Local edema occurs if no neutralizing anitbodies are present, indicating the person is susceptible to diphtheria.

PRECIPITATION REQUIRES THE FORMATION OF A LATTICE BETWEEN SOLUBLE ANTIGEN AND ANTIBODY

Precipitation reactions are serological reactions involving thousands of antigen and antibody molecules cross-linked at multiple determinant sites to form a structure called a **lattice**. The lattices are so huge that particles of precipitate form, and the reaction product is observed visually.

Precipitation tests are performed in either fluid media or gels. In **fluid precipitation**, the antibody and antigen solutions are layered over each other in a thin tube. The molecules then diffuse through the fluid until they reach a zone of equivalence, the ideal concentration for precipitation. A visible mass of particles now forms at the interface or at the bottom of the tube. Fluid precipitation is used frequently in forensic medicine to learn the origin of albumin proteins in bloodstains.

In **gel precipitation**, the diffusion of antigens and antibodies takes place through a semisolid gel, such as **agarose**. (The tests also are called **immunodiffusion tests**.) The Oudin tube technique, described in 1946 by Jacques Oudin, is typical. A plug of gel is placed between solutions of antigen and antibody in a thin tube. As the molecules diffuse through the gel, they eventually reach the zone of equivalence, where they interact and form a visible ring of precipitate. This type of technique is called a **double diffusion assay** because both reactants diffuse. Another application is in the Ouchterlony plate technique, named for Orjan Ouchterlony, who devised it in 1953. In this technique, antigen and antibody solutions are placed in wells cut into agarose in Petri dishes. The plates are incubated, and precipitation lines form at the zone of equivalence (**FIGURE 20.6**). The test is used to detect fungal antigens of *Histoplasma*, *Blastomyces*, and *Coccidioides* (Chapter 15).

Agarose:
a neutral polysaccharide prepared from agar.

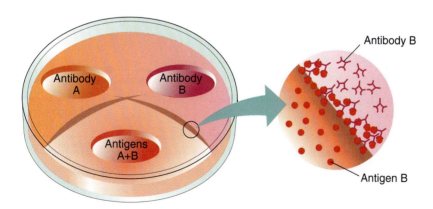

FIGURE 20.6

A Precipitation Test

Wells are cut into a plate of purified agarose. Different known antibodies then are placed into the two upper wells, and a mixture of unknown antigens is placed into the lower well. During incubation, the reactants diffuse outward from the wells, and cloudy lines of precipitate form where the reactants happen to meet. The lines cross each other because each antigen has reacted only with its complementary antibody.

In the procedure known as **immunoelectrophoresis**, the techniques of gel electrophoresis and diffusion are combined for the detection of antigens. A mixture of antigens is placed in a reservoir on an agarose slide, and an electrical field is applied to the ends of the slide. The different antigens then move through the agarose at different rates of speed, depending on their electrical charges. This process is **gel electrophoresis** (FIGURE 20.7). A trough then is cut into the agarose along the same axis, and a known antibody solution is added. During incubation, antigens and antibodies diffuse toward each other and precipitation lines form, as in the Ouchterlony technique.

AGGLUTINATION INVOLVES THE CLUMPING OF ANTIGENS

Agglutination is a serological reaction in which antibodies interact with antigens on the surface of particular objects and cause the objects to clump together, or agglutinate. Agglutination techniques were among the earliest serological reactions adapted to the diagnostic laboratory. For example, until the 1960s, the diagnosis of typhoid fever was based on the agglutination of *Salmonella* cells by antibodies in the patient's serum. This test, called the **Widal test** after its developer Georges Fernand Widal, is now supplemented by more sophisticated procedures performed in plastic microtiter plates.

Agglutination procedures are performed on slides or in tubes. For bacterial agglutination, emulsions of unknown bacteria are added to drops of known antibodies on a slide, and the mixture is observed for clumping. If none occurs, different antibodies are tried until the correct one is found (FIGURE 20.8). This

---

FIGURE 20.7

Immunoelectrophoresis

Ⓐ Gel electrophoresis:
On a gel-coated slide, an antigen sample is placed in a central well. An electrical current is run through the gel to separate antigens by their electrical charge (electrophoresis). Unlike the diagram, the separate antigens cannot be detected visually.

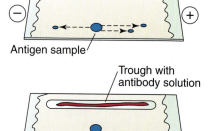

Ⓑ Addition of antibodies:
A trough is made on the slide and a known antibody solution is added.

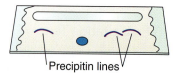

Ⓒ Diffusion of antigens and antibodies:
As antigens and antibodies diffuse toward one another through the gel, precipitin lines are seen where optimal concentrations of antigen and antibodies meet.

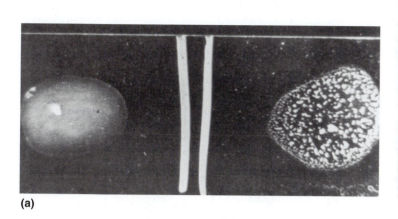

(a)

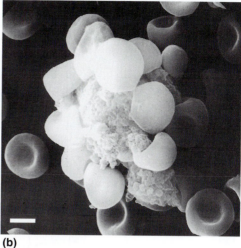

(b)

FIGURE 20.8

Two Views of Agglutination

(a) A slide agglutination procedure. *Salmonella* cells were mixed with *Salmonella* antibodies on the right side of the slide and with antibody-free saline solution on the left side. Clumps of cells are visible on the right side. (b) A scanning electron micrograph of human red blood cells agglutinated by a microcolony of *Mycoplasma pneumoniae*, the cause of primary atypical pneumonia. (Bar = 7 μm.)

process is essentially a trial-and-error method, although the chances for success may be enhanced by using a **polyvalent serum**, that is, one containing a mixture of antibodies. Tube agglutinations may be performed with serum to determine the titer of antibodies present.

Passive agglutination is a modern approach to traditional agglutination methods. Most often, antigens are adsorbed onto the surface of latex spheres or polystyrene particles (**FIGURE 20.9a**). Serum antibodies can be detected rapidly by observing agglutination of the carrier particle. Bacterial infectious agents, such as those caused by *Streptococcus*, can be detected rapidly by mixing the bacterial sample with latex spheres containing streptococcus antibody (**FIGURE 20.9b**). **Hemagglutination** refers to the agglutination of red blood cells. This process is particularly important in the determination of blood types prior to blood transfusion (Chapter 21). In addition, certain viruses, such as measles and mumps viruses, agglutinate red blood cells. Antibodies for these viruses may be detected by a procedure in which the serum is first combined with laboratory-cultivated viruses and then added to the red blood cells. If serum antibodies neutralize the viruses, agglutination fails to occur. This test, called the **hemagglutination inhibition (HAI) test**, is discussed in Chapter 12.

A hemagglutination test called the **Coombs test** is used to detect **Rh antibodies** involved in hemolytic disease of the newborn (discussed in Chapter 21). These Rh antibodies will react with red blood cells (RBCs) bearing the corresponding Rh antigens. In the Coombs test, RBCs having the Rh antigen are combined with the patient's serum, which may or may not have Rh antibodies. After a short incubation period, a sample of **antiglobulin antibodies** are added. These antibodies react with Rh antibodies. If the Rh antibodies were present in the serum, they have now

FIGURE 20.9

Reactions in Passive Agglutination Tests

These tests use antigens or antibodies adsorbed onto the surface of latex spheres. (a) When particles are bound with antigens, agglutination indicates the presence of antibodies, such as the IgM shown here. (b) When particles are bound with antibodies, agglutination indicates the presence of antigens.

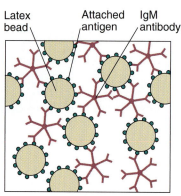

Latex bead   Attached antigen   IgM antibody

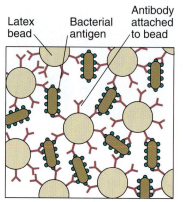

Latex bead   Bacterial antigen   Antibody attached to bead

(a) Reaction in a positive passive agglutination test for antibodies

(b) Reaction in a positive passive agglutination test for antigens.

gathered on the RBC surface, and when the antiglobulin antibodies are added, the RBCs will form a clump, yielding a positive test. However, if there were no Rh antibodies in the serum, no gathering on the RBC surface took place, and when the antiglobulin antibodies are added, no clumping of the RBCs will take place. The absence of clumping indicates a negative test—the patient does not have Rh antibodies in the serum.

FLOCCULATION REQUIRES AGGREGATE FORMATION WITHIN A PRECIPITATE

The **flocculation** test combines the principles of precipitation and agglutination. The antigen exists in a noncellular particulate form that reacts with antibodies to yield large, visible aggregates.

An example of the flocculation test is the Venereal Disease Research Laboratory (VDRL) test used for the rapid screening of patients to detect syphilis. The antigen consists of an alcoholic extract of beef heart called cardiolipin. When diluted with buffer solution, the cardiolipin forms a milky-white precipitate. Serum from a patient then is added. If the serum contains syphilis antibodies, the particles of precipitate react with antibodies and cling together, yielding aggregates. Observation under the low-power objective of the light microscope reveals the extent of aggregate formation and gives a clue to the amount of antibodies present. The **VDRL test** has been in use for many decades and is part of the "blood test" that couples may be required to take before obtaining a marriage license.

To this point . . .

We have surveyed a number of serological reactions used in the diagnostic laboratory to detect interactions between antigens and antibodies. In each case, the reaction is fairly simple and straightforward, and the laboratory technician can usually determine whether an interaction has occurred. For example, particles clump in agglutination, and precipitates form in precipitation. Many of the basic reactions have been adapted by modern technologists to improve on the fundamental theme of the process.

In the final section of this chapter, we shall explore another series of serological reactions and tests. Most of the tests involve a multistep procedure, and the visible manifestation of the reaction usually requires the participation of accessory factors, indicator systems, and specialized equipment. These diagnostic tests are more complicated to perform and require skilled technicians, but their development has ensured the position of laboratory immunology as a key link in the health-care delivery system.

20.3

Other Serological Reactions

Among the other serological reactions are a set of sophisticated procedures that detect antibodies in novel ways, including the use of radioactivity and gene probes. We begin the section, however, with a standard procedure used since the beginning of the 1900s.

COMPLEMENT FIXATION CAN DETECT ANTIBODIES TO A VARIETY OF PATHOGENS

The **complement fixation test** was devised by Jules Bordet and Octave Gengou in 1901. It was later adapted for syphilis by August von Wassermann in 1906, and for many decades it remained a mainstay for syphilis diagnosis. Technologists now use it for detecting antibodies for a variety of viruses, fungi, and bacteria.

The test is performed in two parts. The first part, the **test system**, uses the patient's serum, a preparation of antigen from the suspected pathogen, and complement derived from guinea pigs. The second part, the **indicator system**, requires sheep red blood cells and a preparation called **hemolysin**.

The first step in the test is to heat the patient's serum to destroy any complement present in the serum. Next, carefully measured amounts of antigen and guinea pig complement are added to the serum (FIGURE 20.10). This test system then is incubated at 37°C for 90 minutes. If antibodies specific for the antigen are present in the serum, an antibody-antigen interaction takes place, and the complement is used up, or "fixed." However, there is no visible sign of whether a reaction has occurred.

Now the indicator system (sheep red blood cells and hemolysin) is added to the tube, and the tube is reincubated. If the complement was previously fixed, none will be available to the hemolysin, and lysis of the sheep red blood cells cannot take place. The blood cells therefore would remain intact, and when the tube is centrifuged, the

Hemolysin:
antibodies that will lyse sheep red blood cells in the presence of complement.

FIGURE 20.10

The Complement Fixation Test

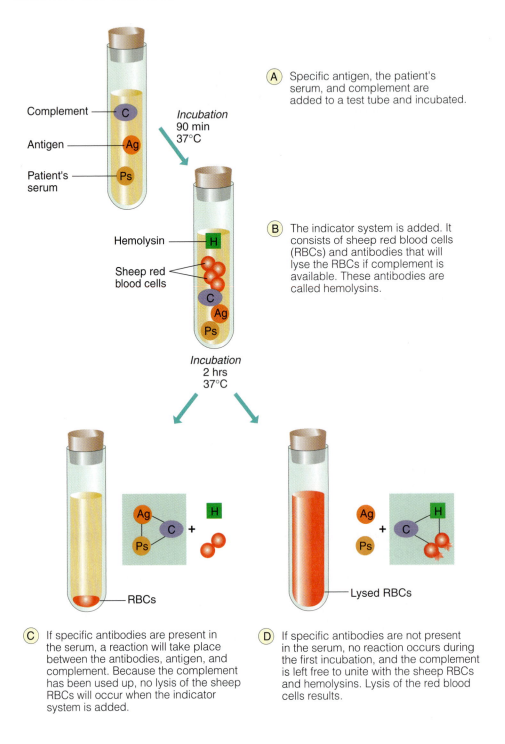

A Specific antigen, the patient's serum, and complement are added to a test tube and incubated.

Complement — C
Antigen — Ag
Patient's serum — Ps

Incubation 90 min 37°C

Hemolysin — H
Sheep red blood cells

B The indicator system is added. It consists of sheep red blood cells (RBCs) and antibodies that will lyse the RBCs if complement is available. These antibodies are called hemolysins.

Incubation 2 hrs 37°C

RBCs

C If specific antibodies are present in the serum, a reaction will take place between the antibodies, antigen, and complement. Because the complement has been used up, no lysis of the sheep RBCs will occur when the indicator system is added.

Lysed RBCs

D If specific antibodies are not present in the serum, no reaction occurs during the first incubation, and the complement is left free to unite with the sheep RBCs and hemolysins. Lysis of the red blood cells results.

technician observes clear fluid with a "button" of blood cells at the bottom. Conclusion: The serum contained antibodies that reacted with the antigen and fixed the complement.

If the complement was not fixed in the test system, it will still be available to the hemolysin, and the hemolysin-complement mixture will lyse the sheep red blood cells. When the tube is centrifuged, the technician sees red fluid, colored by the hemoglobin of the broken blood cells, and no evidence of blood cells at the bottom of the tube. Conclusion: The serum lacked antibodies for the antigen tested.

The complement fixation test is valuable because it may be adapted by varying the antigen. In this way, tests may be conducted for such diverse diseases as encephalitis, Rocky Mountain spotted fever, meningococcal meningitis, and histoplasmosis. The versatility of the test, together with its sensitivity and relative accuracy, have secured its continuing role in diagnostic medicine.

FLUORESCENT ANTIBODY TECHNIQUES CAN IDENTIFY UNKNOWN ANTIGENS (PATHOGENS)

The **fluorescent antibody technique** is a slide test performed by combining particles containing antigens with antibodies and a fluorescent dye (Chapter 3). When the three components react, the dye causes the complex to glow on illumination with ultraviolet light under a fluorescence microscope, as shown in FIGURE 20.11. Two commonly used dyes are fluorescein, which emits an apple-green glow, and rhodamine, which gives off orange-red light.

Fluorescent antibody techniques may be direct or indirect. In the **direct method**, the fluorescent dye is linked to known antibody molecules. The antibodies then are combined with particles that may contain complementary antigens, such as bacteria. If the presumption is correct, the tagged antibodies accumulate on the particle surface, and the particle glows when viewed with fluorescence microscopy. In this way, an unknown antigen or unknown organism can be identified.

The **indirect method** is illustrated by the **FTA-ABS** diagnostic procedure used for detecting syphilis antibodies in the blood of a patient (FIGURE 20.12). A sample of commercially available syphilis spirochetes is placed on a slide, and the slide is then flooded with the patient's serum. Next, a sample of fluorescein-labeled antiglobulin (antihuman) antibodies is added. These are antibodies that unite with human antibodies. They are produced by an animal injected with human antibodies. The slide then is observed using the fluorescence microscope.

FTA-ABS:
fluorescent treponemal antibody absorption test; a diagnostic procedure for syphilis.

FIGURE 20.11

Fluorescent Antibody Staining

Three views of spirochetes in the hindgut of a termite. (a) An unstained area displayed by differential interference contrast microscopy. (b) The same viewing area seen by fluorescence microscopy using rhodamine B as a stain. (c) The same area viewed after staining with fluorescein. (Bar = 10 μm.)

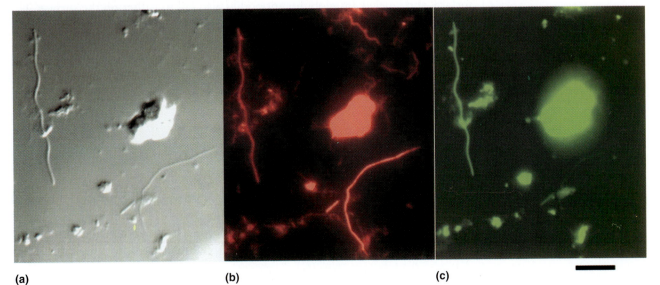

(a)   (b)   (c)

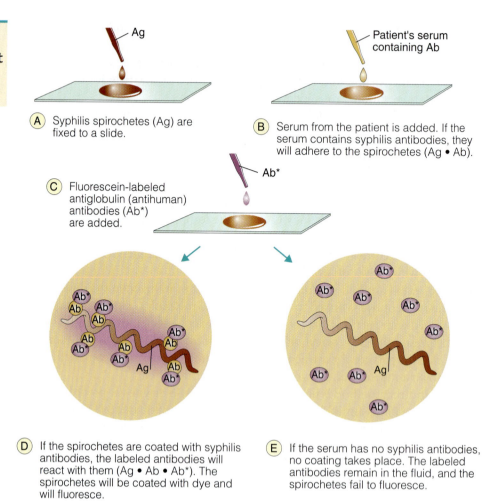

FIGURE 20.12

The Indirect Fluorescent Antibody Technique for Diagnosing Syphilis

(A) Syphilis spirochetes (Ag) are fixed to a slide.

(B) Serum from the patient is added. If the serum contains syphilis antibodies, they will adhere to the spirochetes (Ag • Ab).

(C) Fluorescein-labeled antiglobulin (antihuman) antibodies (Ab*) are added.

(D) If the spirochetes are coated with syphilis antibodies, the labeled antibodies will react with them (Ag • Ab • Ab*). The spirochetes will be coated with dye and will fluoresce.

(E) If the serum has no syphilis antibodies, no coating takes place. The labeled antibodies remain in the fluid, and the spirochetes fail to fluoresce.

The test is interpreted as follows. If the patient's serum contains syphilis antibodies, the antibodies bind to the surfaces of spirochetes, and the labeled antiglobulin antibodies are attracted to them. The spirochetes then glow from the dye. However, if no antibodies are present in the serum, nothing accumulates on the spirochete's surface, and labeled antiglobulin antibodies also fail to gather on the surface. The labeled antibodies remain in the fluid, and the spirochetes do not glow.

Fluorescent antibody techniques are adaptable to a broad variety of antigens and antibodies and are widely used in serology. Antigens may be detected in bacterial smears, cell smears, and viruses fixed to carrier particles. The value of the techniques is enhanced because the materials are sold in kits and are readily available to small laboratories.

RADIOIMMUNOASSAY (RIA) IS USED TO QUANTITATE ANTIGEN-ANTIBODY COMPLEXES

Radioimmunoassay (RIA) is an extremely sensitive serological procedure used to measure the concentration of low-molecular-weight antigens, such as haptens. Since its development in the 1960s, the technique has been adapted for quantitating hepatitis antigens, as well as reproductive hormones, insulin, and certain

Radioimmunoassay: a sensitive serological procedure in which radioactive antigens compete with nonlabeled antibodies for reactive sites on antibody molecules.

drugs. One of its major advantages is that it can detect trillionths of a gram of a substance (MicroFocus 20.7).

The RIA procedure is based on the competition between radioactive-labeled antigens and unlabeled antigens for the reactive sites on antibody molecules. A known amount of the radioactive (labeled) antigens is mixed with a known amount of specific antibodies, and an unknown amount of unlabeled antigens. The antigen-antibody complexes that form during incubation then are separated out, and their radioactivity is determined. By measuring the radioactivity of free antigens remaining in the leftover fluid, one can calculate the percentage of labeled antigen bound to the antibody. This percentage is equivalent to the percentage of unlabeled antigen bound to the antibody because the same proportion of both antigens will find spots on antibody molecules. The concentration of unknown unlabeled antigen then can be determined by reference to a standard curve.

Radioimmunoassay procedures require substantial investment in sophisticated equipment and carry a certain amount of risk because radioactive isotopes are used. For these reasons, the procedure is not widely used in routine serological laboratories. However, immunologists with access to radioimmunoassay have discovered a wealth of information.

The **radioallergosorbent test** (**RAST**) is an extension of the radioimmunoassay. The test may be used, for example, to detect IgE antibodies in the serum of a person possibly allergic to penicillin.

To detect IgE against penicillin, penicillin antigens are attached to a suitable particle. Serum that may contain penicillin IgE is then added. If the antibody is present, it will combine with the penicillin antigens on the surface of the particle. Now another antibody, one that will react with human antibodies, is added. This antiglobulin antibody carries a radioactive label. The entire complex therefore will become radioactive if the antiglobulin antibody combines with the IgE, as FIGURE 20.13 illustrates. By

---

MicroFocus 20.7

SOMETHING SPECIAL FROM A SPECIAL SOMEONE

When the Nobel Prize in Physiology or Medicine was announced on October 14, 1977, the scientific community applauded a special person and a special technique. One of the recipients was Rosalyn Sussman Yalow, a developer of the radioimmunoassay (RIA) technique. This technique is one of the most highly regarded immunological procedures.

RIA has made possible the detection of incredibly small amounts of chemical substances in body fluids. One offshoot was the discovery of certain hormones not previously known to exist in the body. Another was the revelation that individuals receiving insulin injections

produce antibodies against insulin. This finding put in serious doubt the contention that the insulin molecules were too small to be antigenic.

Radioimmunoassay also allows the detection of tumor viruses in the body before the appearance of a tumor. Moreover, it may be utilized to screen for hepatitis B viruses in blood used for transfusions. RIA is said to be sensitive enough to detect trillionths of a gram of a substance. This is equivalent to detecting a lump of sugar in Lake Erie.

Like the process she developed, Rosalyn Yalow is also special. She was educated during a time when opportunities for women were limited, and her work

was done in restricted surroundings at the Veterans Hospital in the Bronx, New York. She was the sixth woman honored by the Nobel Committee and only the second in Physiology or Medicine. (She is also a product of the same New York City neighborhood as Ed Alcamo.) Rosalyn Yalow remains one of the eminent immunologists of our time.

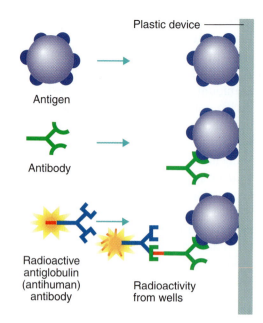

A The antigen to the suspected antibody is bound to a plastic device, such as a plastic plate with wells.

B The test solution, such as urine, is added to the device. The antibody may or may not be present in the solution.

C An antibody that reacts with a human antibody (an antiglobulin antibody) is added. The antiglobulin antibody carries a radioactive label. If radioactivity is detected in the wells, the technician concludes that the solution contained the suspected antibody.

FIGURE 20.13

The Radioallergosorbent Test (RAST)

This technique provides an effective way of detecting very tiny amounts of antibody in a preparation. A known antigen is used. The objective is to determine whether the complementary antibody is present.

contrast, if no IgE was present in the serum, no reaction with the antigen on the particle surface will take place, and the radioactive antibody will not be attracted to the particle. When tested, the particles will not show radioactivity.

The RAST is commonly known as a "**sandwich**" technique. There is no competition for an active site as in RIA, and the type of unknown antibody, as well as its amount, may be learned by determining the amount of radioactivity deposited.

THE ENZYME-LINKED IMMUNOSORBENT ASSAY (ELISA) CAN DETECT ANTIGENS OR ANTIBODIES

The **enzyme-linked immunosorbent assay (ELISA)** has virtually the same sensitivity as radioimmunoassay and the RAST, but does not require expensive equipment or radioactivity. The procedure involves attaching antibodies or antigens to a solid surface and combining (immunosorbing) the coated surfaces with the test material. An enzyme system then is linked to the complex, the remaining enzyme is washed away, and the extent of enzyme activity is measured. This gives an indication that antigens or antibodies are present in the test material.

An application of the ELISA is found in the highly efficient laboratory test used to detect antibodies against the human immunodeficiency virus (HIV) (**FIGURE 20.14**). A serum sample is obtained from the patient and mixed with a solution of plastic or polystyrene beads coated with antigens from HIV. If antibodies to HIV are present in the serum, these "primary antibodies" will adhere to the antigens on the surface of the beads. The beads then are washed and incubated with antiglobulin (antihuman) antibodies chemically tagged with molecules of horseradish peroxidase or a similar

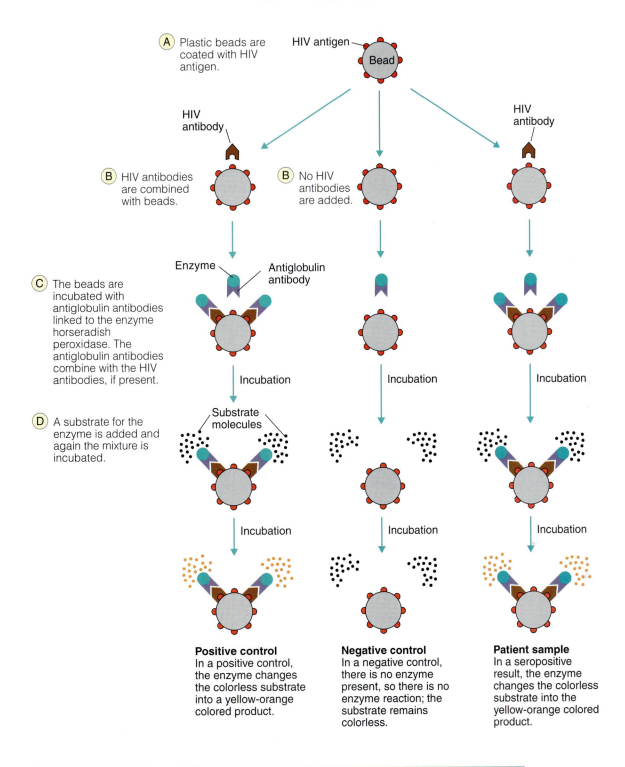

(A) Plastic beads are coated with HIV antigen.

HIV antigen

Bead

HIV antibody

(B) HIV antibodies are combined with beads.

(B) No HIV antibodies are added.

HIV antibody

Enzyme   Antiglobulin antibody

(C) The beads are incubated with antiglobulin antibodies linked to the enzyme horseradish peroxidase. The antiglobulin antibodies combine with the HIV antibodies, if present.

Incubation   Incubation   Incubation

(D) A substrate for the enzyme is added and again the mixture is incubated.

Substrate molecules

Incubation   Incubation   Incubation

Positive control
In a positive control, the enzyme changes the colorless substrate into a yellow-orange colored product.

Negative control
In a negative control, there is no enzyme present, so there is no enzyme reaction; the substrate remains colorless.

Patient sample
In a seropositive result, the enzyme changes the colorless substrate into the yellow-orange colored product.

FIGURE 20.14

The ELISA

The enzyme–linked immunosorbent assay (ELISA) as it is used to detect antibodies (in this case HIV) in a patient's serum. A seronegative result would be similar to the negative control.

enzyme. These antibodies are referred to as the "secondary antibody." The preparation is washed, and a solution of substrate molecules for the peroxidase enzyme is added. Initially the solution is clear, but if enzyme molecules react with the substrate, the solution will become yellow orange in color. The enzyme molecules will be present only if HIV antibodies are present in the serum. If no HIV antibodies are in the serum, no enzyme molecules could concentrate on the bead surface, no change in the substrate molecules could occur, and no color change would be observed.

ELISA procedures may be varied depending on whether one wishes to detect antigens or antibodies. The solid phase may consist of beads, paper disks, or other suitable supporting mechanisms, and alternate enzyme systems such as the alkaline phosphatase system may be used that produce a different color solution in a positive reaction. In addition, the results of the test may be quantified by noting the degree of enzyme-substrate reactions as a measure of the amount of antigen or antibody in the test sample. The availability of inexpensive ELISA kits has brought the procedure into the doctor's office and routine serological laboratory.

Importantly, a positive ELISA for HIV, or for any pathogen being tested, only indicates that antibodies to the pathogen are present. A positive ELISA does not necessarily mean the person has the disease. The presence of antibodies only says the person has been exposed to the pathogen. However, in the case of HIV, a positive ELISA probably does indicate an infection, as no one has ever been known to be cured of an HIV infection. MicroInquiry 20 looks at the ELISA test.

MicroInquiry 20

APPLICATIONS OF IMMUNOLOGY: DISEASE DIAGNOSIS

Ancient Egyptian medical papyri from the fifteenth century B.C. refer to many different disease symptoms and treatments. Some of these symptoms still can be used to identify diseases today. Thus, one of the most traditional "tools" of diagnosis over the centuries is a patient's signs and symptoms. However, there are problems when relying solely on signs and symptoms. Many diseases, at least initially, display common symptoms. For example, the initial symptoms of a Hantavirus infection are very similar to those of the flu. Some diseases do not display symptoms for perhaps weeks or months—or years in the case of AIDS. Yet, it is important to identify these diseases rapidly so appropriate treatment, if possible, can be started.

Serological (blood) tests have been used in the United States since about 1910 both to diagnose and control infectious disease. Today's understanding of immunology has brought newer tests that rely on identifying antibody-mediated (humoral) immune responses; that is, antibody reactions with antigens. Serology laboratories work with serum or blood from patients suspected of having an infectious disease. The lab tests look for the presence of antibodies to known microbial antigens. Serological tests that are seropositive indicate antibodies to the microbe were detected while seronegative means no antibodies were detected in a patient's serum.

Let's use a hypothetical person, named Pat, who "believes" that 12 days ago he may have been exposed to the hepatitis C virus through unprotected sex. Afraid to go to a neighborhood clinic to be tested, Pat goes to a local drugstore and purchases an over-the-counter, FDA-approved, hepatitis C home testing kit (FIGURE A). Using a small spring-loaded device that comes with the kit, Pat pricks his little finger and puts a couple of drops of blood onto a paper strip included with the kit. He fills out the paperwork and mails the paper strip in a prepaid Federal Express® mailer (supplied in the kit) to a specific blood testing facility where the ELISA test is performed. In 10 business days, Pat can call a toll-free number anonymously, identify himself by his unique 14-digit code that came with the testing kit, and ask for his test results. If necessary, he can receive professional post-test counseling and medical referrals.

Two days later, Pat runs into a good friend of his who is a nurse. Pat explains his "predicament" and that he is anxious about having to wait 10 days for the test results. He asks his friend some questions. If you were the nurse, how would you respond to his questions? Answers can be found in Appendix E.

MONOCLONAL ANTIBODIES ARE PRODUCED BY A SINGLE CLONE OF B CELLS

Usually pathogens contain several different antigens. Therefore, when an infection occurs, a different B-cell population is activated for each epitope and different antibodies specific to each of the epitopes are produced. Serum from such a patient would contain **polyclonal antibodies** because they are derived from several different clones of B cells. However, in the laboratory it is possible to produce populations of identical antibodies that bind to only one epitope.

In 1975, Georges Köhler of then–West Germany and Cesar Milstein of Argentina developed a method for the laboratory production of antibodies that recognize only one epitope. The key was using **myelomas**, cancerous tumors that arise from the uncontrolled division of plasma cells. Although these myeloma cells have lost the ability to produce antibodies, they can be grown indefinitely in culture. Köhler and Milstein were able to use myeloma cells to produce **monoclonal antibodies**, which are specific to one epitope.

There are two basic steps to generating monoclonal antibodies. First, a myeloma cell is fused to an activated (antibody secreting) B cell (plasma cell) to form a hybrid cell called a **hybridoma**. Then, each hybridoma is screened for the desired monoclonal antibody and cultured to produce hybridoma clones.

20.1a. What is ELISA and how is it performed?

20.1b. You mention positive and negative controls in your explanation of the ELISA process. Pat asks, "What are positive and negative controls and why are they necessary?"

After your explanations, Pat still has more questions for you to answer.

20.1c. Pat says, "So suppose I am seropositive. Does it mean I have hepatitis C?"

20.1d. Pat says, "Okay, so if I am seronegative, then I must be free of the virus. Right?"

20.1e. If indeed Pat was seronegative, what advice would you give him?

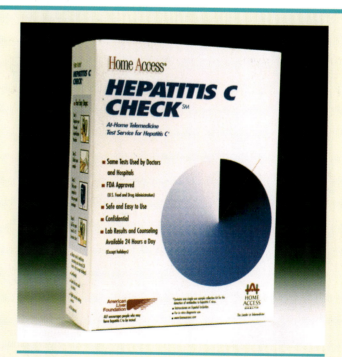

FIGURE A

To produce hybridomas, a mouse is injected with the antigen of interest against which monoclonal antibodies are needed (**FIGURE 20.15**). Injection triggers antibody-mediated (humoral) immunity in the mouse. Plasma cells from the spleen are removed and mixed with myeloma cells. By forcing many cells to fuse together, hybridomas are produced. Hybridomas are placed in a special tissue culture medium. In seven to ten days, hybridomas form small clusters while any unfused cells (B cells or myelomas) cannot survive and die.

Since the antigen originally injected may have several epitopes, different hybridoma clusters may be producing different antibodies. Therefore, individual hybridoma cells are separated from one another and allowed to grow. Each hybridoma cell can multiply indefinitely (myeloma characteristic), producing antibodies recognizing but one epitope (B-cell characteristic).

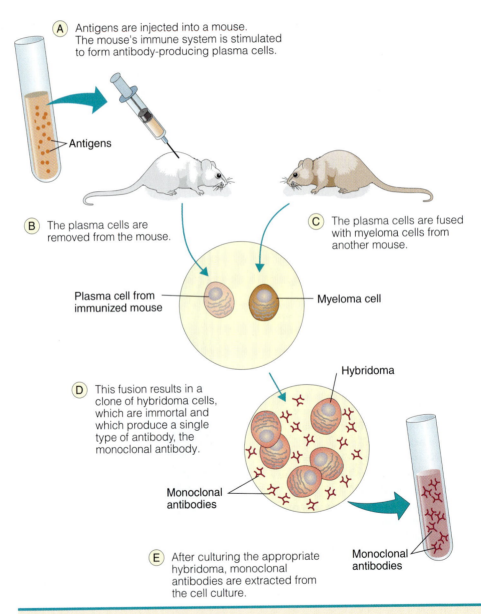

(A) Antigens are injected into a mouse. The mouse's immune system is stimulated to form antibody-producing plasma cells.

Antigens

(B) The plasma cells are removed from the mouse.

(C) The plasma cells are fused with myeloma cells from another mouse.

Plasma cell from immunized mouse

Myeloma cell

(D) This fusion results in a clone of hybridoma cells, which are immortal and which produce a single type of antibody, the monoclonal antibody.

Hybridoma

Monoclonal antibodies

(E) After culturing the appropriate hybridoma, monoclonal antibodies are extracted from the cell culture.

Monoclonal antibodies

FIGURE 20.15

The Production of Monoclonal Antibodies

Each clone of hybridomas is screened for the synthesis of the desired antibody. The hybridoma producing the desired monoclonal antibodies can be propagated (cloned) in tissue culture flasks.

Monoclonal antibodies and the hybridoma technique have been hailed as one of our era's most important methodological advances in biomedicine. In 1984, Köhler and Milstein shared the Nobel Prize in Physiology or Medicine for the development of the monoclonal antibody technique. The antibodies differ from ordinary antibodies because they are far more pure and uniform, and exquisitely sensitized to probe for their antigenic targets. Monoclonal antibodies, for example, have been used to pinpoint the antigens on the surfaces of parasites, thereby enabling researchers to zero in on these antigens for vaccine production. In this regard, they are excellent research tools.

Monoclonal antibodies also may hold the key to the treatment of tumors. For example, scientists have developed a technique in which tumor cells are removed from a patient and injected into a mouse, whereupon the mouse's spleen begins producing tumor antibodies. Spleen cells then are fused with myeloma cells to produce a hybridoma that produces antibodies for that specific tumor. Toxins are attached to the tail (Fc fragment) of the antibodies. When the antibodies are injected back into the patient, they act like "stealth missiles." They react specifically with the tumor cells and the toxin kills the cells without destroying other tissue cells.

Monoclonal antibodies are reproducible because of how they are manufactured; therefore, laboratories throughout the world can use identical antibodies. This allows comparisons of tests and research results that were previously impossible to obtain. Monoclonal antibodies are used also for cleansing bone marrow prior to transplantation, in treating disorders of the immune system, and for an assortment of basic studies and practical approaches to medicine. They represent one of the most elegant expressions of modern biotechnology.

GENE PROBES ARE SINGLE-STRANDED DNA SEGMENTS

Although antibody tests are a valuable resource in the clinical laboratory, a variety of new diagnostic tests permits the identification of an organism and its antigens (MicroFocus 20.8). One new test is based on the use of a DNA fragment called the gene probe and a procedure known as the polymerase chain reaction (PCR). A **gene probe** is a relatively small, single-stranded DNA segment that can hunt for a complementary fragment of DNA within a morass of cellular material, much like a right hand searching for a left hand. When the probe locates its complementary fragment, it emits a signal such as a pulse of radioactivity. If the complementary fragment cannot be found, then no signal is sent. The procedure is remarkable for its accuracy. For example, if we were to line up the 46 human chromosomes as a two-lane highway, the highway would stretch around planet Earth 300 times. A gene probe can locate and unite with a few-mile stretch of this highway. More information on gene probes is presented in Chapter 7.

To use a gene probe effectively, it is valuable to increase the amount of DNA to be searched. The **polymerase chain reaction (PCR)** accomplishes this task. The procedure takes a segment of DNA and reproduces it to a billion copies in a few short hours. The PCR process is a three-step cycle (FIGURE 20.16). Target DNA is mixed with DNA polymerase (the enzyme that synthesizes DNA), short strands of primer DNA, and a mixture of nucleotides. The mixture is then alternately heated and cooled during which time the double-stranded DNA unravels, is duplicated, then reforms the double helix. The process is repeated over and over again in a highly automated PCR

FIGURE 20.16

The Polymerase Chain Reaction

The polymerase chain reaction (PCR) is a method to copy (amplify) a specific DNA sequence. The sequence to be amplified is placed in a test tube with a solution of primer oligonucleotides, a supply of all four DNA nucleotides, and DNA polymerase enzymes that are heat resistant. The primer oligonucleotides are needed to start DNA synthesis. The number of copies of the target sequence doubles in each cycle of PCR, so millions of DNA sequences, identical to the target sequence, can be generated in 30 cycles of PCR.

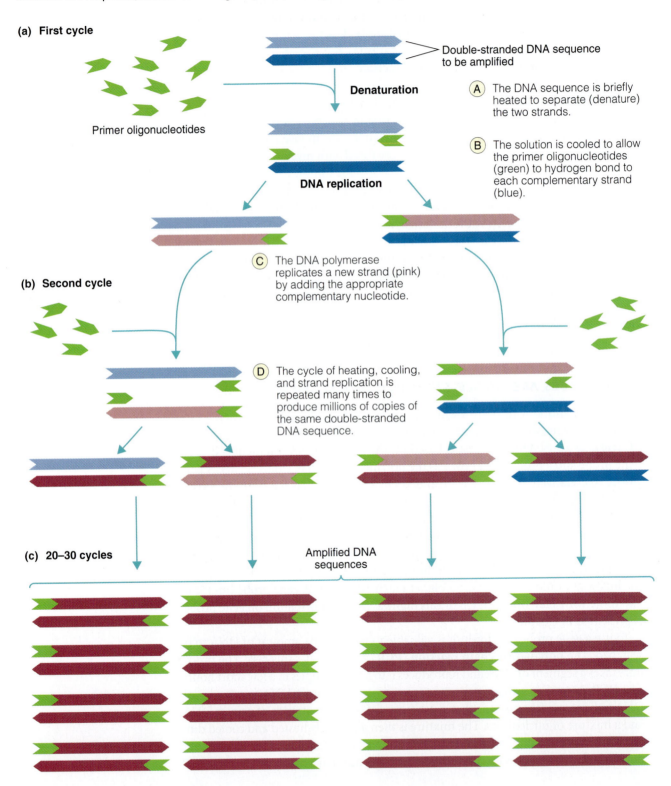

MicroFocus 20.8

CAUGHT IN THE SPOTLIGHT

Current diagnostic tests for tuberculosis can take several weeks to complete because the tubercle bacillus *Mycobacterium tuberculosis* multiplies very slowly, a binary fission taking place every 24 hours or so. While waiting for a definitive diagnosis, physicians must make treatment decisions based on very limited information. It therefore is possible that ineffective drugs may be prescribed during this interval, and that the patient's illness will worsen; also, the patient may transmit the disease to others as the wait goes on.

With help from the firefly, researchers have developed an innovative and imaginative diagnostic test for tuberculosis that could shorten the time interval for detection considerably. Only a few days may be required, and the test could help determine whether that particular strain of *M. tuberculosis* is drug resistant. The new approach relies on the firefly enzyme luciferase to produce a flash of light in living *M. tuberculosis*. The process works this way: A bacteriophage (a bacterial virus) specific for *M. tuber-*

culosis is genetically engineered to carry the gene for luciferase. A sample of phage is then mixed with a culture of bacteria. If the culture contains *M. tuberculosis*, the phage penetrates the bacterium and inserts itself into the bacterial chromosome, carrying the luciferase gene along. The bacterium promptly begins producing luciferase. Now luciferin, a compound attacked by luciferase, is added to the culture together with the high-energy molecule ATP (Chapter 5). If luciferase is present,

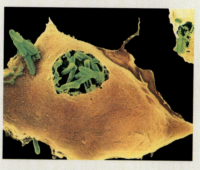

■ *A macrophage (yellow) phagocytizing* Mycobacterium tuberculosis *(green).*

the enzyme breaks down luciferin, and the reaction results in a flash of light. A sensitive instrument detects the light flash, and the culture is confirmed to contain *M. tuberculosis*. The report is made to the physician, and the diagnosis is complete.

To determine drug susceptibility or resistance, the same procedure is used, except a drug is added to the culture. If the bacteria are sensitive to the drug, they die and, quite literally, their "lights go out." If they are resistant, they continue to live, and they produce luciferase—and they give off light.

In the spring of 1993, scientists from New York's Albert Einstein College of Medicine and the University of Pittsburgh reported the test's development in *Science* magazine. As expected, headline writers from numerous publications had a field day as word of the successful test filtered through various journals and newspapers. The well-worn cliché is particularly appropriate in this instance—the future of tuberculosis diagnosis "appears bright."

---

machine, which is the biochemist's equivalent of an office copier. Each cycle takes about five minutes, and each new DNA segment serves as a model for producing many additional copies, which in turn serve as models for producing more copies. Instead of looking for a needle in a haystack, the gene probe now has a huge number of needles.

One place where gene probes and PCR have been useful is in the detection of human immunodeficiency virus (HIV). T lymphocytes are obtained from the patient and disrupted to obtain the cellular DNA. The DNA then is amplified by PCR and the gene probe is added. The probe is a segment of DNA that complements the DNA in the provirus synthesized from the genome of HIV (Chapter 14). If the person is infected with HIV, the probe will locate the proviral DNA, bind to it, and emit radioactivity. An accumulation of radioactivity thus constitutes a positive test. Because the test identifies viral DNA rather than viral antibodies, the physician can be more confident of the patient's health status.

A gene probe test also is available for detecting human papilloma virus (HPV), the virus that causes genital warts (Chapter 12). The test uses a gene probe to detect viral DNA in a sample of tissue obtained from a woman's cervix. Because certain forms of HPV have been linked to cervical tumors, the test has won acceptance as an important preventive technique, and it has been licensed by the FDA. It is commercially available as the ViraPap test.

A similar technique can be used to conduct **water-quality tests** based on the detection of coliform bacteria such as *E. coli* (Chapter 26). Traditionally, *E. coli* had to be cultivated in the laboratory and identified biochemically. With gene probe technology, a sample of water can be filtered, and the bacteria trapped on the filter can be broken open to release their DNA for PCR and gene probe analysis. Not only is the process time saving (many days by the older method, but a few short hours by the newer method), it also is extremely sensitive: A single *E. coli* cell can be detected in a 100-ml sample of water. Moreover, the pathogens transmitted by water, rather than the "indicator" *E. coli*, can be detected by DNA analysis. Thus, the identification of *Salmonella*, *Shigella*, and *Vibrio* species will become more feasible in the future as gene probe analysis becomes more widely accepted.

Gene probe assays are widely available in kit forms for a variety of bacteria (e.g., streptococci, *Haemophilus*, *Listeria*, *Mycobacterium*, and *Neisseria*), as well as for fungi (e.g., *Blastomyces*, *Coccidioides*, and *Histoplasma*). In many cases, the tests are described as "exquisitely accurate," with a high degree of discrimination and reliability as strong as older identification methods. Since first introduced in the 1970s, gene probe tests have been met with periods of unbridled enthusiasm counterbalanced by periods of disappointment. The future value of gene probes will depend in part on the development of ways to minimize false-positive reactions due to contamination, on methods of increasing the sensitivity of tests, and on mechanisms for enhancing the signals from probes bound to their target molecules.

Note to the Student

Ever consider that we are born too soon? Is 9 months in the womb enough? Or would 18 months be preferable?

Absurd, you say? Why, then, is a baby born immunologically "unfinished"—that is, why is its immune system not fully functional until it is roughly 6 months old? And why is it so dependent on its parents that it probably could not lead an independent existence until it is 9 months old? Still not convinced? Then consider a newborn colt or a newly hatched chick. Each is able to walk about and gather food within hours of its birth. Certainly the colt and chick will survive better than a newborn human.

If we are willing to buy into the concept that we are born too soon, the next question is: Why does this happen? The answer, according to the late Stephen Jay Gould and other evolutionary biologists, is the size of our brain. In proportion to the remainder of our body, our brain is larger than any other animal's brain. To have this large brain, we must have a large head. After 9 months, our head can fit through the birth canal, but it could not fit if we stayed inside much longer. So it becomes a matter of give and take. Evolution has given us a large brain (and head), but it also has decreed that we must complete our development outside the comfortable confines of our mother. That development includes immunological development as well as physical development. It also increases our dependence on our parents, and perhaps that's not all so bad—it certainly helps us appreciate the ones who care for us. Thanks, guys.

Summary of Key Concepts

20.1 IMMUNITY TO DISEASE

■ **Naturally Acquired Active Immunity Develops from Exposure to an Infectious Agent.** In this form of immunity, antibodies are produced and lymphocytes activated in response to an infectious agent. Immunity arises from contracting the disease and recovering, or from a subclinical infection.

■ **Artificially Acquired Active Immunity Is Established by Vaccination.** When antibodies and lymphocytes are produced as a result of a vaccination, immunity is established for some length of time. Some first-generation vaccines are being replaced with second- or third-generation vaccines. Often booster shots are required to maintain immunological memory.

■ **Naturally Acquired Passive Immunity Comes from Acquiring Maternal Antibodies.** The passage of antibodies from the mother to fetus, or mother to newborn through colostrum confers immunity for a short period of time.

■ **Artificially Acquired Passive Immunity Is Produced from the Injection of Antisera.** Receiving an antiserum (antibodies) produced in another human or animal confers immunity for a short period of time. Serum sickness can develop if the recipient produces antibodies against the antiserum.

■ **Herd Immunity Results from Effective Vaccination Programs.** Herd immunity results from a vaccination program that lowers the number of susceptible members within a population that can contract a disease. If there are few susceptible individuals, the probability of disease spread is minimal.

■ **Do Vaccines Have Dangerous Side Effects?** Established vaccines are quite safe, although some people do experience side effects that may include mild fever, soreness at the injection site, or malaise after the vaccination. Few people suffer serious consequences.

20.2 SEROLOGICAL REACTIONS

■ **Serological Reactions Have Certain Characteristics.** Serological reactions consist of antigens and antibodies (serum). Successful reactions require the correct dilution (titer) of antigens and antibodies.

■ **Neutralization Involves Antigen-Antibody Reactions.** Neutralization is a serological reaction in which antigens and antibodies neutralize each other. Often there is no visible reaction, so injection into an animal is required to see if the reaction has occurred.

■ **Precipitation Requires the Formation of a Lattice between Soluble Antigen and Antibody.** In a precipitation reaction, antigens and antibodies react to form a matrix that is visible to the naked eye. Different microbial antigens can be detected using this technique.

■ **Agglutination Involves the Clumping of Antigens.** The cross-linking of antigens and antibodies causes the complex to clump. Some infectious agents can be detected readily by this method.

■ **Flocculation Requires Aggregate Formation within a Precipitate.** A flocculation test typically is used to detect syphilis. The presence of syphilis antibodies produces aggregates that can be detected visually.

20.3 OTHER SEROLOGICAL REACTIONS

■ **Complement Fixation Can Detect Antibodies to a Variety of Pathogens.** This serological method involves antigen-antibody complexes that are detected by the fixation (binding) of complement.

■ **Fluorescent Antibody Techniques Can Indentify Unknown Antigens (Pathogens).** The fluorescent antibody technique involves the addition of fluorescently tagged antibodies to a known pathogen to a slide containing unknown antigen (pathogen). If the antibody binds to the antigen, the pathogen will glow (fluoresce) when observed with a fluorescence microscope, and the pathogen is identified.

■ **Radioimmunoassay (RIA) Is Used to Quantitate Antigen-Antibody Complexes.** This serological procedure uses radioactivity to quantitate radioactive antigens by competition with nonradioactive antigens for reactive sites on antibody molecules.

■ **The Enzyme-Linked Immunosorbent Assay (ELISA) Can Detect Antigens or Antibodies.** ELISA can be used to detect if a patient's serum contains antibodies to a specific pathogen or to detect or measure antigens in serum. A positive ELISA is seen as a colored reaction product.

■ **Monoclonal Antibodies Are Produced by a Single Clone of B Cells.** The fusion of a specific B cell with a myeloma cell produces a hybridoma that secretes antibody that recognizes but a single epitope.

■ **Gene Probes Are Single-Stranded DNA Segments.** Gene probes are single-stranded DNA segments that recognize and bind to complementary sequences. Such binding is detected by radioactive means. Such probes can locate or identify specific disease-causing organisms.

Questions for Thought and Discussion

Answers to selected questions can be found in Appendix C.

1. It is estimated that when at least 90 percent of the individuals in a given population have been immunized against a disease, the chances of an epidemic occurring are very slight. The population is said to exhibit "herd immunity," because members of the population (or herd) unknowingly transfer the immunizing agent to other members and eventually immunize the entire population. What are some ways by which the immunizing agent can be transferred?

2. The tendency of women in the present generation is to have children at an older age than in past generations. How might this present an immunological problem for the newborn?

3. A man is found murdered on the front seat of his automobile, and the police observe bloodstains on the floor. It is important to know whether this is the victim's blood, the murderer's blood, or the blood of the victim's dog, which was always with him. However, it also could be fish blood, since the man was an avid fisherman, or blood from the poorly wrapped chicken the man was bringing home from the supermarket. How might the medical examiner proceed?

4. In 1991, scientists first reported success in vaccinating women late in pregnancy to protect their newborns from *Haemophilus* meningitis. The researchers found that levels of meningitis antibodies in the newborns were far above those normally present. Would you favor this approach to protecting newborns? Why or why not?

5. When a child is born in Great Britain, they are assigned a doctor. Two weeks later, a social services worker visits the home, enrolls the child on a national computer registry for immunization, and explains immunization to the parents. When a child is due for an immunization, a notice is automatically sent to the home, and if the child is not brought to the doctor, the nurse goes to the home to learn why. Do you believe a method similar to this can work in the United States to achieve uniform national immunization?

6. For passive immunity, serum containing IgG is routinely used. Why do you suppose IgM is not used, especially since the immunoglobulins are the important components of the primary antibody response? Do you believe that research in this direction would be fruitful?

7. Given a choice, which of the four general types of immunity would it be safest to obtain? Why? Ultimately, which would be the most helpful?

8. One of the hot research items of 1993 was that naked DNA molecules conceivably could be used to immunize an individual. The theory was that DNA from a virus could be made to penetrate to cells such as muscle cells, which would then display that virus's proteins on their surface. What do you suppose would happen next?

9. A complement fixation test is performed with serum from a patient with an active case of syphilis. In the process, however, the technician neglects to add the syphilis antigen to the tube. Would lysis of the sheep red blood cells occur at the test's conclusion? Why?

10. From 1980 to 1989, the incidence of pertussis increased in the United States, and the greatest incidence was found to be in adolescents and adults. Can you think of any reason why adolescents and adults should have been the targets of the bacillus, especially since these individuals were usually considered immune to the disease? Would you be in favor of using the new acellular pertussis vaccine to reimmunize these populations?

11. Suppose the titer of mumps antibodies from your blood was higher than that for your fellow student. What are some of the possible reasons that could have contributed to this? Try to be imaginative on this one.

12. Children between the ages of 5 and 15 are said to pass through the "golden age of resistance" because their resistance to disease is much higher than that of infants and adults. What factors may contribute to this resistance?

13. In 1985, a vaccine for meningococcal meningitis was licensed for use. The vaccine consists of capsular polysaccharides from four different strains of *Neisseria meningitidis*. What form of vaccine does this vaccine represent, and why is it safer to use than older vaccines made from whole meningococci?

14. The ability to keep the human body alive artificially, though brain dead, has stimulated the idea of keeping the organs functioning to produce vaccines for disease treatment. What arguments can be presented for and against this proposition?

15. In 1985, the state of New York dropped its requirement for a VDRL test prior to obtaining a marriage license. Why might you support this action? Can you think of any reasons to oppose it?

Review

On completing the section 20.1 Immunity to Disease, test your comprehension of the section's contents by filling in the following blanks with two terms that answer the description best. Appendix D contains the answers.

1. Two general forms of immunity:
   _____ and _____ .

2. Two types of natural immunity:
   _____ and _____ .

3. Two diseases that MMR is used against:
   _____ and _____ .

4. Two diseases that DPT is used against:
   _____ and _____ .

5. Two types of passive immunity:
   _____ and _____ .

6. Two different types of adjuvants used in vaccines:
   _____ and _____ .

7. Two names for antibody-containing serum:
   _____ and _____ .

8. Two antibody types formed on antigen stimulation:
   _____ and _____ .

9. Two ways newborns have acquired maternal antibodies:
   _____ and _____ .

10. Two characteristics of serum sickness:
    _____ and _____ .

11. Two materials used in second-generation vaccines:
    _____ and _____ .

12. Two diseases for which synthetic vaccines are available or being developed:
    _____ and _____ .

13. Two factors that can determine innate immunity:
    _____ and _____ .

14. Two types of viruses in viral vaccines:
    _____ and _____ .

15. Two bacterial diseases for which toxoids are used:
    _____ and _____ .

16. Two methods for administering vaccines:
    _____ and _____ .

17. Two tracts in which IgA accumulates:
    _____ and _____ .

18. Two viral diseases where passive immunity is used:
    _____ and _____ .

19. Two functions of antibodies in antiserum:
    _____ and _____ .

20. Two bacterial diseases where passive immunity is used:
    _____ and _____ .

Immune Disorders

Allergy differs from most other diseases in that its victims . . .
seldom die and almost as seldom recover.

—Berton Rouché, author of *The Medical Detectives*

DURING THE CLASSICAL GOLDEN AGE of microbiology, investigations were carried out on certain diseases that were not as dangerous as microbial diseases, but nevertheless were a source of great discomfort and inconvenience. One such disease was **hay fever**. In the 1870s, British scientist Charles Harrison Blackley noted that crude pollen placed into the eyes of hay fever sufferers caused swelling of the membranes. Blackley also observed that pollen grains rubbed into a skin scratch produced a local reaction. Some of his critics suggested that the reaction was due to the mechanical injury inflicted by pollen grains, but in 1903, another British investigator, William Philipps Dunbar, supported Blackley's work by showing that saline extracts of pollen grains would cause the same reaction.

Research on hay fever has come a long way since the experiments of Blackley and Dunbar, and, as we shall see in this chapter, the disease is now regarded as a disorder of the immune system. About 35 million Americans currently suffer from hay fever, and thousands more are allergic to foods, cosmetics, leather, or metals. Many people cannot keep pets because of severe sensitivities, and between 2,000 and 4,000 Americans die of asthma annually. All told, an estimated 40 to 50 million people in the United States have some type of allergy.

The common denominator among these problems is a state of increased reactivity known as **hypersensitivity**. First reported in the early 1900s, hypersensitivity stems from activity of the immune system and involves both antibody-mediated and cell-mediated aspects of immunity. It represents a major topic of this chapter and is currently a subject of intense research in immunology.

Hypersensitivity is a multistep phenomenon triggered by exposure to an antigen and consisting of a dormant (latent) stage, during which an individual becomes sensitized, and a reaction following a subsequent exposure to the antigen. The process may involve elements of antibody-mediated immunity or cell-mediated immunity, or sometimes both. The antibody response to the second dose of antigens often occurs within minutes, whereas the cell-mediated response develops over two to three days. For this reason, the terms **immediate hypersensitivity** and **delayed hypersensitivity** traditionally have been used to differentiate two types.

In the early 1970s, P. G. H. Gell and R. A. Coombs proposed another method for classifying hypersensitivities into four types. They classified immediate hypersensitivity into three types: **type I IgE-mediated hypersensitivity**, a process involving IgE, mast cells, basophils, and mediators that induce smooth muscle contraction; **type II cytotoxic hypersensitivity**, which involves IgG, IgM, complement, and the destruction of host cells; and **type III immune complex hypersensitivity**, which involves IgG, IgM, complement, and the formation of antigen-antibody aggregates in the tissues. Gell and Coombs defined delayed hypersensitivity as **type IV cellular hypersensitivity**, in which cytokines and T lymphocytes function. TABLE 21.1 compares these four types.

Realize that for someone who has a hypersensitivity, it is an exaggerated immune defense that causes the problems, not the invading substance. Hypersensitivities and allergies represent immune defenses that are in disorder.

Also included in the broad category of immune disorders are the autoimmune diseases and various immune deficiency diseases as well as the principles of transplantation research and tumor immunology, as FIGURE 21.1 illustrates. We shall survey each of these in the sections that follow.

21.1

Type I IgE-Mediated Hypersensitivity

A type I hypersensitivity can vary from simply a runny nose and watery eyes to a life-threatening reaction. The latter case involves a **systemic anaphylaxis**, a series of events in which chemical substances induce vigorous contractions of the body's smooth muscles. The term is derived from Latin stems that mean "against-protection," a reference to the dangerous nature of the condition.

TYPE I HYPERSENSITIVITY IS INDUCED BY ALLERGENS

Type I hypersensitivity has all the characteristics of an antibody-mediated (humoral) immune response. It begins with the entry of an antigenic substance into the body. This antigen, referred to as an **allergen**, may be any of a wide variety of materials such as plant pollen, certain foods, bee venom, serum proteins, or a drug, such as penicillin. In the case of penicillin, the drug molecule itself is the allergen, but the molecule does not stimulate the immune system until after it has combined with tissue proteins to form an allergenic complex. Doses of antigen as low as 0.001 mg have been known to sensitize a person. Allergists refer to this first dose of antigen as the **sensitizing dose**.

Allergen:
an antigenic substance that induces an allergic reaction.

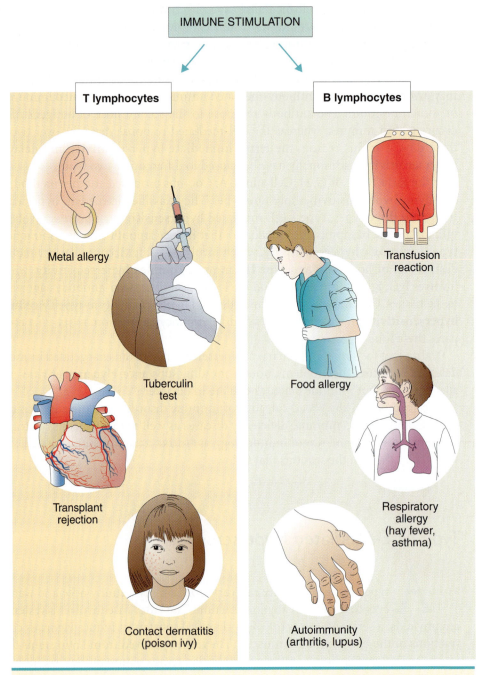

The Various Forms of Immune Disorders

Immune system disorders may be related to T lymphocytes or B lymphocytes. Stimulation of the immune system is the starting point for all these disorders.

The immune system responds to the allergen, and B lymphocytes mature into plasma cells, which produce IgE antibodies (MicroFocus 21.1). This antibody, formerly known as *reagin*, enters the circulation and attaches itself to the surface of mast cells and basophils. **Mast cells** are connective tissue cells numerous in the respiratory and gastrointestinal tracts and near the blood vessels. They measure about 10 μm to 15 μm in diameter and are filled with 500 to 1,500 granules containing histamine and other physiologically active substances. **Basophils** are circulating leukocytes, also rich in granules. They represent about 1 percent of the total leukocyte count in the circulation and measure about 15 μm in diameter. Mast cells and basophils each have over 100,000 receptor sites where IgE antibodies can attach by the Fc tail portion. As IgE accumulates on mast cells and basophils, the individual becomes sensitized, as FIGURE 21.2 illustrates.

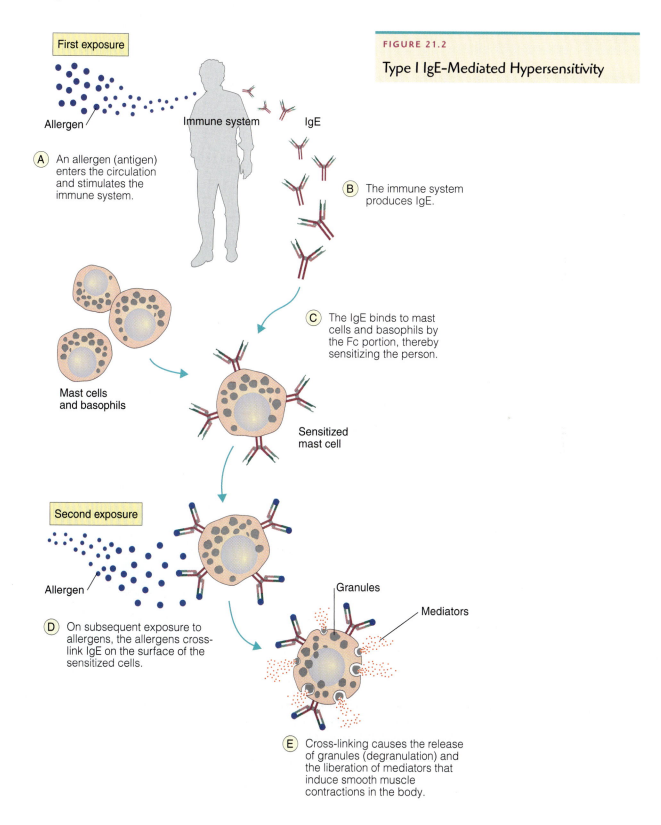

FIGURE 21.2

Type I IgE-Mediated Hypersensitivity

First exposure

Allergen   Immune system   IgE

(A) An allergen (antigen) enters the circulation and stimulates the immune system.

(B) The immune system produces IgE.

(C) The IgE binds to mast cells and basophils by the Fc portion, thereby sensitizing the person.

Mast cells and basophils

Sensitized mast cell

Second exposure

Allergen

(D) On subsequent exposure to allergens, the allergens cross-link IgE on the surface of the sensitized cells.

Granules

Mediators

(E) Cross-linking causes the release of granules (degranulation) and the liberation of mediators that induce smooth muscle contractions in the body.

TABLE 21.1

Overview of the Gell and Coombs Classification of Hypersensitivity Reactions

HYPERSENSITIVITY TYPE	ORIGIN OF HYPERSENSITIVITY	ANTIBODY INVOLVED	CELLS INVOLVED	MEDIATORS INVOLVED
Type I IgE-mediated	B lymphocytes	IgE	Mast cells Basophils	Histamine Serotonin Leukotrienes Prostaglandins
Type II Cytotoxic	B lymphocytes	IgG IgM	RBC WBC Platelets	Complement
Type III Immune complex	B lymphocytes	IgG IgM	Host tissue cells	Complement
Type IV Cellular	T lymphocytes	None	Host tissue cells	Cytokines

MicroFocus 21.1

ITCHES AND ANTIBODIES

In the early 1960s, scientists knew that pollen and other allergens cause mast cells and basophils to release histamine. What they did not know was how allergens induced the cells to spill their contents. The immune system appeared to be involved, but researchers were ignorant of the nature or function of the immune mechanism.

At a hospital in Denver, Colorado, two Japanese doctors, Teruko Ishizaka and her husband, Kimishige, set out to find an antibody that would stimulate the allergic reaction. Neither scientist had any observable allergies, so they decided to use themselves as guinea pigs. When they injected an extract of ragweed pollen under their skin, no reaction took place. But, if they first injected serum from an allergy patient

and followed it with an injection of pollen extract, a raised itchy welt appeared. It was apparent that something in the patient's serum was responsible for the allergy.

The Ishizakas went to the next step. They separated the serum into as many different components as possible and repeated the skin test with each component. When a particular component caused welts, they purified it further and reinjected themselves. After four years of experiments (and lots of itching), they finally isolated their elusive substance. The substance was an antibody. The Ishizakas named it immunoglobulin E (or IgE) because the antibody was directed against antigen E of ragweed pollen. Then, they breathed a sigh of relief that the investigation was over.

TRANSFER OF SENSITIVITY	EVIDENCE OF HYPERSENSITIVITY	SKIN REACTION	EXAMPLES
By serum	30 minutes or less	Urticaria	Systemic anaphylaxis Hay fever Asthma
By serum	5–8 hours	Usually none	Transfusion reactions Hemolytic disease of newborns Thrombocytopenia Agranulocytosis Goodpasture syndrome
By serum	2–8 hours	Usually none	Serum sickness Arthus phenomenon SLE Rheumatic fever LCM Organ rejection
By lymphoid cells	1–3 days	Induration Tissue death	Contact dermatitis Infection allergy Skin graft rejection

Sensitization usually requires a minimum of one week, during which time millions of molecules of IgE attach to thousands of mast cells and basophils. The attachment occurs at the Fc end of the antibody, leaving the Fab ends pointing outward from the cell. Multiple stimuli by allergen molecules may be required to sensitize a person fully. This is why penicillin often must be taken several times before a penicillin allergy manifests itself.

On subsequent exposure to the same allergen, the allergen molecules bind with IgE on the surfaces of sensitized mast cells and basophils, and cross-link IgE antibodies, as shown in FIGURE 21.3. This cross-linking triggers **degranulation**, a release of granule contents at the cell surface. Degranulation requires first an inflow of calcium ions (Ca^{2+}) and a brief stimulation of the enzyme **adenylyl cyclase**, which catalyzes the conversion of ATP to cyclic AMP (cAMP). This leads to the swelling of the granules. Adenylyl cyclase activity then is repressed and there is a sudden decrease in cAMP, which is necessary for degranulation to occur. An understanding of this biochemistry is important because drugs such as epinephrine are given to individuals suffering systemic anaphylaxis (see below). Epinephrine stimulates adenylyl cyclase, increasing the level of cAMP and inhibiting degranulation.

As granules fuse with the plasma membrane, they release a number of mediator substances having substantial pharmacologic activity. The most important preformed mediator of allergic reactions is **histamine**, a derivative of the amino acid histidine. Once in the bloodstream, histamine circulates to the body cells and attaches to histamine receptors called **H-1 receptors**, present on most body cells. The principal effect will be on smooth muscle cells, as we shall see presently. **Heparin, serotonin, bradykinin**, and **tryptase** are other preformed mediators.

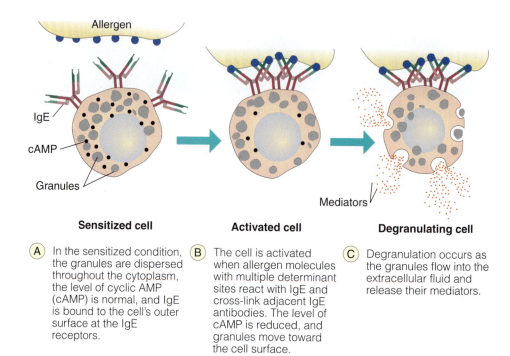

Allergen

IgE

cAMP

Granules

Mediators

Sensitized cell   **Activated cell**   **Degranulating cell**

(A) In the sensitized condition, the granules are dispersed throughout the cytoplasm, the level of cyclic AMP (cAMP) is normal, and IgE is bound to the cell's outer surface at the IgE receptors.

(B) The cell is activated when allergen molecules with multiple determinant sites react with IgE and cross-link adjacent IgE antibodies. The level of cAMP is reduced, and granules move toward the cell surface.

(C) Degranulation occurs as the granules flow into the extracellular fluid and release their mediators.

FIGURE 21.3

The Involvement of Mast Cells and Basophils in Anaphylaxis

Still other mediators must be synthesized after the antigen-IgE reaction. One example is a series of substances called **leukotrienes** (so named because they are derived from *leuko*cytes and have a *triene* (triple) chemical bond). Leukotrienes (once called "slow-reacting substance of anaphylaxis," or SRS-A) result from a complex set of interactions. The inflow of Ca^{2+} activates an enzyme (phospholipase A), which in turn releases from the cell membrane a 20-carbon fatty acid called **arachidonic acid**. The arachidonic acid is acted on by another enzyme and converted to leukotriene A, which is immediately converted to other leukotrienes. The latter, especially leukotriene D4, is extremely potent as a smooth muscle contractor. It also causes leakage in blood vessels and attracts eosinophils to continue the inflammatory reaction.

The second family of synthesized mediators are **prostaglandins**. These substances are well known as human hormones. They also result from enzyme reactions on arachidonic acid. In this case, however, the fatty acid is converted by a different enzyme and various prostaglandins result. One prostaglandin, prostaglandin D2, is a powerful constrictor of the bronchial tubes.

Cytokines also are thought to be involved in the allergic response. Cytokines are produced by most cells and have actions that both stimulate and inhibit inflammation. One cytokine, called **interleukin-4** (IL-4), promotes IgE production by B cells; another, called **interleukin-5** (IL-5), encourages the maturation and activity of eosinophils; and a third, called **tumor necrosis factor alpha** (TNF-α), is released from mast cells and may be responsible for shock in systemic reactions.

SYSTEMIC ANAPHYLAXIS IS A LIFE-THREATENING CONDITION

Systemic anaphylaxis is the most dangerous form of type I hypersensitivity. It involves antigen (allergen) in the bloodstream triggering degranulation of mast cells throughout the body.

The principal activity of the released mediators is to contract smooth muscles in the body. One effect is constriction of the small veins and the expansion of capillary pores, forcing fluid out and into the tissues. The skin becomes swollen around the eyes, wrists, and ankles, a condition called **edema**. The edema is accompanied by a hive-like rash, along with burning and itching in the skin, as the sensory nerves are excited. Contractions also occur in the gastrointestinal tract and bronchial muscles, leading to sharp cramps and shortness of breath, respectively. The individual inhales rapidly without exhaling and traps carbon dioxide in the lungs, an ironic situation in which the lungs are fully inflated but lack oxygen. Death may occur in 10 to 15 minutes as a result of asphyxiation if prompt action is not forthcoming (hence the name "immediate" hypersensitivity). MicroFocus 21.2 describes the emergency treatment that may be given in such a case. Diverse allergies can trigger a systemic anaphylactic reaction, including bee stings, penicillin, antitoxins, and foods such as nuts and seafood.

ATOPIC DISEASES REPRESENT A LOCALIZED ANAPHYLAXIS

Type I hypersensitivity reactions need not result in the whole-body involvement that accompanies anaphylaxis. Indeed, the vast majority of hypersensitivity reactions are accompanied by limited production of IgE and the sensitization of mast cells in localized areas of the body. The result is an **atopic disease**, or **common allergy**. According to public health estimates, up to 50 million Americans (almost 20 percent of the population) have some type of atopic disease.

An example of an atopic disease is **hay fever**, technically referred to as **allergic rhinitis**. This condition develops from springtime inhalations of tree and grass

MicroFocus 21.2

WHEN SYSTEMIC ANAPHYLAXIS STRIKES

Systemic anaphylaxis is a terrifying experience. The skin itches intensely and breaks into hives, the eyes and joints become red and puffy, and the person doubles over with abdominal pains. Breathing becomes difficult, then belabored, and finally is reduced to life-sucking gasps. The symptoms develop within minutes, and the individual usually faints and quickly lapses into a coma.

The key to survival is swift action. Epinephrine (adrenalin) is the highest priority drug. Within minutes of injection, it stabilizes basophils and mast cells to prevent further mediator release. It also dilates the bronchioles to reopen the air passageways, and constricts the capillaries to keep fluid in the circulation.

A smooth muscle relaxant such as aminophylline may be used also. This drug helps dilate the bronchial tubes and pulmonary blood vessels. An antihistamine such as diphenhydramine (Benadryl) may be valuable. This drug competes with histamine for the active sites on smooth muscle receptors, thereby inhibiting the action of histamine. Hydrocortisone may be used to reduce swelling in the tissue, and an expectorant may help clear laryngeal edema.

If the patient does not respond rapidly to the drugs, it may be necessary to insert a tube into the respiratory passageway or perform a tracheostomy. Either must be done quickly, because life now is reduced to a scant few minutes.

pollens, and summer and fall exposures to grass and weed pollens (**FIGURE 21.4**). Fall exposures coincide with the haying season, from which the disease first acquired its name (although there is no fever associated with the condition). Immune stimulation by pollen antigens leads to IgE production, and a sensitization of mast cells follows in the eyes, nose, and upper respiratory tract. Subsequent exposures bring on sneezing, tearing, swollen mucous membranes, and other well-known symptoms. In addition, hay fever symptoms may be caused by house dust, mold spores, dust mites, detergent enzymes, and the particles of animal skin and hair called dander. (Dander itself is not the allergen; the actual allergens are proteins deposited in the dander from the animal's saliva when it grooms itself.) About 35 million Americans suffer from upper respiratory allergic reactions.

Some people experience a **late-phase anaphylaxis**. In this case, it takes several hours for the tissue to become hot, tender, red, and swollen. The mast cells induce this reaction by releasing chemical attractors (called chemotactic factors) that attract other cells to the site to bring about the changes. For example, **eosinophils** exist in unusually high numbers in allergic individuals, and they arrive at the site and release leukotrienes, as well as toxic substances that contribute to tissue damage. **Neutrophils** are normally the phagocytes of the bloodstream, but in allergic reactions, they liberate a number of enzymes that bring about local tissue damage. **Helper T lymphocytes** are involved because they produce interleukin-4 (IL-4), which augments the allergic response. Many of these reactions still are under investigation.

Allergic reactions also are responsible for triggering many cases of **asthma**, which is characterized by wheezing and stressed breathing. Asthma attacks can be caused by airborne allergens such as pollen, mold spores, or products from insects like dust mites. Attacks also can be induced by physical exercise or cold temperatures, without any known allergen being present.

Asthma primarily is an inflammatory disease that occurs in two parts. First, there is an **early response** to allergen exposure. Like hay fever, the degranulation of mast cells releases a variety of chemical mediators. However, the mediators are not

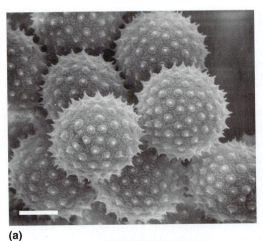 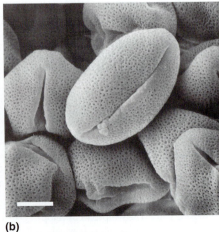

(a)   (b)

FIGURE 21.4

Scanning Electron Micrographs of Two Types of Pollen Grains

(a) Pollen grains of *Sphaeralcea munroana*, a desert plant. (Bar = 10 μm.) (b) Pollen grains from *Penstemon pruinosus*, a flowering plant that grows wild in the mountains of southern California. Antigens in pollen grains such as these stimulate allergic reactions. (Bar = 5 μm.)

released in the nose or eyes, but rather the lower respiratory tract, resulting in bronchoconstriction, vasodilation, and mucus buildup (**FIGURE 21.5**). The synthesis and release of other mediators, including IL-4, IL-5, and TNF-α, result in the recruitment of eosinophils and neutrophils into the lower respiratory tract. These events represent the **late response** because they occur hours after the initial exposure to allergen. The presence of the eosinophils and neutrophils can cause tissue injury and potentially cause blockage of the bronchioles. Bronchodilators can be used to widen the bronchioles. More recently, the use of anti-inflammatory agents such as inhaled steroids or nonsteroidal cromolyn sodium have been prescribed. Cromolyn sodium blocks Ca^{2+} inflow into mast cells and helps prevent degranulation.

Food allergies are accompanied by symptoms in the gastrointestinal (GI) tract, including swollen lips, abdominal cramps, nausea, and diarrhea. The skin may break out in a rash containing **hives**, each hive consisting of a central puffiness, called a wheal, surrounded by a zone of redness known as a flare. Such a rash is called **urticaria**, from the Latin *urtica* for "stinging needle." Allergenic foods include chocolate, strawberries, oranges, cow's milk, and fish (**MicroFocus 21.3**). A dry food such as flour also may cause respiratory allergy.

Eczema (**atopic dermatitis**) is another form of localized anaphylaxis. This is a more prolonged and chronic inflammatory reaction in the skin, most often in children. Serum IgE levels often are elevated and the allergic individual develops a reddened skin rash with periods of intense itching. The localized areas contain increased numbers of eosinophils and helper T cells. The trigger (allergen) for atopic dermatitis is unclear.

WHY DO ONLY SOME PEOPLE HAVE HYPERSENSITIVITIES?

Not everyone suffers from allergies. An interesting avenue of research was opened when it was discovered that the B lymphocytes responsible for IgE and IgA production lie close to one another in the lymphoid tissue, and that the IgA level and its corresponding lymphocytes are greatly reduced in atopic individuals. Immunologists have suggested that in nonallergic individuals, lymphocytes producing IgA shield lymphocytes producing IgE from antigenic stimulation, but that atopic people may lack sufficient IgA-secreting lymphocytes to block the antigens.

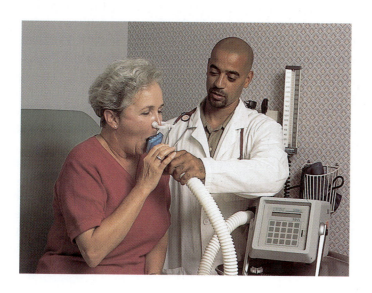

FIGURE 21.5

Testing for Asthma

A pulmonary function test measures the amount of air expelled from the bronchial tubes, as an indication of obstruction. Such a test enables the physician to determine whether the patient is allergic to an airborne antigen.

MicroFocus 21.3

THE PEANUT DILEMMA

Americans love peanuts—salted, unsalted, oil roasted, dry roasted, Spanish, honey crusted, in shells, out of shells, and on and on.

Yet, more than one million Americans fear the peanut because they risk a rather nasty allergic reaction: One can break out in hives, develop a serious headache, experience a racing heartbeat, or double over with intestinal cramps.

And, if that's not bad enough, someone eating peanuts on a plane can release small peanut particles into the air that can cause a reaction in a sensitive person seated nearby. Moreover, a 1997 report in the *New England Journal of Medicine* indicates that peanut-specific antibodies can be transferred from an organ donor to an organ recipient. (In the recipient, the skin reaction is not threatening, but it does complicate matters.) However, a research report in 2003 suggests peanut allergies can be avoided if babies are not given peanuts until their immune systems are mature at age three.

The answer to all those miseries can best be summed up as "V plus V." The first V is for vigilance. Vigilance means avoiding peanuts or peanut butter; but it also means being cautious about egg roll wrappers, chili fillers, and protein extenders in cake mixes, all of which may contain peanuts in one form or another. Drinking liquid charcoal (available in pharmacies) will absorb peanut particles and prevent triggering of increased symptoms. A 2003 report in the *Journal of Allergy and Clinical Immunology* suggests parents keep liquid charcoal at home in case of a peanut allergy in young children. It means vigilance that manufacturers clearly label their products and insistence that peanut-detection tests be performed routinely.

The second V is for vaccine. In 1999, investigators from Johns Hopkins University tested a peanut vaccine and showed that it protects sensitized mice against peanut proteins. The vaccine consists of DNA segments that encode the peanut proteins. Encased in protective molecules and delivered orally, the vaccine decreased the mice's capacity for producing peanut-related IgE. The developers postulated that the vaccine elicits the so-called "blocking antibodies" that bind the peanut antigens before they reach the animal's immune system.

So, does that mean we can expect health officials to distribute vaccine injections where we buy peanut butter, peanut brittle, or beer nuts? Not likely, say the researchers, at least not in the immediate future.

Another theory of atopic disease maintains that allergy results from a breakdown of feedback mechanisms in the immune system. Research findings indicate that B lymphocytes, which synthesize IgE, may be controlled by suppressor T lymphocytes, and regulated in turn by IgE. Under normal conditions, IgE may limit its own production by stimulating suppressor T-cell activity. However, in atopic individuals the mechanism malfunctions, possibly because the T cells are defective, and IgE is produced in massive quantities. Allergic people are known to possess almost 100 times the IgE level of people who do not have allergies. Radioimmunoassay (RAI) techniques and radioallergosorbent tests (RAST) are used to detect the nature and quantity of IgE to specific substances known to cause allergies (Chapter 20).

Some researchers postulate a positive role for the allergic response. They suggest, for example, that sneezing expels respiratory pathogens, and that contractions of the gastrointestinal tract force parasites out of the body. Others have theorized that allergy was once a survival mechanism, and that atopic individuals are the modern generation of people who developed this ability to resist pathogens and passed the trait along. MicroFocus 21.4 describes how hygiene might affect allergic responses.

THERAPIES SOMETIMES CAN CONTROL TYPE I HYPERSENSITIVITIES

First, the best way to avoid hypersensitivities is to identify and avoid contact with the allergen. In other cases, immunotherapy may help control allergies.

A person sensitized to an allergen may undergo **desensitization** therapy to reduce the possibility of anaphylaxis. For example, a person suffering from allergic rhinitis could undergo a procedure that involves injections of tiny but increasing amounts of allergen over a period of hours or weeks, to effect a gradual reduction of granules in sensitized mast cells and basophils. Such treatment prevents a massive degranulation later. One who is sensitive to the immune serum used in disease therapy may need to undergo desensitization before the serum is used in large therapeutic doses.

MicroFocus 21.4

ALLERGIES, DIRT, AND THE HYGIENE HYPOTHESIS

For reasons that are not completely understood, the incidence and severity of asthma—and allergies in general—are increasing in developed nations. In fact, between 1980 and 1994, the prevalence of asthma rose 71 percent in the United States. Today, about 15 million Americans suffer from asthma, including 5 million children and adolescents. Many scientists have suggested the increase is in large part due to our overly clean lifestyle. We use disinfectants for almost everything in the home and antibacterial products have flooded the commercial markets (Chapters 23 and 24). In other words, maintaining overly good hygiene is making us sick. We need to eat dirt! Well, not literally.

The **hygiene hypothesis**, first proposed in 1989, suggests that a lack of early childhood exposure to dirt, bacteria, and other infectious agents can lead to immune system weakness and an increased risk of developing asthma and allergies. In the early 1900s, infants and their immune systems had to battle all sorts of infectious diseases—from typhoid fever and polio to diphtheria and tuberculosis—as well as ones that were more mundane. Such interactions and recovery "pumped up" the immune system and prepared it to act in a controlled manner. Today, most children in developed nations are exposed to far fewer pathogens and their immune systems remain "wimpy," often unable to respond properly to nonpathogenic substances like pollen and cat dander. Their immune systems have not had the proper "basic training."

The hygiene hypothesis has been the subject of debate since 1989. But, new research studies are providing evidence that may make the hygiene hypothesis a theory. In 2003, researchers at the National Jewish Medical and Research Center in Denver reported they had discovered that mice infected with the bacterium *Mycoplasma pneumoniae* had less severe immunological responses when challenged with an allergen. However, if mice were exposed to allergens first, they developed more severe allergic responses. Also, the allergy-producing mediators were at lower levels in mice first exposed to the bacterium. So, early "basic training" of the immune system seems to temper allergic responses.

Two other studies also bolster the hygiene hypothesis. One study used data from the Third National Health and Nutrition Examination Survey conducted from 1988 through 1994 by the Centers for Disease Control and Prevention. The survey included 33,994 American residents ages 1 year to older than 90. The analysis concluded that humans who were seropositive for hepatitis A virus, *Toxoplasma gondii*, and herpes simplex virus type 1—that is, markers for previous microbial exposures—were at a decreased risk of developing hay fever, asthma, and other atopic diseases.

A second study used data collected from 812 European children ages 6 to 13 who either lived on farms or did not live on farms. Using another marker—endotoxin found in dust samples from bedding—the investigators reported that children who did not live on farms were more than twice as likely to have asthma or allergies than children growing up on farms. Presumably, on farms children are exposed to more "immune-strengthening" microbes.

So, all in all, microbial challenges to the immune system as it develops in young children can drive the system to a balanced response to allergens. If the hypothesis proves correct, eating dirt or moving to a farm is not a practical solution, nor is returning to pre-hygiene days. However, a number of environmental factors can help lower incidence of allergic disease early in life. These include the presence of a dog or other pet in the home before birth, attending day care during the first year of life, and simply allowing children to do what comes naturally—play together and get dirty.

TABLE 21.2

Data from Six Cases of Human Anaphylaxis

CASE NO.	SEX	AGE (YR)	AGENT	DOSE (ML)	ROUTE OF ADMINISTRATION
1	F	39	Penicillin	1.5	Intramuscular
2	F	21	Guinea pig hemoglobin	0.2	Subcutaneous
3	M	52	Bee venom	—	Subcutaneous
4	M	45	Penicillin	—	Intramuscular
5	M	56	Hay fever desensitization vaccine	0.0625	Subcutaneous
6	F	38	Penicillin and streptomycin	—	Intramuscular

Another approach to desensitization is to give a series of injections of allergens over a period of weeks. Allergists believe that these exposures cause the immune system to produce IgG antibodies, which circulate and neutralize allergens before they contact sensitized cells. These **blocking antibodies**, as they are called, appear to be an effective device for individuals sensitized to bee stings. They also may be used for people who have food allergies. A promising alternative is to inject Fc fragments of IgE to fill the receptor sites on mast cells and basophils, thereby making the sites unavailable to the person's IgE. TABLE 21.2 summarizes six cases of mainly systemic anaphylaxis in which desensitization procedures were not performed and death resulted.

A novel approach to desensitization is to develop monoclonal antibodies (Chapter 20) that recognize and react with IgE. These **anti-IgE antibodies** can be used to dislodge IgE from the surfaces of mast cells and basophils, thereby disarming the cells and preventing the allergic reaction from occurring. To date, research on these monoclonal antibodies has shown that they can be administered safely and effectively reduce the circulating levels of IgE. The treatment can be used to reduce both the immediate and the late-phase anaphylaxis, especially in asthma patients, as we shall see in the next section.

To this point . . .

We have spent considerable time discussing the process of type I anaphylactic hypersensitivity because it has substantial importance to the millions of people who have allergies. We noted how IgE plays a central role in the process, how mast cells, basophils, and numerous mediators are involved, and how smooth muscle contractions account for the symptoms in anaphylaxis and common allergy. Two forms of desensitization

ESTIMATED TIME FROM CHALLENGE TO DEATH (MIN)	SYMPTOMS AND SIGNS	KNOWN PRIOR EXPOSURE	"ALLERGIC HISTORY"
60	Generalized warmth, tightness of throat, respiratory distress, cyanosis, convulsion, respiratory failure	Yes	"Hives" 2 weeks before death; allergen unknown
16	Headache, wheezing; cyanosis	No	"Asthma"; skin sensitivity to dog hair, kapok, shellfish, ragweed, timothy, orris root, and house dust
20	Unknown	Yes	Severe local reaction to bee sting 20 years previously
60	Dyspnea	Unknown	Unknown
45	Difficulty breathing	Yes	"Hay fever"—injection was 11th in series of weekly desensitization injections
120	Chest pain, cough, collapse, hypotension, cardiac arrest	Unknown	Unknown

were outlined, and the discussion described two theories on why allergies develop. One theory has to do with the blockage of allergens by IgA lymphocytes; the second involves a defect in suppressor T lymphocytes. We also explored some positive roles for the allergic reaction.

We shall continue the discussion of hypersensitivity by studying the salient features of the remaining three types of hypersensitivity. Type II involves a destruction of cells, type III leads to the formation of granular masses called immune complexes, and type IV depends upon the exaggeration of T lymphocyte function. As the chapter progresses, you may note how phenomena of the immune system are helping to enlighten scientists on several disease conditions that were poorly understood in past decades.

21.2

Other Types of Hypersensitivity

Immunological responses involving IgG antibodies or T cells also can lead to adverse hypersensitivity reactions.

TYPE II CYTOTOXIC HYPERSENSITIVITY INVOLVES ANTIBODY-MEDIATED CELL DESTRUCTION

A **cytotoxic hypersensitivity** is a cell-damaging reaction that occurs when IgG reacts with antigens on the surfaces of cells, as FIGURE 21.6 depicts. Complement often is activated and IgM may be involved, but IgE does not participate, nor is there

Type II Cytotoxic Hypersensitivity

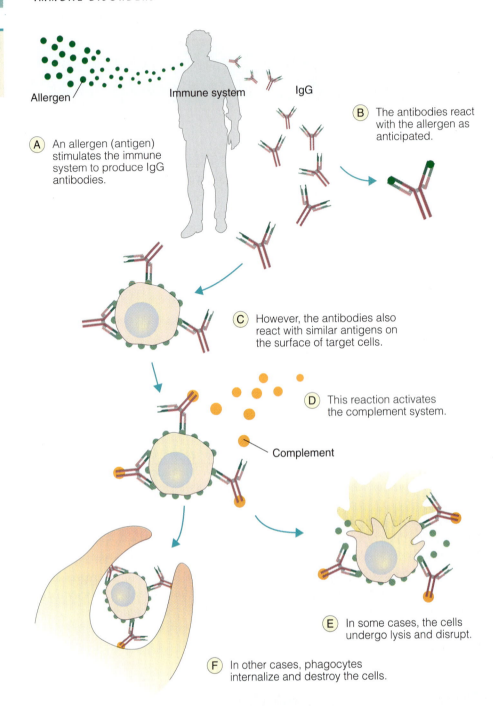

Allergen

Immune system

IgG

(A) An allergen (antigen) stimulates the immune system to produce IgG antibodies.

(B) The antibodies react with the allergen as anticipated.

(C) However, the antibodies also react with similar antigens on the surface of target cells.

(D) This reaction activates the complement system.

Complement

(E) In some cases, the cells undergo lysis and disrupt.

(F) In other cases, phagocytes internalize and destroy the cells.

any degranulation of mast cells. The cells affected in cytotoxic hypersensitivity are known as **target cells.**

A well-known example of cytotoxic hypersensitivity is the **transfusion reaction** arising from the mixing of incompatible blood types. Four major human blood types are known: A, B, AB, and O. Each type is distinguished by unique antigens on the surface of erythrocytes and certain antibodies in the plasma that are directed against antigens not present in the individual's cells (TABLE 21.3). Before a transfusion is attempted, the laboratory technician must determine the **blood type** of all participants so that incompatible types are not mixed. For example, if a person with

TABLE 21.3

Some Characteristics of the Major Blood Groups

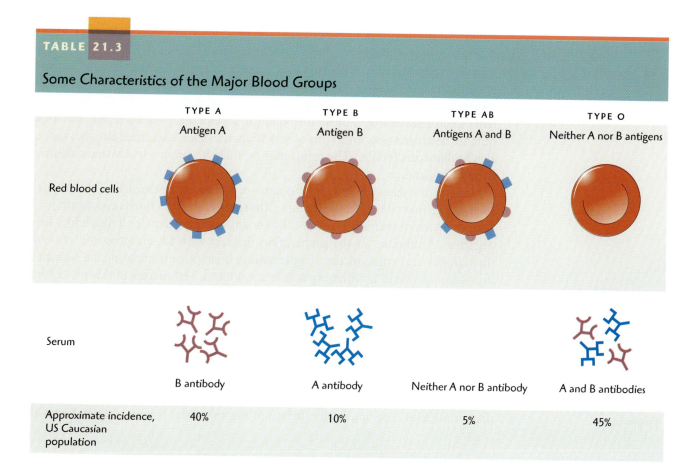

	TYPE A	TYPE B	TYPE AB	TYPE O
	Antigen A	Antigen B	Antigens A and B	Neither A nor B antigens
Red blood cells				
Serum	B antibody	A antibody	Neither A nor B antibody	A and B antibodies
Approximate incidence, US Caucasian population	40%	10%	5%	45%

type A blood donates to a recipient with type O blood, the A antigens on the donor's erythrocytes will react with anti-A antibodies in the recipient's plasma, and the cytotoxic effect will be expressed as agglutination of donor erythrocytes and activation of complement in the recipient's circulatory system. If the conditions are reversed, the donor's A antibodies will react with the recipient's A antigens, although to a lesser degree, because dilution in the recipient's plasma takes place. Most blood banks cross-match the donor's erythrocytes with the recipient's serum, as well as the reverse, to ensure compatibility (MicroFocus 21.5).

Another expression of cytotoxic hypersensitivity is **hemolytic disease of the newborn**, or **Rh disease**. This problem arises from the fact that erythrocytes of approximately 85 percent of Caucasian Americans contain a surface antigen, first described in rhesus monkeys and therefore known as the **Rh antigen**. Such individuals are said to be Rh-positive. The 15 percent who lack the antigen are considered Rh-negative. For African Americans, the figures are 90 percent and 10 percent, respectively. Evidence indicates that the antigen is really a group of antigens that vary among Rh-positive individuals, but we shall consider the group as a single factor for the purposes of discussion.

The ability to produce the Rh antigen is a genetically inherited trait. When an **Rh-negative woman** marries an **Rh-positive man**, there is a 3 to 1 chance (or 75 percent probability) that the trait will be passed to the child, resulting in an Rh-positive child. During the birth process, a woman's circulatory system is exposed to her

Rh antigen:
a group of antigenic substances on the red blood cells of Rh-positive individuals.

child's blood, and if the child is Rh-positive, the Rh antigens enter the woman's blood and stimulate her immune system to produce Rh antibodies (FIGURE 21.7). If a succeeding pregnancy results in another Rh-positive child, IgG antibodies (from memory cell activation) will cross the placenta (along with other antibodies) and enter the fetal circulation. There they will react with Rh antigens on the fetal erythrocytes and cause complement-mediated lysis of the cells. (The fetal circulatory system rapidly releases immature erythroblasts to replace the lysed blood cells, but these cells are destroyed also. From this observation, the disease acquired its older name, **erythroblastosis fetalis**.) The result may be stillbirth or, in a less extreme form, a baby with jaundice.

Modern treatment for hemolytic disease of the newborn consists of the mother receiving an injection of Rh antibodies (**RhoGAM**). The injection is given within 72 hours of delivery of an Rh-positive child (no injection is necessary if the child is Rh-negative). Antibodies in the preparation interact with Rh antigens and remove them from the circulation, thereby preventing them from stimulating the woman's immune system and no memory cells are produced. The success of this procedure has virtually eliminated expectant parents' concerns about disease in their newborns. It should be noted, however, that an Rh-negative woman may produce Rh antibodies as a result of miscarriage or abortion of an Rh-positive fetus, or after a transfusion with Rh-positive blood.

Other examples of cytotoxic hypersensitivity are less familiar. One example, called **thrombocytopenia**, results from antibodies produced against such drugs as aspirin, certain antibiotics, or antihistamines. The antibodies combine with antigens and drug molecules adhering to the surface of thrombocytes (blood platelets), and as

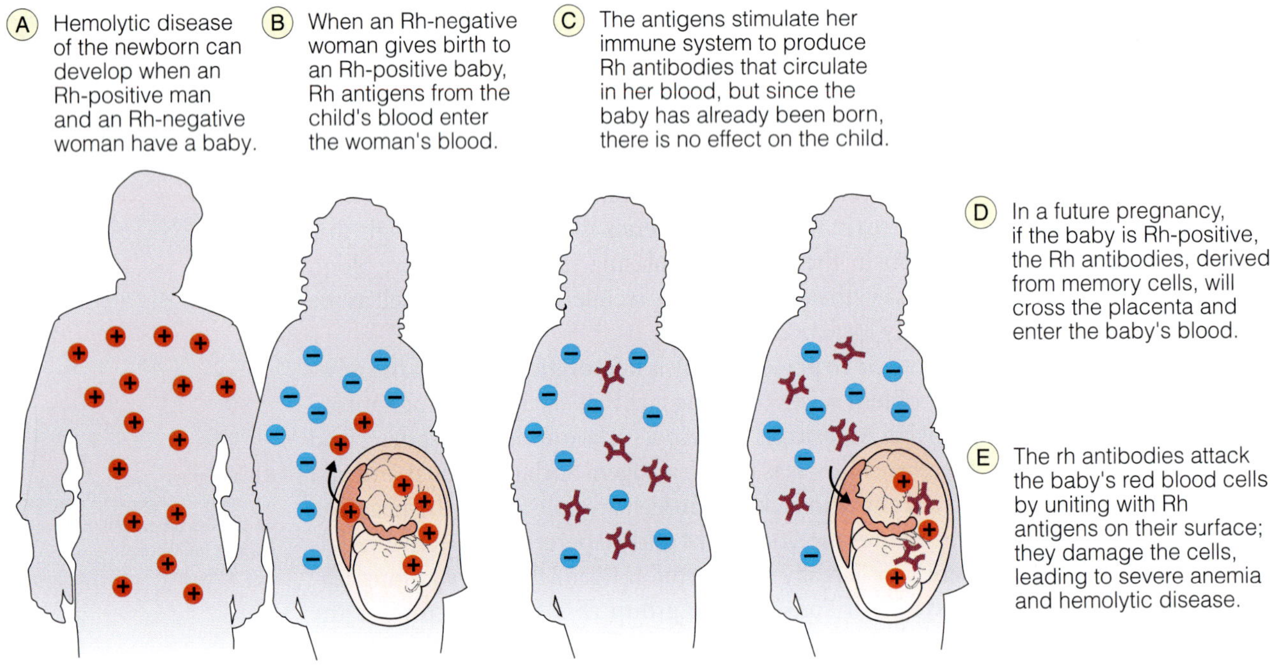

Thrombocytopenia:
a cytotoxic hypersensitivity in which antidrug antibodies attack thrombocytes.

(A) Hemolytic disease of the newborn can develop when an Rh-positive man and an Rh-negative woman have a baby.

(B) When an Rh-negative woman gives birth to an Rh-positive baby, Rh antigens from the child's blood enter the woman's blood.

(C) The antigens stimulate her immune system to produce Rh antibodies that circulate in her blood, but since the baby has already been born, there is no effect on the child.

(D) In a future pregnancy, if the baby is Rh-positive, the Rh antibodies, derived from memory cells, will cross the placenta and enter the baby's blood.

(E) The rh antibodies attack the baby's red blood cells by uniting with Rh antigens on their surface; they damage the cells, leading to severe anemia and hemolytic disease.

FIGURE 21.7

Hemolytic Disease of the Newborn

MicroFocus 21.5

THE KEY TO TRANSFUSIONS

Since earliest times, people believed that blood contained mysterious powers of rejuvenation. The Romans, for example, would rush into the gladiatorial arena to drink the blood of dying gladiators because they thought that blood would restore youth.

Transfusions of blood came into popular use after 1667, when Jean Baptiste Denis, physician to King Louis XIV of France, temporarily restored a dying boy by transfusing lamb's blood into his veins. Transfusions, however, were a mixed blessing: Sometimes they worked, but often they proved fatal. Why this happened perplexed doctors, until Karl Landsteiner provided an answer in 1900.

Landsteiner was an 1891 graduate of the medical school at the University of Vienna. After graduation, he spent five years in Wurtzburg, Germany, working as a chemist with Emil Fischer, the 1902 Nobel laureate for the synthesis of sugars. Landsteiner was more interested in proteins, and he concluded that protein differences could be fruitfully revealed by studying interactions between serum proteins and the components of living cells. In 1897 he became assistant at the Hygienic Institute in Vienna, and in 1900 he applied his method of protein analysis to blood cells.

Landsteiner observed that when erythrocytes from one person were mixed with the serum from another person, the cells sometimes clumped together. In other cases there was no clumping (as if a person's blood was being mixed with its own serum). By meticulous cross-comparisons, Landsteiner concluded that two markers, now called antigens, exist on the surface of the red blood cells. He labeled them with the first two letters of the alphabet, A and B. Landsteiner also surmised that a person's plasma contained antibodies against another person's antigens. These findings formed the basis of the familiar ABO blood groups and explained why some transfusions were successful and others fatal. It now became possible to work out the details for matching up blood groups for safe transfusions.

In World War I, physicians finally recognized the immense value of Landsteiner's work. Over 21 million men were wounded during the war, and for many, a blood transfusion was lifesaving. More than three million Americans now receive safe transfusions annually during surgery, childbirth, or in the treatment of disease. In 1930, Landsteiner was the recipient of the Nobel Prize in Physiology or Medicine. Ten years later, while working with New York physician Alexander S. Weiner, he also discovered the Rh antigen.

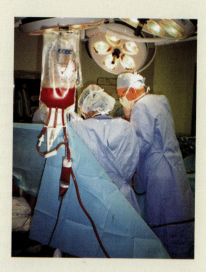

complement is activated, the thrombocytes undergo lysis. The effect is impaired blood clotting, and hemorrhages may occur on the skin and in the mouth. The symptoms subside as the drug is withdrawn. Another condition, referred to as **agranulocytosis**, results from the destruction of neutrophils by antibodies. This problem, also stimulated by drugs, is manifested as a reduced capacity for phagocytosis. In both conditions, antibodies are directed toward the individual's own cells. The term **autoimmune disease** therefore is applied to both phenomena.

A third autoimmune disease is **Goodpasture syndrome**. In this rare disease, antibodies combine with antigens on the membranes of glomeruli in kidneys. Antibody binding activates the complement system and, as the integrity of the membranes is destroyed, blood and proteins leak into the urine. Kidney failure may follow.

Antibodies reacting with antigens on cell surfaces do not always lead to cell destruction, but the reaction may alter the cellular physiology. In **myasthenia gravis**, for example, antibodies react with acetylcholine receptors on membranes covering the muscle fibers. This interaction reduces nerve impulse transfer to the fibers and results in a loss of muscle activity, manifested as weakness and fatigue. In **Graves'**

disease, antibodies unite with receptors on the surfaces of thyroid gland cells, causing an overabundant secretion of **thyroxine**. The patient experiences goiter and a rise in the metabolic rate. A third example of altered cell physiology is **Hashimoto's disease**. This is a condition in which antibodies also attack thyroid gland cells, but the reaction changes their chemistry leading to a thyroxine deficiency. All three diseases are considered autoimmune diseases.

Although cytotoxic hypersensitivity is generally cast in a negative role with deleterious effects on the body, the cytotoxic activity may contribute to the body's resistance to disease. For example, the antigen-antibody interaction occurring on the surface of a parasite leads to destruction of the parasite. The interaction also may encourage chemotaxis or histamine release through the activity of C3a and C5a components of the complement system. Increased phagocytosis and membrane damage from the complement attack complex are other by-products of complement activation. These activities probably account for resistance to many disorders.

TYPE III IMMUNE COMPLEX HYPERSENSITIVITY IS CAUSED BY ANTIGEN-ANTIBODY AGGREGATES

Immune complex hypersensitivity develops when antibodies combine with antigens and form aggregates that accumulate in blood vessels or on tissue surfaces (**FIGURE 21.8**). As complement is activated, the C3a and C5a components increase vascular permeability and exert a chemotactic effect on phagocytic neutrophils, drawing them to the target site. Here the neutrophils release lysosomal enzymes, which cause tissue damage. Local inflammation is common, and fibrin clots may complicate the problem. The antibodies are predominantly IgG, with IgM also found in certain cases.

Serum sickness is a common manifestation of immune complex hypersensitivity (Chapter 20). It develops when the immune system produces IgG against residual proteins in serum preparations. The IgG then reacts with the proteins, and immune complexes gather in the kidney over a period of days. The problem is compounded when IgE, also from the immune system, attaches to mast cells and basophils, thereby inducing a type I anaphylactic hypersensitivity. The sum total of these events is kidney damage, along with hives and swelling in the face, neck, and joints.

Another form of immune complex hypersensitivity is the **Arthus phenomenon**, named for Nicolas Maurice Arthus, the French physiologist who described it in 1903. In this process, excessively large amounts of IgG form complexes with antigens, either in the blood vessels or near the site of antigen entry into the body. Antigens in dust from moldy hay and in dried pigeon feces are known to cause this phenomenon. The names **farmer's lung** and **pigeon fancier's disease** are applied to the conditions, respectively. Thromboses in the blood vessels may lead to oxygen starvation and cell death.

Systemic lupus erythematosus (SLE) is another example of a type III hypersensitivity. SLE, also known as lupus, is an autoimmune disease in which plasma cells produce IgG upon stimulation by nuclear components of disintegrating white blood cells. The terms "autoantigens" and "autoantibodies" apply because the antibodies are formed against the body's own molecules. When the autoantigens and autoantibodies react, immune complexes accumulate in the skin and body organs, and complement is activated. The patient experiences a **butterfly rash**, a facial skin condition across the nose and cheeks (**FIGURE 21.9**). Lesions also form in the heart, kidneys, and blood vessels. In **rheumatoid arthritis**, another autoimmune disease, immune complexes form in the joints. **TABLE 21.4** summarizes immune disorders involving type II and type III hypersensitivities.

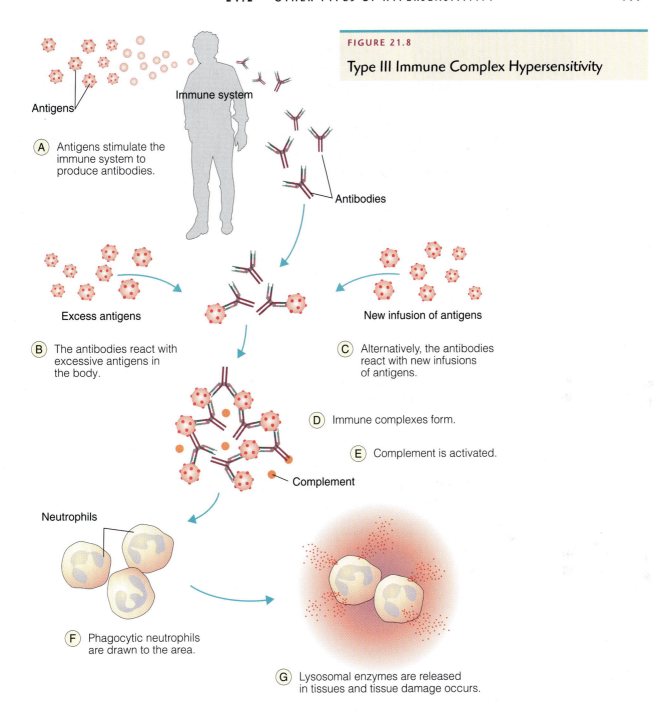

FIGURE 21.8

Type III Immune Complex Hypersensitivity

Antigens

Immune system

(A) Antigens stimulate the immune system to produce antibodies.

Antibodies

Excess antigens

New infusion of antigens

(B) The antibodies react with excessive antigens in the body.

(C) Alternatively, the antibodies react with new infusions of antigens.

(D) Immune complexes form.

(E) Complement is activated.

Complement

Neutrophils

(F) Phagocytic neutrophils are drawn to the area.

(G) Lysosomal enzymes are released in tissues and tissue damage occurs.

Several microbial diseases also are complicated by immune complex formation. For example, the glomerulonephritis and rheumatic fever that follow **streptococcal diseases** (Chapter 8) appear to be consequences of immune complex formation in the kidneys and heart, respectively. In these cases, the deposit of complexes relates to common antigens in streptococci and the tissues. Other immune complex diseases include hemorrhagic shock, which may accompany dengue fever; subacute sclerosing panencephalitis (SSPE), which follows cases of measles; and the slow-forming kidney deposits associated with lymphocytic choriomeningitis (LCM). Research evidence also suggests that Reye's syndrome and Guillain-Barré syndrome may be related to immune complex formation.

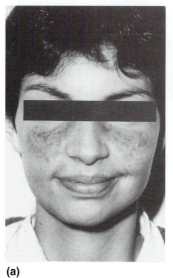
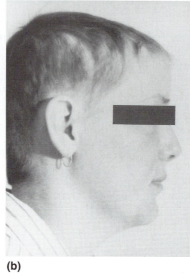
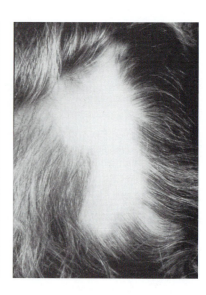

(a)   (b)

FIGURE 21.9

Systemic Lupus Erythematosus

Two patients showing some of the symptoms of systemic lupus erythematosus (SLE). (a) The butterfly rash on the face. Note how the rash extends out to the cheeks like the wings of a butterfly. (b) A young girl displaying the loss of hair that accompanies some cases of SLE.

TABLE 21.4

A Summary of Some Immune Disorders

DISORDER	TYPE OF HYPERSENSITIVITY	TARGET TISSUE	STIMULATING ANTIGEN	EFFECT
Thrombocytopenia	Cytotoxic	Thrombocytes (blood platelets)	Aspirin Antibiotics Antihistamine	Impaired blood clotting Hemorrhages
Agranulocytosis	Cytotoxic	Neutrophils	Drugs	Reduced phagocytosis
Goodpasture syndrome	Cytotoxic	Kidneys	Own antigens (?)	Kidney failure
Myasthenia gravis	Cytotoxic	Membranes of muscle fibers	Not established	Loss of muscle activity
Graves' disease	Cytotoxic	Thyroid gland	Not established	Abundant thyroxine High metabolic rate
Hashimoto's disease	Cytotoxic	Thyroid gland	Not established	Thyroxine deficiency Low metabolic rate
Serum sickness	Immune complex	Kidney	Serum proteins	Kidney failure
Arthus phenomenon	Immune complex	Blood vessels Site of antigen entry	Environmental antigens	Thromboses in blood vessels
Systemic lupus erythematosus	Immune complex	Skin, heart, kidney, blood vessels	Own nucleo-proteins	Butterfly rash Heart, kidney failure
Rheumatoid arthritis	Immune complex	Joints	Own nucleo-proteins (?)	Swollen joints

TYPE IV CELLULAR HYPERSENSITIVITY IS MEDIATED BY ANTIGEN-SPECIFIC T CELLS

Cellular hypersensitivity is an exaggeration of the process of cell-mediated immunity, discussed in Chapter 19. The adjective *cellular* was originally applied because contact between T lymphocytes and antigens was thought to be necessary for the reaction. However, the later identification of cytokines as mediators of the process challenged this assumption. The hypersensitivity cannot be transferred to a normal individual by serum cytokines because their concentration is too low in transfused serum. Transfer is accomplished only with T cells. **TABLE 21.5** compares this **delayed hypersensitivity** with type I immediate hypersensitivity.

Type IV hypersensitivity is a delayed reaction whose maximal effect is not seen until 24 to 72 hours have elapsed (hence the name, delayed hypersensitivity). It is characterized by a thickening and drying of the skin tissue, a process called **induration**, and a surrounding zone of erythema (redness). Two major forms of type IV hypersensitivity are recognized: infection allergy and contact dermatitis.

Infection allergy develops when the immune system responds to certain microbial agents. Effector cells known as **delayed hypersensitivity T lymphocytes** migrate to the antigen site and release cytokines. The cytokines attract phagocytes and encourage phagocytosis (Chapter 19). Sensitized lymphocytes then remain in the tissue and provide immunity to successive episodes of infection. Among the microbial agents that stimulate this type of immunity are the bacterial agents of tuberculosis, leprosy, and brucellosis; the fungi involved in blastomycosis, histoplasmosis, and candidiasis; the viruses of smallpox and mumps; and the chlamydiae of lymphogranuloma.

Infection allergy is demonstrated by injecting into the skin an extract of the microbial agent. As the immune response takes place, the area develops induration and erythema, and fibrin is deposited by activation of the clotting system. An important application of infection allergy is the **tuberculin skin test** for tuberculosis (**FIGURE 21.10**). A purified protein derivative (PPD) of *Mycobacterium tuberculosis* is applied to the skin by intradermal injection (the Mantoux test) or multiple punctures (tines). Individuals sensitized by a previous exposure to *Mycobacterium* species develop a vesicle, erythema, and induration.

Skin tests based on infection allergy are available for many diseases. Immunologists caution, however, that a positive result does not constitute a final diagnosis. They note that sensitivity may have developed from a subclinical exposure to the organisms, from clinical disease years before, or from a former screening test in which the test antigens elicited a T-lymphocyte response.

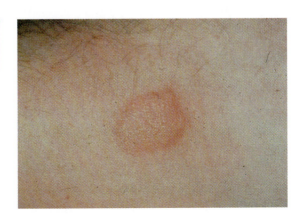

FIGURE 21.10

A Positive Tuberculin Test

The raised induration and zone of inflammation indicate that antigens have reacted with T lymphocytes, probably sensitized by a previous exposure to tubercle bacilli.

TABLE 21.5

Immediate and Delayed Hypersensitivities Compared

	TYPE I IMMEDIATE HYPERSENSITIVITY	TYPE IV DELAYED HYPERSENSITIVITY

Clinical state:	Hay fever
	Asthma
	Urticaria
	Allergic skin conditions
	Serum sickness
	Anaphylactic shock
		Drug allergies
		Infectious allergies
		    Tuberculosis
		    Rheumatic fever
		    Histoplasmosis
		    Trichinosis
		Contact dermatitis

Onset:	Immediate	Delayed
Duration:	Short: hours	Prolonged: days or longer
Allergens:	Pollen Molds House dust Danders Drugs Antibiotics Soluble proteins and carbohydrates Foods	Drugs Antibiotics Microorganisms: bacteria, viruses, fungi, animal parasites Poison ivy and plant oils Plastics and other chemicals Fabrics, furs Cosmetics
Passive transfer of sensitivity:	With serum	With cells or cell fractions of lymphoid series

Contact dermatitis develops after exposure to a broad variety of antigens, such as the allergens in clothing, insecticides, coins, cosmetics, and furs. The offending substances also may include formaldehyde, copper, dyes, bacterial enzymes, and protein fibers. In poison ivy, the allergen has been identified as **urushiol**, a low-molecular-weight chemical on the surface of the leaf. In the body, urushiol complexes with tissue proteins to form allergenic compounds. A poison ivy rash consists of very itchy pinhead-sized blisters usually occurring in a straight row (**FIGURE 21.11**).

The course of contact dermatitis is typical of the type IV reaction. Repeated exposures cause a drying of the skin, with erythema and scaling. Examples are on the scalp when allergenic shampoo is used, on the hands when contact is made with detergent enzymes, and on the wrists when an allergy to watchband chemicals exists. Contact dermatitis also may occur on the face where contact is made with cosmetics, on areas of the skin where chemicals in permanent-press fabrics have accumulated, and on the feet when there is sensitivity to dyes in leather shoes. Factory workers exposed to photographic materials, hair dyes, or sewing materials may experience allergies. The list of possibilities is endless and includes the use of costume jewelry (**MicroFocus 21.6**). Applying a sample of the suspected substance to the skin (a patch test) and leaving it in place for 24 to 48 hours will help pinpoint the source of the allergy. Another method is illustrated in **FIGURE 21.12**. Relief generally consists of avoiding the inciting agents.

Several hypersensitivity reactions also are responses to self-antigens and can cause damage to one's own tissues and organs. Such autoimmune diseases are explored in **MicroInquiry 21**.

FIGURE 21.11

The Poison Ivy Rash

(A) When a sensitized person touches poison ivy, a substance called urushiol stimulates T lymphocytes in the skin; within 24 hours, a type IV reaction takes place.

(B) The reaction is characterized by pinhead-sized blisters that usually occur in a straight row.

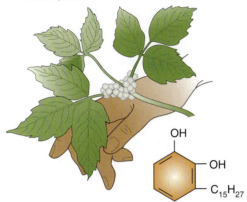

The chemical structure of urushiol

MicroFocus 21.6

PIERCED PROBLEMS

He arrived at the office wearing studs and rings in his ears, eyebrows, nose, navel, and various other body parts. And he was scratching . . . and scratching . . . and scratching.

The doctor looked past the jewelry. She saw nickel. Not the coin, but the metal. Nickel is commonly used to make inexpensive costume jewelry.

Unfortunately, it also provokes contact dermatitis. (Other allergens in costume jewelry include chrome and palladium.)

What to do? The most obvious solution is to remove the jewelry and wait for the rash to fade. (Cortisone cream can be used to reduce the inflammation.) Furthermore, it's a good idea to substitute stainless steel or gold jewelry to pre-

vent a future recurrence. Most sterling silver is safe, but occasionally it may contain nickel or chrome, so a rash may still occur. If doubts persist, a skin test can pinpoint the allergy, and an allergist can offer helpful advice. Then, one's individual expression can continue.

FIGURE 21.12

The Test for Skin Allergies

The forearm is marked and injected with various test allergens. In a few minutes, the allergist notes where a skin wheal has occurred. This individual displays several positive reactions. The reaction can be assessed as slight, mild, moderate, or severe. In other instances, the skin of the back is used.

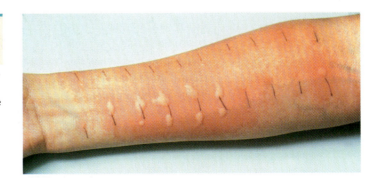

MicroInquiry 21

EXPLORING THE ORIGINS OF AUTOIMMUNITY

One of the key characteristics of the immune system is its ability to recognize self from nonself—that is, to tolerate self antigens while attempting to eliminate nonself antigens. The response to self antigens is called **autoimmunity** and attack on self produces **autoimmune diseases** such as those mentioned in this chapter. Autoimmune diseases are a major threat to the health of all Americans. At least ten million Americans suffer from one or more of 80 illnesses caused by autoimmunity. They are a special threat to women since about 75 percent of the patients are women. The basic question, though, is why and how does the immune system at some point become intolerant to specific self antigens?

First, let's review the normal consequence of an immune response to a for-

eign antigen. Answers can be found in Appendix E.

21a. What cells are involved in the clearance of virus-infected cells from the body?

21b. How do these cells function to eliminate the infected cells?

If the immune system attacks itself, "seeing" some self antigen as "foreign," cytotoxic T cells and NK cells will be involved. Also, activation of B cells can lead to the production of antibodies to the self antigen (autoantibodies). However, it is impossible to clear completely the self antigens because they probably are part of a whole body tissue. The result is a sustained response inflicting tissue damage at the "foreign" site, resulting in an autoimmune disease. So, why were these sites attacked?

Normally tolerance to self is a function of what is called **clonal deletion**, which is a process by which immature B and T cells that recognize self antigens are eliminated. In the thymus, immature T cells are exposed to self antigens. Any cells that bind a self antigen are deleted through apoptosis, a genetically programmed cell death. If a maturing T cell does not bind an antigen, it survives.

21c. Suppose there are "hidden" tissue antigens that are not presented to immature T cells in the thymus. What could happen later in life if those tissue antigens now are recognized by the immune system?

It ends up that there are **immunologically privileged** sites in the body

To this point . . .

We have completed a survey of the four types of hypersensitivity, and we have noted the salient differences among them. In type II cytotoxic hypersensitivity, there is an antigen-antibody interaction on the surface of cells that leads to cell destruction. Transfusion reactions and hemolytic disease of the newborn are symbolic of this cytotoxicity. Several autoimmune diseases, such as thrombocytopenia, myasthenia gravis, and Graves' disease, also result from the cytotoxic reaction. In type III hypersensitivity, immune complex formation takes place as antigens react with antibodies to form masses of granular material. With the activation of complement, a local destruction of body tissue ensues. The effects of serum sickness, the Arthus phenomenon, and systemic lupus erythematosus illustrate how tissue may be affected.

Type IV cellular hypersensitivity is substantially different from the other three types because it involves sensitized T lymphocytes. An exaggeration of cell-mediated immunity is basic to type IV hypersensitivity. Infection allergy and contact dermatitis are two examples of this hypersensitivity. Poison ivy is an example of the latter.

In the final section of this chapter, we shall examine immune deficiency diseases, transplantation immunology, and tumor immunology. The immune deficiency diseases rob the body of its natural defenses against infection and demonstrate the vital role the immune system plays. In transplantation and tumor immunology, we shall encounter some of the most recent findings in immunology, and see how scientists are beginning to understand certain functions of the immune system unrelated to infectious disease. Studies in these areas represent major areas of concentration in modern immunological research.

that are "hidden from" the immune system. Such sites include the brain and the anterior chamber of the eye.

21d. A man suffers trauma to the eye from a herpes infection. Although the viral infection is cured, he later develops a gradual onset of pain, redness, light sensitivity, and blurred vision. His physician identifies it as uveitis, an inflammation of the uvea of the eye. Autoimmunity is suspected. How could the autoimmune disease have originated?

Viruses and bacteria also can lead to autoimmune diseases in another way. Some of these microbes have epitopes that essentially are identical to normal body antigens and both would be recognized by the same antibody. Such molecular mimicry can lead to autoimmunity.

21e. A young girl has rheumatic fever, which is caused by *Streptococcus pyogenes* (Chapter 8). She fully recovers. Some forty years later, she dies from rheumatic heart disease—an autoimmune disease. Her mitral valve fails, which allows the heart to pump blood directly into the lungs causing suffocation. How might *Streptococcus pyogenes* and molecular mimicry be responsible for this autoimmune condition?

An inappropriate expression of major histocompatibility complex (MHC) proteins also can trigger an autoimmune response. Normally, class I MHC are found on most body cells, but class II MHC are found only on antigen presenting cells—macrophages, dendritic cells, and B cells.

21f. A woman suffering from Graves' disease is producing autoantibodies to the acinar (secretory) cells of her thyroid. How could MHC and helper T cells be involved in this autoimmune reaction?

Evidence exists for all these mechanisms, so autoimmunity probably is not caused by a single mechanism. Rather, autoimmunity can arise in different ways. Also, some families show a history of autoimmune disease, suggesting that genetics may play a role.

21.3
Immune Deficiency Diseases

The spectrum of immune deficiency diseases ranges from major abnormalities that are life threatening to relatively minor deficiency states. The latter may be serious in populations where malnutrition and frequent contact with pathogenic organisms are common. Diagnostic techniques for determining immune deficiency diseases include measurements of antibody types, detection tests for B lymphocyte function, and enumeration of T lymphocytes. Assays of complement activity and phagocytosis also are useful in diagnosis.

IMMUNE DEFICIENCIES CAN INVOLVE ANY ASPECT OF THE IMMUNE SYSTEM

An example of an antibody-mediated (humoral) deficiency is **Bruton's agammaglobulinemia**, first described by Ogden C. Bruton in 1952 (MicroFocus 21.7). In this disease, B lymphocytes fail to develop from pre-B cells in the bone marrow. The patient's lymphoid tissues lack mature B cells and plasma cells, all five classes of antibodies are either low in level or absent, and antibody responses to infectious disease are undetectable. Infections from staphylococci, pneumococci, and streptococci are common between the ages of six months and two years. Bruton's agammaglobulinemia is also called X-linked agammaglobulinemia because it is a sex-linked inherited

MicroFocus 21.7

SOMETHING FROM NOTHING

In the 1940s and early 1950s, scientists frequently debated whether antibodies were essential to immunity. Amid this controversy, a pediatrician at Washington, D.C.'s Walter Reed Hospital made a momentous discovery.

The story began when a US Air Force officer's son was admitted to the hospital with acute streptococcal disease. The child's pediatrician, Ogden C. Bruton, used penicillin to control the infection, but two weeks later, the child was back, sick again. Wishing to determine the level of antibodies in the boy's system, Bruton sent a sample of blood for testing on a new machine recently acquired by the hospital.

The next day, the laboratory called Bruton to report that something must be wrong with the machine because it could not detect any antibodies in the blood. Bruton responded by sending over a second sample, but the results were the same: no antibodies. It occurred to Bruton that maybe the trouble was not in the machine but in the blood. Perhaps the blood had no

antibodies. Conceivably, this could account for the recurring infections. Bruton began giving the boy monthly injections of antibodies, with outstanding success.

In 1952, Bruton reported his remarkable findings. His report was a medical bombshell because it established the concept of immune deficiency disease, while helping to solidify the role played by antibodies in resistance to infection.

The story was not finished, however, because certain patients with immune deficiency disease still produced the lymphocytes that immunize against skin grafts. This observation led to the notion that the immune system was actually a dual system—one branch centered in antibodies, a second branch centered in lymphocytes. The sources of antibodies and lymphocytes would be found almost simultaneously, 13 years later.

In 1965, at the meeting of the American Academy of Pediatrics, Robert A. Good's research group was presenting evidence on the importance of the bursa of Fabricius to antibody production. The audience was skeptical, but the evidence

appeared substantial. When the speaker finished, a Philadelphia pediatrician named Angelo DiGeorge stepped to the microphone to add a footnote. DiGeorge described how four children at his hospital were struck repeatedly with severe infections. The children had antibodies in their blood, but curiously, each lacked a thymus. DiGeorge's observation was lost in the shuffle, but a question arose in his mind: Were the children susceptible to skin grafts, or could they possibly be immune?

DiGeorge hurried back to the hospital and devised a set of experiments to determine whether the children could successfully receive skin grafts. After weeks of study, he found they could. The thymus was apparently the key to the production of lymphocytes and hence to graft immunity. Bruton's patient lacked the bursa type of immunity, but DiGeorge's patients lacked the thymus type of immunity. The duality of the immune system was thus strengthened, and a second immune deficiency disease, DiGeorge syndrome, entered the dictionary of microbiology.

trait, much more frequently observed in males than in females. Artificially acquired passive immunity is used to treat infectious disease in patients with this form of immune deficiency.

DiGeorge syndrome is an example of a cell-mediated deficiency in which T lymphocytes fail to develop. The deficiency is linked to failure of the thymus gland to mature in the embryo. Cell-mediated immunity is defective in such individuals, and susceptibility is high to many fungal and protozoal diseases and certain viral diseases. Correction of the defect by grafts of thymus tissue has been reported. Partial thymus insufficiencies exist in some cases.

Combined antibody- and cell-mediated deficiencies are the most dangerous. **Severe combined immunodeficiency disease (SCID)** involves defects in both antibody-mediated and cell-mediated immunity. The B lymphocyte and T lymphocyte areas of the lymph nodes are depleted of lymphocytes, and all immune function is suppressed. For many years immunologists believed that the disease resulted from a defect in stem cells of the bone marrow, but recent evidence indicates an enzyme deficiency causes structural and functional abnormalities in antibody- and cell-mediated immunity.

The longest surviving victim of SCID was a boy known only as David (to protect his privacy). David lived for 12 years inside a sterile, plastic bubble at the Baylor College of Medicine in Houston (**FIGURE 21.13**). On October 21, 1983, he received a bone marrow transplant from his sister as doctors attempted to establish an immune system within his body. Three months later, David left the bubble, but on February 22, 1984, he died as a result of blood cancer traced to a virus apparently brought into his body by the transplant.

Other immune deficiency diseases are linked to the **neutrophils**. In these diseases, the phagocytes fail to engulf and kill microorganisms because of defects in chemotaxis, ingestion, and/or intracellular digestion. For example, in the rare disease known as **Chédiak-Higashi syndrome**, a delayed killing of phagocytized microorganisms is traceable to the inability of lysosomes to release their contents. Another disease, called **Job syndrome**, is accompanied by defective neutrophil chemotaxis. Scientists sometimes call this disease the "lazy-leukocyte syndrome."

Deficiencies in the **complement system** may be life threatening. As noted in Chapter 19, complement is a series of proteins any of which the body may fail to produce. For reasons not currently understood, many patients with complement component deficiencies suffer systemic lupus erythematosus (SLE) or an SLE-like syndrome. Meningococcal and pneumococcal diseases are observed often in patients who lack C3, probably as a result of poor opsonization.

Although **acquired immune deficiency disease (AIDS)** is considered a viral disease (Chapter 14), we shall mention it here also. In patients with AIDS, cell-mediated immunity is severely weakened, while antibody-mediated immunity also may be altered. The number of T lymphocytes is sparse, and a striking imbalance is observed between the helper T lymphocytes (CD4) in "healthy" individuals and in AIDS patients. Usually, the number of CD4 cells is three to four times the number in an AIDS patient (<200 per microliter of blood). However, the exact counts vary greatly between individuals.

FIGURE 21.13

David, the Boy in the Plastic Bubble

David suffered from severe combined immunodeficiency disease (SCID) and lived for 12 years within the sterile confines of a plastic bubble. In 1983, he received a bone marrow transplant from his sister and left the bubble. Unfortunately, he developed a blood cancer and died some months thereafter.

Transplantation Immunology

Modern techniques for the transplantation of tissues and organs trace their origins to Jacques Reverdin, who in 1870 successfully grafted bits of skin to wounded tissues. Enthusiasm for the technique rose after his reports were published, but it waned when doctors found that most transplants were rapidly rejected by the body. Then, in 1954, a kidney was transplanted between identical twins, and again, interest grew. The graft survived for several years, until ultimately it was destroyed by a recurrence of the recipient's original kidney disease. During that time, attempts to transplant kidneys between unrelated individuals were less successful.

THERE ARE FOUR TYPES OF GRAFTS (TRANSPLANTS)

Transplantation technology improved considerably during the next few decades, and today, four types of transplantations, or grafts, are recognized, depending upon the genetic relationship between donor and recipient. A graft taken from one part of the body and transplanted to another part of the same body is called an **autograft**. This graft is never rejected because it is the person's own tissue. A tissue taken from an identical twin and grafted to the other twin is an **isograft**. This, too, is not rejected because the genetic constitutions of identical twins are the same.

Rejection mechanisms become more vigorous as the genetic constitutions of donor and recipient become more varied (MicroFocus 21.8). For instance, grafts between brothers and sisters, or between fraternal twins, may lead to only mild rejection because many of their genes are similar. Grafts between cousins may be rejected more rapidly, and as the relationship becomes more distant, the vigor of rejection increases proportionally. **Allografts**, or grafts between genetically different members of the same species, such as two humans, have variable degrees of success. Most transplants are allografts. **Xenografts**, or grafts between members of different species, such as a monkey and a human, are rarely successful. TABLE 21.6 summarizes this terminology.

Transplanted tissue is rejected by the body if the immune system interprets the tissue as nonself. The rejection mechanisms may take either of two forms. In the first mechanism, cytotoxic T lymphocytes aided by helper T lymphocytes attack and destroy transplanted cells. The process is stimulated by the recognition of foreign MHC proteins (to be discussed next) on the surface of the graft cells. Class II MHC proteins stimulate helper T lymphocytes, while class I MHC proteins are the sites that the cytotoxic cells recognize during their attack (Chapter 19).

A second rejection mechanism involves helper T lymphocytes alone. These lymphocytes are stimulated by class II MHC proteins, after which the lymphocytes release cytokines. The cytokines stimulate phagocytes to enter the graft tissue. The phagocytes secrete lysosomal enzymes, which digest the tissues, leading to a dryness and thickening, as in type IV hypersensitivity. Cell death, or necrosis, follows.

A rejection mechanism of a completely different sort is sometimes observed in bone marrow transplants. In this case, the transplanted marrow may contain immune system cells that form immune products against the host after the host's

MicroFocus 21.8

ACCEPTANCE

Ordinarily a woman will reject a foreign organ, such as a kidney or heart, but she will accept the fetus growing within her womb. This acceptance exists, even though half the fetus's genetic information has come from a "foreigner"—namely, the father. Has her immune system failed?

Apparently not. It seems that the sperm carries an antigenic signal, which induces the woman's immune system to produce a series of so-called blocking antibodies. The blocking antibodies form a type of protective screen that sheaths the fetus and prevents its antigens from stimulating the production of rejection antibodies by the mother. So protected, the fetus grows to term and "escapes" before any immunological damage can be done to its tissues.

But sometimes a rejection in the form of a miscarriage occurs. A number of physicians now believe that at least certain miscarriages have an immunological basis. Their research indicates that the level of blocking antibodies in some pregnant women is too low to protect the fetus, and that the miscarriage is due to the woman's antibodies. Ironically, they have discovered the blocking antibody level may be low because the father's tissue is very similar to the mother's. In such a case, the sperm's antigens elicit a weak antibody response, too low to protect the fetus.

With this knowledge in hand, physicians are now attempting to boost the level of blocking antibodies as a way of preventing miscarriage. They inject white blood cells from the father into the mother, thereby stimulating her immune system to produce antibodies to the cells. These antibodies exhibit the blocking effect. In other experiments, injections of blocking antibodies are administered to augment the woman's normal supply.

Both approaches have been successful in trial experiments, and continuing research has given cause for optimism that fetal rejection will give way to acceptance.

TABLE 21.6

A Summary of Transplantation Terminology

PREFIX	MEANING	COMBINING SUFFIXES	TRANSPLANTATION PARLANCE
Auto-	Self	-graft -genic -antigen -antibody	An autograft is a self-graft, e.g., skin from one site of the patient's body moved to another site on the patient's body
Iso-	Equal; identical with another	"	An isograft is a graft between isogeneic individuals, i.e., between genetically identical individuals such as identical twins
Allo- (homo-)	Similar or like another	"	An allograft is a graft between allogeneic individuals, i.e., between nonidentical members of the same species
Xeno- (hetero-)	Dissimilar or unlike another; foreign	"	A xenograft is a graft between xenogeneics, i.e., between members of different species, as a graft from monkey to human

immune system has been suppressed during transplant therapy. Essentially, the graft is rejecting the host. This phenomenon is called a **graft-versus-host reaction (GVHR)**. It sometimes can lead to fatal consequences in the host body.

THE MAJOR HISTOCOMPATIBILITY COMPLEX IS INVOLVED WITH TRANSPLANT REJECTION

During the 1970s, immunologists determined that the acceptance or rejection of a graft depends largely on a relatively small number of genes called the major histocompatibility complex (MHC) genes. These genes encode a series of cell-surface glycoproteins called **major histocompatibility complex (MHC) proteins**, also known as **human leukocyte antigens (HLAs)**. The 1980 Nobel Prize in Physiology or Medicine was awarded to George Snell, Jean Dausset, and Baraj Benacerraf for their discoveries regarding MHC genes and proteins.

The MHC genes are believed to exist on chromosome 6 in humans. The actual gene complex consists of four clusters of genes, each gene having multiple versions, or alleles. Two individuals chosen at random (a husband and wife, for example) are not expected to have many MHC genes in common. Identical twins, by contrast, have identical MHC genes. MHC proteins are of two types: class I and class II. Class II MHC proteins are important in the recognition of nonself antigens when T lymphocytes combine with macrophages in cell-mediated immunity (Chapter 19). Class I MHC proteins are present on every nucleated cell in the human body and help define the uniqueness of a person's tissue. For this reason, they are important subjects of transplantation immunology.

FIGURE 21.14

Tissue Typing for MHC Proteins

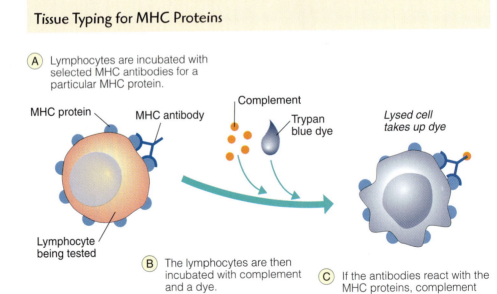

(A) Lymphocytes are incubated with selected MHC antibodies for a particular MHC protein.

MHC protein   MHC antibody   Complement   Trypan blue dye   Lysed cell takes up dye

Lymphocyte being tested

(B) The lymphocytes are then incubated with complement and a dye.

(C) If the antibodies react with the MHC proteins, complement opens pores in the cells and allows the dye to enter. A positive result consists of a stained cell, which indicates that a particular MHC protein is present on the cell surface.

The nature of MHC proteins is a key element in transplant acceptance or rejection: The closer the match between donor and recipient MHC proteins, the greater the chance of a successful transplant. The matching of donor and recipient is performed by **tissue typing** (FIGURE 21.14). In this procedure, the laboratory uses standardized MHC antibodies for particular MHC proteins. Lymphocytes from the donor are incubated with a selected type of MHC antibodies. Complement and a dye, such as trypan blue, are then added. If the selected MHC antibodies react with the MHC proteins of the lymphocytes, the cell becomes permeable and dye enters the cells (living cells do not normally take up the dye). Similar tests then are performed with recipient lymphocytes to determine which MHC proteins are present and how closely the tissues match one another. The blood types of donor and recipient must be identical also.

Studies of the MHC proteins also provide clues about the development of certain immune diseases. In **Graves' disease**, for example, antibodies appear to unite with MHC proteins on the surface of thyroid gland cells and overstimulate the gland. Another example is seen in **ankylosing spondylitis**, a spinal disease in which adjacent vertebrae fuse and cause fixation and stiffness of the spine. The relative risk for this disease is far greater when individuals possess particular MHC proteins. A final possibility is that certain viruses, because of their similarity to MHC proteins, induce antibodies that attack body cells. This may be the source of Guillain-Barré syndrome and Reye's syndrome that sometimes follow occurrences of influenza.

ANTIREJECTION MECHANISMS BRING ABOUT IMMUNOSUPPRESSION

Tissue typing and histocompatibility screening help reduce the rejection mechanism in allograft transfers, but they do not eliminate the mechanism completely. To inhibit rejection, it is necessary to suppress activity of the immune system. One method uses **antimitotic drugs**, which prevent multiplication of lymphocytes in the lymph nodes. The drug **azathioprine** (Imuran) is a nucleic acid antagonist used for this purpose. Another drug is **cyclosporin A**. Originally isolated from strains of fungi as an antimicrobial substance, this drug now is produced synthetically. It appears to suppress cell-mediated immunity without killing T lymphocytes or interfering with antibody formation. Immunologists believe that cyclosporin A prevents the division of helper T lymphocytes by inhibiting transcription of the genes encoding IL-2 and IL-2 receptor.

Another means of diminishing the rejection mechanism is to introduce **antilymphocyte antibodies**. To obtain these antibodies, lymphocytes from the transplant recipient are injected into animals, and later, the animals are bled to obtain the antibody-rich serum. When introduced to the transplant recipient (serum sickness notwithstanding), the antibodies interact with the local lymphocytes, thereby slowing the rejection process and increasing the survival time of the transplant.

Additional methods of reducing rejection include the use of **prednisone**, a steroid hormone that suppresses the inflammatory response, and the bombardment of lymphocyte centers with **radiations** such as X rays. In another experimental treatment, the donor's bone marrow cells are injected into the recipient's bone marrow to induce lymphocytes that will recognize the transplant as self. It should be noted that virtually all treatments leave the recipient in an immunosuppressed condition, thereby increasing the susceptibility to a variety of infectious diseases and cancer. Immunosuppression is, at best, a poor expedient and is viewed as only a stopgap measure until the rejection mechanism can be understood and exploited.

Tumor Immunology

The alteration from a normal cell to a cell with tumor potential is accompanied by many physiological changes, including the change to an immature form, the loss of contact inhibition, and an increased rate of mitosis (Chapter 12). In addition, tumor cells (especially virus-induced ones) contain antigens not found in surrounding cells. Since many of these antigens localize on the cell surface, they might be expected to induce an immune response and make the cancer cells vulnerable to destruction. However, the cells escape the resistance mechanisms of the body and proliferate to form the tumor. As the tumor breaks apart and spreads, a metastasizing cancer develops.

AN IMMUNE RESPONSE CAN BE DIRECTED AGAINST TUMOR CELLS

Host resistance to tumors depends on immune responses directed against the tumor antigens. This process is termed **immune surveillance**. It is considered an ongoing function of cytotoxic T lymphocytes, which recognize the antigens as foreign and destroy the cells containing them in a manner analogous to graft rejection. Another cell thought to function in immune surveillance is the **natural killer (NK) cell**, which is a type of lymphocyte (Chapter 19). NK cells exert a nonspecific, cytotoxic effect on other cells, including tumor cells. They kill cells in the absence of a prior sensitization and without the involvement of antibody. Interferon appears to regulate their activity and can increase it.

Despite immune surveillance, tumors continue to occur in otherwise healthy people as tumor cells escape destruction. Various theories are offered to account for this **immunologic escape**. One possibility is such individuals learn to tolerate tumor cells before immunocompetence has developed, and then they express the tumors later in life. Another theory suggests tumor cells express low levels of immunogenicity; that is, they have MHC molecules needed for immune recognition. It also is possible that tumor antigens elicit antibodies that combine with antigens to form

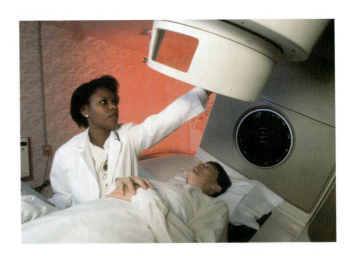

FIGURE 21.15

A Patient Being Treated with Radiation to Reduce a Tumor

Radiation therapy with X rays is used to destroy cancer cells and prevent their spread.

a complex that is endocytased and thus not recognized by cytotoxic T cells. A final theory points to the release by tumor cells of substances that suppress the immune system, such as what happens in **Hodgkin's disease** of the lymph nodes. Although experimental evidence exists for each theory, no one is universally accepted.

IMMUNOTHERAPY ATTEMPTS TO AUGMENT AND ENHANCE NATURAL IMMUNE DEFENSES

Cancer is the second most common cause of death in the industrialized world. Today, most cancer patients are treated with some combination of surgery, radiation, and chemotherapy. Unfortunately, radiation and chemotherapy destroy healthy as well as malignant cells and thus can cause severe side effects (FIGURE 21.15).

One of the most promising new approaches in modern cancer treatment is **immunotherapy**, the development of methods to augment and enhance the body's natural tendency to defend itself against malignant tumors without damaging healthy tissue. Four major immunotherapies are being investigated.

One approach to curing some cancers (including leukemia) is to treat the patient with such high doses of chemotherapy and radiation that the cancer cells are killed. Unfortunately, such high doses also destroy the patient's bone marrow. If the patient is to survive the treatment, he or she must be given a bone marrow transplant, which contains the stem cells from which all blood cells are formed (Chapter 19). The bone marrow can be an autograft (previously purged of any cancer cells it may contain) or an allograft. The latter usually involves bone marrow donated by a family member sharing similar major histocompatibility molecules. In 2002, Steven A. Rosenberg and his associates at the National Cancer Institute reported success with T cells. They reported partial or complete regression of melanomas in 10 of 13 patients who received tumor infiltrating lymphocytes (TILs) composed of CD4 and CD8 T cells identified as efficient melanoma killers.

Cytokines are being investigated and used as **immunostimulants**, nonspecific agents that tune up the body's immune defenses. Interferons are one group of cytokines that are produced naturally by white blood cells in the body (or in the laboratory) in response to infection or inflammation and have been used as a treatment for certain viral diseases (Chapter 12). Interferon-alpha was one of the first cytokines to show an antitumor effect, and it can slow tumor growth, as well as help to activate the immune system. Interferon-alpha has been approved by the FDA and is now commonly used for the treatment of a number of cancers, including certain types of myelomas, leukemias, and melanomas. Interferon-beta and interferon-gamma also are being investigated.

Other cytokines with antitumor activity include IL-2 and TNF-α. IL-2 is an essential factor for lymphocyte multiplication and frequently is used to treat kidney cancer and melanoma. The problem with using cytokines for immunotherapy is that they have side effects, which at the high doses needed to effect cancer therapy are very serious.

Monoclonal antibodies, artificial antibodies produced in the laboratory (Chapter 20), are another potential form of immunotherapy against cancer cells, especially cancers of white blood cells (leukemia, lymphoma, and multiple myeloma). Monoclonal antibodies that seek out the cancer cells can be injected into patients. These antibodies can disrupt cancer cell activities or enhance the immune response against the cancer. An antibody called rituximab (Rituxan) is useful in the treatment of B-cell lymphomas, trastuzumab (Herceptin) is helpful against certain

Hodgkin's disease: an abnormal development of B cells causing enlargement of the lymph nodes; the lymphoma may spread beyond the lymphatic system.

breast cancers, and Alemtuzumab (MabCampath) has produced complete remission of some forms of leukemia.

Researchers also are developing ways of linking cytotoxic drugs, toxins, or radioisotopes to monoclonal antibodies to augment their effectiveness against cancer cells (FIGURE 21.16). In this case, the antibodies would function as a targeted "stealth missile," capable of seeking out a specific target—a cancer cell—without harming the surrounding healthy tissue. Tositumomab (Bexxar) is a monoclonal antibody conjugated with radioactive iodine (^{131}I) used to treat lymphomas.

The fourth avenue of immunotherapy involves the development of cancer vaccines. Although vaccines have revolutionized public health by preventing many serious infectious diseases (Chapter 20), it has been difficult to develop effective vaccines to prevent cancer, or to treat patients who already have cancer. Cancer vaccines typically consist of a source of cancer-associated material (cancer antigen), along with other components, which will stimulate an immune response against the cancer antigen. The challenge has been to find those cancer antigens that will enhance the patient's immune system to fight cancer cells that have the antigen. In malignant lymphoma, a number of laboratory studies have indicated that vaccination using lymphoma-associated proteins can stimulate sufficiently the immune systems of mice to help them resist the development of lymphomas.

As we learn more about how the immune system and immune surveillance operate, immunotherapy should offer great promise as a new component in cancer treatment.

Note to the Student

The body's immunologic responses to transplants and tumors represent an ironic dilemma. Against transplants, the immune system responds vigorously (immunologists attempt to suppress the response); against tumors, the immune system responds weakly (immunologists attempt to enhance the response). Would that the situations could be reversed.

Speaking of immunological responses, animal organs, especially pig kidneys, are being considered as transplant organs for humans. Of the more than 72,000 American patients waiting for a transplant, 47,000 need a kidney. Yet, in 2002 less than 13,000 kidneys were available for transplantation. Overcoming obstacles such as rejection still are an issue with other animals, including pigs, although genetic engineering techniques are being developed to make organ rejection less likely. Indeed, such xenotransplants may become more common as the demand for organs and suitable organ donors continues to increase.

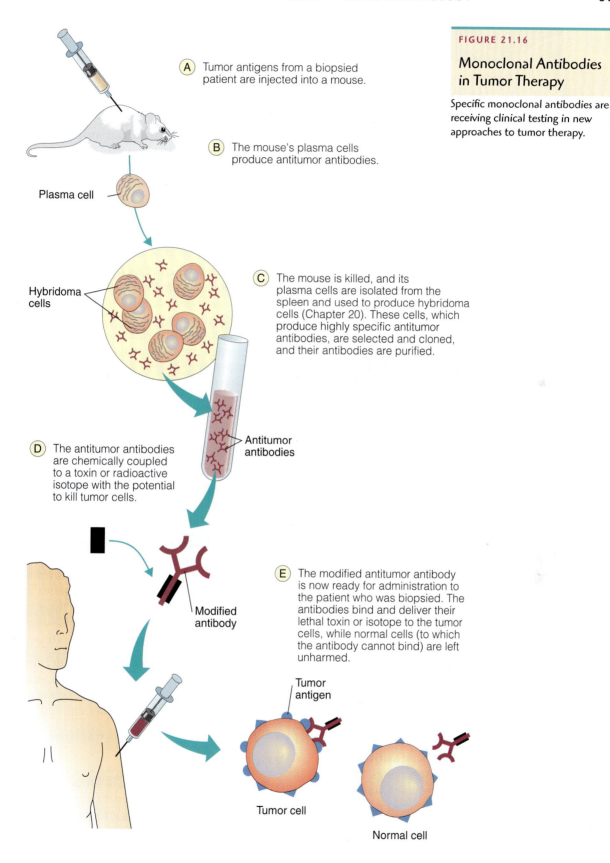

FIGURE 21.16

Monoclonal Antibodies in Tumor Therapy

Specific monoclonal antibodies are receiving clinical testing in new approaches to tumor therapy.

(A) Tumor antigens from a biopsied patient are injected into a mouse.

(B) The mouse's plasma cells produce antitumor antibodies.

Plasma cell

Hybridoma cells

(C) The mouse is killed, and its plasma cells are isolated from the spleen and used to produce hybridoma cells (Chapter 20). These cells, which produce highly specific antitumor antibodies, are selected and cloned, and their antibodies are purified.

Antitumor antibodies

(D) The antitumor antibodies are chemically coupled to a toxin or radioactive isotope with the potential to kill tumor cells.

Modified antibody

(E) The modified antitumor antibody is now ready for administration to the patient who was biopsied. The antibodies bind and deliver their lethal toxin or isotope to the tumor cells, while normal cells (to which the antibody cannot bind) are left unharmed.

Tumor antigen

Tumor cell

Normal cell

Summary of Key Concepts

21.1 TYPE I IgE-MEDIATED HYPERSENSITIVITY

■ **Type I Hypersensitivity Is Induced by Allergens.**
Type I hypersensitivity is caused by IgE antibodies
produced in response to certain antigens called allergens. The antibodies can attach to the surface of mast
cells and basophils and trigger the release of the mediators on subsequent exposure to the same allergen.

■ **Systemic Anaphylaxis Is a Life-Threatening Condition.** Type I hypersensitivities can involve systemic
anaphylaxis, a whole-body reaction in which a series
of mediators induces vigorous and life-threatening
contractions by the smooth muscles of the body.

■ **Atopic Diseases Represent a Localized Anaphylaxis.**
Atopic diseases involve localized anaphylaxis (common allergies), such as hay fever or a food allergy.

■ **Why Do Only Some People Have Hypersensitivities?** Several ideas have been put forth concerning
the nature of allergies. These include the shielding of
IgE-secreting B cells (plasma cells) by IgA-secreting
plasma cells and feedback mechanisms where IgE
limits its own production in nonatopic individuals.

■ **Therapies Sometimes Can Control Type I Hypersensitivities.** Desensitization therapy attempts to
limit the possibility of an anaphylactic reaction
through the injection of tiny amounts of allergen
over time. Therapy also can involve the presence of
IgG antibodies as blocking antibodies.

21.2 OTHER TYPES OF HYPERSENSITIVITY

■ **Type II Cytotoxic Hypersensitivity Involves
Antibody-Mediated Cell Destruction.** In type
II cytotoxic hypersensitivity, the immune system
produces IgG and IgM antibodies, both of which
react with the body's cells and often destroy the latter. The destruction of platelets in thrombocytopenia and neutrophils in agranulocytosis is typical of
immune disorders resulting from this form of
hypersensitivity.

■ **Type III Immune Complex Hypersensitivity Is
Caused by Antigen-Antibody Aggregates.** No cells
are involved in type III hypersensitivity. Rather, the
body's IgG and IgM antibodies interact with dissolved antigen molecules to form visible immune
complexes. The accumulation of immune complexes
in various organs leads to local tissue destruction in

such illnesses as serum sickness, Arthus phenomenon, and systemic lupus erythematosus.

■ **Type IV Cellular Hypersensitivity Is Mediated by
Antigen-Specific T Cells.** Type IV cellular hypersensitivity involves no antibodies, but is an exaggeration
of cell-mediated immunity based in T lymphocytes.
Contact dermatitis is a manifestation of this hypersensitivity.

21.3 IMMUNE DEFICIENCY DISEASES

■ **Immune Deficiencies Can Involve Any Aspect of
the Immune System.** Immune disorders include
immune deficiency diseases in which important cells
of the immune system, such as B or T lymphocytes,
are not formed. Deficiencies in phagocytic cells or
complement components also may be observed.

21.4 TRANSPLANTATION IMMUNOLOGY

■ **There Are Four Types of Grafts (Transplants).** Four
types of grafts or transplants can be performed.
These include autografts, isografts, allografts, and
xenografts. Allografts are the most common. Rejection of grafts or transplants involves cytotoxic T cells
and helper T cells. The graft also can be rejected by
immune cells in the graft that reject the recipient
(graft-versus-host reaction).

■ **The Major Histocompatibility Complex Is Involved
with Transplant Rejection.** The major histocompatibility complex (MHC) proteins are involved in transplant rejection. MHC antigens on all individuals
except identical twins are different, so tissue typing is
important to match as closely as possible donor and
recipient MHC proteins to prevent rejection.

■ **Antirejection Mechanisms Bring About Immunosuppression.** Preventing rejection is strengthened by
using antirejection strategies, including antimitotic
drugs, antilymphocyte antibodies, anti-inflammatory
drugs, or radiation.

21.5 TUMOR IMMUNOLOGY

■ **An Immune Response Can Be Directed against
Tumor Cells.** One of the functions of cytotoxic T
cells and natural killer cells is to survey the body for
abnormal cells (tumors). Such immune surveillance
is not perfect since many people still develop various
types of tumors and cancer.

■ **Immunotherapy Attempts to Augment and Enhance Natural Immune Defenses.** Several strategies are available or being investigated that enhance immune defenses. Such immunotherapies include T-cell transplants, cytokines as immunostimulants, monoclonal antibodies, and cancer vaccines.

Questions for Thought and Discussion

Answers to selected questions can be found in Appendix C.

1. In December 1967, people throughout the world were startled to learn that Christiaan A. Barnard, a South African physician, had made the first successful heart transplant. The patient, Louis Washkansky, survived the rejection mechanism for two weeks before dying of pneumonia. Many scientists consider Barnard's surgery to signal the beginning of modern immunology. What evidence do you think might be offered to support this contention?

2. During war and under emergency conditions a soldier whose blood type is O donates blood to save the life of a fellow soldier with type B blood. The soldier lives, and after the war becomes a police officer. One day he is called to donate blood to a brother officer who has been wounded and finds that it is his old friend from the war. He gladly rolls up his sleeve and prepares for the transfusion. Should it be allowed to proceed? Why?

3. Coming from the anatomy lab, you notice that your hands are red and raw and have begun peeling in several spots. This was your third period of dissection. What is happening to your hands, and what could be causing the condition? How will you solve the problem?

4. Researchers have found that the powder in latex-powdered gloves has the ability to aerate and spread allergy-inducing latex proteins. Ordinarily, the powder helps hurried medical personnel don the gloves in emergency situations. Can you think of any substitute material that can be used to resolve the problem of "sticky gloves"?

5. A scientist observed ten men and noticed the following: All used an excessive amount of antihistamine, all were tall and had blond hair and a mustache, all were avid fishermen, and all enjoyed movies. Then she observed ten more men: All used less antihistamine than the first group, all were short and had black hair and no mustache, all were avid golfers, and all enjoyed fine dining. Which trait in the second group correlates with their tendency to use less antihistamine? What advice would the scientist give to men in the first group?

6. As part of an experiment, one animal is fed a raw egg, while a second animal is injected intravenously with a raw egg. Which animal is in greater danger? Why?

7. "He had a history of nasal congestion, swelling of his eyes and difficulty breathing through his nose. He gave a history of blowing his nose frequently, and the congestion was so severe during the spring he had difficulty running." The person in this description is a certain former president of the United States, and the writer is an allergist from Little Rock, Arkansas. What condition (technically known as allergic rhinitis) is probably being described?

8. When the skin of the feet becomes itchy, red, and scaly, many people assume they have athlete's foot, a fungal disease. However, allergists now know that the irritation could be a contact dermatitis resulting from chemicals in the insoles of sneakers. How might the distinction be made by physical examination of the feet and by laboratory testing?

9. A woman is having the fifth injection in a weekly series of hay fever shots. Shortly after leaving the allergist's office, she develops a flush on her face, itching sensations of the skin, and shortness of breath. She becomes dizzy, then faints. What is taking place in her body, and why has it not happened after the first four injections?

10. You may have noted that brothers and sisters are allowed to be organ donors for one another, but that a person cannot always donate to his or her spouse. Many people feel bad about being unable to help a loved one in time of need. How might you explain to someone in such a situation the basis for becoming an organ donor and why it may be impossible to serve as one?

11. Some years ago there was a commercial for a life insurance company that included this jingle: "There's

nobody else exactly like you; nobody else like you." Why is this concept immunologically correct?

12. The story is told that in 1552, the distinguished physician Jerome Cardan of Pavia was called to England to advise treatment for Archbishop John Hamilton. Hamilton had suffered from asthma for 10 years. Cardan prescribed a carefully controlled diet, plenty of exercise and sleep, and the removal of feathers from the archbishop's mattress. Soon the asthma diminished. Which of Cardan's recommendations was the key to success? Why?

13. Paternity suits are often settled by determining the blood types of parents and child, and then concluding whether the man could be the father of the child. How could determining the major histocompatibility complex be used to replace the blood-typing procedure in paternity suits of the future?

14. In order to prevent hemolytic disease of the newborn, Rh antibodies must be injected into an Rh-negative woman shortly after the birth of an Rh-positive child. Where do you suppose these antibodies were obtained in the past, and what might be their source in the future?

15. The immune system is commonly regarded as one that provides protection against disease. This chapter, however, seems to indicate that the immune system is responsible for numerous afflictions. Even the title is "Immune Disorders." Does this mean that the immune system should be given a new name? On the other hand, is it possible that all these afflictions are actually the result of the body's attempts to protect itself? Finally, why can the phrase "immune disorder" be considered an oxymoron?

http://microbiology.jbpub.com

The site features **eLearning,** an on-line review area that provides quizzes and other tools to help you study for your class. You can also follow useful links for in-depth information, or just find out the latest microbiology news.

Review

This chapter has summarized some of the disorders associated with the immune system. To gauge your understanding, rearrange the scrambled letters to form the correct word for each of the spaces in the statements. The correct answers are in Appendix D.

1. The simple compound _____ is one of the major mediators released during allergy reactions.

   I T M E H I A S N

2. An immune deficiency called _____ syndrome is characterized by the failure of T lymphocytes to develop.

   I D O G G E R E

3. In type IV hypersensitivity, a drying and thickening of the skin known as _____ is an observable symptom.

   A N U I R O N I D T

4. Cases of rheumatoid arthritis are accompanied by immune complex formation in the body's _____.

   N I S T J O

5. In cases of _____ disease, antibody molecules unite with receptors on the surface of thyroid gland cells.

   V S R E A G

6. In a _____ hypersensitivity, antibodies unite with cells and trigger a reaction that results in cell destruction.

   X Y O T C C I T O

7. Hay fever is an example of an _____ disease, one in which a local allergy takes place.

   O C T I A P

8. Immune complex hypersensitivities develop when antibody molecules interact with _____ molecules and form aggregates in the tissues.

   E N I G N A T

9. In people suffering from Bruton's agammaglobulinemia, the lymph nodes are noticeably deficient in _____ cells.

   S L M A A P

10. The skin test for _____ relies on a response by T lymphocytes to PPD placed in the skin tissues.

    U U I C T R L S B E O S

11. Mast cells and _____ are the two principal cells that function in anaphylactic responses.

    S S B I O H P A L

12. During cases of thrombocytopenia, antibodies react with platelets and cause them to undergo _____.

    Y S S L I

13. Rh disease can develop in a fetus if the father's blood type was Rh-_____ and the mother's type was Rh-negative.

    O S I P I E T V

14. The glomerulonephritis that accompanies streptococcal disease may be due to immune complex formation in the _____.

    N I Y E K D

15. A key element in transplant acceptance or rejection is a set of molecules abbreviated as _____ proteins.

    H C M

16. Urticaria is a form of skin _____ that occurs in a person having an allergic reaction.

    A H S R

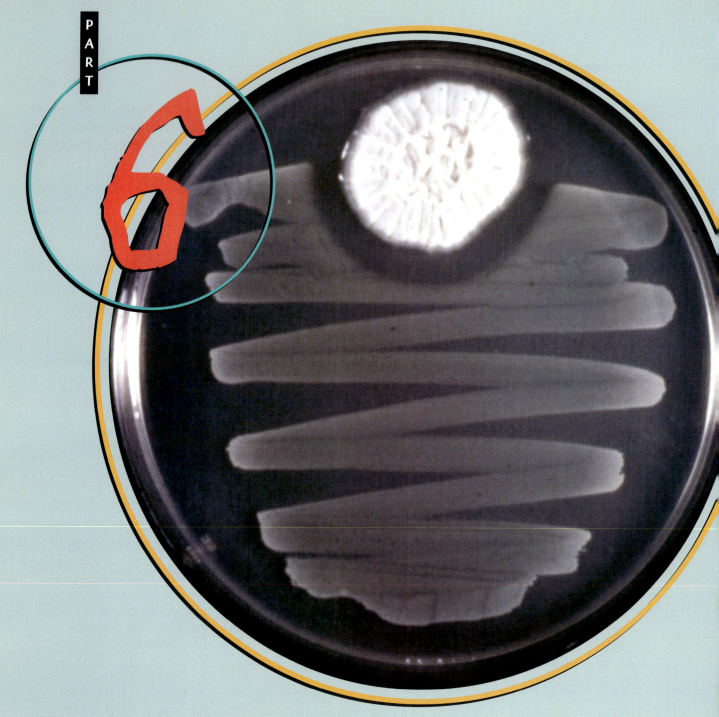

Control of Microorganisms

In the late 1800s and early 1900s, growing support for the germ theory of disease led to a dramatic reduction in the frequency of epidemics. Scientists reasoned that if microorganisms cause disease, then it was possible to control disease by controlling the microorganisms. The idea had been proposed by Pasteur decades before, but not until the turn of the century did it gather momentum and achieve a firm footing in the scientific community.

Two types of control gradually evolved: physical control and chemical control. To achieve physical control, scientists used heat, radiations, filters, and other physical agents to remove microorganisms from instruments, equipment, and fluids. Pasteurization of dairy products was in wide use by 1895, and food preservation methods were gradually updated. To achieve chemical control, doctors used antiseptics and disinfectants in medicine, surgery, and wound treatment. Moreover, municipalities began adding chlorine to their water supplies to protect citizens from waterborne diseases. As control methods became a way of life in public health, the chains of microbial transmission broke down, and disease outbreaks declined.

Little could be done for the patient who already was ill. Here, another type of control was necessary, one that supplemented the body's own natural defenses. That control device would emerge in the 1940s with the discovery and development of chemotherapeutic agents and antibiotics. Now, physicians could do something about diseases they could not hope to treat before. The new drugs and medicines ushered in a second dramatic reduction in the incidence of infectious disease.

The control of microorganisms is an essential factor in maintaining good health. In Part 6, we shall consider a variety of control methods and examine their uses and modes of action. Our survey will begin in Chapter 22 with an exploration of physical methods of control used for objects outside the body. It will continue in Chapter 23, where we examine chemical methods for objects that come in contact with the body (e.g., instruments) and on the body surface itself. Finally, in Chapter 24 we shall move inside the body environment to discuss chemotherapeutic agents and antibiotics. Applied on a broad scale, these control methods remain a major deterrent to infection and disease.

■ *Inhibition of* Staphylococcus aureus *growth by the fungus* Penicillium.

SALES AND RESEARCH

Not all salespeople sell vacuum cleaners or brushes. And not all microbiologists wear white coats and work in ultraclean laboratories. It is quite commonplace for a person trained in microbiology to find a comfortable and enjoyable future with a company that sells and distributes instruments, scientific chemicals, and pharmaceuticals. For example, it is valuable for a sterilizer salesperson to understand the microbiological basis for sterilizing such things as instruments, microbial media, and patient materials. Also, it makes sense for a disinfectant salesperson to realize why microorganisms must be eliminated from a particular surface. Finally, without question, an individual selling new antibiotics should have a strong familiarity with the microorganisms that the antibiotic is intended to eliminate. The bottom line is: Know your product, and know its uses.

Before a product is available for sale, however, it must be developed, and once again, the microbiologist is a member of the team. The microbiologist can appreciate why new methods must be developed for preserving dairy products and for storing foods. The microbiologist can set the direction for developing new sterilizing instruments and have significant input into the development of pharmaceuticals for treating infectious disease. The diagnostic lab depends heavily on instruments produced under the guidance of microbiologists because the lab's objective is to detect microorganisms.

The contemporary health care industry depends heavily on the talents of microbiologists for the development and sale of innovative and novel approaches to diagnosis and treatment. As you might suspect, a healthy dose of chemistry, physics, and mathematics is valuable, depending upon which road one chooses to follow. The ability to tinker with instruments is important to research and development, while strong interpersonal skills are important for the sales phase. What may seem like services and sales really is microbiology at its core.

22

Physical Control of Microorganisms

I had difficulty believing anything like this could exist.

—A university microbiologist recalling his introduction to *Deinococcus radiodurans*, a bacterium that survives one-thousand times the radiation lethal to a human

BY THE BEST ESTIMATES of health officials, the problem began during the first week of July during 1988. Early that week, a person unintentionally contaminated the water while swimming in an indoor pool at a school in Los Angeles, California. The same week, the pool was used by a water polo team for its match against a local opponent. Shortly thereafter, a class of scuba divers used the pool to perfect their skills. A group of elementary schoolchildren also enjoyed the pool during a day-camp field trip.

Beginning in early August, doctors began receiving reports of watery diarrhea, abdominal cramps, and fever. Two members of the scuba class were hospitalized. Eleven patients were tested for stool microorganisms and in seven cases, the protozoan *Cryptosporidium* was located. Indeed, the attack rate was highest for those who were exposed to the water the longest. By the second week of August, 44 individuals were affected out of a total of 60 swimmers.

When CDC epidemiologists arrived at the pool, they inquired about the water-purification methods. They were told that three filters were used to sterilize the water, and chlorine was added to maintain a low bacterial count. Their inspection revealed that the chlorination was adequate, but the filtration left much to be desired. One of the three filters was inoperative, and the flow rate of water through the remaining two filters was only 70 percent of the expected rate. It was clear that the water filtration had not worked.

The episode in Los Angeles illustrates how microorganisms can spread when proper methods of water purification are lacking. In many cases, these purification methods are designed to achieve sterilization. **Sterilization** is the destruction or removal of all life-forms, including viruses. It is an absolute term that cannot be qualified. Thus, one cannot assume "partial sterilization" or "incomplete sterilization," but must consider a material contaminated until it is sterilized. It should be pointed out, however, that material may remain harmful even though it is sterile. For example, a solution of bacterial toxin may contain no living forms but still cause physiological damage in the body.

Although filters and chemicals sometimes may be used to sterilize objects, the principal methods for achieving sterilization are physical agents, such as heat and radiation. These methods are not products of the modern era. They were used by Pasteur, Koch, and microbiologists over a century ago to prevent contamination of their materials and to ensure the accuracy of their work.

Microbiologists recognize that bacterial **endospores** are among the most resistant forms of life (**FIGURE 22.1**). Anthrax spores, for example, have been found to remain viable after 60 years on dry silk threads. The spores are not metabolically inert. Rather, they carry on life processes at minimum rates and possess the necessary enzymes to transform from a dormant state to actively metabolizing vegetative cells (Chapter 4). The extraordinary survival of spores often is attributed to their low water content, which yields a heat-resistant gel-like spore core. Another possible reason is the presence of **dipicolinic acid**, an organic compound that encourages heat resistance by linking to the spore proteins. Destruction of the bacterial spore is one of the principal aims of sterilization methods, especially those involving heat.

However, the bacterial spore is not necessarily the most resistant of all microorganisms. Recent research has focused on the bacterium *Deinococcus radiodurans*, the organism mentioned in the opening quote to this chapter and pictured in

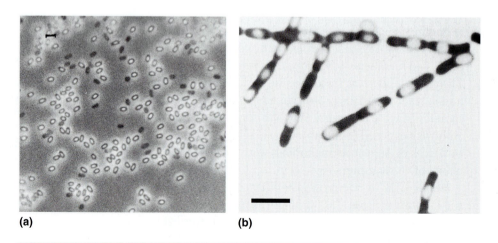

(a)   (b)

FIGURE 22.1

The Objectives of Physical Control

Two views of bacterial spores, the destruction of which is the objective of physical control methods. (a) A phase contrast view of free spores of *Bacillus subtilis*. (Bar =1 μm.) (b) Clear spores within the stained vegetative cells of *Bacillus anthracis*. Bacterial spores are among the most resistant forms of life known to science. (Bar = 10 μm.)

FIGURE 22.2. *D. radiodurans* is notable for its resistance to radiation (the *Guinness Book of World Records* labels it the world's "toughest bacterium"). This resistance is more substantial than for bacterial spores and may derive from the bacterium's unusually high ability to repair its DNA once radiation damage has occurred. In 1998, the complete genome of the organism was deciphered, and researchers began studying the genes for clues about the nature of the extreme radiation resistance. Indeed, the organism is being considered as a tool for detoxifying waste sites that contain radioactive materials, a process called **bioremediation** (Chapter 27).

22.1

Physical Control with Heat

he Citadel is a novel by A. J. Cronin that follows the life of a young British physician, beginning in the 1920s. Early in the story, the physician, Andrew Manson, begins his practice in a small coal-mining town in Wales. Almost immediately, he encounters an epidemic of typhoid fever. When his first patient dies of the disease, Manson becomes terribly distraught. However, he realizes that the epidemic can be halted, and in the next scene, he is tossing all of the patient's bed-sheets, clothing, and personal effects into a huge bonfire.

The killing effect of heat on microorganisms has long been known. Heat is fast, reliable, and relatively inexpensive, and it does not introduce chemicals to a sub-stance, as disinfectants sometimes do. Above growth range temperatures, bio-chemical changes in the cell's organic molecules result in its death. These changes arise from alterations in enzyme molecules or chemical breakdowns of structural

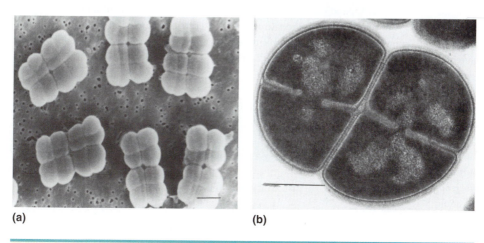

(a)   (b)

FIGURE 22.2

Extreme Radiation Resistance

Deinococcus radiodurans is a bacterium that can withstand a thousand times the radiation lethal to humans. (a) This organism exists in tetrads, as shown in the scanning electron micrograph. (Bar = 2 μm.) (b) The transmission electron micrograph shows the complex cell wall of the four cells in the tetrad. (Bar = 1 μm.)

molecules, especially in cell membranes. Heat also drives off water, and since all organisms depend on water, this loss may be fatal.

The killing rate of heat may be expressed as a function of time and temperature. For example, tubercle bacilli are destroyed in 30 minutes at 58°C, but in only 2 minutes at 65°C, and in a few seconds at 72°C. Each microbial species has a **thermal death time**, the time necessary for killing the population at a given temperature. Each species also has a **thermal death point**, the minimal temperature at which it dies in a given time. These measurements are particularly important in the food industry, where heat is used for preservation (Chapter 25). MicroInquiry 22 examines heat and microbial killing.

When determining the time and temperature for microbial destruction with heat, certain factors bear consideration. One factor is the type of organism to be killed. For example, if materials are to be sterilized, the physical method must be directed at bacterial spores. Milk, however, need not be sterile for consumption, and heat is therefore aimed at the most resistant vegetative cells.

Another factor is the type of material to be treated. Powder is subjected to dry heat rather than moist heat, because moist heat will leave it soggy. Saline solutions, by contrast, can be sterilized with moist heat but are not easily treated with dry heat. Other important factors are the presence of organic matter and the

MicroInquiry 22

EXPLORING HEAT AS AN EFFECTIVE CONTROL METHOD

Is that can of unopened peas in your pantry sterile? Yes, because companies like Green Giant® and the food industry in general have established appropriate procedures for sterilizing commercial foods. Sterilization depends on several factors. Identifying the type(s) of microbes in a product can determine whether the heating process will sterilize or eliminate only potential disease-causing species. In many foods, the microbes usually are not in water but rather in the food material. Microbes in powders or dry materials will require a different length of time to sterilize the product than microbes in organic matter. Environmental conditions also influence the sterilization time. Microbes in acidic or alkaline materials decrease sterilization times while microbes in fats and oils, which slow heat penetration, increase sterilization times. It must be remembered that sterilization times are not precise values. However, by knowing these

factors, heating the product to temperatures above the maximal range for microbial growth will kill microbes rapidly and effectively. Let's explore the factors of time and temperature. Answers can be found in Appendix E.

As we described for the microbial growth curve in Chapter 4, microbial death occurs in an exponential fashion.

Look at TABLE A. The table records the death of a microbial population by heating. Notice that the cells die at a constant rate. In this generalized example, each minute 90 percent of the cells die (10 percent survive). Therefore, if you know the initial number of microorganisms, you can predict the **thermal death time (TDT)**, which is

TABLE A

Microbial Death Rate

TIME (MINUTES)	NUMBER OF CELLS SURVIVING	% KILLED
0	1,000,000	—
10	100,000	90
20	10,000	90
30	1,000	90
40	100	90
50	10	90
60	1	90

acidic or basic nature of the material. Organic matter may prevent heat from reaching microorganisms, while acidity or alkalinity may encourage the lethal action of heat.

THE DIRECT FLAME REPRESENTS A RAPID STERILIZATION METHOD

The use of a **direct flame** can incinerate microbes very rapidly. The flame of the Bunsen burner is employed for a few seconds to sterilize the bacteriological loop before removing a sample from a culture tube and after preparing a smear (FIGURE 22.3). Flaming the tip of the tube also destroys organisms that happen to contact the tip, while burning away lint and dust.

In general, objects must be disposable if a flame is used for sterilization. Disposable hospital gowns and certain plastic apparatus are examples of materials that may be incinerated. In past centuries, the bodies of disease victims were burned to prevent spread of the pestilence. It still is common practice to incinerate the carcasses of cattle that have died of anthrax and to put the contaminated field to the torch because anthrax spores cannot be destroyed adequately by other means. Indeed, British law stipulates that anthrax-contaminated animals may not be autopsied before burning. The 2001 outbreak of foot-and-mouth disease in British cattle

---

defined as the time, at a specified temperature, required to kill a population of microorganisms.

The food industry depends on knowing a microorganism's heat sensitivity when planning the canning or packaging of many foods. One way the industry determines this sensitivity is by using standard curves that take into account the factors mentioned above. FIGURE A represents three such curves, each representing a different bacterial species treated at the same temperature (60°C) in the same food material (curve B represents the plotted data from Table A).

22a. If you had to sterilize this food product that initially contained 10^6 bacteria, how long would it take for each bacterial species? (Hint: $10^0 = 1$)

Another value of importance is the **decimal reduction time (D value)**, which is the time required at a specific temperature to kill 90 percent of the viable organisms. These are the values typically used in the canning industry. D values are usually identified by the temperature used for killing. In Figure A, the temperature was 60°C, so D is written as D_{60}. Look again at Figure A.

22b. Calculate the D_{60} values for the three bacterial populations depicted in curves A, B, and C.

On another day in the canning plant, you need to sterilize a food product. However, the only information you have is a $D_{70} = 12$ minutes for the microorganism in the food product. Assuming the D value is for the volume you have to sterilize and there are 10^8 bacteria in the food product:

22c. At what temperature will you treat the food product?

22d. How long will it take to sterilize the product?

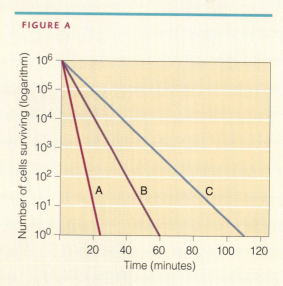

FIGURE A

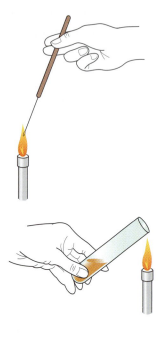

FIGURE 22.3

Use of the Direct Flame as a Sterilizing Agent

Laboratory use of the Bunsen burner. A few seconds in the flame is usually sufficient to effect sterilization.

required the mass incineration of thousands of cattle as a means to stop the spread of the disease (FIGURE 22.4).

THE HOT-AIR OVEN USES DRY HEAT

The **hot-air oven** uses radiating dry heat for sterilization. This type of energy does not penetrate materials easily, and therefore, long periods of exposure to high temperatures are necessary. For example, at a temperature of 160°C (320°F), a period of two hours is required for the destruction of bacterial spores. Higher temperatures are not recommended because the wrapping paper used for equipment tends to char at 180°C. The hot-air method is useful for sterilizing dry powders and water-free oily substances, as well as for many types of glassware, such as pipettes, flasks, and syringes. Dry heat does not corrode sharp instruments as steam often does, nor does it erode the ground glass surfaces of nondisposable syringes.

FIGURE 22.4

Incineration Is an Extreme Form of Heat

Smoke rising from the pyres of burning cows slaughtered to prevent the spread of foot-and-mouth disease. Nearly four million hoofed animals infected with or exposed to foot-and-mouth disease in England in 2001 were incinerated or buried.

The effect of dry heat on microorganisms is equivalent to that of baking. The heat changes microbial proteins by oxidation reactions and creates an arid internal environment, thereby burning microorganisms slowly. It is essential that organic matter such as oil or grease films be removed from the materials, because organic matter insulates against dry heat. Moreover, the time required for heat to reach sterilizing temperatures varies according to the material. This factor must be considered in determining the total exposure time.

BOILING WATER INVOLVES MOIST HEAT

Immersion in boiling water is the first of several moist-heat methods that we shall consider. Moist heat penetrates materials much more rapidly than dry heat because water molecules conduct heat better than air. Lower temperatures and a shorter exposure time are therefore required than for dry heat (FIGURE 22.5).

Moist heat kills microorganisms by denaturing their proteins. **Denaturation** is a change in the chemical or physical property of a protein. It includes structural alterations due to destruction of the chemical bonds holding proteins in a three-dimensional form. As proteins revert to a two-dimensional structure, they coagulate (denature) and become nonfunctional. Egg protein undergoes a similar transformation when it is boiled. (You might find reviewing the chemical structure of proteins in Chapter 2 helpful in understanding this process.) The coagulation of proteins requires less energy than oxidation, and, therefore, less heat need be applied.

Boiling water is not considered a sterilizing agent because the destruction of bacterial spores and the inactivation of viruses cannot always be assured. Under ordinary circumstances, with microorganisms at concentrations of less than 1 million per milliliter, most species of microorganisms can be killed within 10 minutes. Indeed, the process may require only a few seconds. However, fungal spores, protozoal cysts, and large concentrations of hepatitis A viruses require up to 30 minutes' exposure. Bacterial spores often require two hours or more. Because inadequate information exists on the heat tolerance of many species of microorganisms, boiling water is not reliable for sterilization purposes.

If it is imperative that boiling water be used to destroy microorganisms, materials must be thoroughly cleaned to remove traces of organic matter, such as blood or feces. The minimum exposure period should be 30 minutes, except at high altitudes, where it should be increased to compensate for the lower boiling point of water. All materials should be well covered. Washing soda may be added at a 2 percent concentration to increase the efficiency of the process.

THE AUTOCLAVE USES MOIST HEAT AND PRESSURE

Moist heat in the form of pressurized steam is regarded as the most dependable method for the destruction of all forms of life, including bacterial spores. This method is incorporated into a device called the **autoclave**. Over 100 years ago, French and German microbiologists developed the autoclave as an essential component of their laboratories. MicroFocus 22.1 details some of the development.

A basic principle of chemistry is that when the pressure of a gas increases, the temperature of the gas increases proportionally. Because steam is a gas, increasing its pressure in a closed system increases its temperature. As the water molecules in steam become more energized, their penetration increases substantially. This principle is used to reduce cooking time in the home pressure cooker and to reduce sterilizing

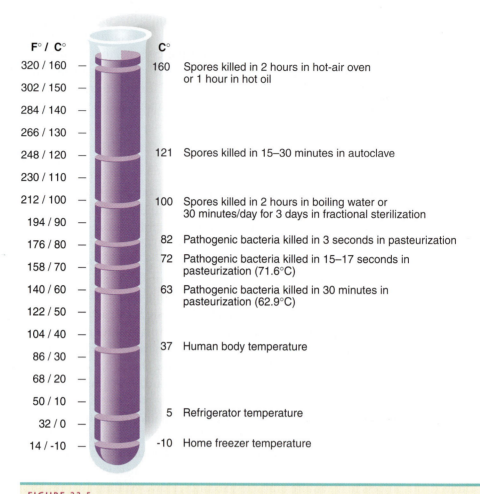

F° / C°		C°	
320 / 160	—	160	Spores killed in 2 hours in hot-air oven or 1 hour in hot oil
302 / 150	—		
284 / 140	—		
266 / 130	—		
248 / 120	—	121	Spores killed in 15–30 minutes in autoclave
230 / 110	—		
212 / 100	—	100	Spores killed in 2 hours in boiling water or 30 minutes/day for 3 days in fractional sterilization
194 / 90	—		
176 / 80	—	82	Pathogenic bacteria killed in 3 seconds in pasteurization
158 / 70	—	72	Pathogenic bacteria killed in 15–17 seconds in pasteurization (71.6°C)
140 / 60	—	63	Pathogenic bacteria killed in 30 minutes in pasteurization (62.9°C)
122 / 50	—		
104 / 40	—		
86 / 30	—	37	Human body temperature
68 / 20	—		
50 / 10	—		
32 / 0	—	5	Refrigerator temperature
14 / -10	—	-10	Home freezer temperature

FIGURE 22.5

Temperature and the Physical Control of Microorganisms

time in the autoclave. It is important to note that the sterilizing agent is the moist heat, not the pressure.

Most autoclaves contain a sterilizing chamber into which articles are placed and a steam jacket where steam is maintained, as shown in **FIGURE 22.6**. As steam flows from the steam jacket into the sterilizing chamber, cool air is forced out and a special valve increases the pressure to 15 pounds/square inch (lb/in²) above normal atmospheric pressure. The temperature rises to 121.5°C, and the superheated water molecules rapidly conduct heat into microorganisms. The time for destruction of the most resistant bacterial spore is now reduced to about 15 minutes. For denser objects, up to 30 minutes of exposure may be required. The conditions must be carefully controlled or serious problems could occur, as **FIGURE 22.7** indicates.

The autoclave is used to control microorganisms in both hospitals and laboratories. It is employed for blankets, bedding, utensils, instruments, intravenous solutions, and a broad variety of other objects. The laboratory technician uses it to sterilize bacteriological media and destroy pathogenic cultures. The autoclave is equally valuable for glassware and metalware, and is among the first instruments ordered when a microbiology laboratory is established.

MicroFocus 22.1

A HEATED CONTROVERSY

Among the last defenders of spontaneous generation was the British physician Harry Carleton Bastian. Louis Pasteur had stated that boiled urine failed to support bacterial growth, but in 1876, Bastian claimed that if the urine were alkaline, microorganisms occasionally would appear. Pasteur repeated Bastian's work and found it correct. This led Pasteur to conclude that certain microorganisms could resist death by boiling. The spores of *Bacillus subtilis*, discovered coincidentally in 1876 by Ferdinand Cohn, were an example.

Pasteur soon realized that he would have to heat his broths at a temperature higher than 100°C to achieve sterilization. He therefore put his pupil and collaborator Charles Chamberland in charge of developing a new sterilizer. Chamberland responded by constructing a pressure steam apparatus patterned after a steam "digester" invented in 1680 by the French physician Denys Papin. The sterilizer resembled a modern pressure cooker. It attained temperatures of 120°C and higher, and established the basis for the modern autoclave. Chamberland also would achieve fame

in later years for his work with porcelain filters.

But, Chamberland's invention was not universally accepted. A German

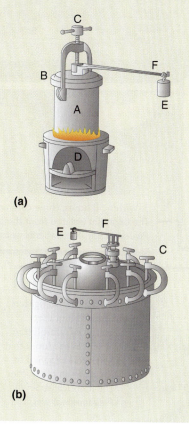

(a)

(b)

group of investigators, led by Robert Koch, criticized the pressurized steam sterilizer because they believed its higher temperatures would destroy critical laboratory media. Instead they preferred an unpressurized steam sterilizer. In 1881, the German group developed a free-flowing steam sterilizer of the type used in tyndallization. In time, however, they came to appreciate the benefits of pressurized steam as a sterilizing agent, so much so that they modified Chamberland's device to an upright model. Ironically, the instrument became known as the Koch autoclave.

■ (a) *Origins of the autoclave. Denis Papin's steam digester, designed in 1680. The digester consisted of a vessel (A) into which food was placed. The lid (B) and screw (C) sealed the vessel. A furnace (D) raised the temperature, and a weight (E) and safety lever (F) controlled the pressure of steam in the vessel. (b) Chamberland's autoclave, built in 1880 according to the principles of Papin's digester. The lid is held in place by screws (C), and a weight (E) and safety lever (F) are used to control pressure. This autoclave is basically similar to a home pressure cooker.*

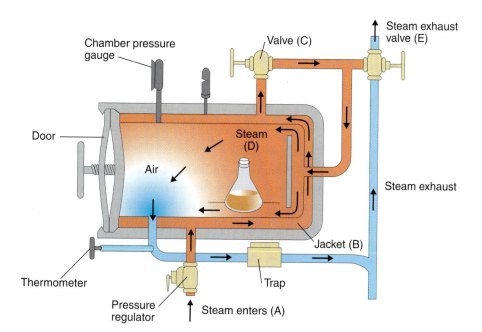

FIGURE 22.6

Operating an Autoclave

Steam enters through the port (A) and passes into the jacket (B). After the air has been exhausted through the vent, a valve (C) opens to admit pressurized steam (D), which circulates among and through the materials, thereby sterilizing them. At the conclusion of the cycle, steam is exhausted through the steam exhaust valve (E).

FIGURE 22.7

An Outbreak of *Pseudomonas* Infection Traced to a Contaminated Solution

This outbreak occurred in a Thailand hospital during 1992. It illustrates the need to carefully monitor the autoclave during use.

TEXTBOOK CASES

1. The problem began when hospital pharmacists prepared bottles of basal salts solution for use in the hospital operating rooms. To sterilize the solutions, the bottles were placed in the autoclave and left to run on its automatic cycle.

2. The bottles were then delivered to surgery to be used to irrigate the eyes of patients undergoing cataract surgery. Some bottles were left unused.

3. Within 30 hours, three cataract patients developed eye inflammations. The organism isolated from the patients was a pathogenic strain of *Pseudomonas*. The patients were treated with antibiotics.

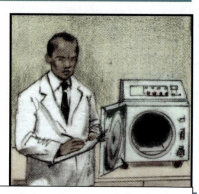

4. Health investigators tested the unused bottles of salt solution as well as the tubes that had been attached to the now-empty bottles. The identical strain of *Pseudomonas* was found.

5. Examining the pharmacy records, investigators noted that the autoclave pressure had reached only 10 to 12 lb/in², rather than the required 15 lb/in². The salts solution apparently was not sterilized.

The autoclave also has certain limitations. For example, some plasticware melts in the high heat, and sharp instruments often become dull. Moreover, many chemicals break down during the sterilization process, and oily substances cannot be treated because they do not mix with water. To gauge the success of sterilization, a strip containing spores of a *Bacillus* species is included with the objects treated (**FIGURE 22.8**). At the conclusion of the cycle, the strip is placed in nutrient broth medium and incubated. If the sterilization process has been successful, no growth will occur, but growth indicates failure.

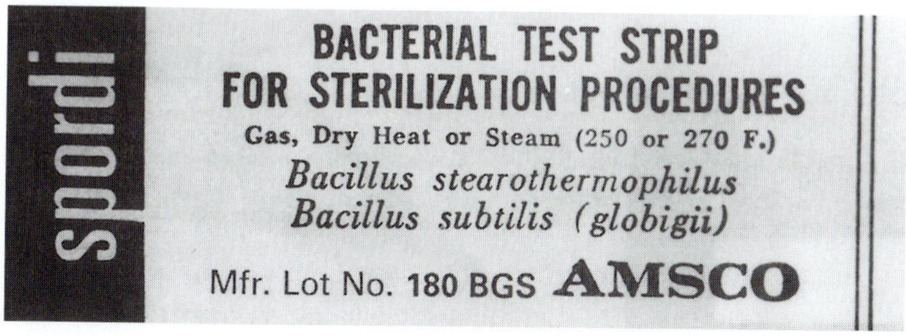

FIGURE 22.8

Testing Effective Sterilization

A bacterial test strip containing *Bacillus stearo-thermophilus* and *Bacillus subtilis*. The strip inside the package is placed into broth medium at the conclusion of the sterilization cycle. If growth fails to appear on incubation, it may be assumed that the sterilization was successful. If growth occurs in the broth, then sterilization may not be assured.

In recent years a new form of autoclave, called the **prevacuum autoclave**, has been developed for sterilization procedures. This machine draws air out of the sterilizing chamber at the beginning of the cycle. Saturated steam then is used at a temperature of 132°C to 134°C at a pressure of 28 to 30 lb/in². The time for sterilization is now reduced to as little as four minutes. A vacuum pump operates at the end of the cycle to remove the steam and dry the load. The major advantages of the prevacuum autoclave are the minimal exposure time for sterilization and the reduced time to complete the cycle.

FRACTIONAL STERILIZATION USES FREE-FLOWING STEAM

In the years before the development of the autoclave, liquids and other objects were sterilized by exposure to free-flowing steam at 100°C for 30 minutes on each of three successive days, with incubation periods between the steaming. The method was called **fractional sterilization** because a fraction of the sterilization was accomplished on each day. It was also called **tyndallization** after its developer, John Tyndall (MicroFocus 22.2).

Sterilization by the fractional method is achieved by an interesting series of events. During the first day's exposure, steam kills virtually all organisms except bacterial spores, and it stimulates spores to germinate to vegetative cells. During overnight incubation, the cells multiply and are killed on the second day. Again, the material is cooled and the few remaining spores germinate, only to be killed on the third day. Although the method usually results in sterilization, occasions arise when several spores fail to germinate. The method also requires that spores be in a suitable medium for germination, such as a broth.

Fractional sterilization has assumed renewed importance in modern microbiology with the development of high-technology instrumentation and new chemical substances. Often, these materials cannot be sterilized at autoclave temperatures, or by long periods of boiling or baking, or with chemicals. An instrument that generates free-flowing steam, such as the Arnold sterilizer, is used in these instances.

PASTEURIZATION DOES NOT STERILIZE

Pasteurization is not the same as sterilization. Its purpose is to reduce the bacterial population of a liquid such as milk and to destroy organisms that may cause spoilage and human disease (FIGURE 22.9). Spores are not affected by pasteurization.

One method for milk pasteurization, called the **holding method**, involves heating at 62.9°C for 30 minutes. Although thermophilic bacteria thrive at this temperature,

MicroFocus 22.2

TEDIOUS BUT WORTHWHILE

While Pasteur and Koch were setting down the foundations of microbiology in Europe, a British physicist named John Tyndall was developing a process for killing the most resistant forms of bacteria.

Tyndall believed that airborne microorganisms were associated with dust particles. In the early 1870s, he devised a wooden chamber with a glass front and glass side windows, and passed a beam of light through the chamber to visualize the dust. A beam of sunlight through a window displays dust particles the same way. Tyndall managed to prepare a sample of dust-free air and showed that it was free of microorganisms, thereby adding credence to Pasteur's theory that microorganisms were present in the air.

In 1876, Tyndall concluded that certain forms of bacteria were more resistant than other forms. His theory was based on observations that samples of old, dried hay were more difficult to sterilize than samples of fresh hay. Though unaware of the existence of spores, he resolved to develop a method to kill the "heat-resistant" bacteria.

Tyndall soon found that extended heating did not work well, but a stop-and-start method seemed useful. He heated hay samples to boiling on five consecutive occasions and allowed the samples to cool to room temperature between heatings. Intervals of 10 to 12 hours between heatings were found most effective in the sterilization process. Tyndall's sterilization method preceded the development of the autoclave by several years. It showed that even the most heat-resistant forms of life could be eliminated, and it made possible the use of sterile broths for verifying the germ theory of disease. Today the method is known as tyndallization, for its developer.

■ Tyndall's apparatus for observing dust particles and determining the presence of microorganisms. Dust-free air contained few microorganisms, and thus the sterile broths open to the air showed little evidence of growth on incubation. When dust was introduced through the tube, however, most broth tubes became cloudy on incubation, indicating that microorganisms were present in the dust.

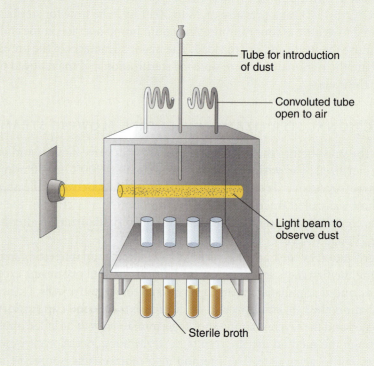

Tube for introduction of dust

Convoluted tube open to air

Light beam to observe dust

Sterile broth

they are of little consequence because they cannot grow at body temperature. For decades, pasteurization has been aimed at destroying *Mycobacterium tuberculosis*, long considered the most heat-resistant bacterium. More recently, however, attention has shifted to destruction of *Coxiella burnetii*, the agent of Q fever (Chapter 8), because this organism has a higher resistance to heat. Since both organisms are eliminated by pasteurization, dairy microbiologists assume that other pathogenic bacteria also are destroyed. Pasteurization also is used to eliminate the *Salmonella* and *Escherichia coli* that can contaminate fruit juices.

Two other methods of pasteurization are the **flash pasteurization** method at 71.6°C for 15 seconds, and the **ultrapasteurization** method at 82°C for 3 seconds. These methods are discussed in Chapter 25.

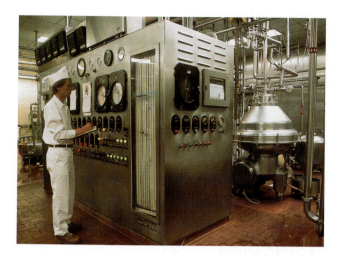

FIGURE 22.9

The Pasteurization of Milk

Milk is pasteurized by passing the liquid through a heat exchanger. The flow rate and temperature are monitored carefully. Following heating, the liquid is rapidly cooled.

HOT OIL CAN STERILIZE INSTRUMENTS

Some dentists and physicians use hot oil at 160°C for the sterilization of instruments. A time period of one hour is usually recommended. Hot oil does not rust metals, and minimal corrosion takes place. However, once sterilization is complete, the instruments must be cleaned and dried for storage, and this step may reintroduce contamination. Silicone is sometimes used as an alternative to oil.

To this point . . .

We have surveyed a number of physical methods for controlling microorganisms that involve heat. These methods are generally aimed at sterilization, a term that denotes the destruction or removal of all life-forms, including bacterial spores. Incineration with a direct flame is the most rapid heating method, but materials must be disposable. Dry heat is used in the hot-air oven where an exposure at 160°C for two hours achieves sterilization. A lower temperature of 100°C is used in boiling water, but sterilization cannot always be assured. The moist heat in boiling water is superior to the dry heat in the hot-air oven because moist heat penetrates better.

The autoclave is regarded as the most dependable instrument for sterilization. Moist heat in the form of steam under pressure is used in this instrument, and sterilization may be achieved in 15 to 30 minutes. The prevacuum autoclave uses a vacuum to draw out air at the beginning of the cycle and steam at the end, thereby reducing the cycle time. Higher pressures and temperatures also are used. We noted how fractional sterilization is a useful tool under certain circumstances, and how pasteurization is not the same as sterilization. Hot oil was mentioned briefly.

We shall now direct our attention to physical methods that do not employ heat. These methods include filtration and the use of ultraviolet light, other radiations, and ultrasonic vibrations. However, the destruction of all organisms may not be as thorough as with heating methods. The chapter will close with a review of certain preservation methods used in foods. Here, too, sterilization may not be achieved, but the number of microorganisms can be reduced substantially.

22.2

Physical Control by Other Methods

Although heat is a valuable physical agent for controlling microorganisms, sometimes it is impractical to use. For example, no one would suggest removing the microbial population from a tabletop by using a Bunsen burner, nor can heat-sensitive solutions be subjected to an autoclave. In instances such as these and numerous others, a heat-free method must be used. This section describes the most-used methods.

FILTRATION TRAPS MICROORGANISMS

Filters came into prominent use in microbiology as interest in viruses grew during the 1890s. Previous to that time, filters were used to trap airborne organisms and sterilize bacteriological media, but now they became essential for separating viruses from other microorganisms. Among the early pioneers of filter technology was Charles Chamberland, an associate of Pasteur. His porcelain filter was important to early virus research, as noted in Chapter 12. Another pioneer was Julius Petri (inventor of the Petri dish), who developed a sand filter to separate bacteria from the air.

The **filter** is a mechanical device for removing microorganisms from a solution. As fluid passes through the filter, organisms are trapped in the pores of the filtering material, as FIGURE 22.10 shows. The solution that drips into the receiving container is decontaminated or, in some cases, sterilized. Filters are used to purify such things as beverages, intravenous solutions, bacteriological media, toxoids, and many pharmaceuticals (FIGURE 22.11).

Several types of filters are used in the microbiology laboratory. **Inorganic filters** are typified by the Seitz filter, which consists of a pad of porcelain or ground glass mounted in a filter flask. **Organic filters** are advantageous because the organic molecules of the filter attract organic components in microorganisms. One example, the Berkefeld filter, uses a substance called **diatomaceous earth**. This material contains the remains of marine algae known as diatoms. Diatoms are unicellular algae that abound in the oceans and provide important foundations for the world's food chains (Chapter 16). Their remains accumulate on the shoreline and are gathered for use in swimming pool and aquarium filters, as well as for microbiological filters used in laboratories.

The **membrane filter** is a third type of filter that has received broad acceptance. It consists of a pad of organic compounds such as cellulose acetate or polycarbonate, mounted in a holding device. This filter is particularly valuable because bacteria multiply and form colonies on the filter pad when the pad is placed on a plate of culture medium. Microbiologists then can count the colonies to determine the number of bacteria originally present. For example, if a 100-ml sample of liquid was filtered and 59 colonies appeared on the pad after incubation, it could be assumed that 59 bacteria were in the sample. FIGURE 22.12 demonstrates the process.

Air also can be filtered to remove microorganisms. The filter generally used is a **high-efficiency particulate air (HEPA) filter**. This apparatus can remove over 99 percent of all particles, including microorganisms with a diameter larger than 0.3 μm. The air entering surgical units and specialized treatment facilities, such as burn units, is filtered to exclude microorganisms. In some hospital wards, such as for respiratory

FIGURE 22.10

The Principle of Filtration

Filtration is used to remove microorganisms from a liquid. The effectiveness of the filter is proportional to the size of its pores. (a) Bacteria-laden liquid is poured into a filter, and a vacuum pump helps pull the liquid through and into the flask below. The bacteria are larger than the pores of the filter, and they become trapped. The liquid dripping into the flask is sterilized if all microbial forms, including viruses, are caught. Otherwise, the liquid will remain contaminated. (b) A view of *Escherichia coli* cells trapped in the pores of a 0.45 μm nylon membrane filter. (Bar = 6 μm.)

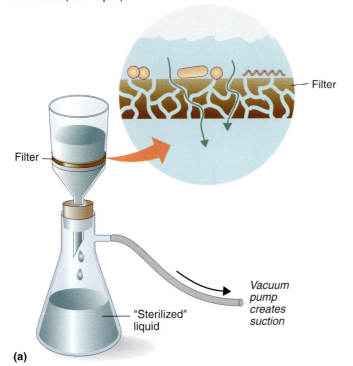

Filter

Filter

"Sterilized" liquid

Vacuum pump creates suction

(a)

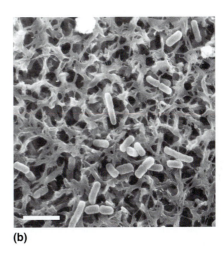

(b)

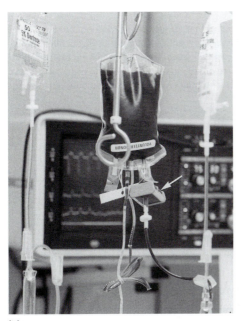

(a)

(b)

FIGURE 22.11

Industrial Fluid Filtration

Two examples of filters used in conjunction with fluids. (a) A filter of woven mesh dacron (arrow) is used to trap clumps of unwanted blood cells that might otherwise enter the recipient's circulation during a transfusion. (b) A cartridge filter removes contaminants from fluids to be used for intravenous injections or for other medical purposes.

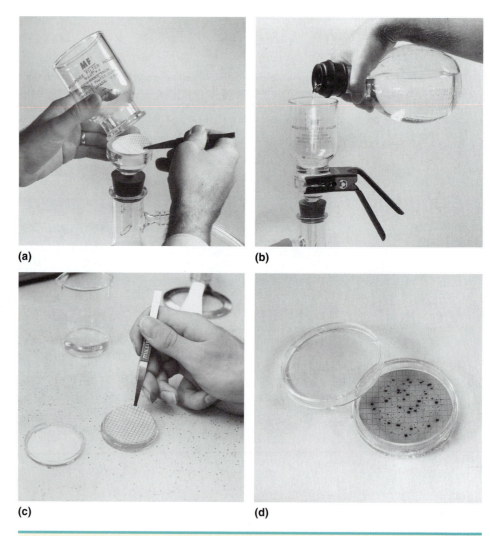

(a)

(b)

(c)

(d)

FIGURE 22.12

The Membrane Filter Technique

(a) The membrane filter consists of a pad of cellulose acetate, or similar material, mounted in a holding device. (b) The holding device is secured by a clamp, and a measured amount of fluid is filtered by pouring it into the cup. The solution runs through to a flask beneath, and bacteria are trapped in the filter material. (c) The filter pad is placed onto a plate of growth medium, and the plate is incubated. (d) After incubation, colonies appear on the surface of the filter pad. The colony count reflects the original number of bacteria in the fluid sample.

diseases, and in certain pharmaceutical filling rooms, the air is recirculated through HEPA filters to ensure its purity.

ULTRAVIOLET LIGHT CAN BE USED TO CONTROL MICROBIAL GROWTH

Visible light is a type of radiant energy detected by the light-sensitive cells of the eye. The wavelength of this energy is between 400 and 800 nanometers (nm). Other types of radiation have wavelengths longer or shorter than that of visible light, and therefore they cannot be detected by the human eye.

One type of radiant energy, **ultraviolet (UV) light**, is useful for controlling microorganisms (MicroFocus 22.3). Ultraviolet light has a wavelength between 100 and 400 nm, with the energy at about 265 nm most destructive to bacteria. When microorganisms are subjected to UV light, cellular DNA absorbs the energy, and adjacent thymine molecules link together. Linked thymine molecules are unable to encode adenine on messenger RNA molecules during the process of protein synthesis, as described in Chapter 6. Moreover, replication of the chromosome prior to binary fission is impaired. The damaged organism can no longer produce critical proteins or reproduce, and it quickly dies.

Ultraviolet light effectively reduces the microbial population where direct exposure takes place. It is used to limit airborne or surface contamination in a hospital room, morgue, pharmacy, toilet facility, or food service operation. It is noteworthy that UV light from the sun may be an important factor in controlling microorganisms in the air and upper layers of the soil, but it may not be effective against all bacterial spores. Ultraviolet light does not penetrate liquids or solids, and it may cause damage in human skin cells.

OTHER TYPES OF RADIATION ALSO CAN STERILIZE MATERIALS

The spectrum of energies (FIGURE 22.13) includes two other forms of radiation useful for destroying microorganisms. These are **X rays** and **gamma rays**. Both have wavelengths shorter than the wavelength of UV light. As X rays and gamma rays pass through microbial molecules, they force electrons out of their shells, thereby creating ions. For this reason, the radiations are called **ionizing radiations**. The ions quickly combine with and destroy proteins and nucleic acids such as DNA, causing cell death. Gram-positive bacteria are more sensitive to ionizing radiations than gram-negative bacteria. Ionizing radiations currently are used to sterilize such heat-sensitive pharmaceuticals as vitamins, hormones, and antibiotics, as well as certain plastics and suture materials.

MicroFocus 22.3

OUT OF HARM'S WAY

It has long been recognized that bacterial spores can resist the effects of ultraviolet light and survive where vegetative bacterial cells quickly die. The fact that spores remain alive for decades in soil exposed to the ultraviolet radiations of the sun is but one manifestation of the spore's resistance.

In the 1980s, scientists caught a glimpse of how this resistance works. Spore formers, it seems, have the ability to produce a certain protein during the early stages of sporulation. The protein, referred to as a small, acid-soluble spore protein (SASP), appears to protect the spore from UV light. In virtually all other bacteria, UV light affects adjacent thymine molecules in the cell's DNA and binds them together to form gnarled, double-looped structures. The disfigured DNA cannot replicate or be repaired. But in spores with SASP, there is no effect on the thymine molecules.

Then, in 1991, new light was shed on the process. Biochemists at the University of Connecticut and Boston University discovered that SASP can bind to DNA and untwist it ever so slightly. This untwisting and change in DNA's geometry apparently makes the DNA resistant to the effects of UV light.

Earth-shattering news? Perhaps not. But that's how science works. We might expect scientific experiments to have profound effects on our lives, but the general rule is that scientific endeavors rarely have an immediate impact. More often, experimental findings (like those discussed here) elicit an "Aha!" from the scientist and from colleagues in the scientific community. Then the scientist goes back to work.

FIGURE 22.13

The Ionizing and Electromagnetic Spectrum of Energies

The complete spectrum is presented at the bottom of the chart, and the ultraviolet and visible sections are expanded at the top. Notice how the bactericidal energies overlap with the UV portion of sunlight. This may account for the destruction of microorganisms in the air and in upper layers of soil.

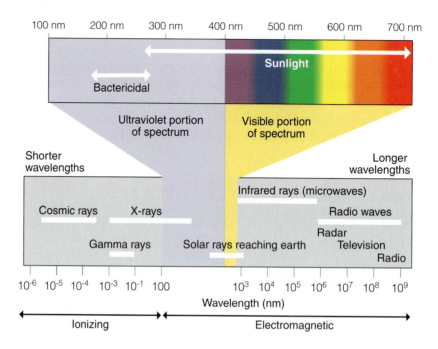

Ionizing radiations also have been approved for controlling microorganisms, and for preserving foods, as noted in MicroFocus 22.4. The approval has generated much controversy, fueled by activists concerned about the safety of factory workers and consumers. First used in 1921 to inactivate *Trichinella spiralis*, the agent of trichinosis (Chapter 17), irradiation now is used as a preservative in more than 40 countries for over 100 food items, including potatoes, onions, cereals, flour, fresh fruit, and poultry (FIGURE 22.14a). The US Food and Drug Administration (FDA) approved cobalt-60 irradiation to preserve poultry in the early 1990s, and in 1997, it extended the approval to preserve red meats such as beef, lamb, and pork. Irradiation is used to prepare many foods for the US astronauts (FIGURE 22.14b), and even was used for sterilizing the mail after the 2001 anthrax terrorist attacks.

Another form of energy, the microwave, has a wavelength longer than that of ultraviolet light and visible light. In a microwave oven, microwaves are absorbed by water molecules. The molecules are set into high-speed motion, and the heat of friction is transferred to foods, which become hot rapidly. Other than the heat generated, there is no specific activity against microorganisms (MicroFocus 22.5).

A final form of radiation we shall consider is light energy. When concentrated by sophisticated devices, light energy forms a **laser beam**. The word *laser* is an acronym for *light amplification by stimulated emission of radiation*. Recent experiments indicate that laser beams can be used to sterilize instruments and the

FIGURE 22.14

Food Irradiation

(a) The FDA has approved irradiation as a preservation method for numerous foods, including many fruits and vegetables, as well as poultry and red meats. (b) Many perishable foods eaten by NASA astronauts are prepared by irradiation.

(a)   (b)

air in operating rooms, as well as a wound surface. Microorganisms are destroyed in a fraction of a second, but the laser beam must reach all parts of the material to effect sterilization.

ULTRASONIC VIBRATIONS DISRUPT CELLS

Ultrasonic vibrations are high-frequency sound waves beyond the range of human hearing. When directed against environmental surfaces, they have little value because air particles deflect and disperse the vibrations. However, when propagated in fluids, ultrasonic vibrations cause the formation of microscopic bubbles or cavities, and the water appears to boil. Some observers call this "cold boiling." The cavities rapidly

MICROWAVES: MYTHS AND FACTS

Myth: Microwaves cook food.

Fact: It's true that microwaves cook food, but in a somewhat indirect manner. Food is composed of a loosely woven nest of large proteins, fats, carbohydrates, and other organic materials. Water molecules float freely among the materials, and, like submicroscopic magnets, they have positive and negative ends. Alternating microwave fields (2.4 billion alternations per second) spin the water molecules about at implausibly high rates, thus creating friction. The heat of friction is transferred to the surrounding food molecules as heat, and the food cooks.

Myth: Microwaves cook from the inside out.

Fact: Not really. Microwaves excite all water molecules throughout the food simultaneously. However, some heat at the food surface is lost to the surrounding air, so the outside of the food cools more quickly than the inside, and the perception is that the inside has been heated more thoroughly.

Myth: There is no solution to soggy, limp pizza heated in a microwave oven.

Fact: Sure there is, but first you have to understand why the pizza gets soggy. In a regular oven, hot water molecules from the dough come to the surface where they meet the hot oven air and evaporate, leaving the pizza crust crisp and the inside moist. In the microwave oven, the surrounding air is cool, so when the hot water molecules reach the surface, they condense to liquid and stay there or soak back into the pizza, creating a soggy mess. The answer to this problem is a metallic film mounted on a piece of cardboard. The metal heats up when exposed to microwaves, and the hot water molecules vaporize when they hit it, thereby forming a dry surface and a crisp pizza.

Myth: Microwaves are useless for sterilization.

Fact: Microwaves have potential as sterilizing agents. One Norwegian company is experimenting with a process that shreds infectious waste, sprays it with water, and exposes it to microwaves until the temperature reaches near-boiling levels. Another company is testing a procedure that sterilizes instruments after sealing them within a vacuum in a glass container. Both methods show promise.

collapse and send out shock waves. Microorganisms in the fluid are quickly disintegrated by the external pressures. The formation and implosion of the cavities are known as **cavitation**. FIGURE 22.15 illustrates this process.

Ultrasonic vibrations are valuable in research for breaking open tissue cells and obtaining their parts for study. A device called the **cavitron** is used by dentists to clean teeth, and ultrasonic machines are available for cleaning dental plates, jewelry, and coins. A major appliance company also has experimented with an ultrasonic washing machine.

As a sterilizing agent, ultrasonic vibrations have received minimal attention because liquid is required and other methods are more efficient. However, many research laboratories use ultrasonic probes for cell disruption and hospitals use ultrasonic devices to clean their instruments. When used with an effective germicide, an ultrasonic device may achieve sterilization, but the current trend is to use ultrasonic vibrations as a cleaning agent and follow the process by sterilization in an autoclave.

TABLE 22.1 summarizes the physical agents used for controlling microorganisms.

PRESERVATION METHODS RETARD SPOILAGE BY MICROORGANISMS

Over the course of many centuries, various physical methods have evolved for controlling microorganisms in food. Though valuable for preventing the spread of

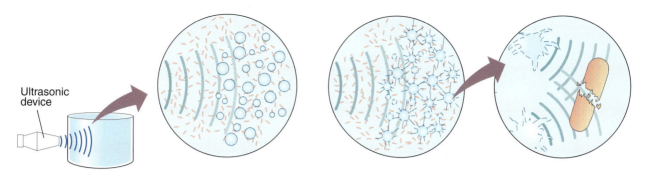

A. High-frequency sound waves cause microscopic bubbles to form in the fluid.

B. As the bubbles collapse, shock waves are created in the fluid.

C. Alternating high- and low-pressure areas impinge upon microorganisms and destroy them.

FIGURE 22.15

How Ultrasonic Vibrations Kill Microorganisms

TABLE 22.1

A Summary of Physical Agents Used to Control Microorganisms

PHYSICAL METHOD	CONDITIONS	INSTRUMENT	OBJECT OF TREATMENT	EXAMPLES OF USES	COMMENT
Direct flame	A few seconds	Flame	All microorganisms	Laboratory instruments	Object must be disposable or heat-resistant
Hot air	160°C for 2 hr	Oven	Bacterial spores	Glassware Powders Oily substances	Not useful for fluid materials
Boiling water	100°C for 10 min 100°C for 2 hr+	— —	Vegetative microorganisms Bacterial spores	Wide variety of objects	Total immersion and precleaning necessary
Pressurized steam	121°C for 15 min at 15 lb/in²	Autoclave	Bacterial spores	Instruments Surgical materials Solutions and media	Broad application in microbiology
Fractional sterilization	30 min/day for 3 successive days	Arnold sterilizer	Bacterial spores	Materials not sterilized by other methods	Long process Sterilization not assured
Pasteurization	62.9°C for 30 min 71.6°C for 15 sec	Pasteurizer	Pathogenic microorganisms	Dairy products	Sterilization not achieved
Hot oil	160°C for 1 hr	—	Bacterial spores	Instruments	Rinsing necessary
Filtration	Entrapment in pores	Berkefeld filter Membrane filter	All microorganisms	Fluids	Many adaptations
Ultraviolet light	265 nm energy	UV light	All microorganisms	Surface and air sterilization	Not useful in fluids
X rays Gamma rays	Short-wave length energy	Generator	All microorganisms	Heat-sensitive materials	Possible toxic chemicals
Ultrasonic vibrations	High-frequency sound waves	Sonicator	All microorganisms	Fluids	Few practical applications

infectious agents, these procedures are used mainly to retard spoilage and prolong the shelf life of foods, rather than for sterilization. Irradiation is an example of a preservation method.

Drying is useful in the preservation of various meats, fish, cereals, and other foods. Since water is necessary for life, it follows that where there is no water, there is virtually no life. Many of the foods in the kitchen pantry typify this principle. One example is discussed in MicroFocus 22.6.

Preservation by **salting** is based upon the principle of osmotic pressure. When food is salted, water diffuses out of microorganisms toward the higher salt concentration and lower water concentration in the surrounding environment. This flow of water, called **osmosis**, leaves the microorganisms to shrivel and die. The same phenomenon occurs in highly sugared foods such as syrups, jams, and jellies. However, fungal contamination may remain at the surface because aerobic molds tolerate high sugar concentrations.

Low temperatures found in the refrigerator and freezer retard spoilage by lowering the metabolic rate of microorganisms and thereby reducing their rate of growth. Spoilage is not totally eliminated in cold foods, however, and many microorganisms remain alive, even at freezer temperatures. These organisms multiply rapidly when food thaws, which is why prompt cooking is recommended.

Note in these examples that there are significant differences between killing microorganisms, holding them in check, and reducing their numbers. The preservation methods are described as **bacteriostatic** because they prevent the further multiplication of bacteria. A more complete discussion of food preservation as it relates to public health is presented in Chapter 25.

MicroFocus 22.6

A DRY AND TWISTED TALE

The kitchen pantry usually contains many foods that resist microbial contamination simply because they are too dry to support life. In many pantries, one of these foods is the pretzel.

According to a major manufacturer, the origin of the pretzel dates back to the early 600s and a monk in southern France or northern Italy. Legend has it that the monk took leftover strips of bread and fashioned them into twisted loops, similar to the arms folded across the chest in prayer. He baked the loops and distributed them to children who performed good deeds, such as learning their prayers well. Each child received a "pretiola" or "little reward." The name was later shortened to pretzel.

It seems unfortunate that anyone would take issue with such an appealing tale, but Tom Burnam has done just that in *The Dictionary of Misinformation*. Burnam maintains that the word *pretzel* is derived from the German word for "branch," and is not Latin for "little reward." He points out that a pretzel looks like the intertwined branches of a tree, and he attempts to punch holes in the legend by noting that folding the arms across the chest is the conventional coffin arrangement, not the position for prayer. Some traditionalists might dispute this.

Whatever the true story, the pretzel was brought to America by European immigrants, especially the Pennsylvania Dutch (really "Deutsch," for German),

who settled in the northeastern states. Today, major pretzel factories are located in Lancaster and Lititz, Pennsylvania. Their products are as dry, twisted, and free of microbial contamination as the original pretzels of centuries ago.

Note to the Student

As the science of microbiology evolved in the late 1800s, physicians were quick to grasp the notion that contaminated instruments, clothing, and similar articles were important to disease transmission. Also, researchers soon realized that contaminated growth media and materials could ruin their experiments and cast doubt on their findings. Physical methods for controlling microorganisms therefore attained acceptance rapidly, and steam and high-pressure sterilizers became commonplace in medical microbiology.

Over 100 years later, things have not changed substantially. The autoclave still occupies a prominent position in the clinical and research laboratory, and most physical methods for sterilization are essentially as they were a century ago. Even tyndallization, tedious as it is, has reemerged for use in decontaminating the products of modern biotechnology. To be sure, there are more blinking lights in today's sophisticated equipment, but it is reassuring to note that yesterday's principles are still valid. Perhaps this is one example of what is implied by the saying, "The more things change, the more they remain the same."

Summary of Key Concepts

22.1 PHYSICAL CONTROL WITH HEAT

The physical methods for controlling microorganisms are generally intended to achieve sterilization. Sterilization is the destruction or removal of all life-forms, with particular reference to the bacterial spore.

- **The Direct Flame Represents a Rapid Sterilization Method.** The direct flame method achieves sterilization in a few seconds. Incineration is an extreme example.

- **The Hot-Air Oven Uses Dry Heat.** Hot-air ovens can achieve sterilization through exposures to hot air at 160°C for two hours. Dry heat produces oxidation reactions in microbial proteins.

- **Boiling Water Involves Moist Heat.** The exposure to boiling water at 100°C for two hours also can result in sterilization, but spore destruction cannot always be assured. Moist heat causes denaturation of proteins.

- **The Autoclave Uses Moist Heat and Pressure.** The autoclave uses pressurized steam at 121°C to sterilize objects in about 15 minutes. It remains one of the most-used instruments in microbiology. A prevacuum sterilizer shortens this time still further using higher temperatures and pressures.

- **Fractional Sterilization Uses Free-Flowing Steam.** Fractional sterilization is another method for sterilization. It involves 30 minutes of exposure to steam on each of three successive days.

- **Pasteurization Does Not Sterilize.** Pasteurization reduces the microbial population in a liquid and is not intended to be a sterilization method. Several variations of the method differ in exposure time and temperature.

- **Hot Oil Can Sterilize Instruments.** Hot oil is used to sterilize some instruments. Sterilization requires treatment at 160°C for one hour.

22.2 PHYSICAL CONTROL BY OTHER METHODS

Certain nonheat methods are used to control microbial populations.

- **Filtration Traps Microorganisms.** Filters consist of various materials that trap microorganisms within the pores of the filter. Inorganic and organic filters have been used but membrane filters are the most common. Air can be filtered using a high-efficiency particulate air (HEPA) filter.

- **Ultraviolet Light Can Be Used to Control Microbial Growth.** Ultraviolet light (UV) is an effective way of killing microorganisms on a dry surface and in the air. UV light causes the formation of thymine dimers in DNA, blocking further growth of the microorganism by inhibiting DNA replication.

- **Other Types of Radiation Also Can Sterilize Materials.** X rays and gamma rays are two forms of ionizing radiation used to sterilize heat-sensitive objects. Irradiation also is used in the food industry to control microorganisms on perishable foods.

- **Ultrasonic Vibrations Disrupt Cells.** Ultrasonic vibrations could be useful also for sterilization, but more typically such sonication devices disrupt cells.

- **Preservation Methods Retard Spoilage by Microorganisms.** For food preservation, drying, salting, and low temperatures can be used to control microorganisms, but sterilization is a virtual impossibility.

Questions for Thought and Discussion

Answers to selected questions can be found in Appendix C.

1. Instead of saying that food has been irradiated, manufacturers indicate that it has been "cold pasteurized." Why do you believe they must use this deception? Do you think it is ethical? What will it take for food manufacturers to avoid the deception and use the correct term? Can you think of any place a euphemism like this one is used in foods?

2. When the local drinking water is believed to be contaminated, area residents are advised to boil their water before drinking. Often, however, they are not told how long to boil it. As a student of microbiology, what might be your recommendation?

3. The label on the container of a product in the dairy case proudly proclaims, "This dairy product is sterilized for your protection." However, a statement in small letters below reads: "Use within 30 days of purchase." Should this statement arouse your suspicion about the sterility of the product? Why?

4. Several days ago, you bought a steak and stored it in the refrigerator. Now you find that the surface of the steak has spots of bacterial contamination. You are certain that the bacteria have not penetrated deeply into the meat. Should you broil the steak, as you had planned, or discard it? Why?

5. In 1997, on approving irradiation as a preservation method for red meat, the acting commissioner of the US Food and Drug Administration indicated that irradiated red meat might cost an extra 3 to 6 cents per pound. Opponents of irradiation point to this extra cost as a deterrent to irradiation. What other deterrents might they offer?

6. Early in the century, a prehistoric woolly mammoth was discovered in the tundra of Siberia. The meat of the animal was so well preserved that it was fed to hunters' dogs. What factors contributed to the meat's preservation?

7. Suppose a liquid needed to be sterilized. What methods could be developed using only the materials found in the average household?

8. Old metropolitan buildings often had hallway chutes in which garbage could be dumped. The garbage would drop into an incinerator and burn. How might this be of value to a budding microbiologist who happened to live in the building?

9. The Scope Shield is a patented device that covers a stethoscope and protects it from disease organisms sticking to these instruments during patient exams. The device is plastic, costs 10 cents, and is meant to be discarded after a single use. Do you think it will be widely used by physicians in the ensuing years? For which organisms will it interrupt transmission?

10. The Bunsen burner was invented in 1855 by the German chemist Robert W. Bunsen. In how many different ways can it be used to sterilize objects and materials under laboratory conditions?

11. The world's oldest known pottery was made in Japan almost 13,000 years ago. With this invention, people now had watertight containers to boil or steam foods. Archaeologists believe that a population explosion soon followed in Japan. One reason is that the Japanese could for the first time avail themselves of leafy vegetables. Can you think of any other reasons?

12. In view of all the sterilization methods we have reviewed in this chapter, why do you think none has been widely adapted to the sterilization of milk? Which, in your opinion, holds the most promise?

13. Why may sunlight be referred to as nature's great sterilizing agent?

14. The word *autoclave* is derived from stems that mean "self-closing." This is a reference to the fact that the chamber closes itself by the pressure of the steam. Would it be correct to equate the words *autoclave* and *sterilizer*? Why?

15. A liquid that has been sterilized may be considered pasteurized, but one that has been pasteurized may not be considered sterilized. Why not?

http://microbiology.jbpub.com

The site features **eLearning,** an on-line review area that provides quizzes and other tools to help you study for your class. You can also follow useful links for in-depth information, or just find out the latest microbiology news.

Review

Use the following syllables to form the term that answers the clue pertaining to sterilization. The number of letters in the term is indicated by the dashes, and the number of syllables in the term is shown by the number in parentheses. Each syllable is used only once. The answers are listed in Appendix D.

A A AU BA BER BRANE BUN CIL CLAVE CLEAN CRO CU DA DALL DE DER DI DRY GAM HOLD I IC IDS IN ING ING ING INS LET LO LUS MA MEM MENTS MI MO NA O OX OS PLAS POW PRES SEN SIS SIS SIS SOL SON SPORE STRU SURE THIR TIC TION TION TO TOMS TOX TRA TRA TU TUR TY TYN UL UL VI WAVES

1. Instrument for sterilization   (3) —— —— —— —— —— —— —— —— ——

2. Type of filter   (2) —— —— —— —— —— —— —— ——

3. Sterilization in an oven   (2) —— —— —— —— —— —— ——

4. Occurs in boiling water   (5) —— —— —— —— —— —— —— —— —— —— —— ——

5. Developed fractional method   (2) —— —— —— —— —— —— ——

6. Preserves meat, fish   (2) —— —— —— —— —— ——

7. High-frequency vibrations   (4) —— —— —— —— —— —— —— —— —— ——

8. Short-wavelength rays   (2) —— —— —— —— ——

9. Most resistant life-form   (1) —— —— —— —— ——

10. Occurs in dry heat   (4) —— —— —— —— —— —— —— —— ——

11. Raised in the autoclave   (2) —— —— —— —— —— —— ——

12. Minutes for tyndallization   (2) —— —— —— —— —— ——

13. Method of pasteurization   (2) —— —— —— —— —— —— ——

14. Sterilized with hot oil   (3) —— —— —— —— —— —— —— —— —— ——

15. Source of an organic filter   (3) —— —— —— —— —— —— —— ——

16. Light for air sterilization   (5) —— —— —— —— —— —— —— —— —— —— ——

17. Melts in the autoclave   (2) —— —— —— —— —— —— ——

18. May remain after filtration   (2) —— —— —— —— —— ——

19. Not penetrated by UV light   (2) —— —— —— —— —— ——

20. Water flow from salting   (3) —— —— —— —— —— —— ——

21. Used to heat water molecules   (3) —— —— —— —— —— —— —— —— ——

22. Genus of spore formers   (3) —— —— —— —— —— —— —— ——

23. Direct flame burner   (2) —— —— —— —— —— ——

24. Essential pretreatment   (2) —— —— —— —— —— ——

25. Prevented by pasteurization   (5) —— —— —— —— —— —— —— —— —— —— ——

23

Chemical Control of Microorganisms

If it is terrifying to think that life may be at the mercy of the multiplication of infinitesimally small creatures, it is also consoling to hope that science will not always remain powerless before such enemies.

—Louis Pasteur (1822–1895)

F OR PERSONAL HYGIENE, washing our hands, taking regular showers or baths, brushing our teeth with fluoride toothpaste, and using an underarm deodorant are common practices we use to control microorganisms on our bodies (**FIGURE 23.1a**). In our homes, we try to keep microbes in check by cooking and refrigerating foods, cleaning our kitchen counters and bathrooms with disinfectant chemicals, and washing our clothes with detergents. All are important ways to maintain a sanitary environment at home.

In our attempt to be hygiene-minded consumers, sometimes we have become excessively "germphobic." The news media regularly report about this scientific study or that survey identifying places in our homes (e.g., toilets, kitchen drains) or environment (e.g., public phones, drinking-water fountains) that represent infectious dangers. Consumer groups distribute pamphlets on "microbial awareness" and numerous companies have responded by manufacturing dozens of household chemical products—some useful, many unnecessary (**FIGURE 23.1b**). These products include everything from waterless hand sanitizers to chemically treated kitchen cutting boards, children's play toys, toilet seats, toothbrushes, and even automobile steering wheels.

Our desire to protect ourselves from microbes also stems from events beyond our doorstep. The news media again report about dangerous disease outbreaks, many of which result from a lack of sanitary controls or a lack of vigilance to maintain those controls. In our

- **23.1 General Principles of Chemical Control**
  Disinfection Can Be Viewed in Different Ways
  Antiseptics and Disinfectants Have Certain Properties
  Antiseptics and Disinfectants Can Be Evaluated for Effectiveness

- **23.2 Important Chemical Agents**
  Halogens Oxidize Proteins
  Phenol and Phenolic Compounds Denature Proteins
  Heavy Metals Interfere with Microbial Metabolism
  Alcohols Denature Proteins and Disrupt Membranes

- **23.3 Other Chemical Agents**
  Aldehydes Inactivate Proteins and Nucleic Acids
  Ethylene Oxide and Other Gases React with Proteins
  Hydrogen Peroxide Damages Cellular Components
  Some Detergents React with Membranes
  Some Dyes Can Interfere with Cell Wall Structure or DNA
  Acids Enhance Chemical Effectiveness

FIGURE 23.1

Controlling Microorganisms

(a) Dental hygiene using toothpaste and mouthwash is just one way we maintain a high state of personal hygiene. (b) Modern household cleaning products are diverse and formulated for every cleaning need to maintain a sanitary condition. (c) Surgeons must thoroughly wash their hands and arms prior to surgery. They are wearing full operating gear, including gowns, face masks, and caps. (d) This African shantytown has slum houses, open sewage, and littered walkways. It is not surprising, in these unsanitary conditions, that diseases such as cholera and typhoid are common here.

(a)

(b)

(c)

(d)

communities, we expect our drinking water to be clean. That goes for our hospitals as well. It is vital to keep microbes under control in an environment where many patients carry infectious diseases and many others are susceptible to them. Nowhere is this more important than in the operating rooms and surgical wards. Here, hospital personnel must maintain scrupulous levels of cleanliness and have surgical instruments that are sterile (**FIGURE 23.1c**). Yet, **nosocomial** (hospital-acquired) **infections** do occur when hygiene barriers are breached.

Microbial control also is a global endeavor. Government and health agencies in many developing nations often lack the means (financial, medical, social) to maintain proper sanitary conditions. These circumstances can result in outbreaks of diseases such as diphtheria, malaria, measles, meningitis, and cholera. Cholera, as an example, tends to be associated with poverty-stricken areas where overcrowding and inadequate sanitation practices generate contaminated water supplies and food (FIGURE 23.1d). Although the disease can be treated effectively, 20,000 people die worldwide every year from cholera. Proper sanitary procedures are essential for the control and prevention of cholera—and most infectious diseases.

We do need to be hygiene conscious. There always will be infectious and pathogenic microbes "out there" representing a potential health menace. If the procedures and methods to control these pathogens fail or are not monitored properly, serious threats to health and well-being may occur. However, as health care providers, nurses, or simply everyday citizens, we need to recognize that the successful control of microorganisms usually requires simple methods and procedures.

In this chapter, we shall examine a variety of chemical methods used for controlling the spread of microorganisms. Our study begins by outlining some general principles and terminology of disinfection practices, and then proceeds to a discussion of the spectrum of antiseptics and disinfectants. Whether for wounds, swimming pools, industrial machinery, or pharmaceutical products, antiseptics and disinfectants are fundamental to public health practices that ensure continued good health. Louis Pasteur would be gratified that science has not remained powerless against microorganisms.

23.1 General Principles of Chemical Control

The notions of sanitation and disinfection are not unique to the modern era. The Bible refers often to cleanliness and prescribes certain dietary laws to prevent consumption of what was believed to be contaminated food. Egyptians used resins and aromatics for embalming even before they had a written language, and ancient peoples burned sulfur for deodorizing and sanitary purposes. Over the centuries, necessity demanded chemicals for food preservation, and spices were used as preservatives as well as masks for foul odors. Indeed, Marco Polo's trips to the Orient for new spices were made out of necessity as well as for adventure.

Medicinal chemicals came into widespread use in the 1800s. As early as 1830, for example, the U.S. Pharmacopoeia listed tincture of iodine as a valuable antiseptic, and soldiers in the Civil War used it in plentiful amounts. In the first decades of that century, people found copper sulfate useful for preventing fungal disease in plants, but unfortunately, this chemical was not well known during Ireland's great potato blight (Chapter 15). Mercury was used sometimes for treating syphilis, as first suggested by Arabian physicians centuries before. Moviegoers probably have noted that American cowboys practiced disinfection by pouring whiskey onto gunshot wounds between drinks.

In the 1860s, a physician at the University of Glasgow named Joseph Lister established the principles of aseptic surgery. Pasteur's references to airborne microorganisms convinced Lister that microorganisms were the cause of wound infections (MicroFocus 23.1). Lister experimented with several chemicals to kill microorganisms

MicroFocus 23.1

SURGERY WITHOUT INFECTION

While Louis Pasteur was speaking and writing about the germ theory of disease, the British physician Joseph Lister was doing his best to reform the practice of surgery.

Lister was an innovative and imaginative individual. He was, for example, among the earliest physicians to put the newly discovered anesthetics to use. By using carefully measured quantities of ether and air, Lister found he could perform operations without torture. But, it distressed him that even though an operation could be painless and successful, the patient might still die of infection. Throughout Great Britain, roughly one out of every two amputations ended in death from "hospital gangrene," "blood poisoning," or other affliction.

Toward the end of 1864, Lister read Pasteur's reports on fermentation and successfully repeated many of Pasteur's experiments. The work convinced Lister that airborne microorganisms were responsible for postsurgical diseases. He decided to test Pasteur's suspicions that microorganisms cause disease.

While searching for an antimicrobial compound, Lister's attention was drawn to a newspaper account describing the use of carbolic acid (phenol) for the treatment of sewage in a town near Glasgow. After exploring the capabilities of this compound in laboratory cultures

and finding it effective, Lister was ready to proceed with human experiments.

The first recorded use of chemical disinfection in a surgical procedure occurred in March 1865 at the Glasgow Royal Infirmary. During surgery to repair a compound fracture, Lister sprayed the air with a fine mist of carbolic acid and soaked his instruments and ligatures in carbolic acid solution. Although the patient subsequently died of infection, Lister remained optimistic. He repeated the experiments, improved the procedures, and finally met with success. In 1867, he reported his results in an article in *The Lancet*, a British medical journal. Lister wrote that his antiseptic methods reduced the mortality

rate in postoperative surgery from 45 percent to 9 percent.

Lister's work was accomplished without a clear knowledge or understanding of pathogenic microorganisms, and in this regard, his achievements merit special note. Though hospital equipment and methods have changed, the principles he established are as valid today as they were over a century ago. Before his death in 1912, he was knighted by the British Crown and is remembered as Sir Joseph Lister.

■ *A painting by Robert Thom depicting Joseph Lister using antiseptic methods in the surgical treatment of a leg wound.*

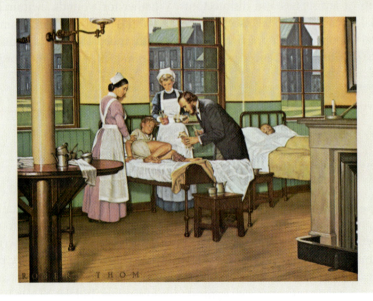

and finally settled on **carbolic acid** (**phenol**). He applied the chemical to wounds and instruments and sprayed it in the air near the operating table. Lister achieved brilliant success and was awarded many accolades as a pioneer microbiologist of his time, although he advocated the same principles recommended by Semmelweis years before.

DISINFECTION CAN BE VIEWED IN DIFFERENT WAYS

The chemical control of microorganisms extends into such diverse areas as hospital environments, food-processing plants, and everyday households. These broad fields have yielded an equally broad terminology that should be explored before we undertake a discussion of individual chemical agents.

The physical agents for controlling microorganisms (Chapter 22) generally are intended to achieve **sterilization**, the destruction of all forms of life, especially bacterial spores. Chemical agents, by contrast, rarely achieve sterilization. Instead, they are expected only to destroy the pathogenic organisms on or in an object. The process of destroying pathogens is called **disinfection** and the object is said to be disinfected. If the object is lifeless, such as a tabletop, the chemical agent is known as a **disinfectant**. However, if the object is living, such as a tissue of the human body, the chemical is an **antiseptic**. FIGURE 23.2 illustrates the fundamental difference between antiseptics and disinfectants. It is important to note that even though a particular chemical may be used as a disinfectant as well as an antiseptic (iodine, for example), the precise formulations are so different that the ability to kill microorganisms differs substantially in the two products.

Antiseptics and disinfectants are usually bactericidal, but occasionally they may be bacteriostatic. A **bactericidal agent** kills microorganisms, while a **bacteriostatic agent** temporarily prevents their further multiplication without necessarily killing them. For example, a bactericidal agent may inactivate the major enzymes of an organism and interfere with its metabolism so that it dies. A bacteriostatic agent, by contrast, disrupts a minor chemical reaction and slows the metabolism, resulting in a longer time between cell divisions. Although a delicate difference sometimes exists between the bactericidal and bacteriostatic nature of a chemical agent, the terms indicate effectiveness in a particular situation.

The word **sepsis** is derived from the Greek *seps*, meaning "putrid." It refers to the contamination of an object by microorganisms and is the stem for *septicemia*, meaning "microbial infection of the blood," and *antiseptic*, which translates to "against infection." It also is the origin of the term **aseptic**, meaning "free of contaminating microorganisms."

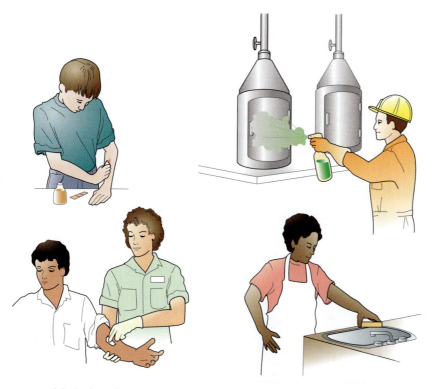

(a) Antiseptic　　　　**(b) Disinfectant**

FIGURE 23.2

Sample Uses of Antiseptics and Disinfectants

(a) Antiseptics are used on body tissues, such as on a wound or before piercing the skin to take blood. (b) A disinfectant is used on inanimate objects, such as a tabletop or equipment used in an industrial process.

Other expressions are associated with chemical control. To **sanitize** an object is to reduce the microbial population to a safe level as determined by local public health standards. For example, in dairy and food-processing plants, the equipment is usually sanitized (and the process is called sanitization). To **degerm** something is merely to remove organisms from its surface. Washing with soap and water degerms the skin surface but has little effect on microorganisms deep in the skin pores.

A final group of terms that deserves mentioning is the -*cidal* agents. Besides the bactericidal agents, there are the **fungicidal** agents, which kill fungi; the **virucidal** agents, for viruses; the **sporicidal** agents, for bacterial spores; and the **germicidal** agents, for various types of microorganisms.

ANTISEPTICS AND DISINFECTANTS HAVE CERTAIN PROPERTIES

To be useful as an antiseptic or disinfectant, a chemical agent must have certain properties, some of which are more desirable than others. The first prerequisite is that it must be able to kill or slow the growth of microorganisms. It also should be nontoxic to animals or humans, especially if it is used as an antiseptic. It should be soluble in water and have a substantial shelf life during which its activity is retained. The agent should be useful in very diluted form and perform its job in a relatively short time. Both factors tend to minimize toxic side effects.

Other characteristics also will contribute to the value of a chemical agent: It should not separate on standing, it should penetrate well, and it should not corrode instruments. The chemical will have a distinct advantage if it does not combine with organic matter such as blood or feces, because the organic matter will bind and "use up" the chemical. Of course, the chemical should be easy to obtain and relatively inexpensive.

Since disinfection is essentially a chemical process, the chemical parameters should be considered when selecting an antiseptic or disinfectant. For example, the temperature at which the disinfection is to take place may be important because a chemical reaction occurring at 37°C (body temperature) may not occur at 25°C (room temperature). Also, a particular chemical may be effective at a certain pH but not another. Moreover, the chemical reaction may be very rapid with one agent and slower with another. Thus, if long-term disinfection is desired, the second agent may be preferable.

Two other considerations are the type of microorganism to be eliminated and the surface treated. For instance, the removal of bacterial spores requires more vigorous treatment than the removal of vegetative cells. Also, a chemical applied to a laboratory bench is considerably different from one used on a wound or for sterilizing an object (**FIGURE 23.3**). It therefore is imperative to distinguish the antiseptic or

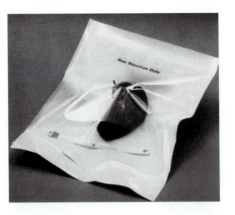

FIGURE 23.3

Packing for Chemical Sterilization

Various chemicals are available for disinfection, antisepsis, and in some cases, sterilization. The chemicals vary with the situation. In this case, a gas mask is to be gas sterilized, and it must be wrapped securely before being exposed to the sterilizing chemical. An antiseptic or disinfectant could not be used to sterilize this object.

disinfectant nature of a chemical before proceeding with its use. Indeed, chemical agents formulated as disinfectants are regulated and registered by the US Environmental Protection Agency (EPA), while chemicals formulated as antiseptics are regulated by the US Food and Drug Administration (FDA).

ANTISEPTICS AND DISINFECTANTS CAN BE EVALUATED FOR EFFECTIVENESS

At the current time, the EPA lists over 8,000 disinfectants for hospital use and thousands more for general use. Evaluating these chemical agents is a tedious process because of the broad diversity of conditions under which they are used.

A standard of effectiveness used for the chemical agents is the **phenol coefficient (PC)**. This is a number that indicates the disinfecting ability of an antiseptic or disinfectant in comparison to phenol under identical conditions (TABLE 23.1). A PC higher than 1 indicates that the chemical is more effective than phenol; a number less than 1 indicates poorer disinfecting ability than phenol. For example, antiseptic A may have a PC of 78.5, while antiseptic B has a PC of 0.28. These numbers are used relative to each other rather than to phenol because phenol is allergenic and irritating to tissues and thus is rarely used in a concentrated form.

The phenol coefficient is determined by a laboratory procedure in which dilutions of phenol and the test chemical are mixed with standardized bacteria, such as *Staphylococcus aureus* and *Salmonella typhi* or other species (FIGURE 23.4). The laboratory technician then determines which dilutions have killed the organisms after a 10-minute exposure but not after a 5-minute exposure. The test has many drawbacks, especially since it is performed in the laboratory rather than in a real-life situation. Nor does it take into account such factors as tissue toxicity, activity in the presence of organic matter, or temperature variations.

A more practical way of determining the value of a chemical agent is by an **in-use test**. For example, swab samples from a floor are taken before and after the application of a disinfectant to determine the level of disinfection. Another method is to dry standardize cultures of bacteria on small stainless steel cylinders and then expose the cylinders to the test chemical. After an established period of time, the bacteria are tested for survival rates. These methods of standardization have value

TABLE 23.1

Phenol Coefficients of Some Common Antiseptics and Disinfectants

CHEMICAL AGENT	STAPHYLOCOCCUS AUREUS	SALMONELLA TYPHI
Phenol	1.0	1.0
Chloramine	133.0	100.0
Tincture of iodine	6.3	5.8
Lysol	5.0	3.2
Mercury chloride	100.0	143.0
Ethyl alcohol	6.3	6.3
Formalin	0.3	0.7
Hydrogen peroxide	—	0.01

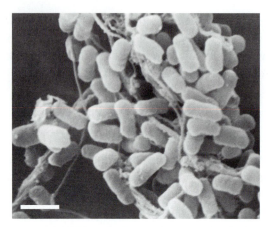

under certain circumstances. However, it is conceivable that a universal test never may be developed, in view of the huge variety of chemical agents available and the numerous conditions under which they are used.

To this point . . .

We have examined how chemicals have been used since ancient times and how increased use occurred in the 1800s as the germ theory of disease was widely propagated. We then outlined some of the important terms applied to chemical agents and distinguished disinfectants used on lifeless objects from antiseptics applied to the skin surface. Sterilization and disinfection also were compared. We explored certain specifications for selecting antiseptics and disinfectants, including characteristics of the chemical agent, the conditions under which it is used, and the type of microorganism it is intended to kill.

The discussion then turned to methods for evaluating the effectiveness of antiseptics and disinfectants, with emphasis on the phenol coefficient method. This method is limited because it is a laboratory test and does not necessarily reflect the conditions under which the chemical is to be used. Some in-use tests were briefly mentioned.

We shall now survey the broad spectrum of chemical agents used for disinfection or as an antiseptic. You will note an equally broad set of applications for these agents. We shall focus on the materials for which they are used and the mechanisms by which they kill microorganisms. Only a few agents in the first group are sterilizing agents.

23.2

Important Chemical Agents

The chemical agents currently in use for controlling microorganisms range from very simple substances, such as halogen ions, to very complex compounds, typified by detergents. Many of these agents in nature have been used for generations (MicroFocus 23.2), while others represent the latest products of chemical companies. In this section, we shall survey several groups of chemical agents and indicate how they are best applied in the chemical control of microorganisms.

MicroFocus 23.2

ANTISEPTICS IN YOUR PANTRY

Today, we live in an age when alternative and herbal medicine claims are always in the news, and these reports have generated a whole industry of health products that make often-unbelievable claims. With regard to "natural products," are there such products that have genuine medicinal and antiseptic properties? It appears so.

Cinnamon. Professor Daniel Y. C. Fung, Professor of Food Science and Food Microbiology at Kansas State University in Manhattan, Kansas, believes cinnamon might be an antiseptic that can control pathogens, at least in fruit beverages. Fung's group added cinnamon to commercially pasteurized apple juice. They then added typical foodborne pathogens (*Salmonella typhimurium, Yersinia enterocolitica*, and *Staphylococcus aureus*) and viruses. After one week of monitoring the juice at refrigerated and room temperatures, the investigators discovered the pathogens were killed more readily in the cinnamon blend than in the cinnamon-free juice. In addition, more bacteria and viruses were killed in the juice at room temperature than when refrigerated.

Garlic. In 1858, Louis Pasteur examined the properties of garlic as an antiseptic. During World War II, when penicillin and sulfa drugs were in short supply, garlic was used as an antiseptic to disinfect open wounds and prevent gangrene. Since then, numerous scientific studies have tried to discover garlic's antiseptic powers.

Many research studies have identified a group of sulfur compounds as one key to garlic's antiseptic properties. When a raw garlic clove is crushed or chewed, the active antiseptic compound is produced. Studies using garlic, at least in the laboratory, suggest that this compound is responsible for combating the microbes causing the common cold, flu, sore throat, sinusitis, and bronchitis. The findings indicate that the compound blocks key enzymes that bacteria and viruses need to invade and damage host cells.

Honey. For the past two decades, Professor Peter Molan, associate professor of biochemistry and director of the honey research unit at the University of Waikato, New Zealand, has been studying the medicinal properties of and uses for honey. Its acidity, between 3.2 and 4.5, is low enough to inhibit many pathogens. Its low water content (15 to 21 percent by weight) means that it osmotically ties up free water and "drains water" from wounds, helping to deprive bacteria of an ideal environment in which to grow. In addition, when honey encounters fluid from a wound, it slowly releases small quantities of hydrogen peroxide that are not damaging to skin tissues. It also speeds wound healing.

If that isn't enough, there also is evidence that honey protects against tooth decay. Professor Molan's group has shown that, in the lab, honey completely inhibits the growth of plaque-forming bacteria, including *Streptococcus mitis, S. sobrinus*, and *Lactobacillus caseii*. Honey cut acid production to almost zero and stopped the bacteria from producing dextran, which is a component of dental plaque. Like its use for wound infections, hydrogen peroxide probably is, in part, responsible for the antimicrobial activity.

But beware! Not all honey is alike. The antibacterial properties of honey depend on the kind of nectar, or plant pollen, that bees use to make honey. At least manuka honey from New Zealand and honeydew from central Europe are thought to contain useful levels of antiseptic potency. Professor Molan is convinced that "honey belongs in the medicine cabinet as well as the pantry."

Sushi. The green, pungent, Japanese horseradish called wasabi may be more than a spicy condiment for sushi. Professor Hedeki Masuda, director of the Material Research and Development Laboratories at Ogawa & Co. Ltd., in Tokyo, Japan, and his colleagues have found that natural chemicals in wasabi, called isothiocyanates, inhibit the growth of *Streptococcus mutans*—one of the bacterial species that cause tooth decay. Researchers tested wasabi's tooth-decay fighting ability in test tubes and found that the substance interferes with the way sugar affects teeth. At this point, these are only test-tube laboratory studies and the results will need to be proven in clinical trials.

Now, if only they could manufacture a wasabi-flavored honey! Well, maybe not.

HALOGENS OXIDIZE PROTEINS

The **halogens** are a group of highly reactive elements whose atoms have seven electrons in the outer shell. Two halogens, chlorine and iodine, are commonly used for disinfection.

Chlorine is available in a gaseous form and as both organic and inorganic compounds. It is widely used in municipal water supplies, where it keeps bacterial populations at low levels. Chlorine combines readily with numerous ions in water; therefore, enough chlorine must be added to ensure that a residue remains for antibacterial activity. In municipal water, the residue of chlorine is usually about 0.2 to 1.0 parts per million (ppm) of free chlorine. One ppm is equivalent to 0.0001 percent, an extremely small amount.

Chlorine also is available as **sodium hypochlorite (NaOCl, Clorox®)** or **calcium hypochlorite (Ca(OCl)$_2$)**. The latter, also known as chlorinated lime, was used by Semmelweis in his studies in Vienna. Hypochlorite compounds release free chlorine in solution. They are typified by the 0.5 percent sodium hypochlorite solution of H. D. Dakin, used extensively for wounds sustained in World Wars I and II. **Dakin's solution** remains popular in Europe, where it is used to treat athlete's foot.

Sodium hypochlorite is used as a bleaching agent in the textile industry; commercially available **bleach** contains about 5 percent of this compound. To disinfect clear water, the Centers for Disease Control and Prevention recommends a half-teaspoon of household chlorine bleach in two gallons of water, with 30 minutes of contact time before consumption. Hypochlorites also are useful in very dilute solutions for disinfecting swimming pools and sanitizing factory equipment (**FIGURE 23.5**).

The **chloramines**, such as chloramine-T, are organic compounds that contain chlorine. These compounds release free chlorine more slowly than hypochlorite solutions and are more stable. They are valuable for general wound antisepsis and root canal therapy.

Chlorine is effective against a broad variety of organisms, including most gram-positive and gram-negative bacteria, and many viruses, fungi, and protozoa. However, it is not sporicidal. In microorganisms, the halogen is believed to cause the release of atomic oxygen, which then combines with and inactivates certain cytoplasmic proteins, such as enzymes. Another theory is that chlorine changes the structure of cell membranes, thus leading to leakage.

The **iodine** atom is slightly larger than the chlorine atom and is more reactive and more germicidal. It is widely found in nature in such plants as marine seaweeds, where it is bound to chemical compounds. Iodine acts by halogenating tyrosine amino acids in protein molecules.

Tincture of iodine, a commonly used antiseptic for wounds, consists of 2 percent iodine and sodium iodide dissolved in ethyl alcohol. For the disinfection of clear water, the CDC recommends five drops of tincture of iodine in one quart of water, with 30 minutes of contact time before consumption. Iodine compounds in different forms are also valuable sanitizers for restaurant equipment and eating utensils.

Iodophors are iodine-detergent complexes that release iodine over a long period of time and have the added advantage of not staining tissues or fabrics. The detergent portion of the complex loosens the organisms from the surface, and the halogen kills them. Some examples of iodophors are Wescodyne, used in preoperative skin preparations; Ioprep, for presurgical scrubbing; Iosan, for dairy sanitation; and Betadine, for local wounds. Iodophors also may be combined with nondetergent

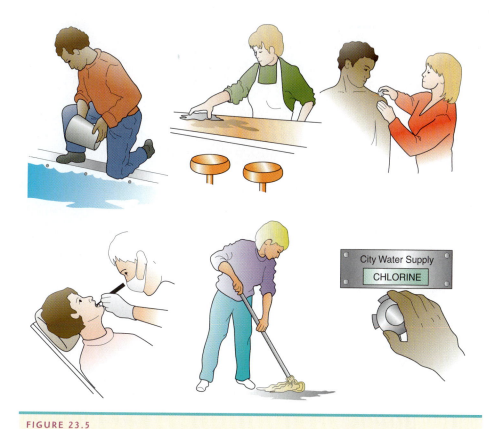

FIGURE 23.5

Some Practical Applications of Disinfection with Chlorine Compounds

carrier molecules. The best known carrier is **povidone**, which stabilizes the iodine and releases it slowly. However, compounds like these are not self-sterilizing (**FIGURE 23.6**). In 1989, for example, four cases of peritoneal *Burkholderia* (*Pseudomonas*) *cepacia* infection were related to a contaminated povidone-iodine product.

PHENOL AND PHENOLIC COMPOUNDS DENATURE PROTEINS

Phenol and phenolic compounds (phenolics) have played a key role in disinfection practices since Joseph Lister used them in the 1860s. Phenol remains the standard against which other antiseptics and disinfectants are evaluated in the phenol coefficient test. It is active against gram-positive bacteria, but its activity is reduced in the presence of organic matter. Biochemists believe that phenol and its derivatives act by denaturing proteins, especially in the cell membrane.

Phenol is expensive, has a pungent odor, and is caustic to the skin; therefore, the role of phenol as an antiseptic has diminished. However, phenol derivatives called **cresols** have greater germicidal activity and lower toxicity than the parent compound (**FIGURE 23.7**). Mixtures of ortho-, meta-, and para-cresol (creosote) are used commercially as wood preservatives for railroad ties, fenceposts, and telephone poles. Another phenol derivative, **hexylresorcinol**, is used in a mouthwash and topical antiseptic (ST37) and in throat lozenges (Sucrets). It has the added advantage of reducing surface tension, thereby loosening bacteria from the tissue and allowing greater penetration of the germicidal agent.

Cresol:
phenol derivative containing methyl groups.

FIGURE 23.6

When Disinfectants Fail to Work

A scanning electron micrograph of *Burkholderia cepacia*. This species was found embedded in the interior surface of a pipe located in a manufacturing plant where iodine disinfectants were produced. Investigators located the bacteria in the pipe after a series of nosocomial diseases were traced to a contaminated iodine product made at the plant. (Bar = 1 μm.)

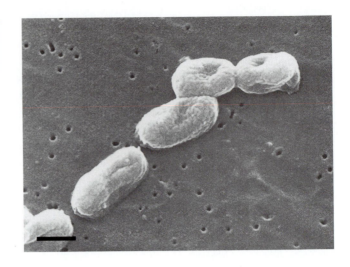

Phenol

Bisphenol

Hexylresorcinol

Para-cresol

Hexachlorophene

Meta-cresol

Ortho-cresol

Orthophenylphenol

Chlorhexidine

FIGURE 23.7

Phenol Derivatives

The chemical structures of some important derivatives of phenol used in disinfection and antisepsis.

Combinations of two phenol molecules called **bisphenols** are prominent in modern disinfection and antisepsis. **Orthophenylphenol**, for example, is used in Lysol, Osyl, Staphene, and Amphyl. Another bisphenol, **hexachlorophene**, was used extensively during the 1950s and 1960s in toothpaste (Ipana), underarm deodorant (Mum), and bath soap (Dial). One product, **pHisoHex**, combined hexachlorophene with a pH-balanced detergent cream. Pediatricians recommended it to retard staphylococcal infections of the scalp and umbilical stump, and for general cleansing of the newborn. However, a late 1960s study indicated that excessive amounts could be absorbed through the skin and cause neurological damage, and hexachlorophene was subsequently removed from over-the-counter products. The product pHisoHex is still available, but only by prescription.

An important bisphenol relative is **chlorhexidine**. This compound is used as a surgical scrub, hand wash, and superficial skin wound cleanser. A 4 percent chlorhexidine solution in isopropyl alcohol is commercially available as Hibiclens. The chemical is believed to act on the cell membrane of gram-positive and gram-negative bacteria. Chlorhexidine in a concentration of 0.2 percent also is the most extensively tested and most effective antiplaque and antigingivitis agent. However, evidence indicates that bacteria may grow within it (**FIGURE 23.8**).

A bisphenol in widespread use is **triclosan**, a broad-spectrum antimicrobial agent that destroys bacteria by disrupting cell membranes (and possibly, cell walls) by blocking the synthesis of lipids. Triclosan, known commercially as Irgasan and Ster-Zac, is fairly mild and nontoxic, and it is effective against pathogenic bacteria (but only partially against viruses and fungi). The chemical is included in antibacterial soaps, lotions, mouthwashes, toothpastes, toys, food trays, underwear, kitchen sponges, utensils, and cutting boards. Products developed by Microban, Inc. incorporate triclosan into the plastic and synthetic fibers of many of these items. The negative side to triclosan use is that bacteria may develop resistance to the chemical, just as they have developed resistance to antibiotics. Indeed, a ninth-grade student presenting her experiment at a Massachusetts science fair was among the first to draw attention to this possibility.

HEAVY METALS INTERFERE WITH MICROBIAL METABOLISM

The term *oligodynamic action,* meaning "small power," expresses the activity of heavy metals, such as mercury, silver, and copper, on microorganisms. The elements are called **heavy metals** because of their large atomic weights and complex electron configurations.

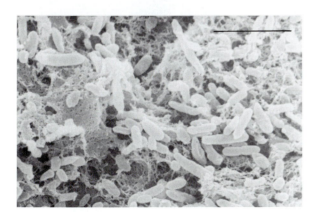

FIGURE 23.8

Contamination

A scanning electron micrograph of the inner surface of a plastic chlorhexidine bottle contaminated with *Serratia marcescens.* Epidemiologists located these bacilli following an outbreak of *S. marcescens* infections in a hospital. The study indicates that *S. marcescens* survives in chlorhexidine and may be transferred in the disinfectant. (Bar = 5 µm.)

Heavy metals are very reactive with proteins, particularly at the protein's sulfhydryl groups (—SH), and they are believed to bind protein molecules together by forming bridges between the groups. Because many of the proteins involved are enzymes, cellular metabolism is disrupted, and the microorganism dies. However, heavy metals are not sporicidal.

Mercury is one of the traditional heavy-metal antiseptics, having been used as **mercuric chloride (HgCl$_2$)** by the Greeks and Romans for treating skin diseases. However, mercury is very toxic to the host, and the antimicrobial activity of mercury is reduced when other organic matter is present. In such products as Mercurochrome, Merthiolate, and Metaphen, mercury is combined with carrier compounds and is less toxic when applied to the skin, especially after surgical incisions. Another mercury derivative is **thimerosal**, previously used as a preservative in vaccines (Chapter 20). As of 1999, the US Public Health Service recommended its removal and replacement as a safety measure.

Copper is active against chlorophyll-containing organisms and is a potent inhibitor of algae. As **copper sulfate (CuSO$_4$)**, it is incorporated into algicides and is used in swimming pools and municipal water supplies. Copper sulfate also is mixed with lime to form the bluish-white Bordeaux mixture used since 1882 to control the growth of fungi (MicroFocus 23.3).

Silver in the form of **silver nitrate (AgNO$_3$)** is useful as an antiseptic and disinfectant. For example, one drop of a 1 percent silver nitrate solution may be placed in the eyes of newborns to protect against infection by *Neisseria gonorrhoeae*. This gram-negative diplococcus can cause blindness if contracted by a

MicroFocus 23.3

FROM OUT OF THE BLUE

One October day in 1882, Professor Alexis Millardet of the University of Bordeaux was strolling through a vineyard in Medoc, France. Downy mildew of grapes, a fungal disease, had been widespread that year, and Millardet was surprised to note that the grapevines beside his path were rather healthy. By contrast, those most everywhere else were diseased. He paused to examine the leaves and found them covered with a bluish-white deposit.

Millardet inquired of the owner what the deposit might be. The owner responded that it was customary to spray the vines with a mixture of copper sulfate (blue) and lime (white) to make them look poisonous. This would quickly deter any would-be thieves from helping themselves to the grapes.

Millardet's curiosity was aroused. It was clear that the pathway vines were greener and healthier than their counterparts farther off in the field. He therefore devised a set of experiments to test the antifungal properties of the copper sulfate and lime mixture. Within three years, he concluded that the combination of chemicals was an effective deterrent to mildew, and the now famous "Bordeaux mixture" came into being. Today the mixture is one of the most widely used fungicides in all the world.

The story has an ironic twist. While mildew was ravaging French vines, other Mediterranean countries hurriedly planted their own vines, anticipating the collapse of the French wine industry. But the collapse never came. The Bordeaux mixture saved French

vines and left the neophyte grape growers with lots of grapes but nowhere to sell them.

newborn during passage through the birth canal (Chapter 11). In 1884, Karl S. F. Credé first used silver nitrate to prevent gonococcal eye disease; many states now require that Credé's method be followed (although an erythromycin or tetracycline ointment is often substituted because silver nitrate can cause irritation). Silver compounds also are used to treat suturing threads.

ALCOHOLS DENATURE PROTEINS AND DISRUPT MEMBRANES

Alcohols are effective skin antiseptics and valuable disinfectants for medical instruments. For practical use, the preferred alcohol is ethyl alcohol. **Ethyl alcohol (ethanol)** is active against vegetative bacterial cells, including the tubercle bacillus, but it has no effect on spores. It denatures proteins and dissolves lipids, an action that may lead to cell membrane disintegration. Ethyl alcohol also is a strong dehydrating agent.

Because ethyl alcohol reacts readily with any organic matter, medical instruments and thermometers must be thoroughly cleaned before exposure. Usually, a 50 to 80 percent alcohol solution is recommended because water prevents rapid evaporation and assists penetration into the tissues. A 10-minute immersion in 70 percent ethyl alcohol is generally sufficient to disinfect a thermometer or delicate instrument. Ethyl alcohol is used in many popular hand sanitizers.

Alcohol is used to preserve cosmetics, and to treat skin before a venipuncture or injection. It mechanically removes bacteria from the skin and dissolves lipids. **Isopropyl alcohol**, or rubbing alcohol, has high bactericidal activity in concentrations as high as 99 percent. Methyl alcohol (methanol) is toxic to the tissues and is used infrequently.

To this point . . .

We have discussed four major groups of chemical agents used for disinfectant and antiseptic purposes. Halogen compounds, which include chlorine and iodine, are effective against most microorganisms. The halogens can be used either in elemental forms or as derivatives, such as chloramines and iodophors. Phenol rarely is used in modern disinfection practices, but the phenolic compounds, such as bisphenols, are valuable skin cleansers. The heavy metals include mercury, copper, and silver. Silver nitrate often is used to prevent the transmission of gonorrhea to newborns. Alcohols are valuable for the disinfection of instruments, especially as 70 percent ethyl alcohol.

Through the discussions of these chemical agents you may have noted certain general considerations that apply to all. For example, the chemical agent will usually combine with any organic matter present, so that cleanliness is an important prerequisite to disinfection. Also, the chemical agents react with a wide spectrum of organisms, not just one type. Moreover, they are generally useless against bacterial spores. Virtually no chemical agent is useful within the human body, but most are employed to interrupt the spread of organisms outside the body.

In the final section of this chapter, we shall discuss three chemical agents used for sterilization purposes. These agents are sporicidal if sufficient time is given for them to act and proper conditions are established. We also shall mention a group of other agents that are used on the skin surface and in wounds.

23·3

Other Chemical Agents

The chemical agents we discussed in previous sections are used as disinfectants and antiseptics. In addition, there are some chemicals that can be used for sterilization purposes, especially for modern high-technology equipment (as well as the mundane but essential Petri dish). Three such agents are considered next.

ALDEHYDES INACTIVATE PROTEINS AND NUCLEIC ACIDS

Formaldehyde is a gas at high temperatures and a solid at room temperatures. When 37 grams of the solid are suspended in 100 ml of water, a solution called **formalin** results. For over a century, formalin was used in embalming fluid for anatomical specimens (though rarely used any more) and by morticians, as well as for disinfecting purposes (MicroFocus 23.4). In microbiology, formalin is used for inactivating viruses in certain vaccines and producing toxoids from toxins.

In the gaseous form, formaldehyde is expelled into a closed chamber where it is a sterilizing agent for surgical equipment, hospital gowns, and medical instruments. However, penetration is poor, and the surface must be exposed to the gas for up to 12 hours for effective sterilization. Instruments can be sterilized by placing them in a 20 percent solution of formaldehyde in 70 percent alcohol for 18 hours. Formaldehyde, however, leaves a residue, and instruments must be rinsed

MicroFocus 23.4

FOR PEOPLE WHO WEAR SHOES

Here's some good news for people who can't bear to part with those extraordinarily comfortable but oh-so-smelly shoes: You can get rid of the smell and enjoy many more years with them. But, before you do anything, you should understand what's going on.

The first thing you should know is that shoes become smelly when odors accumulate in the material of the shoe. The odors come from gases that bacteria produce while growing in the material. It seems that airborne cocci, recently identified as members of the genus *Micrococcus*, thrive in the sweat made by feet (about a gallon per week, in some people). This sweat is absorbed by the

shoe's material. The bacteria break down the sweat's organic components and produce sulfur compounds not unlike the hydrogen sulfide in a swamp or landfill. These sulfur compounds gather in the material and produce the odor we turn up our nose at.

So what to do? First, try to rotate your shoes as often as possible, and let them air out as long as feasible between wearings. (Some of the gas will dissipate.) Try to wear cotton, silk, or other natural fiber socks rather than synthetic materials, because bacteria thrive better in synthetic materials. This is because synthetics retain more heat, increase sweating, and limit evaporation. And use a foot powder to absorb

sweat to make life difficult for the micrococci.

Now, for those old shoes. Buy some formaldehyde at a local pharmacy, and try wiping the insides of the shoes with it. Be sure to follow all the precautions that come with the formaldehyde because it can be poisonous. You can also roll up rags, stuff them inside the shoes, and soak them with the formaldehyde. Place the shoes in a closed place, such as a box, and leave them for a few days in an airy environment where the fumes can disperse. Be sure to dry out the shoes before wearing them again. And if it works, score another one for the disinfectants.

before use. Many allergic individuals develop a contact dermatitis to this compound (Chapter 21).

Formaldehyde is an **alkylating agent**. It reacts with amino and hydroxyl groups of nucleic acids and proteins, and with carboxyl and sulfhydryl groups in proteins by inserting between them a small carbon fragment (an alkyl group) and forming bridges (**FIGURE 23.9a**). This insertion changes the structures of the molecules and interferes with an organism's biochemistry, thereby leading to death.

Glutaraldehyde has become one of the most effective chemical liquids for sterilization purposes. This small molecule destroys vegetative cells within 10 to 30 minutes and spores in 10 hours. Glutaraldehyde is an alkylating agent, usually employed as a 2 percent solution. To use it for sterilization purposes, materials have to be precleaned, immersed for 10 hours, rinsed thoroughly with sterile water, dried in a special cabinet with sterile air, and stored in a sterile container to ensure that the material remains sterile. If any of these parameters is altered, the materials may be disinfected but may not be considered sterile.

Because the activity of glutaraldehyde is not greatly reduced by organic matter, the chemical is recommended for use on surgical instruments where residual blood may be present. In addition, glutaraldehyde does not damage delicate objects, and therefore it can be used to sterilize optical equipment, such as the fiber-optic endoscopes used for arthroscopic surgery. It gives off irritating fumes, however, and instruments must be rinsed thoroughly in sterile water. At pH 7.5, glutaraldehyde kills *S. aureus* in five minutes and *Mycobacterium tuberculosis* in ten minutes.

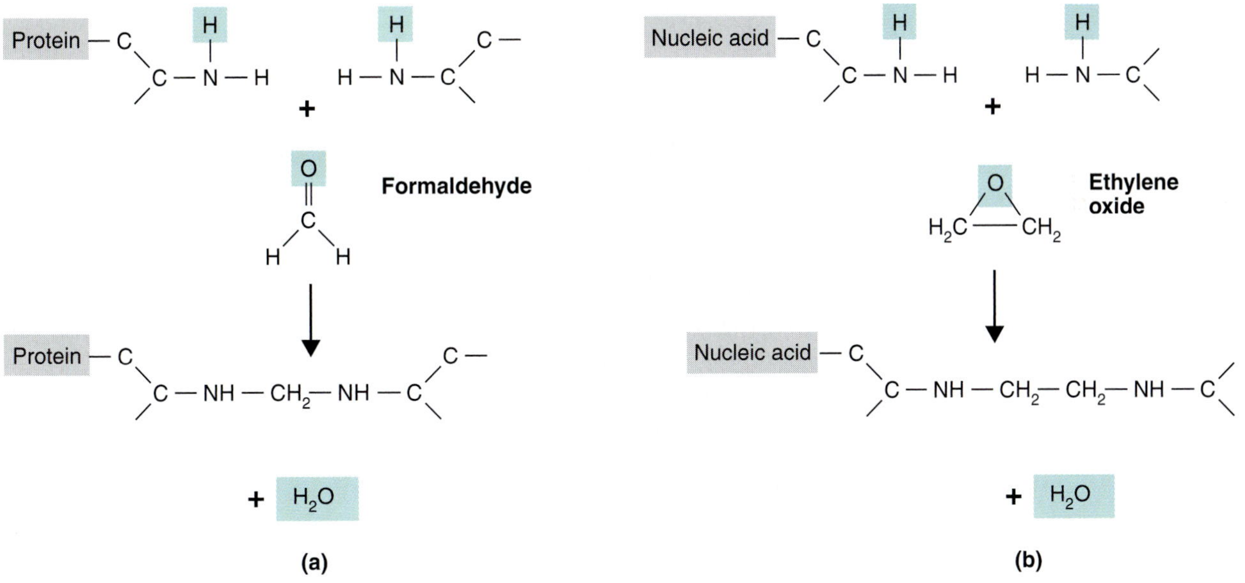

FIGURE 23.9

Alkylating Agents

The chemical reactions that take place between alkylating agents and other molecules. (a) Formaldehyde reacts with amino groups on protein molecules and forms bridges between adjacent groups. (b) Ethylene oxide reacts with the amino groups on the nitrogenous bases of nucleic acid molecules and forms bridges between adjacent groups. The effect in both cases is to alter the structure of the molecules and change the biochemistry of the microorganisms.

ETHYLENE OXIDE AND OTHER GASES REACT WITH PROTEINS

The development of plastics for use in microbiology required a suitable method for sterilizing these heat-sensitive materials. In the 1950s, research scientists discovered the antimicrobial abilities of **ethylene oxide (EtO)** that essentially made the plastic Petri dish and plastic syringe possible.

Ethylene oxide is a small molecule with excellent penetration capacity and sporicidal ability (**FIGURE 23.9b**). However, it is both carcinogenic (cancer causing) and highly explosive. Its explosiveness is reduced by mixture with Freon gas in Cryoxide or carbon dioxide gas in Carboxide, but its toxicity remains a problem for those who work with it. The gas is released into a tightly sealed chamber where it circulates for up to four hours with carefully controlled humidity (**FIGURE 23.10**). The chamber then must be flushed with inert gas for 8 to 12 hours to ensure that all traces of EtO are removed, otherwise the chemical will cause "cold burns" on contact with the skin.

Ethylene oxide is used to sterilize paper, leather, wood, metal, and rubber products, as well as plastics. In hospitals, it is used to sterilize catheters, artificial heart valves, heart-lung machine components, and optical equipment. The National Aeronautics and Space Administration (NASA) uses the gas for sterilization of interplanetary space capsules. Ethylene oxide chambers have become chemical counterparts of autoclaves for sterilization procedures. Often they are called gas autoclaves.

Chlorine dioxide has properties very similar to chloride gas and sodium hypochlorite but produces nontoxic by-products and is not a carcinogen. Chlorine dioxide can be used as a gas or liquid. In a gaseous form, with proper containment and humid-

(a)

(b)

FIGURE 23.10

Sterilization with Ethylene Oxide

(a) A bank of ethylene oxide sterilizers used to achieve "gas sterilization" or "cold sterilization" of laboratory materials. (b) Materials are wheeled into the sterilizer and subjected to a multihour process, during which the most resistant microorganisms are destroyed. Plastic Petri dishes and other plastic items are sterilized by this method.

ity, a 15-hour fumigation can be used to sanitize air ducts, food and meat processing plants, and hospital areas. In a liquid form, it can be used to decontaminate hospital and office equipment.

Sulfur dioxide is a heavy, colorless, poisonous gas with a pungent, irritating odor familiar as the smell of a just-struck match. To prevent wine spoilage, European and American winemakers use sulfur dioxide to inhibit the growth of molds and bacteria that otherwise tend to cause spoilage. Sulfur dioxide also stops oxidation (browning) and preserves the wine's natural flavor. Therefore, to make a consistently stable wine, many winemakers add sulfur dioxide, which exists as sulfites in the wine.

Sulfites also are used as a preservative and anti-browning agent in many other beverages and foods. These include dried fruits, sugar, jams, baked goods, pizza dough, frozen and dehydrated potatoes, processed vegetables, cheeses, and many prescription drugs. Approximately 10 percent of people with asthma, as well as some non-asthmatics, are sensitive to ingested sulfites. Foods containing sulfur dioxide therefore should be used with caution.

HYDROGEN PEROXIDE DAMAGES CELLULAR COMPONENTS

Hydrogen peroxide (H_2O_2) has been used as a rinse in wounds, scrapes, and abrasions. However, H_2O_2 applied to such areas foams and effervesces, as **catalase** in the tissue breaks down hydrogen peroxide to oxygen and water. The furious bubbling may remove microorganisms mechanically, but its action is short-lived. Anaerobic bacteria are sensitive to hydrogen peroxide because the sudden release of oxygen gas inhibits their growth. H_2O_2 is more effective on inanimate objects. Although bacteria may generate catalase, they are swamped by the amount of H_2O_2 applied. Hydrogen peroxide decomposition also results in a reactive form of oxygen—the superoxide radical—highly toxic to microorganisms.

New forms of H_2O_2, such as Super D hydrogen peroxide, are more stable than traditional forms and therefore do not decompose spontaneously. Such inanimate materials as soft contact lenses, utensils, heat-sensitive plastics, and food-processing equipment can be disinfected within 30 minutes. Sterilization also can be achieved after six hours of exposure to a 6 percent solution. Microbiologists currently are researching the use of hydrogen peroxide for destroying bacteria in milk and milk products.

SOME DETERGENTS REACT WITH MEMBRANES

A **soap** is a chemical compound of fatty acids combined with potassium or sodium hydroxide. The pH of the compound is usually about 8.0, and some microbial destruction is due to the alkaline conditions it establishes on the skin. However, the major activity of soap is as a degerming agent for the mechanical removal of microorganisms from the skin surface.

Soaps are **wetting agents**—that is, they emulsify and solubilize particles clinging to a surface. The surface tension also is reduced by the soap. In addition, soaps remove skin oils, further reducing the surface tension and increasing the cleaning action.

Detergents are synthetic chemicals acting as strong wetting agents and surface tension reducers. Since they are actively attracted to the phosphate groups of cellular membranes, they also alter the membranes and encourage leakage from the cytoplasm. When used to clean cutting boards (MicroFocus 23.5), they can reduce the possibility of transmitting contaminants.

MicroFocus 23.5

CUTTING BOARD WARS

He: "I'll get out a cutting board to cut up the salad."

She: "You might want to use the plastic one instead of the wood one."

He: "Why's that?"

She: "Because bacteria get caught in the grooves of the wood."

He: "But I read somewhere that the wood draws moisture and bacteria so deep into the grooves they can't reach food. The article said that plastic lets bacteria stay close to the surface so they get on the food."

She: "Well I heard that plastic cleans better than wood, so it's safer."

He: "Yeah, but wood has antibacterial powers that plastic doesn't have."

She: "Maybe so, but plastic doesn't have all those grooves and scars where bacteria can hide."

He: "How about this: I read about some guy in California who found that Salmonella *infections are more likely if a household uses a plastic board.*"

She: "Well, I don't know about anybody in California, but I do know that plastic is easier to dry thoroughly. Wood stays moist, and that lets bacteria stay alive."

He: "How about we forget the salad and go out to McDonald's?"

She: "How about Burger King?"

The **anionic detergents** yield negatively charged ions in solution. These detergents are somewhat active against gram-positive bacteria, but the negative charges of bacteria usually repel them and limit their use to common laundry products. Anionic detergents such as Triton W-30 and Duponol find value as iodophors.

Other detergents are **cationic**. These derivatives of ammonium chloride contain four organic radicals in place of the four hydrogens, and at least one radical is a long-chain alkyl group (**FIGURE 23.11**). The positively charged ammonium group is counterbalanced by a negatively charged chloride ion. Such compounds often are called **quaternary ammonium compounds** or, simply, quats. They react with cell membranes and can destroy some bacterial and enveloped viruses.

The cationic detergents have rather long, complex names, such as **benzalkonium chloride** in Zephiran and **cetylpyridium chloride** in Ceepryn. Other detergents are used in Phemerol and Diaparene. Cationic detergents are bacteriostatic on a broad range of bacteria, especially gram-positive bacteria, and are relatively stable, with little odor. They are used as sanitizing agents for industrial equipment and food utensils; as skin antiseptics; as disinfectants in mouthwashes and in storage solutions for contact lenses; and as disinfectants for use on hospital walls and floors. Their use as disinfectants for food-preparation surfaces can help reduce contamination incidents (**FIGURE 23.12**). Mixing with soap, however, reduces their activity, and certain gram-negative bacteria, such as *Burkholderia (Pseudomonas) cepacia*, can grow in them.

SOME DYES CAN INTERFERE WITH CELL WALL STRUCTURE OR DNA

Dyes are useful in microbiology as staining reagents and in laboratory media, where they help select out certain organisms from a mixture. A group of dyes called **triphenylmethane dyes** is useful as antiseptics against species of *Bacillus* and *Staphylococcus*. The group includes malachite green and crystal violet. Crystal violet has been used traditionally as **gentian violet** for trench mouth (Chapter 11), and for *Candida albicans* infections such as thrush (Chapter 15). Interference with cell wall construction appears to be the mode of activity. The dye is bactericidal at very weak dilutions of less than 1:10,000.

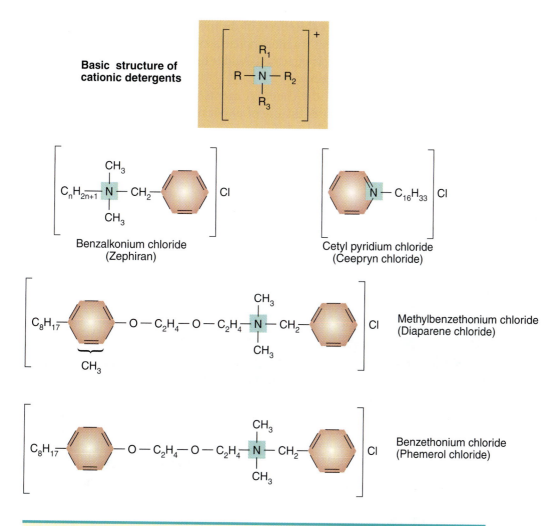

FIGURE 23.11

Cationic Detergents

The chemical structures of some important cationic detergents used in disinfection and antisepsis. Note that a long chain of carbon atoms, called an alkyl group, is included in each molecule, and that nitrogen is bonded to four radicals.

A second group of dyes, the **acridine dyes**, includes acriflavine and proflavine. Both dyes are used as antiseptics for staphylococcal infections in wounds. They apparently act by combining directly with DNA, thereby halting RNA synthesis.

ACIDS ENHANCE CHEMICAL EFFECTIVENESS

Certain acids are useful as antiseptics or disinfectants. The popular ones include **benzoic, salicylic,** and **undecylenic** acids for tinea infections of the skin. These infections are caused by various species of fungi, as noted in Chapter 15.

Organic acids are particularly valuable as food preservatives. Lactic and acetic acids, for example, are important preservatives in sour foods such as cheeses, sauerkraut, and pickled products. Propionic acid is added to bakery products to keep microbial populations low. Chapter 25 discusses these preservatives in more detail.

In general, acid enhances the effects of disinfectants and antiseptics and makes them more soluble. Also, heat is a more potent sterilizing agent if acid conditions are present, because hydrolysis is increased. Acid is therefore a valuable adjunct to disinfection, albeit in an indirect way.

TABLE 23.2 summarizes the chemical agents used in controlling microorganisms. MicroInquiry 23 explores the use of physical and chemical agents in the 2001 anthrax incidents.

TEXTBOOK CASES

FIGURE 23.12

A Case of *Plesiomonas* and *Salmonella* Infection

This incident occurred over a five-day period in 1996. An unusually high number of people became ill as a result of food contamination traced to contaminated water.

1. On June 19 and 20, it rained heavily in Livingston County, New York. The rain caused soil to run off from a local poultry farm into a stream adjoining the farm. The bacteria *Plesiomonas shigelloides* and *Salmonella* Hartford were probably in the soil, and they entered the water supply.

2. The stream ran into town and passed several feet over a well dug on the grounds of a local convenience store about 500 yards from the farm. The owner also sold pizza and catered neighborhood parties.

3. Two days later, on June 23, the caterer prepared food for a party. He made macaroni salad, potato salad, and a mixed green salad. He used the well water to wash and mix ingredients, and to clean his kitchen utensils. In doing so, he probably introduced contamination.

4. That evening, the caterer delivered the salads to the party along with a variety of other dishes and baked goods. Attending were 189 people who were celebrating the wedding anniversary of the host and hostess.

5. By June 24, 30 party guests were ill with diarrhea, nausea, abdominal cramps, and, in some cases, vomiting. Follow-up studies revealed that 43 of 56 people who ate macaroni salad became ill, and 36 of 49 guests who had potato salad were also sick. A total of 60 of the 189 attendees suffered illness.

6. Public health investigators examined the caterer's premises and were drawn to the well. The chlorinator was not operating, and there was no filtration mechanism. Lab tests revealed *Plesiomonas* and *Salmonella* in the water. They concluded that well water was the probable culprit in the outbreak.

TABLE 23.2

Summary of Chemical Agents Used to Control Microorganisms

CHEMICAL AGENT	ANTISEPTIC OR DISINFECTANT	MECHANISM OF ACTIVITY	APPLICATIONS	LIMITATIONS	ANTIMICROBIAL SPECTRUM
Chlorine	Chlorine gas Sodium hypochlorite Chloramines	Protein oxidation Membrane leakage	Water treatment Skin antisepsis Equipment spraying Food processing	Inactivated by organic matter Objectionable taste, odor	Broad variety of bacteria, fungi, protozoa, viruses
Iodine	Tincture of iodine Iodophors	Halogenates tyrosine in proteins	Skin antisepsis Food processing Preoperative preparation	Inactivated by organic matter Objectionable taste, odor	Broad variety of bacteria, fungi, protozoa, viruses
Phenol and derivative	Cresols Trichlosan Hexachlorophene Hexylresorcinol Chlorhexidine	Coagulates proteins Disrupts cell membranes	General preservatives Skin antisepsis with detergent	Toxic to tissues Disagreeable odor	Gram-positive bacteria Some fungi
Mercury	Mercuric chloride Merthiolate Metaphen	Combines with —SH groups in proteins	Skin antiseptics Disinfectants	Inactivated by organic matter Toxic to tissues Slow acting	Broad variety of bacteria, fungi, protozoa, viruses
Copper	Copper sulfate	Combines with proteins	Algicide in swimming pools Municipal water supplies	Inactivated by organic matter	Algae Some fungi
Silver	Silver nitrate	Binds proteins	Skin antiseptic Eyes of newborns	Skin irritation	Organisms in burned tissue Gonococci
Alcohol	70% ethyl alcohol	Denatures proteins Dissolves lipids Dehydrating agent	Instrument disinfectant Skin antiseptic	Precleaning necessary Skin irritation	Vegetative bacterial cells, fungi, protozoa, viruses
Formaldehyde	Formaldehyde gas Formalin	Reacts with functional groups in proteins and nucleic acids	Embalming Vaccine production Gaseous sterilant	Poor penetration Allergenic Toxic to tissues Neutralized by organic matter	Broad variety of bacteria, fungi, protozoa, viruses
Glutaraldehyde	Glutaraldehyde	Reacts with functional groups in proteins and nucleic acids	Sterilization of surgical supplies	Unstable Toxic to skin	All microorganisms, including spores
Ethylene oxide	Ethylene oxide gas	Reacts with functional groups in proteins and nucleic acids	Sterilization of instruments, equipment, heat-sensitive objects	Explosive Toxic to skin Requires constant humidity	All microorganisms, including spores
Chlorine dioxide	Chlorine dioxide gas	Reacts with functional groups in proteins and nucleic acids	Sanitizes equipment, rooms, buildings	Explosive	All microorganisms, including spores
Hydrogen peroxide	Hydrogen peroxide	Creates aerobic environment Oxidizes protein groups	Wound treatment	Limited use	Anaerobic bacteria
Cationic detergents	Commercial detergents	Dissolve lipids in cell membranes	Industrial sanitization Skin antiseptic Disinfectant	Neutralized by soap	Broad variety of microorganisms

Summary of Chemical Agents Used to Control Microorganisms (*continued*)

CHEMICAL AGENT	ANTISEPTIC OR DISINFECTANT	MECHANISM OF ACTIVITY	APPLICATIONS	LIMITATIONS	ANTIMICROBIAL SPECTRUM
Triphenyl-methane dyes	Malachite green Crystal violet	React with cytoplasmic components	Wounds Skin infection	Residual stain	Staphylococci Some fungi Gram-positive bacteria
Acridine dyes	Acriflavine Proflavine	React with cytoplasmic components	Skin infection	Residual stain	Staphylococci Gram-positive bacteria
Acids	Benzoic acid Salicylic acid Undecylinic acid Lactic and propionic acids	Alter pH	Skin infections Food preservatives	Skin irritation	Many bacteria and fungi

MicroInquiry 23 DECONTAMINATION OF ANTHRAX-CONTAMINATED MAIL AND OFFICE BUILDINGS

This chapter has examined the chemical procedures and methods used to control the numbers of microorganisms on inanimate and living objects. These control measures usually involve a level of sanitation, although some procedures may sterilize. Some examples were given for their use in the home and workplace. However, what about real instances where large-scale and extensive decontamination has to be carried out?

In October 2001, the United States experienced a bioterrorist attack. The perpetrator(s) used anthrax spores as the bioterror agent. Four anthrax-contaminated letters were sent through the mail on the same day, addressed to NBC newscaster Tom Brokaw, the *New York Post*, and to two United States Senators, Senator Patrick Leahy and Senate Majority Leader Tom Daschle (**FIGURE A**). The Centers for Disease Control and Prevention (CDC) confirmed that

anthrax spores from at least the Daschle letter contaminated the Hart Senate Office Building and several post office sorting facilities in Trenton, New Jersey (from where the letters were mailed), and Washington, D.C., areas. This resulted in the closing of the Senate building and the postal sorting facilities.

With the Senate building and postal sorting facilities closed, the CDC, the Environmental Protection Agency

(EPA), other governmental agencies, and commercial companies had to devise and implement a strategy to decontaminate these buildings and the mail sorting machines (**FIGURE B**). Let's see how well you could come up with such strategies based on your knowledge of physical and chemical control methods. Answers can be found in Appendix E.

FIGURE A

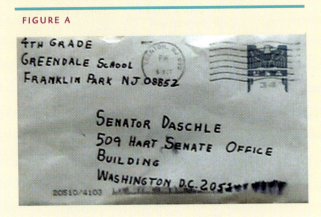

In this chapter, we have focused on a broad variety of antiseptics and disinfectants but few, you will note, are of significant value on the body surface and virtually none are used to treat internal disease. Most chemical agents are used for equipment, instruments, materials, and other related purposes. The major reason so few are available as antiseptics is simple: Chemicals will often do more harm to the human body than to the micro-organisms they are intended to kill.

Witness the spectacular rise and fall of hexachlorophene during the 1960s and 1970s. This compound's antimicrobial capabilities, residual activity, and lack of side effects appeared too good to be true. Eventually, though, it was found to be hazardous to the tissues. Similarly, many thousands of possible antiseptics have fallen by the wayside. If you have a background in chemistry, then you know of the dazzling array of available chemicals. Yet how many are useful as antiseptics?

Assignment 1: Your expertise has been asked in determining the best way to sanitize the mail to protect the mail facilities against further anthrax incidents.

23a. Based on your knowledge and reading on physical and chemical control agents, suggest a way that could be put into effect that

FIGURE B

would safeguard the mail. (Assume that money is no object!)

Assignment 2: As an "expert" on microbial control procedures, you also have been called in to advise the governmental agencies as to the best way to decontaminate the Hart Senate Office Building and the post office sorting facilities that were contaminated with *Bacillus anthracis* endospores. These are large, multi-room facilities with many pieces of furniture and instruments, including computers, copy machines, and mail sorting machines.

23b. Make a list of the most practical physical and/or chemical agents that could be used

to accomplish the decontamination task. The governmental officials also will want a plan describing how best to use these agents for decontamination.

Assignment 3: At a town meeting, a concerned citizen asks you how the government has decontaminated letters that might have been cross-contaminated with endospores from the Daschle letter as it passed through the postal sorting facilities.

23c. Suggest how such mail could be decontaminated.

23d. What could a common citizen do to decontaminate himself or herself if a letter potentially containing *B. anthracis* endospores was received.

Summary of Key Concepts

23.1 GENERAL PRINCIPLES OF CHEMICAL CONTROL

■ **Disinfection Can Be Viewed in Different Ways.** Chemical agents are effectively used to control the growth of microorganisms, even though they do not achieve sterilization. Instead, they usually destroy pathogenic microorganisms, a process called disinfection. A chemical agent used on a living object, such as the body surface, is an antiseptic; one used on a nonliving object, such as a tabletop, is a disinfectant. Most chemicals are bacteriostatic agents, although a few gases can be bactericidal.

■ **Antiseptics and Disinfectants Have Certain Properties.** Both antiseptics and disinfectants are selected according to certain criteria, including an ability to kill microorganisms or interfere with their metabolism. They are best if they are nontoxic to animals and humans, water soluble, have a long shelf life, and are effective at dilute concentrations.

■ **Antiseptics and Disinfectants Can Be Evaluated for Effectiveness.** The phenol coefficient test can be used to evaluate antiseptics and disinfectants. Chemicals are contrasted based on their effectiveness compared to phenol. In-use tests are more practical for everyday applications of antiseptics and disinfectants.

23.2 IMPORTANT CHEMICAL AGENTS

■ **Halogens Oxidize Proteins.** Halogens, such as chlorine and iodine, are useful for water disinfection, wound antisepsis, and for various forms of sanitation.

■ **Phenol and Phenolic Compounds Denature Proteins.** Phenol derivatives, such as hexachlorophene, are valuable skin antiseptics and active ingredients in presurgical scrubs. The chemical agents react with most types of organic matter (including microorganisms), so precleaning is necessary prior to disinfection.

■ **Heavy Metals Interfere with Microbial Metabolism.** Heavy metals, such as silver in silver nitrate and copper in copper sulfate are useful as antiseptics and disinfectants, respectively.

■ **Alcohols Denature Proteins and Disrupt Membranes.** Alcohol, which is commonly used in a 70 percent solution of ethyl alcohol, is an effective skin antiseptic. Precleaning and complete immersion are required for alcohol use.

23.3 OTHER CHEMICAL AGENTS

■ **Aldehydes Inactivate Proteins and Nucleic Acids.** Formaldehyde and glutaraldehyde are examples of alkylating agents that unite with amino and hydroxyl groups in proteins and nucleic acids to alter the biochemistry of microorganisms.

■ **Ethylene Oxide and Other Gases React with Proteins.** Ethylene oxide gas is carcinogenic and highly explosive. Under controlled conditions, the gas is an effective sterilant for plasticware. Chlorine dioxide gas can be used to sanitize air ducts, food and meat processing plants, and hospital areas. In a liquid form, it can be used to decontaminate hospital and office equipment. Another gas, sulfur dioxide, prevents spoilage.

■ **Hydrogen Peroxide Damages Cellular Components.** Hydrogen peroxide acts by releasing oxygen to cause an effervescing cleansing action. It is better as a disinfectant than an antiseptic.

■ **Some Detergents React with Membranes.** Soaps and anionic detergents are effective degerming agents. Quats are more effective as a disinfectant than as an antiseptic.

■ **Some Dyes Can Interfere with Cell Wall Structure or DNA.** Triphenylmethane dyes interfere with cell wall structure while acridine dyes interfere with DNA structure.

■ **Acids Enhance Chemical Effectiveness.** Acids are used as preservatives in foods and enhance the chemical properties of disinfectants.

Questions for Thought and Discussion

Answers to selected questions can be found in Appendix C.

1. Suppose you were in charge of a clinical laboratory where instruments are routinely disinfected and equipment is sanitized. A salesperson from a disinfectant company stops in to spur your interest in a new chemical agent. What questions might you ask the salesperson about the product?

2. Statistics compiled by the CDC indicate that between 5 and 10 percent of hospitalized patients (between 1.75 and 3.5 million patients per year) acquire infections during their stay. A significant percentage of these situations could be prevented, officials maintain, if hospital workers took more care in washing their hands. What reasons, do you suppose, are offered by workers for their negligence?

3. While on a camping trip, you find that a luxury hotel has been built near the stream where you once swam and from which you drank freely. Fearing contamination of the stream, you decide that before drinking the water, some form of disinfection would be wise. The nearest town has only a grocery store, pharmacy, and post office. What might you purchase? Why?

4. European manufacturers have included chlorhexidine in their toothpastes and mouthwashes for many years, but there has been resistance to this practice in the United States. Would you favor or oppose such a move? Why?

5. A study reported in 1992 in the *New England Journal of Medicine* recounted the reluctance of patients to ask their physician a vitally important question just before their examination began. Can you guess what that question is? (It is not "How much do you charge?")

6. A portable room humidifier can incubate and disseminate infectious microorganisms. If a friend asked for your recommendations on disinfecting the humidifier, what might you suggest?

7. With over 11 million children currently attending day-care centers in the United States, the possibilities for disease transmission among children has mounted considerably. Under what circumstances may antiseptics and disinfectants be used to preclude the spread of microorganisms?

8. Researchers at Virginia Polytechnic Institute have shown that apples infected with gram-negative rods can be made safe for consumption by dipping the apples in a mixture of vinegar and hydrogen peroxide, both available at grocery stores. How does each component in the mixture work?

9. In 1912, J. W. Churchman coined the word *bacteriostasis* to indicate that certain dyes were inhibitory rather than destructive to bacteria. What does the word mean today, and how is it applied to the germicides?

10. A recently published brochure called *Operation Clean Hands* lists several instances where individuals should wash their hands thoroughly. One list entitled "Before you" includes "prepare food" and "insert contact lenses." A second list entitled "After you" includes "change a diaper" and "play with an animal." How many items can you add to each list?

11. A student has finished his work in the laboratory and is preparing to leave. He remembers the instructor's precautions to wash and disinfect his hands before leaving. However, he cannot remember whether to wash first then disinfect, or to disinfect then wash. What advice might you give?

12. Before taking a blood sample from the finger, the blood bank technician commonly rubs the skin with a pad soaked in alcohol. Many people think that this procedure sterilizes the skin. Are they correct? Why?

13. A Clorox advertisement published in 1993 carries the following message: "Raw foods like chicken can carry germs and bacteria that cause salmonella sickness. It's important to kill the bacteria on any surface raw foods touch with a little Clorox. Soap and water won't do the trick." Immediately above the statement was a photograph of a raw chicken, a green pepper, three carrots, and a red Bermuda onion. At the bottom of the page a box describes "a little Clorox" as a solution made by mixing a sink full of water with ⅛ cup of Regular Clorox Liquid Bleach. How many things can you find wrong with this advertisement? Suppose you were the company microbiologist. What would you say in your version of the advertisement?

14. Suppose you had just removed the thermometer from the mouth of your sick child and confirmed

your suspicion of fever. Before checking the temperature of the next child, how would you treat the thermometer to disinfect it?

15. The water in your home aquarium always seems to resemble pea soup, but your friend's is crystal clear.

Not wanting to appear stupid, you avoid asking him his secret. But one day, in a moment of desperation, you break down and ask, whereupon he knowledgeably points to a few pennies among the gravel. What is the secret of the pennies?

Review

The chemical agents are a broad and diverse group, as this chapter has demonstrated. To test your knowledge of the chapter contents, match the chemical agent on the right to the statement on the left by placing the correct letter in the available space. A letter may be used once, more than once, or not at all. The answers are listed in Appendix D.

_____ 1. The halogen in bleach

_____ 2. Sterilizes heat-sensitive materials

_____ 3. Used to prevent gonococcal eye disease

_____ 4. Part of chlorhexidine molecule

_____ 5. Oxygen retards anaerobic bacteria

_____ 6. Seventy percent concentration recommended

_____ 7. Active ingredient in Betadine

_____ 8. Quaternary compounds, or quats

_____ 9. Can induce a contact dermatitis

_____ 10. Valuable food preservative

_____ 11. Often used as a tincture

_____ 12. Rinse for wounds and scrapes

_____ 13. Example of a heavy metal

_____ 14. Enhances antimicrobial activity of heat

_____ 15. Two molecules in hexachlorophene

_____ 16. Aids mechanical removal of organisms

_____ 17. Exerts an oligodynamic action

_____ 18. Used by Joseph Lister

_____ 19. Used for plastic Petri dishes

_____ 20. Benzoic and salicylic for tinea

_____ 21. Found in Zephiran and Diaparene

_____ 22. Used to purify waters

_____ 23 Derivatives of ammonium compounds

_____ 24. Active ingredient in Dakin's solution

_____ 25. Broken down by catalase

A. iodine

B. ethylene oxide

C. hydrogen peroxide

D. ethyl alcohol

E. acid

F. chlorine

G. glutaraldehyde

H. soap

I. phenol

J. silver

K. cationic detergent

L. dye

M. anionic detergent

N. formaldehyde

http://microbiology.jbpub.com

The site features **eLearning,** an on-line review area that provides quizzes and other tools to help you study for your class. You can also follow useful links for in-depth information, or just find out the latest microbiology news.

24 Chemotherapeutic Agents and Antibiotics

During the third and fourth years of [medical] school we also began to learn something that worried us all, although it was not much talked about. On the wards of the great Boston teaching hospitals . . . it gradually dawned on us that we could do nothing to change the course of the great majority of the diseases we were so busy analyzing, that medicine, for all its facade as a learned profession, was in real life a profoundly ignorant occupation.

—Lewis Thomas describing 1933 medicine in the preantibiotic era

For CENTURIES, PHYSICANS BELIEVED that heroic measures were necessary to save patients from the ravages of infectious disease. They prescribed frightening courses of purges and bloodlettings, enormous doses of strange chemical concoctions, blood-curdling ice water baths, deadly starvations, and other drastic remedies. These treatments probably complicated an already bad situation by reducing the natural body defenses to the point of exhaustion.

A revolution in medicine took place about 1825 when a group of doctors in Boston and London experimented to see what would happen if such treatments were withheld from patients. Surprisingly, they found that the survival rate was essentially the same—and, in some cases, better. Over the next few decades, the lessons from their experiments spread, and as the worst features of heroic therapy disappeared, doctors adopted a more conservative, nonmeddling approach to disease. It became the doctor's job to diagnose the illness, explain it to the family, predict what would happen, and then stand by and care for the patient within the limits of what was known.

When the germ theory of disease emerged in the late 1800s, insights about microorganisms added considerably to the understanding of disease and increased the storehouse of knowledge available to the doctor. However, it did not change the fact that little, if anything, could be done for the infected patient (FIGURE 24.1). Tuberculosis continued to kill

one out of every seven people; and streptococcal disease was a fatal experience, as were pneumococcal pneumonia and meningococcal meningitis.

Then, in the 1940s, chemotherapeutic agents and antibiotics burst on the scene, and another revolution in medicine began. Doctors were astonished to learn that they could kill microorganisms in the body without doing substantial harm to the body itself. Medicine had a period of powerful, decisive growth, as doctors found they could successfully alter the course of infectious disease. The chemotherapeutic agents and antibiotics effected a radical change in medicine and charted a new course that has been followed to the present day.

In this chapter, we shall discuss the antimicrobial drugs that have become mainstays of our health-care delivery system. Some have been known for generations (MicroFocus 24.1), but most are of recent vintage. We shall explore their discovery and examine their uses, while noting the important side effects attributed to many of them. When Louis Pasteur performed his experiments over 100 years ago, he implied that microorganisms could be destroyed and that some day, a way would be found to successfully treat many diseases. Only since the 1940s has Pasteur's prophecy become reality.

24.1 Chemotherapeutic Agents

Chemotherapeutic agents are chemical substances used within the body for therapeutic purposes. The term generally implies a chemical that has been synthesized by chemists or produced by a modification of a preexisting chemical. By contrast, an **antibiotic** is a product of the metabolism of a microorganism. We shall maintain that distinction in this chapter, although many antibiotics are currently produced by synthetic or semisynthetic means and are more correctly "chemotherapeutic agents." Our discussion of chemotherapeutic agents will begin with a brief review of their development.

MicroFocus 24.1

THE FEVER TREE

Rarely had a tree caused such a stir in Europe. In the 1500s, Spaniards returning from the New World told of its magical powers for malaria patients, and before long, the tree was dubbed "the fever tree." The tall evergreen grew only on the eastern slopes of the Andes Mountains. According to legend, the Countess of Chinchón, wife of the Spanish ambassador to Peru, developed malaria in 1638 and agreed to be treated with its bark. When she recovered, she spread news of the tree throughout Europe, and a century later, Linnaeus named it *Cinchona* after her.

For the next two centuries, cinchona bark remained a staple for malaria treatment. Peruvian Indians called the bark *quina-quina* (bark of bark), and the

term *quinine* gradually evolved. In 1820, two French chemists, Pierre Pelletier and Joseph Caventou, extracted pure quinine from the bark and increased its availability still further. The ensuing rush

to stockpile the chemical led to a rapid decline in the supply of cinchona trees from Peru, but Dutch farmers made new plantings in Indonesia, where the climate was similar. The island of Java eventually became the primary source of quinine for the world.

During World War II, Southeast Asia came under Japanese domination, and the supply of quinine to the West was drastically reduced. Scientists synthesized quinine shortly thereafter, but production costs were prohibitive. Finally, two useful substitutes were synthesized in chloroquine and primaquine. Today, as resistance to these drugs is increasingly observed in malarial parasites, scientists once again are looking to the fever tree to help control malaria.

THE HISTORY OF CHEMOTHERAPY STARTS WITH PAUL EHRLICH

In the drive to control and cure infectious disease, the efforts of microbiologists in the early 1900s were primarily directed toward enhancing the body's natural defenses. Sera containing antibodies lessened the impact of diphtheria, typhoid fever, and tetanus; and effective antibody-inducing vaccines for smallpox and rabies (and later, diphtheria and tetanus) reduced the incidence of these diseases.

Among the leaders in the effort to control disease was an imaginative investigator named Paul Ehrlich. Ehrlich envisioned antibody molecules as "magic bullets" that seek out and destroy disease organisms in the tissues without harming the tissues. His experiments in stain technology indicated that certain dyes also had antimicrobial qualities, and by the early 1900s, his attention had turned to magic bullets of a purely chemical nature.

One of Ehrlich's collaborators was the Japanese investigator Sahachiro Hata. Hata wished to perform research on chemical control of the syphilis spirochete *Treponema pallidum*, and Ehrlich was happy to oblige. Previously, Ehrlich and his staff had synthesized hundreds of arsenic-phenol compounds, and Hata set to work testing them for antimicrobial qualities. After months of painstaking study, Hata's attention focused on **arsphenamine**, compound #606 in the series. Hata and Ehrlich (FIGURE 24.2) successfully tested arsphenamine against *T. pallidum* in animals and human subjects, and in 1910, they made a derivative of the drug available to doctors for use against syphilis. Arsphenamine, the first modern chemotherapeutic agent, was given the common name **salvarsan** because it offered *salv*ation from syphilis and contained *ars*enic.

Salvarsan met with mixed success during the ensuing years. Its value against syphilis was without question, but local reactions at the injection site, and indiscriminate use

by some physicians, brought adverse publicity. Moreover, some church officials had used the threat of syphilis as a deterrent to immoral behavior, and they were less than enthusiastic about salvarsan's therapeutic effect. Ehrlich's death in 1915, together with the general ignorance of organic chemistry and the impending World War I, further eroded enthusiasm for chemotherapy. Instead, interest strengthened in serum and vaccine therapy for war-related diseases.

Significant advances in chemotherapy would not occur for another 20 years. During this interval, German chemists continued to synthesize and manufacture dyes for fabrics and other industries, and they routinely tested their new products for antimicrobial qualities. Among these products was a red dye, **prontosil**, synthesized in 1932.

Prontosil had no apparent effect on bacteria in culture. However, things were different in animals. When Gerhard Domagk tested prontosil in animals, he found a pronounced inhibitory effect on staphylococci, streptococci, and other gram-positive bacteria. In February 1935, Domagk injected the dye into his daughter Hildegard, who was gravely ill with septicemia. (She had pricked her finger with a needle, and blood infection followed rapidly.) Hildegard's condition gradually improved, and, to many historians, her recovery set into motion the age of modern chemotherapy. For his discovery, Gerhard Domagk was awarded the 1939 Nobel Prize in Physiology or Medicine (*in absentia*, however, because Chancellor Adolf Hitler prohibited him from accepting it).

The next great leap occurred in 1935. A group at the Pasteur Institute, headed by Jacques and Therese Tréfouël, isolated the active principle in prontosil. They found it to be **sulfanilamide (SFA)**, a substance first synthesized by Paul Gelmo in 1908. Sulfanilamide was highly active against gram-positive bacteria, and it quickly became a mainstay for treating wound-related diseases during World War II.

SULFANILAMIDE AND OTHER SULFONAMIDES INTERFERE WITH METABOLIC REACTIONS

Sulfanilamide was the first of a group of chemotherapeutic agents known as **sulfonamides**. In 1940, the British investigators D. D. Woods and E. M. Fildes proposed a mechanism of action for sulfanilamide and other sulfonamides, and provided insights about how these substances interfere with the metabolism of bacteria without damaging body tissues. The mechanism came to be known as **competitive inhibition** (FIGURE 24.3).

Certain bacteria synthesize an important molecule called **folic acid** for use in nucleic acid production. Humans cannot synthesize folic acid and must consume it in foods or vitamin capsules. However, bacteria possess the necessary enzyme to manufacture folic acid. Indeed, they are incapable of absorbing folic acid from the surrounding environment.

In the production of folic acid, the bacterial enzyme joins together three important components, one of which is **para-aminobenzoic acid (PABA)**. This molecule is similar to sulfanilamide in chemical structure (Figure 24.3). Therefore, if the environment contains large amounts of sulfanilamide, the enzyme selects the sulfanilamide molecule instead of the PABA molecule for use in folic acid production. Once combined with the enzyme, the sulfanilamide binds tightly, and by the process of competitive inhibition interferes with the enzyme, thus making it unavailable for folic acid synthesis. As the production of folic acid is reduced, nucleic acid synthesis ceases, and the bacteria die.

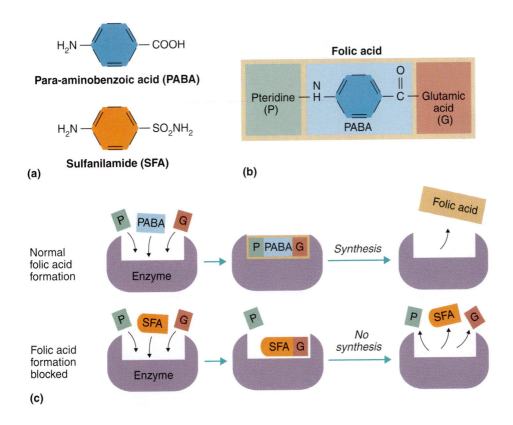

FIGURE 24.3

The Disruption of Folic Acid Synthesis by Competitive Inhibition

(a) The chemical structures of para-aminobenzoic acid (PABA) and sulfanilamide (SFA) are very similar. (b) Folic acid is made up of three components: pteridine (P), PABA, and glutamic acid (G). (c) In the normal synthesis of folic acid, a bacterial enzyme joins the three components to form folic acid. However, in competitive inhibition, the enzyme takes up SFA because it is abundant. The SFA assumes the position normally reserved for PABA, and folic acid cannot form. Without folic acid, nucleic acid metabolism is interrupted and the bacterium dies.

Sulfamethoxazole

Isoniazid

Modern sulfonamides are typified by **sulfamethoxazole**. Doctors prescribe this drug against gram-positive bacteria, as well as for urinary tract infections due to gram-negative rods. Frequently the drug is combined with **trimethoprim**, an agent that inhibits another step in folic acid synthesis. Commercially the drug combination is known as Bactrim. It is frequently used to treat *Pneumocystis* pneumonia occurring in AIDS patients (Chapter 15). Another common sulfonamide, **sulfisoxazole acetyl**, is marketed as Gantrisin cream for vaginal infections caused by gram-negative bacteria. In some patients, a drug allergy to sulfonamides develops, with a skin rash, gastrointestinal distress, or type II cytotoxic hypersensitivity (Chapter 21).

OTHER CHEMOTHERAPEUTIC AGENTS HAVE A VARIETY OF ACTIONS

The discovery and development of sulfanilamide led to the development of numerous other chemotherapeutic agents, many of which are currently in wide use. One example is the antituberculosis drug **isoniazid** (isonicotinic acid hydrazide, or INH). Biochemists believe that isoniazid interferes with cell wall synthesis in *Mycobacterium* species by inhibiting the production of mycolic acid, a component of the wall (**MicroFocus 24.2**). Isoniazid often is combined in therapy with such drugs as rifampin and ethambutol. **Ethambutol** is a synthetic, well-absorbed drug that is tuberculocidal. It also interferes with wall synthesis. Visual disturbances as a side effect limit its use to the treatment of tuberculosis.

MicroFocus 24.2

THE TROJAN HORSE

In Greek mythology, the Trojan Horse was a colossal statue of a horse in which Greek soldiers hid to gain access to the city of Troy. Now, scientists have uncovered a similar plot used by the drug isoniazid to kill *Mycobacterium tuberculosis*, the bacterium that causes tuberculosis.

What happens is this: Isoniazid, a small, synthetic molecule, enters the bacterial cytoplasm as a benign, nontoxic chemical substance no different than any nutrient passing through the cell wall and membrane. Then the drug reveals its true identity. A common enzyme called catalase activates the isoniazid and converts it to a toxic form. The activated drug, now toxic, attacks a protein used by the bacterium to synthesize mycolic acid, a key component of its cell wall. Without a strong cell wall, the bacterium is left vulnerable to

destructive elements in the environment, and it quickly dies.

In the Greek tale, the city of Troy fell to invaders. But bacteria don't read books, and in some cases, the bacterium fights back and resists the drug. It stops producing catalase by switching off the gene that encodes the enzyme. Thus, the isoniazid remains inactivated. The action also leaves the bacterium in a tenuous position because catalase breaks down hydrogen peroxide, a corrosive compound normally produced during its metabolism.

To resolve this dilemma, a second bacterial gene encodes a second enzyme (alkyl hydroperoxidase), which takes over the job of catalase and destroys the hydrogen peroxide.

In mythology, the Greeks carried the day and won the city of Troy. And in the early halcyon days of antibiotic use,

isoniazid was a prime weapon in the fight against tuberculosis. In 1998, an estimated 8 million people developed tuberculosis worldwide, and about 3 million died. Tubercle bacilli are learning to resist the Trojan Horse. Let's hope the isoniazid has another trick up its sleeve.

Three other chemotherapeutic drugs, all inhibitory to *Mycobacterium* species, deserve a brief mention. The first two are **pyrazinamide (PZA)** and **para-aminosalicylic acid (PAS)**, both used for treating tuberculosis. Their action is unknown. The third agent is diaminodiphenylsulfone, or **dapsone**, which also blocks PABA metabolism. This drug is used to treat leprosy.

Another chemotherapeutic agent, a quinolone called **nalidixic acid**, blocks DNA synthesis in certain gram-negative bacteria (*Proteus, Klebsiella, Enterobacter, Escherichia coli*) that cause urinary tract infections. Synthetic derivatives of nalidixic acid called **fluoroquinolones** also are used in urinary tract infections as well as for gonorrhea and chlamydia, and for intestinal tract infections due to gram-negative bacteria. Examples of the fluoroquinolone drugs are **ciprofloxacin** (Cipro), which became well known during the 2001 anthrax bioterror incidents, **enoxacin**, and **norfloxacin**.

Still another chemotherapeutic agent is **nitrofurantoin**, a drug actively excreted in the urine for treating urogenital infections. **Metronidazole** (Flagyl) has been used for decades against trichomoniasis, amoebiasis, and giardiasis. In cells, it forms unstable complexes that bind to DNA and inhibit replication. However, evidence that the drug causes tumors in mice has prompted physicians to use caution when prescribing it.

The treatment of malaria has long depended upon the use of **quinine**. When the tree bark used in its production became unavailable during World War II, researchers quickly set to work to develop two alternatives: chloroquine and primaquine. **Chloroquine** is effective for terminating malaria attacks; **primaquine** destroys the malaria parasites outside red blood cells. Both drugs bind to and alter DNA.

Ciprofloxacin

Quinine

To this point . . .

We have explored some of the events that led to the development of chemotherapeutic agents, first by Paul Ehrlich in the early 1900s and then by Gerhard Domagk in the 1930s. The landmark work of these investigators established the principle that chemical agents could be effective in the treatment of established diseases and encouraged other researchers to synthesize new compounds.

The focus next shifted to sulfanilamide. We described the mechanism of action of this compound and illustrated how a chemotherapeutic agent could interfere with an important metabolic process within a microorganism. Though sulfanilamide is rarely used anymore, a number of modern derivatives, such as sulfamethoxazole, often are prescribed. We also mentioned several other chemotherapeutic agents and their uses in order to see the spectrum of these drugs. A variety of gram-positive, gram-negative, and acid-fast bacteria, as well as several protozoa and fungi, can be controlled with chemotherapeutic agents.

We now shall turn our attention to the antibiotics. These are naturally occurring products of the metabolism of microorganisms. As in the previous section, we shall examine the experiments leading to the discovery of antibiotics and then proceed to a discussion of the important groups of antibiotics. Because antibiotics are a key to successful recovery from a large number of diseases, many of the names in this section should be familiar. Antiviral agents are discussed in Chapter 12 and will be considered only briefly here.

24.2

Antibiotics

The word antibiotic is derived from *antibiosis*, which literally means "against life." In 1889, the French researcher Paul Vuillemin coined the term to describe a substance he isolated some years earlier from *Pseudomonas aeruginosa*. The substance, called pyocyanin, inhibits the growth of other bacteria in culture—but it was too toxic to be useful in disease therapy. Vuillemin's term has survived as **antibiotics**, which now are considered to be chemical products or derivatives of microorganisms that are inhibitory to other microorganisms (Chapter 1).

Scientists are uncertain as to how the ability to produce antibiotics arose in some microorganisms, but it is conceivable that random genetic mutations were responsible. Clearly, the ability to produce an antibiotic conferred an extraordinary evolutionary advantage on the possessor in the struggle for survival. In this section we will discuss the sources of antibiotics, their modes of action, and side effects, and how they are used by physicians to control infectious disease. Our study will begin with Fleming's discovery of penicillin and the events that followed.

THE DISCOVERY OF ANTIBIOTICS BEGAN WITH FLEMING

One of the first to postulate the existence and value of antibiotics was the British microbiologist Alexander Fleming (**FIGURE 24.4**). During his early years, Fleming

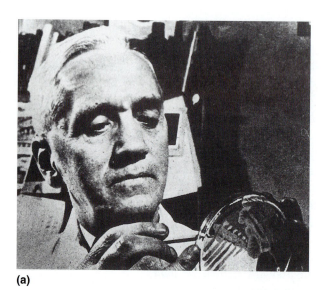

(a)

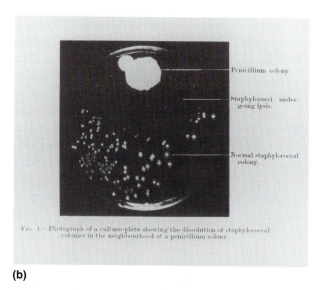

(b)

FIGURE 24.4

Penicillin and Its Discoverer

(a) Alexander Fleming, the British microbiologist who reported the existence of penicillin in 1928 but was unable to purify it for use as a therapeutic agent. (b) The actual photograph of Fleming's culture plate, originally published in the *British Journal of Experimental Pathology* in 1929. The photograph, taken by Fleming, shows how staphylococci in the region of the *Penicillium* colony have been killed (they are "undergoing lysis") by some unknown substance produced by the mold. Fleming called the substance penicillin. Ten years later, penicillin would be rediscovered and developed as the first modern antibiotic.

experienced the excitement of the classical Golden Age of microbiology and spoke up for the therapeutic value of salvarsan. In a series of experiments in 1921, he described **lysozyme**, the nonspecific enzyme that breaks down cell walls in gram-positive bacteria. MicroFocus 24.3 describes an ironic incident in Fleming's life.

The discovery of antibiotics is an elegant expression of Pasteur's dictum, "Chance favors the prepared mind." In 1928, Fleming was performing research on staphylococci at St. Mary's Hospital in London. Before going on vacation, he inoculated staphylococci onto plates of nutrient agar, and on his return, he noted that one plate was contaminated by a green mold. His interest was piqued by the failure of staphylococci to grow near the mold. Fleming isolated the mold, identified it as a species of *Penicillium*, and found that it produces a substance that kills gram-positive organisms. Though he failed to isolate the elusive substance, he named it **penicillin**. (A specimen of the *Penicillium* species in a Petri dish autographed by Fleming was sold at auction in 1998 for $13,121.)

Fleming was not the first to note the antibacterial qualities of *Penicillium* species. Joseph Lister had observed a similar phenomenon in 1871; John Tyndall did likewise in 1876; and a French medical student, Ernest Duchesne, wrote a research paper on the subject in 1896. Now, in 1928, Fleming proposed that penicillin could be used to eliminate gram-positive bacteria from mixed cultures. Further, he unsuccessfully tried the filtered broth on infected wound tissue. At the time, vaccines and sera were viewed as essential to disease therapy, and Fleming's request for financial support went unheeded. Moreover, biochemistry was not sufficiently advanced to make complex separations possible, and funds for research were limited since the Great Depression had begun. Fleming's discovery soon was forgotten.

In 1935, Gerhard Domagk's dramatic announcement of the antimicrobial effects of prontosil fueled speculation that chemicals could be used to fight disease in the body. Then, in 1939, Rene Dubos of New York's Rockefeller Institute reported that soil bacteria produce antibacterial substances. By that time, a group at England's Oxford University, led by pathologist Howard Florey and biochemist Ernst Boris Chain (FIGURE 24.5), had reisolated Fleming's penicillin and were conducting trials with highly purified samples. An article in *The Lancet* in 1940 detailed their success. However, England was involved in World War II, so a group of American companies developed the techniques for the large-scale production of penicillin and made the drug available for commercial use (MicroFocus 24.4). Fleming, Florey, and

MicroFocus 24.3

COINCIDENCE OR FATE?

In August 1961, the *Bacteriological News*, a publication of the American Society for Microbiology, carried the following story.

In the 1870s, the son of a British nobleman became mired in a bog and was in danger of losing his life. Then a Scotsman happened along. The Scotsman waded into the bog and pulled the boy free.

When the nobleman learned of the deed, he offered money to the Scotsman, but the Scotsman politely refused. Instead, the Scotsman pointed out that he also had a son and perhaps the nobleman would be willing to educate the boy. The nobleman agreed, and the bargain was struck.

The Scotsman's son was Alexander Fleming. After some years, Fleming attended St. Mary's Hospital School of Medicine, and in the course of his research, he discovered penicillin. Meanwhile, the nobleman's son also was rising to a prominent position in British politics. During World War II, he was stricken with pneumonia but was treated with penicillin and survived. His name was Winston Churchill.

FIGURE 24.5

The Developers of Penicillin

In 1940, Alexander Fleming (left, standing) learned that Howard Florey and Ernst Boris Chain (center and right, standing) had reisolated penicillin and were conducting tests on laboratory mice. He decided to visit them at the Sir William Dunn School of Pathology in Oxford, England. This painting by Robert Thom recalls the meeting of the three scientists who would share the Nobel Prize five years later. Norman Heatley, another member of the research group, is at the lower right.

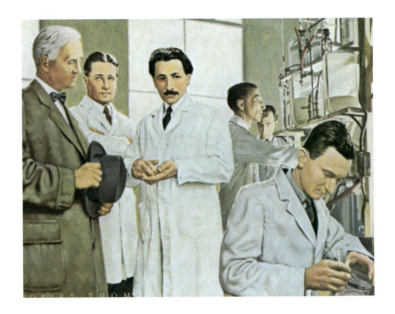

Chain shared the 1945 Nobel Prize in Physiology or Medicine for the discovery and development of penicillin.

PENICILLIN INHIBITS BACTERIAL CELL WALL SYNTHESIS

Since the 1940s, penicillin has remained the most widely used antibiotic because of its low cost and thousands of derivatives. **Penicillin G**, or benzylpenicillin, is currently the most popular penicillin antibiotic and is usually the one intended when doctors prescribe "penicillin." Other types are penicillin F and penicillin V, all with the same basic structure of a **beta-lactam nucleus** and several attached groups (FIGURE 24.6).

The penicillins are active against a variety of gram-positive bacteria, including staphylococci, streptococci, clostridia, and pneumococci (FIGURE 24.7). In higher concentrations, they also are inhibitory to the gram-negative diplococci that cause gonorrhea and meningitis, and they are useful against syphilis spirochetes. Peni-

Beta-lactam nucleus:
a distinctive chemical group central to the penicillin molecule.

MicroFocus 24.4

TRANSPORTING A TREASURE

Their timing could not have been worse. Howard Florey, Ernest B. Chain, Norman Heatley, and others of the team had rediscovered penicillin, refined it, and proven it useful in infected patients. But it was 1939, and German bombs were falling on London. This was no time for research into new drugs and medicines.

There was hope, however. Researchers in the United States were willing to attempt the industrial production of penicillin, so the British scientists would move their lab across the ocean. There were many problems, to be sure, but one was particularly interesting—how to transport the vital *Penicillium* cultures. If the molds were to fall into enemy hands or if the enemy were to learn the secret of penicillin, all their work would be wasted. Then Heatley made a suggestion: They would rub the mold on the inside linings of their coats, deposit the mold spores there, and transport the

Penicillium cultures across the ocean that way.

And so they did. On arrival in the United States, they set to work to reisolate the mold from their coat linings, and they began the laborious task of manufacturing penicillin. One of the great ironies of medicine is that virtually all the world's penicillin-producing mold has been derived from those few spores in the linings of the British coats. Few coats in history have yielded so noble a bounty.

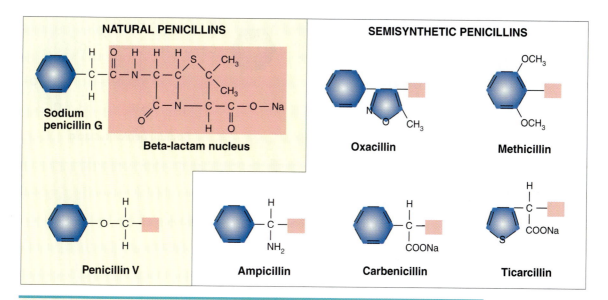

FIGURE 24.6

Some Members of the Penicillin Group of Antibiotics

The beta-lactam nucleus is common to all the penicillins. Different penicillins are formed by varying the side group on the molecule.

cillin functions during the synthesis of the bacterial cell wall. It blocks the cross-linking of carbohydrates in the peptidoglycan layer during wall formation. This results in such a weak wall that internal pressure causes the cell to swell and burst. Penicillin is therefore bactericidal in rapidly multiplying bacteria (as in an infection). Where bacteria are multiplying slowly or are dormant, the drug may have only a bacteriostatic effect, or no effect at all.

Over the years, two major drawbacks to the use of penicillin have surfaced. The first is the **anaphylactic reaction** occurring in allergic individuals (Chapter 21). This allergy applies to all compounds related to penicillin. Swelling around the eyes or wrists, flushed or itchy skin, shortness of breath, and a series of hives are signals that sensitivity exists and that penicillin therapy should cease immediately.

The second disadvantage is the evolution of penicillin-resistant bacteria. These organisms produce **penicillinase** (also called **beta-lactamase**), an enzyme that converts penicillin into harmless penicilloic acid (**FIGURE 24.8**). It is probable that the ability to produce penicillinase always has existed in certain bacterial mutants, but that the ability manifests itself when the organisms are confronted with the drug. Thus, a process of natural selectivity takes place, and the rapid multiplication of penicillinase-producing bacteria yields organisms over which penicillin has no effect. Recent years, for example, have witnessed an increase in penicillinase-producing *Neisseria gonorrhoeae* (PPNG), with the result that penicillin is now less useful for gonorrhea treatment. (Antibiotic resistance is discussed in depth later in this chapter.)

FIGURE 24.7

The Action of Penicillin on Bacteria

The *Penicillium* mold in the center of this petri dish has inhibited the growth of *Staphylococcus* colonies around the mold.

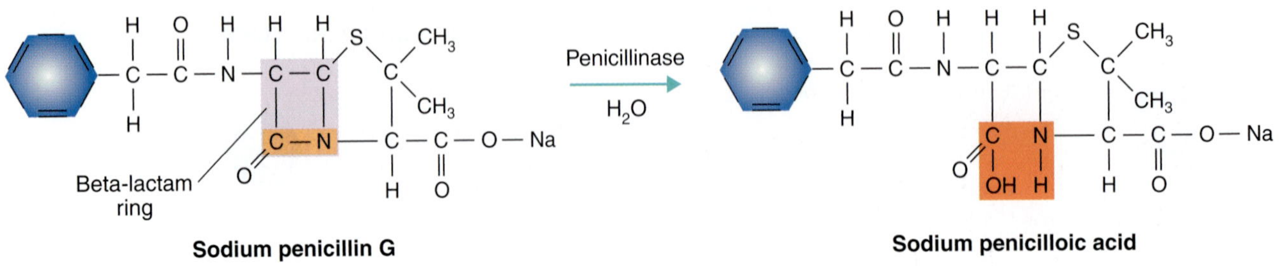

Sodium penicillin G

Beta-lactam ring

Penicillinase
H_2O

Sodium penicilloic acid

FIGURE 24.8

The Action of Penicillinase on Sodium Penicillin G

The enzyme converts penicillin to harmless penicilloic acid by opening the beta-lactam ring and inserting a hydroxyl group to the carbon and a hydrogen to the nitrogen.

A LARGE GROUP OF SEMISYNTHETIC PENICILLINS HAS BEEN DEVELOPED

In the late 1950s, the beta-lactam nucleus of the penicillin molecule was identified and synthesized, and scientists found they could attach various groups to this nucleus and create new penicillins. In the following years, thousands of **semisynthetic penicillins** emerged (Figure 24.6).

Ampicillin exemplifies a semisynthetic penicillin. It is less active against gram-positive cocci than penicillin G, but is valuable against several gram-negative rods as well as gonococci and meningococci. The drug resists stomach acid and is absorbed from the intestine after oral consumption. **Amoxicillin**, a chemical relative of ampicillin, also is acid stable and has the added advantage of not binding to food as many antibiotics do. Because ampicillin and amoxicillin are excreted into the urine, they are used to treat urinary tract infections.

Another semisynthetic penicillin, **carbenicillin**, is used primarily for infections of the urinary tract. Other semisynthetic penicillins include **methicillin, nafcillin, piperacillin,** and **oxacillin**. Still another is **ticarcillin**, a penicillin derivative often combined with **clavulanic acid** (the combination is called Timentin) for use against organisms resistant to other penicillins. The clavulanic acid inactivates penicillinase and thus overcomes the resistance. None of these drugs may be prescribed where allergy to the parent drug exists, and many have been implicated in gastrointestinal disturbances and kidney and liver damage.

Semisynthetic penicillin: a natural beta-lactam nucleus with a synthetic side chain.

CEPHALOSPORINS ALSO INHIBIT CELL WALL SYNTHESIS

While evaluating seawater samples along the coast of Sardinia in 1945, an Italian microbiologist named Giuseppe Brotzu observed a striking difference in the amount of *E. coli* in two adjoining areas. Subsequently he discovered that a fungus, *Cephalosporium acremonium*, was producing an antibacterial substance in the water. The substance, named cephalosporin C, was later isolated and characterized by scientists, and eventually it formed the basis for a family of antibiotics known as **cephalosporins**.

Cephalosporins generally are arranged in four groups, or "generations." **First-generation** cephalosporins are variably absorbed from the intestines and are useful against gram-positive cocci and certain gram-negative rods (FIGURE 24.9).

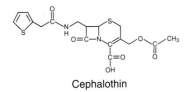

Cephalothin

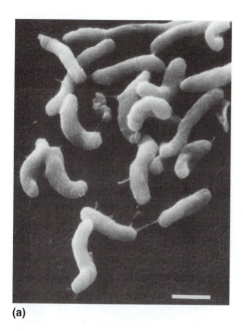

(a)

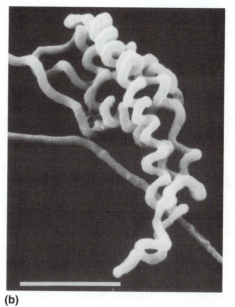

(b)

FIGURE 24.9

The Effects of Cephalexin on *Vibrio cholerae*

(a) A scanning electron micrograph of control cells grown in a medium free of antibiotic. The cells exhibit the typical vibrio shape with a short curve and an incomplete spiral. (Bar = 1 μm.) (b) Experimental cells treated with 3.13 μg of cephalexin per ml. The cells have elongated and formed right-handed spirals that are complete. (Bar = 5 μm.)

They include cephalexin (Keflex) and cephalothin (Keflin). **Second-generation** drugs are active against gram-positive cocci as well as numerous gram-negative rods (e.g., *Haemophilus influenzae*). They include cefaclor, cefoxitin, and cefuroxime (Zinacef). The **third-generation** cephalosporins are used primarily against gram-negative rods (e.g., *Pseudomonas aeruginosa*) and for treating diseases of the central nervous system. Cefotaxime (Claforan), ceftriaxone (Rocephin), and ceftazidime (Fortaz) are in the group. The **fourth generation** cephalosporins (e.g., cefepime) have improved activity against gram-negative bacteria that cause urinary tract infections.

Cephalosporins resemble penicillins in chemical structure, except that the beta-lactam nucleus has a slightly different composition. They are used as alternatives to penicillin where resistance is encountered, or in cases where penicillin allergy exists. Side effects appear to be minimal, but allergic reactions have been reported, and **thrombophlebitis** can occur. The drugs function by interfering with cell wall synthesis in bacteria. Cephalosporins are resistant to penicillinases but are sensitive to another group of beta-lactamases.

Thrombophlebitis:
a venous inflammation with an associated blood clot.

AMINOGLYCOSIDES INHIBIT PROTEIN SYNTHESIS

The **aminoglycosides** are a group of antibiotic compounds in which amino groups are bonded to carbohydrate molecules (glycosides) that are bonded to other carbohydrate molecules. All aminoglycosides attach irreversibly to bacterial ribosomes, thereby blocking the reading of the genetic code on messenger RNA molecules. Since oral absorption is negligible, the antibiotics must be administered by injection. Their use has declined in recent years with the introduction of second- and third-generation cephalosporins, and with the introduction and development of quinolone drugs, such as the fluoroquinolones.

In 1943, the first aminoglycoside was discovered by researchers led by Selman A. Waksman, a soil microbiologist at Rutgers University. Waksman's group isolated an antibacterial substance from a mold-like bacterium named *Streptomyces griseus* and named the substance **streptomycin** (MicroFocus 24.5). At the time, the discovery was sensational because streptomycin was useful against tuberculosis and

Streptomycin

MicroFocus 24.5

SERENDIPITY

They met by chance on a ship sailing from France to the United States: Rene Dubos (doo-bo'), a 23-year-old French student interested in soil science, and Selman A. Waksman, a professor of soil microbiology at Rutgers University in New Jersey. The year was 1924. For the next two decades, their lives would intertwine as each carved out a niche in modern microbiology.

As they chatted aboard the ship, Waksman suggested that Dubos come to Rutgers to earn a doctorate in microbiology. Dubos took the advice, and by 1927, he had his Ph.D. and a job at Rockefeller Institute in New York City. There he discovered a bacterial enzyme that destroys the capsules of pneumococci and hastens their death.

But it was 1931, and Dubos's work stalled because biochemistry was still in a state of infancy. Soon, Domagk's work on prontosil burst on the scene, and Dubos began searching for ways to destroy whole organisms, not just capsules. He isolated a soil bacterium, *Bacillus brevis*, and in 1939 he extracted from it an antibiotic called tyrothricin. Further extractions yielded a second antibiotic, gramicidin. Both substances killed a variety of gram-positive bacteria, but both were too toxic for use in the body.

Nevertheless, Dubos's discoveries encouraged Florey and Chain to continue their work with penicillin and showed that antimicrobial substances could be obtained from soil bacteria.

The lesson also had an impact on Selman Waksman. Over the years he had followed his former pupil's research, and in 1939 he began testing soil bacteria for antimicrobial compounds. The work was long, systematic, and plodding. In 1940, Waksman's group isolated the toxic antibiotic actinomycin, and in 1942, they found another antibiotic, streptothricin. But before this second drug could be thoroughly evaluated, streptomycin emerged.

The saga of streptomycin began in August 1943. Working with Albert Shatz and Elizabeth Bugie, Waksman isolated the mold-like bacillus *Streptomyces griseus* from the throat of a chicken. The bacillus produced streptomycin, an antibiotic with extraordinary capabilities. Preliminary studies showed its effectiveness against tubercle bacilli, and exhaustive tests at the Mayo Clinic in 1944 confirmed the results. Merck and Company soon began industrial production of the antibiotic, and within a decade, 26 companies throughout the world were manufacturing it. In 1952,

Waksman was awarded the Nobel Prize in Physiology or Medicine for his accomplishment.

Serendipity is a word derived from Horace Walpole's fairy tale *The Three Princes of Serendip*. In the tale, desirable things happen by accident or chance. Many antibiotics are the products of serendipity, but even more fundamental are the serendipitous meetings of two people such as once happened on a ship traveling from France to the United States.

■ *A scanning electron micrograph of* Streptomyces griseus, *the organism isolated by Waksman, and one of the first species of bacteria to yield an antibiotic.* (Bar = 5 μm.)

---

numerous diseases caused by gram-negative bacteria. Since then it has been largely replaced by safer drugs, but streptomycin is still prescribed on occasion for such diseases as tuberculosis. The major side effect of therapy is damage to the auditory branch of the nerve extending from the inner ear. Deafness may result.

Gentamicin, a still-useful aminoglycoside, is administered for serious infections caused by gram-negative bacteria, especially those causing urinary tract infections. The antibiotic is produced by species of *Micromonospora*, a bacterium related to *Streptomyces*. Damage to the kidney and the hearing mechanism has been reported.

Neomycin and kanamycin are two older antibiotics of the aminoglycoside group, having been isolated from *Streptomyces* species in 1949 and 1957, respectively. **Neomycin** sometimes is used to control intestinal infections because it is poorly absorbed, and it is prescribed as an ointment for bacterial conjunctivitis. Commercially, it is available in combination with polymyxin B and bacitracin as Neosporin. **Kanamycin** is used primarily against gram-negative bacteria in wounded tissue. A

Gentamicin

derivative of kanamycin called **amikacin** is used for controlling numerous nosocomial diseases and for intestinal and urinary tract diseases. Physicians use **tobramycin** against *Pseudomonas* species, and **paromomycin** against *Entamoeba histolytica*. Both antibiotics are derived from *Streptomyces* species. In 1998, the FDA approved an aerosolized version of tobramycin (TOBI) for treating respiratory infections in patients with cystic fibrosis.

CHLORAMPHENICOL ALSO INHIBITS TRANSLATION

Chloramphenicol was the first broad-spectrum antibiotic discovered. Its isolation in 1947 by John Ehrlich, Paul Burkholder, and David Gotlieb was hailed as a milestone in microbiology because the drug was capable of inhibiting a wide variety of gram-positive and gram-negative bacteria, as well as several species of rickettsiae and fungi. During the next 30 years, however, physicians tempered their enthusiasm for chloramphenicol as side effects became apparent and new drugs appeared. Nevertheless, chloramphenicol still retains its importance in the treatment of many diseases.

Chloramphenicol is a small molecule that passes into the tissues, where it interferes with protein synthesis in microorganisms (**FIGURE 24.10**). It diffuses into the nervous system and is thus useful in treating meningitis. It also remains the drug of choice in the treatment of typhoid fever and is an alternative to tetracycline for typhus fever and Rocky Mountain spotted fever. Originally isolated from the waste products of *Streptomyces venezuelae*, chloramphenicol became the first synthetic antibiotic when scientists at the Parke-Davis Company manufactured it from raw materials (its trade name is Chloromycetin). The drug has two major side effects: In the bone marrow, it prevents hemoglobin incorporation into the red blood cells, causing a condition called **aplastic anemia**; and it accumulates in the blood of newborns, causing a toxic reaction and sudden breakdown of the cardiovascular system known as the **gray syndrome**. For these reasons, chloramphenicol is not used to treat minor infections.

Chloramphenicol

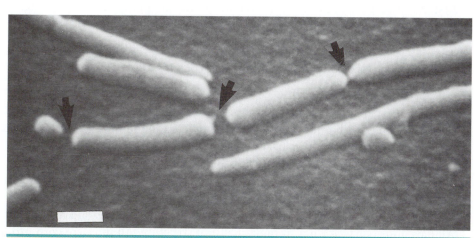

FIGURE 24.10

The Effect of Chloramphenicol on Bacteria

Escherichia coli was treated with chloramphenicol and photographed with the scanning electron microscope 20 minutes after the exposure. Note the formation of "minicells" and their thread-like attachments to the parent cells. Phenomena such as these are not usually found during normal cell division. They reflect the interference of the antibiotic with the organism's metabolism. (Bar = 0.5 μm.)

To this point . . .

We have surveyed four major groups of antibiotics and have noted their sources, modes of activity, and uses. We mentioned the side effects associated with each one, because they are a major consideration in determining which drug to prescribe. Penicillin and penicillin derivatives are used primarily for gram-positive bacteria and have their activity at the cell wall of microorganisms. Penicillinase production and patient allergy limit their use. Cephalosporin antibiotics also function at the cell wall. The aminoglycoside group consists of numerous antibiotics that interfere with protein synthesis; physicians prescribe them for diseases caused by gram-negative bacteria. Some are used on the skin, while others must be injected. Chloramphenicol is a broad-spectrum antibiotic with numerous uses but potentially lethal side effects.

As the discussion progressed, you may have noted how the antibiotics are products of microorganisms, especially Penicillium, Cephalosporium, *and* Streptomyces *species. Many of the original products can be modified to form semisynthetic antibiotics. In the case of chloramphenicol, the antibiotic is produced totally by synthetic means. This is why antibiotics often are placed under the umbrella of chemotherapeutic agents.*

We shall continue our discussion by examining the tetracycline antibiotics and a miscellaneous group of other drugs, including several that are used against fungi.

24.3

Other Antibiotics

Some antibiotics, such as the penicillins, have withstood the test of time and have remained valuable adjuncts in the therapy for infectious diseases. Another "old-timer" is the tetracycline group of antibiotics that we discuss first in this section. Some of the newer antibiotics are presented in the paragraphs that follow.

TETRACYCLINES ARE ANOTHER GROUP OF DRUGS THAT AFFECT PROTEIN SYNTHESIS

In 1948, scientists at Lederle Laboratories discovered chlortetracycline, the first of the tetracycline antibiotics. This finding completed the initial quartet of "wonder drugs": penicillin, streptomycin, chloramphenicol, and tetracycline.

Modern **tetracyclines** are a group of **broad-spectrum antibiotics** with a range of activity similar to chloramphenicols. They include naturally occurring **chlortetracycline** and **oxytetracycline** isolated from species of *Streptomyces* (**FIGURE 24.11**), and the semisynthetic **tetracycline, doxycycline, methacycline,** and **minocycline**. All have four benzene rings in their chemical structure. All interfere with protein synthesis in microorganisms by binding to ribosomes.

Tetracycline antibiotics may be taken orally, a factor that led to their indiscriminate use in the 1950s and 1960s. The antibiotics were consumed in huge quantities by tens of millions, and in some people, the normal microbiota of the intestine was destroyed. With these natural controls eliminated, fungi such as *Candida albicans* flourished. Patients then had to take an antifungal antibiotic such as nystatin, but

Broad-spectrum antibiotic: one that is effective against a wide range of bacteria, rickettsiae, chlamydiae, and fungi.

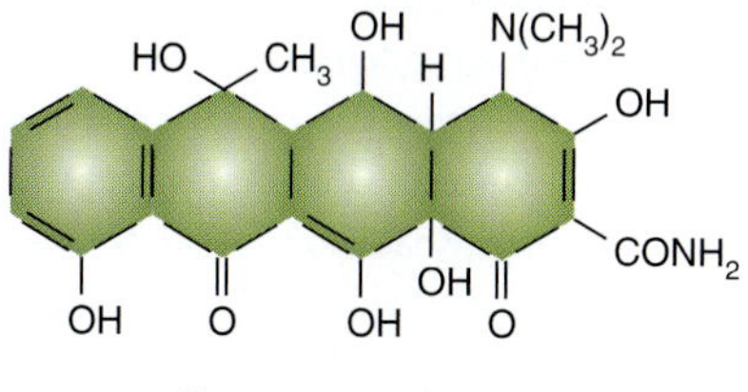

FIGURE 24.11

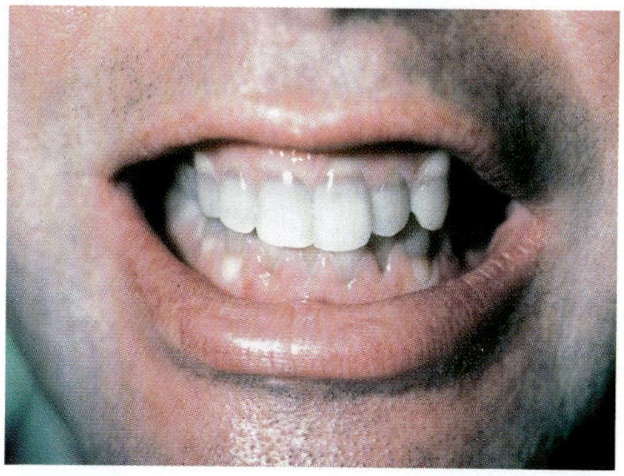

Oxytetracycline

(a) The chemical structure of oxytetracycline. Note the presence of four benzene rings, which characterizes all the tetracycline antibiotics. (b) The staining of teeth associated with tetracycline use.

because this drug was sometimes toxic, the preferred course was to replace the intestinal bacteria by consuming large quantities of bacteria-laden yogurt. Tetracyclines also cause a yellow-gray-brown discoloration of teeth and stunted bones in children. These problems are minimized by restricting use of the antibiotic in pregnant women and children through the teen years.

Despite these side effects, tetracyclines remain the drugs of choice for most rickettsial and chlamydial diseases, including the STD chlamydia (Chapter 11). They are used against a wide range of gram-negative bacteria, and they are valuable for treating primary atypical pneumonia, syphilis, gonorrhea, pneumococcal pneumonia, and certain protozoal diseases. Although resistances have occurred, newer tetracyclines such as minocycline (Minocin) and doxycycline (Vibramycin) appear to circumvent these. Evidence indicates that tetracycline may have been present in the food of ancient people, as MicroFocus 24.6 points out.

OTHER ANTIBIOTICS HAVE A VARIETY OF EFFECTS

A miscellaneous group of antibiotics merits brief discussion because the drugs in this group are commonly used in modern therapy.

Erythromycin is a clinically important antibiotic in the group called the macrolides. **Macrolides** consist of large carbon rings attached to unusual carbohydrate molecules. In the 1970s, researchers discovered that erythromycin was effective for treating primary atypical pneumonia and Legionnaires' disease. The antibiotic is a *Streptomyces* product and a protein synthesis inhibitor. It is recommended for use against gram-positive bacteria in patients with penicillin allergy and against both *Neisseria* and *Chlamydia* species that can affect the eyes of newborns. Although it has few side effects, it sometimes interferes with gastrointestinal functions, presumably by combining with receptor sites used by body hormones to control nutrient transport in the GI tract.

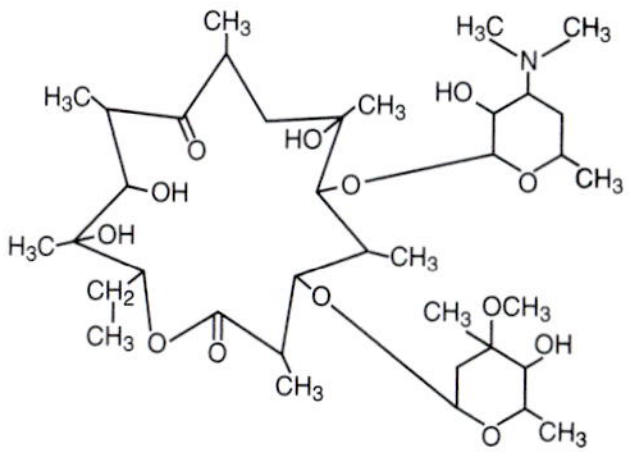

Another macrolide antibiotic is **clarithromycin**, a semisynthetic drug. Clarithromycin (Biaxin) acts by binding to ribosomes to inhibit protein synthesis in gram-negative bacteria, as well as the same gram-positive bacteria inhibited by

Erythromycin

MicroFocus 24.6

WONDER BREAD

In September 1980, a chance observation led to the discovery that antibiotics were protecting humans from disease long before anyone suspected.

The remarkable find was made by Debra L. Martin, a graduate student at Detroit's Henry Ford Hospital. After preparing thin bone sections for microscopic observation, Martin placed her slides under a fluorescent microscope because none other was available at the time. When illuminated with ultraviolet light, the sections glowed with a peculiar yellow-green color. Her colleagues identified the glow as that of the antibiotic tetracycline.

These were no ordinary bone sections. Rather, they were from the mummified

remains of Nubian people excavated along the floodplain of the Nile River. Anthropologists from the University of Massachusetts, led by George Armelagos, had previously established that the Nubian population was remarkably free of infectious disease, and now Martin's discovery gave a possible reason why.

Streptomyces species are very common in desert soil, and the anthropologists postulated that bacteria may have contaminated the grain bins and deposited tetracycline. Bread made from the antibiotic-rich grain then conferred freedom from disease. The theory was strengthened when the amount of tetracycline in the ancient bone was shown to be equivalent to that in therapeutic doses used in medicine.

Another practice, reported in 1944, indicates that modern people were more deliberate in their use of contaminated bread. A doctor traveling in Europe noted that a loaf of moldy bread hung in the kitchens of many homes. He inquired about it and was told that when a wound or abrasion was sustained, a sliver of the bread was mixed with water to form a paste; the paste was then applied to the skin. A wound so treated was less likely to become infected. Presumably the modern bread, like the ancient bread, contained a chemical that would be recognized today as an antibiotic.

erythromycin. Still another macrolide called **azithromycin** (Zithromax) has a similar mode of action and spectrum of activity. Both antibiotics are dangerous to fetal tissue and should not be taken by pregnant women.

Vancomycin, a cell wall inhibitor, is a product of a *Streptomyces* species. It is administered by intravenous injection against diseases caused by gram-positive bacteria, especially severe staphylococcal diseases where penicillin allergy or bacterial resistance is found, as discussed later in the chapter. It also is used against *Clostridium* species, and against *Enterococcus* species (enterococci) that cause mild intestinal diseases. (In recent years, vancomycin-resistant strains of enterococci termed VRE have emerged.) Its major side effects are damage to the ears and kidneys; the drug is not routinely prescribed for trivial conditions.

As drug resistance has developed and spread among staphylococci, the choice of antibiotics has gradually diminished, and vancomycin has emerged as a key treatment in therapy. Unfortunately, resistance to vancomycin also has been observed and substitute drugs have been sought. One possibility is Synercid, the trade name for the two-drug combination of **quinupristin** and **dalfopristin**. Synercid is effective against resistant strains of *Staphylococcus aureus* and *Streptococcus pneumoniae*. It is the first of the **streptogramin** class of antibiotics to be approved. Both components interfere with protein synthesis, but the interference is enhanced when they are used together.

Rifampin is a semisynthetic drug prescribed (in combinations with isoniazid and ethambutol) for tuberculosis and leprosy patients. It also is administered to carriers of *Neisseria* and *Haemophilus* species that cause meningitis and as a prophylactic when exposure has occurred. It acts by interfering with RNA synthesis in bacteria. Rifampin therapy may cause the urine, feces, tears, and other body secre-

Rifampin

tions to assume an orange-red color and may cause liver damage. The drug is administered orally and is well absorbed. A related drug called **rifapentine** (Priftin) has been approved as a treatment for tuberculosis.

Clindamycin and its parent drug, **lincomycin**, are alternatives in cases where penicillin resistance is encountered. Both are active against gram-positive bacteria, including several anaerobic species (e.g., *Bacteroides* species). Use of the antibiotics is limited to serious infections, however, because the drugs eliminate competing organisms from the intestine and permit *Clostridium difficile* to overgrow the area. The clostridial toxins then may induce a condition called **pseudomembranous colitis**, in which membranous lesions cover the intestinal wall (Chapter 11).

Both bacitracin and polymyxin B are polypeptide antibiotics produced by *Bacillus* species. These antibiotics are generally restricted to use on the skin because internally they may cause kidney damage and are poorly absorbed from the intestine. **Bacitracin** is available in pharmaceutical skin ointments and is effective against gram-positive bacteria such as staphylococci. **Polymyxin B** is valuable against *Pseudomonas aeruginosa* and other gram-negative bacilli, particularly those that cause superficial infections in wounds, abrasions, and burns. The two antibiotics often are combined with neomycin in Neosporin. Bacitracin inhibits cell wall synthesis, while polymyxin B injures bacterial membranes.

Bacitracin

Spectinomycin is a *Streptomyces* product that shares chemical properties with the aminoglycosides. The antibiotic came into prominence in the 1970s for use against gonorrhea caused by penicillin-resistant gonococci. It is given by intramuscular injection and appears to interfere with protein synthesis.

Monobactams are a group of antibiotics first synthesized by researchers at Squibb Laboratories in the early 1980s. The core of monobactam antibiotics is a beta-lactam nucleus isolated from *Chromobacter violaceum*, a purple-pigmented bacterium. One antibiotic, **moxalactam**, is active against a broad variety of gram-negative bacteria, especially those involved in nosocomial diseases and bacterial meningitis. Resistance to penicillinase appears to be high, but interference with platelet function and severe bleeding have limited its use. Another set of beta-lactam drugs are the **carbapenems**. The representative of this group called **imipenem** (Primaxin) is effective against a variety of gram-positive bacteria and gram-negative rods, as well as anaerobes (e.g., *Bacteroides fragilis*). It is prescribed often in cases where resistances occur, and it appears to have minimal side effects.

Imipenem

FIGURE 24.12 summarizes the sites of activity for the bacterial antibiotics.

ANTIFUNGAL ANTIBIOTICS CAN PRODUCE TOXIC SIDE EFFECTS

Fungal diseases pose a special problem for the medical mycologist because there are few drugs available for treatment. For infections of the intestine, vagina, or oral cavity due to *Candida albicans*, physicians often prescribe **nystatin** (MicroFocus 24.7). This product of *Streptomyces* is commercially available as Mycostatin or Achrostatin and is sold in ointment, cream, or suppository form. It acts by changing the permeability of the cell membrane by combining with fungal sterols. Often it is combined with antibacterial antibiotics to retard *Candida* overgrowth of the intestines during the treatment of bacterial diseases. Another antifungal product of *Streptomyces*, **candicidin**, has similar uses but is not as widely prescribed.

Griseofulvin is an antibiotic used for fungal infections of the skin, hair, and nails, such as ringworm and athlete's foot. Griseofulvin interferes with mitosis and causes the tips of molds to curl. It is a product of a *Penicillium* species and is taken orally.

Griseofulvin

FIGURE 24.12

The Sites of Activity in a Bacterial Cell for Various Antibiotics

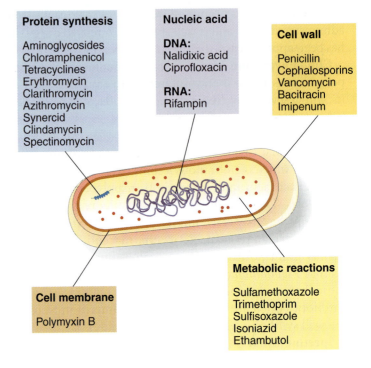

Protein synthesis

Aminoglycosides
Chloramphenicol
Tetracyclines
Erythromycin
Clarithromycin
Azithromycin
Synercid
Clindamycin
Spectinomycin

Nucleic acid

DNA:
Nalidixic acid
Ciprofloxacin

RNA:
Rifampin

Cell wall

Penicillin
Cephalosporins
Vancomycin
Bacitracin
Imipenum

Metabolic reactions

Sulfamethoxazole
Trimethoprim
Sulfisoxazole
Isoniazid
Ethambutol

Cell membrane

Polymyxin B

MicroFocus 24.7

PERSISTENCE

The discovery of nystatin stems from the research of two scientists, Rachael F. Brown and Elizabeth Hazen. Their work was accomplished through persistent effort and the intelligent use of basic scientific principles. Sound research led them to a naturally occurring substance that would prove valuable in fighting fungal disease.

Rachael Brown was an organic chemist with a Ph.D. from the University of Chicago. Elizabeth Hazen was a mycologist. In the mid-1940s, the two scientists were employed at the New York State Department of Health where their interest was piqued by increasing reports concerning antibiotics. Penicillin was in widespread use by that time, and streptomycin had been discovered by Waksman in 1942. Both antibiotics were useful against bacteria, but none had yet been developed for fungal disease. Brown and Hazen would try to fill that gap.

In the 1940s, scientists knew that soil-borne bacteria of the genus *Streptomyces* were potential sources of antibiotics. Hazen therefore collected soil samples from various places and tested the waste products of soilborne microorganisms to determine whether they could inhibit fungal growth. A bacterium from soil on a farm owned by a certain Henry Nourse was especially promising. The organism was apparently unknown before Hazen isolated it, and she therefore named it *Streptomyces noursei* after the farmer. Now it was Brown's turn. Using her skills in chemistry, she isolated, purified, and characterized the active ingredient in the waste product. With Hazen, she demonstrated that minuscule amounts of the active principle were extraordinarily inhibitory to fungi. Hazen and Brown named the ingredient nystatin, for New York State.

Nystatin was introduced to the scientific community at the 1949 meeting of the National Academy of Sciences. Two years later, a patent was issued for production, and E. R. Squibb received exclusive license to manufacture the antibiotic. Before long, nystatin became a key treatment for various forms of candidiasis, and when the Arno River flooded Florence, Italy, in the 1970s, nystatin was used to combat fungi attacking the art treasures. Nystatin also has commercial value for preventing spoilage in foods, especially bananas, and it is used in surgery to preclude fungal infection.

For Brown and Hazen, the accolades were many, including several honorary degrees and awards. Students at Mount Holyoke College currently vie for the Rachael Brown Fellowship, and students at Mississippi University for Women are eligible for the Elizabeth Hazen Scholarship, both named for alumnae of the respective colleges.

For serious systemic fungal infections, the drug of choice is **amphotericin B**. This antibiotic degrades the cell membranes of fungal cells, and is effective for treating serious diseases (Chapter 15). However, it causes a wide variety of side effects and therefore is used only in progressive and potentially fatal cases.

Other antifungal antibiotics are synthetic compounds. One example, **flucytosine**, is converted in fungal cells to an inhibitor that interrupts nucleic acid synthesis. The drug is used primarily with amphotericin B in systemic diseases. Another example, the **imidazoles**, include clotrimazole, miconazole, itraconazole, and ketoconazole. These compounds interfere with sterol synthesis in fungal cell membranes. **Clotrimazole** (Gyne-Lotrimin) is used topically for *Candida* skin infections, while the other drugs are used topically as well as internally for systemic diseases. Side effects are uncommon. **Miconazole** is commercially available as Micatin for athlete's foot and Monistat for yeast infections. **Itraconazole** is sold as Sporanox for athlete's foot. TABLE 24.1 summarizes the chemotherapeutic and antibiotic agents currently in use.

Miconazole

ANTIVIRAL AGENTS INHIBIT THE VIRAL REPLICATION CYCLE

The inhibition and destruction of viruses present unique challenges for drug researchers because viruses are extraordinarily simple. They do not have the complex structures or the cell walls and membranes of other microorganisms, and they engage in few physiological activities other than replication. Thus, viruses have few places where researchers may focus their attention. Nevertheless, a number of antiviral agents have emerged in recent years. They are discussed in depth in Chapter 12, but we shall review them here briefly.

Most antiviral agents seek to interrupt an aspect of the viral replication cycle. For example, **amantadine** and **zanamivir** prevent the attachment of influenza viruses to host cell membranes. **Acyclovir** and **ganciclovir** are erroneously incorporated into viral DNA during the replication cycles of herpes simplex viruses and cytomegaloviruses, respectively. **Azidothymidine (AZT)** and **ribavirin** act in a similar way when DNA is synthesized using viral RNA as a template.

Acyclovir

Two newer classes of antiviral agents are inhibitors of enzymes used by viruses. The first class includes the anti-HIV drugs **nevirapine** and **delavirdine**. These drugs bind to and inhibit reverse transcriptase when HIV is replicating. In the second class are **saquinavir** and **ritonavir**. They bind to protease, the enzyme needed to form the viral capsid.

The **interferons** continue to attract attention because they are naturally produced by body cells (Chapter 12). Now synthesized by genetic engineering methods, the interferons provoke host cells to produce antiviral proteins, and they stimulate natural killer cells into action. The FDA has approved interferons for treating hepatitis B, genital warts, and a type of leukemia.

Azidothymidine
(AZT)

NEW APPROACHES TO ANTIBIOTIC THERAPY ARE NEEDED AS RESISTANCE INCREASES

As reports of antibiotic resistance and harmful side effects appear in the literature, scientists hasten their search for new approaches to antibiotic therapy. One approach centers on the **two-component signaling pathway** found in bacterial cells but not in animal cells. When infectious bacteria reach the human lung, for example, an enzyme called a kinase is activated. The kinase then signals a second

TABLE 24.1

A Summary of Major Chemotherapeutic Agents and Antibiotics

CHEMOTHERAPEUTIC AGENT OR ANTIBIOTIC	SOURCE	ANTIMICROBIAL SPECTRUM	ACTIVITY IMPEDED	SIDE EFFECTS
Sulfonamides Sulfanilamide Sulfamethoxazole	Synthetic	Broad spectrum	Folic acid metabolism	Kidney and liver damage Allergic reactions
Isoniazid	Synthetic	Tubercle bacilli	Cell wall synthesis	Liver damage
Metronidazole	Synthetic	*Trichomonas vaginalis*	Cell metabolism	Tumors in mice
Chloroquine and primaquine	Synthetic	*Plasmodium* species	Cell metabolism	Eye damage
Fluoroquinolones Ciprofloxacin Enoxacin	Synthetic	Broad spectrum	DNA synthesis	Few reported
Penicillins Penicillin G Ampicillin Amoxicillin Nafcillin Oxacillin	*Penicillium notatum* and *Penicillium chrysogenum* Some semisynthetic Some synthetic	Broad spectrum, especially gram-positive bacteria	Cell wall synthesis	Allergic reactions Selection of penicillinase-producing strains
Cephalosporins Cephalothin Cephalexin	*Cephalosporium* species Some semisynthetic	Gram-positive bacteria Broad spectrum	Cell wall synthesis	Occasional allergic reactions
Aminoglycosides Gentamicin Neomycin Amikacin Streptomycin	*Micromonospora* species *Streptomyces* species	Broad spectrum, especially gram-negative bacteria	Protein synthesis	Hearing defects Kidney damage
Chloramphenicol	*Streptomyces venezuelae*	Broad spectrum, especially typhoid bacilli	Protein synthesis	Aplastic anemia Gray syndrome
Tetracyclines Chlortetracycline Oxytetracycline Doxycycline Minocycline	*Streptomyces* species Some semisynthetic	Broad spectrum, rickettsiae, chlamydiae, gram-negative bacteria	Protein synthesis	Destruction of natural flora Discoloration of teeth Stunted bones
Macrolides Erythromycin Clarithromycin Azithromycin	*Streptomyces erythraeus* Some semisynthetic	Gram-positive bacteria, *Mycoplasma* Broad spectrum	Protein synthesis	Gastrointestinal distress
Vancomycin	*Streptomyces orientalis*	Gram-positive bacteria, especially staphylococci	Cell wall synthesis	Ear and kidney damage
Rifampin Rifapentine	*Streptomyces mediterranei* Semisynthetic	Tubercle bacilli Gram-negative bacteria	RNA synthesis	Liver damage
Clindamycin Lincomycin	*Streptomyces lincolnesis*	Gram-positive bacteria	Protein synthesis	Pseudomembranous colitis
Bacitracin	*Bacillus subtilis*	Gram-positive bacteria, especially staphylococci	Cell wall synthesis	Kidney damage
Polymyxin	*Bacillus polymyxa*	Gram-negative bacteria, especially in wounds	Cell membrane function	Kidney damage
Spectinomycin	*Streptomyces spectabilis*	Gonococci	Protein synthesis	Few reported

CHEMOTHERAPEUTIC AGENT OR ANTIBIOTIC	SOURCE	ANTIMICROBIAL SPECTRUM	ACTIVITY IMPEDED	SIDE EFFECTS
Moxalactam Imipenem	*Chromobacter violaceum*	Broad spectrum	Protein synthesis (?)	Few reported
Nystatin	*Streptomyces noursei*	Fungi, especially *Candida albicans*	Cell membrane function	Few reported
Griseofulvin	*Penicillium janczewski*	Fungi, especially in superficial infections	Nucleic acid synthesis	Occasional allergic reactions
Amphotericin B	*Streptomyces nodosus*	Fungi, especially in systemic infections	Cell membrane function	Fever Gastrointestinal distress
Imidazoles Clotrimazole Ketoconazole	Synthetic	Fungi, especially in superficial infections	Inhibit sterol synthesis	Few reported

protein called a **transcription factor**, which binds to the bacterial DNA and "turns on" the genes that will encourage infection. Scientists believe they can develop a chemical compound to prevent kinase activation and thereby interrupt the infection process.

Another approach is to focus on the lipid of the outer membrane in gram-negative bacteria. This membrane, discussed in Chapter 4, includes a unique lipopolysaccharide (LPS) with several unusual carbohydrates that could be targets for new drugs. Indeed, researchers already have shown that bacteria with defective or disorganized LPS are susceptible to complement-mediated antibody activity. Furthermore, neutralizing the LPS would limit the release of endotoxins associated with LPS.

Interfering with regulatory systems might conceivably be another focus of research on antibiotic action. Almost all bacteria, for instance, use an enzyme called **DNA adenine methylase** to coat DNA with methyl groups and regulate DNA replication and repair. Researchers have significantly reduced the virulence of a *Salmonella* species by disabling the gene that encodes the enzyme. They now hope to develop a chemical compound to neutralize the enzyme (rather than the gene), thereby pinpointing a protein found only in the bacterium. At this point, however, they are not completely sure of the enzyme's activity.

Discovering a new class of antibiotic compounds might be another trail to follow. A number of researchers are investigating **peptide antibiotics**, a group of small proteins isolated from the immune system of various animals. The peptides are found in innumerable animal species; they guard against microbial entry at the body surface (e.g., the tongue of a cow, the stomach of a shark, the skin of a human). The antibiotics appear to disrupt cell membranes in bacteria, and their positive charge enhances binding to the negatively charged bacterial membrane. However, none have yet entered the pipeline as potential drugs, one reason being the estimated $300 million necessary to bring a new antibiotic to market. MicroInquiry 24 looks at what is involved in bringing a drug to the pharmacy shelf.

MicroInquiry 24

TESTING DRUGS—CLINICAL TRIALS

Before any antibiotic or chemotherapeutic agent can be used in therapy, it must be tested to ensure its safety and efficacy. The process runs from basic biomedical laboratory studies to approval of the product to improve health care. On average it takes over eight years to study the drug in the laboratory (pre-clinical testing), test it in animals, and finally run human clinical trials.

In this MicroInquiry, we will examine a condensed version of the steps through which a typical antibiotic would pass. Some questions asked are followed by the answer so you can progress further in your analysis. Please try to answer these questions before progressing through the scenario. Also, methods are used that you have studied in this and previous chapters. A real pharmaceutical company might have other, more sophisticated methods available. Answers can be found in Appendix E.

The scenario. You are head of a drug research group with a pharmaceutical company that has isolated a chemical compound (let's call it FM04) that is believed to have antibacterial properties and thus potential chemotherapeutic benefit. As head, you must move this drug through the development pipeline from testing to clinical trials, evaluating at each step whether to proceed with further testing.

1. Preclinical testing. Many experiments need to be done before the chemical compound can be tested in humans. Since FM04 is "believed" to have antibacterial properties,

24.1a. What would be the first experimental tests to be carried out?

You would experimentally test the drug on bacterial cells in culture to ascertain its relative strength and potency as an antibacterial agent. This testing might include the agar disk diffu-

sion method described in this chapter. **FIGURE A** shows three diffusion disk plates, each plated with a different bacterial species (*Escherichia coli*, *Staphylococcus aureus*, and *Pseudomonas aeruginosa*) and disks with different concentrations of FM04 (0, 1, 10, 100 μg).

24.1b. What can you conclude from these disk diffusion studies? Should drug testing continue? Explain.

Let's assume the bacterial studies look promising and several months of additional studies confirm the antibacterial properties. FM04 now would be tested on human cells in culture to see if there are any toxic effects on metabolism and growth. **TABLE A** presents the results using the same drug concentrations from the bacterial studies. (The drug was prepared as a liquid stock solution to give the final drug concentration in culture.)

24.1c. What would you conclude from the studies of human cells in culture? Should drug testing continue? Explain.

24.1d. What was the point of including a solution with 0 μg of the drug?

2. Animal testing. When drugs are injected or ingested into the whole body, concentrations required for a

desired chemotherapeutic effect may be very different from the experimental results with cells. Higher concentrations often are required and these may have toxic side effects. In animal testing, efforts should be made to use as few animals as possible and they should be subjected to humane and proper care. Let's assume the research studies of FM04 on mice used oral drug concentrations of 1, 10, 100, and 1,000 mg. **TABLE B** presents the simplified results.

24.2a. What would you conclude from the mouse studies? Should drug testing continue? Explain.

Since the results look positive and higher doses can be administered without serious toxic side effects (except at 1,000 mg), additional studies would be carried out to determine how much of the drug is actually absorbed into the blood, how it is degraded and excreted in the animal, and if there are any toxic breakdown products produced. These studies could take a few years to complete. Again, for the sake of the exercise, let's say the studies with FM04 remain promising.

Although mice are quite similar to humans physiologically, they are not identical, so human clinical trials need to be done. This is the reason the US Food

---

TABLE A

Adverse Cellular Effects Based on FM04 Drug Concentration

FINAL DRUG CONCENTRATION (μg/CULTURE)	ADVERSE CELLULAR EFFECTS
0	None
1	None
10	None
100	Abnormal cells and cell death

FIGURE A

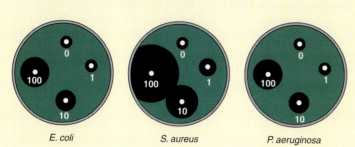

E. coli   S. aureus   P. aeruginosa

and Drug Administration (FDA) requires all new drugs to be tested through a *three-phase clinical trial period.*

First, as head of the research team, you must provide all the preclinical and animal testing results as part of an FDA *New Drug Application* (NDA) before human tests can begin. The application must spell out how the clinical investigations will be conducted. As head of the investigational team, you must design the protocol for these human trials.

24.2b. How would you conduct the human tests to see if FM04 has positive therapeutic effects as an antibiotic in humans?

Briefly, there are many factors to take into consideration in designing the protocol.

(*i*) How will the drug be administered? Let's assume it will be oral in the form of pills.

(*ii*) You need to recruit groups of patients (volunteers) of similar age, weight, and health status (healthy as well as patients with the infectious condition to which the drug may be therapeutic). Note that the bacterial studies suggested FM04 had its most potent effect on *Staphylococcus aureus*, a gram-positive species. Perhaps patients with a staph infection would be recruited.

(*iii*) Eventually, volunteers (patients) need to be split into two groups, one that receives the drug and one that receives a placebo (an inactive compound that looks like the FM04 pill).

(*iv*) The clinical studies should be carried out as "double blind" studies. In these studies, neither the patients nor the investigators know which patients are getting the drug or placebo. This prevents patients and investigators from "wishful thinking" as to the outcome of the studies.

In actual clinical studies there are even more factors to consider, but we will assume that FM04 wins approval for clinical testing.

3. Clinical trials. In *Phase 1 studies,* a small number (20 to 100) of healthy patients are treated with the test drug at different concentrations (doses) to see if there are any adverse side effects. These studies normally take at least several months. If unfavorable side effects are minimal, *Phase 2* trials begin.

Phase 2 studies use a larger population of patients (several hundred) who have the disease the drug is designed to treat. Here, it is especially important to split the patients into the test and control groups since the major purpose of these trials is to study drug effectiveness. There needs to be a control group to accurately contrast effectiveness. These trials can take anywhere from several months to two years to complete. Provided there are no serious side effects or toxic reactions, or a lack of effectiveness, *Phase 3* trials begin. There is an ethical question that can arise at this stage in the clinical trials, especially if the drug shows signs of reversing an illness that might be life threatening. Even medical experts disagree as to the answer.

24.3 Is it ethical to give ill patients placebos when effective treatment appears available?

TABLE B

Toxicity Results on 50 Mice Given Different Concentrations of FM04.

DRUG CONCENTRATION (mg)	NUMBER OF MICE	ADVERSE REACTIONS
0	10	None
1	10	None
10	10	None
100	10	None
1,000	10	Tremors and seizures in 8 mice

Phase 3 studies include several hundred to a few thousand patients. The trials can take several years to complete because the purpose is to evaluate safety, dosage, and effectiveness of the drug.

If the clinical trials are positive, as head you can recommend the pharmaceutical manufacturer apply for an FDA license. The license application is reviewed by an internal FDA committee that examines all clinical data, proposed labeling, and manufacturing procedures. The most important questions that need to be addressed are:

(i) Were the clinical studies well controlled to provide "substantial evidence of effectiveness"?

(ii) Did the clinical study results demonstrate that the drug was safe; that is, do the benefits outweigh the risks?

Based on its findings, the committee then makes its recommendation to the FDA, which decides if licensing is to be given. Note that very few drugs (perhaps 5 in 5,000 tested compounds) actually make it to human clinical trials. Then, only about 1 of those 5 are found to be safe and effective enough to reach the pharmacy.

To this point . . .

In this section, we considered a few additional antibiotics that are useful in fighting infectious diseases. The tetracyclines were another group of antibiotics that affect proteins synthesis. These broad-spectrum antibiotics can be taken orally, but side effects can cause tooth discoloration and stunted bones in children. Three macrolides—naturally produced erythromycin and semisynthetic clarithromycin and azithromycin—inhibit ribosome function.

Vancomycin is a cell-wall synthesis inhibitor of gram-positive bacteria, although vancomycin-resistant strains are on the increase. Rifampin, an RNA synthesis inhibitor, is part of tuberculosis therapy. The polypeptide antibiotics bacitracin and polymyxin B are used on superficial skin wounds. We finished by mentioning the monobactams, such as moxalactam and imipenem.

We also described some antifungal antibiotics, including nystatin, which interacts with fungal sterols to change membrane permeability. Griseofulvin is used for superficial infections such as ringworm, while amphotericin B is used on systemic infections. Synthetic agents included the imidazoles, such as clotrimazole and miconazole. A review of antiviral agents included those agents that interfere with the viral replication cycle.

The section ended with a discussion of new approaches that can help alleviate the problem of increasing resistance to antibiotics. Our survey will conclude with a description of laboratory tests used to determine the effectiveness of antimicrobial agents under experimental conditions. Also, we will discover the mechanisms that give rise to antibiotic resistance and discuss the misuses and abuses of antibiotics.

Antibiotic Assays and Resistance

The substantial variety of antibiotics and chemotherapeutic agents developed since the 1930s (MicroFocus 24.8) necessitates that the physician determine which one is best for the patient under the circumstances of the infection. Accordingly, an antibiotic sensitivity assay is performed. The assay also helps determine whether a microorganism is resistant to a particular antibiotic, a problem that has developed into a major concern of modern medicine.

THERE ARE SEVERAL ANTIBIOTIC SUSCEPTIBILITY ASSAYS

Antibiotic susceptibility assays are used to study the inhibition of a test organism by one or more antibiotics or chemotherapeutic agents. Two general methods are in common use: the tube dilution method, and the agar disk diffusion method.

The **tube dilution method** determines the smallest amount of antibiotic necessary to destroy a population of a test organism. This amount is known as the **minimum inhibitory concentration (MIC)**. To determine the MIC, the microbiologist prepares a set of tubes with different concentrations of a particular antibiotic. The tubes then are inoculated with an identical population of the test organism, incubated, and examined for the growth of bacteria. The extent of growth diminishes as the concentration of antibiotic increases, and eventually an antibiotic concentration is observed at which growth fails to occur. This is the MIC.

The second method, the **agar disk diffusion method**, operates on the principle that antibiotics will diffuse from a paper disk or small cylinder into an agar medium containing test organisms. This method is shown in FIGURE 24.13a. Inhibition is observed as a failure of the organism to grow in the region of the antibiotic. A common application of the agar disk diffusion method is the **Kirby-Bauer test**, named after W. M. Kirby and A. W. Bauer, who developed it in the 1960s. This procedure

MicroFocus 24.8

TRANSITION

The late Lewis Thomas was one of the remarkable individuals of the twentieth century. He was a well-known researcher, physician, professor, administrator, and writer. One of his notable books is a collection of essays entitled *The Youngest Science: Notes of a Medicine-Watcher* (1983). In an essay called "1933 Medicine," Thomas gives us an idea of what it was like to be a physician during the transition into the age of antibiotics:

"The two great hazards to life were tuberculosis and syphilis. These were feared by everyone in the same way that cancer is feared today. There was nothing to be done for tuberculosis except to wait it out, hoping that the body's own defense mechanisms would eventually hold the tubercle bacilli in check. . . . Then came the explosive news of sulfanilamide, and the start of the real revolution in medicine. I remember with astonishment when the first cases of

pneumococcal and streptococcal septicemia were treated in Boston in 1937. The phenomenon was almost beyond belief. Here were moribund patients, who would surely have died without treatment, improving in their appearance within a matter of hours of being given the medicine and feeling entirely well within the next day or so."

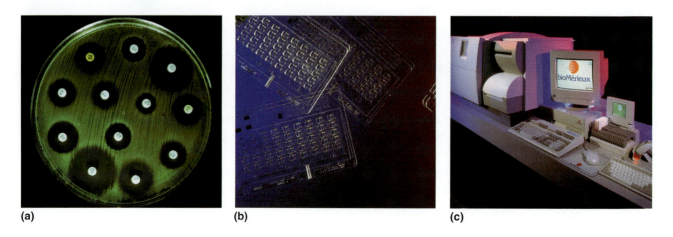

(a)   (b)   (c)

FIGURE 24.13

Antibiotic Susceptibility Testing

Two methods for determining an organism's antibiotic susceptibility are shown. (a) The more traditional method is the agar disk diffusion method. Bacteria are inoculated onto a plate of agar medium containing disks of various antibiotics. The plate is incubated, and the presence of a clear zone of inhibition shows susceptibility to a particular antibiotic. The absence of a zone or presence of a small zone indicates resistance. In a modern automated system, bacteria are automatically inoculated into wells on a card (b), each well containing a different antibiotic. The card is then incubated, and the presence or absence of growth is assayed by a computer (c). A printout relates the organism's susceptibility (no growth) to the antibiotic, or its resistance (growth).

determines the susceptibility of a microorganism to a series of antibiotics and is performed according to standards established by the FDA. A more sophisticated procedure is noted in FIGURE 24.13b, c.

THERE ARE FOUR MECHANISMS OF ANTIBIOTIC RESISTANCE

During the past 25 years, an alarming number of bacterial strains have evolved with resistance to chemotherapeutic agents and antibiotics. Public health microbiologists note that resistant organisms are increasingly responsible for human diseases of the intestinal tract, lungs, skin, and urinary tract. Those in intensive care units and burn wards are particularly vulnerable, as are infants, the elderly, and the infirm. Common diseases like bacterial pneumonia, tuberculosis, streptococcal sore throat, and gonorrhea, which a few years ago succumbed to a single dose of antibiotics, are now among the most difficult to treat.

One of the major concerns of public health officials is the bacterium *Staphylococcus aureus*. Capable of causing staphylococcal septicemia, pneumonia, endocarditis, and meningitis, *S. aureus* is involved in over 250,000 infections per year, primarily in hospitals and nursing homes. Over the years, strains of *S. aureus* have developed resistance to penicillin and numerous other drugs, and the term **MRSA**, for **multidrug-resistant *Staphylococcus aureus***, has often been used in hospital settings. Through those years, vancomycin remained a viable alternative for treating MRSA infections.

Then, in 1997, an MRSA strain evolved with intermediate (partial) vancomycin resistance; scientists named it **VISA**, for **vancomycin intermediately resistant *Staphylococcus aureus***. Although researchers have found useful alternatives in drug combinations, they are grappling with the possibility that nothing will be left in the antimicrobial arsenal to treat patients infected by this strain of staphylococci. Indeed, in 2001, reports of **vancomycin-resistant *S. aureus*** (**VRSA**) was reported in many countries, indicating global spread.

Microorganisms in the environment have been involved in chemical warfare with their neighbors for hundreds of millions of years. To counteract the production of antibiotics by some microbes, others developed self-protective (resistance) mechanisms to the natural antibiotics. Thus, it is not surprising that microbes such as bacteria quickly develop resistance to the natural antibiotics used to combat infectious diseases.

Resistance to antibiotics can develop through four major routes.

Altered metabolic pathway. Resistance to sulfonamides may develop when the drug fails to bind with enzymes that synthesize folic acid because the enzyme's structure has changed. Moreover, drug resistance may be due to an altered metabolic pathway in the microorganism, a pathway that bypasses the reaction normally inhibited by the drug.

Antibiotic inactivation. Resistance can arise from the microorganism's ability to enzymatically inactivate the antibiotic. The production of penicillinases (beta-lactamases) by penicillin-resistant gonococci is an example. By breaking the beta-lactam ring, penicillin- or cephalosporin-resistant bacteria have a mechanism to prevent blockage of cell wall synthesis.

The aminoglycosides normally block mRNA translation on bacterial ribosomes. Aminoglycoside-resistant bacteria have developed ways to enzymatically modify aminoglycosides so they cannot bind to ribosomes. The enzymatic modifications include phosphorylation, acetylation, or adenylation of the antibiotic.

Reduced permeability/active export of antibiotics. Another resistance mechanism involves keeping the concentrations of antibiotic needed to affect inhibition below what is required. For example, changes in membrane permeability in penicillin-resistant *Pseudomonas* prevent the antibiotic from entering the cytoplasm. Bacteria such as *E. coli* and *S. aureus*, which are resistant to tetracyclines, actively export (pump out) the drug. Cytoplasmic and membrane proteins in these bacteria act as pumps that remove the antibiotic before it can have an effect in the cytoplasm.

Target modification. A fourth route leading to microbial resistance involves altering the drug target. Some streptomycin-resistant bacteria can modify the structure of their ribosomes so that the antibiotic cannot bind to the ribosome and protein synthesis is not inhibited. Other targets include the RNA polymerase and enzymes involved in DNA replication.

ANTIBIOTICS HAVE BEEN MISUSED AND ABUSED

Resistance may expand in a bacterial population during the normal course of events (**FIGURE 24.14**), but antibiotic misuse and abuse encourages the emergence of resistant forms. For example, drug companies promote antibiotics heavily, patients pressure doctors for quick cures, and physicians sometimes write prescriptions without ordering costly tests to pinpoint the patient's illness. In addition, people may diagnose their own illness and take leftover antibiotics from their medicine chests for ailments where antibiotics are useless. Moreover, many people fail to complete their prescription, and some organisms remain alive to evolve to resistant forms by gathering "genetic debris" from the local area. And the survivors proliferate well because they face reduced competition from susceptible organisms.

Hospitals are another forcing ground for the emergence of resistant bacteria. In many cases, physicians use unnecessarily large doses of antibiotics to prevent

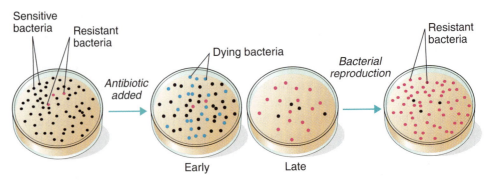

Sensitive bacteria
Resistant bacteria
Antibiotic added
Dying bacteria
Early   Late
Bacterial reproduction
Resistant bacteria

(A) Some members of the population have genes for drug resistance as a part of their genomes. In the absence of antibiotic, this resistance is not used.

(B) When the population of bacteria is treated with an antibiotic, the sensitive members gradually die off and the resistant ones thrive as the resistance genes become operative.

(C) The antibiotic-resistant bacteria now multiply and fill the environment with their enormous numbers. This antibiotic can no longer be used to control this population.

FIGURE 24.14

The Expansion of Drug Resistance

Drug resistance can occur widely in a population of bacteria as a result of natural selection during exposure to an antibiotic.

infection during and following surgery. This increases the possibility that resistant strains will overgrow susceptible strains and subsequently spread to other patients, thereby causing nosocomial disease. Antibiotic-resistant *E. coli, P. aeruginosa, Serratia marcescens*, and *Proteus* species now are widely encountered causes of illness in hospital settings.

Antibiotics also are abused in developing countries where they often are available without prescription, even though they have toxic side effects. Some countries permit the over-the-counter sale of potent antibiotics, and large doses encourage resistance to develop. Between 1968 and 1971, some 12,000 people died in Guatemala from shigellosis attributed to antibiotic-resistant *Shigella dysenteriae*.

Moreover, the problem of antibiotic abuse is widespread in livestock feeds. An astonishing 40 percent of all the antibiotics produced in the United States finds its way into animal feeds to check disease and promote growth. By killing off less hardy bacteria, chronic low doses of antibiotics create ideal growth environments for resistant strains. Transferred to humans through meat, these resistant organisms may cause intractable illnesses.

Allied to the problem of antibiotic resistance is the concern for transfer of the resistance. Researchers have demonstrated that plasmids and transposons account for the movement of antibiotic-resistant genes among bacteria (Chapter 7). Thus, the resistance in a relatively harmless bacterium may be passed to a pathogenic bacterium where the potential for disease is then increased by resistance to standardized treatment.

Dealing with the emerging antibiotic resistance is a major problem confronting contemporary researchers. One approach may lie in methods to boost the immune system, as noted in Chapter 19. Another approach may involve the use of bacteriophages (Chapter 12) to inhibit certain species of bacteria. Still another may involve innovative drugs and medicines that overcome the resistance. For example, biochemists have demonstrated that it is possible to prevent a bacterium from pumping an antibiotic out of its cytoplasm. Curbing the overuse of antibiotics through public and physician education and governmental regulation is a useful approach as well (although this approach will be particularly difficult, since the antibiotics market is currently worth about $23 billion). Increasing the research and development of effective vaccines and rapid diagnostic methods also will cut down on antibiotic needs. Indeed, all the principles that apply to the control of infection apply equally well to controlling the spread of antibiotic-resistant organisms.

Antibiotics traditionally have been known as miracle drugs, but there is a growing body of evidence that they are becoming overworked miracles. Some researchers suggest that antibiotics should be controlled as strictly as narcotics. The antibiotic roulette that is currently taking place should be a matter of discussion to all individuals concerned about infectious disease, be they scientist or student.

Note to the Student

In the last 100 years, there were two periods in which the incidence of disease declined sharply. The first period took place in the early 1900s. At this time an understanding of the disease process led to numerous social measures, such as water purification, careful food production, insect control, milk pasteurization, and patient isolation. Sanitary practices such as these made it possible to prevent virulent microorganisms from reaching their human targets.

The second period began in the 1940s with the development of antibiotics, and blossomed in the years thereafter when physicians found they could treat established cases of disease. Major health gains were made as serious illnesses came under control.

An outgrowth of these events has been the belief by many people that science can cure any infectious disease. A shot of this, a tablet of that, and then perfect health. Right? Unfortunately not.

Scientists may show us how to avoid infectious microorganisms and doctors may be able to control certain diseases with antibiotics, but the ultimate body defense depends upon the immune system and other natural measures of resistance. Used correctly, the antibiotics provide that extra something needed by natural defenses to overcome pathogenic microorganisms. The antibiotics supplement natural defenses; they do not replace them.

The great advances in chemotherapy should be viewed with caution. Antibiotics undoubtedly have relieved much misery and suffering, but they are not the cure-all some people perceive them to be. In the end, it is well to remember that good health comes from within, not from without.

Summary of Key Concepts

Chemotherapeutic agents and antibiotics work with the body's natural defenses to stop the growth of bacteria and other microorganisms in the body.

24.1 CHEMOTHERAPEUTIC AGENTS

- **The History of Chemotherapy Starts with Paul Ehrlich.** Ehrlich saw antibiotics as a chemotherapeutic approach to alleviating disease. One of his students isolated the first chemotherapeutic agent, salvarsan. Domagk showed that prontosil could be used to treat bacterial infections, setting in motion the age of modern chemotherapy.

- **Sulfanilamide and Other Sulfonamides Interfere with Metabolic Reactions.** One group of chemotherapeutics, the sulfonamides, interferes with the production of folic acid in bacteria. Modern sulfonamides include sulfamethoxazole and trimethoprim.

- **Other Chemotherapeutic Agents Have a Variety of Actions.** Other agents have various uses and various chemical compositions. Isoniazid and ethambutol are antituberculosis drugs that affect wall synthesis. Ciprofloxacin and drugs like metronidazole and chloroquine interact with DNA in different ways to inhibit replication.

24.2 ANTIBIOTICS

- **The Discovery of Antibiotics Began with Fleming.** The discovery of natural antibiotics started with Fleming's discovery of penicillin, which was purified and prepared for chemotherapy by Florey and Chain.

- **Penicillin Inhibits Bacterial Cell Wall Synthesis.** Penicillin was first obtained from the green mold *Penicillium*. The antibiotic interferes with cell wall synthesis in gram-positive bacteria. It can cause an anaphylactic reaction in sensitive individuals.

- **A Large Group of Semisynthetic Penicillins Has Been Developed.** Many other penicillin derivatives, designed to counteract drug resistance, are available in numerous synthetic and semisynthetic forms such as ampicillin and amoxicillin.

- **Cephalosporins Also Inhibit Cell Wall Synthesis.** Where penicillin allergy or resistance is encountered, a cephalosporin antibiotic can be used for therapy. Certain cephalosporin drugs are first-choice antibiotics, and a wide variety of these drugs is currently in use.

- **Aminoglycosides Inhibit Protein Synthesis.** For gram-negative bacteria, the aminoglycoside antibiotics can be employed. Gentamicin, neomycin, and kanamycin are useful members of the aminoglycoside group that bind to and inhibit ribosomes.

- **Chloramphenicol Also Inhibits Translation.** Chloramphenicol is a broad-spectrum antibiotic valuable against both gram-positive and gram-negative bacteria, but serious side effects limit its use.

24.3 OTHER ANTIBIOTICS

- **Tetracyclines Are Another Group of Drugs that Affect Protein Synthesis.** Less severe side effects accompany tetracycline use, and this antibiotic is recommended against gram-negative bacteria as well as rickettsiae and chlamydiae.

- **Other Antibiotics Have a Variety of Effects.** A variety of other antibiotics are used for various infectious diseases. This group includes the macrolides, which inhibit protein synthesis; vancomycin and bacitracin, which inhibit cell wall synthesis; and rifampin, which blocks DNA transcription.

- **Antifungal Antibiotics Can Produce Toxic Side Effects.** Certain antibiotics, such as nystatin, amphotericin B and the imidazoles, are valuable against fungal infections. Amphotericin B has serious side effects when used to treat systemic fungal diseases.

- **Antiviral Agents Inhibit the Viral Replication Cycle.** Antiviral agents include those that block attachment (amantadine), replication (acyclovir, AZT, ribavirin), and biosynthesis (saquinavir, ritonavir).

- **New Approaches to Antibiotic Therapy Are Needed as Resistance Increases.** New approaches include two-component signaling pathways, drugs that target the gram-negative lipopolysaccharide, methylase-neutralizing drugs, and peptide antibiotics.

24.4 ANTIBIOTIC ASSAYS AND RESISTANCE

- **There Are Several Antibiotic Susceptibility Assays.** The ability of an antibiotic to affect bacterial growth can be determined by a tube dilution assay that measures the minimal inhibitory concentration needed to destroy microbial cells. The agar disk diffusion method determines the susceptibility of a microbe to a series of antibiotics.

■ **There Are Four Mechanisms of Antibiotic Resistance.** Bacteria have developed resistance to antibiotics and chemotherapeutic agents by altering metabolic pathways, inactivating antibiotics, reducing permeability/active export of antibiotics, and modifying the drug target.

■ **Antibiotics Have Been Misused and Abused.** A major problem attending antibiotic use is the development of antibiotic-resistant strains of microorganisms. Arising from any of several sources such as changes in microbial biochemistry, antibiotic resistance threatens to put an end to the cures of infectious disease that have come to be expected in contemporary medicine.

Questions for Thought and Discussion

Answers to selected questions can be found in Appendix C.

1. In 1877, Pasteur and his assistant Joubert observed that anthrax bacilli grew vigorously in sterile urine but failed to grow when the urine was contaminated with other bacilli. What was happening?

2. Historians report that 2,500 years ago, the Chinese learned to treat superficial infections such as boils by applying moldy soybean curds to the skin. Can you suggest what this implies?

3. During World War II, American soldiers were required to carry a full canteen of water. If they found it necessary to sprinkle a sulfa drug on a wound, they were instructed to drink the complete contents of the canteen. Why do you think this was necessary?

4. One of the novel approaches to treating gum disease is to impregnate tiny vinyl bands with antibiotic, stretch them across the teeth, and push them beneath the gumline. Presumably, the antibiotic would kill bacteria that form pockets of infection in the gums. What might be the advantages and disadvantages of this therapeutic device?

5. In May 1953, Edmund Hillary and Tenzing Norgay reached the summit of Mount Everest, the world's highest mountain. Since that time over 150 other mountaineers have reached the summit, and groups have gone to Nepal from all over the world on expeditions. The arrival of "civilization" has brought a drastic change to the lifestyle of Nepal's Sherpa mountain people. For example, half of all Sherpas used to die before the age of 20, but with antibiotics available for disease, the population has grown from 9 million to 15 million in three decades. Medical enthusiasts are proud of this increase in the life expectancy, but population ecologists see a bleaker side. What do you suspect they foresee, and what

does this tell you about the impact that antibiotics have on a culture?

6. It was a crude remark, but during a discussion on the side effects of antibiotics, a student blurted out: "Better red than dead!" What antibiotic do you think was being discussed?

7. Most naturally occurring antibiotics appear to be products of the soil bacteria belonging to the genus *Streptomyces*. Can you draw any connection between the habitat of these organisms and their ability to produce antibiotics?

8. Bacitracin derives part of its name from Margaret Tracy, a patient from whom doctors isolated the *Bacillus subtilis* used in antibiotic production. Lincomycin is so named because it was first obtained from a bacterium isolated in Lincoln, Nebraska. From your knowledge of prefixes and suffixes, can you guess how other antibiotics in this chapter got their names?

9. Why would a synthetically produced antibiotic be more advantageous than a naturally occurring antibiotic? Why would it be less advantageous?

10. In 1994, researchers surveyed a number of emergency departments in urban hospitals to determine the top 20 drugs prescribed most often. Tylenol placed first. The top-rated antibiotic (fourth place) was amoxicillin. Why do you think this antibiotic was prescribed so often?

11. Of the thousands and thousands of types of organisms screened for antibiotics since 1940, only five genera appear capable of producing these chemicals. Does this strike you as unusual? What factors might eliminate potentially useful antibiotics?

12. History shows that over and over, creativity is a communal act, not an individual one. How does the 1945 Nobel Prize to Fleming, Florey, and Chain reflect this notion?

13. Ecologists tell us that by upsetting the balances in nature, a group of organisms may emerge with unusual characteristics. For instance, the continual use of rodenticides in the 1960s and 1970s permitted a variety of pesticide-resistant "superrat" to emerge in certain American cities. How does this principle relate to the appearance of PPNG?

14. Is an antibiotic that cannot be absorbed from the gastrointestinal tract necessarily useless? How about one that is rapidly expelled from the blood into the urine? Why?

15. The antibiotic issue can be argued from two perspectives. Some people contend that because of side effects and microbial resistance, antibiotics will eventually be abandoned in medicine. Others see the future development of a superantibiotic, a type of "miracle drug." What arguments can you offer for either view? Which direction in medicine would you support?

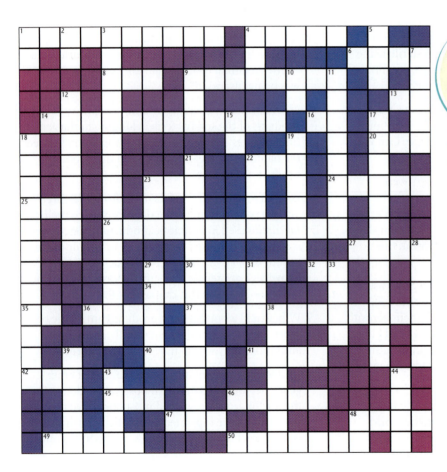

Review

On completing this chapter, you are invited to test your knowledge of its contents by completing the following crossword puzzle. The solution is in Appendix D.

■ ACROSS

1. Has a beta-lactam nucleus; most familiar antibiotic.
4. Four carbon _____ exist in certain antibiotic molecules.
6. Penicillinase also is known as _____-lactamase.
8. Erythromycin has relatively _____ activity against gram-negatives.
9. Causes body secretions to assume a red-orange color.
13. An important antifungal drug is _____-conazole.
14. Drug of choice for rickettsial and chlamydial infections.
16. Country (intls) where penicillin developed after British discovery.
20. Number of major drawbacks to penicillin use.
22. An example of an aminoglycoside antibiotic is _____-tamicin.
23. Organ treated with neomycin to relieve infection.
24. Age group where tetracycline used for acne.

25. Unit (abbr) to establish useful dose of antibiotic.
26. Macrolide antibiotic used as a penicillin substitute.
27. An antibiotic and its receptor are similar to a _____ and key.
30. Body organ possibly damaged by antibiotic use.
32. Bacterium (intls) from which bacitracin isolated.
34. The drug recommended for tuberculosis is _____-niazid.
35. Disease (intls) treated with INH and rifampin.
36. Number of benzene rings in tetracycline molecule.
37. Antibiotic whose major side effect is damage to the auditory mechanism.
40. A desirable trait for antibiotics is _____ excretion to the urine.
41. Type of microorganism that produces certain antibiotics.
42. Organ affected by excessive streptomycin use.
45. A _____-day period of antibiotic use may be required.
46. Method for applying a topical antibiotic.
47. Number of beta-lactam nuclei in a penicillin molecule.

48. Body part where fungus infection may require antibiotic.
49. Bacterium (abbr) treated with penicillin or vancomycin.
50. Genus of fungus susceptible to nystatin therapy.

■ **DOWN**

1. Acidity level (abbr) that can influence antibiotic activity.
2. Penicillin has _____ effect on viruses.
3. Broad-spectrum antibiotic that interferes with protein synthesis.
4. Nucleic acid whose synthesis is inhibited by rifampin.
5. Many _____ cephalosporin antibiotics are available today.
7. Type of chemical group in gentamicin and neomycin.
9. Cell (abbr) having no hemoglobin following chloramphenicol use.
10. Fungus (intls) from which penicillin first isolated.
11. Antifungal drug to treat yeast infection.
12. Used in the eye for conjunctivitis.
15. After reaction with penicillinase, penicillin becomes _____-active.
17. Genus of antibiotic-producing bacteria.

18. Type of bacteria treated with aminoglycosides.
19. Method of introducing cephalosporins to body.
21. Used against penicillin-resistant bacteria.
22. Bacterial characteristic that often determines antibiotic use.
23. Where penicillin allergy exists, _____-thromycin is useful.
28. Function can be affected by antibiotic use.
29. Microorganism unaffected by antibiotics.
31. Swollen during penicillin allergy.
33. Ampicillin and amoxicillin are _____-synthetic drugs.
38. Organic compound whose synthesis is inhibited by certain antibiotics.
39. Urinary _____ infections are treated with certain antibiotics.
43. Antibiotic synthesis is a multi-_____ process.
44. Part of body on which polymyxin is used.
48. Salt (abbr) of antibiotic that is often used.

Microbiology and Public Health

Public health has many facets. Several—such as mental health, nutritional health, and environmental health—lie outside the realm of this text. Other facets, however, are concerned with communicable disease and therefore fall within the scope of microbiology. Immunization, for example, is a public health measure because it creates a barrier within the individual that lessens susceptibility to infection. Case findings and antibiotic treatments are also under the umbrella of public health because they are used to limit the spread of communicable disease.

In Part 7, we shall be concerned with sanitation, a facet of public health designed to prevent microorganisms from reaching the body. The word *sanitation* is derived from the Latin *sanitas*, meaning "health." Sanitation came into full flower in the mid-1800s, and it is a relatively recent phenomenon. Before that time, living conditions in some Western European and American cities were almost indescribably grim: Garbage and dead animals littered the streets; human feces and sewage stagnated in open sewers; rivers were used for washing, drinking, and excreting; and filth was rampant.

The belief that filth was a catalyst to disease fueled the sanitary movement of the mid-1800s. As the Industrial Revolution sparked great population increases in the cities, health problems mounted, and sanitary reformers spoke up for effective sewage treatment, water purification, and food preservation. The germ theory of disease, developed in the 1870s, strengthened the movement and justified its actions because communicable disease could now be blamed on microorganisms in food and water. As sanitary reformers merged with bacteriologists, they issued loud calls for government support of public health and stimulated the development of sophisticated methods for dealing with food and water, which we shall study in Chapters 25 and 26.

Not all microorganisms are bad. Indeed, some contribute mightily to public health. For example, certain microorganisms are used to produce a broad assortment of foods and dairy products that make up a regular part of our diet. In addition, microorganisms play a dominant role in numerous cycles of elements in the environment and forge a link between what is useless and what is useful to other living things. Moreover, scientists employ microorganisms on industrial scales to synthesize a variety of products that we could not obtain otherwise. We shall see examples of these contributions in Chapters 25, 26, and 27.

■ *Microbiological testing of cooked and fresh foods is a daily routine in the food processing industry.*

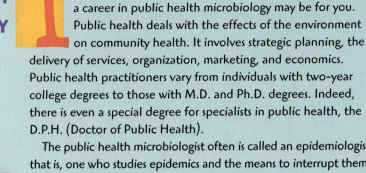

PUBLIC HEALTH MICROBIOLOGY

If you enjoy the adventure and excitement of travel, then a career in public health microbiology may be for you. Public health deals with the effects of the environment on community health. It involves strategic planning, the delivery of services, organization, marketing, and economics. Public health practitioners vary from individuals with two-year college degrees to those with M.D. and Ph.D. degrees. Indeed, there is even a special degree for specialists in public health, the D.P.H. (Doctor of Public Health).

The public health microbiologist often is called an epidemiologist, that is, one who studies epidemics and the means to interrupt them. This individual is concerned with serious foodborne and waterborne diseases of the community, such as cholera and typhoid fever, and with worldwide diseases, such as AIDS. Even the less serious diseases, such as chickenpox and measles, are studied because large-scale outbreaks can exact heavy human tolls. Public health also takes into account the new and emerging epidemics occurring in various parts of the world. Public health microbiologists, sometimes called "disease detectives," were the doctors and researchers who arrived to study recent outbreaks of Ebola fever in Africa.

Public health microbiologists also are involved with the sanitary treatment of water supplies, and they constantly are on the lookout to ensure conformity with public health standards. They oversee the proper disposal of sewage and industrial wastes and work to ensure the safety of milk, dairy products, and other foods we consume. The spread of disease by insects, rodents, and wildlife also is in the domain of the public health microbiologist working locally and in distant parts of the globe. Public health microbiologists attempt to prevent and control the spread of contagious diseases in the urban environment as well as in the forest, open savanna, mountains, and deserts. There is virtually no place on Earth that the public health microbiologist will not put his or her talents to use.

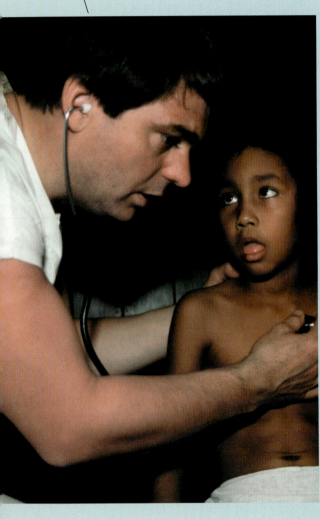

Microbiology of Foods

A dangerous quarter-pounder is as un-American as a murderous Mickey Mouse.

—A *Newsweek* writer commenting on the possibility of bacterial contamination in hamburger meat

THE IDEA WAS APPEALING and the price was right: a patty melt sandwich and a soft drink for lunch. The rye bread was toasted, the hamburgers were stacked and waiting to be cooked, the American cheese slices were lined up next to the grill, and the aroma from the sauteed onions was irresistible. It was October 1983, at the Skewer Inn, a restaurant at the Northwoods Mall in Peoria, Illinois. The stage was set for the third largest recorded outbreak of botulism in US history.

Between October 14 and 16, numerous people stopped by the now-defunct restaurant and enjoyed patty melt sandwiches. Soon, however, 36 individuals began experiencing the paralyzing signs of botulism. They suffered blurred vision, difficulty swallowing and chewing, and labored breathing. One by one they called their doctors, and within a week, all were hospitalized. Twelve patients had to be placed on respirators. After many anxious hours, all but one recovered.

Investigators from the Centers for Disease Control and Prevention (CDC) arrived in Peoria shortly thereafter. They obtained detailed food histories from patients and from others who ate at the restaurant during the same three-day period. First, they identified patty melt sandwiches as the probable cause (24 of 28 patients interviewed specifically recalled eating the sandwiches); then they began a search to pinpoint the item that might be responsible. The data pointed to the onions. Investigators isolated *Clostridium botulinum* spores from the skins of fresh onions at the restaurant, and learned that once sauteed, the onions were left uncovered on the warm stove for hours. Furthermore, the onions were not reheated before serving. Spores had probably germinated within anaerobic mounds of warm onions and deposited their deadly toxins.

Incidents like this one highlight how most foods, even cooked foods, provide excellent conditions for the growth of microorganisms and the depositing of their toxins. The organic matter in food is plentiful, the water content is usually sufficient, and the pH is either neutral or only slightly acidic. To the food manufacturer or restaurant owner, the growth of microorganisms may spell economic loss or a reputation for bad business. To the consumer, it may mean illness or, in some cases, death.

The primary focus of this chapter will be to examine the types of microorganisms that contaminate various foods and to point out the consequences of contamination. We also will examine food spoilage and the preservation methods used to prevent spoilage. Toward the end of the chapter, we shall see that some forms of microbial growth in food are actually desirable because they lead to numerous fermented foods we consume regularly.

25.1 Food Spoilage

Food spoilage has been a continuing problem ever since humans first discovered they could produce more food than they could eat in a single meal. Schoolchildren are taught that Marco Polo traveled to China in the thirteenth century to obtain spices and explore new trade routes. What many fail to realize is that spices were more than just a luxury at that time. They were essential for improving the smell and taste of spoiled food. Refrigeration was virtually unknown, and canning was yet to be invented.

Food is considered **spoiled** when it has been altered from the expected form. Usually, the food has an unpleasant appearance, aroma, and taste. Sometimes, however, these signs may be difficult to detect, such as when staphylococci deposit enterotoxins in food or when too few bacteria grow to cause a perceptible change. The consumption of toxins or microorganisms may cause a number of food poisonings or infections, including those noted in Chapter 9.

Contaminating microorganisms enter foods from a variety of sources. Airborne organisms, for example, fall onto fruits and vegetables, then penetrate the product through an abrasion of the skin or rind. Crops carry soilborne bacteria to the processing plant. Shellfish concentrate organisms by straining contaminated water and catching the organisms in their filtering apparatus. And rodents and arthropods transport microorganisms on their feet and body parts as they move about among foods.

Human handling of foods also provides a source of contamination. For example, bacteria from an animal's intestine contaminate meat handled carelessly by a butcher. Of even more concern are raw vegetables such as those obtained at salad bars. In March 1983, for instance, 107 Maryland residents contracted shigellosis (Chapter 9) after eating from the salad bar in the cafeteria at a military hospital. Earlier that year, 123 cases of hepatitis A were diagnosed in patients who dined at a salad bar restaurant in Lubbock, Texas. In both cases, investigators believed that food handlers were the source of pathogenic microorganisms.

SEVERAL CONDITIONS DETERMINE IF SPOILAGE CAN OCCUR

Since food is basically a culture medium for microorganisms, the chemical and physical properties of food have a significant bearing on the type of microorganisms growing on or in it (FIGURE 25.1). Water, for instance, is a prerequisite for life, and therefore the food must be moist, with a minimum water content of 18 to 20 percent. Microorganisms do not grow in foods such as dried beans, rice, and flour because of their low water content.

Another important factor is **pH**. Most foods fall into the slightly acidic range on the pH scale, and numerous species of bacteria multiply under these conditions. In foods with a pH of 5.0 or below, acid-loving molds often replace the bacteria. Citrus fruits, for example, generally escape bacterial spoilage but yield to mold contamination.

A third property of a food is its physical structure. A raw steak, for example, is not likely to spoil quickly because microorganisms cannot penetrate the meat easily. However, an uncooked hamburger can deteriorate rapidly, since microorganisms exist within the loosely packed ground meat as well as on the surface.

The chemical composition of the food may be another determining factor in the type of spoilage possible. Fruits, for instance, support organisms that metabolize carbohydrates, whereas meats attract protein decomposers. Starch-hydrolyzing bacteria

Moist foods   Neutral foods   Unrefrigerated (25°C)   Ground or sliced meat

Foods that spoil quickly

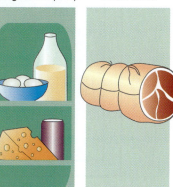

Dry foods   Acidic foods   Refrigerated (5°C)   Whole meat

Foods that resist spoilage

FIGURE 25.1

Food Spoilage

These drawings show the conditions under which foods are likely to spoil quickly or resist spoilage.

and molds often are found on potatoes, corn, and rice products. The presence of certain vitamins encourages particular microorganisms to proliferate, while the absence of other vitamins provides natural resistance to decay.

Oxygen and temperature are other considerations. Vacuum-sealed cans of food do not support the growth of aerobic bacteria, nor do vegetables or most bakery products support anaerobes. Similarly, the refrigerator is usually too cold for the growth of human pathogens, but the warm hold of a ship or a humid, hot warehouse storeroom is an environment conducive to the growth of these pathogens. It is common knowledge that contamination is more likely in cooked food at warm temperatures than in refrigerated cooked food (FIGURE 25.2).

FIGURE 25.2

A Case of *Vibrio parahaemolyticus* Food Poisoning

This outbreak occurred in Port Allen, Louisiana, in June 1978. *Vibrio parahaemolyticus* was subsequently found in the leftover shrimp, as well as in a major number of stool specimens from patients.

TEXTBOOK CASES

1. On the morning of June 21, a large amount of shrimp was cooked for a dinner by bringing the water to a "rolling boil." The gas was then turned off.

2. The shrimp were repackaged in the boxes in which they came, and covered with aluminum foil to keep them warm.

3. The shrimp were transported 40 miles in an unrefrigerated truck to the site of the dinner. The food was held unrefrigerated until 7:30 p.m., when served. A total of almost 8 hours had passed since preparation. Microorganisms proliferated during this extended period.

4. A group of 1,700 people attended the dinner that evening. Shrimp were served to all as an appetizer in shrimp cocktail.

5. About 16 hours later, 1,100 of the 1,700 guests reported diarrhea, cramps, nausea, and vomiting. The diagnosis was food poisoning.

The food industry recognizes three groups of foods loosely defined on the basis of their chemical and physical properties. **Highly perishable** foods are those that spoil rapidly. They include poultry, eggs, meats, most vegetables and fruits, and dairy products. Foods such as nutmeats, potatoes, and some apples are considered **semiperishable**, because they spoil less quickly. Orange juice is usually considered semiperishable because of its high acidity, but in 1999, a major outbreak of *Salmonella* infection was linked to unpasteurized orange juice distributed to 15 states by an Arizona producer. **Nonperishable** foods are stored often in the kitchen pantry. Included in this group are cereals, rice, dried beans, macaroni and pasta products, flour, and sugar. **FIGURE 25.3** summarizes the three groups.

THE CHEMISTRY OF SPOILAGE PRODUCES SPECIFIC PRODUCTS

Spoilage in foods often is due to the naturally occurring chemistry of contaminating microorganisms. Yeasts, for instance, live in apple juice and convert the carbohydrate into **ethyl alcohol**, a product that gives spoiled juice an alcoholic taste. Certain bacteria convert food proteins into amino acids, then break down the amino acids into foul-smelling end products. Cysteine digestion, for example, yields hydrogen sulfide, which imparts a rotten egg smell to food. Tryptophan digestion yields **indole** and **skatole**, which give food a fecal odor.

Two other possible products of the microbial metabolism of carbohydrates are acid, which causes food to become sour, and gas, which causes sealed cans to swell. Moreover, when fats break down to fatty acids as in spoiled butter, a rancid odor or taste may evolve. Capsule production by bacteria causes food to become slimy, and pigment production imparts color. In numerous historical incidents, the red pigment from *Serratia marcescens* in bread has been interpreted as a sign of blood (**MicroFocus 25.1**).

Some foods resist spoilage naturally because they contain antimicrobial chemicals. The white of an egg has **lysozyme**, an enzyme that digests the cell wall of grampositive bacteria. Garlic contains certain compounds that inhibit many bacteria (MicroFocus 23.2). Oil of cloves appears to have healing tendencies, while radish and onion extracts both retard bacterial growth.

Highly perishable foods   Semiperishable foods   Nonperishable foods

FIGURE 25.3

The Perishability of Foods

Examples of highly perishable, semiperishable, and nonperishable foods. The physical and chemical properties of these foods are reliable indicators of their rate of perishability.

MicroFocus 25.1

THE BLOOD OF HISTORY

Because of its characteristic blood-red pigment and propensity for contaminating bread, *Serratia marcescens* has had a notable place in history. For example, the dark, damp environments of medieval churches provided optimal conditions for growth on sacramental wafers used in Holy Communion. At times, the appearance of "blood" on the Host was construed to be a miracle. One such event happened in 1264 when "blood dripped" on a priest's robe. The event was later commemorated by Raphael in his fresco *The Mass of Bolsena*. Unfortunately, religious fanatics used such episodes to institute persecutions because they believed that heretical acts caused the "blood" to flow.

It was not until 1819 that Bartholemeo Bizio, an Italian pharmacist, demonstrated that the bloody miracles were caused by a living organism. Bizio thought the organism was a fungus. He named it *Serratia* after Serafino Serrati, a countryman whom he considered the inventor of the steamboat. The name *marcescens* came from the Latin word for decaying, a reference to the decaying of bread.

In 332 B.C., Alexander the Great and his army of Macedonians laid siege to the city of Tyre in what is now Lebanon. The siege was not going well. Then one morning, blood-red spots appeared on several pieces of bread. At first it was thought to be an evil omen, but a soothsayer named Aristander pointed out that the "blood" was coming from within the bread. This suggested that blood would be spilled within Tyre and that the city would fall. Alexander's troops were buoyed by this interpretation and with renewed confidence they charged headlong into battle and captured the city. The victory opened the Middle East to the Macedonians. Their march did not stop until they reached India.

MEATS AND FISH CAN BECOME CONTAMINATED IN SEVERAL WAYS

Meats and fish originally are free of contamination because the muscle tissues of living animals normally are sterile. Spoilage organisms enter during handling, processing, packaging, and storage. For example, if a piece of meat is ground for hamburger, microorganisms from the surface accumulate in the teeth of the grinder along with other dustborne organisms. Bacteria from preparers' hands or from a sneeze compound the problem. The grinder may be cleaned well and refrigerated, but rarely is it sterilized. The importance of foodborne infection in ground meat is pointed up by the 1993 outbreak of hemolytic diarrhea traced to hamburger meat contaminated by *Escherichia coli* O157:H7. Over 500 patrons of Jack-in-the-Box restaurants were involved. A 1998 outbreak of the same disease necessitated a recall of 25 million pounds of hamburger meat (Chapter 9).

There also is the problem of the so-called "choke points," or places where animals come together and epidemics spread. In the United States, for instance, about 9,000 farms produce calves for beef; the animals then are sent to about 46,000 feedlots for developing, and then, to about 80 plants for slaughter. Microorganisms can spread at any of these points. Moreover, animal feeds often contain the entrails of poultry as added protein sources, and another opportunity for bacteria to spread is presented.

Processed meats, such as luncheon meats, sausages, and frankfurters, represent special hazards because they are handled often. A 1999 listeriosis outbreak in Oklahoma was linked to hot dogs and deli meats (Chapter 9). Also, natural sausage and salami casings made from animal intestines may contain residual bacteria, especially botulism spores. When preparers pack such casings tightly with meat, *Clostridium botulinum* may multiply and produce its powerful toxins. As early as the 1820s, people recognized the symptoms of "sausage poisoning" and coined the name botulism, which translates loosely to sausage.

Organ meats, such as livers, kidneys, and sweetbreads (thymus and pancreas), are less compact tissue than muscle and thus spoil more quickly. Moreover, because the organs contain many natural filtering tissues, bacteria tend to be trapped here. Foods like these therefore should be cooked as soon as possible after purchase.

The extent of contamination in meats often consists of a harmless "greening" seen on the surface of a steak. This discoloration is commonly due to the gram-positive rod *Lactobacillus*, or the gram-positive coccus *Leuconostoc*. The green color results from pigment alteration in the meat and represents no hazard to the consumer. Slime formation on the outer casings and souring in frankfurters, bologna, or processed meats also may result from these organisms as well as from the *Streptococcus* species.

Microbiologists often can trace spoilage in fish to the water from which it is taken or held. Fish tissues deteriorate rapidly and the filleting of fish on a bloodstained block encourages contamination. It also is interesting to note that bacteria in fish are naturally adapted to the cold environment that fish live in and thus, cooling will not affect them so thoroughly; freezing is preferred. Shellfish are of particular concern because they commonly obtain their food by filtering particles from the water. Clams, oysters, and mussels therefore concentrate such pathogens as hepatitis A viruses, typhoid bacilli, cholera vibrios, or amoebic cysts. Many cases of cholera have been related to raw oysters.

POULTRY AND EGGS CAN SPOIL QUICKLY

Contamination in poultry and eggs may reflect human contamination (FIGURE 25.4), but it usually stems from microorganisms that have infected the bird. Members of the genus *Salmonella*, with over 2,400 pathogenic serotypes, may cause diseases in chickens and turkeys, then pass to consumers via poultry and egg products. For example, an outbreak of **salmonellosis** in New York was traced to contaminated eggs used to make cheese lasagna and stuffed shells. Processed foods such as chicken pot pies, whole egg custard, mayonnaise, egg nog, and egg salad also may be sources of salmonellosis. **Psittacosis** (Chapter 8) is another problem because *Chlamydia psittaci* infects poultry. In one instance, 27 employees in a turkey-processing plant contracted psittacosis while preparing consumer products from diseased turkeys.

Eggs are normally sterile when laid, but the outer waxy membrane, as well as the shell and inner shell membrane, may be penetrated by bacteria after several hours. *Proteus* species cause **black rot** in eggs as hydrogen sulfide gas accumulates from the breakdown of cysteine. This gives eggs a rotten odor. Other spoilages in eggs include **green rot** from *Pseudomonas* species and **red rot** from the growth of *Serratia marcescens*. Green rot results from a fluorescent pigment that causes the egg to glow when placed in front of an ultraviolet light. Red rot gives the yolk a blood-red appearance.

The primary focus of contamination in an egg is the yolk rather than the white. This is because of the nutritious quality of the yolk and because the white has a pH of approximately 9.0. Also, the lysozyme in egg white is inhibitory to gram-positive bacteria. However, *Salmonella enteritidis* (FIGURE 25.5) is a common problem.

BREADS AND BAKERY PRODUCTS CAN SUPPORT BACTERIAL AND FUNGAL GROWTH

In the production of bakery products, ingredients such as flour, eggs, and sugar are generally the sources of spoilage organisms. Although most contaminants are killed during baking, some bacterial and mold spores survive because bread is heated internally to 100°C for only about nine minutes. Members of the spore-forming

FIGURE 25.4

A Case of *Staphylococcus aureus* Food Poisoning

This incident took place in Sussex County, Delaware, between March 8 and 10, 1979. Analysis showed that *S. aureus* phage type 95 was present in the nasal swab from the person who ground the chicken, as well as in the meat grinder and in the leftover chicken salad.

1. On March 8, 1979, chicken was cooked for chicken salad for a wedding reception to be held two days later.

2. The chicken was deboned and refrigerated overnight in a large washtub.

3. On March 9, the chicken was ground in a meat grinder by a person who had toxigenic *Staphylococcus aureus* in his nose.

4. The chicken was then mixed with celery, onions, and mayonnaise and refrigerated overnight in the same washtub.

5. On March 10, the chicken salad was delivered, unrefrigerated, to the wedding reception. A total of seven hours passed before it was consumed.

6. Some hours later, 64 of the 107 guests at the wedding experienced nausea, diarrhea, intestinal cramps, and other symptoms of food poisoning.

FIGURE 25.5

Salmonella enteritidis

Salmonella enteritidis is one of the most widely found contaminants in poultry and egg products. (Bar = 5 μm.)

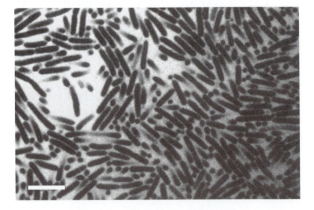

genus *Bacillus* commonly survive, and as they proliferate, their capsular material accumulates, giving the bread a soft cheesy texture with long, stringy threads. The bread is said to be ropy.

Cream fillings and toppings in bakery products provide excellent chemical and physical conditions for bacterial growth. For example, custards made with whole eggs may be contaminated with *Salmonella* species, and whipped cream may contain dairy organisms such as species of *Lactobacillus* and *Streptococcus*. The acid produced by these bacteria results in a sour taste. High sugar environments of chocolate toppings and sweet icings support the growth of fungi. Most bakery products should be refrigerated during warm summer months.

SOME GRAINS ARE SUSCEPTIBLE TO SPOILAGE

Two types of grain spoilage are important in public health microbiology. The first type of spoilage is caused by the ascomycete *Aspergillus flavus*. This mold produces **aflatoxins**, a series of toxins that accumulate in stored grains such as wheat as well as peanuts, soybeans, and corn. Scientists have implicated aflatoxins in liver and colon cancers in humans. The toxins are consumed in grain products, as well as meat from animals that feed on contaminated grain.

The second type of grain spoilage is caused by *Claviceps purpurea*, the cause of **ergot poisoning** (ergotism). Rye plants are particularly susceptible to this type of spoilage (MicroFocus 15.8), but wheat and barley grains may be affected also. The toxins deposited by *C. purpurea* may induce convulsions and hallucinations when consumed. The drug LSD is derived from the toxin.

MILK AND DAIRY PRODUCTS SOMETIMES SOUR

Milk is an extremely nutritious food. It is an aqueous solution of proteins, fats, and carbohydrates that contains numerous vitamins and minerals. Milk has a pH of about 7.0 and is an excellent growth medium for humans and animals, as well as microorganisms.

About 87 percent of the substance of milk is water. Another 2.5 percent is a protein called **casein**, actually a mixture of three long chains of amino acids suspended in fluid. A second protein in milk is **lactalbumin**. This protein forms the surface skin when milk is heated to boiling. Lactalbumin is a whey protein, one that remains in the clear fluid (the whey) after the casein curdles during milk spoilage or fermentation. Carbohydrates make up about 5 percent of the milk. The major carbohydrate is **lactose**, sometimes referred to as milk sugar (*lactus* is Latin for "milk"). Rarely found elsewhere, lactose is a disaccharide that can be digested by relatively few bacteria, and these are usually harmless. The last major component of milk is **butterfat**, a mixture of fats that can be churned into butter. Butterfat comprises about 4 percent of the milk and is removed in the preparation of nonfat (skim) milk or low-fat milk. When bacterial enzymes digest fats into fatty acids, the milk develops a sour taste and becomes unfit to drink.

A common type of milk spoilage often takes place in the kitchen refrigerator or dairy case at the supermarket. Here, *Lactobacillus* or *Streptococcus* species multiply slowly and ferment the lactose in milk. Soon, large quantities of lactic and acetic acids accumulate. Enough acid may develop to change the structure of the protein and cause it to solidify as a curd. Dairy microbiologists refer to such an acidic curd as a **sour curd**. The lactobacilli and streptococci causing it usually have survived the pasteurization process.

Sweet curdling in milk may result when enzymes from species of *Bacillus*, *Proteus*, *Micrococcus*, or certain other bacteria attack the casein. As weak hydrogen bonds break, casein loses its three-dimensional structure and curdles. The reaction is said to be sweet because little acid production occurs. It is an essential step in the production of cheese. The clear liquid is **whey**, an aqueous solution of lactose, minerals, vitamins, lactalbumin, and other milk components. Whey is used to make processed cheeses and "cheese foods."

Milk also may be contaminated by gram-negative rods of the coliform group of bacteria, including *E. coli* and *Enterobacter aerogenes*. These bacteria produce acid and gas from lactose. The acid curdles the protein, and the gas forces the curds apart, sometimes so violently that they explode out of the container. The result is a **stormy fermentation**. *Clostridium* species also cause this reaction.

Ropiness in milk is similar to that in bread. It develops from capsule-producing organisms, such as *Alcaligenes*, *Klebsiella*, and *Enterobacter*. These gram-negative rods multiply in milk, even at low temperatures, and deposit gummy material that appears as stringy threads and slime.

Another form of spoilage is caused by species of *Pseudomonas* and *Achromobacter* that produce the enzyme **lipase**. This enzyme attacks butterfats in milk and digests them into glycerol and fatty acids, giving milk a sour taste and a putrid smell. A similar problem may develop in butter.

Additional types of milk spoilage result from the red pigment deposited in dairy products by *Serratia marcescens*, the blue or green pigment of *Pseudomonas* species, and the gray rot caused by certain *Clostridium* species. In gray rot, hydrogen sulfide (H_2S) from cysteine imparts a rotten-egg smell to milk, and the H_2S reacts with minerals to yield a gray or black sulfide compound. Spoilage due to wild yeasts is usually characterized by a pink, orange, or yellow coloration in the milk. Acid conditions stimulate mold decay in cheese products.

Milk normally is sterile in the udder of the cow, but contamination occurs as it enters the ducts leading from the udder, as well as from other unlikely sources (**MicroFocus 25.2**). The colostrum, or first milk, is laden with organisms. Species of soilborne *Lactobacillus* and *Streptococcus* are acquired during the passage, together with various coliform bacteria from dust, manure, and polluted water. The milk may derive other organisms from dairy plant equipment and unsanitary handling of dairy products by plant employees.

MicroFocus 25.2

GOOD AND NOT SO GOOD

In parts of Great Britain, milkmen still visit homes regularly and deliver bottles of fresh milk. That's good. Magpies, crows, and other birds arrive shortly thereafter and use their strong beaks to peck through the foil caps of the bottles and take a drink. That's not so good.

The birds use the milk to feed their young broods in the nest. That's good. But while taking a drink, they transmit *Campylobacter* species to the milk. That's not so good.

By having their milk delivered, British families save a trip to the market. That's good. Unfortunately, they also spend extra time in the bathroom suffering the misery of diarrhea. And that's not so good.

The moral of the story? "Bewildered Brits better beware bacteria-bearing birds."

To this point . . .

We have discussed the problem of food spoilage and have indicated how various conditions may contribute to the type of spoilage observed. Among these conditions are the amount of water present, the pH of the food product, the physical structure of the food and its chemical composition, and the oxygen content and temperature of the environment in which the food is found. We also noted how the chemistry of spoilage leads to the deposit of certain end products in foods. For example, hydrogen sulfide production from cysteine gives a rotten smell to eggs, and pigment production imparts color to foods.

The discussion then turned to various foods and the spoilage that takes place within them. We examined the sources and types of spoilage within meats and fish, poultry and eggs, breads and bakery products, grains, and milk and dairy products. In each case, some examples of the microbial flora were noted.

We now will move on to the topic of food preservation. This section will outline various methods for preventing microorganisms from reaching food and causing spoilage. Food preservation also is an essential prerequisite to sanitation and public health because preservation methods halt the spread of infectious microorganisms. We will explore several traditional methods of preservation, and a number of contemporary methods used in food technology.

25.2

Food Preservation

Centuries ago, humans battled the elements to keep a steady supply of food at hand. Sometimes there was a short growing season; at other times, locusts descended on their crops; at still other times, they underestimated their needs and had to cope with scarcity. However, experience taught humans they could overcome these difficult times by preserving foods. Among the earliest methods was drying vegetables and strips of meat and fish in the sun. Foods also could be preserved by salting, smoking, and fermenting. Individuals could now trek far from their native habitat, and soon they took to the sea and moved overland to explore new lands.

The next great advance did not come until the mid-1700s. In 1767, Lazaro Spallanzani attempted to disprove spontaneous generation by showing that beef broth would remain unspoiled after being subjected to heat (Chapter 1). Nicholas Appert applied this principle to a variety of foods (MicroFocus 25.3). Neither Appert nor any of his contemporaries was quite sure why food was being preserved, but it was clear that the spoilage could be retarded by prolonged heating. The significance of microorganisms as agents of spoilage awaited Pasteur's classic experiments with wine several generations later.

Through the centuries, preservation methods have had a common objective: to reduce the microbial population and maintain it at a low level until food can be consumed. Modern preservation methods are mere extensions of these principles. Though today's methods are sophisticated and technologically dynamic, advances in preservation processes are counterbalanced by the great volumes of food that must be preserved and the complexity of food products. Thus, the problems that early humans

MicroFocus 25.3

TO FEED AN ARMY

Part of Napoleon's genius was understanding the finer points of warfare, including how to feed an army. Recognizing that thousands of men-at-arms were a glut on the countryside, he broke up his army into smaller units that foraged on their own as they moved. When the time for battle neared, he reassembled his forces and engaged the enemy.

The shortcomings of this system became painfully clear when Napoleon crossed into northern Italy in 1800 and engaged the Austrians at the Battle of Marengo. Aware that the French army was scattered about, the Austrians

charged before Napoleon could bring his forces together. Disaster was averted only when units arrived on the flanks to repel the Austrians. Napoleon had learned an important lesson: The next time he went to war, food would go with him.

Included in Napoleon's plan to resurrect France was a ministry that encouraged industry by offering prizes for imaginative inventions. A winemaker named Nicholas Appert attracted attention with his process of preserving food. Appert placed fruits, vegetables, soups, and stews in thick bottles, then boiled the bottles for several hours. He used

wax and cork to seal the bottles and wine cages to prevent inadvertent opening of the bottle. By 1805, Appert had set up a bottling industry outside Paris and had a thriving business.

The Ministry of Industry encouraged Appert to publish his methods and submit samples of bottled foods for government testing. The French navy took numerous bottles on long voyages and reported excellent food preservation. In 1810, Appert was awarded 12,000 francs for his invention. Two years later, Napoleon assembled hundreds of cannons, thousands of men, and countless bottles of food, and marched off to war with Russia.

faced do not differ fundamentally from those confronting modern food technologists. In this section, we shall examine the methods of food preservation currently in use. Since the methods depend on physical and chemical control practices, a review of Chapters 22 and 23 might be useful.

HEAT DENATURES PROTEINS

Heat kills microorganisms by changing the physical and chemical properties of their proteins. In a moist heat environment, proteins are denatured and lose their three-dimensional structure, reverting to a different three-dimensional form or a two-dimensional form (Chapter 2). As structural proteins and enzymes undergo this change, the organisms die.

The most useful application of heat is in the process of canning. Shortly after Appert established the use of heat in preservation, an English engineer named Bryan Donkin substituted iron cans coated with tin for Appert's bottles. Soon he was supplying canned meat to the British navy. In the United States, the tin can was virtually ignored until the Civil War period. In the years thereafter, mass production began and soon the tin can became the symbol of prepackaged convenience.

Modern canning processes are complex (**FIGURE 25.6**). Machines wash, sort, and grade the food product, then subject it to steam heat for three to five minutes. This last process, called **blanching**, destroys many enzymes in the food product and prevents any further cellular metabolism from taking place. Then the food is peeled and cored, and its diseased sections are removed. Canning comes next, after which the air is evacuated and placed in a pressured steam sterilizer similar to an autoclave at a temperature of 121°C or lower, depending on the pH, density, and heat penetration rate.

The sterilizing process is designed to eliminate the most resistant bacterial spores, especially those of the genera *Bacillus* and *Clostridium*. However, the process is con-

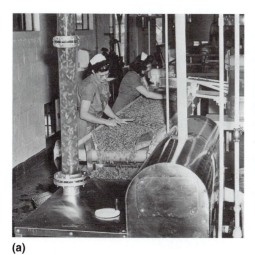
(a)

(b)

FIGURE 25.6

Two Steps in the Industrial Canning Process

(a) Initial inspection of green beans is made after cutting and sieve sizing. The beans are washed in the apparatus in the foreground and then conveyed through sanitary glass piping (at the left) to the blanching machine. (b) A continuous cycle orbital cooker used by modern vegetable processors to reduce cooking time of the product.

sidered **commercial sterilization**, which is not as rigorous as true sterilization, and some spores may survive. Moreover, should a machine error lead to improper heating temperatures, a small hole allow airborne bacteria to enter, or a proper seal not form, contamination may result.

Contamination of canned food is commonly due to facultative or anaerobic bacteria that produce gas and cause the ends of the can to bulge. Food microbiologists call a can a flipper if the bulge can be flattened easily. It is a springer if pushing the bulge pushes out the opposite end of the can. A soft swell occurs when both ends bulge. If neither end can be pushed in because of the large amount of gas, a hard swell is present. The organisms often responsible for gas production are *Clostridium* species as well as coliform bacteria, a group of gram-negative non–spore-forming rods that ferment lactose to acid and gas. Contamination is usually obvious since the spoiled product generally has a putrid odor.

Growth of acid-producing bacteria presents a different problem because spoilage cannot be discerned from the can's shape. Food has a flat-sour taste from the acid and has probably been contaminated by a *Bacillus* species, a coliform, or another acid-producing bacterium that survived the heating.

The process of **pasteurization** was developed by Louis Pasteur in the 1850s to eliminate bacteria in wines. His method was first applied to milk in Denmark about 1870, and by 1895 the process was widely employed. Although the primary object of pasteurization is to eliminate pathogenic bacteria from milk, the process also lowers the total number of bacteria and thereby reduces the chance of spoilage (Chapter 22). The more traditional method involves heating the milk in a large bulk tank at 62.9°C (145°F) for 30 minutes. This is the **holding method**, also known as the **LTLT method** for "low temperature, long time." Machines stir the milk constantly during the pasteurization to ensure uniform heating, and cool it quickly when the heating is completed. More concentrated products, such as cream, often are heated at the higher temperature of 69.5°C (155°F) to ensure successful pasteurization. An innovative method of egg pasteurization recently has been introduced, as **MicroFocus 25.4** explains.

The more modern method of pasteurization is called the **flash method**. In this process, raw milk is first warmed using the heat of previously pasteurized milk. Machines then pass the milk through a hot cylinder at 71.6°C (161°F) for a period of 15 to 17 seconds. Next, the milk is cooled rapidly, in part by transferring its heat to the incoming milk. This is the **HTST method**, for "high temperature, short time." It is useful for high-quality raw milk, in which the bacterial count is consistently low. A new method called **ultrapasteurization** is used in some dairy plants. In this process, milk and milk products are subjected to heat at 82°C (180°F) for 3 seconds. Following pasteurization, a set of laboratory tests is performed to assay the quality of the milk and the success of pasteurization (**TABLE 25.1**).

The bacteria that survive pasteurization may be involved in spoilage. *Streptococcus lactis*, for instance, grows slowly in refrigerated milk, and when its numbers reach 20 million per milliliter, enough lactic acid has been produced to make the milk sour. Organisms that survive the heat of pasteurization are described as **thermoduric**. Pasteurization is virtually useless against **thermophilic** bacteria, since they grow naturally at 60°C to 70°C. These organisms generally do not grow at refrigerator temperatures or cause human disease because conditions are too cool. Pasteurization also has no effect on spores.

Although milk in the United States is normally pasteurized and refrigerated, exposure to steam at 140°C for 3 seconds can sterilize it. This **ultra-high temper-**

MicroFocus 25.4

"BRING ON THE CAESAR SALAD!"

Some people have it Hawaiian-style with pineapple and ham; some have it Italian-style with pasta and tomatoes; and some enjoy it Southwestern-style with roasted chili peppers. Some toss it with grilled chicken or flank steak or grilled calamari or Cajun scallops. Regardless of the addition, however, the salad remains the same—Caesar salad.

For history buffs, the first Caesar salad is reported to have evolved on July 4, 1924, in the mind of Caesar Cardini, the proprietor of a restaurant (Caesar's Place) in Tijuana, Mexico. Cardini was desperate for a fill-in during a particularly busy day, so he threw together some Romaine lettuce, Parmesan cheese, lemon, garlic, oil, and raw eggs. His customers were enchanted.

Over the years, the reputations of the salad and its inventor grew. There was only one problem, however—the eggs. For true Caesar salad, raw eggs are used to add creaminess to the dressing. But that became a problem when increasing cases of *Salmonella* infections were traced to raw eggs. Eggless Caesar dressings (with heavy cream) were tried, but it just wasn't the same.

Caesar salad aficionados, take heart—the pasteurized egg is on the way. Purdue University microbiologists have found that *Salmonella* can be eliminated by heating eggs in hot water or a microwave oven, then maintaining them at 134°F in a hot-air oven for one hour. A New Hampshire company (Pasteurized Eggs, LP) is touting the benefits of its new machine for destroying eggborne *Salmonella*. The machine heats eggs slowly, then directs them to successive baths in water ranging from 62°C to 72°C. Indeed, the process works so well that the US Department of Agriculture (USDA) has issued a new stamp certifying that eggs treated by the company meet standards for egg pasteurization established by the Food and Drug Administration (FDA).

Agricultural officials estimate that almost 50 billion eggs are produced for American consumption each year, and that over 2 million are infected with *Salmonella*. In the years ahead, that second number should dwindle considerably as egg pasteurization becomes standard practice (as milk pasteurization already has). Then it will be safe to sample the cookie dough, or to have eggs over easy, or to enjoy Caesar salad the way it was meant to be enjoyed.

TABLE 25.1

Laboratory Tests Used to Assay the Quality of Milk

TEST	PURPOSE	IMPORTANCE
Phosphatase test	Determines whether phosphatase is present. Phosphatase is an enzyme normally destroyed during pasteurization.	To determine whether sufficient heat was used during pasteurization. If phosphatase is present, pathogens also might be present.
Reductase test	Estimates the number of bacteria in milk. The rate at which methylene blue is reduced to its colorless form is proportional to the number of bacteria present in a milk sample.	High-quality milk contains so few bacteria that a standard concentration of methylene blue will not be reduced in 6 hours. Low-quality milk has many bacteria, and methylene blue is reduced in 2 hours or less.
Standard plate count	Determines the total number of viable bacteria per ml of milk. Diluted milk is mixed with nutrient agar and incubated 48 hours; colonies are counted and the number of bacteria in the original sample is calculated.	The number per milliliter may not exceed 100,000 in raw milk before being pooled with other milk. It may not exceed 20,000 after pasteurization.
Test for coliforms	Determines the number of viable coliform bacteria per ml of milk. Similar to the standard plate count except that special media for coliform bacteria are used.	A positive coliform test indicates contamination with fecal material. The number per ml may not exceed established standards.
Test for pathogens	Detects the presence of pathogens. Methods depend on the pathogens suspected.	Helps locate the source of infectious agents that may be present in milk.

ature (**UHT**) results in milk (e.g., Parmalat) with an indefinite shelf life as long as the container remains sealed. Small containers of coffee cream often are prepared this way.

LOW TEMPERATURES SLOW MICROBIAL GROWTH

By lowering the environmental temperature, one can reduce the rate of enzyme activity in a microorganism and thus lower the rate of growth and reproduction. This principle underlies the process of refrigeration and freezing. Although the organisms are not killed, their numbers are kept low and spoilage is minimized. Ironically, the food is preserved by preserving the microorganisms.

Well before contemporary humans developed refrigerators, the Greeks and Romans had partially solved the problem of keeping things cold. They simply dug a snow cellar in the basements of their homes, lined the cellar with logs, insulated it with heavy layers of straw, and packed it densely with snow delivered from far-off mountaintops ("The iceman cometh!"). The compressed snow turned to a block of ice, and foods would remain unspoiled for long periods when left in this makeshift refrigerator. The modern refrigerator at 5°C (41°F) provides a suitable environment for preserving food without destroying appearance, taste, or cellular integrity (FIGURE 25.7). However, **psychrotrophic** microorganisms survive and cause green meat surfaces, rotten eggs, moldy fruits, and sour milk. Pathogens such as *Listeria monocytogenes* and *Yersinia enterocolitica* also grow at low temperatures.

FIGURE 25.7

Important Temperature Considerations in Food Microbiology

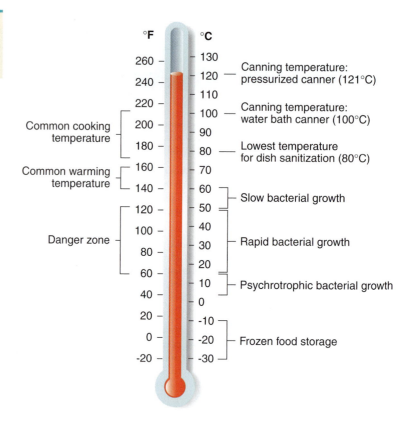

FIGURE 25.8

Food Contamination

A scanning electron micrograph of unidentified flagellated bacteria growing on the skin of a chicken carcass. Bacteria such as these contaminate frozen food and grow to large numbers when the food is thawed and held for long periods before cooking. (Bar 5 = μm.)

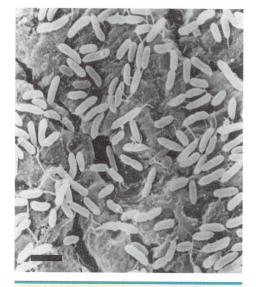

When food is placed in the freezer at −10°C (14°F), ice crystals form rapidly. These crystals tear and shred microorganisms and kill a significant number. However, many microorganisms survive, and the ice crystals are equally destructive to food cells. Therefore, when the food thaws, bacteria multiply quickly. Organisms such as staphylococci produce substantial amounts of enterotoxins, and *Salmonella* serotypes and streptococci and other bacteria grow to large numbers (FIGURE 25.8). Rapid thawing and cooking are recommended. Moreover, food should not be refrozen, because during thawing and refreezing, bacteria deposit sufficient enterotoxin to cause food poisoning the next time the food is thawed. Microwave cooking, which requires minimal thawing, may eliminate some of these problems.

Deep freezing at −60°C results in smaller ice crystals, and although the physical damage to microorganisms is less severe, their biochemical activity is reduced considerably. Small ice crystals do not damage food cells as severely as do the larger crystals formed at higher temperatures. Some food producers blanch their product before deep freezing, a process that further reduces the number of microorganisms.

A major problem in freezing is **freezer burn**, which may occur over long periods of time as food dries out from moisture evaporation. Another disadvantage is that the energy cost of freezing is considerable. Nevertheless, freezing has been a mainstay of preservation since

Clarence Birdseye first offered frozen foods for retail purchase in the 1920s. Approximately 33 percent of all preserved food in the United States is frozen.

DRYING REMOVES WATER

The advantage of drying foods is best expressed by the phrase "Where there is no water, there is no natural life." Indeed, dry foods cannot support microbial growth. In past centuries, people used the sun for drying, but modern technologists have developed sophisticated machinery for this purpose. For example, the spray dryer expels a fine mist of liquid such as coffee into a barrel cylinder containing hot air. The water evaporates quickly, and the coffee powder falls to the bottom of the cylinder.

Another machine for drying is the heated drum. Machines pour liquids such as soup on the drum, and the water evaporates rapidly, leaving dried soup to be scraped off the drum. A third machine uses a belt heater that exposes liquids such as milk to a stream of hot air. The air evaporates any water and leaves dried milk solids. Unfortunately, spore-forming and capsule-producing bacteria are problems because they resist drying.

During the past 20 years, freeze-drying, or **lyophilization**, has emerged as a valuable preservation method, although it has a long history (**MicroFocus 25.5**). In this process, food is deep frozen, and then a vacuum pump draws off the water in a machine like the one pictured in **FIGURE 25.9**. (Water passes from its solid phase [ice] to its gaseous phase [water vapor] without passing through its liquid phase [water].) The dry product is sealed in foil and easily reconstituted with water. Hikers and campers find considerable value in freeze-dried food because of its light weight and durability. However, there is a disquieting note: Lyophilization also is a useful method for storing, transporting, and preserving bacterial cultures.

OSMOTIC PRESSURE CAN HELP PRESERVE FOODS

When living cells are immersed in large quantities of a compound such as salt or sugar, water diffuses out of cells through cell membranes and into the surrounding environment, where it dilutes the high concentration of the compound. This diffusion of water is called **osmosis**, and the tendency of water to move across a membrane is termed **osmotic pressure**.

Osmotic pressure can be used to preserve foods because water flows out of microorganisms as well as food cells. For example, in highly salted or sugared foods, microorganisms dehydrate, shrink, and die. Jams, jellies, fruits, maple syrups, honey, and similar products typify foods preserved by high sugar concentrations. Salted foods include ham, cod, bacon, and beef, as well as certain vegetables such as sauerkraut, which has the added benefit of large quantities of acid. It should be noted, however, that staphylococci tolerate salt and may survive to cause staphylococcal food poisoning.

CHEMICAL PRESERVATIVES ARE FOUND IN OR ARE ADDED TO MANY FOODS

For a chemical preservative to be useful in foods, it must be inhibitory to microorganisms while easily broken down and eliminated by the human body without side effects. These requirements have limited the number of available chemicals to a select few.

MicroFocus 25.5

IT STARTED WITH "STOMPED POTATOES"

Freeze-drying, or lyophilization, is the removal (sublimation) of water from many substances without going through a liquid state. Such freeze-dried foods last longer than other preserved foods and are very light, which makes them perfect for everyone from backpackers to astronauts. However, the origins of freeze-dried foods go back to the ancient Incas of Peru.

The basic process of freeze-drying food was known to the Peruvian Incas that lived in the high altitudes of the Andes Mountains of South America. The Incas stored some of their food crops, including potatoes, on the mountain heights. First, most of the moisture was crushed out of the potatoes by stomping on them with their feet. Then, the cold mountain temperatures froze the stomped potatoes and the remaining water inside slowly vaporized under the low air pressure of the high altitudes in the Andes. Today, a similar procedure still is used in Peru for a dried potato product called chuno. It is a light powder that retains most of the nutritional properties of the original potatoes but, being freeze-dried, chuno can be stored for up to four years.

On the commercial market, the Nestle Company was asked by Brazil in 1938 to help find a solution to save their coffee bean surpluses. Nestle developed a process to convert the coffee surpluses into a freeze-dried powder that could be stored for long periods. The result was Nescafe® coffee, which was first introduced in Switzerland. However, it was during World War II that freeze-drying developed into an industrial process. Blood plasma and penicillin were needed for the armed forces and it was discovered that freeze-drying was the best way to preserve and transport these materials.

In 1964, Nestle developed an improved method of producing instant coffee by freeze-drying. Today, there are several other coffees, such as Tasters Choice®, that use a similar process of preservation. In the late 1970s, freeze-drying was used for taxidermy, food preservation, museum conservation, and pharmaceutical production.

Today, more than 400 commercial food products are preserved by freeze-drying. In the medical field, freeze-drying is the method of choice for extending and preserving the shelf life of enzymes, antibodies, vaccines, pharmaceuticals, blood fractions, and diagnostics. The benefits of all freeze-dried products is that they rehydrate easily and quickly because of the porous structure created by the freeze-drying process.

But it all started with the Incas stomping on their potatoes.

FIGURE 25.9

An Industrial Model Lyophilizer

This model removes 500 pounds of product moisture in 24 hours of freeze-drying. Using vacuum and heat, water is drawn off from ice without passing through the liquid phase to produce the freeze-dried product.

A major group of chemical preservatives are organic acids, such as sorbic acid, benzoic acid, and propionic acid. Microbiologists believe that these compounds damage microbial membranes and interfere with the uptake of certain essential organic substances such as amino acids. Chemicals are used primarily against molds and yeasts, but their acidity also is a deterrent to bacterial growth.

Sorbic acid, which came into use in 1955, is added to syrups, salad dressings, jellies, and certain cakes. **Benzoic acid**, the first chemical (1908) to be approved by the Food and Drug Administration (FDA), protects beverages, catsup, margarine, and apple cider. **Propionic acid** is incorporated in wrappings for butter and cheese, and is added to breads and bakery products, where it inhibits the ropiness commonly due to *Bacillus* species and prevents the growth of fungi. Other natural acids in food add flavor while serving as preservatives. Examples are **lactic acid** in sauerkraut and yogurt, and **acetic acid** in vinegar.

The process of **smoking** with hickory or other woods accomplishes the dual purposes of drying food and depositing chemical preservatives. By-products of smoke, such as aldehydes, acids, and certain phenol compounds, effectively inhibit microbial growth for long periods of time. Smoked fish and meats have been staples of the diet for many centuries.

Sulfur dioxide has gained popularity as a preservative for dried fruits, molasses, and juice concentrates. Used in either gas or liquid form, the chemical retards color changes on the fruit surface and adds to the aesthetic quality of the product. Another gas, **ethylene oxide**, is employed for the preservation of spices, nuts, and dried fruits, especially those packaged in cellophane bags. This same gas is used for the chemical sterilization of packaged Petri dishes and other plastic devices.

Some foods contain their own natural preservatives. Examples are the antimicrobial substances in garlic, the lysozyme in egg white, and the benzoic acid of cranberries.

Sorbic acid:
an organic acid originally obtained from the mountain ash.

Benzoic acid:
a phenyl-containing organic acid.

Propionic acid:
a colorless organic acid.

RADIATION CAN STERILIZE FOODS

Though much of the public is apprehensive about foods exposed to radiation, various forms of radiation are used to sterilize foods. Taste and appearance have been preserved by freezing food in liquid nitrogen and exhausting oxygen from the package before irradiation. Meat storage facilities use **ultraviolet (UV) light** to reduce surface contamination, and water can be treated with UV light when chlorine is not useful.

Gamma rays are used to extend the shelf life of fruits, vegetables, fish, and poultry from several days to several weeks. This form of radiation also increases the distances fresh food can be transported and significantly extends the storage time for food in the home. Gamma rays are high-frequency forms of electromagnetic energy emitted by a radioactive isotope called **cobalt-60**. The radiations are not radioactive, and they cannot cause food to become radioactive. They kill microorganisms by reacting with and destroying microbial DNA (Chapter 22). Opponents to their use point out, however, that the radiations also break chemical bonds in foods and cause new ones to form, thereby raising the possibility of new and toxic chemical compounds.

Interest in gamma radiation for food preservation grew during the 1950s under President Eisenhower's Atoms for Peace program. For the next quarter-century, the FDA conducted extensive tests to determine whether the process was safe. In March 1981, the FDA approved radiated foods such as spices, condiments, fruits, and vegetables for sale to American consumers. Gamma radiation of pork to prevent trichinosis won approval in 1985, and irradiated strawberries made it to market in 1992. Irradiation of red meat was approved in 1997, as further explored in Chapter 22.

FOODBORNE DISEASE CAN RESULT FROM AN INFECTION OR INTOXIFICATION

Food may be a mechanical vector for infectious microorganisms or a culture medium for growth. People then are affected by the organism or the toxin it has produced in the food. In the former case, a food infection is established; in the latter, a food poisoning or intoxication occurs.

Food infections are typified by typhoid fever, salmonellosis, cholera, and shigellosis, all of which are of bacterial origin. The protozoal infections amoebiasis, balantidiasis, and giardiasis represent foodborne diseases. Viral infections are exemplified by hepatitis A. **Food intoxications** include botulism, staphylococcal food poisoning, and clostridial food poisoning. Since full discussions of these diseases are presented elsewhere in this text (Chapters 9, 14, and 18), we shall not examine them individually here.

In the United States, public health microbiologists estimate that between 2 and 10 million people are affected by foodborne disease annually. Many episodes require medical attention, but the vast majority of patients recover rapidly without serious complications. In many cases, the incident might have been avoided by taking some basic precautions. For example, unrefrigerated foods are a prime source of staphylococci and *Salmonella* serotypes (MicroFocus 25.6), and perishable groceries such as meats and dairy products should not be allowed to warm up while other errands are performed. Also, a thermometer should be used to ensure that the refrigerator temperature is below 5°C (40°F) at all times.

Another way to avoid foodborne disease is to cover any skin boils while working with foods, since boils are a common source of staphylococci. The hands always should be washed thoroughly before and after handling raw vegetables or salad fixings to avoid cross-contamination of other foods. It is wise to cook meat from a frozen or partly frozen state; if this is impossible, the meat should be thawed in the

---

MicroFocus 25.6

FOWL PLAY

It was the annual company picnic and the softball game was finally over. Now the serious eating could begin. There were salads, stuffed eggs, barbecued chickens, lots of desserts, and a picnic table overflowing with goodies. Unfortunately, there was also an unwelcome guest at the picnic: a serotype of *Salmonella*. During the next three days, over half the attendees would suffer abdominal cramps, diarrhea, headaches, and fever. The common thread was the

barbecued chicken—all the sick attendees had eaten it.

Public health inspectors questioned the woman in charge of the chickens. She led inspectors back to a local supermarket, where the barbecued chickens were sold "ready to eat." It seemed that store employees cooked the birds, then put them back in the original trays where the uncooked birds had been stored. Investigators found evidence of *Salmonella* in the trays. Further questioning

revealed that the woman had bought the chickens in the morning and stored them for the next seven hours in the trunk of her car, believing they would remain cool. They did not. In fact, they reached incubator temperatures quickly and by dinner time, at 7:00 P.M., they were teeming with *Salmonella*.

The lessons are clear: Keep meats cold; store them in clean, fresh trays after cooking; and play softball *after* dinner, not before.

refrigerator. Cutting boards should be cleaned with hot, soapy water after use, and old cutting boards with cracks and pits should be discarded. MicroInquiry 25 looks at how good are your sanitary precautions.

Studies indicate that leftovers are implicated in most outbreaks of foodborne disease (FIGURE 25.10). It is therefore important to refrigerate leftovers promptly and keep them no more than a few days. Thorough reheating of leftovers, preferably to boiling, also reduces the possibility of illness.

Many instances of foodborne disease occur during the summer months, when foods are taken on picnics where they cannot be refrigerated. As a general principle,

FIGURE 25.10

A Case of Botulism Due to Leftover Food

This incident happened in California in August 1984. Because the stew was discarded, it could not be tested for *Clostridium botulinum*. However, the 16-hour interval during which the stew remained at room temperature was sufficient for the germination of clostridial spores. The first man was unaffected because he ate the stew immediately after cooking.

1. In August 1984, a man prepared stew from fresh ingredients including meat, unpeeled potatoes, and carrots.

2. After simmering it for 45 minutes, the man ate some of the stew and left the remainder on the stove overnight.

3. Sixteen hours later, the man's roommate tasted the stew previous to heating it and noted a sour taste. He decided not to eat it and subsequently threw it away.

4. Forty hours after tasting the stew, the man developed signs of botulism. He was hospitalized, and type A botulism toxin was detected in his serum. He was treated with antitoxin, and recovered.

MicroInquiry 25

KEEPING MICROORGANISMS UNDER CONTROL

How knowledgeable are you concerning food safety in your home or apartment? If you are, do you actually practice what you know? Take the following quiz (honestly) and then we will see where microbes lurk and where you need to eliminate or keep them under control. Answers can be found in Appendix E.

25.1. Your refrigerator should be kept at what temperature?
- a. 0°C (32°F)
- b. 5°C (41°F)
- c. 10°C (50°F)
- d. Do not know.

25.2. Frozen fish, meat, and poultry products should be defrosted by
- a. setting them on the counter for several hours.
- b. microwaving.
- c. placing them in the refrigerator.

25.3. After cutting raw fish, meat, or poultry on a cutting board, you can safely
- a. reuse the board as is.
- b. wipe the board with a damp cloth and reuse it.
- c. wash the board with soapy hot water and sanitize it with a mild bleach and reuse it.

25.4. After handling raw fish, meat, or poultry, how do you clean your hands?
- a. Wipe them on a towel.
- b. Rinse them under hot, cold, or warm tap water.
- c. Wash them with soap and warm water.

25.5. When was the last time you sanitized your kitchen sink drain and garbage disposal?
- a. Last night
- b. Several weeks ago
- c. Several months ago
- d. Cannot remember

dairy foods, such as custards, cream pies, pastries, and deli salads should be excluded from the picnic menu. For outdoor barbecues, one dish should be used for carrying hamburgers to the grill and another dish for serving them. Many of these principles apply equally well to fall and winter tailgate parties.

Over 90 percent of **botulism** outbreaks reported to the CDC are traced to home-canned food. To prevent this sometimes fatal foodborne disease, health officials urge that homemakers use the pressure method to can foods. Reliable canning instructions should be obtained and followed stringently. Foods suspected of contamination should not be tasted to confirm the suspicion, but should be discarded immediately. When in doubt, boiling the food for a minimum of 10 minutes and thoroughly washing the utensil used to stir the food are recommended. Bulging or leaking cans must be discarded in a way that will not endanger other people or animals.

Milk is an unusually good vehicle for the transmission of pathogenic microorganisms because its fat content protects organisms from stomach acid, and, being a fluid, it remains in the stomach a relatively short period of time. The diseases of cows transmitted by milk include bovine tuberculosis, brucellosis, and Q fever (Chapter 9). Since the 1980s, several outbreaks of milkborne disease have been linked to *Salmonella* serotypes. In 1984, for example, 16 cases of **salmonellosis** due to *Salmonella typhimurium* occurred in nuns at a convent in Kentucky. A failure in milk pasteurization accounted for the episode.

Another milkborne organism of significance is *Campylobacter jejuni*, the cause of **campylobacteriosis** (Chapter 9). In 1984, this gram-negative curved rod was isolated from nine kindergarten children and adults who sampled raw milk at a bottling plant in southern California while on a school field trip. Scientists estimate that *Campylobacter jejuni* is present in the intestinal tracts of about 40 percent of dairy cattle.

25.6. Leftover cooked food should be

   a. cooled to room temperature before being put in the refrigerator.

   b. put in the refrigerator immediately after the food is served.

   c. left at room temperature overnight or longer.

25.7. How do you clean your kitchen counter surfaces that are exposed to raw foods?

   a. Use hot water and soap, then a bleach solution.

   b. Use hot water and soap.

   c. Use warm water.

25.8. How was the last hamburger you ate cooked?

   a. Rare

   b. Medium

   c. Well-done

25.9. How often do you clean or replace your kitchen sponge or dishcloth?

   a. Daily

   b. Every few weeks

   c. Every few months

25.10. Normally dishes in your home are cleaned

   a. by an automatic dishwasher and air-dried.

   b. after several hours soaking and then washed with soap in the same water.

   c. right away with hot water and soap in the sink and then air-dried.

So, how did you do?

Public health microbiologists seek to limit milkborne disease by inspecting food and dairy plants regularly, and making recommendations on improved sanitary practices. In the field, sick animals are treated with antibiotics and immunized. The success of mass brucellosis immunizations in the 1970s showed their effectiveness as a public health measure. New methods for immunizing poultry are explored in Chapter 9.

HACCP SYSTEMS ATTEMPT TO IDENTIFY POTENTIAL CONTAMINATION POINTS

Fueled by consumer awareness, the entire food industry has been placed under a food-safety spotlight. Among the most important food safety systems is **Hazard Analysis and Critical Control Point (HACCP)**, a set of scientifically based safety regulations now federally enforced in the seafood, meat, and poultry industries. In an HACCP system, manufacturers identify individual processing points that could affect the safety of a product. These points are called **critical control points**, or **CCPs** (in the jargon of food technology). The CCPs are supervised to ensure that any hazards associated with the operation are contained or, preferably, eliminated (the key concept is prevention). When all possible hazards are controlled at the CCPs, the safety of the product can be assumed without further testing or inspection.

HACCP systems are overseen by the FDA and the US Department of Agriculture (USDA). The systems are developed by each food establishment and tailored to its individual product, processing, or distribution conditions. The standard regulations require food processors to monitor and control eight key sanitation areas: (1) the safety of water that contacts food or used to make ice; (2) the condition and cleanliness of utensils, gloves, outer garments, and other food contact surfaces; (3) the prevention of cross-contamination from raw products and unsanitary objects to foods; (4) the maintenance of hand-washing and toilet facilities; (5) the protection of foods and food surfaces from adulteration with lubricants, fuel, pesticides, sanitizing agents, and other contaminants; (6) the proper labeling, storage, and use of toxic compounds; (7) the control of employee health conditions that could result in food contamination; and (8) the exclusion of pests from the food plant.

The HACCP system is a risk-reduction system originally developed in the 1960s for foods used in space travel ("space foods"), but not applied to the food industry until the 1990s. It places the responsibility for food safety on the shoulders of industry, but it also focuses consumer attention on food handling and safety issues. Several well-publicized foodborne disease outbreaks and product recalls have raised questions about the safety and quality of foods, and HACCP systems will attempt to restore consumer confidence. Improved epidemiological investigations and increased surveillance will add to that confidence in the ensuing years. Indeed, January 26, 1998, was an important date for consumers. On that day, HACCP systems began at 312 of the largest meat and poultry processing plants in the United States (FIGURE 25.11); by 2000, very small plants were using HACCP. The FDA established HACCP for the seafood industry in 1995 and all juice industries were using HACCP by 2004. The FDA currently is considering HACCP for other food industries, both domestic and imported.

FIGURE 25.11

Meat Inspections

A meat inspector checks beef prior to sale. The long-standing "sniff-and-poke" method of inspection is being replaced by a newly instituted HACCP system of ensuring meat safety.

To this point . . .

We have focused on the topic of food preservation and have outlined the various ways in which preservation can be achieved. One of the most widespread methods of preservation is by using heat in the canning process. Another method is by employing low temperatures in the refrigerator and freezer. A third is by drying food through various industrial modes, and a fourth is by drawing fluid out of microorganisms by osmotic pressure.

Two additional methods of preservation are chemical preservatives and radiation. Chemicals currently used include sorbic, benzoic, and propionic acid, as well as sulfur dioxide and various chemicals in wood smoke. Gamma rays are a type of radiation shown to be an excellent food preservative under experimental conditions. We pointed out some advantages to the use of gamma rays, but extensive testing still needs to be completed, and consumer acceptance remains an obstacle to adoption. Prevention of illness from food and dairy products can be effected by consumers, however, and we noted several methods.

The discussion up to this point has cast microorganisms in a negative role as spoilers of food and objects of preservation methods. But microorganisms also may play a positive role in the food and dairy product industries because their chemical activities result in numerous food products. We shall discuss briefly some of these products in the chapter's final section.

25.3

Foods from Microorganisms

Over the centuries, social customs and traditions have brought acceptance of a variety of foods produced by microorganisms. Some individuals regard these foods as "spoiled," but to many people, the food is "fermented" (FIGURE 25.12).

MANY FOODS ARE FERMENTED PRODUCTS

Fermented foods have three things in common: They are less vulnerable to extensive spoilage than unfermented foods; they are less likely to be vectors of foodborne illness than unfermented foods; and they have been accepted by the cultures in which they were developed (indeed, in some cases, they are considered delicacies). It is conceivable that ancient peoples were first attracted to the preservative qualities of fermented foods and coincidentally learned to appreciate their tastes. Here, we will mention a few typical fermented food products.

Sauerkraut (German for "sour cabbage") is not only a well-preserved and tasty form of cabbage, but also nutritionally sound. For instance, the vitamin C content of sauerkraut is equivalent to that of citrus fruits, and sauerkraut often was taken on British sea voyages to prevent scurvy because citrus fruits were too expensive.

Sauerkraut is prepared commercially by adding salt to shredded cabbage and packing the cabbage tightly to encourage anaerobic conditions. The first organisms to multiply are species of *Leuconostoc*, a gram-positive coccus found naturally in the

Fermentation:
in an industrial setting, any commercial process, whether carried out by an aerobic or anaerobic process.

cabbage. These bacteria ferment carbohydrates in the plant cells and produce acetic and lactic acids. After some days, the acids lower the pH of the cabbage to about 3.5. Species of *Lactobacillus* then proliferate, and the additional lactic acid they produce by fermentation further reduces the pH to about 2.0. The salt helps retard mold contamination while drawing juices out of the plant cells. A compound called **diacetyl** (the flavoring agent in butter) is produced by *Leuconostoc*, adding aroma and flavor.

In the United States, "pickle" is practically synonymous with "pickled cucumber." Over 37 types of dill, sour, and sweet pickles have been categorized, but essentially the fermentations are similar. Cucumbers are placed in a salt solution of 8 percent or higher, at which point the cucumber changes color from bright green to olive green. Next comes curing.

Three groups of microorganisms are important to the fermentation and curing of cucumbers. *Enterobacter aerogenes*, a gram-negative rod, produces large amounts of CO_2, which takes up all the air space and establishes anaerobic conditions. *Lactobacillus* and *Leuconostoc* species then dominate and form abundant amounts of acid, which softens the tissues and sours the cucumbers (**FIGURE 25.13**). Finally, certain yeasts grow and establish many flavors associated with ripe pickles. Dill, garlic, and other herbs and spices are added to finish the product. Pickled pepper, tomatoes, and other vegetables undergo a somewhat similar process. Most pickles are heat pasteurized or further acidified to increase their shelf life, but "kosher-style" pickles are given no further treatment.

Vinegar is a fermented food that traditionally has been made by the spontaneous souring of wine. Indeed, the word is derived from the French *vinaigre*, which means "sour wine."

A widely used method for industrial vinegar production follows a procedure first devised in Germany in the early 1800s. Yeasts ferment the fruit juice to alcohol until the alcohol concentration is about 10 to 20 percent. Machines then spray the alcoholic juice into a tank containing the bacterium *Acetobacter aceti* growing on the surface of wood shavings, gravel, or other substrate. As alcohol percolates through, bacterial enzymes convert it to acetaldehyde and then acetic acid. The vinegar recir-

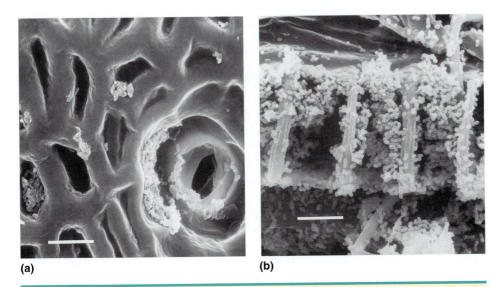

(a)   **(b)**

FIGURE 25.13

A Cucumber and Its Bacterial Flora

(a) A scanning electron micrograph of *Lactobacillus plantarum* on the surface of the cucumber. The opening at the center right is a stoma, through which gases pass for the cucumber's metabolism. Note the accumulation of lactobacilli at this site. (Bar = 10 μm.) (b) A longitudinal section through the vascular tissue of a brined cucumber showing *Leuconostoc* species along the tubular walls. (Bar = 10 μm.)

culates several times before collection at the bottom of the tank. Residual alcohol evaporates in the heat, and the product usually has an acetic acid content of about 3 to 5 percent. The flavor of vinegar is determined by oils, sugars, and other compounds produced by the bacteria, plus the residue of organic compounds in the wine from which it was made.

Several other fermented foods also are worthy of note. One example, **soy sauce**, is made from roasted soybeans and wheat inoculated with the fungus *Aspergillus oryzae* and allowed to stand for three days (**FIGURE 25.14**). The fungus-covered product, called **koji**, then is added to a solution of salt and microorganisms, and aged for about a year. During this time, lactobacilli produce acid, and yeasts produce small amounts of alcohol. Together with the fungal products, the acid and alcohol determine the flavor. The liquid pressed from the mixture is soy sauce.

Another example is fermented sausages. These are produced generally as dry or semidry products and include pepperoni from Italy, thuringer from Germany, and polsa from Sweden. Curing and seasoning agents first are added to ground meats, followed by stuffing into casings and incubation at warm temperatures. Mixed acids produced from carbohydrates in the meat give the sausage its unique flavor and aroma.

Cocoa and coffee also owe their flavor in part to microorganisms. Cocoa derives some of its taste from the microbial fermentation that helps remove cocoa beans from the pulp covering them in the pod. Likewise, coffee is believed to obtain some of its flavor from the fermentation of coffee berries when the beans are soaked in water to loosen the berry skins before roasting.

The influence of microorganisms on animal nutrition is evident in the animal food called **silage**. Silage is made in huge, cylindrical silos that commonly stand

adjacent to barns. The farmer packs the silo with corn, grain stalks, grass, potatoes, and virtually anything that can be fermented. During storage, various bacteria ferment the plant carbohydrates in a process similar to that for sauerkraut and pickles. Fermentation yields a broad mixture of acids, aldehydes, and ketones, as well as diacetyl, which is relished by cattle. After several weeks, the well-preserved plant material is sweet smelling, succulent, and nutritious to animals. It also is inexpensive to the farmer.

SOUR MILK PRODUCTS ARE THE RESULT OF FERMENTATION

Over the centuries, fermented milk products have assumed a key place in our diet. The sour milks are typical examples of fermented milk products. **Buttermilk** is made by adding starter cultures of *Streptococcus cremoris* and *Leuconostoc citrovorum* to vats of skim milk. The *Streptococcus* ferments lactose to lactic and acetic acids, and the *Leuconostoc* continues the fermentation to yield various aldehydes and ketones, and the compound diacetyl. These substances, especially diacetyl, give buttermilk its flavor, aroma, and acidity. For **sour cream**, the same process is used, except that pasteurized light cream is the starting point.

Acidophilus milk is produced in much the same way as buttermilk, except that the skim milk is inoculated with *Lactobacillus acidophilus*. This bacterium is a normal member of the human intestinal flora. Many health-care practitioners believe that the lactobacilli in acidophilus milk augment naturally occurring lactobacilli and help the digestion process, while keeping molds in check. A newer type of acidophilus milk, **sweet acidophilus milk**, lacks the sour taste. It is prepared by adding *Lactobacillus acidophilus* to pasteurized milk.

Yogurt is a form of sour milk (**MicroFocus 25.7**), made by adding dry milk solids to boiled milk to achieve a custard-like consistency. The two starter cultures are *Streptococcus thermophilus* and *Lactobacillus bulgaricus*. Yogurt is sometimes called Bulgarian milk because it was popular among peasants in Bulgaria and other Balkan countries. After World War I, many Americans became health conscious, and Elie Metchnikoff (the discoverer of phagocytosis) took note of the longevity of Bulgarian peasants and attributed it to the yogurt they drank. Metchnikoff believed that the streptococci and lactobacilli in yogurt assume residence in the intestine and replace organisms that contribute to aging. Although the aging theory has largely been dis-

MicroFocus 25.7

MAKING YOUR OWN YOGURT

The popularity of yogurt as a nutritious low-calorie food has prompted many to try their hand at making it at home. Here is a recipe that works well.

Heat one quart of milk to about 170°F (77°C), stirring often and using a thermometer to check the temperature. This heating will evaporate some of the liquid and reduce the bacterial popula-

tion. Let the milk cool to about 130°F (about 55°C), then add one cup of powdered milk and one-third cup of unflavored commercial yogurt. Mix thoroughly and pour into small containers with lids. Styrofoam coffee cups may be used.

For the incubation step, you will need a small cooler of the type used for picnics. Fill the cooler with several

inches of water at 130°F (55°C). Now place the containers in the cooler, close the lid tightly, and let the containers stand for about six to eight hours. (A large pan of hot water in the oven also works well, or the cups can be wrapped with hot towels.) During this time, the bacteria will multiply and the yogurt will thicken. Refrigerate, add fresh or frozen fruit, and enjoy.

counted, many microbiologists hold that bacteria in yogurt support good health much as the bacteria in acidophilus milk do.

CHEESE PRODUCTION INVOLVES A VARIETY OF MICROORGANISMS

Cheese production begins when the casein curdles out of milk (**FIGURE 25.15**). Usually this accompanies a souring of the milk by streptococci, but the process may be accelerated by adding **rennin**, an enzyme obtained from the stomach lining of a calf. The milk curd is essentially an unripened cheese. It may be marketed as cottage cheese, or pot cheese. Cream cheese also is unripened cheese with a butterfat content of up to 20 percent.

To prepare ripened cheese, the milk curds are washed, pressed, sometimes cooked, and cut to the desired shape. Often the curds are salted to add flavor, control moisture, and prevent contamination by molds. If **Swiss cheese** is to be made, two types of bacteria grow within the cheese: *Lactobacillus* species, which ferment the lactose to lactic acid; and *Propionibacterium* species, which produce organic compounds and carbon dioxide, which seeks out weak spots in the curd and accumulates as holes, or eyes (**FIGURE 25.16**). **Cheddar cheese** is scalded at a lower temperature than Swiss. Cheddar and Swiss are examples of cheese ripened internally by bacteria. Provolone, Edam, and Gouda are others.

Another group of cheeses are somewhat softer in texture, a characteristic deriving from the partial breakdown of the protein curds by microbial enzymes. Growth takes place primarily at the surface of these cheeses, and the products tend to be pungent. Within the group of soft cheeses are Muenster, Port du Salut, and Limberger. Yeasts and species of the gram-positive rod *Brevibacterium* are among the surface flora. The rind of the cheese is derived from microbial pigments.

The mold-ripened cheeses are represented by Camembert and Roquefort. **Camembert cheese** is made by dipping salted curds into *Penicillium camemberti* spores to stimulate a surface growth. The fungus grows on the outside of the curd and digests the proteins, thus softening the curd. **Roquefort** is a blue-veined cheese produced by *Penicillium roqueforti*. The mold penetrates cracks within the curd, creating the distinctive veins within the cheese. Most people, however, would rather remain blissfully ignorant of this fact.

(a) **(b)** **(c)** **(d)** **(e)** **(f)**

FIGURE 25.15

Steps in the Production of Swiss Cheese

(a) The milk is mixed with rennet and heated in a large kettle. After the curds form, the cheesemaker cuts the curds into small pieces using a series of copper wires called a cheese harp. This speeds the expulsion of the whey and ensures uniform heating of the curds by increasing the surface area. (b) About 2½ hours after adding the rennet, a square piece of cheesecloth composed of coarse-weave hemp is drawn down along the wall of the vat and passed under the curds. (c) The four corners of the cheesecloth are tied together, and the curds are removed by a block and tackle. (d) The curds are deposited on a wooden hoop, and the surface is kneaded lightly with the palms of the hands. (e) Press boards are placed on top of the cheese mass until no more whey emerges. This process, which usually requires a day, forms the familiar wheels of cheese. (f) The cheese is then transported to a warm cellar, where it is salted and set aside to ripen. Bacteria ferment the cheese for about two weeks at a temperature of 50°C to 57°C and a relative humidity of 90 percent. The cheese is then inspected, graded, and sold for consumption.

FIGURE 25.16

An Array of Cheeses Produced by Microorganisms

Note to the Student

Since 1925, only five deaths from botulism have been attributed to commercially canned food in the United States. During this period, almost 100 billion cans of food were produced for sale to consumers.

These figures are a testament to the high standards achieved by the canning industry. They represent an achievement of which we consumers can be justifiably proud. I say "we consumers" because we are the ones who understand that foods can be a vehicle for disease, and we refuse to tolerate a manufacturer's ignorance. Working through our representatives in government agencies, we exact heavy penalties from companies whose products are tainted. Witness the 25 million pounds of hamburger meat recalled in 1988 (Chapter 9) and the 36,000 pounds of ham recalled in 1983 (this chapter). Both incidents were due to the possibility of microbial contamination.

The next time you shop at the supermarket, stop and take note of the broad variety of foods we consume, and consider that we buy and eat these foods with full confidence that none will make us ill. It is a confidence that is not shared by peoples in many parts of the world.

http://microbiology.jbpub.com

The site features **eLearning,** an on-line review area that provides quizzes and other tools to help you study for your class. You can also follow useful links for in-depth information, or just find out the latest microbiology news.

Summary of Key Concepts

The microbiology of foods is concerned with the spoilage that occurs in foods, with methods for preserving foods, and with the activities of microorganisms in the formation of certain foods.

25.1 FOOD SPOILAGE

- **Several Conditions Determine if Spoilage Can Occur.** Food spoilage has been a continuing problem since ancient times. Certain conditions, such as water content, pH, physical structure, and chemical composition determine the extent of food spoilage because they influence the growth of microorganisms.

- **The Chemistry of Spoilage Produces Specific Products.** Products of spoilage include alcohol, foul-smelling products such as indole and skatole, and acids and gases. Capsules and pigments also are involved with spoilage.

- **Meats and Fish Can Be Contaminated in Several Ways.** In meats and fish, spoilage organisms enter during processing; in fish, water often is the source.

- **Poultry and Eggs Can Spoil Quickly.** In poultry, the spoilage may reflect human contamination, but often it is due to members of the genus *Salmonella*, which infect the bird.

- **Breads and Bakery Products Can Support Bacterial and Fungal Growth.** Bakery ingredients generally bring contaminants to bread and other bakery products.

- **Some Grains Are Susceptible to Spoilage.** Grains, including peanuts and rye may be spoiled by toxin-producing fungi.

- **Milk and Dairy Products Sometimes Sour.** Dairy products are spoiled by microorganisms surviving pasteurization, such as curd, capsule, and pigment producers.

25.2 FOOD PRESERVATION

- **Heat Denatures Proteins.** To preserve food from spoilage, a number of methods are used including heat. The most useful application of heat is in the process of canning, while pasteurization is used for milk and other liquid products.

- **Low Temperatures Slow Microbial Growth.** Low temperatures are achieved in the refrigerator and freezer. Low temperatures do not kill all microor-

ganisms contaminating a refrigerated or frozen food product.

- **Drying Removes Water.** Drying is useful because water is an absolute necessity for life, and dried foods therefore are unable to support microbial life. An ever-increasing list of foods preserved by freeze-drying is appearing on the market.

- **Osmotic Pressure Can Help Preserve Foods.** Foods high in sugar or salt preserve foods by drawing water out of microorganisms contaminating the food product.

- **Chemical Preservatives Are Found in or Are Added to Many Foods.** Various chemicals, such as propionic, sorbic, and benzoic acids, are used to preserve foods.

- **Radiation Can Sterilize Foods.** Ultraviolet light and gamma rays typify the radiations used in processing certain foods.

- **Foodborne Disease Can Result from an Infection or Intoxication.** Individuals can contract a foodborne disease either through an infection or by ingesting a microbial toxin.

- **HACCP Systems Attempt to Identify Potential Contamination Points.** The Hazard Analysis and Critical Control Point (HACCP) are a set of scientifically based safety regulations that attempt to identify food processing points that are subject to contamination.

25.3 FOODS FROM MICROORGANISMS

- **Many Foods Are Fermented Products.** Certain foods "spoiled" by microorganisms have come to be accepted as the norm. Among these microbial products are sauerkraut, vinegar, and fermented sausages. The spoilage in these foods causes no harm to consumers, and the foods reflect the helpful activities that microorganisms perform to add to the quality of our lives. Other foods involving fermentation products include soy sauce, sausages, cocoa, and coffee.

- **Sour Milk Products Are the Result of Fermentation.** Buttermilk, sour cream, and yogurt are products where microbial action produces a fermented milk product.

- **Cheese Production Involves a Variety of Microorganisms.** Numerous dairy products, such as cheeses, also are products of microorganisms.

Questions for Thought and Discussion

Answers to selected questions can be found in Appendix C.

1. A writer in a food technology magazine once suggested that refrigerators be fitted with ultraviolet lights to reduce the level of microbial contamination in foods. Would you support this idea?

2. In 1997, the USDA approved the spraying of steam followed by vacuuming for the removal of microorganisms from the surface of meats. What are the advantages of this method over the previously used hand trimming? Can you think of any other useful methods for decontaminating meats?

3. Chicken and salad are two items on the dinner menu at home, and you are put in charge of preparing both. You have a cutting board and knife for slicing up the salad items and cutting the chicken into pieces. Which task should you perform first? Why? What other precautions might you take to ensure that the meal is not remembered for the wrong reason?

4. Tradition has it that Peruvian Incas of the Andes Mountains preserved their potatoes and other foodstuffs by placing them for several weeks on high mountainsides exposed to the air. Which method of food preservation were they practicing?

5. On January 12, 1996, the author opened a container of sour cream that had become lost in the back of the refrigerator some nine months before (the expiration date listed on the bottom was April 27, 1995). The sour cream appeared satisfactory, and there was no unusual smell. The author proceeded to spoon it onto a baked potato and dig in. What factors might have contributed to the sour cream's preservation so long after the expiration date?

6. It is a hot Saturday morning in July. You get into your car at 9:00 A.M. with the following list of chores: Pick up the custard eclairs for tonight's dinner party, drop off clothes at the cleaners, buy the ground beef for tomorrow's barbecue, deliver the kids to the Little League baseball game, pick up a broiler at the poultry farm. Microbiologically speaking, what sequence should you follow?

7. Suppose you had the choice of purchasing "yogurt made with pasteurized milk" or "pasteurized yogurt." Which would you choose? Why? What are the "active cultures" in a cup of yogurt?

8. How would you answer a child who asks, "Who puts the holes in Swiss cheese?" Why might blue (Roquefort) cheese pose a possible threat to someone who has an acute allergy to penicillin?

9. It is 5:30 P.M. and you arrive on campus for your evening college class. You stop off at the cafeteria for a bite to eat. Which foods might you be inclined to avoid purchasing?

10. To avoid *Salmonella* infection when preparing eggs for breakfast, the operative phrase is "scramble or gamble." How many foods can you name that use uncooked or undercooked eggs and that can represent a health hazard?

11. The local fish market occasionally receives an unusually large order of fish, and it offers a special sale to customers. Whole fish are piled high on a bed of ice, and signs are put out advertising a one-time-only reduced price. However, smart consumers know that it might be better to pass up the sale, especially if they cannot pick out the individual fish they want to purchase. Why? If you could not resist the temptation to buy the sale fish, which ones would you purchase?

12. Which principles of preservation ensure that each of the following remains uncontaminated on the pantry shelf: vinegar, olive oil, brown sugar, tea bags, spaghetti, hot cocoa mix, pancake syrup, soy sauce, rice?

13. On Saturday, a man buys a steak and a pound of liver and places them in the refrigerator. On Monday, he must decide which to cook for dinner. Microbiologically, which is the better choice? Why?

14. Certain fruits, such as oranges and cantaloupes, are peeled before they are eaten. For this reason, many people believe it unnecessary to wash them. Why might they be wrong?

15. Foods from tropical nations such as Mexico tend to be very spicy, with lots of hot peppers, spices, garlic, and lemon juice. By contrast, foods from cooler countries such as Norway and Sweden tend to be much less spicy. Why do you think this pattern has evolved over the ages?

16. You and a friend are going to an orchard to pick apples with the intention of pressing them in your new cider mill. However, you are aware of the recent outbreaks of infection with *E. coli* O157:H7 that

were traced to fresh apple cider. What precautions can you take to ensure a "healthy" experience?

17. A standard set of recommendations and regulations exists for individuals who work in restaurants and cook food for customers (i.e., food handlers). However, very few regulations exist for individuals who pick fruits or vegetables in the fields. How many recent incidents can you identify where fresh-picked fruits or vegetables were linked to infectious disease? What regulations would you recommend for such workers?

Review

On completing this chapter on food microbiology, test your knowledge of its contents by using the following syllables to compose the term that answers the clue. Each term is a genus of microorganism important in food microbiology. The number of letters in the genus is indicated by the dashes, and the number of syllables in the genus is shown by the number in parentheses. Each syllable is used only once, and the answers are listed in Appendix D.

A A A AS AS BA BA BAC BAC CE CEPS CHLA CIL CIL CIL CLAV CLO CLO CO COC CUS DI DO EN GIL I I I LA LAC LEU LUS LUS LUS MO MON MY NEL NOS O PER PRO PSEU RA SAL SER STREP STRID STRID TE TER TER TER TI TO TO TO TOC UM UM US

1. Common poultry contaminant (4) _ _ _ _ _ _ _ _ _ _ _ _

2. Discolors meat surface (5) _ _ _ _ _ _ _ _ _ _ _ _ _ _ _

3. Destroyed in canning (4) _ _ _ _ _ _ _ _ _ _ _

4. Red pigment in bread (4) _ _ _ _ _ _ _ _ _

5. Causes psittacosis (4) _ _ _ _ _ _ _ _ _ _

6. Black rot in eggs (3) _ _ _ _ _ _ _ _

7. Sours dairy products (4) _ _ _ _ _ _ _ _ _ _ _ _

8. Spore-forming contaminant (3) _ _ _ _ _ _ _ _

9. Used to make vinegar (5) _ _ _ _ _ _ _ _ _ _ _ _

10. Sauerkraut producer (4) _ _ _ _ _ _ _ _ _ _ _

11. Causes ergot poisoning (3) _ _ _ _ _ _ _ _ _

12. Produces potent exotoxins (4) _ _ _ _ _ _ _ _ _ _ _

13. Green rot in foods (4) _ _ _ _ _ _ _ _ _ _ _

14. Used to cure cucumbers (5) _ _ _ _ _ _ _ _ _ _ _ _ _

15. Aflatoxins in grains (4) _ _ _ _ _ _ _ _ _ _ _

Environmental Microbiology

Milwaukee may happen again.

—Microbiologist Rita Colwell describing the urgency to develop safe drinking water standards to avoid repeating the epidemic that struck Milwaukee, Wisconsin, in 1993

I N THE EARLY 1800s, the steam engine and its product, the Industrial Revolution, brought crowds of rural inhabitants to European cities. To accommodate the rising tide, row houses and apartment blocks were hastily erected, and owners of existing houses took in tenants. Not surprisingly, the bills of mortality from typhoid fever, cholera, tuberculosis, dysentery, and other diseases mounted in alarming proportions.

As the death rates rose, a few activists spoke up for reform. Among them was an English lawyer and journalist named Edwin Chadwick. Chadwick subscribed to the then-novel idea that humans could shape their environment and could eliminate diseases of filth by doing away with filth. In 1842, he published a landmark report indicating that poverty-stricken laborers suffered a far higher incidence of disease than people from middle or upper classes. Chadwick attributed the difference to the abominable living conditions of workers, and he declared that most of their diseases were preventable. His report established the basis for the Great Sanitary Movement, a wave of reform that began in Europe and spread to developed countries.

Chadwick was not a medical man, but his ideas captured the imagination of both scientists and social reformers. He proposed that sewers be constructed using smooth ceramic pipes, and that enough water be flushed through the system to carry waste to some distant depository. In order to work, the system required the installation of new water and sewer pipes, the development of powerful pumps to bring water into homes, and the elimination of older sewage systems. He foresaw intrusions upon private property to permit water mains and extensive construction to allow straight sewer pipes. The cost would be formidable.

Chadwick's vision eventually came to reality, but it might have taken decades longer without the intervention of **cholera**. In 1849, a cholera epidemic broke out in London and terrified so many people that public opinion began to form in favor of Chadwick's proposal. Another epidemic occurred in 1853, during which John Snow proved that water was involved in transmission of the disease (Chapter 18). In both outbreaks, the disease reached the affluent as well as the poor, and the mortality rate exceeded 50 percent. Construction of the sewer system began shortly thereafter, with John Simon, London's first Medical Officer of Health, in command of the project.

The proverbial "icing on the cake" came in 1892 when a devastating epidemic of cholera erupted in Hamburg, Germany. For the most part, Hamburg drew its water directly from the polluted Elbe River. Adjacent to Hamburg lay Altona, a city where the German government had previously installed a water filtration plant. Altona remained free of cholera. The contrast was sharpened further by a street that divided Hamburg and Altona. On the Hamburg side of the street, multiple cases of cholera broke out; across the street, none occurred. Chadwick and his fellow sanitarians could not have imagined a more clear-cut demonstration of the importance of water purification and sewage treatment.

Dealing with water pollution and treating sewage are but two of the myriad activities that involve microbiology in public health. Concerns about water pollution and sewage treatment are environmental in scope. Other environmental issues involve the place of microbiology in biogeochemical cycles, whose microorganisms are critical to the recycling of nitrogen, sulfur, and carbon. Thus, the focus in environmental microbiology is to study microorganisms as they affect the natural environment.

26.1

Water Pollution

For purposes of simplification, scientists classify water into two major types: groundwater and surface water. **Groundwater** originates from deep wells and subterranean springs, and, because of the filtering action of soil, deep sand, and rock, it is virtually free of microorganisms. As the water flows up along channels, contaminants may enter it and alter its quality. **Surface water** is found in lakes, streams, and shallow wells. Its microbial population may reflect the air through which rain has passed, the meat-packing plant near which a stream flows, or the sewage-treatment facility located along a riverbank.

Certain terms are significant in water microbiology. For example, water is considered **contaminated** when it contains a chemical or biological poison, or an infectious agent. In water that is **polluted**, the same conditions apply, but the poison or agent is obvious. Polluted water carries an unpleasant taste, smell, or appearance. **Potability**, by contrast, refers to the drinkability of water. Potable water is fit for consumption, while unpotable water is unfit.

UNPOLLUTED AND POLLUTED WATER CONTAIN VERY DIFFERENT MICROBIAL POPULATIONS

A body of unpolluted water, such as a mountain lake or stream, is usually low in organic nutrients, and thus, only a limited number of bacteria are present, perhaps a few thousand per milliliter. Most bacteria are soil organisms that have run off into the water during a rainfall. An example are the **actinomycetes**, a group of mold-like bacteria that give a musty odor to soil. Other inhabitants include yeasts, and bacterial and mold spores. **Cellulose digesters** such as members of the genus *Cellulomonas* also are found. These bacteria digest cellulose in plant cell walls. **Autotrophic bacteria** also are common, and free-living protozoa such as *Paramecium*, *Euglena*, *Tetrahymena*, and *Amoeba* abound (**FIGURE 26.1a**).

A polluted body of water, such as a polluted lake or river, presents a totally different picture (**FIGURE 26.1b**). The water contains large amounts of organic matter from sewage, feces, and industrial sources, and the population of microorganisms is usually **heterotrophic**. A major type of bacteria in polluted water is **coliform bacteria**, a group of gram-negative, non–spore-forming bacilli usually found in the human intestine. Coliform bacteria ferment lactose to acid and gas. Included in this group are *Escherichia coli* and species of *Enterobacter*. Noncoliform bacteria also common in polluted water include *Streptococcus*, *Proteus*, and *Pseudomonas* species.

In polluted water, microorganisms contribute to a chain of events that drastically alters the ecology of the environment, as **FIGURE 26.2** illustrates. When phosphates accumulate in the water, algae bloom. The algae supply nutrients to microorganisms, which multiply rapidly and use up the available oxygen. Soon, other protozoa, small fish, crustaceans, and plants die and accumulate on the bottom. Anaerobic species of bacteria such as *Desulfovibrio* and *Clostridium* then thrive in the mud and produce

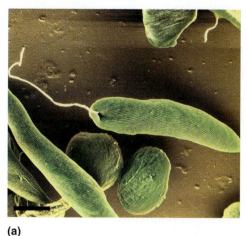

(a)

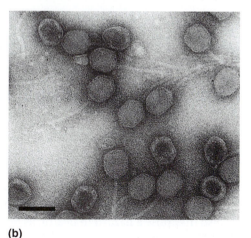

(b)

FIGURE 26.1

Microorganisms in Water Environments

(a) A scanning electron micrograph of the protozoan *Euglena*, a common inhabitant of unpolluted water. (Bar = 10 μm.) (b) A transmission electron micrograph of bacteriophages that replicate in *Vibrio parahaemolyticus*. Bacteriophages like these are commonly found in sewage-contaminated polluted water. (Bar = 70 nm.)

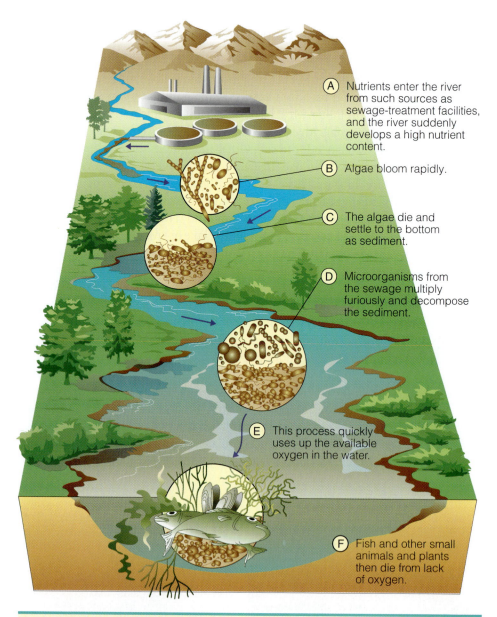

(A) Nutrients enter the river from such sources as sewage-treatment facilities, and the river suddenly develops a high nutrient content.

(B) Algae bloom rapidly.

(C) The algae die and settle to the bottom as sediment.

(D) Microorganisms from the sewage multiply furiously and decompose the sediment.

(E) This process quickly uses up the available oxygen in the water.

(F) Fish and other small animals and plants then die from lack of oxygen.

FIGURE 26.2

The Death of a River

gases that give the water a stench reminiscent of rotten eggs. The organisms in mud also may pose a hazard to health, as **MicroFocus 26.1** points out.

The marine environment of the oceans illustrates another view of microbial populations in water. In the high salt concentration of ocean water, **halophilic**, or salt-loving, microorganisms survive. In addition, the organisms must be **psychrophilic** since it is very cold below the surface. Those at the bottom must withstand great pressure and are therefore **barophilic**, or pressure loving. The organisms in these environments pose no threat to humans because they cannot grow in the body tissues.

Marine microorganisms are vital to ecological cycles because they form the foundations of many food chains. For example, marine algae such as **diatoms** (**FIGURE 26.3**) and organisms such as the **dinoflagellates** capture the sun's energy and, using carbon dioxide, convert the energy to chemical energy in carbohydrates. The microorganisms then are consumed by other animals in the food chain. Dinoflagellates have made the news in recent years because certain species of *Gonyaulax* and *Gymnodinitum* are responsible for the red tide.

Most marine microorganisms are found along the shoreline, or **littoral zone**, because this is where nutrients are plentiful. Certain unusual types of microorganisms have been found on the ocean floor in the **benthic zone** and even at the bottom of six-mile-deep trenches, the **abyssal zone**. Two types of protozoa, the foraminiferans and radiolarians, are of special interest to oil companies because these protozoa were dominant species during formation of the oil fields, and their fossils serve as markers for oil-bearing layers of rock (Chapter 16).

THERE ARE THREE TYPES OF WATER POLLUTION

Water is vital to such industries as food processing, meat packing, and paper manufacturing. Water also is used extensively in pharmaceutical plants and mines, and for cooling purposes in power-generating units. It irrigates agricultural lands and

FIGURE 26.3

Marine Microorganisms

A phase contrast micrograph of a collection of diatoms. Note the broad variety of shapes and sizes of microorganisms in this group. Diatoms trap the sun's energy in photosynthesis and use it to form carbohydrates that are passed on to other marine organisms as food. (Bar = 2 μm.)

FIGURE 26.4

A Coliform Bacterium

A scanning electron micrograph of *E. coli* on the microvilli of an animal's small intestine. *E. coli* is commonly found in water that is biologically polluted. The bacillus represents the coliform group of bacteria and often is used as an indicator of bacterial pollution of water. (Bar = 10 μm.)

provides the focus for many recreational facilities. Uses such as these, however, commonly add to contamination and pollution of water.

Physical pollution of water occurs when particulate matter such as sand or soil makes the water cloudy, or when cyanobacteria bloom during midsummer and their remains give water the consistency of pea soup. **Chemical pollution** results from the introduction of inorganic and organic waste to the water. For example, water passing out of a mine contains large amounts of copper or iron. Other chemical pollutants in water include phosphates and nitrates from laundry detergents, as well as acids such as sulfuric acid.

The third type of pollution, **biological pollution**, is the main concern of our discussion. This type of pollution develops from microorganisms that enter water from human waste, food-processing and meat-packing plants, medical facilities, and similar sources (**FIGURE 26.4** shows a common pollutant). Normally, water can handle biological material because heterotrophic microorganisms digest organic matter to carbon dioxide, water, and useful ions (phosphates, nitrates, and sulfates). With the rapid movement of the water, aeration is constant, and waste is soon diluted and eliminated. However, when water stagnates or is overloaded with waste, it cannot deal with biological material and becomes polluted.

A critical measurement in polluted water is the **biochemical oxygen demand**, or **BOD**. This refers to the amount of oxygen that microorganisms require to decompose the organic matter in water. As the number of microorganisms increases, the demand for oxygen increases proportionally. In the laboratory, the BOD is determined by measuring the dissolved oxygen content of water immediately after collection and then after incubation at 20°C for five days. The difference in oxygen content represents the amount used up by microorganisms in the water sample. Results are generally expressed as parts per million (ppm), with a BOD of several hundred ppm usually considered high.

DISEASES CAN BE TRANSMITTED BY WATER

Water consumption may be the vehicle for transfer of a broad variety of intestinal diseases, including bacterial diseases such as typhoid fever, cholera, shigellosis, and others described in Chapter 9. Waterborne epidemics of these diseases, however, are rare because of continual surveillance. Many illnesses transmitted by drinking water are due to less familiar bacteria such as species of *Legionella* (**FIGURE 26.5**), *Yersinia*, and *Campylobacter*, and toxin-producing strains of *E. coli*. Septicemia, necrotizing fasciitis, and gangrene may develop from the initial infection. An emerging pathogen associated with contaminated water is *Vibrio vulnificus*, a gram-negative bacterium that can cause serious intestinal illness and septicemia in individuals with preexisting liver disease or compromised immune systems. In the 10-year period preceding 1993, 125 persons became infected with *V. vulnificus* and 44 died. Raw oyster consumption was implicated in the majority of deaths.

Several species of marine bacteria also contribute to wound infections when contaminated seawater has contacted the exposed tissue. Among the marine pathogens

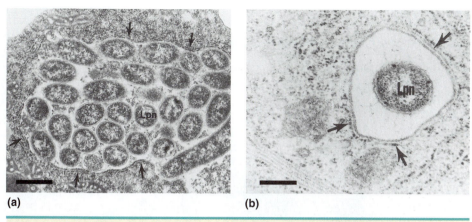

(a) **(b)**

FIGURE 26.5

Legionella in the Environment

Legionella pneumophila is the gram-negative rod responsible for Legionnaires' disease. It infects the respiratory passages when it is inhaled in airborne droplets of water. In nature, the bacterium lives within the cytoplasm of the waterborne protozoan *Hartmanella*. (a) Numerous bacterial rods (labeled Lpn) are seen in a transmission electron micrograph of the cytoplasm of a *Hartmanella* cell. (Bar = 0.6 μm.) (b) A close-up of a cross section of the bacterium (Lpn) is seen in the cytoplasm. The bacterium exists within a clear body known as a vacuole. The arrows point out the multilayered bordering membranes of the vacuole. (Bar = 0.2 μm.)

is *Erysipelothrix rhusiopathiae*, a gram-negative pleomorphic rod that causes **erysipeloid**. Infections usually occur on the hands or feet and are accompanied by a bright-red, well-demarcated lesion that burns or itches. Another pathogen is *Mycobacterium marinum*, an acid-fast rod that causes a small papular lesion at the wound site, particularly near the knuckles of the hand. The lesion often progresses to a nodule, and infection may spread via the lymphatic system. Fish handlers, seafood workers, and aquatic sports enthusiasts may suffer infection from these organisms, and antibiotics are used to limit the growth of both. The marine bacterium, *Vibrio vulnificus* (mentioned above) can cause wound infections with gangrene and necrotizing fasciitis; a fourth bacterium, *Aeromonas hydrophila*, is discussed in Chapter 9.

Viral diseases transmitted by water include hepatitis A, rotavirus disease, gastroenteritis due to Coxsackie or Norwalk virus, and in rare instances, polio. These diseases are generally related to fecal contamination of water. Many **protozoa** form cysts that survive for long periods in water. For this reason, water may be a vehicle for the transfer of *Entamoeba histolytica* and *Giardia lamblia*. A notable outbreak of *Cryptosporidium* infection occurred in 1993 when the municipal water supply of Milwaukee, Wisconsin, became contaminated (Chapter 16).

Three **dinoflagellates** bear mention because of their involvement in human poisonings. The first, *Gonyaulax catanella*, produces a toxin that may cause muscular paralysis and death from asphyxiation. The toxin is ingested from shellfish that feed on the dinoflagellate. The second dinoflagellate is *Gambierdiscus toxicus* (**FIGURE 26.6a**). This marine microorganism is consumed by small fish that concentrate the toxin and pass it to larger fish, such as sea bass and red snapper. Human consumption of the fish leads to neurological and muscular intoxication and a condition called **ciguatera fish poisoning** (from *cigua* for "poisonous snail," originally thought to be the cause).

The third dinoflagellate is *Pfiesteria piscicida* (**FIGURE 26.6b**). This organism was observed in heavy concentrations in 1997 in waters near Maryland's Chesapeake

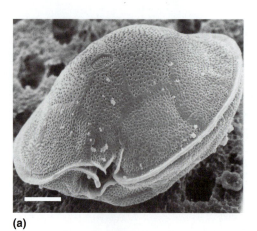
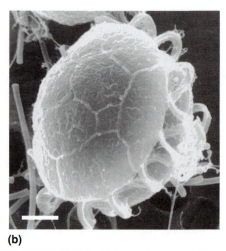

(a)   **(b)**

FIGURE 26.6

The Dinoflagellates of Human Poisonings

(a) *Gambierdiscus toxicus*, a cause of ciguatera fish poisoning. (Bar = 15 μm.) (b) *Pfiesteria shumwayae*, a species related to the *Pfiesteria* linked to human illness in 1997. (Bar = 10 μm.)

Bay. It was linked to massive fish kills and human illnesses characterized by skin rash, memory loss, and difficult breathing. Researchers believe that agricultural runoff was the source of the extremely complex dinoflagellate (over 20 stages in its life cycle have been observed, including amoeboid and flagellated forms). Eutrophication in the waters may have encouraged the protozoal bloom. At least two *Pfiesteria* toxins are being studied, and the role of a secondary fungus invader is also being investigated, as MicroFocus 16.1 explains. The dinoflagellate was named for Lois Pfiester, a University of Oklahoma marine microbiologist.

To this point . . .

We have explored the microbial floras of various water environments to illustrate the differences that exist among them. In unpolluted and polluted waters, the amount and type of nutrients are important determining factors in the microbial population, while in marine environments, the salt, temperature, and pressure conditions regulate the population.

We then focused on water pollution. We compared the physical, chemical, and biological sources of pollution, and we discussed the biochemical oxygen demand (BOD) as a way of determining the extent of biological pollution. We also outlined the types of diseases that could be transmitted by water, and referred to bacterial, viral, and protozoal diseases. Three dinoflagellates were noted because their poisons can cause human illness.

In the next section, we will examine methods used for water purification and sewage treatment where the chain of transmission of waterborne diseases can be broken. We shall see how water is prepared for drinking purposes and how microorganisms can be used effectively to digest organic matter in sewage. The discussion then will pass to the laboratory, where the water-quality bacteriologist seeks to detect water pollution by a series of laboratory tests. The monitoring of water supplies by bacteriologists is an essential feature of environmental health for all of us.

26.2

The Treatment of Water and Sewage

Some years ago, health care workers in Africa asked villagers to name their single greatest need. The answer was almost unanimous: "Water." A startling survey by the World Health Organization indicates that three of every four humans alive today do not have enough water to drink, or, if water is available, the supplies are contaminated.

Water is unfit to drink when it contains human sewage, animal waste, or disease-causing microorganisms (FIGURE 26.7). However, the situation can be reversed through the proper management of water resources. Water-purification procedures prevent pathogenic microorganisms from reaching the body, while sewage-treatment processes remove pathogens from body waste products. In this section, we will examine how these are accomplished.

WATER PURIFICATION IS A THREE-STEP PROCESS

Three basic steps are included in the preparation of water for drinking: sedimentation, filtration, and chlorination. In the **sedimentation** step, leaves, particles of sand and gravel, and other materials from the soil are removed in large reservoirs or settling tanks. Chemicals such as aluminum sulfate (alum) or iron sulfate are dropped as a powder onto water and they form jelly-like masses of coagulated material called **flocs**. The flocs fall through the water and cling to organic particles and microorganisms, dragging a major portion to the bottom sediment in the process of **flocculation**. MicroFocus 26.2 describes a new, environmentally safe flocculant.

The **filtration** step is next. Although different types of filtering material are available, most filters use a layer of sand and gravel to trap microorganisms. A **slow sand filter**, containing fine particles of sand several feet deep, is efficient for smaller-scale operations. Within the sand, a layer of microorganisms acts as a supplementary filter. This layer is called a **schmutzdecke**, or dirty layer. A slow sand filter may purify

FIGURE 26.7

From Water to the Intestine

A mass of waterborne *Giardia lamblia* along the intestinal walls of an infected animal. This protozoan has a flat shape, with numerous flagella. A disk-like sucker apparatus is seen on the ventral surface of several organisms. Since the 1970s, *G. lamblia* has been recognized as the most widespread protozoal cause of human intestinal disease in the United States. Most cases are related to consumption of contaminated water. (Bar = 5 μm.)

MicroFocus 26.2

PURIFYING WATER WITH THE "MIRACLE TREE"

In developed nations, water purification usually uses chemical powders such as aluminum sulfate or iron sulfate for the flocculation process. Many developing nations do not have such chemicals available nor can they readily afford to purchase them. Is there something else that could be used for the flocculation step in water purification? Yes, there is. It's the "miracle tree." Scientifically known as *Moringa oleifera*, this tropical tree was given the name "miracle tree" because of the many uses the tree has—including an environmentally friendly way to purify water.

M. oleifera survives in arid areas and produces elongated seedpods. The local people use the leaves and roots for food, the wood is used in building, and some parts of the tree are used in traditional medicines. High-grade oil

for lubrication and cosmetics is produced from the seeds. However, the seeds also have another use—purifying water. In those parts of Africa and India where the trees grow, people grind up the seeds, and add the powder to cloudy water to precipitate all the solid particles. Clean drinking water results.

Ian Marison, research director at École Polytechnique Fédérale de Lausanne in Switzerland and his colleagues have examined the ground-up seed residue and have isolated a charged peptide. When they add this peptide to cloudy water, within two minutes the water goes from cloudy to clear.

And that's not all. The peptide also has bactericidal properties and can disinfect heavily contaminated water. According to Marison, it even works against some drug-resistant strains of

Staphylococcus, Streptococcus, and *Legionella,* and is effective against some nonbacterial waterborne pathogens. In fact, Marison's group has discovered that if they tweak the peptide's structure, they can increase the antimicrobial effect of the peptide.

This is certainly good news for developing nations where the trees mainly grow. However, if commercially produced, the plant flocculant also would be useful to developed nations since the chemical flocculants often have safety and environmental problems.

As Marison says, "It's a biological, biodegradable, sustainable resource. *Moringa* grows where there is very little water; it grows very, very fast and it costs almost nothing." It truly is a "miracle tree."

---

over 3 million gallons of water per acre per day. To clean the filter, the top layer is removed and replaced with fresh sand.

A **rapid sand filter** contains coarser particles of gravel. A schmutzdecke does not develop in this filter, but the rate of filtration is much higher, with over 200 million gallons purified per acre per day. This type of filter is commonly used in municipal water systems. It must be cleaned more often than the slow sand filter, a process accomplished by forcing water back through the filter by mechanical pressure. Both slow and rapid sand filters remove approximately 99 percent of the microorganisms from water.

The final step is **chlorination**, in which chlorine gas is added to the water. Chlorine is an active oxidizing agent that reacts with any organic matter in water (Chapter 23). It is important, therefore, to continue adding chlorine until a residue is present. A residue of 0.2 to 1.0 parts of chlorine per million (ppm) of water often is the standard used. Under these conditions, most remaining microorganisms die within 30 minutes. FIGURE 26.8 illustrates the steps in the water-purification process.

At this point, some communities **soften** water by removing magnesium, calcium, and other salts. Softened water mixes more easily with soap, and soap curds do not form. Water also may be fluoridated to help prevent tooth decay. Scientists believe that fluoride strengthens tooth enamel and makes it more resistant to the acid produced by anaerobic bacteria in the mouth.

SEWAGE TREATMENT CAN BE A MULTISTEP PROCESS

Systems for the treatment of human waste range from the primitive outhouse, which is nothing more than a hole in the ground, to the sophisticated sewage-

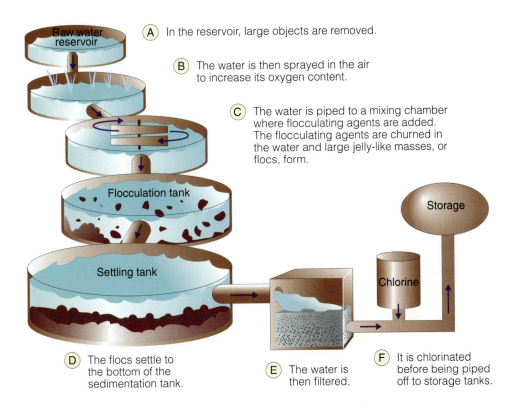

A. In the reservoir, large objects are removed.

B. The water is then sprayed in the air to increase its oxygen content.

C. The water is piped to a mixing chamber where flocculating agents are added. The flocculating agents are churned in the water and large jelly-like masses, or flocs, form.

Storage

Raw water reservoir

Flocculation tank

Settling tank

Chlorine

D. The flocs settle to the bottom of the sedimentation tank.

E. The water is then filtered.

F. It is chlorinated before being piped off to storage tanks.

FIGURE 26.8

Steps in the Purification of Municipal Water Supplies

treatment facilities used by many large cities. All operate under the same basic principle: Water is separated from the waste, and the solid matter is broken down by microorganisms to simple compounds for return to the soil and water.

In many homes, human waste is emptied into underground **cesspools**. These are concrete cylindrical rings with pores in the wall. Water passes into the soil through the bottom and pores of the cesspool, while solid waste accumulates on the bottom. Microorganisms, especially anaerobic bacteria, digest the solid matter into soluble products that enter the soil and enrich it. Some hardware stores sell enzymes and dried bacteria, usually *Bacillus subtilis* spores, to accelerate digestion.

Some homes have a **septic tank**, an enclosed concrete box that collects waste from the house. Organic matter accumulates on the bottom of the tank, while water rises to the outlet pipe and flows to a distribution box. The water is then separated into pipes that empty into the surrounding soil. Since digested organic matter is not absorbed into the ground, the septic tank must be pumped out regularly.

Sewers are at least as old as the Cloaca Maxima of Roman times. Until the mid-1800s, however, sewers were simply elongated cesspools with overflow pipes at one end. They collected filth and had to be pumped out regularly. Finally, in 1842, Edwin Chadwick's report raised the possibility that sewage spreads disease, and soon thereafter a movement (fueled by a cholera outbreak) sprang up to sanitize European cities.

Small towns collect sewage into large ponds called **oxidation lagoons**. Here, the sewage is left undisturbed for up to three months. During that time, aerobic bacteria

Oxidation lagoon:
a large pond into which sewage is piped for natural digestion of organic matter.

digest organic matter in the water, while anaerobic organisms break down sedimented material. Under controlled conditions, the waste may be totally converted to simple salts such as carbonates, nitrates, phosphates, and sulfates. At the cycle's conclusion, the bacteria die naturally, the water clarifies, and the pond may be emptied into a nearby river or stream.

Large municipalities rely on a mechanized sewage-treatment facility to handle the massive amounts of waste and garbage generated daily (**FIGURE 26.9**). The first step in the process, **pretreatment**, involves grit and insoluble waste removal. Next comes **primary treatment**, in which raw sewage is piped into huge open tanks for organic waste removal. This waste, called **sludge**, is passed into sludge tanks for further treat-

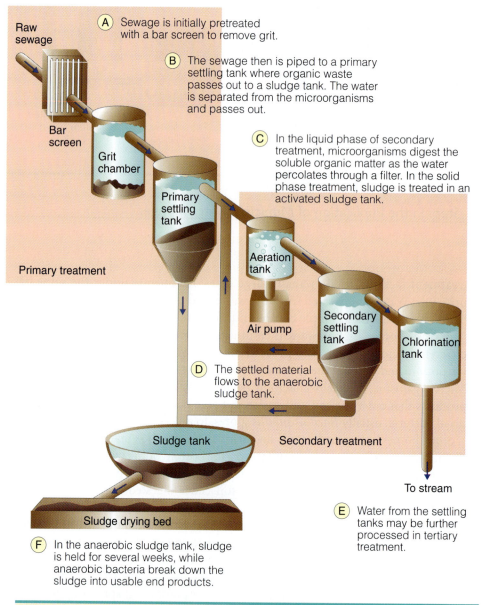

A Sewage is initially pretreated with a bar screen to remove grit.

B The sewage then is piped to a primary settling tank where organic waste passes out to a sludge tank. The water is separated from the microorganisms and passes out.

C In the liquid phase of secondary treatment, microorganisms digest the soluble organic matter as the water percolates through a filter. In the solid phase treatment, sludge is treated in an activated sludge tank.

D The settled material flows to the anaerobic sludge tank.

E Water from the settling tanks may be further processed in tertiary treatment.

F In the anaerobic sludge tank, sludge is held for several weeks, while anaerobic bacteria break down the sludge into usable end products.

Raw sewage
Bar screen
Grit chamber
Primary settling tank
Primary treatment
Aeration tank
Air pump
Secondary settling tank
Chlorination tank
Sludge tank
Secondary treatment
Sludge drying bed
To stream

FIGURE 26.9

A Sewage-Treatment Facility

ment. Flocculating materials, such as aluminum and iron sulfate, then are added to the raw sewage to drag microorganisms and debris to the bottom.

The **secondary treatment** of sewage has two phases, a liquid phase and a solid phase. The **liquid phase** involves aeration of the water portion to encourage aerobic growth of microorganisms. As they grow, microorganisms digest proteins into simple amino acids, carbohydrates into simple sugars, and fats into fatty acids and glycerol. Acids and alcohols also are produced, and carbon dioxide evolves. The water then is passed through a clarifier and filter to remove the microorganisms and remaining organic matter, after which it flows into a stream or river.

The **solid phase** of secondary treatment is carried on in a **sludge tank**. Within the tank, microbial growth is encouraged by either an aerobic or an anaerobic process. In the aerobic process, compressed air is forced into the sludge, and the suspended particles form tiny gelatinous masses swarming with microorganisms, which thrive on the organic matter. *Zoogloea ramigera*, a gram-negative rod, produces the slime to which other microorganisms attach and congregate. The **activated sludge**, as it is termed, gathers to itself much of the microorganisms, organic material, color, and smell of the sewage. The activated sludge is drawn off and dried.

In the anaerobic method of sludge digestion, sewage is held in the tank for up to 30 days while the sludge ferments. Gases such as methane, carbon dioxide, and nitrogen are derived from this process. The methane may be captured and used to run the machinery of the sewage facility. Other gases, such as ammonia and hydrogen sulfide, are of value to chemical industries. The digested sludge, together with the dried activated sludge, may be used as agricultural fertilizer since it contains many valuable salts, or the sludge may be carted to landfill sites or offshore dumping grounds.

The separation of solid sludge leaves a certain amount of water that may be further processed in **tertiary treatment** by purifying the water. Sedimentation is followed by filtration and chlorination, after which the water is placed back into circulation and made available to consumers. In many municipalities, it is also important to remove salts such as phosphates from the water because they may spark blooms of algae.

Agricultural waste poses still another problem for sanitary microbiologists. Most animal waste currently is handled through systems in which manure is piped into clay-lined oxidation lagoons and allowed to remain while bacteria decompose the waste. The accumulation of ammonia nitrogen from urine is a problem, however, as is the buildup of phosphorus in the soil. Moreover, the stench tends to be overpowering, except during the late summer when photosynthetic bacteria thrive and turn the water a deep purple. Microbiologists have isolated a species of *Rhodobacter* from the purple water, and they have shown that the bacteria can break down many odoriferous (odor-causing) compounds, including volatile fatty acids and phenols. In the future, seeding such bacteria in the lagoons may help increase their efficiency and minimize their noxious odors.

BIOFILMS ARE PREVALENT IN THE ENVIRONMENT

For decades, microbiologists have focused on free-floating bacteria growing in laboratory cultures. In recent years, however, they have come to realize the importance of bacteria living in biofilms. A **biofilm** is an immobilized population of bacteria (or other microorganisms) caught in a sticky web of tangled polysaccharide fibers adhering to a surface. Biofilms develop on virtually all surfaces in contact with a watery environment (an example is depicted in **FIGURE 26.10**). Examples of such environments are the surfaces of catheter tubes, teeth, aquatic plants or animals, the urinary tract, water pipes, and stones. Contact with a fluid environment ensures a plentiful

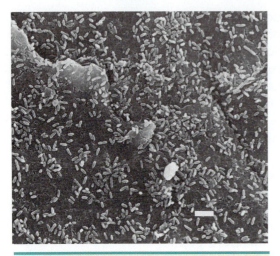

FIGURE 26.10

Biofilm Contamination

A scanning electron micrograph of a biofilm of *Pseudomonas aeruginosa* adhering to material used in urinary catheters in hospital settings. Experiments indicated that large numbers of these cells remained alive after exposure to high concentrations of tobramycin, an antibiotic to which the bacillus is normally susceptible. (Bar = 5 μm.)

supply of nutrients, and as the bacteria grow, they secrete the polysaccharides characteristic of the slimy layer.

Bacteria behave much differently in a biofilm than in a culture tube. Within their microcosm, for example, aerobic bacteria coexist with anaerobic bacteria as they share passageways and interact metabolically. In this state, they form communities that store nutrients, benefit from each other's metabolic by-products, and resist predators such as protozoa and bacteriophages. By contrast, **planktonic bacteria** live as individuals and do not enjoy these benefits. They tend to be of smaller size and have a lower metabolic rate of respiration. Thus, the members of a biofilm often are morphologically and physiologically distinct from their free-floating neighbors.

Apparently they communicate. Researchers have studied *Pseudomonas aeruginosa* in their mushroom-like biofilms and noted that at least two extracellular signals exist: one for cell-to-cell communication and one for cell-density determination. The latter signal governs the pattern of the so-called quorum sensing.

Biofilms carry both positive and negative connotations. Among the benefits of biofilms are their uses in **bioremediation**, where biofilm populations are used to degrade toxic wastes and other synthetic products of industry. They also are employed in the oil industry to fill empty spaces after oil has been pumped out. In natural settings, they degrade organic compounds, thus retarding pollutant buildup.

On the negative side, concern continues to mount about the tendency of biofilms to form on contact lens surfaces and cause eye infections. Moreover, biofilms form on catheters and medical implements (e.g., artificial hearts), and biofilms are important considerations in the development of dental caries and urinary tract infections (both Chapter 11). The concern is particularly acute because the bacteria in biofilms resist phagocytosis by white blood cells, display enhanced resistance to antibiotics, and withstand the destructive action of disinfectants and antiseptics.

In the industrial setting, biofilms pose problems when their sulfate-reducing members anaerobically convert sulfur compounds to hydrogen sulfide. The latter corrodes water pipes, especially those that carry seawater (where sulfate abounds). When they contaminate computer chips, biofilms act as conductors and thereby interfere with electronic signals. Indeed, one researcher has called biofilms the "venereal disease of industry." Some scientists estimate that in nature, 99 percent of all microbial activities occur in biofilms.

THE BACTERIOLOGICAL ANALYSIS OF WATER TESTS FOR INDICATOR ORGANISMS

Many methods are available for detecting the bacterial contamination of water, and various ones are selected according to the resources of the testing laboratory. Since it is impossible to test for all pathogenic microorganisms, water-quality bacteriologists have adopted the practice of testing for certain indicator bacteria normally found in the human intestinal tract. If these bacteria are present, fecal contamination has probably taken place.

Among the most frequently used indicator organisms are the coliform bacteria. **Coliform bacteria** are normally found in the intestinal tracts of humans and many warm-blooded animals. They can survive for extensive periods of time in the environment, and they are relatively easy to cultivate in the laboratory. *E. coli* is the most important indicator organism within the group.

The **membrane filter technique** is a popular laboratory test in water microbiology because it is straightforward and can be used in the field. A technician places a specially designed collecting bottle in the water and takes a 100-ml sample (FIGURE 26.11). The water then is passed through a membrane filter, and the filter pad is transferred to a plate of bacteriological medium, as outlined in Chapter 22. Bacteria trapped in the filter will form colonies, and by counting the colonies, the technician may determine the original number of bacteria in the sample.

Another method for testing water is the **standard plate count (SPC) technique**. Samples of water are diluted in sterile buffer solution, and carefully measured amounts are pipetted into Petri dishes. Agar medium is added, and the plates are set aside at incubation temperatures. A count of the colonies multiplied by the reciprocal of the dilution (the dilution factor) yields the total number of bacteria per ml of the original sample.

A third test is a statistical evaluation called the **most probable number (MPN) test**. In this procedure, a technician inoculates water in 10-ml, 1-ml, and 0.1-ml amounts into lactose broth tubes. The tubes are incubated and coliform organisms are identified by their production of gas from lactose. Referring to an MPN table, a statistical range of the number of coliform bacteria is determined by observing how many broth tubes showed gas. MicroInquiry 26 gives you a chance to carry out a standard qualitative water analysis.

DNA-based analysis also can be used to conduct water-quality tests based on the detection of indicator bacteria such as *E. coli*. With DNA technology, a sample of water is filtered, and the bacteria trapped on the filter are broken open to release their

MPN test:
a statistical evaluation of the number of coliform bacteria in a water sample.

FIGURE 26.11

Collection of Water for Analysis

DNA for amplification by the polymerase chain reaction (Chapter 20). Analysis for *E. coli* genes then is performed using **DNA probes** (Chapter 20). Not only does the process save time, it is extremely sensitive; for instance, a single *E. coli* cell can be detected in a 100-ml sample of water. Moreover, the pathogens transmitted by water (rather than the indicator organism) can be detected by DNA analysis. The identification of *Salmonella*, *Shigella*, and *Vibrio* species will become more feasible in the future as probe analyses become more widely accepted. The same principles hold for identifying viruses and virtually all other microorganisms.

Ensuring the safety of potable waters is one of the high priorities of public health officials. The use of DNA probes and amplification methods represents a revolutionary and exciting era in water-quality testing. Where the previous procedures required many days of waiting, the DNA-based procedures often are complete within a few short hours. Quickly determining whether a health risk exists allows the introduction of health measures when they can benefit the most people.

To this point . . .

We have given close scrutiny to water pollution and the methods used in water purification and sewage treatment. We noted the three basic steps in the preparation of water for drinking: sedimentation, to remove bulky objects; filtration, to eliminate the vast majority of microorganisms; and chlorination, to destroy the last remnants of microbial life. We also saw how sewage treatment may involve something as simple as the outhouse, cesspool, or septic tank, or a highly sophisticated operation such as is used in metropolitan centers. Primary treatment involves screening the sewage, while secondary treatment is more concerned with the microbial decomposition of organic waste. Sludge tanks are used in the secondary treatment. Tertiary treatment also is practiced in some communities.

We then surveyed some of the tests used by bacteriologists to determine the fitness of water for consumption. We emphasized coliform bacteria as indicators of water pollution. The membrane filter technique, standard plate count, and most probable number procedure were outlined to illustrate the range of available tests. We also explored the DNA-based analyses that will become more commonplace in the future.

In the next section of this chapter, we shall shift gears and examine the important positions occupied by microorganisms in the cycles of elements. We will use three examples—the carbon, sulfur, and nitrogen cycles—to show how microorganisms contribute to the quality of life. Working in their countless numbers in the environment, microorganisms effect a series of chemical changes fundamental to all life processes on Earth.

26.3

The Cycles of Elements in the Environment

The thought of microorganisms usually conjures a negative reaction because of their disease implication, and the contents of this chapter have undoubtedly supported that notion. It would be unwise, however, to neglect the positive role of microorganisms in the environment, because it is a substantial one. We have alluded to their role in the decay of organic matter, and we shall expand the idea by briefly examining the place of microorganisms in three vital cycles of nature: the carbon, sulfur, and nitrogen cycles. Indeed, the number of microorganisms in the environment is almost beyond belief. In 1998, microbiologists at the University of Georgia took a census of the various environments (i.e., ocean, air, soil, subsurface, and within other living things), and they calculated the number of microorganisms inhabiting the Earth to be an astounding 5 million trillion trillion, or enough to cover the top three feet of France!

THE CARBON CYCLE IS INFLUENCED BY MICROORGANISMS

Planet Earth is composed of numerous elements, among which is a defined amount of **carbon** that must constantly be recycled to allow the formation of organic compounds, of which all living things are made. **Photosynthetic organisms** take carbon in the form of carbon dioxide and convert it into carbohydrates using the sun's energy and chlorophyll pigments (Chapter 5). The vast jungles of the world, the grassy plains of the temperate zones, and the plants of the oceans show the results of this process. Photosynthetic organisms, in turn, are consumed by grazing animals, fish, and humans, who use some of the carbohydrates for energy and convert the remainder to cell parts. To be sure, some carbon dioxide is released back to the atmosphere in cellular respiration, but a major portion of the carbon is returned to the ground when the animal or plant dies (FIGURE 26.12).

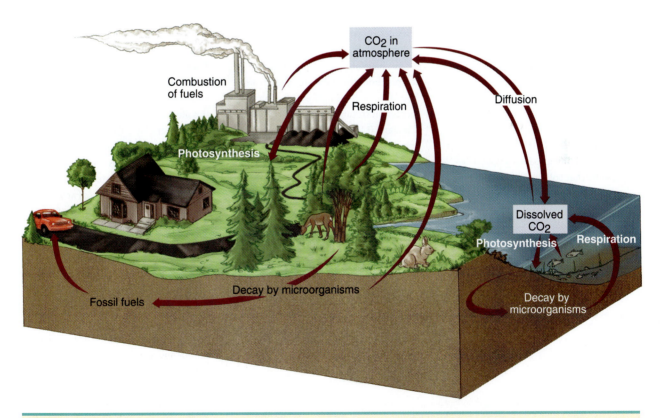

FIGURE 26.12

A Simplified Carbon Cycle

Photosynthesis represents the major method for incorporating carbon dioxide to organic matter, and cellular respiration accounts for its return to the atmosphere. Microorganisms are crucial to all decay in soil and ocean sediments.

DOING A STANDARD QUALITATIVE WATER ANALYSIS

One of the traditional methods of analyzing bacteriological water quality is through the **most probable number (MPN) test**. In this analysis, the detection of coliform bacteria (gram-negative enteric bacteria) involves up to three sets of tests, called the presumptive, confirmed, and completed tests.

The presumptive test is designed to detect coliform bacteria in a water sample. Of the enteric bacteria, only the coliform bacteria use lactose as a carbon and energy source. Therefore, a lactose broth medium represents a selective medium for coliforms. If coliforms are present, they ferment the lactose and produce acid and carbon dioxide gas. The presence of acid is seen by a change in the broth color (red to yellow) due to the presence of a pH indicator and gas production will be trapped in a small, inverted tube within the lactose broth tube (**FIGURE A**).

Three sets of five lactose broth tubes each are used. Set 1 tubes contain 10 ml of double-strength lactose broth, while set 2 tubes contain 1 ml and set 3 tubes contain 0.1 ml of single-strength lactose broth. Each tube then is inoculated with a known volume of the water sample. If any of the tubes produce gas, you can presume that coliform bacteria are present in the original water sample.

If a presumptive test indicates the presence of coliforms, you can determine the MPN by noting how many of the tubes in each group have produced acid and gas. Then, using a standard MPN table, you can obtain a statistical estimate for the number of coliforms present (**TABLE A**). For example, if three of the 10 ml tubes, two of the 1 ml tubes, and zero of the 0.1 ml tubes had positive results, the presumptive test would be read as 3-2-0, indicating approximately 14 coliforms per 100 ml of water (95 percent confidence that there were actually between 6 and 36 bacteria present).

Scenario: Working for the state health department, your job is to regularly analyze lake water from a public recreation area for coliform bacteria that would indicate fecal pollution. The health department needs to ensure that the water contains no appreciable number of coliform bacteria that otherwise could make people sick after swimming or playing in the water. Let's explore the presumptive test and then mention a few details about the confirmed and completed tests. Answers can be found in Appendix E.

You have obtained six lake-water samples from the public recreation area. You set up a presumptive test using 15 lactose broth tubes (5 each with 10 ml, 1 ml, and 0.1 ml of broth) for each of the six water samples. Using sterile pipettes, you transfer 10 ml aliquots into each of the 10 ml tubes, 1 ml aliquots into the 1 ml tubes, and 0.1 ml aliquots into the 0.1 ml tubes. All the tubes are incubated at 37°C for 24 hours.

26.1. Why do the 10 ml tubes contain double-strength lactose broth?

26.2. Why were sterile pipettes used? Is this absolutely necessary for this test?

Here are the results for the six samples:

SAMPLE NUMBER	POSITIVE TUBES		
	10 ML	1 ML	0.1 ML
1	0	0	1
2	0	0	0
3	0	1	0
4	5	4	5
5	0	2	0
6	2	0	1

26.3. What would you conclude from these presumptive test results?

The confirmed test is designed to confirm the presence of coliform bacteria in positive or uncertain presumptive test results.

26.4. Why is it necessary to confirm a positive presumptive test?

In a confirmed test, a sample from each identified lactose broth tube is streaked onto an eosin-methylene blue (EMB) agar plate. This is a selective and differential growth medium. It is selective because it inhibits the growth of gram-positive (non-enteric) bacteria and is differential because of any enteric bacteria present, only colonies of *Escherichia coli*, the prime indicator of fecal pollution, produce dark centers with a green metallic sheen.

26.5. Which lake-water samples need to be confirmed? What was your reasoning for selecting these samples?

If needed, a confirmed test can be run for any samples that gave a doubtful result in the confirmed test.

26.6. Suppose that only sample 4 gives a positive confirmed test; as the water analysis expert what would be your recommendation to the state department of health regarding this recreational lake?

FIGURE A

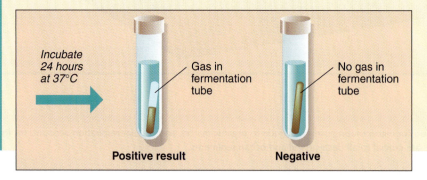

Incubate 24 hours at 37°C

Gas in fermentation tube

No gas in fermentation tube

Positive result   Negative

TABLE A	The MPN index per 100 ml for combinations of positive and negative presumptive test results when five 10-ml, five 1-ml, and five 0.1-ml portions of sample are used.

NUMBER OF TUBES WITH POSITIVE RESULTS						NUMBER OF TUBES WITH POSITIVE RESULTS					
FIVE OF 10 ML EACH	FIVE OF 1 ML EACH	FIVE OF 0.1 ML EACH	MPN INDEX PER 100 ML	CONFIDENCE LIMIT LOW	HIGH	FIVE OF 10 ML EACH	FIVE OF 1 ML EACH	FIVE OF 0.1 ML EACH	MPN INDEX PER 100 ML	CONFIDENCE LIMIT LOW	HIGH
0	0	0	<1.8	—	6.8	4	2	1	26	9.8	70
0	0	1	1.8	0.09	6.8	4	3	0	27	9.9	70
0	1	0	1.8	0.09	6.9	4	3	1	33	10	70
0	2	0	3.7	0.7	10	4	4	0	34	14	100
1	0	0	2	0.1	10	5	0	0	23	6.8	70
1	0	1	4	0.7	10	5	0	1	31	10	70
1	1	0	4	0.7	12	5	0	2	43	14	100
1	1	1	6.1	1.8	15	5	1	0	33	10	100
1	2	0	6.1	1.8	15	5	1	1	46	14	120
2	0	0	4.5	0.79	15	5	1	2	63	22	150
2	0	1	6.8	1.8	15	5	2	0	49	15	150
2	1	0	6.8	1.8	17	5	2	1	70	22	170
2	1	1	9.2	3.4	22	5	2	2	94	34	230
2	2	0	9.3	3.4	22	5	3	0	79	22	220
2	3	0	12	4.1	26	5	3	1	110	34	250
3	0	0	7.8	2.1	22	5	3	2	140	52	400
3	0	1	11	3.5	23	5	3	3	180	70	400
3	1	0	11	3.5	26	5	4	0	130	36	400
3	1	1	14	5.6	36	5	4	1	170	58	400
3	2	0	14	5.7	36	5	4	2	220	70	440
3	2	1	17	6.8	40	5	4	3	280	100	710
3	3	0	17	6.8	40	5	4	5	430	150	1100
4	0	0	13	4.1	35	5	5	0	240	70	710
4	0	1	17	5.9	36	5	5	1	350	100	1100
4	1	0	17	6	40	5	5	2	540	150	1700
4	1	1	21	6.8	42	5	5	3	920	220	2600
4	1	2	26	9.8	70	5	5	4	1600	400	4600
4	2	0	22	6.8	50	5	5	5	≥1600	700	—

It is here that the microorganisms exert their influence, for they are the primary **decomposers** of dead organic matter. Working in their countless billions in the water and soil, bacteria, fungi, and other microorganisms consume the organic substances and release carbon dioxide for reuse by the plants. This activity results from the concerted action of a huge variety of microorganisms, each with its own nutritional pattern of protein, carbohydrate, or lipid digestion (FIGURE 26.13). Without microorganisms, the Earth would be a veritable garbage dump of animal waste, dead plants, and organic debris accumulating in implausible amounts.

Microorganisms accomplish a similar goal in compost, where they decompose manure and other natural materials and convert them to compost for crop fertilization. Although both conventional and organic agriculture use manure as part of regular farm soil fertilization programs, only certified organic farmers are required to have a plan detailing the methods for building soil fertility using raw or aged manure. Furthermore, certified organic farmers are prohibited from using raw manure for at least 60 days prior to harvesting crops for human consumption.

But, there is more. Microorganisms also break down the carbon-based chemicals produced by industrial processes, including herbicides, pesticides, and plastics. In addition, they produce methane, or natural gas, from organic matter and are probably responsible for the conversion of plants to petroleum and coal deep within the recesses of the Earth. Moreover, many microorganisms trap CO_2 from the atmosphere and form carbohydrates to supplement the results of photosynthesis. In these activities, the microorganisms represent a fundamental underpinning of organic creation.

THE SULFUR CYCLE RECYCLES SULFATE MOLECULES

The sulfur cycle may be defined in more specific terms than the carbon cycle. **Sulfur** is a key constituent of such amino acids as cysteine and methionine, which are important components of proteins. Proteins are deposited in water and soil as living

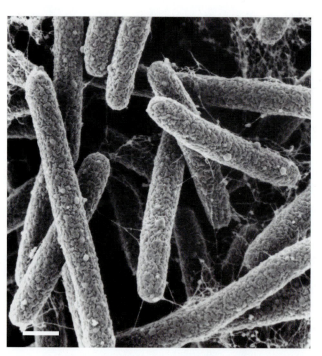

FIGURE 26.13

A Soil Bacterium

A scanning electron micrograph of the soil bacterium *Myxococcus xanthus*. This bacterium belongs to a group that displays gliding motility, complex social interactions, and under certain circumstances, a resistant spore called the myxospore. In this view, the fibrils connecting the cells at the surface can be seen. (Bar = 1 μm.)

organisms die, and bacteria decompose the proteins and break down the sulfur-containing amino acids to yield various compounds, including hydrogen sulfide. Sulfur also may be released in the form of sulfate molecules commonly found in organic matter. Anaerobic bacteria, such as those of the genus *Desulfovibrio*, subsequently convert sulfate molecules to hydrogen sulfide.

The next set of conversions involves several genera of bacteria, including members of the genera *Thiobacillus*, *Beggiatoa*, and *Thiothrix*. These bacteria release sulfur from hydrogen sulfide during their metabolism and convert it into sulfate. The sulfate now is available to plants, where it is incorporated into the sulfur-containing amino acids. Consumption by animals and humans completes the cycle.

THE NITROGEN CYCLE IS DEPENDENT ON MICROORGANISMS

The cyclic transformation of nitrogen is of paramount importance to life on Earth. **Nitrogen** is an essential element in nucleic acids and amino acids. Although it is the most common gas in the atmosphere (about 80 percent of air), animals cannot use nitrogen in its gaseous form, nor can any but a few species of plants. The animals and plants thus require the assistance of microorganisms to trap the nitrogen. Among the first to recognize this relationship was Martinus Beijerinck, who did his work at the end of the 1800s (MicroFocus 26.3).

The nitrogen cycle begins with the deposit of dead plants and animals in the soil. In addition, nitrogen reaches the soil in urea contained in urine. A process of digestion and putrefaction by soil bacteria and other microorganisms follows, thus yielding a mixture of amino acids (FIGURE 26.14). Amino acids are broken down further by microbial metabolism, and the ammonia that accumulates may be used directly by plants.

MicroFocus 26.3

OF LUMPS AND BUMPS

The man from the Delft laboratory had an audacious proposal to the assembled farmers: "Don't plant your crops in the same field as last year," he said. "Leave the field alone for the next two years; let it lie fallow." The year was 1887. The man was Martinus Willem Beijerinck (bi'jer-ink). The country was the Netherlands. And, because agricultural land was at a premium, the proposal was revolutionary.

Beijerinck was a local bacteriologist. While his medical colleagues were investigating the germ theory of disease and its implications, Beijerinck was out in the fields. He was observing that fields are very productive when the land has just been cleared and freshly planted. He was noting that fields yield bountiful crops when the farmer is away for a couple of

years. Now he thought he had the answer: great populations of bacteria.

Beijerinck was an expert on plants, but he also had something that most other botanists lacked: a solid background in chemistry. He was of the opinion that nitrogen is essential for plant growth, but he had no idea how nitrogen bridges the gap between atmosphere and plant. Then it dawned on him that bacteria were the bridge. And the little lumps and bumps were the key. Time and again he observed great hordes of bacteria in the little lumps and bumps ("nodules") on plant roots. He didn't see the nodules as often on tended crops, but they always seemed to be on wild plants growing in untended fields.

Beijerinck performed the laboratory experiments that strengthened his

views: He took bacteria from the nodules and inoculated them to seedlings of various plants. In many cases, the plants developed nodules, and when he planted the seedlings, the nitrogen content of the soil rose dramatically. So, his advice that 1887 day was straightforward: Leave the field alone for a spell; plant elsewhere; let the wild plants thrive; and when the field is finally planted, the crop yield will be worth the wait.

Indeed, he was right. Modern farmers know that every now and then, it is important to let a field lie fallow and "refresh" itself. To the macrobiologist, the lumps and bumps on the roots of wild plants are the important parts. To the microbiologist, it's what's inside that counts. Either way, we all benefit.

FIGURE 26.14

A Simplified Nitrogen Cycle

Plant and animal protein and metabolic wastes are decomposed by bacteria into ammonia. The ammonia may be used by plants, or it may be converted by *Nitrosomonas* and *Nitrobacter* species to nitrate, which also is used by plants. Some nitrate is broken down to atmospheric nitrogen. This nitrogen is returned to the leguminous plants by nitrogen-fixing microorganisms as nitrate, which is converted to ammonia. Animals consume the plants to obtain proteins that contain the nitrogen.

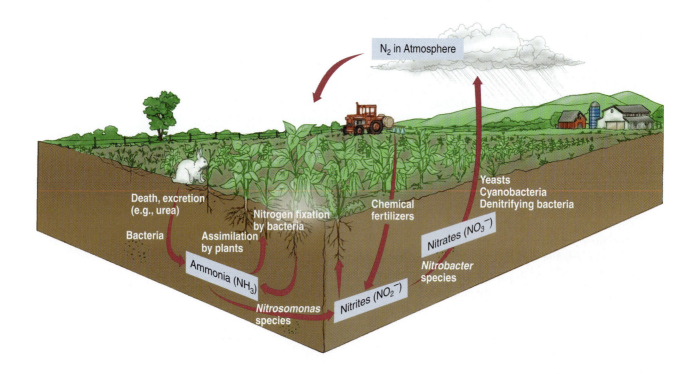

Next, **mineralization** takes place. In this process, complex organic compounds are converted to inorganic compounds and additional ammonia. Much of the ammonia is converted to nitrite ions by *Nitrosomonas* species, a group of **nitrifying bacteria** that are aerobic gram-negative rods. In the process, the bacteria obtain energy for their metabolic needs. The nitrite ions then are converted to nitrate ions by species of *Nitrobacter*, another group of nitrifying bacteria that are aerobic gram-negative rods, which obtain energy from the process. Nitrate is a crossroads compound: it can be used by plants for their nutritional needs, or it can be liberated as atmospheric nitrogen by certain microorganisms, including the denitrifying bacteria.

For the nitrogen released to the atmosphere, a reverse trip back to living things is an absolute necessity for life to continue as we know it. The process is called **nitrogen fixation**. Once again microorganisms in water and soil play a key role because they possess the enzyme systems that trap atmospheric nitrogen and convert it to compounds useful to plants (**FIGURE 26.15**). In nitrogen fixation, gaseous nitrogen is incorporated to ammonia, which fertilizes plants.

Two general types of microorganisms are involved in nitrogen fixation: free-living species and symbiotic species. **Free-living species** include bacteria of the genera *Bacillus, Clostridium, Pseudomonas, Spirillum*, and *Azotobacter*, as well as types of cyanobacteria and certain yeasts. Generally, the free-living species fix nitrogen dur-

Nitrogen fixation:
the chemical process by which atmospheric nitrogen is incorporated to organic compounds.

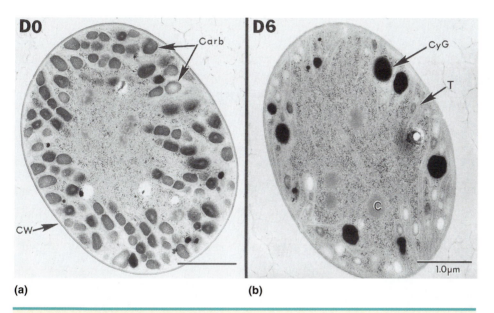

(a)   **(b)**

FIGURE 26.15

Metabolic Cycling in Cyanobacteria

Scanning electron micrographs of the cyanobacterium *Cyanothece*, showing the cycling of its metabolic functions. (Bars = 1.0 μm.) (a) During light periods (D0), nitrogen fixation is at a minimum. At this time, a series of carbohydrate granules (Carb) appears in the cytoplasm of the cell, indicating that photosynthesis is taking place. The granules are plentiful near the cell wall (CW). (b) When the cell enters a dark period (D6), nitrogen fixation occurs at a high rate, but photosynthesis slows. As the figure shows, most of the carbohydrate granules have disappeared, possibly because their contents are used as an energy source to fuel the reactions of nitrogen fixation. An empty granule (C) and some thylakoid membrane (T) can be seen. Also visible is a granule containing a cyanobacterial pigment called cyanophycin (CyG).

ing their growth cycles. The nitrogen-fixing ability of these species cannot be overemphasized.

Symbiotic species of nitrogen-fixing microorganisms live in association with plants that bear their seeds in pods. These plants, known as **legumes**, include peas, beans, soybeans, alfalfa, peanuts, and clover. Species of gram-negative rods known as *Rhizobium* infect the roots of the plants and live within swellings, or nodules, in the roots. Although complex factors are involved, the central theme of the mutualistic relationship is that *Rhizobium* fixes nitrogen and makes nitrogen compounds available to the plant while taking energy-rich carbon compounds in return. The bulk of the nitrogen compounds accumulates when *Rhizobium* cells die. Legumes then use the compounds to construct amino acids and, ultimately, protein. Animals consume the soybeans, alfalfa, and other legumes and convert plant protein to animal protein, thereby completing the cycle.

Humans have long recognized that soil fertility can be maintained by rotating crops and including a legume. The explanation lies in the ability of rhizobia to fix nitrogen within the nodules of legumes (**FIGURE 26.16**). So much nitrogen is captured, in fact, that the net amount of nitrogen in the soil actually increases after a crop of legumes has been grown. When cultivating legumes, there is no need to add nitrogen fertilizer to the soil. In addition, when crops such as clover or alfalfa are plowed under, they markedly enrich the soil's nitrogen content. Thus, humans are indebted to microorganisms for such edible plants as peas and beans, as well as for the indirect products of nitrogen fixation—namely, steaks, hamburgers, and milk.

Symbiosis: a close and permanent association between two organisms.

(a)

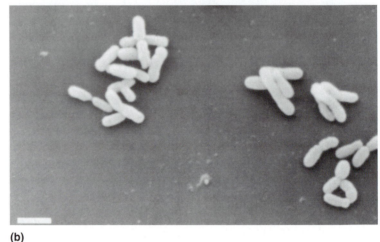

(b)

FIGURE 26.16

Nitrogen Fixation

(a) Nodules on the roots of a cowpea, a legume plant. Species of *Rhizobium* live within the nodules and fix nitrogen to nitrogen-containing compounds. When the bacteria die, the compounds are utilized by the legume to synthesize amino acids. (b) A scanning electron micrograph of a *Rhizobium* species. (Bar = 2 μm.)

Note to the Student

In many chapters of this text, we have studied the negative role of microorganisms in disease, and I would not blame you for becoming paranoid about taking a breath of air, bite of food, or drink of water.

However, there is another side to the story of microorganisms, a side that reflects the positive roles they play in our lives. These roles are highlighted by the foods and dairy products manufactured by microorganisms (Chapter 25), and by the industrial products derived from their growth (Chapter 27).

The positive role of microorganisms is exemplified further by their activity in the treatment of sewage. Through a complex network of processes, microorganisms transform the devilish cocktail of sewage into simple compounds that can be handled by the environment. Working with quiet competence, they break down the vile mixture of human and animal feces, urine, hair, oily filth from roads, bloody effluent from slaughterhouses, and as ugly a profusion of grot as can be imagined. The mammoth and unceasing task is accomplished unfailingly and efficiently.

Nor does it end here. In the carbon, nitrogen, and sulfur cycles, the microorganisms convert the basic elements of life on Earth to usable forms and replenish the soil to nourish all living things. By far, the great majority of microorganisms are engaged in constructive, cooperative, healthy, and wholesome activities. It is good to remember that microorganisms represent the most efficient way of breaking down organic matter, and, in the general scheme of things, this happens after death, not before.

Summary of Key Concepts

Environmental microbiology is concerned in large measure with the pollution bought on by microorganisms.

26.1 WATER POLLUTION

■ **Unpolluted and Polluted Water Contain Very Different Microbial Populations.** An unpolluted water environment is inhabited by limited numbers of soil bacteria, but a polluted environment contains an enormous variety of heterotrophic organisms from sewage, feces, and industrial sources. Coliform bacteria, the gram-negative rods of human and animal intestinal tracts, abound in polluted water. By contrast, a marine environment has halophilic microorganisms, as well as psychrophilic and barophilic organisms.

■ **There Are Three Types of Water Pollution.** Of the three types of pollution, biological pollution is of primary interest to the water microbiologist. The biochemical oxygen demand (BOD) is a measure of the amount of biological pollution. A high BOD indicates the presence of large numbers of organisms needing oxygen for decomposition processes.

■ **Diseases Can Be Transmitted by Water.** Concern exists for a variety of bacterial, viral, and protozoal diseases transmitted by water. Most of the diseases were described in Chapter 9.

26.2 THE TREATMENT OF WATER AND SEWAGE

■ **Water Purification Is a Three-Step Process.** To prepare water for drinking purposes, municipalities employ various levels of water purification, including sedimentation, filtration, and chlorination.

■ **Sewage Treatment Can Be a Multistep Process.** Sewage also can be treated by different steps according to the needs of the municipalities. Cesspools and septic tanks are used for local treatment, and variations of oxidation lagoons and secondary and tertiary treatments are used on larger scales.

■ **Biofilms Are Prevalent in the Environment.** As much as 99 percent of microbial activities in nature may occur in biofilms. The immobilized populations adhere to many different surfaces. They can be used in such processes as bioremediation and oil spill clean up. Biofilms also can cause disease and be quite resistant to chemicals and antibiotics.

■ **The Bacteriological Analysis of Water Tests for Indicator Organisms.** To test the effectiveness of purification procedures, several bacteriological tests are available, including the membrane filter technique, the standard plate count, the most probable number test, and a number of DNA-based analyses.

26.3 THE CYCLES OF ELEMENTS IN THE ENVIRONMENT

■ **The Carbon Cycle Is Influenced by Microorganisms.** In the carbon cycle, bacteria and fungi are essential to the breakdown of organic matter and the release of carbon back to the atmosphere for recycling.

■ **The Sulfur Cycle Recycles Sulfate Molecules.** Anaerobic bacteria are important to the sulfur cycle by producing and degrading sulfate.

■ **The Nitrogen Cycle Is Dependent on Microorganisms.** In the nitrogen cycle, many microorganisms release nitrogen from urea, amino acids, and nitrogenous organic matter. Many types of nitrogen-based conversions are performed by microorganisms, and bacteria are essential in the steps of nitrogen fixation where nitrogen is brought back into the cycle. Indeed, the processes performed by nitrogen-fixing bacteria are so essential that life as we know it probably would not exist without bacterial intervention.

Questions for Thought and Discussion

Answers to selected questions can be found in Appendix C.

1. When sewers were constructed in New York City in the early 1900s, engineers decided to join storm sewers carrying water from the streets together with sanitary sewers bringing waste from the homes. The result was one gigantic sewer system. In retrospect, was this a good idea? Why?

2. In the 1970s, a popular bumper sticker read: "Have you thanked a green plant today?" The reference was to photosynthesis taking place in plants. Suppose you saw this bumper sticker: "Have you thanked a microorganism today?" What might the owner of the car have in mind?

3. The English scientist John Harrington is credited with the invention of the first functional water closet (toilet) for the disposal of human waste. Why was this a significant advance in sanitation and water microbiology?

4. A student notes in her microbiology class that a particular species of bacteria actively dissolves fats, greases, and oils. Her mind stirs, and she wonders whether such an organism could be used to unclog the cesspool that collects waste from her house. What do you think she is considering? Will it work?

5. The water in a particular bay is relatively free of bacteria in the wintertime but generally polluted with bacteria in the summer. One reason is that summer boaters illegally empty their holding tanks into the bay before docking. Another is that fishermen clean their catch along the docks and dump the refuse into the bay. How many other reasons can you suggest for the summertime pollution?

6. The victims of disease, both animals and people, are buried underground, yet the soil is generally free of pathogenic organisms. Why?

7. In August 1989, two men were tragically killed while working in an industrial septic tank during cleaning. The cause of death was listed as methane poisoning. What was the source of the methane, and how was it related to the septic tank?

8. What information might you offer to dispute the following four adages common among campers and hikers? (1) Water in streams is safe to drink if there are no humans or large animals upstream. (2) Melted ice and snow is safer than running water. (3) Water gurgling directly out of the ground or running out from behind rocks is safe to drink. (4) Rapidly moving water is germ free.

9. In 1978, Legionnaires' disease broke out in the garment district of New York City and was given front-page treatment in the press. During that same period, the coliform count rose dramatically in water in the Murray Hill section of the city, but this was given minor coverage in the papers. Many public health officials believed that the Murray Hill problem posed the more substantial threat. Do you agree? Why?

10. Your family is building a new home and has the choice of installing a cesspool or a septic tank. Which might you be inclined to choose? Why?

11. In his classic book *Rats, Lice, and History,* Hans Zinsser writes: "As soon as a state ceases to be mainly agricultural, sanitary knowledge becomes indispensable for its maintenance." What evidence can you provide to support this statement? Can you think of any evidence to refute it?

12. Some years ago, the syndicated columnist Erma Bombeck wrote a humorous book entitled *The Grass Is Always Greener Over the Septic Tank.* (The title was an adaptation of the expression "The grass is always greener on the other side of the fence.") Indeed, the grass is often greener over the septic tank. Why is this so? How can you locate your home's cesspool or septic tank in the days following a winter snowfall?

13. The author of a biology textbook writes: "Because the microorganisms are not observed as easily as the plants and animals, we tend to forget about them, or to think only of the harmful ones . . . and thus overlook the others, many of which are indispensable to our continued existence." How do the carbon, sulfur, and nitrogen cycles support this outlook?

14. A park in a local community has two swimming pools: an Olympic-sized pool for swimmers, and a small wading pool for toddlers. In which pool does the greater potential for disease transmission exist, and what precautions may be taken to limit the transmission of microorganisms?

15. Clamming and mussel gathering are often prohibited in contaminated waters even though fishing is permitted. What is the reason for this apparent discrepancy?

Review

Using your knowledge of environmental microbiology, consider each characteristic and the three possible choices below. In the space, place the letter or letters of the most appropriate choice(s). The answers are listed in Appendix D.

_____ 1. Genus (genera) of coliform bacteria
   a. *Escherichia*
   b. *Staphylococcus*
   c. *Enterobacter*

_____ 2. Where halophilic bacteria live
   a. lake
   b. ocean
   c. mountain stream

_____ 3. Used as markers for oil drilling
   a. bacteriophages
   b. radiolaria
   c. foraminifera

_____ 4. Waterborne microbial disease(s)
   a. hepatitis A
   b. amoebiasis
   c. hepatitis B

_____ 5. Step(s) in water purification
   a. filtration
   b. chlorination
   c. sedimentation

_____ 6. Found in polluted water
   a. *Proteus* species
   b. *E. coli*
   c. AIDS virus

_____ 7. Test(s) for oxygen consumption in water
   a. SPC
   b. BOD
   c. MPN

_____ 8. Type(s) of pollution when microorganisms are present
   a. biological
   b. physical
   c. chemical

_____ 9. Toxin-producing dinoflagellate(s)
   a. *Entamoeba*
   b. *Gambierdiscus*
   c. *Gonyaulax*

_____ 10. Produce(s) flocs in water
   a. iron sulfate
   b. copper sulfate
   c. aluminum sulfate

_____ 11. Needed to perform the standard plate count
   a. Petri dishes
   b. pipettes
   c. agar medium

_____ 12. Possible cause(s) of red tide
   a. *Gonyaulax*
   b. *Streptococcus*
   c. *Gymnodinium*

_____ 13. Found in anaerobic mud at lake bottom
   a. *Giardia*
   b. diatoms
   c. *Clostridium*

_____ 14. Release(s) carbon in carbon cycle
   a. viruses
   b. fungi
   c. bacteria

_____ 15. Function(s) in nitrogen cycle
   a. *Thiobacillus*
   b. *Nitrobacter*
   c. *Thiothrix*

27 Industrial Microbiology and Biotechnology

Never underestimate the power of the microbe.

—Microbiologist Jackson W. Foster of the University of Texas

HUMANS PROBABLY DISCOVERED alcoholic beverages by accident. It is conceivable that sunlight warmed some sort of fallen grape or other fruit and accelerated the fermentation of its juices by yeasts. Humans must have sampled this "spoiled" fruit with curiosity, and if the taste of the aromatic concoction was not especially pleasing, the euphoric feeling that followed probably brought them back for more. By trial and error, humans discovered the important factors in fermentation and soon learned to control the process. In doing so, they became the first industrial microbiologists.

However, industrial microbiology is not restricted to alcoholic fermentations. It includes bread baking and cheese production, as well as the synthesis of organic compounds, antibiotics, insecticides, and the myriad products of genetic engineering. Basically, industrial microbiology refers to the uses of microorganisms in commercial enterprises.

Many products of industrial microbiology contribute to public health as aids to nutrition. Other products are used to interrupt the spread of disease. Still others hold promise for improving the quality of life in the years ahead. As we shall see in this chapter, industrial microbiology is an extremely diversified field, in which inexpensive raw materials are converted to valuable commodities through the metabolism of microorganisms.

Microorganisms in Industry

Certain properties make microorganisms well suited for industrial processes. Microorganisms not only possess a broad variety of enzymes to make an array of chemical conversions possible, they also have a relatively high metabolic activity that allows conversions to take place rapidly. In addition, they have a large surface area for the quick absorption of nutrients and release of end products. Moreover, they usually multiply at a high rate, as evidenced by the 20-minute generation time for *Escherichia coli* under ideal conditions.

In the industrial process, microorganisms act like chemical factories. To be effective, they should liberate a large amount of a single product that can be isolated and purified efficiently. The organisms should be easy to maintain and cultivate, and should have genetic stability with infrequent mutations. Their value is enhanced if they can grow on an inexpensive, readily available medium that is a by-product of other industrial processes. For example, a large amount of whey is produced in cheese manufacturing, and microorganisms that convert whey components to lactic acid add to the overall profit of the cheese industry. FIGURE 27.1 displays some possible conversions from a single metabolic substance.

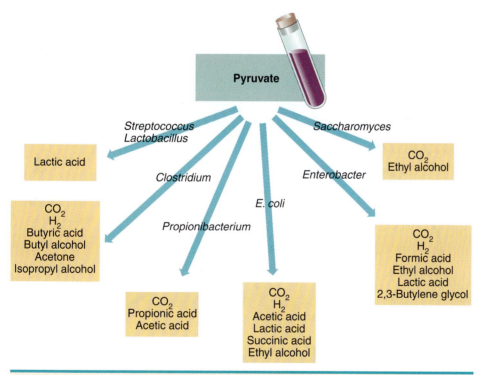

FIGURE 27.1

The Products of Fermentation

Industrial microbiology includes the metabolic conversions of organic compounds to other organic compounds by microorganisms. These conversions are called fermentations because they take place in the absence of oxygen. The process begins with the digestion of glucose (and other carbohydrates) by glycolysis, as explained in Chapter 5. The product of glycolysis is pyruvate, the starting point for the conversions shown here.

MICROORGANISMS PRODUCE MANY USEFUL ORGANIC COMPOUNDS

Microorganisms are used in industry to produce a variety of organic compounds, including acids, growth stimulants, and enzymes. In some cases, the production results from an apparent accident in nature in which an organism manufactures many thousands of times the amount necessary for its own metabolism.

One of the first organic acids to be made in bulk by microorganisms was **citric acid**. Manufacturers use this organic compound in soft drinks, candies, inks, engraving materials, and a variety of pharmaceuticals, such as anticoagulants and effervescent tablets (e.g., Alka-Seltzer). The organism most widely used in citric acid production is the mold *Aspergillus niger*. Microbiologists inoculate the mold into a medium of cornmeal, molasses, salts, and inorganic nitrogen in huge shallow pans or fermentation tanks. The absence of a Krebs cycle enzyme in the mold prevents the metabolism of citric acid (citrate) into the next component of the cycle, and the citric acid accumulates in the medium. **FIGURE 27.2** outlines this chemistry.

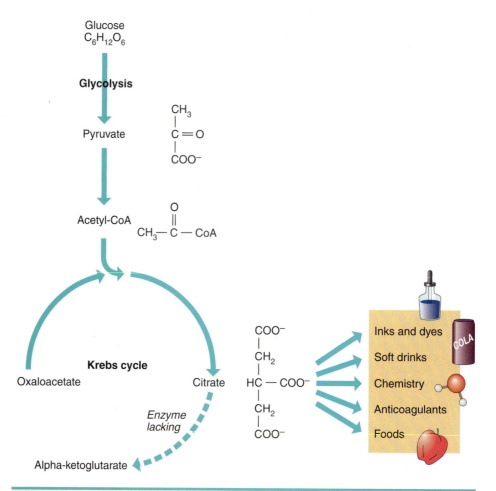

FIGURE 27.2

The Chemistry of Citrate Production

Aspergillus niger is grown in a mixture of nutrients, where it digests glucose into pyruvic acid. The pyruvic acid is then converted to acetyl-CoA, which condenses with oxaloacetic acid in the Krebs cycle to yield citric acid. However, the chemistry goes no further, because the next enzyme in the cycle is absent. Citric acid therefore accumulates and is isolated for use in various products, as shown.

Another important microbial product is **lactic acid**, a compound employed to preserve foods, finish fabrics, prepare hides for leather, and dissolve lacquers. Lactic acid is commonly produced by bacterial activity on the whey portion of milk. *Lactobacillus bulgaricus* is widely used in the fermentation because it produces only lactic acid from lactose.

Gluconic acid (gluconate), another valuable organic acid, is useful in medicine as a carrier for calcium, because gluconic acid is easily metabolized in the body, leaving a store of calcium for distribution. This acid is produced from carbohydrates by *A. niger* and species of the bacterium *Gluconobacter* cultivated in fermentation tanks. Calcium gluconate also is added to the feed of laying hens to provide calcium that strengthens the eggshells.

When the amount of amino acid produced by a microorganism exceeds the need, the remainder is excreted into the environment. Such is the case with **glutamic acid** (glutamate) produced by certain species of *Micrococcus, Arthrobacter*, and *Brevibacterium*. Glutamic acid is a valuable food supplement for humans and animals, and its sodium salt, monosodium glutamate, is used in food preparations.

In the production of lysine, another amino acid, two organisms are involved. *E. coli* is first cultivated in a medium of glycerol, corn steep liquor, and other ingredients, and the compound diaminopimelic acid (DAP) accumulates. Several days later, *Enterobacter aerogenes* is added to the mixture. This organism produces an enzyme that removes the carboxyl group from DAP to produce the lysine used in breads, breakfast cereals, and other foods.

Two important vitamins, riboflavin (vitamin B_2) and cyanocobalamin (vitamin B_{12}), also are products of microbial growth. **Riboflavin** is a product of *Ashbya gossypii*, a mold that produces 20,000 times the amount it needs for its metabolism. **Cyanocobalamin** is produced by selected species of *Pseudomonas, Propionibacterium*, and *Streptomyces* grown in a cobalt-supplemented medium. The vitamin prevents pernicious anemia in humans and is used in bread, flour, cereal products, and animal feeds.

MICROORGANISMS ALSO PRODUCE IMPORTANT ENZYMES AND OTHER PRODUCTS

The production of microbial enzymes for commercial exploitation has been an important industry since the emergence of industrial microbiology. Currently, over two dozen types of microbial enzymes are in use, and several others are in the research or developmental stage (FIGURE 27.3). Industrial enzymes have reached an annual market of $1.6 billion.

Among the important microbial enzymes used commercially are amylase, pectinase, and several proteases (TABLE 27.1). **Amylase** is produced by the mold *Aspergillus oryzae*. It is used as a spot remover in laundry presoaks, as an adhesive, and in baking, where it digests starch to glucose. **Pectinase**, a product of a *Clostridium* species, is employed to ret flax for linen. In this process, manufacturers mix the flax plant with pectinase to decompose the pectin "cement" that holds cellulose fibers together. The cellulose fibers then are spun into linen. MicroFocus 27.1 describes a more traditional process for retting. Pectinase also is used to clarify fruit juices.

Proteases are a group of protein-digesting enzymes produced by *Bacillus subtilis, Aspergillus oryzae*, and other microorganisms. Certain proteases are used for bating hides in leather manufacturing, a process in which organic tissue is removed from the skin to yield a finer texture and grain. Other proteases find value as liquid glues, laundry presoaks, meat tenderizers, drain openers, and spot removers.

FIGURE 27.3

Enzyme Production, a Mainstay of the Industrial Microbiology Process

A chemist extracts an enzyme from a bacterium found in ship-worms that is a powerful new stain remover for laundry detergents.

One of the most appreciated but lesser known uses of a microbial enzyme is in making soft-centered chocolates. **Invertase**, an enzyme from yeast, is mixed with flavoring agents and solid sucrose, and then covered with chocolate. The enzyme converts some of the sucrose to liquid glucose and fructose, forming the soft center of the chocolate. In medical microbiology, doctors use another microbial enzyme, **streptokinase**, to break down blood clots formed during a heart attack. Still another enzyme, **hyaluronidase**, is used to facilitate the absorption of fluids injected under the skin. The microbial roles for these enzymes are discussed in Chapter 18.

Gibberellins are a series of plant hormones that promote growth by stimulating cell elongation in the stem. Botanists use the hormones to hasten seed germination and flowering, and agriculturalists find them valuable for setting blooms in the plant. This increases the yield of fruit and, in the case of grapes, enhances their size.

TABLE 27.1

Some Enzymes and Their Microbial Sources Used in Commercial Markets

COMMERCIAL MARKET	ENZYME	SOURCE	USE
Dairy	Proteases	Bacteria and molds	Hydrolysis of whey proteins; coagulant in cheese production
	Lactase	Molds and yeasts	Hydrolysis of lactose to produce lactose-free milk
Brewing	Cellulases	Molds	Liquefaction and clarification processes
	Amylases and proteases	Bacteria and molds	
Wine and juices	Pectinases	Bacteria and molds	Increase juice yield and clarification
Meat	Proteases	Bacteria and molds	Meat tenderizing
Confectionary	Invertase	Yeasts	Liquefaction of sucrose
Textiles	Pectinase	Bacteria and molds	Flax retting
Pulp and paper	Xylanases, hemicellulases, lipases	Molds and yeasts	Cleaner and more efficient pulp and paper processing
Detergents	Proteases, amylases, lipases, cellulases	Bacteria and molds	Better cleansing of laundry and dishes

MicroFocus 27.1

IT SMELLED BAD, BUT IT WORKED

In past centuries, industrial pectinase was not available for retting flax, nor was protease available for bating hides. Nevertheless, the processes were carried on efficiently and successfully.

The retting process began by bundling flax plants and drying them in stacks. The stacks were then placed in a long trench several feet deep, covered with water, and weighted down with stones to exclude as much air as possible. After a few days, the water turned black, and an unmistakable stench signaled that retting was taking place. Two weeks later the flax was so soft and pliable that the fibers could be removed easily by pounding with wooden blocks. Today's microbiologists point out that *Clostridium* species were probably producing pectinase in the trenches.

The method for treating hides was equally messy. Skins were mixed with dog or fowl manure and set aside to cure. Fragments of tissue and hair gradually dissolved in the muck, and soon the hide became soft and pliable. Apparently, the proteases from fecal bacteria were responsible for the digestion. Bating hides was another smelly process, to be sure, but like the method for retting, it was usually reliable.

Gibberellins are produced during the metabolism of the fungus *Gibberella fujikuroi* and may be extracted from these organisms for commercial use (MicroFocus 27.2).

Although most natural food flavoring ingredients are produced by traditional processes from plant origins, new biotechnology methods have made it possible to produce novel flavoring ingredients by converting relatively cheap starting materials into higher-value flavor and aroma additives. The latter are used in foods, beverages, cosmetics, and other consumer items. An example are the fruit, peach, and coconut flavoring agents called **lactones**. Although lactones can be generated from long-chain fatty acids from sweet potatoes, the fatty acids occur in limited quantity and are expensive to modify. To circumvent this problem, microbiologists use species of *Mucor* and other fungi to convert medium-chain fatty acids to compounds that other microorganisms can easily transform to lactones. Another example are

MicroFocus 27.2

FOOLISH SEEDLINGS

During the 1890s, Japanese rice growers noticed that elongated seedlings sometimes appeared among their normal-sized seedlings in rice paddies. Although the elongated seedlings demonstrated vigorous early growth, the plants died before reaching maturity. After a time, growers began calling the condition "foolish seedling disease."

Little was known about the disease until 1926, when a Japanese botanist named Eiichi Kurosawa discovered an ascomycete fungus, *Gibberella fujikuroi*, growing on the plants. He isolated the fungus, transferred it to healthy seedlings, and found that they too developed elongated stems. In further studies, he reproduced the phenomenon with an extract made from the fungus and with samples of culture media in which the fungus grew. Kurosawa concluded that some chemical was involved and named the chemical gibberellin, after the fungus.

Gibberellin was isolated and identified by Japanese biochemists during the 1930s, but because of World War II, Western scientists did not learn of the work until 1950. That year, scientists from the United States and Great Britain read the scientific papers from Japan and began their own research. By 1956, they had isolated gibberellin from bean seeds and showed that it was really a mixture of plant substances. Today, 57 different gibberellins have been separated. The chemicals are known to be hormones that have dramatic effects on cell division, seed maturation, stem elongation, and numerous other activities in plants. Their function in the fungus *Gibberella*, however, still remains uncertain.

methylketones, which confer strong cheese-associated flavors in dairy products. The flavoring agents are derived industrially from *Penicillium roqueforti* incubated in lipase-treated milk fats.

In addition to the major products we have surveyed, microorganisms provide a number of specialized materials. Typical of the miscellaneous microbial products is **alginate**, a sticky substance used as a thickener in ice cream, soups, and other foods. Another product of microbial origin is **perfume**. Musk oil, for example, is prepared from ustilagic acid, a product of the mold *Ustilago zeae*, which, ironically, causes smut disease (Chapter 15). Moreover, there are numerous pharmaceutical products derived from the ergot poisons of the mold *Claviceps purpurea*. These derivatives are prescribed to induce labor, treat menstrual disorders, and control migraine headaches.

To this point . . .

We have opened the study of industrial microbiology by outlining the properties of microorganisms that make them assets to manufacturing processes and by surveying the techniques used in industrial operations. We then discussed some of the organic compounds and products produced by microorganisms, emphasizing specific groups of compounds.

Among the important organic acids produced by microorganisms are citric acid, lactic acid, gluconic acid, and acetic acid. Microorganisms also are used to produce amino acids such as glutamic acid and lysine, and vitamins such as riboflavin and cyanocobalamin. The important enzymes from microorganisms include amylase, pectinase, and protease. Several additional classes of organic compounds can be more economically derived by microbial activity than by chemical synthesis. The text highlighted the panorama of applications for these organic substances.

In the next section, we shall discuss additional products from microorganisms. We shall begin with alcoholic beverages and examine the fermentation processes for beer, wine, and distilled spirits. These are among the most developed of all industrial processes, so much so that fermentation is considered by many people to be an art.

27.2

Alcoholic Beverages

The fermentation of beer, wine, and other alcoholic beverages is one of the most venerable and universal of human domestic activities (**FIGURE 27.4**). The origin of beer fermentation, for example, has been traced as far back as 4000 B.C., when legend tells us that Osiris, the god of agriculture, taught Egyptians the art of brewing once they had learned how to farm the land. Wine production apparently has an equally long history because archaeologists have discovered evidence of grape cultivation in the Nile Valley during the same period. Furthermore, scientists have found evidence of wine in jars excavated from an Iranian site 7,000 years old. Thus, it is conceivable that the Egyptians, Sumerians, Assyrians, and other Near East peoples were among the earliest consumers, if not connoisseurs, of alcoholic beverages.

FIGURE 27.4

Ancient Wine Making

A Greek vase from the sixth century B.C. showing grapes being crushed for wine by two satyrs. In mythology, satyrs were part man, part goat creatures (note the horns, tails, and feet), who attended Dionysus, the Greek god of wine. Bacchus was the Roman counterpart of Dionysus.

BEER IS PRODUCED BY THE FERMENTATION OF MALTED BARLEY

As early as 3400 B.C., a tax was placed on beer in the ancient Egyptian city of Memphis on the Nile (MicroFocus 27.3). The Greeks later brought the art of brewing to Western Europe, and the Romans refined it. Indeed, the main drink of Caesar's legions was beer. During the Middle Ages, monasteries were the centers of brewing, and by the 1200s, breweries and taverns were commonplace in Great Britain. However, centuries passed before beer made its appearance in cans. That auspicious event took place in the United States in Newton, New Jersey, in 1935. The six-pack was a logical successor.

The word beer is derived from the Anglo-Saxon *baere*, meaning "barley." Thus, beer is traditionally a product of yeast fermentations of barley grains. However, yeasts are unable to digest barley starch, and therefore it must be predigested for them (FIGURE 27.5). This is accomplished in the process of **malting**, where barley grains are steeped in water while naturally occurring enzymes digest the starch to simpler carbohydrates, principally maltose (malt sugar).

At the brewery, the malt is ground with water to achieve further digestion of the starch. This process, called **mashing**, often includes corn as a starch supplement. Brewers then remove the liquid portion, or **wort**, and boil it to inactivate the enzymes. Dried petals of the vine *Humulus lupulus*, called **hops**, are added to the wort, giving it flavor, color, and stability. Hops also prevent contamination of the wort, because the leaves contain at least two antimicrobial substances. At this point, the fluid is filtered, and yeast is added in large quantities.

The yeast usually employed in beer fermentation is one of two species of *Saccharomyces* developed for centuries by brewers. One species, *S. cerevisiae*, gives a uniform dark cloudiness to beer and is carried to the top of the fermentation vat by foaming carbon dioxide. This yeast therefore is called a **top yeast**. It is used primarily in English-type brews such as ale and stout. The second species, *S. carlsbergensis*, ferments the malt more slowly and produces a lighter, clearer beer, having less alcohol. This yeast sediments and is thus called a **bottom yeast**. Its product is pilsener or lager beer. Almost three-quarters of the world's beer is lager beer.

A normal fermentation requires approximately seven days in a fermentation tank. The young beer then is transferred to vats for secondary aging, or **lagering**, which may take an additional six months. If the beer is intended for canning or bottling, it is pasteurized at 140°F for 55 minutes to kill the yeasts, or filtered through

MicroFocus 27.3

THIS HQT'S FOR YOU!

Beer, called hqt by the ancient Egyptians, was a very important drink. They often used beer in religious ceremonies and, since water could be a source of illness, they served beer at mealtime to both adults and children. In fact, archaeologists have discovered that workmen at the great pyramids had five types of beer that they drank three times a day. It was the staple drink of the poor (wages often were paid in beer), it was a drink of the pharaohs, and a drink offered to the gods.

Because of the prevalence of beer in Egyptian life, many Egyptologists have studied beer residue from ancient Egyptian vessels. The traditional view held that their beer was made by crumbling lightly baked well-leavened bread into water. After straining through a vat, the water was allowed to ferment because of the yeast from the bread. It then was flavored with date juice or honey, because the straining method would not give much flavor.

In 1996, these traditional views were challenged. Delwen Samuel, an archaeobotanist at the University of Cambridge, examined just what cereals and grains the Egyptians used to brew beer by looking at 2,000-year-old beer residues using a scanning electron microscope. Her findings suggested that the ancient Egyptians used barley to make malt and a type of wheat, called emmer, instead of hops. She says they heated the mixture and then added yeast and uncooked malt to the cooked malt. After adding the second batch of malt, the mixture was allowed to ferment. In her analysis, Samuel says she found no traces of flavorings.

Samuel and her colleagues tried brewing the beer using the recipe derived by the analysis. They brewed it at a modern brewery and found the beer to be fruity and sweet because it lacked the bitterness of hops. The beer was reported to have an alcoholic content of between 5 and 6 percent. Samuel gave her recipe to Scottish and Newcastle Breweries that then made a limited edition, 1,000-bottle batch of "King Tut ale." It was sold at Harrods department store for $100 per bottle, the proceeds going toward further research into Egyptian beer making.

In 2002, a major Japanese brewery claimed to have recreated a 4,000-year-old Egyptian beer by following a recipe from ancient hieroglyphics in Egyptian tomb paintings. Kirin Brewery Company Ltd., Japan's second largest beer maker, said the eight gallons of brewed beer was dark brown, contained no froth, and had a strong sour taste and an alcohol content of about 10 percent. The company does not plan to sell the beer commercially; it was developed for research purposes only!

a membrane filter. Some yeast is used to seed new wort, and the remainder may be dried for animal feed or pressed to tablets for human consumption. The alcoholic content of beer is approximately 4 percent.

Although commonly referred to as rice wine, the Oriental beverage **sake** is more like a rice beer. It is produced by allowing *Aspergillus oryzae* to convert rice starch to fermentable sugar. *Saccharomyces* species then ferment the sugar until the alcohol level is about 14 percent. The sake is now ready for consumption.

WINE IS PRODUCED BY THE FERMENTATION OF FRUIT OR PLANT EXTRACTS

During the Middle Ages and the centuries thereafter, wine was called *aqua vitae*, the "water of life." The title was appropriate because wine was one of the few safe things to drink. Indeed, until the late 1800s, safe drinking water was virtually non-existent in the Western world (note that the Bible makes no references to water for drinking purposes). Wine, by contrast, was generally free of pathogens due to its acidity and alcohol content, and it provided a few minerals and vitamins to the diet, while serving as a pain reliever. Of course, it also bred a society that was somewhat inebriated most of the time.

Essentially, all wines are derived from the natural conversion of grape or other fruit sugars to ethyl alcohol by the enzymes in *Saccharomyces*. Wild yeasts naturally

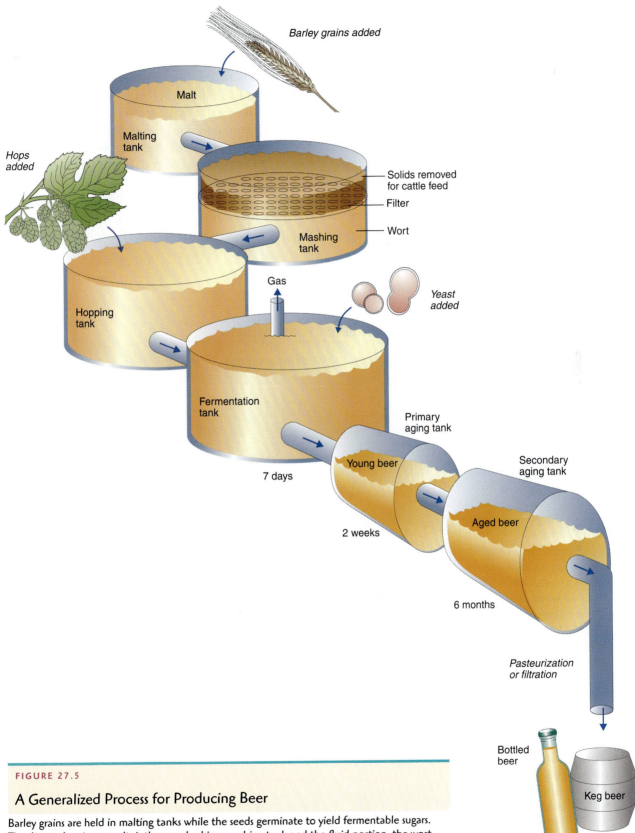

FIGURE 27.5

A Generalized Process for Producing Beer

Barley grains are held in malting tanks while the seeds germinate to yield fermentable sugars. The digested grain, or malt, is then mashed in a mashing tank and the fluid portion, the wort, is removed. Hops are added to the wort in the next step, followed by the yeast growth and alcohol production in fermentation. The young beer is aged in primary and secondary aging tanks. When it is ready for consumption, it is transferred to kegs, bottles, or cans.

occurring on the grapes may be used, but in the United States, the general trend is to use controlled cultures of yeast, usually *S. cerevisiae* variety *ellipsoideus*.

Wine may be made from fruit, fruit juice, or plant extracts such as dandelions. Among the grapes, the species *Vitis vinifera* is recognized as the highest-quality fruit. The winemaking process begins with crushing to produce the juice, or **must**. For **red wine**, black grapes are used, including skins and sometimes, the stems (MicroFocus 27.4). **White wine**, by contrast, is made from black or white grapes without their skins or stems. Yeasts begin to multiply immediately and initiate the fermentations. Anaerobic conditions soon are established as carbon dioxide evolves and takes up all the air space. The mixture froths with CO_2 and "ferments" in the original sense of the word.

Alcohol production requires only a few days, but the aging process in wooden casks may go on for weeks or months. During this time wine develops its unique flavor, aroma, and bouquet. These result from the array of alcohols, acids, alde-

MicroFocus 27.4

THE BENEFITS OF RED WINE

In 1992, scientists from the Bordeaux region of France did a population and epidemiological study and noted that French and other Mediterranean peoples ate large amounts of fatty foods, yet suffered a relatively low incidence of coronary artery disease. A 1996 report by the television news magazine program *60 Minutes* also pointed out that fatty meats, creams, butters, and sauces had little apparent effect on French hearts.

This so-called "French Paradox" was due, in part, to the apparent medicinal properties of wine, especially red wine. In 1996, researchers said that phenol-based compounds in red wine inhibit the oxidation of low-density lipoproteins (LDLs) in the blood, and by doing so, prevent the buildup of cholesterol and blood platelets in the arteries. The scientists pointed out that red wine contains more phenolics than white wine, and far more than beer. The phenolics also are present in other foods (e.g., raisins and onions), but not in the quantity found in the skins of grapes used for red wine, which concentrate even more after fermentation has taken place.

But that's not all. Scientists have known for decades that putting organisms

on a calorie-restricted diet dramatically reduces the incidence of age-related illnesses such as cancer, osteoporosis, and heart disease. In 2003, David Sinclair, an assistant professor of pathology at Harvard Medical School, and his colleagues said that an easier way to live longer might be simply to drink red wine.

A compound known as resveratrol, which belongs to the same family of phenols mentioned above, naturally exists in grapes and red wine. This compound was shown to extend the life span

of yeast cells by up to 80 percent. The molecule apparently mimics the life-extending effects of calorie restriction. Sinclair said he hoped resveratrol would prove to prolong life in other organisms, including humans. He believes that his study might help explain why moderate consumption of red wine has been linked to lower incidence of heart disease and why resveratrol prevents cancer in mice. He also noted that red wines that come from harsher growing areas, such as Spain, Chile, Argentina and Australia, contain higher levels of resveratrol than those wines produced where grapes are not highly stressed or dehydrated (France and the United States).

However, red wine is not for everyone. Indeed, alcohol should be avoided by pregnant women, people taking medications, those under the legal drinking age, and anyone with a family history of alcoholism. For these and for anyone else wishing to stay away from alcohol, alcohol-free red wines are appearing in the marketplace. They offer the opportunity to take advantage of a natural health ingredient while enjoying a glass of nature's bounty. "Ah, a glass of wine, thou, a healthy heart—and more years to contemplate it."

hydes, and other organic compounds produced by the yeast during aging. (Wine is estimated to have thousands of components, most of which have not yet been identified.) Soil and climate conditions (the *terroir*, in French) also contribute to the wine because they determine what organic compounds are present in the grape. The type of yeast and nature of wood derivatives from the fermentation casks are other determining factors. Thus, there are "vintage years" and "poor years." FIGURE 27.6 shows some of the steps in the wine-making process.

The broad variety of available wines result from modifications of the basic fermentation process. In **dry wines**, for example, most or all of the sugar is metabolized, while in **sweet wines**, fermentation is stopped while there is residual sugar. **Sparkling wines**, including champagne, sparkle because of a second fermentation taking place inside the bottle. For a sweet **sauterne**, vintners enhance the sugar content of grapes by a controlled infection with the mold *Botrytis cinerea*. The mold literally sucks water out of the grapes, thereby increasing the sugar concentration.

The strongest natural wines measure about 15 percent alcohol because yeasts cannot tolerate alcohol above this level. Most table wines average about 10 to 12 percent alcohol, with **fortified wines** reaching 22 percent alcohol. In fortified wines, brandy or other spirits are added to produce such wines as Port, Sherry, and Madeira. For mass production, wine is pasteurized to increase its shelf life, filtered, and bottled.

(a)

(b)

(c)

FIGURE 27.6

The Large-Scale Production of Wine

(a) A view of industrial model wine presses. The press on the right is open for loading with crushed grapes. Once loaded, a rubber bag is inflated in the center of the press to gently squeeze the grapes against the inside walls of the press. This extracts as much juice as possible from crushed grapes without breaking the seeds. (b) Large, steel tanks in which the juice ferments. (c) Wooden casks of wine aging in a cool cellar. The wine "ages" in casks for months or years.

DISTILLED SPIRITS ARE PRODUCED BY CONCENTRATING THE ALCOHOL FROM FERMENTATION

Distilled spirits contain considerably more alcohol than beer or wine. Each is designated with a **proof number**, which is twice the percentage of the alcohol content. For example, a 90 proof product contains 45 percent alcohol.

The production of distilled spirits begins like a wine fermentation. A raw product is fermented by *Saccharomyces* species, then aged, and finally matured in casks. At this point, the process diverges as manufacturers concentrate the alcohol by a distillation apparatus using heat and vacuum. Next, they mature the product in wooden casks to introduce unique flavors from various chemicals, such as aldehydes and volatile acids. Finally, the alcohol is standardized by diluting it with water before bottling.

Four basic types of distilled spirits are produced: brandy, whiskey, rum, and neutral spirits. **Brandy** is made from fruit or fruit juice, while **rum** is produced from molasses. **Whiskey** is a product of various malted cereal grains, such as scotch from barley, rye from rye grain, and bourbon from corn. The final type, **neutral spirits**, includes vodka, which is made from potato starch and left unflavored, and gin, which is flavored with the oils of juniper berries.

To this point . . .

We continued our study of industrial microbiology with a discussion of alcoholic beverages that are the result of microbial fermentations. We went through the process whereby beer is produced by fermentation of the wort and wine is produced by the fermentation of fruit or plant extracts. Distilled spirits are also a fermentation product, but here the spirits come from the distillation of the alcohol produced.

We now will move on with a brief discussion of antibiotics and insecticides, and then explore the developing concept of bioremediation. We will close with a survey of genetic engineering as an industrial process. Modern microbiologists are extremely optimistic about the future of genetic engineering. It is fitting, therefore, that we end our study of the microorganisms on this upbeat note.

27·3 Other Microbial Products

In addition to the products we have discussed, microorganisms are the sources of antibiotics and a number of valuable insecticides. Moreover, they are the producers of enzymes that break down natural and synthetic wastes in bioremediation. They are the biological factories for the genetic engineering technology that has revolutionized industrial microbiology. In the final section of this text, we shall study the methods for antibiotic and insecticide production and bioremediation, and we discuss some details of the genetic engineering process.

ANTIBIOTICS ARE PRODUCED COMMERCIALLY

Penicillin was the first antibiotic to be produced on an industrial scale. In 1941, Robert D. Coghill of the Fermentation Division of the USDA made the suggestion that the deep-tank method used to produce vitamins might be applied to penicillin. In the ensuing months, he offered several modifications to stimulate the growth of *Penicillium notatum* and increase the penicillin yield. For example, corn steep liquor in the culture medium increased the output 20 times, and the substitution of lactose for glucose made penicillin production still more efficient. Moreover, the search for a higher-yielding producer of the drug led researchers to *Penicillium chrysogenum*, a mold isolated from a rotten cantaloupe from a Peoria, Illinois, supermarket. Treatment with ultraviolet light resulted in a mutant with still higher penicillin yields. By 1943, the United States was producing enough penicillin for the Allied forces, and by 1945, sufficient amounts were available for the civilian population.

To the present time, over 8,000 antibiotic substances have been described and approximately 100 such drugs are available to the medical practitioner. Although most antibiotics are produced by species of *Streptomyces* (**FIGURE 27.7**), a significant number are products of *Penicillium* or *Bacillus* species. The worldwide production of antibiotics exceeded 25,000 tons in 1990, and two-thirds were penicillins.

Antibiotic production is carried on in huge, aerated tanks of stainless steel similar to those used in brewing. A typical tank may hold 30,000 gallons of medium. Older methods employed enormous mats of fungi or actinomycetes on the surface of the tank. Newer technology, however, employs small fragments of submerged

(a)

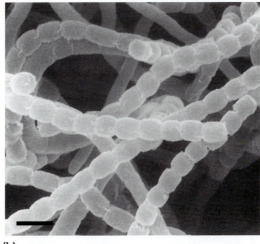

(b)

FIGURE 27.7

An Antibiotic Producer

Many modern antibiotics are produced by species of the soilborne rod *Streptomyces*. (a) A colony of *Streptomyces griseus*, the organism from which Selman Waksman isolated streptomycin in the 1940s. (b) A scanning electron micrograph of *S. griseus* grown on nutrient agar. This view displays the long chains of cells that characterize this organism. (Bar = 2 μm.)

hyphae or cells, rotated and agitated in the medium with a constant stream of oxygen. After several weeks of growth, the microorganisms are removed, and the antibiotic is extracted from the medium for further conversion to the desired product. The remaining brown mash of microorganisms may be dried and sold as an animal feed additive. Another alternative is to process it for use as human food.

SOME MICROBIAL PRODUCTS CAN BE USED TO CONTROL INSECTS

To be useful as an insecticide, a microorganism should be relatively specific for an insect pest and should act rapidly. It should be stable in the environment and easily dispensed, as well as inexpensive to produce. It helps if its odor is pleasant.

In the early part of this century, a scientist named G. S. Berliner found that sporulating cells of a *Bacillus* species were inhibitory to moth larvae. Berliner named the organism *Bacillus thuringiensis* after the German state Thuringia where he lived. The bacillus remained in relative obscurity until recent years, when scientists learned that *B. thuringiensis* produces toxic crystals in older cells during the process of sporulation (**FIGURE 27.8**). The toxic substance, an alkaline protein, is deposited on leaves and ingested by caterpillars (the larval forms of butterflies, moths, and related insects). In the caterpillar gut, the insecticide lyses the cells of the gut wall, possibly by inhibiting ATP phosphorylase or by forming pores in the microvilli membranes. As gut liquid diffuses between the cells, the larvae experience paralysis, and bacterial invasion soon follows.

Bacillus thuringiensis (Bt) appears to be harmless to plants and other animals. It is produced by harvesting bacteria at the onset of sporulation and drying them into a commercially available dusting powder. The product is useful on tomato hornworms and gypsy moth caterpillars. Its success encouraged further research and led to the discovery of a new strain called *B. thuringiensis israelensis* (Bti), first isolated in Israel.

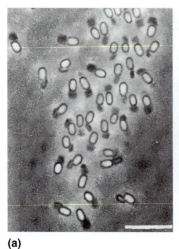  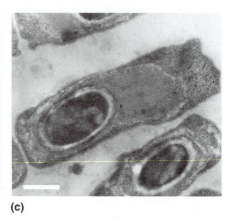

(a)   (b)   (c)

FIGURE 27.8

The Crystals and Spores of *Bacillus thuringiensis*

Photographs of a strain of the gram-positive rod *Bacillus thuringiensis* showing its spores and the crystals associated with the insecticidal ability of this organism. (a) A phase-contrast photomicrograph of *B. thuringiensis* spores attached to the insecticidal crystals. (Bar = 5 μm.) (b) A scanning electron micrograph of the spore-crystal complex after isolation from the cells. (Bar = 0.5 μm.) (c) A transmission electron micrograph of a sporulating cell that contains a crystal-spore complex surrounded by a containing membrane called the exosporium. (Bar = 0.5 μm.)

However, bacteria sprayed onto plants can soon wash off, so the protective effect of the insecticide can be limited. DNA technologists have realized that long-term protection can be provided by inserting the genes for toxin production directly into plants. To date, they have identified and isolated the gene for toxin production and have spliced it into plant genes using the Ti plasmid of *Agrobacterium tumefaciens* as the carrier (or vector), as noted in Chapter 7.

The first seeds with inserted *B. thuringiensis* (Bt) genes were added to the soil in 1996. Genetically engineered corn, cotton, seed potatoes, and other crops were planted on over three million acres in the United States. The crops displayed increased resistance to pests, but scientists noted an emerging resistance to Bt toxin in certain caterpillars. The resistance was thought to be related to a recessive gene passed along by typical Mendelian patterns. To prevent the development of resistance, scientists suggested planting adjacent fields without insecticide to dilute the gene. However, research reported in 1999 suggests that the resistance gene is dominant in certain pests, a factor that would work against the management strategy.

Another approach to gene-related plant resistance has been used in corn infested with corn borers. Researchers isolated the *B. thuringiensis* gene for toxin production and spliced it into a bacterium that lives harmlessly with corn plants. The researchers then forced a colony of gene-altered bacteria into corn seeds. When the seeds were sown, the toxin-producing bacteria flourished along with the plant, and when an insect ate the plant, it consumed the toxin. This approach is advantageous because only the insect attacking the plant is subjected to the toxin.

Another useful *Bacillus* species is *Bacillus sphaericus*, which kills at least two species of mosquitoes that ingest its poison. To increase the bacterium's efficiency, researchers inserted two of its genes to the bacterium *Asticcacaulis excentris* and achieved insecticidal activity against the mosquitoes that transmit malaria, filariasis, and St. Louis encephalitis. Using the gene carrier *A. excentris* is advantageous because it is easier to grow in large quantity; it tolerates sunlight better than *B. sphaericus*; and it floats in water, where mosquitoes feed (the heavier *Bacillus* species sinks because of its spores).

In 1941, *Bacillus popilliae* was introduced as a control measure for Japanese beetles. The bacillus infects beetle larvae and causes **milky spore disease**, so named because the blood of the larvae becomes milky white. The larvae eventually die, and the bacillus spores remain in the soil to infect other larvae.

Many bacteriologists also are investigating the insecticidal abilities of the gram-negative rod *Photorhabdus luminescens*. Its toxin, known as Pht, attacks the gut lining of larvae (as Bt does), and it is contained in large cytoplasmic crystals (also, as Bt); however, its spectrum of activity is wider than for Bt and includes numerous caterpillars as well as cockroaches. The Pht genes have been isolated, and efforts are underway to introduce them to plant cells. Normally, *P. luminescens* lives in the intestines of soilborne nematodes. The latter invade insect tissues in the soil, and the bacteria-derived toxin kills the insect. The bacteria also produce luciferase, an enzyme that induces a light-generating reaction and causes the nematode to glow. The significance of this reaction is currently unknown.

Viruses also show promise as pest-control devices, partly because they are more selective in their activity than bacteria. Once released in the field, the viruses spread naturally. It also is possible to harvest infected insects, grind them up, and use them to disseminate the virus to new locations. Among the insects successfully controlled with viruses are the cotton bollworm, cabbage looper, and alfalfa caterpillar. Chapter 7 explains experiments of this type.

Researchers also can develop insecticides by using a toxin from the venom of a scorpion. The toxin paralyzes the larvae of moths and other lepidopteran insects. It is attached to a **baculovirus**, a virus with a high affinity for lepidopteran tissues. Then, the virus is sprayed on lettuce and cotton plants infested with moth larvae. At the conclusion of the field trial, the plot is sprayed with 1 percent bleach to destroy any remaining viruses.

Viral genes also have been used to protect grapevines. In 1993, French biotechnologists announced the successful incorporation of genes from the **grape fan-leaf virus (GFLV)** to champagne grapevines. This virus is transmitted by a nematode and is endemic in the soils of many French regions. It causes malformation of the plant's leaf ("fan-leaf") and induces the plant to lose chlorophyll and become yellow. To protect the vines, researchers inserted the genes for viral capsids to *A. tumefaciens* and infected the plants with this bacterium. Soon the cells were producing viral capsid proteins, and they became resistant to the virus.

Even a fungus is being employed in the pesticide wars. California researchers have used *Lagenidium giganteum* to protect against crop-damaging mosquitoes in soybeans and rice, and in mosquito-infested nonagricultural settings such as wetlands. The fungus forces its spores into mosquito larvae, which die in a day or two. Marketed as Laginex, the fungal preparation has been approved for certain uses by the US Environmental Protection Agency (EPA).

OTHER PRODUCTS ALSO ARE BEING DEVELOPED

Industrial products of microbiology technology continue to emerge as the years unfold. In 1998, for instance, British scientists announced that they were using long strands of *Bacillus subtilis* to guide the assembly of such inorganic structures as fibers of magnetite, cadmium sulfide, and titanium dioxide. The researchers take advantage of a mutant whose cells fail to separate after binary fission, and soon elongate to form entangled chains and filaments that dry to a single hair-like structure almost a yard in length. The strands are used as templates (models), that is, a type of organic scaffold to which the inorganic materials cling. Such strands enable chemists to design materials in the micrometer range.

Another development allows chemists to obtain the dye **indigo** from leaves of a plant called woad. British biochemists discovered that a species of *Clostridium* ferments the carbohydrates in woad leaves and converts the indigo to a state where it is useful as a dye. This chemistry is valuable because the currently used synthetic versions of the dye tend to pollute waterways, while the naturally occurring dye is nontoxic. Denim garments of the future are prime candidates for the new dye.

Then, there is the *bacterial cement* that South Dakota researchers have used to fix cracks in concrete blocks in the laboratory. Biochemists begin with urea-digesting strains of *Bacillus pasteurii* and *Sporosarcina ureae*, both found in the soil. They mix the two organisms with sand and add urea (the major waste product of urine), then add some calcium chloride. The bacteria digest the urea and produce ammonia, which reacts with water to form ammonium hydroxide. The latter converts calcium chloride to calcium carbonate, which crystallizes as limestone. Soon, the gaps between the sand have filled in with the solidifying limestone (the bacteria die within the solid material), and the bacterial cement has set. One observer has wryly noted that this is truly "a concrete use for bacteria."

BIOREMEDIATION HELPS CLEAN UP POLLUTION NATURALLY

Recruiting bacteria and other microorganisms to break down synthetic waste is an immensely appealing idea. It signals a willingness to work with nature and adapt to its sophisticated sanitation systems, rather than trying to reinvent them. Putting microorganisms to work in this manner is the crux of **bioremediation**.

Although the term is recent and somewhat imposing, the concept of bioremediation is not new. In the 1800s, for example, night-soil men would, for a small fee, travel from house to house collecting sewage and excrement. After making their rounds, they would scatter their collections on fields, to be broken down by naturally occurring soil bacteria. Although modern waste-disposal systems have replaced the night-soil men, a new concern is the plethora of environmental pollutants contaminating the land. Bioremediation seeks to exploit microorganisms to degrade these pollutants.

The advantages of bioremediation were displayed following a major oil spill from the tanker *Exxon Valdez* in 1987 along the Alaska coastline (FIGURE 27.9). Previous studies showed that where oil is spilled, the bacteria that degrade oil (e.g., *Pseudomonas* species) are already present, and all technologists need to do is encourage their growth. Thus, when the oil spill occurred, technologists "fertilized" the oil-soaked water with nitrogen sources (e.g., urea), phosphorus compounds (e.g., laureth phosphate), and other mineral nutrients to modify the environment and stimulate the growth of indigenous (naturally occurring) microorganisms. Areas treated this way were cleared of oil significantly faster than nonremediated shorelines. Indeed, the oil degraded five times faster when microorganisms were put to work.

Bioremediation also can be applied to help eliminate **polychlorinated biphenyls (PCBs)** from the environment. PCBs were used widely in industrial and electrical machinery before their threat to environmental quality was realized. The compounds contain numerous chlorine atoms and chlorine-containing groups, and researchers have discovered that certain anaerobic bacteria remove the atoms and groups and reduce the compounds to smaller molecules. Aerobic bacteria now take over and reduce the molecular size still further. Field demonstrations in New York's

FIGURE 27.9

The *Exxon Valdez* Oil Spill

In 1987, the oil tanker *Exxon Valdez* ran aground on the shoreline of Alaska. Numerous species of bacteria demonstrated their value in oil digesting during the ensuing cleanup efforts. This was one of the first large-scale attempts to use microorganisms in bioremediation.

Hudson River have shown the value of the combination anaerobic-aerobic degradation; where once there was an accumulation of PCBs, now carbon dioxide, water, and hydrogen chloride have evolved.

Many years ago, **trichloroethylene (TCE)** was a much-used cleaning agent and solvent. At the time, scientists did not realize that TCE would diffuse through the soil and contaminate underground wells and water reservoirs (aquifers). To combat the problem, scientists are exploiting bacteria that grow on methane to degrade the TCE. During their metabolism, the bacteria produce a methane-digesting enzyme that coincidentally breaks down TCE. Technologists pump methane and other nutrients into the TCE-contaminated water, and as the bacteria grow, they digest the TCE as well as the methane. The deliberate enhancement of microbial growth yields an environmental cleanup.

Among the big news stories of 1999 was a "superbug" able to withstand 3,000 times more radiation than humans. The bacterium, a tetracoccus called *Deinococcus radiodurans*, was found in a tin of irradiation-sterilized ground beef. Researchers are hoping to use it in the daunting task of cleaning up thousands of toxic-waste sites that include radioactive materials such as plutonium and uranium. Genetic engineering methods have produced a *D. radiodurans* strain that degrades ionic mercury compounds common to these sites. The strain uses a gene cluster from *E. coli* to reduce mercury to the less toxic form found in thermometers.

During the 1940s through the 1960s, a major component of weaponry was the explosive compound **2,4,6-trinitrotoluene (TNT)**. Like other synthetic wastes, this compound has contaminated the soil from residues deposited around weapons plants. Scientists have found that they can reduce the level of contamination by encouraging bacterial growth with molasses. In a pilot study, researchers mixed

MicroInquiry 27

WORKING WITH MICROORGANISMS

As we approach the end of this text, you now should have a deep appreciation for the diverse roles microorganisms can have. Certainly, their roles go beyond causing human disease. In this last MicroInquiry, let's see some additional ways that microbes are being put to beneficial use. Answers can be found in Appendix E.

OIL EXPLORATION THROUGH MICROBIAL CHEMISTRY

For decades, we have been hearing about the impending oil shortage. Indeed, in the new millennium, although a shortage has yet to occur, it is becoming harder for oil companies to find deposits of light ("sweet") crude oil, which is what is mostly refined. This leaves behind about 60 percent of the world's reserves that exist as a very thick and sticky form called heavy ("sour") crude oil. Much of it remains below ground because it is very difficult and costly to recover and refine into the liquid gasoline that you put in the automobile.

27.1a. Outline a microbial way that this sour crude oil could be recovered more reasonably?

Another problem with heavy crude oil is that it contains large amounts of potential contaminants, including sulfur compounds and heavy metals, which could be spread into the atmosphere through refining. This would reverse the trend to cut sulfur emissions in the air. Add to this the high cost of refining such sulfur-rich crude and you have a dilemma.

27.1b. How can this heavy crude be more economically and safely extracted and refined?

PROSPECTING WITH MICROBES

The technology of metal extraction has been around for a long time. More than 2,000 years ago, the Romans used iron, copper, lead, gold, silver, and alloys such as bronze and brass for tools, weapons, coins, and jewelry. Unknown to them, microbes had solubilized metal ores, which the Roman metallurgists collected from mines in Spain and Wales.

The twentieth century saw an explosion in the production of such high-grade ores that, like crude oil, are becoming depleted. With this expansion of metal production, besides disfig-

water with TNT-laced soil and added molasses at regular intervals. In a matter of weeks, the TNT concentration plummeted.

Plants apparently are able to conscript bacteria for the task of environmental cleanup: Following the Gulf War of 1991, oily devastation remained in much of the Arabian Desert. Within four years, however, plant life returned, aided in large measure by *Arthrobacter* species. When researchers dug into oil-soaked desert, they found healthy plant roots surrounded by reservoirs of these oil-degrading gram-negative rods.

For many years, the "haul-and-bury" technique was the prevailing method for disposing of synthetic waste. As the public becomes increasingly intolerant of that approach, the importance of bioremediation will become more apparent. Technologists are testing microorganisms for their ability to degrade flame-retardants, phenols, chemical warfare agents, and numerous other waste products of industry. Indeed, one prominent researcher has called bioremediation "a field with its own mass and momentum."

MicroInquiry 27 explores a few more examples with microorganisms.

INDUSTRIAL GENETIC ENGINEERING CONTINUES TO MAKE ADVANCES

In 1973, Herbert Boyer, of the University of California at San Francisco, and Stanley Cohen, of Stanford University, performed the first practical experiment in genetic engineering. Working with *E. coli*, they removed the genes for kanamycin resistance and spliced them to a plasmid that already carried genes for tetracycline resistance. The plasmids then were mixed with *E. coli* cells not resistant to either antibiotic. Finally, the cells were streaked on plates of culture medium containing both

---

uring the land, comes the environmental pollution from the smelting process that produces sulfur dioxide in the atmosphere and toxic chemicals in the water and soil.

27.2 Outline a biotechnological scheme to provide for a continued supply of metals?

SUPERFUND AND BIOREMEDIATION

Historically and still in use today are chemical and physical methods, such as excavation and incineration, to degrade many of the toxic materials found at Superfund sites. However, the Environmental Protection Agency (EPA) has tested the feasibility of bioremediation at several of these sites.

Creosote contamination: Between 1973 and 1985, a lumber company in southern Missouri operated a wood treating facility that preserved railroad ties with a creosote/diesel fuel mixture. These operations contaminated the soil at the site with polynuclear aromatic hydrocarbons (PAHs), which were major components of the creosote/diesel mixture. After the facility was shut down and designated as a Superfund site in 1987, the EPA oversaw construction and operation of a land treatment unit to remove the PAH-contaminated soils at the site. EPA studies indicated that indigenous microorganisms that could digest aromatic hydrocarbons existed in the soil.

27.3 Outline a process whereby excavated contaminated soil could be decontaminated.

Petrochemical wastes: In Texas, an industrial waste disposal facility contained an estimated 70 million gallons of wastes from area petrochemical companies that were disposed of on site between 1966 and 1971. Contaminants in the lagoon sediments included PAHs, chlorinated organics, and metals.

27.4 Outline a process where by lagoon sediments could be decontaminated.

kanamycin and tetracycline. The bacteria that grew were ones that took up the plasmids and were now resistant to both antibiotics. Feats like these launched the modern era of biotechnology. As one observer later noted: "Biotechnology used to be BBC (before Boyer Cohen). Now, it is ABC (after Boyer Cohen)."

Genetic engineering based in plasmid technology has been hailed as the beginning of modern industrial microbiology. **Plasmids** are ultramicroscopic ringlets of double-stranded DNA that exist apart from the chromosome (**FIGURE 27.10**). A single plasmid may contain between 2 and 250 genes. The significance of **plasmid technology** lies in the fact that plasmids can be spliced with fragments of DNA from unrelated organisms and inserted into host organisms. The host organisms then produce the protein whose genetic message is carried by the foreign DNA. The mechanics of this reengineering process are explored in Chapter 7. Some contemporary products already obtained by plasmid technology include interferon, insulin, a vaccine for hoof-and-mouth disease, human growth hormone, and urokinase.

When microbiologists first developed the art of plasmid technology in the 1970s, the likely choice for the prototype "bacterial factory" was *E. coli*. Nonpathogenic strains had long been used as test organisms in the laboratory, and the genetics of *E. coli* were well understood. In the 1980s, however, attention shifted to *Bacillus subtilis* and yeasts as host organisms. Advocates of *B. subtilis* point out that this gram-positive bacillus normally secretes the proteins it makes, while *E. coli* retains them. Also, *B. subtilis* is not regarded as a human pathogen, in contrast to *E. coli*, nor does it contain endotoxins in its cell wall. Yeast supporters point to the traditional role of their organism in fermentation processes and, therefore, the public acceptance of an organism without disease potential. One industrial firm has reengineered a *Saccharomyces* species to produce a synthetic vaccine for hepatitis B, while a second firm has used yeast to obtain rennin, the enzyme used to make cheese.

The implications of genetic engineering are far-reaching and thought provoking. One group of biotechnologists has focused on the antibiotic-producing genes of plasmids and is attempting to remove the genes from *Streptomyces* species and insert them into more rapidly growing organisms. An allied group is trying to amplify the number of plasmids in antibiotic producers in order to obtain a higher yield of product.

FIGURE 27.10

Plasmids

A transmission electron micrograph of bacterial plasmids. The large plasmid is from the intestinal organism *Bacteroides fragilis*. The smaller ones have been isolated from *Escherichia coli*. Both large and small plasmids contain genes that encode antibiotic resistance.

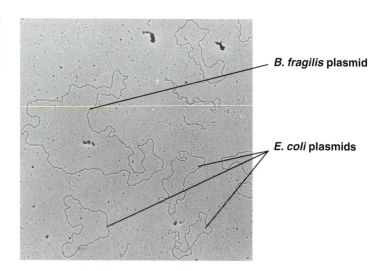

B. fragilis plasmid

E. coli plasmids

In the 1960s, a "Green Revolution" took place in which high-yielding supergrains were exported to poor countries throughout the world to encourage self-sufficiency. However, the program stalled when stocks of essential petroleum fertilizers fell to high oil prices. Agriculturalists now hope for a second Green Revolution sparked by genetic engineering (FIGURE 27.11). They foresee the day when the genes for **nitrogen fixation** can be extracted from bacteria such as *Rhizobium* and inserted into grain plants such as wheat, rye, and barley. The most optimistic planners look to the future and foresee fertilizers becoming obsolete, plants using microbial toxins to drive off insects, grains growing in salty water, and crops living for weeks without water.

Another dream of biotechnologists is using **replacement organisms** to interrupt disease cycles in nature. For example, a **transgenic** snail that resists invasion of *Schistosoma* species conceivably could interrupt the life cycle of the parasite that causes schistosomiasis (Chapter 17). A slightly different strategy is being employed by DNA technologists who are attempting to produce transgenic bollworms by inserting a gene that activates a "suicide gene" in offspring of the cotton bollworm. Work continues to progress on the transgenic mosquito. Researchers have identified the critical genes that allow anopheline mosquitoes to harbor and transmit malarial parasites. Hopefully, those genes can be altered, and mutated insects can be produced. Released in large numbers, the mosquitoes could dilute or overwhelm native mosquito populations and break the chain of disease transmission.

Other research projects in genetic engineering have equally important goals. Biochemists at breweries are attempting to insert the starch-digesting enzyme **amylase** into yeasts to eliminate the malting process in beer production. At chemical companies, scientists are seeking to incorporate genes for **cellulase** activity into microorganisms, to make cellulose a useful source of glucose. For the 20,000 hemophiliacs in the United States, new hope dawned when a genetic engineering company announced that it had isolated and cloned the genes for **Factor VIII**, an

Transgenic:
referring to an organism that contains a gene from another organism that is incorporated in a stable condition.

FIGURE 27.11

On the Horizon

An artist's fanciful conception of a futuristic superplant produced by genetic engineering.

essential blood-clotting protein missing in people who have hemophilia. The next step was to splice the genes to bacteria and other cells to produce a safe product.

Nor does it end here. Research in genetic engineering holds promise for new strategies for cancer prevention, new diagnostic procedures for microbial diseases and genetic abnormalities, new methods to correct genetic disorders, new hormones, antibiotics, and vaccines, and a generally improved quality of life. The discoveries and insights made possible by genetic engineering have been described as breathtaking. In the future, we can expect startling developments in medicine, agriculture, and the pharmaceutical and chemical industries. Indeed, it is an exciting time to be a microbiologist.

Note to the Student

It should be clear from this chapter that the microorganisms make a substantial contribution to the quality of life. Rather than gush with enthusiasm and rhetoric on the positive roles they play, I would prefer to paraphrase several concepts of applied microbiology set down by the late industrial microbiologist David Perlman of the University of Wisconsin. In a 1980 publication, Perlman wrote:

1. The microorganism is always right, your friend, and a sensitive partner;
2. There are no stupid microorganisms;
3. Microorganisms can and will do anything;
4. Microorganisms are smarter, wiser, and more energetic than chemists, engineers, and others; and
5. If you take care of your microbial friends, they will take care of your future.

Summary of Key Concepts

27.1 MICROORGANISMS IN INDUSTRY

Microorganisms occupy an important place in industry as the producers of many important products, such as organic compounds, alcoholic beverages, antibiotics, and insecticides.

- **Microorganisms Produce Many Useful Organic Compounds.** Among the organic compounds synthesized on an industrial scale are organic acids (such as citric, lactic, and gluconic acids), along with various amino acids and vitamins.

- **Microorganisms Also Produce Important Enzymes and Other Products.** Microbes also produce a variety of enzymes such as amylase, pectinase, and protease. Microorganisms produce other key products and save much expense and tedium for the chemist. In addition, the plant hormones known as gib-

berellins can be produced on an industrial scale by microorganisms. Other microbes are needed for flavoring in the cheese industry.

27.2 ALCOHOLIC BEVERAGES

Beer, wine, and spirits are among the products of yeast cell fermentation.

- **Beer Is Produced by the Fermentation of Malted Barley.** Yeasts of the genus *Saccharomyces* ferment barley grains to produce beer under anaerobic conditions. The process requires an aging step to develop the full flavor of the product.

- **Wine Is Produced by the Fermentation of Fruit or Plant Extracts.** Yeasts of the genus *Saccharomyces* ferment grape juice to wine under anaerobic conditions. This process also requires an aging step to develop the full flavor of the product.

■ **Distilled Spirits Are Produced by Concentrating the Alcohol from Fermentation.** To produce spirits, alcohol produced from fermentation is distilled off to produce brandy, rum, whiskey, or neutral spirits.

27.3 OTHER MICROBIAL PRODUCTS

■ **Antibiotics Are Produced Commercially.** Though antibiotics were originally a microbial product, most of these drugs are synthetically produced today.

■ **Some Microbial Products Can Be Used to Control Insects.** By contrast, the spores of *Bacillus thuringiensis* and *B. popilliae* are used in the live form as insecticides for plants. Both are successfully employed against the caterpillars of various insect pests. In addition, scientists are investigating the use of bacterial insecticides, such as the toxin produced by *Photorhabdus*. Scorpion toxins attached to insect viruses are being developed as potent insecticides.

■ **Other Products Also Are Being Developed.** Microorganisms have been used to guide the assembly of inorganic fibers, to obtain indigo dye from plants, and to cement cracks in concrete blocks.

■ **Bioremediation Helps Clean Up Pollution Naturally.** Bioremediation is still another innovative use for industrial microorganisms. In this process, naturally occurring microorganisms are encouraged to grow in a polluted environment and break down the pollutants. Bioremediation has been used successfully to degrade the oil in oil spills and to help eliminate polychlorinated biphenyls and trichloroethylene from the environment. Researchers are hoping to use the process in the daunting task of cleaning toxic waste sites that contain radioactive materials.

■ **Industrial Genetic Engineering Continues to Make Advances.** The future of industrial microbiology will center on genetic engineering and plasmid technology. By inserting foreign genes into vector organisms such as bacteria and yeasts, biotechnologists can produce rare proteins on an industrial scale and provide treatments for such diseases as diabetes, hemophilia, and cancer. Genetic engineers predict plants will be developed with inborn resistance to disease, animals will be engineered to produce human proteins, and novel treatments will be designed to cure or prevent specific illnesses. Medicine, agriculture, and industry eagerly anticipate the fruits of modern biotechnology that are made possible with microorganisms.

Questions for Thought and Discussion

Answers to selected questions can be found in Appendix C.

1. The poet John Donne once wrote: "No man is an island, entirely of itself." This maxim applies not only to humans, but to all living things in the natural world. What are some roles the microorganisms play in the interrelationships among living things?

2. Certain beer companies have developed strains of yeasts that break down more of the carbohydrate in barley malt than traditional yeasts. What do you think their product is called?

3. In industrial microbiology, it is extremely helpful if the leftover product of one process can be fermented by microorganisms to produce a valuable product. Can you think of any leftover products that might be useful in industry?

4. When the *Mayflower* set sail for the New World, its intended destination was Virginia. Instead, it landed at Plymouth, Massachusetts, because, as one diarist put it, "We could not now take time for further search or consideration, our victuals being much spent, especially our beer." What do these last few words tell you about the Pilgrims?

5. Bioremediation holds the key to solving numerous types of enviornmental problems in the future. Yet, the process has been used for generations without our realizing it. How many instances can you point out where bioremediation is currently in use?

6. The discovery of the extremely high resistance of *Deinococcus radiodurans* to radioactivity has prompted hopes that this organism can be used in bioremediation. Suppose you were to go on a hunt for novel organisms that could be used for other environmental cleanups. Where might you look?

7. A product called Dipel contains *Bacillus thuringiensis*. It is used widely in the Northeast for destruction of the gypsy moth caterpillar. What dangers might result from extensive application of this product?

8. In Europe, in the month of March, it is customary to have a festival where the primary drink is bock beer. Bock beer is a very dark, lager beer that represents the dregs of the brewery. The word *bock* is German for "ram," the sign of Aries, which begins in March. How does this festival coincide with the brewing process, and why does it occur in March?

9. In several places in this text, we have noted how apparently harmless organisms have been discovered later to be dangerous in humans. Yeasts have been consumed in breads and alcoholic beverages for centuries, and still no pathogenic signs have been observed. Can you postulate why?

10. Certain bacteria produce many thousands of times their required amount of specific vitamins. Some biologists suggest that this makes little sense because the excess is wasted. Can you suggest a reason for this apparent overproduction in nature?

11. A biotechnologist suggests that one day it may be possible to engineer certain bacteria to produce antibiotics and then to feed the bacteria to diseased people. The bacteria would then serve as antibiotic producers within the body. Would you favor research of this type?

12. It was noted in this chapter that *Aspergillus oryzae* is used as a commercial source of amylase, and that the same organism is essential for the production of sake. How are the two processes related?

13. One of the side effects of the use of Bt toxin has been a reduction in the population of Monarch butterflies. Why do you think this may have happened, and what might you as a concerned citizen do about it?

14. The editors of *The Economist*, a British magazine, have referred to genetic engineering as "one of the biggest industrial opportunities of the late twentieth century." What evidence can you offer to support this contention, and why might you be inclined to reject it?

15. How many times in the last 24 hours have you had the opportunity to use or consume the industrial product of a microorganism?

Review

Many different microorganisms find value in industrial microbiology. To test your knowledge of these organisms, match the microorganism on the right to the characteristic on the left by placing the correct letter in the space. A letter may be used once, more than once, or not at all. The answers are listed in Appendix D.

_____ 1. Used to ferment grapes.

_____ 2. Produces penicillin.

_____ 3. Used in riboflavin production.

_____ 4. Produces many antibiotics.

_____ 5. Enhances grape sugar content.

_____ 6. Lactic acid producer.

_____ 7. Used in lysine production.

_____ 8. Biological insecticide.

_____ 9. Source of musk oil.

_____ 10. Used for citric acid production.

_____ 11. Fungal source of proteases.

_____ 12. Synthesizes pectinase.

_____ 13. Produces amylase.

_____ 14. Produces plant growth hormones.

_____ 15. Yeast for making beer.

_____ 16. Ferments rice starch to sake.

_____ 17. Used for genetic engineering.

_____ 18. Causes milky spore disease.

_____ 19. Degrades ionic mercury compounds.

_____ 20. Gluconic acid from carbohydrates.

A. *Aspergillus niger*

B. *Aspergillus oryzae*

C. *Bacillus subtilis*

D. *Bacillus popilliae*

E. *Clostridium* species

F. *Agaricus phalloides*

G. *Saccharomyces* species

H. *Ashbya gossypii*

I. *Staphylococcus aureus*

J. *Enterobacter aerogenes*

K. *Botrytis cinera*

L. *Claviceps purpurea*

M. *Streptomyces* species

N. *Penicillium chrysogenum*

O. *Lactobacillus bulgaricus*

P. *Ustilago zeae*

Q. *Acetobacter aceti*

R. *Gibberella fujikuroi*

S. *Deinococcus radiodurans*

T. *Rhizopus nigricans*

Appendix A: Metric Measurement

	FUNDAMENTAL UNIT	QUANTITY	SYMBOL	NUMERICAL UNIT	SCIENTIFIC NOTATION
Length	meter (m)				
		kilometer	km	1,000 m	10^3
		centimeter	cm	0.01 m	10^{-2}
		millimeter	mm	0.001 m	10^{-3}
		micrometer	μm	0.000001 m	10^{-6}
		nanometer	nm	0.000000001 m	10^{-9}
Volume (liquids)	liter (l)				
		milliliter	ml	0.001 l	10^{-3}
		microliter	μl	0.000001 l	10^{-6}
Mass	gram (g)				
		kilogram	kg	1,000 g	10^3
		milligram	mg	0.001 g	10^{-3}
		microgram	μg	0.000001 g	10^{-6}

Appendix B: Temperature Conversion Chart

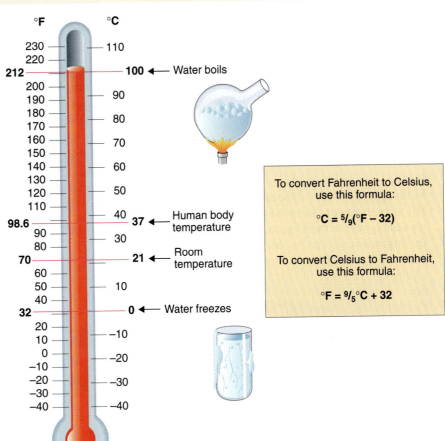

°F   °C

230 — 110
220 —
212 — **100** ← Water boils
200 —
190 — — 90
180 — — 80
170 —
160 — — 70
150 —
140 — — 60
130 —
120 — — 50
110 —
        — 40
98.6 — **37** ← Human body temperature
90 — — 30
80 —
70 — **21** ← Room temperature
60 —
50 — — 10
40 —
32 — **0** ← Water freezes
20 —
10 — — −10
0 —
−10 — — −20
−20 —
−30 — — −30
−40 — — −40

To convert Fahrenheit to Celsius, use this formula:

$$°C = \frac{5}{9}(°F - 32)$$

To convert Celsius to Fahrenheit, use this formula:

$$°F = \frac{9}{5}°C + 32$$

Appendix C: Answers to Selected Questions for Thought and Discussion

CHAPTER 1

3. Since measles is caused by a virus, and because the measles viruses cannot be seen or cultivated by methods that work for bacteria, your work would probably result in a series of dead ends and your frustrations would mount quickly.

6. Spontaneous generation would hold that the microorganisms of infectious disease arise spontaneously in the body and are not obtained from outside the body. Thus, we could do little to interrupt the spread of microorganisms, and all the antimicrobial processes we employ (water purification, sanitation, care in food preparation) would be useless. The germ theory holds that disease agents enter the body from the outside. This means that transmission can be interrupted, and it brings an element of control to epidemics.

9. Proving that the microorganisms did not arise spontaneously in the hamburger meat could be shown by obtaining another fresh sample of hamburger meat. After separating it into two portions, each portion could be placed in a small glass dish, covered with aluminum foil, and sterilized in some way (see Chapter 22 for sterilization methods). Then, one portion would be uncovered and exposed to the air while the other would remain covered. Only the uncovered portion should harbor microorganisms after some period.

12. Pasteur was something of a revolutionary in the way he thought and attacked the medical establishment. His germ theory of disease was severely at odds with prevailing ways of thinking, and his persistent hammering away at contemporary dogmas (e.g., the nature of fermentation, the cause and transmission of infectious disease, and the ability to interrupt epidemics) initiated a revolution that has carried through to the present day. In 1910, the French honored their "rebel" in connection with Mexican rebels. Pasteur was also a great patriot, as were the Mexicans.

15. *E. coli* is easier to grow, costs less to maintain, and requires less space than using animals.

CHAPTER 2

2. Detergents act on the lipid portions of bacterial membranes. They dissolve the membranes and cause leakage from the cytoplasm and subsequent death to the cell.

4. Wöhler showed that organic molecules could be synthesized, in effect demonstrating that living things could be synthetically produced. This finding stood in stark contrast to the pre-1800s belief that living things were something special and not within the ability of scientists to create. It also brought chemistry into the realm of biology, and made biologists begin thinking like chemists.

6. The organic matter of an individual consists of proteins, lipids, carbohydrates, and nucleic acids. Oxygen is a key component of each type of these groups of organic compounds. Also, there are large volumes of water in the body. It is not surprising, therefore, that a 120-pound individual contains 78 pounds of oxygen.

11. Carbon atoms have four electrons in their outer shell. They cannot lose their electrons easily, nor can they attract electrons from other atoms. Hence, they enter into covalent bonds, sharing their four electrons with four other atoms or radicals. This leads to enormous variations of chemical combination.

15. Carbohydrates have hydroxyl groups; lipids have hydroxyl and carboxyl groups; proteins have amino and carboxyl groups; and nucleic acids have hydroxyl and phosphate groups.

CHAPTER 3

2. The correct form is: "The famous bacterium *E. coli*."

5. One reason would be the size difference. Once a cell reaches a certain size, the surface-to-volume ratio makes it difficult to depend on diffusion to move nutrients and other materials through the cell cytoplasm. Thus, eukaryotic cells need an endomembrane system, energy systems, and a cytoskeleton. Bacteria, being smaller, can rely on diffusion due to their larger surface-to-volume ratio.

9. You will receive an order of mushrooms because the Italian *fungi* (Italian pronunciation "foon-gee") translates to "mushroom."

13. Contrary to the belief of many, oil does not increase the magnification of the light microscope. It merely allows the gathering of enough light to make using the instrument possible. The magnification would be the same without the oil, but the resolution would be much reduced.

15. A class of artists or pseudoartists could have a field day with this project. Imagine all the shapes taken by protozoa, bacteria, fungi, algae, and the rest, and a rather diverse group of "patrons" would emerge. I am reminded of the famous bar scene in *Star Wars*.

CHAPTER 4

1. Various advantages derive from being able to form a spore, capsule, or flagellum. The answers may help you to see that the bacterium's life is not dramatically different than that of other creatures.

3. If bacillus is being used as a genus, it should be capitalized and italicized. However, often the print news media omit the italics. If bacillus is lowercase, it refers to the bacterial shape.

6. Boiling may kill the majority of bacteria, but bacterial spores survive two hours or more of boiling. It is wrong to believe the water is sterile after a few minutes of boiling. It has been disinfected, however, and is perfectly safe to drink under normal circumstances.

10. In the pre–electron microscope days, the chapter on bacterial anatomy would probably be very brief. Certainly there would be no discussion of pili, plasmids, magnetosomes, ribosomes, or inclusions.

14. Bacteria may produce toxins that interfere with metabolic patterns in the body. Thus, the physical growth of bacteria may not be a necessary prerequisite to disease, but consuming food that contains toxins may be dangerous.

CHAPTER 5

1. The cartoon in question appeared in a syndicated newspaper and was authored by Athelstan Spilhaus, former president of the AIBS. Spilhaus probably conjectured that the cellulose of newspapers would be broken down to units of glucose. The glucose would be metabolized by bacteria into metabolic intermediates, which would then be converted to amino acids by adding amino groups to carbon skeletons in a reversal of deamination. The amino acids are fed to cattle, which would combine them to produce proteins.

6. One reason ATP is not supplied to the growth medium is that it probably could not be absorbed into the cytoplasm of a bacterial cell. You should speculate on other possible reasons.

8. The cyanobacterium is a photoautotroph while *E. coli* is a chemoheterotroph. Therefore, the carbon dioxide gas produced by cellular respiration can be used by the cyanobacteria for photosynthesis. The gas will accumulate in the *E. coli* flask since photosynthesis does not occur.

10. You should recognize the *flavin* part of riboflavin and guess that it is used to make flavin adenine dinucleotide (FAD). Here is an example of how recognizing word stems helps one to make an intelligent guess. Without FAD, oxidative phosphorylation grinds to a halt, energy production ceases, and death is imminent.

15. Essentially, the answer summarizes the chapter by showing how carbohydrates, fats, and proteins are digested and synthesized, and how the processes relate to one another. It is an excellent exercise in metabolism.

CHAPTER 6

1. The ribosome is the machine that translates the language of genes (nucleotides) into the language of proteins (amino acids). It brings together the mRNA that has the nucleotide sequence for a protein and the tRNAs that carry the amino acids for that protein.

4. Since there is no histidine added to the agar growth medium and *Salmonella* is his⁻, theoretically no bacterial colonies should form. Since some do, one must presume that among all the bacteria, a very small number of cells generated a spontaneous mutation that converted the histidine gene back to a functional his⁺ form.

7. A mutation is a permanent change in an organism's DNA. Therefore, a new, beneficial mutation that is transmitted to other cells or organisms (prokaryotic or eukaryotic) can change the makeup of the genetic content and those cells or organisms then produce a disproportionate number of new cells or organisms because of natural selection. Although mutations are rare, the impact on evolution can be very great.

12. One way would be to use a positive selection plating technique. A soil sample could be collected from a hazardous waste site known to contain benzene. The soil would be plated on a nutrient agar master plate. After 48 hours of growth, colonies could be replica-plated to another nutrient agar plate and to another plate containing a defined medium with benzene added as the sole carbon source. Only bacteria that

can metabolize benzene will grow on the benzene-containing plate.

15. A normal promoter sequence is the place where the RNA polymerase binds to start transcription of a gene (or operon). More than likely, the deleted segment will not permit attachment of the RNA polymerase, so that the gene (or operon) cannot be transcribed. It also is possible that the mutation causes a less efficient binding of the RNA polymerase to the promoter, such that the gene is transcribed less often than normal.

CHAPTER 7

1. Some ingenuity is required to explain genetic transformation as observed by Griffith in 1928. You might begin by assuming a mistake was made and guessing how Griffith eliminated this possibility. From there, you will have to let your intuition run overtime.

5. Perhaps the organism was one that had existed previously but as a harmless species. A recombination process may have provided it with the genes to assume a pathogenic mode and thus cause disease. It also might have undergone, over time, a series of mutations that conferred disease-causing abilities.

8. Although the terms *reproduction* and *recombination* are sometimes confused, it is good to remember that reproduction leads to "more" cells, while recombination leads to "different" cells.

10. The plasmid only has a restriction cut site for EcoR1. Therefore, mixing plasmid with EcoR1 will produce a linear plasmid, while mixing the plasmid with Pvu1 will cause no change.

13. (a) *Legionella pneumophila* is the causative agent of Legionnaires' disease in humans, which kills many people every year. (b) By knowing the bacterium's DNA sequence, perhaps sites sensitive to antibiotics can be identified or a vaccine can be produced.

CHAPTER 8

1. Probably the most important factor was the introduction of the vaccine in 1987. Other factors might have been a strong surveillance system, the willingness of doctors to report cases, the establishment of a case definition to ensure that reports were accurate, using tests to assess continued vaccine effectiveness, and the willingness of public health departments to spend the necessary funds to help the eradication effort. Can you suggest any other public health measures that might have been used?

3. This question assumes that a pathogen is in an advantageous condition, a situation with which you might take issue. Perhaps the virus places the bacillus

in a difficult position because pathogenicity is not always desirable, especially after the host dies. On the other hand, the virus permits active growth in the tissues, a situation not possible without the toxin.

4. This is a tough one, and I wouldn't be surprised if you didn't guess correctly. At the turn of the century, people believed that microorganisms hid in corners, so they built this hospital with rounded edges, carefully avoiding any corners. It was actually one of the first attempts at preventive medicine.

12. It would probably be a good idea to change the name to avoid confusion with the influenza virus. One choice is *Haemophilus meningitidis*. Perhaps you can come up with a more novel name.

14. Modern microbiological discoveries are the fruits of labor of many investigators, so it is unlikely that one or two will be immortalized in a common bacterial name. However, part of the scientific name may honor an investigator. For example, *Borrelia burgdorferi*, the Lyme disease spirochete, is named for Willy Burgdorfer, the investigator who isolated the spirochete from the tick in the early 1980s.

CHAPTER 9

7. The laboratory instructor intended to show how soilborne organisms contaminate food and cause the can to swell. The colleague pointed out that *Clostridium botulinum* would be cultivated from the soil and suggested students should not be exposed to this organism.

9. Chickens and other forms of poultry may be infected with *Salmonella* serotypes. Evidence of this problem is often difficult to ascertain, and poultry manufacturers may unknowingly send infected chickens off to market. If the chicken is cut up on a carving board, the *Salmonella* cells can be deposited there and picked up by the salad ingredients. Old wooden carving boards are a particular problem because of numerous cracks and fissures. Thorough cooking kills the *Salmonella* in the poultry, but the salad is eaten raw and is a source of live bacteria. In this situation, the salad should be prepared before the chicken is cut up if the same carving board is used.

10. The public health official is probably asking a bit much to expect that doctors will recognize botulism symptoms. With about 50 cases per year, botulism is quite rare in the United States, and doctors who deal with botulism will probably be seeing their first and last case. Usually there is little to distinguish botulism from paralysis-associated disease such as a stroke. When the eating history of the patient is examined, however, the possibility of botulism arises.

11. The disease was brucellosis. All recovered and returned to work. The health department also made several recommendations to prevent further outbreaks, including the use of rubber gloves, face shields, and negative air pressure on the floor.

12. My knee-jerk reaction would be to select the frozen hamburger because there would be less opportunity for staphylococci or other bacteria to grow during the past two days. However, if you cooked the hamburgers for equal amounts of time, the frozen one would absorb less heat than the fresh one, and fewer internal bacteria would be destroyed. Therefore, the fresh one might be safer.

CHAPTER 10

2. The medical upheaval was bubonic plague. The town marks the farthest reach of the vast graveyard that the suburbs of London became after the plague.

9. Good-quality boots, preferably hip length, would be a wise suggestion.

12. The numerous soilborne diseases discussed in the chapter provide ample possibilities. Certainly tetanus and gas gangrene must be prime candidates for serious disease, but the list can also include anthrax, melioidosis, and leptospirosis. Since each disease has a different pathology, you should explore how each can be deadly in warfare.

13. When leaves and debris pile at the curbside, soilborne arthropods flourish, including ticks, which commonly occur on grass and leaves. A child's chance playing in the leaves may bring it in contact with the ticks, and any of the three diseases indicated may follow.

15. Lice transmit typhus. Therefore, by reducing the lice population via head shaving, the people also reduced the opportunity for transfer of the rickettsiae of typhus. Accumulations of lice eggs are called nits. In past generations, nits were so common in the hair that people would habitually pick them out, a practice that resulted in the common expression "nit-picking."

CHAPTER 11

3. Impetigo is a skin disease often caused by staphylococci and occurring most commonly among children. Children tend to have more contact with one another during the summer months than during any other time of the year, and the skin is often unclothed at this time.

5. During douching, fluid is forcibly expelled into the vaginal tract to cleanse it. The fluid can force microorganisms in the vaginal tract up into the uterus and/or Fallopian tubes, where infection can occur. Pelvic inflammatory disease (PID) may result.

8. I can see pluses and minuses in this method. What do you think?

11. Organizations that help control the spread of leprosy point out that "leprosy" is a pejorative term and should be replaced by "Hansen's disease." This will probably take many years to accomplish, in view of the long history of leprosy. You might wish to focus on the modern methods for the detection and treatment of leprosy, and indicate that many diseases pose a greater threat to life. Thousands of leprosy patients are treated at centers throughout the United States, often on an outpatient basis.

15. The disease is dental caries. You may be surprised that caries has a disease connotation, but a close look reveals that dental caries fulfills the prerequisites for an infectious disease.

CHAPTER 12

2. The concept certainly seems to be true. Then again, are viruses "organisms" or replicating entities that possess one but not all the properties of a living thing? And is "much" of their behavior or "all" of their behavior directed toward reproduction? And is it really "reproduction" as we generally mean the term? What do you think? Can you rephrase the concept to better suit the virus?

8. For bacterial disease, it is treatment; for viral disease, it is prevention. Think of all the vaccines available for various viral diseases, and compare that number with the limited number available for bacterial diseases.

11. It's simple—"prion" sounds better than "proin."

13. Dmitri Ivanowsky did not discover viruses; he merely showed that something smaller than a bacterium was capable of causing disease. At the time it was a momentous discovery, because bacteria were believed to be the ultimate in simplicity. Ivanowsky pointed out that a *filterable* virus caused tobacco mosaic disease. His key word was *filterable*, to show that whatever the agent was, it was able to pass through a filter that trapped bacteria.

15. Reproduction has a connotation in biology that implies the generation of new individuals by asexual or sexual processes. Although new individuals are generated in viral replication, the process is neither asexual nor sexual, but a completely separate process seen nowhere else in biology. Hence, virologists prefer "replicate" to "reproduce."

CHAPTER 13

2. Here is what transpired: Twice-daily reports were collected from delegations on whether any members had measles symptoms. All visitors to medical stations were observed for measles symptoms. Letters were sent to all participants, volunteers, and staff, advising them of the situation and the control measures, signs, and symptoms. Daily telephone calls were made to all local hospital emergency rooms. State health departments of the competition participants were notified of the outbreak. Can you think of anything the epidemiologists missed? (P.S. No additional cases occurred.)

5. Children are easier to reach at 15 months for immunization, but after 20 years or so, the immunity has probably worn thin. By contrast, teenagers may be harder to reach and certify for immunization, but the level of immunity will be much higher. Additional discussion can lead from here. Avoiding common cold viruses effectively can be accomplished by several means: Keep hands away from the eyes; wash hands often; use disposable tissues, rather than handkerchiefs, to cover coughs; and avoid people with colds. You should add to this list with other imaginative aproaches.

11. The evidence points to chickenpox. (You may notice that I use "chickenpox" as one word. This is for consistency with smallpox and cowpox, both of which are conventionally written as single words.)

12. It will certainly lead to a smelly shoe. It will also stop people from coming too close to you, which will decrease your possibility of coming in contact with disease organisms. Chapter 23 has a box on garlic that you might enjoy reading.

13. The incident helps prove that the agents of disease are transmissible. Pasteur and his supporters were right on target.

CHAPTER 14

1. In 1994, 1,700 residents of New York certainly did not experience rabies. The number most likely refers to the number of people who received rabies immunizations because of a possible exposure to rabies viruses from an animal. There was, in fact, only one case of rabies in all of New York that year.

4. Care to be a part of history? Then watch the continuing battle to eradicate polio from the world. Four strategies have been employed: (a) maintaining high vaccine coverage among children with at least three doses of oral vaccine, (b) developing sensitive systems of surveillance, (c) administering supplementary doses of vaccine during National Immunization Days, (d) instituting "mopping-up" vaccine campaigns in high-risk areas where wild poliovirus is believed to exist.

9. Incidents such as these are the bane of laboratory instructors. The dangers that come to mind include AIDS and hepatitis B, but a case can be made for numerous other dangers such as staphylococcal disease. At this juncture, you might like to consider whether the pretreatment with alcohol is an effective safeguard against disease transfer.

11. Certainly, you should accept the offer. In fact if you work in any profession where blood is encountered (health-care or laboratory worker, mortician, police officer, corrections officer, firefighter), you should consider obtaining a hepatitis B immunization. Ask around. Your hospital, doctor's office, school, or other employer may offer it as a perk.

13. Disease reflects a competition between host and parasite. If the parasite overcomes the body defenses, disease ensues; but if the body defenses overcome the parasite, the latter is driven away. In each case, the parasite and defenses compete until one wins out. AIDS is dramatically different. HIV eliminates body defenses by destroying portions of the body's immune system. Left defenseless, the body is subjected to marauding parasites in the form of opportunistic microorganisms. Few other diseases take this approach.

CHAPTER 15

3. There are several possible reasons for the small number of intestinal diseases of fungal origin. Perhaps the spores are destroyed by stomach acid, intestinal nutrients for fungal growth are lacking, receptor sites for tissue attachment are not present, or the oxygen supply is limited. Discussions like this are valuable because they help you focus on the requirements for infection to take place.

7. This student has an intriguing idea. The chitin will encourage chitin-digesting bacteria to emerge, and she may then isolate them to develop a natural fungicide.

8. Lime raised the pH of Mr. A's soil and prevented fungal disease in the turf. The mushrooms in Mr. B's lawn in June are a signal that fungi can grow in an acidic environment, and the brown spots are proof. Mr B should invest in lime next year.

9. Tests were performed for various fungi. The results pointed to *Histoplasma capsulatum* and histoplasmosis. All were treated and recovered.

11. According to public health epidemiologists, the outbreak of coccidioidomycosis was most likely due to dust made airborne by the earthquake. The disease is not transmitted from person to person, but from the inhalation of airborne *Coccidioides immitis.*

CHAPTER 16

2. The British found they could improve the taste of quinine by mixing it with gin. The drink came to be their "gin and tonic."

6. It will probably depend on whether the residents have suffered from intestinal disease due to protozoa. The residents of Milwaukee, Wisconsin, would probably favor a proposal, but other taxpayers would probably reject it until they experienced such an epidemic.

7. This is quite a bold statement, but it reflects the fact that dirty diapers are as capable of transmitting infectious disease as the open sewers of generations ago. Day-care center workers should take special note.

9. As the borders of Rome expanded, Romans ventured into far-off lands where mosquitoes thrived and where the malaria parasite was prevalent. Infected Romans probably brought the disease back to Rome, and the mosquito populations of the aqueducts and pools propagated the parasite and spread the disease.

11. The disease was probably trypanosomiasis (sleeping sickness).

CHAPTER 17

2. On a global scale, diseases such as these "impede national and individual development, make fertile land inhospitable, impair intellectual and physical growth, and exact a huge cost in treatment and control programs." Solutions are straightforward: Develop new drugs, vaccines, diagnostic tests, and control methods. Can you suggest any novel approaches?

6. This question illustrates how knowing the stems of scientific words helps decipher their meaning. *Diphyllobothrium latum* is a very thin, broad tapeworm. It is associated with insanitary water, which the worm enters via human feces.

10. The thread-like body may have been an eyeworm, and the warmth of the fire may have brought it to the surface. Figure 17.14 in this chapter portrays an eyeworm in the eye.

13. Improved sanitation methods would reduce mosquito populations and decrease the possibility of transmitting *Wuchereria*. By contrast, clearing forest lands for agriculture and installing drainage ditches would encourage mosquito populations and increase the possibility for filariasis. Now it's your turn to continue the comparisons.

15. The concept of studying parasitology to appreciate the web of life is intriguing and worthy of note. Hundreds of thousands of individuals are infected with multicellular parasites. The relationship is benign in huge numbers of cases but parasitical in many others, as this chapter shows. Perhaps a geographical summary of the parasites might help you appreciate how worldwide the parasites are.

CHAPTER 18

1. Consider how many people have handled a stamp and the conditions under which it has been stored before it comes to you. Then think about whether you would want to place it in your mouth.

4. The salad bar remains one of the most common possibilities for disease transmission in the restaurant business. You should discuss how to limit microbial transmission by the chilling of foods, the selection of foods to be offered at the salad bar, the thorough washing of foods, cautions to patrons, the turnover of foods, the disinfection of salad bar areas, dust control, and other means.

7. A "tetanus shot" is a preparation of tetanus toxoid to induce the immune system to produce tetanus antitoxins. These antitoxins would protect against tetanus toxins, since tetanus spores had probably entered the wound from the soil.

12. Can you imagine how many people have handled a dollar bill before it gets into your hands? Wetting the fingers before touching the dollar bill brings whatever was on the bill into your mouth.

14. Few people realize how many bacteria are present under the fingernails. When the fingers are brought to the mouth, infection can take place. In a day-care center (or any situation), where fecal matter or blood or tissue is contacted, it is a good idea to brush under the fingernails regularly.

CHAPTER 19

2. Begin with the chemical signals that stimulate phagocytosis, and go from there. Remember that the heart and other organs are constantly "at work" but that the immune system lies at rest until it receives a signal to act. The signal can be an epitope; other signals are MHC molecules, interleukins, and antibody molecules.

6. Here is another intriguing possibility for microbiology research. The obvious advantage is freedom from tooth decay, but many problems must be solved. Is *S. mutans* the only agent of caries? Could the immune system be stimulated to produce enough IgA for protection? Would the IgA be delivered to and survive in the oral cavity? Would new organisms emerge as caries agents? Could IgA be used in a mouthwash or toothpaste? See MicroFocus 11.5 for more details on such vaccines.

7. The cockroach does appear to have an immune system, and the system apparently is a factor in its ability to survive. Far from being immunologically primitive, as suspected, the arthropods are able to produce antibody-like proteins. The next steps would be to identify and characterize these substances. How would you continue the experiments?

11. It's not necessarily unusual. How many times does an organism emit signals that stimulate the "wrong" response.

15. You are encouraged to exercise imagination on your trip and consider the perils of phagocytosis, mechanical barriers, lysozyme, T lymphocytes, and other protective measures in the body. An appreciation for resistance mechanisms should evolve from this discussion.

CHAPTER 20

1. In herd immunity, the immunizing agent can be spread in numerous ways. For example, day-care workers changing the diaper of a child recently immunized against polio may pick up the viruses on their fingertips and inoculate themselves and other children they contact. Body secretions, such as saliva and respiratory droplets, may spread measles, mumps, or rubella viruses from children who recently received the MMR vaccine. These examples may serve as a starting point for the discussion.

4. The approach would seem favorable. In effect, the woman is being used as an "immunological factory" to produce antibodies for her newborn as well as herself. I cannot think of any negative effects other than the normal hazard of using a vaccine.

5. It would seem to be a good idea, but it will take lots of money to implement. And the willingness for the American taxpayer to part with the funds will depend in part on how serious the threat of disease is perceived to be. You should continue the debate from here.

7. With some insight and imagination, a case can be made for each type of immunity as being safest to obtain. Similarly, reasons may be offered for each type as being most helpful. Discussions such as these put immunity into perspective and help you make choices supported by what you have learned. Ultimately, I would think that the individual situation would dictate the choice.

14. Keeping a body alive for vaccine production is certainly possible, but the ethical implications must be considered. Strong arguments can be made pro and con. You may wish to set up a debate panel on a topic such as this one.

CHAPTER 21

3. Formaldehyde used for preservation of the animals is probably causing a contact dermatitis. The sensitivity began to develop during the first two exposures, and now, during the third exposure, it is manifesting itself. Rubber gloves would be helpful for protecting the hands.

7. The "he" is William J. Clinton; the condition is hay fever.

8. Athlete's foot usually occurs between the toes, while contact dermatitis occurs where the foot comes in contact with the shoe. Laboratory testing will reveal fungal cells if the condition is athlete's foot, but no cells if it is dermatitis. A patch test can be performed by the allergist to determine if a contact allergy exists.

11. The major histocompatibility complex codes for histocompatibility molecules, which exist on the cell surface. Over 50 versions of each gene have already been discovered, and the number of gene combinations and histocompatibility molecules is enormous. Immunologically, there is probably no one else like you.

15. It is interesting to question whether the immune system is actually protecting the body during immune disorders. Allergies can be interpreted as protective mechanisms for ridding the body of antigens, and the theory can be extended to other types of hypersensitivity, as well as to transplants and tumors. Autoimmune diseases stand in stark contrast because the body appears to be attacking itself. An oxymoron is two terms that do not fit together (the dictionary definition is "two mutually exclusive juxtaposed words"). "Immune disorder" appears to be an oxymoron because *immune* means "free of" and a disorder is a problem. Therefore, how can you have a problem that you don't have? (For another oxymoron, see the MicroFocus on the enormous bacterium in Chapter 3.)

CHAPTER 22

3. A suspicious person might inquire what will happen within 30 days. Will spontaneous generation take place? Is it possible that the contents were sterilized at the manufacturing plant, but that the porous container is now permitting airborne microorganisms to enter? If so, then the product was once sterilized but is now contaminated, and evidence of contamination will appear by the expiration date.

4. Broiling of meat is done at extremely high heat. This heat would kill any surface organisms rapidly, and the meat would be safe for consumption.

10. In the laboratory, the Bunsen burner is used to sterilize loops, needles, and the tips of tubes and flasks. Some imagination might be needed to identify other uses.

12. Milk's taste apparently changes as the temperature becomes too high. Also, the protein will coagulate. The butterfat is another problem, because it will not pass through a filter. Recall that milk is a suspension rather than a true solution. A fruitful discussion could take place on the economic implications of a useful sterilization method for milk.

14. The autoclave is undoubtedly the first word to come to mind when we think of "sterilizer." However, the autoclave is merely a pressurized steam apparatus that can be used to achieve sterilization only under specific conditions. Unless these conditions are established in the machine, sterilization is not assured. Therefore, it would be incorrect to believe that "autoclave" and "sterilizer" are synonymous words.

CHAPTER 23

1. In selecting a laboratory disinfectant or sanitizing agent, it would be advisable to ask whether the agent was bactericidal or bacteriostatic under the desired conditions. Inquiry should also be made about its toxicity, solubility in water, shelf life, use in diluted form, penetrating ability, corrosiveness, temperature and pH of use, and cost. Many other avenues of inquiry are noted in the chapter.

4. Chlorhexidine is often used by dentists during treatments, but this chemical has not yet made its way into toothpastes. The upside is that it may reduce periodontal problems by reducing bacterial populations. The downside is that over a long period of time, the chemical may prove toxic to the tissues. Can you add to the list?

7. Some ways of reducing the spread of microorganisms in day-care centers are the disinfection of tabletops used for diaper changes, regular furniture and toy disinfection, and hand washing and the use of an antiseptic after each diaper change. You should suggest other methods that are practical and useful.

9. *Bacteriostatic* is an adjective derived from the noun *bacteriostasis*. The adjective means essentially the same as the noun coined by Churchman; it applies to an agent that inhibits the further growth of bacteria without necessarily destroying them.

13. This advertisement was a nightmare. Some questions that occurred to me: Why wipe down or dunk a raw chicken? It's the inside that's a problem, not the outside—the cooking heat will take care of the outside. Why clean a raw pepper with Clorox? The skin is so smooth that ordinary water will probably wash away anything that's a problem. Why worry about the outsides of carrots? I always thought you peeled carrots before cooking them. Clean a red onion? Gimme a break! Exactly how big is a sink full? And why worry about being so precise about the amount of Clorox (⅛ cup) when you have no idea of the sink size? Can you spot the other problems? How about ". . . germs and bacteria"; "salmonella . . ."; "*like* chicken . . ."?

CHAPTER 24

1. Pasteur and Joubert were probably observing the effects of an antibiotic produced by the contaminating microorganisms. It is clear that the concept of antibiosis was known over 100 years ago. You might enjoy speculating whether Pasteur and Joubert envisioned a therapeutic compound in the mixture of bacteria.

4. Sounds like an interesting idea. What do you think?

5. The arrival of antibiotics and modern medical practices in Nepal typifies how medical advances can disrupt the lives of a population. With more mouths to feed, forests had to be cleared, and great pressure was placed in preparing the land for crops. Living space soon became a premium, and natural resources, such as water supplies, were tapped to their limit. Greater populations also meant greater sanitation problems and, consequently, more opportunity for the spread of disease. Discussions such as these help us understand the negative aspects of medical advances.

8. Here are some origins of antibiotic names: Erythromycin is named after *Streptomyces erythraeus*, a red-pigmented actinomycete that produces it; tetracycline is named for the four benzene rings found in the molecule; and nystatin is named for New York

State because it was discovered at the Department of Health laboratory in New York. The use of a dictionary, along with some imagination, should help reveal the origins of other antibiotic names in this chapter.

15. The antibiotic issue is one that can be argued for hours. Perhaps this might be a good question for a debate panel. Side effects, the emergence of antibiotic-resistant bacteria, the effect on human population growth, and human dependence on antibiotics would stimulate the anti-antibiotic side. The misery and death from disease, rising food costs, job impact in the pharmaceutical industry, and generally better quality of life might be discussed by the pro-antibiotic side.

CHAPTER 25

3. The correct sequence would be to prepare the salad before the chicken. *Salmonella* serotypes may be present in the chicken, and cross-contamination to the salad may take place. Cooking will eliminate *Salmonella* in chicken, but salad is eaten raw and is potentially dangerous. You should explore other precautions. A thorough cleaning of the knife and cutting board is a good place to begin.

5. The sour cream was preserved primarily by the acid present, and no symptoms of food poisoning or any other intestinal illness developed in the following days. The incident points up that the expiration date is merely a statistical guess on when the product will look or taste bad. It is not necessarily the date on which the product will begin posing a peril to health.

10. Some examples to ponder: Custard egg products, ice cream, fresh mayonnaise, meat loaf mixtures, many sauces, prepared cheese products, and on and on.

13. The liver would probably be the better choice because it will spoil more rapidly than the steak. Liver is an organ meat, with a looser tissue consistency and richer blood supply than muscle tissue. Contamination is therefore more probable in the liver.

17. Individuals who pick fruits and vegetables in the fields should be vaccinated against any diseases they could transmit. Toilet facilities should be made available to them, and they should be discouraged from working if they are ill. The water used to wash the produce should be purified.

CHAPTER 26

1. This idea of connecting storm and sanitary sewers was a poor one because it resulted in a huge body of water where storm runoff polluted sanitary waste, and vice versa. Separation probably would have

made handling the waste considerably easier. However, you might present arguments for connecting the systems, the most obvious argument being the economic benefit.

5. Summertime pollution can be traced to many factors in addition to those cited: Animals are more active during the summer, and they tend to drink from and defecate in the water; summer thunderstorms cause substantial soil runoff of organisms from animal remains; bathers introduce intestinal organisms to the water; algae invade the water, inducing fish kills and bacterial growth during decomposition; and the heat, especially in a stagnant bay, leads to bacterial proliferation.

9. The presence of high coliform counts indicated that the water had been contaminated with sewage and that typhoid fever, bacterial dysentery, amoebiasis, hepatitis, or other serious diseases were imminent. Certainly the danger was equivalent to that posed by Legionnaires' disease.

12. The grass is greener over the septic tank because the organic-rich water flowing from it provides a natural lawn fertilizer. In the wintertime, the hot water from a cesspool or septic tank warms the soil and melts the snow over it. To locate a cesspool or septic tank in climates that have snow, one need only watch to see where the snow melts first.

13. The carbon, sulfur, and nitrogen cycles provide irrefutable evidence of the roles that microorganisms play in enhancing the quality of life. It is instructive to try to imagine a world where no microorganisms are available to recycle these elements.

CHAPTER 27

1. By the time you reach Chapter 27, you should have a firm understanding of how microorganisms contribute to the quality of life. Examples abound in this chapter. Combined with examples from Chapters 25 and 26, they are a formidable list that should help you see the positive roles of microorganisms in our lives.

3. You should explore a variety of industrial processes and identify useful end products. Some examples might be waste from meat-packing plants and petroleum processes, wood by-products from paper manufacturing, residue from food-canning operations, and sewage-treatment end products.

9. This is an intriguing question similar to that proposed for lactobacilli and streptococci in yogurt (Chapter 25). You might speculate about what factors might be necessary to establish pathogenicity in the yeast. The question also illustrates how we consume

microorganisms regularly without a thought of dangerous consequences.

10. Why so much vitamin? Perhaps a defect in the enzyme system prevents the metabolism of the vitamin to the next step. Perhaps the organisms live in a symbiotic or synergistic relationship with other organisms that utilize the vitamin. As Pearlman suggests in the Note to the Student, "There are no stupid microorganisms."

15. The list of microbial products should be formidable and should include food, drink, and numerous industial products.

Appendix D: Answers to Review Questions

CHAPTER 1

```
M  C  G  .  .  .  .  F  L  E  A  .  .  .  R
V  I  R  U  S  E  S  .  .  D  .  R  .  .  U  .  S
A  B  L  .  .  A  N  I  M  A  L  C  U  L  E  S
S  W  A  N  .  .  P  .  P  .  N  .  .  E  .  T
M  .  E  .  R  O  U  X  .  H  .  C  O  L  O  N  Y
A  .  R  E  D  I  .  C  E  .  T  .  E  .  .  S
T  .  .  D  .  C  E  .  H  .  P  .  .  T
W  E  L  C  H  .  K  .  S  E  M  M  E  L  W  E  I  S
O  .  .  A  .  E  .  R  .  B  A  .  C
.  S  M  I  T  H  .  S  O  I  L  .  R  .  T  K
F  .  P  .  T  .  A  .  A  .  W  I  N  E
R  .  A  .  S  H  I  G  A  .  .  N  .  R  O  D  S
A  .  L  .  .  R  .  B  R  U  C  E
C  E  L  L  .  P  .  B  .  O  .  S  .  R
N  O  .  A  .  V  A  N  L  E  E  U  W  E  N  H  O  E  K
S  U  N  .  E  .  S  .  .  .  O  .  S  O
T  .  Z  .  Y  .  T  .  .  O  .  .  W  .  S  C
O  A  G  E  .  E  H  R  L  I  C  H  .  .  .  H
R  .  .  N  .  U  .  .  .  L  .  A
T  O  X  I  N  .  R  E  E  D  .  T  E  T  A  N  U  S
```

CHAPTER 2

1. IONIC 2. ORGANIC 3. HYDROGEN
4. GLUCOSE 5. FATS 6. ENZYMES
7. NUCLEOTIDES 8. PRIMARY 9. DEHYDRA-
TION 10. ACIDITY 11. ELEMENT 12. ISOTOPES
13. CARBOXYL 14. MALTOSE 15. HYDROGEN

CHAPTER 3

1. G 2. T 3. K 4. L 5. X 6. U 7. N 8. W 9. D 10. A
11. V 12. C 13. P 14. F 15. O 16. I 17. B 18. J 19. Z
20. Q

CHAPTER 4

```
.  B  .  D  B  S  .  V  .  D  N  A
S  A  R  C  I  N  A  .  S  I  Z  E  .  .  S  A  L  I  V  A
L  .  .  P  .  C  .  .  B  .  X  .  .  I
I  .  F  L  U  I  D  .  R  .  T  R  A  N  S  P  O  R  T
M  .  F  O  .  L  .  I  .  R  .  .  I
E  .  O  .  G  L  Y  C  O  C  A  L  Y  X  .  D  .  .  D
.  .  U  .  U  .  .  N  .  .  .  .  I
C  H  R  O  M  O  S  O  M  E  .  D  N  A  .  F  .  S
.  .  U  .  .  A  .  .  .  N  .  C  L  E
P  E  P  T  I  D  O  G  L  Y  C  A  N  .  O  A  .  A
E  .  A  .  .  N  .  A  .  .  C  G  .  S
N  .  N  M  .  .  E  .  S  P  I  R  O  C  H  E  T  E
P  I  L  U  S  .  F  T  .  .  S  .  O  U  .  L
C  .  .  I  .  O  .  .  U  .  D  S  .  L
R  I  B  O  S  O  M  E  S  .  M  L  .  P  .  .  U
L  .  P  .  B  .  O  .  .  E  .  P  L  A  S  M  I  D
G  L  U  E  .  G  R  A  M  .  S  .  .  A  .  E  .  Y
I  .  U  .  .  I  .  E  .  K  .  W  .  Q  .  M  O  V  E
O  N  E  .  .  N  A  G  .  .  I  .  B  .  U
.  T  W  O  .  .  .  N  U  C  L  E  O  I  D
```

CHAPTER 5

1. metabolism; digestion; catabolism 2. protein;
speed up; substrate 3. glycolysis; energy; ATP
4. fermentation; oxygen; glucose; alcohol 5. electrons;
cytochromes; energy; ATP 6. pyruvate; carbon;
carbon dioxide; NAD 7. amino; amino; deamination
8. carbon dioxide; photosynthesis; glucose 9. chemical reactions; carbohydrates; *Nitrosomonas*
10. fatty acids; beta oxidation; two-carbon units;
cellular respiration

CHAPTER 6

1a. —ATGGCGATAGTTAAACCCGGAGGGTGA—
1b. 8 1c. 8; met-ala-ile-val-lys-pro-gly-gly 1d. 27
2a. DNA; thymine (T) bases are found only in DNA.
2b. met-cys-tyr-gln-asn-phe-asn-ala; silent
2c. met-cys-tyr-gln; nonsense

CHAPTER 7

1. PLASMIDS 2. PALINDROME 3. COMPETENCE
4. BACTERIOPHAGE 5. ENDONUCLEASE
6. PNEUMOCOCCUS 7. PILI 8. LYSOGENY
9. LIGASE 10. INTERFERON 11. GRIFFITH
12. CONJUGATION 13. VIRULENT 14. GENOME
15. PROPHAGE

CHAPTER 8

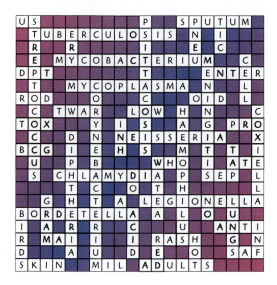

CHAPTER 9

1. ANTITOXIN 2. CHOLERA 3. UNDULANT
4. DYSENTERY 5. ROSE 6. STAPHYLOCOCCAL
7. CHICKENS 8. CARRIERS 9. NEGATIVE
10. CEREUS 11. CAMPYLOBACTER 12. ABOR-
TION 13. PARALYSIS 14. NOSE 15. ACID
16. FLUID 17. RARE 18. VIBRIO 19. SALMO-
NELLA 20. HEAT 21. ANAEROBIC 22. STOOL
23. SEAFOOD 24. MILK 25. ANIMALS

CHAPTER 10

1. G 2. B 3. K 4. L 5. O 6. N 7. C 8. J 9. M 10. G
11. A 12. N 13. L 14. F 15. I 16. A 17. G 18. E
19. M 20. C 21. L 22. F 23. B 24. M 25. G

CHAPTER 11

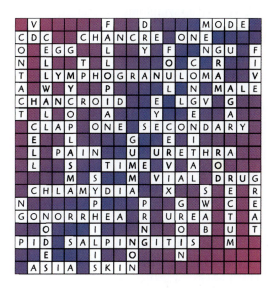

CHAPTER 12

1. CAPSID 2. HELIX 3. BACTERIOPHAGE
4. ANTIBODIES 5. NEGRI 6. ACYCLOVIR
7. INTERFERON 8. ATTENUATED 9. VIROID
10. TUMOR 11. ULTRAVIOLET 12. ONCOGENE
13. RECEPTOR 14. ICOSAHEDRON 15. GENOME
16. ENDERS 17. VIRION 18. LYSOGENY
19. AMANTADINE 20. ENVELOPE 21. STANLEY
22. FORMALDEHYDE 23. PRION 24. CAP-
SOMERES 25. INACTIVATED

CHAPTER 13

1. RNA; icosahedral; nose; mild 2. contact; blisters;
emotional stress; acyclovir 3. in crops; teardrops;
shingles; red; painful 4. measles; a skin rash; airborne
droplets 5. Reye's; testes; orchitis; brain; SSPE
6. measles; Koplik spots; mouth; red rash 7. fifth,
B19; parvovirus; DNA virus 8. airborne droplets;
mumps; salivary 9. smallpox; cowpox; Jenner
10. influenza; vaccine; bacteria 11. RNA; lungs; children; clump together; syncytia 12. coronavirus;
enveloped; person-to-person 13. DNA; inclusions;
common colds; eye; meninges 14. icosahedral; million; thin; weeks; stress 15. infectious; transplacental passage; newborns; rubella; herpes simplex
16. attenuated; children; long-term; measles;
mumps; German measles

CHAPTER 14

1. F (RNA) 2. F (no symptoms) 3. True
4. True 5. F (intestine) 6. True 7. F (rabies) 8. True
9. F (T lymphocyte) 10. True 11. True 12. F (gastroenteritis) 13. F (eyes, retina) 14. True 15. True
16. F (New York City) 17. F (polio) 18. F (retrovirus)
19. F (Coxsackie) 20. True 21. F (liver) 22. F (inactivated viruses) 23. F (rotavirus) 24. True 25. True
26. F (infectious mononucleosis) 27. F (high) 28. F
(hantaviruses) 29. F (ctyomegalovirus) 30. F (arm or
shoulder muscle)

CHAPTER 15

1. I 2. Q 3. W 4. M 5. K 6. U 7. A 8. S 9. C 10. E
11. P 12. S 13. U 14. G 15. O 16. B 17. P 18. H
19. M 20. S 21. Z 22. M 23. K 24. P 25. N

CHAPTER 16

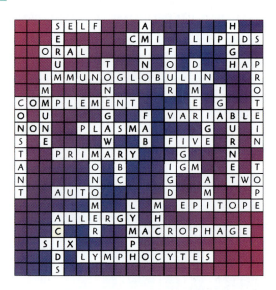

CHAPTER 17

1. a 2. c 3. a, c 4. c 5. a, b, c 6. b 7. a, c 8. a 9. b
10. a, b, c 11. a 12. a, b 13. c 14. b, c 15. c

CHAPTER 18

1. F (endemic) 2. True 3. True 4. F (infection)
5. F (liver) 6. True 7. F (low) 8. F (symptom)
9. F (mechanical) 10. True 11. F (carrier) 12. F
(antitoxins) 13. True 14. F (coagulase) 15. True
16. True 17. F (direct) 18. True 19. F (acute)
20. F (incubation)

CHAPTER 19

CHAPTER 20

1. innate, acquired 2. active; passive 3. measles;
mumps; rubella 4. diphtheria; pertussis, tetanus
5. natural; artificial 6. alum; mineral oil 7. hyperimmune serum; antiserum 8. IgG; IgM; IgA, IgE
9. transplacental passage; breast feeding 10. hive-like
rash; labored breathing; swollen joints 11. capsule
polysaccharides, pilus protein 12. hepatitis B; foot-and-mouth disease 13. genetic factors; physiological
factors; biochemical factors 14. inactivated; attenuated
15. diphtheria; tetanus 16. injection; oral consumption; nasal spray 17. gastrointestinal; respiratory 18.
hepatitis A; chickenpox 19. therapy; prophylaxis 20.
tetanus; diphtheria

CHAPTER 21

1. HISTAMINE 2. DIGEORGE 3. INDURATION
4. JOINTS 5. GRAVES 6. CYTOTOXIC 7. ATOPIC
8. ANTIGEN 9. PLASMA 10. TUBERCULOSIS
11. BASOPHILS 12. LYSIS 13. POSITIVE
14. KIDNEY 15. MHC 16. RASH

CHAPTER 22

1. AUTOCLAVE 2. MEMBRANE 3. POWDER
4. DENATURATION 5. TYNDALL 6. DRYING
7. ULTRASONIC 8. GAMMA 9. SPORE
10. OXIDATION 11. PRESSURE 12. THIRTY
13. HOLDING 14. INSTRUMENTS 15. DIATOMS
16. ULTRAVIOLET 17. PLASTIC 18. TOXINS

19. SOLIDS 20. OSMOSIS 21. MICROWAVES
22. BACILLUS 23. BUNSEN 24. CLEANING
25. TUBERCULOSIS

CHAPTER 23

1. F 2. B 3. J 4. I 5. C 6. D 7. A 8. K 9. N 10. E
11. A 12. C 13. J 14. E 15. I 16. C 17. J 18. I 19. B
20. E 21. K 22. F 23. K 24. F 25. C

CHAPTER 24

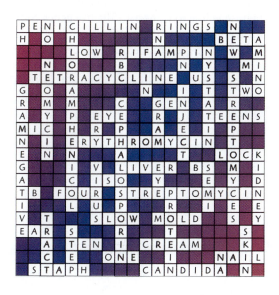

CHAPTER 25

1. SALMONELLA 2. LACTOBACILLUS
3. CLOSTRIDIUM 4. SERRATIA 5. CHLAMYDIA
6. PROTEUS 7. STREPTOCOCCUS 8. BACILLUS
9. ACETOBACTER 10. LEUCONOSTOC 11. CLAV-
ICEPS 12. CLOSTRIDIUM 13. PSEUDOMONAS
14. ENTEROBACTER 15. ASPERGILLUS

CHAPTER 26

1. a, c 2. b 3. b, c 4. a, b 5. a, b, c 6. a, b 7. b 8. a 9. b,
c 10. a, b 11. a, b, c 12. a, c 13. c 14. b, c 15. b

CHAPTER 27

1. G 2. N 3. H 4. M 5. K 6. O 7. J 8. D 9. P 10. A
11. B 12. E 13. B 14. R 15. G 16. B 17. C 18. D
19. S 20. A

Appendix E: MicroInquiry Answers

MICROINQUIRY 1

Self-explanatory.

MICROINQUIRY 2

2.1a. Uracil is the base specific to DNA and thymine is the base specific to DNA.

2.1b. Radioactively label uracil with isotope of one element (say ^{3}H) and radioactively label thymine with the isotope of another element (^{14}C).

2.1c. You cannot use sulfur (^{35}S) to label nucleic acids because this element is only found in a few amino acids of proteins.

2.1d. You should have concluded that the chemical does affect growing bacterial cells. Specifically, it affects protein synthesis because as time passes, the radioactive label in proteins levels off as shown in the figure. This means that proteins are no longer being made and, therefore, the ^{35}S cannot be incorporated. DNA and RNA are not affected since there is continued incorporation of radioactive isotope into new DNA and RNA molecules.

MICROINQUIRY 3

Self-explanatory.

MICROINQUIRY 4

4.1. Self-explanatory.

4.2. Self-explanatory.

4.3. Only *Staphylococcus* colonies will appear on the agar plate. However, only *S. aureus* colonies would have a yellow halo surrounding the colonies, indicating the cells fermented mannitol and produced acid. Other nonfermenting species of *Staphylococcus* produce no yellow halos—the agar remains red.

MICROINQUIRY 5

Self-explanatory.

MICROINQUIRY 6

6.1. When tryptophan is absent from the growth medium, the repressor protein of the *trp* operon fails to bind to the operator. The RNA polymerase is free to transcribe the five genes coding for tryptophan synthesis. When tryptophan is added to the growth medium, the cells preferentially use what is available rather than expend energy to make their own amino acids. So, in the cell tryptophan binds to the repressor protein, which can now bind to the operator and block transcription by the RNA polymerase (Figure B). In this example, tryptophan is said to be a co-repressor because it is needed to repress or "turn off" gene transcription.

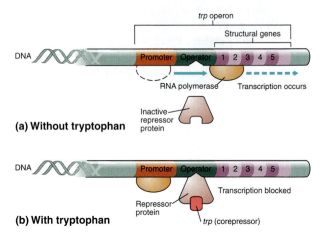

MICROINQUIRY 7

7.1. Bacterial colonies without a plasmid and those containing the recombinant plasmid will not grow on (a) because they lack tetracycline resistance genes. Only the bacterial cells with intact plasmids (*tet*^{R}) grow. So, we have identified those clones that lack the recombinant plasmid. Bacterial colonies lacking plasmids will not grow on medium (b) because they are sensitive to ampicillin. Colonies containing the recombinant plasmid and intact plasmid will grow because they contain the *amp*^{R} gene. Colonies with recombinant (insulin-containing) plasmids are identified because they grow on ampicillin-containing but not on tetracycline-containing growth media.

MICROINQUIRY 8

8.1a.   Often a tubercle is visible in a chest X-ray.

8.1b.   Identification of *M. tuberculosis* by acid-fast staining from sputum usually is sufficient for a diagnosis of TB because these bacteria are not normally found in respiratory secretions.

8.1c.   Skin becomes thick with a raised red welt developing within 48 to 72 hours.

8.1d.   A tuberculin skin test does not necessarily mean the person has tuberculosis. It simply indicates the person has been exposed to *Mycobacterium tuberculosis*, been immunized recently, or had a previous tuberculin skin test.

8.1e.   Tuberculosis; *Mycobacterium tuberculosis*.

81f.   Most likely from his former roommate.

8.1g.   Tuberculosis is more virulent in HIV-infected patients because their immune system is compromised from fighting the HIV infection.

8.1h.   The patient was on INH for an extended period because of his HIV infection. With a lowered immune system, the patient's immune system will not mount a strong response to the TB infection. Drugs, such as INH, are the best hope for control and recovery.

8.2a.   *Corynebacterium diphtheriae*.

8.2b.   Dead tissue accumulates with mucus, white blood cells, and fibrous material (pseudomembrane) in the throat. Respiratory blockage can result, which can lead to death.

8.2c.   Having the proper vaccinations, in this case the DTaP vaccine.

8.2d.   Treatment requires both antibiotics for the bacteria and antitoxins to neutralize the diphtheria toxins.

8.3a.   *Streptococcus pneumoniae*.

8.3b.   The patient's underlying medical conditions (heavy smoker and alcoholic) place him at high risk for the disease.

8.3c.   Vaccination with the pneumococcal 23-valent vaccine.

8.3d.   Polysaccharide capsule and toxin production.

8.3e.   Alpha-hemolytic.

MICROINQUIRY 9

9.1a.   *Campylobacter jejuni*.

9.1b.   Drinking raw milk or contaminated water; consumption of contaminated food, especially chicken.

9.1c.   Foods, especially chicken, contaminated during processing or preparation.

9.1d.   Guillain-Barré syndrome where immune system attacks nerves.

9.1e.   Erythromycin.

9.2a.   *Salmonella typhi*, *Vibrio cholerae*, and *Escherichia coli*.

9.2b.   All gram-positive bacteria.

9.2c.   Typhoid fever.

9.2d.   Contaminated food or water; raw milk.

9.2e.   Don't drink raw milk; drink purified water; eat cooked foods.

9.3a.   *Listeria monocytogenes*.

9.3b.   One of the forms of listeriosis is listeric meningitis, which is characterized by headache, stiff neck, and delirium.

9.3c.   Penicillin or tetracycline.

9.3d.   They should avoid foods (deli meats, soft cheeses) that potentially could be contaminated with the bacterium. People with a strong immune system can fight off the bacterium.

9.3e.   Psychrotrophic refers to the ability of the bacterium to grow at refrigerator temperatures. Therefore, refrigeration does not stop growth of *Listeria* in contaminated refrigerated foods, like feta cheese.

9.4a.   *Escherichia coli* O157:H7.

9.4b.   Bloody diarrhea and hamburger meat that had been in the refrigerator for "some time."

9.4c.   Other *E. coli* strains because they lack the bloody diarrhea symptom.

9.4d.   It can lead to complications called hemorrhagic colitis and hemolytic uremic syndrome.

9.4e   On MacConkey agar, *Escherichia coli* O157:H7 produces white colonies while other strains produce red or pink colonies.

MICROINQUIRY 10

10.1a.   Headache, rash on thighs, fever, hiking near Seattle.

10.1b.   Lyme disease; *Borrelia burgdorferi*.

10.1c. Tick bite.

10.1d. Without treatment, infection can lead to an early disseminated stage characterized by meningitis, facial palsy, and peripheral nerve disorders. Joint and muscle pain also can occur.

10.1e. Wear protective clothing and avoid tick-infested areas. If bitten by a tick, remove whole tick and wash the skin wound with soap and water, and apply antiseptic.

10.2a. *Clostridium tetani*; tetanus.

10.2b. Sedatives and muscle relaxants; penicillin and tetanus antitoxin.

10.2c. No, because symptoms develop rapidly and the patient could die waiting for lab results.

10.2d. If the spores entered via a foot wound, they will germinate and growth will produce dead tissue.

10.2e. See 10.2d.

10.3a. *Rickettsia rickettsii* and *Ehrlichia chaffeensis*.

10.3b. *R. rickettsii*. A pink rash on palms and feet; tick bite; geography (South Carolina).

10.3c. Lyme disease, relapsing fever, tularemia, and ehrlichiosis.

10.3d. Weil-Felix test.

10.3e. *E. chaffeensis* does not usually cause a rash and it is associated with a lowered WBC count, not a raised count.

10.4a. Gas gangrene; *Clostridium perfringens*.

10.4b. Anaerobic conditions, as the bacterium is an anaerobe.

10.4c. Debridement.

10.4d. Use of a hyperbaric chamber, which introduces oxygen gas into necrotic tissues. Oxygen gas will kill *C. perfringens* cells.

MICROINQUIRY 11

11.1a. *E. coli*, *Proteus mirabilis*, *Enterobacter faecalis*, *Klebsiella pneumoniae*, *Pseudomonas aeruginosa*, *Staphylococcus aureus*.

11.1b. Due to anatomical differences. The female urethra is relatively short; proximity of urethra to anus.

11.1c. Infections of the urethra, bladder, and possibly the kidneys.

11.1d. Avoiding tight-fitting clothes and urinating as soon after intercourse as possible.

11.1e. Biofilms are a key factor. Bacteria are protected from immune attack by a slimy coating. The coating also makes antibiotic therapy difficult since the drug does not penetrate the biofilm.

11.2a. *Chlamydia trachomatis* and *Neisseria gonorrhoeae*.

11.2b. *C. trachomatis* is an obligate pathogen and only grows in tissue culture. The reproductive cycle is described on page 383.

11.2c. Laboratory tests include the use of fluorescent antibodies on a cervical swab or an immunoassay test.

11.2d. A *C. trachomatis* infection can be spread to others through sexual intercourse.

11.2e. Recovery from a gonorrhea infection does not generate lifelong immunity because of the weak immune response to the bacterium.

11.3a. *Pasteurella multocida*.

11.3b. Cat bite, swollen and painful wrist, tenderness at the site, small puncture wound, and small abscess. Gram-negative rods.

11.3c. The pharynx.

11.3d. Limiting contact with the animal and washing the site thoroughly.

11.4a. Gonorrhea; *Neisseria gonorrhoeae*.

11.4b. Gonorrhea can be spread by sexual intercourse. Complications in females include salpingitis and pelvic inflammatory disease.

11.4c. HIV, syphilis, and chlamydial infections, because they all are sexually transmitted diseases.

11.4d. Although these are additional signs of gonorrhea in males, the infection has not affected these tissues.

11.4e. Ceftriaxone or cefixime.

MICROINQUIRY 12

12.1a. 10 hours. During this time, the viral nucleic acid is being replicated and new capsid parts are being synthesized.

12.1b. 15 hours. During this time, the viral nucleic acid is being replicated, new capsid parts are being synthesized, and nucleic acid and capsids are being assembled into new virions (maturation stage).

12.1c. 10^8 virions are released from 10^5 cells in culture. Therefore, the burst size is 10^3 virions per infected cell.

12.1d. The low number represents virus particles that were not washed away after the 60 minute viral inoculation.

MICROINQUIRY 13

13.1a. Antigenic drift. The new H1N2 strain appears to have resulted from the reassortment of the genes of the currently circulating influenza A(H1N1) and A(H3N2) subtypes.

13.1b. The hemagglutinin protein of the A(H1N2) virus is similar to that of the currently circulating A(H1N1) viruses, and the neuraminidase protein is similar to that of the current A(H3N2) viruses.

13.1c. Because the current influenza vaccine contains strains with both H1 and N2 proteins similar to those in the new strain, the current vaccine should provide good protection against the new A(H1N2) virus. No unusual levels of disease have been associated with this virus and, at this time, it is uncertain if the A(H1N2) virus will persist and circulate widely.

13.2a. Antigenic shift. Reports indicated that H7N7 infections had occurred among pigs and humans in the Netherlands, and among birds in Belgium and Germany.

13.2b. While it is unusual for people to get influenza infections directly from animals, sporadic human infections and limited outbreaks caused by avian influenza A viruses, including H7N7, have been reported. When such infections are identified, public health authorities monitor the situation closely.

13.2c. Because influenza H7N7 viruses do not commonly infect humans, there is little or no antibody protection against these viruses in the human population. If an avian or other animal influenza virus is able to infect people, cause illness, and spread efficiently from person to person, an influenza pandemic could begin. The vast majority (79) of these people had conjunctivitis, and six of those with conjunctivitis reported influenza-like illness symptoms (e.g., fever, cough, muscle aches).

MICROINQUIRY 14

14.1a. Hepatitis C; hepatitis C virus.

14.1b. History of intravenous drug abuse, nausea, vomiting, fever, headache, abdominal pain, and slight jaundice. Also, negative HBsAg and serum antibodies to HAV.

14.1c. This is indicative of liver damage, which would be the infection site for the virus.

14.1d. Being a drug user, contaminated needles are the probable source.

14.1e. Most likely hepatitis B and HIV, both of which can be transmitted by contaminated needles.

14.2a. Infectious mononucleosis; Epstein-Barr virus.

14.2b. Some of the symptoms (sore throat, fever) could have been due to a strep infection.

14.2c. The elevated B lymphocyte count is because the Epstein-Barr virus infects B lymphocytes and the immune system is attempting to replace the damaged cells.

14.2d. Heart defects, paralysis of the face, and rupture of the spleen.

14.2e. Burkitt's lymphoma. The malarial parasite may stimulate tumor development.

14.3a. Hepatitis A and E.

14.3b. Hepatitis A because of the jaundice, short incubation period, initial symptoms, and travel history.

14.3c. A transmission route could be from contaminated water or food eaten at local restaurants.

14.3d. Hepatitis A antibodies in serum.

14.3e. Administering hepatitis A immune globulin.

14.4a. Dengue fever.

14.4b. Clues included fever, backache, headache, bone and joint pain, and eye pain. Her trip to Bangladesh (where Dengue is prevalent) and mosquito bites also were clues.

14.4c. Four strains of the virus are known.

14.4d. She has a viral disease for which antibiotics are useless.

14.4e. Returning to Southeast Asia puts her at risk of being infected by another strain of dengue virus, which could lead to a severe or deadly illness called dengue hemorrhagic fever.

MICROINQUIRY 15

Historically, fungi were thought to be related to the plants. By analyzing the characteristics supplied in this inquiry, fungi share many more characteristics with animals than they do with plants. This suggests both probably evolved from a common protistan ancestor in the distant past.

MICROINQUIRY 16

16.1. From the information given, the most likely vehicle of transmission of *Giardia lamblia* was the taco ingredients. Although all of the ill persons ate the commercially prepared salsa, salsa was unlikely to have transmitted *Giardia* cysts because the cysts would not remain viable after the pasteurization and canning processes.

Two explanations for the contamination are possible. First, if the water system was contaminated and the lettuce and tomatoes could have been contaminated when washed. Because the lettuce, tomatoes, and onions were all cut on the same board, cross-contamination could have occurred. However, because plumbing changes and outlet flushing were made before completion of the epidemiologic investigation, this hypothesis could not be tested.

A second explanation is that the woman who prepared the lettuce and tomatoes was infected and excreting *Giardia* cysts that contaminated the vegetables during preparation. However, this mode is less likely because this woman had acute onset of diarrhea 10 days after the meal, suggesting a new infection at that time.

Therefore, the case remains unsolved, but more than likely, the water system was contaminated. *Giardia* normally is transmitted via contaminated water.

MICROINQUIRY 17

Self-explanatory.

MICROINQUIRY 18

18.1a. Approximately 110,000 cases in 1993 and 45,000 cases in 2001.

18.1b. Prevalence: 1993 = 0.09%; 2001 = 0.13%.

18.2a. Continuous common source epidemic. Plateau indicates person-to-person transmission.

18.2b. Possibly three plateaus, suggesting spread within different communities or groups of individuals.

18.3a. Highest reported cases (≥15) in Northeast states and upper Midwest. Fewer cases (1–14) in Eastern and Pacific states.

18.3b. Perhaps the infected host is newly introduced in this area. Another possibility is the area recently has become an area that is frequently visited by hikers or outdoors people. Depending on how the data are reported, the cases could have been the result of visiting another endemic area in the United States, but not reported until returning home.

18.4a. Highest number of cases in infants less than 1 year old, with higher incidence in 10–14 and 15–19 age groups.

18.4b. Vaccinations would eliminate the high numbers of infant cases, and booster shots would reduce the cases in all age groups.

MICROINQUIRY 19

19.1a. The alternate complement pathway can be stimulated by pathogen surfaces while the classical pathway is stimulated by antigen-antibody complexes. Membrane attack complexes would be of significance only with nonencapsulated bacteria since these complement proteins assemble into membrane complexes on the cell membrane of the bacterium.

19.1b. The alternate pathway is activated spontaneously by the bacterial surface and can assist in an elevated phagocytic response by phagocytes. Membrane attack complexes can stimulate the destruction of nonencapsulated bacteria.

19.2a. Without a thymus, T cells cannot mature or function in an immune response. Without helper T cells and cytotoxic T cells, cell-mediated immunity would not exist. Some B cells could mature if T-independent antigens are present.

19.2b. HAV infects liver cells, which could not be eliminated without cytotoxic T-cell activity.

19.3a. C3 is critical for all three of the complement's effector functions: inflammatory response, opsonization and phagocytosis, and membrane attack complex formation.

19.3b. A C3 deficiency cannot trigger any of the functions mentioned above, so the patient will experience more bacterial infections.

19.4a. It could be due to a genetic abnormality that prevents movement out of the blood or migration of the cells from the blood to the skin.

19.4b. Neutrophils are a type of granulocyte and are critical to the second line of defense against pathogen infection. Being first at the infection site, they help control and eliminate the infection. Antibodies do not enter a localized infection site and cytotoxic T cells eliminate infected cells.

19.5a. Lymphoid progenitors are the lineage of white blood cells that give rise to the lymphocytes, including T and B cells. Without these cells and an effective cell-mediated and antibody-mediated immunity, the individual will be susceptible to many infectious diseases.

19.5b. In reticular dysgenesis, no white cells or platelets are produced, so any trauma or infection will be lethal to the fetus.

19.6a. T-cell proliferation is dependent on stimulation by IL-2. Without IL-2 receptors, T cells cannot become activated.

19.6b. Based on 19.6a, there is no clonal expansion of the T-cell populations that would be specific to any viral, bacterial, or fungal infection.

MICROINQUIRY 20

20.1. ELISA is a serological test for the presence of antibodies or antigens in serum. In your case, Pat, the hepatitis C kit will assay for the presence or absence of hepatitis C antibodies in the blood sample you sent.

20.2. To make sure all the reagents are functioning properly, positive and negative controls are run along with your blood sample. A positive control contains antibodies specific to hepatitis C antigen so a positive color reaction will be produced. A negative control lacks hepatitis C antibodies, so no color reaction will develop.

20.3. If your blood sample tests positive, it means you are seropositive. That is, your blood contains anti-HCV antibodies.

20.4. A *seronegative* result would mean you are "probably" not infected with HCV. However, you still should be retested in six months because sometimes it can take that long for the immune system to produce enough antibodies to be detected by ELISA.

20.5. Do another HCV test in six months. Even if that one is positive, lab errors or some other infection may mimic an HCV antigen. If that ELISA test proves positive, another type of test, called a *Western blot*, is run that detects the actual presence of HCV in your blood. If all tests come back positive, then I am afraid you are infected with the microbe because when ELISA is used with the Western blot, the results are more than 99.9 percent accurate.

MICROINQUIRY 21

21.1. Cytotoxic T cells and natural killer (NK) cells.

21.2. After recognizing and binding to virus-infected cells by the class I MHC/peptide complex on the infected cell surfaces, the cytotoxic T cells release granzymes and perforins that kill and destroy the infected cells.

21.3. T cells will "see" the antigens as foreign and will trigger cell-mediated and antibody-mediated responses.

21.4. Being a privileged site, self antigens never were presented to immature T cells. Therefore, the herpes infection may have produced some type of tissue trauma that allowed immune system access to the eye, resulting in the activation of T cells.

21.5. After years of normal wear and tear on the heart and heart valves, self antigens were exposed that were recognized as foreign (mimic *S. pyogenes* antigens). Damage caused by chronic immune system attack eventually produced mitral valve failure.

21.6. If class II MHC molecules were inappropriately expressed on these cells, then there is a possibility that helper T cells will be activated to peptides derived from the thyroid cells and presented with the class II MHC molecules. Helper T-cell activation would activate cytotoxic T cells, and B-cell maturation would produce plasma cells secreting the autoantibodies.

MICROINQUIRY 22

22.1. A = 90 percent kill every 4 minutes, so in 24 minutes there theoretically would be one cell remaining alive. Another 4 minutes should kill the last cell, so 28 minutes would be required. Likewise, B = 70 minutes and C = 126 minutes, both one time interval beyond the point where one cell remains alive.

22.2. A = 4 min.; B = 10 min.; C = 18 min.

22.3. 70°C.

22.4. 108 minutes.

MICROINQUIRY 23

23.1. A physical method is the only way and UV irradiation would be the method of choice to ensure that mail and parcels were not damaged.

23.2. Chemical disinfectants (bleach, phenol solutions) could be used for open surface areas. However, due to the large size of the facilities a physical gas would be more efficient since machinery and furniture also are contaminated with spores. The best and safest chemical gas would be chlorine dioxide.

23.3a. See the answer for question 1.

23.3b. Thoroughly wash skin with soap and water. Wash clothes in normal fashion, using bleach if possible. Wipe down all physical surfaces with bleach solution.

MICROINQUIRY 24

24.1a. Self-explanatory.

24.1b. The drug appears most effective against *S. aureus* as seen by the large rings where growth was inhibited. The drug looks promising against *S. aureus*.

24.1c. The drug looks good at 10 µg but not at 100 µg. Need to do intermediate concentration tests.

24.1d. The control was to make sure there wasn't something in the solution used to dissolve the drug that could cause an adverse cellular effect. Drug testing should continue.

24.2. A concentration somewhere between 100 mg and 1,000 mg was toxic to the animals. Need to test intermediate concentrations to narrow down the toxicity range.

24.3. Self-explanatory.

24.4. There are pros and cons to this decision. For example, without a control group, it might not be possible to accurately evaluate the efficacy of the drug. On the other hand, if early drug trials show obvious benefit, the control group could be given the drug. Discuss with classmates.

MICROINQUIRY 25

25.1. **Your refrigerator should be kept at 5° C (41° F).** Although this temperature will not kill most microbes, it will slow their growth, making it less likely you might spread or could get sick from food stored there.

25.2. Freezing also does not kill all microbes. So, foods like frozen chicken should not be set on the counter to defrost. You risk illness as bacteria grow rapidly at room temperature. **Microwaving the food following package directions or thawing it in the refrigerator will keep any microbes from growing.** Incidentally, that kitchen sponge you used to clean up spilled food juices should be rinsed and sanitized by microwaving it 20 seconds.

25.3. Raw foods can be contaminated with several disease-causing microbes that can cling to a cutting board. So, **clean that board with soap and hot water and then sanitize it with a mild bleach solution** (or put it in the dishwasher).

25.4. **Wash your hands with soap and warm water for about 20 seconds** so you don't deposit harmful bacteria on other surfaces. Also, keep tabby away from kitchen counters and where food is being prepared. Pets carry and spread diseases to foods.

25.5. Your sink drain and garbage disposal can harbor several species of viruses and bacteria. Some sinks can contain more bacteria than in a flushed toilet. So, **every week you should sanitize your drain: pour a solution of 1 teaspoon of chlorine bleach in 1 quart of water down the drain.**

25.6. In the oven, the high temperatures used in cooking kill microbes. However, **cooled leftovers should be refrigerated within two hours after cooking.** Once refrigerated, they are safe to eat for three to five days. Read discussion on page 929.

25.7. Did you clean your kitchen counters after preparing raw food? **Hot, soapy water and a dilute chlorine bleach are recommended.** Hot water and soap alone do not get rid of all possible bacteria.

25.8. Meats should be cooked until there is no red seen and the juices run clear. **Well-done meats that reach 160°F kill foodborne microbes.** Meats and other improperly cooked foods can retain foodborne microbes.

25.9. Most kitchen sponges contain some bacteria that can make people sick. In one study, 20 percent of sponges and dishrags collected from many of the 1,000 kitchens tested in five American cities contained *Salmonella* bacteria, which can cause food poisoning, typhoid fever, and gastrointestinal diseases. **So, microwave that damp kitchen sponge for 20 seconds and wash dishrags.** Also, change to a new sponge often or use a germ-resistant sponge.

25.10. Now that dinner is over, what to do with the dishes? Don't soak them for several hours because the soaking water becomes "nutrient broth" for bacteria. **Either wash them in the dishwasher and air-dry them, or wash them within two hours in hot, soapy water and let them air-dry.**

MICROINQUIRY 26

26.1. Provides a richer growth medium, ensuring the growth of coliforms, if present.

26.2. Do not want to introduce any contaminating bacteria. It is unlikely that pipettes would contain coliforms.

26.3. Sample 4 appears to contain high numbers of coliforms. For this sample, the presumptive test read 5–4–5, indicating an MPN of 430 per 100 ml (confidence limits of 150–1,100 organisms).

26.4. Double-check on the presumptive test results.

26.5. Sample 4 and possibly sample 6 (MPN 14).

26.6. Close off the area of the lake where sample 4 was collected until further testing indicates the water again is safe. Also, investigate the cause for the high coliform numbers.

MICROINQUIRY 27

27.1a. Bacteria are a possible answer. Identify or genetically engineer bacterial species that can break down the thick crude into lighter forms.

27.1b. They are bacteria that are known to produce enzymes that break down sulfur compounds into water-soluble products while other bacteria can remove heavy metal contaminants. Since each oil field is slightly different in terms of the heavy crude, different "bacterial cocktails" may be needed. In some cases, this probably would require genetic modifications to give the microbes the precise genes to deal with crude-oil digestion. However, if the right mix of bacteria can be assembled, the result could be a lighter, cleaner crude for refining.

27.2. Chemolithotrophic ("rock-eating") bacteria represent groups that survive by using inorganic compounds (minerals) in rocks as their source of nutrients and energy. Bacteria associated with iron pyrite break the minerals into acidic solutions of iron that then can be used to dissolve out usable forms of copper. Also, using natural bacteria or genetically engineered ones, it may become possible to convert low-grade ores into a sustainable yield of metals for the future.

27.3. Contaminated soil was excavated and only mineral nutrients had to be added for growth since indigenous microorganisms already existed in the soil. Such biostimulation was performed from December 1989 through September 1991, and approximately 16,000 tons of soil were treated. Through bioremediation, PAH concentrations were reduced by 70 percent.

27.4. From January 1992 through November 1993, approximately 300,000 tons of sediment and subsoil from the lagoon were treated using the indigenous microorganisms that were biostimulated with added oxygen and growth nutrients. The bioremediation process achieved the specified soil cleanup goals for the contaminants within 11 months of treatment. For example, benzene levels were reduced from 608.0 mg/kg to 4.4 mg/kg.

Glossary

This glossary contains concise definitions for the major microbiological terms and concepts only. If a term is not present, please refer to the index.

A

abscess A circumscribed pus-filled lesion characteristic of staphylococcal skin disease.

abyssal zone The environment at the bottom of oceanic trenches.

acid-fast technique A process in which certain bacteria resist decolorization with acid alcohol after staining with a primary dye.

acidophile A microorganism that grows at acidic pHs.

acquired immune deficiency syndrome (AIDS) A serious viral disease caused by human immunodeficiency virus (HIV) in which the T lymphocytes are destroyed and opportunistic illnesses occur in the patient.

acme period The phase of a disease during which specific symptoms occur and the disease is at its height.

actinomycete A soil bacterium that exhibits fungus-like properties when cultivated in the laboratory.

actinomycosis A fungus-like bacterial disease caused by *Actinomyces israelii*, characterized by draining sinuses of the face and other organs.

acute disease A disease that develops rapidly, exhibits substantial symptoms, and then comes to a climax.

acute necrotizing ulcerative gingivitis (ANUG) A bacterial infection of the mucous membranes in the oral cavity; sometimes called Vincent's angina or trench mouth.

acyclovir A drug used as a topical ointment to treat herpes simplex and injected for herpes encephalitis.

adenosine triphosphate (ATP) A molecule that carries energy for use in chemical reactions; 1 mole of ATP liberates 7300 calories of energy when digested to adenosine diphosphate (ADP) and phosphate ions.

adenovirus An icosahedral DNA virus involved in respiratory infections, viral meningitis, and viral conjunctivitis.

adhesin A protein in bacterial pili that assists in attachment to the surface molecules of cells; a binding molecule that encourages parasites to hold firmly to the tissues of their host.

adjuvant A substance attached to a vaccine component that increases the efficiency of a vaccine.

aerotolerant A bacterium not inhibited by O_2.

aflatoxin A toxin produced by *Aspergillus flavus* that may pass from contaminated grain, milk, or meat products to humans and may induce tumors.

agar A derivative of marine seaweed used as a solidifying agent in many microbiological media.

agglutination A type of antigen-antibody reaction that results in visible clumps of organisms or other material.

agranulocytosis The destruction of neutrophils (granulocytes) resulting from the reaction of antibodies with antigens on the neutrophil surface; a type II hypersensitivity reaction.

AIDS *See* acquired immune deficiency syndrome.

alga An organism in the kingom Protista that performs photosynthesis; algae differ structurally from mosses, ferns, and seed plants.

allergen An antigenic substance that stimulates an allergic reaction in the body.

alloantigen A type of antigen that exists in certain but not all members of a given species; examples are the A, B, and Rh antigens in humans.

allograft A tissue graft between two members of the same species, such as between two humans.

alpha-hemolytic streptococci Streptococci that partially destroy red blood cells; when cultivated in blood agar, an olive green color forms around colonies of these streptococci.

amantadine A synthetic drug thought to block the penetration of influenza viruses to host cells.

Ames test A diagnostic procedure used to detect potential cancer agents by their ability to cause mutations in *Salmonella* cells.

aminoglycosides A group of antibiotics that contain amino groups bonded to carbohydrate groups; examples are gentamicin, streptomycin, and neomycin.

amoebiasis A protozoal disease of the intestine caused by *Entamoeba histolytica*, transmitted by food and water; characterized by intestinal ulcers and the involvement of multiple internal organs.

amoxicillin A semisynthetic penicillin antibiotic related to ampicillin.

amphitrichous The presence of flagella at the opposite poles of a bacterial cell.

amphotericin B An antifungal drug used to treat serious fungal diseases, such as cryptococcosis, histoplasmosis, and blastomycosis.

ampicillin A semisynthetic penicillin derivative active against gram-positive bacteria and certain gram-negative bacteria.

anabolism A chemical process involving the synthesis of organic compounds; usually an energy-requiring process.

anaerobe Referring to an organism that does not require free oxygen.

analog A compound closely related to a naturally occurring compound.

anaphylatoxin A chemical substance that initiates a series of events leading to smooth muscle contraction.

anaphylaxis A life-threatening allergic reaction in which a series of mediators cause contractions of smooth muscle throughout the body.

anionic detergent A detergent that yields negatively charged ions in solution.

anthrax A serious bacterial disease caused by *Bacillus anthracis*, characterized by severe blood hemorrhaging. Also called woolsorter's disease.

antibiotic An antimicrobial substance produced by molds and bacteria that inhibit other microorganisms.

antibody A highly specific protein molecule produced by plasma cells in the immune system in response to a specific chemical substance; antibodies function in antibody-mediated immunity.

antibody-mediated immunity (AMI) A type of immunity arising from the activity of antibodies directed against antigens in the tissues. Also called humoral immunity.

anticodon A three-base sequence on the tRNA molecule that binds to the codon on the mRNA molecule during translation.

antigen Any chemical substance that elicits a response by the body's immune system.

antigenic determinant A section of an antigen molecule that stimulates a specific response and to which the response is directed; often consists of several amino acids or monosaccharides. Also called epitope.

antigenic variation A process in which chemical changes occur periodically in antigens; known to take place in influenza viruses.

antiglobulin antibody An antibody that reacts with a human antibody.

antiseptic A chemical used to kill pathogenic microorganisms on a living object, such as the surface of the human body.

antiserum Serum that is rich in a particular type or types of antibody.

antitoxin An antibody that circulates in the bloodstream and provides protection against toxins by neutralizing them.

aplastic anemia A side effect of chloramphenicol therapy in which red blood cells are produced with little or no hemoglobin.

apicomplexan A protozoan containing a number of organelles at one end of the cell used for host penetration; no motion is observed in adult forms of the organisms. Also called sporozoa.

arboviral encephalitis A type of encephalitis caused by a viral agent, transmitted by an arthropod.

Archaea The domain of living things that excludes the Eubacteria and Eukarya.

archaebacteria A group of bacteria believed to be of ancient origin.

arsphenamine An arsenic-phenol compound synthesized by Ehrlich and Hata for use against syphilis spirochetes; the first modern chemotherapeutic agent.

Arthropoda A large phylum of animals having jointed appendages and a segmented body; includes insects, such as lice, mosquitoes, and fleas, and spiderlike arachnids, such as ticks and mites.

arthrospore A fungal spore formed by fragmentation of the hypha.

artificially acquired active immunity Immunity resulting from the production of antibodies in response to antigens in a vaccine or toxoid.

artificially acquired passive immunity Immunity resulting from an exposure to or injection of antibodies.

ascomycete A member of the phylum Ascomycota.

Ascomycota A phylum of fungi whose members have septate hyphae and form ascospores within saclike asci, among other notable characteristics.

ascospore A sexually produced spore formed by members of the fungal phylum Ascomycota.

ascus A saclike structure that contains ascospores; formed by the ascomycetes group of fungi.

aseptic Free of microorganisms.

aseptic meningitis A type of meningitis in which no bacterium or other agent can be readily identified; usually refers to meningitis caused by a virus.

aspergilloma A dense, round ball of mycelium often occurring in the lungs and commonly caused by *Aspergillus fumigatus*.

asthma A period of wheezing and stressed breathing resulting from a type I hypersensitivity reaction taking place in the respiratory tract.

asymptomatic Without symptoms.

athlete's foot A fungal disease of the webs of the toes caused by various species of fungi.

atopic disease A type I hypersensitivity reaction, characterized by limited production of IgE and a localized reaction in the body. Also called common allergy.

attenuated virus A weakened variant of a virus that emerges during successive transfers of the virus in tissue cultures; used in immunizations because of its reduced virulence. Sometimes called live virus.

autoantigens A person's own proteins and other organic compounds that elicit a specific response in the body.

autoclave A laboratory instrument that sterilizes microbiological materials by means of steam under pressure.

autograft Tissue taken from one part of the body and grafted to another.

autoimmune disease A disease in which antibodies react with an individual's own chemical substances and cells.

autotroph An organism that uses CO_2 as a carbon source.

auxotroph An organism that lacks a nutritional need. *See* **prototroph**.

avirulent organism An organism that is normally without pathogenicity.

axial filament A microscopic fiber located along cell walls in certain species of spirochetes; contractions of the filaments yield undulating motion in the cell.

azathioprine An antimitotic drug used to treat cancer.

azithromycin A macrolide antibiotic that inhibits protein synthesis in Gram-negative and Gram-positive bacteria.

 B

B19 disease An alternate name for fifth disease.

babesiosis A tickborne protozoal disease of the red blood cells caused by *Babesia microti*, characterized by periods of high fever, anemia, and blood clotting.

bacille Calmette Guérin (BCG) A strain of attenuated *Mycobacterium bovis* used for immunization against tuberculosis and, on occasion, leprosy.

bacillus A bacterial rod.

bacitracin An antibiotic derived from a *Bacillus* species, effective against Gram-positive bacteria when used topically.

bacterial dysentery An alternate name for shigellosis.

bactericidal agent An agent that kills bacteria.

bacteriochlorophyll A pigment located in bacterial membrane systems that upon excitement by light, loses electrons and initiates photosynthetic reactions.

bacteriocins A group of bacterial proteins toxic to other bacteria.

bacteriophage A type of virus that attacks and replicates within bacteria.

bacteriorhodopsin A photosynthetic pigment found in the archaea.

bacteriostatic agent An agent that prevents the multiplication of bacteria without killing them.

baculovirus A virus with a high affinity for lepidopteran tissues; used as a carrier for genes that encode pesticides.

balantidiasis A waterborne and foodborne protozoal disease of the intestine caused by *Balantidium* coli, characterized by mild to severe diarrhea.

Bang's disease An alternate name for brucellosis.

barophilic microorganism A microorganism that lives under conditions of high pressure.

bartonellosis A relatively mild sandfly-transmitted disease of the blood, caused by *Bartonella bacilliformis*, characterized by fever and a skin rash.

basal body A structure at the base of a bacterial flagellum having a central rod and set of enclosing rings.

basidiomycete A member of the phylum Basidiomycota.

Basidiomycota A phylum of fungi whose members have septate hyphae and form basidiospores on supportive basidia, among other notable characteristics.

basidiospore A sexually produced spore formed by members of the basidiomycetes group of fungi.

basidium A clublike structure that contains basidiospores; formed by the basidiomycetes group of fungi.

basophil A type of leukocyte that functions in allergic reactions and, possibly, other reactions not yet established.

bejel A mild syphilislike disease occurring in remote parts of the world, caused by a species of *Treponema*.

benthic zone The environment of the ocean floor.

Bergey's Manual (of Systemic Bacteriology) The official manual of bacteriology that lists the names and characteristics of the known bacteria and presents a classification scheme for these organisms.

beta-hemolytic streptococci Streptococci that destroy red blood cells completely; when cultivated in blood agar, a clearing forms around the colonies of the streptococci.

beta-lactam nucleus The chemical group central to all penicillin antibiotics.

beta-lactamase An alternative name for penicillinase, the enzyme that converts penicillin to penicilloic acid.

bilharziasis An alternate name for schistosomiasis.

binary fission An asexual process by which a cell divides to form two new cells; specifics of the process vary among organisms.

binomial system The system of nomenclature that uses the genus and specific epithet to refer to organisms.

biochemical oxygen demand (BOD) A number referring to the amount of oxygen utilized by the microorganisms in a sample of water during a 5-day period of incubation.

biofilm A complex community of bacteria attached to a surface, such as a catheter or industrial pipeline.

bioinformatics The use of computers and statistical techniques to manage biological information.

biological vector An infected living organism that transmits disease agents.

bioremediation The use of naturally occurring microorganisms to degrade toxic wastes and other synthetic products of industry.

biotechnology The commercial application of genetic engineering using living organisms.

bipolar staining A characteristic of *Yersinia* and *Francisella* species in which stain gathers at the poles of the cells, yielding the appearance of safety pins.

bisphenol A combination of two phenol molecules used in disinfection.

blackwater fever An alternate name for malaria.

blastomycosis A fungal disease caused by Blastomyces dermatiditis, characterized by infection of the lungs and other systemic organs. Also called Gilchrist's disease.

blastospore A fungal spore formed by budding.

B lymphocyte A lymphocyte formed in the fetal bone marrow in humans; responsible for antibody-mediated immunity. Also called B cell.

boil A raised pus-filled lesion.

botulism A foodborne disease of the nervous system characterized by paralysis.

boutonneuse fever A tickborne rickettsial disease due to *Rickettsia conori*, characterized by fever and a skin rash.

bovine spongiform encephalopathy (BSE) A prion-caused disease of cows in which infected brain tissue becomes soft and spongy; accompanied by neurological defects. Also called mad cow disease.

bradykinin A peptide that functions in type I hypersensitivity reactions by contracting smooth muscles.

breakbone fever A term applied to dengue fever, reflecting sensations that the bones are breaking.

broad-spectrum antibiotic An antibiotic useful for treating many groups of microorganisms, including gram-positive and gram-negative bacteria, rickettsiae, fungi, and protozoa. *See also* narrow-spectrum antibiotic.

bronchopneumonia A disease characterized by scattered patches of pneumonia, especially in the bronchial tree.

brucellosis A foodborne bacterial disease of the blood caused by *Brucella* species and accompanied in humans by blood involvement and undulating fever; in large animals, a disease of the reproductive organs. Also known as Malta fever, Bang's disease, and undulant fever; called contagious abortion in animals.

Bruton's agammaglobulinemia An immune deficiency disease in which the body fails to produce plasma cells from B lymphocytes.

bubo A swelling of the lymph nodes.

budding An asexual process of reproduction in fungi, in which a new cell forms at the border of the parent cell and then breaks free to live independently.

Burkitt's lymphoma A type of lymphoid cancer occurring in the connective tissues of the jaw.

C

calicivirus An RNA virus appearing as a six-pointed star with abundant calcium in the capsid; caliciviruses include the Norwalk-elke viruses.

campylobacteriosis A foodborne and waterborne bacterial disease of the intestine caused by *Campylobacter jejuni*, characterized by diarrhea.

cancer A disease characterized by the radiating spread of cells that reproduce at an uncontrolled rate.

candidiasis An infection due to *Candida albicans* that manifests as a yeast infection of the intestine, the vaginal tract, or the mucous membranes of the mouth (thrush).

capsid The surrounding layer of protein that encloses the genome of a virus.

capsomere A protein subunit of the capsid on the surface of a virus; the number of capsomeres varies among different viruses.

capsule A layer of polysaccharides and small proteins that adheres to the surface of certain bacteria; serves as a buffer between the cell and its environment.

carbenicillin A semisynthetic penicillin antibiotic used to treat urinary tract infections caused by certain Gram-negative bacteria.

carbuncle An enlarged abscess formed from the union of several smaller abscesses.

carcinogen A cancer-causing substance.

cardiolipin An alcoholic extract of beef heart used as an antigen in the VDRL test.

carrier An individual who has recovered from a disease but retains live organisms in the body and continues to shed them.

Carrion's disease An alternate name for bartonellosis.

casein A mixture of three long chains of amino acids; the major protein in milk.

catabolism A chemical process in which organic compounds are digested; usually an energy-liberating process.

cationic detergent A detergent that yields positively charged ions in solution.

cat-scratch disease An infectious disease of the lymph channels accompanied by lymph node swellings. Also called cat-scratch fever.

cavitation The formation and implosion of bubbles in a liquid in which ultrasonic vibrations are propagated.

cefoxitin A semisynthetic derivative of a *Streptomyces* species used as a penicillin substitute where penicillin resistance is encountered.

cell-mediated immunity (CMI) Immunity arising from the activity of cytotoxic T-lymphocytes on or near the body cells. Also called tissue immunity.

cell membrane A thin bilayer of phospholipids and proteins that surrounds the bacterial cell cytoplasm.

cell wall A carbohydrate-containing structure surrounding fungal, algal, and most bacterial cells.

cephalosporin A group of antibiotics derived from the mold *Cephalosporium* and used against Gram-positive bacteria and certain Gram-negative bacteria.

cercaria A tadpolelike intermediary stage in the life cycle of a fluke.

cestode A type of flatworm, commonly known as a tapeworm.

Chagas' disease An alternate name for American trypano-somiasis.

chancre A circular, purplish hard ulcer with a raised margin that occurs during primary syphilis.

chancroid A sexually transmitted bacterial disease of the external genital organs caused by *Haemophilus ducreyi*, characterized by soft, spreading chancres and swollen lymph nodes. Also called soft chancre.

Chediak-Higashi syndrome An immune disorder characterized by delayed killing of phagocytized microorganisms.

chemically defined medium A medium in which the nature and quantity of each component are identified. Also called synthetic medium.

chemoautotroph An organism that derives energy from chemical reactions and uses the energy to synthesize nutrients from carbon dioxide.

chemoheterotroph An organism that derives energy from chemical reactions and uses the energy to synthesize nutrients from carbon compounds other than carbon dioxide.

chemotactic factor A lymphokine that draws phagocytes to the antigen site in cell-mediated immunity.

chemotaxis A movement toward a chemical attractant.

chemotherapeutic agent A synthetic or semisynthetic chemical compound that is inhibitory to microorganisms and is used in the body for therapeutic purposes; includes antibiotics synthetically manufactured.

chickenpox A communicable skin disease caused by a DNA icosahedral virus, characterized by teardrop-shaped, highly infectious, skin lesions. Also called varicella.

chitin A polymer of acetylglucosamine units found in the cell walls of fungi that provides rigidity.

chlamydia A sexually transmitted bacterial disease of the urethra and reproductive organs caused by *Chlamydia trachomatis*; accompanied by urinary tract symptoms and possible pelvic inflammatory disease, leading to sterility.

chlamydiae A group of very small bacteria visible only with the electron microscope and cultivated within living tissue medium.

chlamydial ophthalmia An infection of the eye tissues caused by *Chlamydia trachomatis*; may occur in newborns from exposure to chlamydiae during birth.

chlamydial pneumonia An influenzalike disease of the lungs caused by *Chlamydia pneumoniae*.

chloramine A chlorine derivative formed by adding a chlorine to an amino group on a carrier molecule.

chloramphenicol A broad-spectrum antibiotic derived from a *Streptomyces* species, used to treat typhoid fever and various other diseases; interferes with protein synthesis.

chlorhexidine A bisphenol compound widely used as an antiseptic and disinfectant.

chlorophyll A pigmented molecule that functions in photosynthesis; exists free in the cytoplasm of prokaryotes and within the chloroplasts of eukaryotes.

chloroquine A synthetic drug used to treat malaria in the clinical phase.

chocolate agar A bacteriological medium consisting of a nutritious base and whole blood; the medium is heated to disrupt the blood cells and release the hemoglobin.

cholera A foodborne and waterborne bacterial disease of the intestine caused by *Vibrio cholerae*, characterized by massive diarrhea, fluid and electrolyte imbalance, and severe dehydration.

chronic disease A disease that develops slowly, tends to linger for a long time, and requires a long convalescence.

cilium A hairlike appendage on some eukaryotic cells; cilia assist motion in certain protozoa and are used for filtering air by respiratory epithelial cells in humans.

Ciliophora A group of protozoa whose members move by means of cilia.

clarithromycin A macrolide antibiotic that inhibits protein synthesis in Gram-negative and Gram-positive bacteria.

clindamycin An antibiotic used as a penicillin substitute and for certain anaerobic bacterial diseases.

clinical disease A disease in which the symptoms are apparent.

clonal selection theory A theory that shows how certain lymphocytes are selected out from the mixed population of B lymphocytes when stimulated by processed antigens.

clone A collection, or colony, of identical cells arising from a single cell.

clostridial myonecrosis An alternate name for gas gangrene, referring to the death of muscle cells following an invasion of certain *Clostridium* species.

clotrimazole An imidazole drug used in the treatment of fungal diseases, such as candidiasis.

coagulase An enzyme that catalyzes the formation of a fibrin clot; produced by virulent staphylococci.

coccidioidomycosis A fungal disease of the lungs caused by *Coccidioides immitis*, characterized by cough, malaise, and other respiratory symptoms.

coccus A spherical bacterium.

codon A three-base sequence on the mRNA molecule that specifies a particular amino acid in the protein molecule.

coenocytic Referring to a fungus containing no septa (crosswalls) and multinucleate hyphae.

coenzyme An organic molecule that forms the nonprotein part of an enzyme molecule.

cold agglutinin screening test (CAST) A laboratory procedure in which *Mycoplasma* antibodies agglutinate human red blood cells at cold temperatures.

cold sore A herpes-induced blister that may occur on the lips, gums, nose, and adjacent areas.

coliform bacteria Gram-negative nonsporeforming bacilli usually found in the human and animal intestine; they ferment lactose to acid and gas.

colony A visible mass of microorganisms of one type.

Colorado tick fever A tickborne disease caused by an RNA virus, occurring in the western United States; characterized by high fever and joint pains.

colostrum The first milk secreted from the mammary glands of animals or humans.

commensalism A close and permanent association between two populations of organisms in which only one population benefits.

common allergy An allergic reaction taking place in a localized area of the body. Also called atopic disease.

communicable disease A disease that is transmissible among various hosts.

comparative genomics The comparison of DNA sequences between organisms.

competent cell A bacterium that can take up DNA in the recombination process of transformation.

complement A group of proteins that functions in a cascading series of reactions during the response by the body to certain antigens; the complement cascade is stimulated by antigen-antibody activity.

congenital rubella syndrome A condition in which rubella viruses pass across the placenta of an infected woman and cause damage in the fetus.

conidium An asexually produced fungal spore formed on a supportive structure without an enclosing sac.

conidiophore The supportive structure on which conidia form.

conjugation A type of bacterial recombination in which genetic material passes from a live donor cell into a live recipient cell during a period of contact.

conjunctivitis A general term for disease of the conjunctiva, the thin mucous membrane that covers the cornea and forms the inner eyelid. Also called pinkeye.

contact dermatitis A type IV hypersensitivity in which the immune system responds to allergens such as clothing materials, metals, and insecticides; the reaction is usually characterized by an induration.

contagious abortion An alternate name for brucellosis in animals.

contagious disease A communicable disease whose agent passes with particular ease among hosts.

continuous flow technique An industrial process in which medium is continually added to a fermentation tank to replace that which has been used.

convalescent serum Antibody-rich serum obtained from a convalescing patient.

Coombs test An antibody test used to detect rh antibodies involved in hemolytic disease of the newborn.

copepod A small aquatic arthropod that serves as an intermediary host for the fish tapeworm.

covalent bond A chemical bond created by the sharing of electrons between atoms or molecules.

cowpox A mild viral disease of the skin tissues that occurs in humans and bovine species. Also called vaccinia.

Coxsackie virus An RNA virus of the entero-virus group transmitted by food and water and involved in intestinal disease; also, the cause of disease of the heart muscle (myocarditis) and the chest wall (pleurodynia).

cresol A derivative of phenol that contains one or more methyl groups on the benzene ring.

croup A collective name for upper respiratory tract infections often caused by adenoviruses, characterized by hoarse coughing.

cryptococcosis A fungal disease of the lungs and spinal cord that occurs as an opportunistic disease in immune-compromised individuals, such as AIDS patients.

cryptosporidiosis A waterborne protozoal disease of the intestine caused by *Cryptosporidium coccidi* and *C. parvum*, characterized by intense diarrhea; often occurs in immune-compromised individuals, such as AIDS patients.

cyanobacterium A pigmented bacterium occurring in unicellular and filamentous forms; formerly called blue-green alga.

cyclosporiasis A protozoal disease of the intestine, characterized by vomiting, nausea, and watery diarrhea.

cyclosporin A A drug that suppresses cell-mediated immunity and encourages the body's acceptance of transplanted tissue.

cyst A dormant and very resistant form of a microorganism, such as a protozoan or multicellular parasite.

cystic fibrosis A genetic disease characterized by a buildup of sticky mucus in the respiratory tract and often accompanied by bacterial infection.

cytokine A glycoprotein secreted by helper and other types of T lymphocytes; stimulates the activity of immune system cells. Also known as lymphokine.

cytomegalovirus (CMV) An icosahedral DNA virus that causes infected cells to enlarge.

cytomegalovirus (CMV) disease A viral disease of multiple organs accompanied by nonspecific symptoms an malaise; transmissible to the fetus of a pregnant woman; an opportunistic disease in AIDS patients.

cytotoxic hypersensitivity A cell-damaging or cell-destroying hypersensitivity that develops when IgG reacts with antigens on the surfaces of cells.

D

dalton A unit of weight equal to the mass of one hydrogen atom; used to measure molecular weights.

dalfopristin A streptogramin antibiotic used with quinupristin to inhibit reproduction in *Staphylococcus aureus*.

dander Particles of animal skin, hair, or feathers that may contain materials that cause allergic reactions.

dapsone A chemotherapeutic agent used to treat leprosy patients.

Darling's disease An alternative name for histoplasmosis.

deamination A biochemical process in which amino groups are enzymatically removed from amino acids and hydroxyl groups are substituted, thereby allowing the carbon skeleton to be used for energy purposes.

death phase The final portion of a bacterial growth curve in which environmental factors overwhelm the population and induce death. Also called the decline phase.

degerm To mechanically remove organisms from a surface.

dehydration synthesis A process of bonding two molecules together by removing the products of water and joining the open bonds.

delavirdine An antiviral drug that binds to and inhibits reverse transcriptase; used against HIV in AIDS patients.

denaturation A process in which proteins change from the tertiary structure to the secondary structure; heat and certain chemicals may induce denaturation.

dendritic cell A cell having long fingerlike extensions, found within all tissues; they phagocytize infected cells.

dengue fever A viral disease transmitted by the *Aedes aegypti* mosquito, characterized by bone-breaking sensations. Also called breakbone fever.

deoxyribonucleic acid (DNA) The genetic material of all cells and many viruses.

dermatophytosis A fungal disease of the skin tissues.

desensitization A process in which minute doses of antigens are used to remove antibodies from the body tissues to prevent a later allergic reaction.

desquamation Peeling of the skin of the fingertips and toes; associated with toxic shock syndrome and Kawasaki disease.

deuteromycete A member of the phylum Deuteromycota.

Deuteromycota A phylum of fungi whose members have no known sexual cycle of reproduction.

diarrhea Excessive loss of fluid from the gastrointestinal tract.

diatomaceous earth Filtering material composed of the remains of diatoms.

diatoms A group of eukaryotic marine algae that perform photosynthesis.

differential medium A growth medium in which different species of microorganisms can be distinguished.

DiGeorge syndrome An immunodeficiency disease in which the thymus fails to develop, thereby leading to a reduced number of T lymphocytes.

dimorphic fungus A fungus that takes a yeast form in the human body, and a hyphal form when cultivated in the laboratory. Also called biphasic fungus.

dinoflagellates A group of photosynthetic marine algae that form one of the foundations of the food chain in the ocean.

diphtheria A bacterial disease of the respiratory tract caused by toxin-producing *Corynebacterium diphtheriae*, characterized by tissue destruction and the accumulation of pseudomembranes.

dipicolinic acid An organic substance that helps stabilize the proteins in a bacterial spore, thereby increasing spore resistance.

diplococcus A pair of cocci.

disease Any change from the general state of good health.

disinfectant A chemical used to kill pathogenic microorganisms on a lifeless object such as a tabletop.

DNA ligase An enzyme that binds together DNA fragments; important in genetic engineering experiments.

DNA polymerase An enzyme that catalyzes DNA replication by combining complementary nucleotides to an existing strand.

double helix The spiral staircase arrangement of the chromosome, in which two strands of DNA sit opposite each other, with the nitrogenous bases forming the rungs of the staircase.

Downey cell A swollen lymphocyte with foamy cytoplasm and many vacuoles that develops as a result of infection with infectious mononucleosis viruses.

droplets Airborne particles of mucus and sputum from the respiratory tract that contain disease organisms.

dysentery A condition marked by frequent, watery stools, often with blood and mucus.

E

Eaton agent An alternate name for *Mycoplasma pneumoniae*.

Ebola virus An RNA virus existing in Africa that causes a type of hemorrhagic fever.

echovirus An RNA virus transmitted by food and water and involved in diseases of the intestine and the skin.

edema A swelling of the tissues brought about by an accumulation of fluid.

ehrlichiosis A tickborne rickettsial disease caused by *Ehrlichia chaffeensis*, characterized by fever, headache, and malaise.

electrophoresis A laboratory technique involving the movement of charged organic molecules through an electrical field; used to separate DNA fragments in diagnostic procedures.

elementary body An infectious form of a chlamydia in the early stage of reproduction.

elephantiasis A condition of swelling and distortion of the tissues, especially the legs and scrotum, commonly caused by infection with *Wuchereria bancrofti*. Also called filariasis.

encephalitis Inflammation of the tissue of the brain or infection of the brain.

endemic typhus A relatively mild fleaborne disease of the blood caused by *Rickettsia typhi*, characterized by fever and a skin rash. Also called tabardillo.

endogenous disease A disease caused by organisms commonly found within the body.

endonuclease An enzyme that cleaves a DNA molecule at the sugar-phosphate bond; used in genetic engineering techniques.

endospore An extremely resistant dormant cell produced by some gram-positive bacteria.

endotoxin A metabolic poison produced chiefly by Gram-negative bacteria; endotoxins are part of the bacterial cell wall and consequently are released on cell disintegration; composed of lipid-polysaccharide-peptide complexes.

enteritis A synonym for gastrointestinal illness; commonly, an alternate name for salmonellosis.

enterotoxin A toxin that is active in the gastrointestinal tract of the host.

enterovirus A virus that infects intestinal cells.

envelope The flexible membrane of protein and lipid that surrounds many types of viruses.

enzyme A reusable protein molecule that brings about a chemical change while remaining unchanged itself; the molecule may include a nonprotein part.

enzyme-linked immunosorbent assay (ELISA) A serological test in which an enzyme system is used to detect test material linked to antigens or antibodies on a solid surface.

eosinophil A type of leukocyte whose functions are not clearly established but may involve phagocytosis.

epidemic parotitis An alternate name for mumps.

epidemic typhus A relatively serious louseborne disease of the blood caused by *Rickettsia prowazecki*, characterized by high fever and a skin rash beginning on the body trunk.

episome A plasmid attached to the chromosome of a bacterium.

Epstein-Barr (EB) virus An icosahedral DNA virus thought to cause infectious mononucleosis; also associated with Burkitt's lymphoma.

ergot disease A toxemia transferred to humans from rye grain; the toxin is produced by the fungus *Claviceps purpurea.*

erythema A zone of redness in the skin due to an accumulation of blood.

erythema chronicum migrans (ECM) An expanding circular red rash that occurs on the skin of patients with Lyme disease.

erythema infectiosum An alternate name for fifth disease.

erythrogenic toxin A streptococcal toxin that leads to the rash in scarlet fever.

erythromycin An antibiotic derived from a *Streptomyces* species; used against Gram-positive bacteria, mycoplasmas, and certain other organisms.

estivo-autumnal malaria A type of malaria in which the attacks occur at widely spaced intervals.

Eubacteria The domain of living things that includes all organisms not classified as Archaea or Eukarya.

Eukarya The domain of living things encompassing all eukaryotic organisms.

eukaryote A relatively complex organism whose cells contain organelles as well as a nucleus with multiple chromosomes and a nuclear membrane envelope; reproduction involves mitosis. *See also* prokaryote.

exanthem A maculopapular rash occurring on the skin surface.

exotoxin A metabolic poison produced chiefly by Gram-positive bacteria; exotoxins are released to the environment on production; composed of protein and affect various organs and systems of the body.

F

F factor A DNA plasmid in the cytoplasm of an F⁺ bacterial cell that may pass to a recipient bacterial cell in conjugation and change the recipient into an F⁺ cell.

Fab fragment The portion of an antibody molecule that combines with the determinant site of the antigen.

facultative Referring to an organism that grows in the presence or absence of oxygen.

farmer's lung A condition that results from a type III immune complex hypersensitivity following exposure to antigens.

Fc fragment The portion of an antibody molecule that combines with phagocytes, viral receptor sites, and complement.

fermentation Anaerobic respiration in which intermediaries in the process are used as electron acceptors; also refers to the industrial use of microorganisms.

fifth disease A communicable skin disease caused by a DNA virus and accompanied by a fiery red rash especially on the checks of the face. Also called B19 disease and erythema infectiosum.

filariform larva A hairlike intermediate form of the hookworm.

fimbria A short, hairlike structure used by bacteria for attachment; sometimes used as an alternative term for pilus.

flaccid paralysis Paralysis in which the limbs have little tone and become flabby.

flagellum A long, hairlike appendage composed of protein and responsible for motion in microorganisms; found in some bacteria, protozoa, and algae.

flare A spreading zone of redness around a wheal that occurs during an allergic reaction.

flash method A process of pasteurization in which milk is heated at 71.6°C for 15 to 17 seconds and then cooled rapidly. Also known as the HTST method, for "high temperature, short time."

flatworm A common name for a member of the phylum Platyhelminthes.

flavin adenine dinucleotide (FAD) A coenzyme that functions in electron transport during oxidative phosphorylation.

flocculation The formation of jellylike masses of coagulated material in the water-purification process; also, a serological reaction in which particulate antigens react with antibodies to form visible aggregates of material.

flucytosine An antifungal drug that interrupts nucleic acid synthesis in cells.

fluid mosaic model The model for the cell membrane where proteins "float" within or on a bilayer of phospholipid.

fluke A flatworm belonging to the phylum Platyhelminthes. Also known as trematode.

folic acid The organic compound in bacteria whose synthesis is blocked by sulfonamide drugs.

fomite An inanimate object, such as clothing or a utensil, that carries disease organisms.

foraminiferan A shell-containing amoeboid protozoan having a chalky skeleton with windowlike openings between sections of the shell.

formalin A solution of formaldehyde used as embalming fluid, in the inactivation of viruses, and as a disinfectant.

Friedländer's bacillus An alternate name for *Klebsiella pneumoniae*.

fruiting body The general name for an asexual reproductive structure of a fungus.

functional genomics The identification of gene function from a gene sequence.

functional group A group of atoms on hydrocarbons that function in chemical reactions.

fungemia The dissemination of fungi through the circulatory system.

Fungi One of the five kingdoms in the Whittaker classification of living things, composed of the molds and yeasts.

fungicidal agent An agent that kills fungi.

fusospirochetal disease An alternate name for trench mouth.

 G

gamma globulin A general term for antibody-rich serum.

gangrene A physiological process in which the enzymes from wounded tissue digest the surrounding layer of cells, inducing a spreading death to the tissue cells.

gas gangrene A serious soilborne bacterial disease caused by *Clostridium perfringens*, characterized by gas accumulation in the muscle tissues and gangrene; often called wet gangrene.

gastroenteritis Infection of the intestinal tract often due to a virus.

gene A segment of a DNA molecule that provides the biochemical information for protein synthesis and inherited traits.

generalized transduction A type of transduction in which the prophage accidentally incorporates bacterial DNA into its own DNA while replicating during the lytic cycle; the bacterial DNA is then carried into the next cell by the virus. *See also* specialized transduction.

generation time The time interval between bacterial divisions; varies among bacteria and can be as brief as 20 minutes.

genetic engineering The use of bacterial and microbial genetics to isolate, manipulate, recombine, and express genes.

genome The complete set of genes in a virus or an organism.

genomics The study of an organism's DNA.

gentamicin An aminoglycoside antibiotic often used to treat infections caused by Gram-negative bacteria.

genus A rank in the classification system of organisms composed of two or more species; a collection of genera constitute a family.

germ theory of disease The theory that holds that microorganisms are responsible for infectious diseases.

germicidal agent An agent that kills microorganisms.

giardiasis A foodborne and waterborne protozoal disease of the intestine caused by *Giardia lamblia*, characterized by mild to severe diarrhea.

gibberellins A series of plant hormones that promote growth by stimulating cell elongation in the stem.

Gilchrist's disease An alternate name for blastomycosis.

glomerulonephritis A complication of streptococcal disease involving the inflammation of blood vessels in the kidneys due to reactions between antigens and antibodies.

glycocalyx A layer of capsule or slime around certain bacteria that assists in attachment to a surface and imparts resistance.

glycolysis A series of enzyme-catalyzed chemical reactions in which glucose is broken down into two molecules of pyruvate with a net gain of two ATP molecules.

gonococcal ophthalmia Infection of the eye by *Neisseria gonorrhoeae*; may occur in newborns exposed during birth.

gonococcus A colloquial name for *Neisseria gonorrhoeae*.

gonorrhea A sexually transmitted bacterial disease of the urethra and reproductive organs caused by *Neisseria gonorrhoeae*; accompanied by urinary tract symptoms and possible pelvic inflammatory disease, leading to sterility.

Goodpasture syndrome A type II hypersensitivity reaction in which antibodies are directed against antigens on the membranes of kidney cells, often resulting in kidney failure.

graft-versus-host reaction (GVHR) A phenomenon in which a tissue graft produces immune substances against the host.

Gram stain technique A technique used for differentiating bacteria into two groups, gram-positive and gram-negative, depending upon their ability to retain a crystal violet iodine complex on treatment with an alcohol solution.

granuloma inguinale A sexually transmitted bacterial disease of the external genital organs caused by *Calymmatobacterium granulomatis*, characterized by bleeding ulcers and swollen lymph nodes.

Graves' disease A type II hypersensitivity reaction in which antibodies react with receptors on thyroid gland cells, resulting in an overabundant secretion of thyroxine.

gray syndrome A side effect of chloramphenicol therapy, characterized by a sudden breakdown of the cardiovascular system.

green rot Spoilage in eggs due to the production of a green, fluorescent pigment by *Pseudomonas* species.

griseofulvin An antifungal drug used against tinea infections of the skin, hair, and nails.

Guillain-Barré syndrome (GBS) A complication of influenza and chickenpox, characterized by nerve damage and poliolike paralysis.

Guinea worm The common name for *Dracunculus medinensis*, a multicellular roundworm parasite of the human blood and subcutaneous tissues.

gumma A soft, granular lesion that forms in the cardiovascular and/or nervous systems during tertiary syphilis.

H

halogen A chemical element whose atoms have seven electrons in their outer shell; examples are iodine and chlorine.

halophile An organism that lives in environments with high concentrations of salt.

Hansen's disease An alternate name for leprosy.

hapten A small molecule that combines with tissue proteins or polysaccharides to form an antigen.

Hashimoto's disease A type II hypersensitivity reaction in which antibodies react with thyroid gland cells, leading to a deficiency of thyroxine.

hay fever A type I hypersensitivity reaction resulting from the inhalation of tree and grass pollens.

Hazard Analysis Critical Control Point (HACCP) A set of federally enforced regulations to ensure the dietary safety of seafood, meat, and poultry.

helix A figure resembling a tightly wound coil such as a corkscrew or spring; one of the major shapes of viral capsids.

helminth A term referring to a multicellular parasite; includes roundworms and flatworms.

helper T lymphocyte A T lymphocyte that enhances the activity of B lymphocytes.

hemagglutination The agglutination of red blood cells.

hemagglutinin An enzyme on the surface spikes of certain influenza viruses that enables the viruses to bind to red blood cells.

hemoglobin The red oxygen-carrying pigment in erythrocytes.

hemolysin An enzyme that dissolves red blood cells; produced by streptococci, staphylococci, gas gangrene, bacilli, and other microorganisms.

hemolytic disease of the newborn A disease in which Rh antibodies from a pregnant woman combine with Rh antigens on the surface of fetal erythrocytes and destroy them; a type II hypersensitivity reaction. Also known as Rh disease and erythroblastosis fetalis.

hemolytic uremic syndrome (HUS) A disease characterized by the formation of hemorrhages in the kidney and often resulting in kidney failure; associated with *E. coli* O157:H7.

hemorrhagic colitis Bloody diarrhea associated with infection by *E. coli* O157:H7.

hemorrhagic fever Any of a series of viral diseases characterized by high fever and hemorrhagic lesions of the throat and internal organs.

hepatitis A A foodborne and waterborne disease of the liver caused by a highly resistant RNA virus, characterized by jaundice, abdominal pain, and degeneration of the liver tissue.

hepatitis B A bloodborne disease of the liver caused by a fragile DNA virus, characterized by jaundice, abdominal pain, and degeneration of the liver tissue.

hepatitis B core antigen (HBcAG) An antigen located in the inner lipoprotein coat enclosing the DNA of a hepatitis B virus.

hepatitis B surface antigen (HBsAg) An antigen located in the outer surface coat of a hepatitis B virus; previously known as the Australia antigen.

hepatitis C A bloodborne disease of the liver caused by a virus, characterized by jaundice, abdominal pain, and degeneration of the liver tissue.

hepatocarcinoma Cancer of the liver tissue.

hermaphroditic organism An organism that possesses both male and female reproductive organs.

herpes simplex A viral disease of the skin and nervous system, often characterized by blisterlike sores.

heterophile antigen An antigen that occurs in apparently unrelated species of organisms.

heterotroph An organism that feeds on preformed organic matter and obtains energy from this matter.

hexachlorophene A bisphenol compound containing six chlorine atoms; used in disinfectants and antiseptics.

histamine A mediator in type I hypersensitivity reactions; released from the granules in mast cells and basophils, causing the contraction of smooth muscles.

histocompatibility molecule A molecule on the surface of animal tissue cells involved in transplant acceptance and rejection.

histoplasmosis A fungal disease of the lungs and other systemic organs caused by *Histoplasma capsulatum*, often occurring in immune-compromised individuals, such as AIDS patients.

holding method A process of pasteurization in which milk is heated at 62.9°C for 30 minutes. Also known as the LTLT method, for "low temperature, long time."

host An organism on or in which a microorganism lives and grows.

human granulocytic ehrlichiosis (HGE) A tickborne form of ehrlichiosis in which the causative agent infects the body's neutrophils.

human monocytic ehrlichiosis (HME) A tickborne form of ehrlichiosis in which the causative agent infects the body's monocytes.

humoral immunity Immunity arising from the activity of antibodies directed against antigens in the tissues. Also called antibody-mediated immunity.

Hutchinson's triad Deafness, impaired vision, and notched, peg-shaped teeth; may accompany congenital syphilis.

hyaluronidase An enzyme that digests hyaluronic acid and thereby permits the penetration of parasites through the tissues; known as the spreading factor.

hydatid cyst A thick-walled body formed in the human liver by *Echinococcus granulosus.*

hydrogen bond A weak chemical bond that forms between oppositely charged poles of adjacent molecules.

hydrophobia "Fear of water," an emotional condition arising from the inability to swallow as a consequence of rabies.

hyperimmune serum Serum that contains a higher than normal amount of a particular antibody.

hypha A microscopic filament of cells that represents the basic unit of a fungus.

I

icosahedron A symmetrical figure composed of 20 triangular faces and 12 points; one of the major shapes taken by the viral capsid.

idoxuridine The drug 5-iodo-2-deoxyuridine (IDU) that replaces thymine in DNA molecules; useful in treating herpes simplex.

imidazoles A group of antifungal drugs that interfere with sterol synthesis in fungal cell membranes; examples are miconazole and ketoconazole.

imipenem A beta-lactam antibiotic used again Gram-positive bacteria.

immune adherence phenomenon Increased phagocytosis of a cell resulting from the attachment of certain complexes to the cell surface.

immune complex hypersensitivity Type III hypersensitivity, in which antigens combine with antibodies to form aggregates that are deposited in blood vessels or on tissue surfaces.

immune complex An aggregate of antigen-antibody material that is deposited in body tissues; characteristic of type III hypersensitivity reactions.

immunoelectrophoresis A laboratory technique in which antigen molecules move through an electric field and then diffuse to meet antibody molecules to form a precipitation line; used in diagnostic procedures.

immunoglobulin An alternate term for antibody.

imperfect fungi Fungi that are known to multiply only by asexual processes.

impetigo contagiosum An infectious skin disease usually caused by staphylococci or streptococci, characterized by abscesses and boils.

inactivated virus A weakened virus that results from treatment with physical or chemical agents; used in immunizating agents because of its reduced virulence; sometimes called a dead virus.

inclusion body A granulelike storage body that forms in the bacterial cytoplasm.

incubation period The time that elapses between the entry of a parasite to the host and the appearance of symptoms. Also called period of incubation.

induced mutation A mutation arising from a mutagenic agent used under controlled laboratory conditions.

induration A thickening and drying of the skin tissue that occurs in type IV hypersensitivity reactions.

infection The relationship between two organisms and the competition for supremacy that takes place between them.

infection allergy A type IV hypersensitivity reaction in which the immune system responds to the presence of certain microbial agents.

infectious mononucleosis A disease of the white blood cells caused by an icosahedral DNA virus, characterized by sore throat, mild fever, and malaise.

inflammation A nonspecific defensive response to injury; usually characterized by red color from blood accumulation, warmth from the heat of blood, swelling from fluid accumulation, and pain from injury to local nerves.

influenza A disease of the lungs caused by a helical RNA virus, characterized by cough and malaise.

initial body A noninfectious form of chlamydia in the later stage of reproduction.

innate immunity An inborn capacity for resisting disease.

insect An arthropod having six legs; examples are fleas, mosquitoes, and lice.

insertion sequence A segment of DNA that forms a copy of itself, after which the copy moves into areas of gene activity to interrupt the genetic coding sequence.

interferon An antiviral protein produced by body cells on exposure to viruses; interferons trigger the production of a second protein that binds to mRNA coded by the virus, thereby inhibiting viral replication.

interleukin A lymphokine produced by white blood cells that acts on other white blood cells; important in immune processes.

intergenic transfer The transfer of DNA between different genera.

intermediary host The host in which the larval or other intermediary stage of a multicellular parasite is found.

invasiveness The ability of a parasite to invade the tissues of the host and cause structural damage to the tissues.

iodophor A complex of iodine and detergents that releases iodine over a long period of time; used as an antiseptic and disinfectant.

ionizing radiation A type of radiation such as gamma rays and X rays, that causes the formation of ions.

isograft Tissue taken from an identical twin and grafted to the other twin.

isomers Molecules with the same molecular formula but different structural formulas.

isoniazid A chemotherapeutic agent effective against the tubercle bacillus.

isopropyl alcohol A two-carbon alcohol compound widely used as a disinfectant; commonly known as rubbing alcohol.

isotope Variant of an atom in which the number of neutrons differs.

J

jaundice A condition in which bile seeps into the circulatory system, causing the complexion to have a dull yellow color.

Job syndrome An immune disorder characterized by defective chemotaxis between phagocyte and microorganism.

K

kala-azar An alternate name for leishmaniasis.

kanamycin An aminoglycoside antibiotic derived from a *Streptomyces* species; used to treat infections caused by Gram-negative bacteria.

Kaposi's sarcoma A type of skin cancer that affects immune-compromised individuals, such as AIDS patients.

kappa factor A nucleic acid particle produced by species of *Paramecium*; appear responsible for the production of toxins by these protozoa.

Kawasaki disease A disease of undetermined origin characterized by a skin rash and desquamation and often involving the cardiovascular system.

keratitis Infection of the cornea of the eye.

keratoconjunctivitis Eye inflammation accompanied by infection of the cornea and conjunctiva; the condition is characterized by tearing, swelling, and sensitivity to light.

ketoconazole An imidazole drug often used in the treatment of various fungal diseases.

Kirby-Bauer test An agar diffusion test used to determine the antibiotic concentration effective against a test organism.

Koch's postulates A set of procedures by which a specific organism can be related to a specific disease.

Koch-Weeks bacillus An alternate name for *Haemophilus aegyptius.*

Koplik spots Red patches with white central lesions that form on the gums and walls of the pharynx during the early stages of measles.

Krebs cycle A cyclic series of enzyme-catalyzed reactions in which carbon from acetyl-CoA is released as carbon dioxide; the reactions also yield protons and high-energy electrons that are transported among coenzymes and cytochromes as their energy is released.

kuru A disease characterized by slow degeneration of the brain tissue, possibly caused by a prion.

L

lactalbumin A protein occurring in minor quantities in cow's milk.

lactose A milk sugar composed of one molecule of glucose and one molecule of galactose.

lag phase A portion of a bacterial growth curve encompassing the first few hours of the population's history; minimal reproduction occurs.

larva A preadult immature stage in the life cycle of certain eukaryotic organisms; also, a tiny worm in the life cycle of a multicellular parasite.

Lassa fever A hemorrhagic fever disease of the blood caused by an RNA virus, characterized by severe throat lesions.

lateral (horizontal) gene transfer The movement of DNA segments between cells of the same generation.

lecithinase A toxin produced by gas gangrene bacilli that dissolves the membranes of tissue cells.

legionellosis An alternate name for Legionnaires' disease.

Legionnaires' disease A bacterial disease of the lungs caused by *Legionella pneumophila*, characterized by pneumonia, fever, and malaise. Also called legionellosis.

legume A plant that bears seeds in pods; examples are beans, soybeans, and alfalfa.

leishmaniasis A protozoal disease of the white blood cells caused by *Leishmania* species, transmitted by sandflies; characterized by fever, sores, and emaciation; occurs in visceral and cutaneous forms. Also called kala-azar.

leproma A tumorlike growth on the skin associated with leprosy.

lepromin test A skin test used in the screening and diagnosis of leprosy.

leprosy A bacterial disease transmitted by contact and caused by *Mycobacterium leprae*; characterized by destruction of the skin tissues and the peripheral nerves, leading to local anesthesia.

leptospirosis A soilborne bacterial disease caused by *Leptospira interrogans*, characterized by mild fever.

leukemia Cancer of the white blood cells.

leukocidin A bacterial enzyme that destroys phagocytes, thereby preventing phagocytosis.

leukocyte An alternate name for white blood cell.

leukopenia A condition characterized by a drop in the normal number of white blood cells.

leukotriene A substance that acts as a mediator during type I hypersensitivity reactions; formed from arachidonic acid after the antigen-antibody reaction has taken place.

lichen An organism composed of a fungal mycelium within which are embedded photosynthetic algae.

lincomycin An antibiotic used as a penicillin substitute for diseases caused by Gram-positive bacteria.

lipase A fat-digesting enzyme produced by certain contaminating bacteria in milk.

lipid An organic compound that dissolves in organic solvents; composed of carbon, hydrogen, and oxygen; lipids include fats and are used in energy metabolism and structural compounds.

lipopolysaccharide (LPS) A molecule composed of lipid and polysaccharide, found in the outer membrane of gram-negative bacteria, where it functions as an endotoxin.

Lipschütz body An inclusion that forms in the nucleus of cells infected with herpesviruses.

listeriosis A soilborne and foodborne bacterial disease caused by *Listeria monocytogenes* and accompanied by mild symptoms, except in pregnant women where miscarriage may occur.

littoral zone The environment along the shoreline of an ocean.

lobar pneumonia Pneumonia that involves an entire side or lobe of the lung.

local disease A disease restricted to a single area of the body, usually the skin.

lockjaw A common name for tetanus, based on the spasms of the jaw muscles.

logarithmic (log) phase The second portion of a bacterial growth curve, in which active growth leads to a rapid rise in numbers of the population.

long incubation hepatitis An alternate name for hepatitis B.

lophotrichous The presence of two or more flagella at one end of a bacterial cell.

louse A type of insect of the genus *Pediculus* that transmits epidemic typhus and other diseases.

lumpy jaw A common name for actinomycosis in the jaw, characterized by hard nodules in the tissue.

Lyme disease A serious arthropodborne bacterial disease transmitted by ticks, caused by *Borrelia burgdorferi*; characterized by a skin rash (ECM) and malaise, and later a swelling and degeneration of the large joints.

lymph node A bean-shaped organ located along lymph vessels, involved in the immune response; the major cells of lymph nodes are phagocytes and lymphocytes.

lymphocyte A type of leukocyte that functions in the immune system.

lymphocytic choriomeningitis A disease of the brain tissue possibly caused by a virus, characterized by headache, drowsiness, and stupor, as well as large numbers of lymphocytes in the meninges.

lymphogranuloma venereum A sexually transmitted bacterial disease of the external genital organs, caused by *Chlamydia trachomatis* and characterized by swollen lymph nodes.

lymphokine A glycoprotein that increases the efficiency of immune system reactions, such as at the antigen site. Also called cytokine.

lymphoid progenitors Primitive cells that arise from stem cells, modified to form B lymphocytes or T lymphocytes.

lyophilization A process in which food or other material is deep frozen, after which its liquid is drawn off by a vacuum.

lysis The rupture of a cell and the loss of cell contents.

lysogenic bacterium A bacterium that carries a prophage.

lysogeny The process by which a virus remains in the cell cytoplasm as a fragment of DNA or attaches to the chromosome, but fails to replicate in or destroy the cell.

lysosome A microscopic organelle found in most eukaryotic cells; contains digestive enzymes.

lysozyme An enzyme found in tears and saliva that digests the peptidoglycan of gram-positive bacteria and leads to their destruction.

lytic cycle A process by which a virus replicates within a host cell and ultimately destroys the host cell.

M

Machupo A type of viral hemorrhagic fever occurring primarily in South America.

mad cow disease A common name for bovine spongiform encephalopathy.

macrophage A large cell derived from monocytes and found within the tissues; macrophages actively phagocytize foreign bodies and comprise the reticuloendothelial system.

macule A pink-red skin spot associated with infectious disease.

maculopapular rash A rash consisting of pink-red spots that later become dark red before fading; occurs in rickettsial diseases.

Madura foot Substantial swelling of the tissues of the foot due to *Nocardia asteroides*.

magnetosome A cytoplasmic body in certain bacteria that assists orientation to the environment by aligning with the magnetic field.

major histocompatibility complex (MHC) A set of genes that controls the expression of MHC proteins; involved in transplant rejection.

major histocompatibility (MHC) protein Any of a set of proteins at the surface of all body cells that identify the uniqueness of the individual.

malaria A serious protozoal disease of the red blood cells caused by *Plasmodium* species, transmitted by mosquitoes, and characterized by periods of high fever, anemia, and blood clotting.

mannitol salt agar An enriched medium that encourages the growth of staphylococci; contains a high percentage of salt, a feature that makes it inhibitory to most other organisms.

Marburg disease A viral disease caused by an RNA virus and characterized by fever and hemorrhaging.

mast cell A connective tissue cell to which IgE fixes in type I hypersensitivity reactions; mast cells degranulate and release histamine during allergic attacks.

Mastigophora A group of protozoa whose members move by means of flagella.

maternal antibodies Type IgG antibodies that cross the placenta from the maternal to the fetal circulation and protect the newborn for the first few months of life.

measles A communicable respiratory disease, caused by an RNA helical virus, characterized by respiratory symptoms and a blushlike skin rash. Also called rubeola.

mechanical vector A living organism, or an object, that transmits disease agents on its surface.

melioidosis A soilborne bacterial disease, caused by *Pseudomonas pseudomallei*; characterized by lung abscesses.

memory cell A cell derived from B lymphocytes or T lymphocytes that reacts rapidly upon the future recurrence of antigens in the tissues.

meninges The covering layers of the brain and spinal cord; the three meninges are the dura mater, arachnoid, and pia mater.

meningitis A general term for inflammation of the meninges due to any of several bacteria, fungi, viruses, or protozoa.

meningococcemia A type of endotoxic shock caused by *Neisseria meningitidis*.

meningococcus A common name for *Neisseria meningitidis*.

merozoite A stage in the life cycle of *Plasmodium* species; parasites invade the red blood cells in this form.

mesophile An organism that grows in temperature ranges of 20°C to 40°C.

messenger RNA (mRNA) An RNA transcript containing the information for synthesizing a specific polypeptide.

metabolism The sum of all biochemical processes taking place in a living cell.

metacercaria An encysted intermediary stage in the life cycle of a fluke.

metachromatic granule A phosphate-storing granule that stains deeply with methylene blue; commonly found in *Corynebacterium diphtheriae*. Also called volutin granule.

methanogen A bacterium that lives on simple compounds in anaerobic environments and produces methane during its metabolism.

metronidazole A chemotherapeutic agent used to treat trichomoniasis and other protozoal diseases.

miasma An ill-defined entity generally referring to an altered chemical quality of the atmosphere; believed to cause disease before the establishment of the germ theory of disease.

miconazole An imidazole drug used in the treatment of topical and systemic fungal diseases.

microaerophile Referring to an organism that grows best in an oxygen-reduced environment.

microbiota The populations of microorganisms that colonize various parts of the human body without causing disease.

microfilaria An intermediate eellike form of *Wuchereria bancrofti*.

micrometer (μm) A unit of measurement equivalent to one millionth of a meter that is commonly used in measuring the size of microorganisms.

microorganism A microscopic form of life including bacteria, viruses, fungi, protozoa, and some multicellular parasites.

miliary tuberculosis Tuberculosis that spreads through the body.

miracidium A ciliated larva representing an intermediary stage in the life cycle of a fluke.

mite A spiderlike arthropod that transmits such diseases as rickettsialpox and tsutsugamushi.

mold A type of fungus that consists of chains of cells and appears as a fuzzy mass in culture.

molecular epidemiology The study of the sources, causes, and mode of transmission of diseases by using molecular diagnostic techniques.

molecular taxonomy The systematized arrangement of related organisms based on molecular characteristics, such as ribosomal RNA nucleotide sequences.

molecule The smallest part of a compound that retains the properties of the compound.

molluscum body A cytoplasmic inclusion that occurs in cells infected with the viruses of molluscum contagiosum.

molluscum contagiosum A viral skin disease caused by a DNA virus and accompanied by firm, waxy, wartlike lesions with a depressed center.

Monera One of the five kingdoms in the Whittaker classification of living things, composed of prokaryotes; also called Prokaryotae.

monoclonal antibody A type of antibody produced by a clone of hybridoma cells, consisting of antigen-stimulated plasma cells fused to myeloma cells.

monocyte A leukocyte with a large bean-shaped nucleus; functions in phagocytosis.

monotrichous The presence of a single flagellum on a bacterial cell.

most probable number (MPN) A laboratory test in which a statistical evaluation is used to estimate the number of bacteria in a sample of fluid; often employed in determinations of coliform bacteria in water.

moxalactam A monobactam antibiotic used against Gram-negative bacteria.

M protein A protein that enhances the pathogenicity of streptococci by allowing organisms to resist phagocytosis and adhere firmly to tissue.

multidrug-resistant *Staphylococcus aureus* (MRSA) A strain of *S. aureus* displaying resistance to a multitude of available antibiotics.

mumps A communicable disease caused by an icosahedral DNA virus, characterized by swelling of the parotid and other salivary glands.

mutation A change in the characteristic of an organism arising from a permanent alteration of a DNA sequence.

mutualism A close and permanent association between two populations of organisms in which both benefit from the association.

myasthenia gravis A type II hypersensitivity reaction in which antibodies react with acetylcholine receptors on the membranes of muscle fibers.

mycelium A mass of hyphae from which most fungi are built.

mycologist One who studies fungi.

mycology The study of fungi.

mycoplasmas A group of tiny submicroscopic bacteria that lacks cell walls and is visible only with an electron microscope.

mycoplasmal pneumonia An airborne primary pneumonia transmitted by droplets, caused by *Mycoplasma pneumoniae*; characterized by lung tissue destruction and respiratory symptoms. Also called primary atypical pneumonia and walking pneumonia.

mycoplasmal urethritis A sexually transmitted disease caused by *Mycoplasma hominis* and characterized by urinary tract discomfort.

mycotoxin A fungal toxin.

myocarditis Infection of the heart muscle, often due to Coxsackie virus.

N

nalidixic acid A chemotherapeutic agent that blocks protein synthesis in certain gram-negative bacteria that cause urinary tract infections.

nanometer (nm) A unit of measurement equivalent to one billionth of a meter; the unit is abbreviated as nm and is often used in measuring viruses and the wavelength of energy.

narrow-spectrum antibiotic An antibiotic that is useful for a restricted group of microorganisms. *See also* broad-spectrum antibiotic.

natural killer (NK) cell A type of defensive body cell that attacks and destroys cancer cells and infected cells without the involvement of the immune system.

naturally acquired active immunity Immunity resulting from the immune system's response to disease antigens.

naturally acquired passive immunity Immunity resulting from the passage of antibodies to the fetus via the placenta or the milk of a nursing mother. Also called congenital immunity.

necrosis Cell death.

negative stain technique A staining process that results in clear bacteria on a stained background when viewed with the light microscope.

Negri body A cytoplasmic inclusion that occurs in brain cells infected with rabies viruses.

nematode A common name for the roundworm, a member of the phylum Nematoda, or Nemathelminthes.

neomycin An aminoglycoside antibiotic derived from a *Streptomyces* species; used topically for infections caused by Gram-negative bacteria, especially in the eye.

neoplasm An uncontrolled growth of cells; often called a tumor.

neuraminidase An enzyme on the surface spikes of certain viruses that dissolves the cell membrane when the virus attaches to a cell during replication.

neurotoxin A toxin that is active in the nervous system of the host.

neutralization A type of antigen-antibody reaction in which the activity taking place between reactants is not visible.

neutrophil A type of polymorphonuclear cell that functions chiefly as a phagocyte.

nevirapine An antiviral drug that binds to and inhibits the action of reverse transcriptase; used against HIV.

nicotinamide adenine dinucleotide (NAD) A coenzyme that transports electrons during oxidative phosphorylation and fermentation reactions.

night soil Human feces sometimes used as an agricultural fertilizer.

nitrogen fixation A general term for the chemical process in which organisms trap atmospheric nitrogen and use it to form organic compounds.

nitrogenous base Any of five nitrogen-containing compounds found in nucleic acids, including adenine, guanine, cytosine, thymine, and uracil.

noncommunicable disease A disease whose causative agent is acquired from the environment and is not easily transmitted to the next host.

normal flora *See* microbiota.

Norovirus A virus transmitted by food and water and involved in intestinal diseases.

nosocomial disease A disease acquired during an individual's stay at a hospital.

nucleocapsid The combination of genome and capsid in a virus.

nucleoid The chromosomal region of a bacterium.

nucleotide A component of a nucleic acid consisting of a carbohydrate molecule, a phosphate group, and a nitrogenous base.

nutrient agar A common bacteriological growth medium consisting of beef extract, peptone, water, and agar.

nutrient broth A common bacteriological growth medium consisting of beef extract, peptone, and water.

nystatin An antifungal drug effective against *Candida albicans*.

O

obligate anaerobe An organism that grows only in the absence of O_2.

Okazaki fragments Segments of DNA resulting from discontinuous DNA replication.

oncofetal antigen An antigen in tumor cells thought to be expressed during differentiation of the tissues in the embryonic stage.

oncogene A region of DNA in human cells thought to induce uncontrolled growth of the cell if permitted to function.

oncology The study of tumors and cancers.

oocyst An oval body in the reproduction cycle of certain protozoa that develops by a complex series of asexual and sexual processes.

oospore A sexually produced spore that is formed by members of the oomycetes group of fungi.

operon The unit of bacterial DNA that controls and expresses a particular set of structural genes.

ophthalmia Severe inflammation of the eye.

opportunist An organism that invades the tissues when body defenses are suppressed.

opsonins Antibodies or complement components that encourage phagocytosis.

opsonization Enhanced phagocytosis due to the activity of antibodies or complement.

orchitis A condition caused by the mumps virus in which the virus damages the testes.

organelle A membrane-enclosed compartment in eukaryotic cells; the site of various cellular functions.

Oriental sore A cutaneous form of leishmaniasis characterized by skin sores.

ornithosis An alternate name for psittacosis.

osmosis The flow of water from a region of low concentration of chemical substance through a semipermeable membrane to a region of high concentration of the same chemical substance.

oxidase An enzyme that catalyzes oxidation-reduction reactions.

oxidation A chemical change in which electrons are lost by an atom; also, the union of oxygen with a chemical substance.

oxidative phosphorylation A series of sequential steps in which energy is released from electrons as they pass among coenzymes and cytochromes, and ultimately, to oxygen; the energy is used to combine phosphate ions with ADP molecules to form ATP molecules.

P

pandemic A worldwide epidemic.

papilloma A tumor of the skin tissue.

papilloma virus An icosahedral DNA virus and one of the causes of skin warts.

papule A pink pimple on the skin.

parainfluenza A disorder caused by an RNA helical virus, characterized by mild upper respiratory illness.

parasite A type of heterotrophic organism that feeds on live organic matter such as another organism.

parasitemia The spread of protozoa and multicellular worms through the circulatory system.

parasitism A close and permanent association between two organisms in which one feeds on the other and may cause injury to the other organism.

parasitology The discipline of biology concerned with pathogenic protozoa.

paromomycin A drug used in the treatment of amoebiasis and balantidiasis.

paroxysm A sudden intensification of symptoms, such as a severe bout of coughing.

parvovirus An icosahedral DNA virus that causes disease in dogs and fifth disease in humans.

passive agglutination An immunological procedure in which antigen molecules are adsorbed to the surface of latex spheres or other carriers that agglutinate when combined with antibodies.

pasteurellosis A bacterial disease transmitted by animal bites, caused by *Pasteurella multocida*; characterized by abscess formation at the bite site.

pasteurization A heating process that destroys pathogenic bacteria in a fluid such as milk and lowers the overall number of bacteria in the fluid.

pathogen A type of parasite that causes disease in the host organism.

pathogenicity The ability of a parasite to gain entry to a host and bring about a physiological or anatomical change interpreted as disease.

Paul-Bunnell test A diagnostic procedure for infectious mononucleosis in which a sample of a patient's serum is combined with sheep red blood cells; the cells agglutinate in a positive test.

pebrine A protozoal disease of silkworms studied by Louis Pasteur.

pellicle A rigid covering layer of certain protozoa composed in part of chitinlike material.

pelvic inflammatory disease (PID) A disease of the pelvic organs; often a complication of a sexually transmitted disease.

penicillin Any of a group of antibiotics derived from *Penicillium* species or produced synthetically; effective against gram-positive bacteria and several gram-negative bacteria by interfering with cell wall synthesis.

penicillinase An enzyme produced by certain microorganisms that converts penicillin to penicilloic acid and thereby confers resistance against penicillin.

peptidoglycan A complex molecule of the bacterial cell wall composed of alternating units of *N*-acetylglucosamine and *N*-acetylmuramic acid crosslinked by short peptides.

perforin A protein secreted by cytotoxic T lymphocytes to dissolve the cell membrane attacked during cell-mediated immunity.

period of convalescence The phase of a disease during which the body's systems return to normal.

period of decline The phase of a disease during which symptoms subside.

peritrichous The presence of flagella over the entire surface of the bacterial cell.

pertussis A bacterial disease of the upper respiratory tract in which an accumulation of mucus causes a narrowing of the tubes and a characteristic "whoop" on inhalation, thus the common name, whooping cough.

pH An abbreviation for the negative logarithm of the amount of hydrogen ions in 1 liter of solution.

pH scale A scale that extends from 1 to 14 and indicates the degree of acidity or alkalinity of a solution.

phage *See* **bacteriophage**.

phagocyte A cell that performs phagocytosis.

phagocytosis A process in which solid particles are taken into the cell; important in nutritional processes and in defense against disease.

phagosome A vesicle that contains particles of phagocytized material.

phenol coefficient A number that indicates the effectiveness of an antiseptic or disinfectant compared to phenol.

phosphatase An enzyme normally found in milk; it is destroyed by pasteurization processes, and its absence indicates that the pasteurization has been successful.

photoautotroph An organism that uses light energy to synthesize nutrients from carbon dioxide.

photoheterotroph An organism that uses light energy to synthesize nutrients from carbon compounds other than carbon dioxide.

photophobia Sensitivity to bright light.

photosynthesis A biochemical process in which light energy is converted to chemical energy and used in carbohydrate synthesis.

picornavirus A small virus containing RNA in its genome.

pigeon fancier's disease A condition that develops from a type III hypersensitivity reaction following exposure to antigens from a pigeon.

pilin A protein subunit found in the bacterial pilus.

pilus One of many short, hairlike appendages of bacteria that anchor the cell to a surface; pili are also involved in conjugations between bacteria.

pinkeye A disease of the conjunctival membranes of the eye usually caused by bacteria or viruses and accompanied by red, swollen eyes. Also called conjunctivitis.

pinta A mild syphilislike disease of the skin occurring in remote parts of the world, caused by a species of *Treponema*.

pinworm The common name for *Enterobius vermicularis*, a multicellular roundworm parasite that infects the intestine.

plague A serious arthropodborne bacterial disease transmitted by the flea, caused by *Yersinia pestis* and characterized by severe blood hemorrhaging.

plaque A clear area on a lawn of bacteria where viruses have destroyed the bacteria; also, the gummy layer of gelatinous material consisting of bacteria and organic matter on the teeth.

plasma The fluid portion of blood remaining after the cells have been removed; serum plus the clotting agents.

plasma cell The cell derived from B lymphocytes; plasma cells produce antibodies.

plasma membrane The phospholipid bilayer with proteins that surrounds the eukaryotic cell cytoplasm.

plasmid A small, closed-loop molecule of DNA apart from the chromosome; plasmids carry genes for drug resistance and pilus formation and are used in genetic engineering techniques.

pleomorphic organism An organism that occurs in a variety of shapes.

pleurodynia Infection of the chest wall commonly due to Coxsackie virus.

pneumococcus A common name for *Streptococcus pneumoniae*.

***Pneumocystis* pneumonia (PCP)** A lung infection caused by *Pneumocystis jiroveci*, characterized by consolidation of the lung and suffocation; occurs primarily in immune-compromised individuals, such as AIDS patients. Also called pneumocystosis.

pneumonia An infectious disease of the lower respiratory tract.

pneumonic plague A form of plague in which *Yersinia pestis* invades the lung tissues and causes severe respiratory symptoms.

polymorphonuclear cell A type of white blood cell with a multilobed nucleus.

polyvalent serum Serum that contains a mixture of antibodies.

Pontiac fever A form of Legionnaires' disease.

porin A protein in the outer membrane of Gram-negative bacteria that acts as channels for the passage of biomolecules.

portal of entry The site at which a parasite enters the host.

portal of exit The site at which a parasite leaves the host.

Pott's disease Tuberculosis of the spine.

pox Pitted scars remaining on the skin of individuals who have recoverered from smallpox.

precipitation A type of antigen-antibody reaction in which thousands of molecules of antigen and antibody cross-link to form particles of precipitate.

prednisone A steroid hormone that suppresses the inflammatory response; used to retard transplant rejection.

primaquine A synthetic drug used to treat malaria in the dormant phase.

primary antibody response The initial response by the immune system to an antigen, characterized by an outpouring of IgM. *See also* secondary antibody response.

primary atypical pneumonia (PAP) An airborne bacterial disease of the lungs caused by *Mycoplasma pneumoniae*, characterized by degeneration of the lung tissue. Also called mycoplasmal pneumonia and walking pneumonia.

primary disease A disease that develops in an otherwise healthy individual.

prion An infectious particle of protein, possibly involved in human diseases of the brain.

proctitis Infection of the rectum.

prodromal phase The phase of a disease during which general symptoms occur in the body.

proglottid One of a series of segments that make up the body of a tapeworm.

prokaryote A microorganism composed of single cells having a single chromosome but no intracellular organelles, nucleus, or nuclear membrane envelope; reproduction does not involve mitosis. *See also* eukaryote.

prontosil A red dye found by Domagk to have significant antimicrobial activity when tested in live animals, and from which sulfanilamide was later isolated.

prophage The viral DNA of a temperate phage inserted into the bacterial DNA.

prophylactic serum Antibody-rich serum used to protect against the development of a disease.

prostaglandins Substances resulting from interactions involving arachidonic acid and acting as mediators in type I hypersensitivity reactions.

protein A chain of amino acids used as a structural material or enzyme in living cells; all proteins have primary and secondary structures, and some have a tertiary structure.

Protista One of the five kingdoms in the Whittaker classification of living things, composed of the protozoa and unicellular algae; also, the term used by Haeckel to denote single-celled organisms of microscopic size.

proto-oncogene A region of DNA in the chromosome of human cells; they are altered by carcinogens into oncogenes that transform cells.

prototroph An organism that contains all its nutritional needs. *See* **auxotroph**.

pruritus Itching sensations.

pseudomembrane An accumulation of mucus, leukocytes, bacteria, and dead tissue in the respiratory passages of diphtheria patients.

pseudomembranous colitis A condition of the wall of the small intestine, characterized by membranous lesions; believed to be caused by *Clostridium difficile*.

pseudopod One of many projections of the cell membrane that allow movement in members of the Sarcodina group of protozoa.

psittacosis An airborne bacterial disease of the lung caused by *Chlamydia psittaci*, characterized by respiratory discomfort and influenzalike symptoms. Also called ornithosis.

psychrophile An organism that lives at cold temperature ranges of 0°C to 20°C.

psychrotrophic Referring to an organism that normally lives at medium temperatures but tolerates and grows at cold temperatures.

ptomaine A nitrogen compound having a strong odor; once thought responsible for food poisoning.

pure culture An accumulation or colony of microorganisms of one type.

pus A mixture of serum, dead tissue cells, leukocytes, and bacteria that accumulates at the site of infection.

Q

Q fever A rickettsial disease characterized by flulike symptoms.

quartan malaria A type of malaria in which attacks occur at 4-day intervals.

quinine A drug derived from the bark of the *Cinchona* tree; used to treat malaria.

quinupristin A streptogramin antibiotic used with quinupristin to inhibit reproduction in *Staphylococcus aureus.*

R

R factor A small, circular DNA molecule that occurs frequently in bacteria and carries genes for drug resistance. Also called R plasmid.

rabbit fever An alternate name for tularemia.

rabies A serious nervous system disease due to a helical RNA virus, transmitted by an animal bite; characterized by the destruction of brain tissue, leading to paralysis and death.

radioallergosorbent test (RAST) A type of radioimmunoassay in which antigens for the unknown antibody are attached to matrix particles.

radioimmunoassay (RIA) An immunological procedure that uses radioactive-tagged antigens to determine the identity and amount of antibodies in a sample.

rat-bite fever A bacterial disease of the skin and blood due to either *Aspergillum minor* or *Streptobacillus moniliformis* and transmitted by a rat bite.

reagin An alternate name for the IgE that stimulates anaphylaxis in the body.

recombinant DNA molecule A DNA molecule containing DNA from two different sources.

recombination An alteration in the genetic information in a microorganism arising from the acquisition of DNA; in bacteria, recombination occurs by transformation, conjugation, and transduction.

red rot Spoilage in milk and eggs due to the production of red pigment by *Serratia marcescens.*

redia An intermediary stage in the life cycle of a fluke.

relapsing fever A serious arthropodborne bacterial disease transmitted by ticks and lice, caused by *Borrelia recurrentis*; characterized by periods of fever.

rennin An enzyme that accelerates the curdling of protein in milk.

reservoir A human or other animal that retains disease organisms in the body but has not experienced disease and shows no evidence of illness.

resolving power The numerical value of a lens system indicating the size of the smallest object that can be seen clearly when using that system.

respiration Any biochemical process in which energy is liberated; respiration may occur in the presence of oxygen (aerobic respiration) or in its absence (anaerobic respiration).

respiratory syncytial (RS) disease An airborne viral disease occurring in young children, characterized by pneumonia in the lung tissues.

respiratory syncytial (RS) virus A helical RNA virus that causes lung infections primarily in young children.

restriction enzyme A type of endonuclease that splits open a DNA molecule at a specific restricted point; important in genetic engineering techniques.

retrovirus An RNA virus that uses reverse transcriptase to synthesize DNA using RNA as a template.

reverse transcriptase An enzyme that synthesizes a DNA molecule from the code supplied by an RNA molecule.

Reye's syndrome A complication of influenza and chickenpox, characterized by vomiting and convulsions as well as liver and brain damage.

rhabditiform larva An elongated intermediate form of the hookworm.

rheumatic fever A complication of streptococcal disease in which damage to the heart valves develops from the reactions between antigens and antibodies.

rheumatoid arthritis An autoimmune disease characterized by immune complex formation in the joints.

rhinovirus An icosahedral RNA that causes diseases of the upper respiratory tract commonly known as head colds.

ribonucleic acid The nucleic acid involved in protein synthesis and gene control. Also the genetic information in some viruses.

ribosomal RNA (rRNA) An RNA transcript that forms part of the ribosome's structure.

ribosome A cellular component of RNA and protein that participates in protein synthesis.

rice-water stool A colorless, watery diarrhea containing particles of intestinal tissue in cholera patients.

rickettsiae A group of small bacteria generally transmitted by arthropods; most rickettsiae are cultivated only within living tissue medium.

rickettsialpox A relatively mild miteborne disease of the blood caused by *Rickettsia akari*, characterized by fever and a skin rash.

rifampin An antibiotic prescribed for tuberculosis and leprosy patients and for carriers of *Neisseria* and *Haemophilus* species.

rifapentine A drug related to rifampin, used to treat tuberculosis.

Rift Valley fever A disease caused by an RNA virus, occurring primarily in the Rift Valley of Africa, characterized by intense fever and joint pains.

ringworm A fungal disease caused by numerous species of fungus, characterized by scaly, circular patches of infection, usually on the head; ringworm of the feet is athlete's foot; ringworm of the groin, nails, face, and other body regions also occurs.

ritonavir An antiviral drug that inhibits the activity of protease in the formation of the viral capsid; belongs to a group known as protease inhibitors.

Ritter's disease A type of impetigo contagiosum of the skin caused by *Staphylococcus aureus*.

Rocky Mountain spotted fever A relatively serious tickborne disease of the blood caused by *Rickettsia rickettsii*, characterized by high fever and a skin rash beginning on the body extremities and proceeding toward the body trunk.

rolling circle mechanism A type of DNA replication in which a strand of DNA "rolls off" the loop and serves as a template for the synthesis of a complementary strand of DNA.

roseola A condition marked by fever and body rash.

rose spots Bright red skin spots associated with diseases such as typhoid fever and relapsing fever.

rotavirus An RNA virus transmitted by food and water, involved in intestinal diseases.

rubella A communicable skin disease caused by an icosahedral RNA virus, characterized by mild respiratory symptoms and a measleslike rash; can cause damage in the fetus if contracted by a pregnant woman.

rubeola An alternate name for measles.

S

Sabia virus An arenavirus that causes hemorrhagic illnesses.

Sabin vaccine A type of polio vaccine prepared with attenuated viruses; the vaccine is taken orally.

Sabouraud dextrose agar A growth medium for fungi.

saddleback fever A condition characterized by fluctuations of fever.

Salk vaccine A type of polio vaccine prepared with viruses inactivated with formaldehyde; the vaccine is injected into the body.

salpingitis Blockage of the Fallopian tubes; a possible complication of a sexually transmitted disease.

sandfly fever A sandfly-transmitted viral disease occurring in Mediterranean regions, characterized by high fever and joint pains.

sanitize To reduce microbial populations to a safe level as determined by public health standards.

saprobe A type of heterotrophic organism that feeds on dead organic matter, such as rotting wood or compost; formerly called a saprophyte.

saquinavir An antiviral drug that inhibits the activity of protease in the formation of the viral capsid; belongs to a group known as protease inhibitors.

sarcina A cubelike packet of eight cocci.

Sarcodina A group of protozoa whose members move by means of pseudopodia.

sarcoma A tumor of the connective tissues.

scalded skin syndrome A staphylococcal skin disease in infants characterized by a red, wrinkled surface with a sandpaper texture.

scarification A method of inoculating an immunizing agent by scratching the skin.

scarlatina A mild case of scarlet fever.

Schick test A skin test used to determine the effectiveness of diphtheria immunization.

schistosomiasis A waterborne blood disease caused by *Schistosoma* species and characterized by fever, chills, and liver damage.

schmutzdecke A slimy layer of microorganisms that develops in a slow sand filter.

sclerotium A hard purple body that forms in grains contaminated with *Claviceps purpurea*.

scolex The head region of a tapeworm where the attachment organ is located.

scrapie A disease of sheep in which nerve damage causes the animal to scratch the skin.

scrofula A bacterial disease of the lymph nodes of the neck caused by *Mycobacterium scrofulaceum*.

secondary anamnestic response A vigorous immune response stimulated by a second or subsequent entry of antigens to the body.

secondary antibody response The second response by the immune system to an antigen; characterized by an outpouring of IgG. *See also* primary antibody response.

secondary disease A disease that develops in a weakened individual.

Seitz filter A filter composed of a pad of porcelain or ground glass.

selective medium A growth medium that contains ingredients to inhibit certain microorganisms while encouraging the growth of others.

septicemia A generalized bacterial infection of the bloodstream due to any of several organisms, including streptococci and staphylococci; once known as blood poisoning.

septum A cross-wall in the hypha of a fungus.

serological reaction An antigen-antibody reaction studied under laboratory conditions and involving serum.

serology A branch of immunology that studies serological reactions.

serotonin A derivative of tryptophan that functions in type I hypersensitivity reactions to contract smooth muscles.

serotype A rank of classification below the species level based on an organism's reaction with antibodies in serum; used for several bacteria, especially *Salmonella*.

serum The fluid portion of the blood consisting of water, minerals, salts, proteins, and other organic substances; contains no clotting agents.

serum sickness A type of hypersensitivity reaction in which the body responds to proteins contained in foreign serum.

severe combined immunodeficiency disease (SCID) An immune disease in which the lymph nodes lack both B lymphocytes and T lymphocytes.

shigellosis A foodborne and waterborne bacterial disease of the intestine caused by *Shigella* species, characterized by extensive diarrhea, often with blood and mucus.

shingles An alternate name for herpes zoster; a condition of the nerves and skin due to varicella-zoster virus.

short incubation hepatitis An alternate name for hepatitis A.

silage A type of animal feed produced by fermenting grains and other plants in silos, the huge cylindrical structures that often stand next to barns.

simple stain technique The use of a cationic dye to contrast cells.

slime layer A thinner, flowing, and less tightly bound form of a glycocalyx.

slapped-cheek disease An alternate name for fifth disease.

smallpox An extinct viral disease of the skin and body organs caused by a complex DNA virus and accompanied by bleeding skin pustules, disfigurements, and multiple organ involvements. Also called variola.

smut A fungal disease of agricultural crops, so named because of the sooty black appearance of infected plants.

sodium hypochlorite (NaOCl) A derivative of chlorine used in disinfection practices; also known as bleach.

sodoku An alternate name for rat-bite fever caused by a spirillum.

soft chancre An alternate name for chancroid.

sorbitol An alcoholic carbohydrate fermented slowly by *E. coli* O157:H7 and used in the medium to identify this organism.

specialized transduction A transduction in which the prophage carries some bacterial genes when it breaks free from the chromosome; the bacterial genes are then replicated and carried into the next cell by the virus. *See also* generalized transduction.

species The fundamental rank in the classification system of organisms.

species immunity Immunity present in one species of organisms but not another species.

specific immunologic tolerance A phenomenon in which a person's own proteins and polysaccharides contact and inactivate cells that might later respond to them immunologically (as "nonself").

spectinomycin An antibiotic used as a substitute for penicillin in cases of gonorrhea that are caused by penicillinase-producing *Neisseria gonorrhoeae*.

spherule A stage in the life cycle of the fungus *Coccidioides immitis*.

spike A functional projection of the viral envelope.

spirillum A bacterium characterized by twisted or curved rods, generally with a rigid cell wall and flagella.

spirochete A twisted bacterial rod with a flexible cell wall containing axial filaments for motility.

spontaneous generation The doctrine that nonliving matter could spontaneously give rise to living things.

spontaneous mutation A mutation that arises from chance events in the environment.

sporangiospore An asexually produced fungal spore formed within a sporangium.

sporangium A protective sac that contains asexually produced fungal spores.

spore A highly resistant structure formed from vegetative cells in several genera of bacteria, including *Bacillus* and *Clostridium*; also, a reproductive structure formed by a fungus.

sporicidal agent An agent that kills bacterial spores.

sporocyst An intermediary stage in the life cycle of a fluke; usually occurs in the snail.

sporotrichosis A soilborne fungal disease of the lymph channels caused by *Sporothrix schenkii*, characterized by knotlike growths under the skin surface and occasional skin lesions.

Sporozoa A group of protozoa whose members have no means of locomotion in the adult form.

sporozoite A stage in the life cycle of *Plasmodium* species; the parasite enters the human body in this form.

sputum Thick, expectorated matter from the lower respiratory tract.

staphylococcus A bacterium characterized by spheres in a grapelike cluster.

stationary phase The third portion of a bacterial growth curve in which the reproductive and death rates of cells are equal.

stem cell A primordial cell of bone marrow from which hematopoietic and lymphopoietic cells develop.

sterilization The removal of all life forms, especially bacterial spores.

stormy fermentation Fermentation and curdling of milk accompanied by gas accumulation that forces the curds apart.

streptobacillus A chain of bacterial rods.

streptococcus A chain of bacterial cocci.

streptokinase (strep´to-ki´nās) An enzyme that dissolves fibrin clots; produced by virulent streptococci.

streptomycin An antibiotic derived from *Streptomyces griseus* that is effective against gram-negative bacteria and the tubercle bacillus; interferes with protein synthesis.

subacute sclerosing panencephalitis (SSPE) A rare complication of measles that affects the brain, characterized by a loss of cortical functions and a decrease in cognitive skills.

subclinical disease A disease in which there are few or inapparent symptoms.

substrate The substance(s) upon which an enzyme acts.

subunit vaccine A vaccine that contains parts of microorganisms, such as capsular polysaccharides or purified pili.

sulfamethoxazole A sulfonamide compound used to treat vaginal infections, conjunctivitis, and other diseases.

sulfanilamides A group of sulfur-containing compounds used as antimicrobial agents.

sulfur granule A small, hard granule found in tissue infected with *Actinomyces israelii*; they resemble the sulfur granules used in a pharmacy.

superantigen An antigen that stimulates an immune response without any prior processing.

suppressor T lymphocyte A T lymphocyte that interferes with the activity of the immune system.

swimmer's itch Dermatitis of the skin caused by the body's reaction to certain species of schistosomes.

symbiosis An interrelationship between two populations of organisms where there is a close and permanent association for mutual benefit.

syncytium A giant tissue cell in culture formed by the fusion of cells infected with respiratory syncytial viruses.

syndrome A collection of symptoms.

synergism A close and permanent association between two populations of organisms such that the populations are able to accomplish together what neither could otherwise accomplish alone.

synthetic vaccine A vaccine that contains chemically synthesized parts of microorganisms, such as proteins normally found in viral capsids.

syphilis A sexually transmitted bacterial disease of multiple organs, caused by *Treponema pallidum* and occurring in three stages, characterized by extensive skin lesions, multiple organ involvements, and complications of the nervous and cardiovascular systems.

systemic disease A disease that has disseminated to the deeper organs and systems of the body.

systemic lupus erythematosus (SLE) An autoimmune disease in which antibodies form against nuclear components of the individual's cells and then unite with the antigens to form immune complexes in the skin and body organs.

T

tabardillo An alternate name for endemic typhus.

tapeworm A ribbonlike flatworm belonging to the phylum Platyhelminthes; also called cestode.

taxonomy The science dealing with the systematized arrangements of related living things in categories.

teichoic acid A polysaccharide found in the cell wall of gram-positive bacteria.

temperate phage A bacteriophage that enters a bacterium but does not replicate; the phage DNA may remain in the bacterial cytoplasm or attach to the bacterial chromosome.

tertian malaria A type of malaria in which the attacks occur every 48 hours.

tetanospasmin An exotoxin produced by *Clostridium tetani*; acts at synapses, thereby stimulating muscle contractions.

tetanus A soilborne bacterial disease of the muscles and nerves, caused by toxins produced by *Clostridium tetani*; characterized by uncontrolled muscle spasms.

tetracyclines A group of antibiotics, each of which is characterized by four benzene rings with attached side groups; used for many diseases caused by gram-negative bacteria, rickettsiae, and chlamydiae; inhibit protein synthesis in bacteria.

T-dependent antigen An antigen that requires the assistance of T lymphocytes to stimulate antibody-mediated immunity.

therapeutic serum Antibody-rich serum used to treat a specified condition.

thermal death point The temperature required to kill an organism in a given length of time.

thermal death time The length of time required to kill an organism at a given temperature.

thermoacidophile A prokaryote living under high temperature conditions.

thermoduric organism An organism that normally lives at medium temperatures but may tolerate and live at high temperatures.

thermophile An organism that lives at high temperature ranges of 40°C to 90°C.

thioglycollate broth A bacteriological medium that contains thioglycollic acid; the latter binds oxygen from the atmosphere and creates an environment suitable for anaerobic growth.

three-domain system The system of classification established by Carl Woese and his coworkers based on their studies of ribosomal RNA.

thrombocytopenia A reduced count of blood platelets (thrombocytes) resulting from the reaction of antibodies with antigens on the platelet surface; a type II hypersensitivity reaction.

thrush Oral candidiasis.

thymus A flat, bilobed organ that lies in the neck below the thyroid; T lymphocytes are modified in the thymus.

ticarcillin A semisynthetic penicillin antibiotic used against gram-positive bacteria.

tincture of iodine An iodine solution consisting of 2 percent iodine plus iodide in ethyl alcohol.

T-independent antigen An antigen that does not require the assistance of T lymphocytes to stimulate antibody-mediated immunity; found in bacterial flagella and capsules.

tinea Any of a group of fungal infections of the skin, including tinea pedis (athlete's foot) and tinea capitis (ringworm of the scalp).

tissue immunity An alternate term for cell-mediated immunity.

titer The most dilute concentration of antibody that will yield a positive reaction with a specific antigen; a method of expressing the amount of antibody in a sample of serum.

T lymphocyte A lymphocyte modified in the thymus gland, associated with cell-mediated immunity. Also called T cell.

topoisomerase An enzyme that prevents excessive coiling of DNA molecules.

TORCH An acronym for four diseases that pass from the mother to the unborn child: toxoplasmosis, rubella, cytomegalovirus disease, and herpes simplex; the O stands for other diseases.

toxic shock syndrome A bacterial disease of the blood caused by toxins produced by *Staphylococcus aureus*; characterized by shock and circulatory collapse.

toxin A poisonous substance produced by a species of microorganism; bacterial toxins are classified as exotoxins or endotoxins.

toxoid An immunizing agent produced from an exotoxin that elicits antitoxin production by the body.

toxoplasmosis A protozoal disease of the blood caused by *Toxoplasma gondii*, transmitted by contact with cats or the consumption of beef, characterized by malaise and nonspecific symptoms; a serious disease in immune-compromised patients; may cause fetal injury in pregnant women.

trachoma A contact disease of the eye caused by *Chlamydia trachomatis*; characterized by tiny, pale nodules on the conjunctiva, and possibly leading to blindness.

transcription The biochemical process in which RNA is synthesized according to a code supplied by the bases of a gene in the DNA molecule.

transduction A type of bacterial recombination in which a virus transports fragments of DNA from a donor cell to a recipient cell.

transfer RNA (tRNA) A molecule of RNA that unites with amino acids and transports them to the ribosome in protein synthesis.

transformation A type of bacterial recombination in which competent bacteria acquire fragments of DNA from disintegrated donor cells and incorporate the DNA into their chromosomes.

transgenic Referring to organisms containing DNA from another source.

translation The biochemical process in which the code on the mRNA molecule is translated into a sequence of amino acids in the protein molecule.

transposon A segment of DNA that moves from one site on a DNA molecule to another site, carrying information for protein synthesis. Also known as jumping gene.

traveler's diarrhea A foodborne and waterborne bacterial disease of the intestine often caused by *Escherichia coli*, characterized by diarrhea.

trematode A flatworm, commonly known as a fluke.

trench fever A louseborne rickettsial disease due to *Rochalimaea quintana*, characterized by fever and a skin rash. Also called His-Werner disease.

trichinosis A foodborne disease of the muscles caused by the parasite *Trichinella spiralis*, contracted from poorly cooked pork, characterized by muscle pains.

triclosan A bisphenol antiseptic incorporated into a wide variety of household products.

trichocyst An organelle of *Paramecium* species that discharges filaments used to trap the organism's prey.

trichomoniasis A protozoal disease of the reproductive tract caused by *Trichomonas vaginalis*, transmitted by sexual contact, characterized by discomfort in the urinary and lower reproductive tracts.

trivalent vaccine A vaccine consisting of three components, each of which stimulates immunity.

trophozoite The feeding form of a microorganism, such as a protozoan.

trypanosomiasis A protozoal disease of the blood caused by *Trypanosoma* species, transmitted by arthropods, and characterized by periods of high fever and coma. Also called sleeping sickness.

tsutsugamushi A relatively mild miteborne disease of the blood caused by *Rickettsia tsutsugamushi*, characterized by fever and a skin rash. Also called scrub typhus.

tubercle A hard nodule that develops in tissue infected with *Mycobacterium tuberculosis*.

tuberculin test A skin test used for the early detection of tuberculosis; performed by applying purified protein derivative to the skin and noting a thickening of the skin with a raised vesicle in a few days.

tuberculosis An airborne bacterial disease of the lungs caused by *Mycobacterium tuberculosis*, characterized by degeneration of the lung tissue and spreading to other organs.

tularemia A mild arthropodborne bacterial disease, transmitted by the tick and by contact, caused by *Francisella tularensis*; characterized by mild fever and vague symptoms of malaise.

tumor An abnormal functionless mass of cells.

tyndallization A sterilization method in which materials are heated in free-flowing steam for 30 minutes on each of three successive days.

typhoid fever A foodborne and waterborne bacterial disease of the intestine and blood caused by *Salmonella typhi*; characterized by intestinal ulcers, high fever, and red skin spots.

U

ultrapasteurization A pasteurization process in which milk is heated at 82°C for 3 seconds.

undulant fever An alternate name for brucellosis in humans, due to the undulating nature of the fever.

ureaplasmal urethritis A sexually transmitted bacterial disease of the urinary and genital organs caused by *Ureaplasma urealyticum*, characterized by urinary tract symptoms.

urticaria A hivelike rash of the skin.

urushiol A chemical substance on the leaves of certain plants that induces poison ivy.

V

vaccination Originally, the process by which Jenner introduced vaccinia (cowpox) to volunteers; currently, any immunization procedure involving an injection.

vaccine A preparation of modified microorganisms, treated toxins, or parts of microorganisms used for immunization purposes.

vaccinia An alternate name for cowpox, a mild form of smallpox.

vaginitis A general term for infection of the vagina.

vancomycin An antibiotic used in treating diseases caused by gram-positive bacteria, especially staphylococci.

variant Creutzfeldt-Jakob disease (vCJD) A nervous disorder in humans believed to be caused by a prion and possibly related to bovine spongiform encephalopathy.

varicella An alternate name for chickenpox; means "little vessel," a reference to small chickenpox lesions.

varicella-zoster (VZ) virus The virus that causes varicella (chickenpox) and herpes zoster (shingles).

variola An alternate name for smallpox.

VDRL test A precipitation test used in the detection of syphilis antibodies.

vector A living organism that transmits the agents of disease.

vertical gene transfer The passing of genes from one cell generation to the next.

vesicle A fluid-filled skin lesion, such as that occurring in chickenpox.

vibrio A form of bacterium occurring as a curved rod; resembles a comma.

vidarabine A drug useful in treating herpes zoster and herpes encephalitis.

Vincent's angina Bacterial infection of the tonsil and soft palate areas due to several species of bacteria; a form of acute necrotizing ulcerative gingivitis.

viremia The spread of viruses through the circulatory system.

virion A completely assembled virus outside its host cell.

viroid A tiny fragment of nucleic acid associated with certain plant diseases.

virucidal agent An agent that inactivates viruses.

virulence The degree of pathogenicity of a parasite.

virulent phage A bacteriophage that replicates within a bacterium and destroys the bacterium.

virus A particle of nucleic acid (either DNA or RNA) surrounded by a protein sheath, and, in some cases, a membranous envelope; neither prokaryotic nor eukaryotic; highly infectious.

W

walking pneumonia A colloquial expression for a mild case of pneumonia.

wart A small, usually benign skin growth commonly due to a virus.

Waterhouse-Friderichsen syndrome The formation of hemorrhagic lesions in the adrenal glands; thought to be an allergic reaction with immune complex formation; associated with meningococcal meningitis.

Weil-Felix test A diagnostic procedure for rickettsial diseases, performed by combining serum from a patient and *Proteus* OX-19 on a slide; in a positive test, the *Proteus* cells clump together.

wetting agent An agent that emulsifies and solubilizes particles clinging to a surface; an example is soap.

wheal An enlarged, hivelike zone of puffiness on the skin, often due to an allergic reaction.

whey The clear liquid remaining after protein has curdled out of milk.

whipworm A common name for *Trichuris trichiura*.

whooping cough An alternate name for pertussis.

Widal test An immunological procedure used for diagnostic purposes in which *Salmonella* antibodies agglutinate *Salmonella* antigens.

woolsorter's disease An alternate name for anthrax, derived from the fact that people who work with wool inhale the spores of the causative agent.

X

xenograft A tissue graft between members of different species, such as between an animal and a human.

Y

yaws A mild syphilislike disease occurring in remote parts of the world, caused by a species of *Treponema*.

yeast A type of fungus that is unicellular and resembles bacterial colonies when grown in culture.

yellow fever A highly fatal viral disease transmitted by the *Aedes aegypti* mosquito, characterized by intense fever and jaundice.

Z

zoonosis An animal disease that may be transmitted to humans.

zygomycete A member of the phylum Zygomycota.

Zygomycota A phylum of fungi whose members have coenocytic hyphae and form zygospores, among other notable characteristics.

zygospore A sexually produced spore formed by members of the Zygomycota fungi.

Index

W

Photograph Acknowledgments

Note: Photographs from *ASM News, Journal of Bacteriology, Journal of Clinical Microbiology, Infection and Immunity, Applied and Environmental Microbiology,* and *Antimicrobial Agents and Chemotherapy* are used with permission from the American Society for Microbiology.

CDC = Centers for Disease Control and Prevention, Public Health Service, U.S. Department of Health and Human Services, Atlanta, Georgia.

COVER: Dennis Kunkel Microscopy, Inc.

PART 1 OPENER: ©Arthur Siegelman/Visuals Unlimited.

PART 1 MICROBIOLOGY PATHWAYS: Courtesy of the CDC.

CHAPTER 1

Box 1.1: Bettman/Corbis. 1.1a: ©James King-Holmes/Photo Researchers, Inc. 1.1b: © Vincent Yu/AP Photo. 1.2a,b: Library of Congress, LC-USZ62-95187. 1.3a: Courtesy of Pfizer, Inc. Box 1.3: National Library of Medicine. Box 1.4: National Library of Medicine. 1.4: Courtesy Royal Society, London. 1.5a: Bettman/Corbis. Box 1.6: AP/World Wide Photos. 1.6: Courtesy of Pfizer, Inc. 1.7: ©SPL/Photo Researchers, Inc. 1.8: Courtesy of the Public Health Foundation/CDC. 1.9a: National Library of Medicine. 1.9b: Courtesy of Pfizer, Inc. 1.10a: Courtesy of National Library of Medicine. 1.10b: Bettmann/Corbis. 1.11: Courtesy of the CDC. 1.12: Courtesy of Institut Pasteur, Paris. 1.14a: ©Professor P. Motta and T. Naguro/Photo Researchers, Inc. 1.14b: ©Dr. Dennis Kunkel/Visuals Unlimited. 1.15a: Courtesy of Pfizer, Inc. 1.15b: ©St. Mary's Hospital Medical School/Photo Researchers, Inc. 1.16: Courtesy of Pfizer, Inc. 1.17: Reproduced with permission of the New York State Department of Health. 1.18: ©PhotoDisc. 1.19: Bacterial Biofilm in a Chronic Respiratory Tract Infection. ©Hiroyuki Kobayashi. Licensed for use, ASM MicrobeLibrary.org. 1.20a: Courtesy of Tom Loughlin, Office of Response and Restoration, NOAA. 1.20b: Courtesy of the Exxon Valdez Oil Spill Trustee Council/NOAA. 1.20c: Courtesy of Edgar Kleindinst, NMFS Wood Hole Laboratory/NOAA.

CHAPTER 2

Box 2.2: Courtesy of Danisco Animal Nutrition. Box 2.3: Courtesy of L. Tao, J.M. Tanzer, and T.J. MacAlister. Bicarbonate and potassium regulation of the shape of *Streptococcus mutans* NCTC 10449S. *J. Bacteriol.* 169:2543 (1987).

CHAPTER 3

Box 3.1: ©PhotoDisc. 3.2: From R.E. Mueller, et al. *AEM* 47:715 (1984). 3.3a: ©Dr. Kari Lounatmaa/Photo Researchers, Inc. 3.3b: Courtesy of Dr. Hans Paerl and Dr. Pia Moisander, University of North Carolina at Chapel Hill, Institute of Marine *Science*s. 3.4c: From John, Cole, & Marciano-Cabral. *AEM* 47:12–14 (1984). 3.4a: Scanning micrograph by L. Tetley, from K. Vickeman, *Nature* 273:613 (1978). 3.4b: Courtesy of Robert L. Owen, *Gastroenterology* 76:759–769 (1979). Box. 3.5: ©A. Tesks, Woods Hole Oceanographic Institution. 3.6a: ©Andrew Syred/Photo Researchers, Inc. 3.6b: Courtesy of Jacques Descloitres, MODIS Land Rapid Response Team, NASA/GSFC. 3.7a: Courtesy of E.M. Peterson, R.J. Hawlay, and R.A. Calerone. 3.7c: USDA/ARS Information Staff. 3.7b: ©Dr. C. W. Mims/The University of Georgia. 3.9a: From *The Hundred Greatest Men,* New York: Appleton & Company, 1885. Courtesy of the General Libraries, University of Texas at Austin. 3.9b: Courtesy of Wolfgang Peter (www.anthroposophie.net). 3.11: Courtesy of Bill Wiegand/University of Illinois. 3.14a: Courtesy of Carl Zeiss MicroImaging, Inc. 3.16d: ©Jack Bostrack/Visuals Unlimited. 3.17a,b,c: Visuals Unlimited/David M. Phillips. 3.18: ©Dr. Peter Lewis, The University of Newcastle, Australia. 3.19a: Courtesy of LEO Electron Microscopy. 3.20a,b: Courtesy of N. Gotoh, from *J. Bacteriol.* 171:983 (1989). MicroInquiry 3.A: Courtesy of the CDC. MicroInquiry 3.B: ©CNRI/Photo Researchers, Inc.

PART 2 OPENER: ©George Haling/Photo Researchers, Inc.

PART 2 MICROBIOLOGY PATHWAYS: USDA.

CHAPTER 4

4.2a: Courtesy of D.L. Shungu, J.B. Cornett, and G.D. Schockman, *J. Bacteriol.* 138:601 (1979). 4.2c: Courtesy of K. Amako. 4.2b: Courtesy of Dr. M. Matsuhashi, from Yamada et al., *J. Bacteriol.* 129:1513–1517 (1977), Figure 2A. Box 4.2: John Mitchell/Photo Researchers, Inc. Box 4.3: Courtesy of Dennis A. Bazylinski, Iowa State University. 4.3: D.L. Balkwill, D. Maratean, and R.P. Blakemore, from *J. Bacteriol.* 141:1399–1408 (1980). 4.MF4: ©Reuters/Landov. 4.5c: Photograph by Gary Gaard. Courtesy of Dr. A. Kelman (Department of Plant Pathology, University of Wisconsin-Madison). 4.6: ©Dr. Dennis Kunkel/Visuals Unlimited. Box 4.6: Courtesy of John Pacy, King's College London. 4.7a: Courtesy of Elliot Juni, Department of Microbiology and Immunology, The University of Michigan. 4.7b: Courtesy of S.D. Acres, VIDO, University of Saskatchewan, and J.W. Costeron, University of Calgary. From *Infect. Immun.* 37(3):1170 (1982). 4.10: With permission of Victor Lorian, M.D., Bronx Lebanon Hospital Center, Bronx, NY. 4.11: Courtesy of E. Kellenberger, from *J. Bacteriol.* 173:3149 (1992). 4.12a: Courtesy of Dr. C. Robinow, from *The Bacteria,* Vol. 1, Academic Press, New York (1960), Fig. 7, p. 214. 4.12b: Courtesy Gerhardt, Pankrants, and Scherrer, *Appl. Environ. Microbiol.* 32:438–439 (1976). 4.12c: Courtesy of Drs. Yoshii, J. Tokunaga, and J. Tawara, *Atlas of Scanning Electron Microscopy in Microbiology,* IGAKU-SHOIN, Ltd., 1976. 4.14b: ©Lee Simon/Photo Researchers, Inc. 4.17a: Courtesy of P. Scherer, from *ASM NEWS* 56:569 (1990). 4.17b: Courtesy of Drs. Z. Yoshii, J. Tokunaga, and J. Tawara, *Atlas of Scanning Electron Microscopy in Microbiology,* IGAKU-SHOIN, Ltd., 1976. 4.20b: Jack Bostrack/Visuals Unlimited. 4.21e: John Durham/Science Photo Library/Photo Researchers, Inc.

CHAPTER 5

5.1b: "The *Candida boidinii* Peroxisomal Membrane Protein Pmp30 Has a Role in Peroxisomal Proliferation and Is Functionally Homologous to Pmp27 from Saccharomyces cerevisiae" by Y. Sakai, PA. Marshall, et al. *J. Bacteriol.* 177(23): 6773-6781 (Dec. 1995), ASM fig. 4a&b, page 6778. 5.1a: ©Stem Jems/Photo Researchers, Inc. 5.13: ©Dr. Dennis Kunkel/Visuals Unlimited. 5.15a: Courtesy of H.W. Jannasch, from *Appl. Environ. Microbiol.* 41:528–538 (1981). 5.15b: ©Woods Hole Oceanographic Institution.

CHAPTER 6

6.2a: H. Potter-D. Dressler/Visuals Unlimited. Box 6.2: ©Vittorio Luzzati/ Photo Researchers, Inc. Box 6.6: Corbis. 6.7: Reprinted with permission from O. L. Miller, B. A. Hamkalo, and C. A. Thomas. 1977. *Science* 169: 392. Copyright 1977 American Association for the Advancement of Science. 6.9a,b: Courtesy of C. Nombela, from *J. Bacteriol.* 172:2384 (1990).

CHAPTER 7

Box 7.1: AbleStock. 7.4: ©Dr. Dennis Kunkel/Visuals Unlimited. 7.6a: Courtesy of the CDC. 7.6b: Oliver Meckes/Photo Researchers, Inc. Box 7.6: ©PhotoDisc. Box 7.8: Courtesy of The Wellcome Trust Sanger Institute—Photography by Richard Summers. 7.13:©Prof. Dr. K.O. Stetter and Dr. Reinhard Rachel, University of Regensburg, Germany. 7.14: Courtesy of Michael J. Daly.

PART 3 OPENER: ©Dr. Dennis Kunkel/Visuals Unlimited.

PART 3 MICROBIOLOGY PATHWAYS: ©PhotoDisc.

CHAPTER 8

8.1: Courtesy of John Pacy, King's College London. Box 8.1: Chinch Gryniewicz/Ecoscene/Corbis. 8.2a: Courtesy of L. Tao, from *J. Bacteriol.* 169:2543 (1987). 8.2b: Courtesy of U. Fluckiju and V.A. Fischetti, from *Infect. Immun.* 66:974–979 (1998). 8.2c: Courtesy of V. Fischetti, from *ASM News* 57:619 (1991) 8.3: Ken Greer/Visuals Unlimited. 8.4b: Courtesy Professor Mumtaz Virji and Dr. David J.P. Ferguson from cover of *ASM News* 64(7), 1998. Box 8.5: Courtesy Stephanie Torta. Box. 8.6: Courtesy of the California Tuberculosis Controllers Association. 8.7: "Portrait of a Woman," by Franz Hals. The Detroit Institute of Arts, City of Detroit Purchase. Box 8.7: AP/Wide World Photos. 8.8: ©SPL/Photo Researchers, Inc. 8.9a: Courtesy of Gourley, Leach, and Howard, *J. Gen. Microbiol.* 81:475 (1974), with permission from ASM. 8.9b: Courtesy of Respiratory Disease Laboratory Section, DCDP, from *Appl. Environ. Microbiol.* 47(3):467 (1984). 8.10: ©Dr. Kari Lounatmaa/Photo Researchers, Inc. 8.11a: S. Ito and A.R. Flesher, Harvard Medical School. 8.11b: Courtesy of J. Kwaik, from *Appl. Environ. Microbiol.* 62(6): 2022–2028 (1996), fig. 1E p. 2025. 8.13: Courtesy of Dr. Edgar Ribi and William R. Brown, Ribi ImmunoChem Research, Inc., Hamilton, Montana. 8.15: Courtesy of Arking, Appelt, and Balin, from *Science News* 154: 325 (1998).

CHAPTER 9

Box 9.1: Bettmann/Corbis. 9.1: "Characterization of a Region of the IncH12 Plasmid R478 Which Protects *Escherichia coli* from Toxic Effects Specified by Components of the Tellurite, Phage and Colicin Resistance Cluster," by K.F. Whelan, R.K. Sherburne and D.E. Taylor, *J. Bacteriol.* Vol. 179(1): 63–71 (Jan. 1997), ASM fig. 6c, page 69. 9.3a: Courtesy of Drs. Z. Yoshii, J. Tokunaga, and J. Tawara, *Atlas of Scanning Electron Microscopy in Microbiology,* IGAKU-SHOIN, Ltd., 1976. 9.3b:Courtesy of the CDC. Box 9.3: ©PhotoDisc. Box 9.4: USDA-ARS Information Staff. 9.5a: Courtesy of D.J. Thomas and T.A. McMeekin, University of Tasmania, Tasmania, Australia. 9.5b: Courtesy of Dr. W.L. Dentler. 9.6b: Corbis. 9.6a: ©Dr. Dennis Kunkel/Visuals Unlimited. 9.7: From B. Kenny, S. Ellis, A.D. Leard, J. Warawa, H.. Mellor, and M.A. Jepson, Co-ordinate regulation of distinct host cell signalling pathways by malfunctional enteropathogenic *E. coli* (EPEC) effector molecules. *Mol. Microbiol.* 44:4 (2002). © Blackwell Publishing. 9.10a,b: Courtesy of Diane E. Taylor, from *J. Bacteriol.* 164(1):338–343 (1985).

9.11: Courtesy of Tilney, from *Trends in Microbiol.* 1(1) (1993). 9.12: USDA-ARS Information Staff. 9.13: Courtesy of Chai, from *Appl. Environ. Microbiol.* 62:1300–1305 (1996). 9.14: Courtesy of V. Miller, from *ASM News* 58:26 (1992) with permission of American Society for Microbiology.

CHAPTER 10

Box 10.1: S & G Press Agency, Ltd. 10.1a,b: Courtesy of Drs. Z. Yoshii, J. Tokunaga, and J. Tawara, *Atlas of Scanning Electron Microscopy in Microbiology.* IGAKU-SHOIN, Ltd., 1976. 10.2: Charles W. Stratton/Visuals Unlimited. 10.3: Courtesy of The Royal College of Surgeons, Edinburgh. 10.4: Courtesy of the CDC. 10.5a,b: Courtesy of Drs. Z. Yoshii, J. Tokunaga, and J. Tawara, *Atlas of Scanning Electron Microscopy in Microbiology.* IGAKU-SHOIN, Ltd., 1976. 10.6: Bettman/Corbis. Box. 10.7: Grant Heilman/Grant Heilman Photography. 10.8a,b: Courtesy of Armed Forces Institute of Pathology Neg. Nos. N-85837-2 and AN1147-1A, respectively. 10.9: Courtesy of the CDC. 10.10a: Courtesy of A. Garon, Rocky Mountain Laboratory, Hamilton, Montana. 10.10b: Courtesy of J. Radolf, from *J. Bacteriol.* 173:8004 (1991). 10.11: Science VU/Visuals Unlimited. 10.14a: Courtesy of Timothy J. Lysyk/Agriculture and Agri-Food Canada. 10.14b: Oliver Meckes/Photo Researchers, Inc. 10.14c: John Burbidge/SPL/Photo Researchers, Inc. 10.15 insert: Armed Forces Institute of Pathology. 10.16: Armed Forces Institute of Pathology.

CHAPTER 11

11.1a: Corbis. 11.1b: Ken Greer/Visuals Unlimited. 11.1c: Courtesy of the CDC. Box 11.2: Courtesy of National Hansen's Disease Center, Carville, LA. 11.2a:©Science Source/Photo Researchers, Inc. 11.2b: Courtesy of the CDC. 11.4a.b: Courtesy of Drs. Z. Yoshii, J. Tokunaga, and J. Tawara, *Atlas of Scanning Electron Microscopy in Microbiology.* IGAKU-SHOIN, Ltd., 1976. 11.4c: Biophoto Associates/Photo Researchers, Inc. Box 11.4: Ken Cavanagh/Photo Researchers, Inc. 11.5b: Courtesy of Dr. L.A. Page. From *Bergey's Manual of Determinative Bacteriology*, 8th ed., Williams and Wilkins Co.: Baltimore (1974). 11.7: Courtesy of D. Krause from *J. Bacteriol.* 173:1041 (1991). 11.8a,b: Courtesy of American Leprosy Mission, 1 Broadway, Elmwood Park, NJ. 11.9b: Courtesy of the CDC. 11.10 insert: Courtesy of the CDC. 11.11: Courtesy of World Health Organization–Prevention and Blindness Programme. 11.12: David M. Phillips/Visuals Unlimited. 11.13a,b: Armed Forces Institute of Pathology. 11.14b: Courtesy of the Naval Dental Research Institute, Naval Training Center, Great Lakes, Illinois. 11.15: Courtesy of Gobel, from *J. Clin. Microbiol.* 36: 1399 (1998). 11.16a: ©Andrew Syred/Photo Researchers, Inc. 11.16b: Courtesy M. Robert Belas, Center of Marine Biotechnology, Maryland Biotechnology Institute. 11.17: Courtesy of Z. Samara, Rabin Medical Center, Israel. 11.19: NNIS data.

PART 4 OPENER: ©Dr. Linda Stannard, Uct/Photo Researchers, Inc.

PART 4 MICROBIOLOGY PATHWAYS: Courtesy Anne Simon.

CHAPTER 12

12.1a: Courtesy of National Library of Medicine. 12.1b,c: Courtesy of the American Type Culture Collection. 12.3: Courtesy of C.K.Y. Fong, from *J. Clin. Microbiol.* 30:1612 (1992). Box 12.4: PhotoDisc. 12.4b Courtesy of Dr. Glenn Howard, Pall Biomedical Products Corporation, Glen Cove, NY. 12.7: Courtesy of the CDC. 12.8a,b,c: Courtesy of Dr. Sara Miller, Duke University Medical Center. 12.9b: Courtesy of Paul R. Chipman, Petr G. Leiman, and Michael G. Rossmann of Purdue University, and Vadim V. Mesyanzhinov of the Moscow Institute of Bioorganic Chemistry. 12.11: Courtesy of Drs. Z. Yoshii, J. Tokunaga, and J. Tawara, *Atlas of Scanning Electron Microscopy in Microbiology.* IGAKU-SHOIN, Ltd., 1976. 12.17: Courtesy of Fisher Scientific Company. 12.24a: Courtesy Stanley Prusiner/Institute for Neurodegenerative Diseases/University of California at San Francisco.

CHAPTER 13

Box 13.1: ©Eye of Science/Photo Researchers, Inc. 13.1a: Courtesy of the CDC. 13.1b: Bettmann/CDC. 13.4: Courtesy of Dr. Sara Miller, Duke University Medical Center. 13.5a,b: Courtesy of Sailen Barik et al. University of South Alabama. Box 13.7: Courtesy of Gertrude B. Elion. 13.8: ©Dr. Steve Patterson/Photo Researchers, Inc. 13.9a: St. Bartholomew's Hospital/SPL/Photo Researchers, Inc. 13.9b: Corbis. 13.10: Courtesy of Spotswood L. Spruance, from *J. Clin. Microbiol.* 22(3):366–368 (1985). 13.12a: Biofoto Associates/Photo Researchers, Inc. 13.12b: Science Photo Library/Photo Researchers, Inc. 13.13: Courtesy of J. Vreeswijk, from *J. Clin. Microbiol.* 30:2487 (1992). 13.16: Dr. P. Marazzi/SPL/Photo Researchers, Inc. 13.17b: Hans Gelderblom/Visuals Unlimited. 13.17c: Courtesy of V. Murphy, from *ASM News* 57:579 (1991). 13.18a,b: Courtesy of the CDC.

CHAPTER 14

Box 14.1: Courtesy of Parke-Davis, Division of Warner-Lambert Company, Morris Plains, NJ. 14.2a: London School of Hygiene and Tropical Medicine/SPL/Photo Researchers, Inc. 14.2b: CDC Photo by James Gathany. 14.4: Courtesy of Dr. E.H. Cook, Jr., Hepatitis and Viral Enteritis Division, CDC, Phoenix, AZ. 14.6a,b: Courtesy of Fred B. Williams, Jr., U.S. Environmental Protection Agency, Cincinnati, OH 45268. Box 14.9: S. Nagandra/Photo Researchers, Inc. 14.9b: Scott Camazine/Photo Researchers, Inc. 14.9a: Courtesy of the CDC. 14.10: Courtesy of the Pan American Health Organization. Source: Pan American Health Organization (PAHO). Available at: http://www.paho.org. Accessed July 2003. 14.15: Centers for Disease Control. Available at:

http://www.cdc.gov/ncidod/dvrd/rabies/Epidemiology/Epidemiology.htm. Accessed July 2003. 14.16: Courtesy of Drs. F.A. Murphy and A.K. Harrison, in *Rhabdoviruses*, D.H.L. Bishop, ed., Vol. I, pp. 66–106, CRC Press, 1979. 14.18: Courtesy of the CDC. 14.21: Courtesy of the CDC. 14.22: ©Dr. Lisa Stannard, UCT/Photo Researchers, Inc.

CHAPTER 15
15.1a: Courtesy of John Pitt of CSIRO Food *Science*, Australia. 15.1b: Courtesy of M.A. Lachance. 15.2: Courtesy of C. Nombela, from *J. Bacteriol.* 172:2384 (1990). 15.4: "Disseminated Invasive Infection Due to *Metarrhizium anisopliae* in an Immunocompromised Child," by D. Burgner, G. Eagles, M. Burgess, P. Procopis, M. Rogers, D. Muir, R.Pritchard, A. Hocking and M. Priest. *J. Clin. Microbiol.* 36: 1146–1150 (Apr.1998) ASM Fig. 1a, page 1147. Box 15.5a,b: Courtesy of Gary Strobel, Montana State University. 15.5: Courtesy of Drs. L.F. Ellis and S.W. Queener. Reproduced by permission of the National Research Council of Canada, from the *Canadian Journal of Microbiology,* Vol. 21, pp. 1982–1996. Box 15.6: Courtesy Fleishmann's Yeast. 15.6a,b: Courtesy of Drs. Z. Yoshii, J. Tokunaga, and J. Tawara, *Atlas of Scanning Electron Microscopy in Microbiology.* IGAKU-SHOIN, Ltd., 1976. Box 15.7: Courtesy Block Drug Company, Inc. Box 15.8: USDA-ARS Information Staff. 15.8: Courtesy of Northern Regional Research Center, Peoria, IL 61604. 15.9a: Visuals Unlimited. 15.9b: Richard Gross. 15.9c: Jack Bostrack/Visuals Unlimited. 15.10: Courtesy of K.J. Kwon-Chung, from *J. Clin. Microbiol.* 30:3290 (1992). 15.11a,b: Courtesy of T. Sewall, from *ASM News* 54:657 (1988). 15.12b: Richard Gross. 15.15a: Courtesy of Joseph Schlitz Brewery. 15.15b: David M. Phillips/Visuals Unlimited. 15.16a: Stuart Levitz, Boston University School of Medicine from *ASM News* Vol. 64(12): 693–698, Fig. 2. 15.16b: Courtesy of M.A. Kydstra, from *Infect. Immun.* 16:129 (1977). 15.17: Copyright © 1987 American Society of Clinical Pathologists. Reprinted with permission. Image courtesy of the CDC. 15.18a,b: Courtesy of the CDC. 15.20a,b: "Case of Onychomycosis Caused by Microsporum racemosum" by P. Garcia-Martos, J. Gene, et al. *J. Clin. Microbiol.* 37(1): 258–260 (Jan.1999) ASM fig. 1, page 258, fig. 2a, page 259. 15.22a: ©E. Gueho–CNRI/Photo Researchers, Inc.15.22b: Courtesy of the CDC.

CHAPTER 16
Box 16.1: AP/Wide World Photos. 16.1: Courtesy of D. White, from *ASM News* 54:583 (1988). 16.3: From John, Cole, & Marciano-Cabral *AEM* 47:12–14 (1984). Box 16.4: Portrait of Charles Darwin, 1840 by George Richmond (1809-96). Down House, Downe, Kent, UK/Bridgeman Art Library. 16.4a–f: Courtesy of Carolina Biological Supply Company. 16.5: ©John Walsh/Photo Researchers, Inc. 16.6: Diana Laulaien-Schein. Box 16.7: ©PhotoDisc. 16.8a,b: Courtesy of Govinda S. Visvesvara, Division of Parasitic Diseases, CDC. 16.10a–c: Courtesy of Robert L. Owen and the editors of *Gastroenterology* 76:759–769 (1979). 16.11a,b: Courtesy of C. Chetty, A. Armstrong, and M. Miller, Abbott Laboratories. 16.12a: Courtesy of the Laboratory of *Parasitology*, University of Pennsylvania School of Veterinary Medicine. 16.12b: ©SPL/Photo Researchers, Inc. 16.13: Courtesy of T. Minnick, from *ASM News* 51:239 (1985). 16.14: Courtesy of the CDC. 16.16a,b: Courtesy of Chiappino, Nichols, and O'Connor, *J. Protozool.* 31:228 (1984). 16.19: Courtesy of Saul Tzipori, from *Microbiological Review* 47(1):84–96 (March 1983). 16.20a: ©PhotoDisc. 16.20b: Courtesy of Michael Arrowood, Ph.D., Division of Parasitic Diseases, CDC.

CHAPTER 17
17.1a: Courtesy of Armed Forces Institute of Pathology, Neg. No. 69-77765-2. 17.1b: Courtesy of Dr. Clive E. Bennet, Jr., *Parasitology* 61:892 (1975). 17.4a–c: Courtesy of Dr. J. Jujino. 17.7a,b: Courtesy of B. Schieven, from *ASM News* 57:566 (1991). 17.8: Courtesy of Dr. K.A. Wright, Jr., *J. Parasitology* 65(3):441 (1979). 17.10a,b: Courtesy of Armed Forces Institute of Pathology, Neg. Nos. 69-3584 and 60-3583, repsectively. 17.12 insert: Courtesy of Armed Forces Institute of Pathology, Neg. No. 74-11349. 17.13: R. Umesh Chandran, WHO/SPL/Photo Researchers, Inc. 17.14a,b: Courtesy of Armed Forces Institute of Pathology, Neg. Nos. 73-6654 and 75-11789-4, respecitively.

PART 5 OPENER: ©P. Webster/Custom Medical Stock Photo.

PART 5 MICROBIOLOGY PATHWAYS: ©PhotoDisc.

CHAPTER 18
Box 18.1: © Photodisc. 18.1: Courtesy of Drs. Z. Yoshii, J. Tokunaga, and J. Tawara, *Atlas of Scanning Electron Microscopy of Microbiology*, IGAKU-SHOIN, Ltd., 1976. 18.4a: From Ksiazek et al, A novel coronavirus associated with severe acute respiratory syndrome, *N Engl J Med*, 348:1966, 2003. Copyright © 2004 Massachusetts Medical Society. All right reserved. 18.4b: ©Logical Images/Custom Medical Stock Photo. 18.4c: ©*Science* Source/Photo Researchers, Inc. Box 18.7: Courtesy Paul Ewald, Amherst College. Box 18.8: London School of Hygiene & Tropical Medicine /SPL/Photo Researchers, Inc. 18.10a,b: Courtesy of Drs. D.C. Savage and R.V.H. Blumershine, *Infect. Immun.* 10:240–250 (1974). 18.14a: Courtesy of J.A. Dowsett and J. Reid, from *Mycologia* 71:329 (1979). 18.14b: Courtesy of D. Corwin and S. Falkow, Rocky Mountain Laboratory, NIAID, NIH, Hamilton, Montana. 18.15: Courtesy of B. Finlay, from *ASM News* 58:486 (1992). 18.16: "Reciprocal Exchange of Minor Components of Type and 1 and F1C Fimbriae Results in Hybrid Organelles with Changed Receptor. 19.3 Specificities, " by P.

Klemm, G. Christiansen, B. Kreft, R. Marre and H. Bergmans *J. Bacteriol.* Vol. 176(8): 2227–2234 (Apr.1994) ASM fig. 2a&b, page 2231. 18.17: Courtesy of Jose Galan, SUNY, Stony Brook. 18.18a–c: Courtesy of Tilney, from *Trends in Microbiology* 1(1) (1993). 18.20: "Tobramycin Resistance of Pseudomonas aeruginosa Cells Growing as a Biofilm on Urinary Catheter Material," by J.C. Nickel, I. Ruseska, J.B. Wright, and J.W. Closterton *Antimicrobial Agents and Chemotherapy* 27(4): 619–624 (Apr.1985) ASM fig. 2, page 620, fig. 3, page 621.18.21: Courtesy of Howard J. Faden.

CHAPTER 19
Box 19.1: Courtesy of David Pfennig. 19.2: Courtesy of Drs. D.F. Bainton and M.G. Farquhar. 19.9: Juergen Berger/Max-Planck Institute/Photo Researchers, Inc. 19.12: SPL/Photo Researchers, Inc.

CHAPTER 20
MicroInquiry 20.A: Courtesy of Home Access Health Corporation. Box 20.1: Corbis. 20.3: Courtesy of G. Christiansen, *J. Bacteriol.* 176:2227–2234 (1994). 20.4: Courtesy of the CDC. Box 20.7: Corbis. Box 20.8: Photo Researchers, Inc. 20.8a: Courtesy of the CDC.20.8b: From Razin et al., *Infect. Immun.* 30:538–546 (1980). 20.11: "Phylogeny of Not-Yet-Cultured Spirochetes from Termite Guts," by B.J. Paster, F.E. Dewhirst, S.M. Cooke, V.Fussing, L.K. Poulsen and J.A. Breznak *Appli. Environ. Microbiol.* Vol. 62(2): 347–352. (Feb. 1996) ASM fig. 2 a-c, page 350.

CHAPTER 21
Box 21.1: Ed Reschke. Box 21.3: USDA-ARS Information Staff. 21.4a,b: Photos by Scott T. Clay-Poole. Box 21.5: Visuals Unlimited. 21.5: David M. Grossman/Photo Researchers, Inc. Box 21.8: ©PhotoDisc. 21.9a,b: Courtesy of Dr. Robert A. Marcus. 21.10: Ken Greer/Visuals Unlimited. 21.12: SIU/Photo Researchers, Inc. 21.13: Corbis. 21.15: Larry Mulvehill/Photo Researchers, Inc.

PART 6 OPENER: Christine L. Case/Skyline College.

PART 6 MICROBIOLOGY PATHWAYS: ©PhotoDisc.

CHAPTER 22
22.1a: Courtesy of J. Moran, from *J. Bacteriol.* Vol. 176: 2003–2012. (Apr. 1994) ASM fig. 4, page 2007. 22.1b: Courtesy of the CDC. 22.2a,b: "Salt-Mediated Multicell Formation in *Deinococcus radiodurans*," by F.I. Chou and S. T. Tan *J. Bacteriol.* Vol. 173(10): 3184–3190 (May,1991) ASM fig. 1e, page 3185, 2A, page 3187. 22.3: Lester V. Bergman/Corbis. 22.4: ©Simon Fraser/Photo Researchers, Inc. Box 22.5: Courtesy Panasonic Consumer Products. Box 22.6: Snack Food Association. 22.8: Courtesy of AMSCO/American Sterilizer Company, Erie, PA 16512. 22.9: Larry Lefever/Grant Heilman Photography. 22.10b: Courtesy of Glenn Howard, Pall Biomedical Products Corporation, Glen Cove, NY. 22.11a,b: Courtesy of Pall Biomedical Products Corporation, Glen Cove, NY. 22.12a–d: Courtesy of Millipore Corporation, Bedford, MA. 22.14a: Nordion/Visuals Unlimited. 22.14b: Courtesy of US Army Natick Soldier Center.

CHAPTER 23
MicroInquiry 23.A: ©Ted Horowitz/Corbis. MicroInquiry 23.B: Courtesy of the US Census Bureau, Public Information Office. Box 23.1: Courtesy of Parke-Davis, Division of Warner-Lambert Company, Morris Plains, NJ. 23.1a,b ©Jones and Bartlett Publishers. 23.1c ©D.Y. Riess MD/Alamy Images. 23.1d ©Flat Earth/FotoSearch. Box 23.2: ©PhotoDisc Box 23.3: ©PhotoDisc. 23.3: Courtesy of the CDC. 23.4: Courtesy C.J. Thomas and T.A. McMeekin, University of Tasmania, Tasmania, Australia. 23.6: CDC Photo by Janice Carr. 23.8: Courtesy of T.J. Marrie and J.W. Costerton, from *Appl. Environ. Microbiol.* 42(6):1093 (1981). 23.10a,b: Courtesy of American Sterilizer Company.

CHAPTER 24
Box 24.1: Courtesy of Forest and Kim Starr/Haleakala Field Station/USGS/BRD/PIERC. 24.1: The Granger Collection, New York. Box 24.2: Vanni Archive/Corbis. 24.2: Courtesy of Pfizer, Inc. 24.4a,b: Courtesy of the National Library of Medicine. Box 24.5: Courtesy of S. Horinuchi, from *J. Bacteriol.* 172:3003 (1990) with permission of American Society for Microbiology. 24.5: Courtesy of Pfizer, Inc. 24.7: James Webb, Phototake. 24.9a,b: Courtesy of H. Konishi, from *J. Bacteriol.* 168:1476 (1986). 24.10: Courtesy of Dr. D.R. Zusman. 24.11b: Kenneth E. Greer/Visuals Unlimited. 24.13a: Miles/Visuals Unlimited. 24.13b,c: Courtesy BioMerieux, Inc.

PART 7 OPENER: Science Photo Library/Photo Researches, Inc.

PART 7 MICROBIOLOGY PATHWAYS: Corbis

CHAPTER 25
Box 25.2: Corbis. 25.5: "Bacterial Plasmolysis as a Physical Indicator of Viability," by D.R. Korber, A. Choi, G.M. Wolfaardt and D.E. Caldewell. *Appl. Environ. Microbiol.* 62(11): 3939–3947 (Nov. 1996) ASM fig. 6a, page 3944. 25.6a,b: Courtesy of the National Food Processors Association, Washington, DC. 25.8: Courtesy of C.J. Thomas and R.A. McMeekin, University of Tasmania, Tasmania, Australia, from *Appl. Environ. Microbiol.* 41(2):492–503.

25.9: Robert Longuehaye, NIBSC/SPL/Photo Researchers, Inc. 25.11:USDA. 25.12: Courtesy of Paul Gustafson. 25.13a,b:Courtesy of M.A. Dawschel and H.P. Flemming, USDA-ARS, Raleigh, NC. 25.14:Courtesy of Kikkoman Foods, Inc., Walworth, WI. 25.15a–f: Courtesy of the Switzerland Cheese Association. 25.16: Courtesy Dr. Ed Alcamo.

CHAPTER 26

26.1a: Biophoto Associates/Photo Researchers, Inc. 26.1b: Courtesy of Tomio Kawata. 26.3: Courtesy of Dr. Neil Sullivan, University of Sourthern California/NOAA Corps Collection. 26.4: Photo provided by A.C. Scott, Central Veterinary Laboratory, Wexbridge, U.K., British Crown Copyright 1982. 26.5a,b: Courtesy of J. Kwaik, from *Appl. Environ. Microbiol.* 62(6): 2022–2028 (1996), fig. 1D, page 2024, fig. 1F, page 2025. 26.6a: Courtesy, Florida Fish & Wildlife Conservation Commission. 26.6b: AP/Wide World Photos. 26.7: Courtesy of Robert L. Owen, et al., from *Gastroenterology* 76: 757–769 (1979). 26.10: "Resistance of *Pseudomonas aeruginosa* Cells Growing as a Biofilm on Urinary Catheter Material," by J.C. Nickel, I. Ruseska, J.B. Wright, and J.W. Closterton *Antimicrobial Agents and Chemotherapy* 27(4): 619–624 (Apr.1985) ASM fig. 2, page 620, fig. 3, page 621. 26.11: Courtesy of Scott Bauer/USDA. 26.13: Allan R.J. Eaglesham, Boyce Thompson Institute. 26.15: Courtesy of Louis Sherman, used with permission of the ASM. 26.16a: Courtesy of Harold Evans. 26.16b: "Generation of Buds, Swellings, and Branches instead of Filaments after Blocking the Cell Cycle of *Rhizobium meliloti*," by J. N. Latch and William Margolin. *J. Bacteriol.* Vol. 179(7):2373–2381 (Apr.1997) ASM fig. 4E, page 2378.

CHAPTER 27

27.3: USDA-ARS Information Staff. Box 27.4: ©PhotoDisc. 27.4: Photo ACL, Bruxelles. Prière de verser 1000 FB au compte 000-0343629-55 du Patrimoine des Muscès royaux d'art et d'histoire. 10 parc du Cinquantenaire B-1040 Bruxelles. 27.6a–c: Courtesy of Wine Institute of California. 27.7a: Courtesy of Pfizer, Inc. 27.7b: "Cloning and Characterization of a Gene Involved in Aerial Mycelium Formation in *Streptomyces griseus*," by N. Kudo, M. Kimura, T. Beppu, and S. Horinuchi *J. Bacteriol.* Vol. 177(22): 6401–6410 (Nov.1995) ASM fig.2a, page 6404. 27.8a–c: "Characterization of a Novel Strain of *Bacillus thuringiensis*," by J. E. Lopez-Mesa & J. E. Ibarra *Appl. Environ. Microbiol.* Vol. 62(4): 1306–1310 (Apr.1996) ASM Fig. 1 a-c, page 1307. 27.9: Courtesy of the Exxon Valdez Oil Spill Trust Council/NOAA. 27.10: Courtesy of the CDC.

Leuconostoc citrovorum   lü-kū-nos'tok sit-rō-vôr'um

Listeria monocytogenes   lis-te'rē-ä mo-nō-sī-tô'je-nēz

Loa loa   lō'ä lō'ä

Metarrhizium   met-är-ē'sē-um

Methanobacterium   meth-a-nō-bak-tėr'ē-um

Methanococcus jannaschii   meth-a-nō-kok'kus jan-nä'shē-ē

Micrococcus luteus   mī-krō-kok'kus lū'tē-us

Micromonospora   mī-krō-mō-nos'pōr-ä

Microsporum racemosum   mī-krō-spô'rum ras-ē-mōs'um

Moriga oleifera   mor-e'-ga ō-lif'ėr-ä

Mucor   mū'kôr

Mycobacterium abscessus   mī-kō-bak-ti'rē-um ab-ses'us

M. avium-intracellulare   ā'vē-um-in'trä-cel-ū-lä-rē

M. bovis   bō'vis

M. cheloni   kē-lō'ē

M. haemophilum   hē-mo'fil-um

M. leprae   lep'rī

M. marinum   mãr'in-um

M. scrofulaceum   skrof-ū-lä'sē-um

M. tuberculosis   tü-bėr-kū-lō'sis

Mycoplasma genitalium   mī-kō-plaz'mä jen'i-tä-lē-um

M. hominis   ho'mi-nis

M. pneumoniae   nu-mō'nē-ī

Myxococcus xanthus   micks-ō-kok'kus zan'thus

Naegleria fowleri   nī-gle'rē-ä fou'lėr-ē

Nannizzia   nan'nė-zė-ä

Nanoarchaeum equitans   na-nō-ärk-ē-um ē-kwi-tänz

Necator americanus   ne-kā'tôr ä-me-ri-ka'nus

Neisseria gonorrhoeae   nī-se'rē-ä go-nôr-rē'ī

N. meningitidis   me-nin ji'ti-dis

Neurospora   nū-ros'pōr-ä

Nitrobacter   nī-trō-bak'tėr

Nitrosomonas   nī-trō-sō-mō'näs

Nocardia asteroides   nō-kär'dē-a as-tėr-oi'dēz

Ornithodoros   ôr-nith-ō'dô-rōs

Panagrellus   pa-nä-grē'lus

Paragonimus westermani   pãr-ä-gōn'e-mus we-stėr-ma'nē

Paramecium   pãr-ä-mē'sē-um

Pasteurella multocida   pas-tyėr-el'lä mul-tō'si-dä

Pediculus humanus   ped-ik'ū-lus hū-man'us

Pelagibacter ubique   pe-lag-i-bak'tėr ū'bi-kwē

Penicillium camemberti   pen-i-sil'lē-um kam-am-bėr'tē

P. chrysogenum   krī-so'gen-um

P. janczewski   jānk-zū'skē

P. notatum   nō-tä'tum

P. roqueforti   rō-kō-fôr'tē

Penstemon pruinosus   pens'ste-mun prū-in-ō'sus

Pfiesteria piscicida   fes-ter'ē-ä pis-si-sē'dä

P. shumwayae   shum-wā'nē

Phlebotomus   fle-bot'ō-mus

Photorhabdus luminescens   fō-tō-rab'dus lü-mi-nes'senz

Phytophthora infestans   fī-tof'thô-rä in-fes'tans

Picrophilus   pik-rä'fil-us

Plasmodium falciparum   plaz-mō'dē-um fal-sip'är-um

P. malariae   mä-lā'rē-ī

P. ovale   ō'va'lē

P. vivax   vī'vaks

Plesiomonas shigelloides   ple-sē-ō-mō'näs shi-gel-loi'dēs

Pneumocystis carinii   nü-mō-sis'tis kär-i'nē-ī (or kar-i'nē-ē)

P. jiroveci   jėr-ō-vek'ē

Porphyromonas gingivalis   pôr'fī-rō-mō-näs jin-ji-val'is

Propionibacterium acnes   prō-pē-on'ē-bak-ti-rē-um ak'nēz

Proteus mirabilis   prō'tē-us mi-ra'bi-lis

Providencia stuartii   prō-vi-den'sē-ä stū-är'tē-ē

Pseudomassaria   sū-dō-mas'sär-ē-ä

Pseudomonas aeruginosa   sū-dō-mō'näs ã-rü ji-nō'sä

P. cepacia   se-pā'sē-ä

P. marginalis   mär-gin-al'is

P. pseudomallei   sū-dō-mal'lē-ī

P. putida   pyū-tē'dä

Pseudoterranova decipiens   sū-dō-ter-rä-nō-vä dē-si-pē-enz

Rhizobium   rī-zō'bē-um

Rhizopus stolonifer   rī'zo-pus stō-lon-i-fėr

Rhodnius   rōd'nē-us

Rhodobacter   rō-dō-bac'tėr

Rhodococcus rhodnii   rō-dō-kok'kus rōd-nē-ē

Rickettsia akari   ri-ket'sē-ä ä-kär'ī

R. conorii   con-ōr'ē-ē

R. prowazekii   prou-wa-ze'kē-ē

R. rickettsii   ri-ket'sē-ē

R. tsutsugamushi   tsü-tsü-gäm-ü'shē

R. typhi   tī'fē

Rochalimaea quintana   rōk-ä-li-mē'ä kwin-tä'nä

Saccharomyces carlsbergensis   sak-ä-rō-mī'sēs kä-rls-bėr-gen'sis

S. cerevisiae   se-ri-vis'ē-ī

S. ellipsoideus   ē-lip-soi'dē-us

Salmonella enteritidis   säl-mōn-el'lä en-tėr-i'tī-dis

S. heidelberg   hī'del-berg